VNR METRIC HANDBOOK OF ARCHITECTURAL STANDARDS

VNR METRIC HANDBOOK OF ARCHITECTURAL STANDARDS

Patricia Tutt
and
David Adler

VNR VAN NOSTRAND REINHOLD COMPANY
NEW YORK CINCINNATI TORONTO LONDON MELBOURNE

Library of Congress Catalog Card Number 79-9183

ISBN 0-442-25189-0 cloth

ISBN 0-442-24902-0 paper

Printed in Great Britain.

Published by Van Nostrand Reinhold Company
A division of Litton Educational Publishing, Inc.
135 West 50th Street, New York, NY 10020, U.S.A.

16 15 14 13 12 11 10 9 8 7 6 5 4 3 2 1

Library of Congress Cataloging in Publication Data

Main entry under title:

VNR metric handbook of architectural standards.

Edition of 1969 by L. Fairweather published under title: VNR metric handbook.
Bibliography: p.
Includes index.
1. Building—Tables, calculations, etc.
2. Metric system—Tables, calculations, etc. I. Tutt, Patricia. II. Adler, David. III. Fairweather, Leslie. VNR metric handbook.
TH151.F34 1979 690'.021'2 79-9183
ISBN 0-442-25189-0
ISBN 0-442-24902-0 pbk.

Contents

Preface

With the completion of the change to metric in the United Kingdom construction industry, the orginal purpose of the *AJ Metric Handbook* has been fulfilled. However, in the eleven years since it first appeared as three special issues of the AJ it has become an essential boardside reference. It was therefore decided to carry out a full revision: expanding its functions of providing architects with sketch design information and assistance to students. The number of building types has been doubled, and all the sections revised by experts in the appropriate fields. In particular there are new sections in the growth areas of transport, industrial and leisure buildings. The functional aspects such as thermal comfort and structures have been greatly expanded, and a new section added on landscaping.

It must be appreciated that no one book can cover everything on these subjects. The *New Metric Handbook* can be seen as a digest of the Architectural Press publications. At the end of most sections will be found a bibliography including books that are essential further reading.

While every care been taken to ensure the accuracy of the information at the time of compilation, it is inevitable that there will be errors and changes. The editors would be grateful to be informed when these are detected, and also for suggestions for additions and improvements. However, neither they, the authors nor the publisher can accept responsiblity for loss or damage resulting from inaccuracies.

David Adler
September 1979

Acknowledgements

The Architectural Press and the editors wish to acknowledge the help and information provided by many organisations, government bodies and individuals during the preparation of this volume. In these acknowledgements are listed the sources of information given in the text, tables and illustrations. Any omissions are unintentional and the publishers will gladly make any further acknowledgements necessary when the book is reprinted.

Dr B. V. Akerblom, ***Standing and sitting*** p 165
Jeremy Alexander p 489
Architects' Co-Partnership, Essex University p 266
Ariel Hotel, Heathrow p 322
American Society of Heating, Refrigerating and Air Conditioning Engineers, p 389
London Borough of Barnet, Copthall p 217
Eric Berglund, *Bord* p 165
British Metal Window Association, p 358 ff
BPB Industries Ltd p 356
British Standards Institution (major references)

BS4	p 348 ff	BS4011	p 20
BS340	p 355	BS4229	p 354
BS350	p 495 ff	BS4483	p 352
BS368	p 353	BS4484	p 13
BS416	p 363	BS4514	p 364
BS437	p 363	BS4568	p 363
BS449	p 450 ff	BS4576	p 364
BS460	p 363	BS4607	p 363
BS849	p 354	BS4787	p 359
BS990	p 358	BS4848	p 348
BS1178	p 354	BS5254, 5255	p 364
BS1192	p 10 ff	BS5328	p 466
BS1474	p 354	BS5555	p 4 ff
BS1972	p 364	CP3: Chap IV: Part 3	p 117
BS2028: 1364	p 353	CP3: Chap V: Part 1	p 446
BS2655	p 481 ff	CP111	p 454
BS2871	p 361	CP112	p 457
BS3284	p 364	CP114	p 465
BS3505	p 364	CP305	p 336 ff
BS3867	p362 ff	CP2004	p 478
BS3893	p 124	DD51	p 19
BS3936	p 372	PD5686	pp 5–8
		PD6031	p 11

British Steel Corporation p 346 ff
British Woodwork Manufacturers' Association windows p 357, kitchen units p 361
Building Design Partnership, pp 126, 265, 267
Building Research Establishment, pp 126 and 400
Casson, Conder & Ptrs p 271
Central Directorate of Housing and Building, The Hague p 1
Chartered Institution of Building Services pp 291, 382, 412
Cheshire County Council, p 39
Chicopee Motor Inn p 322
Brian Clouston & Ptrs p 368
Cockpit Theatre, Marylebone p 186
Constrado p 472 ff
Costain Homes Ltd p 318
D. J. Cottam p 170
Crucible Theatre, Sheffield p 185
Department of Education and Science p 253 ff
Department of the Environment/Department of Transport (material reproduced with the permission of the Controller of Her Majesty's Stationery Office):

DB1	p 334
DB6	pp 1, 310
DB16	p 304
DB24	pp 305, 306
DB32	p 34 ff

Technical Memorandum H9/76 p 35
Roads in Urban Areas p 35 etc
Circular 82/69 p 313
Sunlight and Daylight p 421
Various p 300 etc
Department of Health and Social Security p 139 ff
Henry Dreyfuss p 23 etc
Encylopedia Britannica (1875 ed) p 399
FISA (International Rowing Federation) p 212
Foster Associates p 138
Frederick Gibberd & Ptrs p 369
Brenda Goddard (various)
Selwyn Goldsmith p 23 ff
Greater London Council pp 312, 484
London Borough of Hammersmith and Fulham p 39
Haymarket Theatre, Leicester p 186
Her Majesty's Stationery Office:
Petrol (Consolidation) Act 1928, Technical Appendix p 47
Building Regulations 1976 pp 110, 439 ff
Paper 2, Wind Effects on Buildings p 407
Guidance Note: Sound Insulation p 436
Home Office, Guide to Safety at Sports Grounds 1973 p 221
International Air Transport Association p 60 ff
King Edward's Hospital Fund for London p 144
Professor A. Kira p 336 ff
Dr L. Knopp p 195 ff
Lancashire County Council p 44
Lilleshall Hall National Sports Centre p 234
Metrication Board p 18
Moorwood-Vulcan Ltd p 172
National Coal Board p 399
National House Builders' Council p 315 ff
National Playing Fields Association p 214
National Water Sports Centre p 212
E. Neufert p 30
New County Window Cleaning Co p 489
Pilkingtons Ltd p 357
Porters Field Riding School, Leyton p 335
Powell and Moya p 287
Queen Elizabeth College, London p 306
Robert Matthew, Johnson-Marshall & Ptrs p 265
Roche & Dinkeloo p 83
Royal Garden Hotel, London p 332
Royal Institute of British Architects p 14
Sports Council p 214 ff
Stotts of Oldham p 169
Texan Engineering Station/B. H. Evans p 407
Thorndike Theatre, Leatherhead p 185
Trust Houses Forte p 172
Universities Grants Commission/DES p 274 ff
University of Surrey, Guildford p 265
Volvo p 83
Peter Warwick, Architect p 318
L. G. Wood p 408
plus numerous governing bodies for the various sports and pastimes covered in sections 25 to 28.
The different authors and the editors would also like to thank most sincerely all the people who helped them in the preparation of their contributions by typing, checking and soothing fevered brows. Without this vast multitude, the task would have been impossible!

1 Introduction

Contents

1 THE HANDBOOK

The original purpose of this handbook, first published at the time of the changeover to the metric system in Great Britain, was to offer those engaged in the planning, design and construction of buildings a grounding in the principles of the metric system, with a reasonably comprehensive spread of basic data presented in metric terms.

The first edition (March 1969) was published in response to the British Standards Institution document *Change to the metric system in the construction industry— questionnaire* (PD 5767:1966), which included AJ information sheets and design guides in the proposed list of essential reference publications of an 'industrial' nature which would require metrication under the BSI programme for the metric change.

The first edition followed current BSI ruling and used the comma as the decimal marker. This decision was later rescinded and the decimal point was officially approved. The second edition incorporated this change.

1.01 This edition

The metric system is now well established in Great Britain with the exception of a few specialised industries, although there are still considerable problems in co-ordination where rationalised metric and rounded up imperial sizes are used together.

The continuing demand for the handbook reflects the regular use designers make of the design data it provides. This edition recognises that by extending design guidance and including many new sections on building types and design parameters not previously covered.

1.02 Layout

The first part of the book gives a description and explanation of the system itself, while the bulk of the book is concerned with presenting basic metric data for a wide range of building types and for environmental and structural design, complemented by common conversion factors and tables, a bibliography and list of organisations.

1.03 Design data

This handbook puts at the architect's disposal a unique collection of basic metric design data which will remain, as far as possible, valid and unaffected by forthcoming publications.

Selected anthropometric and ergonomic data form the core of the information provided and it is quite unlikely that any minor dimensional variance with other sources will materially affect the validity of the design data. A comparative study of international sources will confirm the existence of marginal differences between the dimensional values given in various publications; slight deviations are of no consequence in the wider context of overall space requirements.

It is not the intention of this handbook to set up or give rise to any dimensional standards not compatible with prevailing trends and dimensional co-ordination principles, but where in the authors' opinion the situation requires a firm decision on some aspect not covered by precedence this decision has been taken after careful scrutiny of the wider issues involved. In most instances the international field of technical metric information has been consulted and cross-checked to compare the validity of the design data given.

1.04 Space requirements

In evaluation of space requirements a sensible dimensional approach was aimed at similar to that established by the *Mobleringsplaner* published by Staters Byggeforskningsinstitut, Copenhagen; *Woningbouw* published by the Bouwcentrusm, Rotterdam and *Voorschriften en wenken voor het ontwepen van woningen 1965* published by the Central Directorate of Housing and Building, The Hague, **1.1**. This approach was also accepted by the Ministry of Housing and Local Government in its Design Bulletin 6 *Space in the home* (metric edition), **1.2**,

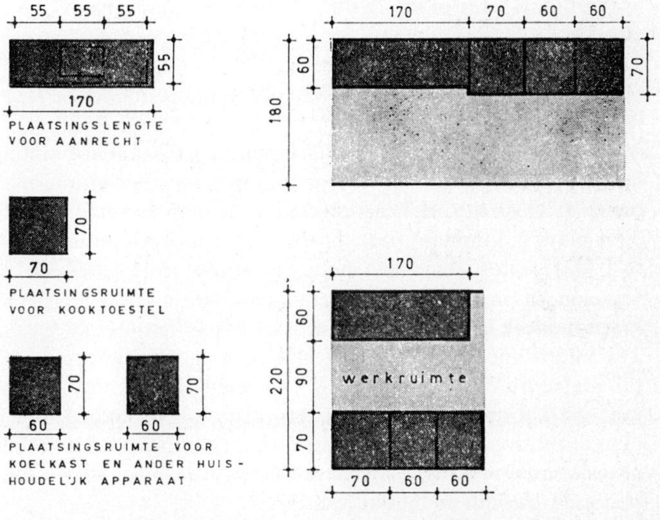

1.1 *Dimensional approach of 'Voorschriften' (cm)*

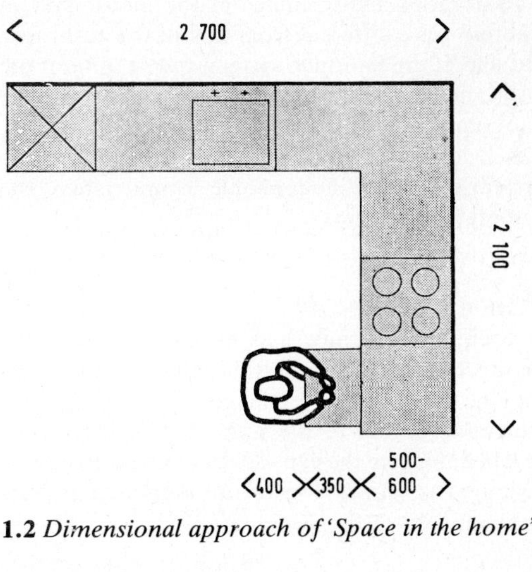

1.2 *Dimensional approach of 'Space in the home' (mm)*

and is generally followed by Danish, Swedish, German and Norwegian architects and designers.
Metric sizes indicated throughout this handbook do not necessarily imply dimensional standards for actual objects or furniture shown, but refer generally to the space required to accommodate them within the overall design.

2 HISTORICAL BACKGROUND

The scientific evolution of the eighteenth and nineteenth centuries, and the industrial revolution of the nineteenth century, accelerated the movement towards unification of existing systems of weights and measures—all by name apparently interchangeable but in practice unrelated.
Academic efforts to introduce the decimal system of measurement began in the sixteenth century sponsored by astronomers and physicists. The search for definable and stable standards of weights and measures culminated in 1790 in France in the proposal to establish a metre as the new decimal unit of length expressed in geodetic terms and defined as the length of a simple pendulum with a swing of one second at sea level on latitude 45 deg.
Although adopted by the French in the same year its final official recognition did not take place in France until 1840. In the next stage of evolution the metre, while still relating to geodetic factors, became associated with a more constant and physically measurable standard. The metre was defined as the distance between two points or lines and marked accordingly on a metal bar of platinum and iridium (90 per cent Pt and 10 per cent Ir)—the length of the marked unit representing one ten-millionth part of the earth's meridional quadrant.

2.01 Final definition of a metre

The growth of science and technology necessitated further refinements in the basic formula of a metre and the final stage of evolution brought about replacement of the standard metal bar as an international prototype unit by a new definition of a metre in which the physical determination of the wavelength of light plays a dominant part. The most recent pronouncement in 1960 states that the metre is equal to 1 650 763·73 wavelengths in vacuum of the orange radiation ($2p_{10}$–$5d_5$) of a krypton atom of mass 86.
The importance of creating alternative means of determining the metre by the formula of light wavelengths lies in the fact that the natural standard of length can be reproduced independently anywhere in the world without having to refer to the international prototype metre in Sèvres in France, where the platinum-iridium metal bar is kept.

2.02 Architecture not radically affected

The mathematical ratios and formulae of the metric system form a background quite different from that of the anthropometric relationship of the foot/inch system, but in spite of the differences it is unlikely that architecture will be greatly affected by the changeover. The judicious use of proportions, ratios and dimensionally co-ordinated sizes available within the metric system will finally help people to understand the simplicity of operations it offers.

3 MAKING THE CHANGE NOW

For those in countries only now making the change to the metric system this handbook should still fulfil its original purpose in guiding the designer through his country's changeover period and introducing him to the principles of the metric system. He will benefit from the experience of Great Britain in making the change, as reflected in the accumulated amendments to this handbook.

3.01 The new dimensional environment

The breakaway from the foot/inch convention will perhaps not come easily to those with a deep-rooted attachment to the duodecimal system. All that is going to happen, however, is the change from one convention of measurement to another, to gain simplicity of operation and standardisation in line with 90 per cent of the world's population. A practical and emotionally neutral atmosphere will help the architect to approach this new system with a more responsive mind, thus making the learning process easier.

3.02 Some practical problems

The primary obstacle facing every architect undertaking his first metric design will be his lack of confidence in his own dimensional judgement. The only remedy will be to learn the new data as quickly as possible, in the new units, without any translation.
There will be, however, a few problems:

- technical literature will not change immediately or for purposes of economy will change to metric units only, confusing those without the metricated technical back-up. Some manufacturers may publish data showing both imperial and metric units, but experience in Great Britain has shown that prolonged use of both sets of units has little advantage
- further difficulties will lie in assessing what is and is not available in rationalised metric sizes ie those metricised in relation to the dimensional coordination principle (see section 4). Industry must give the practising architect, qs and engineer a comprehensive and accurate timetable showing when metric products will be on the market and in what quantity they will be available
- publications showing equivalent or rounded-off sizes may represent an intermediate stage prior to the determination of final rationalised sizes. This will cause confusion on the drawing board and on site. Definite standards of overall thickness and size etc of materials should be established as soon as possible.

3.03 Rounded and rationalised dimensions

The following example will show what the architect may be confronted with in the process of the changeover. The thickness of sheet material in its equivalent metric form, published by manufacturers as $^1/_8$ in = 3·17 mm, $^1/_4$ in = 6·35 mm, $^3/_8$ in = 9·52 mm, $^1/_2$ in = 12·7 mm, $^3/_4$ in = 19·05 mm, 1 in = 25·4 mm, may, in the process of translation, be rounded off and appear as $^1/_8$ in = 3·2 mm, $^1/_4$ in = 6·4 mm, $^3/_8$ in = 9·5 mm, $^1/_2$ in = 12·7 mm, $^3/_4$ in = 19·1 mm, or 19 mm, 1 in = 25·4 mm. The final rationalised metric thicknesses of materials may take the form of increments of 3 mm, 6 mm, 10 mm 12 mm and 25 mm. From this process it is obvious that rounding-off may not give the final answer and that a logical rationalisation must be achieved to arrive at the right size for the product. The fractional values involved in the examples quoted will hardly matter in scaled drawings as they are of minute thickness, but cumulatively they cannot be ignored.

3.04 Importance of the millimetre

The value of a millimetre should not be underestimated. Many architects imagine that one millimetre is simply the thickness of a pencil line. In fact 1·5 mm equals $^1/_{16}$ in, or $^1/_8$ in equals 3 mm—both dimensions important enough not to be ignored especially in assessing the dimensional values of tolerances and gaps for jointing solutions.

3.05 Conversion of statutory regulations

A major feature of the changeover will be the conversion of building regulations into the metric system. Work on dimensional co-ordination and controlling dimensions must precede the final issue of metricated controls and problems may arise in the interim.

4 REFERENCES

BRITISH STANDARDS INSTITUTION
PD 5686: 1978 The use of SI units
PD 6030: 1967 Programme for the change to the metric system in the construction industry
PD 6031: 1968 Use of the metric system in the construction industry (2nd edn)
DD 51: 1977 Guidance on dimensional co-ordination in buildings
MINISTRY OF HOUSING AND LOCAL GOVERNMENT AND WELSH OFFICE Metrication of house building, Circular 1/68
MINISTRY OF HOUSING AND LOCAL GOVERNMENT
Public Health Act 1961: The Building Regulations 1965: Metric equivalents of dimensions
SCOTTISH DEVELOPMENT DEPARTMENT Metric space standards, Circular 27/68
A J Elder *Guide to the Building Regulations 1976,* London, Architectural Press, 1979 (6th edn, 1st edn 1968)
METRICATION BOARD
Going metric, quarterly
How to write metric: a style guide to teaching and using SI units, 50p.

2 Basic metric system and SI units

CI/SfB (F7)
CI/SfB (1976 revised) (F7)

Contents

1 DEVELOPMENT OF ORIGINAL METRIC SYSTEM

The original metric system of measures based on a decimal system of units went through several refinements, with sponsors varying between astronomers, geodetic mathematicians, physicists and scientists. The progress of technology necessitated revision of the MKS system (metre, kilogram, second) defined in 1900, and discoveries in the electro-magnetic field resulted in adding the fourth basic unit of ampere, the unit of electric current, to the existing system of units of mechanics. The MKSA (or Giorgi system) as the new metric system was known, lasted till 1960 when again the growth of thermodynamics, nuclear physics and electronic science caused further refinement of the metric system by adding the degree kelvin as the unit of temperature and the candela as the unit of luminous intensity. This final set of six basic units became known as 'Système International d'Unités' known in all languages as SI units. A seventh unit the mole (amount of substance) was added later, and the unit of temperature was amended to the kelvin.

2 SI BASE UNITS

The six SI units which will most likely be the limit of the major units affecting the architect in day-to-day practice, are shown in table I.

Table I Six base units of the SI System

Quantity	Name of unit	Unit symbol
Length	metre	m
Mass	kilogram	kg
Time	second	s
Electric current	ampere	A
Thermodynamic temperature	kelvin*	k
Luminous intensity	candela	cd

* The degree Celsius (°C) will be used in practice, see para 6.01.

3 DERIVED SI UNITS

From the six units given in table I a number of other units have been derived which are necessary for the calculation of structural forces, electrical and thermal values and other related properties. These are given in table II.

4 DECIMAL MULTIPLES OF SI UNITS

A range of multiples and submultiples of SI units is given in table III.
Table IV gives a comprehensive list of SI units and a selection of recommended decimal multiples and submultiples of the SI units together with other units or other names of units which may be used (from PD 5686).

Table II Derived SI units

Physical quantity	SI unit	Unit symbol
Force	newton	N = kg m/s²
Pressure, stress	pascal	Pa = N/m²
Work, energy, quantity of heat	joule	J = Nm
Power	watt	W = J/s
Electric charge	coulomb	C = As
Electrical potential	volt	V = W/A
Electric capacitance	farad	F = C/V
Electric resistance	ohm	Ω = V/A
Electric conductance	siemens	S = I/Ω
Frequency	hertz*	Hz = I/s
Magnetic flux	weber	Wb = Vs
Magnetic flux density	tesla	T = Wb/m²
Inductance	henry	H = W/A
Luminous flux	lumen	lm = cd sr†
Illuminance	lux	lx = lm/m²

* The term 'cycle per second' (c/s) has been used in the UK.
† sr is the symbol for steradian.

Table III A range of multiples and submultiples of SI units

Multiples and submultiples	Prefix	Symbol
10^{18}	exa	E
10^{15}	peta	P
10^{12}	tera	T
10^{9}	giga	G
10^{6}	mega	M
10^{3}	kilo	k
10^{2}	hecto	h
10^{1}	deca	da
10^{-1}	deci	d
10^{-2}	centi	c
10^{-3}	milli	m
10^{-6}	micro	u
10^{-9}	nano	n
10^{-12}	pico	p
10^{-15}	femto	f
10^{-18}	atto	a

5 THE UNIT OF FORCE

From the units listed in table II it will be seen that all work on stress and pressure contains the metric unit of force: the newton. A great deal has been written about the newton and the newton per square metre but the meaning of these terms is not always explained against the old conventions of pound-force and pound-force per square inch respectively.

5.01 Effect of gravity

Until the International Organisation for Standardisation (ISO) accepted the newton as a new unit of force, metric countries were using kilogram or kilogram-force units, while Britain was employing pounds or pound-force units. The problem in using these units lies in the fact that the kilogram as a unit refers to a quantity of matter subject to a gravitational acceleration of 9·81 m/s². The use of such a unit in spheres where gravitational acceleration plays no part at all proved cumbersome and at times inconsistent; the introduction of the SI system cleared away this ambiguity. The newton as a new unit of force has *no* 9·81 m/s² encumbrance of gravitational force and its use simplifies the application of the newton for quantities such as pressure, stress, work and power.

Table IV List of SI units and a selection of recommended decimal multiples and sub-multiples of the SI units together with other units or other names of units which may be used (from PD 5686)
An asterisk against a unit means that the unit may be used in the UK but is not yet included in the ISO draft recommendation.

Item No in ISO/R31	Quantity	SI unit	Selection of recommended decimal multiples and sub-multiples of SI unit	Other units or other names of units which may be used	Remarks
Space and time					
1–1.1	plane angle	rad (radian)	mrad µrad	degree (. .°), $1° = \frac{\pi}{180}$ rad minute (. .′), $1' = \frac{1}{60}°$ second (. .″), $1'' = \frac{1}{60}'$ grade (. .^{G}), $1^G = \frac{\pi}{200}$ rad	
1–2.1	solid angle	sr (steradian)			
1–3.1 . . 7	length	m (metre)	km cm mm µm nm	*International nautical mile (1 n mile = 1852 m)	
1–4.1	area	m^2	km^2 dm^2 cm^2 mm^2	hectare (ha), 1 ha = 10^4 m^2 are (a), 1 a = 10^2 m^2	
1–5.1	volume	m^3	dm^3 cm^3 mm^3	hectolitre(hl), 1 hl = 10^{-1} m^3 litre (l). 1 l = 10^{-3} m^3 = 1 dm^3 centilitre (cl), 1cl = 10^{-5} m^3 millilitre (ml), 1 ml = 10^{-6} m^3 = 1 cm^3	In 1964 the Conférence Générale des Poids et Mesures adopted the name litre (l) as the synonym for cubic decimetre (dm^3) but discouraged the use of the name litre for precision measurements
1–6.1	time	s (second)	ks ms µs ns	day (d). 1 d= 24 h hour (h), 1 h = 60 min minute (min), 1 min = 60 s	Other units such as week, month and year (a) are in common use
1–8.1	angular velocity	rad/s			
1–10.1	velocity	m/s		kilometre per hour (km/h) 1 km/h = 0·277 778 m/s *knot (kn) 1 kn = 1 n mile/h = 0·514 444 m/s	
1–11.1	acceleration	m/s^2			
Periodic and related phenomena					
2–3.1	frequency	Hz (hertz)	THz GHz MHz kHz		
2–3.2	rotational frequency	1/s		min^{-1}	
Mechanics					
3–1.1	mass	kg (kilogram)	Mg g mg µg	tonne (t), 1 t = 10^3 kg	
3–2.1	density (mass density)	kg/m^3	Mg/m^3 or kg/dm^3 or g/cm^3	1 t/m^3 or 1 kg/1 g/ml g/l	For litre (l) see item 1–5.1
3–5.1	momentum	kg m/s			

Table IV List of SI units and a selection of recommended decimal multiples and sub-multiples of the SI units together with other units or other names of units which may be used (from PD 5686) *continued*
An asterisk against a unit means that the unit may be used in the UK but is not yet included in the ISO draft recommendation.

Item No in ISO/R31	Quantity	SI unit	Selection of recommended decimal multiples and sub-multiples of SI unit	Other units or other names of units which may be used	Remarks
Mechanics (continued)					
3–6.1	moment of momentum, angular momentum	kg m²/s			
3–7.1	moment of inertia	kg m²			
3–8.1	force	N (newton)	MN kN mN		
3–10.1	moment of force	N m	MN m kN m		
3–11.1 3–11.2	pressure and stress	Pa (pascal) or N/m²	GPa MPa or N/mm² kPa	bar = 10⁵ Pa mbar	
3–19.1	viscosity (dynamic)	Pas	mPa s	centipoise (cP) 1 cP = 1 mPa s	
3–20.1	kinematic viscosity	m²/s	mm²/s	centistokes (cSt) 1 cSt = 1 mm²/sec	
3–21.1	surface tension	N/m	mN/m		
3–22.1	energy work	J (joule)	TJ GJ MJ kJ mJ	kilowatt hour (kW h) 1 kW h = 3·6 × 10⁶ J = 3·6 MJ electronvolt (eV) 1 eV = (1·60210 ± 0·00007) × 10⁻¹⁹ J	The units W h, kW h, MW h, GW h and TW h are used in the electrical industry The units keV, MeV and GeV are used in accelerator technology
3–23.1	power	W (watt)	GW MW kW mW μW		
Heat					
4–1.1	thermodynamic temperature	K (kelvin)			
4–2.1	Celsius temperature	°C			
4–1.1 4–2.1	temperature interval	K			1°C = 1 K
4–3.1	linear expansion coefficient	1/K			
4–4.1	heat, quantity of heat	J	TJ GJ MJ kJ mJ		
4–5.1	heat flow rate	W	kW		
4–7.1	thermal conductivity	W/(m K)			

Table IV List of SI units and a selection of recommended decimal multiples and sub-multiples of the SI units together with other units or other names of units which may be used (from PD 5686) *continued*
An asterisk against a unit means that the unit may be used in the UK but is not yet included in the ISO draft recommendation

Item No in ISO/R31	Quantity	SI unit	Selection of recommended decimal multiples and sub-multiples of SI unit	Other units or other names of units which may be used	Remarks
Heat (continued)					
4–8.1	coefficient of heat transfer	W/(m² K)			
4–10.1	heat capacity	J/K	kJ/K		
4–11.1	specific heat capacity	J/kg K	kJ/kg K		
4–13.1	entropy	J/K	kJ/K		
4–14.1	specific entropy	J/(kg K)	kJ/(kg K)		
4–16.1	specific energy	J/kg	MJ/kg kJ/kg		
4–18.1	specific latent heat	J/kg	MJ/kg kJ/kg		
Electricity and magnetism (see notes 1, 2 and 3 at end of this table)					
5–1.1	electric current	A (ampere)	kA mA μA nA pA		
5–2.1	electric charge, quantity of electricity	C (coulomb)	kC μC nC pC		1 Ah = 3·6 kC
5–3.1	volume density of charge, charge density	C/m³	C/mm³ MC/m³ or C/cm³ kC/m³ mc/m³ μc/m³		
5–4.1	surface density of charge	C/m²	MC/m² or C/mm² C/cm² kC/m² MC/m² μC/m²		
5–5.1	electric field strength	V/m	MV/m kV/m or V/mm V/cm mV/m μv/m		
5–6.1 5–6.2 5–6.3	electric potential potential difference electromotive force	V (volt)	MV kV mV μV		
5–7.1	displacement	C/m²	C/cm² kC/m² mC/m² μC/m²		

Table IV List of SI units and a selection of recommended decimal multiples and sub-multiples of the SI units together with other units or other names of units which may be used (from PD 5686) *continued*
An asterisk against a unit means that the unit may be used in the UK but is not yet included in the ISO draft recommendation.

Item No in ISO/R31	Quantity	SI unit	Selection of recommended decimal multiples and sub-multiples of SI unit	Other units or other names of units which may be used	Remarks
Electricity and magnetism (continued)					
5–11.1	capacitance	F (farad)	mF μF nF pF		
5–41.1	resistance	Ω (ohm)	GΩ MΩ kΩ mΩ μΩ		
5–42.1	conductance	S (siemens)	kS mS μS		1 S = 1/Ω
5–43.1	resistivity	Ω m	GΩ m MΩ m kΩ m Ω cm mΩ m μΩ m nΩ m		μΩ cm = 10^{-2} Ω m $\frac{\Omega\ mm^2}{m} = 10^{-6}\ \Omega\ m$ = μΩ m are also used
5–44.1	conductivity	S/m	MS/m kS/m		
5–45.1	reluctance	1/H			
5–46.1	permeance	H (henry)			

NOTE 1. In electricity and magnetism the SI units assume the rationalized form of the equations between the quantities. See ISO/R 31, Part V.
NOTE 2. THE IEC has not considered the arrangement and the content of the list. In order to give guidance to ISO, IEC/TC 24 provided the list of multiples and sub-multiples used here, but without division into columns.
NOTE 3. This list is a selection only from that given in PD 5686.

5.02 Objections to the newton

Before SI units were accepted some metric countries introduced the term kilopond for the kilogram-force unit. The newton supersedes both. There is however quite a body of opinion among the structural and civil engineering profession which does not favour use of the SI system in structural and civil engineering calculations. The objection is not centred on the use of the metric system as a whole but round the use of SI units, with particular hostility towards the newton as a unit of force: structural engineers prefer the use of kilogram-force (kgf) or kilopond. The argument revolves around the practicability of using newtons in calculation of stress, pressure and other derived values. Introduction of the newton causes figures in common use to be high, with many digits, and therefore more difficult to memorise and manipulate. In this, the structural and civil engineer is worse off than the architects, qs or builder who, at the worst, could fall back on existing information from metric countries. The structural engineer cannot do this because the metric countries are still trying to introduce the new SI units among their engineers, and all *their* data is in the old metric units. There is as yet no international standardisation agreed upon multiples and submultiples of newtons required for structural calculations.

5.03 Use of multiple units

The structural engineers' real fear appears to be the physical problems involved in working compound calculations resulting in large figures with a lack of confidence about which multiple or submultiple of the unit to place at the end. One can understand their apprehension when looking at the following comparative figures:

1 lbf/in²	=	6895 N/m²
Pressure of 100 lbf/in²	=	7·04 kgf/cm²
	=	689·48 N/m² × 10³

The last figure can be expressed as 689 480 N/m² or in a more manageable way as 689·5 kN/m². The newton and the newton per square metre (especially the latter) are too small for practical purposes and it is generally understood that multiples of the basic units will be accepted for all practical purposes rather than the basic units themselves.

A slight preference has been expressed for writing multiples of pressure and stress units in the form N/mm² (newtons per square millimetre) rather than MN/m² (meganewtons per square metre). It is recommended that this practice should be adopted but regarded as subject to revision in the light of experience. This recommendation helps a great deal towards future rationalisation but the architect must be aware that it is by no means standardised. Use of kN/m² for load and N/mm² for stress has been endorsed by the Institution of Civil Engineers and Institute of Structural Engineers.

See also section 45.

Use of the pascal as the basic unit of stress (equal to N/m²) is expected to become universal within the next few years. The practical unit for structural purposes will be the MPa.

6 UNITS FOR HEATING, VENTILATING, ELECTRICAL AND MECHANICAL ENGINEERING

The position of the services engineering professions is one of complete acceptance of the metric system, including the SI units. The CIBS prepared a report which concluded: 'Adoption of the SI system of units would help international relationships. Particularly it would help this industry with its sales to Europe and its association with international professional commercial bodies.' Finally in 'Conclusions and recommendations' it states:

'When a change to the metric system is necessary it is recommended that the heating and ventilating industry should support the adoption of the SI system. No system intermediate between SI and traditional systems should be contemplated'.

6.01 Temperature

The kelvin belongs to the group of SI base units as a quantitative unit of thermodynamic temperature. The units of kelvin and degree Celsius (centigrade) temperature interval are identical. A temperature expressed in degrees Celsius is equal to the temperature expressed in kelvins less 273·15. What this means in simple language is that each grade, degree or interval of the Celsius scale is in magnitude the same as a kelvin, but in relation to the zero degree Celsius, the zero point of the kelvin scale slides down 273·15 degrees (point of absolute zero), making 0K (ie zero) = −273·15°C or 0°C = +273·15°K.

For customary temperatures and temperature intervals, the degree Celsius (°C) will be used.

6.02 Energy units

For calculations of atmospheric pressure, water pressure, steam pressure, force, energy and power the newton and the pascal are the basic units. The significant changes are in the calculations of heat values. The familiar Btu unit and even the old metric kilocalorie (kcal) disappear, giving way to the new unit of quantity of heat, the joule (J). The customary horse-power (hp) unit has changed to kilowatts (kW). The thermal transmittance (U value) has become in SI units watt per square metre per kelvin ($W/(m^2K)$), instead of the familiar $U = Btu/ft^2h$ deg F.

See also section 41.

6.03 Lighting units

In lighting, the unit of illumination, lumen per square foot has given way to the metric unit of lux ($1\ lx = 1\ lm/m^2$).

See also section 43.

6.04 Sound units

Sound insulation values have not changed through the introduction of the metric system with the exception of the unit of frequency which became the hertz (Hz) unit instead of the British cycle per second (c/s) with 1 Hz = 1 cycle per second (c/s). The decibel unit (dB) used in acoustic and sound insulation calculations remains unaltered.

See also section 44.

7 REFERENCES

BRITISH STANDARDS INSTITUTION PD 5686: 1978 The use of SI units

PD 6031: 1968 The use of the metric system in the construction industry (2nd edn)

BS 350: Part 1: 1959 Conversion factors and tables

BS 350: Part 2: 1962

Supplement No 1 (1967) to BS 30: Part 2: 1962.

Chester H. Page, Paul Vigaureux (eds) *SI—The International System of Units*, National Physical Laboratory, HMSO, 3rd edn, 1977

3 Notation and drawing office practice

CI/SfB (A3t)
CI/SfB (1976 revised) (A3t)

Contents

1 NOTATION

1.01 Decimal marker

The decimal marker on the line is the standard decimal point in the UK; the mid-point position, such as in this publication, is also commonly used. When the value to be expressed is less than unity it should always be preceded by zero (eg 0.6 not .6). Whole numbers may be expressed without a decimal marker. The appropriate number of decimal places should be chosen depending on the circumstances in which the resulting value is to be used.

Thousand marker

No thousand marker should be used but where legibility needs to be improved, a space can be left in large groups of digits at every thousand point. Where only four digits are used a space between the first digit and the others is usually not desirable (eg 15 000, 1500). The comma should never be used either thousand marker or decimal marker in the UK, although it is in common use as a decimal marker in other metric countries and will be encountered when referring to foreign books and magazines.

1.02 Use of SI units

Centimetres or millimetres

Anyone familiar with the metric system since childhood will experience no difficulty in using a mixture of millimetres, centimetres or metres and will recognise at a glance plan dimensions or sizes of products from dimensions indicated by figures alone, without m, cm or mm alongside—but only experience, and the convention associated with the given material, product or scale of drawing tells the reader how to interpret these mixed values. Such a situation should not be entertained in the UK or any other countries now changing to the metric system: the methods of communication must be as direct and unambiguous as possible. In accepting the metre and millimetre exclusively for linear measurement the *AJ Metric Handbook* conforms to international agreement and to subsequent ratifications of that agreement by the BSI in:
PD 6031:1968 (2nd edition) *The use of the metric system in the construction industry* and
BS 1192:1969 *Recommendations for building drawing practice* (metric units).
A standardised approach to notation makes it unnecessary to waste time on checking, even mentally, whether dimensions are in centimetres or in millimetres, with consequent chances of error due to mistaken identity.

1.03 Specifying imperial and metric sizes

Where it is necessary to give equivalent dimensions, it should be done thus:
1 On metric projects, using imperial sized products and components, give metric equivalent sizes, usually to nearest mm, in brackets—eg 4 in (102 mm).
On imperial projects using metric sized products and components: give imperial equivalent sizes, usually to nearest 1/16 in, in brackets—eg 100 mm (3 15/16 in). When both sizes are not required, and where no ambiguity can arise, the first size may be discarded and the second size in brackets retained. Thus the use of brackets indicates that on a metric project an imperial size is required and vice versa. The use of this convention should be noted in the drawing. Examples:

	Sizes of product/component	
Metric projects	**Imperial**	**Metric**
Normally	4 in (102 mm)	100 mm
Where no ambiguity can arise	(102 mm)	100 mm
Imperial projects		
Normally	4 in	100 mm (3 15/16 in)
Where no ambiguity can arise		(3 15/16 in)

Figure **3.1** shows the difference between imperial and metric modular sizes. From this it will be seen that they are not interchangeable, eg the difference between 3 m and its modular equivalent of 10 ft is nearly 2 in or 50 mm.

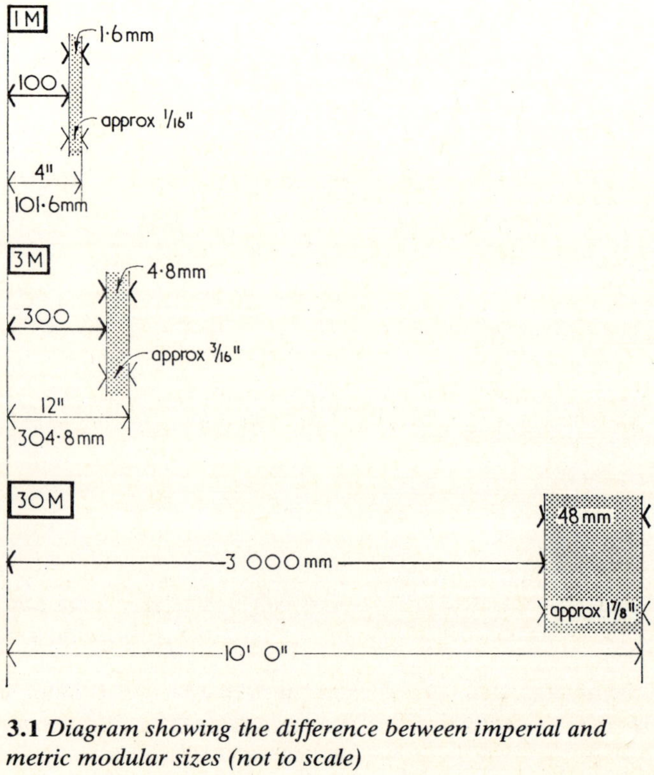

3.1 *Diagram showing the difference between imperial and metric modular sizes (not to scale)*

1.04 Linear measurements on drawings

When both metres and millimetres are used on drawings it will be less confusing if the dimension is always written to three places of decimals, ie 3·450. No unit symbol need be shown unless a lesser number of decimal places is used; ie 3·450 or 3·45 m or, under some circumstances, 3·5 m, are all correct. The first of these (3·450) is to be preferred. 3450 mm can also be used.

Where no ambiguity can arise symbols may be discarded, according to the following three rules (see table I):

1 Whole number indicates millimetres.

2 Decimalised expressions to three places of decimals indicate metres (and also, by implication, millimetres).

3 All other dimensions must be followed by the unit symbol.

Rule 1 could safely be broken in certain circumstances, eg in a large-scale detail drawing where all dimensions are in whole number millimetres, except for the occasional one which may be a whole number plus a decimal fraction. There could be no confusion over the unit size in relation to the other dimensions and it would not seem necessary to add mm after the dimension because it was not a whole number.

Table I Examples of the different ways of expressing linear measurements on drawings (from PD 6031 revised)

Millimetres	Metres		Kilometres
3·5 mm			
3			
30			
300	0·300	0·3 m, 0·30 m	
3000	3·000	3 m, 3·0 m, 3·00 m	
3300	3·300	3·3 m, 3·30 m	
3330	3·330	3·33 m	
3333	3·333		
	3000·000	3000 m	3 km

2 PAPER SIZES

The International A-series of paper sizes has been adopted for all drawings and written material.

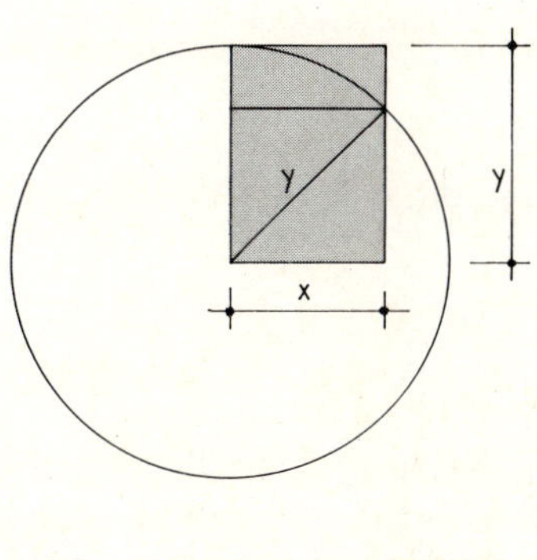

3.2 *Derivation of the rectangle A0, which has a surface area of 1 m²*

2.01 Sizes in the A-series

The A range is derived from a rectangle A0, **3.2**, of area 1 m² with sides x and y such that x:y = 1 : $\sqrt{2}$ (ie x = 841 mm; y = 1189 mm).

The other sizes in the series are derived downwards by progressively halving the size above across its larger dimension. The proportions of the sizes remain constant, **3.3**.

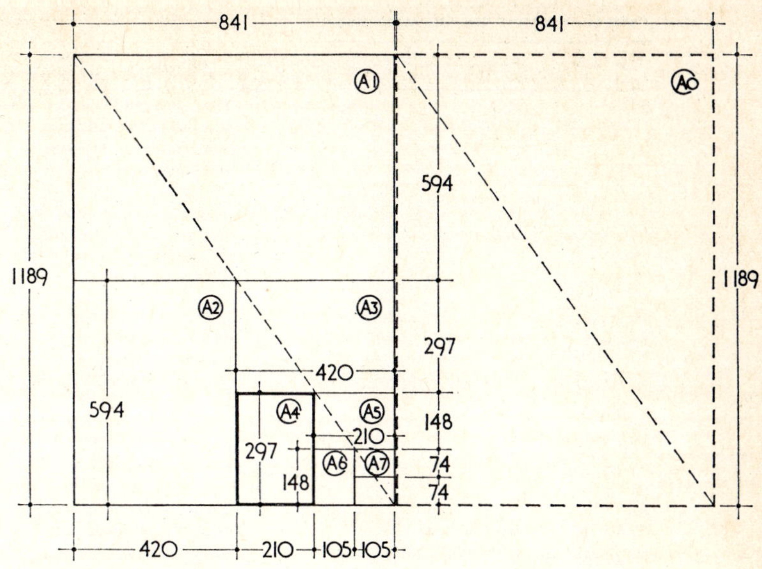

3.3 *A-sizes retain the same proportion (1:$\sqrt{2}$), each size being half the size above*

2.02 Trimmed sizes and tolerances

The A formats are trimmed sizes and therefore exact; stubs of tear off books, index tabs etc, are always additional to the A dimensions. Printers purchase their paper in sizes allowing for the following tolerances of the trimmed sizes:

- for dimensions up to and including 150 mm, ±1·5 mm
- for dimensions greater than 150 mm up to and including 600 mm, ±2 mm.
- for dimensions greater than 600 mm, ±3 mm.

Recommended methods of folding the larger A sized prints are given in **3.4**.

A size	mm
A0	841x1189
A1	594x841
A2	420x594
A3	297x420
A4	210x297
A5	148x210
A6	105x148
A7	74x105
A8	52x74
A9	37x52
A10	26x37

measurements represent trimmed sizes

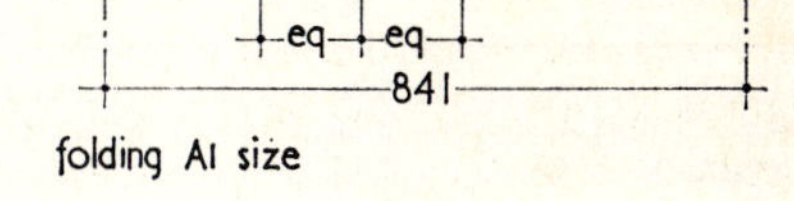

3.4 *A-series of paper sizes*

2.03 Drawing boards

As traditional types of working drawing are superseded it is probable that the range of paper sizes will be restricted to A1, A2, A3 and A4. Only rarely will it be necessary to use A0. The traditional sizes of drawing boards can accommodate the likely range of paper sizes as indicated in table II.

New drawing boards are now manufactured to fit A size paper, while vertical and horizontal filing cabinets and chests have internal dimensions approximately corresponding to the board sizes listed in table III.

Table II Paper and drawing board sizes

Paper size	*Metric board*	*Traditional board*	*Comments*
AO 1189 × 841	AO 1270 × 920	Antiquarian 1372 × 813	Tight vertical fit on antiquarian board; size limited to very big jobs and surveys
A1 841 × 594	A1 920 × 650	Double elephant 1092 × 737	Largest size in common use; paper must be shifted on double elephant board if used vertically
A2 594 × 420	A2 650 × 470	Imperial 813 × 584	If drawing has largest paper dimension vertical, traditional board size must be double elephant
A3 420 × 297	—	Half imperial 594 × 405	A3 drawings are normally executed on an A2 or Imperial board
A4 297 × 210	Special boards with built-in rules available	Half imperial	A4 drawings are normally executed on an A2 or Imperial board

Table III Nominal sizes of drawing boards for use with parallel motion or drafting machines attached

Type of board	**size**	**width mm**	**length mm**	**thickness**
Parallel motion unit only or parallelogram type drafting machine	A1	730	920	22
	A0	920	1270	22
	A0a*	920	1370	22
	2A0†	1250	1750	22

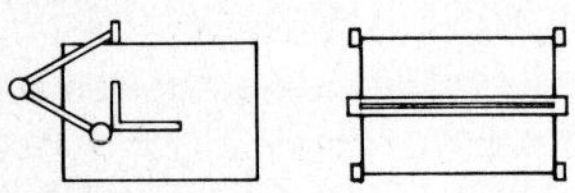

Type of board	**size**	**width mm**	**length mm**	**thickness**
Track or trolley type drafting machine requiring additional 'parking' area to one side	A1 extended	650	1100	22
	A0 extended	920	1500	22
	A0a extended*	920	1600	22

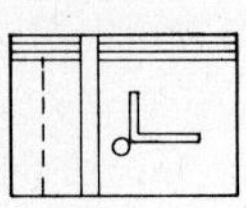

Type of board	**size**	**width mm**	**length mm**	**thickness**
Parallel motion unit with drafting head requiring additional 'parking' area at bottom of board	A1 deep	730	920	22
	A0 deep	1000	1270	22
	A0a deep*	1000	1370	22

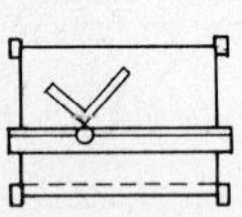

* A0a size temporary, to accommodate continued use of antiquarian sheets (762 × 1346) during the change to ISO 'A' sizes.
† This size has four battens at the rear instead of the usual two.

3 ORDNANCE SURVEY MAPS

3.01 Metrication

The first Ordnance Survey maps based completely on metric measurements were published in 1969. The gradual changeover concentrated initially on the large-scale OS maps, including the six-inch-to-the-mile series. Metrication of the popular one-inch and smaller scale maps followed behind.

• 1:1250 and 1:2500 maps

The International Organisation for Standardisation (ISO) recommended the use of 1:1000 and 1:2000 ratios for Block Plans—but because of large costs and labour involved, Great Britain decided not to change immediately to these rational metric scales, but to retain for at least 10 years the existing ratios of 1:1250 and 1:2500.

Method of conversion

In the meantime the new and revised 1:1250 and 1:2500 maps are being converted to metric on the following basis: heights of bench marks will be shown to 0·01 m accuracy and spot heights to 0·1 m accuracy. On the 1:2500 maps areas of parcels of land will be given to the accuracy of 0·001 hectare, and will also be given in acres as at present. The first metric maps at those scales appeared in October 1969 but it will be many years before all 1:1250 and 1:2500 maps (there are some 150 000 of them) are converted to metric form.

Bench marks and levels

The *Recommendations for Building Drawing Practice* contained in BS 1192:1969 (metric units) state that although Ordnance Survey bench marks are taken in metres to an accuracy of 0·01 m, the levels given in BS 1192:1969 are to 0·001 m. Metric levelling staffs may be read by interpolation to an accuracy of up to 10 mm (ie staff gradations will be in 10 mm increments and accuracy will depend on skill in reading and interpolating values of less than 10 mm. See para **6.04**). Temporary bench marks will usually be established for particular projects. Also, some levels will be obtained by direct measurement and three places of decimals (0·001 m) is consistent with the recommended notation for linear dimensions.

• Six-inch and 1:25 000 scales

The six-inch (1:10 560) scale is being replaced by the 1:10 000 with metric contours. The contour interval is 10 metres in the more mountainous areas and 5 metres in the remainder of the country. The first sheets at the 1:10 000 scale were published in December 1969, but it will be some time before the country is covered with a homogeneous series at this scale.

Publication of the 1:25 000 map with metric contours follows the 1:10 000 maps. The contour intervals will be consistent over the whole of the 1:25 000 sheets.

4 BENCH MARK LISTS

The height of bench marks are given in both metres and feet in Bench Mark Lists obtainable from Ordnance Survey Headquarters, Romsey Road, Maybush, Southampton SO9 4DH.

5 LEVELS

Levels record the distance of a position above or below a defined datum.

5.01 Datum

As a minus sign is easily misread, a suitable fixed point should be taken as TBM (temporary bench mark) such that all other levels are positive. Thus datum should be clearly indicated or described on the drawings and all levels and vertical dimensions should be expressed in metres to three places of decimals above this datum. Particularly on large jobs, it is usually necessary to relate the job datum to the Ordnance Survey datum. It is important to state clearly which Ordnance datum is being used, eg Newlyn, Liverpool or otherwise.

5.02 Levels on plan

It is important to differentiate on site layout drawings between existing levels and intended levels, thus:

Existing level:	× 58·210
Intended level:	× 60·255 (boxed)

The exact position to which the level applies should be indicated by 'x'.
Finished floor levels should be indicated by the letters FFL followed by the figures of the level, thus:
FFL 12·335

5.03 Levels on section and elevation
The same method should be used as for levels on plan except that the level should be projected beyond the drawing with an arrowhead indicating the appropriate line, as in **3.5**.

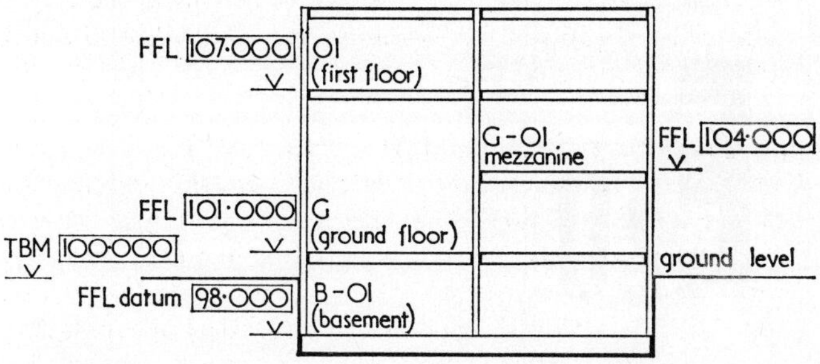

3.5 *Method of indicating levels on sections and elevations*

6 INSTRUMENTS FOR METRIC MEASUREMENT
Measuring instruments for constructional works. Metric graduation and figuring of instruments for linear measurement. The following notes are based on BS 4484.

6.01 Folding rules and rods, laths, and pocket tape rules
Lengths of instruments will be as follows:
(*a*) folding rules: 1 m
(*b*) laths: 1 m, 1·5 m or 2 m
(*c*) folding and multi-purpose rods: 2 m
(*d*) pocket tape rules: 1 m, 2 m, 3 m, or 5 m.

The form of graduation is shown in **3.6**. The instruments are graduated in millimetres along one edge with 5 m and 10 m graduation marks. Along the other edge the millimetre-graduations are omitted.

6.02 Steel and synthetic tapes
Lengths are 10 m, 20 m, or 30 m long. Etched steel bands are available in 30 m and 50 m lengths.
Tapes are graduated at intervals of 100 mm, 10 mm (with the 50 mm centre graduation mark 'arrowed') and 5 mm. The first and last metre of the tape are further subdivided into minor graduation marks at 1 mm intervals (see **3.7**). Note that synthetic material tapes however are not subdivided into millimetres over the first and last metre.

6.03 Chains
Studded steel band chains are in lengths of 20 metres, divided by brass studs at every 200 mm position and figured at every five metres. The first and last metre are further divided into 10 mm intervals by smaller brass studs with a small washer or other identification at half metre intervals. The markings appear on both sides of the band.
Land chains are also in lengths of 20 metres, made up of links, which from centre to centre of each middle connecting link, measure 200 mm. Tally markers are attached to the middle connecting ring at every whole-metre position. Red markers are used for 5 m positions, with raised numerals; yellow markers of a different shape and with no markings are used for the rest, **3.8**.

6.04 Levelling staffs
Lengths are 3 m, 4 m or 5 m long with a reading face not less than 38 mm wide.
Graduation marks are 10 mm deep, spaced at 10 mm intervals. At every 100 mm the graduation marks offset to the left and right of centre, **3.9**. The outside edges of the lower three graduation marks join together to form an 'E' shape. Different colours distinguish graduation marks in alternate metres.
Staffs are figured at every 100 mm interval with metre numbers (small numerals) followed by the decimal point and first decimal part of the metre (large numerals).

6.05 Ranging rods
Lengths are 2 m, 2·5 m or 3 m painted in either 200 mm or 500 mm bands alternating red and white. A rod of 2 m length painted in 200 mm bands is shown in **3.10**.

7 METRIC SCALES

7.01 Metric ratios
The metric changeover introduces a range of new scales with ratios similar but not quite the same as foot/inch units. The internationally agreed and recommended range of scales for use with the metric system is reflected in the ease with which

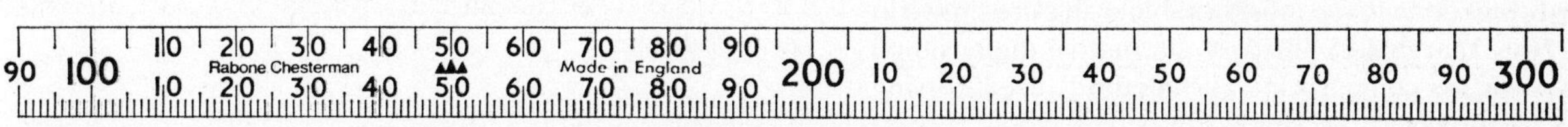

3.6 *Graduation markings for folding rules and rods, laths and pocket tape rules*

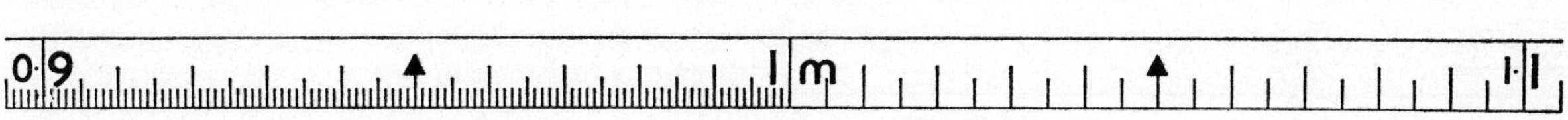

3.7 *Graduation markings for steel tapes*

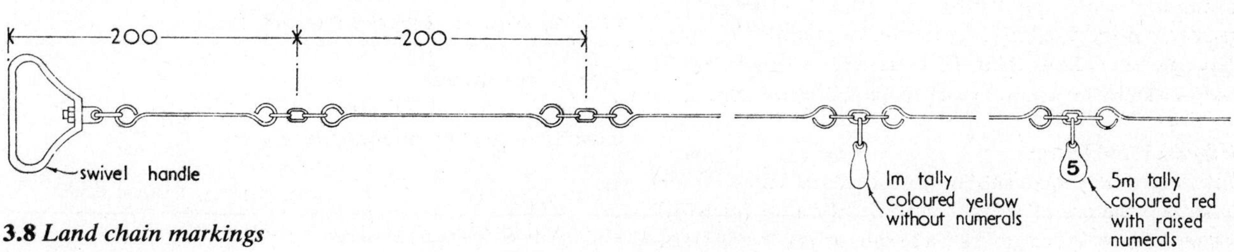

3.8 *Land chain markings*

Table IV Preferred scales for use with metric system

Use	**Scale**	**Nearest foot/inch scales**
Maps	1:1 000 000	1:1 000 000
	1:500 000	1:625 000
	1:200 000	1:250 000
	1:100 000	1:126 720 (1/2 in to 1 mile)
Town surveys	1:50 000	1:63 360 (1 in to 1 mile)
	1:20 000	1:25 000
	1:10 000	1:10 560 (6 in to 1 mile)
	1:5000	—
	†1:2500	1:2500
	1:2000	—
Block plan	†1:2500	1:2500
	1:2000	—
	†1:1250	1:1250
	1:1000	—
Location drawings		
Site plan	1:500	1:500
	1:200	1:192 (1/16 in to 1 ft)
General location	1:200	1:192 (1/16 in to 1 ft)
	1:100	1:96 (1/8 in to 1 ft)
	1:50	1:48 (1/4 in to 1 ft)
Ranges	1:100	1:96 (1/8 in to 1 ft)
	1:50	1:48 (1/4 in to 1 ft)
	1:20	1:24 (1/2 in to 1 ft)
Component drawings		
Assembly	1:20	1:24 (1/2 in to 1 ft)
	1:10	1:12 (1 in to 1 ft)
	1:5	1:4 (3 in to 1 ft)
Details	1:10	1:12 (1 in to 1 ft)
	1:5	1:4 (3 in to 1 ft)
	1:1	1:1 (full size)

* Draft ISO Recommendation No. 1427, 'Architectural and building drawings. Presentation of drawings. Scales'.
† The traditional scales 1:2500 and 1:1250 are included in addition to the rational scales of 1:2000 and 1:1000 because the cost to Ordnance Survey of changing their existing practice makes it impractical for them to adopt the new scales within the foreseeable future.

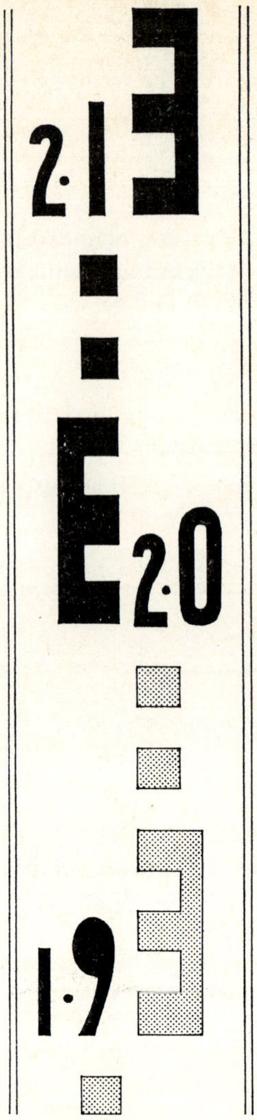

3.9 *Levelling staff marked in 10 mm increments*

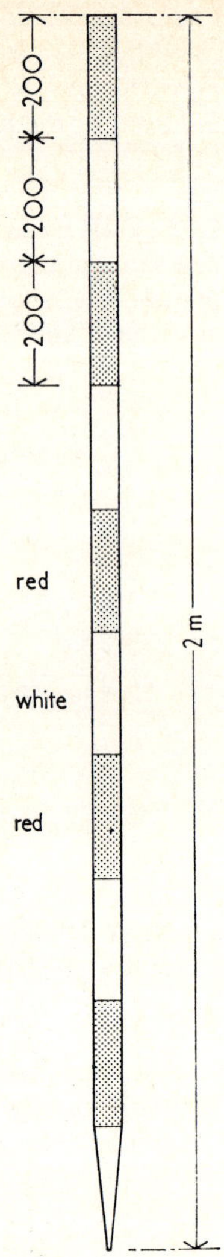

3.10 *Metric ranging rod*

metric ratios can be operated, **3.11**. The decimal subdivision of the scale 1:1 and 1:100 allows for almost all drawing to be done with one metric scale. However, to acquire metric knowledge quickly, it is suggested that the recommended ratios of the metric scales should be used as indicated in table IV.

7.02 Types of drawings

Types of drawings done to the most suitable scales are shown in **3.12** to **18**. Note that in **3.16** and **3.17** alternative dimensional units are shown for comparison. The method of expressing dimensions as shown in the shaded drawings is *not* recommended.

Table V RIBA metric scale

Description		**Metric ratios** Full size (1:1)	1:5	1:10	1:20	1:50	1:100	1:200	1:1250	1:2500	Sizes
RIBA Metric Scale For use in fully metric work	Edge 1	m*		m			m				
Uses metric ratios and gives dimensions in metric	Edge 2				m			m			300 mm 150 mm
	Edge 3		m			m					
	Edge 4								m	m	(Also in plastic)

m Gives dimensions in metric. * Full size (1:1) is read off the 1:100 edge, which is calibrated millimetre dividings.

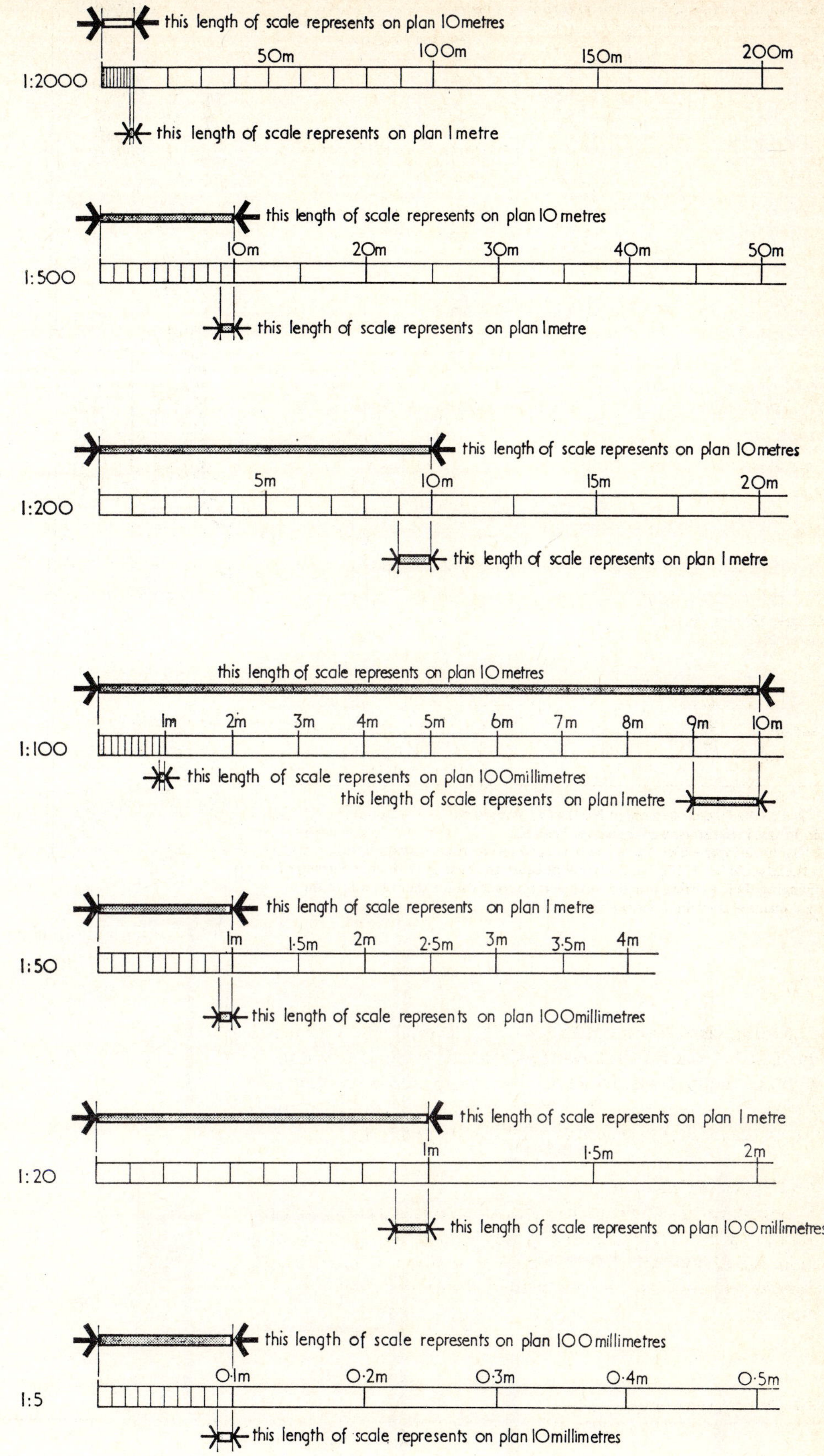

3.11 *Representation of metric lengths (to scale). This drawing may be used to check the correct interpretation of a scale*

Scales on drawings

The scale should be stated on every drawing; drawings which are intended to be read by the non-specialist (eg sketch drawings), or which are to be microfilmed or published, should in addition have a drawn scale. Where two or more scales are used on the same sheet these should be particularly clearly indicated.

7.03 RIBA metric scales

A number of metric scales are now on the market. Details of the scale being sold by the RIBA are given in table V.

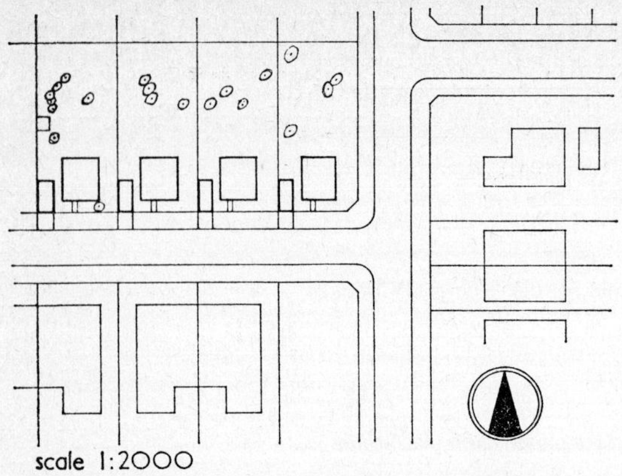

3.12 *Layout plan (note that during the metrication of the Ordnance Survey 1:2500 scale may continue to be used for some time)*

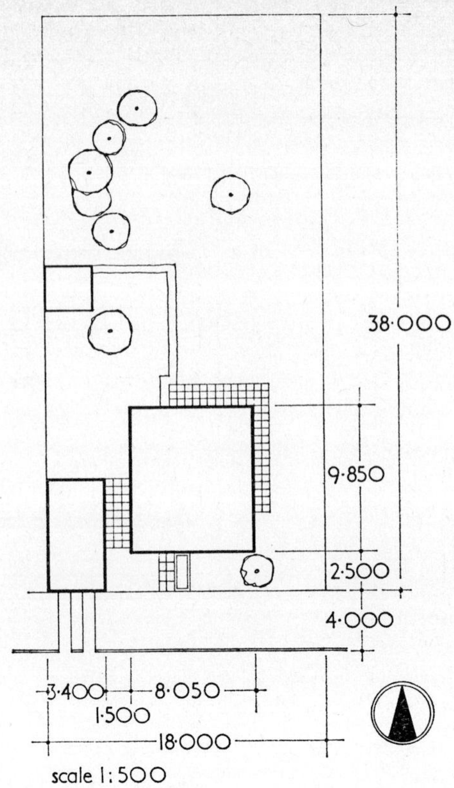

3.13 *Site plan*

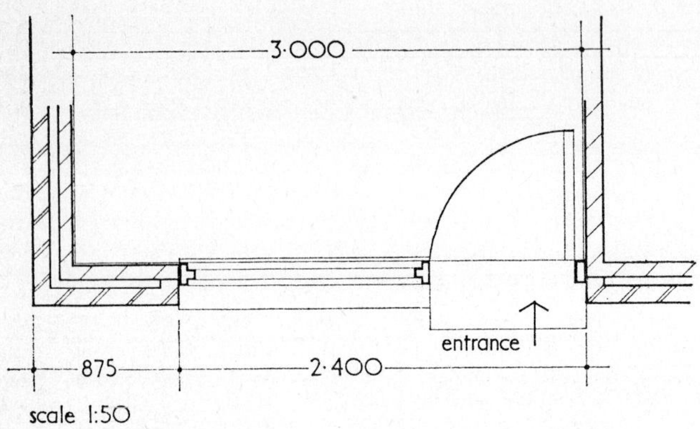

3.15 *Location drawing*

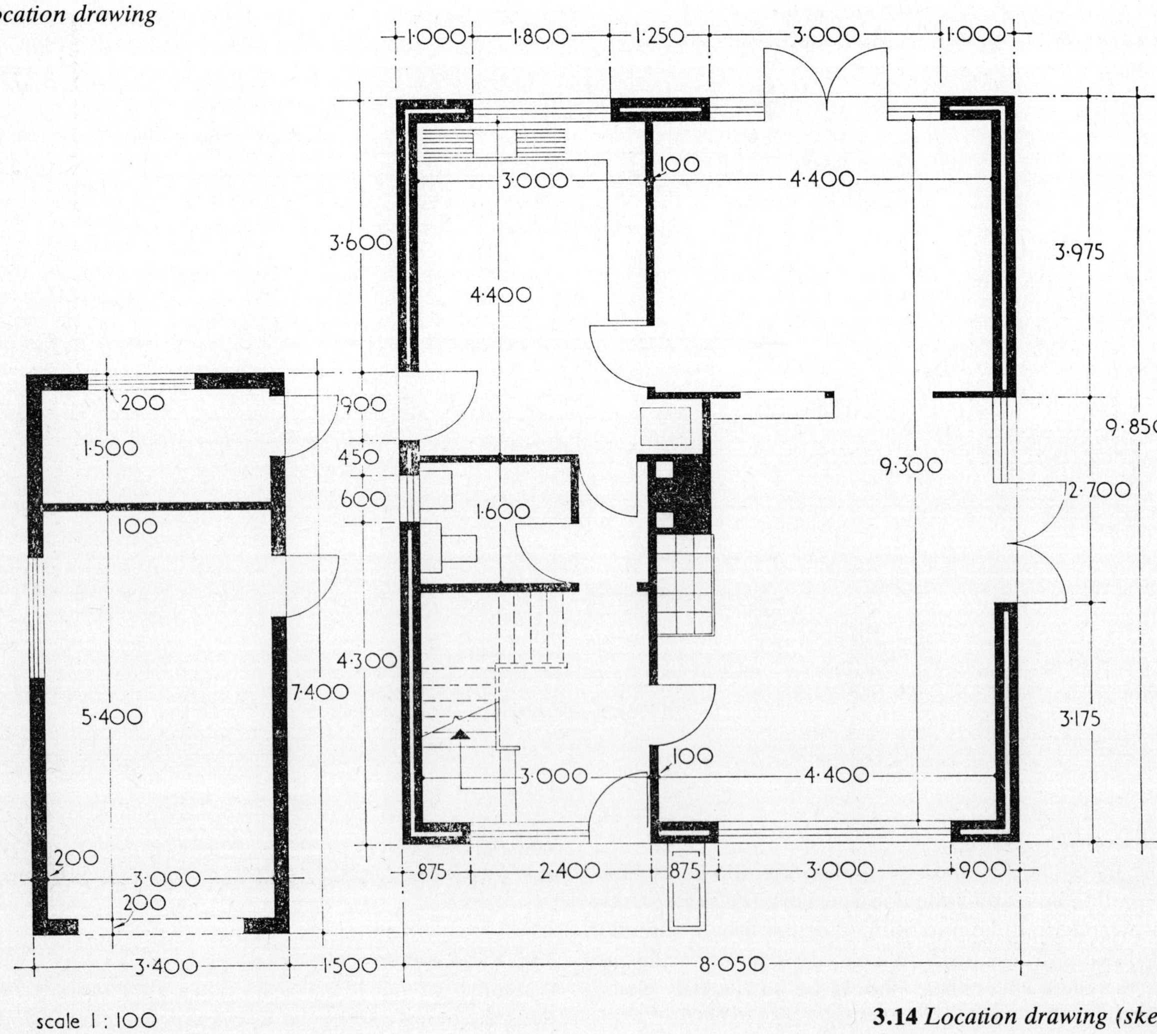

3.14 *Location drawing (sketch plan)*

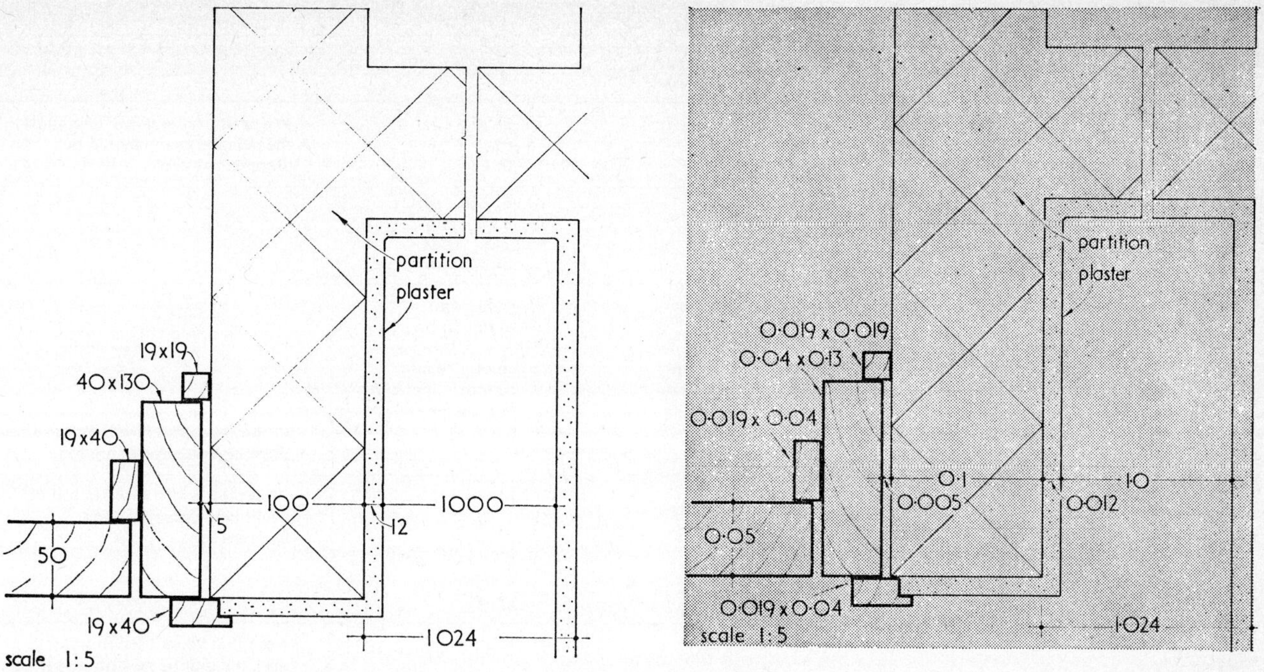

3.16 *Assembly detail drawing (shaded version not recommended)*

Note: *These illustrations are reproduced at the scales indicated. For the larger scale drawings dimensions should be shown in millimetres and not in metres (as in the shaded drawings). There can be no confusion about unit size as the dimensions, in this context, could not be other than millimetres.*

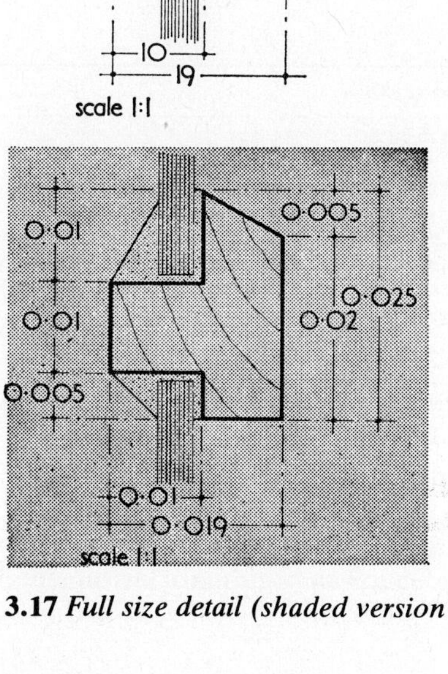

3.17 *Full size detail (shaded version not recommended)*

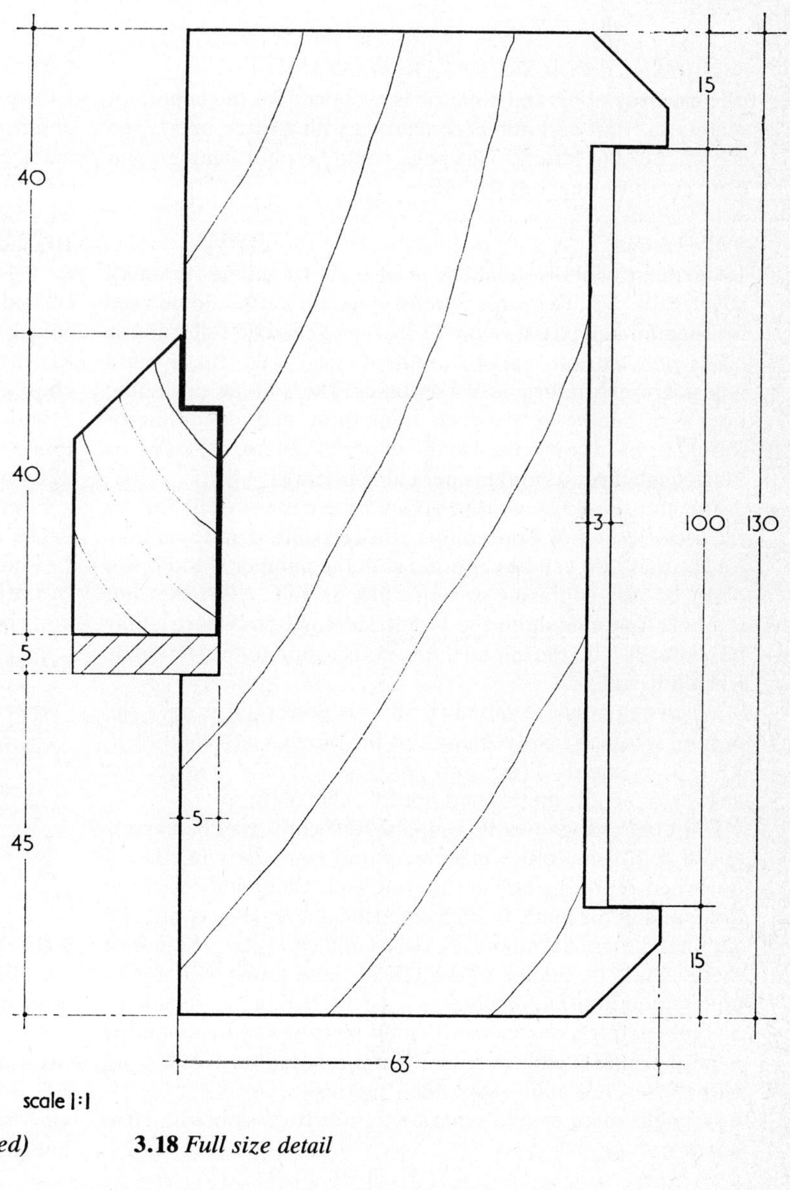

3.18 *Full size detail*

Table VI Summary of symbols and notation

Quantity	Description	Correct unit symbol	Acceptable alternatives	Incorrect use	Notes
Numerical values		0·1 0·01 0·001		·1 ·01 ·001	When the value is *less* than unity, the decimal point should be *preceded* by zero
Length	metre millimetre	m mm		m. M meter m.m. mm. MM M.M. milli-metre	
Area	square metre	m²	sq m	m.sq sm sq.m sq m.	
Volume	cubic metre cubic millimetre	m³ mm³	cu m cu mm	cu.m. m.cu. cu.mm. mm.cub. mm.cu.	
	litre (liquid volume)	l, ltr		l. lit.	Preferably write *litre* in full to avoid 'l' being taken for figure 'one'
Mass (weight)	tonne	t		ton	Preferably write *tonne* in full to avoid being mistaken for imperial ton
	kilogram	kg		Kg kG kg. kilogramme	
	gram	g		g. G.	
Force	newton	N		N. n	Note that when used in written text, the unit of newton is spelled out in full and begins with a lower case letter 'n'. When used as unit symbol, in calculations or in a formula it is then expressed as capital letter 'N'

8 GUIDANCE FOR SECRETARIAL STAFF

The use of symbols and notation is explained for the benefit of secretarial staff and others concerned with written and typed documents and letters. This page could be photo-copied and distributed throughout the office.

8.01 Symbols

1 The main symbols should be used as shown on the summary sheet, table VI. The same symbol ie m mm kg should be used for singular and plural values (1 kg, 10 kg) and no full stops or other punctuation marks should be used after the symbol unless it occurs at the end of a sentence. The 'solidus' or sloping line as a separator between numerator and denominator should be used as it is used in the imperial system, ie 3kg/m³ or 3 kg/cu m (three kilograms per cubic metre).

2 The unit should be written in full if there is any doubt about the use of a symbol. For example, the recognised unit symbol l for the unit litre can be confused with the number 1 and it will often be less confusing to write litre in full. Also, the unit symbol t for tonne should be avoided in situations where it may be confused with the imperial ton, and the unit tonne should be written in full.

3 When symbols are raised to various powers, it is *only* the symbol which is involved and not the number attached to it. Thus 3 m³ equals 3 (m)³ and *not* 3 m × 3 m × 3 m (ie the answer is 3 cubic metres and not 27 cubic metres).

4 Until the changeover is complete, difficulty may be experienced with some office machinery and computers in current use when reproducing the squaring and cubing indices m² or mm², and m³ or mm³. In such cases the abbreviations 'sq' and 'cu' may be used temporarily (sq m and cu m) but the use of these should be limited to the UK as these terms may not be understood internationally.

5 Hyphenating a unit is wrong (milli-metres) and hyphenating a unit because it will overrun a margin should be avoided (eg millimetres, not milli—next line—metres).

6 A single space should separate figures from symbols: 10 m not 10m.

8.02 Notation

1 Care should be taken when expressing sizes of units to keep descriptions consistent within themselves and not mix different units, eg 1500 mm × 600 mm × 25 mm thick and *not* 1·5 m × 600 mm × 25 mm thick. Occasionally, as with timber sizes, it may be necessary to mix the units, eg 100 mm × 75 mm × 10 m long, but in any event, the latter description is preferred to: 0·100 × 0·075 × 10 m. An exception to this rule is when certain mathematical formulae must contain similar units for calculation purposes in which case these figures may be written as 0·1 × 0·075 × 10 (if all in metres) or 100 × 75 × 10 000 (if all in millimetres).

2 It is important to distinguish clearly between the (metric) tonne and the imperial ton. The tonne is equivalent to 2204·6 lb whilst the ton is equal to 2240 lb—a difference of 1·6 per cent.

3 The interval of temperature should be referred to as degree Celsius (°C) and not as centigrade. The word centigrade is used by the continental metric countries as a measure of plane angle and equals 1/10 000th part of a right angle.

Examples

Correct use	Incorrect use
33 m	33 m 33 m.
10·100 m	10 m 100 mm*
33 mm	33 mm 33 mm
50.750 kg	50 kg 750 g

* NOTE. Some metric values are expressed differently in certain metric countries. The value of 10·100 m for example could mean ten thousand one hundred metres and *not* ten metres one hundred millimetres, as in UK.

9 REFERENCES

PD 6031:1968 The use of the metric system in the construction industry (2nd ed)

BS 1192:1969 Recommendations for building drawing practice (metric units)

BS 4484:1969 Measuring instruments for constructional works. Part 1 Metric graduation and figuring of instruments for linear measurement

4 Dimensional co-ordination

CI/SfB (F4j)
CI/SfB (1976 revised) (F43)

Contents

1 INTRODUCTION

This section is based on the BSI *Draft for development: Guidance on dimensional co-ordination, DD 51:1977.* UK readers should consult this document for more detailed information.

1.01 Definition

Dimensional co-ordination (DC) is the application of a range of related dimensions to the sizing of building components and assemblies, and the buildings incorporating them. The principles of DC are to assist in the co-ordination of the many parts that go to make up the total construction, but which are supplied from widely separated sources. At an international level, 100 mm is accepted as the basic module (often referred to by the letter 'M').

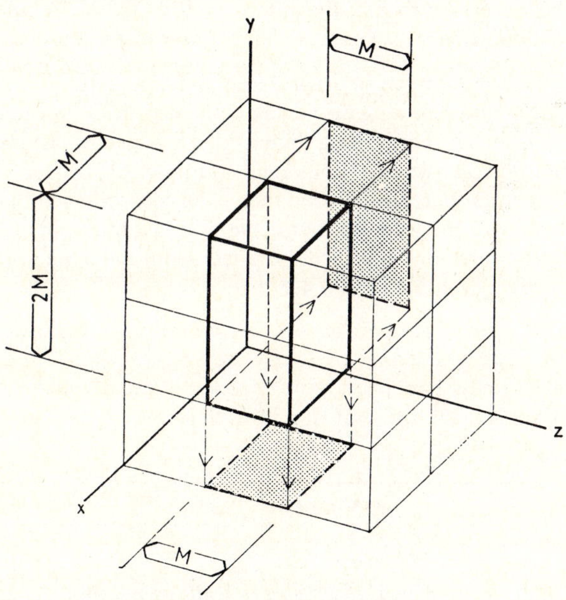

4.1 *Three-dimensional grids of basic modules*

Dimensional co-ordination relies on establishment of rectangular and three-dimensional grids of basic modules into which components can be introduced in an interrelated pattern of sizes, **4.1**. The modular grid network delineates the space *into* which the component fits. This is the most important facet of dimensional co-ordination—the component must always be *undersized* in relation to the space grid.

In the engineering world the piston and cylinder principle establishes the size relationship between dimensional space grid and component, **4.2**. The size of the cylinder must allow

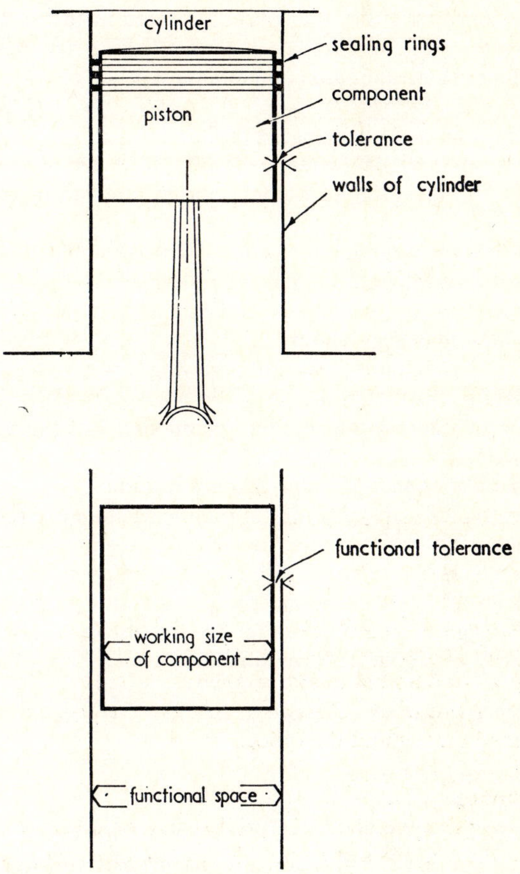

4.2 *The piston and cylinder principle*

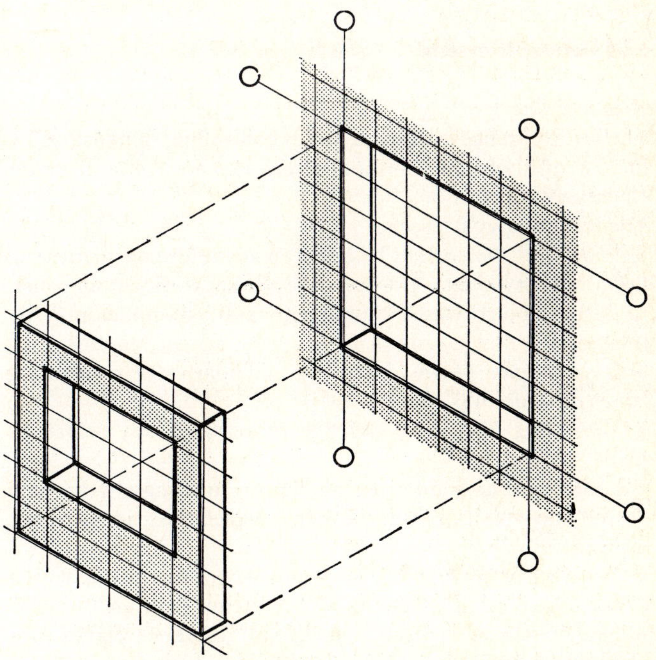

4.3 *Fitting a component into a dimensionally co-ordinated grid*

for the right degree of accuracy and tolerance to enable the piston to move up and down.
The degree of accuracy to be allowed for in the building process is related to the economics of jointing and adequate space must be allowed for size of component plus joint. Transgressing the rules of locating components within the allotted space contained by grid lines will cause considerable difficulty in site assembly.
The basic arrangement of components within the grid layout shows them fitting into the spaces allocated to them: dimensionally they are co-ordinated, thus allowing the designer maximum use of standard components, **4.3**.

1.02 The aims and advantages of DC

The basic aims of DC (as defined in BS 4011:1966) are:
- to obtain maximum economy in the production of components
- to reduce the manufacture of non-standard units
- to avoid wasteful cutting on site.

Advantages to designers may include:
- reduction in design labour
- reduced production of working drawings by the use of standard details
- choice of interrelated standard components at the various price levels.

Potential advantages to manufacturers include:
- more effective use of labour in producing standard lines
- reduction in the stocking, invoicing and other operations connected with numerous differently sized products. There should also be advantages to contractors, not only through better design of components for fit, but through increasing familiarity with standard components.

2 BASIC ELEMENTS OF DC

2.01 Preference for size

The preferred increments are:
first preference (multimodule) multiples of 300 mm
second preference (basic module) multiples of 100 mm
third preference (submodule) multiples of 50 mm up to 300 mm
fourth preference (submodule) multiples of 25 mm up to 300 mm.

2.02 Reference system

Grid and line

The DC reference system identifies controlling dimensions by the use of a grid on plans and a series of horizontal lines on elevations and sections.
The terminology is precise:
- *Controlling dimensions* lie between key reference planes (eg floor to floor height). They provide a framework within which to design and to which components and assemblies may be related.
- *Key reference planes* define the boundaries of controlling zones or structural axes.
- *Controlling lines* on a drawing represent a key reference plane.
- *Axial controlling lines* are shown on drawings by a chain dotted line with a circle at the end, in which the grid reference is given.
- *Face controlling lines* are shown by a continuous line with a circle at the end in which the grid reference is given.
- *Zones* between vertical or horizontal reference planes provide spaces for one or more components which do not necessarily fill the space. Provided that use of associated components is not inhibited, a building component (or group of components) may extend beyond the zone boundary, as may trims and finishes.

3 LOCATING COMPONENTS BY GRID

3.01 Types of grid

A structural grid (axial controlling lines) **4.4**, is established physically by the contractor on site and serves as the main reference in construction. It becomes subject to setting out deviations which affect the spaces required for assemblies of components.

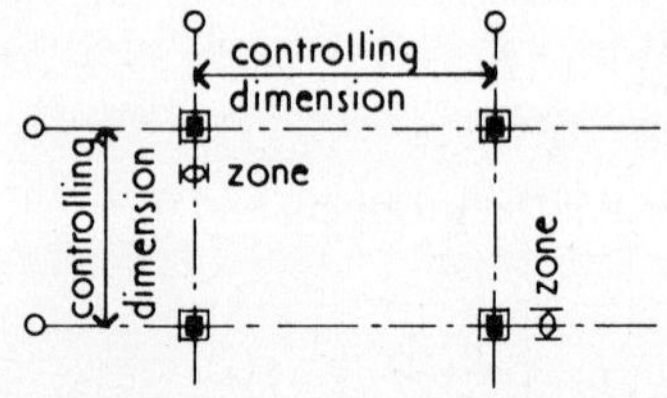

4.4 *Axial control*

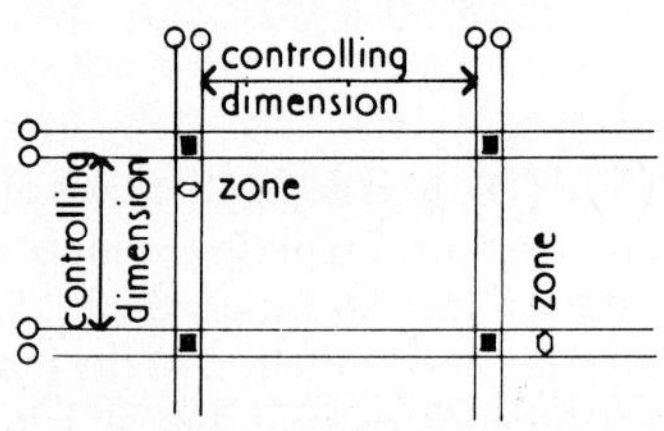

4.5 *Facial control*

A planning grid (face controlling lines) **4.5**, locates non structural elements.

3.02 Relation between structural and planning grids

Structural and planning grids may coincide but do not necessarily do so. The controlling dimensions for spacing structural elements on plan on axial lines are in multiples of 300 (table I). If a 300 mm square grid is used then axial controlling lines will coincide with the grid **4.6**, but if the grid is a multiple of 300 then the controlling lines will be offset from the axial grid by 300 or by a multiple of 300 **4.7**.

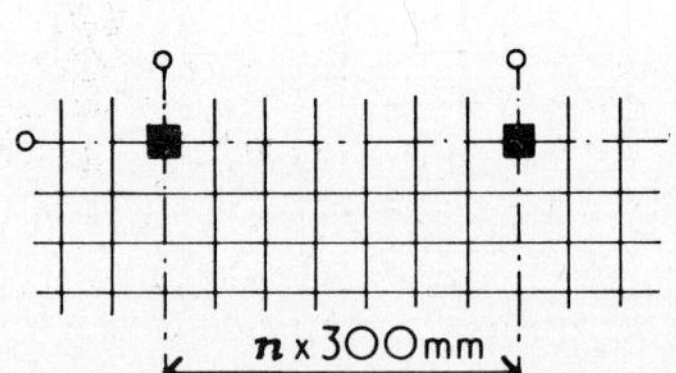

4.6 *Uninterrupted grid: loadbearing walls or columns on axial lines*

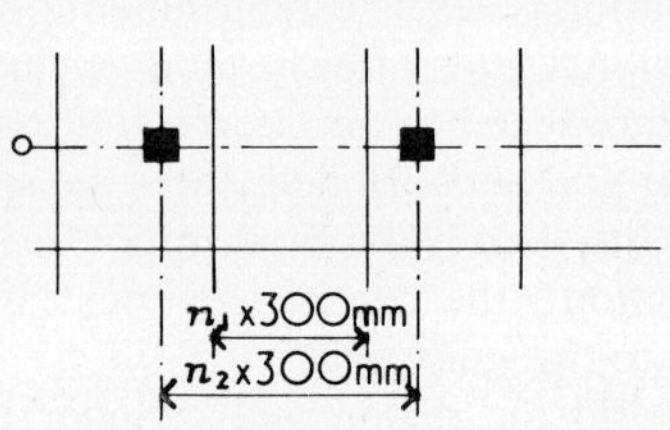

4.7 *Controlling lines offset from the grid*

3.03 Relating zones to a 300 mm grid

If widths of structural zones are multiples of 300 mm, the grid is continuous, **4.8**. If the zone is not a multiple of 300, however, the grid is interrupted by the dimension of that zone, **4.9**. This is referred to as a neutral zone.

- *A neutral zone* is a zone that does not conform to the recommended dimensions given in table I.

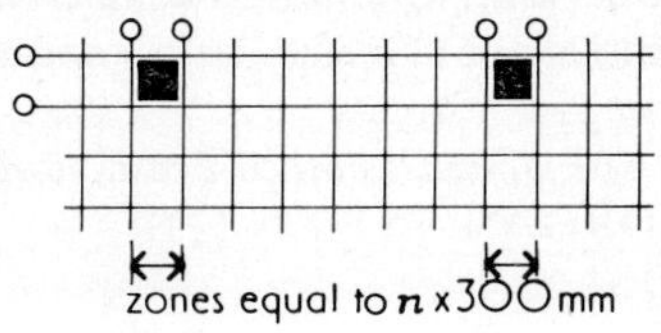

4.8 *Continuous grid*

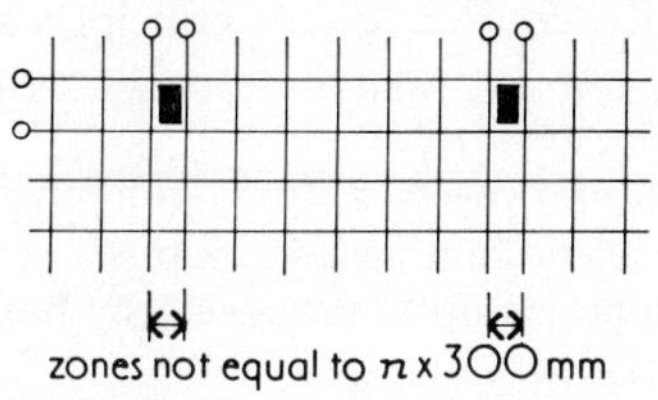

4.9 *Interrupted grid and neutral zones*

3.04 Key reference planes

Key reference planes **4.10** should generally occur at:

- finished floor level
- finished suspended ceiling level
- finished wall surface.

Sizes of zones indicated by key reference planes should be selected from table I. Where controlling or reference lines bound floor or roof soffits, deflection should be allowed for in the zone.

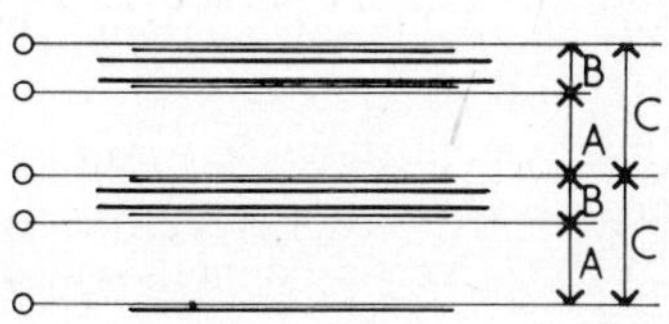

4.10 *Vertical control;* A *being floor to ceiling height controlling dimension;* B *floor and roof zone;* C *floor to floor and floor to roof height controlling dimension*

4 SIZE OF COMPONENTS

4.01 Co-ordinating and work sizes

Controlling dimensions are co-ordinating sizes:

- *Co-ordinating sizes* **4.11** make allowance for fitting and jointing. They represent the overlaid grid which does not usually coincide with actual junction lines on the face of the building. They are indicated by open arrowheads.
- *Work sizes* are the specified manufactured sizes (within permissible deviations). They are indicated by closed arrowheads.

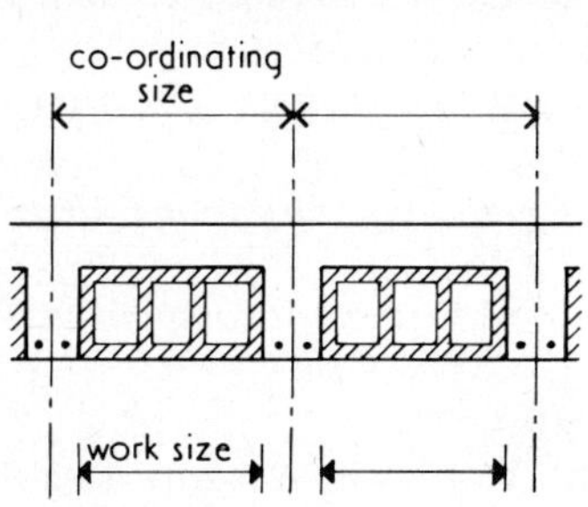

4.11 *Co-ordinating and work sizes*

Table I Sizing of zones and heights

Range mm	Multiples of size mm
Horizontal controlling dimensions	
Widths of zones for columns and loadbearing walls	
100 to 600	300 (first preference) 100 (second preference)
Spacing of zones for columns and loadbearing walls	
From 900[1]	300
Vertical controlling dimensions	
Floor to ceiling heights	
2300[2] to 3000	100
3000 to 6600	300
over 6600	600
Heights of zones for floors and roofs	
100 to 600[3]	100
over 600	300
Floor to floor (and roof) heights	
2700[4] to 8400	300
over 8400	600
Changes in level	
300 to 2400	300
above 2400	600

1 Housing use 800*
2 Farm buildings may use 1500 and 1800
Domestic garages may use 2100
Housing may use 2350*
3 Housing may use 250*
4 Housing may use 2600*
* All special sizes for housing are under review

4.02 Tolerance and fit

Joint sizes are critical. There are graphical aids* to help reconcile all the factors affecting tolerance, such as

- expansion and contraction
- variability in manufactured size
- satisfactory joint clearance range
- variations in setting out dimensions, adjacent components etc
- number of components in an assembly
- variations in interpretation of work size from a given co-ordinating size.

4.03 Degree of accuracy

Designers should identify where fit is critical and where it is not, or they must assess:

- where standard sizes are appropriate and readily available
- if some components can be made to order without a significant cost penalty
- whether cutting is acceptable (and the effect on performance)
- the likely order of assembly.

5 BOUNDARY CONDITIONS

Some assembly and support conditions may necessitate variations in elements to allow for:

- an extended floor slab beyond the clear span to gain a bearing on a wall
- reduction in size to permit the application of a finish
- an increased height of positioning to allow for building directly off the floor slab or extending through a suspended ceiling to reach the soffit of the floor slab.

These allowances (termed 'boundary conditions') should be in multiples of 25 mm. They may be uneconomic to produce, limiting the applications of the product to which they apply.

6 COMMUNICATION

The representation of the dimensional co-ordination framework should be consistent on all drawings. On general

* *Graphical aids for tolerances and fits: handbook for manufacturers, designers and builders,* Building Research Establishment Report, London, HMSO, 1974

location drawings a grid representing 300 mm (or a multiple of 300) may be used. Assembly details may use grids of 300 or 100 mm.

Reference lines
Reference lines or grids should be thin, to distinguish them from other, particularly constructional, lines.

Gridded paper and scales
Table II gives the recommended range of scales for each type of drawing related to appropriate paper grid sizes. Scale and the increment represented by the grid should be indicated on all gridded sheets.

Table II Choice of scales and grids

Type of drawing	Scale	Paper grid size (mm)	Increment represented (mm)
Block plan	1:2000 1:1000	Not applicable	
Site plan	1:500 1:200	Not applicable	
Sketch	1:200	6	1200
	1:100	3*	300
General location	1:100	6	600
		3*	300
	1:50	6	300
		2*	100
Component ranges	1:100	6	600
		3*	300
	1:50	6	300
		2*	100
	1:20	15	300
		5	100
Component details	1:10	10	100
		5	50
	1:5	20	100
		10	50
		5	25
	1:1	100	100
		50	50
		25	25
Assembly	1:20	15	300
		5	100
	1:10	10	100
		5	50
	1:5	20	100
		10	50
		5	25

* These sizes are below the limits for hand drawn grids.

Dimension lines
Different types of dimensions should be distinguished by the type of arrowhead **4.11**.
Running dimensions should be set off from a datum **4.12**.

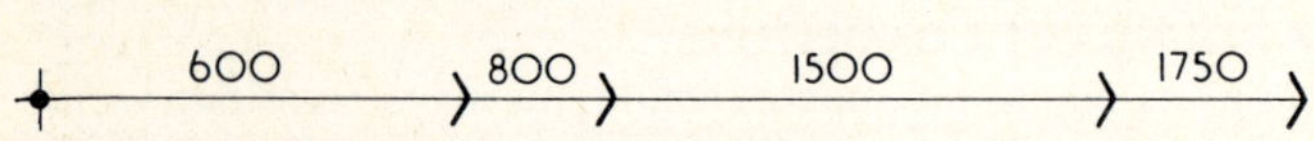

4.12 *Running dimensions. (There is some debate whether the symbol at the datum should be the dot or an arrowhead)*

Assembly details
Assembly details should show components in their context, ie in relation to the adjoining element, with details of the joint.

7 DIMENSIONALLY CO-ORDINATED PRODUCTS

Section 5 of DD 51 lists British Standards where products are dimensionally co-ordinated. Many appear in section 39, Materials, of this handbook.

8 REFERENCES

BRITISH STANDARDS INSTITUTION
BS 2900: 1970 *Recommendations for the co-ordination of dimensions in building.* Glossary of terms
BS 4011: 1966 *Recommendations for the co-ordination of dimensions in building. Co-ordinating sizes for building components and assemblies*
BS 4330: 1968 *Recommendations for the co-ordination of dimensions in building. Controlling dimensions*
BS 5606: 1978 *Accuracy in building*
PD 6432: *Dimensional co-ordination in building. Arrangement of building components and assemblies within functional groups*, Part 1: 1969 Functional Groups 1, 2, 3 and 4. Part 2: 1969 Functional Group 5
DD51: 1976 *Guidance on dimensional co-ordination*
INTERNATIONAL ORGANISATION FOR STANDARDISATION
ISO 1791:1973 Modular co-ordination—vocabulary
ISO 1006:1973 Modular co-ordination—basic module
ISO 2848:1974 Modular co-ordination—principles and rules
ISO 1040:1973 Modular co-ordination—multimodules for horizontal co-ordinating dimensions
ISO R 1790:1970 Modular co-ordination—reference lines of horizontal controlling co-ordinating dimensions
ISO 1789:1973 Modular co-ordination—storey heights and room heights for residential buildings
ISO 2776: 1974 *Modular co-ordination—co-ordinating sizes for door-sets— external and internal*
PUBLIC SECTOR: INTERDEPARTMENTAL CONSTRUCTION DEVELOPMENT GROUP (formerly Interdepartmental Component Co-ordination Group)
DC 10 *Recommended dimensions of basic spaces for selected building components and assemblies used in educational, health, housing and office buildings*, 1969.
DC11 *Recommended dimensions of basic spaces for selected engineering service and drainage components used in educational, health, housing and office buildings*, 1973
DC12 *Dimensional co-ordination for building. Designing with components.* An appreciation of the problems of fit that arise and techniques that may be used in solving them, 1972
PUBLIC SECTOR: HOUSING
Department of the Environment (housing wing) and former Ministry of Housing and Local Government
MHLG Circular 31/67 (1967) Vertical dimensional standards in housing
MHLG Design bulletin 16 *Co-ordination of components in housing. Metric dimensional framework,* 1968
MHLG Circular 69/69 (1969) Metric house shells
DOE(H) *Housing development notes,* 1972, *Housing on a dimensional framework*
PUBLIC SECTOR: GENERAL AND RESEARCH
Department of the Environment (works wing) and former Ministry of Public Building and Works
Metrication in the construction industry No 1: Metric in practice, background and general principles, SI units dimensional co-ordination, 1969
Department of the Environment (Building Research Establishment)
Graphical aids for tolerances and fits. Handbook for manufacturers, designers and builders, Building Research Establishment Report, 1974
PUBLIC SECTOR: DEPARTMENT OF EDUCATION AND SCIENCE
Building bulletin 24 *Controlling dimensions for educational buildings*, 1964
Building bulletin 42 *Co-ordination of components for education buildings*, 1968
PUBLIC SECTOR: HEALTH
Department of Health and Social Security Health Service Design note 5 *Co-ordination of components for health buildings and space data (metric)*, 1969

5 Anthropometric data

Selwyn Goldsmith

Selwyn Goldsmith is the author of Designing for the disabled

CI/SfB (E2d)
CI/SfB (1976 revised) (U41)

Contents

1 PRINCIPLES OF ANTHROPOMETRICS

The body and reach characteristics of people have a direct influence on design. Average (mean) dimensions of some of the more important of these characteristics for men and women are shown in **5.1** and **5.2**.

Although in certain situations it is appropriate to use the average as a criterion it must be emphasised that averages should be treated with caution. Where account is taken of average dimensions only, the likelihood is that in any specific circumstance only about half of the population under consideration will be satisfied. To ensure that the broad range of the population is accommodated account must be taken of people whose dimensional characteristics deviate from the mean. In tables I and II and in **5.3**, **5.4** and **5.5** values for each characteristic are given for the 5th percentile, ie the position at or below which measures for 5 per cent of the total population are found, and for the 95th percentile, the position at or below which measures for 95 per cent are found. It is not always economic or practicable to cover 100 per cent of the population by catering for people at the extremes, and attempts to do so can compromise the convenience of solutions for the broad range of normal people. It may not for example be possible to obtain a solution to a specific design problem which is equally efficient for a typical ambulant person and a person in a wheelchair.

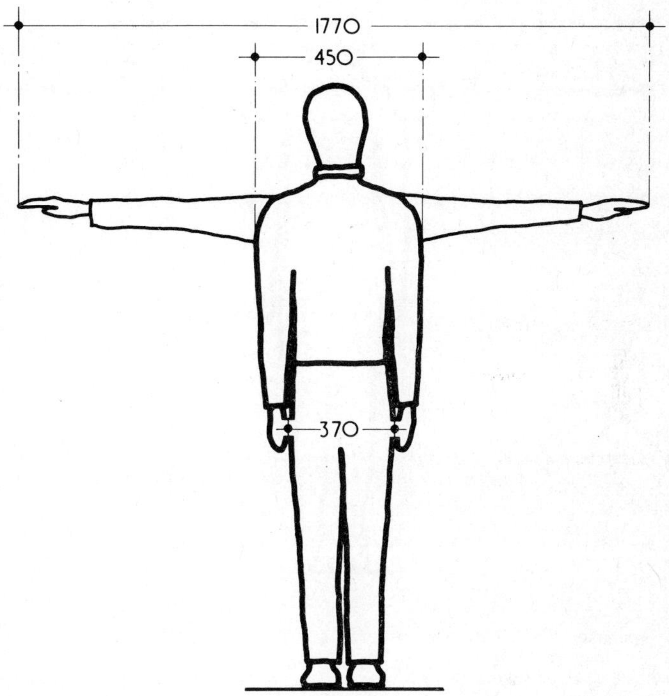

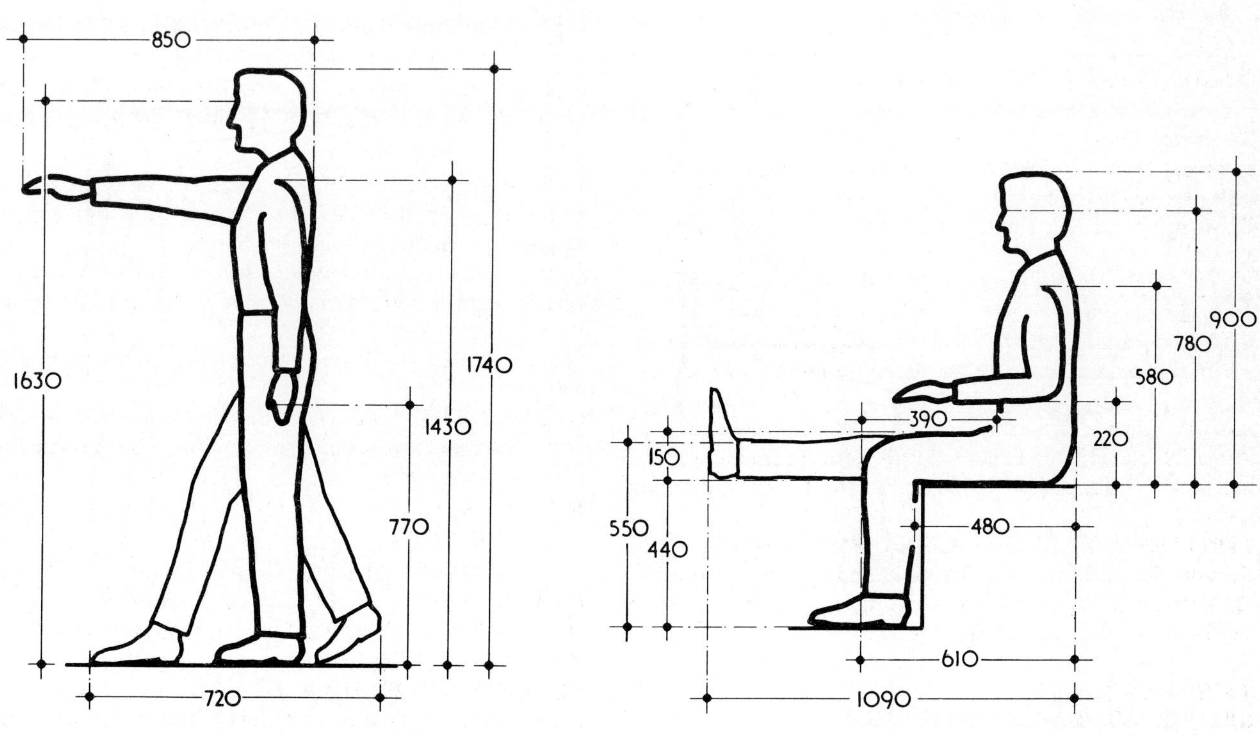

5.1 *Mean average (50th percentile) dimensions of adult British males*

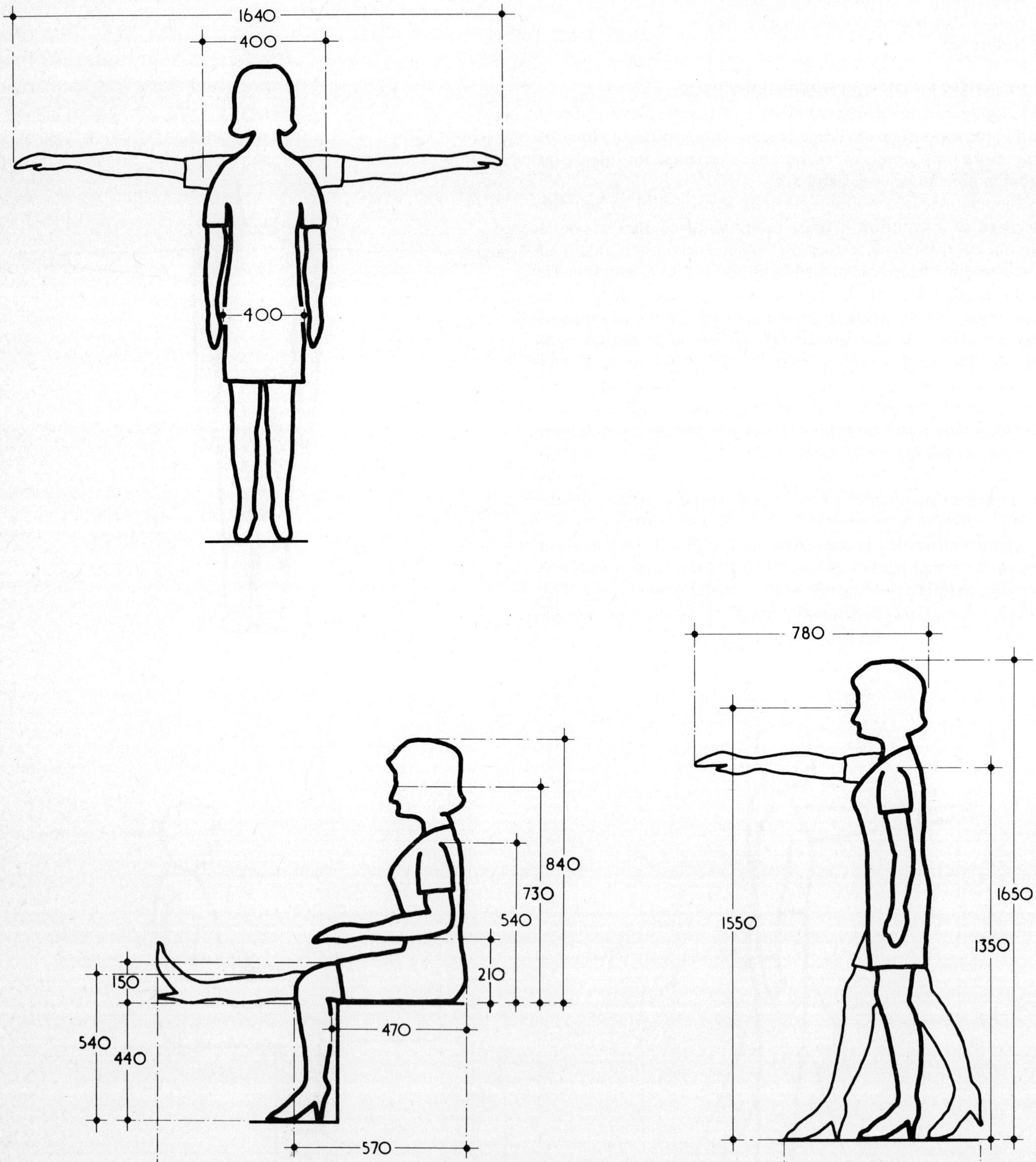

5.2 *Mean average (50th percentile) dimensions of adult British females*

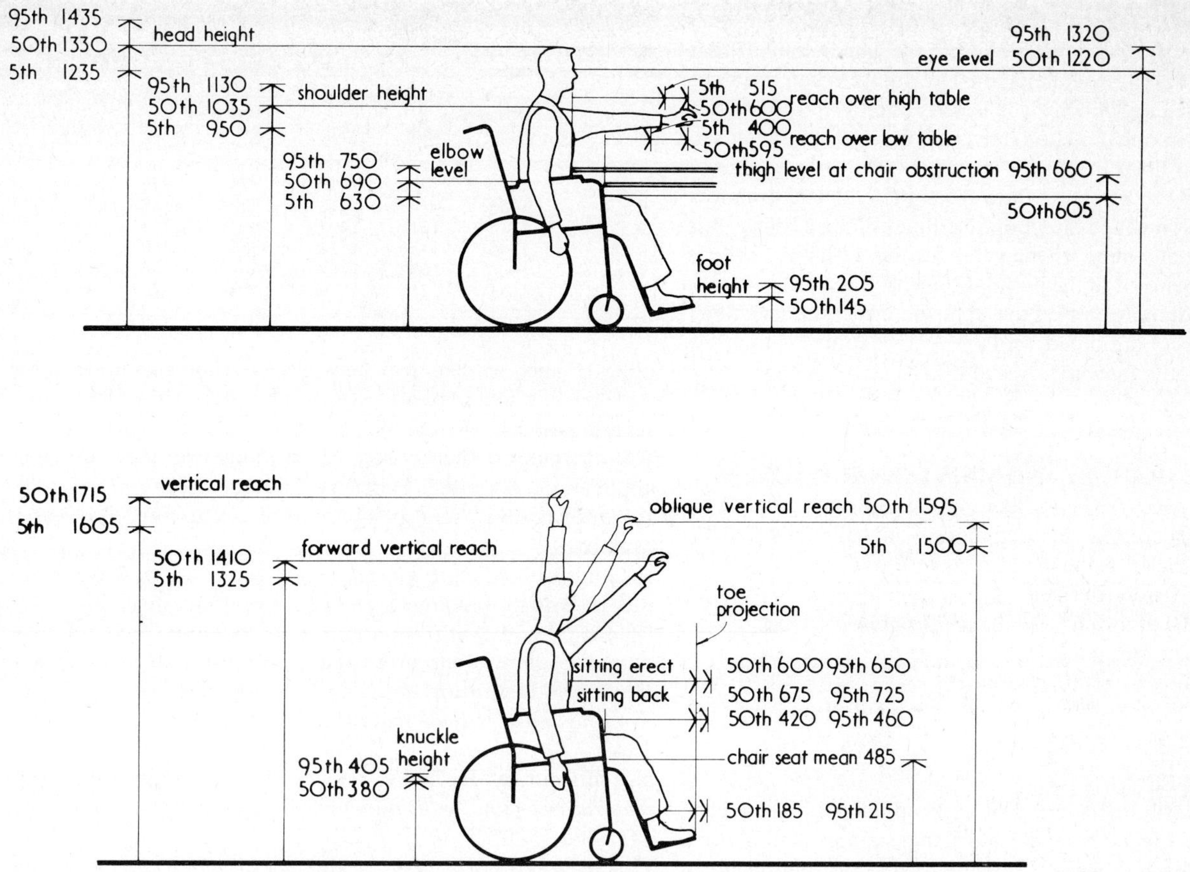

5.3 *Dimensions of different percentiles of adult male wheelchair users. These dimensions and those in* **5.4** *relate to people who use standard wheelchairs and who have no major impairment of upper limbs*

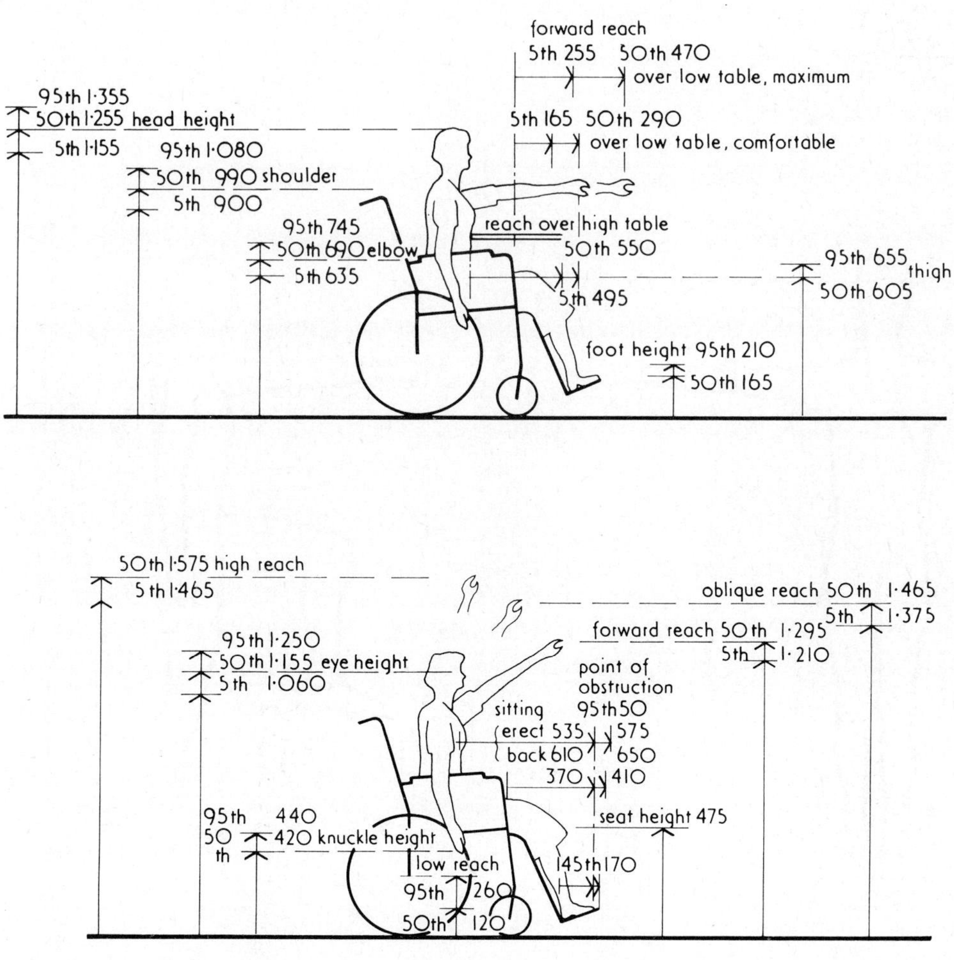

5.4 *Dimensions of adult female wheelchair users*

2 APPLICATION OF ANTHROPOMETRIC DATA

As an example, it can be observed from the data in table II that 95 per cent of men are taller than 1628 mm. When data are applied to design problems it is usually found that there is a limiting factor in one direction only, eg if the problem relates to obstructions at head height the measures of short people are not significant. In applying data from table II the architect should inquire in which direction the dimension is critical. It is not the case that whenever the value for the 95th percentile is observed 95 per cent of the population will be accommodated; if the critical dimension is in the opposite direction only 5 per cent will be accommodated, and the correct course is to apply the 5th percentile instead.

3 BODY CLEARANCES AND MAINTENANCE ACCESS

Typical allowances for body clearances are given in **5.6** to **5.11**. Space requirements for maintenance are given in **5.12** to **5.19**. Minimum entries for one man are as follows:
330 mm to 450 mm: difficult
450 mm to 610 mm: fair
610 mm to 920 mm: good
(See also section 6: Internal circulation.)

4 REFERENCES

References, including sources of anthropometric data reproduced here, are listed below. For original sources of data used in table II reference should be made to AJ information sheet 1185. Any apparent inconsistencies between dimensions in table II, such as distances between eyes and top of head for standing and seated people, are because data were drawn from several studies.

AJ information sheet 1185 Anthropometric data and their application, AJ, 1963, February 13
AJ information sheet 1194 Internal circulation, AJ, 1963, March 20
S. Goldsmith *Designing for the disabled,* London, Royal Institute of British Architects, 1977 (3rd revised edn) £20
BS 4467:1969. Anthropometric and ergonomic recommendations for dimensions in designing for the elderly, London, 1969, The Institution
S. Goldsmith, Wheelchair housing, AJ, 25-6-75, pp 1320–1348.
N. Diffrient, A. R. Tilley, J. C. Batdagjy, *Humanscale 1/2/3, pictorial selectors with guidance,* Cambridge, Massachusetts, MIT Press, 1974.
Henry Dreyfuss *The measure of man: human factors in design,* Whitney, 1967.

Table I Estimated heights, in millimetres, of children at various ages

Age	Boys			Girls		
	Percentiles			Percentiles		
	5th	50th	95th	5th	50th	95th
3	879	942	1005	876	930	984
6	1068	1143	1218	1059	1138	1217
9	1215	1311	1407	1204	1300	1396
12	1345	1458	1571	1355	1468	1581
15	1504	1633	1762	1507	1603	1699
18	1651	1755	1859	1534	1626	1718

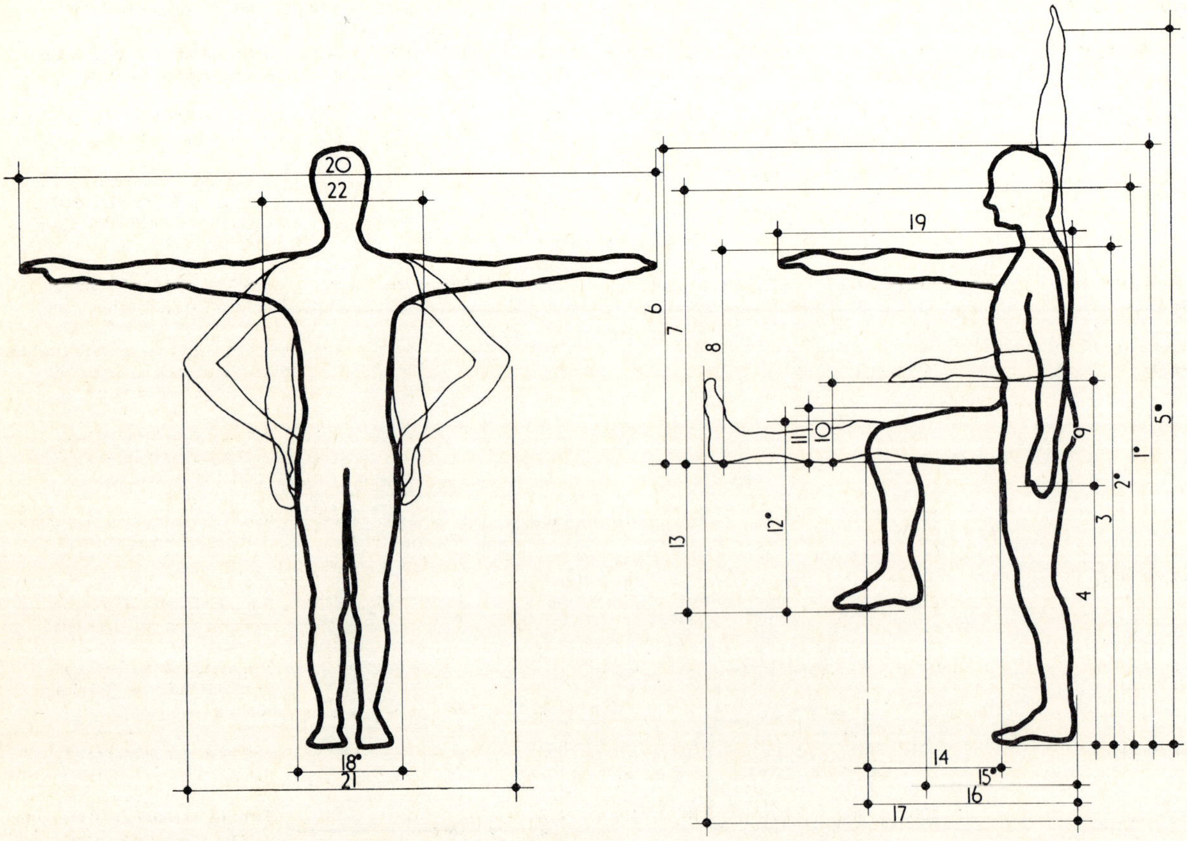

5.5 *Key dimensions listed in table II. To assist interpretation these figures are shown unclothed, though allowances have been made in table II for the wearing of clothes and shoes. Dimension references marked ● are most commonly used.*

Table II Estimated dimensions, in millimetres, of body and reach characteristics of the British population

Because in nearly all situations to which the architect applies anthropometric data users will be clothed, the data in this table includes allowances for clothing and shoes. The allowances for footwear are 28 mm for men, 40 mm for women and 31 mm for elderly women. The allowances for clothing, affecting most of the dimensions from item 6 on, range according to circumstance from 3 mm to 20 mm. In situations where clothes are not worn, eg bathrooms and shower rooms, appropriate deductions should be made.

Key dimension	**MEN aged 18 to 40**			**WOMEN aged 18 to 40**			**ELDERLY women aged 60 to 90**			**Examples of applications to design problems**
	Percentiles			**Percentiles**			**Percentiles**			
	5th	**50th**	**95th**	**5th**	**50th**	**95th**	**5th**	**50th**	**95th**	
STANDING										
1 Stature	**1628**	**1737**	**1846**	**1518**	**1647**	**1742**	**1454**	**1558**	**1662**	95th: Minimum floor to roof clearance; allow for headgear, say 100 mm, in appropriate situations.
2 Eye height	**1524**	**1633**	**1742**	**1427**	**1546**	**1643**	**1338**	**1441**	**1544**	50th: Height of visual devices, transoms, notices, etc.
3 Shoulder height	1328	1428	1528	1254	1353	1440	1195	1288	1375	5th: Height for maximum forward reach.
4 Hand (knuckle) height	703	770	837	703	741	787	653	732	800	95th: Maximum height of grasp points for lifting.
5 Reach upwards	**1972**	**2118**	**2264**	**1821**	**1970**	**2097**	**1710**	**1852**	**1994**	5th: Maximum height of controls: subtract 40 mm to allow for full grasp.
SITTING										
6 Height above seat level	841	900	959	770	839	882	739	798	857	95th: Minimum seat to roof clearance; allow for headgear (men 75 mm, women 100 mm) in appropriate situations.
7 Eye height above seat level	726	785	844	681	735	765	621	684	740	50th: Height of visual devices above seat level.
8 Shoulder height above seat level	537	587	637	494	544	580	470	529	579	5th: Height above seat level for maximum forward reach.
9 Forearm length	307	328	343	277	305	325	—	—	—	50th: Easy reach forward at table height.
10 Elbow above seat level	178	224	270	190	211	218	143	193	243	50th: Height above seat of armrests or desk tops.
11 Thigh clearance	124	149	174	121	146	171	93	131	169	95th: Space under tables.
12 Top of knees, height above floor	**506**	**552**	**598**	**502**	**535**	**571**	**460**	**510**	**545**	95th: Clearance under tables above floor or footrest.
13 Underside thigh, height above floor	402	435	468	401	436	459	366	404	442	50th: Height of seat above floor or footrest.
14 Front of abdomen to front of knees, distance	336	386	436	—	—	—	—	—	—	95th: Minimum forward clearance at thigh level from front of body or from obstruction, eg desk top.
15 Rear of buttocks to back of calf, distance	**436**	**478**	**520**	**430**	**465**	**498**	**424**	**470**	**516**	5th: Length of seat surface from backrest to front edge.
16 Rear of buttocks to front of knees, distance	568	614	660	526	572	608	520	579	638	95th: Minimum forward clearance from seat back at height for highest seating posture.
17 Extended leg length	998	1090	1182	890	964	1025	890	967	1025	5th (less than): Maximum distance of foot controls, foot-rest etc from seat back.
18 Seat width	**328**	**366**	**404**	**360**	**405**	**441**	**321**	**388**	**455**	95th: Width of seats, minimum distance between armrests.
SITTING AND STANDING										
19 Forward reach	773	848	923	714	782	833	665	736	807	5th: Maximum comfortable forward reach at shoulder level.
20 Sideways reach	1634	1768	1902	1516	1643	1760	—	—	—	5th: Limits of lateral finger tip reach; subtract 130 mm to allow for full grasp.
21 Width over elbows akimbo	881	955	1041	820	881	952	—	—	—	95th: Lateral clearance in work space.
22 Shoulder width	420	462	504	376	405	428	381	431	481	95th: Minimum lateral clearance in work space above waist.

NOTE: Dashes indicate that no data were given in the sources used, or that data were unreliable.
Figures printed in bold type are the most commonly used dimensions (see also **5.4**)

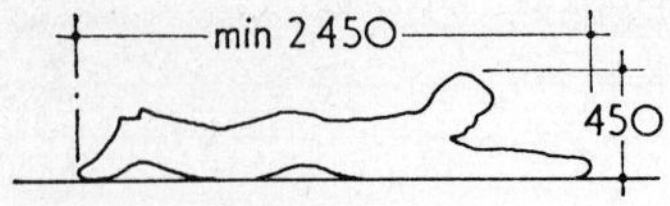

5.6 *Body clearance: prone*

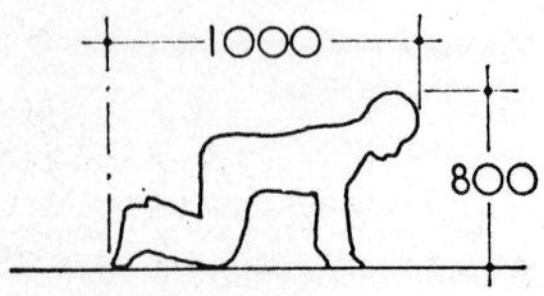

5.7 *Body clearance: crawl*

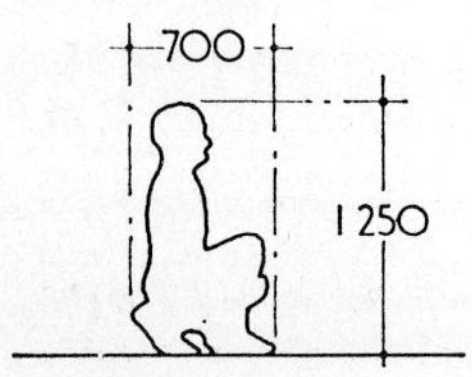

5.8 *Body clearance: squat*

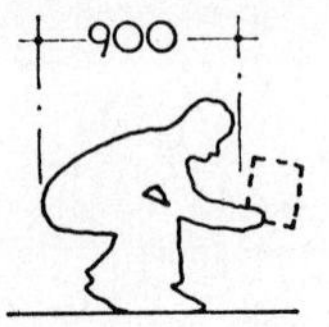

5.9 *Body clearance: stoop*

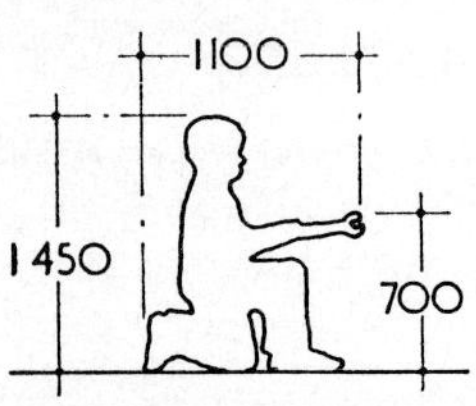

5.10 *Body clearance: kneel*

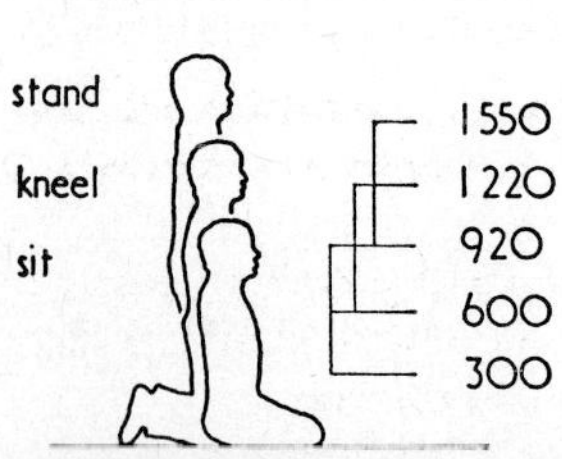

5.11 *Body clearance: maintenance reach levels*

5.12 *Service access; crawlway*

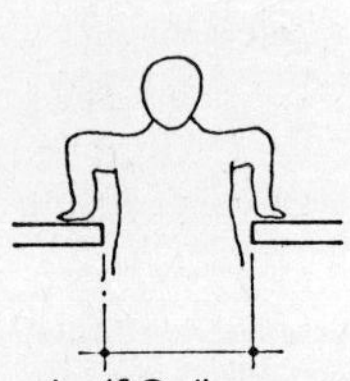

5.13 *Service access: hatch*

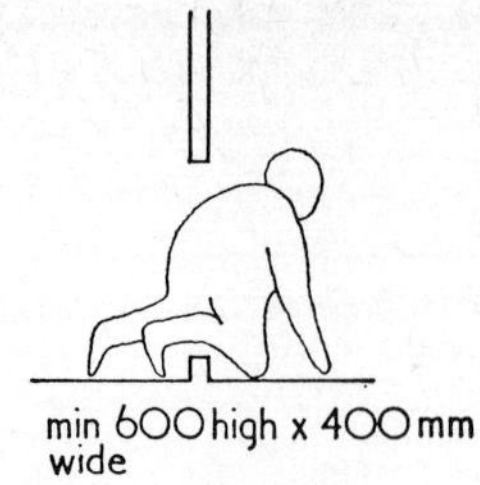

5.14 *Service access: access panel*

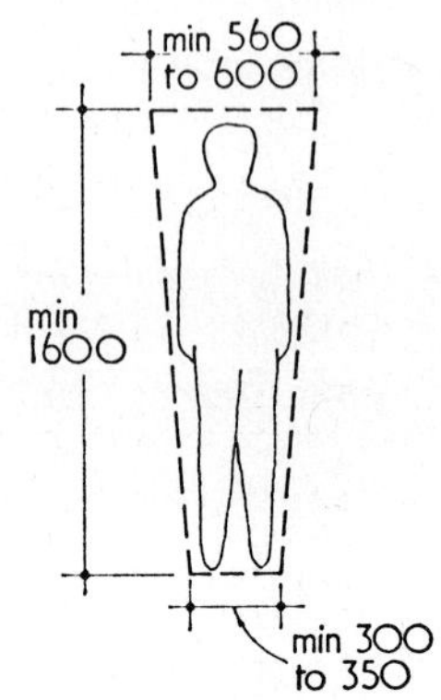

5.15 *Service access: catwalk*

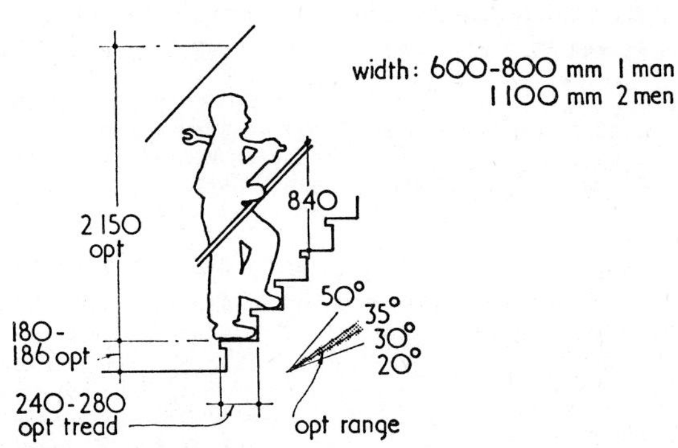

5.16 *Service access: stairs*

min entries for one man (mm):
330-450 difficult
450-610 fair
610-920 good

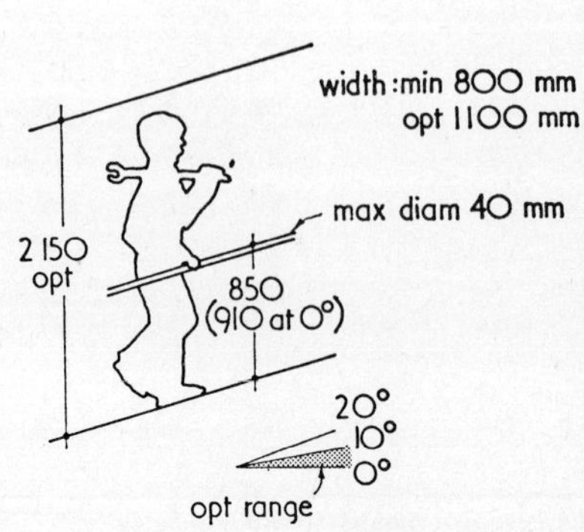

5.17 *Service access: ramps*

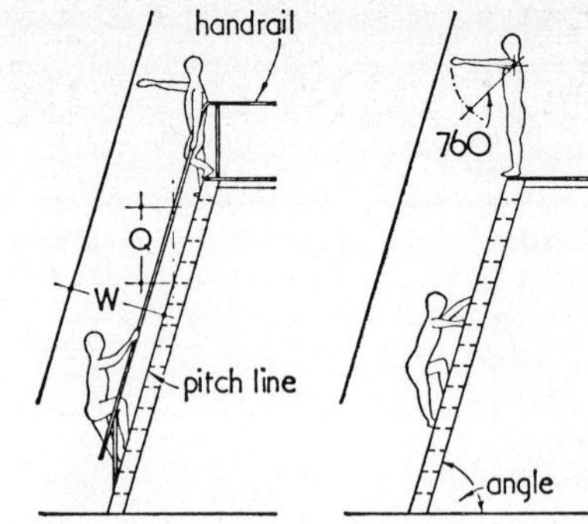

recommended for angles 50° to 75°
handrails are required on both sides if risers are not left open or if there are no side walls
widths: 500mm to 600mm with handrails
600mm min between side walls

angle	W(mm)	Q(mm)
50°-55°	1 620-1 570	880
57°-60°	1500-1450	900
63°-66°	1 370-1 320	910
69°-72°	1 270-1 200	920
74°-77°	1 150-1 050	950

recommended riser 180mm to 250mm
tread 75 mm to 150 mm
45mm diam max for handrail

5.18 *Service access: step ladders*

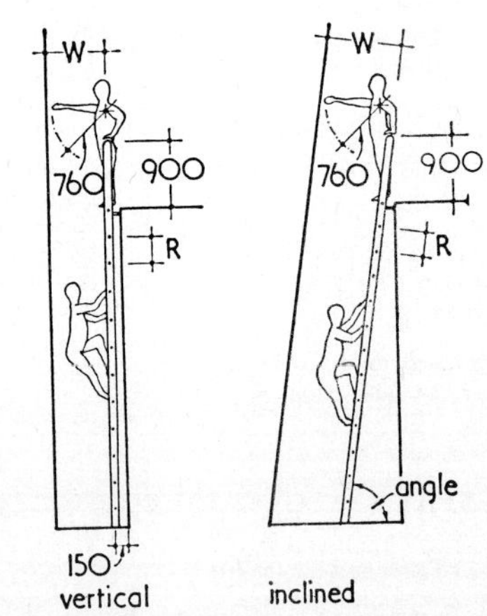

generally suitable for vertical movements from 75° to 90°
ladder frame should extend 900mm above platform
widths: 380mm min, 450mm desirable
600mm min between side walls
150mm toe space

angle	R(mm)	W(mm)
75·0°	330	1 150
78·0°	335	1 050
80·5°	340	1 000
83·0°	350	950
85·0°	360	900
87·5°	370	850
90·0°	380max 300min	800

provide back guard over 6 000 mm high

5.19 *Service access: rung ladders*

6 Internal circulation

CI/SfB 91
CI/SfB (1976 revised) 91

Contents

1 INTRODUCTION

1.01 Fire safety

Many aspects of internal circulation are dependent on regulations and controls arising from fire safety considerations. These are all covered in section 41.

1.02 Lifts and escalators

Criteria relevant to lifts and escalators is given in section 46.

2 SPACE ALLOWANCES IN BUILDINGS

As a guide to assessing circulation, the areas listed in table I may be used. These areas are overall space requirements for both the activity and its associated circulation.

Table I Minimum areas per person in various types of buildings

Occupancy	Area per person (m^2)
Assembly halls (closely seated)	0·46 m^2 (based on movable seats, usually armless, 450 mm centre to centre; with fixed seating at 500 mm centre to centre will increase to about 0·6 m^2)
Dance halls	0·55 m^2 to 0·9 m^2
Restaurants (dining-area)	0·9 m^2 to 1·1 m^2
Retail shops and showrooms	4·6 m^2 to 7·0 m^2 (including upper floors of department stores except special sales areas)
Department stores, bazaars or bargain sales areas	0·9 m^2 (including counters, etc) 0·46 m^2 (gangway areas only)
Offices	9·3 m^2 (excluding stairs and lavatories)
Factories	7 m^2

3 SPACE REQUIRED FOR WAITING AREAS

Table II gives areas to be allowed for specific circulation areas.

Table II Area per person to be allowed in various circulation areas

Occupancy	Area per person (m^2)
Overall allowance for public areas in public-handling buildings	2·3 to 2·8
Waiting areas, allowing 50 per cent seating, 50 per cent standing without baggage, allowing cross-flows (eg airport lounge)	1·1 to 1·4
Waiting areas, 25 per cent seating, 75 per cent standing, without serious cross-flows (eg waiting rooms, single access)	0·65 to 0·9
Waiting areas, 100 per cent standing, no cross-flows (eg lift lobby)	0·5 to 0·65
Circulating people in corridors, reduced to halt by obstruction	0·2
Standing people under very crowded conditions—acceptable temporary densities	Lift car capacities: 0·2 m^2 (four-person car); 0·3 m^2 (thirty-three-person car)

4 FLOW CAPACITIES OF CORRIDORS AND STAIRCASES

Average space allowances per person in circulation areas may be based on the following:

General design purposes	0·8 m^2
People moving at over 1·3 m/s (good walking pace)	3·7 m^2
People moving at 0·4 to 0·9 m/s (shuffle)	0·27 to 0·37 m^2
People at standstill caused by obstruction	0·2 m^2

Dimensions of stairs for ambulant disabled and elderly people are shown in **6.1**. A formula for suitable relationship of riser to going for most normal staircases is 2R + G = 600 to 630 mm where R = riser; G = going.
R should be max 190 mm
G should be min 250 mm

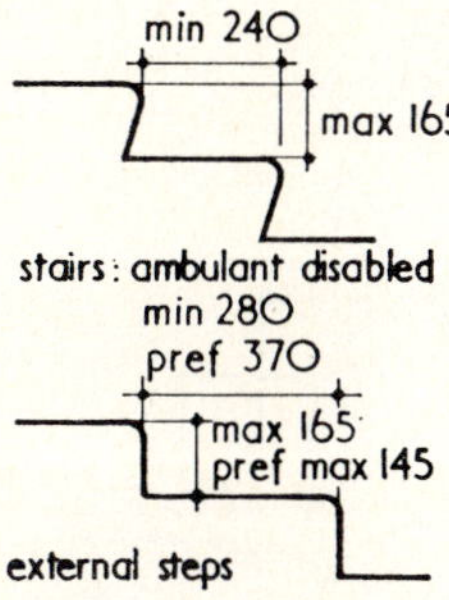

6.1 *Stairs for ambulant disabled and elderly people*

5 CIRCULATION

Some common sizes affecting circulation are shown in **6.2** and certain corridor widths in **6.3**. German examples are shown in **6.4** for comparison.
Common obstructions which may affect circulation flow are shown in **6.5** (See also sections 9 to 37 of this handbook for circulation allowances in particular building types).
Circulation areas for disabled people in wheelchairs and on crutches are given in **6.6** to **6.15** (from *Designing for the disabled).*

6 REFERENCES

AJ information sheet 1194 Internal circulation. AJ, 1963, March 20
AJ information sheet 1195 Sizes of lifts and lift lobbies. AJ 1963, March 20
AJ information sheet 1053 Staircases: required sizes 1. AJ, 1962, January 10
S. Goldsmith, *Designing for the disabled,* London, Royal Institute of British Architects, 1977 (3rd revised edn) £20
E. Neufert *Bauentwurfslehre*, Berlin, Ullstein, 1967
Manuale dell' architetto. Rome, Consiglio Nazionale delle Ricerche, 1962 (3rd edn)

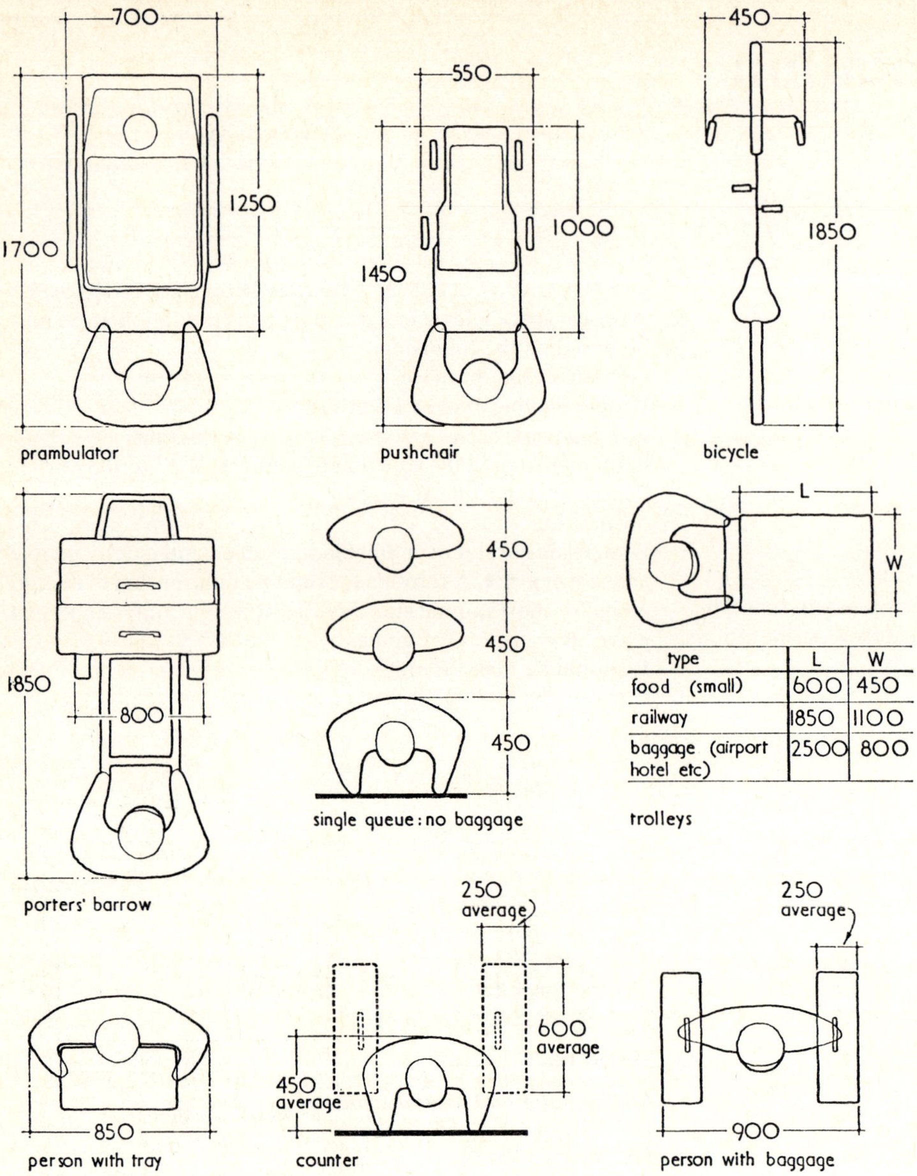

type	L	W
food (small)	600	450
railway	1850	1100
baggage (airport hotel etc)	2500	800

6.2 *Circulation: common sizes*

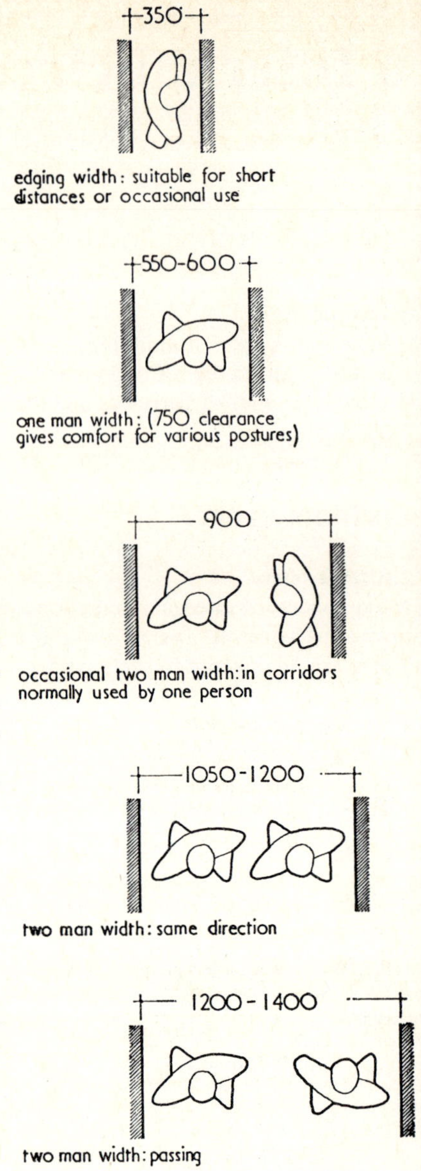

6.3 *Circulation: corridor widths*

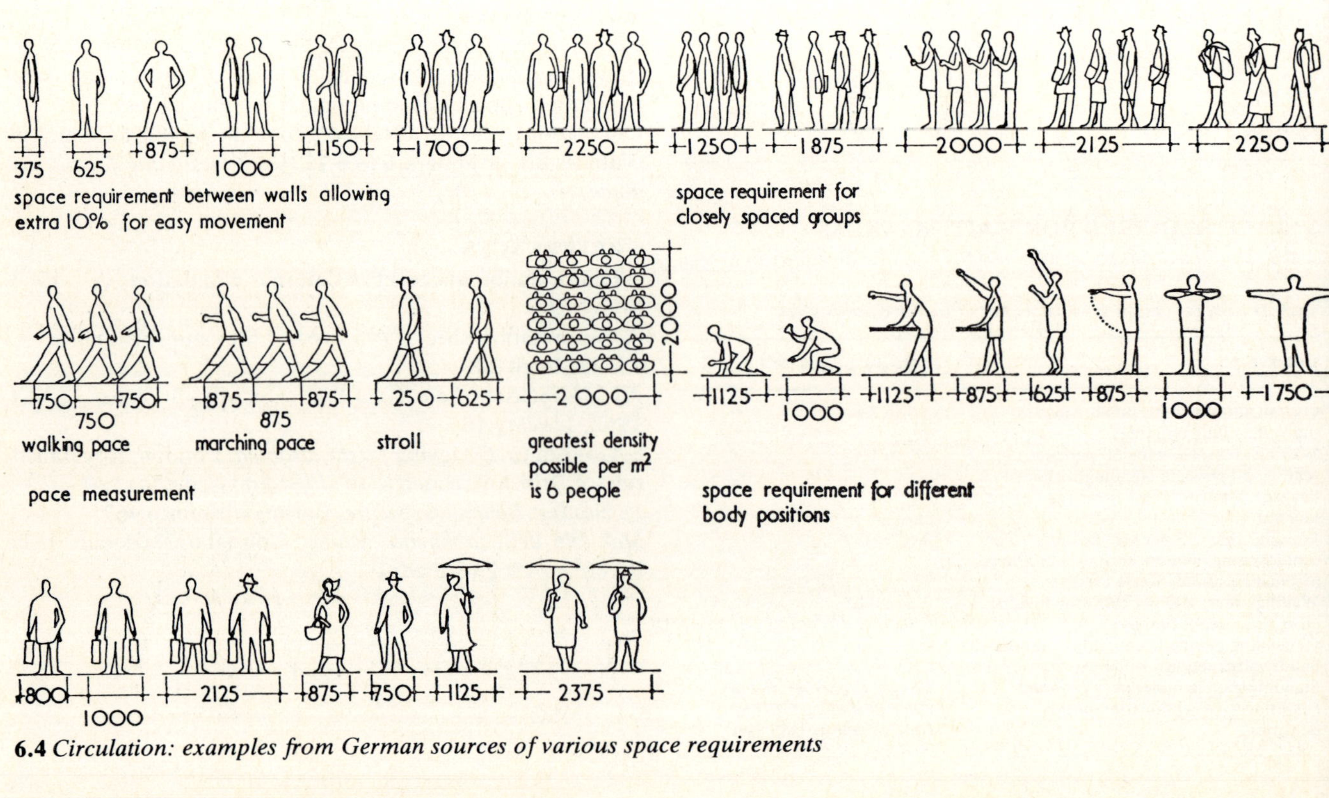

6.4 *Circulation: examples from German sources of various space requirements*

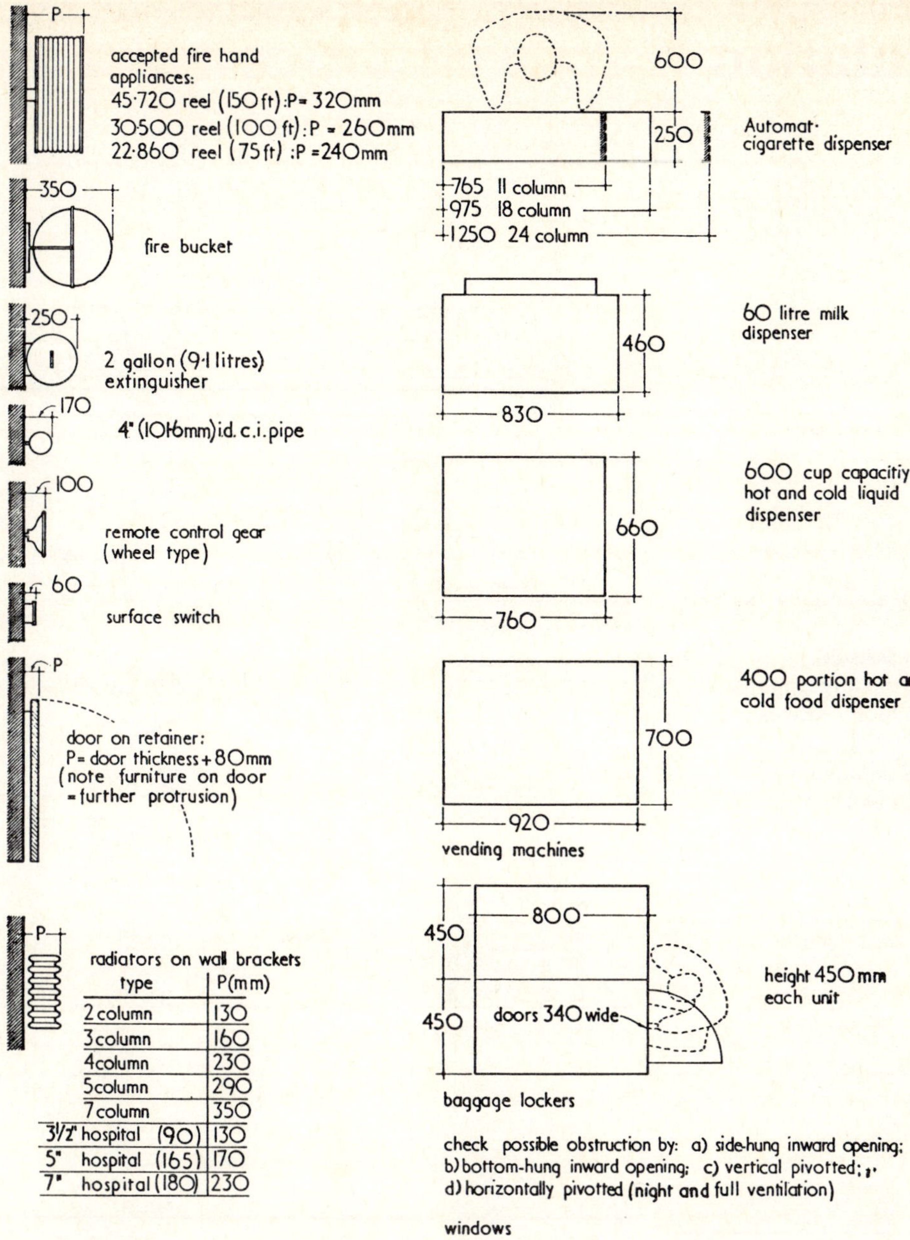

type	P(mm)
2 column	130
3 column	160
4 column	230
5 column	290
7 column	350
3½" hospital (90)	130
5" hospital (165)	170
7" hospital (180)	230

6.5 *Circulation: common obstructions*

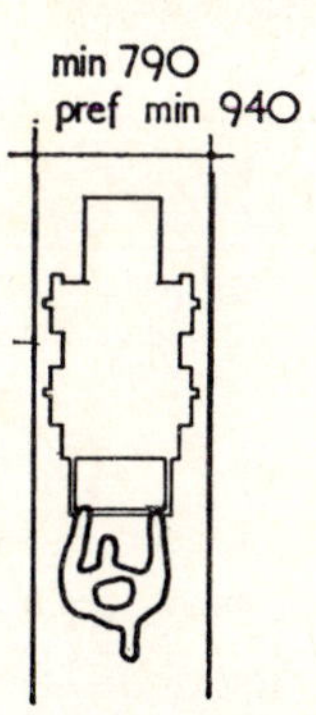

6.6 *Large wheelchairs: forward movement*

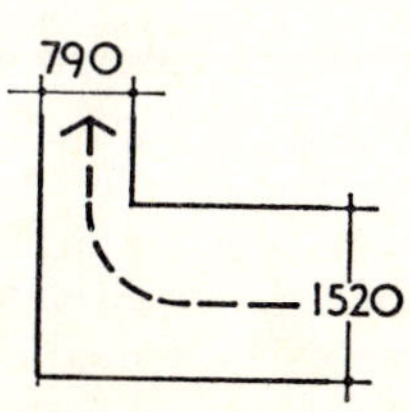

6.7 *Large wheelchairs: forward turn through 90deg*

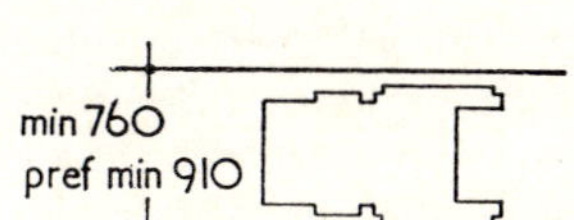

6.8 *Small wheelchairs: forward movement*

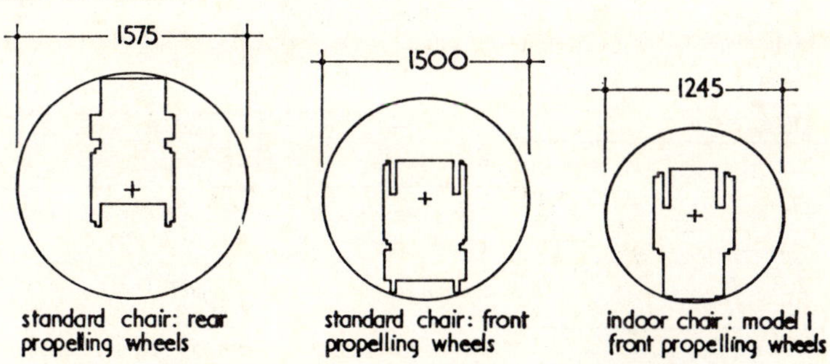

6.9 *Small wheelchairs: comparative turning space requirements*

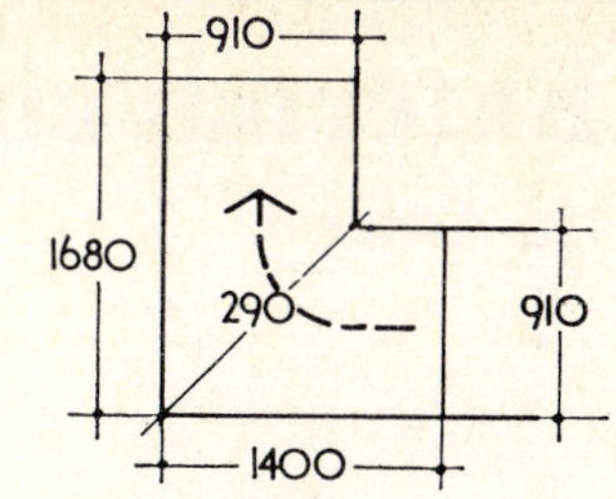

6.10 *Small wheelchairs: forward turn through 90deg preferred minimum space*

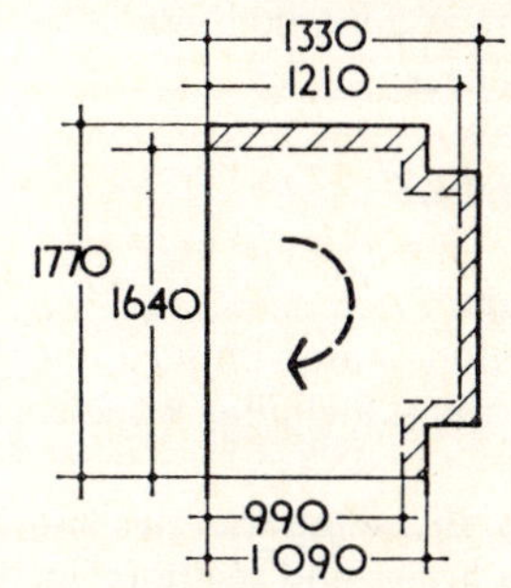

6.11 *Small wheelchairs: turn through 180deg*

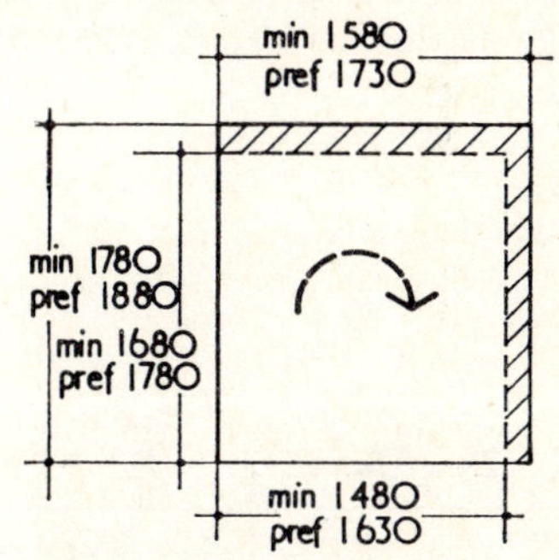

6.12 *Small wheelchairs: turn through 180deg for hemiplegics and so on*

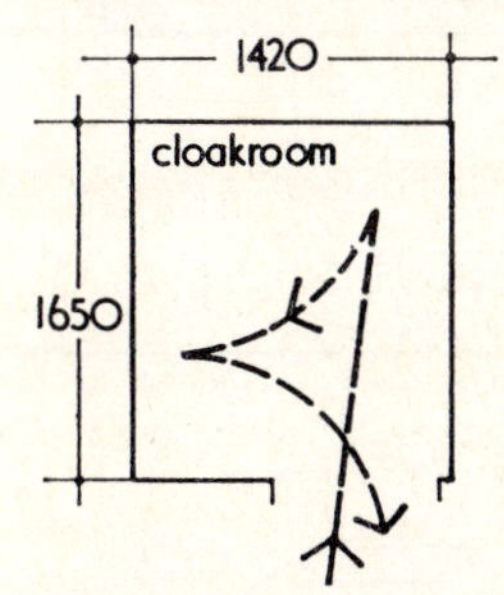

6.13 *Small wheelchairs: three-point turn in cloakroom*

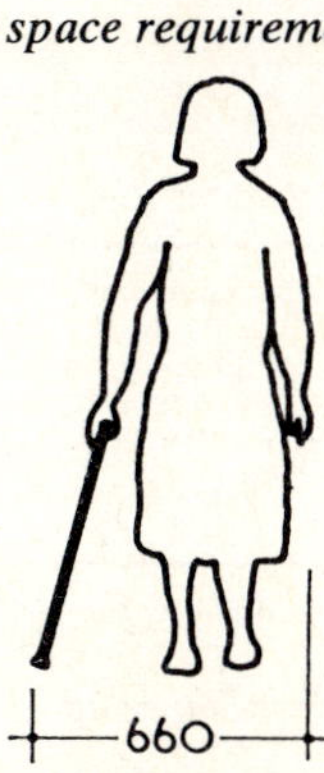

6.14 *Stick user*

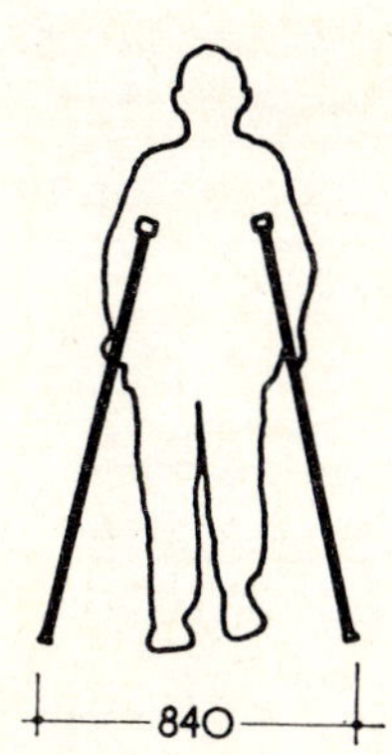

6.15 *Crutch user*

7 External circulation

David Adler

CI/SfB 12
CI/SfB (1976 revised) 128

Contents

1 VEHICLES

1.01 Scope

This section deals with facilities for:

1 Pedestrians, including wheelchairs, prams and juvenile vehicles.
2 Cycles, including tricycles and motor assisted.
3 Motor-cycles and scooters, including the larger mopeds.
4 Automobiles: cars and small vans up to 2½ tonnes unladen weight.
5 Commercial vehicles up to 10 tonnes unladen weight.
6 Public Service Vehicles (PSVs): buses and coaches.
7 'Juggernauts', large commercial vehicles at present maximum 32·5t, but on the Continent up to 40t. This class includes those with draw-bar trailers (see fig **7.1**).

Standards for class 6 vehicles (PSVs) are largely covered in section 9: Bus and coach facilities.

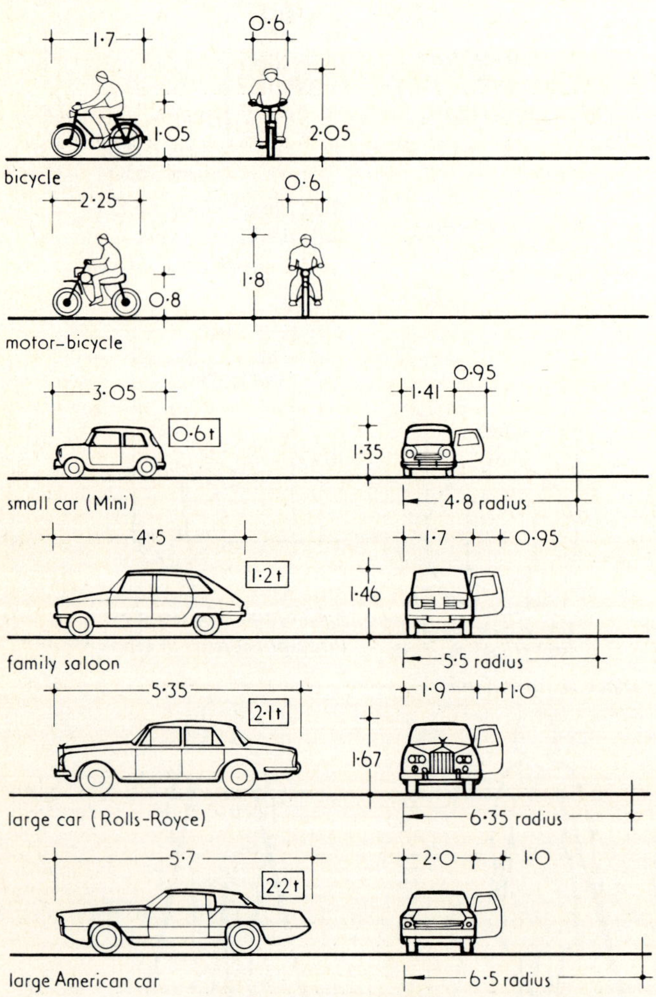

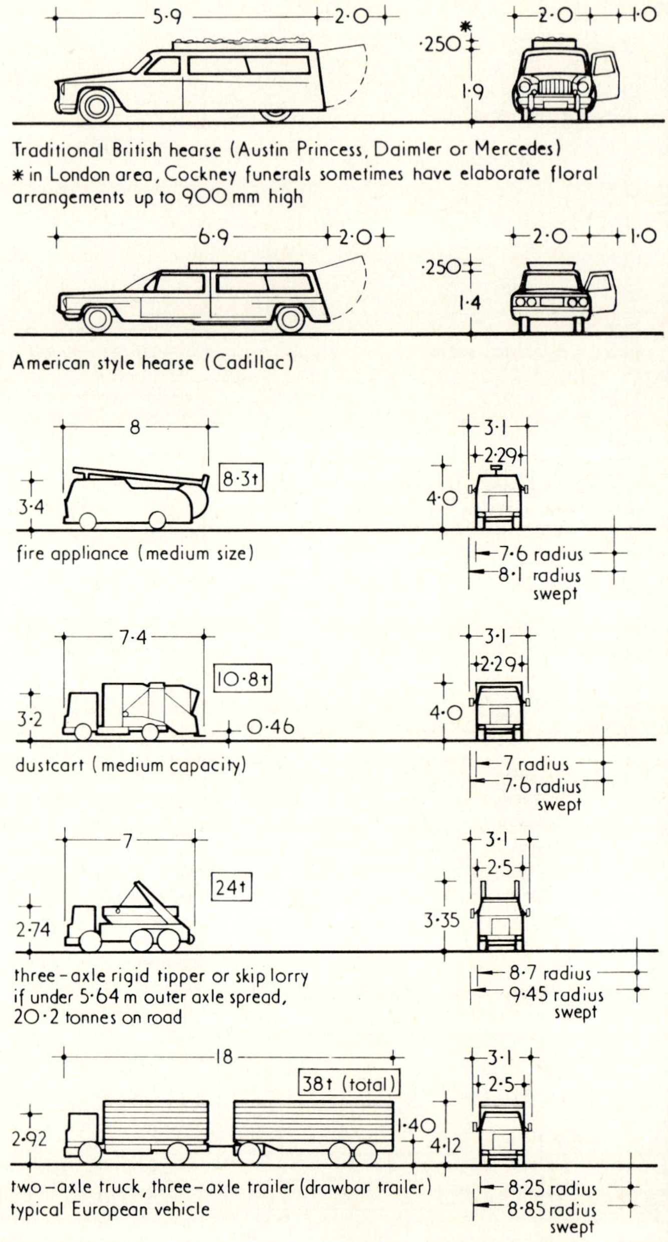

7.1 *Dimensions, weights and turning circles of typical road vehicles*

1.02 Dimensions

The dimensions of some examples of each class are given in **7.1**. In any specific case, the manufacturer's data should be consulted.

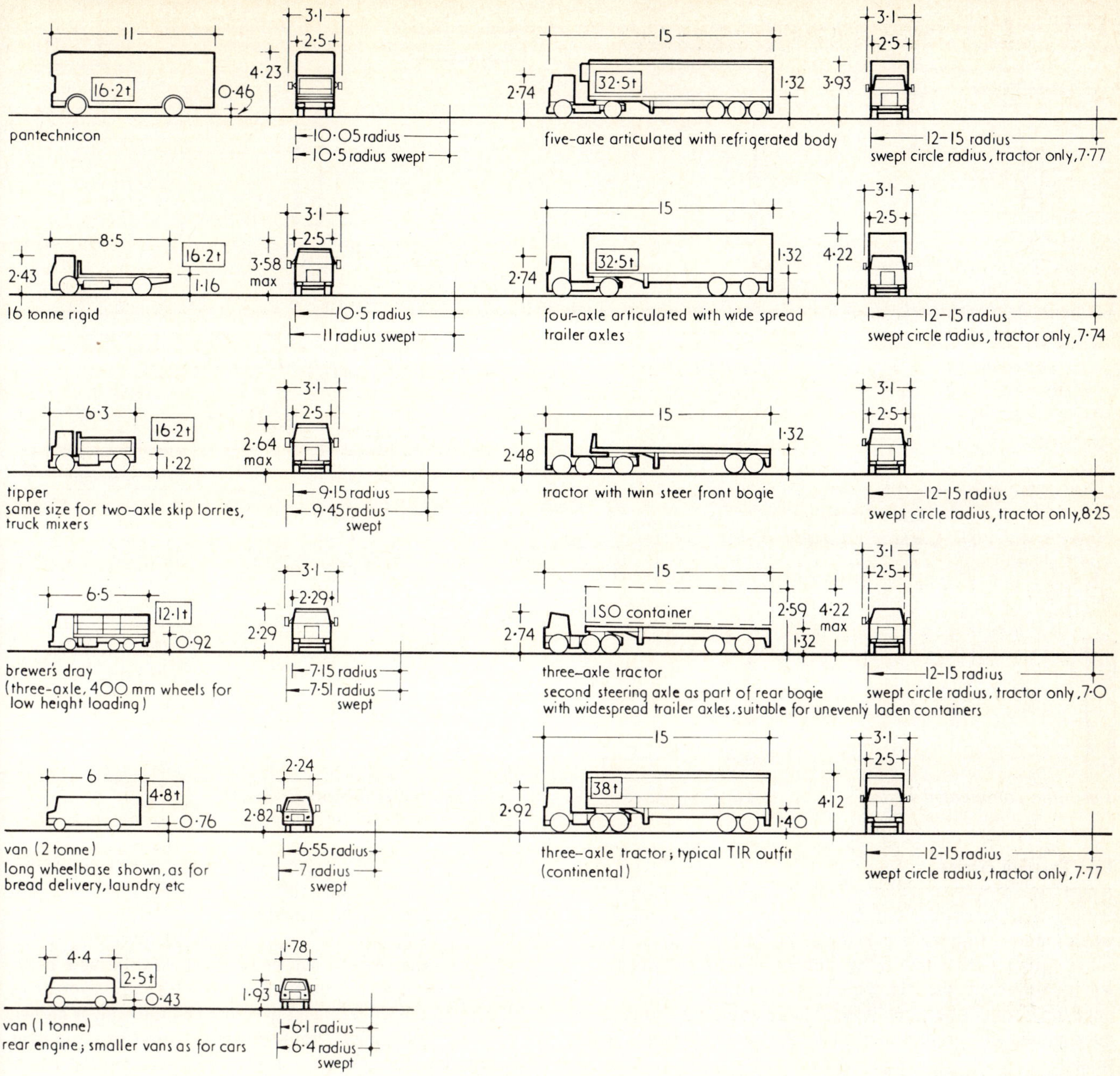

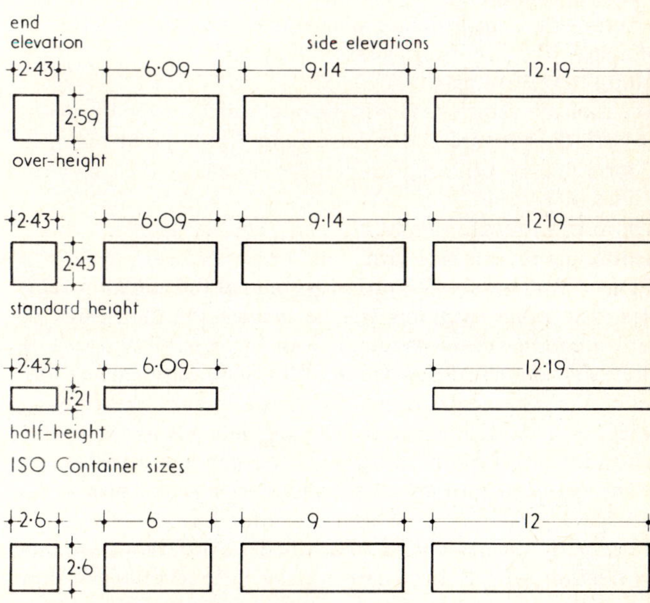

7.2 *Dimensions of standard containers*

Unit construction

In the field of the larger commercial vehicles, unit construction is now almost universally employed. In this system a given chassis can be fitted with a variety of body shells for specific purposes and loads, mainly of standard dimensions. The body can be changed at will, permitting one body to be loaded while the chassis is on the road with another body delivering goods elsewhere.

A particular example is the standard container, which is used on lorries, ships, railways or even as a storage unit in the open **7.2**. As this system was first developed in the USA, the standard dimensions are imperial, but the German railways developed a parallel version.

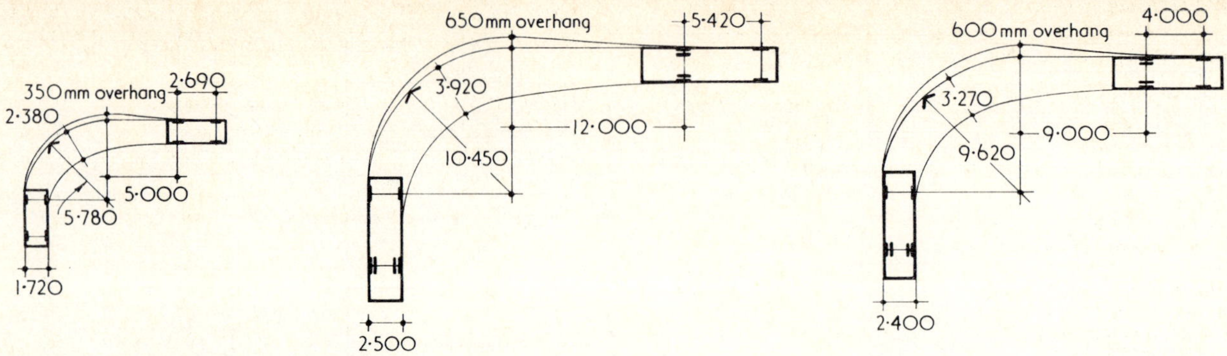

7.3 *Geometric characteristics of typical vehicles turning through 90°:*

a *private car* **b** *pantechnicon* **c** *refuse collection vehicle*

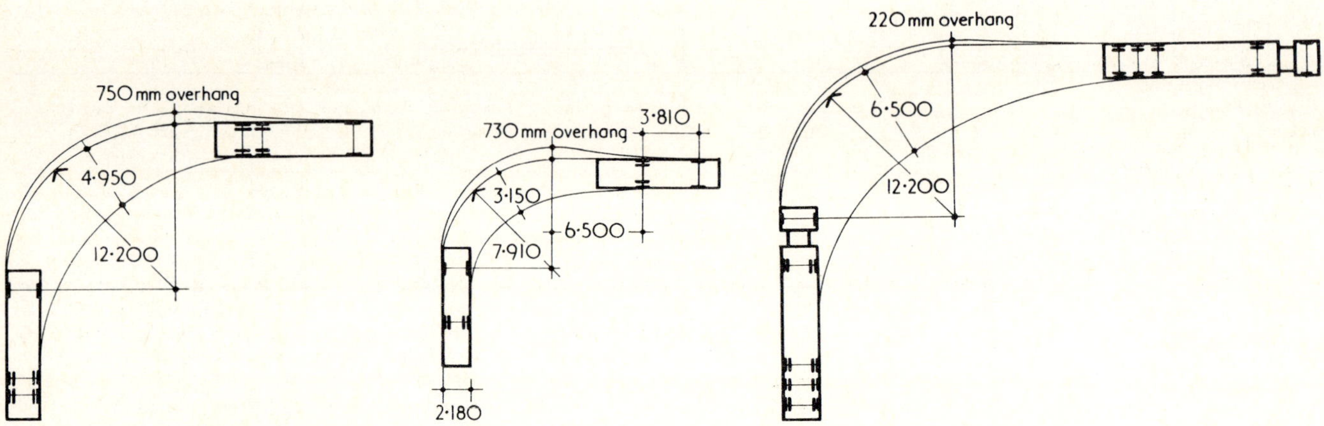

d *medium commercial vehicle* **e** *fire appliance* **f** *largest commercial vehicle*

1.03 Turning circles

Apart from the physical dimensions, it is necessary to know the critical characteristics of the vehicle in motion, particularly when manœuvring while parking or preparing to load. These characteristics are complicated, and usually the manufacturer will quote solely the diameter of the turning circle, either between kerbs or between walls. Manœuvring diagrams have been published for various vehicles for the following operations:

- turning through 90°
- causing the vehicle to face in the opposite direction by means of a 360° turn in forward gear
- ditto in reverse gear
- causing the vehicle to face in the opposite direction by means of both the forward and reverse gears (three-point turn), in T-form
- ditto in Y-form
- ditto in a forward side turn
- ditto in a reverse side turn.

Figure **7.3** shows the 90° turns for some of the common vehicles. The other diagrams will be needed for the design of turn-rounds in culs-de-sac etc. The use of the published turning circle sizes is not sufficient for the following additional factors:

- The distance required for the driver to turn the steering wheel from straight ahead to full lock, depends on the speed, which for the purposes of fig **7.3** is between 8 and 16 km/h.
- The radius of turn differs between a right-hand turn and a left-hand.
- The path traversed by the rear wheels is different from that by the front wheels. In a commercial vehicle travelling at slow speed, the rear wheel follows a smaller arc to the front wheels, the amount depending largely on the distance between the axles. The divergence between the arcs of the wheels on the same side of the vehicle is termed the 'cut in', and value of this determines the total track width of the turning vehicle, always greater than when on the straight.
- While few vehicles have a measurable side overhang of the body beyond the wheel track, many have considerable overhang at front and rear. This is important at the front: the extra width beyond the wheel tracks described by the body is known as 'cut-out'. Allowance should be made for front and rear overhang when designing turn-rounds etc by having no vertical obstructions within 1·2 m of the carriage-way edge.

2 ROADS IN GENERAL

2.01 Hierarchy

The broad hierarchy of roads is:

- motorways and trunk roads
- distributors (primary, district and local)
- access roads.

2.02 Definitions

The following definitions are important:

carriageway: the area of road surface dedicated to vehicles

carriageway width: the distance between the kerbs forming the carriageway edges

dual carriageway: a road with a central reservation, each separate carriageway carrying traffic in reverse direction

lane: a width of carriageway capable of carrying a single line of vehicles, usually delineated with white-painted dashed lines on the carriageway surface

lane width: since the maximum vehicle width permitted is 2·5 m, and the minimum clearance between parallel vehicles is 0·5 m, the minimum lane width is 3 m. However, vehicles travelling at speed require greater clearance and large vehicles

Table I Acceptable gradient

Motorways and primary distributor through carriageways		4%
in hilly districts		5%
consider climbing lanes for heavy vehicles if length of gradient exceeds:		
	500 m at	3%
	350 m at	4%
	250 m at	5%
	200 m at	6%
slip roads		5%
where circumstances make unavoidable		8%
where a preponderance of commercial vehicles		4%
Other distributors and access roads		7%
(although existing roads are up to 12%)		
at junctions and their approaches		5%
Ramps to lorry loading bays		
(along centre-line of curved ramps)		10%
Ramps to car parking garages		10%
if a transition at 1/2 of ramp gradient is provided for 2·4 m at each end		20%
a ramp rising to road level from basement car park should not exceed 5% for 4 m from back of pavement (to ensure clear vision at entry to public road).		
Cycle tracks	up	5%
	down	6·5%
Ramps for disabled people in wheelchairs (maximum length 9 m)		8·5%
Pedestrian ramps		10% (optimum 8·5%)
a stepped ramp with 115 mm step at 1·725 m intervals will have an effective gradient of		16·5% (optimum 14%)
the effective capacity of a ramp is		
20 persons/minute/300 mm width up to		5%
14 persons/minute/300 mm width over		5%

need greater widths on curves, so faster roads have wider lanes

cycle track: a special 'road' dedicated for the sole use of vehicles in class 2 above

footway: an area of road devoted solely for the uses indicated in class 1 above, and running alongside the vehicular carriageway. In Britain the footway is also called the 'pavement', in the USA the 'sidewalk'

footpath: a class 1 facility not forming part of a road.

Table III Reduction allowance in flows for heavy vehicle content (based on Technical Memorandum H9/76)

	Total reduction in design flow level (veh/h)		
Heavy vehicle content (per cent)	**Motorway and dual carriageway all purpose road**	**10 m wide and above carriageway with two way traffic**	**Under 10 m wide carriageway with two way traffic**
	per lane	*per carriageway*	*per carriageway*
0–15	0	0	0
15–20	100	150	100
20–25	150	225	150

Table II Design flows of urban roads (based on Technical Memorandum H9/76 published by DOE)

Description		**Urban motorway**	**All purpose road, no frontage access, no standing vehicles, negligible cross traffic**	**All purpose road, frontage development, side roads, pedestrian crossings, bus stops, waiting restrictions throughout day, loading restrictions at peak hours**
Effective width of carriageway in metres (excluding refuges or central reservation)				
Carriageway with one lane in each direction:	*Peak hourly flow, veh/h, both directions of flow**			
6·1				1100
6·75				1400
7·3			2000	1700
9				2200
10			3000	2500
Undivided carriageways: 2 lanes in each direction:	*Peak hourly flow, veh/h one direction of flow*			
12·3			2550	1700
13·5			2800	1900
14·6			3050	2100
3 lanes in each direction: 18				2700
Dual carriageways: 2 lanes in each direction:	*Peak hourly flow, veh/h one direction of flow*			
6·75 × 2			2950†	
7·3 × 2		3600	3200†	
3 lanes in each direction: 11 × 2		5700	4800†	
One-way roads:	*Peak hourly flow, veh/h*			
6·1				1800
6·75			2950	2000
7·3			3200	2200
9				2850
10				3250
11			4800	3550

* 60/40 directional split can be assumed
† Includes division by line of refuges as well as central reservation

2.03 Maximum recommended gradients

The maximum tolerable gradients depend to an extent on the geography of the area. A climb acceptable in South Wales will be very steep in East Anglia. Generally, commercial vehicles cannot negotiate a gradient of 12 per cent or more in freezing conditions unless some method of de-icing (such as salting, gritting or road surface heating by buried cables) is used. The figures in table I should be used purely as a guide.

2.04 Capacities of roads

Table II gives the peak capacities of urban roads in vehicles per hour, assuming a heavy vehicle content not more than 15 per cent. For greater heavy vehicle contents the reductions in table III apply. From these tables the required road specification can be found from the predicted traffic volume. If this is unknown, table IV can be used instead.

2.05 Access roads

In general, architects are normally only concerned with access roads which are the essential link between the distributor road network and the facilities for parking and loading/unloading associated with the various types of building. They fall into one of the following types:

- *Residential:* serving housing estates. These roads are designed to carry cars and the types of small commercial vehicles used by tradesmen. They must also cater for the weekly or twice-weekly refuse collection vehicle, which must be able to reach within 25 m (or in some cases 15 m) of each house, and to turn round reasonably easily at the ends of culs-de-sac. Pantechnicons are expected at rare intervals, so that provided access and return are possible, they do not have to be particularly easy. There must be rapid acccss for fire-fighting appliances, although there is normally no need to provide substantial turn-round areas for the return journey.
- *Commercial:* serving shopping or administrative centres. These roads must be designed to carry the larger vehicles expected to use them, and the traffic flow predictions are essential in ensuring they are adequate for the purpose.
- *Industrial:* serving factory estates, warehouses etc. Since the effectiveness of these enterprises may well depend on rapid turn-round of vehicles, these roads must be generously designed with the largest commercial vehicles in mind.

Table IV Recommended carriageway widths

Road type	Recommended carriageway width (m) between kerbs or edge lines	
Primary distributor	One-way, four lanes	14·6
	Overall width for divided carriageway, two lanes each way with central refuges	14·6
	Two-way, four lanes total, no refuges	13·5
	One-way, three lanes	11
	Two-way, three lanes (recommended only for tidal flow)	9
	One-way, two lanes	7·3
District distributor	One-way, two lanes	7·3
	One-way, two lanes if the proportion of heavy commercial traffic is fairly low	6·75
	Two-way, two lanes	7·3
Local distributor & access road in industrial district	Two-way, two lanes	7·3
Local distributor & access road in commercial district	Two-way, two lanes	6·75
	Minimum two-way, two lane back service road used occasionally by heavy vehicles	5
Local distributor in residential district	Two-way, two lanes used by heavy vehicles minimum	6
Access road in residential district	see text and fig **7.5**	
	Where all vehicles are required to be able to pass each other	5·5
	Where a wide car can pass a pantechnicon	4·8
	Where two wide cars can pass each other, but a pantechnicon can only pass a cyclist	4·1
	Where a single track only is provided, as for a one-way system, or where passing places are used	
	for all vehicles	3
	for cars only (drives)	2·75
Rural roads	One-way, four lanes	14·6
	One-way, three lanes	11
	Two-way, three lanes	10
	One-way, two lanes	7·3
	Two-way, two lanes	7·3
	Motorway slip road	6
	Minimum for two-way, two lanes	5·5
	Minimum at junctions	4·5
	Single-track between passing places	3·5
	Overall at passing place	6

3 ROADS IN RESIDENTIAL AREAS

3.01 Environmental areas

This section focuses mainly on roads in residential areas. However, the principles are the same in industrial and business areas, only the details will differ.

An 'environmental area' is surrounded by distributor roads from which access to the properties within is gained solely via the access roads within it. The access road network is designed with the following in mind:

- road access to within 25 m (or 15 m in certain cases) of each house
- road access to all private garages, whether within curtilages or in garage courts; and to all parking areas
- through traffic from one distributor road to another, or to another part of the same road (avoiding a traffic blockage) is either impossible, or severely discouraged
- necessary tradesmen, eg 'milk-rounds', calling at all or most properties in sequence, are not unreasonably diverted
- in general, traffic is not allowed to proceed too fast, but visibility at all times is at least the stopping distance for the possible (not the legal) speed limit.

Figure **7.4** illustrates typical access road layouts.

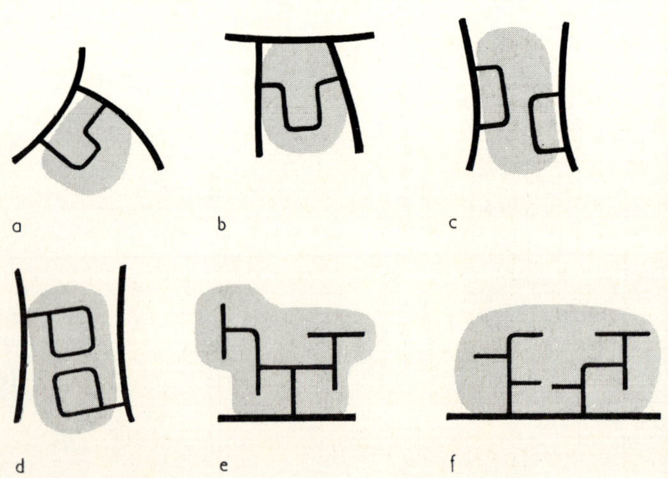

7.4 a *and* **b** *are through roads, so tortuous as to discourage the through traffic.*
7.4 c *and* **d** *are non-through systems, but avoid the necessity for hammer-head turn-rounds.*
7.4 e *is a cul-de-sac system, but will have a substantial traffic flow at the entrance.*
7.4 f *is to be preferred on this count, although both systems pose problems for the 'milk-round'.*

3.02 Types of access road

Access roads in residential areas are of three types:

- *Major access roads (or transitional)*. These are short lengths of road connecting a distributor road with the minor access road network, the latter at a T-junction. They are normally 6 m wide, have no direct access to property along their length, and serve from 200 to 400 dwellings.
- *Minor access or collector roads*. These form the backbone of the network, will serve up to 200 dwellings and be 5·5 m wide with only one footway. Occasionally a single-track 'car way' 2·75 m wide is used for access to about 50 dwellings, in conjunction with a separate footpath system.
- *Shared private drives, mews courts, garage courts and housing squares*. Generally these facilities serve up to 20 dwellings, and are designed for joint pedestrian/vehicle use with hard surfaces, no upstand kerbs and no footways. Access to them from the collector roads is marked by some device such as a short ramp or rough surface material, with the purpose of slowing down the traffic.

3.03 Designed controls

Apart from unwanted through traffic, the major problem in access roads is to ensure that the carriageways and verges are not used for parking or for loading and unloading. Legal restraints require enforcement which is expensive and intermittent. It is therefore up to the designer to provide physical restraints. One way is to make the road as narrow as possible, although this is not always effective on a two-way carriageway unless there are frequent central islands, which require lighting for the sake of safety.

Figure **7.5** shows the effect of narrowing a road from the normal minimum two-lane dimension of 5·5 m. Sub-standard widths of road should be straight, and should have passing places at intervals depending on the frequency of use by larger vehicles. They are only appropriate where there is a great preponderance of automobile traffic such as in residential areas and car park accesses. They can cause serious problems on the occasion of a genuine breakdown.

A second solution to the 'wild-cat' parking and loading problem is to ensure sufficient areas for the purpose as near as possible to the buildings served. This may not always be practical, or even desirable.

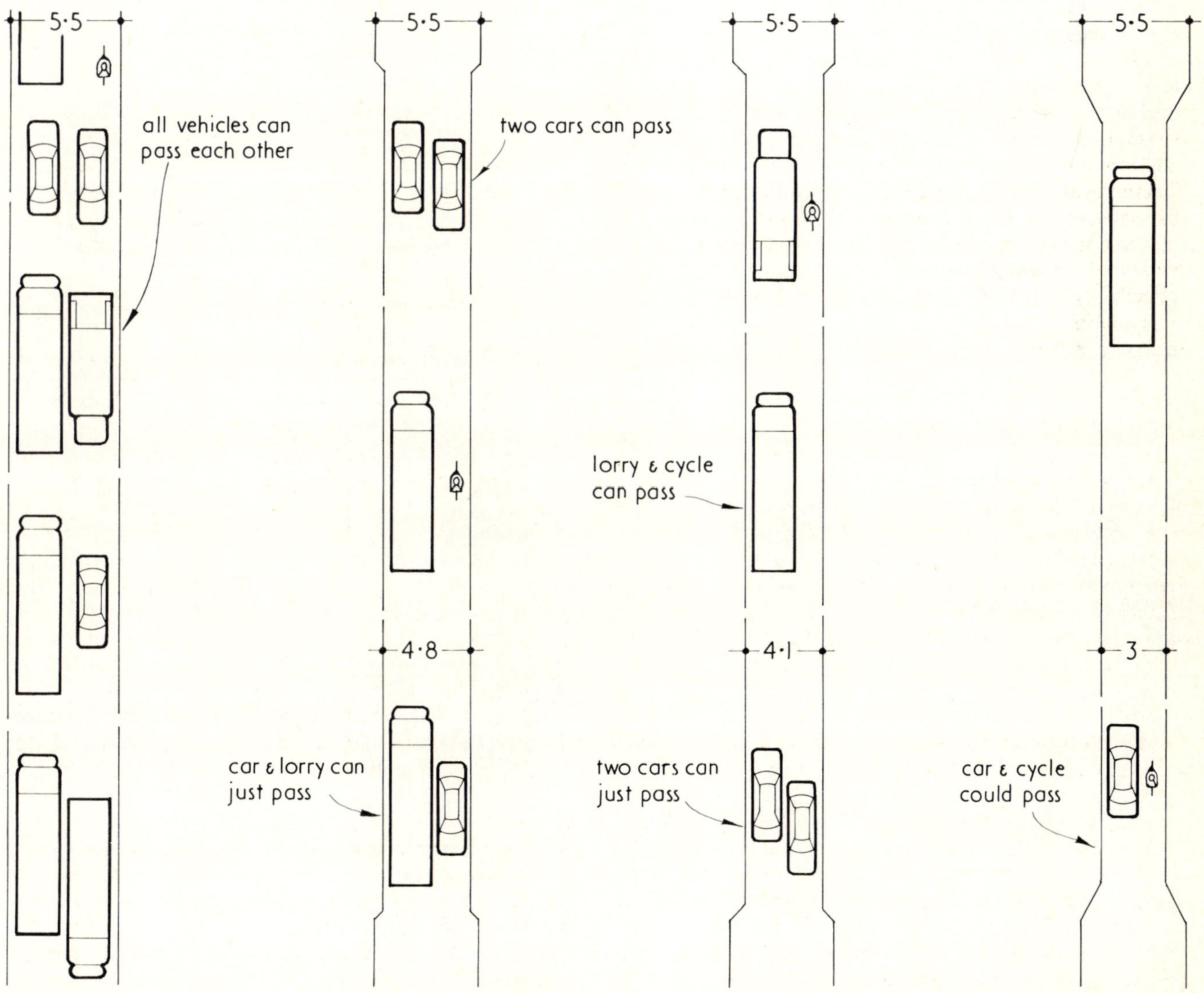

7.5 *Characteristics of various carriageway widths on two-way roads*

Table V Outside turning radius of front axle (m)

Minimum radius			15		30		45		60		75–400		400 +		
R	X	Y	X	Y	X	Y	X	Y	X	Y	X	Y	X	Y	
10·45	3·92	4·57	3·44	3·89	2·96	3·19	2·80	2·95	2·73	2·84	2·68	2·77	2·53	2·54	Pantechnicon
9·62	3·27	3·87	2·94	3·33	2·66	2·85	2·58	2·71	2·53	2·63	2·50	2·58	2·42	2·43	Refuse vehicle
7·91	3·15	3·88	2·67	3·06	2·42	2·61	2·34	2·47	2·30	2·40	2·27	2·35	2·19	2·20	Fire appliance
5·78	2·38	2·73	1·96	2·10	1·84	1·91	1·80	1·85	1·78	1·81	1·76	1·78	1·73	1·74	Private car

X = Maximum width of wheel path
Y = Maximum width of wheel path plus overhang

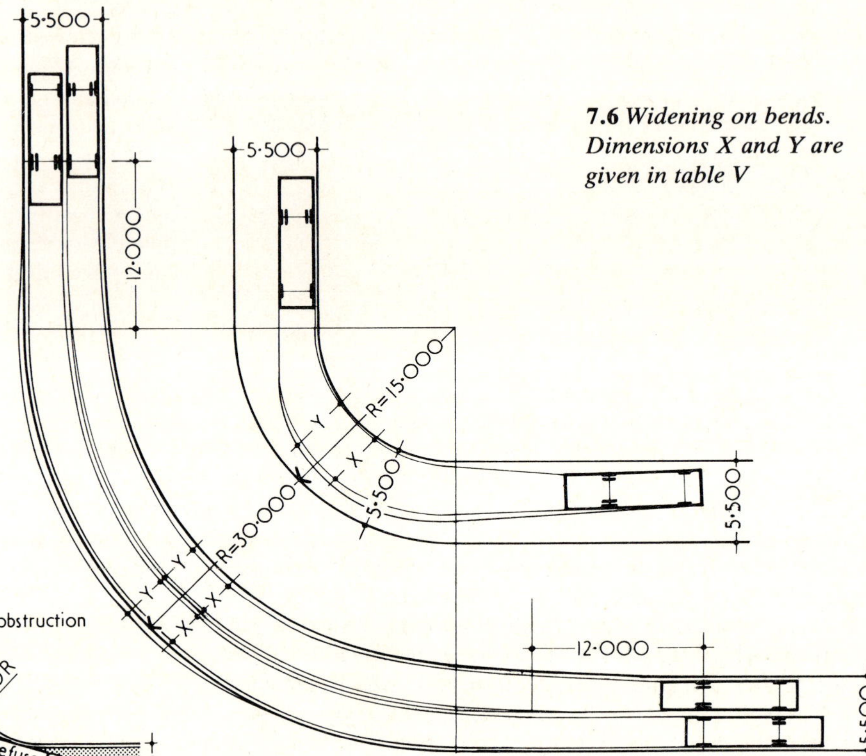

7.6 *Widening on bends. Dimensions X and Y are given in table V*

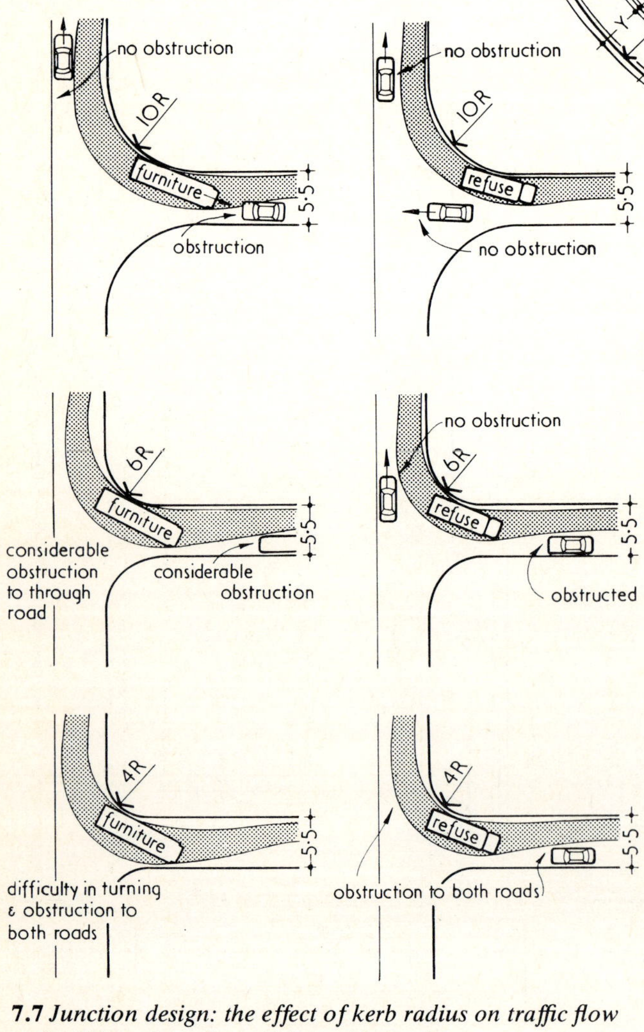

7.7 *Junction design: the effect of kerb radius on traffic flow at the T-junction of two 5·5 m wide roads*

3.04 Curves

As mentioned in para **1.03**, when a vehicle travels round a curve the width of road it occupies is greater than the track width on the straight. Table V combined with fig. **7.6** indicates the magnitude of this, and the normal width of carriageway is increased on curves to compensate.

3.05 Corners

Since the internal radius of turn of a commercial vehicle is about 8 m, it will be seen that a kerb radius of 10 m will be needed for such vehicles to maintain a constant distance from the kerb while turning the corner, while allowing some spare for the distance covered while turning the steering wheel. A kerb radius of 10 m in all cases would mean large areas of carriageway at junctions, and would be inappropriate in scale in many places, particularly in residential areas. Where traffic volumes are low there is no reason why the occasional larger vehicle should not encroach on the opposite side of the road, provided that clear visibility is maintained so that vehicles affected by the manoeuvre can take avoiding action in time. Figure **7.7** illustrates the effects of using radii of 10, 6 and 4 m.

3.06 Visibility and stopping distance

It is an axiom of road design that the driver should be able to see a distance at least as far as the distance he requires to stop. If the object he sees is also a moving vehicle, the sight distance must allow both vehicles to stop before colliding.

Table VI Recommended standards in junction design

Junction type		Radius (m)	Minimum Junction Spacing (m)		Sightlines (m)	
Road A	Road B	R	Adjacent	Opposite	X	Y
Local distributor	Any other road	10	80	40	5	60 in 30 mph zone 80 in 40 mph zone 100 in 50 mph zone
Minor access road	Major access road	6			2·4	40
Minor access road	Minor access road	6	25	8	2·4	40
Minor access road	entrance to mews or garage court	4·2	25	8	2·4	40
Single track road	entrance to mews or garage court	8 & 5 Offset	25	8	Junctions must be intervisible	
Mews or garage court	entrance to mews or garage court	4·2			2·4	10

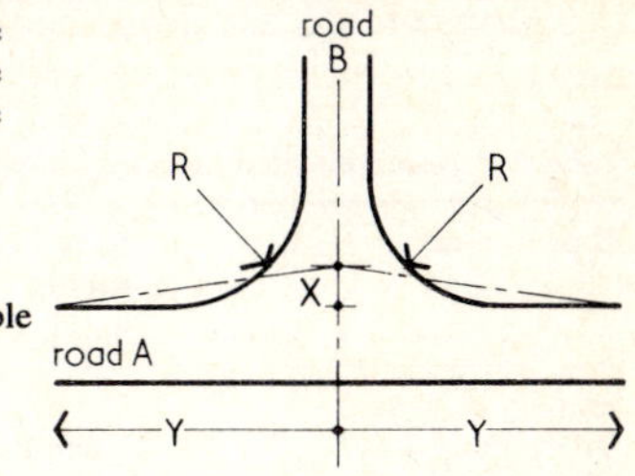

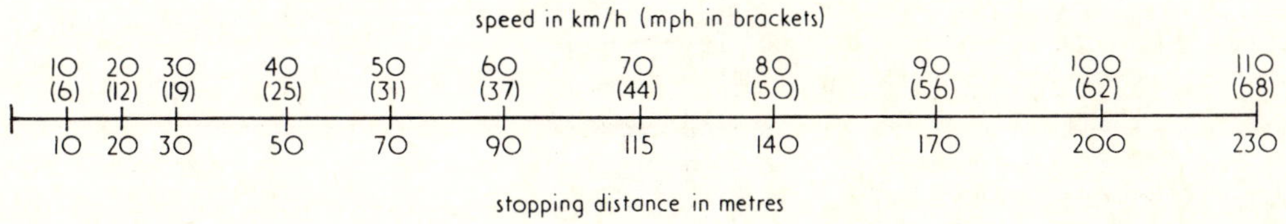

7.8 *Design stopping distances*

Figure **7.8** gives the design stopping distances for speeds up to 110 km/h (approx 70 mph). These distances are about 2¹/₄ times the stopping distances given in the Highway Code for vehicles with good brakes in ideal conditions. This is to allow for reduced brake performance, poor weather conditions and impaired visibility.

When emerging from a side road onto a through road, the vehicle driver must be able to see a vehicle on the through road a distance away of that vehicle's stopping distance. When crossing a footway, the driver should be able to see 2·4 m along the footway. Where small children are to be expected this visibility should be to within 600 mm high from the ground; but where there are no children (such as in industrial areas) 105 mm will be sufficient. Figures **7.9** and **7.10** indicate the areas that must be free of obstruction, and table VI gives some recommended standards in residential areas.

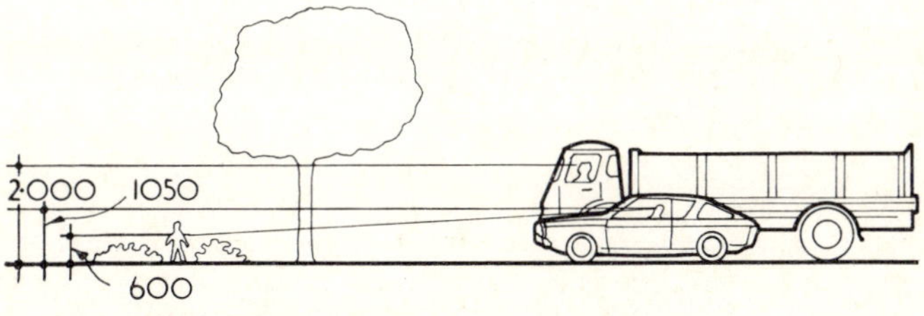

7.9 *Required heights for unobstructed visibility*

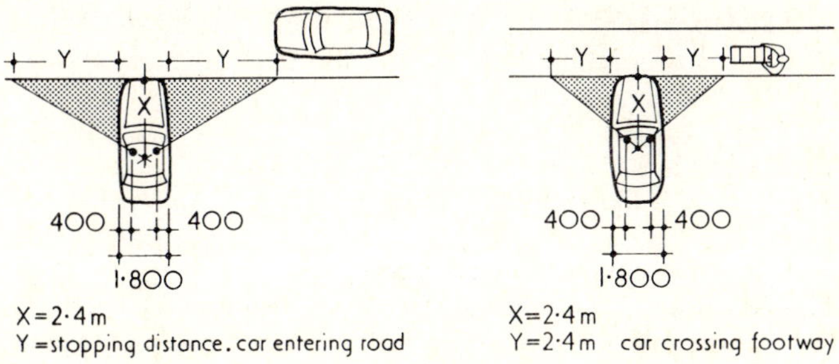

7.10 *Required distances for unobstructed visibility*

3.07 Turn-round areas

Where more conventional arrangements are used in turn-round areas in mews courts and housing squares, the minimum standards in fig **7.11** can be employed. Some local authorities, however, lay down more generous minimum standards to accommodate refuse collection and fire-fighting vehicles. In cases of doubt use the specimen vehicle track diagrams, **7.3**.

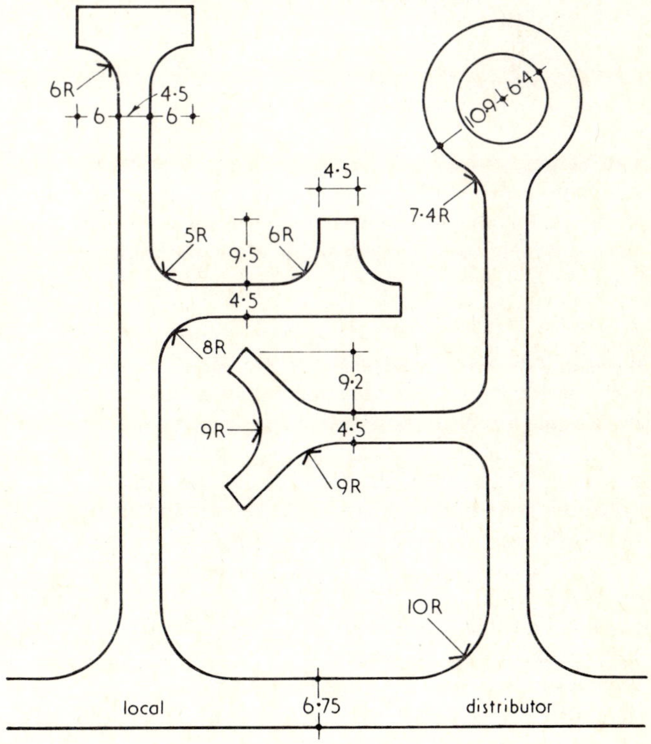

7.11 *Typical recommended dimensions for use in urban areas*

3.08 Footways and footpaths

Fig **7.12** and table VII indicate the standards required for footways and footpaths. There is no reason why these should be constant in width, and in fact desirable that there should be occasions of interest along them—seats, trees, small play areas etc.

Table VII Recommended footway widths

Type of road	Recommended minimum footway widths m
Primary distributor:	
Urban motorway	No footways
All purpose road	3·00*
District distributor	3·00 in principal business and industrial districts* 2·50 in residential districts*
Local distributor	3·00 in principal business and industrial districts* 2·00 in residential districts*
Access roads	Principal means of access: 3·00 in principal business districts* 2·00 in industrial districts* residential districts: see diagrams **7.12** 3·50–4·50 adjoining shopping frontages Secondary means of access: 1·00 verge instead of footway on roads in principal business and industrial districts 0·60 verge instead of footway on roads in residential districts

* If no footway is required provide verge at least 1 m wide. Where slabs are used widths are net paved areas (excluding kerbs)

3.09 Cycle tracks

See table VIII for recommended widths. It is not normally desirable for cycleways to be combined with pedestrian facilities. Where cycle tracks are provided with subways under roads etc, the minimum headroom should be 2·3 m.

Table VIII Cycle tracks* and cycle ways

Type of traffic	Standard width m	Minimum width m
One-way traffic	2·75	1·80
Two-way traffic	3·60	—

* Where no cycle track is provided it is sometimes compensated for by widening the inside traffic lane to 4·25 m

3.10 Verges

Where no footway is provided, a soft verge of 1 m should be provided for the accommodation of services (water, gas, electricity, telephone cables) and to allow for vehicular overhang.

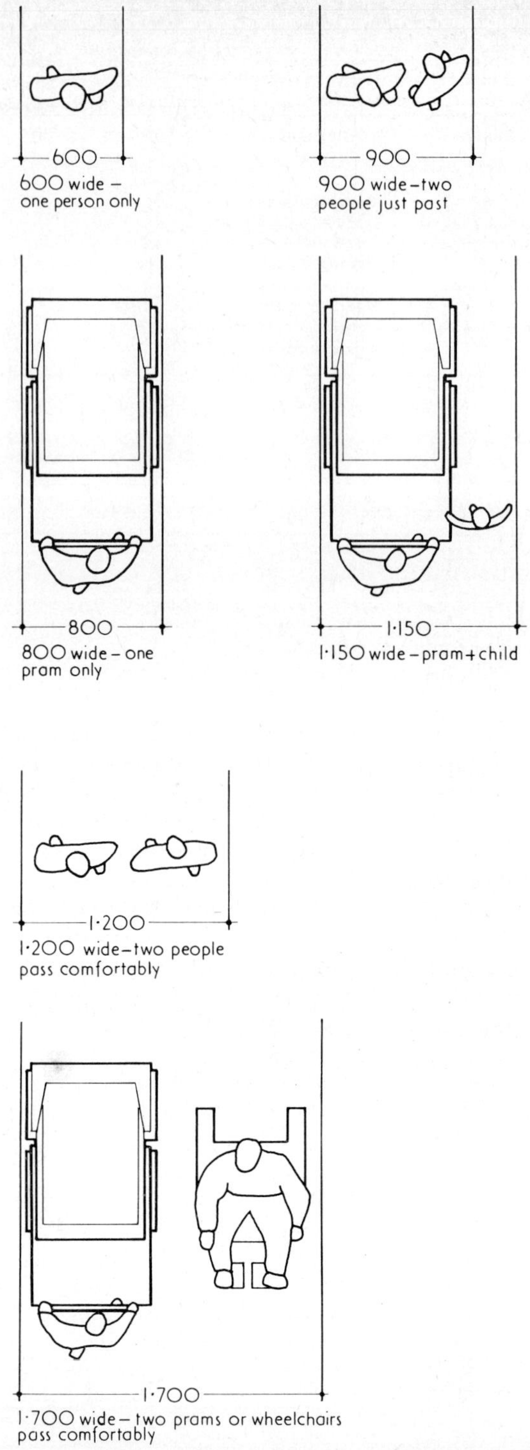

7.12 *Characteristics of various footway widths*

4 PARKING AND LOADING/UNLOADING

Facilities for parking and loading/unloading are often grossly inadequate for the purpose. Table IX covers recommendations for the scale of the provision, but each specific case should be examined to determine the expected requirements.

Once the scale of the provision has been found, the arrangement will depend on the size and shape of the available space, and on the type of vehicle expected. Figures **7.13** to **7.22** give examples of various facilities, but again, these should be taken purely as a guide.

Figure **7.23** shows various types of multi-storey garage. No dimensions are shown as these can vary widely depending on the site. An additional type (not illustrated) incorporates a mechanical stacking system operated by attendants. In practice, this rarely shows any advantage over conventional types.

Figures **7.24** to **7.27** show dimensions of railway carriages and wagons.

5 REFERENCES

Peter Falconer and Jolyon Drury *Handbook of industrial storage and distribution*, London, Architectural Press, 1973

DOE *Roads in urban areas*, 1966 (metric supplement 1974)

DOE *Residential roads and footpaths*, 1977

Cheshire County Council *Design aid, housing, roads*, September 1976

Lancashire County Council *Car parking standards*, 1976

G. Brock *Road width requirements of commercial vehicles when cornering*, Transport and Road Research Laboratory, DOE, 1973

Essex County Council *A design guide for residential areas*, December 1973

DOE Design bulletin 12, *Cars in housing 2*, 1971

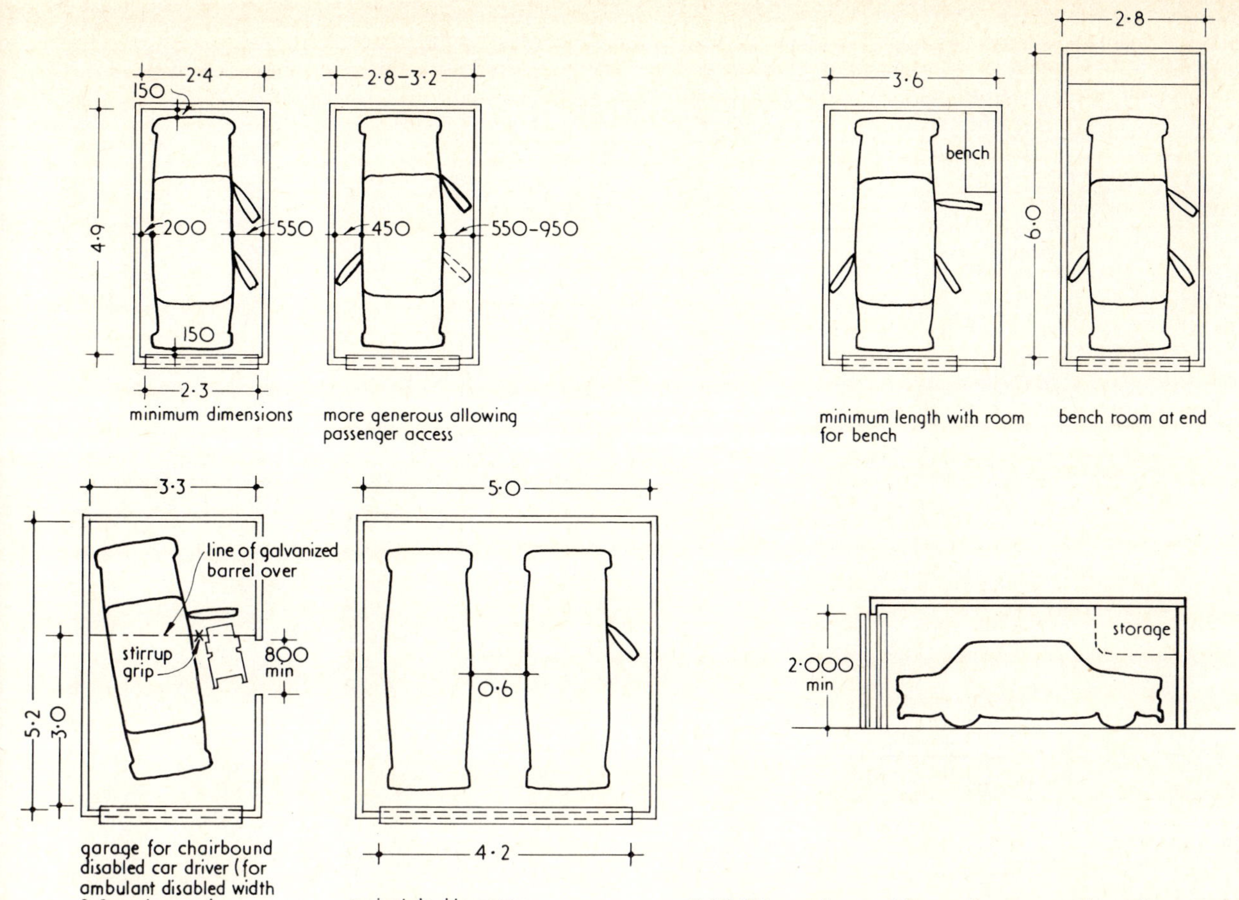

7.13 *Dimensions of domestic garages for private cars*

7.14 *Basic parking dimensions*
Standard parking bay (stall) 4·8 × 2·4, allow 24 m² per car
Can contain most European cars
Areas include half the clear zone, but no access gangways

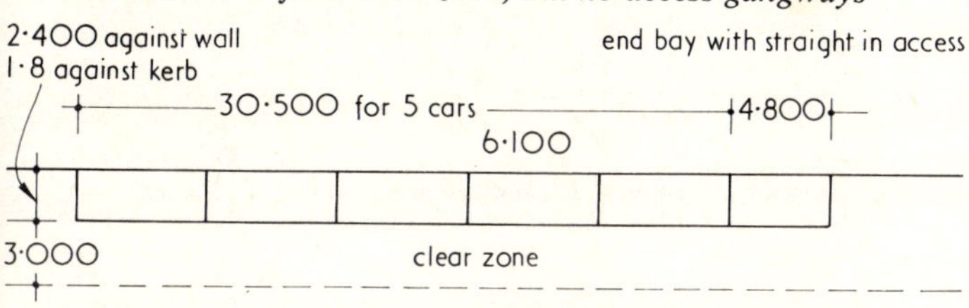

a *in-line parking, area per car: 20·1 m² against kerb*
23·8 m² against wall

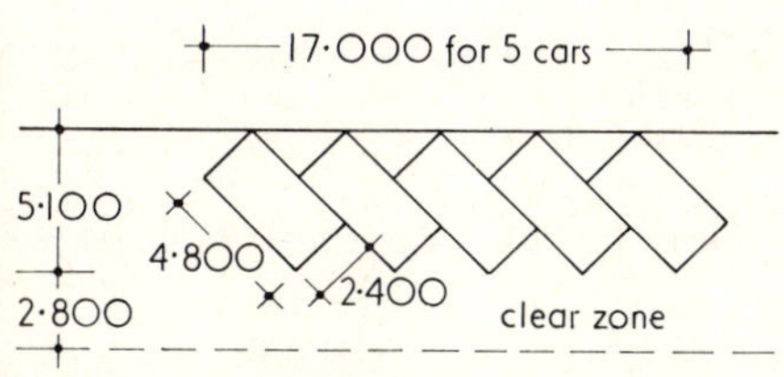

b *echelon parking (45°), other angles can be used*
area per car: 22·1 m²
19·2 m² where interlocking in adjacent rows

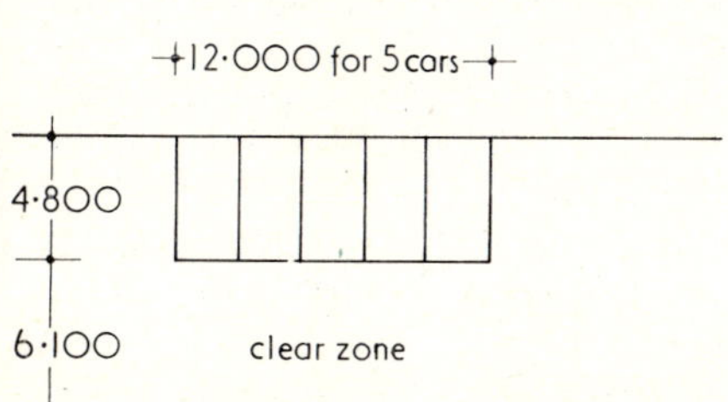

c *head-on parking, area per car: 18·8 m²*

7.15 *Basic parking dimensions*
Standard parking bay (stall) 5·8 × 2·8, allow 33 m² per car
To accommodate American and large European cars
Areas include clear zone, but no access gangways

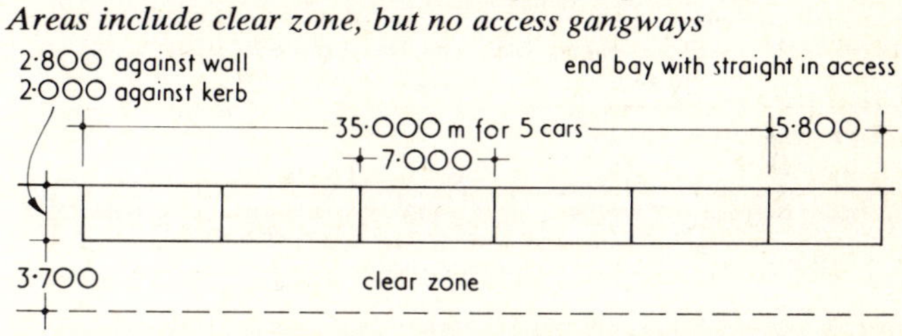

a *in-line parking, area per car: 27·0 m² against kerb*
32·6 m² against wall

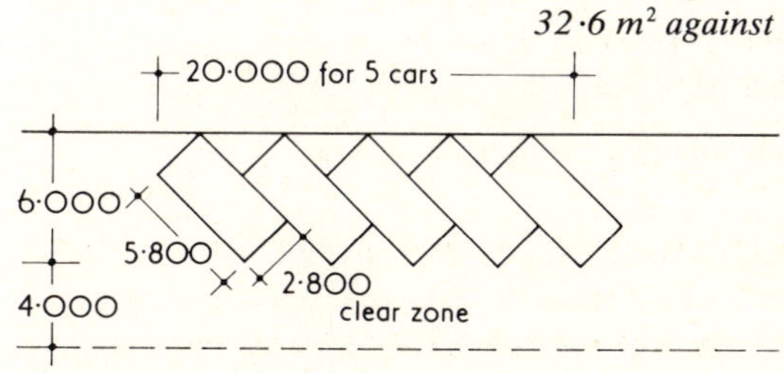

b *echelon parking (45°), other angles can be used*
area per car: 32·0 m²
28·0 m² where interlocking in adjacent rows

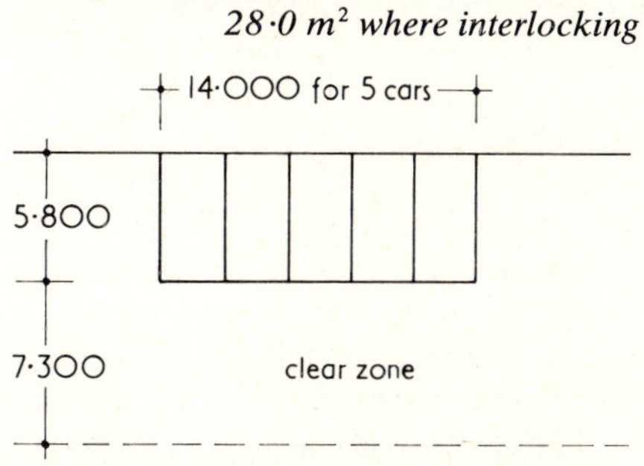

c *head-on parking, area per car 26·5 m²*

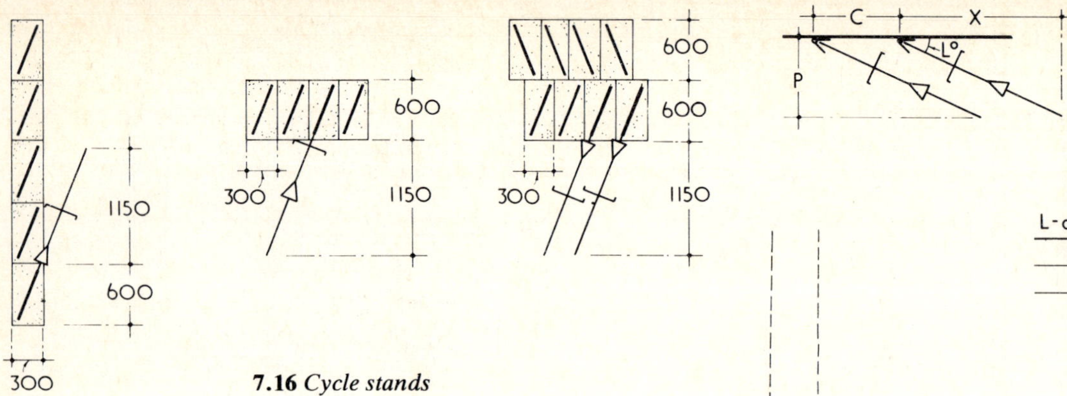

L-degrees	C(mm)	X(mm)	P(mm)
25	840	1600	800
45	600	1450	1450
90	640	—	1900

7.16 *Cycle stands*

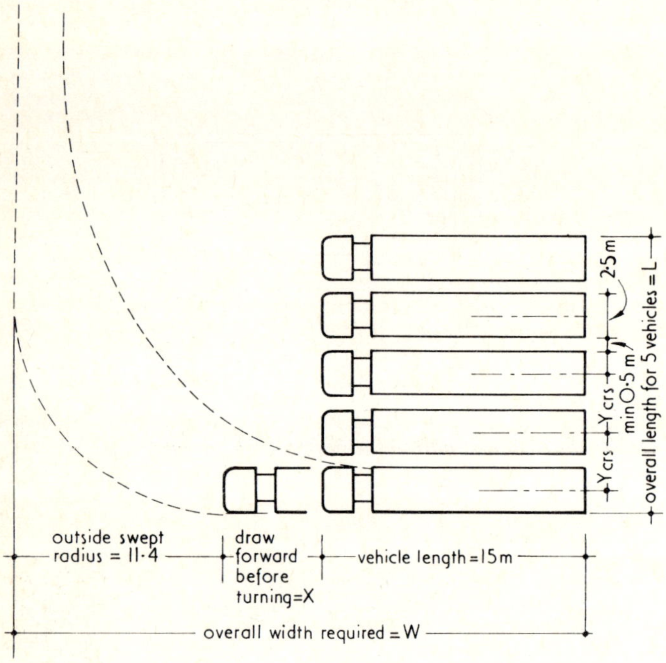

X draw forward	Y centres	W o/a width	L o/a length for 5	Area per vehicle (m^2)
1	5·0	27·4	22·5	123
2	4·4	28·4	20·1	114
3	4·0	29·4	18·5	109
4	3·7	30·4	17·3	105
5	3·4	31·4	16·1	101
6	3·0	32·4	14·5	94

7.17 *Lorry parking and loading bays—head-on; for the largest vehicles*

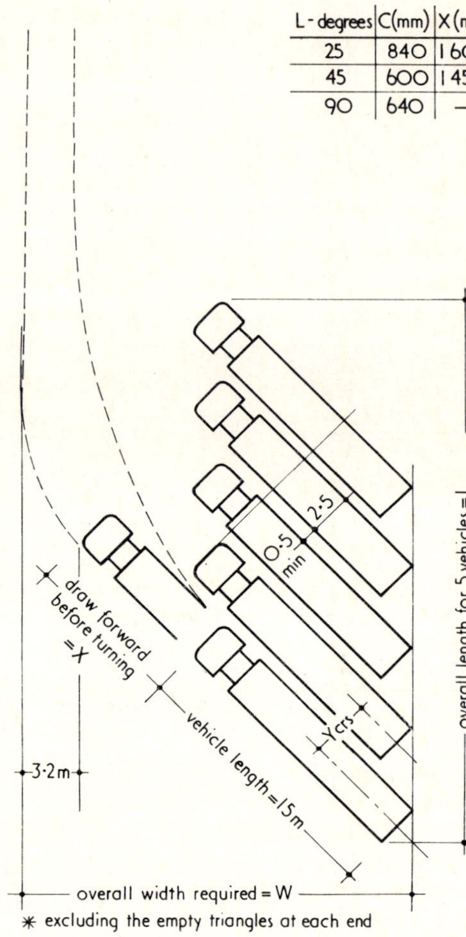

X draw forward	Y centres	W o/a width	L o/a length for 5	Area per vehicle (m^2) gross	 net
4	4·8	18·4	39·5	145	113
5	4·5	19·1	37·8	144	111
6	4·2	19·8	36·1	144	108
7	3·9	20·5	34·4	141	105
8	3·6	21·2	32·7	139	101
9	3·4	21·9	31·6	138	100
10	3·2	22·6	30·5	138	98
11	3·1	23·4	29·9	140	99
12	3·0	24·1	29·3	141	99

7.18 *Lorry parking and loading bays—diagonal (45°); for the largest vehicles*

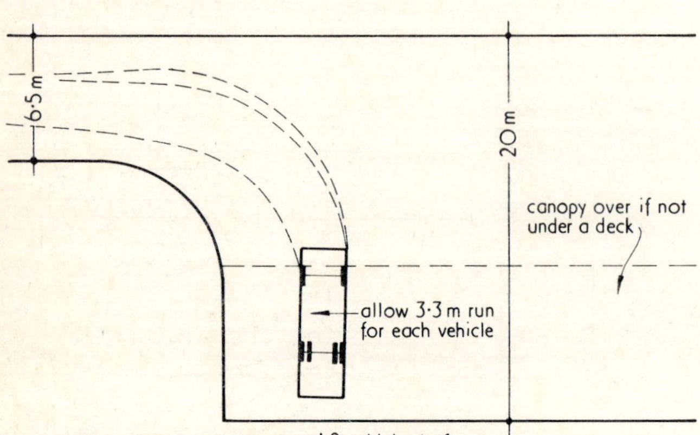

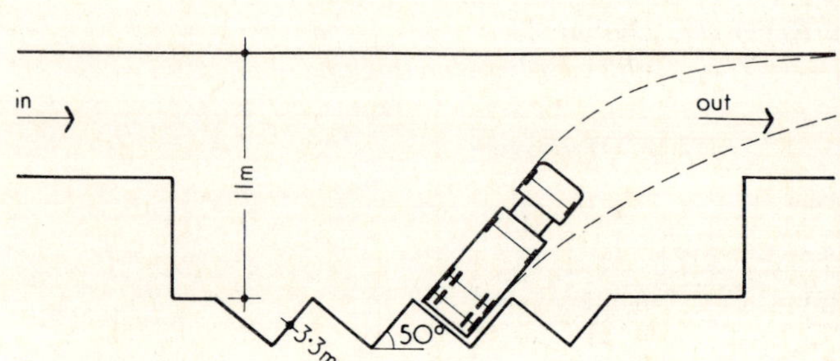

7.20 *Minimal loading docks. Appropriate for: 1 limited number of vehicles per day 2 extremely high land costs (as in city centres) or 3 other physical restraints*

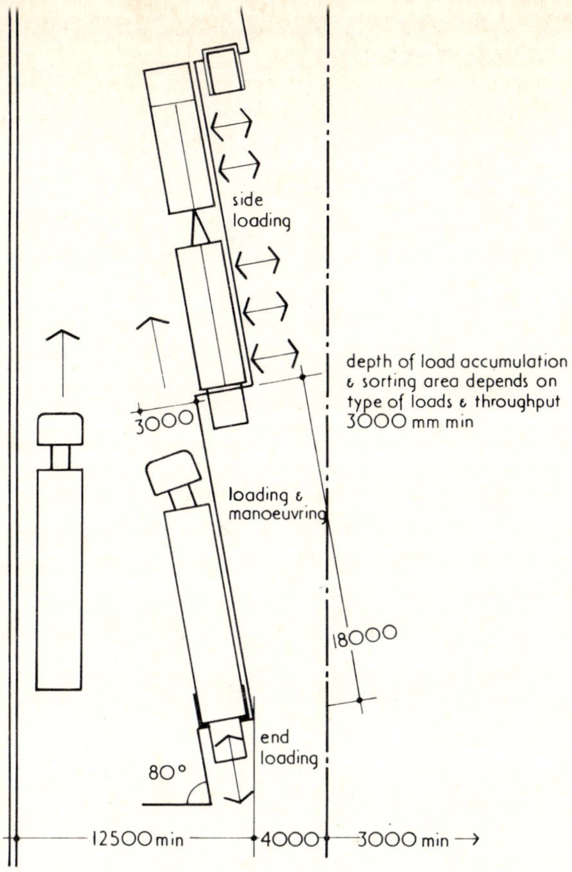

7.19 *Finger dock—where manoeuvring depth is limited and side loading is required as well as end loading. Very fast turn-round times are possible although capacity is small*

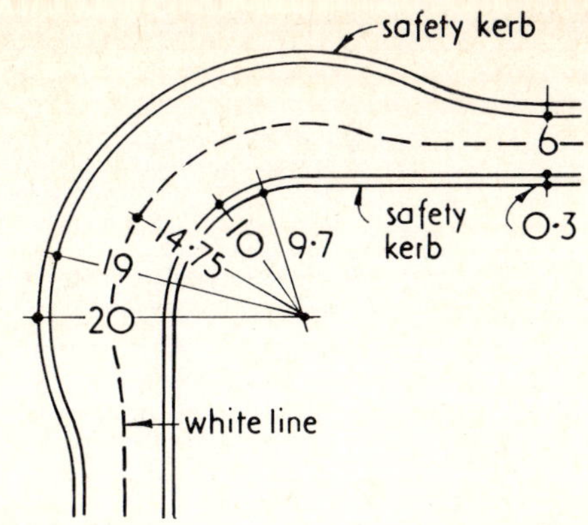

7.21 *Design for a ramp on a sharp curve, such as in access to shopping centre loading dock. Maximum gradients:*
10% on straight
7% on inner kerb

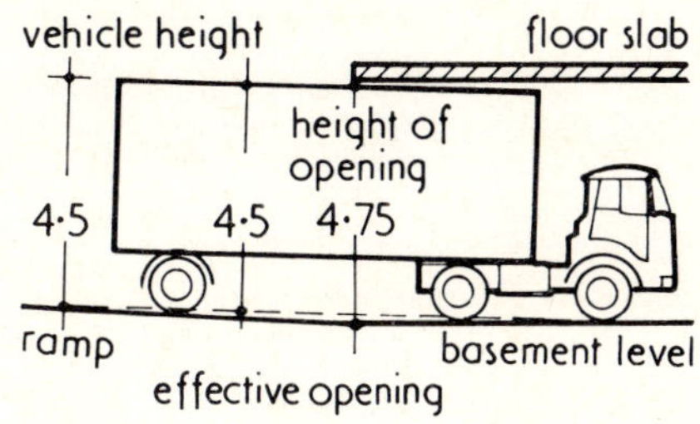

7.22 *Headroom criteria for covered loading docks*

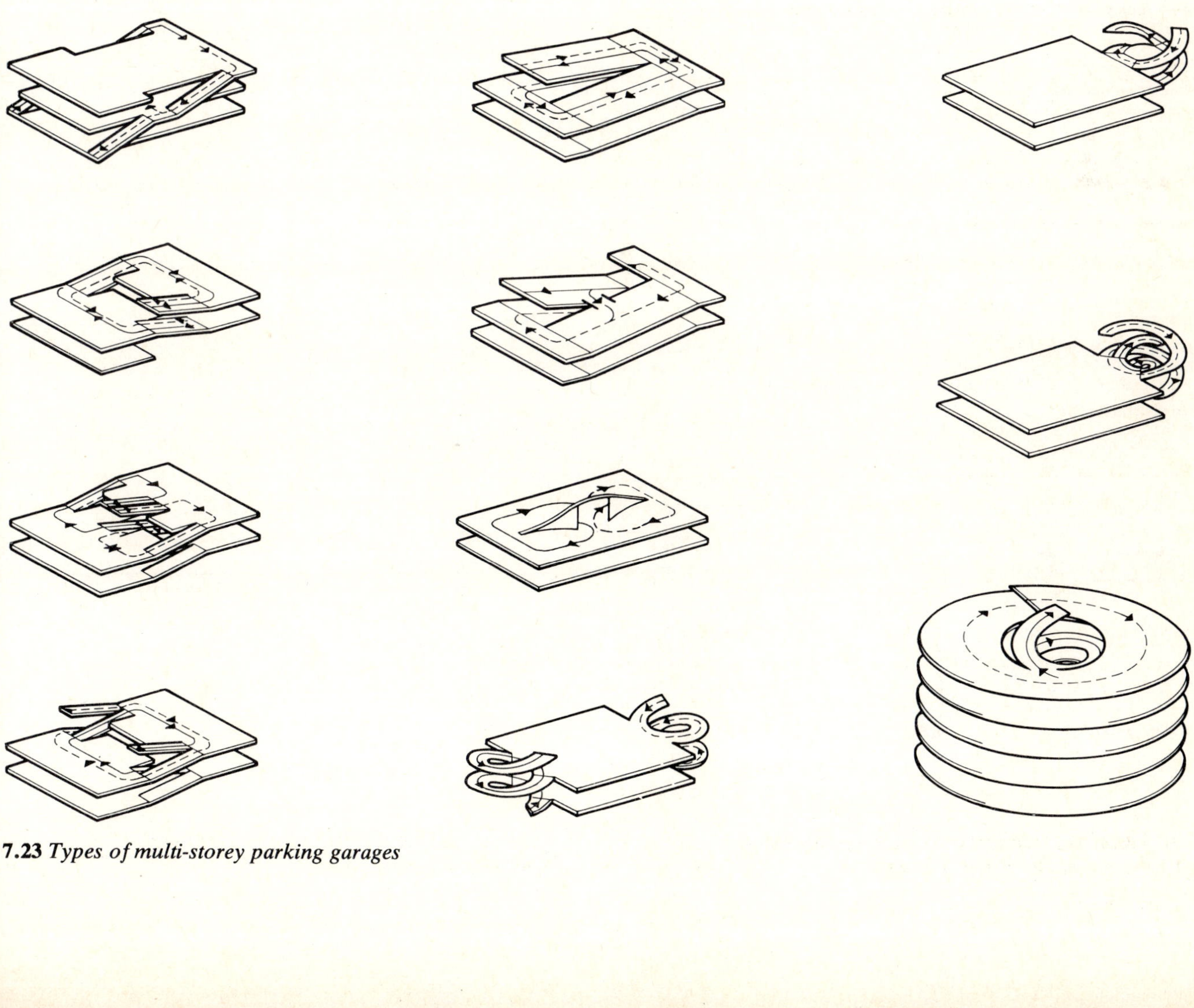

7.23 *Types of multi-storey parking garages*

Table IX Parking and loading/unloading requirements

Type of building	Parking provision	Loading/unloading provision
Normal housing	Residents: one garage space for each occupancy, (preferably within the curtilage) Visitors: where houses are served directly from a road, driveways provide a minimum of one car space within curtilage of each Where visitors cannot park within curtilage, one off-street space per four dwellings	Refuse collection vehicle within 25 m of each disposal point (dustbin position). Some authorities require vehicle within 15 m. Where communal containers (paladins) are used, maximum distance 9 m Furniture removal vehicle as near as possible, not further than 25 m
Minimum-cost housing	Space should be provided, if not laid out, to allow for one resident's or visitor's parking space per dwelling, provided public transport is available	As above
Old people's housing	One garage space per two dwellings	As above
Sheltered housing	Resident and non-resident staff: one car space per two members present at peak period Visitors: use empty staff places, but provide one additional place per five dwellings	As above, plus provision for special passenger vehicle with tail lift, etc Minimum provision for daily loading/unloading 50 m²
Shops	Staff: one car space (preferably in enclosed yard behind shop) for each 100 m² gross floor area or, if known, one space per managerial staff plus one for every four other staff Customers: one space for each 25 m² gross floor area. In superstores with gross floor area exceeding 2000 m² allow one space per 10 m². (Not appropriate when goods sold are obviously bulky, eg carpets, boats)	See diagrams of loading bays. General minima as follows: Gross floor space not exceeding: — Minimum space required: 500 m² — 50 m² 1000 — 100 2000 — 150 each additional 1000 m² — 50 m²
Banks	Staff: one space for each managerial or executive staff, plus one per four others Customers: one space per 10 m² of net public floor space in banking hall	Minimum 25 m²
Offices	Staff: one space for each 25 m² of gross floor area, or one space for each managerial and executive staff, plus one space per four others Visitors: 10% of staff parking provision	General minima: Gross floor space not exceeding: — Minimum space required: 100 m² — 50 m² 500 — 100 1000 — 150 each additional 1000 m² — 25 m²
Production buildings (factories)	Staff: one car space per 50 m² of gross floor area Visitors: 10% of staff parking provision	See loading bay diagrams. Provision to be commensurate with expected traffic General minima as follows: Gross floor space not exceeding: — Minimum space required: 100 m² — 70 m² 250 — 140 500 — 170 1000 — 200 2000 — 300 each additional 1000 m² — 50 m²
Storage buildings (warehouses)	Staff: one space per each 200 m² of gross floor space	
Hotels, motels and public houses	Resident staff: one space per household Non-resident staff: one space for each three staff members employed at peak period Resident guests: one space per bedroom Bar customers: one space for each 4 m² of net public space in bars Occasional diners: no additional provision required If conferences are held in the hotel, space required should be assessed separately at one space for each five seats	General minima as follows: Gross floor space not exceeding: — Minimum space required: 500 m² — 140 m² 1000 — 170 2000 — 200 each additional 1000 m² — 25
Restaurants and cafés	Resident staff: one space per household Non-resident staff: one space per three members employed at peak period Diners: one space for each two seats in dining area (For transport cafés, the space should be a lorry space of 45 m², and the arrangement should be such that vehicles can enter and leave without reversing)	General minima as follows: Dining floor space not exceeding: — Minimum space required: 100 m² — 50 m² 250 — 75 500 — 100
Licensed clubs	Resident staff: one space per household Non-resident staff: one space for each three members employed at peak period Performers: one space for each solo performer and/or group expected at peak Patrons: one space per two seats, or one space per 4 m² net public floor space	Minimum 50 m²

Table IX Parking and loading/unloading requirements (continued)

Type of building	Parking provision	Loading/unloading provision
Dance halls and discotheques	Staff: one space per three members at peak period Performers: three spaces Patrons: one space per 10 m^2 of net public floor space	Minimum 50 m^2
Cinemas	Staff: one space per three members at peak period Patrons: one space per 5 seats	Minimum 50 m^2 Space required within site by main entrance for two cars to pick up and set down patrons
Theatres	Staff: one space per three members at peak period Performers: one space per 10 m^2 of gross dressing room accommodation Patrons: one space for each three seats	Minimum 100 m^2 Space required within site by main entrance for two cars to pick up and set down patrons
Bingo halls	Staff: one space for each three members at peak period Patrons: one space for each ten seats	Minimum 50 m^2 Space required within site by main entrance for two cars to pick up and set down patrons
Swimming baths	Staff: one space for every two members normally present Patrons: one space per 10 m^2 pool area	Minimum 50 m^2
Sports facilities and playing fields	Staff: one space per three members normally present Players: one space for each two players able to use the facility simultaneously, provided public transport is reasonably close. Otherwise two spaces for each three players Spectators: provide only if more than three times the number of players	Minimum 50 m^2
Marinas	Staff: one space per three members normally present Boat-users: two spaces for each three mooring-berths. (If other facilities are included, eg restaurant, shop etc, provide additional spaces at 50% normal provision for each additional facility)	Minimum 50 m^2
Community centres and assembly halls	Staff: one space for each three members normally present Patrons: one space for every five seats for which the building is licensed	Minimum 50 m^2
Places of worship	Worshippers: one space per ten seats in space for worship	Minimum 50 m^2 Space provided within site close to main entrance for two cars to set down and pick up worshippers
Museums and public art galleries	Staff: one space per two members normally on duty Visitors: one space per 30 m^2 of public display space	Minimum 50 m^2
Public libraries	Staff: one space per three members normally on duty Borrowers: one space for each 500 adult ticket holders with a minimum of three spaces. If there are separate reference facilities, provide additional spaces at one for each ten seats	Minimum 50 m^2 If used as a base for a mobile library, provide another 50 m^2 to park this
Hospitals	Staff: one space for each doctor and surgeon, plus one space for each three others Outpatients and visitors: one space for each three beds	General minima as follows: Gross floor space not exceeding: — Minimum space required: 1000 m^2 — 200 m^2 2000 — 300 4000 — 400 6000 — 500 every additional 2000 m^2 — 100 m^2
Health centres, surgeries, clinics	Staff: one space per doctor etc one space per two members of staff other than doctors etc employed at peak period Patients: two spaces per consulting room	Sufficient for requirements specified, including if necessary space for special vehicle for non-ambulant patients
Special schools, day-care centres and adult training centres	Staff: one space for each two members normally present Attenders: in many cases these will be transported to the centre. For certain centres for the physically handicapped, allow one space for special or adapted self-drive vehicle per four attenders	Minimum 30 m^2 Accommodation for special passenger vehicle Space provided within the site for cars and/or buses to set down and pick up
Nursery and primary schools	Staff: one space per two members normally present Visitors: two spaces Hard surface play area used for parking on open days etc.	Minimum 30 m^2

Table IX Parking and loading/unloading requirements (continued)

Type of building	Parking provision	Loading/unloading provision
Secondary schools	Staff: one space per two members normally present Visitors: schools with up to 1000 pupils—four spaces, larger schools—eight spaces. Hard surface play area used for parking on occasion	Minimum 50 m² Space provided within site for school buses to set down and pick up
Sixth form colleges	Staff: one space per two members normally present Visitors: colleges with up to 1000 pupils—five spaces, larger colleges—ten Hard surface play area used for parking on occasion	Minimum 50 m²
Further education colleges and retraining centres	Staff: one space for each member normally present Students and visitors: one space for each three students normally present	Minimum 50 m²

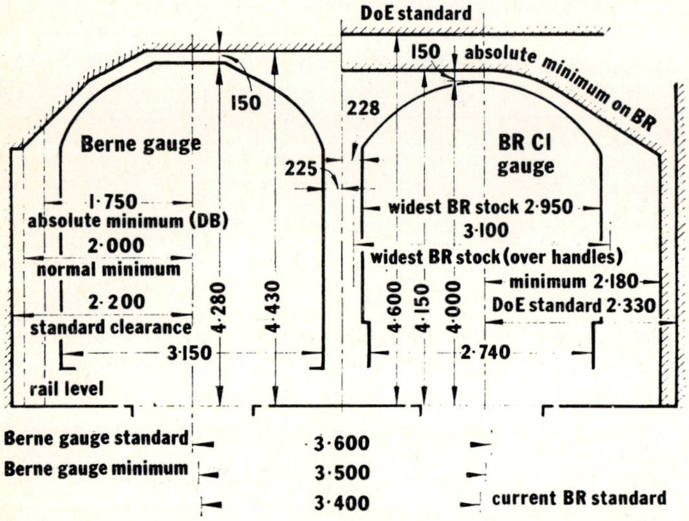

7.24 *Cross-section comparing typical British and European (Berne gauge) standards for rolling stock giving controlling dimensions for use in railway planning: design elements (bridges, platforms, etc) must be clear of these profiles*

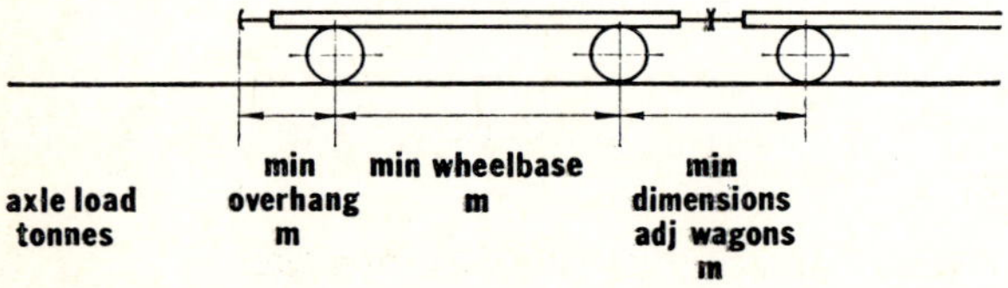

axle load tonnes	min overhang m	min wheelbase m	min dimensions adj wagons m
17·7	1·06	3·65	2·13
25·4	1·52	4·5	3·04

7.25 *Two-axle wagon dimensions*

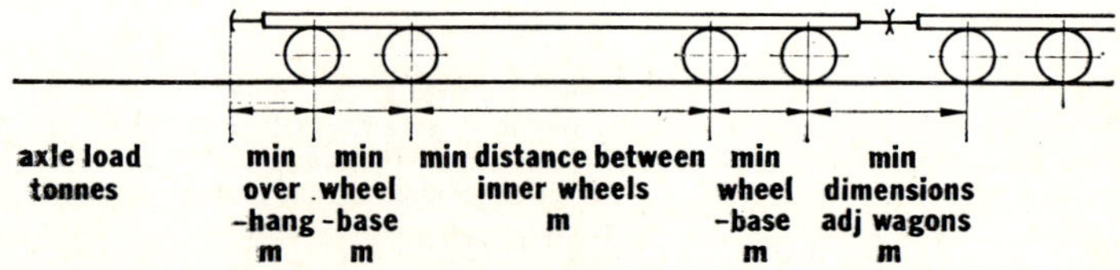

axle load tonnes	min over-hang m	min wheel-base m	min distance between inner wheels m	min wheel-base m	min dimensions adj wagons m
14·2	1·37	1·8	4·87	1·8	2·74
25·4	1·5	2·0	9·1	2·0	3·04

7.26 *Bogie wagon dimensions*

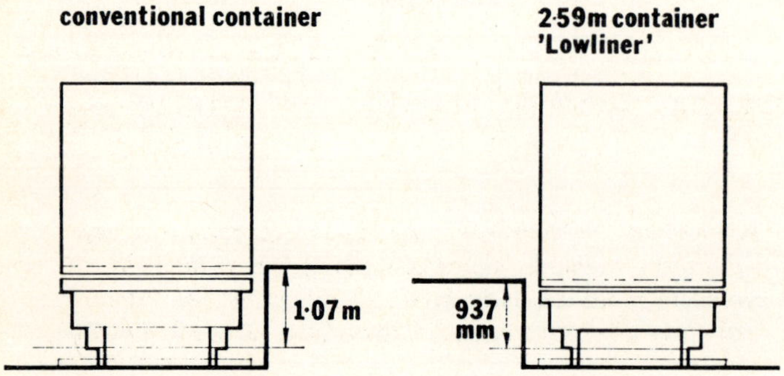

7.27 *Container wagon floors have different heights, and may also vary up to 100 mm in laden and unladen condition*

8 Vehicle servicing

John Carter
John Carter is an architect in practice

CI/SfB 224
CI/SfB (1976 revised) 127

Contents

1 INTRODUCTION

1.01 Types of service station

There are three types. These are:

- company owned and operated (known in the trade as 'co-co', 'company' being the petroleum company). There is only a small number of this type
- dealer-owned and operated. The petroleum company may finance the construction, extension or improvement of this type of station in return for an agreement that the dealer sells only their petrol. He is usually free to sell other products, such as lubricating oils, by other companies. About seventy-five per cent of all service stations are of this type
- company owned but dealer operated.

The two company owned types of service station account for less than a quarter of all stations, but for nearly half of all petrol sales.

1.02 Commercial motivation

Selling image

Petroleum companies set great store by their visual image: motivation is totally commercial and all design decisions are influenced by this.

Sales and profit

There are approximately 30 000 service stations in Great Britain. Sales vary between 1500 000 gal (661 443 litres) and 800 000 gal (3636800 litres) per year. In early 1976, about 185 000 gal (841010 litres) per year was regarded as the break even point. But petrol sales alone offer a very small profit margin. Other, more profitable, sales and services are usually essential, ranging from repair and maintenance to the sale of confectionery. Some services benefit others, eg DOT testing in itself, is not profitable, but offers a repair and maintenance spin off.

2 PUMPS

2.01 Pump capacity per hour

A pump is capable of dispensing up to 150 gal (682 litres) per hour to an unbroken queue of customers, assuming it can dispense all grades (ie is a multigrade pump). Single grade pumps generally have a slower output because customers have a smaller choice of pumps and must either wait longer or manoeuvre to another pump dispensing the same grade.

2.02 Multi-grade petrol

Multigrade petrol is usually mixed at the pump eg 4 star plus 2 star. 5 star is the exception. It is more expensive to refine and therefore uneconomic to mix with a lower grade. It is sold in separate pumps at present, but it is government policy that it will eventually be withdrawn from the market.

4 star is the best seller, with 60 to 70 per cent of all sales.

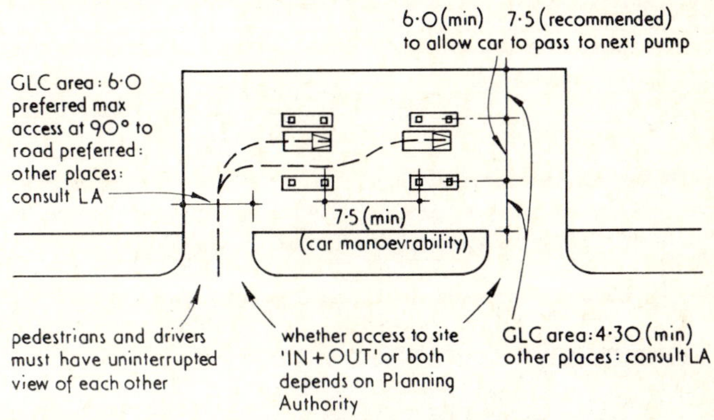

8.1 *Layout of typical petrol filling station*
Petroleum inspectorate require petrol outlet (service hose, tank filler pipe, vent pipe, etc) to be not less than:
4·3 m from property boundary (but waiver is usual in difficult cases)
4·3 m from 'non-intrinsically-safe' electrical appliance (ie where sparking is possible), and
12·0 m from any flame or non-flameproofed electrical appliance

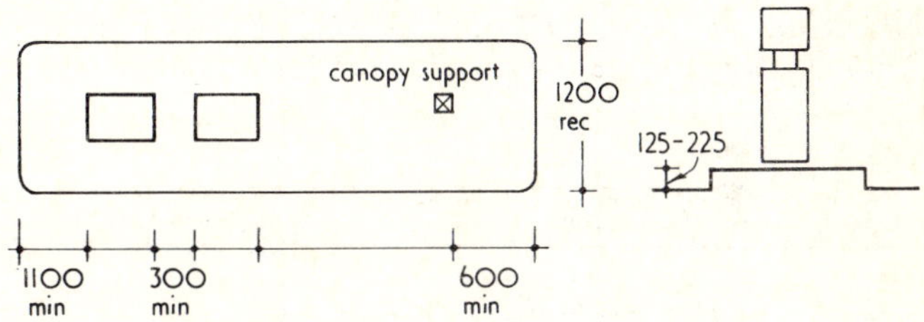

8.2 *Pump island*
kerb: usually precast concrete with rounded angles
kiosk: not advisable on island unless electrical appliance (eg heater, cash register) can be 1·2 m min from pump

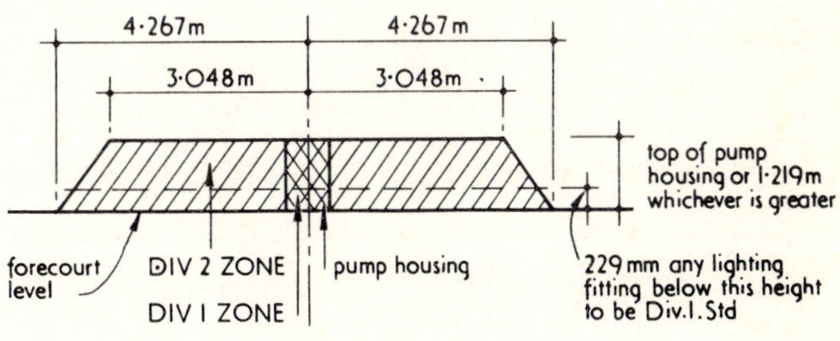

8.3 *Elevation of forecourt showing classification of areas:*
Division 1—where a dangerous atmosphere is likely in normal conditions, and where only flameproof electrical fittings to CP 1003: part 1:1964 and BS 1259:1958 can be used
Division 2—where a dangerous atmosphere is likely only in very abnormal conditions, and where only non-sparking electrical fittings to CP 1003:part 3:1967 (sections 2 and 3) can be used

2.03 Layout of pumps

As much as 90 per cent of one day's sales may occur in the peak period hour (literally one hour's trading). The actual timing of this period will vary between stations.

To meet this demand, an efficient circulation pattern, with no inbuilt delays or obstructions is essential, **8.1**. The multigrade pump will ensure the most efficient movement through the forecourt. If single grade pumps are used they should be accessible from all pump lanes so that customers do not need to change lanes to get the grade they want. Lanes should be sufficiently wide (7·5 m) to allow waiting cars to move past cars being served. A typical pump island is shown in **8.2**.

The requirements of the Model Code of Principles[1] with regard to the safety zoning of forecourts are given in **8.3**.

The number of pumps provided will depend primarily on expected sales volume but should allow for pump servicing time which, on older, less reliable pumps, averages 25 per cent.

2.04 Self or staff service

At present about 20 per cent of all sales are self-service, a proportion that is expected to increase only slowly.

3 PETROL STORAGE AND DRAINAGE

3.01 Petrol storage tanks

Tanks may be sited on or off the forecourt and vehicle manoeuvring space but siting should allow for almost daily lifting of inspection covers to check petrol stocks (by dipstick). Heavy duty covers will be required in vehicle areas.

Common sizes of petrol tank are given in table I. These tanks may be compartmented to hold different petrol grades.

Table I Approximate sizes of petrol storage tanks

Capacity litres	Diameter (mm)	Overall length (mm)
2270	1370	1750
2730	1370	2060
3410	1370	2600
4550	1370	3350
5680	1530	3430
6820	1830	2900
9100	1830	3800
11370	2000	4040
13640	2130	4120
22730	2290	5950

Choice of tank size is based on 4½ days forecast of sales plus one road tanker load (25 000 litres), eg for a station selling 45 460 litres per week,

4½ days supply	29 500	
road tanker load	25 000	
total storage	54 500	litres

The proportions of each grade of petrol normally stored are:

5 star	approx	10 per cent
4 star	approx	65 per cent
3 star	approx	15 per cent
2 star	approx	10 per cent

3.02 Tank installation

Criteria affecting the installation of petrol tanks are given in **8.4**. Note the interdependence of tank diameter, depth and distance from the pump. More detailed recommendations are given in the Model Code of Principles.[1]

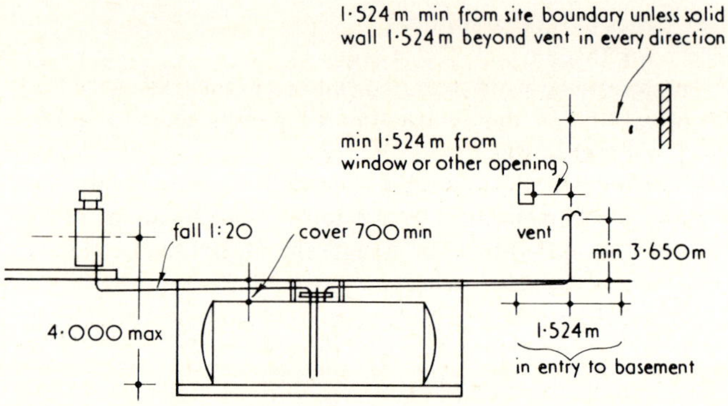

8.4 *Installation of petrol storage tank*

The 4·0 m maximum suction lift is to prevent gravitation and evaporation of the petrol and allows for the continued use of old, worn pumps with less suction than new ones.

Petroleum regulations do not require encasement of the tank, but it is advisable, particularly beneath wheel loads (up to 11·5 tonnes) and to protect tanks against corrosive ground water. Encasement in sand in a brick pit is hardly less expensive than mass concrete, but the tank is easier to remove.

The tanks must be held down during concreting (preferably by strapping them to the concrete base). They must be pressure tested by the controlling authority both before and after encasement.

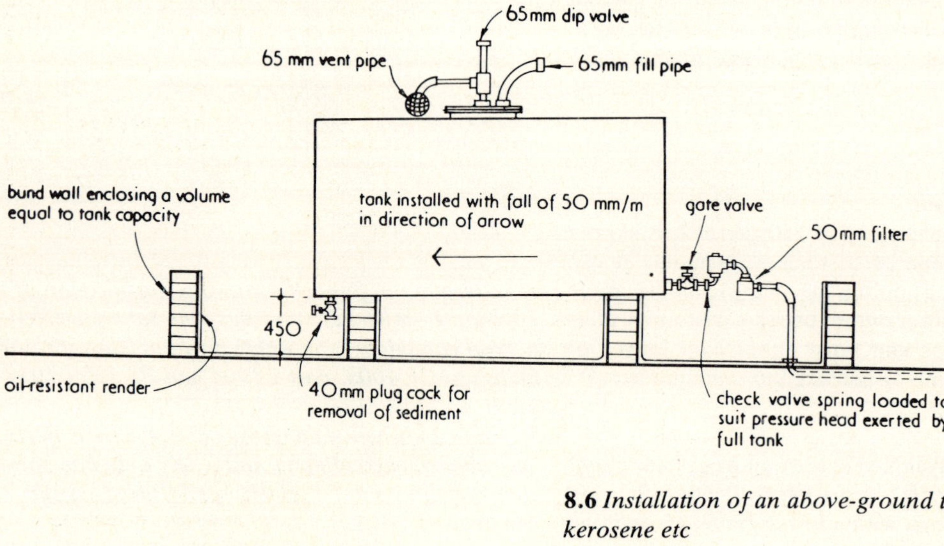

8.6 *Installation of an above-ground tank for diesel oil, kerosene etc*

3.03 Manholes

Petrol must not escape from manholes above or below ground level and rainwater should not percolate into them.
There are three alternatives:

- raising the inspection cover above the general surface level, **8.5a**
- surrounding the cover with a channel, **8.5b**
- using a sealed cover.

Heavy duty covers must be used if the manhole is in the carriageway.

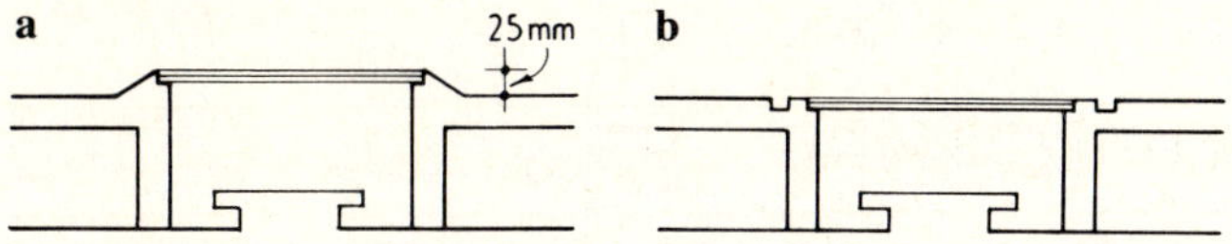

8.5 *Methods of installing inspection covers to prevent ingress of water:*
a *raising above general level*
b *surrounding with a sunken channel*

3.04 Other tanks

Tanks may also be required for kerosene, auto diesel fuel, gas oil (duty free diesel fuel for stationary engines or farm vehicles) and also for used sump oil. These are usually rectangular tanks, kept above ground, **8.6**.
The controlling authority may request a 'bund' wall surrounding the tank to contain spillage of the entire contents in the event of a leak. The normal capacity of tanks above ground is about 2700 litres.

3.05 Interceptor chamber

Interceptor chambers are used to comply with the requirement that no petrol enters the mains drainage system.
The chamber should be between the main sewer and the last surface water collecting manhole. Petrol is lighter than water and will float to the top and evaporate up through the vent manifold. The three chambers give it three channels in which to do this, **8.7**.

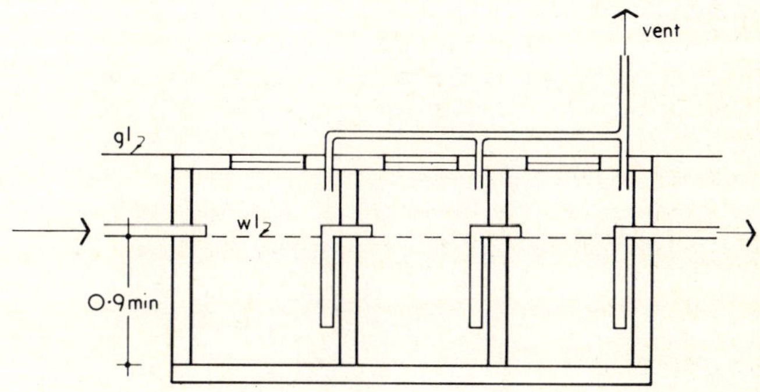

8.7 *Petrol interceptor chamber*

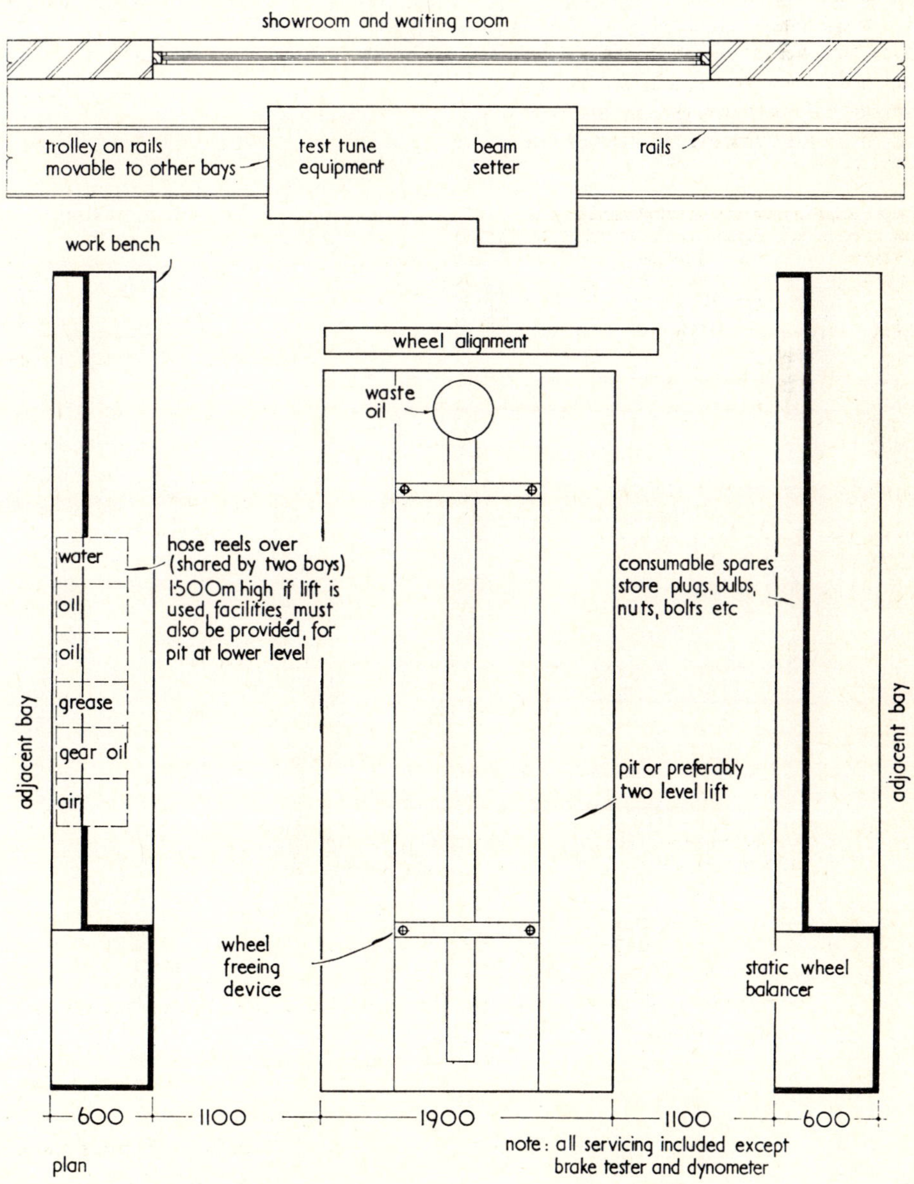

8.8 *'One-stop' service bay for lubrication and mechanical services*

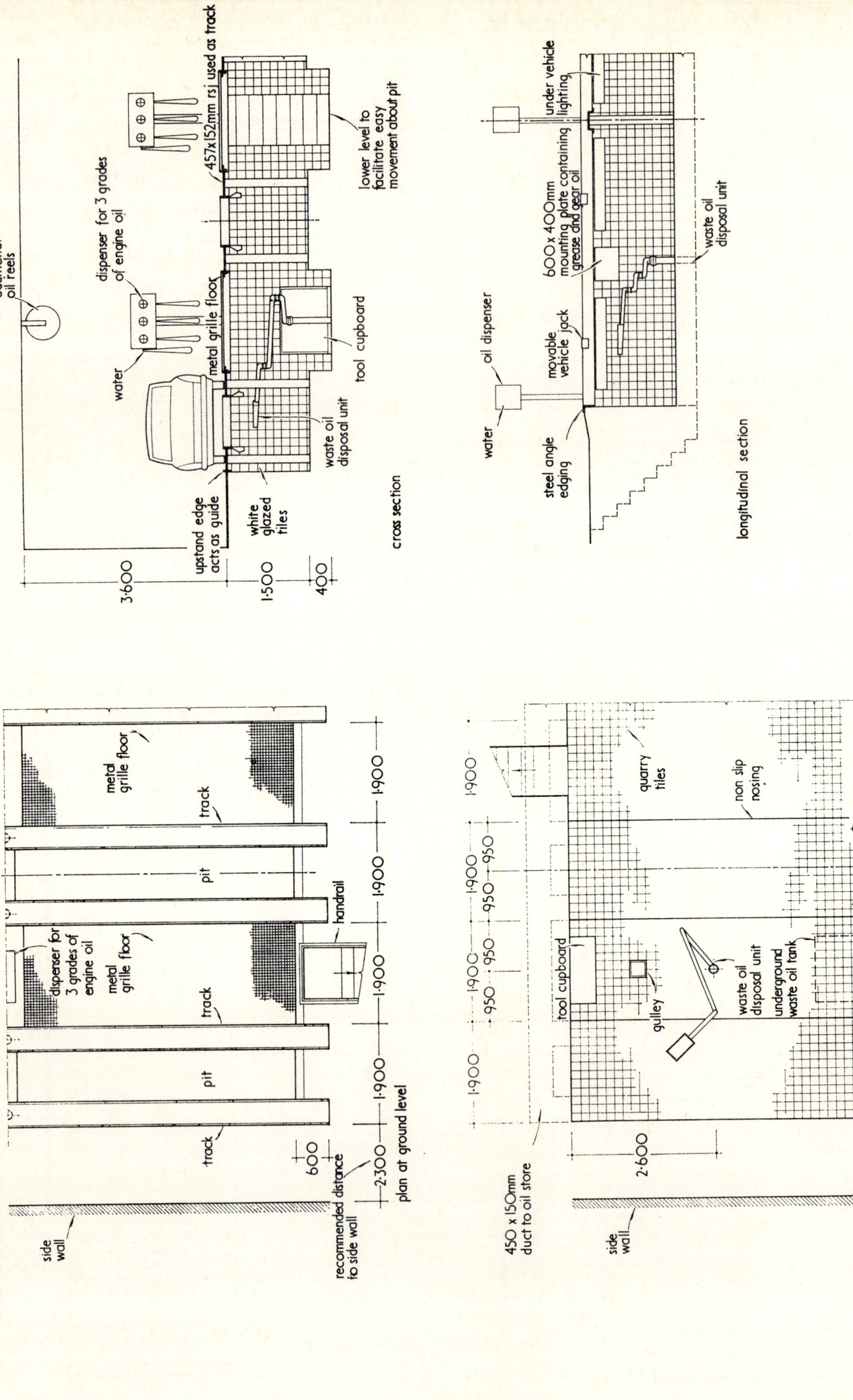

8.9 *Service pits for private vehicles*

The interceptor chamber must be sited where local authority silt collection vehicles can get at it. Vents must be brought above ground level before being connected to the manifold and main vent stack.

Local authorities usually require that the chamber should have a 60 gallon (275 litres) capacity, but there are smaller preformed plastic or GRP interceptors on the market which some authorities will accept.

The chamber should not be rendered on the inside. Render tends to fall off and block the pipes.

4 VEHICLE SERVICING

4.01 Vehicle servicing bays

The Conditions of Appointment of authorised examiners (under section 43 of the Road Traffic Act 1972) include appendices 'Requirements for premises and equipment'.

These are very detailed and should be consulted.[2] Regulations are now more stringent with the aim of reducing the number of licensed testing stations.

The DOT test requirements give seven recommended alternative layouts of hoist, brake tester and headlight tester. Note that the roller brake tester requiring a floor pit is shortly to become mandatory.

At least three servicing bays are needed to be economically viable. Bays for testing roadworthiness require a space of 9·0 m × 4·00 m.

A typical service bay is shown in **8.8**.

4.02 Hoist or pit

Some companies do not install pits (except of course for commercial vehicles) because of the danger of accidental ignition of petrol soaked mechanics' overalls. Smoking is forbidden but mechanics do smoke. Hydraulic or mechanical hoists are preferred but they require 4·8 m min ceiling or truss tie headroom whereas pits only require 3·3 m. Service pits for private cars are shown in **8.9**, commercial vehicles in **8.10**.

4.03 Car wash

A typical layout for an automatic car wash is shown in **8.11**.

4.04 Heating and ventilation

An extract vent system in the servicing area must not draw air from less than 1·2 m above floor level, to avoid drawing petrol fumes into the system and through electric fan motors. In any case the equipment should be approved non-sparking apparatus.

5 OTHER FACILITIES

A service station may need some or all of the following, depending on size and type of trade:

- *Office:* essential for all but the smallest station, to keep the safe containing takings, DOT test certificate blanks etc. It is used to interview credit customers in confidence and to make up wage packets.
- *Toilets:* at a small station, staff and customers can share these.
- *Staff rest room:* for relaxing, making tea.
- *Shop:* to sell motor accessories and/or confectionery, soft drinks and other impulse goods. Counters need space for self-service console if self-service, or cash register. Shop width should be not less than 3·0 m; area between 25·0 and 60·0 m² with storage space for goods, off the shop area. There is usually no special reason why these other facilities should be attached to servicing or wash bays.
- *Security:* design and detailing should make forced access by thieves very difficult in any part not visible from the road. This generally means rear wall, possibly flank walls and roof (eg rooflights).

6 REFERENCES

[1] Petroleum (Consolidation) Act 1928. Model code of principles of construction and licensing conditions (part 1) for the storage of cans, drums and other receptacles, petrol filling stations, HMSO 1968, reprinted 1971, 21 p.

[2] DOE Statutory vehicle testing scheme, *Appointment of authorised examiners, Vehicle inspection division.* Department of the Environment.

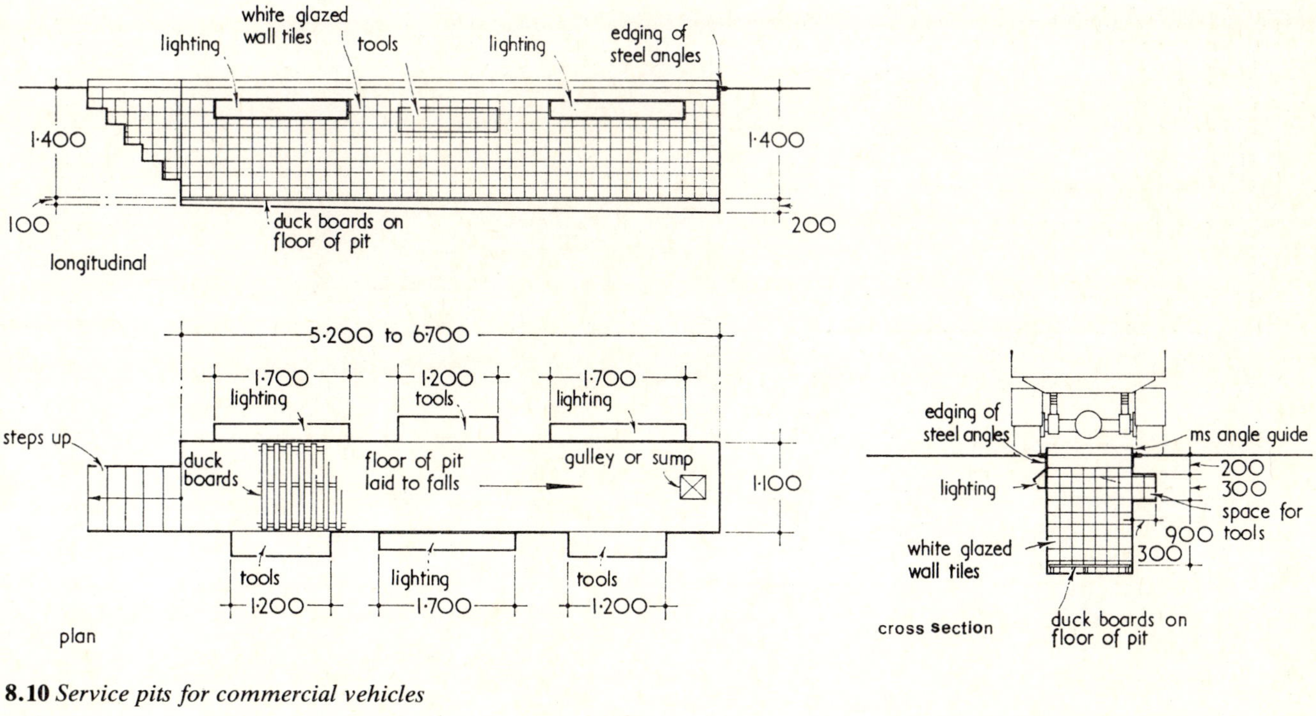

8.10 *Service pits for commercial vehicles*

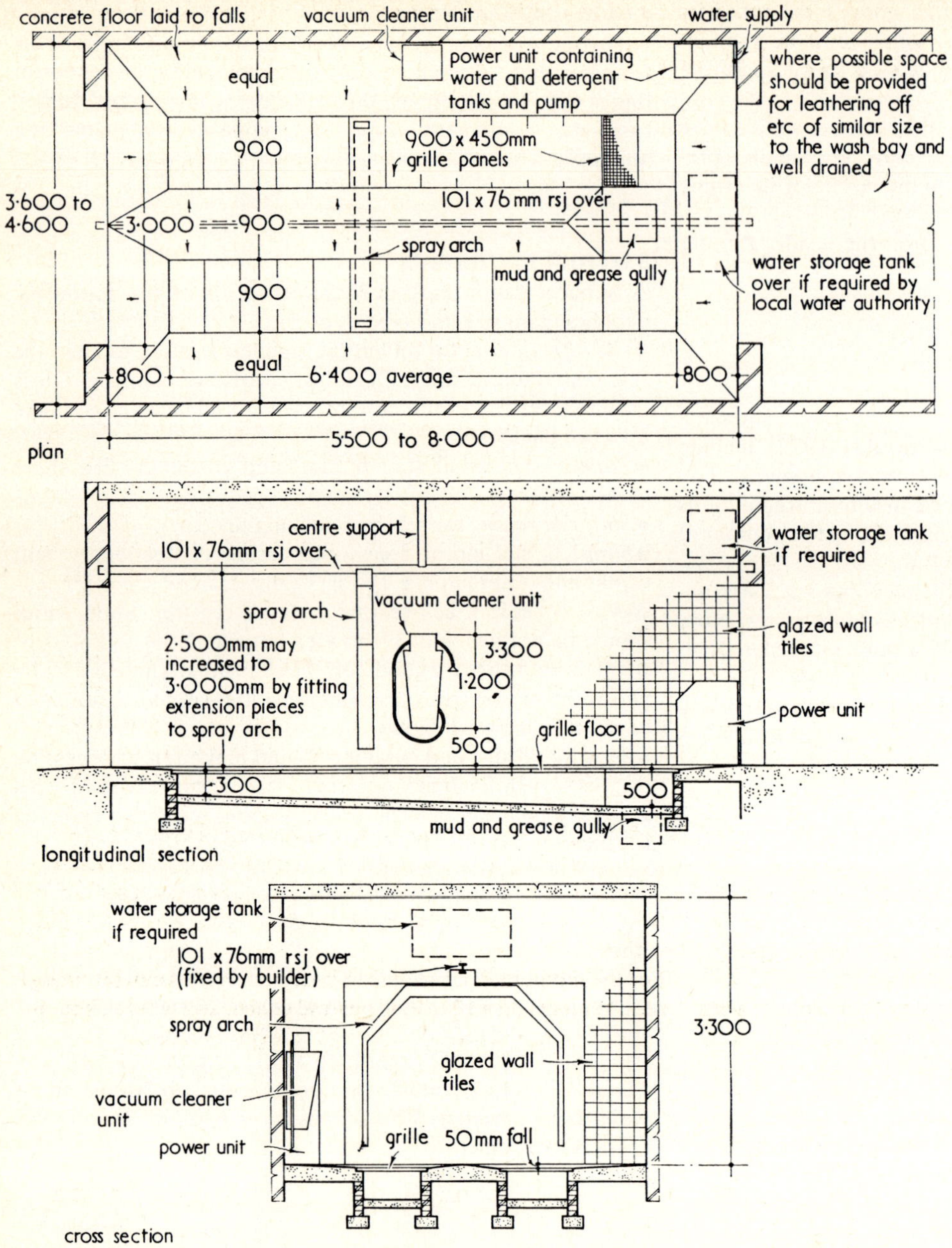

8.11 *Small washing bay with spray arch for automatic operation*

9 Bus and coach facilities

David Knipe

Formerly Assistant Regional Architect, Southern region, National Bus Company; now in private practice

CI/SfB 221
CI/SfB (1976 revised) 124

Contents

1 INTRODUCTION

This section is not exhaustive in all the aspects of designing for buses and coaches. It does not, for example, deal with planning for the fuelling, washing, chassis-cleaning and brake-testing of vehicles; nor does it discuss the selection of materials for and the design of roads, barriers and shelters.

It is, however, intended to be a guide to the understanding of the requirements of bus and coach operators, together with their vehicles, staff and passengers.

2 THE VEHICLES

2.01 Vehicle size

Present legislation in the British Isles restricts rigid road vehicles to a maximum 12 m length. Articulated vehicles may be longer, but, as yet, apart from isolated experiments, no such buses or coaches operate in Britain.

The 12 m long coach is, however, a reality, particularly with respect to continental touring coaches. Buses have not quite reached that proportion, though from **9.1** and **9.2** it will be seen that an 11·4 m long and a 4·4 m high bus is a combined single- and double-deck design criteria.

Articulated buses

In Europe articulated buses, **9.3**, are in full operation, though it should be said that they do so at the moment during peak travelling periods and where roads are of generous proportions. These buses are 17 m long and restricted to a single-deck design. Europe is not alone in operating them for they can also be seen in various parts of America, Canada and Australia.

Trends

It is difficult to predict in what direction, if any, bus and coach design will move. Experiments have been carried out with buses driven by electric motors and powered by a large traction battery pack towed in a trailer behind a standard bus adapted for the purpose. Again, in Europe, a limited number of services are operating using electric buses.

With the possibility of innovation in battery design and specification, particularly to its size and weight, this may ultimately dispense with the use of a trailer and result in buses similar in shape, size and styling to the diesel ones which we are familiar with.

For the moment, it is prudent to base one's minimum planning requirements on a 12 m long bus and coach, and where circumstances permit make allowance for longer vehicles.

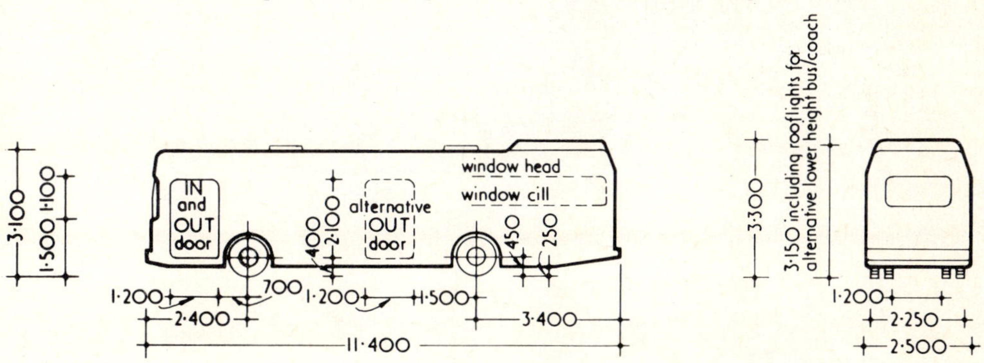

9.1 *Single decker bus*

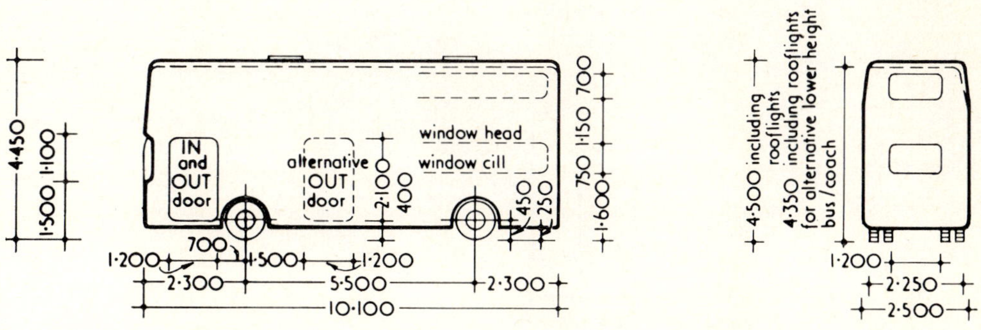

9.2 *Double decker bus*

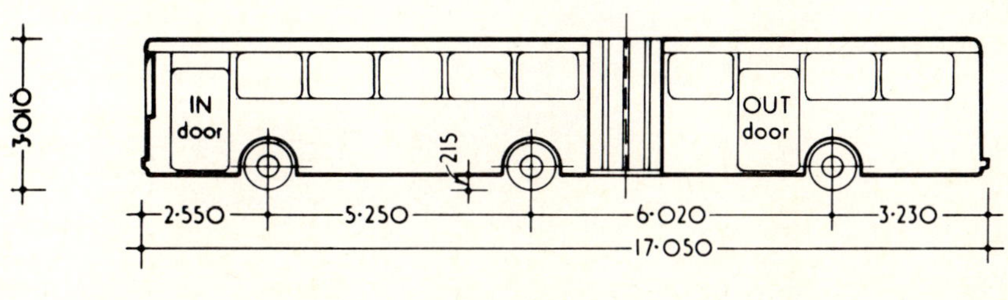

9.3 *Articulated bus, common in Europe*

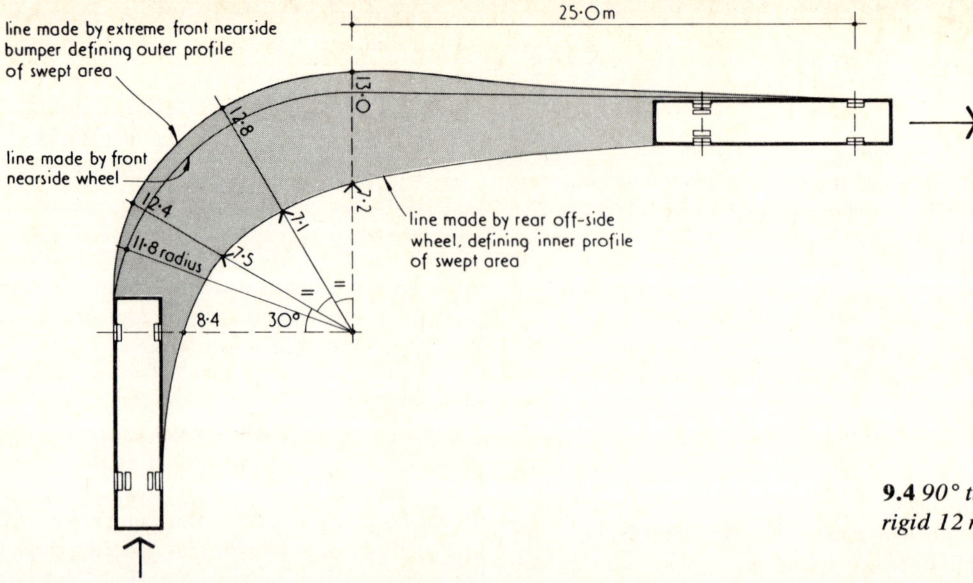

9.4 *90° turning pattern for rigid 12 m vehicle*

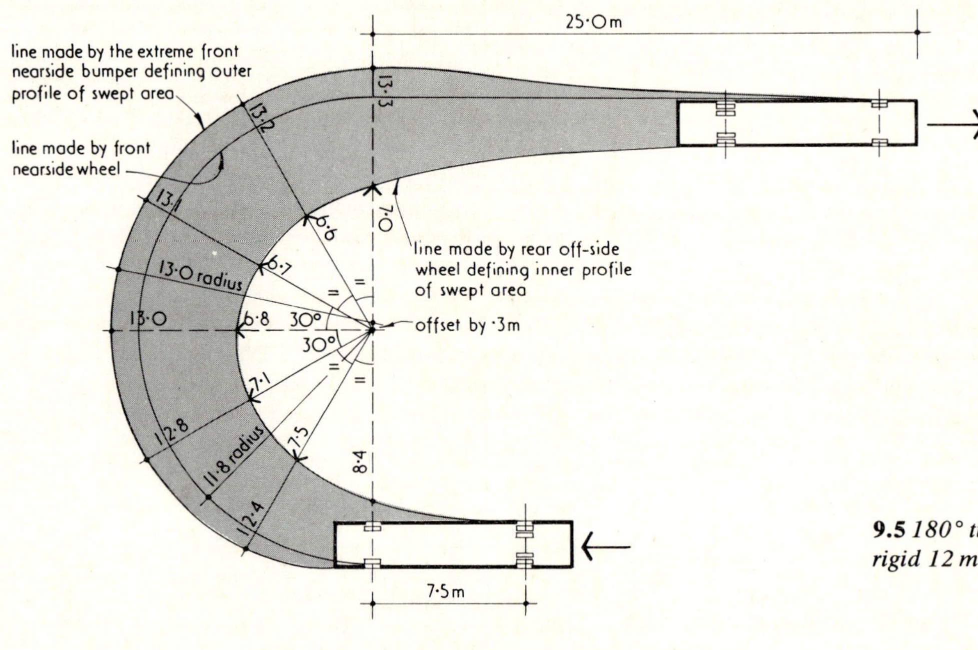

9.5 *180° turning pattern for rigid 12 m vehicle*

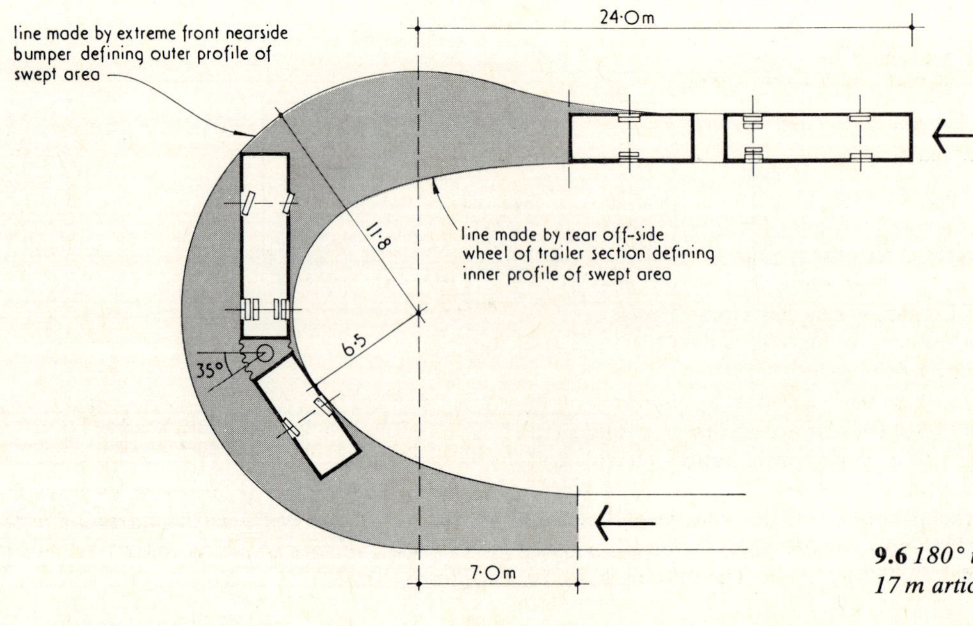

9.6 *180° turning pattern for 17 m articulated vehicle*

3 VEHICLES' MOVEMENTS AND PARKING

3.01 Buses and coaches

Turning patterns

The patterns made by a bus or coach, assuming average driving ability in turning through 90° and 180° angles, **9.4**, **9.5**, show that the movements are not simply circular. This is because the front wheels are steered whereas the rear wheels are virtually dragged around a corner.

With the front wheels turning in a radius of about 11·8 m, the overhang at the front of the vehicle does not turn in a similar circular movement until about 90° of the turn is complete, while the rear overhang and wheels are obliged to move in a spiral shape.

In addition the vehicle requires considerable distance to manoeuvre into and out of its full turning circle. The return of the vehicle into a straight line of motion does not occur until a point well after the front wheels have been straightened.

Swept area

The area contained by the curves made by the outer point of the front overhang and the track of the inner rear wheels in motion is known as the 'swept area'. It is not easy to set out these curves geometrically and so a rough guide method is illustrated on the diagrams.

Turning circle

A much simpler way of ensuring sufficient space for turns is to plan based on a 26 m turning circle, which compares with, say, a 10 m turning circle of an average length saloon car.

Changing lock

When a layout requires the vehicle to carry out a snaking movement it must be remembered that time and distance is required to change from one wheel lock to another. For vehicles moving slowly, a distance of 6 to 10 m should be allowed to change from full lock to full lock.

3.02 Articulated vehicles

Although considerably longer than the conventional bus or coach, the articulated vehicle has a more compact turning circle at 23·6 m, **9.6**. The distance required to straighten out after completing a turn is also nominally less.

This is achieved by a linkage between the front steering wheels of the front cabin, and the single set of wheels of the trailer unit so that the trailer unit 'tracks' the front section.

The turntable between the two sections limits the angle between them to 35° and when this angle is reached when reversing, to avoid a 'jack-knifing' situation a buzzer sounds and the brakes are applied automatically. These are then only released by engaging a forward gear.

3.03 Parking

The layout for parking will depend upon the purpose.

If one is considering a bus and coach park where any one of the vehicles wishes to enter or leave at any time, then **9.7** illustrates the guideline. The width of the bays could be increased to 4 m to give a greater circulation space for passengers between vehicles if the use of the park is predominantly for setting down and collecting coach parties.

If, however, the purpose is for a bus or coach operator to park a maximum number of vehicles in an enclosed or open site, where passengers are no longer a consideration, then a different approach is required, **9.8**.

Often, in these circumstances, vehicles can be parked in a pre-determined order and then the order reversed, in principle, to move all vehicles away again. The parking bays can

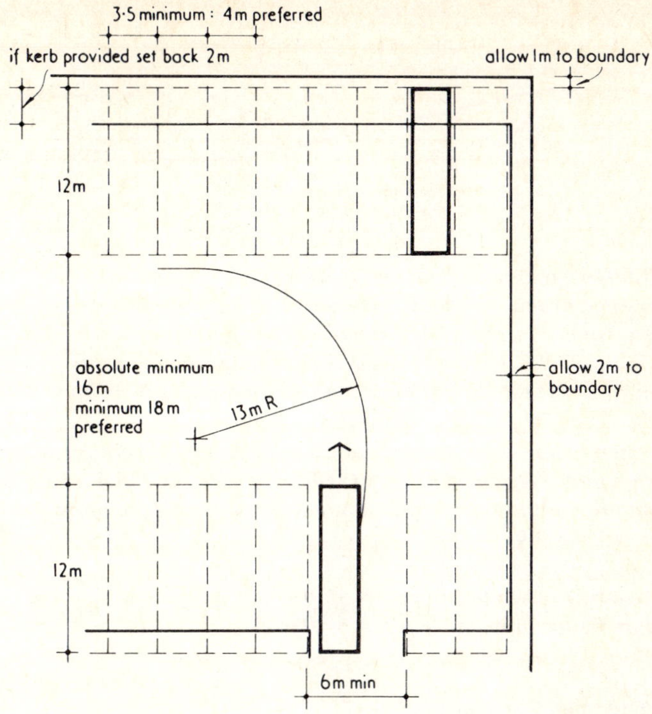

9.7 *Coach park for random arrival and departure of vehicles. The larger bay size (4 m) is necessary if coach parties enter and leave the coaches in the park*

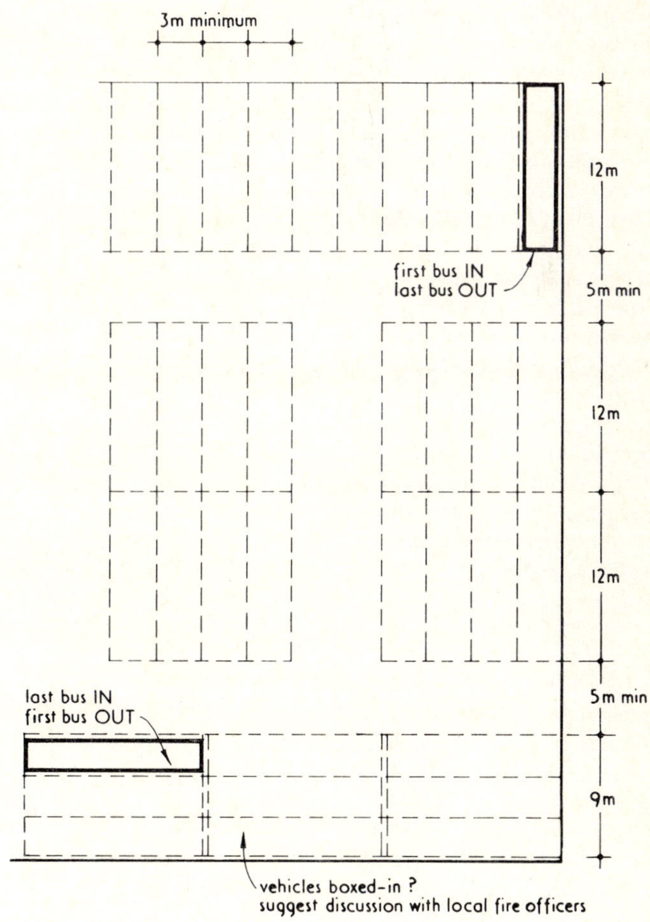

9.8 *Bus garaging layout for where the buses are parked in a pre-determined order to get the maximum number of buses in the space available subject to limitations imposed by the fire officer*

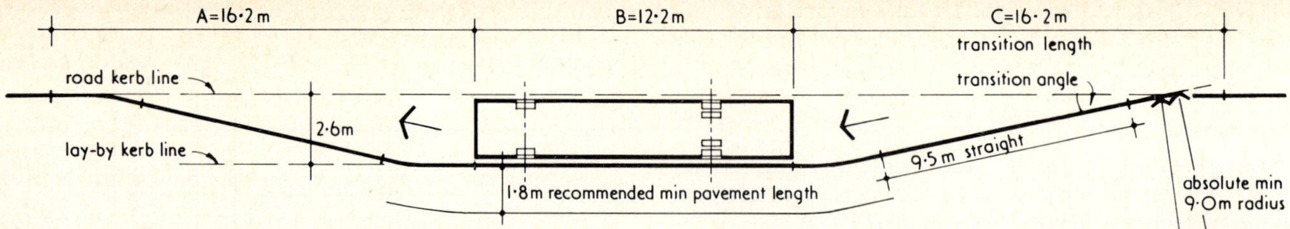

9.9 *Design criteria for a lay-by with one bus stop, assuming normal urban speeds of approach. The transition length of 16·2 m is considered the minimum requirement for a 12 m rigid vehicle. A three bus stop is considered the desirable maximum in a lay-by (ie maximum comfortable boarding distances for passengers to walk).*
When setting out alternative layouts for lay-bys, overall length should be calculated as A + (x) B + C, where x is the number of buses to be accommodated. So, using the transition lengths in **9.9**:
Lay-by for one bus stop = 44·6 m
Lay-by for two bus stops = 56·8 m
Lay-by for three bus stops = 69·0 m etc,

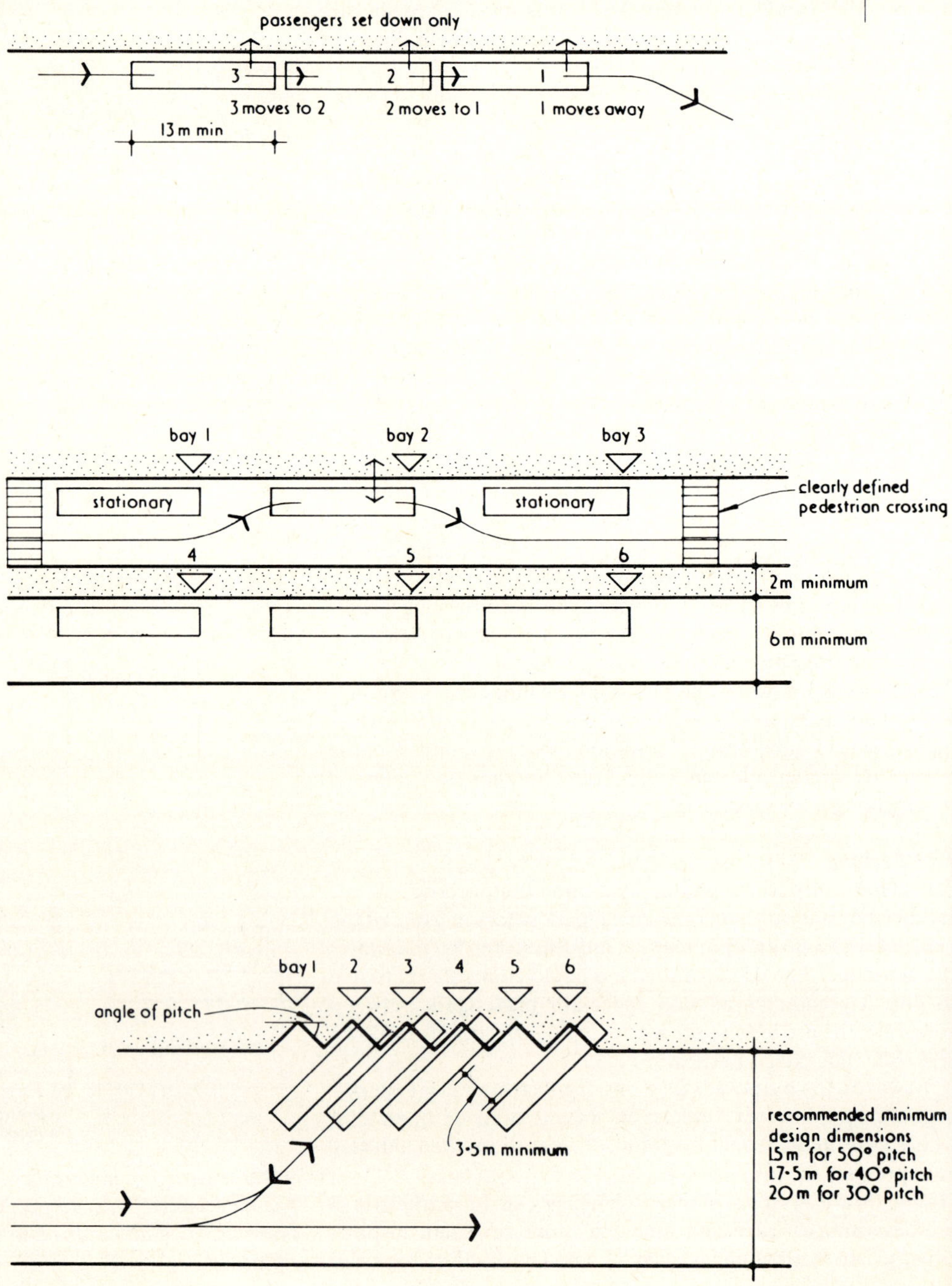

9.10 *Vehicle manoeuvres used in approaching parking bays:*
a *Shunting is used where a vehicle only sets down passengers onto their concourse before moving away to park or to a bay position for collecting passengers. This manoeuvre avoids waiting to occupy a predetermined bay and effectively reduces journey time.*
b *Drive through bays are fixed bay positions for setting down and/or collecting passengers. They are in a line, so a vehicle often has to approach the bay between two stationary vehicles. In practice it is often necessary to have isolated islands for additional bays with the inevitable conflict of passenger and vehicle circulation.*
c *'Saw-tooth' layouts have fixed bay positions for setting down and/or collecting passengers with the profile of the concourse made into a saw-tooth (sometimes referred to as echelon) pattern. In theory the angle of pitch between the vehicle front and the axis of the concourse can be anything from 1° to 90°. In practice, however, it usually falls between 20° and 50°. The vehicle arrives coming forward and departs going backwards, thus reducing the conflict between passenger and vehicle circulation, but demanding extra care to be taken when reversing out of the bays*

then be reduced to 3·0 m width. It would be prudent, however, not to box in too many vehicles and to allow for aisles of at least 5 m width to enable a fire engine to circulate throughout and give sufficient working space for firemen to operate in an emergency.

4 BUS STOP LAY-BYS

The basic bus station is of course a bus stop or stops by the side of a road. Modern road systems are designed for the bus to move over from the main traffic flow into a lay-by arrangement.

The layout in **9.9** illustrates the principle of design for a single bus stop lay-by, assuming normal urban speeds and minimum design criteria, and notes the method of calculating lay-by sizes for additional bus stops under similar conditions.

The transition angle is made less and the transition length made more as the approach speeds increase. In extreme cases, as for example urban motorways, it may be advisable to provide a slip-road for exclusive use by buses, rather than a lay-by of unwarranted length. In this case the lane width should be at least 5·8 m wide to allow buses entering and leaving to overtake those stationary.

5 BUS AND COACH STATIONS

5.01 Location

Where a concentration of bus stops is required in one particular location, it is preferable to provide a bus station, which could be defined as an area away from the general traffic flow giving buses and coaches freedom of movement to set down and collect passengers in safety and comfort.

Its location would normally be in or close to the centre of town, and ideally near to the railway station in order to promote easy and efficient passenger interchange. They may also be located next to places of high intensity public use, eg busy airports, ports and conference centres.

5.02 Factors affecting size of station

Stations will vary in size governed by the following basic points, apart from the obvious physical restraints of the site:

• *The number of bays to be incorporated* (the term 'bay' is used in connection with stations instead of the term 'bus stop').

This is determined by the number of bus and coach services to be operated from the station, and by how practical it is, related to the local timetable, to use an individual bay for a variety of service routes.

• *The vehicle manoeuvre selected to approach the bays.* Three basic types of manoeuvre are used, namely 'shunting', 'drive-through', and 'saw-tooth', which are described and illustrated in **9.10**. The 'saw-tooth' is further explored in **9.11** and **9.12**.

The choice of manoeuvre will be influenced by the size and proportions of the site available, the bus operators' present and anticipated needs, and in particular the preference of their staff. Some will accept the saw tooth arrangement while others prefer the drive through.

The area of the site is further added to by the requirement of 'lay over'. This is that vehicles having set down their passengers, but which are not required to collect passengers are parked on the station until needed again. The layout for this should be based on the requirements for parking, as described in section **9.2**, but preferably in such a manner that no vehicle is boxed-in by another, and of course positioned so as not to interfere with other bus movements.

In some cases economy of space can be achieved, again dependent upon local timetables, by using spare bays for lay-over purposes.

• *The facilities to be provided for passengers.* Provision for passengers will depend upon anticipated intensity of use and possibly existing amenities. If, for example, there are already public toilets, a bus and coach information centre and cafés nearby, then these may not be required on the station concourse. However, waiting-room facilities will probably be required with somebody on hand to give information and supervision. In more comprehensive schemes, in addition to a waiting room, a buffet and public toilets, one may plan for kiosks and enquiry, booking, left luggage and lost property offices.

• *The facilities to be provided for staff.* There will invariably be an inspector or inspectors in a station who, as well as assisting passengers, is primarily concerned with supervising the comings and goings of vehicles, their drivers and conductors.

If there is a depot near to the station then most staff facilities will be provided there. However, if the depot is some distance away, it will be necessary to provide canteen and toilets for them on the station site, so that during breaks and between working shifts they do not need to get back to the depot until they return their vehicle for long-term parking.

Should the depot be even more remote, it will be necessary to provide all facilities at the station site and only basic amenities at the depot. In this case, as well as the canteen and toilets, a recreation area, locker rooms and 'pay-in' facilities should be provided. The latter is an office area where drivers/conductors check, then hand over monies taken as fares, which in turn are checked and accounted for by clerical staff.

• *The facilities for bus maintenance.* It will be appreciated that the proper inspection, repair and servicing of buses and coaches is an integral part of a bus operator's responsibility.

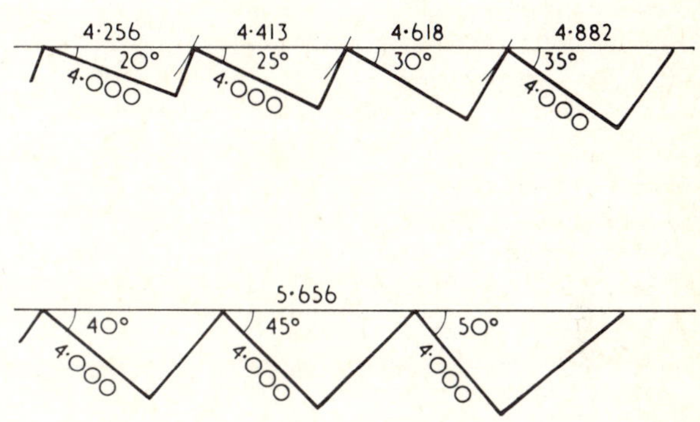

9.11 *As the angle of pitch in saw-tooth bays increases so does the distance between each bay*

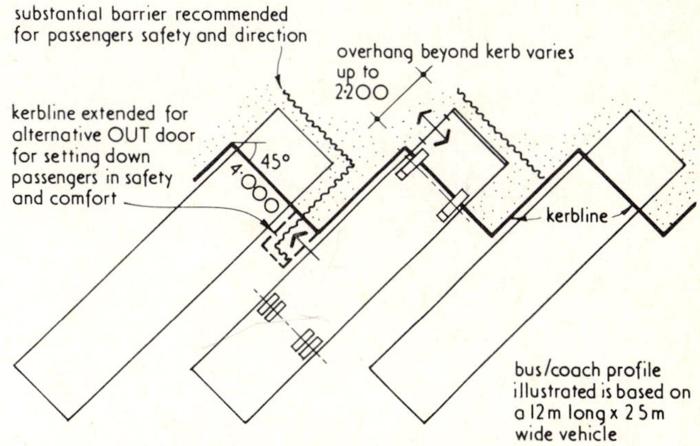

9.12 *Passenger safety and control are particularly important when detailing saw-toothed bays*

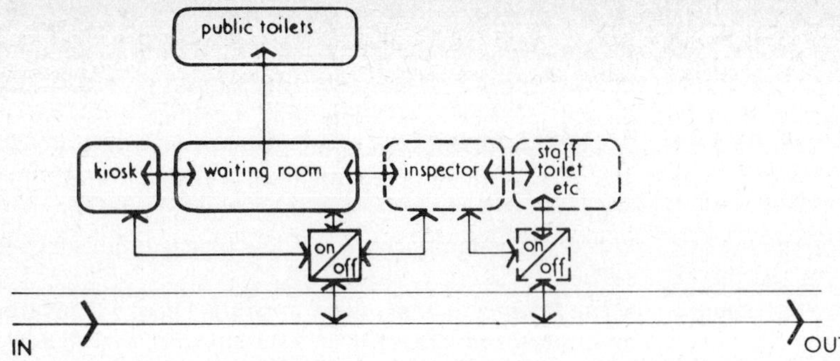

9.13 *Accommodation layout for a bus station in a small town*

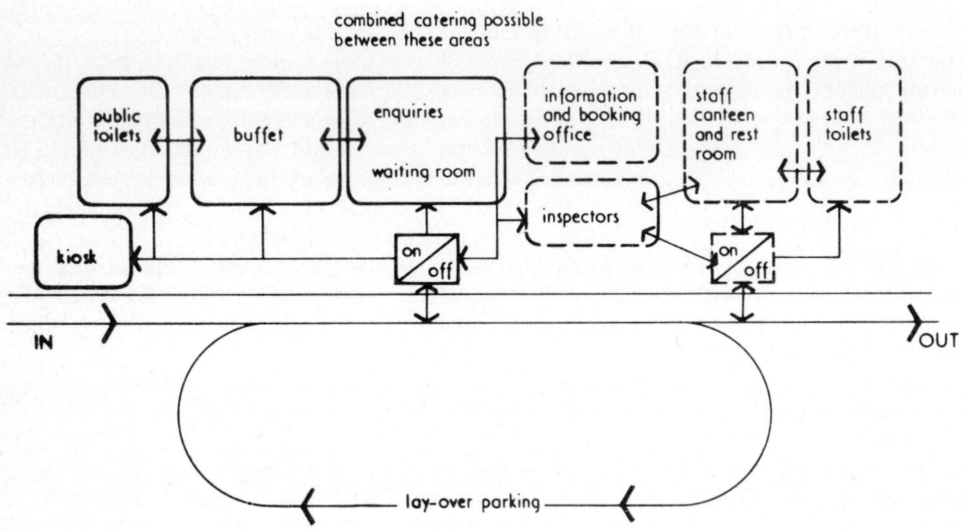

9.14 *Layout for a medium-sized town with both terminal and in transit services*

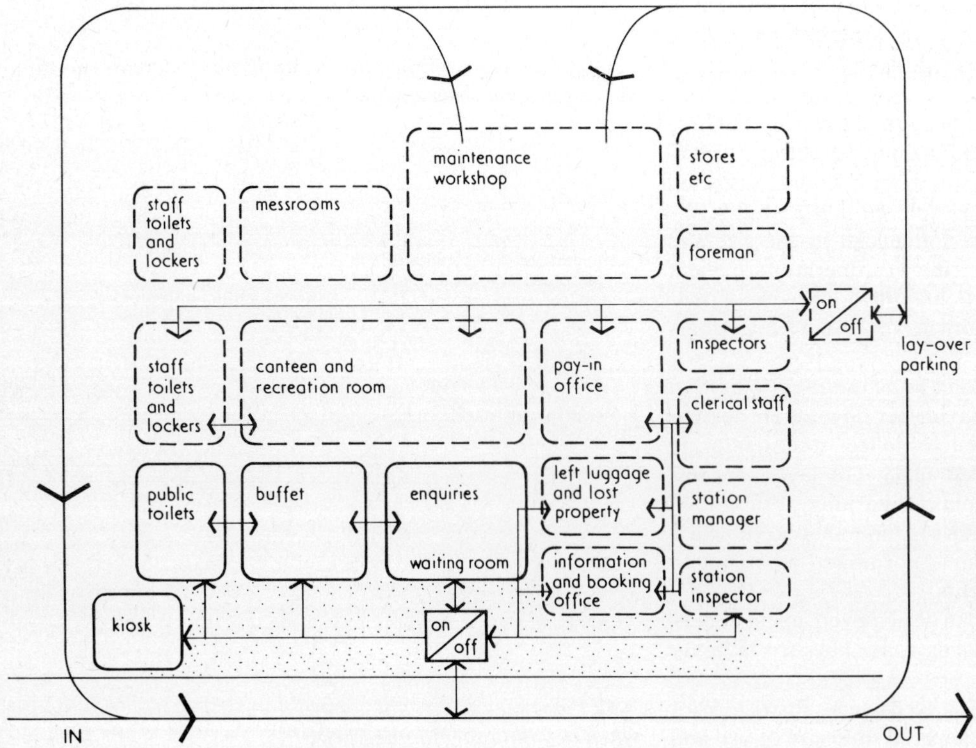

9.15 *Layout for a bus station in a large new town*

Normally such work would be carried out at a local depot, with a repair workshop together with fuelling, washing and garaging facilities.

The provision of some or all these facilities within a station complex is unusual, but by no means unique. For a new town bus station or for a station in a traffic congested township, where it will be difficult and time consuming to drive to and from the station and depot, the inclusion of at least a workshop would be advantageous.

Having established the accommodation to be provided on the station site, the problem is then to combine them in well-planned arrangement. As a guide, their relationship one to another is illustrated by **9.13**, which might be a station for a small town with buses and coaches passing through; **9.14**, which might be for a medium-sized town with both terminal and in-transit services, and lastly **9.15** which might be for a large new town with similar services.

5.03 Joint company use

If two or more bus and coach companies are to operate from the same station, this can lead to the different types of vehicle manoeuvre being planned for on the same site. Figure **9.16**, which is based on the proposals for a new station within a centre-town commercial development in the south-east of England, illustrates such a situation. The more local and predominant company favoured the saw-tooth layout, while the other preferred the drive-through arrangement. From it one can see how full use has been made of a restricted site, and how the conflict between passengers' and vehicles' circulation has been reduced as far as practical.

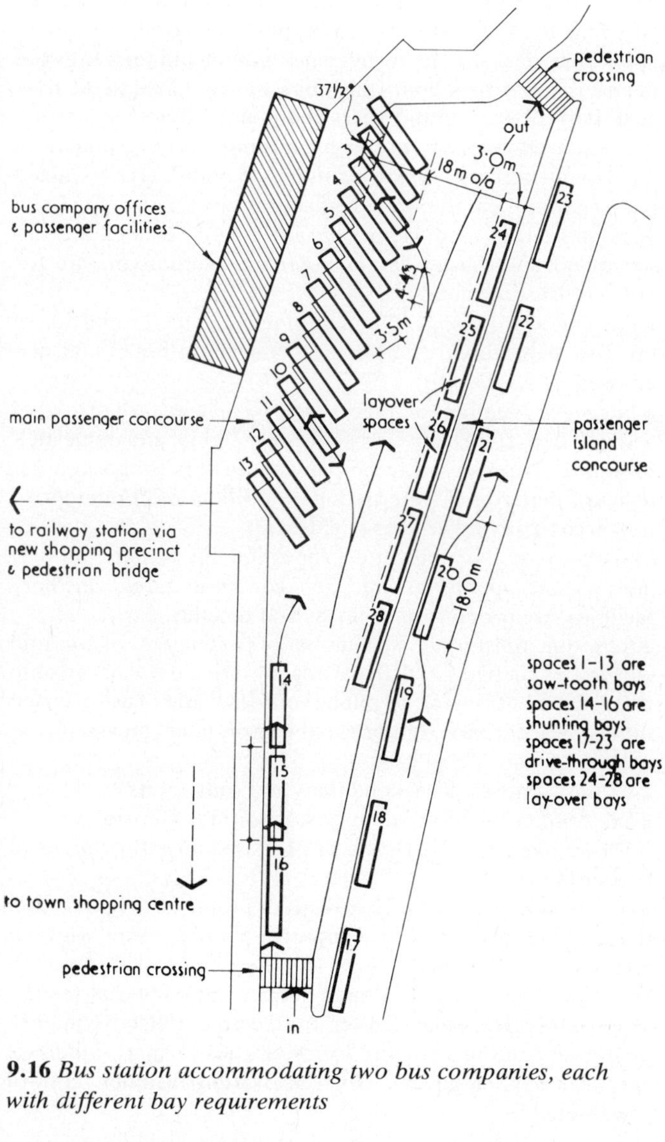

9.16 *Bus station accommodating two bus companies, each with different bay requirements*

10 Air passenger terminals

Gollins, Melvin Ward

CI/SfB 241
CI/SfB (1976 revised) 144

Contents

1 INTRODUCTION

1.01 New building type

Although established as a special building form by 1939, the major development of the airport terminal and its associated buildings can only compass 30 years. An extraordinary proliferation of building types and solutions have appeared in this period, stimulated by unprecedented rates of growth in the industry. The most-up-to-date terminals became obsolete or required extensive alteration or extension in a few years, and solutions which appeared attractive on paper were found difficult to operate or update as changes occurred.

1.02 Alternative forms

There are two basic approaches to terminal planning, which have a fundamental effect on the building form and detail layout, arising from the position of the aircraft in relation to the passenger building.

- The most widely utilised brings the aircraft against or as near to the building as possible and connection is made between them by methods varying in sophistication from walking to complex mechanical airbridges. The obvious problem with terminals of this layout is that their scale is directed mainly by the aircraft, which have steadily increased in size over the development period, and which require considerable manoeuvring space about the building perimeter.
- The alternative is to place aircraft remote from the terminal, connecting them by means of transportation systems. One of the earliest and most successful of this type is Dulles Airport, Washington, where passenger transport vehicles (PTVs) are provided to make the connection from aircraft to building.

1.03 Standards

Although a large number of airports and their buildings have been completed, there has not evolved a solution which can be applied as standard. It would appear that each has its own particular characteristics and external requirements that have dictated the solution. Even the basic principles and criteria which are contained in this section must be applied with caution as universal standards. (In particular, this is true of airport size: most of the information given relates to larger international terminals.) Consequently, the following sections deal mainly with checklists and requirements which need further investigation according to the demands of the airport authority or airline.

Although more basic standards will be established in the years to come, the field is still open for further ideas and experiments. That the complexity will continue to increase is unfortunately also a possibility, as witnessed by the recent necessary introduction of much stricter security measures.

2 CHECKLIST OF BASIC DATA REQUIREMENTS

2.01 Sizing the terminal and related facilities

Requirements will vary considerably with particular requirements. The relevant airport authority and airline should give guidance, but for very large terminals the use of sophisticated mathematical models may prove necessary. For initial planning purposes, determine the peak hourly passenger throughput, total arrivals and departures, and allow between 18 m² and 23 m² per person. This results in an approximate gross area including offices and all support facilities for an international terminal. For domestic, allow 60 to 70 per cent of this figure. For more accurate determination of areas, a particular passenger throughput is chosen from experience to form the basis of calculations. This is known as the standard busy rate (SBR) and is the thirtieth most busy hour of the year, usually taken as approximately 80 per cent of the peak hourly throughput. To determine the peak, SBR and other factors affecting terminal planning, the following basic data will be required, and where applicable, data should be given per annum and for a selected peak week and peak hour (the 'busy' hour) related to the base year, current year, plus 5 years and plus 10 years.

- *Aircraft movements (passenger aircraft only):* Number of movements/aircraft types including 'Combi' types—mixed passenger and cargo/number of stands per type/average and peak passenger load factors/proportion of scheduled, non-scheduled, and charter/details of general aviation (private aircraft) if using terminal.
- *Passenger movements:* Number of movements/proportion of arrivals, departures, transfer and transit (international and domestic).
- *Baggage:* Number of pieces per passenger/proportion of arrivals, departures and transfer (international and domestic).
- *Visitors:* Number of accompanying visitors/proportion for arrivals, departures (international and domestic)/number of non-accompanying visitors (sightseers).
- *Employees:* Number and proportion for airport, aircraft and airline operations and for concessionaires, ancillary facilities, etc/proportion of males and females.
- *Landside transportation:* Number of passengers, visitors and employees arriving and departing by private vehicles (note ratio of owner drivers); by public vehicles (ratio of use of coach or bus, hire car, taxi, rail or rapid transit, helicopter, etc).

2.02 Basic data on local conditions and constraints

- *Site conditions:* topography/soil conditions/climate.
- *Master plan:* check existing or proposed airport or regional master plans.
- *Airside constraints:* check existing or proposed height restrictions (glide paths, radar contours, control tower viewing angles, etc)/siting restrictions (relationship and distance to runways, taxiways, public safety zones, etc)/material restrictions (pilot glare, radar 'clutter' from certain reflective materials)/noise contours/existing fixed facilities (airport buildings, service areas, cargo and general aviation facilities, control tower, etc).

• *Landside constraints:* check existing or proposed—road and public transportation facilities/adjacent commercial, industrial, residential developments, etc.

2.03 Basic economic data

• *Cost planning:* method of financing and income generation.

• *Labour:* labour availability and costs (current and projected for both terminal construction and operation).

• *Materials and equipment:* note any local anomalies regarding cost, availability and maintenance.

2.04 Basic legislative and statutory requirements

The following refers to UK practice. Local requirements, including environmental and political factors, should be determined.

• *Authorities:* The local airport authority or BAA/Civil Aviation Authority (CAA)/HM Customs and Excise/HM Immigration/Port Health Authority/the local authority and advisory bodies—International Civil Aviation Organisation (ICAO)/ International Air Transport Association (IATA).

• *Basic legislation:* Airports Authority Act/Civil Aviation Act/ Offices, Shops and Railway Premises Act/Factories Act/ Town and Country Planning Act/Fire Precautions Act/Building Regulations 1976.

3 CHECKLIST OF BASIC PLANNING OBJECTIVES

3.01 Land transport

The main aim is to provide fast, efficient public and private transport facilities between the terminal and transport networks/centres of population and between separate terminals. Each interchange with each mode of transport should be in close proximity to the terminal and routes should be easily intelligible to the first time user. Note particular importance of kerbside length and car parking capacity (see paras **5.01**, **5.02** and **5.03**).

3.02 Passenger convenience and safety

The passenger is primarily concerned with moving between one mode of transport (land) and another (air) with the minimum of inconvenience and in the shortest time.

• *Main aims:* The route should involve as short a walking distance as possible, preferably unencumbered by baggage; be easy to follow, especially for the first-time user; have the minimum number of changes in level and direction; be protected from the weather and be safe (protected from vehicles and aircraft).

• Note particular importance of facilities for young or invalid passengers, late departing passengers and their baggage, meeting or leaving visitors, access to amenities en route and organisation of necessary control points to cause minimum delay.

3.03 Security and its effect upon flow planning

• *National security:* Main aims—to control the movement of unauthorised persons and commodities into or out of a country (international terminals only). Passengers are technically considered to have entered or left a country at a point within the terminal known as the airside landside barrier. All passenger access to or from aircraft is prohibited except by crossing this barrier at points controlled by government agencies.

Control over departures is generally less comprehensive than over arrivals, applies to passengers only (permitting baggage to be checked in independently at any convenient location, landside or airside) and involves two barrier controls, immigration and customs.

Control over arrivals applies to passengers and their baggage necessitating the following sequential organisation before crossing the barrier: health control (located as close to arrivals aircraft as practicable to minimise risk of infection); immigration control; baggage reclaim; customs control (Note: crossing of the airside-landside barrier by employees and terminal facilities serving both airside-landside areas require special security controls. Note also that domestic passengers and passengers from countries having reciprocal agreements may not require processing as above and should have separate and secure routes to by-pass these controls).

• *Terminal and airline security:* Main aims—to prevent weapons (including explosives) from being carried on board aircraft or through the terminal either directly (on the person or in the baggage of arriving and departing passengers) or indirectly (by transfer from the person or baggage of arrivals, transit or transfer passengers to departure passengers prior to boarding). To achieve this, consider:

• Complete and secure separation of arrival and departure routes at all stages from gate position to airside-landside barrier/airport controlled security checks on passengers and hand luggage at transit-transfer crossover points to departure areas; and on departing and arriving passengers at the airside-landside barrier/security checks on luggage prior to distribution to aircraft and to baggage reclaim facilities/optional airline checks at gate positions.

• Where separation of routes as above cannot be achieved, it is generally impracticable to mount airport controlled security checks on passengers and hand luggage at all arrival gate positions, therefore protection is restricted to the most vulnerable areas only (the public concourse and departure lounge). The concourse is protected by a check on arriving passengers and their baggage at the airside-landside barrier; the departure lounge by separating it from the arrival areas and by checking all passengers and their hand luggage on entry to it—ie at the airside-landside barrier and at transit-transfer entry points. Note—this will not prevent weapons transfer in other airside areas (and consequent risk to aircraft), therefore provision should be made for airlines to install their own security facilities at the departure gate positions, if required.

3.04 Baggage handling

• *For departing passengers:* to free them of heavy baggage at the earliest opportunity, to weigh, identify, sort, security check and dispatch the baggage to the appropriate aircraft safely, reliably and in synchronisation with the flight departure time. The system must also accept occasional baggage arriving too late for normal processing.

• *For arriving passengers:* the system must make provision for unloading baggage from aircraft, dispatching to the terminal, sorting, security checking (if required) and distribution to a baggage reclaim area for presentation to and convenient identification and retrieval by passengers. This must be achieved safely, reliably and in synchronisation with the movement of passengers from aircraft to baggage claim. (For passenger convenience, baggage reclaim should be located as close to the landside transportation facilities as practicable—for international arrivals, immediately prior to the customs check at the airside-landside barrier.)

• *For transfer passengers:* provision must be made for baggage to be either sorted and transferred direct across the apron or to follow the arriving baggage to the sorting point for transfer to departure.

3.05 Extensibility and adaptability

Technological, organisational and economic developments can, and invariably will effect considerable changes in terminal operations. These cannot be forecast with accuracy, but within the constraints of economic feasibility an inbuilt capacity for extensibility and adaptability should be aimed for, capable of implementation without disruption to normal terminal

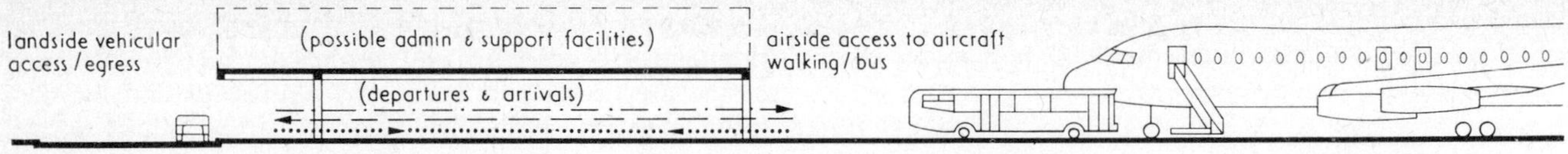

10.1 *Forms of typical terminals shown by cross-sections:*

a *Single level terminal, generally applicable to small or domestic terminals*
Arrival and departure routes split horizontally as flow plan diagram, **10.3**

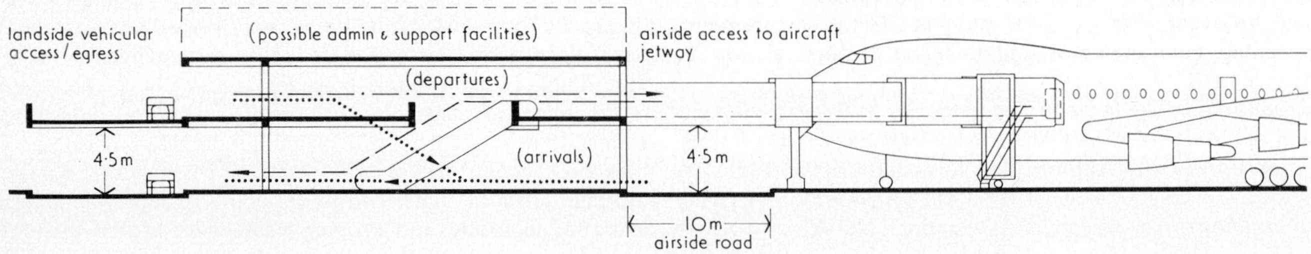

b *Two-level terminal—jetway type*

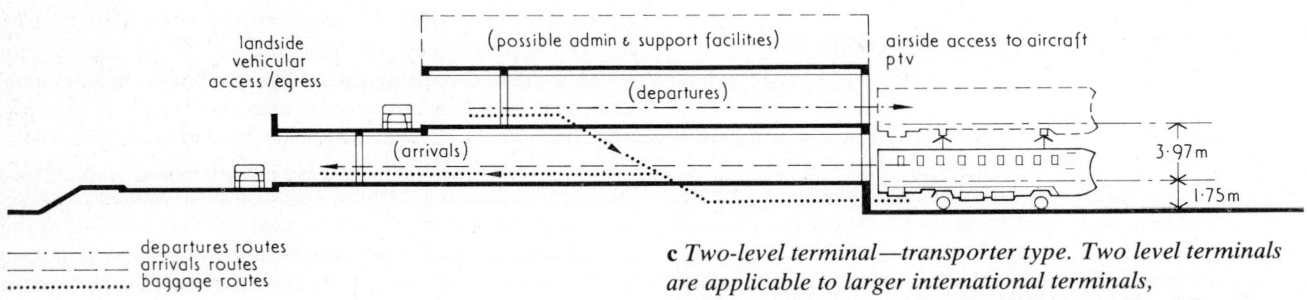

c *Two-level terminal—transporter type. Two level terminals are applicable to larger international terminals, domestic/international routes split horizontally while arrival/departure routes split vertically.*
Note possible variants of one and a half and three level types

operations. Areas of particular concern: car parking kerbside/ concourse/preflight assembly/baggage reclaim and sort/ aircraft parking stand arrangements/small incremental increases in the number of stands/security control techniques.

3.06 Basic administrative requirements

• *Airport management:* Main aims—to reconcile all the conflicting requirements of users and operators within a cost effective design. Particular concerns: capital cost/running cost (note importance of making maximum effective use of staff)/user appeal (to airline operators, passengers and visitors—note importance of passenger convenience, general amenities, VIP facilities, etc)/income generation (in addition to landing fees note increasing importance of concessionaires—a major source of revenue).

• *Airline operators:* Main aims—to provide a fast, efficient and attractive service to customers which is both competitive and economic. Particular concerns: passenger convenience/ passenger facilities/aircraft security, maintenance and servicing/maximum utilisation of staff and equipment/ minimum turn round time of aircraft on the ground. Note individual airline requirements and international recommendations set by IATA.

• *Government agencies:* The main agencies are health, immigration, customs and police. Main aims—to administer and enforce government legislation affecting international travel (see para **3.03**), general law enforcement, traffic control, etc.

• *Aircraft operations controllers:* Main aims—to control local aircraft movements both in flight and on the ground/to provide facilities and equipment for the maintenance of security and safety of aircraft in the air and on the ground to national (CAA, FAA, etc) and international (ICAO) standards.

4 CHECKLIST OF BASIC PLANNING CONCEPTS

4.01 Basic terminal types and configurations

The principal concepts employed in planning terminal design are shown **10.1**, but note that in practice considerable variations upon these are common.

The four basic terminal configurations are shown in **10.2**. These relate to three basic terminal types:

• *centralised* (where all passenger routes converge on and diverge from a centralised passenger processing and amenities facility)

• *decentralised* (where passenger routes pass through a number of separate processing amenity facilities, therefore generally more applicable to domestic operations)

• *Unit terminal* (where two or more quite separate terminals of the centralised or decentralised type are provided to serve specific functions, eg individual airlines, international only, domestic only, etc).

4.02 Basic terminal flow patterns

The basic vehicular/passenger flow patterns for a typical international/domestic terminal are shown in **10.3**.

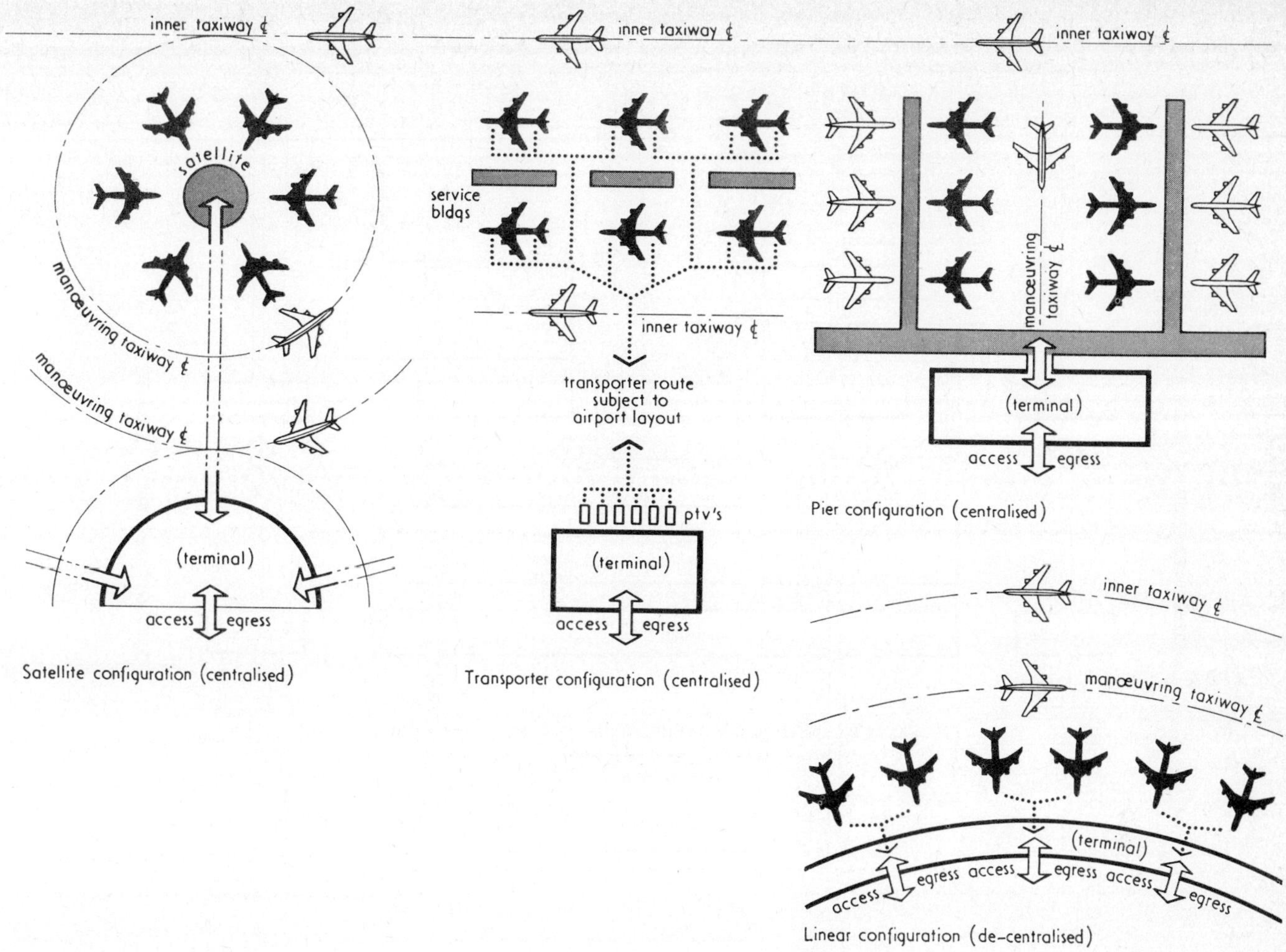

10.2 *The four basic terminal configurations. Selected examples are:*
Satellite: Charles de Gaulle, Paris, France
Transporter: Dulles, Washington, USA
Pier: Heathrow, London, UK
Linear: Dallas/Fort Worth, USA
Unit terminals: Kennedy, New York, USA

Note hybrid variants:
Pier/Satellite: Frankfurt-am-Main, Germany
Pier/Satellite/Centralised/Decentralised: Tampa, USA
Pier/Transporter: Lambert, St Louis, USA
Linear/Pier/Satellite: San Francisco, USA
Linear/Pier: Boston/Logan, USA

5 CHECKLIST OF DETAILED DESIGN REQUIREMENTS—PASSENGER AND VISITOR FACILITIES (LANDSIDE)

5.01 Transport between terminal and centres of population

• *Private vehicles:* Consider—vehicle circulation (one-way: airport entrance direct to car parking and kerbside(s) thence to exit); to and from car parking and kerbside(s)/separation and control of airside-landside vehicular routes/filter, passing lanes adjacent kerbside(s)/effect of right hand–left hand drive on terminal layout (driver's door to kerbside; departures set down upstream of arrivals pick up)/note local vehicle dimensions and percentage mix/fuelling point.

• *Car parking:* Consider—extensibility/weather protection (cars and persons)/check-in facilities/local parking standards/ segregation; by user (passengers-arrivals and departures, visitors, employees); by usage/(long-term and short-term parking)/location.

Adjacent to terminal on opposite side of access road (note the problem of pedestrian/vehicle segregation and effect of mezzanine bridge link between two-level access road on terminal floor heights); adjacent to terminal ends on same side of access road (note problem of terminal extensibility, walking distances); above the terminal (note expense, ramping, extensibility, vertical access to concourses); underground (note expense, extensibility, ramping, orientation problems). Calculations: for average private/public transport mix in a European terminal take 70 to 90 per cent of peak hour pass movement (total arrivals and departures), assume one car per person to give total parking requirements. Note local differences from the above norm.

• *Public transport:* Consider—tracked systems (note terminus and track requirements, proximity to terminal, protected passenger access)/bus-coach (note operators' requirements—airline; charter company; bus company;—kerbside loading-unloading requirements, parking and if large numbers involved consider separate terminus)/taxi-hire cars (note taxi ranks, upstream arrivals kerbside, drivers' facilities, hire car parking and office close to terminal)/helicopter (note special transportation and security requirements between landing point and terminal if crossing of the airside-landside barrier involved). Consider also—city centre baggage check-in (provision for baggage to cross the airside-landside barrier independently of passengers)/fuelling point for public vehicles.

5.02 Transport within airport environs

• *Inter-terminal transport systems (ITTS):* where more than one terminal, consider—location of transport (airside—note conflict with airside operations, cost related to limited usage; landside—note double processing and orientation problems

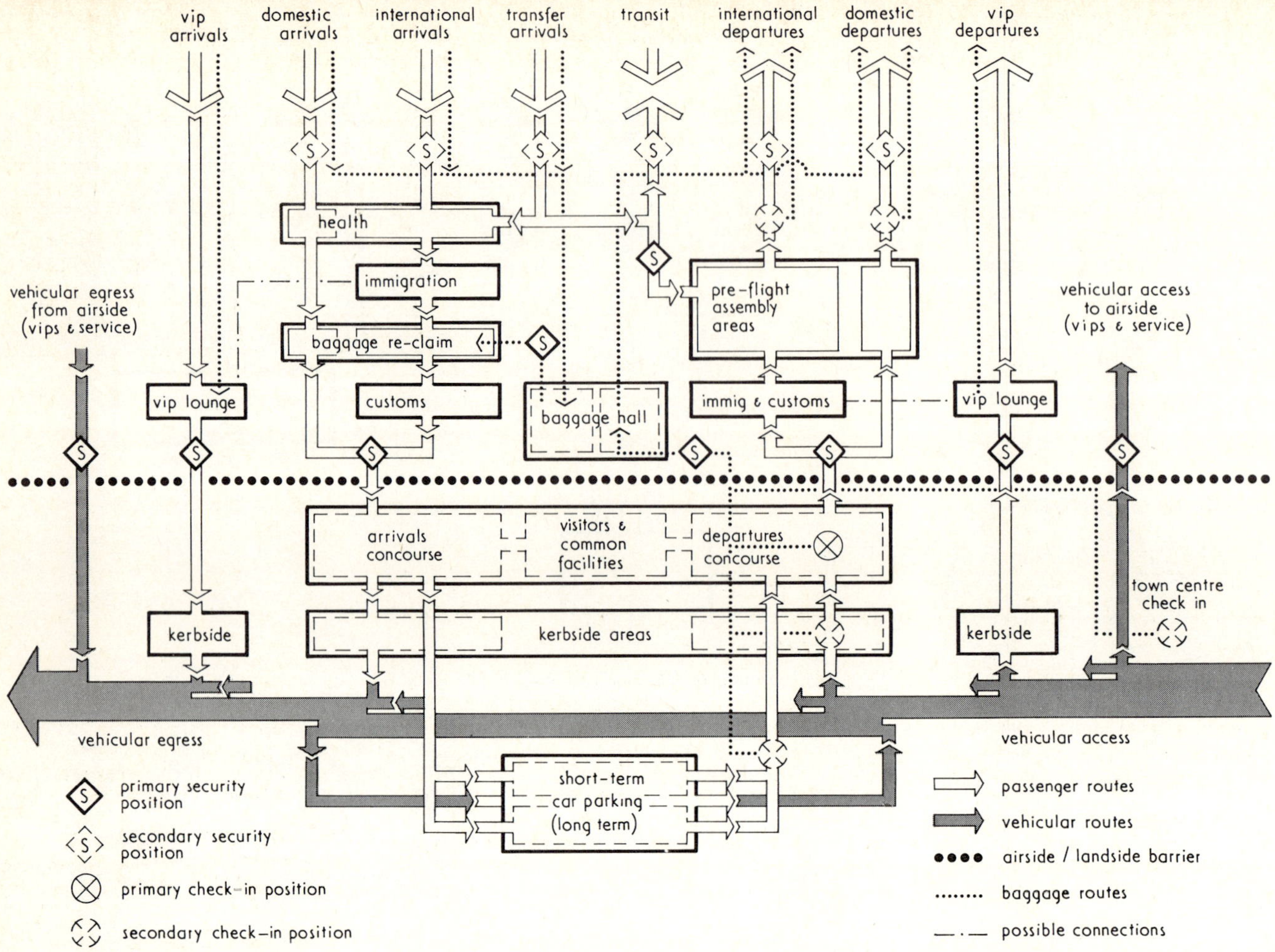

10.3 *Flow plan diagram*

for transfer passengers)/modes—walking, bus, coach, taxi or tracked systems (note waiting and boarding facilities).

• *Miscellaneous transport requirements:* consider—transport to and from remote parking areas, offices, hotels, conference facilities, etc, for passengers, visitors and employees.

5.03 Kerbside areas

Consider—weather protection/baggage check-in facilities/ baggage trolley and wheelchair availability/safety for vehicle loading-unloading (note some coaches have offside, nearside and rear baggage compartments. Avoid multiple set down lanes and pedestrian islands).

• *Calculations:* Allow minimum 0·1 m length per passenger × peak hour passenger total for each kerbside, arrivals and departures.

5.04 Concourse areas

• *Passenger facilities—departures:* Consider—general concourse area/baggage check-in facilities (see para 5.05)/airline counters and offices/flight insurance concessionnaire/cashier's desk/porters' room and amenities/baggage trolley and wheelchair availability/departures information board/common facilities.

• *Passenger facilities—arrivals:* Consider—general concourse area (preferably with separate visitors' area)/airline counters and offices/taxi and hotel reservation facilities (desk or direct telephone)/entertainments and tourist information agencies/ porters' room and amenities/baggage trolley and wheelchair availability/message board/arrivals information board/customs enquiry office/common facilities (see below).

• *Common facilities (departures and arrivals):* Consider—24 hour banking and currency exchange/public toilets/catering (restaurant, buffet, bar, coffee lounge/public telephones/post office and telegraph facilities/airline and club lounges/ VIP–CIP facilities (see para 6.04)/nursery/general sitting and waiting areas/first aid-medical facilities/car rental-return office/lost property room/place of worship/newsreel and TV rooms/airport information desk/luggage lockers (subject to security)/writing desks/public address system, clocks and direction signs/concessionnaires (hairdresser, newsagent, tobacconist, chemist, florist, shoe repair and shine, souvenirs, vending machines, advertising, etc).

• *Area calculation:* For approximate total area of both concourses including common facilities, excluding check-in and main restaurant), allow 2·5 m²–3·0 m² per passenger per peak hour total. Assume average visitor/passenger ratio in western countries of 0·5/1 (long haul), 0·2/1 (short haul), less than 0·2/1 (domestic); in Africa or far east ratio can exceed 4 or 6/1. Where separate concourses are provided, make departures 1·5 × larger than arrivals. Provide seating at 1 per 2 passengers per peak hour total.

• *Visitors' facilities:* In addition to the above facilities consider—visitors' viewing galleries and lounge(s) (note possible security controls and revenue generating potential).

5.05 Check-in facilities

This is the point at which departing passengers have their flight and generally their seat numbers confirmed, their baggage weighed and distributed to aircraft. Consider—location (landside—in town centre, car park, kerbside or departures con-

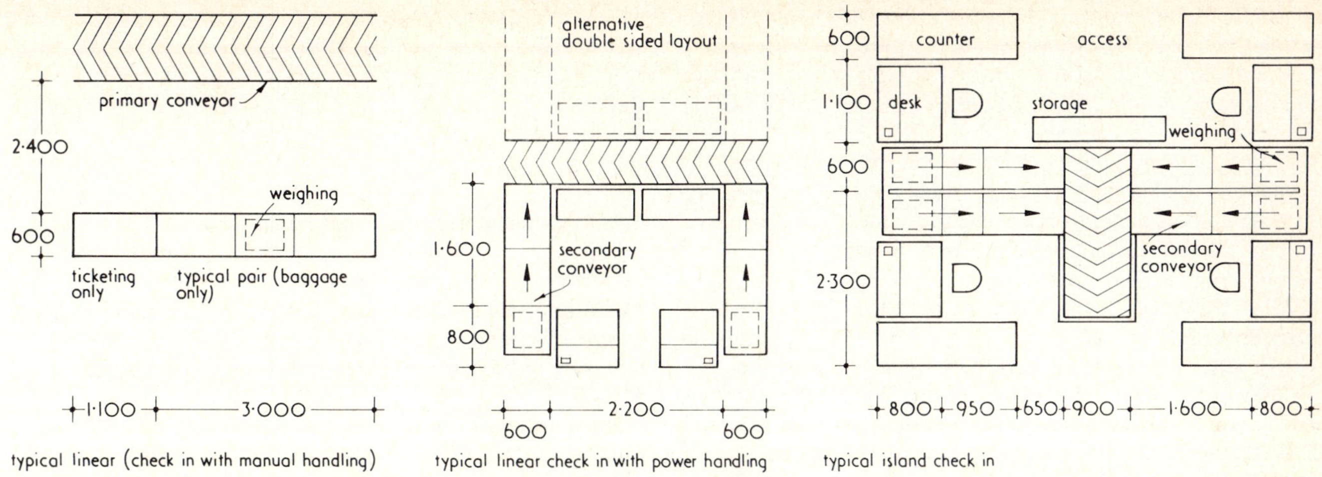

10.4 *Check-in installations*

course; airside—at departures gate or holdroom)/separation of airline information and sales counters from check-in to maximise throughput/individual air line requirements (document storage, power and telephone, electronic data processing facilities with links to other check-in positions and central control, coupon conveyors, baggage weighing methods)/flight number and airline sign boards/provision for out-of-gauge baggage/staff offices. Note greater facilities provision needed for long haul processing than for short haul or domestic. Average processing times, short haul/charter—1 minute, long haul—1½ minutes. Note the main desk arrangements, **10.4**: linear groups with single or double conveyor and single or double sided access island groups with 'pass through' facility (4 to 6 desks per group). Calculations: allow 1·1 m²–1·2 m² per passenger per peak hour departures total. Increase if check-in separate from concourse and for long haul and charter flights. Decrease if for short haul or domestic and where common or single airline check-in provided.

6 CHECKLIST OF DETAILED DESIGN REQUIREMENTS—PASSENGER AMENITIES AND PROCESSING FACILITIES (AIRSIDE)

6.01 Departing passengers

- *Airport security control,* **10.5** (see paras **3.03** and **7.03**).
- *Immigration and customs control:* (generally a combined facility with immigration control and call-upon customs control when required). Consider: Immigration desks and associated offices, detention room(s) and amenities/customs desks and associated offices, secure store and amenities/streaming of nationals from others. Calculations: For international terminal allow 0·2 m² per passenger per peak hour, departures total (excluding support facilities) for immigration. For customs, currency control (if required) allow 1 desk (1·5 m²) per 1000 passengers per peak hour total.
- *Pre-flight assembly areas—departure lounge:* Consider—general waiting-seating areas/separation of international-domestic/concessionnaires (restaurant, buffet, bar, coffee lounge, duty free shop and store, book stall, other shops), public toilets/departure indicator board/airline information desks/VIP-CIP facilities (see para **6.04**)/security control on transit-transfer access/public telephones (if permitted). Calculations: Assuming 20 minutes average passenger waiting time, allow 0·7 m² per passenger, per peak hour departures total (including waiting, and support facilities). Increase to 1·0 m² per passenger per peak hour total if delayed passengers held in lounge. Reduce area if hold rooms provided. Provide 1 seat per passenger per peak hour total in eating and waiting areas.

10.5 *Hand baggage and passenger security control installation.*
Shown here in plan and elevation is a three-part system: metal detection, X-ray and explosives detection. These can be integrated or separated. From two to five operators are required. The gateway should not be closer than 1 m to mains cables or metal objects such as pipes, frames or structure. When two gateways are within 3 m of each other, the openings should be at right-angles

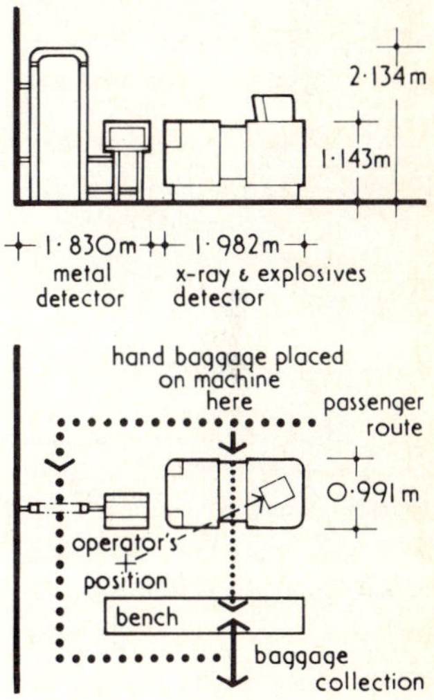

Allow 0·22 m² per passenger per peak hour total for duty free shops.

- *Pre-flight assembly areas—hold rooms:* (supplementary to departure lounge for handling high capacity aircraft. Located adjacent to aircraft boarding gates). Consider—general waiting-seating areas/separation of arrivals—departures/late check-in and baggage handling facility/divisibility and extensibility (to accommodate segregated boarding procedures, larger passenger load factors, alternative aircraft stand arrangements)/airline information desk/facility for airline security control (note no public toilets provided after security control)/departure indicator board. Calculations: approximate area = total aircraft capacity × 1·0 m² (excluding toilets). For seating, assume 80 per cent load factor for largest aircraft served and provide seating for 50% of this figure.

6.02 Arriving passengers

- *Health control:* Consider—specific local requirements/pre-control waiting and hold areas/public toilets/health desks/examination room(s), doctor's room(s)/nurses' room(s)/X-ray/vaccination/quarantine room(s)/staff amenities/secure access to ambulance bay. Calculations: for international terminal allow 0·08 m² per passenger per peak hour arrivals total (including support facilities).
- *Immigration control:* Consider—waiting area/immigration desks/offices/interview room(s)/detention room(s)/security officers' room(s)/dark room/store/staff amenities. Calculations: allow 60 per cent of peak hour arrivals passenger total ×

1·0 m² (including pre-immigration waiting and channel area).

• *Baggage reclaim:* Consider—separation of domestic—international/pre-reclaim waiting area with toilets and seating/baggage reclaim hall/flight indicator boards/out-of-gauge facility/unclaimed baggage store/bonded store/staff amenities/baggage handling equipment. See **10.6** for details of commonly used systems (note possibility also of bringing luggage containers direct from aircraft to reclaim for self-retrieval by passengers). Calculations: pre-reclaim waiting area: allow 30 per cent of arrivals peak hour rate × 1 m². Seating for 10 per cent of passengers. Baggage reclaim: allow 80 per cent of arrivals peak hour rate × 1 m². Baggage retrieval (continuous belts): allow 30 m reclaim periphery per 150 passengers.

• *Customs control:* 'red-green' channelling (declare, non-declare)/customs desks and/or benches/duty room(s)/search room(s)—male and female/interview room(s)/goods search room(s)/lock-up/cash office/general offices/customs enquiry office (accessible from landside arrivals concourse)/staff amenities/aircrew controls. Calculations: assume red/green system. Allow 0·4 m²–0·5 m² per passenger per peak hour arrivals total (including examination and search rooms excluding offices, duty rooms, lock-up).

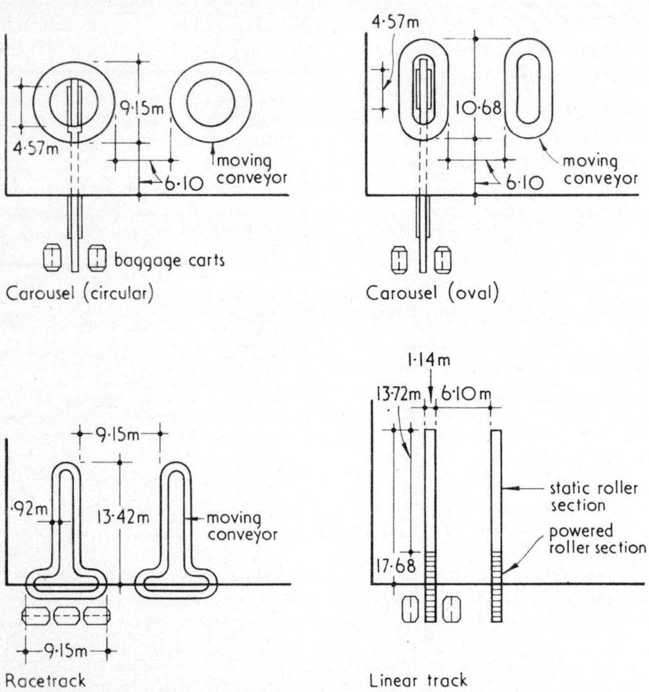

10.6 *Four types of baggage reclaim installations*

6.03 Transit and transfer passengers

Consider—airline and/or transit-transfer information desks close to arrival gates/segregated transit passenger facilities/ access to pre-flight assembly areas (note security controls—paras **3.03** and **7.03**/note interline transfer passenger requirements.

6.04 VIP/CIP passengers

Note: Facilities for CIPs (commercially important passengers) are provided by the airlines. Consider—separate lounge, bar and general amenities in pre-flight assembly areas. Facilities for VIPs are provided by the airport operators and will vary with local custom, size of airport, etc. Consider—facilities related to status, eg heads of state (direct access by car to–from aircraft or via separate building with independent aircraft-vehicular access and facilities). Other VIPs (via separate routes and facilities within main terminal to–from aircraft parked at normal gate positions. Consider—separate vehicular access and baggage handling/separation of VIPs from entourage/private facilities (lounges, catering, place of worship, etc)/press and interview room(s)/communications facilities/security control on press admission to airside areas, safety of VIPs.)

6.05 Methods of passenger movement through the terminal

• *Pedestrian movements* (see para **3.02**). Consider also—the alternative routes by ramp or lift instead of staircases and escalators (maximum ramp gradient 1:12)/maximum recommended walking distances (IATA) can be used as a guide.

Kerbside to baggage check-in/reclaim	average 20 m
Furthest car park space to baggage check-in and reclaim	300 m
Furthest gate to baggage check-in and reclaim	330 m
Gate to aircraft	50 m

• *Mechanised movements:* Consider—escalators in preference to staircases if level changes unavoidable/lifts as a supplement for invalid passengers and for servicing only/passenger conveyors for horizontal travel distances in excess of 100 m (note these can also accommodate changes of level and can extend the walking distances given above). Note too the possible availability of a high-speed conveyor, **10.7**, tracked systems and electric golf-type vehicles.

6.06 Methods of connection to aircraft

• *Open apron approach* (bus and/or walking across the apron). Consider—passenger safety and convenience (weather protection, possible changes of level—down steps to tarmac—up steps to aircraft, risk from aircraft blast and vehicles, time to load-unload bus and walk).

• *Passenger transport vehicle (PTV)*/mobile lounge/transporter, **10.8**. Current PTVs provide space for holding 100 plus passengers, the whole of which can be raised or lowered (to accommodate different aircraft sill heights and terminal floor levels) and transported to aircraft at remote parking positions. Consider relationship to airport-airline operations/effect upon terminal planning (see para **4.02**). Note: latest designs incorporate scissors lifts or telescopic rams to allow the PTV to pass under the 4·5 m clearance recommended for airside structures.

• *Loading bridge* ('air-jetty'/'air-bridge'/jetway), **10.9**. Provide a direct connection at first floor level between aircraft doors and gate positions for use with pier, satellite and linear configurations (see para **4.01**). Consider—maximum gradient of bridge to sill height of aircraft 1:10 over short distance (sill height DC9–2·39 m; Boeing 747—4·88 m)/minimum clearance below bridge for airside vehicles—4·5 m/bridges preferably provided to portside doors only to leave starboard side clear for aircraft servicing/wide bodied aircraft generally require two bridge connections. (Note overwing bridge available but seldom recommended.)

7 CHECKLIST OF DETAILED DESIGN REQUIREMENTS—GENERAL TERMINAL FACILITIES

7.01 Baggage handling system

See para **3.04** for main aims of system. This is the single most important mechanism in a terminal and as it can involve very complex automated handling equipment, specialist advice should be sought. The mechanics of a typical relatively simple system are as follows

• At the departures check-in point baggage is weighed, labelled and transferred to the primary conveyor (see para **5.05**) for distribution via an X-ray security check (if required) to the baggage hall (directly if check-in within terminal, indirectly if

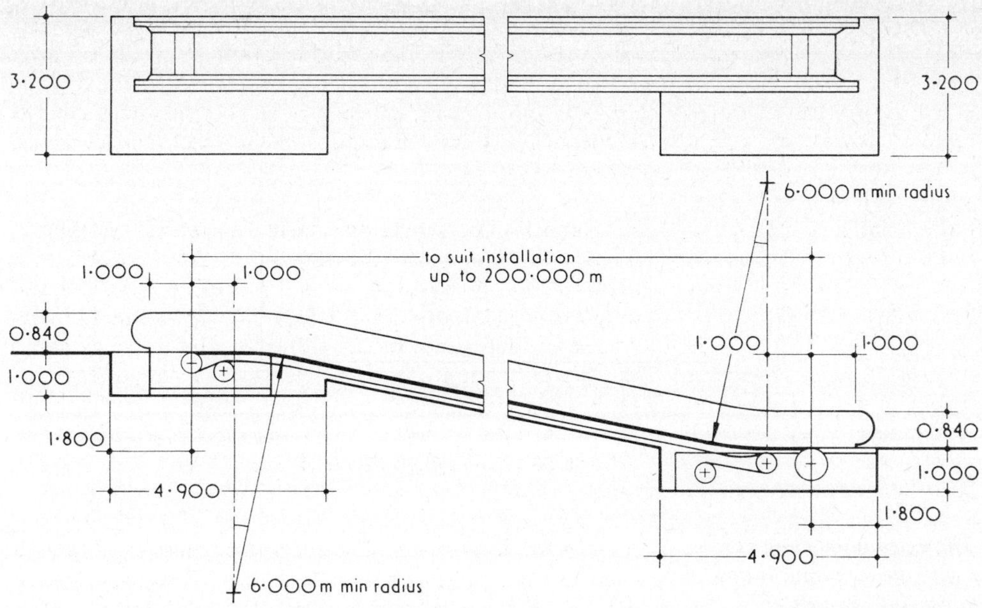

10.7 *Mechanised passenger conveyor systems:*
a *Typical one-speed passenger conveyor; may be flat or up to 12° for prams, trolleys etc or 15° for special installations. Other systems available permit 'valley and hill' profiles, or surface laying of conveyor on drive motor onto existing floors. Capacity of illustrated system is 7200 persons per hour. Systems are available up to 8000 pph. Speed range is 0·45 to 0·6 m/s. Tread widths 1000 to 1400 mm*

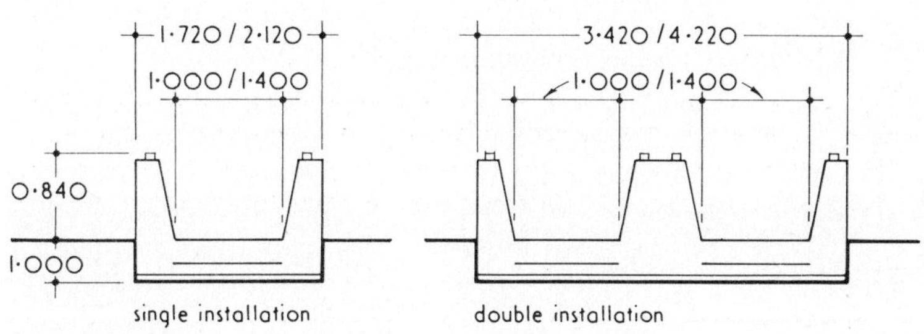

b *Cross-sections*

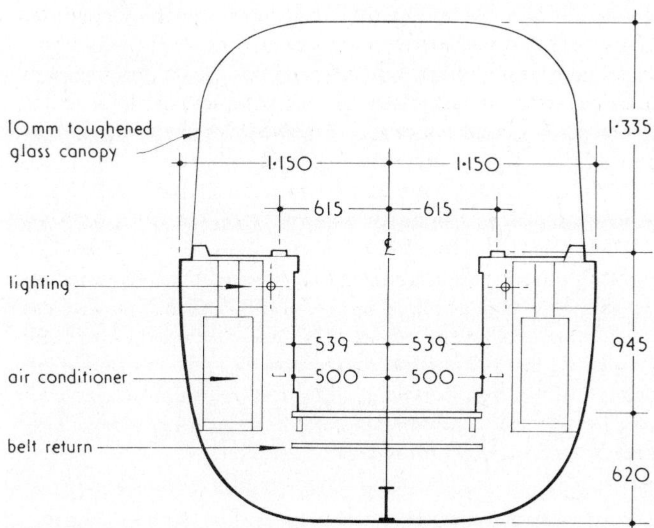

c *Section through passenger conveyor at Charles de Gaulle Airport, Paris*

remote). The primary conveyor may feed direct onto a straight line or circulating accumulator for sorting, or have provision for multiple sorting by visual or electronic identification of labels onto a series of secondary straight line feeds related to specific flights. Baggage is then normally transferred manually to wheeled container dollies or baggage carts, **10.10** for grouping into trains for towing to the appropriate aircraft parking stand. Containers will be hoisted direct, baggage carts will require unloading onto mobile conveyors, for stowage within the aircraft baggage hold. For arrivals, a separate but basically similar process carries the baggage to the reclaim area (see para **6.02**). In developing the system, consider—individual airline preferences versus advantages of single consolidated system/degree of automation related to need and cost effectiveness (note fail-safe requirement)/alternative baggage reclaim facility using self-selection direct from containers/synchronisation of system with passenger processing times/flexibility and extensibility of system/safety and reliability (aim to reduce number of handling operations, changes in level and direction, and apron congestion/provide for weather protection and accessibility to all parts of the system for rapid maintenance in use/storage facilities for early departures and unclaimed or unaccompanied baggage/security against loss in transit/facilities for out-of-gauge and late-arriving departure baggage/facilities for transfer baggage processing/container and/or cart storage.

7.02 Catering

Main catering facilities are related to public concourse areas (landside) but additional facilities required for airside departure lounge, VIP and CIP areas (see paras **5.43**, **6.13**, **6.40**) and for employees. Consider:

- alternative facilities for delayed passengers (transport off airport to hotels, etc)
- meal service on board aircraft
- serving in-flight type meals within terminal
- common servicing of airside-landside facilities (subject to

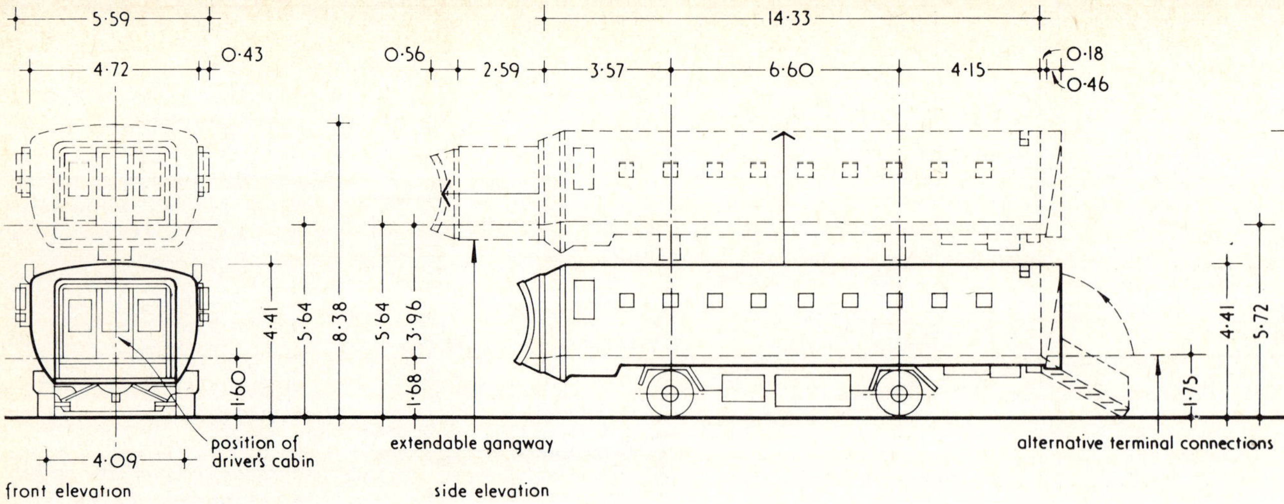

10.8 *Passenger transport vehicle. The extendable gangway connects either with the aircraft or with the terminal. It has hydraulic vertical travel to adapt to varying aircraft sill heights. Passenger capacity is 90 seated plus 60 standing. Normal maximum gross weight 25 310 kg, crush load weight 36 877 kg. Maximum speed 30 km/hr, maximum turning radius over front corner 16·76 m*

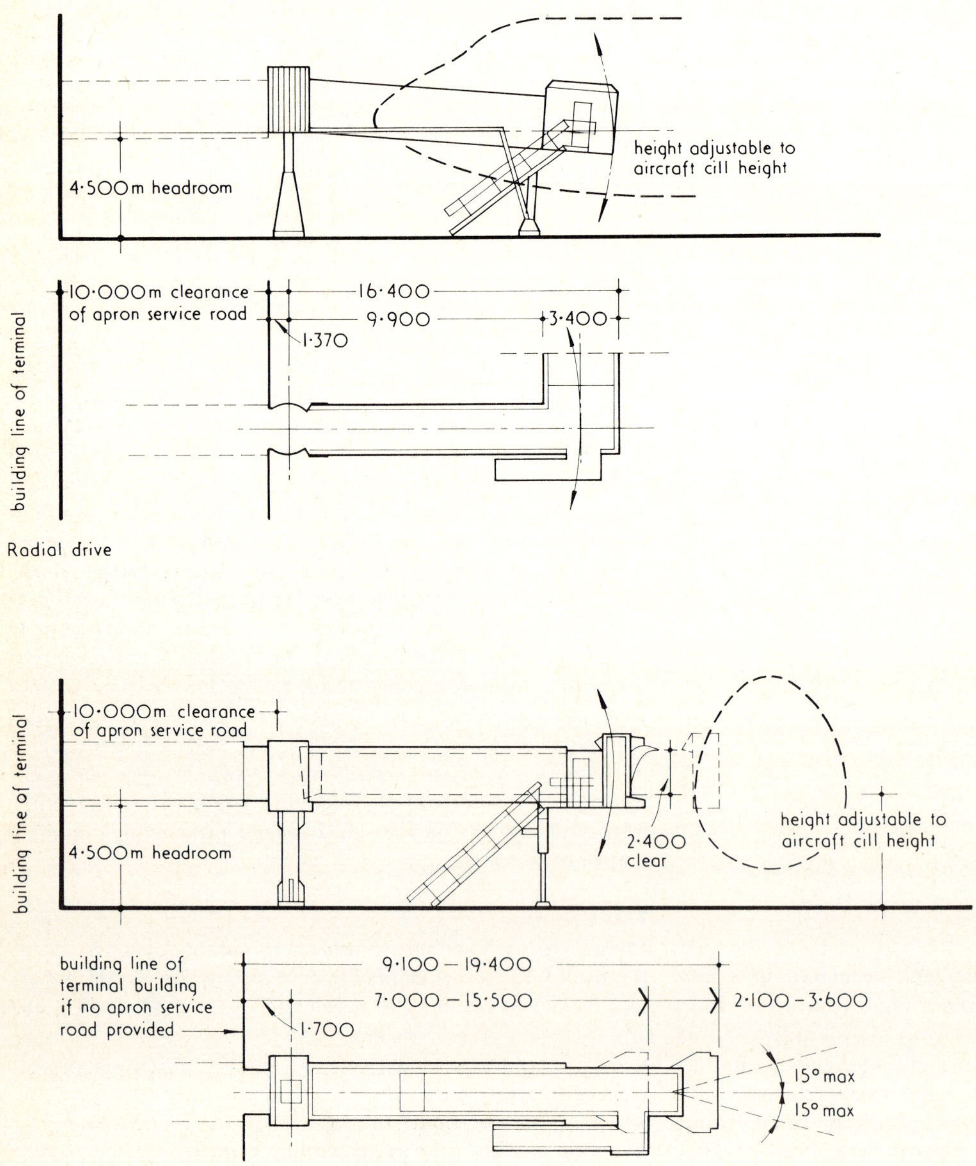

10.9 *Loading bridge types: plans and elevations*

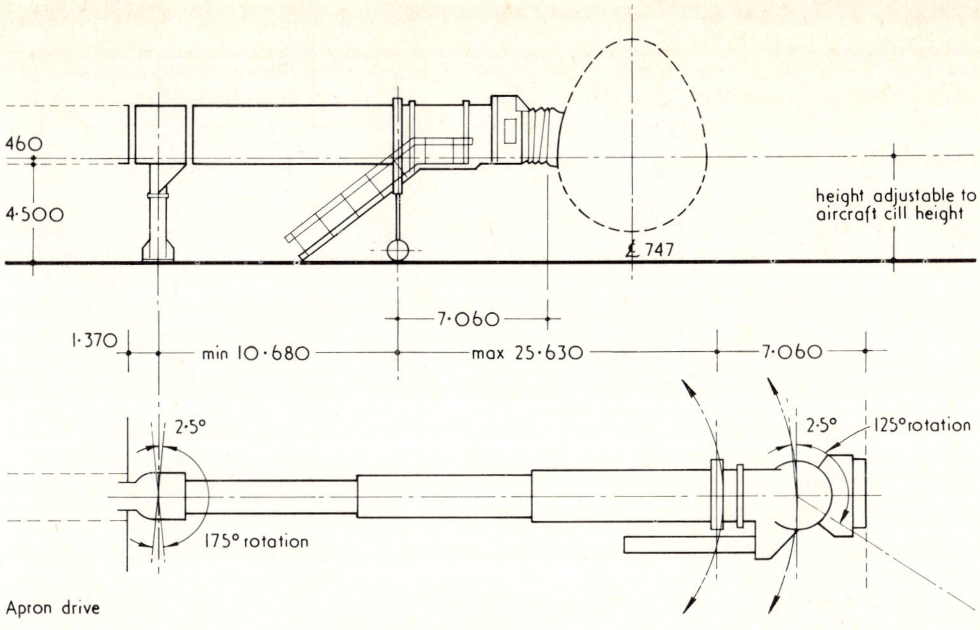

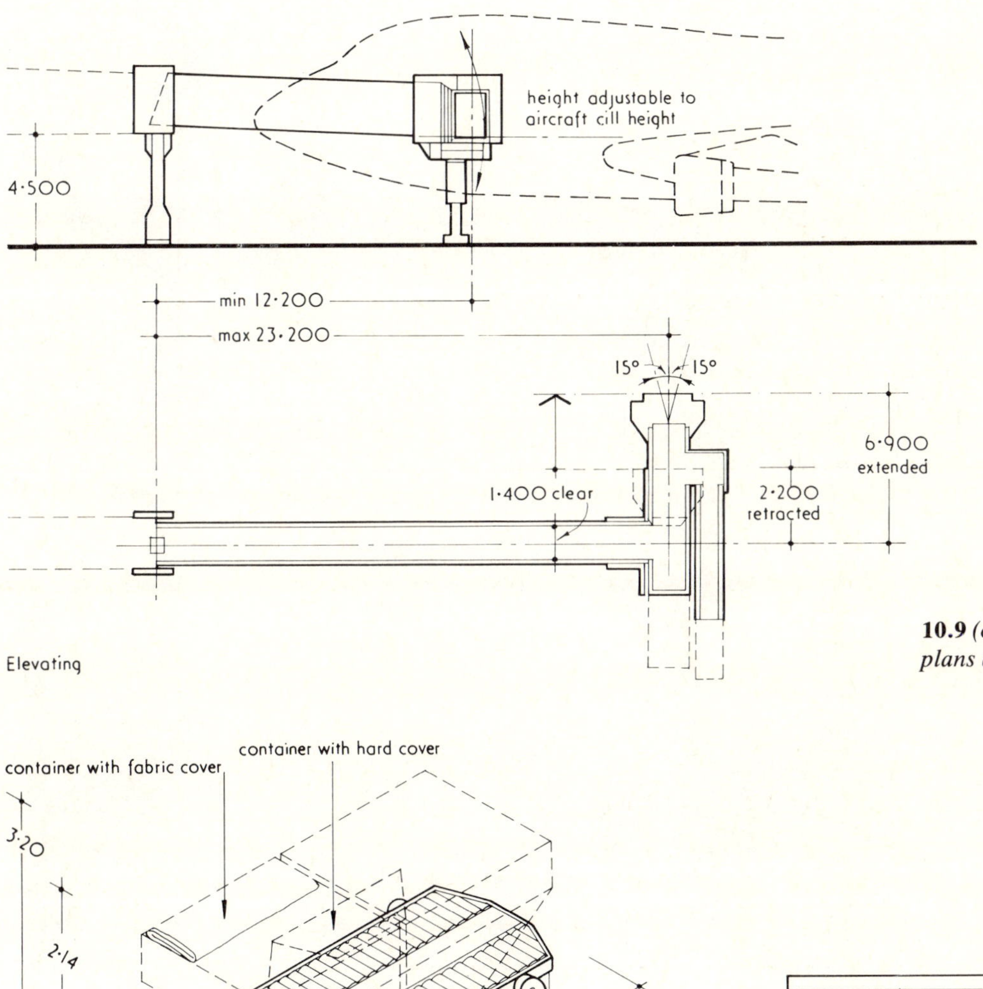

10.9 *(continued) Loading bridge types: plans and elevations*

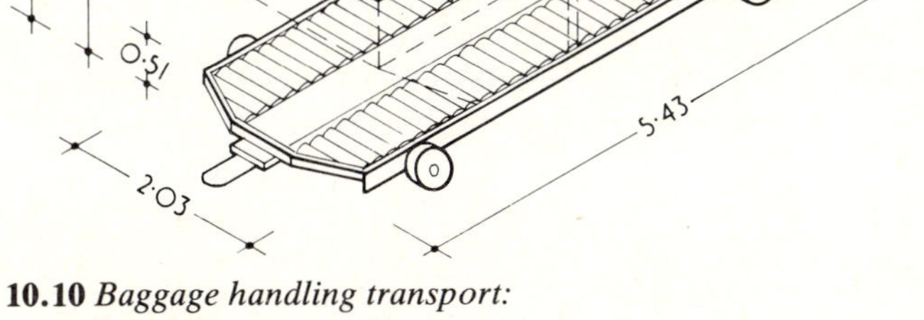

10.10 *Baggage handling transport:*
a *Double container dolly*

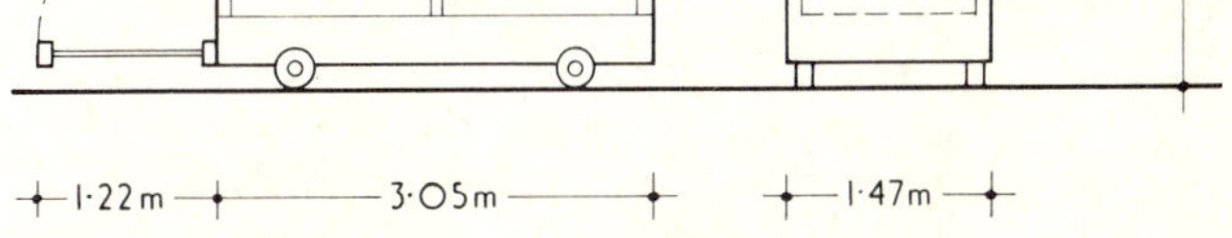

b *Baggage cart*

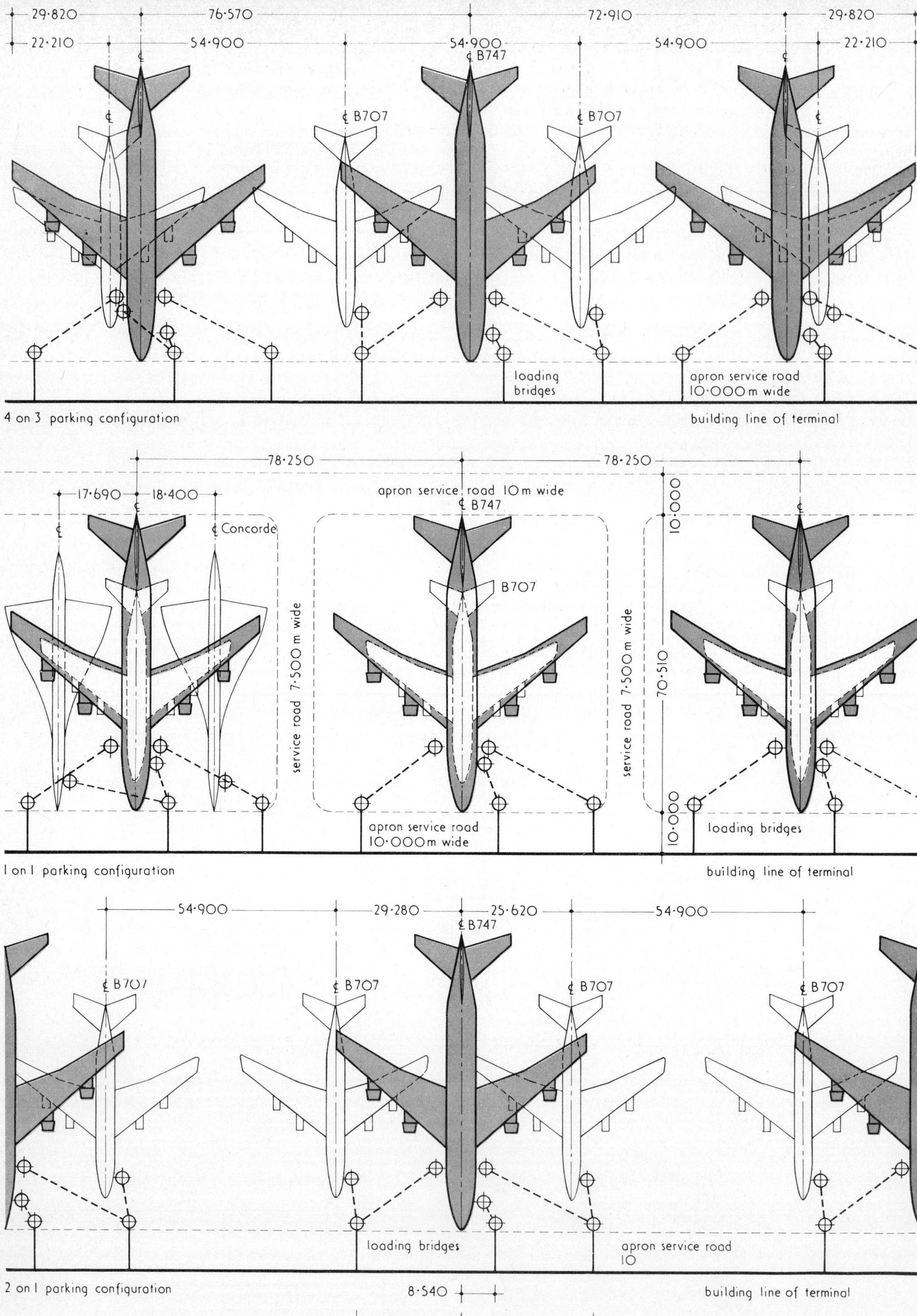

10.11 *Typical aircraft parking stand arrangements*

security control)/delivery, receiving and storage areas/liquor stores/food preparation, kitchen and serving facilities/staff amenities (dining, changing, lockers, toilets, showers).

Note also special requirements of flight catering services ('uplift' catering) by airlines or concessionnaires, within the terminal or in separate facilities depending upon the number of meals served.

Calculations: assume 70 per cent of total passengers, visitors and staff use facilities in ratio; 65 per cent use buffet/bar (1·3 m² per person); 25 per cent use grill/coffee shop (1·6 m² per person); 10 per cent use restaurant (1·4 m² per person). Increase by 50–60 per cent to include service and staff facilities. Increase area if delay catering provided.

7.03 Security

See para **3.03** for main aims. Note additional requirements: to deny access to persons carrying weapons into areas overlooking airside operations (may require airport controlled security checks on visitors entering viewing galleries, car parking areas, etc)/to prevent the loss of property and damage to facilities and equipment within the terminal and in all airside-landside areas (note: protection of goods in transit; vulnerability of aircraft, control tower, etc: need for extension of airside-landside barrier outside terminal and security checks at all crossing points). See below for details of typical passenger and hand baggage security devices.

7.04 Administrative and general terminal support facilities

• *Airport administration and general management facilities:* Consider—offices/conference rooms/overnight accommodation/airport police facilities (offices, detention rooms and amenities)/porters' rooms/engineering stores and workshops/ maintenance stores/cleaners' rooms/general store rooms/ baggage loaders' facilities (lockers, rest rooms and amenities/ employees' canteen.

• *Airline operators' facilities:* Consider—offices in addition to those within the concourse areas/employees and aircrew accommodation and amenities (rest-rooms, changing, canteen, etc/computer facilities/operations and duty room(s).

7.05 Building services

In addition to normal building services. consider—air-conditioning/sound–proofing/public address systems/flight information boards control systems/telephone (internal, public and private)/telex and computer facilities/pneumatic tube systems/sewage disposal (if not provided)/electricity sub-stations and standby power generators. Note that 24 hour occupation of building will increase demand on all services, particularly electrics and water storage.

8 CHECKLIST OF DETAILED DESIGN REQUIREMENTS—AIRCRAFT AND APRON OPERATIONS

8.01 Aircraft parking stand arrangements

• 1 on 1 / 2 on 1 / 4 on 3 stand arrangements, **10.11**.

8.02 Aircraft servicing arrangements

Figure **10.12** is a typical servicing arrangement for a Boeing 747.

8.03 Aircraft fuelling operations

• *Fuel store requirements.* Consider—storage capacity related to numbers of aircraft served (current and projected) and to proximity or otherwise to refinery facilities/types of fuel required/location (dependent upon safety regulations, and environmental and security considerations).

• *Fuel distribution methods.* Consider—bowsers (decreasing use due to large fuel demands of wide bodied aircraft)/fuel pits (fuel piped from storage to connection point adjacent aircraft stand then pumped by mobile unit into aircraft)/hydrants (fuel

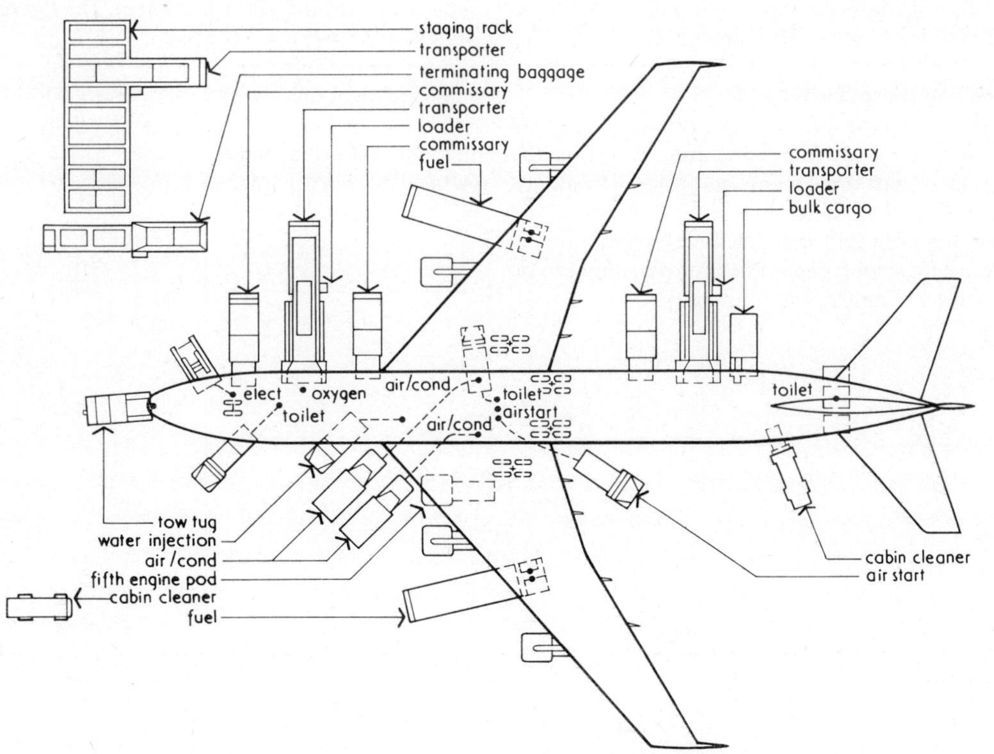

10.12 *Servicing arrangements for turnaround for passenger model Boeing 747–100/200B + C. Under normal conditions external electric power, airstart and air conditioning are not required when the auxiliary power unit is used. The forward oxygen fill point is needed for model B747–100/200B, the aft fill point is for B747–200C.*

pumped under pressure direct from storage to connection point adjacent aircraft stand then metered by mobile unit into aircraft).

8.04 Aircraft maintenance facilities

Requirements will vary according to airport and airlines local requirements (airports serving as 'home base' to airlines require greater provisions). Consider—categories of maintenance, eg major maintenance including complete overhaul; overhaul on major components only—engines, landing gear, etc; light maintenance—wheel changes, adjustments to equipment, etc; very light maintenance (apron service—refuelling, cabin servicing, catering, etc). Consider also—hangar facilities/ancillary airline offices/training facilities.

8.05 Air traffic control (ATC) facilities

• *Control tower.* Main aims—to provide visual control of take offs, landings and movements of all aircraft on the ground (ground movements control may be supplemented by separate control points or by local ground radar and/or automatic taxiway guidance systems). Size and facilities required will depend upon local requirements.

Consider—height (related to distance from runways and angles of vision over local obstructions)/location adjacent terminal but note possible constraint upon terminal extensibility.

• *Aeronautical information service (AIS)*/meteorological/flight clearance facilities. Consider—general planning (location adjacent control tower, common waiting area giving access to each facility)/AIS requirements (general office with clerical, copying and teleprinter facilities; senior officer's room; sleeping cubicle; storage areas)/meteorological requirements (general office with clerical, cartography, wind indicator and barograph facilities; copying; signals; teleprinter and facsimile facilities; library; senior officer's room; sleeping cubicle; storage areas)/flight clearance requirements (general office with teleprinter and link to control tower; storage and amenities; aircrew briefing room).

• *Miscellaneous ATC facilities.* Consider—radio communication equipment room/radar(s)/equipment repair and testing workshops/staff training facilities/staff amenities.

8.06 Fire fighting and rescue facilities

Type and number of facilities required relates to ICAO (Annex 14)/CAA (UK) standards (categories based on the forecast number of aircraft movements and take-off weight). Consider location (unobstructed views of, and access to, all airport areas)/all weather rapid response times—3 minutes to crash site anywhere on airport (note need for personnel to be accommodated close to vehicles)/staff amenities.

8.07 Miscellaneous facilities

Consider—customs facilities for aircrew and aircraft control (facilities to board aircraft, offices, crew examination room(s), search room(s), amenities)/vehicle maintenance facilities (motor transport depots, workshops)/apron staff facilities (offices and amenities).

9 REFERENCES AND BIBLIOGRAPHY

9.01 General information

Edward G. Blankenship *The airport*, London, Pall Mall, 1974

John Vulliamy Airports: passenger and cargo terminals, in *Planning: buildings for habitation, commerce and industry*, London, Newnes-Butterworths (9th edn)

9.02 Basic technical information

The apron and terminal building—Planning manual, 3 Vols, Springfield, Virginia, National Technical Information Service, 1975

INSTITUTION OF CIVIL ENGINEERS' PUBLICATIONS AND PAPERS

World airports—the way ahead, 1966; *Airports for the future*, 1967; *Engineering the world's passenger terminals*, 1966; *The challenging future*, see 'Passenger buildings' by B. J. Mayes—Session VII Paper 18, Proceedings of the 5th World Airports Conference 1976; *The capacity of airports—planning considerations,* paper by J. D. Perret Ref. 7372/1.2.72

IATA PUBLICATIONS

Airport Terminals Reference Manual and *Airport Terminals Bulletins*

ICAO PUBLICATIONS

International Standards and Recommended Practices for Aerodromes Annex 14 and *Airport Master Planning, Sizing of Departure Lounges in Airport Terminal Buildings*, Proceedings of Am. Soc. Civil Engineers 1969, 311–331, USA

Research Studies on Airport Terminal Design by Univ. of Loughborough, UK, 1975

Adis Kanafani and Hanan A. Kivett *The Planning of Passenger Handling Systems*, USA 1972

CAA publications (UK)

FAA publications (USA)

9.03 Specialist magazines

Airport Forum

Airports International

11 Air cargo terminals

Jolyon Drury

CI/SfB 241
CI/SfB (1976 revised) 144

Jolyon Drury is a consultant architect advising on the design of production, storage and distribution facilities

Contents

1 INTRODUCTION

1.01 Wide bodied aircraft

The phenomenal growth of the air cargo industry in the late 60s has now stabilised to about 10% increase p.a., stimulated by additional capacity as a result of the change to predominantly wide-bodied aircraft like the Boeing 747, the Douglas DC10, the Lockheed 1011 and the A 300B Airbus, on major routes. The Boeing 747, in passenger form, can carry as much cargo in its belly-hold as the pure cargo 707 could formerly on the main deck: in cargo form it can lift over 100 tons of freight across the Atlantic. Wide-bodied aircraft construction dictates the carriage of belly-hold cargo in containers, and in other forms of unit load on the main decks. The belly-hold containers are interchangeable within the wide-bodied types while conventional air-cargo pallets, as used in the main decks of the 707/DC8 type freighters, can be carried on the wide-bodied jets' main deck and in certain positions in the belly-hold as well. The Boeing 747 freighter stands out in being able to accept square-section containers to the same ISO dimension as shipping containers on the main deck, loaded through the nose or side door; most other types of air-portable unit load device are contoured to the fuselage profile. This new load-carrying flexibility where an air-cargo pallet can start its journey in the hold of a wide-bodied short haul feeder, like the A 300B Airbus, can be transferred to the main deck of a Boeing 747 for a transcontinental haul, and again be transferred to the main deck of a 707/DC8 type aircraft for onward travel to inland destinations, will inevitably be reflected in the design demands of the cargo buildings throughout the route. Cargo aircraft characteristics are given in table I.

Table I: Cargo aircraft characteristics: all accept palletised freight

Type	Payload (approx)	Remarks
Jet-prop		
Canadair CL44	30 tons	Swing tail, still popular
Britannia	16 tons	Charter cargo
Argosy	12 tons	Nose load, feeder routes
Merchantman	19 tons	British Airways only
Electra: similar to Merchantman: now Thirdworld charter		
Jet cargo aircraft		
Douglas DC9	17 tons	In both pure cargo and combination versions
Douglas DC8–63	46 tons	The most efficient narrow-bodied cargo aircraft
Douglas DC10	71 tons	Some on charter work have self-loading pallet lifts
Boeing 727–200 series	23 tons	
Boeing 737–200 series	16 tons	
Boeing 707–320C	42 tons	Very popular
Boeing 747F	120 tons	Can accept ISO containers on the main deck

1.02 Anticipating change

The evolution from bulk multi-piece cargo, handled at the airport, to consolidate off-airport into unit-load devices was rapid and appeared to take some airlines and airport authorities by surprise: in the early 70s there was confusion and much wasted equipment. One major lesson learnt is that air-cargo facilities, large or small, must be built for flexibility both to accept the systems imposed by rapidly evolving aircraft types, improvements in materials handling technology and changes in a highly competitive market place, which, on some routes, will involve very rapid growth.

1.03 Air cargo terminal design

There is no standard design for an air-cargo centre but there are basic principles that have to be fulfilled. The terminal design will be affected by:

- whether it is an airline-sponsored project or for multiple occupancy from an airport authority
- the characteristics of the route (ie predominently passenger aircraft or predominently cargo or cargo charter)
- the economic characteristics of the area (eg whether it is an importing, developing country)
- the characteristics of the aircraft, in turn generated by the route demand but also prestige
- the position of the terminal in the overall cargo route structure for the region (ie whether it is a transhipment point, an international terminal, an originating point for internal and/or international flights)

1.04 Planning for change

The basic diagram, **11.1**, shows the functional elements of an air-cargo building; the adjoining diagrams illustrate, in approximate terms, how the proportion of the functional areas carry a different weighting for different service demands, **11.2**. For example, if a developing country, such as in the Middle East, is involved; for the first phase the majority of space will be devoted to inward cargo, a large proportion of which will be awaiting collection (as agents in those countries tend to use the facilities as warehousing). But in a major international interchange route, where wide-bodied freighters predominate, the majority of cargo can be expected to pass rapidly through as unit loads. Equally, in terms of building design generated by equipment demands, the developed routes increasingly involve sophisticated unit-load handling whereas in developing countries the simpler the handling method the better: these factors will affect height/area considerations. But routes, economies and mechanical sophistication will continue to develop and unless the building is designed for a limited life, it must be able to accept mechanical handling systems or expansion without the total disruption of the terminal. For example, in newly emerging countries, where cargo is developing through passenger services and where finance is limited, an initially simple building might serve as both the passenger and

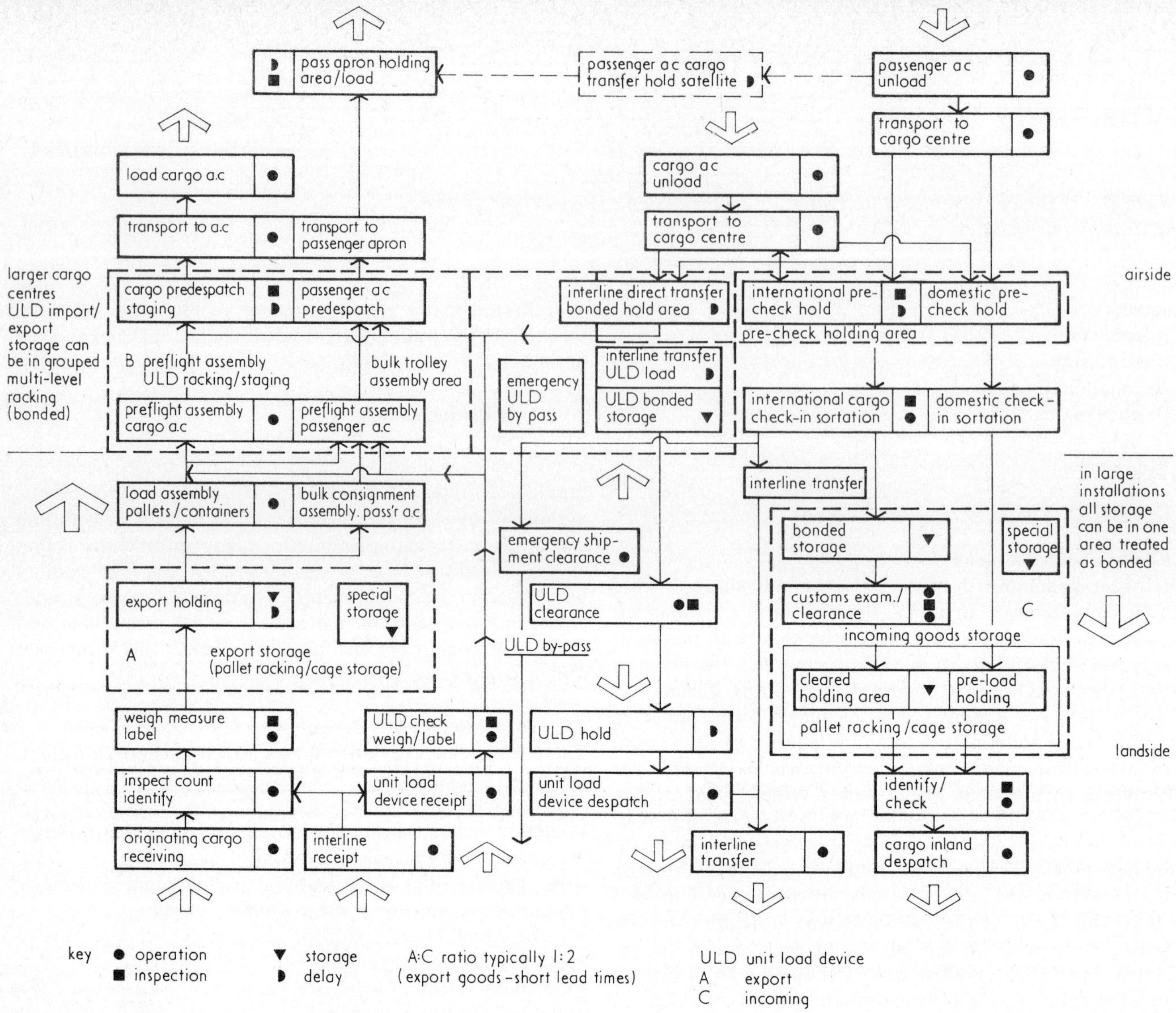

11.1 *Typical cargo terminal system diagram (for international and domestic flights)*

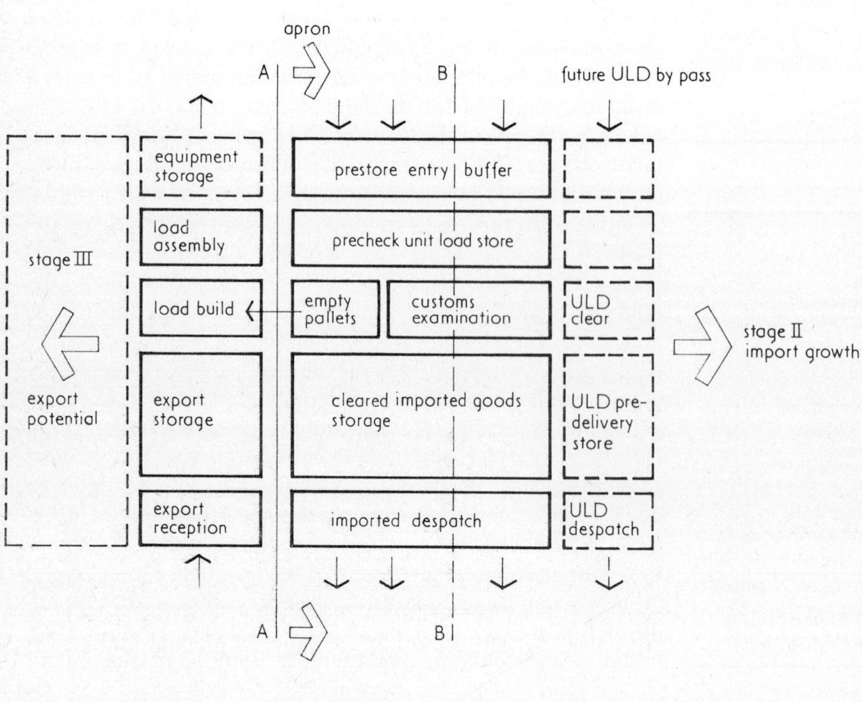

11.2 *Differential growth of elements of cargo centre:* **a** *Developing country terminal. Emphasis on imported material, initially all ULDs (unit load devices) are broken down on the airport (security). As development of country and agency networks grow, ULDs allowed off the airport and export trade may grow. Flexibility is implied, as when country becomes sufficiently developed the emphasis import/export (AA/BB) may change roles: ie notional barrier import/export may need to change. NB change I and II likely to combine expansion and re-use of existing premises*

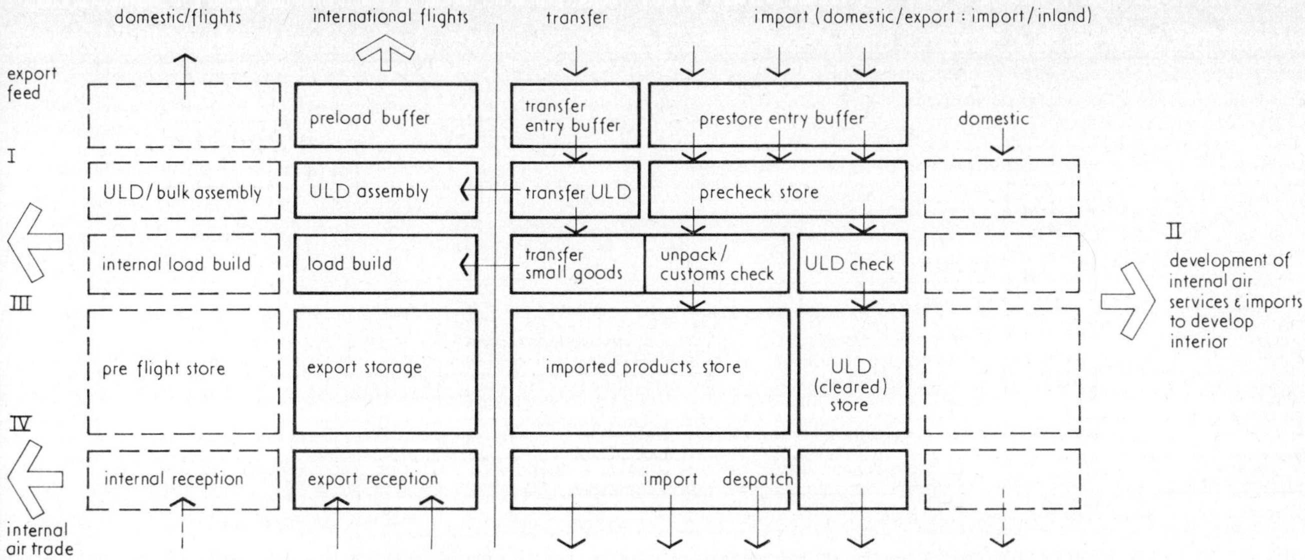

***11.2b** Developed country, routes opening up, becomes airline route centre. As previously shown, exports equal imports. Growth in domestic air feeder services, I to aid imports; II to develop interior; III internal flights feed export traffic and IV wholly internal traffic*

cargo terminal. As the volume of cargo and passengers increases, it can then develop into a passenger terminal and a new cargo facility can be constructed, **11.3**. On a developed but growing route, a terminal constructed for cargo from passenger aircraft and some narrow-bodied freighters may need to accept a substantial increase in volume, with larger unit loads and larger loads per aircraft. A trend that must be considered, too, is the ability of the 747 and future cargo aircraft to accept ISO containers; this will allow pre-loaded, customs-sealed, multi-modal containers to bypass terminal facilities and have access direct to the aircraft: some of the larger air-cargo installations may soon more resemble a maritime container port than an air-cargo centre as we know it.

2 BASIC FACILITY

2.01 Activity areas

There are four main activity zones:
- landside access: vehicle loading/discharge
- landside terminal operation
- airside terminal operation
- airside access: aircraft loading/discharge.

2.02 Design factors

Principal factors controlling design are outlined in table II.

2.03 Siting

The main considerations in selecting a site are:
- good access to major road links (an air-cargo centre is a central collection/despatch point)
- easy access to apron (for containers, service vehicles, individual loads, security)
- minimum taxiing distance from runways
- access to aircraft maintenance areas
- expansion potential/planning flexibility. Building, landside parking, apron, administration. Each should be able to expand without prejudicing the operation of the others
- access to passenger apron. Particularly relevant where wide-bodied aircraft involved: two thirds of cargo by volume travels in passenger aircraft, two thirds by weight in freighters. On developing routes, new problem is wide-bodied 'combination' aircraft which can carry a mixed passenger/cargo load on the main deck as well as cargo in the belly-hold; potentially more than 250 passengers and 12 six-ton pallets of cargo can be carried on a Boeing 747C.

There is a conflict in siting a cargo terminal. Authorities want to segregate heavy goods traffic from passenger vehicles on the landside and keep cargo aircraft areas separate from passenger aprons, **11.3**; but with rapid turn-round times, easy access from cargo centre to passenger aprons is essential. 'Combination' aircraft have to use the passenger apron facilities, **11.4**: with this quantity of cargo and belly-hold cargo as well, the passenger aprons, already crowded with service equipment, become very congested with cargo and its handling equipment. Significantly, all the cargo travelling across the North Atlantic could be contained within the belly-holds of the passenger aircraft services at present. Some of the major cargo areas such

Table II: Primary design factors: whether for a particular airline or planned for leasing

Identification of user	Airline(s) or multiple occupancy	Where information from
Cargo forecast: quantities	International/domestic/ transfer. Peak months/ seasons, expansion detail	Airlines, relevant ministry, IATA
Route characteristics	Proportion import/ export; daily peaks, eg to meet curfew at airport or at receipt airports	Airlines
Aircraft factors	Turn-round times, aircraft types, ie unit load demand	Airlines/IATA/ ICAO
Data control and plans for later installation	Parallel with cargo movement: changes to electronic data processing can fundamentally alter handling characteristics	Airlines, airport authority
Demand for flexibility	Service (charter cargo). Aircraft type (trade up to wide-bodied). Number of users (increasing airline interest). Type of users (individual airlines, consortia, airline agents). Cargo type (small goods, indivisible loads, hazardous, perishable, valuable, live, mail)	Airlines/IATA
Labour characteristics		Visit the location

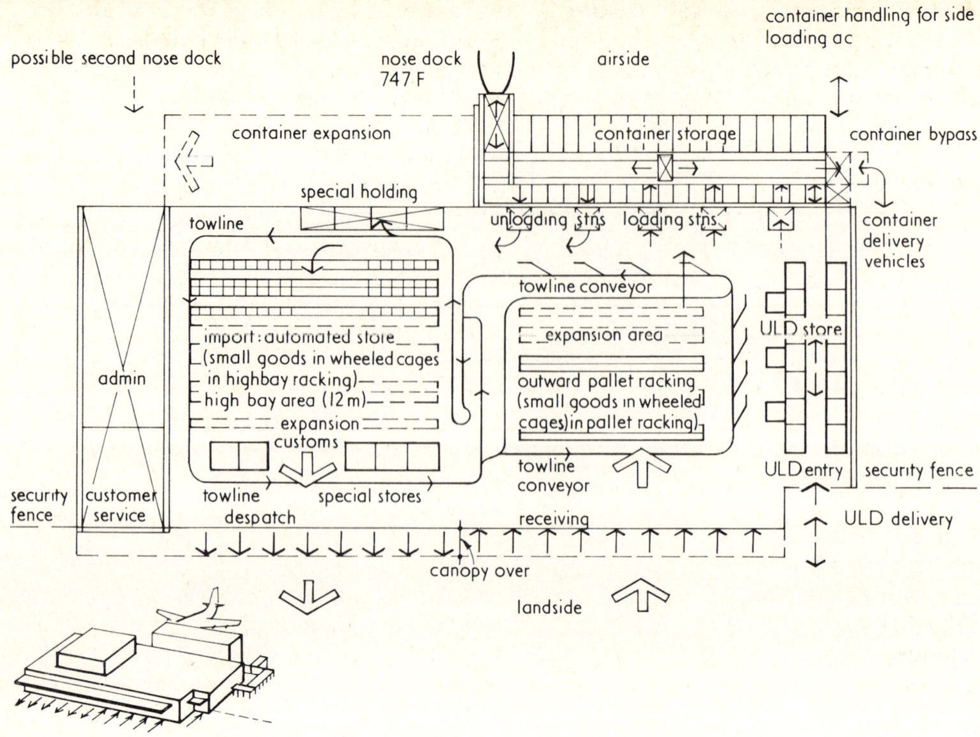

11.3 Typical airline specific cargo centre for major routes: emphasis on container growth (ie packed/cleared off-airport). Not to scale. NB larger volume devoted to incoming cargo due to demand for on-airport breakdown (lack of off-airport clearance facilities)

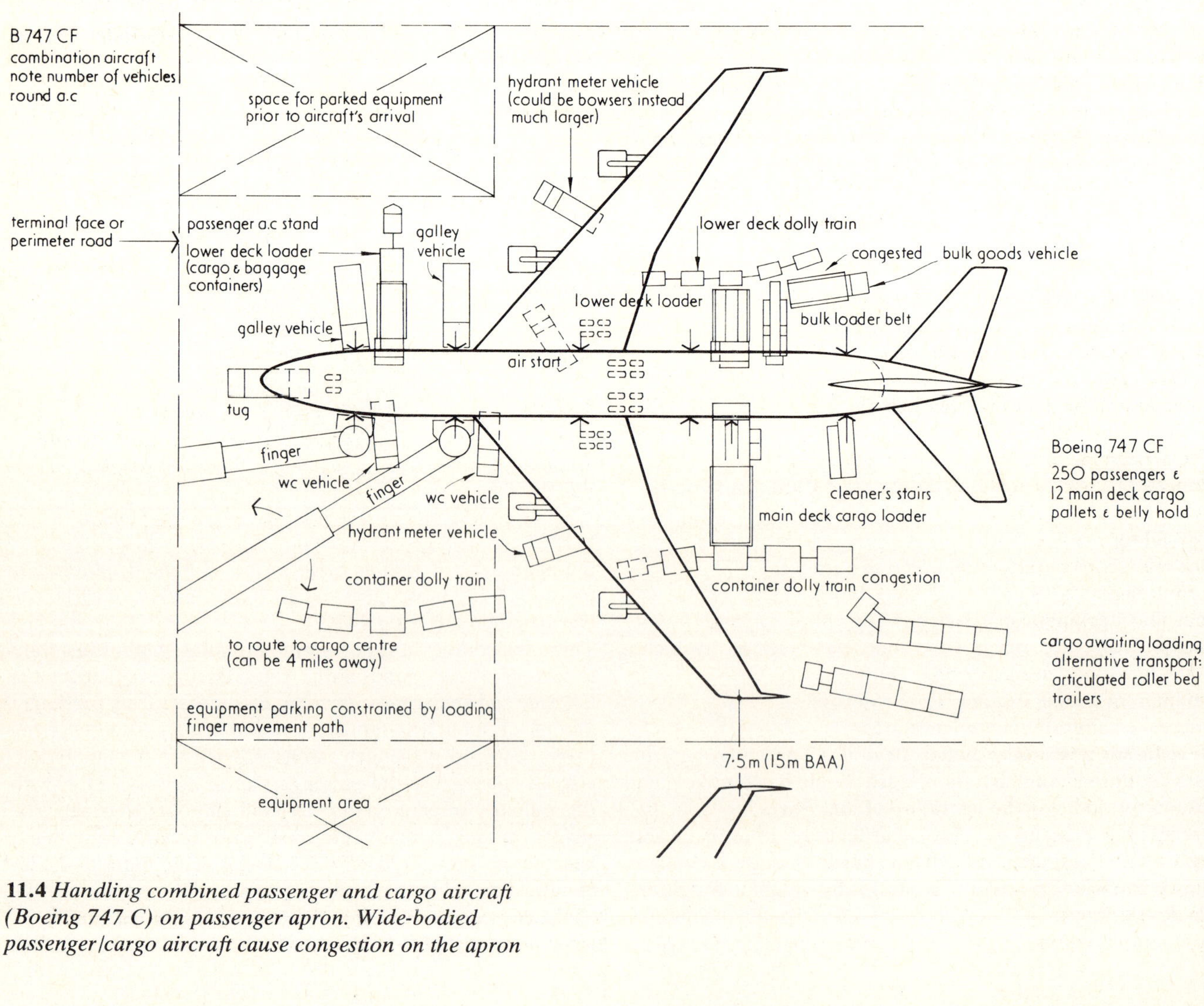

11.4 Handling combined passenger and cargo aircraft (Boeing 747 C) on passenger apron. Wide-bodied passenger/cargo aircraft cause congestion on the apron

as Heathrow and John F. Kennedy airport at New York are up to four miles away from the passenger terminals with poor access: serious consideration must be given to providing cargo satellites in passenger areas especially for forward cargo and interline transfers.

3 ACCOMMODATION SCHEDULE

This will differ for each of the types of cargo-centre growth formerly illustrated, but a basic list includes:

3.01 Handling building

Export (E)
E1 Heavy vehicle parking/manoeuvring
E2 Unloading dock
E3 Identification/sorting area
E4 Weighing/measuring/labelling area
E5 Pre-load storage
E6 Domestic pre-flight assembly
E7 International pre-flight assembly
E8 Unit load device, loading/trolley load
E9 Despatch staging/flight grouping
Note: E6 and E7 can be the same.

Import (I)
I1 Pre-check holding area
I2 Check-in sorting area
I3 Domestic unload
I4 Bonded (international) unload
I5 Domestic pre-collection area
I6 Bonded pre-check storage
I7 Customs examination area
I8 Post-check store
I9 Inter-line transfer store
I10 Transfer cargo bond
I11 Pre-load sort/check area
I12 Heavy-vehicle loading/manoeuvring area
Note: I5 and I8 can be the same area; with electronic data processing I6 and I8 can be the same area; E1 and I12 are normally a common area. E5 and I8 in ratio of 1:2 of exports to imports: import storage likely to grow at increasing ratio, especially in developing countries.

3.02 Accommodation required at warehouse floor level

This will include:
- cold store for vaccines/perishables
- deep freeze for airline stores or special cargo
- strongroom for bullion/documents
- human remains store
- livestock area (IATA livestock regulations)
- dangerous goods store (IATA dangerous goods regulations)
- radio-active material store
- airmail secure store
- aircraft spares/tool storage
- airline flight stores
- machine parking/maintenance space, both in and out of building
- empty unit load device storage area
- rubbish collection area.

3.03 Additional accommodation

Also required are:
- administration offices, attached but not intruding into the working area
- traffic office, adjacent to the vehicle door
- public counter
- teleprinter room
- reprographic room
- manager's office
- staff offices for sales, administration
- traffic planners
- weight and balance office
- customs offices
- warehouse staff amenities, inclusive wc, lockers etc
- clerical staff amenities
- customs staff amenities
- a canteen which can be split up for three sections above or for all of them
- and possibly agents offices can be useful in a developing area
- aircrew facilities
- a rest room
- meteorology room
- public health and immigration area
- first aid room
- transformer room
- standby power plant.

4 BASIC PLANNING

4.01 Planning ahead

When working within a long-range master plan, details would be obtained from the airport authority. Master plan should include apron constraints for developing aircraft types, buildings must not infringe on site lines, flight paths or interfere with electronic aids.

4.02 Basic shape

Consider—
- *Type of terminal* (proportion of import/export; demands of handling equipment). IATA research shows maximum rectangular ratio of 1/1·5; a square is ideal to minimise the perimeter. If more than 1/1·5, then bottlenecks can occur in the handling.

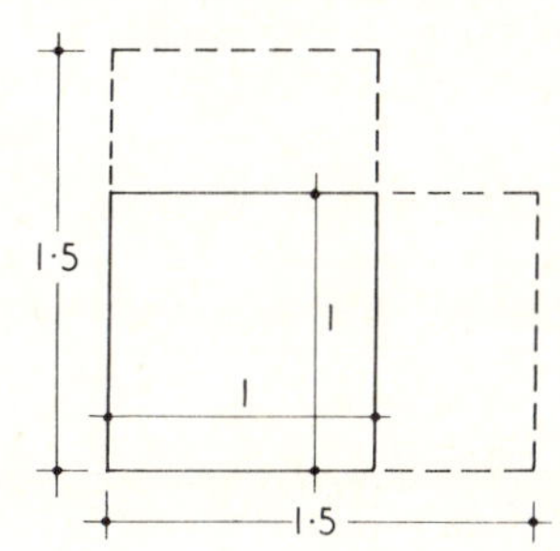

11.5 *Diagram illustrating cargo units for leasing. Operators/airlines with different space requirements. Ideally, flexibility in both axes, but in depth is more important than in width.*
a *Cargo terminal plan/perimeter ratio should fall within 1:1·5 to avoid congestion in planning and handling*

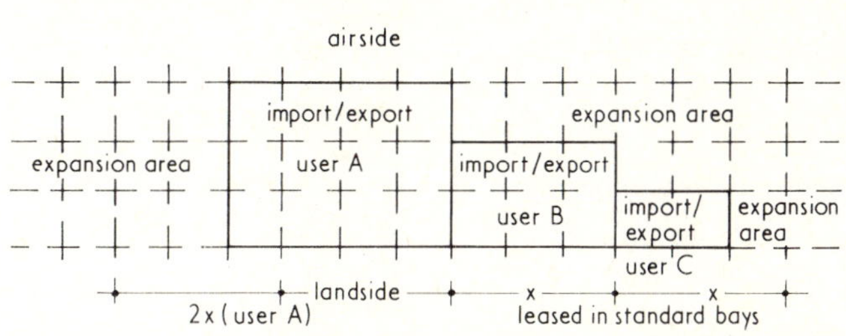

b *Column grid notional only, determined by space demand, local constructional conditions and cost*

• *Multiple occupancy.* Restricted depth is more difficult in planning terms than restricted width. Ideally in blocks of two with gaps between them. Land is often at a premium on airport sites **11.5**. Shape is often dictated by the number of airside/ landside stands. Calculate aircraft stands on an absolute peak, truck stands on eighty per cent peak.

4.03 Critical heights

Structural grid height, to be commensurate with handling technology and to accept changing techniques. Keep administration building out of warehouse or as a mezzanine but be careful of columns under it as these often occur where the door area is required onto the loading dock. Consider clear height for pallet racking for air-pallets up to three high, **11.6**. Flat floor is essential; there is a potential conflict on the height of the landside loading dock which usually needs to be between 1200 mm and 1500 mm above ground level.

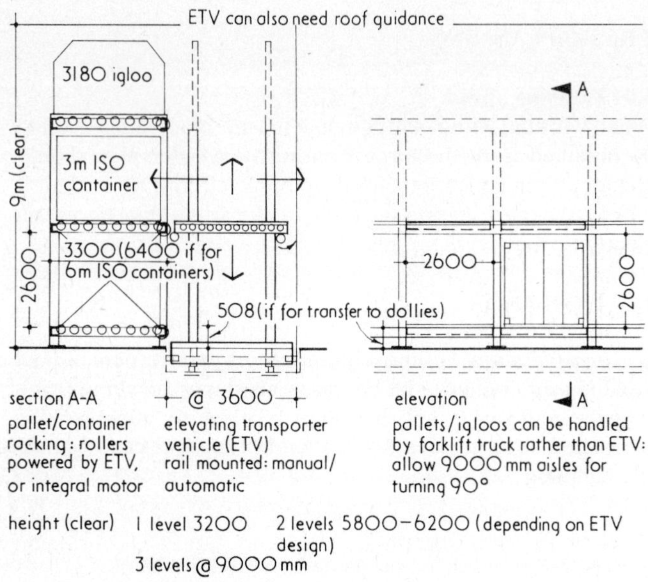

11.6 *Racking in air cargo buildings for storing igloos, air cargo pallets and ISO containers where ETVs are used (elevating transfer vehicles)*

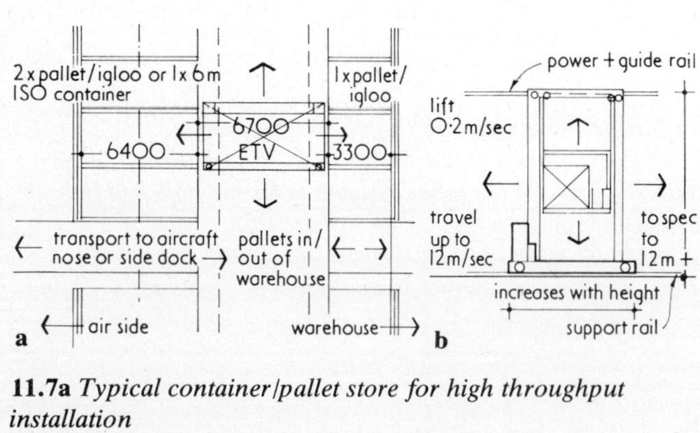

11.7a *Typical container/pallet store for high throughput installation*
b *Typical ETV (elevating transfer vehicle)*

4.04 Local conditions

Consider—local labour/technology/climatic demands, for instance intense heat, sunlight, or rain.

4.05 Legislation

This will differ between countries but principles have been laid down by the ICAO. Advice can be obtained from the country's aviation authority, customs, immigration and port health authorites, and from police, fire and local authority sources. Normally standard controls are needed such as building construction control, fire prevention permission, insurance company permission, and planning authority clearance. In Great Britain it is from the Department of Trade and Industry (Civil Aviation Act 1949, Air Navigation Order 1972).

Legislation referred to in Great Britain is:

Airports Authority Act 1965
Civil Aviation Act 1949, plus 1968, 1971, and 1972 amendments
Offices, Shops and Railway Premises Act 1963
Factories Act 1963
Town and Country Planning Act 1971
Fire Precautions Act 1971
Building Regulations 1976 and 1978 amendment
Health and Safety at Work Act 1974

Further information is available from the ICAO and the IATA.

4.06 Specific legislation

Compartmentation; means of escape; fire prevention, detection, alarm, control; materials combustibility; cladding; insulation; fire-fighting methods; noise control; adjoining area protection; labour legislation, eg the handling of plant design.

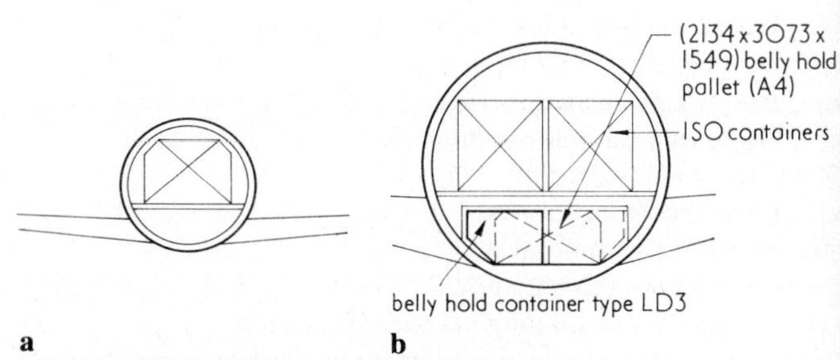

11.8 *Containers, pallets etc in the aircraft*
a *Narrow-bodied aircraft (B707-320, DC8-63)*
b *Wide-bodied aircraft (B747-F)*

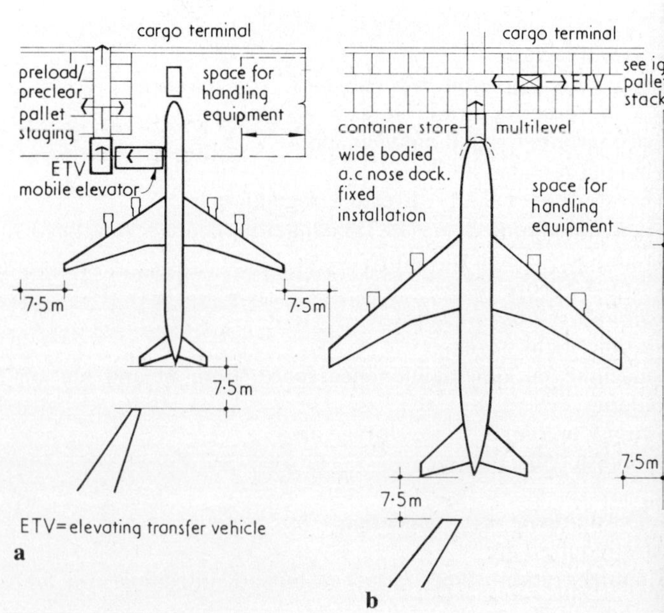

11.9 *Loading and unloading the aircraft. NB British Airports Authority now demand 15 m between wingtips*
a *Narrow body dock*
b *Wide body dock*

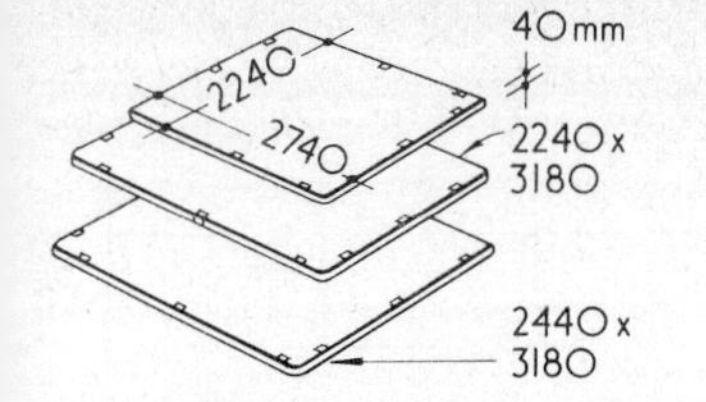

a *Standard air cargo pallets*

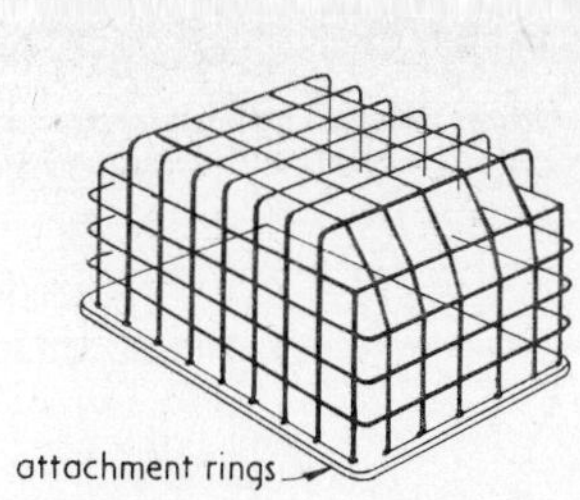

b *Net over air cargo pallets (note contour). Hanging storage required for net*

c *Non-structural igloo plus holding-down net. Do not stack*

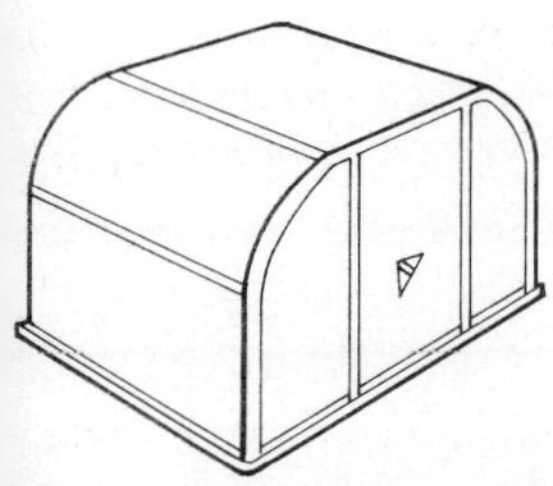

d *Structural igloo. Do not stack*

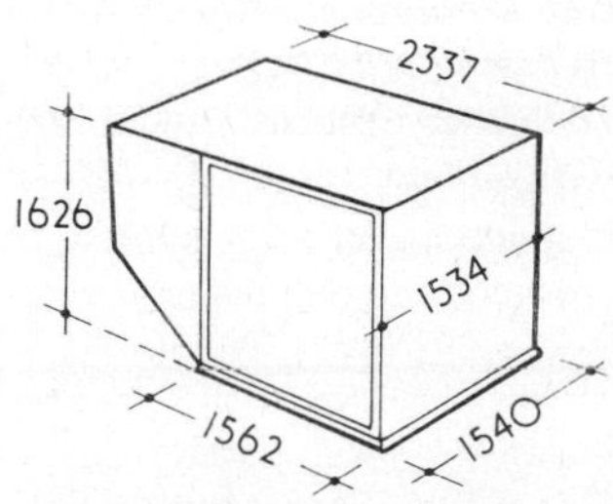

e *Belly-hold container for wide-bodied jets*

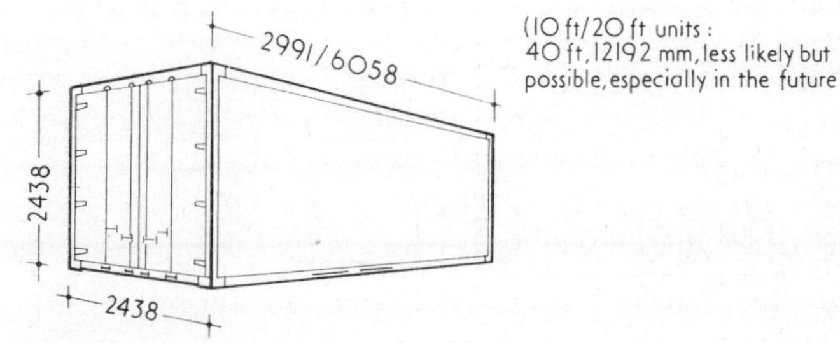

f *ISO container (B747F) with special light construction plus flat base for aircraft. Gross weight (20 ft) 13 000 kg*

11.10 *Airportable ULDs (unit load devices). Pallets/igloos common to most cargo aircraft types, eg B707, DC8. h for DC10F = 2120*

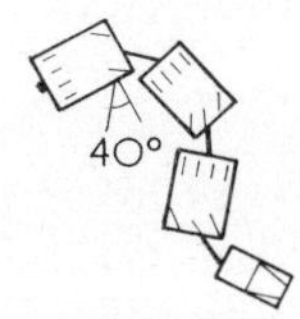

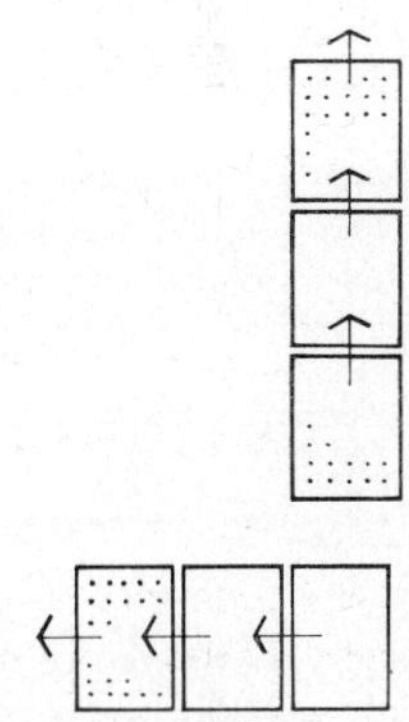

11.11 *Non-powered apron transport:*
a *Dollies can make tight (40°) turns. Maximum train is four dollies*

b *Dollies used as conveyors in either axis when close-coupled*

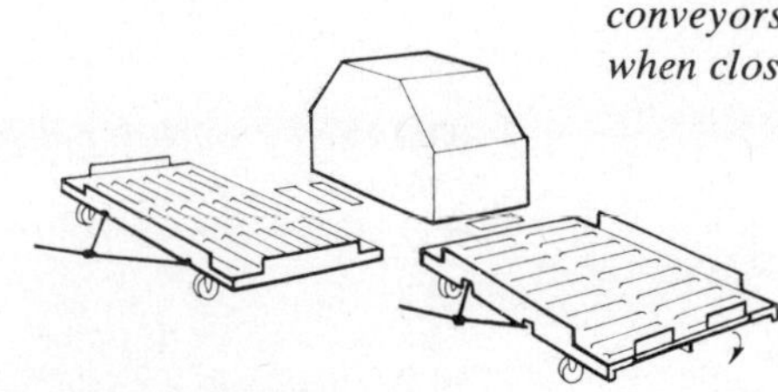

c *Lateral transfer dolly and through end transfer dolly*

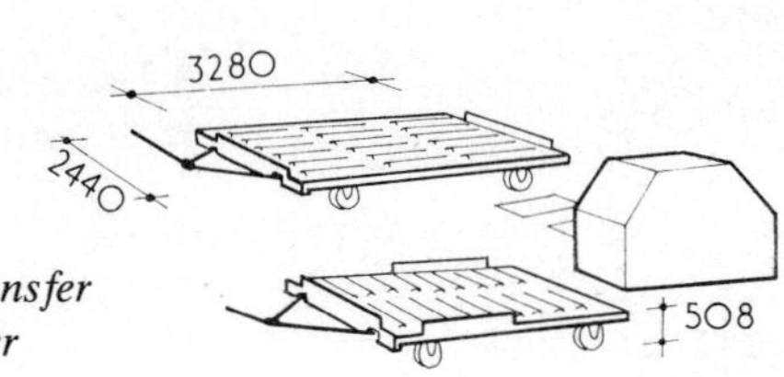

d *Through lateral transfer dolly and end transfer dolly*

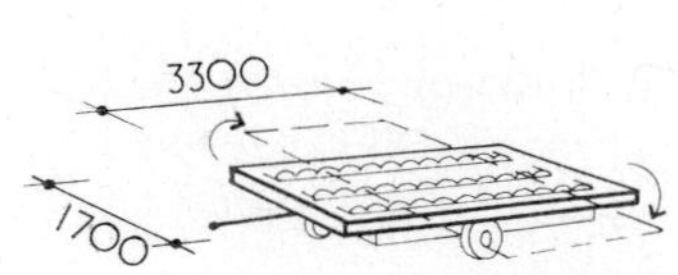

e *Rotating roller table dolly for belly hold container (LD1/3)*

4.07 Area demands

Again there are no hard and fast rules.

Consider—

- peak flows: daily by flight departure, seasonal
- the effect of predominantly small goods
- the effect of predominantly unit loads
- arrival/despatch patterns, both landside and airside
- landside evening peaks, weekend peaks
- airside, the effect of peaks from curfews

Note: A curfew is the time when, to prevent excessive aircraft noise, an airport will neither allow any further aircraft in or out. This generates take-off and landing peaks.

5 DETAIL PLANNING

The planning of an air-cargo building requires the interaction of a large number of factors: flexibility to accept change is essential. This guide, **11.8** to **11.20** can only act as a skeleton for outline and feasibility purposes. More detailed information will be found in the references.

6 REFERENCES

Airport terminals reference manual, IATA (International Air Transport Association)
Airport handling manual, IATA
Restricted Articles Regulations, IATA
Live Animals Regulations, IATA
Quick reference guide to unit load device interchange, IATA
Peter Falconer and Jolyon Drury *Building and Planning for Industrial Storage and Distribution,* London, Architectural Press, 1973
Aircraft manufacturer's literature

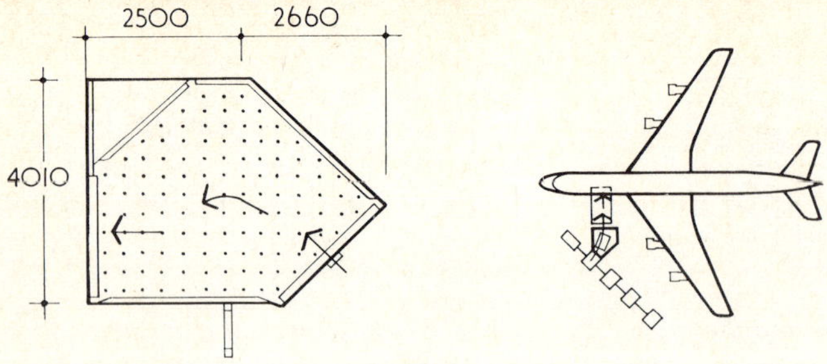

11.12 *Angle transfer ball platforms can accelerate handling by eliminating tight dolly turns and conflict with aircraft and other handling plant*

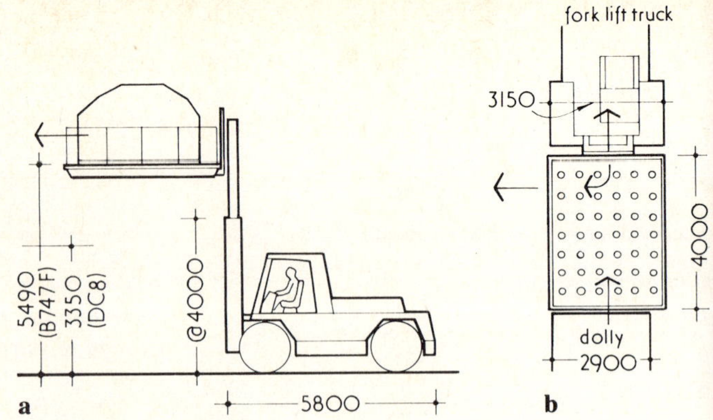

11.15 *Loading aircraft by fork-lift truck (typical for developing country).*
a *16 000 kg capacity fork-lift truck lifting slave pallet for igloos/pallets*
b *Collects pallet from dolly at 90° from aircraft*
Turning circle diameter 12 m

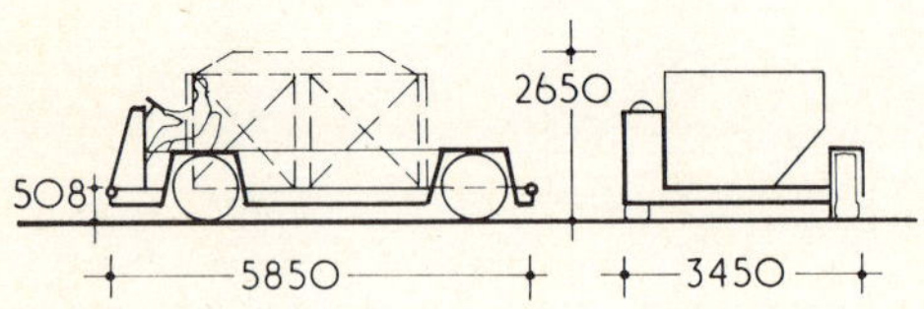

11.13 *Powered apron transport:*
a *Self-propelled pallet transporter carries 1 no 2240 × 3180 pallet or 2 no LD3 belly-hold containers*
Weight laden 9800 kg
Turning circle diameter 16 m

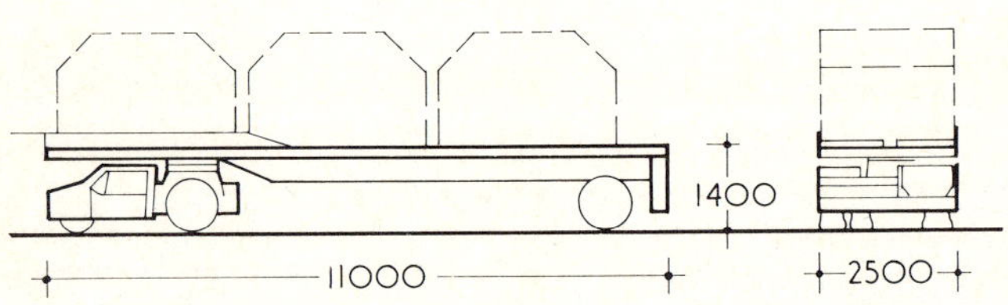

b *Low profile articulated pallet and container transporter*
Turning circle diameter approximately 22 m

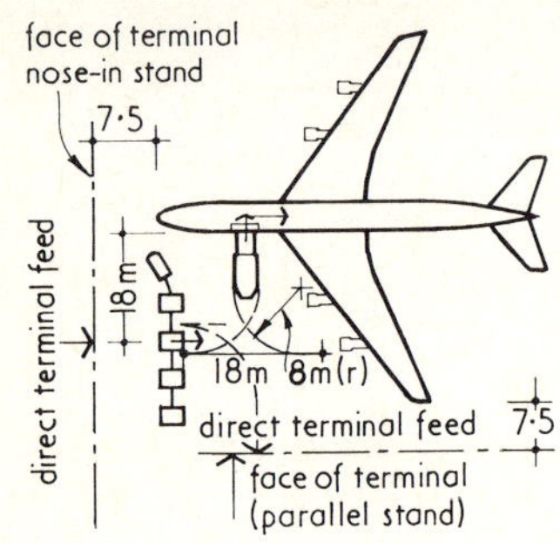

11.16 *Loading aircraft by fork-lift truck (not to scale). Typical for developing country*

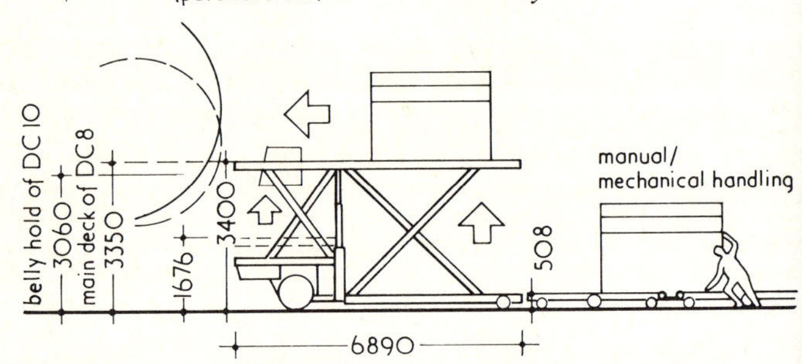

11.17 *Pallet loader for narrow-bodied aircraft main deck or wide-bodied aircraft belly hold. Igloos on pallet dollies: width 2980 mm, turning circle diameter 13 850 mm, weight 7720 kg without pallet*

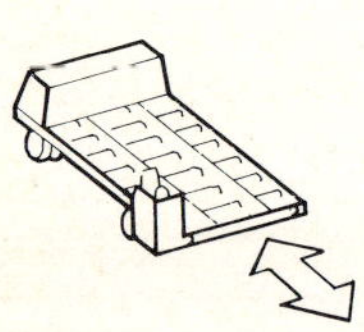

11.14 *Self-propelled pallet transporters*
a *One-way transfer (shuttle car)*
Used as a short distance bridge between aircraft loader and staging

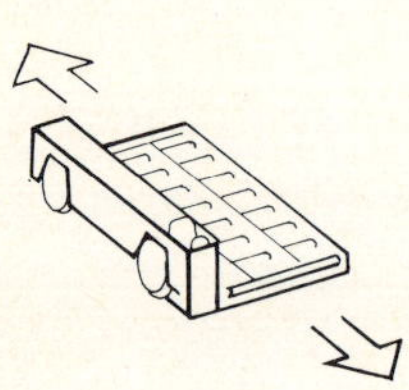

b *Two-way transfer. The most common transporter (all have powered roller decks)*

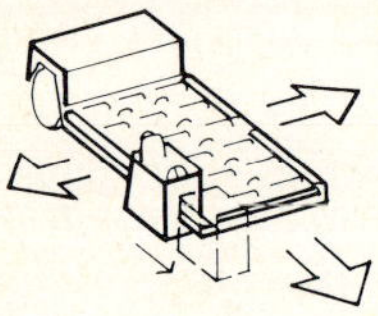

c *Three-way transfer. Used where space is at a premium. Cab moves forward for side transfer*

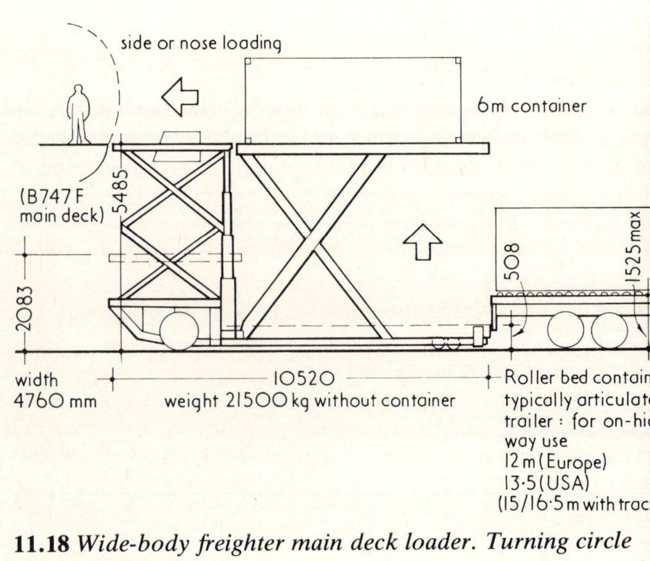

11.18 *Wide-body freighter main deck loader. Turning circle diameter 21 000 mm*

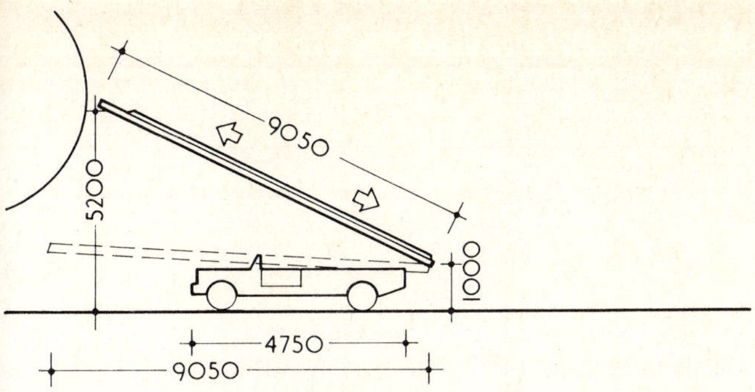

11.19 *Belt loader for mail or bulk cargo. Width 2000 mm, turning circle diameter 12 190 mm, belt width 800 mm offset, weight 2900 kg*

11.20 *Tractors for towing aircraft:*

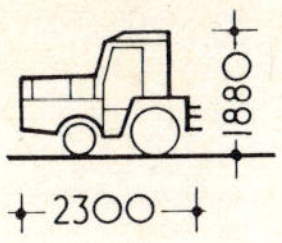

a *Diesel tow-tractor (external). Width 1180 mm, turning circle diameter 4830 mm, weight 1814 kg*

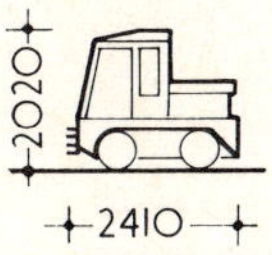

b *Electric tow tractor (internal and on aprons). Width 1184 mm, turning circle diameter 4268 mm, weight 2300 kg*

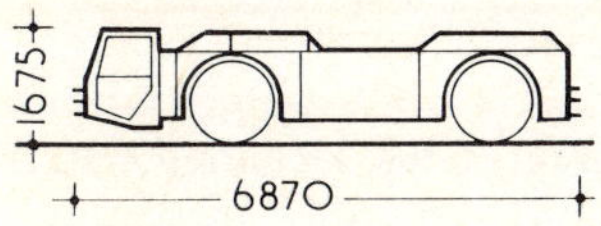

c *Diesel tow-tractor for heavy aircraft, push-out tractor. Width 2920 mm, weight 45 455 kg*

12 Factories

Jolyon Drury

CI/SfB 27
CI/SfB (1976 revised) 282

Contents

1 INTRODUCTION

1.01 Functions of a factory

A factory is a complex network of functions, including materials storage, component manufacture, assembly, interprocess storage, packaging, despatch and transport interface all of which must work together, **12.1**.

1.02 History of factory development

The history of factory development has been one of continually changing requirements following improvements in production equipment, mechanical handling and motive power associated with shifts of philosophy from individual craftsmanship to soulless production line. The latest innovation is rapidly outdated, and the buildings designed to accommodate it are often obsolete before they are commissioned. The following will, in the future, probably be considered landmarks in factory development:

- computer factory for IBM, Havant (1968), Arup Associates.
- diesel engine factory for Cummins, Darlington, Roche and Dinkeloo, **12.2**
- cigarette factory for Players, Nottingham, Arup Associates.
- car assembly plant for Volvo, Kalmar, **12.3**.

2 CLASSIFICATION OF PRODUCTION BUILDING TYPES

Factories can be classified as light, medium, heavy or bulk process industry.

2.01 Light industries

These include:

- high precision work in laboratory-like conditions, **12.4** and **12.5**
- small-scale craft workshops as are now being encouraged in both urban and rural areas similar to **12.8**.

Design will depend on circumstances, but will tend to approximate to laboratory or office type design.

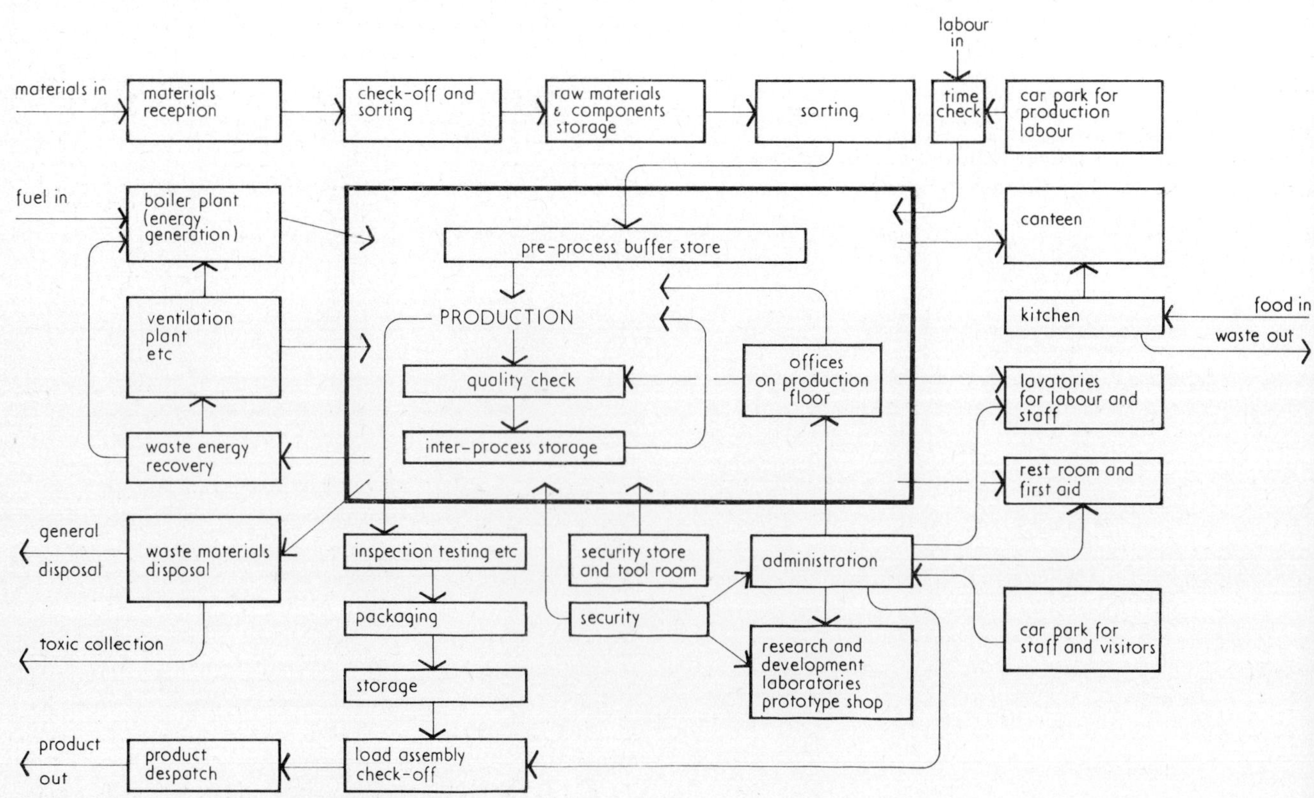

12.1 *Relationship diagram for a typical factory*

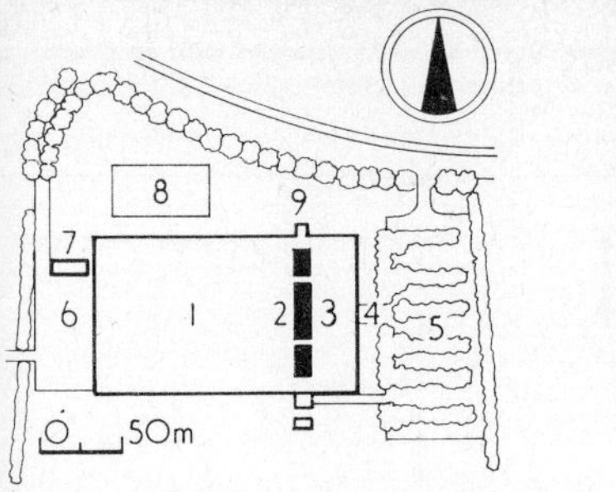

a *Site plan:*
1 *factory area*
2 *brick-enclosed core*
3 *office area*
4 *main entrance*
5 *car parks*
6 *factory yard for incoming and outgoing goods*
7 *stores building*
8 *pool*
9 *chimney*

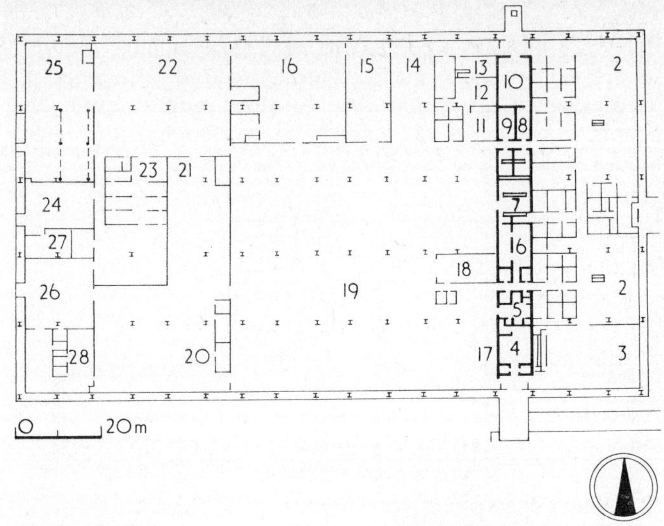

12.2 *Cummins' diesel engine factory at Darlington by Roche and Dinkeloo*

b *Factory building plan:*
1 *entrance hall and reception*
2 *offices*
3 *canteen*
4 *kitchen*
5 *first-aid post*
6 *training room*
7 *lavatory*
8 *electric power plant*
9 *machinery*
10 *boiler room*
11 *calibration*
12 *metallurgical laboratory*
13 *chemical laboratory*
14 *servicing*
15 *stores room for assembly components*
16 *assembly*
17 *tool stores*
18 *tools issue*
19 *main hall*
20 *raw materials*
21 *spares stores*
22 *stores of finished products*
23 *test cells*
24 *test stand*
25 *goods out*
26 *goods in*
27 *dynometer*
28 *inspection of incoming goods*

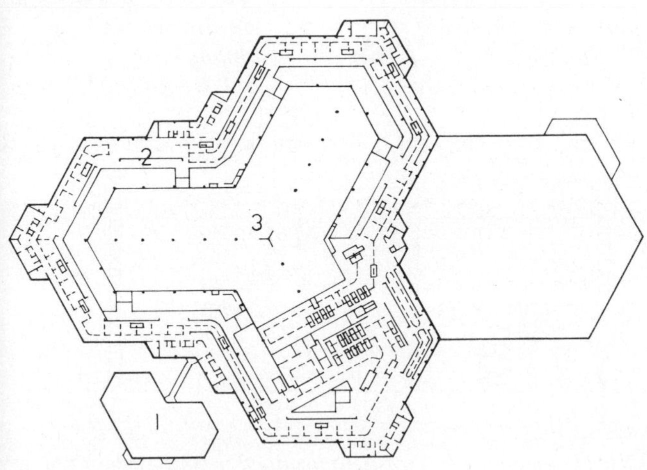

a *upper level*

12.3 *Volvo car assembly plant at Kalmar:*
Team assembly areas are disposed around the perimeter on each level, automatically fed with materials

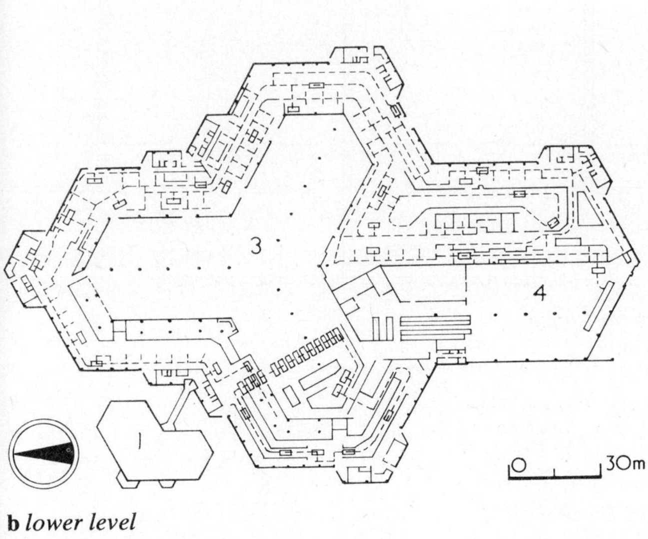

b *lower level*

1 *offices*
2 *assembly*
3 *parts storage*
4 *loading*

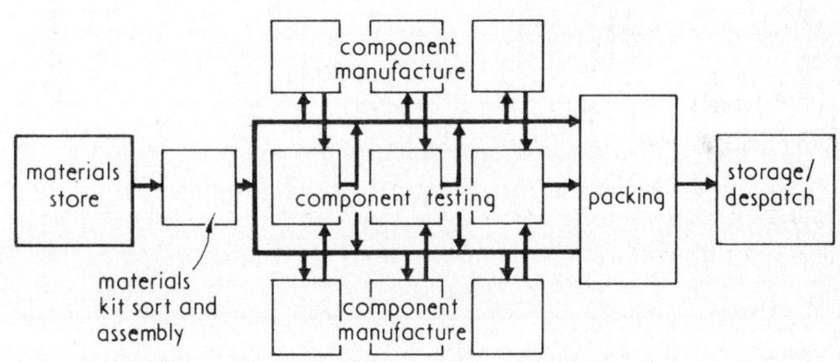

12.4a *Typical process flow diagram for light production and assembly such as small electronic components' manufacture, and similar high technology processes. 'Kit sort' refers to the making up of kits of components for assemblers*

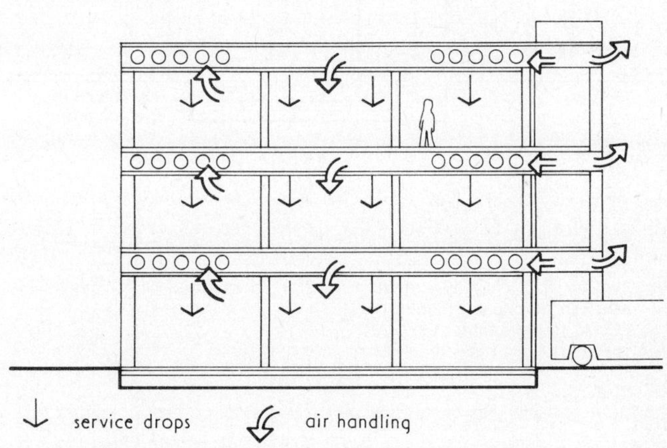

12.4b *Section through typical factories for light, high technology production;*
Multi-storey construction, as new or conversion of existing building: could be flatted units

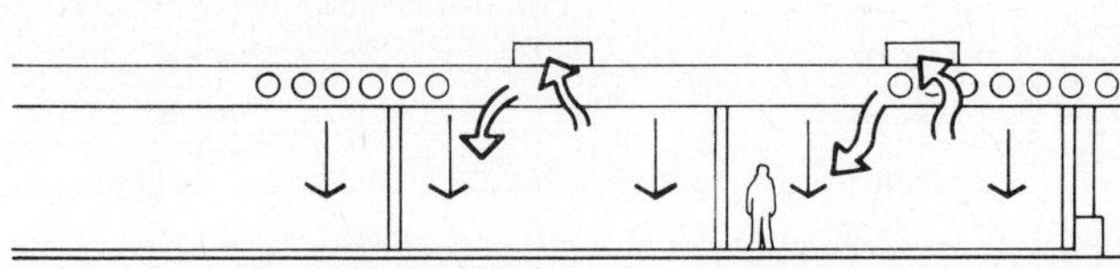

12.5 *Light production and assembly: single storey for small scale and high technology assembly. High degree of service freedom in roof zone*

2.02 Heavy industries

Industries such as steel-making and ship-building require spaces (not necessarily enclosed) designed around the work or the mechanical plant, **12.6**. Traditionally it is difficult to build adaptable structures, **12.7a** and **b**, but modern handling techniques enable 'loose fit' buildings to be designed.

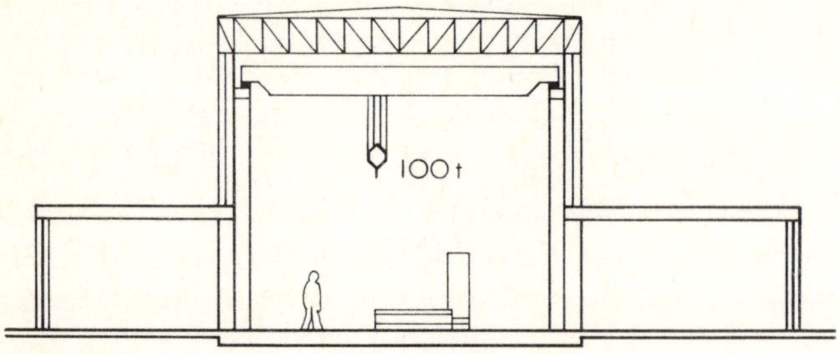

12.7 *Heavy engineering workshops:*
a *Section, and*

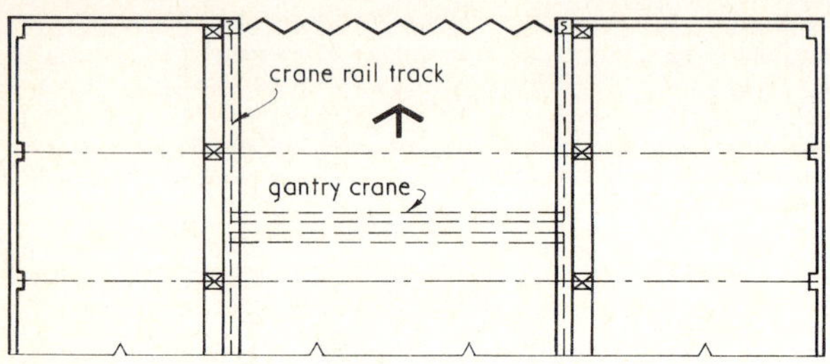

b *Part plan of traditional type. Heavy-duty gantry cranes move the workpiece to the appropriate machine tools and assembly areas*

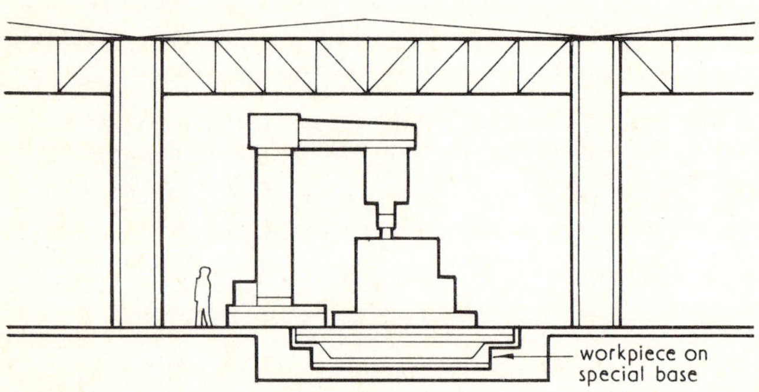

c *Section, and*

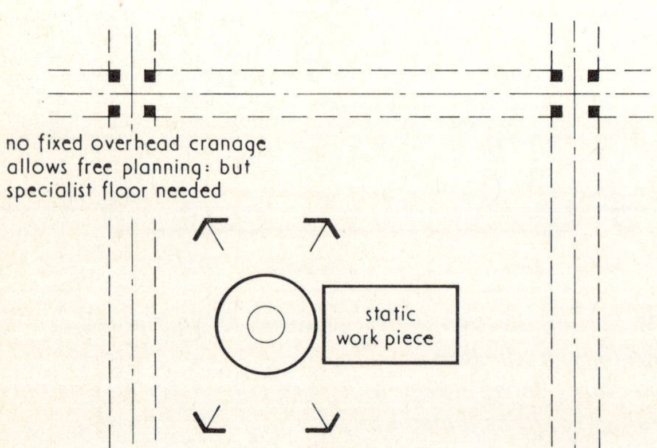

d *Part plan of recently developed workshop where large workpieces remain static, being built up on special bases that are likely to be employed for transport and installation. Machine tools and components are brought to the workpiece, air-cushion techniques are widely used*

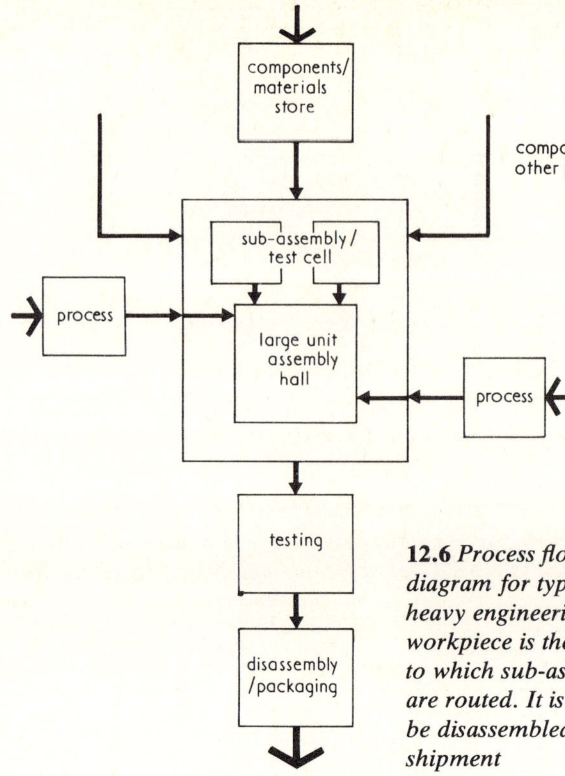

12.6 *Process flow diagram for typical heavy engineering. The workpiece is the centre to which sub-assemblies are routed. It is likely to be disassembled for shipment*

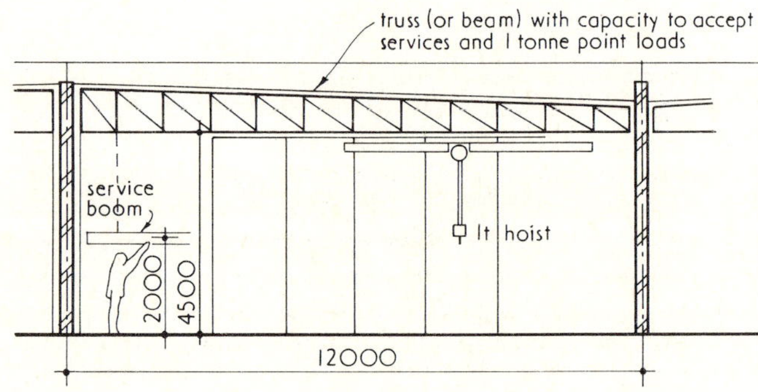

12.8 *Typical 'nursery' for light production and assembly, low technology, may be built speculatively*
a *Section through unit*

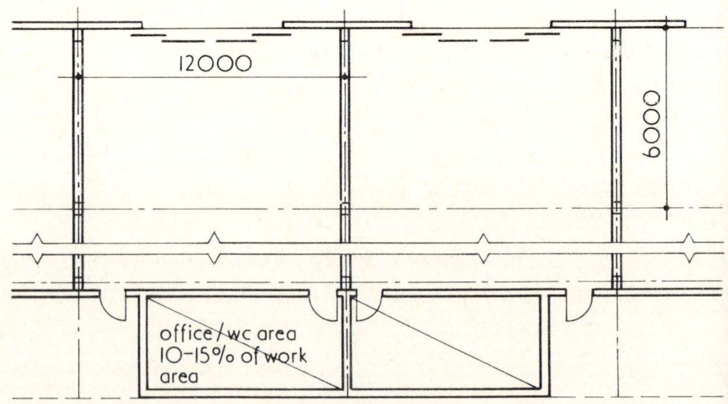

b *Part plan*

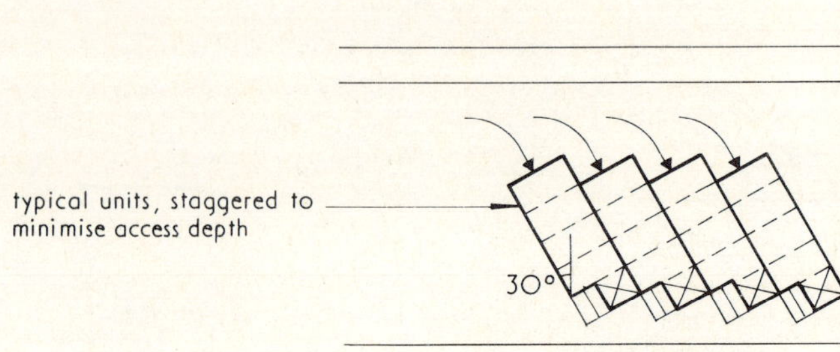

c *Units staggered in plan to reduce site depth required*

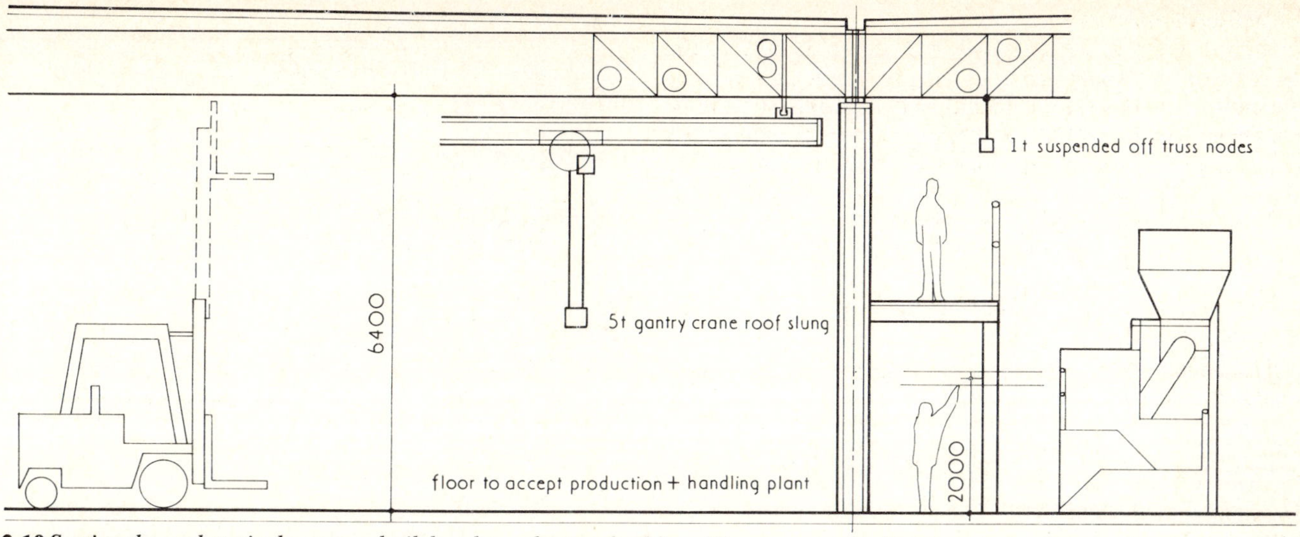

12.10 *Section through typical purpose-built batch production building. The spans, typically 18 × 12 m and trussed roof construction are selected for cheap and rapid adaptation to a variety of uses. Floor loading 25 kN/m²*

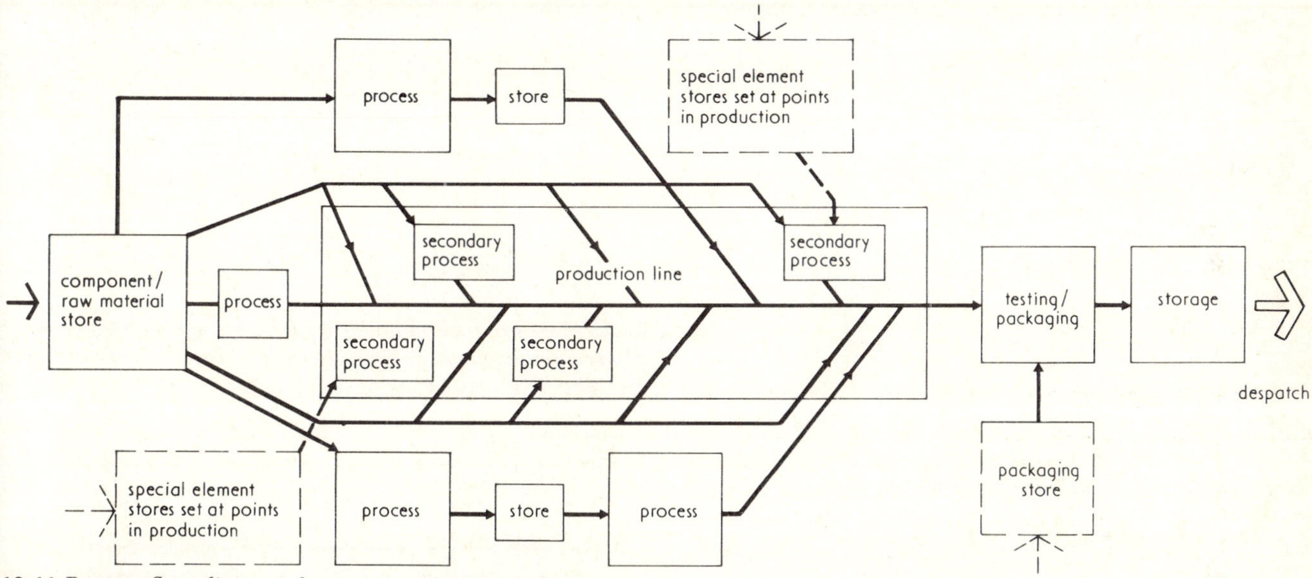

12.11 *Process flow diagram for mass production and assembly. This applies to high volume line assembly as in the motor industry, with some components being built into sub-assemblies before final assembly on the main line*

2.03 Medium industries

The greatest need for careful and thoughtful design is in this field. These industries can be subdivided into:

- *light-medium:* small-scale engineering and assembly, clothing factories, paint shops, **12.8**
- *general-medium:* batch production of components for other factories, medium-sized printing, **12.9**, **12.10**
- *heavy-medium:* industries requiring intensive use of buildings and services as in mass production, **12.11**, **12.12**.

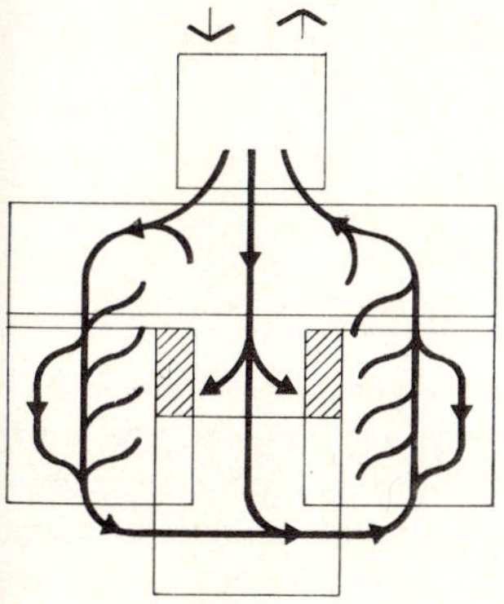

12.12 *Materials flow for mass production does not have to impose a predominantly linear building form. Group assembly 'cells' may feed onto a circulatory route, allowing personnel and services to be grouped into specifically equipped zones*

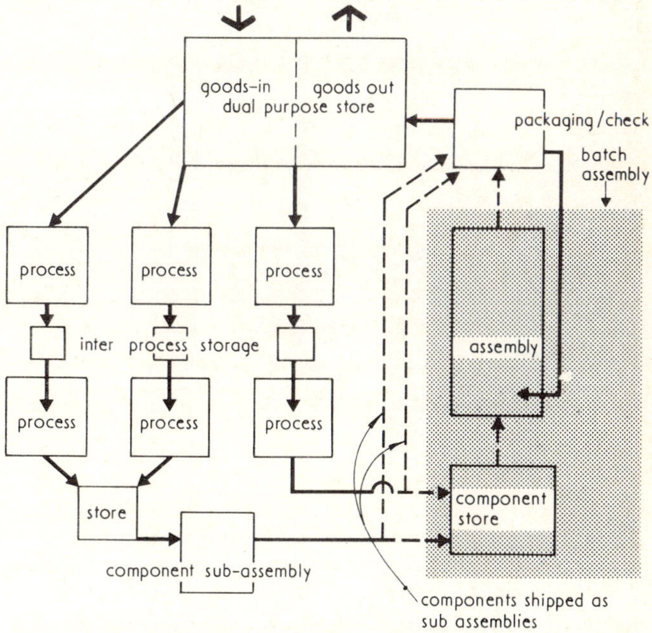

12.9 *Process flow diagram for batch production and assembly. Sometimes involves the assembly and shipping out of complete sub-assemblies, more commonly the production and despatch of batches of discrete components*

3 ADAPTABILITY

3.01 Design for change

The industrial building usually has to change all or part of its use several times during the payback period. Adaptability must therefore be built-in: a minimal first cost will soon be negated by the expense of fitting new processes or working methods into an inherently unsuitable building.

In the recent past, factories were either designed rigidly around a specific process; or else speculatively to a mean specification, resulting in buildings that are unsuitable for many of the modern processes.

Adaptability must allow:

- change of process to avoid obsolescence
- change of process and product following change of ownership.

Changes will normally only be within the broad groupings of building types given in **2.03**.

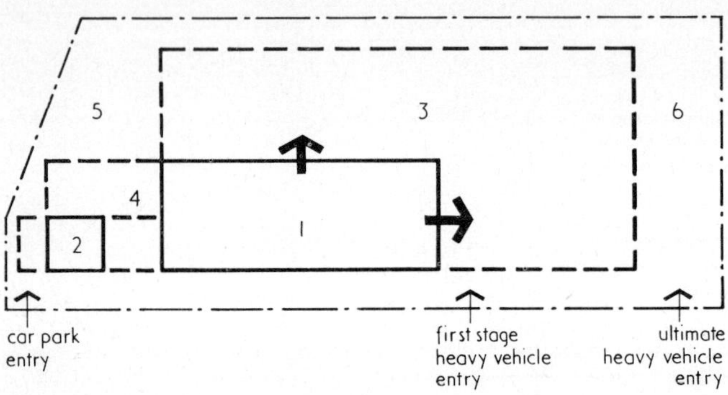

1 *first stage factory*
2 *first stage office*
3 *factory expansion*
4 *various options for office expansion*
5 *car park*
6 *heavy vehicle area*

12.13 *Small or medium-size factory development, with a free-standing office building. The uneven boundary increases the possibility of conflict when the factory and offices expand simultaneously, and restricts commensurate expansion of car parking*

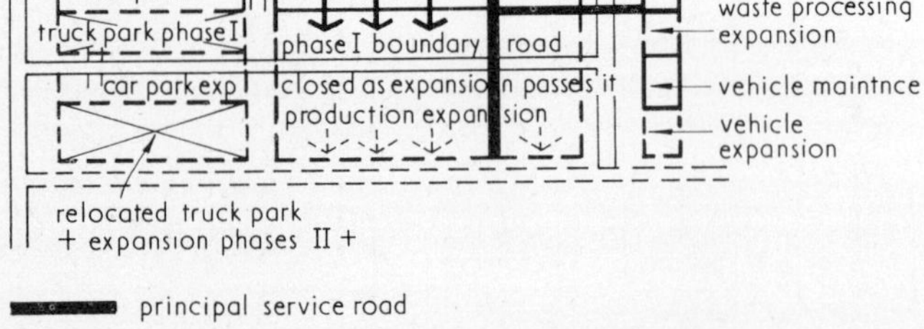

12.14 *A large plant with segregated development zones:*

- *The factory and associated car and truck parking. When the factory expands the truck park becomes the expanded car park and a new truck park is constructed adjacent to despatch*
- *The administration block and associated car parking, separated from manufacturing by landscaping*
- *The ancillary area, incorporating individual growth provision for each element within the zone boundary*

3.02 Design for extension

Apart from alterations within the envelope, there may also be requirements for extension; and the design should anticipate this **12.13**, **12.14**.

4 WORKING METHODS

4.01 Alternative methods

The alternative methods of work organisation are:

- linear assembly
- team technology.

While the latter is a more recent introduction, there is no indication that it will completely supplant the former. Consequently, production buildings must be able to accommodate either; or even both in different areas **12.15**, **12.16**.

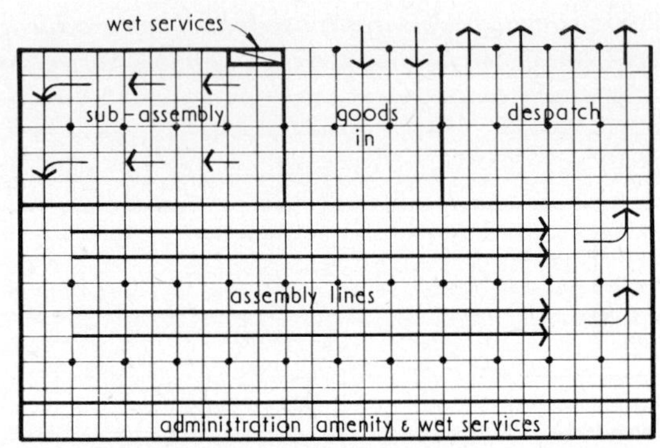

12.15 *Mass production buildings have to accept changes in production technology. This plan shows a conventional line assembly that may be adapted to the form in* **12.16**

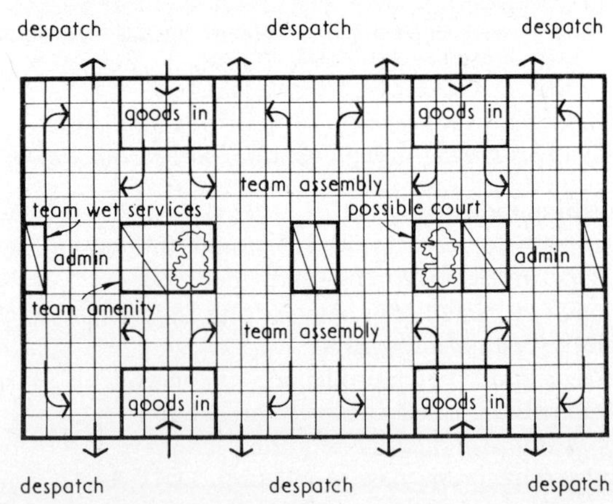

12.16 *The factory can change to team assembly due to new product. Note localisation of amenity and wet service areas to identify with teams. Chance of opening courts adjacent to amenity areas, though these may change position as production demands*

4.02 Linear assembly

In this method, machines are arranged along work travel routes. At each station components are added, until the work has been completely assembled and finished. Supplies of components and materials are needed at each station; and waste must be removed.

4.03 Team technology

This appears to restore to the labour force a feeling of responsibility and achievement. The machines are arranged in groups, and all or a substantial part of the work is assembled within the group. There is a need for storage of materials and components. The main planning requirements are for unrestrictive space and strong floors to enable the machines to be relocated at will, with adaptable overhead services systems. Storage and assembly spaces should be interchangeable.

5 MACHINE SIZES

The sizes of typical machines for light and medium duty industries are shown in **12.17** and **12.22**.

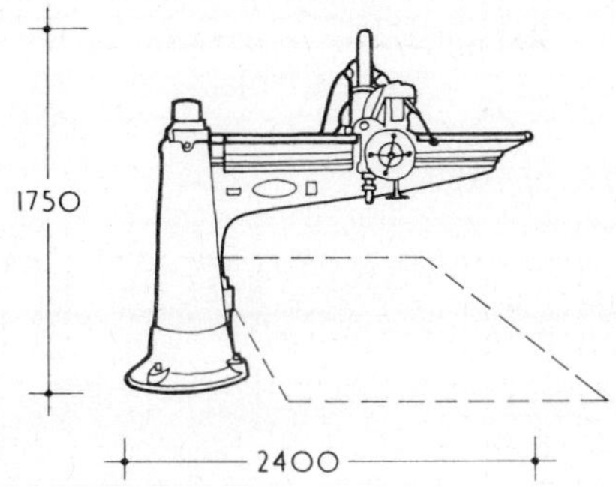

12.17 *Drilling machines:*
a *Plate drill*

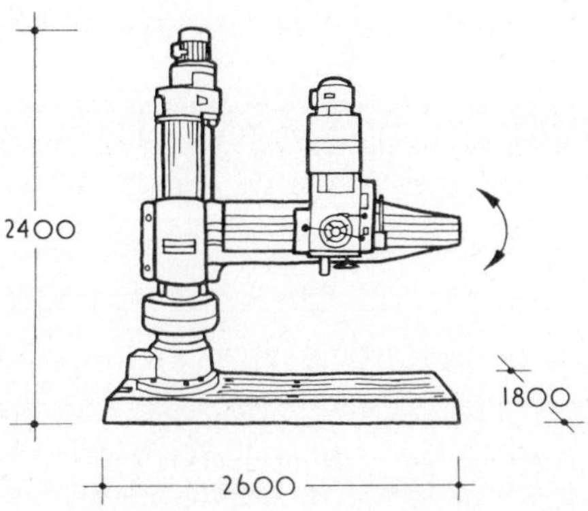

b *Radial drill*

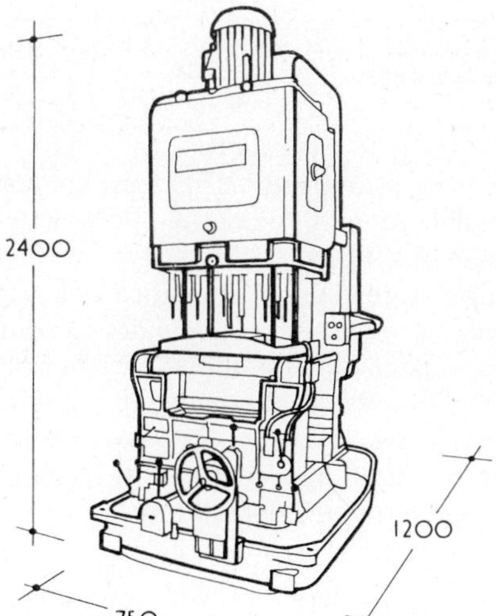

c *Adjustable multi-drill*

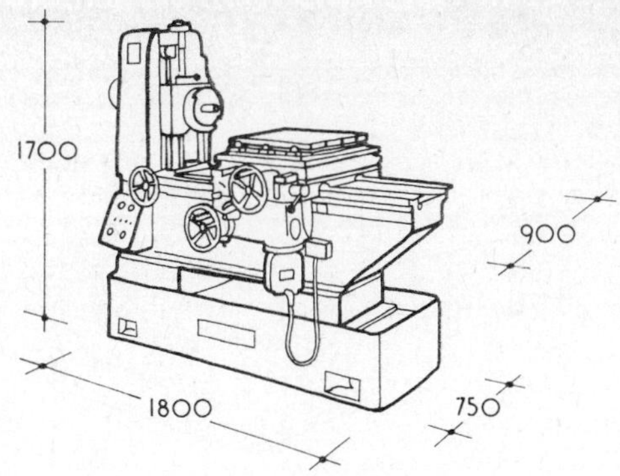

12.18 *Lathes:*
a *General purpose chuck lathe*

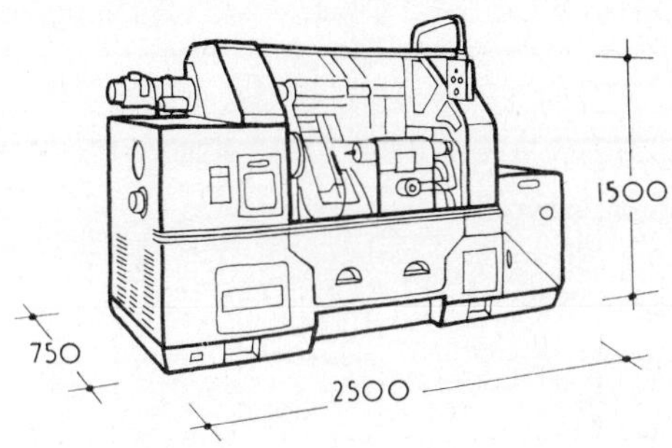

b *Hydraulic copying lathe*

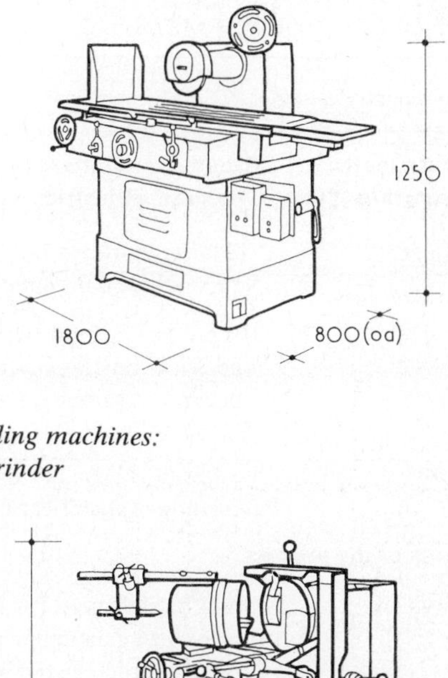

12.19 *Grinding machines:*
a *Surface grinder*

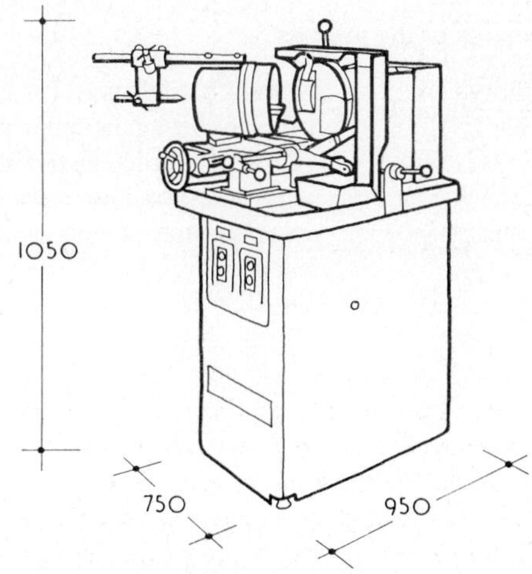

b *Twist drill grinding machine*

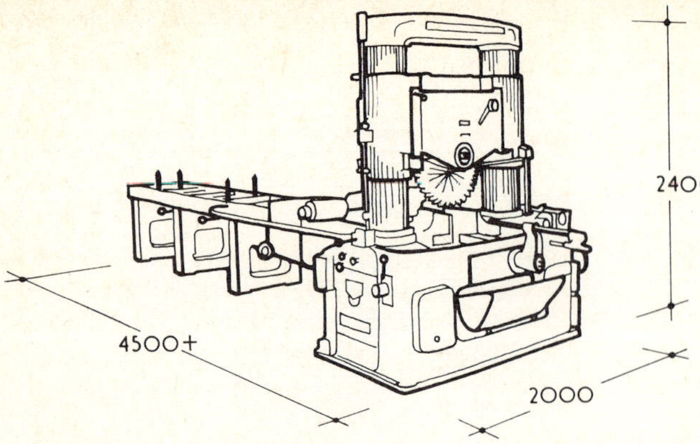

12.20 *Cold sawing machine*

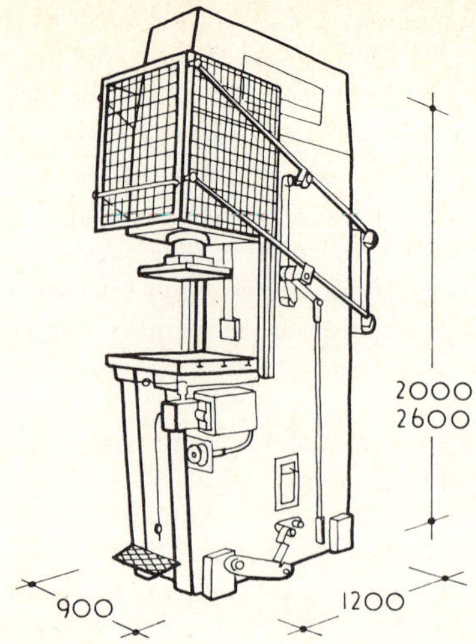

12.21 *Hydraulic pedal press*

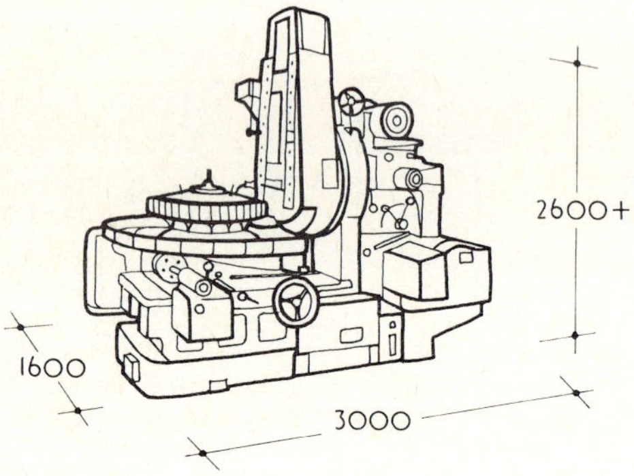

12.22 *Gear cutting machine.*

The majority of machine tools do not exceed 7·5 kN/m²

6 OUTLINE SPECIFICATION OF A TYPICAL MULTI-STRATEGY FACTORY (Updated version of AJ Information Sheet 1378, 6.4.66)

Scope

Type of industries for which appropriate	Buildings of this type are suitable for most manufacturing functions, excluding 'light', 'heavy' and 'process' industries.
Size of project	Total area of production space can vary widely. Average size of all projects is 2500 m², so most are smaller. This specification is suitable for projects from about 1000 m² upwards.
Type of project	40 per cent of industrial projects are adaptations and extensions of existing premises. This specification sets out the general requirements of those projects, or parts of projects, free from special restraints.

Criterion	**Performance specification**	**Design notes**
Requirements of the process		
Adaptability	Should be designed for general purpose use and not around a particular process. General purpose characteristics should be maintained wherever possible, eg in stores and production warehousing as well as in production space itself.	Building positioned on site leaving maximum possible room for extension, preferably in two directions, **12.14**. Single-storey building designed as a large open space. Standardised, mainly dry construction, easily extended or modified. Framework able to carry a variety of alternative roof and wall claddings, services and handling equipment. Those external walls not on or near site boundaries designed for easy demolition.
Plan shape	Probably not critical except where linear flow-processes employed. Rectangular form maximises usable area, facilitates extension.	Rectangular plan form with ratio of long to short sides between 1:1 (minimises internal travel distances where no particular traffic routes are dictated by process) and say, 3:1 (average 2:1).

Criterion	Performance specification	Design notes
Requirements of the process (continued)		
Physical environment	Process requirements will not usually be critical: workplace environment and energy efficiency very important.	See under 'environmental requirements of labour force'. In general the production process will not require special dust-free conditions, nor will it create a dusty or especially dirty atmosphere. If there are toxic or corrosive hazards within the general production space, these should be isolated by local compartmentation and extraction equipment. High standards of cleanliness (eg very exact avoidance of foreign matter) or hygiene (eg avoidance of bacterial contamination) for some high technology factories.
Structural dimensions	Exact plan dimensions rarely critical, except where flow processes employed. Aim should be to optimise convenience for production layouts provided by open space, eg the convenience of stanchions for locating small equipment, switches etc, balanced against the potential adaptability: freedom for service drops and the location of equipment against the cost of greater spans and the loss of overhead craneage.	Span 18 m; bay spacing 12 m or even 18 m (which would permit production line to be turned at right-angles if needed). These are proven dimensions in USA but they are greater than those found in many British factory buildings and (excluding 18 m square bays) are unlikely to increase costs significantly over smaller spans.
	Internal clear height probably most critical dimension, for once built can only be modified with difficulty. Height needed for high stacking, overhead equipment, possibly facility to install mezzanines (for works offices, lavatories, control gear, extension of production space etc) overhead conveyors etc. Space for services needed above clear height level.	Internal clear height minimum 6 m. Main vehicle entrance doors (ground level loading) 5 m.
Provision for services	Facility to take any production service (water, steam, gas, electrical power, etc) to any point within production area with minimum disturbance to building, and therefore production.	Production and building services carried in roof space above level indicated by 'clear height', with vertical droppers as required to machine positions. This eliminates overhead craneage, but allows monorail hoists and conveyors. Roof structure designed appropriately. Drainage used to be below floor level, although alternative more costly but flexible arrangements are preferable. A permanent grid of drainage runs beneath the floor (a minimum of say, one run in the middle of each 18 m span) will minimise disturbance.
Provision for movement of materials and equipment	It should be possible for the production engineer to use the type of material-handling equipment best suited to his product and production methods. Use of fork-lift trucks or similar wheeled materials-handling equipment will be general; overhead conveyors may be used. Cranes more usual in engineering than other industries. Heaviest floor loading is likely to result from wheels of fork-lift trucks (36 kN) and point loads from stacked storage cages and from pallet racking.	Separate foundations will be provided for any special or heavy equipment, especially that which vibrates. Wherever possible, the upper surface of such foundations will be at or below finished floor level. Much equipment is now 'stuck-down' to the floor. Conventionally, an RC floor slab with integral granolithic finish is used, although deterioration of the floor finish is a common problem in industrial buildings. Durable floors can be obtained, but they require a suitable base, good workmanship and close supervision. Particular finishes may be needed to resist attack from acids or oils used in certain processes.

Criterion	Performance specification	Design notes
Requirements of the process (continued)		
Support for production loads	There are two opposed points of view about supports for such production loads as conveyors, local hoists and other overhead equipment. One is that since production loads cannot be predetermined, they should not be allowed to bear on the building structure, and should be carried either on the plant, or on a separate structure, as and when this becomes necessary. This can lead to substructures inhibiting floor area and future flexibility. Although initially more expensive the preferred alternative is to design the roof structure to carry a general minimum of local loads, and to provide the facility to suspend conveyors etc at will.	Design assumptions might be that bottom boom of trusses (assumed spaced at 3 to 3·6 m centres) carry uniformly distributed load of 8kN/m run, and a point load of 10 kN on any panel point at say, 3 m centres. Structural supports for heavier loads are then provided on an ad hoc basis by the production engineer.
Environmental requirements of the labour force		
Visual environment	Practically all visual tasks will be met by illumination levels within the range of 200 to 750 lux; illumination in the middle of the range will be most common. Limiting values of glare index (as IES Code) are likely to be within the range 22–28. Colour schemes should be designed both to assist the distribution of light and to minimise fatigue.	Either daylight or 'windowless' design. If daylight design, a monitor roof shape is a useful compromise between even light level and energy conservation. View windows in external walls. Fluorescent lighting installation arranged in regular pattern over whole production floor to give 300 to 500 lux consistent illumination level E $\frac{E_{min}}{E_{max}}$ must be at least 0·7 wired in three phases to reduce flicker, and in trunking for simple replacement. Point luminaires may be used in areas of higher headroom, or to provide a high and even intensity. Reflecting surfaces decorated with colours of high reflectivity (eg underside of roofs: Munsell value 9), but care that glare from surfaces does not disturb machine operators, eg fork-lift truck drivers.
	Natural light design levels: Warehouse, packing, large assembly, heavy forging, casting, saw mills Daylight Factor 2 per cent (say 10–15 per cent floor area) 300–500 lux Bench and machine work, fine casting, motor repair, general office work, average general purpose lighting, Daylight Factor 4–5 per cent (say 12–15 per cent floor area) 500 lux Drawing work, medium assembly, weaving, small typesetting, Daylight Factor 6 per cent (say 15–20 per cent floor area) 500–750 lux Small inspection and assembly, small bench and machine work, 1000 lux + Daylight Factor 10 per cent.	For 10 per cent and over use PSALI (permanent supplementary artificial lighting installation). For a general purpose building and for resale the design level should not be below a Daylight Factor of 5 per cent. The method of achieving this must be checked against insulation regulations.
Thermal environment	Optimum values of temperature, air movement, etc will depend largely upon nature of work—whether for example it is sedentary or active. Main environmental problem will be to avoid uncomfortable heat in summer. Minimum temperatures: heavy work 10°; light work 13°; sedentary 16°	For most light industry plant should be able to provide air temperature of 18–21°C. Minimum value of thermal insulation for roof and walls U = 0·7 W/m²°C Mechanical ventilation, at least in factories of average or greater size. Air-change rate (fresh air supply) minimum 5 litres/second/person

Criterion	Performance specification	Design notes
Environmental requirements of the labour force (continued)		
Acoustic environment	Production processes highly variable in noise output. Control by encapsulating machinery and by using interspersed storage stacks.	Thermal insulation material can give a measure of acoustic control, particularly in providing absorption.
Fire protection	Some industries are regarded as having 'abnormal' fire risk because of the process or materials used; building design will be affected by requirements for additional compartmentation. Generally, fire hazard is classed as 'moderate' to 'low'. The general requirement of fire safety, of a maximum division of the production area into self-contained fire-resisting compartments, is at variance with the general production need for open space, and should be carefully considered.	Fire division walls may be required to obtain acceptable insurance rate. Areas will depend on process, etc. 'Fire curtains' in roof space. Fire vents in roof surface of total area not less than 1 per cent of floor area. Avoidance of combustible materials in sheeted claddings. Sprinklers are also being increasingly required by insurance companies, both over the process and in the roof depth to protect services.
Explosion hazard	Not normally considered critical, but can be accommodated with blow-out panels, or placing part of process outside the main building.	
Building economics	The cost of using a factory building is an important element in the long-term cost of manufacturing. Nevertheless, without adequate justification, few managements are prepared to pay more than the minimum to obtain their essential specification, one reason being that investment in plant, equipment, perhaps labour is likely to show a higher return than investment in buildings. The current British tax structure discourages spending more on initial building with the aim of reducing maintenance costs, but this is a false economy (see paras **3.01**, **3.02**).	A 'basic' specification: concrete floor slab; exposed structural framework and services; simple finishes, such as painted steelwork, untreated concrete, fairfaced brickwork; self-finished insulating materials forming roof lining.

7 NON-PRODUCTION ACCOMMODATION

7.01 Offices

There is a tendency for administrative and production space to be interchangeable.

Two types of offices will be required in close conjunction with the production space:

- foreman's desk space in sight and proximity of work supervised. This is formed from easily demountable components to allow for rapid relocation. Sometimes to avoid floor obstructions this accommodation is raised to mezzanine levels where visibility is improved
- executive offices for the local administrative staff, or the company headquarters where these are not elsewhere. This type of accommodation is designed in accordance with section 16, Offices, and will depend on the numbers to be accommodated. As a rough guide, allow 10–15 per cent of the production floor area, or 5 m² per person.

7.02 Lavatories

For sanitary accommodation see section 20, Sanitary installations. A first aid facility is normally provided in conjunction with this.

7.03 Canteens

Staff are not allowed to eat in dirty or dusty surroundings. If the process demands a clean environment the reverse may apply, and the importation of food into the working area may need to be discouraged. Canteens are therefore nearly always now provided. See section 20, Catering, for details of design.

8 BIBLIOGRAPHY

Architects' Journal Handbook of factory design, J. Drury, 1977–8

Plant layout and materials handling, Pemberton, Macmillan.

Industrial buildings and factories, O. W. Grube.

Insulation Act, 1972.

Factories Act, 1961.

Offices, Shops and Railway Premises Act, 1963, HMSO.

Health and Safety at Work Act, 1974, HMSO.

Health and Safety Booklets, HMSO.

13 Industrial storage buildings

Jolyon Drury

CI/SfB 28
CI/SfB (1976 revised) 284

Contents

1 INTRODUCTION

Few industrial storage buildings are designed to make a profit (steel stockholders and cash and carry stores are exceptions); the majority perform the function of a valve or pipeline, limiting the supply of a product to suit demand, to stabilise prices and allow steady and economic manufacture within fluctuating market conditions. Industrial storage is therefore a service at a cost that must be minimised.

The payback period most frequently chosen for such a building is 25 years. During that time, it is likely that the storage method will need to change at least three times, and that the type of goods handled will change even more frequently. Flexibility for expansion and manner of use are therefore important design considerations.

2 IDENTIFICATION OF WAREHOUSE AND STORAGE TYPES

The three main types are:

- *Transit* Between manufacture and the market, **13.1**.
- *Distribution* Similar to a transit unit, but accepts a wide variety of goods from a number of manufacturers, sorts them into orders and distributes them to a number of outlets, **13.2**. A components warehouse for a factory performs a similar function.
- *Repository* A warehouse used for stockholding, either as a service (eg a furniture repository) or within a company (eg a cold store), **13.3**.

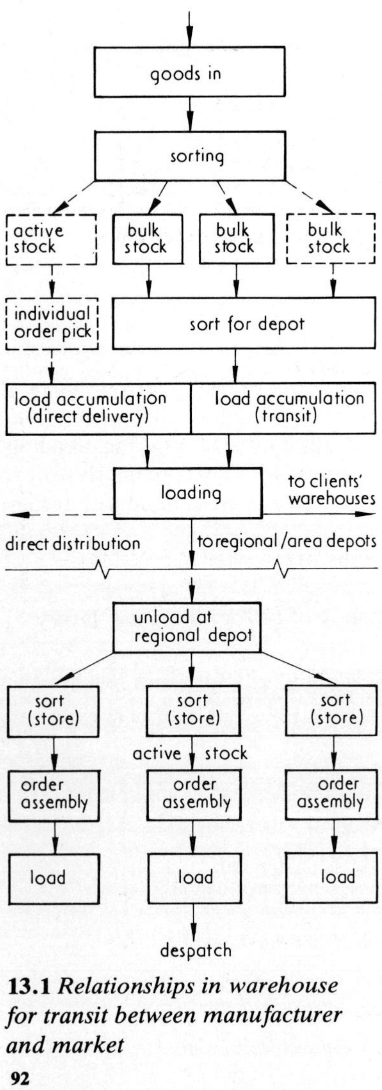

13.1 *Relationships in warehouse for transit between manufacturer and market*

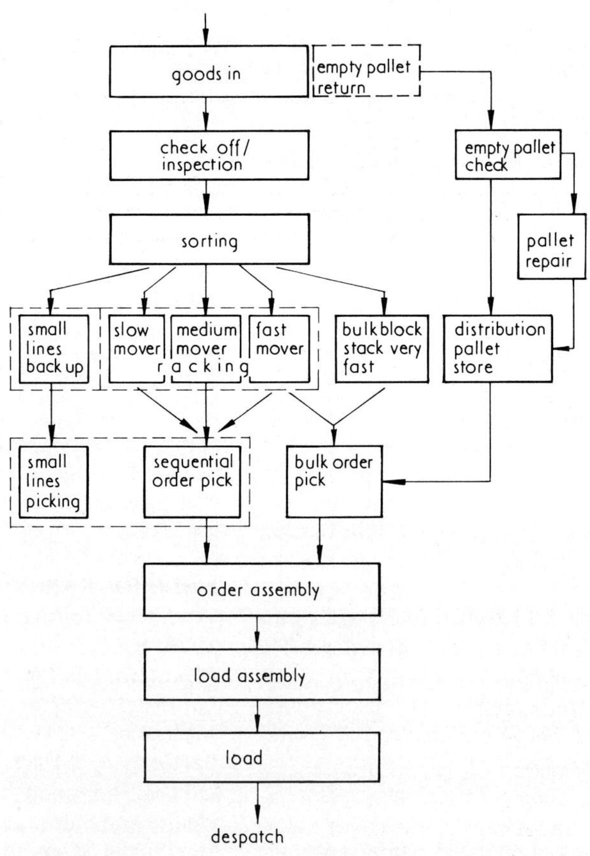

13.2 *Relationships in distribution warehouse*

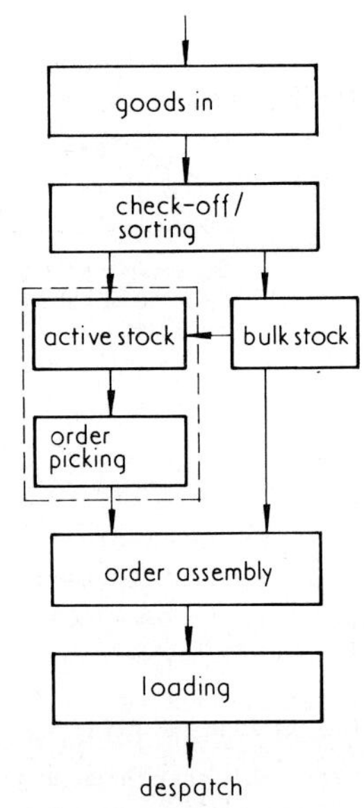

13.3 *Relationships in a stock-holding warehouse*
The bulk stock area is dominant

3 PRELIMINARY DECISIONS

The initial decision about what type of building is required will involve a choice between these three types, dependent on the client organisation's needs. Such a study is generally undertaken in co-operation with a specialist consultant. Other factors to be considered at the pre-design stage are:

1 The orientation of the loading bays and the heavy vehicle marshalling areas. Future expansion must be taken into account.
2 The orientation of the goods sorting and load accumulation areas which must be related to the disposition of the storage area, ie block stacks or racking and loading bays.
3 Will the required bulk of the building be acceptable in terms of planning consent?
4 Are the existing roads suitable to meet increased demand?
5 Is there public transport for operatives?
6 Are there night operating restrictions which will entail special features to muffle night noise?

Can this be catered for by any design measures/configurations?

4 HEIGHT, AREA AND TYPE OF HANDLING SYSTEM

The most economical way of gaining volume for storage is to use height, **13.4** (table I), but this affects the choice of the handling system to be employed. Consider:

- The type of unit load to be handled and the physical characteristic of the goods—crushability, durability, the type of unit loads that will be assembled after sorting (table II).
- The speed of turnover. This will determine what storage method is the most efficient.

Table I Typical internal clear heights for storage areas

Minimum clear internal height* (m)	Type of storage
5–5.5	Minimum cost low rise block stacking warehouse. Suitable for light industrial factory use
7.5	Minimum for any industrial storage building combining racking and block stacking
12+	When turret trucks are used
15–30	Fully automatic, computer controlled warehouses and stacker cranes are to be used

* Clearances for structural members, sprinklers, lighting must be added to obtain overall height of buildings

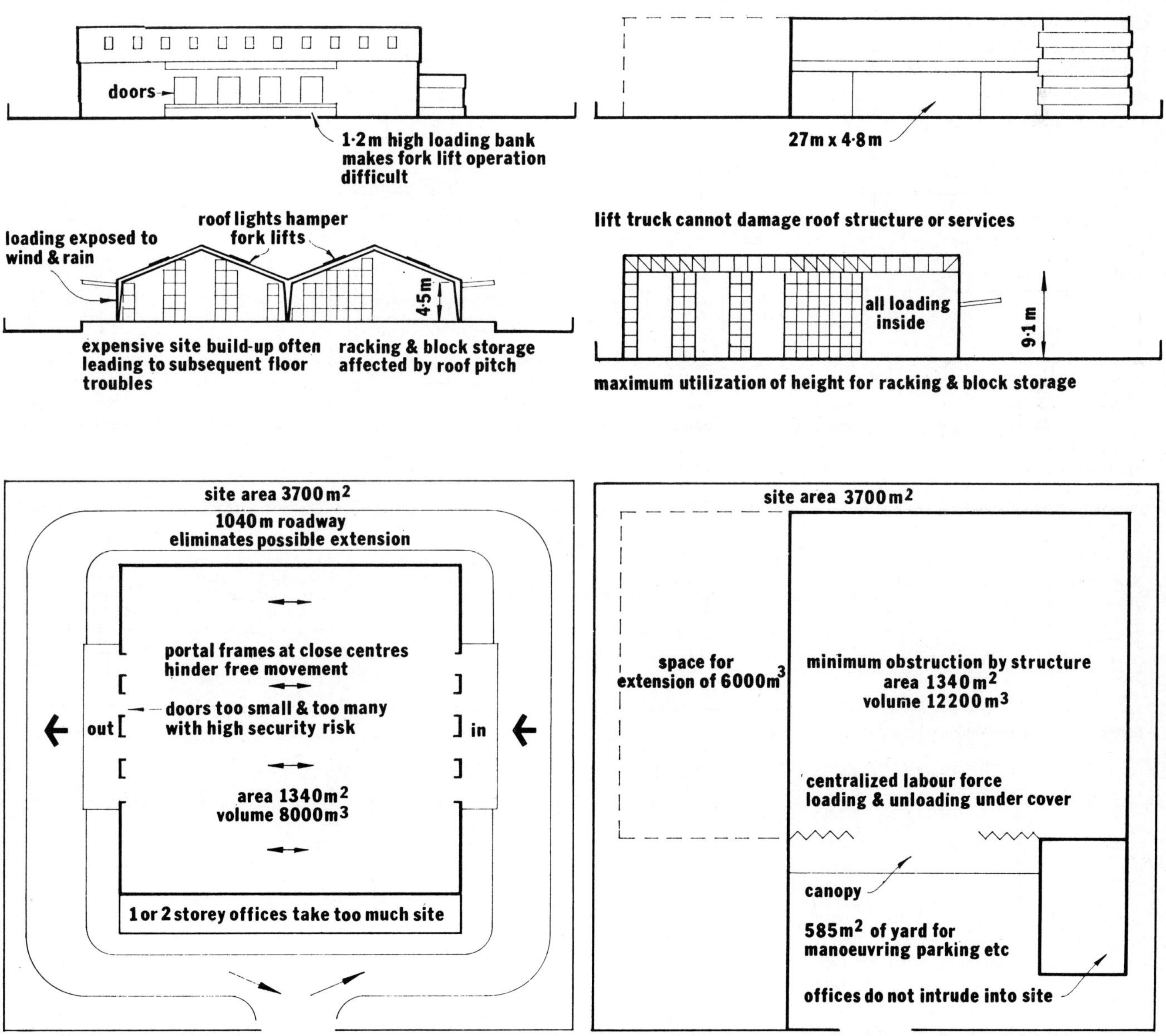

13.4 *Comparison between typical alternative structures:*
a *Traditional portal frame*

b *Modern 'big box'. This provides a much more flexible solution at lower cost when operating efficiency is assessed.*

Table II Classification of materials for handling and storage as unit loads

Description	Examples	Storage method
Materials not strong enough to withstand crushing—not suitable as integral unit load	Automobile components, made-up textiles, electrical appliance components, manufacturing chemists' sundries, light engineering products, glassware	On pallet in rack
Materials strong enough to withstand crushing—suitable for unit loads	Casks and drums, sawn and machined timber, sheet materials	On pallet, or self-palletised and block stowed
Irregular shaped materials, strong in themselves suitably packed into unit loads	Goods in cases, crates or cartons	On post pallets and stacked, on pallets in rack or self-palletised
Bagged materials which form a flat surface under load	Grain, powder, and similar	On pallet and block stowed
Bagged materials which do not form a flat surface under load or will not take pressure	Forgings, moulded or machined parts, nuts and bolts	On pallet in rack
Large irregular loose materials	Moulded plastics; sheet metal pressings	On post pallets and stacked
Small irregular loose materials	Machined and moulded parts, pressings, forgings	In cage pallets and stacked
Materials hot from production processes	Castings and forgings	On post pallets and stacked
Materials too long to be handled other than by side loader or boom	Steel sections, tubes, timber	Horizontally in tube or bar racks
Materials strong enough to withstand crushing but subject to damage	Partly machined automotive parts, painted finished materials, books	Steel box pallets with special partitions
Perishable goods	Frozen meat, vegetables, drink	Cartons, soft packs pallets, box pallets, etc

5 STORAGE METHOD

Storage methods (see tables III to V) include:

1 Very fast throughput involving a limited number of products: block stacking, **13.5**, rather than racking. First in, first out, or first in last out configuration, depending on the shelf life of the goods.

2 A wider variety of goods, but still with fast turnover: drive-in racking, **13.6**, or 'live' (roll-through) storage, **13.7**. Pallets are

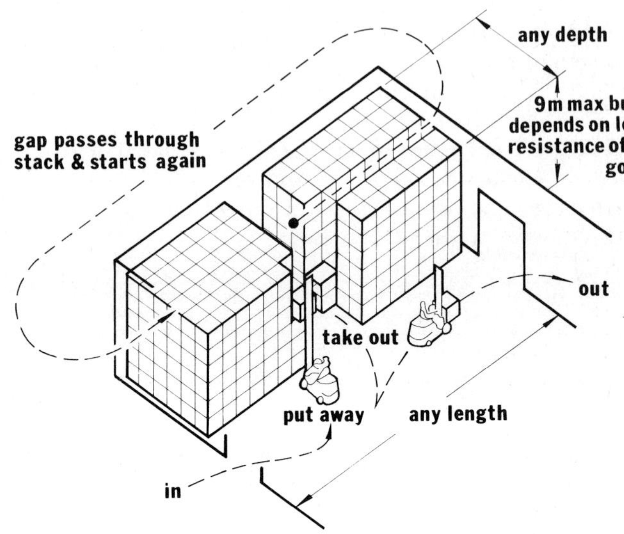

13.5 *Method of block stacking for stock rotation. Where cartons are being stacked on pallets, a height of three pallets is the normal maximum*

Table III Mechanical handling

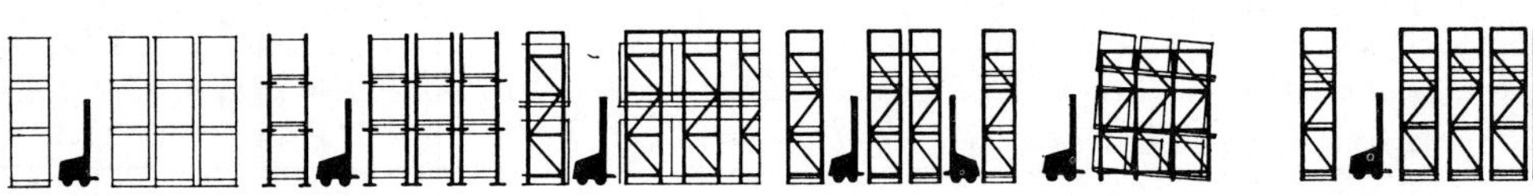

	Block stacking	Post pallets	Drive-in racking	Beam pallet racking	Gravity live storage	Powered mobile racking
Cubic space utilisation %	100	90	65	35–50	80	80
Effective use of installation capacity %	75	75	75	100	70	100
Accessibility of unit load %	10	10	30	100	30	100
Order picking %	1	30	30	100	30	100
Speed of throughput	Fastest	Good	Poor	Good	Good	Quite good
Load crushing	Bad	Nil	Nil	Nil	Some	Nil
Stability of load	Poor	Fair	Good	Good	Fair	Good
Ease of relocation	Not applicable	Not applicable	Fair	Good	Difficult	Difficult
Speed of installation	Not applicable	Not applicable	Good	Fastest	Fair	Slowest
Rotation of stock	Poor	Poor	Poor	Good	Excellent	Good

Table IV Manual handling

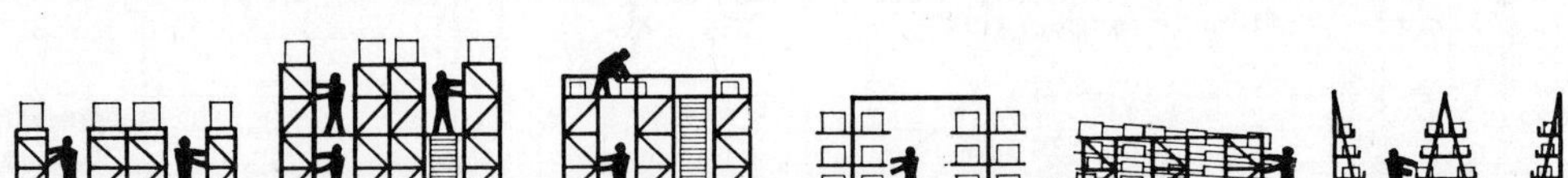

	Long span shelving	Tiered shelving	Raised storage area	Cantilever shelving	Lightweight live storage	Fir tree racking
Cubic space utilisation %	45	45	80	50	65	25
Effective use in installation capacity %	95	95	50	100	70	70
Accessibility of goods	Good	Good	Poor	Good	Excellent	Good
Ease of relocation	Good	Fair	Difficult	Fair	Very difficult	Best
Load range kN/m²	2–9·5	2–9·5	2·8–11	2–4·7	Up to 0·2 kN per m run of track	2·6–4·4 kN per arm
Speed of picking	Good	Fair	Poor	Good	Very good	Good
Speed of installation	Very good	Good	Fair	Fair	Slowest	Fastest
Rotation of stock	Very good	Good	Poor	Very good	Excellent	Very good

Table V Load Mounting

Load mounting	Type of load Heavy unstable load	Flat cards/ sheets	Sacked/ bagged loads	Small unit loads	Drums Reels Barrels	Coils	Casks	Bales	Textile Raw materials
Special cradle with/ without pallet	*								
Standard pallet		*	*	*	*	*	*	*	
Flat board pallet + decking supports		*	*			*		*	
Direct mounting on timber panels		*	*	*		*	*	*	*
Drum supports					*				
Post pallets—cage/bin			*	*		*	*		*
Coil supports					*	*			
Skips/skeps with skids								*	

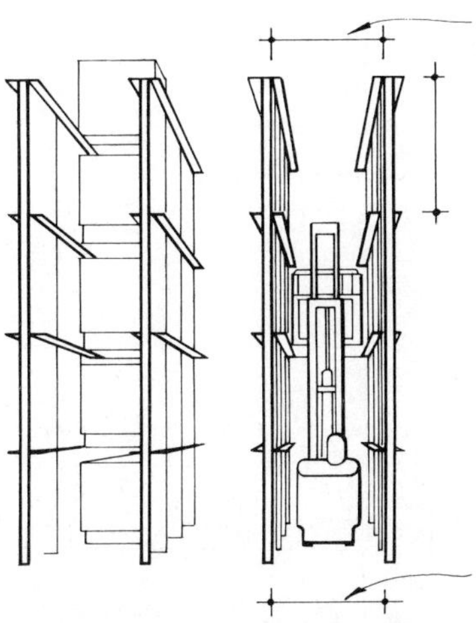

13.6 *Drive-in racking for fork-lift. A maximum depth of six pallets, with fluorescent lighting in the racking structure. Four pallet depth is preferable*

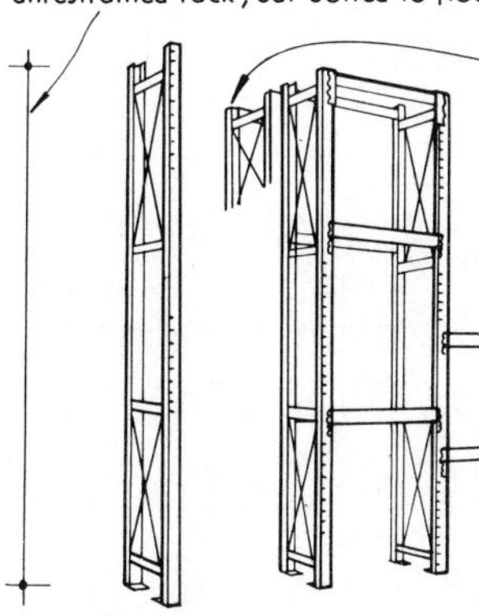

13.8 *Pallet racking*

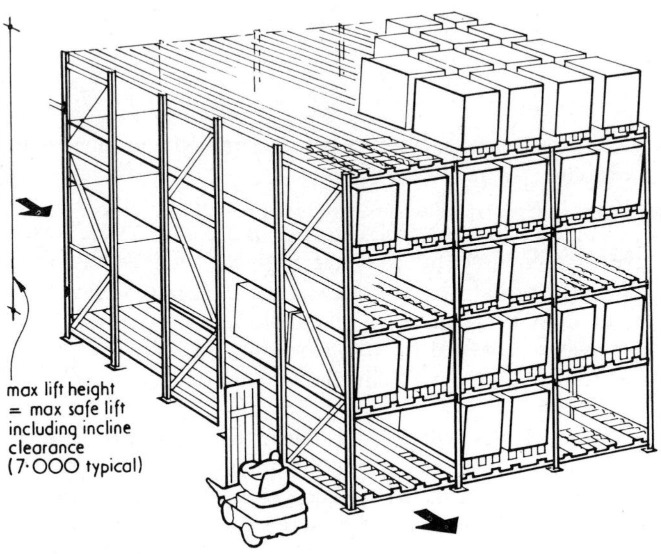

13.7 *Roll-through racking*

placed into racking up to four positions deep, with the pallets' edges resting on runners attached to the rack's uprights. First in, last out.

Live racking involves inclined storage lanes. For heavy pallets and shock-sensitive goods, braking and separating equipment can be incorporated.

3 Pallet racking, **13.8**, **13.9**. For a wide variety of goods, the speed of throughput decreases. Pallet racking is the solution with a large variety of products, brands or pack sizes. Each pallet is normally allotted a unique position in the racking.

6 DISPOSITION OF THE RACKING

There are two common alternatives:

- the rack is oriented at 90° to the order assembly areas, with the fast turnover stock in the bays nearest to it; or
- one complete racking face is oriented along one side of the order assembly area and reserved for very fast moving stock.

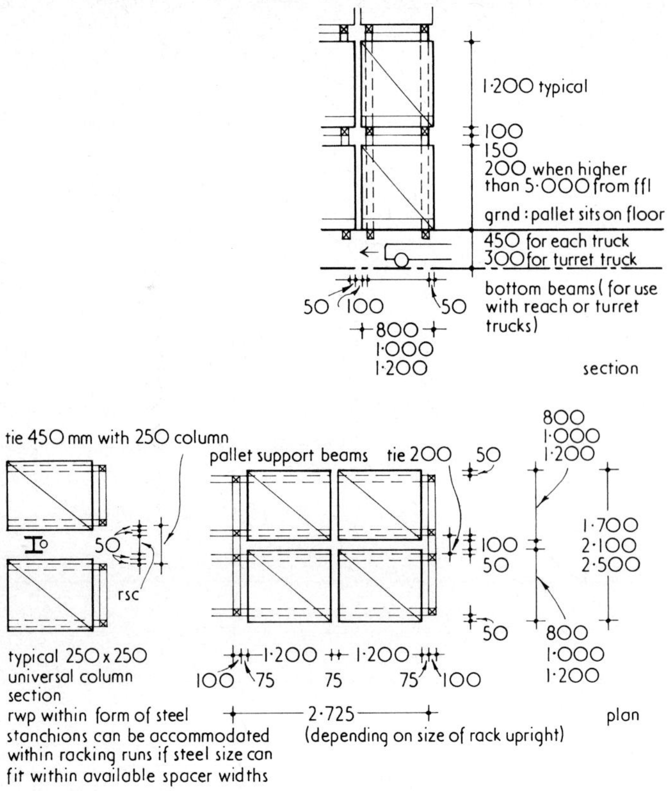

13.9 *Construction of pallet racking*

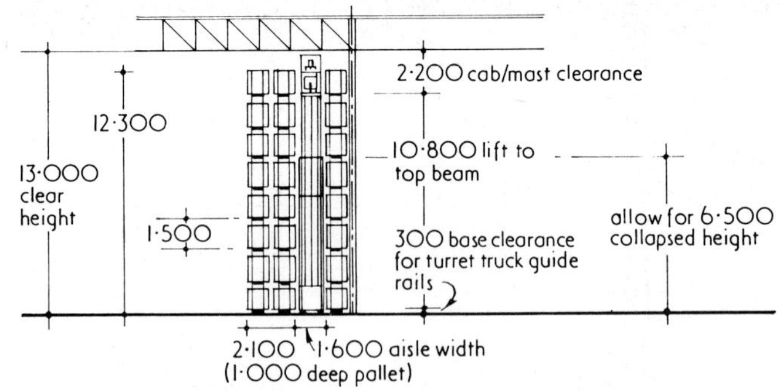

13.12 *Section through warehouse for turret truck operation. Floor tolerance ± 3 mm in 3 m run*

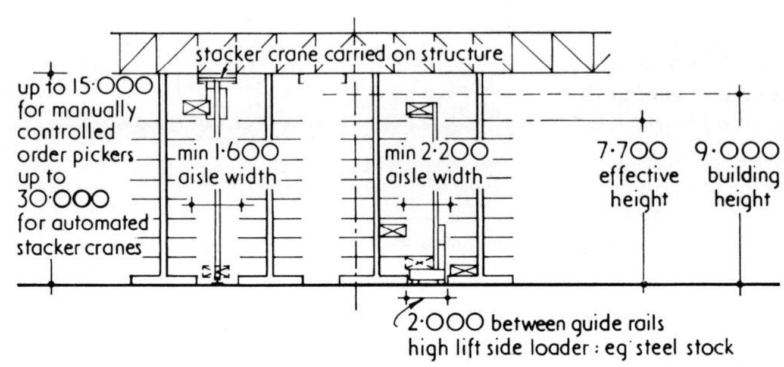

13.13 *Section through warehouse for stacker crane handling (left) and steel stockholding with side loader (right)*

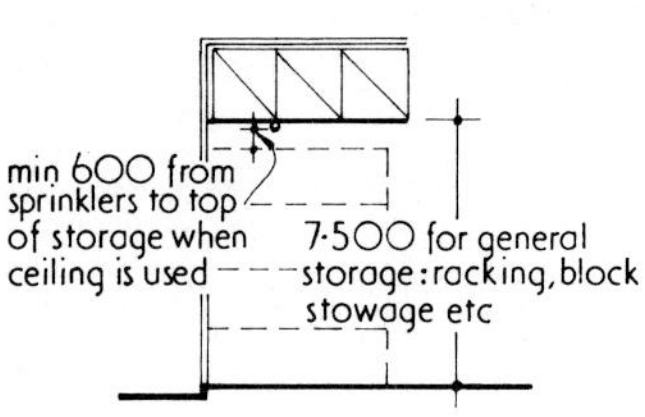

13.10 *Section through small warehouse for fork-lift operation*

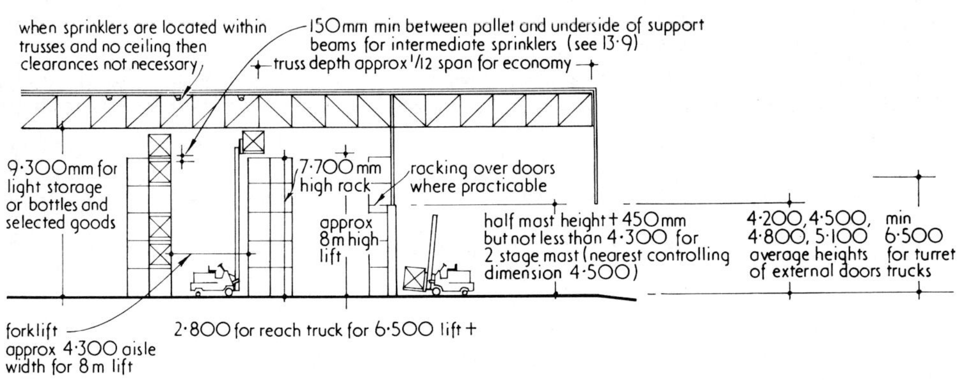

13.11 *Section through large warehouse for fork-lift or reach truck operation*

7 RELATIONSHIP OF STORAGE METHOD, MECHANICAL HANDLING EQUIPMENT AND BUILDING HEIGHT

The effect of handling equipment on warehouse section is shown in **13.10** to **13.13**. These factors depend on site conditions:

1 For very constricted sites where a large volume of goods needs to be held high bay, automated warehouses can prove the most economical solution. Such warehouses have been built up to 30 m high, the racking being used as the roof and wall cladding supporting structure. Handling machines run on fixed tracks, **13.13**, **13.14**.

2 For medium and large-scale installations where full automation is not justified, storage areas up to 12 m high allow free standing racking (bolted to the floor) with aisle widths marginally wider than the largest pallet, **13.15**. 'Turret trucks' used in this type of plant are free path machines based on forklift technology, **13.16**.

3 Where the cost of high bay stacking and high lift machinery is not justified, forklifts and reach trucks are used **13.17**. Reach trucks are suitable for conventional pallet weights (1 to 1·5 tonnes) over flat floors. They can lift to 7 m and operate in aisles of about 2·5 m. A forklift truck can carry heavier loads but requires aisles of 3·2 to 4 m width, **13.18**. Heavier trucks are required to lift greater heights and tend to require a greater aisle width.

4 Mobile racking where pallet racking is mounted on mobile bases and rests face to face may be suitable where storage is to be installed in an existing structure or where the site is limited in area and the turnover of products comparatively low. It is costly to install and the floor slab has to accept double the normal distributed load.

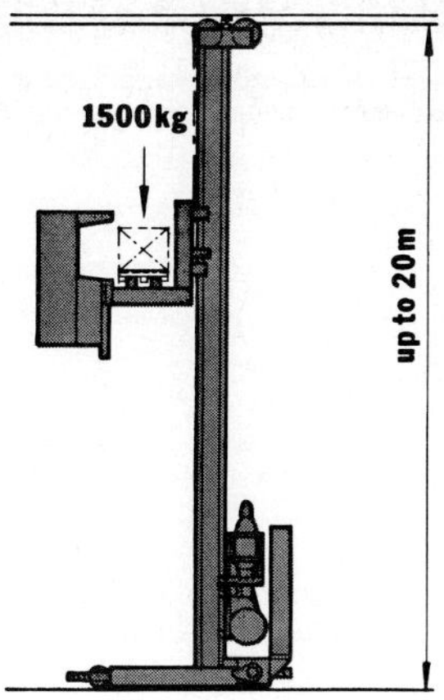

a

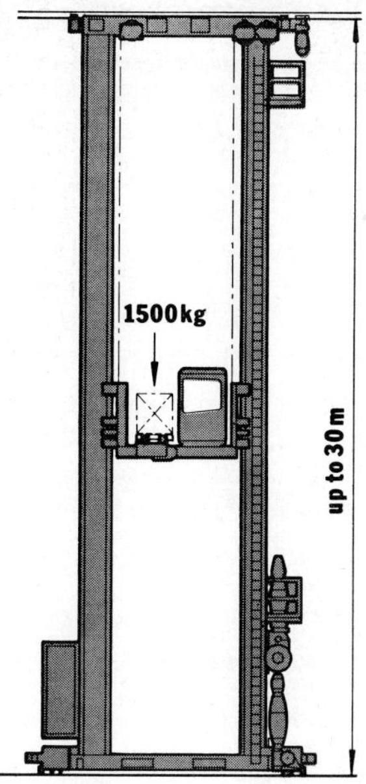

b

13.14 *Dimensions of:*
a *Order picker*
b *Stacker crane*

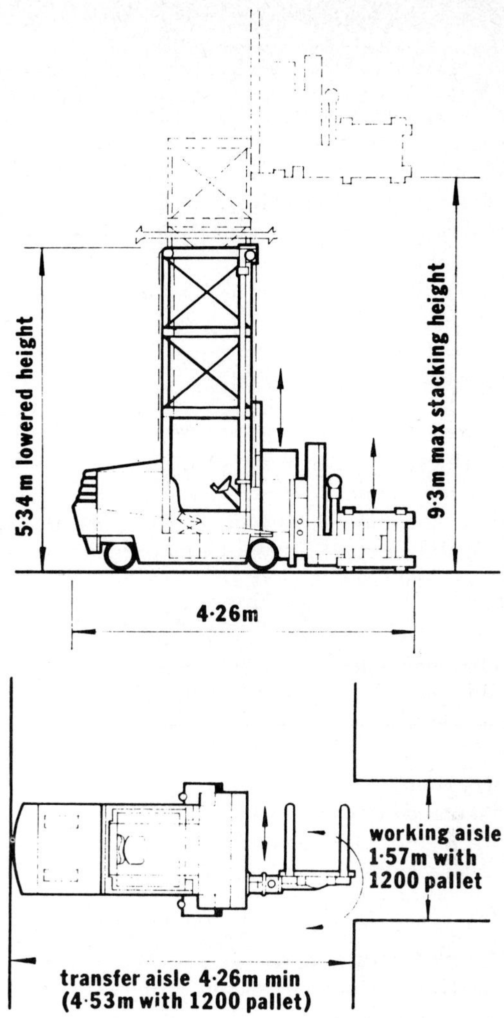

13.15 *(right) Free path stacker/order picker with elevating cab, fixed mast and rotating fork. The four-post mast gives extra stability. Out of the aisle can also be used as a fork-lift truck. The free lift on the fork carriage also allows differential movement between the pallet and the picking platform. Minimum building height 2·2 m above top lifting level*

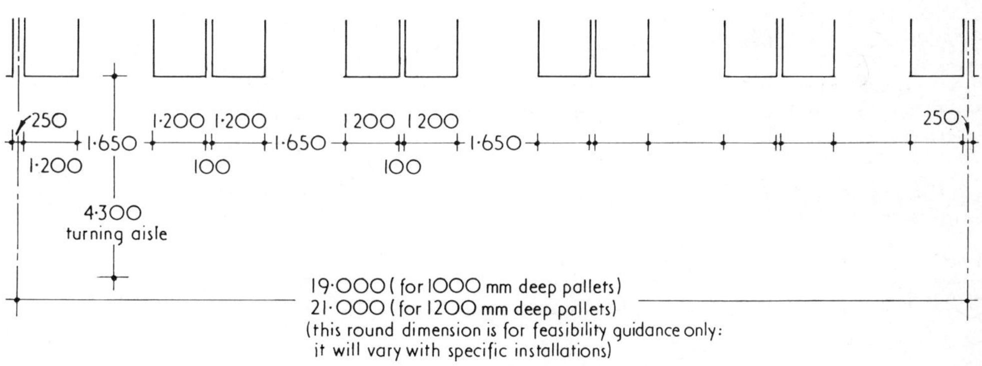

13.16 *Relationship to structure of turret truck aisles*

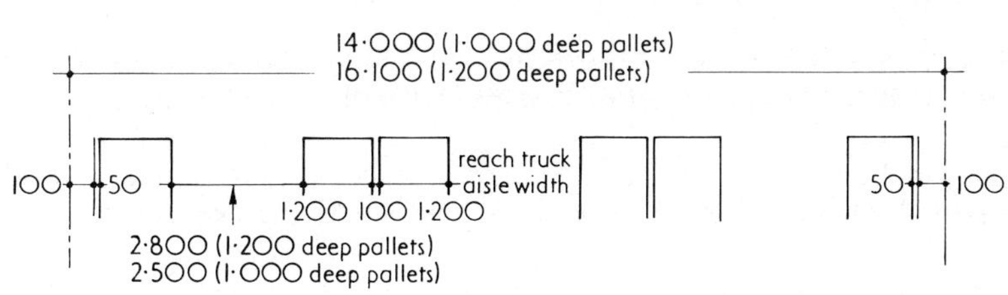

13.17 *Relationship to structure of reach truck aisles*

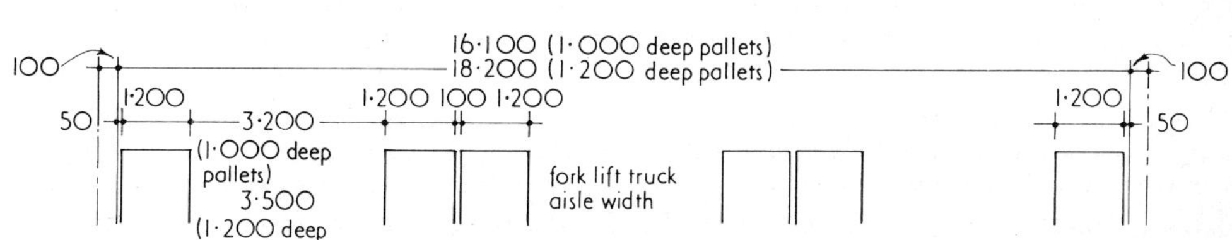

13.18 *Relationship to structure of fork-lift truck aisles. Note: 16 100 mm span is common to fork-lift and reach truck requirements*

8 OUTLINE SPECIFICATION

8.01 Storage area

Pitched roofs, though strong on first cost, waste storage volume and run the risk of being damaged by handling equipment. Three factors favour the flat or low pitch roof type:

- the column pitch can be wide, **13.17**, **13.18**.
- they are more adaptable to a change of use or changes dictated by new processes
- they are more suitable for the installation of services such as cooled air.

8.02 Order picking and assembly

Space demanded will vary with the type of business involved and the method of order assembly, in turn generated by the method of despatch and transport. For instance a brewery warehouse may despatch whole pallet loads, **13.7**, but a pharmaceutical warehouse may handle and assemble a very large number of small items. Therefore it may require a large area for order assembly, **13.19** to **13.21**.

8.03 Loading bay and load accumulation area

The loading bay is the critical link between the storage and distribution system (table VI), **13.22**. It usually combines inward and despatch movements. It must provide sufficient space for:

- incoming goods to be checked off
- empty unit load devices to be removed
- despatch loads to be accumulated (table VII).

A full vehicle length (12 m) should be allowed as the zone behind the loading dock.

8.04 Office and amenity areas

Large warehouses can employ more than 100 order picking staff (mainly female) each shift. Extensive washing and changing facilities will be required. Also space for operatives to rest and smoke outside the storage area.

Table VI Bay widths for loading and unloading of vehicles

No of vehicles at one time	Recommended width of bay (m)	Minimum width of bay (m)	Minimum depth of bay (within building) (m)
Side loading within building			
1	11	10	
2	18	16	18–19
3	24	22	
More than 3	Add 6·5 per vehicle	Add 6 per vehicle	
End loading			
Per vehicle	4–5	3·75	16

Table VII How to calculate approximate numbers of roll pallets in despatch assembly area

	Load platform size Length m	Width m	Number of roll pallets	Weight of load and vehicle Tonnes
Rigid vehicles	5·1	2·3	18	11
	6	2·3	21	13
	6·9	2·3	24 (28)	13–14 (30 gtw)
	7·5	2·3	27 (31)	15 (32 gtw)
Artic vehicles	10·2	2·3	36	21–23
	11	2·3	39	23–24
	11·75	2·3	42	29
	12·6	2·3	45	32

Note 1: Figures in brackets are roll pallets for drawbar trailers, and gross weight of vehicle with drawbar trailer

Note 2: These 720 × 800 mm roll pallets are placed three abreast in the trucks. Other types are larger, and are packed two units across: adjust the numbers accordingly

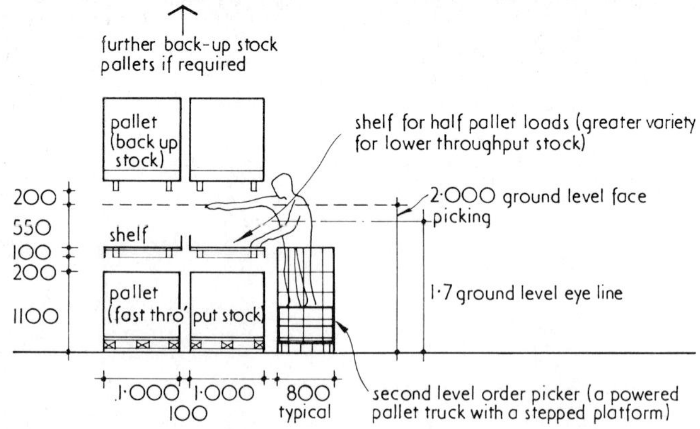

13.19 *Second level order picking, typically used for food distribution and supermarket replenishment. The operative fills a roll pallet or cage from the pallet on the floor and the shelf above it*

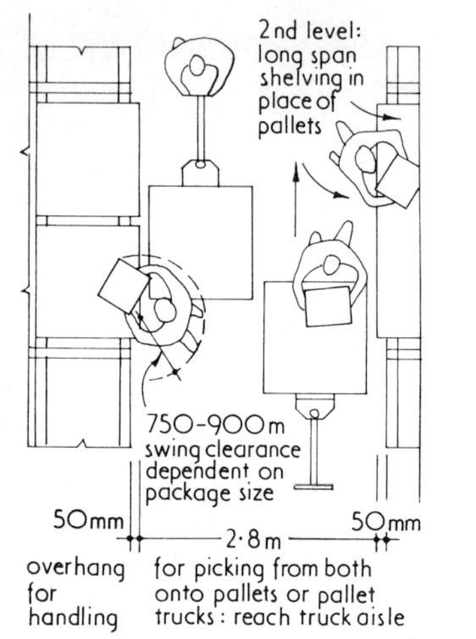

13.20 *Reach truck aisle for second level order picking*

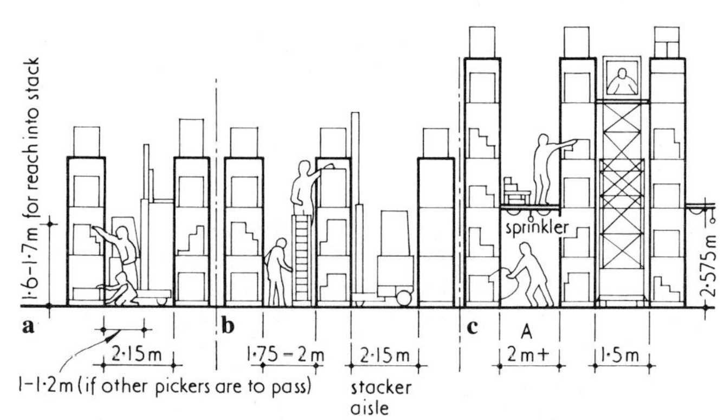

13.21 *Stacker aisles for order picking:*
a *Pulling from lower levels—replenished by stacker truck* **(13.25)**
b *Alternating pick-up and replenishment aisles*
c *Multi-level alternative aisles, replenished by turret truck*

8.05 Equipment maintenance areas

Most mechanical handling equipment for internal use is battery powered electric. The batteries need charging at night or after shifts of about 12 hours. Requirements for maintenance areas are:

- a distilled water supply
- 1 tonne hoisting tackle for removing batteries
- fume extraction
- acid resistant floor.

Major services and repairs tend to be done off site.

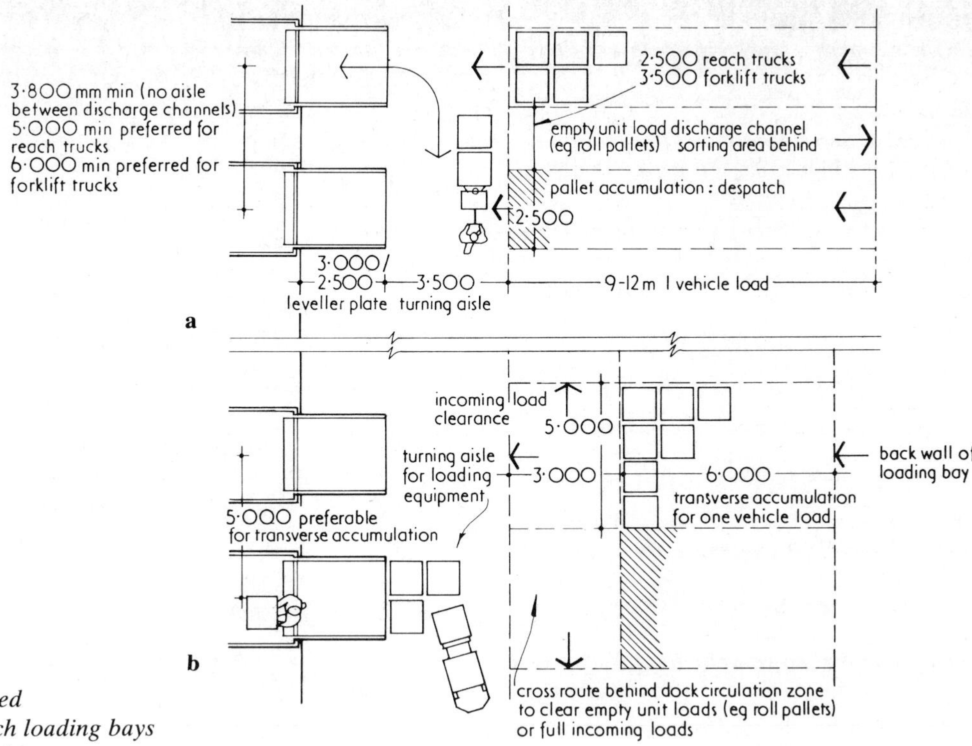

13.22 *Combined arrival/despatch loading bays*
a *Where available*
b *Where depth is limited*

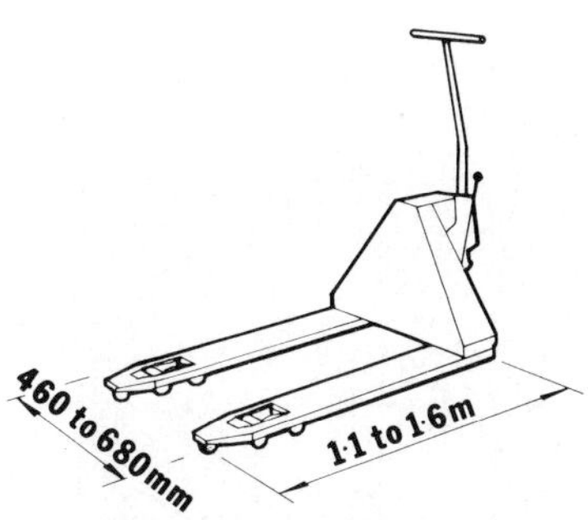

13.23 *Manual pallet truck. For use inside the warehouse building up orders, loading vehicles on raised docks or with tail-lifts, general pallet handling. Increasingly used in retail premises for handling bulk goods. Capacity up to 1500 kg generally and for short distance travel. Operatives soon tire when pushing heavy loads any distance. Forklengths available 0·8 to 1·6 m, widths from 460 to 680 mm. Heights: lowered 83 mm, raised 203 mm. Pallet width should be 150 mm over fork (typical length is 1·06 m for a 1·2 m pallet). Where gangways are narrow and stability is important, a heavy truck should be used with maximum width between forks. This device will turn in its own length but needs additional clearance for overhangs. Normally it requires level floors to operate satisfactorily, but large wheels in nylon or with solid rubber tyres plus articulating axles are available for use in older buildings; although instability may occur. Steel wheels are available but are less popular. Where loading ramps are used, pallet trucks with brakes should be used. Adaptors are available for use as a stillage truck.*

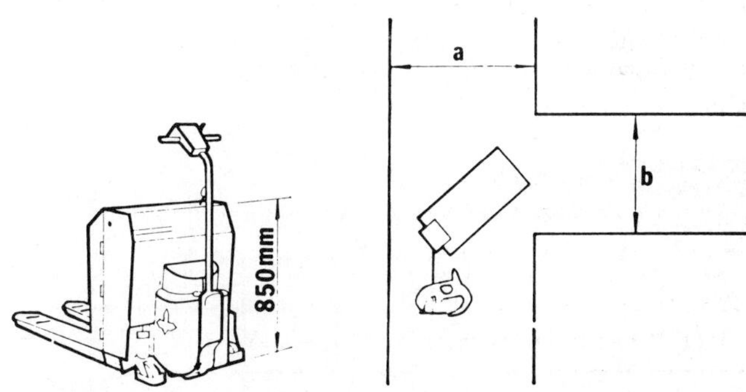

13.24 *Powered pallet truck. For internal transfer, loading vehicles on docks, order build-up, transporting roll pallets to load assembly position. For use with all types of pallet and cages. Capacity 1800 to 3000 kg, forklengths 0·75 to 1·8 m, speeds up to 3·6 km/h running light, widths up to 850 mm, usually 760 mm. Long forks available to carry three roll pallets at once. Special forks for drums and paper rolls. Will turn in its own length but needs additional clearance for overhangs. Some have 200° turn on the single power steering wheel. Aisle width depends on forklength:*
a *(90° stacking aisle) = 1840 mm (truck + 1 m pallet)*
b *(intersecting aisle) = 1570 mm.*
Turning circle 1·78 m radius with 960 mm long forks. This device requires level floors and a three-phase charging point. It can manage ramps up to 10%. Some larger capacity units can also be ridden on, and can tow non-powered pallet trucks if long distances are involved

9 SECURITY

Warehouses are, by definition, prone to theft. Most thefts are carried out during working hours. This can be minimised by ensuring that:

- there is no direct access from loading bays to the warehouse, especially through the order picking zone, without supervision
- access from office accommodation to the warehouse should be visible from the office area
- changing rooms, showers (necessary in cold stores) and WCs should not have direct access from the warehouse, and equally, should not be accessible from outside. Visiting drivers should have segregated WC facilities
- if small, valuable goods are involved, a search room may be required
- operatives' parking should be well separated from heavy vehicles' parking and away from the loading area.

10 HANDLING EQUIPMENT

Some typical handling equipment is shown in **13.23** to **13.27**.

11 BIBLIOGRAPHY

Peter Falconer and Jolyon Drury, *Industrial storage and distribution* (AJ handbook of industrial storage), London, Architectural Press, 1975

Insulation Act 1972, HMSO

Factories Act 1961, HMSO

Offices, Shops and Railway Premises Act 1963, HMSO

Health and Safety at Work etc Act 1974, HMSO

Fire Offices Committee: *Rules for automatic sprinkler installations*, 29th edition, revised 1973

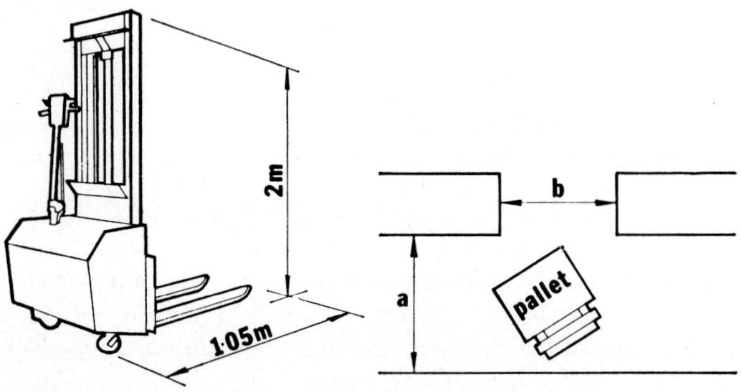

13.25 *Power travel and lift pedestrian-controlled stacker truck. When travelling the pallet rests on the stacker frame which has travel wheels. Power lifting is independent of the travel frame, and is directly into the rack. Only suitable for short travel distances. Lifting range up to 3·6 m. Can be supplied with attachments. Capacity up to 1500 kg at 600 mm centres, straddle width 0·86 to 1·3 m, travel speed up to 4·8 km/h laden. Will turn with full load on 2·1 m aisle*

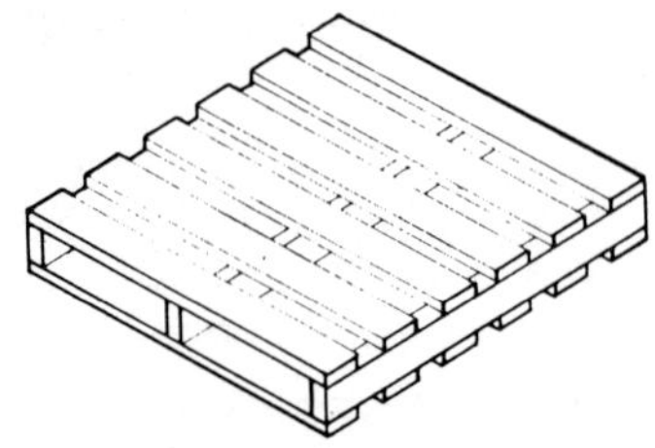

13.26a *Two-way entry pallet*

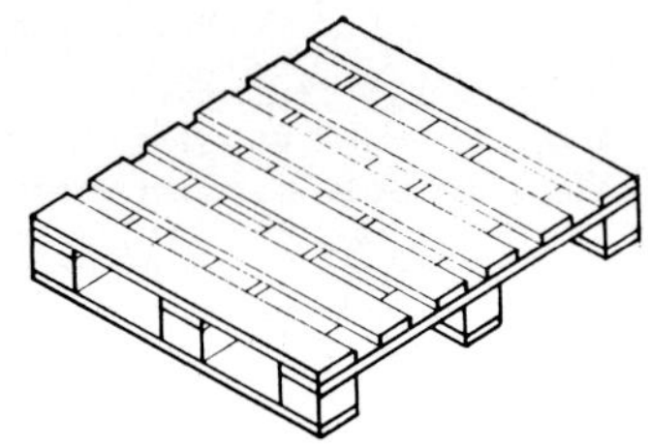

b *Four-way entry pallet*

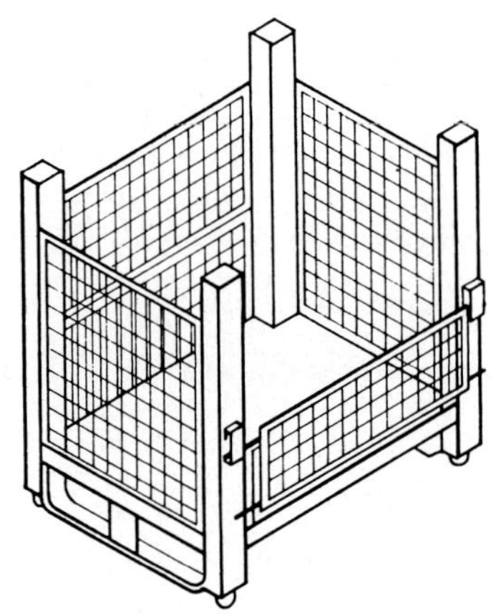

c *Post pallet*

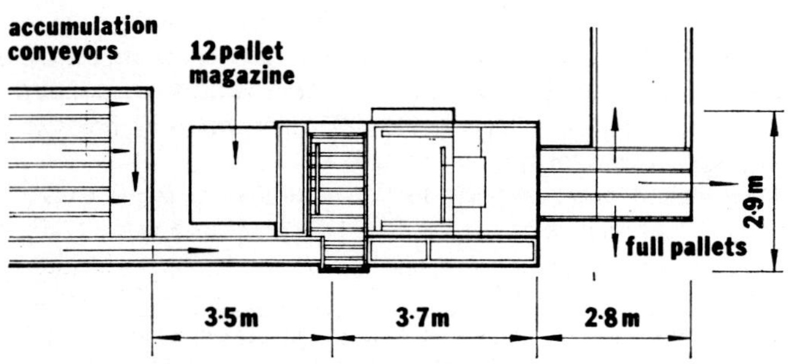

13.27 *Plan of typical palletising machine. Top right is buffer track required for slower shrink wrapper*

14 Farm buildings

John Weller

CI/SfB 26
CI/SfB (1976 revised) 263

John Weller is an architect specialising in rural work. He was consultant editor for the AJ special issue, Rural settlement and land use, AJ 21.1.76

Contents

1 INTRODUCTION

1.01 The agricultural economy

Agriculture, superficially, is stable. Farming assets in 1974 were assessed as over £21 000M and only 8 per cent of this represented borrowed capital. One-fifth of the assets belonged to tenants. This creates a sense of landlord/tenant partnership. Similarly, since 1900, there has been a rapid growth in owner-occupied farms, from 10 to 55 per cent. But, throughout farming, stability is insecure due to national fiscal policies and ownership rights.

By 1977, UK agricultural turnover was in excess of £6000M. Investment in farm buildings, inclusive of servicing, was some £300M. To the farmer, buildings are 'fixed equipment', thus, like other machinery, they are part of the overheads to an enterprise, which is usually defined in terms of the end product.

1.02 Planning

Buildings, irrespective of the enterprise, should be planned in terms of their functions for storage, processing or production. Food, like other industrial processes, should be designed for materials handling and flowline production. Superimposing linear buildings within or over traditional courtyard forms is both a visual and tactical problem.

Stock housing produces effluents. Farm waste management is an essential part of the building design and wastes should normally be recycled.

1.03 Building functions

Depending on managerial philosophy, building functions may be specialist, semi-specialist or flexible in their form. Farmers tend to equate flexibility with general-purpose layouts and with low capital investments. This tends to be a false equation. The loss of quality control, often difficult to evaluate, makes most 'cheap umbrellas' poor performers for specific end products.

The demand for flexibility reflects two things—lack of political confidence in stable markets and rapidity of technical change. UK food production is becoming the responsibility of the EEC (via CAP, the Common Agricultural Policy). This aims the create market stability. Technical change is liable to continue. However, any expansion of power demand may become more selective.

1.04 Storage requirements

In simple terms, most storage requirements are those of containers, whether cylinders, bins or bunkers. Wide span portals are suitable for some layouts for cattle, bulk storage and implements. Compact and insulated 'boxes' of low profile are best for calves, pigs or poultry. These may be prefabricated or cross-wall, the latter mainly being self-build. They may include total or partial environmental control. In contrast, 'kennels' are cheaply framed, semi-open, mono-pitch structures suitable for some cattle and pig layouts. Most building types are available in either prefabricated or package deal form. Performance specifications are rare. Overall costs are lower than for most industrial processes.

1.05 Lifespan of buildings

Most pre-1960 buildings are inefficient for modern production and most traditional buildings are redundant. A few are suitable for casual storage, administration, isolation units, or spare boxes. The issue of redundancy is not easy to resolve. However, tourism, recreation and craft work are all growth activities in rural areas. A tenth of all farms have some tourist income. In upland areas, it can become a major source of income. Farm planning should take into account possible alternative uses for buildings and land.

The normal economic life for farm buildings is 10 years, though some are depreciated, like money, over 5 years. This can be a major design constraint. Some estates may permit a longer term of 20 to 60 years, especially for 'umbrella' enclosures. Grants are available for all except plastic, cheap tents and for factory farms (ie without supporting land). EEC grants are more generous but require carefully prepared development proposals.

1.06 Appearance

Farm building appearance, especially since many are exempt from control and since most are cheap compared to other building types, is a contentious issue. Simple forms, good colour, defined planes, and co-ordinated fittings can make many buildings acceptable in appearance, especially with careful siting and landscaping. However, roof surfaces and pitches are likely to contrast, possibly conflict, with vernacular buildings and, near rising land, become dominant. Components tend to be poor and unrelated in design to the basic structure. Surrounds to buildings, including yards, tanks, fences etc are often more unsightly than the buildings.

1.07 Criteria

Farm management, in relation to resources of land area and terrain, climate, soil, capital etc, is such that nearly every farm building problem is different. This is true in spite of prefabrication and package deals. The lack of common criteria, except perhaps for poultry, is one of the main weaknesses which retards real improvement of building standards—this as well as lack of long-term confidence. In many enterprises, it is difficult to establish a good design brief, but the basic layout, **14.1**,

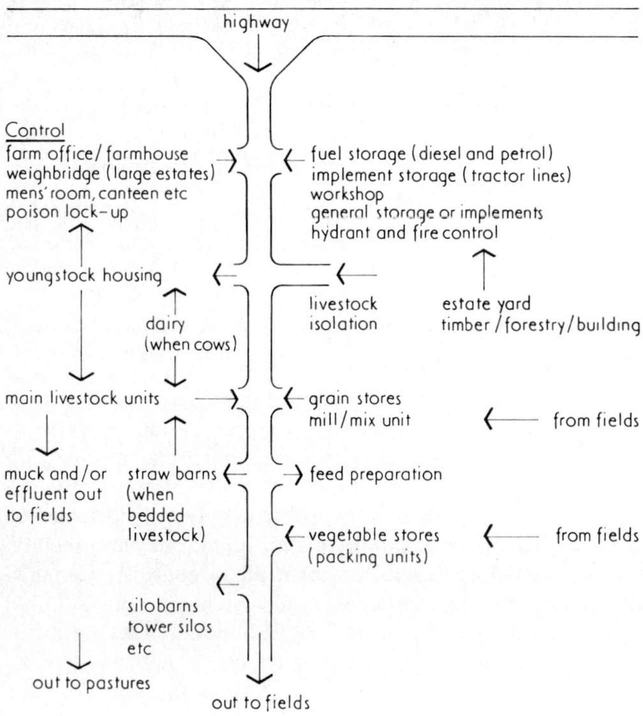

14.1 *Basic layout, mixed arable and stock farm. Although the arrangement shown has been stylised, in fact farms are usually linear to the main service road*

shows the general relationship between the elements of the farm and the main service road.

14.2 shows an example of a typical farm.

2 FARM ANIMALS

Average sizes and weights of animals are shown in **14.3**. Width of animal given is normal trough space allowed (ie about ²/₃ of overall width). Length given is normal standing (not fully extended).

3 FARM MACHINERY

Average sizes and weights of tractors and other machinery are given in **14.4**.

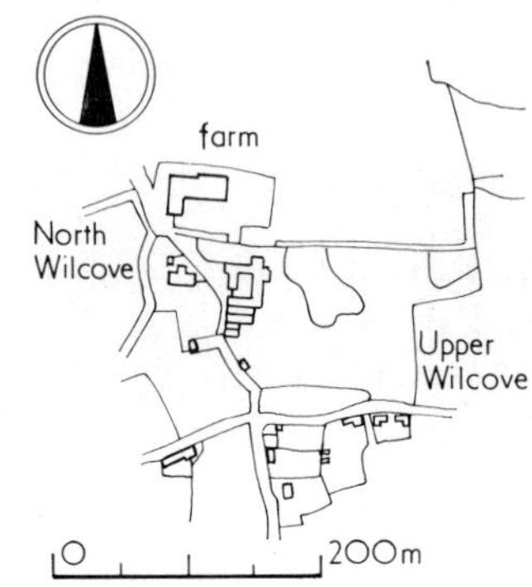

14.2 *Typical farm: Wilcove*
a *Site plan*

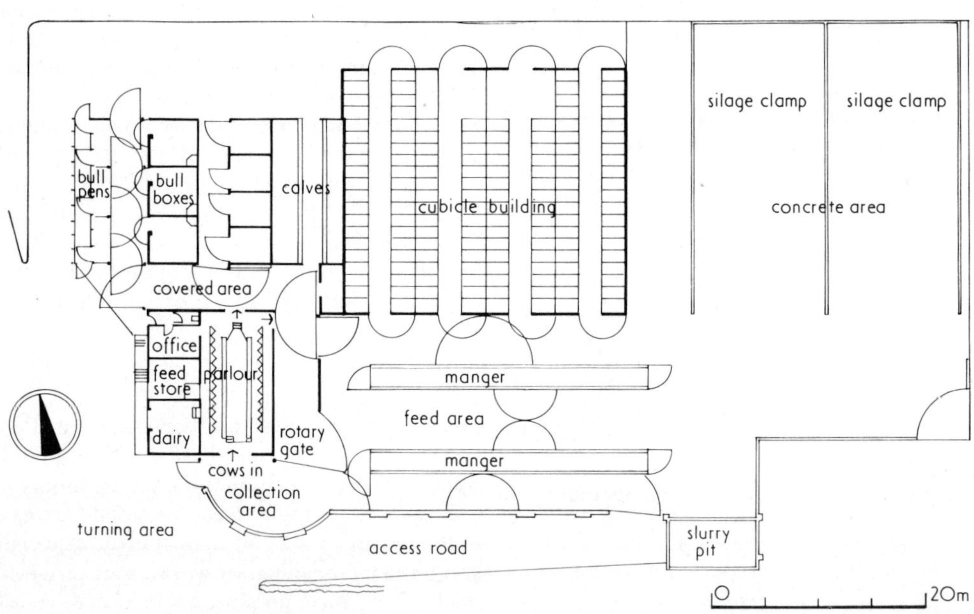

b *Plan*

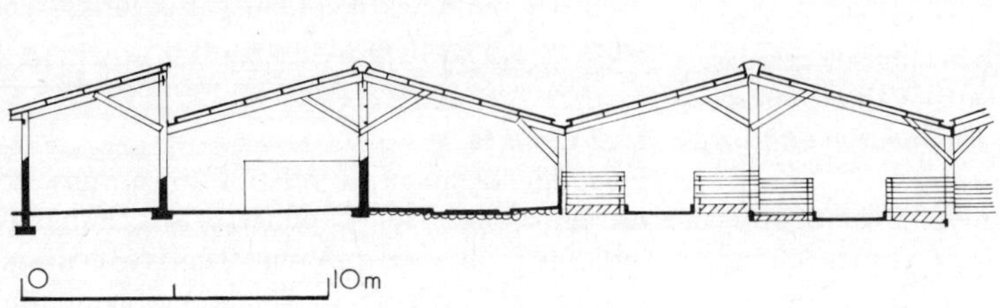

c *Part cross-section*

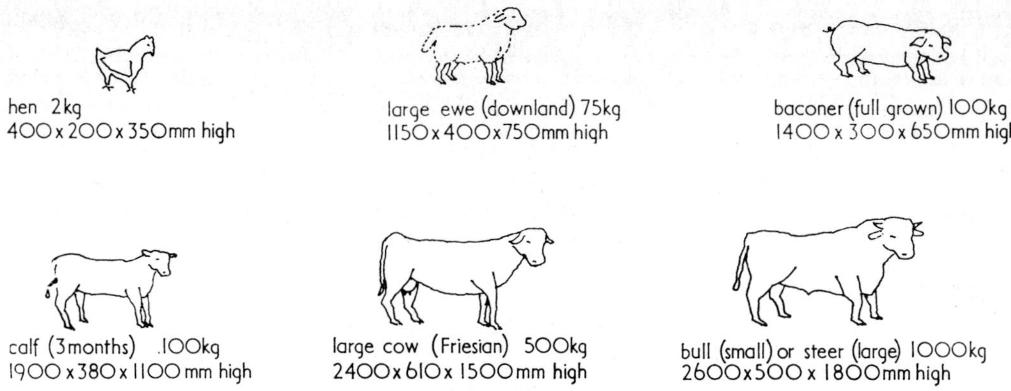

14.3 *Farm animals: average sizes and weights*

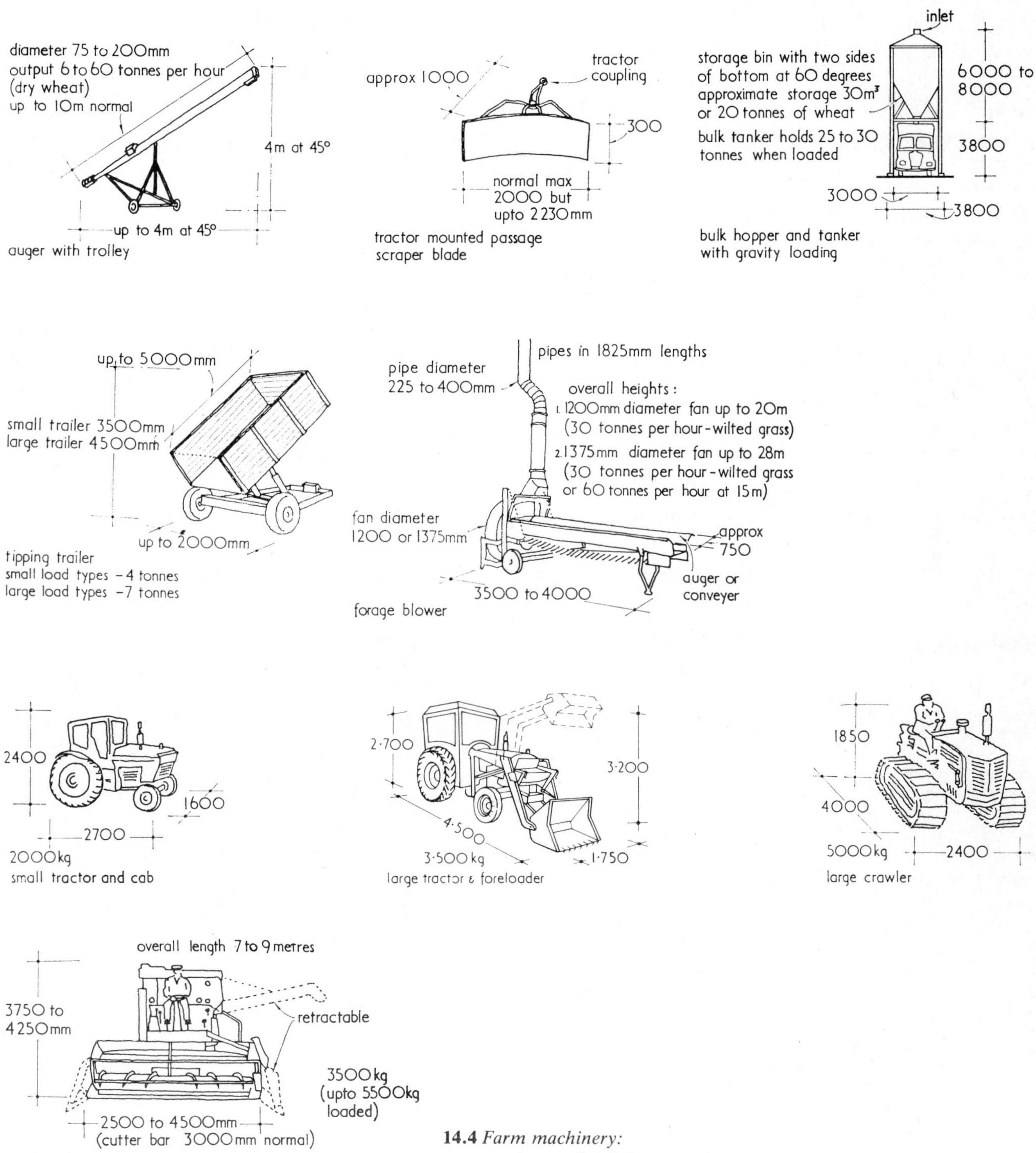

14.4 *Farm machinery: average sizes and weights*

4 DAIRY CATTLE HOUSING

Buildings suitable for 120 cattle unit including parlour complex are shown in **14.5** to **14.9**. Allow approximate building area for cubicle houses (16·7 m) of up to 27 m wide × 55 m long plus 10 m turn area at one end plus 4 m road. A 'kennel' has same basic dimensions but the roof is lower and is held by the cubicle division and the passage is not completely roofed, as **14.17**. Rotary milking parlours are shown in **14.10** to **14.12**, **14.13** showing the milking layout for a 150 to 200 cow unit, including a rotary parlour.

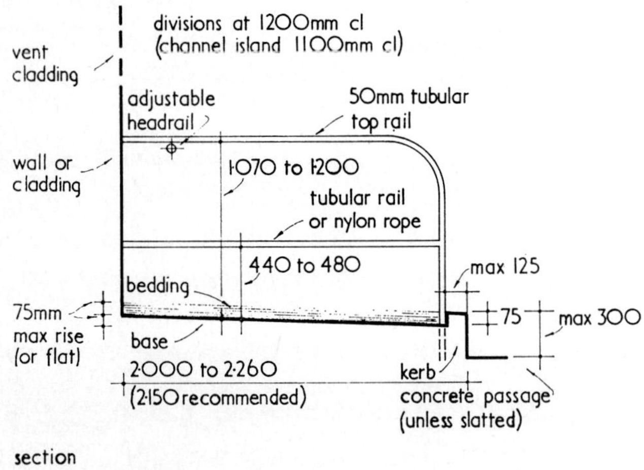

14.5 *Cubicle division: basic dimensions for Friesian cows*

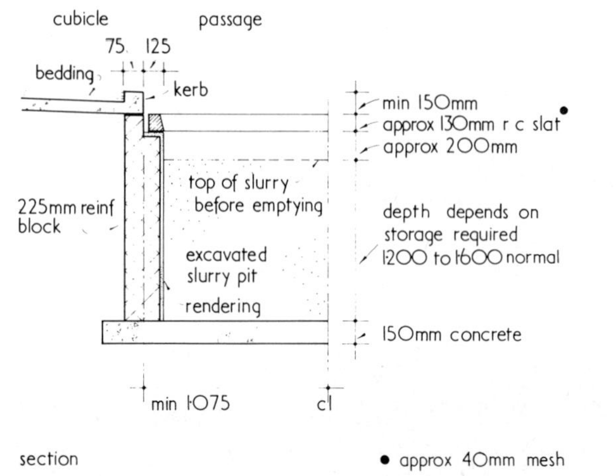

14.6 *Cubicle division: enlarged detail of passage if slatted*

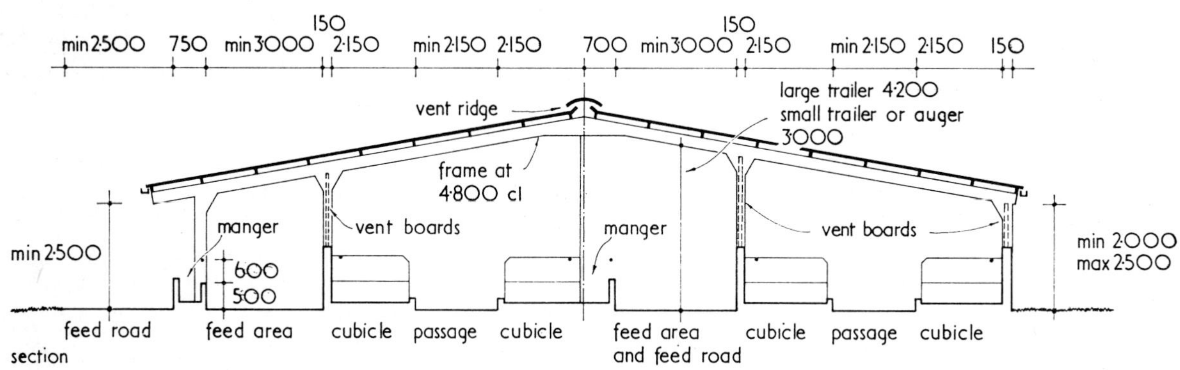

14.7 *Cubicle house: alternative sections showing perimeter feeding to left of centre line, centre feeding to right*

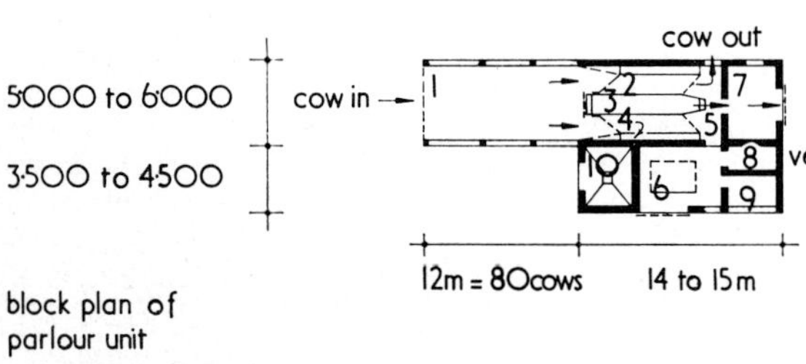

1 *collecting yard: allow 1·2 m² per cow*
2 *standings: allow 910 mm run per cow × 6, plus 1370 mm for slope*
3 *pit: for 6 unit parlour allow 7·5 m (approx) total length*
4 *mangers: see section, one hopper per 2 cows*
5 *exit area: allow up to 1·5 m beyond end of pit*
6 *dairy with bulk tank: allow approx 4 × 6 m*
7 *hold box: about 3·5 m wide*
8 *motor-room: about 3·5 × 1·3 m with louvre panel*
9 *office: with fittings*
10 *feed store: about 18–20 m² with auger sump in floor*

14.8 *Block plan of parlour unit:*

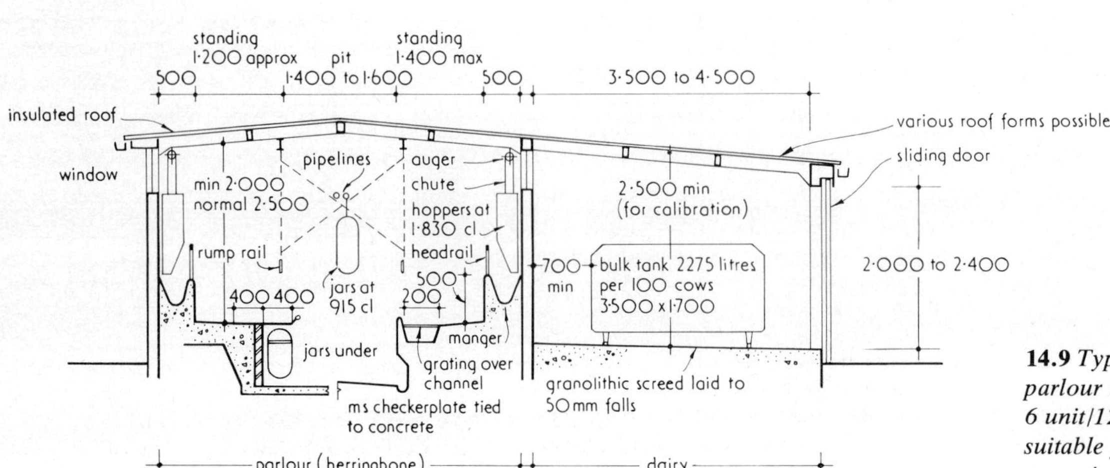

14.9 *Typical section through parlour unit,* **14.8** *This is a 6 unit/12 standing parlour suitable for one-man operation*

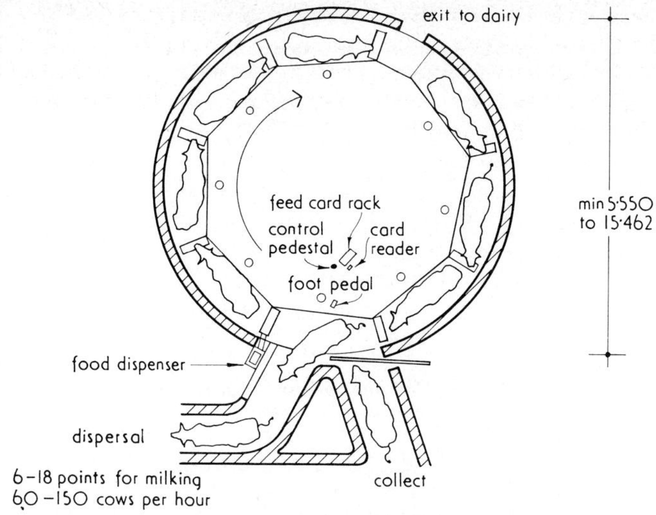

14.10 *Rotary (tandem or carousel) milking parlour*

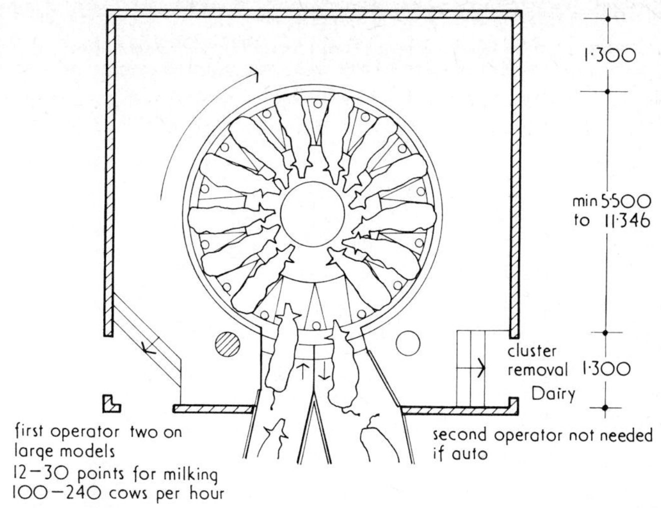

14.11 *Rotary (abreast or turnstile) milking parlour*

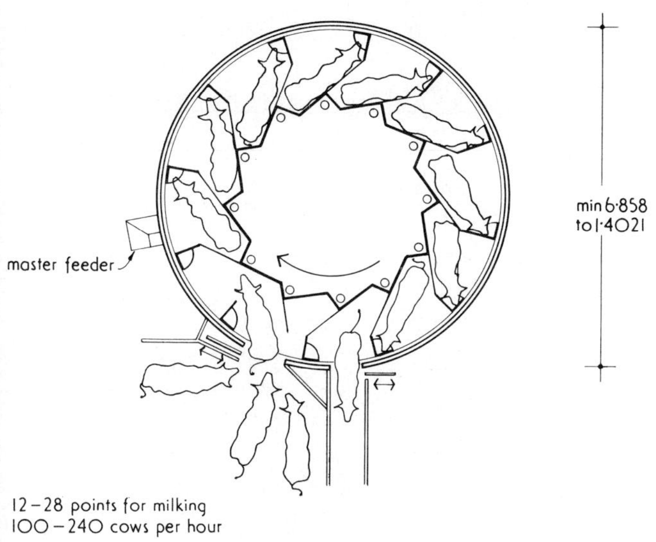

14.12 *Rotary (herringbone) milking parlour*

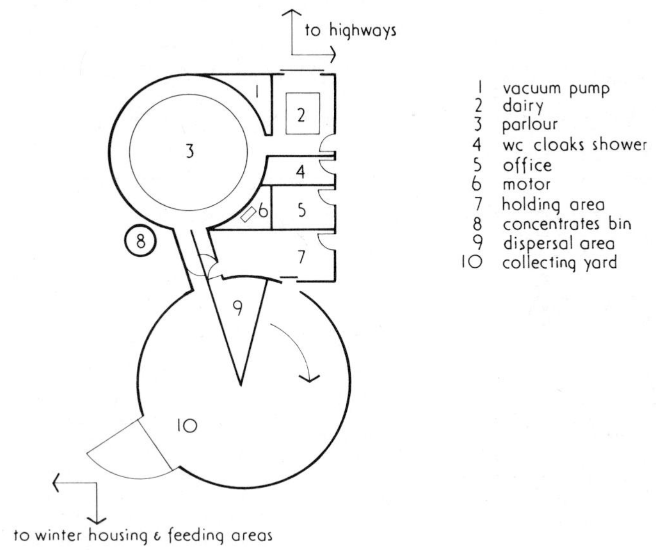

14.13 *Layout for a 150–200 cow rotary milking parlour unit*
All rotating platforms are 760–840 mm above operator's floor level

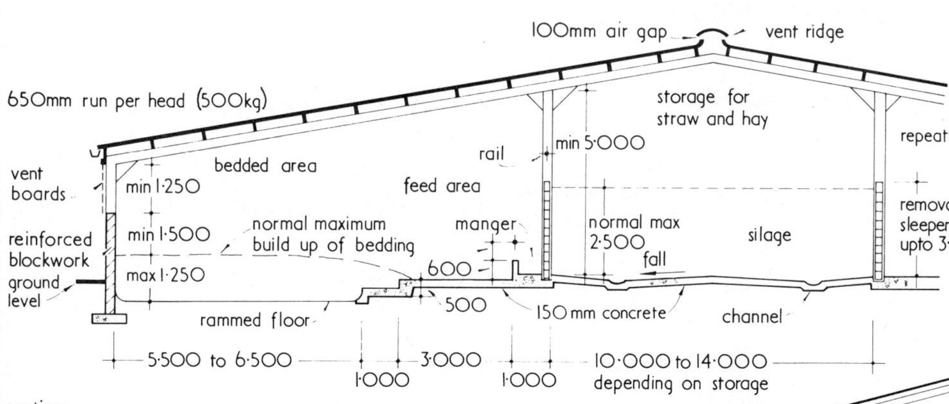

14.14 *Strawed yard for beef cattle with easy feeding*

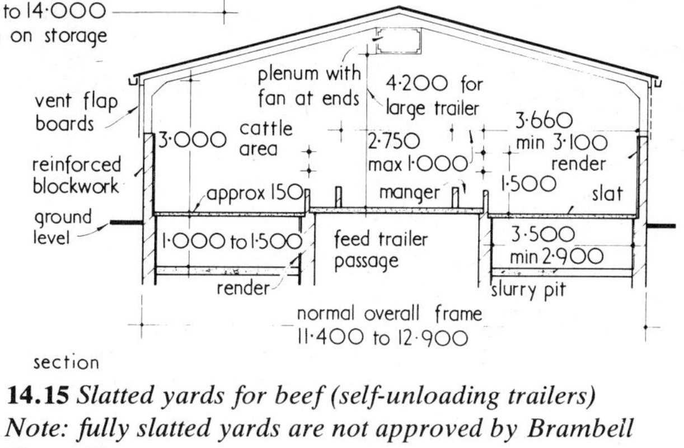

14.15 *Slatted yards for beef (self-unloading trailers)*
Note: fully slatted yards are not approved by Brambell Committee

5 BEEF CATTLE AND CALF HOUSING

Strawed and slatted yards for beef cattle are shown in **14.14** and **14.15**. A typical calf house is illustrated in **14.16**. **14.17** shows a section of kennels suitable for beef or dairy cattle and **14.18** a 'general purpose' strawed yard for cattle (700 mm/head for manger for adults, 500 mm for yearlings).

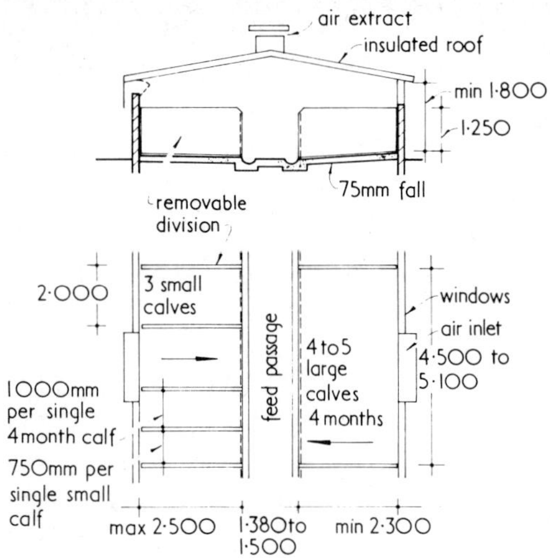

14.16 *Typical calf house*

6 SHEEP HOUSING

A typical section of sheep housing is shown in **14.19**. A dipping tank suitable for large breeds is shown in **14.20**.

7 PIG HOUSING

Three types of fattening house are shown in **14.21** to **14.23**, and two types of farrowing house in **14.24** and **14.25**.

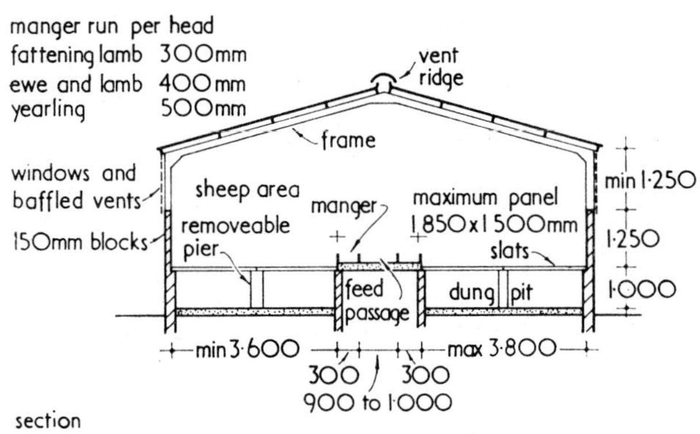

14.19 *Typical sheep housing*

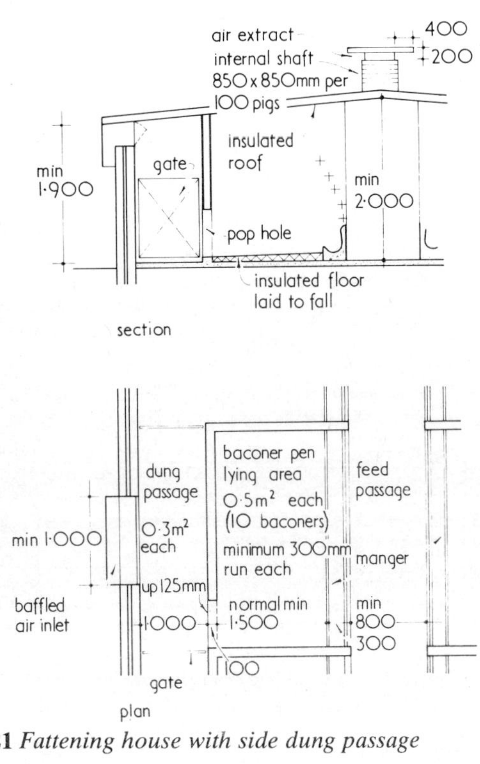

14.21 *Fattening house with side dung passage*

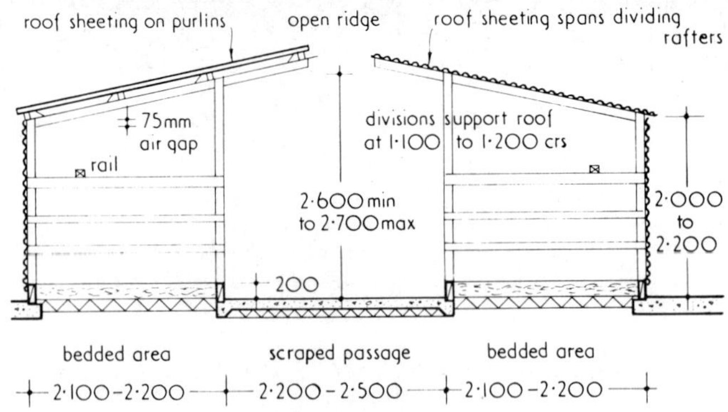

14.17 *Kennel suitable for beef or dairy cattle: section*

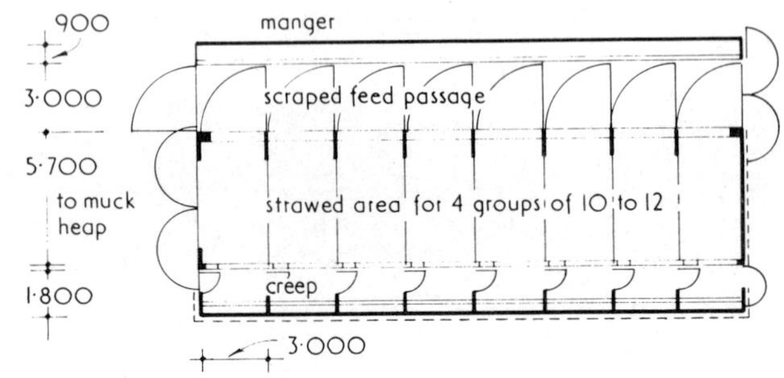

14.18 *General purpose strawed yard for cattle: plan*

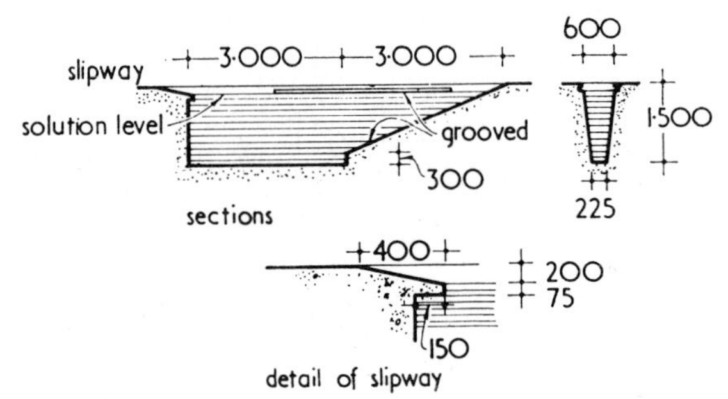

14.20 *Dipping tanks for large breeds: ewes. Allow 2·25 litres solution per head*

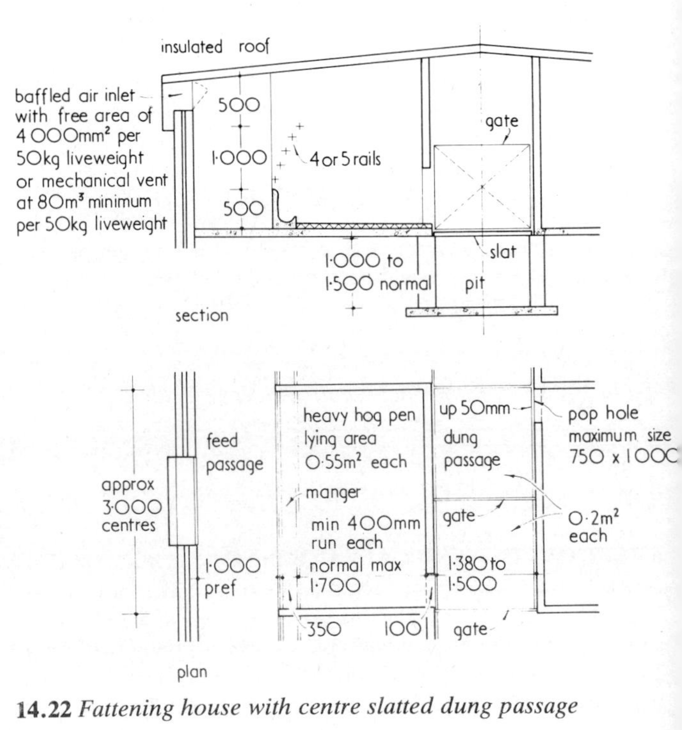

14.22 *Fattening house with centre slatted dung passage*

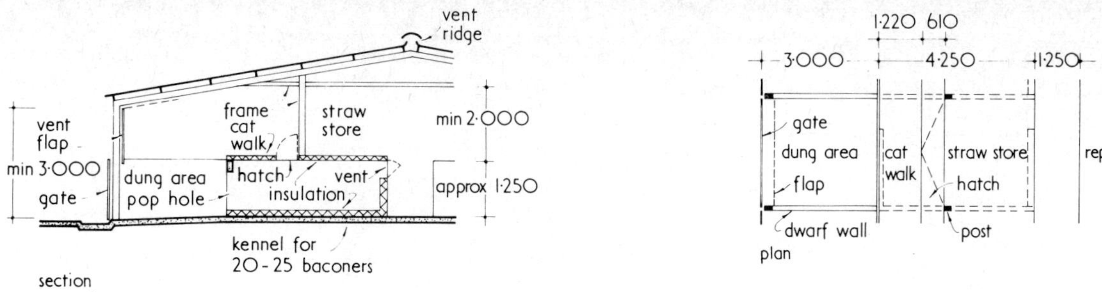

14.23 *Fattening house with strawed system and floor feeding*

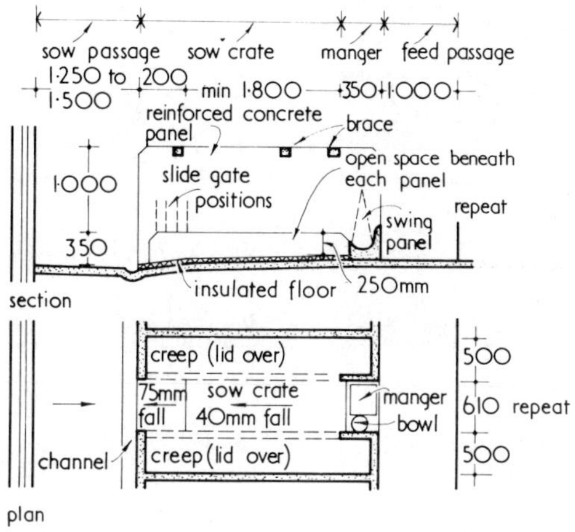

14.24 *Farrowing house: crate (permanent)*

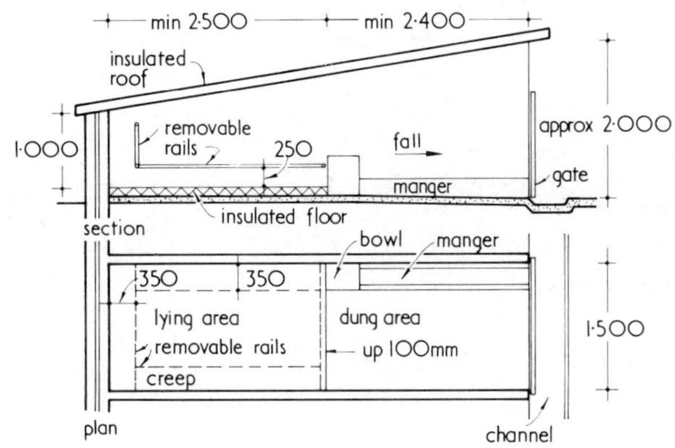

14.25 *Farrowing house: Soleri (open front)*

8 POULTRY HOUSING

Rearing, fattening and egg houses are shown in **14.26** to **14.29**, turkeys in **14.30**.

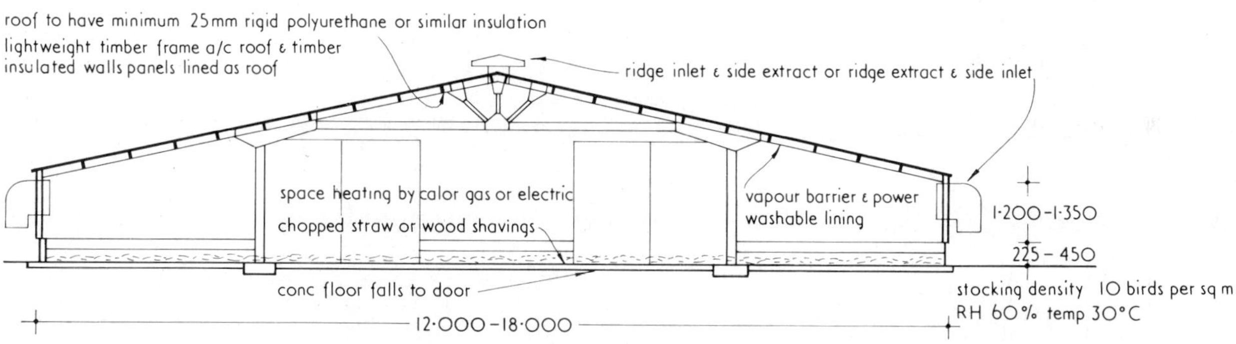

14.26 *Section through broiler and rearing house*

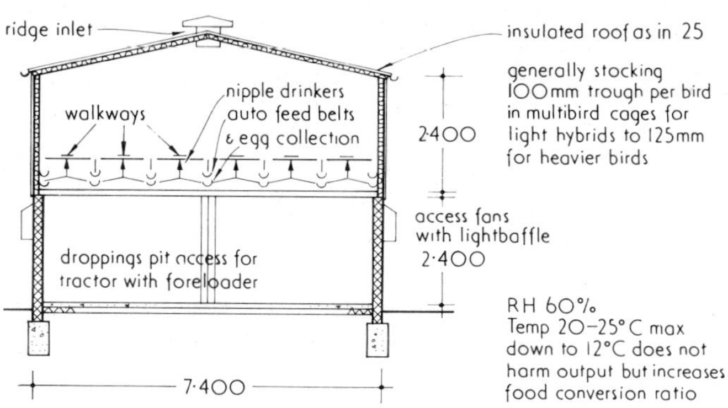

14.27 *Flat deck deep pit battery house*

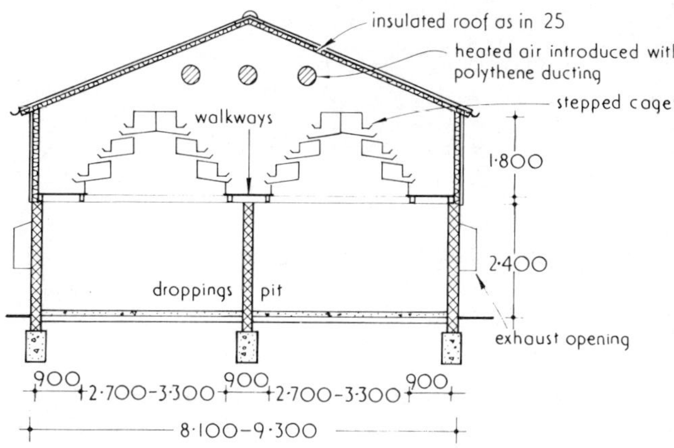

14.28 *California cage deep pit battery house*

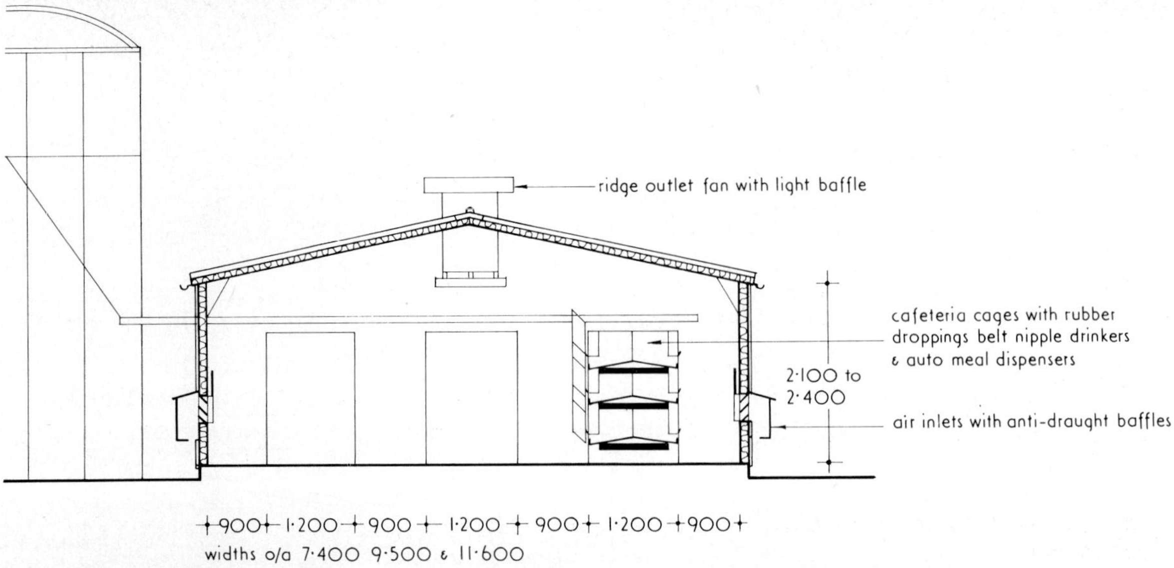

14.29 *Cafeteria cage battery house:*
a *Section*

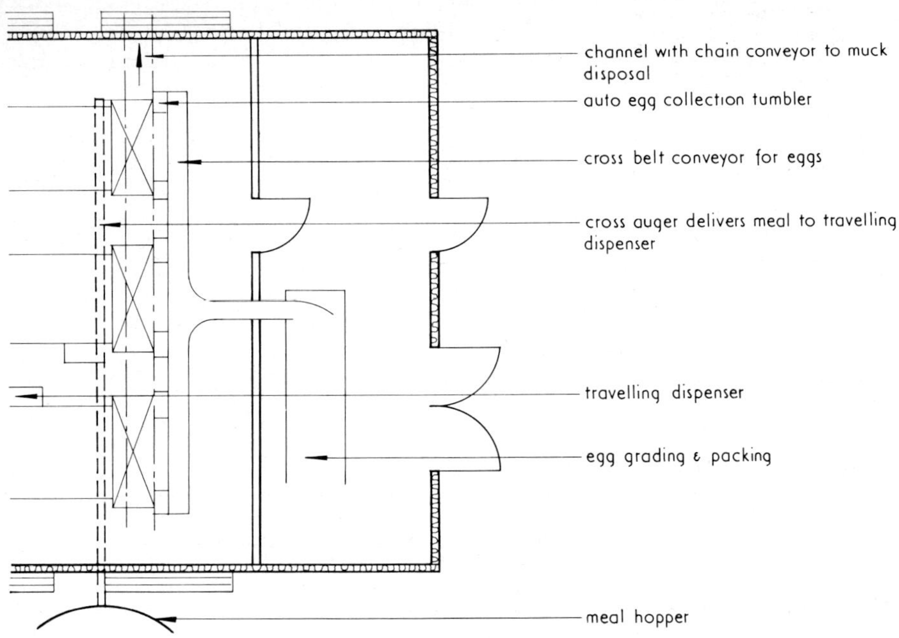

b *Plan of end of house showing gear*

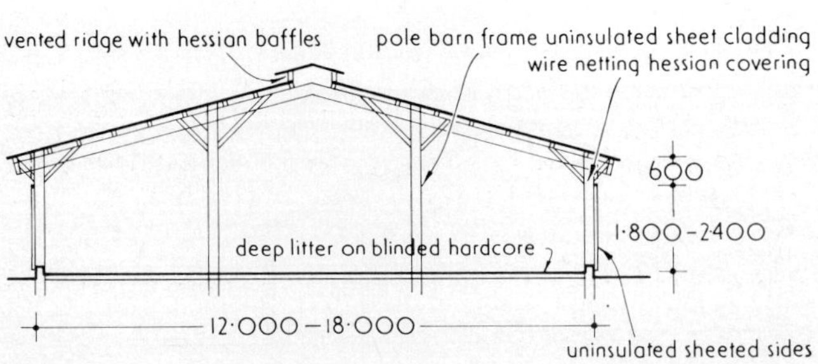

14.30 *Pole barn to fatten turkeys. 30 kg/m² stocking density*

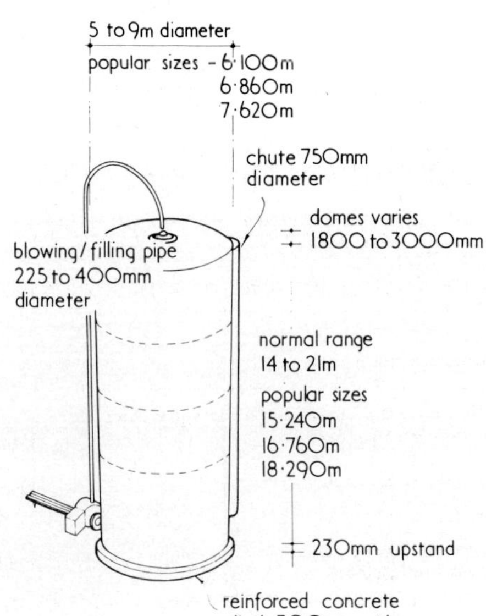

14.31 *Tower silo (wilted grass: 40–50% dry matter)*
Note: wet grain is normally stored in towers of under 12 m height × 6 m diameter approx

9 CROP STORAGE AND EFFLUENT PRODUCED

Some typical feed and produce stores are shown in **14.31** to **14.38**. Tables I and II indicate the scope of manure likely to be produced and its manurial values.

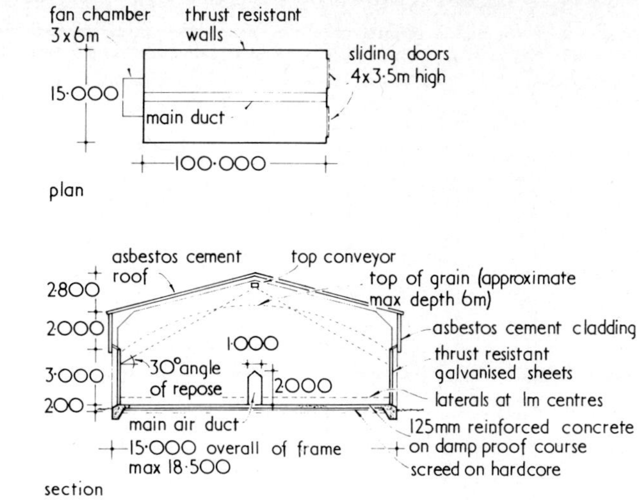

14.32 *Food storage: grain. Plan shows lateral system and allows approx 1200 tonnes storage*

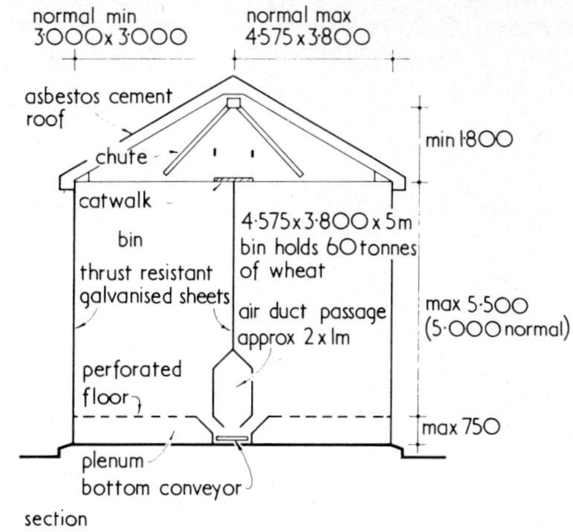

14.33 *Grain drying/storage: nest of bins (square or rectangular) with roof*

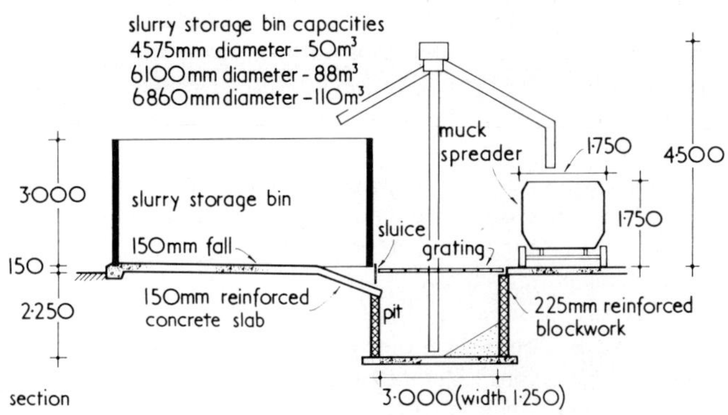

14.34 *Slurry storage above ground*

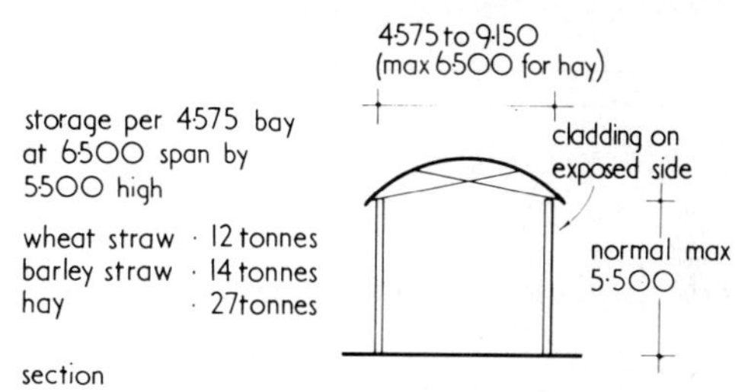

14.35 *Bale storage: Dutch barn*

Table I Average production of effluent*

		Production per head per week Weight kg	Output litres	Volume m³	Total solids kg	BOD kg	BOD population equivalent
Man	adult	75	10	0·01	0·57	0·41	1·0
Cow	dairy	450	250	0·25	21·20	4·20	10·2
,,	large dairy	550	380	0·38	32·22	6·13	14·8
Calf	3-month	100	200	0·20	19·05	2·54	6·2
Pig	porker	50	38	0·04	3·00	1·20	2·0
,,	baconer	95	51	0·05	3·50	1·40	3·4
,,	wet-fed	95	100	0·10	3·50	1·40	3·4
,,	farrow sow	110	75	0·08	3·60	1·45	3·6
Poultry	adult layer	2·25	3·75	0·005	1·27	0·09	0·13
Sheep	adult ewe	75	35	0·04	3·81	0·70	1·7
Silage†	30% dry matter	tonne	2·20	0·001	–	–	–
,,	20% ,, ,,	tonne	37·00	0·04	–	–	–

* This chart was first prepared by the author for the Institution of Water Pollution Control's Symposium on Farm Wastes, January 1970, and was based partly on data provided by J. R. Simpson

† Silage effluent production is given as the average per week during the first month after ensiling per tonne stored, thereafter dropping sharply to almost nothing after the second month

Table II Effluent manurial values*

	Approximate percentage composition Nitrogen N		Phosphoric acid P_2O_5		Potash K_2O		Moisture H_2O
	dung	urine	dung	urine	dung	urine	
Cow	0·32	0·95	0·21	0·03	0·16	0·95	75–80
Pig	0·60	0·30	0·50	0·10	0·40	1·00	70–75
Poultry	1·50	–	1·50	–	0·80	–	50–60
Sheep	0·70	–	0·20	–	0·70	–	65–70

* Manurial values in relation to breed, feeding policy, age, etc, are unknown and, therefore, wide variations around these figures can be experienced: loss of manurial values due to storage conditions, leaching out, period stored, etc, are also unknown

From *Farm buildings*, vol 2, J. B. Weller

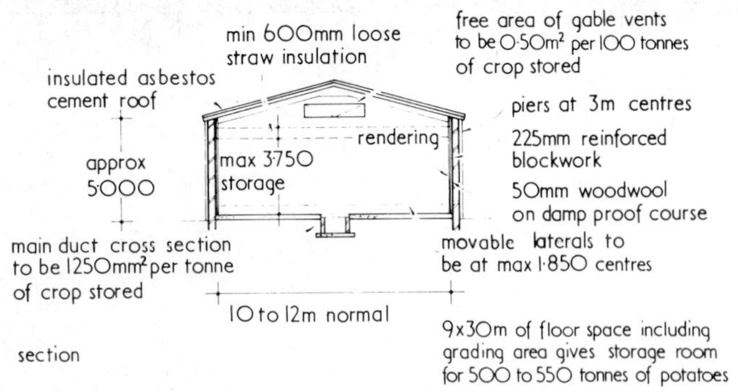

14.36 *Floor storage: potatoes*

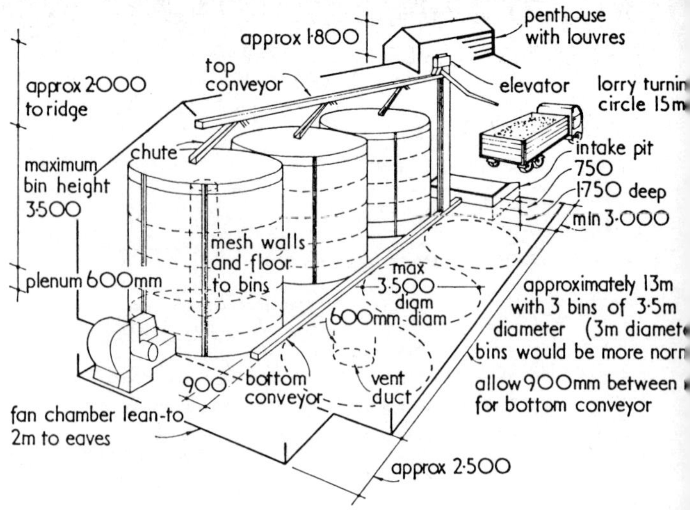

14.38 *Onion store*

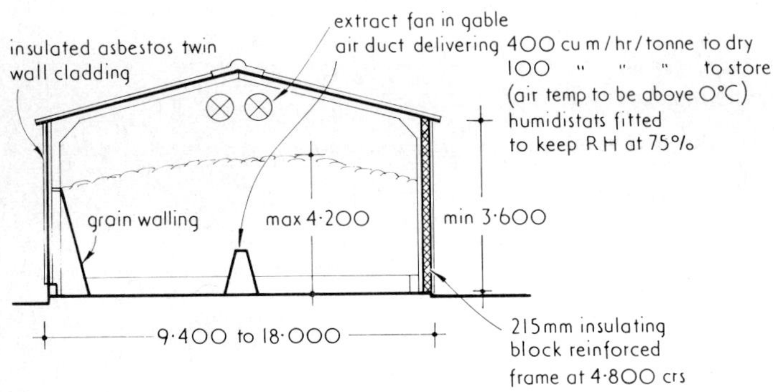

14.37 *Grain drying/storage: radial flow bins in barn*

10 BUILDING LEGISLATION

• *The Building Regulations 1976*

Agricultural buildings are mainly classed as 'buildings partially exempt from the provisions of these regulations' as described in Class 6 of Schedule 2. The parts with which the buildings should comply are:

A10 and **A11**	Depositing notices
B	Fitness of materials
C and **D**	Structural stability and moisture penetration if the capacity of the building exceeds 100 m³
E	Fire, excepting E15
K3	Preservation of zones of open space
L	Boilers and flues

• *The Town and Country Planning Act 1971*

A large number of farm buildings are classed as permitted development in the Town and Country General Development Order 1973 (SI 1973 No 31) Schedule 1, Class VI as below:

Description and development	Conditions
Agricultural buildings, works and uses **1** The carrying out on agricultural land having an area of more than one acre and comprised in an agricultural unit of building or engineering operations requisite for the use of that land for the purposes of agriculture (other than the placing on the land of structures not designed for those purposes or the provision and alteration of dwellings), so long as:	**a** the ground covered by a building erected pursuant to this permission does not, either by itself or after the addition thereto of the ground area covered by any existing building or buildings (other than a dwelling-house) with the same unit erected or in course of erection within the preceding two years and wholly or partly within 90 m of the nearest part of the said building, exceed 465 m² **b** the height of any buildings or works does not exceed 3 m in the case of a building or works within 3 km of the perimeter of an aerodrome nor 12 metres high in any other case **c** no part of any buildings (other than moveable structures) or works is within 25 m of the metalled portion of a trunk or classified road.

Article 3, para 3 of the above statutory instrument states:

1 *This permission* . . . shall not authorise any development which involves the formation, laying out or material widening of a means of access to an existing highway which is a trunk or classified road, or creates an obstruction to the view of persons using any highway used by vehicular traffic at or near any bend, corner, junction or intersection so as to be likely to cause danger to such persons.

• *Other relevant legislation*

The Milk and Dairies Regulations 1959
Rivers (Prevention of Pollution) Acts 1951 and 1961
Public Health Acts 1936 and 1961
Water Resources Act 1963
Agriculture (Miscellaneous Provisions) Act 1968, Part 1 Livestock
Agriculture (Safety, Health and Welfare) Act 1956
Agriculture (Stationary Machinery) Regulations 1959
Civil Amenities Act 1967
Clean Air Acts 1956 and 1968
Countryside Act 1968
Factories Act 1961.

15 Crown courts

Anthony Hooper

CI/SfB 317
CI/SfB (revised) 317

Anthony Hooper is an architect who has been involved in the design of Crown Courts while working for the DOE

Contents

1 THE CROWN COURT

The Crown Court was established in 1972 to replace the Assize and Quarter Sessions as the higher criminal court, as a result of a report by the Royal Commission set up in 1969 to study the courts system in Britain, under the chairmanship of Lord Beeching.

The Crown Court is under the jurisdiction of the Lord Chancellor's office and the building provision is the responsibility of the Property Services Agency.

Lower criminal courts (magistrates' courts) are the responsibility of the local authorities, under the Home Office.

The Crown Court sits in a courtroom, the design of which has been the subject of a study by a working party set up following the publication of the Beeching report.

It is the intention that Crown courtrooms should be as environmentally perfect as possible for the purpose they serve and not impressive, awe-inspiring edifices as in the past. They should be comfortable, quiet and dignified but bland and not distracting in finish and detail. Movement should be easy between different parts of the courtroom without changes in level except where necessary to ensure good sight lines. Lighting levels and colour must be such as to ensure that there is correct colour rendering, that all participants, exhibits and written evidence can be seen clearly without strain or dazzle. Elsewhere lighting should be of a lower intensity. The acoustics and noise levels should ensure that the proceedings can be heard in all parts of the courtroom without amplification (except for nervous witnesses) while at the same time ensuring that movement by the public, press and others does not distract or annoy the participants, and also so that tape recordings of the proceedings can be made.

2 THE COURTROOM AS AN ELEMENT WITHIN THE COURTHOUSE

The courtroom is the primary workspace in a courthouse, which is developed around it. The need to segregate judge, jury, defendant and others in the courtroom and within the courthouse generally is of paramount importance but it is not intended to detail these segregated circulations here but to explain how the occupants of a courtroom come together and to touch on why they need to be segregated.

3 RELATIONSHIPS WITHIN THE COURTROOM

There are four main elements in Crown Court cases—judge, jury, witness and counsel (barristers and stitors). The defendant does not take part except as a witness.

There must be direct confrontation between the jury and the witness, **15.1**.

Judge, jury, witness and counsel must be closely related, see

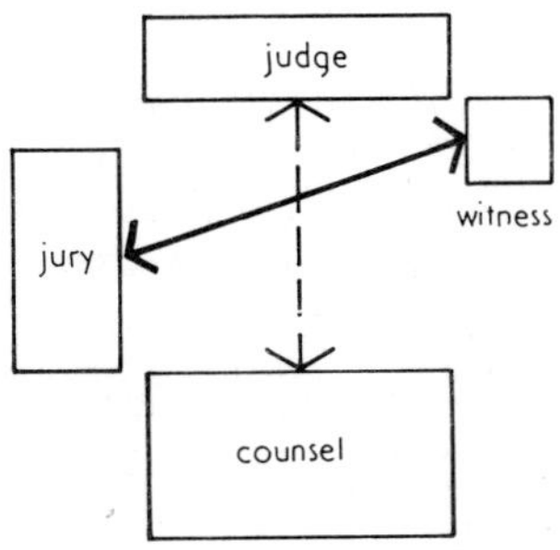

15.1 *Relationship between jury and witness*

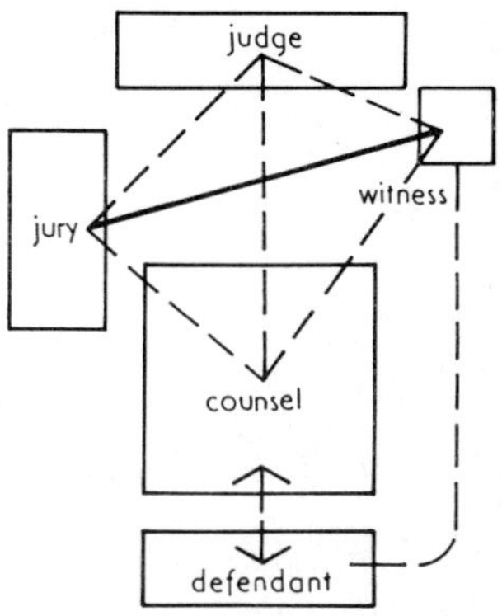

15.2 *Relationships of participants*

and hear each other clearly at all times without mechanical aids and without excessive turning from side to side, **15.2** and **15.19**. The basic positioning of the occupants is shown in **15.3**. The theory behind these relationships can be summarised thus:

- The judge, **15.4** presides over the courtroom and should be able to observe the whole courtroom and to see clearly the principal participants as well as the defendant in the dock, and when called, the antecedents and probation officers, **15.18**.
- The court clerk, **15.5** administers the case and needs to keep a watching eye over the court. He often advises the judge and he should be able to stand up and speak to the judge without being overheard.
- The exhibit tables, **15.6** are in the centre area of the court and used to display exhibits put forward for evidence.
- Counsel are barristers and solicitors, **15.7** who represent the defendant or prosecution. They need to be able to see the jury, judge and witness to whom they address their remarks. The barrister at each end of the front bench should be able to keep every jury member and the witness on the stand within about a 90° angle to obviate too much turning from side to side and to

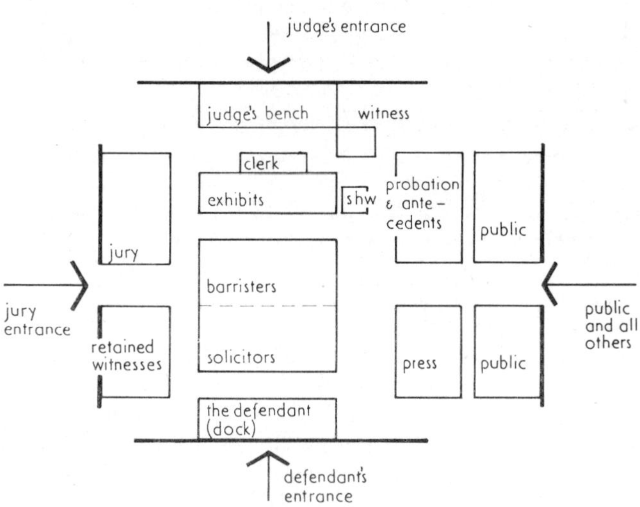

15.3 *The basic components of a Crown courtroom*

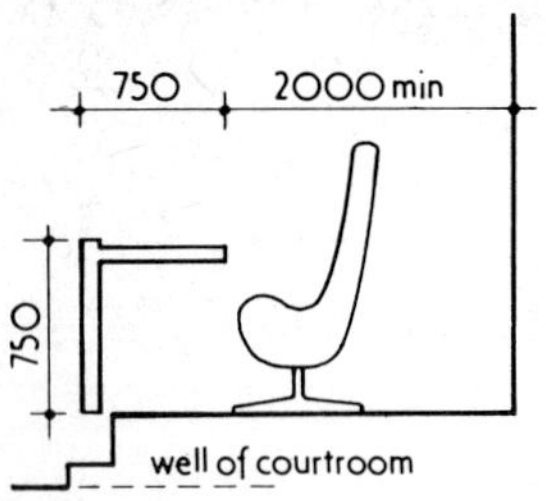

15.4 *Judge's bench*

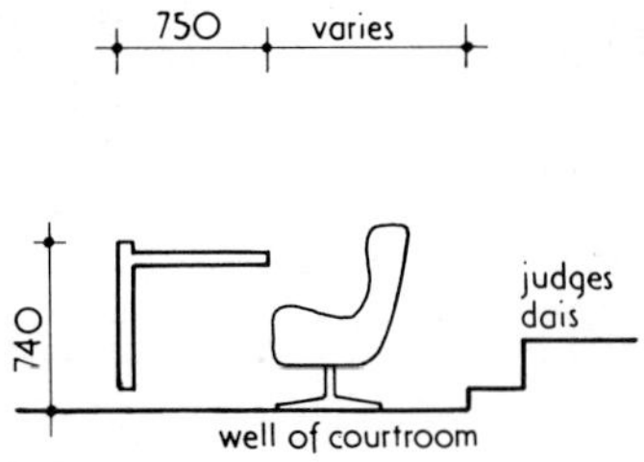

15.5 *Court clerk*

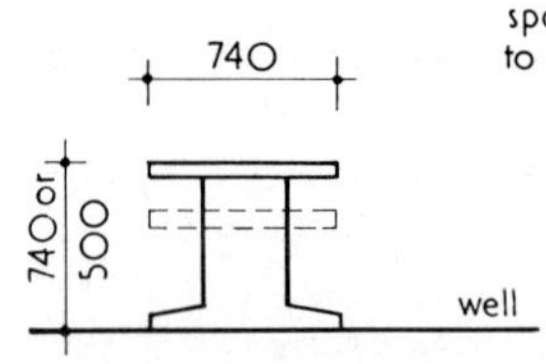

15.6 *Exhibits table*
Note: height 740 mm when used as counsel bench, 500 mm when used as exhibits table

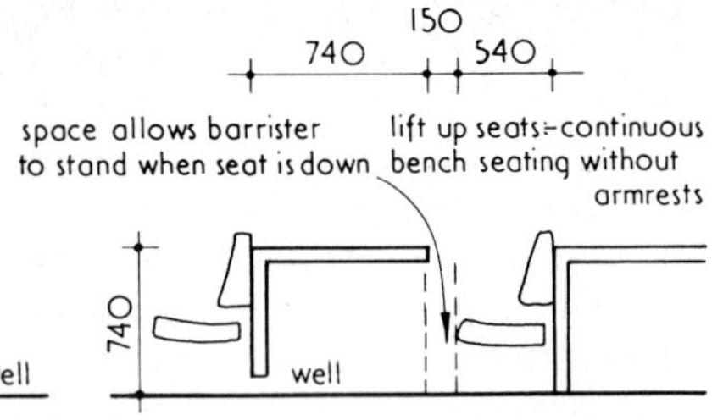

15.7 *Barristers and solicitors*

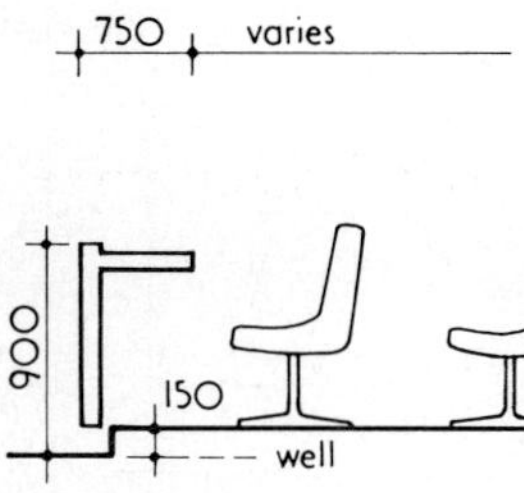

15.8 *Defendant (dock)*

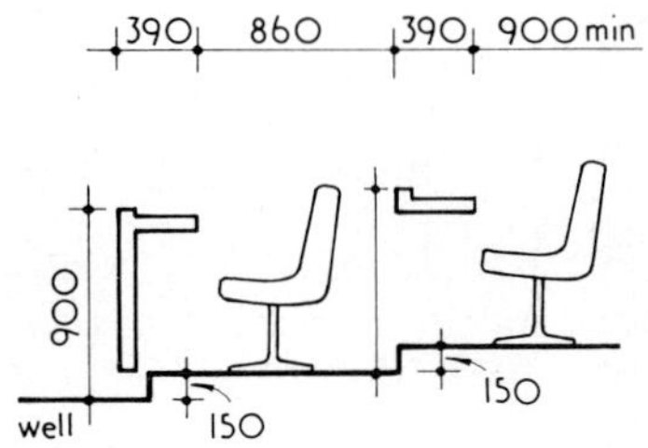

15.9 *Jury*

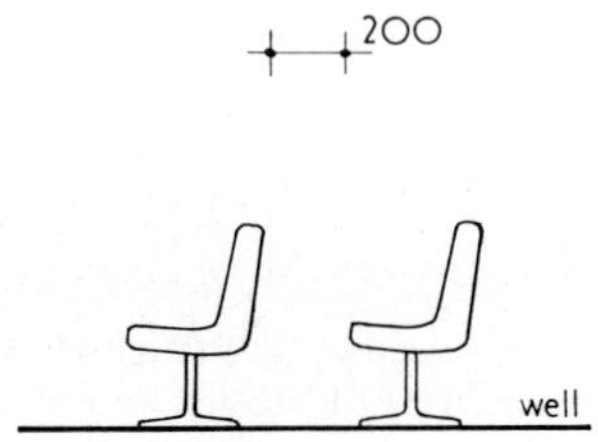

15.10 *Retained witness*

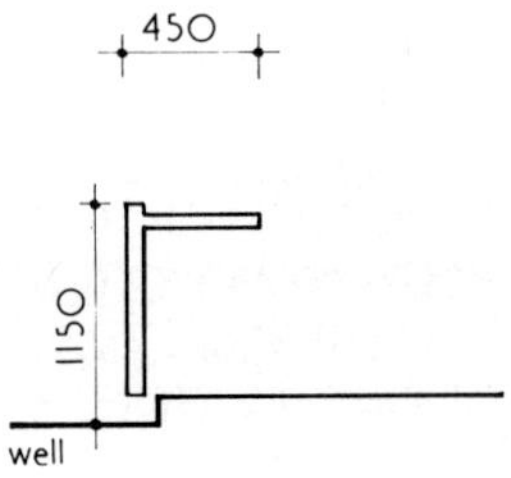

15.11 *Witnesses*

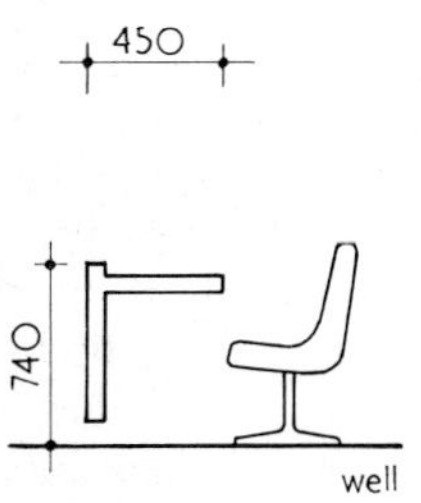

15.12 *Shorthand writer. The table should be movable*

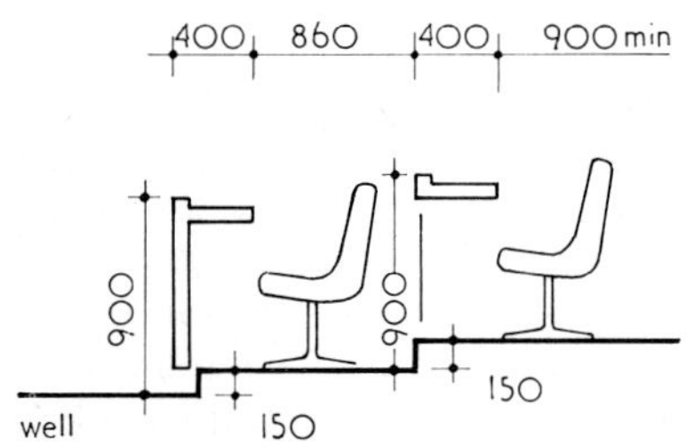

15.13 *Probation and antecedents*

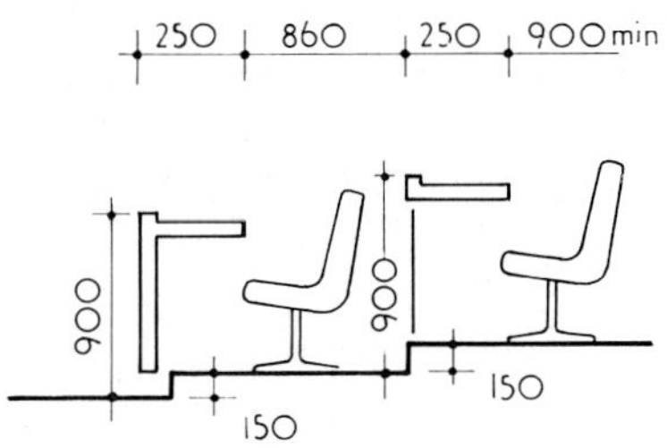

15.14 *Press*

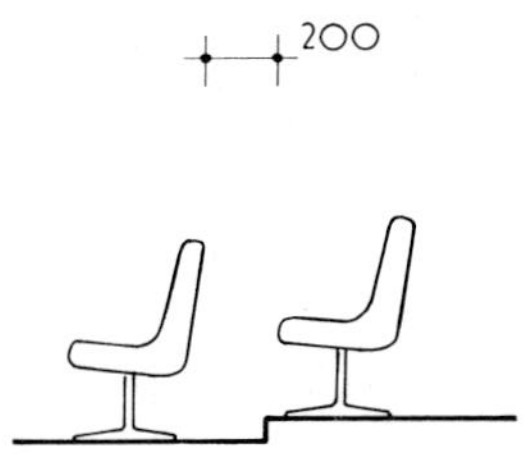

15.15 *Public*

ensure that the judge, the other main party, shall have at least a part face view. The counsel benches are wide enough to hold the large, and numerous, documents and books that are often in use.

• The defendant is assumed to be innocent until proved guilty and current practice is to reduce the prisoner appearance of the dock, **15.8**, by lowering the barriers enclosing it as much as possible compatible with security. Defendants sit in separate fixed seats and if they are thought to be a security risk a prison officer will sit on a seat immediately behind, or at each side of him. Each defendant must have a writing surface and the dock is controlled by a dock officer. The dock is situated behind and near to the counsel benches.

• The jury, **15.9** comprises 12 members of the public whose duty it is to reach a verdict based on the evidence presented. The jury sit opposite the witness stand and must be able to see the defendant in the dock, as well as the judge and counsel. They must have a writing surface and a place to put documents.

• If the judge directs that a witness should be retained they wait in seats within the courtroom, **15.10** where they are not regarded as main participants.

• The witness waits outside the courtroom and when called gives evidence from a stand, **15.11** near to the judge's bench. The stand faces the jury who must be able to observe the face of the witness. The witness is questioned by the barristers and occasionally by the judge.

• The shorthand writer keeps a transcript of the trial and consequently must be able to see and hear everyone who speaks. The shorthand writer's table, **15.12** is movable to enable it to be moved to the other side of the courtroom if required.

• Probation and antecedents officers, **15.13** give their evidence from their seats, after the jury have reached a verdict. The evidence is used to assist the judge when he passes sentence.

• The press, **15.14** are not party to the proceedings but they should be so situated as to allow them to see the participants.

• The public, **15.15** comprise every other person in the courtroom, including the press and those dismissed from the case, who are not participants and are allowed in courtrooms to see that 'justice is done'. They are placed to one side of the body of the court.

4 ACCESS AND EGRESS

There are four entrances to a Crown courtroom and these should be kept separate. Immediate access to the courtroom should be as indicated in **15.3**, but obviously vertical or horizontal circulation methods can be used to maintain segregation elsewhere. The segregation should take account of the following:

a The judge should enter the courtroom from circulation which is restricted to judges' use. Leading off this circulation will be his retiring room where he goes to consider the evidence, confer with counsel and prepare his summing up.

b The jury should enter the courtroom directly from a waiting area outside the room. This waiting area will have access to 'jury restricted' circulation off which will be located jury retiring rooms, and the same circulation will give access to the jury dining room and assembly area which is entered off the public circulation.

c Defendants in custody enter the courtroom directly into the dock from a defendants' waiting area. Access to the waiting area is from segregated circulation, via stairs or lifts, which leads into custody or cell accommodation. Defendants not in custody enter by the public entrance.

d The public (in this context everyone except those on the judges' benches, the jury benches and in the dock) will enter the courtroom from a public concourse. Off this concourse will be access to:

- public dining, lounge and toilet facilities
- barristers' and solicitors' suites
- witness waiting rooms
- press room
- staff offices
- court offices.

This requirement to segregate courtroom users is of prime importance. Fire escape from a courtroom can be via any of the doors, except the dock door, but attention has to be given to aisle widths, length of seat rows, emergency lighting, exit signs etc.

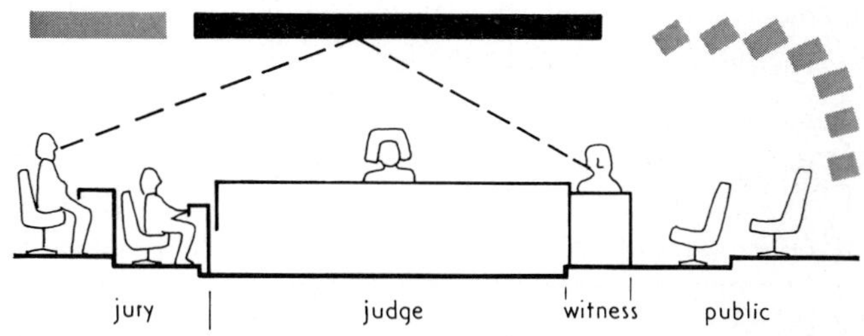

15.16, 15.17 *Required acoustic treatment*

5 THE ENVIRONMENT WITHIN THE COURTHOUSE

5.01 Ventilation

To assist in the smooth running of the court, to help keep participants interested and comfortable and to avoid the distractions of the outside it is recommended that the courtrooms are internal rooms with air-conditioning. The air-conditioning should be fed from the ceiling but the plant should be placed remote from the room to prevent the noise of the plant from entering the courtroom.

5.02 Lighting

Lighting should be bland and should not cause distraction. Fittings should preferably be fully recessed and spotlights should never be used. Lights should be positioned and selected to give good illumination on bench tops and not produce glare. The centre well of the courtroom should have the most intense level of lighting with a lower level over the public and dock. The light source should be carefully selected to give reasonable facial modelling and accurate colour rendering. In order to reduce glare and contrast, bench tops and furniture generally should be of high reflectance ie light colours—likewise for the walls and ceiling.

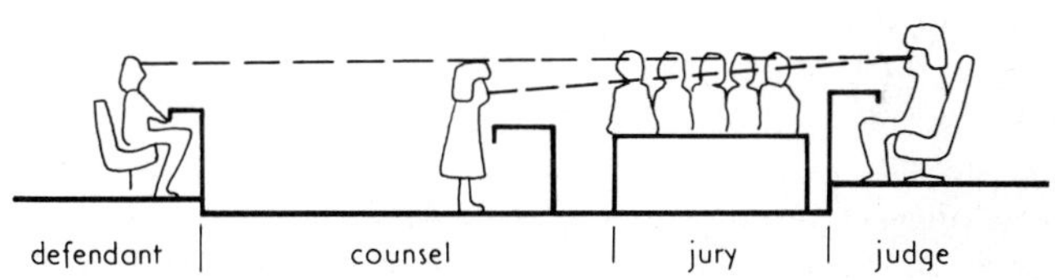

15.18, 15.19 *Sightlines*

5.03 Acoustics

Good acoustics are vital to enable the participants to hear clearly, without strain. There should be reflective surfaces on the ceiling over the main well of the room, on the side walls and behind the judge, **15.16** and **15.17**. There should be absorbant surfaces around the edges of the ceiling and behind the public area. The floor should be carpeted all over and the entrance doors from the public concourses should be lobbied to keep out extraneous noise. An NR rating of 25+2 dB is recommended and a reverberation period of not more than 0·5 to 0·75 secs should be achieved.

5.04 Finishes

Floors should be carpeted and soft upholstery used for all seating. Bench tops must be hard wearing surfaces suitable for writing on. Furniture should be hard finishes, easy to maintain and not easily damaged. Walls can be plastered with suitable sound absorbent panel type finishes where required. The ceiling can be tiled in a manner suited to meet acoustic requirements.

6 BIBLIOGRAPHY

Consultative memorandum—Court design, Joint Working Party—LCO, HO, DOE, 1972

Design study no 2 Magistrates' courts, WP report, GLC, 1969

Design study 1977 Magistrates' courthouses, GLC, 1977

Article:

C. Theodore Larson *Future shock hits American courthouse: opportunities and parameters for design, AIA Journal,* July 1975

The planning of court buildings, Home Office, HMSO, 1965, metric edition 1969

Report of the interdepartmental committee on the business of the criminal courts, Home Office, Lord Chancellor's Office, HMSO, February 1961, reprinted 1968

Royal commission on assizes and quarter sessions (the Beeching Report), 1966–9, HMSO

Sherriff Courts. Recommendation on design and accommodation standards, Joint Working Party—PSA Scotland: Scottish Courts Administration Crown Office 1973

16 Offices

Hugh Ellwood

Hugh Ellwood is an architect partner of Building Design Partnership, Preston

CI/SfB 32
CI/SfB (1976 revised) 32

Contents

1 INTRODUCTION

This section is set out to follow the parallel development of the brief and design where the designer considers all aspects at the appropriate time and returns to investigate and develop them further at later stages.

Different elements of the building have different lifespans:

- *The shell:* The structure, core and skin which are intended to last as long as the building is useful, perhaps for 40 years.
- *The scenery:* Partitions, furniture and furnishings related to the life of a tenancy or a particular stage of an organisation's development, lasting perhaps 5 to 7 years.
- *The set:* The constant re-arrangement of office 'scenery' to meet changing short-term needs with the time scale sometimes measured in weeks.
- *The services:* Basic installations such as lifts or mechanical services will last 20 to 25 years. Services at a small scale such as light fittings and electrical points will change with the scenery and must be capable of accommodating the shortest-term requirements of the set.

2 INCEPTION AND PHYSICAL SCALE

To assess whether the project is a feasible proposition and that legislation will permit the required development on the particular site, the type of building and scale of accommodation must be established. The architect must therefore assess the client's business activities and staff numbers to arrive at an overall floor area.

For a rough, but realistic, total figure a minimum of 9·3 m² to 11·6 m² net should be allowed for each person, increased to a minimum of 14 m² per person if eating and lounge facilities are included. To allow for circulation and support functions, this net area should be assumed to be nearly 80 per cent of the total gross floor area. These standards for the normal office are summarised in table I.

Table II gives some figures for open-plan offices taken from an unpublished DOE internal report of 1971. Here the net office space was found to be 66 per cent of the gross for general offices and 69 per cent for drawing offices.

However, the above figures may be found to be too crude, particularly as space standards vary widely. As more information becomes available the architect can establish more accurate space standards as shown in table III.

3 FEASIBILITY AND GENERAL STRATEGY

In assessing how the client's requirements can be accommodated within the constraints of the site and relevant legislation, the brief should be seen not as a rigid set of instructions, but as a statement of intent which will develop with the design process. At this stage only sufficient statement of intent is required to establish the overall strategy and to decide quickly on basic architectural elements.

The first step is to determine the type of business organisation **16.1** which can be described by the degree to which work places are distinguished by area, number of pieces of furniture and

Table I Space standards at feasibility stage

Absolute minimum area per person	3·7 m² to 4·2 m² (including aisles, filing cabinets and desk space). But this is too low for individual offices
Minimum area for reasonable conditions	4·2 m² to 6·0 m² (14 m³ to 17 m³). But allow more for individual offices with single occupation
Requirements of Offices, Shops and Railway Premises Act 1963	3·7 m² minimum floor area per person 11·3 m³ minimum room capacity per person
Additional space for visitors	Allow minimum 1·8 m² extra for visitors
Recommended allowance for completely integrated office with all activities under one roof	Average of 9·3 m² to 11·6 m² per person excluding circulation, wcs and so on
Completely integrated office plus eating and lounge facilities	Minimum 14·0 m² per person
Proportion of total floor area to be aimed at for office working space	Nearly 80 per cent of gross internal floor area

Table II Results from measurements of space allocation in a number of open-plan offices. General office space is compared with the special case of drawing office space*

Category of planned open office	Net work area expressed as % total net area	Net area for support expressed as % total net area	Total net area per person	Net work area per person
General	69%	31%	8·9 m²	6·32 m²
Drawing	72%	28%	11·24 m²	8·08 m²
All	70%	30%	9·66 m²	6·97 m²

*The drawing office is included just to make the point of how a specific (and sometimes marginal office activity) alters the ratios

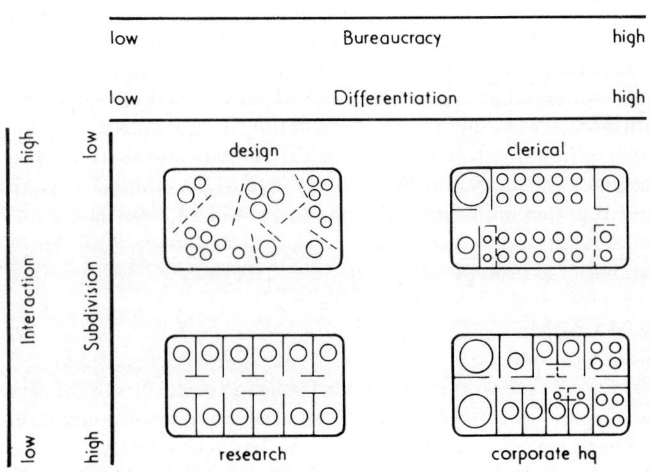

16.1 *Bureaucracy is related to and explains differentiation, and interaction explains subdivision*

Table III Calculating office area requirements

Action	Example 1 Accumulating area allowance per person—m²	Number of personnel	Accumulating total area required	Example 2 Accumulating area allowance per person—m²	Number of personnel	Accumulating total area required
1 Take sum of workplace areas	10 m² per person	862	8620 m²	5 m² per person	525	2625 m²
2 Add 10% for access to workplaces (circulation within departments)	10 m² + 10% = 11 m² per person	862	9482 m²	5 m² + 10% = 5·5 m² per person	525	2887 m²
3 Add area for special, within-department facilities; eg meeting and conference rooms, display or exhibition areas, etc	11 m² + 2·5 m² (say) = 13·5 m² per person *1	882 (additional 20 serving special facs)	11 907 m²	None provided	—	—
4 Add 15% for inter-departmental circulation	13·5 m² + 15% = 15·5 m² per person	882	13671 m²	5·5 m² + 15% = 6·3 m² per person	525	3318 m²
5 Add area for support facilities to office areas—filing registries, archives, vending machines, etc	15·5 m² + 3·3 m² (say) = 18·8 m² per person *2	882	16581 m²	6·3 m² + 2·2 m² (say) = 8·5 m² per person	525	4462 m²
6 Add areas for special facilities—computer rooms, restaurants, etc. Provision varies so much from company to company that no general rules apply. Areas must be determined by preparing rough layouts						
Special facility A	18·8 m² + 1·5 m² = 20·3 m² per person *3	882	17904 m²	None provided		—
Special facility B	20·3 m² + 1·3 m² = 21·6 m² per person *4	882	19051 m²	8· m² + 0·75 m² = 9·25 m² per person	525	4856 m²
Net usable area (NUA) (all the above)	21·6 m² per person	882	19051 m²	9·25 m² per person	525	4856 m²
Gross overall area (NUA + core and columns, etc—allow 18%–20%)	21·6 m² + 20% = 25·9 m² per person	882	22844 m²	9·25 m² + 20% = 11·1 m² per person	525	5827 m²

NB
Area allowances in items 1–4 above are based on the numbers of people to be accommodated. Area allowances in item 5 and 6 are based on the requirements of machinery and plant

Example 1
A multi-national corporation having a high proportion of high status personnel and providing high quality space standards and a high level of special facilities

Example 2:
A largely clerical organisation providing minimum space standards and few special facilities

equipment, position within the building shell and other physical means, and the degree to which workplaces are partitioned. Examples of the relationship between building shell and organisations are given in table IV.

4 OUTLINE PROPOSALS—THE SHELL

4.01 Legislation

The architect must interpret regulations which affect the form and internal layout of the building shell such as density; light and overshadowing; fire regulations (means of escape). The legislative controls are listed in table V.

4.02 Means of escape

The requirements for means of escape are summarised as follows:

Maximum travel distance with escape possible in one direction only—12·2 m except for on the ground and first floors when it may be 30·5 m if the following requirements are met:

- opening windows in all offices 840 × 535 mm wide minimum
- lower level of opening light not more than 3800 mm above ground and 915 mm above FFL
- ground beneath window free of observation for 1830 mm from building.

Maximum travel distance with escape possible in alternative direction—46 m with no point in an office more than 12·2 m to nearest exit door unless second exit door provided.

Maximum distance between two adjacent exits from a storey—61 m.

Fire fighting stair—at least one fire-fighting stair is required in buildings with floors over 18·3 m above ground level which should:

- be continuous throughout building
- have access at ground level direct to open air
- have openable windows at each landing level
- have permanent ventilation at the top of the enclosure of minimum 5 per cent of enclosed area
- have protected and ventilated lobby at each floor.

Staircase width

Single staircase serving gross floor area of less than 230 m²–765 mm wide. Single staircase serving gross floor area of more than 230 m²–1070 mm wide.

Two staircases, one floor only above ground level, gross area less than 1860 m²–1070 mm wide. Two staircases, one floor only above ground level, for every additional 280 m² add 152 mm. For multi-staircase buildings, see table VI.

4.03 Hygiene facilities

(See also section 38, Sanitary installations.)

- *Water closets: men*

The legal minimum for men (Offices, Shops and Railway Premises Act 1963, sections 9 and 10) is shown in table VII.

- *Water closets: women*

For women (and for men where urinals are not provided), the legal minimum is shown in table VIII.

- *Wash basins*

The Offices, Shops and Railways Premises Act 1963 requires washing facilities to be provided on the scale shown in table IX (where there is no exemption from the requirement concerning running water). Section 10 requires hot and cold water, soap and towels or other means of drying.

Table IV Building shell and business organisation

	Design office	Advertising agency	Top management	Clerical office
	Intensely interactive project-based groups in loose touch with each other. Serviced by normal support functions. Visitors at all levels. Partners in close touch. Concentrated work with occasional confidentiality	Isolated work groups, co-ordinative work. Two kinds of groups; working group competes for the services of the other. Usual support services. Directors not involved in day-to-day work; concerned with clients	Isolated executives with secretarial and PA support. Confidential and contemplative work. Visitors	Large supervised groups—paper and/or machine intensive. Highly inter-active groups. No public entry
Cellular	Unsuitable because of group size and informal interaction. Separate rooms breed isolation	Unsuitable because of group sizes though could be made to work	Suitable; permits confidentiality and suitable reception of visitors. Special arrangements required for board room	Unsuitable for large groups—isolation incompatible with supervision requirements.
Group space	More suitable than cellular, though still not ideal	Most suitable. Groups prefer this degree of territorial definition	Possibly suitable but spaces may be too large	Possibly suitable—depends on space and group size, though similar problems as cellular
Open plan	Inappropriate for method of working, management style and occasional need for privacy (meetings)	Inappropriate for method of working and management style	Unsuitable, incompatible with status management style and requirement for confidentiality	Suitable, accepts group size fluctuations, supervision and high interaction
Landscaped	Probably very suitable. Group identity and territorial definition sustained—with day-to-day rearrangement potential. Allows easy access for visitors	Not very suitable. Frenetic and competitive mode of working disturbs others	Unsuitable. Top management too exposed	As open-plan but supervision restricted

Table V Qualifying legislation

		Subject	Act/Regulation	Explanatory text	With whom negotiated
General legislation	a	Detail town planning consent, 'approval of reserved matter'	Town and Country Planning Act 1971	Local and national design policies (eg as to tall buildings)	Local planning authority
	b	(Sunlight and daylight guides, not legislation)		DOE *Sunlight and daylight* HMSO 1971	Local planning authority
	c	Established rights of light			
	d	Public health generally (safety and health)	Public Health Acts 1936 et seq.		Adjoining owners
	e	Provision of sanitary accommodation and adequate environment in work places	Offices, Shops and Railway Premises Act 1963	*General Guide (to Offices, Shops and Railway Premises Act 1963)* HMSO	Local authority public health officers Medical officer of health (Dept of Health and Social Security)
Legislation operating outside the GLC area	f	Generally	Building Regulations 1976 The Building Standards (Scotland) (Consolidation) Regulations 1971	Building Regulations	Building inspector
	g	Nature of enclosures with reference to protection from fire	Building Regulations 1976, E7	Explanatory bulletin, appendix to building regulations	Building inspector
	h	Means of escape in case of fire	Offices, Shops and Railway Premises Act 1963	Department of Employment *Means of escape in case of fire in offices, shops and railway premises* Health and Safety at Work series No 40 HMSO 1973	Fire brigade
Legislation operating within the GLC area	i	Generally			District surveyor
	j	Nature of enclosures with reference to protection from fire	London Building (Constructional) By-laws 1972		District surveyor
	k	Minimum amount of windows	London Building (Constructional) By-laws 1972, Part XIV		District surveyor
	l	Means of escape in case of fire. Tall and large buildings	London Building Acts (Amendment) Act 1939, section 20	GLC *Code of Practice, Means of escape in case of fire, Section* 20, GLC 1976	GLC architects department in collaboration with London fire brigade
	m	Means of escape in case of fire other than section 20 'buildings', above)	London Building Acts (Amendment) Act 1939, section 35	GLC *Code of Practice, Means of escape in case of fire, Section* 34, GLC 1976	GLC architects department

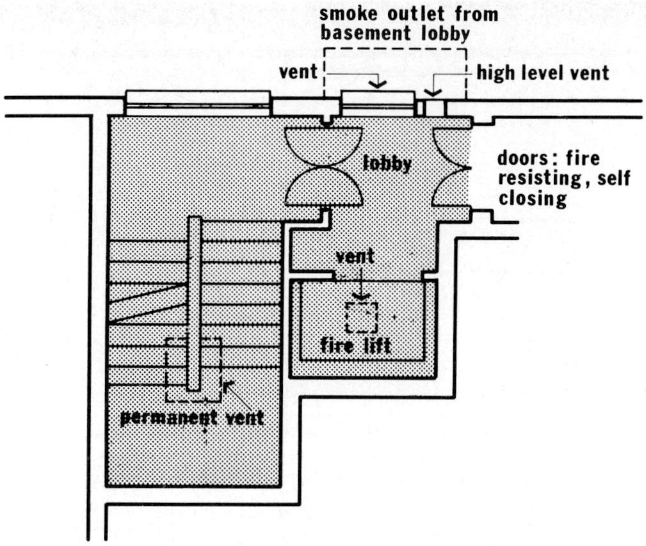

In buildings over 18·3 m high some staircases should be constructed as fire-fighting staircases with smoke outlets, vents and fire-resisting, self-closing doors

Table VI Minimum staircase width for multi-staircase buildings with more than two storeys above ground level (from CP3: chapter IV, part 3)

Gross floor area of the storey (calculated at 9·3 m² per person) not exceeding m²	Number of persons in the storey	Minimum width for each staircase		
		2 staircases mm	3 staircases mm	4 staircases mm
230	25	765	765	765
930	100	1070	1070	1070
1070	115	1220	1070	1070
1210	130	1370	1070	1070
1350	145	1525	1070	1070
1490	160	1680	1070	1070
1630	175	1830	1070	1070
1860	200		1070	1070
2140	230		1220	1070
2420	260		1370	1070
2700	290		1525	1070
2800	300		1525	1070
2980	320		1680	1070
3210	345		1680	1220
3260	350		1680	1220
3630	390			1370
4050	435			1525
4470	480			1680
4890	525			1830

Table VII Legal minimum of water closets for men

Number of men	Number of water closets	Number of urinal stalls
1–15	1	1
16–20	1	1
21–30	2	2
31–45	2	2
46–60	3	3
61–75	3	3
76–90	4	4
91–100	4	4
Over 100	4	plus 1 closet for every 25 persons (or fractions of 25) in excess of 100. Every fourth additional closet may be replaced by a urinal

Table VIII Legal minimum of water closets for women

Numbers regularly employed at any time	Number of water closets
1–15	1
16–30	2
31–50	3
51–75	4
76–100	5
Over 100	5 plus one additional closet for 25 persons in excess of 100 (a fraction of 25 counts as 25)

Table IX Scale of washing facilities to be provided

Numbers regularly employed at any one time (or where separate accommodation is required for each sex, the number of males or females)	Number of washbasins to be provided
1–15	1
16–30	2
31–50	3
51–75	4
76–100	5
Over 100	5 plus 1 basin for every 25 persons (or fraction of 25) in excess of 100

5 THE BASIC SHELL

The form of the office shell is determined by the location of the core and major circulation routes and the depth of space. The effect of these determinants is shown in figures **16.2** to **16.12**.

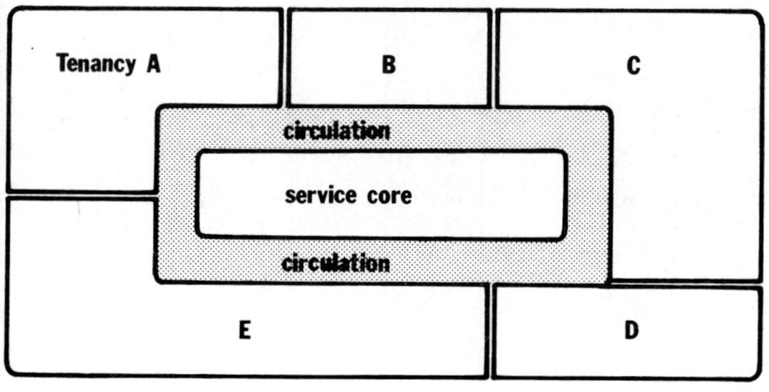

16.2a *In a tower, the core helps to subdivide shell into separate tenancies*

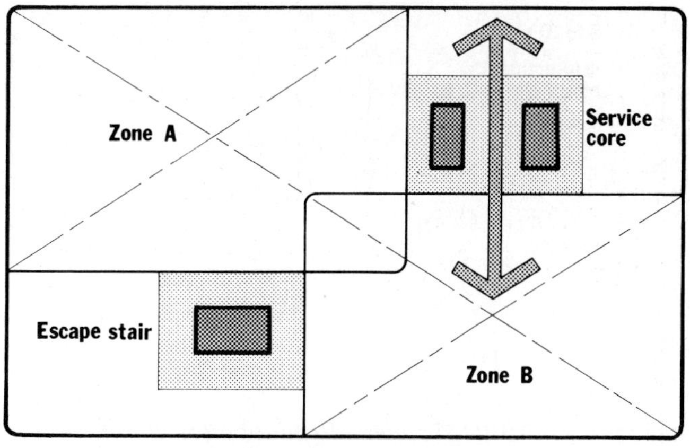

16.2b *Internal service cores can inhibit the effective use of a deep building as an open office*

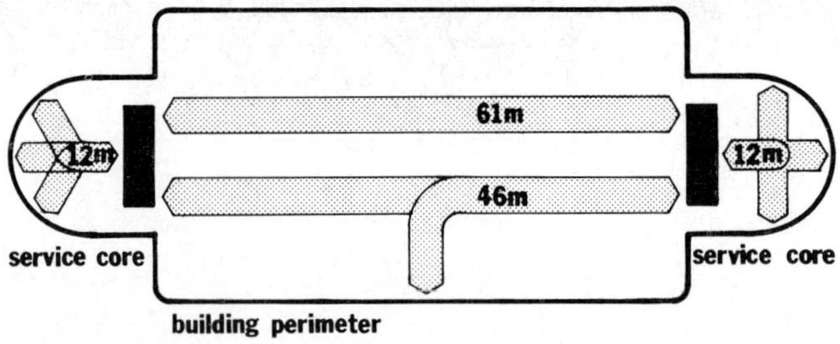

16.3 *Escape routes and distances: 12 m, 46 m and 61 m for different conditions (from BS CP 3: Chapter 4)*

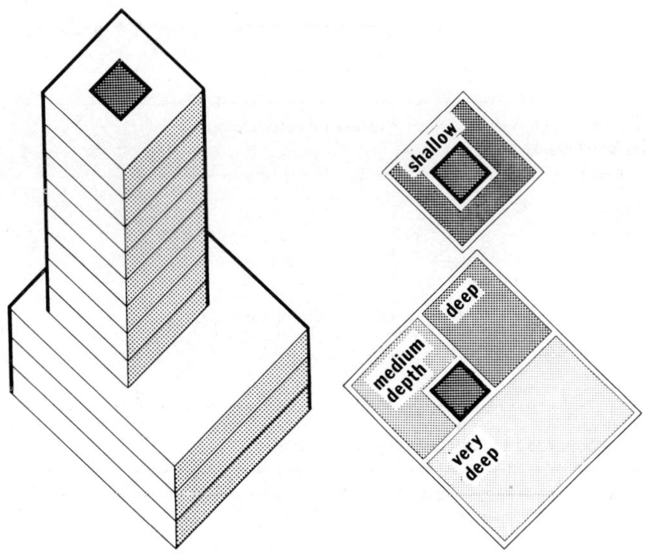

16.4a, b *Illustration of how different types of office space with different kinds of use can be provided in one office building by the simple device of core placed asymmetrically in the podium*

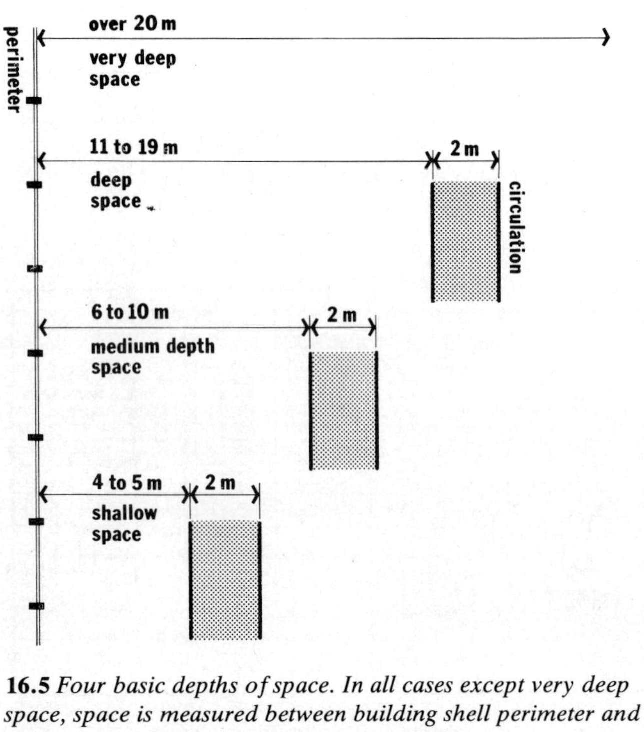

16.5 *Four basic depths of space. In all cases except very deep space, space is measured between building shell perimeter and primary circulation across uninterrupted office space. Circulation is assumed to be 2 m deep*

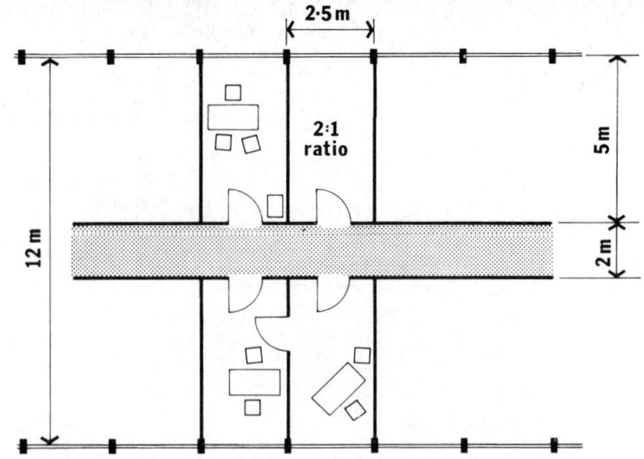

16.6 *Shallow floors often depend upon perimeter ventilation and permit reasonably shared rooms. If 2:1 is accepted as a suitable proportion and if the minimum acceptable width for a single room is 2·5 m, then a shell depth of 12 m results*

16.7 *Shallow space, although essentially linear, can be provided in a variety of office shell forms:*

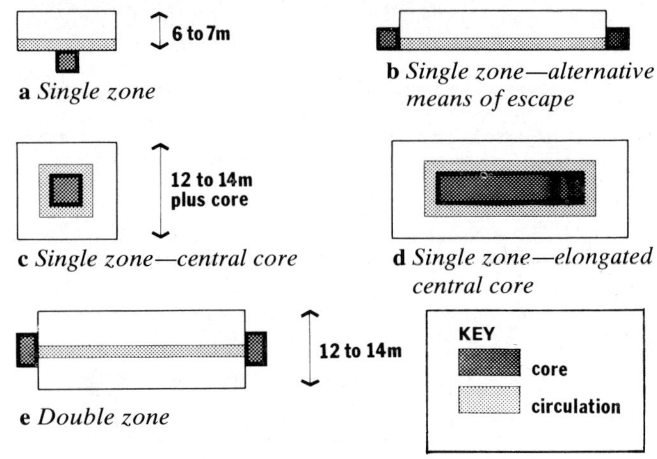

a *Single zone*

b *Single zone—alternative means of escape*

c *Single zone—central core*

d *Single zone—elongated central core*

e *Double zone*

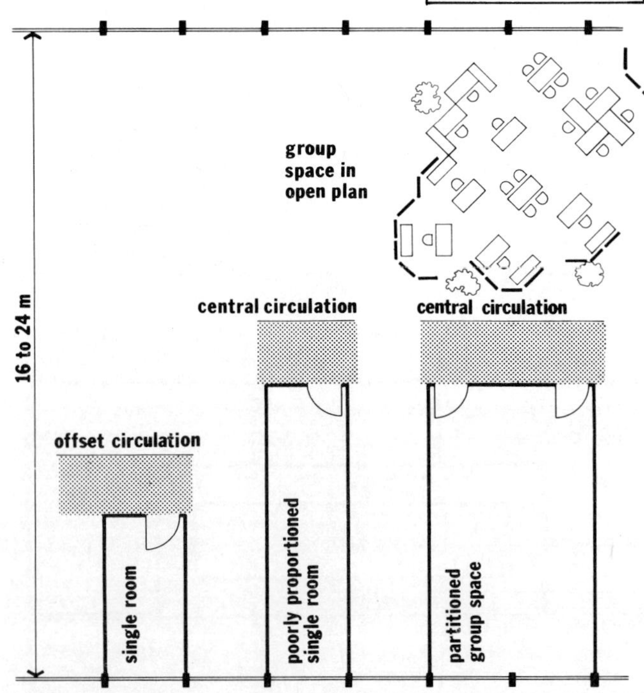

16.8 *In office shells of 16 m to 24 m depth (double zone, medium-depth space) acceptable single rooms can only be provided with an offset corridor. The use of an offset corridor greatly enhances the adaptability of shells of this depth because it enables more than one depth of office space to be provided*

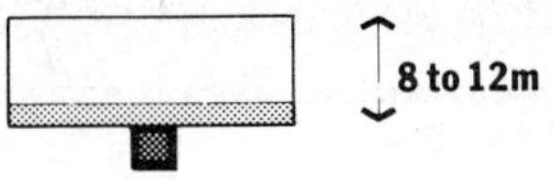

a *Single zone*

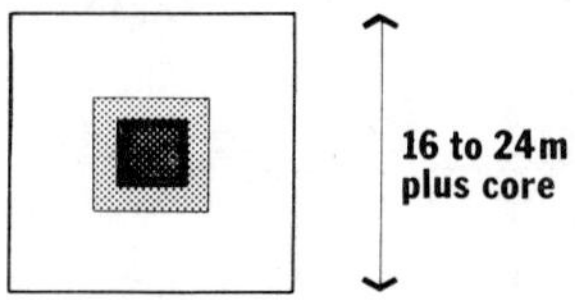

b *Single zone—central core*

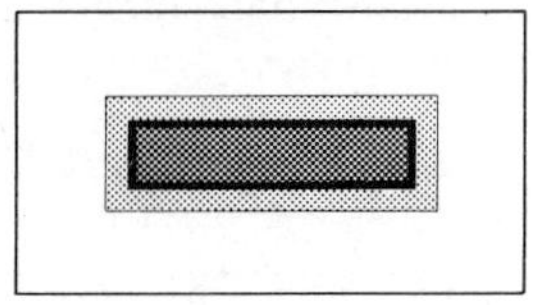

c *Single zone—elongated central core*

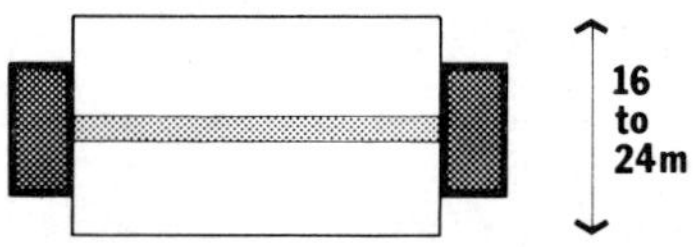

d *Double zone—fixed central circulation*

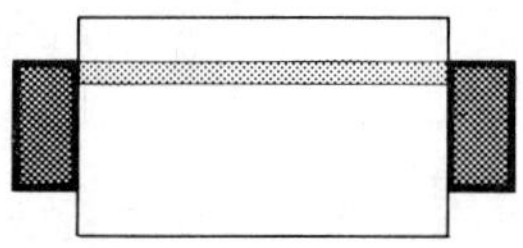

e *Double zone—asymmetrical position of circulation providing shallow and medium-depth space*

16.9 *Medium-depth space can be provided in a range of office building shells. If, as in* **e**, *the position of circulation is not fixed (eg by central columns) both shallow and medium-depth space can be provided in a shell that can also provide two zones of medium-depth space*

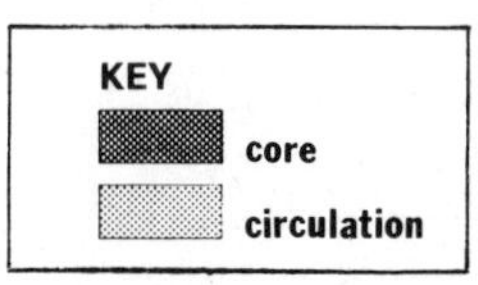

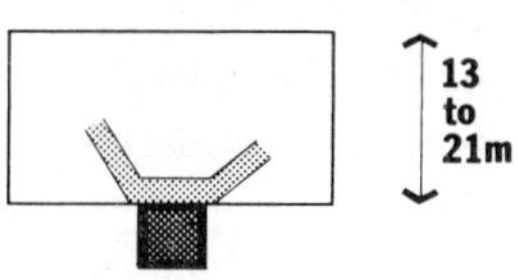

a *Single zone*

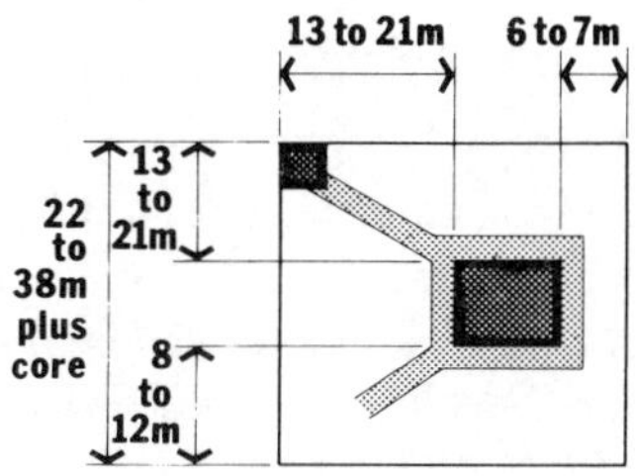

b *Single zone—central core*

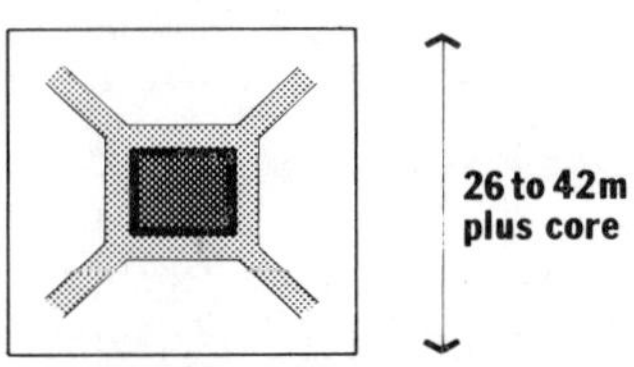

c *Single zone—asymmetrical core*

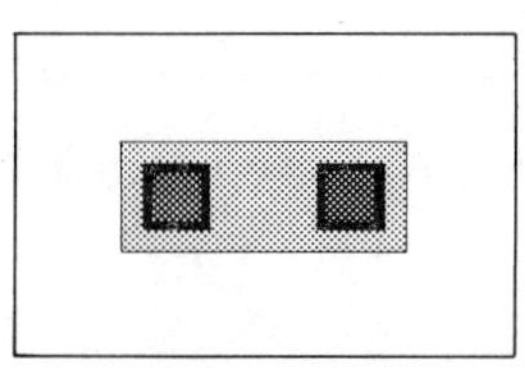

d *Double zone—split core*

16.11 *Some examples of deep space. In* **c** *the asymmetrical placing of the core creates a mixture of deep, medium and shallow space.*

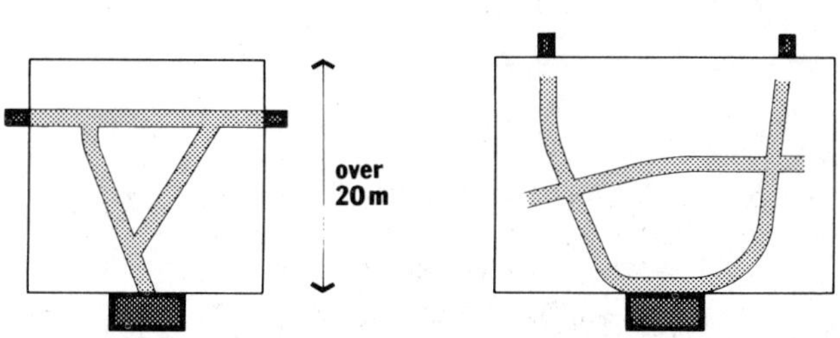

16.12a, b *In very deep space so many choices are available for locating circulation that the concept of zones of space defined by circulation is no longer useful*

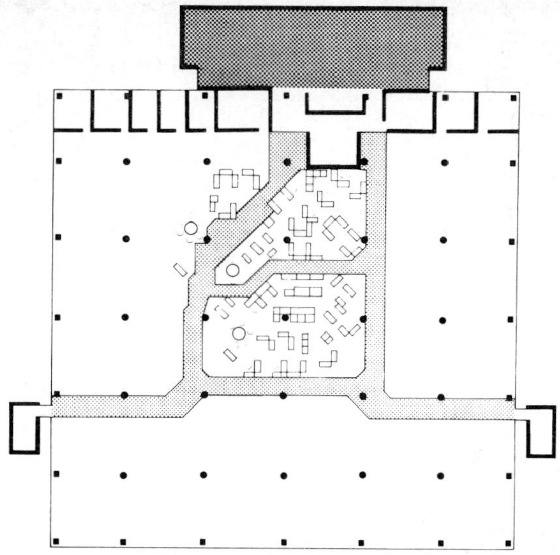

16.10 *The consequences of increasing the proportion of enclosed offices at the perimeter of a deep office space are shown as follows:*
a *(8·4% of the area) hardly affects the quality of the open space*

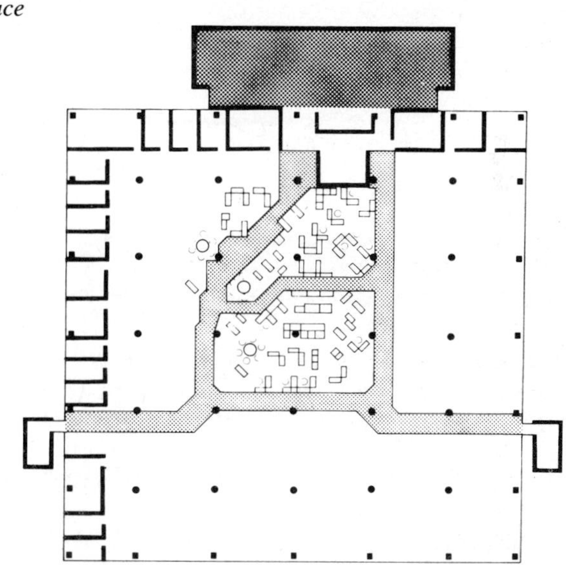

b *(16·8% of the area) has begun to reduce the aspect from some workplaces*

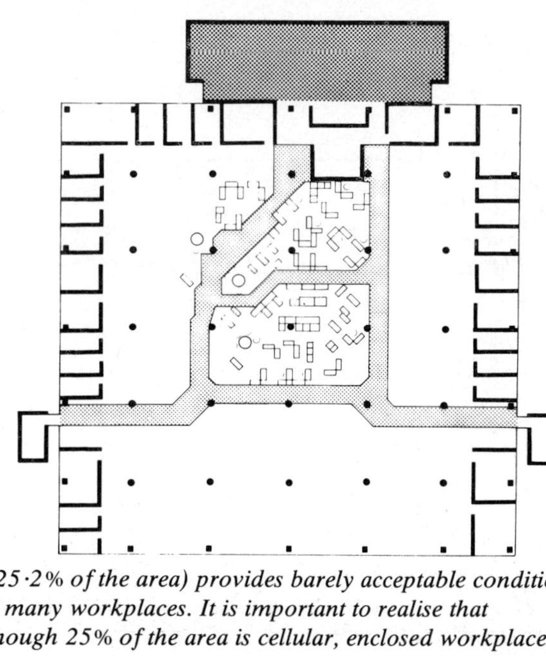

c *(25·2% of the area) provides barely acceptable conditions for many workplaces. It is important to realise that although 25% of the area is cellular, enclosed workplaces are provided for only approximately 10% of the workforce*

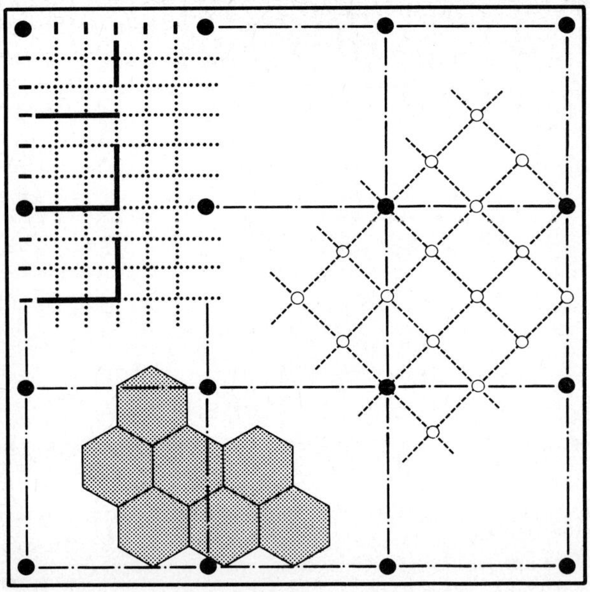

16.13 *Four kinds of grid: structural, constructional, servicing and planning*

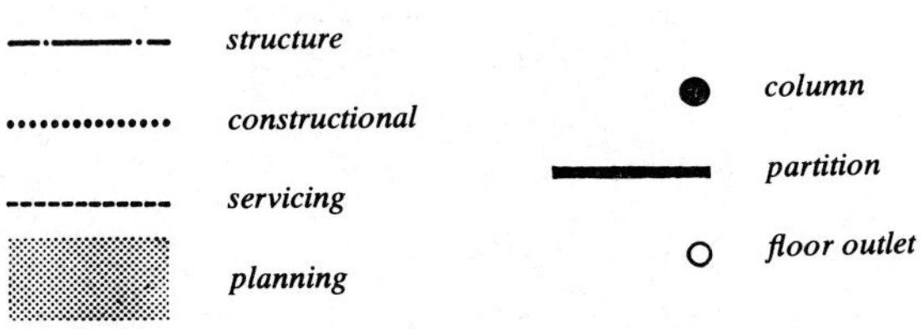

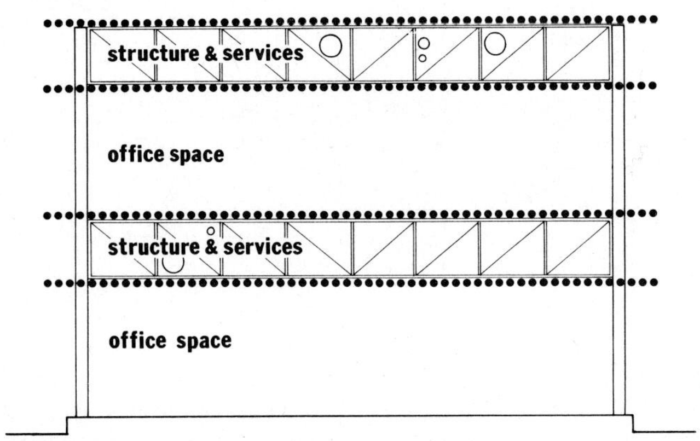

16.14 *Grids cannot be considered in plan alone—the section must be considered*

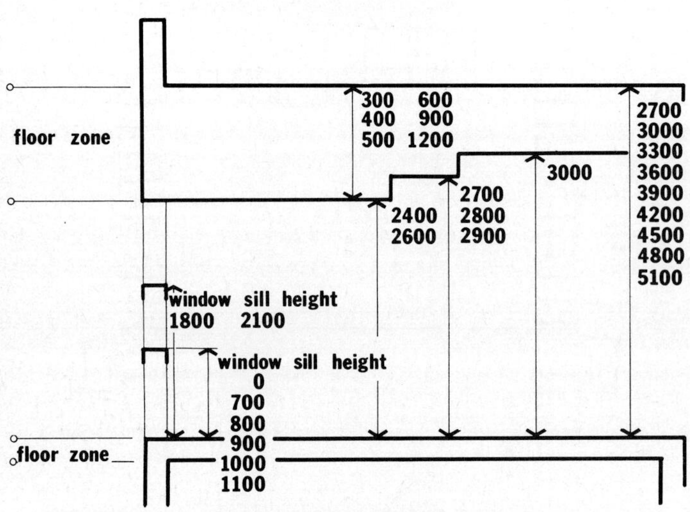

16.15 *Ceiling-height dimensions related to floor-to-floor heights and floor thicknesses. Window sill height to be considered for visual comfort of workers sitting at desks*

6 STRUCTURE AND SERVICES

In determining the shell form, the architect must respect a number of important dimensional disciplines which can be co-ordinated and related by the use of grids, **16.13**, **16.14**. Examples of controlling dimensions are given in **16.15** to **16.22**.

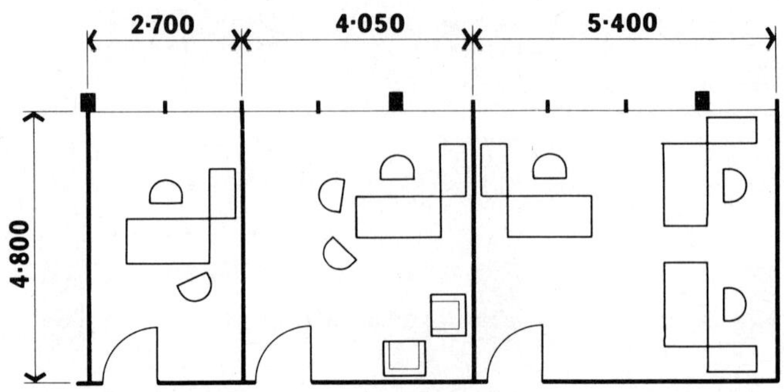

16.16 *Room arrangements. Furniture arrangements in rooms ranging from width of 2·700 m for one person to 5·400 m for three people. The increment is determined by the 1·350 m mullion spacing*

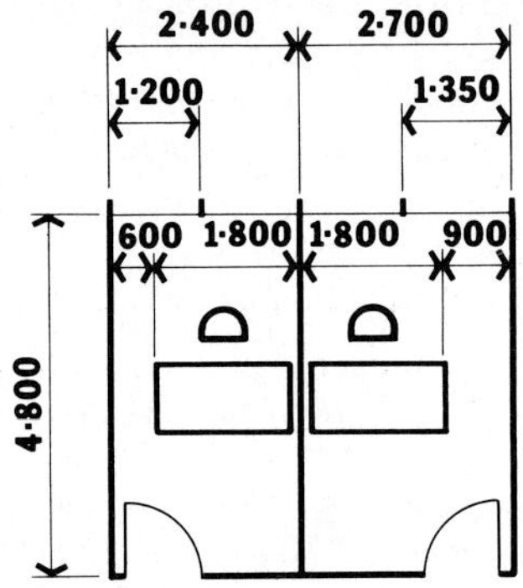

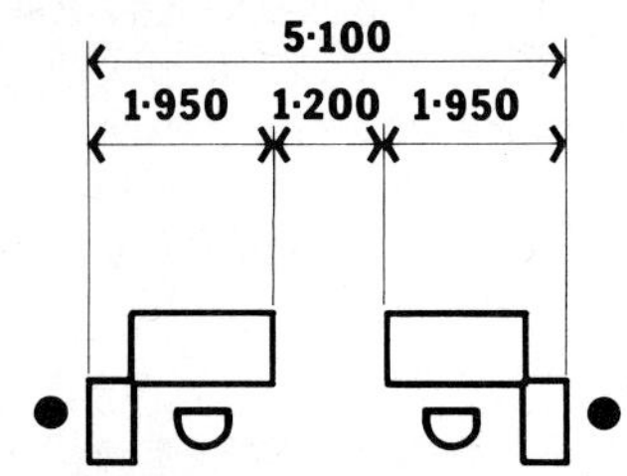

16.18 *How one of the common units of open layout—workplace, aisle and workplace—relates to the spacing of columns*

16.17 *Minimum usable width for a room—comparison of the effect of a 1800 mm desk in two rooms: one 2400 mm, the other 2700 mm wide*

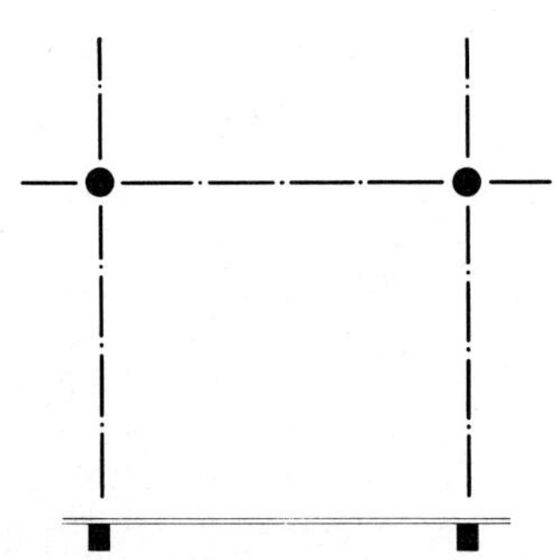

16.19a *Columns outside the envelope*

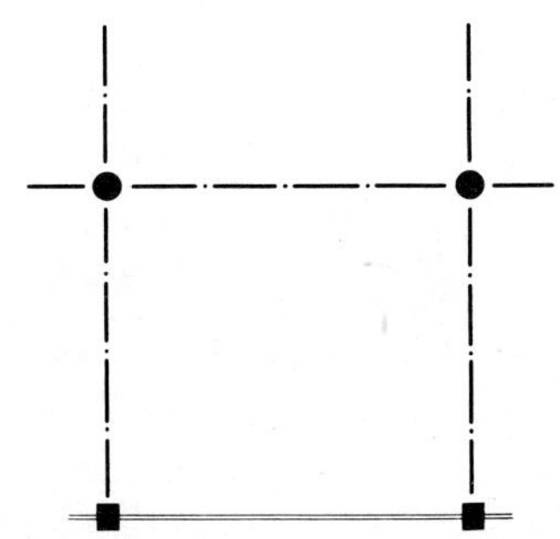

16.19b *Columns within the thickness of the envelope*

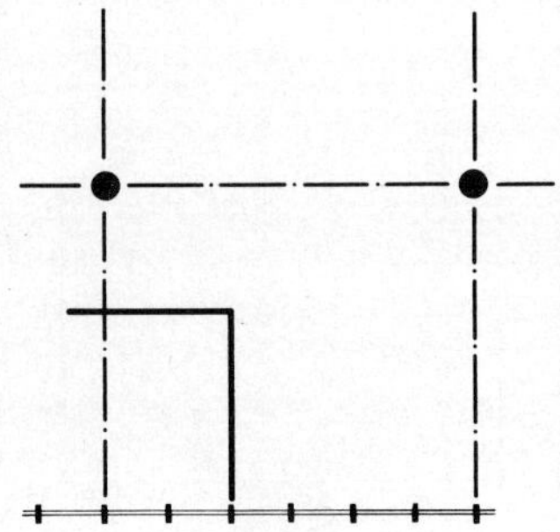

16.19c *Structural mullions—spacing of mullions determines where partitions can be placed*

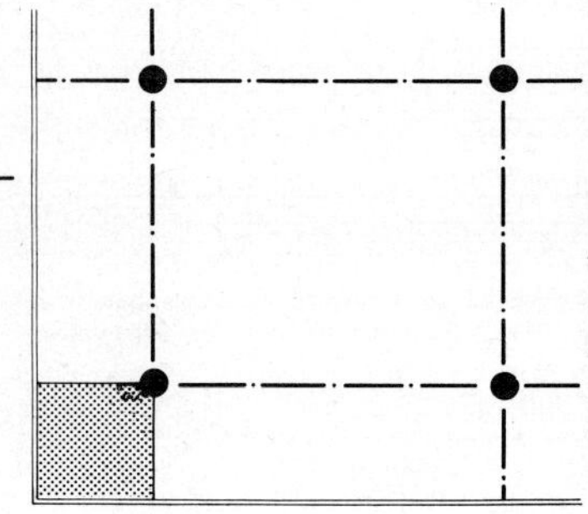

16.19d *Cantilever structure—columns within office space*

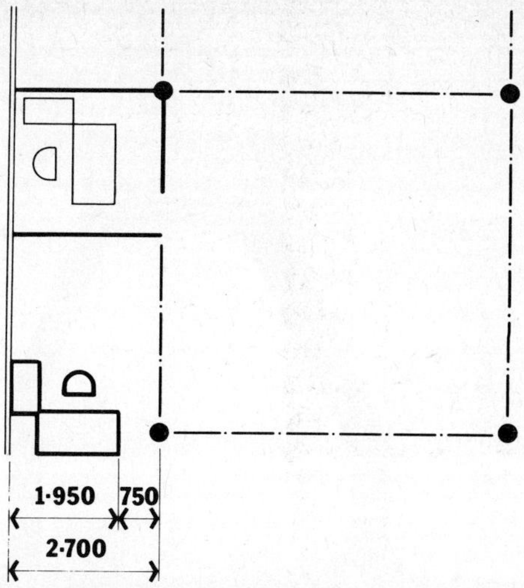

a

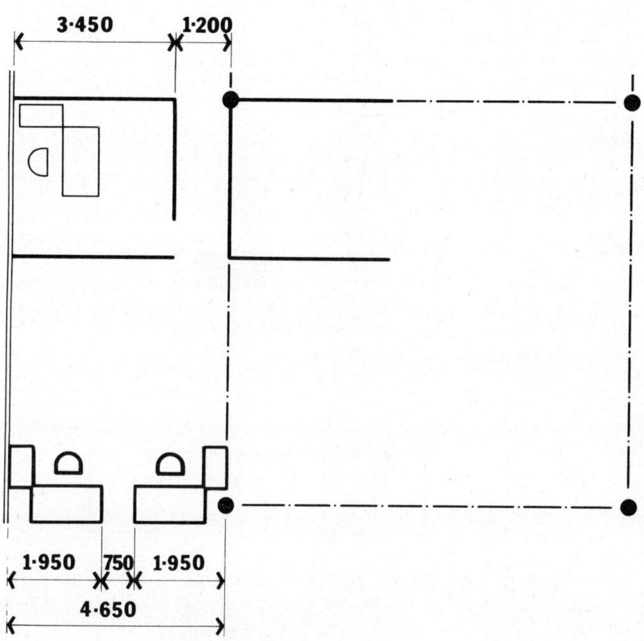

b

16.20 *Planning office layouts with cantilever condition—cantilever should allow either* **a** *minimum room depth or workplace and aisle between column and perimeter. After that condition has been satisfied, the next step* **b** *is to accommodate two workplaces and an aisle, or a room and a corridor: this step means a considerable increase in the cantilever*

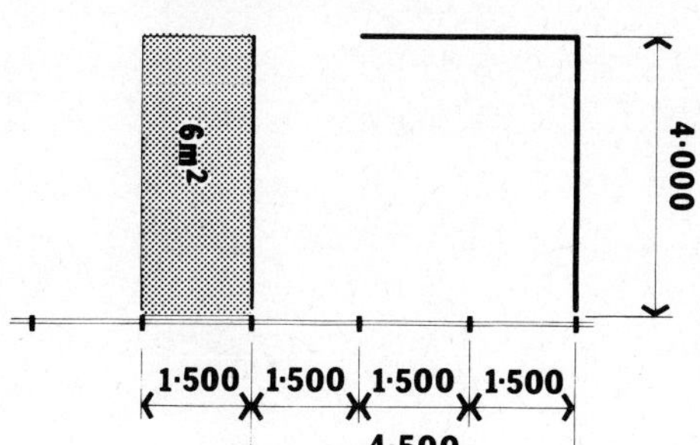

16.21 *Room increasing in increments of 6 m²—mullion modules may produce room sizes which can be increased or decreased only in big increments*

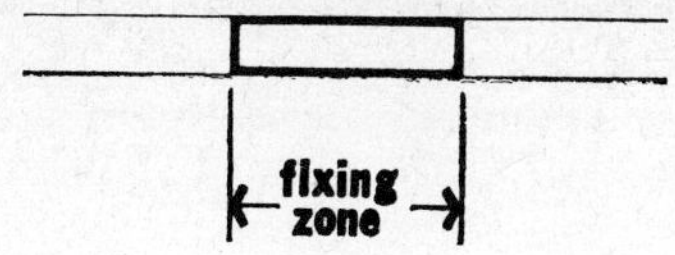

16.22 *Loadbearing external walls: zones of freedom in locating partitions*

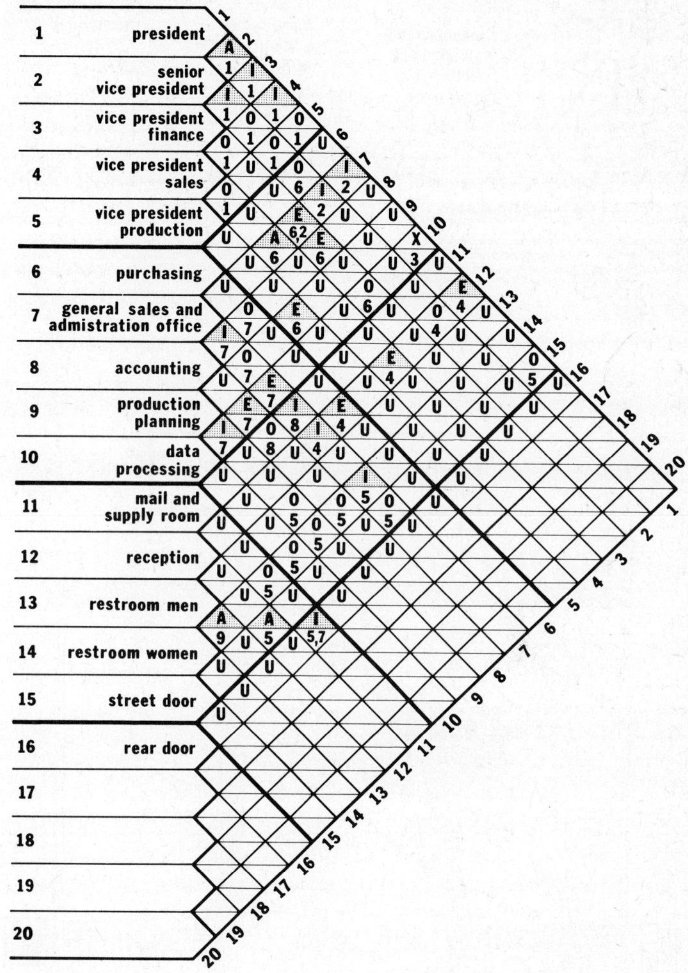

reason

reasons which govern closeness value

code	reason
1	personal contacts
2	use of steno pool
3	noise
4	number of visitors
5	convenience
6	supervisory control
7	movement of paper
8	use of supplies
9	share same utilities

importance

closeness rating

value	closeness
A	absolutely necessary
E	especially important
I	important
O	average satisfies
U	unimportant
X	undesirable

16.23 *Relationship chart for a small firm*

7 THE PHYSICAL LAYOUT

7.01 Relationships

Having established the relationships between groups and departments, **16.23**, **16.24**, the architect will analyse the constraints of the building shell, **16.25**, and will produce the layout which shows groups in relationship to each other as in **16.26**. This will then be worked up into the final layout as in **16.27**, with every workplace and indeed every piece of furniture positioned.

Detailed arrangements will vary with each organisation but there are rules for guidance at two levels: those for planning the whole layout and those for specific furniture arrangements.

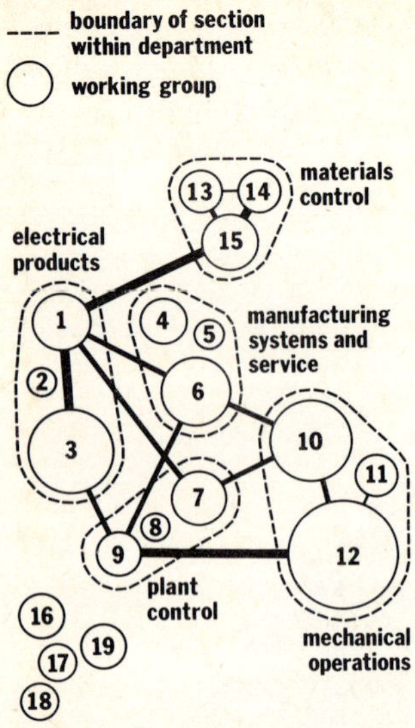

1 Product control
2 Purchasing
3 Engineering electrical production
4 Plant engineering
5 New facilities planning
6 Manufacturing information systems
7 Financial operators
8 Internal control
9 Financial planning and analysis
10 Procurement
11 Plant control
12 Engineering mechanical operations
13 Invoice control
14 Staff
15 Order schedules

16.24 *Adjacency diagram depicting interaction between workgroups within a department*

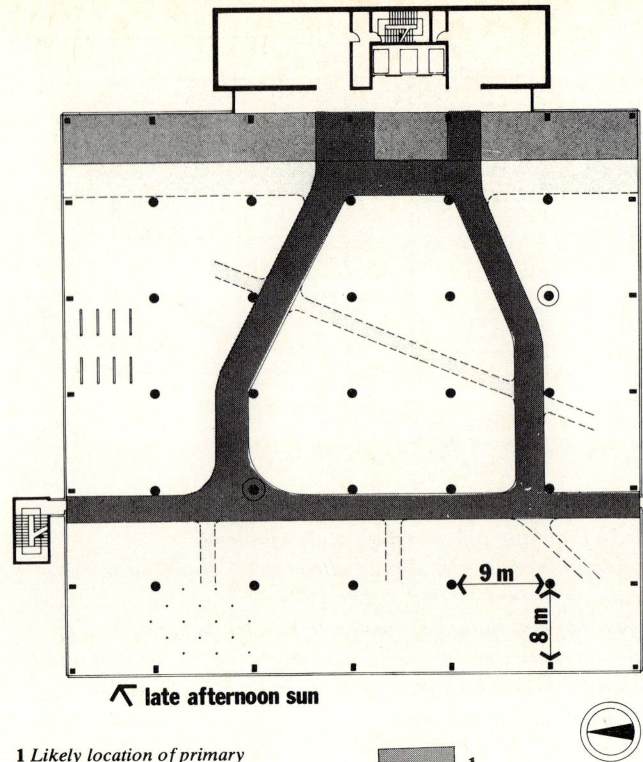

1 Likely location of primary circulation
2 Wet column: good location for wet areas and vending machines
3 Power and telephone grid outlets in floor (2·2 m centres)
4 Lighting 8 ft tubes at 5 ft centres
5 Circulation
6 Offices

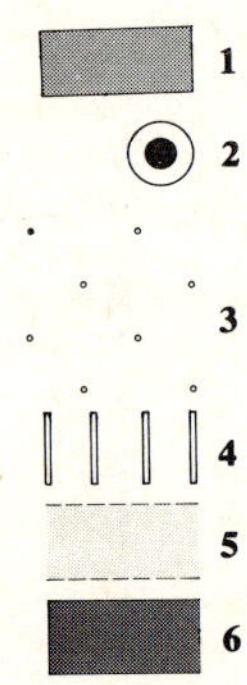

16.25 *Analysis of the constraints the building may impose on the organisation*

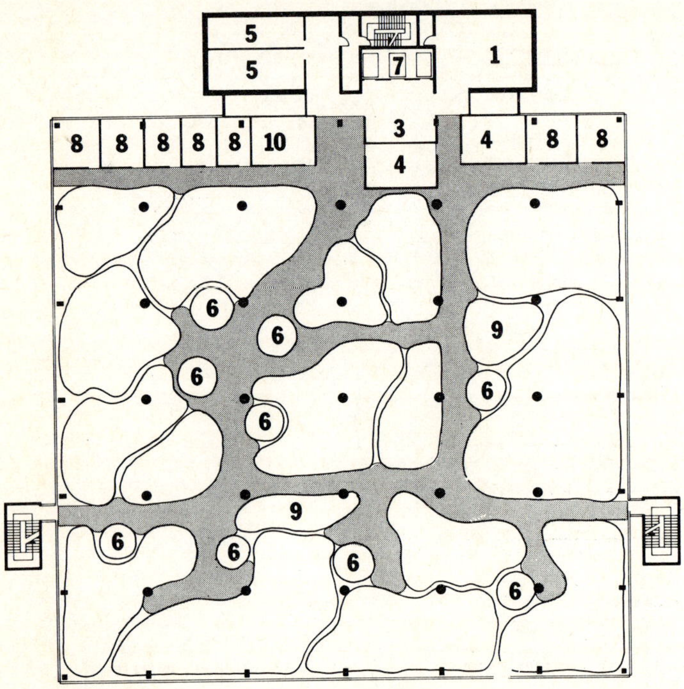

1 Photocopy
2 Circulation
3 Reception
4 Conference
5 Lavatories
6 Meeting areas
7 Lifts
8 Office
9 Break areas
10 Storage

16.26 *Layout of working groups*

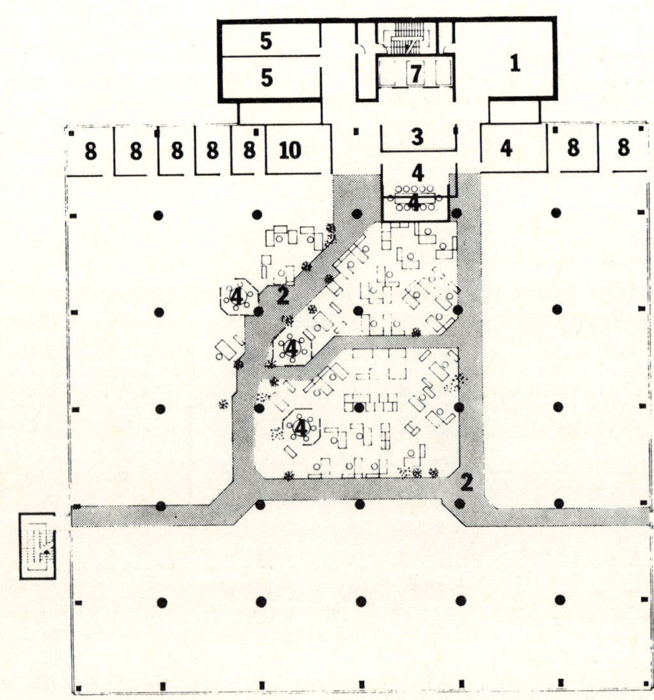

16.27 *Final layout, showing some furniture—refer to* **16.26** *for key*

7.02 Rules for planning the whole layout

Working groups will be arranged to follow patterns of work flow and interaction. Careful distinction is to be made between:

- primary circulation which generally links access and egress points and the major groups (not less than 2 m wide—increasing with the volume of traffic)
- secondary circulation which connects groups not adjacent to primary routes with the primary routes (not less than 1·5 m wide) and
- tertiary circulation which is the circulation within the working groups (not less than 750 mm).

Group territory will be well defined and there will be no cross circulation between groups. All routes will be kept 'clean' and defined by planters and screens.

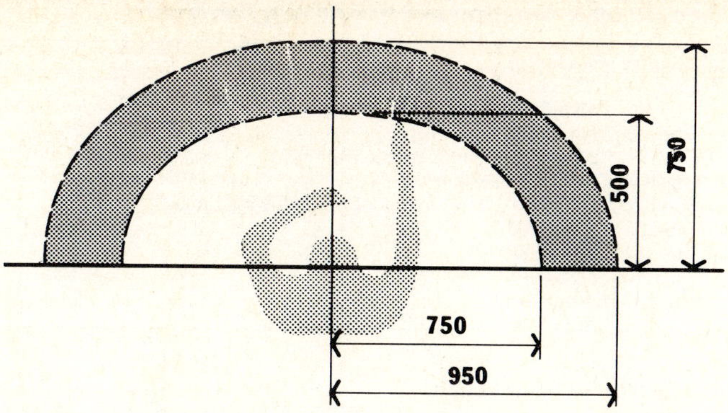

16.28 *Average reach of person sitting at desk. To reach outer area, the user has to bend but not stand up*

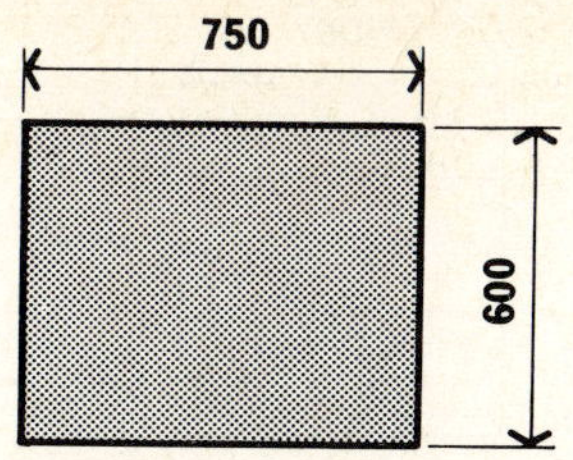

16.29a *Basic space for writing and typing*

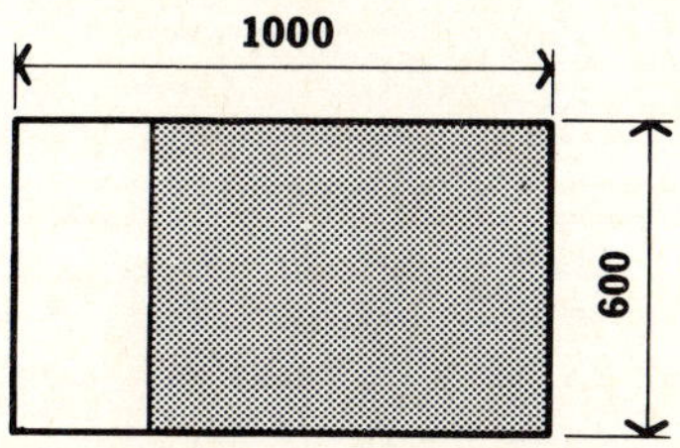

16.29b *With space for paper on one side*

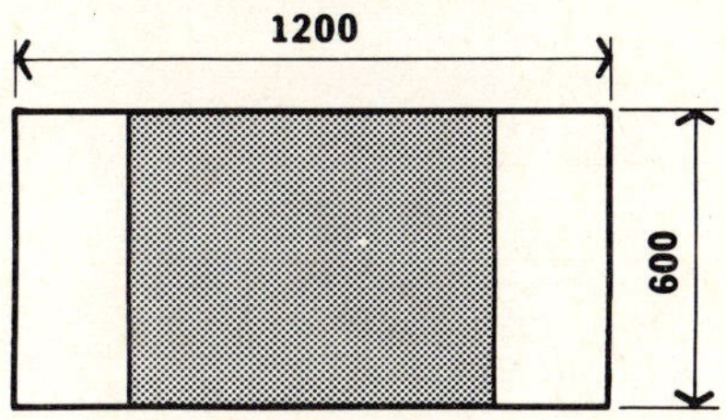

16.29c *With space for paper on both sides*

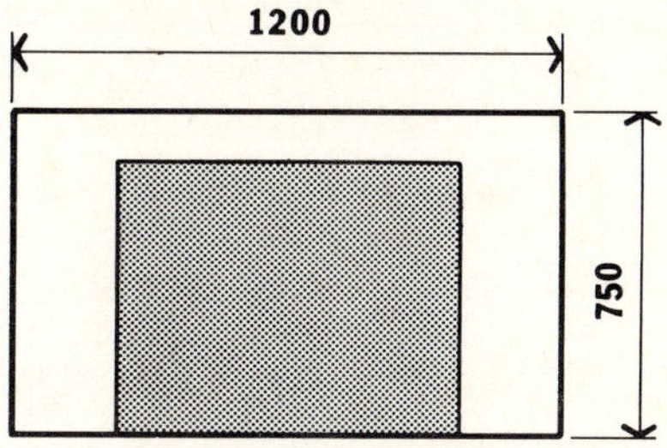

16.29d *Paper plus space for pens and telephones*

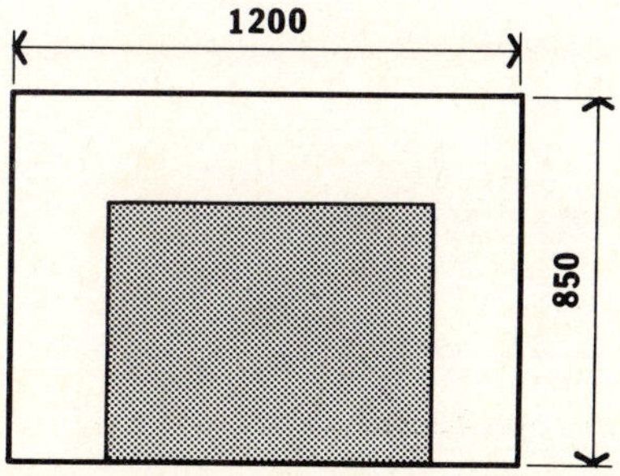

16.29e *Generous amount of space for paper*

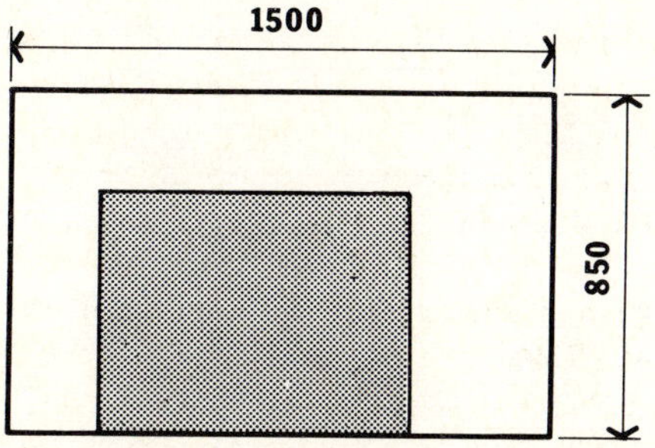

16.29f *Space for papers plus area for references*

7.03 Rules for specific furniture arrangement

Generally;

- no person will directly face a circulation route, another person or storage furniture
- each person should be able to see who is coming
- functions with the highest visitor contact will be placed closest to primary or secondary routes
- planting should always satisfy a particular functional need such as a landmark, screening, or as a holding element on circulation routes to block or direct traffic.

Table X Possible desk sizes based on figures 16.29 and 16.30

Executive/manager	1500 × 750 mm plus extension 750 × 500 mm
Clerk	1300 × 700 mm or 1000 × 700 mm plus extension 1000 × 500 mm
Secretary/typist	1300 × 700 mm plus extension 1000 × 500 mm
Typist	1200 × 700 mm

Table XI Office desk sizes
BS 3893:1965, Specification for Office Desks, Tables and Seating, is currently under revision. The following gives the generally available metric sizes

Type	Length	Depth	Height*
	2000	1000	
Executive	1800	825	650–750
		900	
General purpose	1500	750	650–750
Double pedestal	1350		
General purpose	1200	750	650–750
Single pedestal	1000	450	

* Satisfactory working height varies from 635 mm for a small woman to 760 mm for a large man. Many desks allow for adjustment to height. Where typewriters or desk-top computers are used the desk top should be 50 mm lower or incorporate a well at this level.

Table XII Office table sizes

Lengths	Widths	Height
2000	900	710–760
1800	750	
1500	600	
1200		
1000		
900		

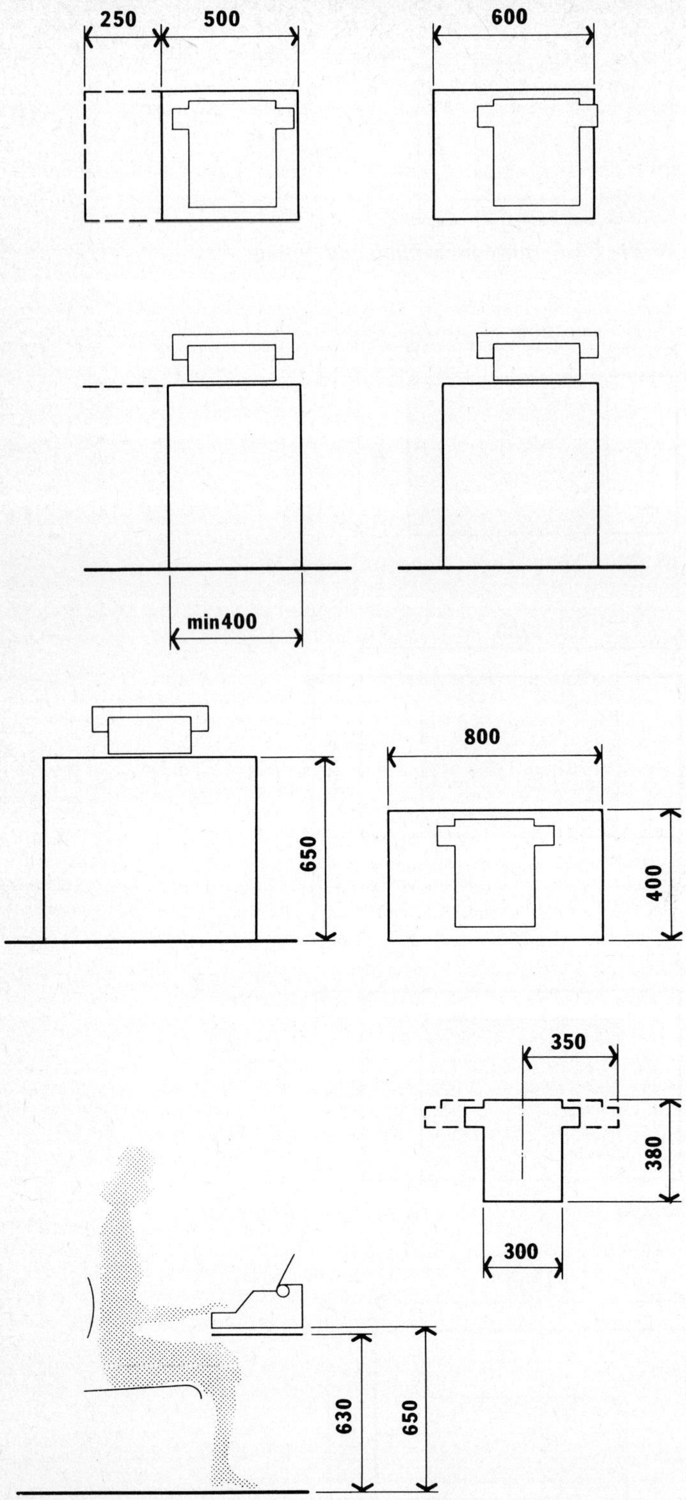

16.30 *Desks suitable for typing (Swedish)*

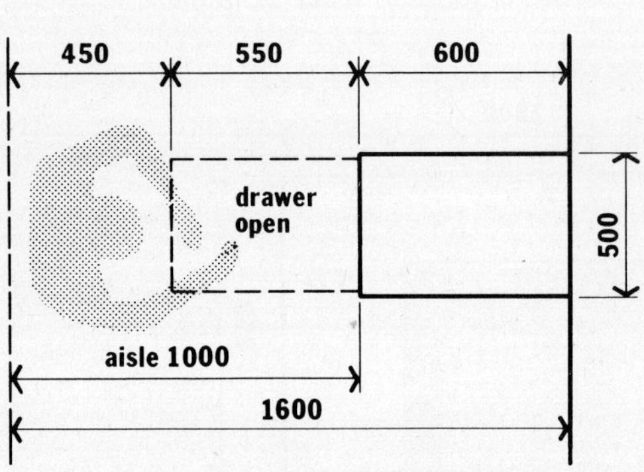

16.31 *Space requirements of drawer filing cabinet*

Table XIII Office chairs (translated from BS 3893) Sizes in mm

a Executive and clerical

Heights:	432 desirable for fixed chairs (430)
	432–508 for adjustable chairs (430 to 510)
Widths:	minimum 406 (400)
Depths:	356–470; recommended 381 (380)

b Typists and machine operators

Adjustable through 102 (100) in one of the following ranges:
406 to 508 (400 to 500)
483 to 584 (480 to 580)
559 to 660 (560 to 660)
635 to 737 (640 to 740)

8 OFFICE FURNITURE

8.01 Desks, tables and chairs

Average reach of a person sitting at a desk is given in **16.28**, and typical desk top requirements in **16.29** and table X. Desks for typing are given in **16.30**. These are Swedish standards based on A4 paper.

The British Standard Specifications for office desks, tables and seating are covered in BS 3893:1965 which is under radical revision. Tables XI and XII and XIII indicate interim sizes available from office equipment suppliers.

8.02 Filing cabinets

Space requirements for drawer cabinets and lateral filing are shown in **16.31** and **16.32**, with circulation requirements in **16.33**. See also minimum space requirements listed in table XIV.

8.03 Space requirements

Space required per employee will depend on type of work; use of equipment or machinery; degree of privacy; and storage needs.

8.04 Work spaces for individuals

Spaces required for different desk and table layouts are shown in **16.34** to **16.39** and calculated as in para 8.05

8.05 Desk spacings and layouts

If desk		= 1500 × 750 mm	
	x	= 2400 mm	and area = 3·96 m²
with file	y	= 2850 mm	and area = 4·71 m²
If desk		= 1200 m × 750 m	
	x	= 2100 mm	and area = 3·47 m²
with file	y	= 2550 mm	and area = 4·20 m²
If desk and table		= 1500 × 750 mm	
	x	= 2400 mm	and area = 5·76 m²
with file	y	= 2850 mm	and area = 6·84 m²
If desk and table		= 1200 × 750 mm	
	x	= 2100 mm	and area = 5·04 m²
with file	y	= 2550 mm	and area = 6.12 m²
If desk		= 1500 × 750 mm	
	x	= 3150 mm	and area = 5·20 m²
If desk		= 1200 × 750 mm	
	x	= 2850 mm	and area = 4·70 m²
If desk		= 1500 × 750 mm	
	x	= 3600 mm	and area = 5·94 m²
If desk		= 1200 × 750 mm	
	x	= 3300 mm	and area = 5·45 m²
If desk		= 1500 × 750 mm	
	x	= 3225 mm	and area = 5·33 m²
If desk		= 1200 × 750 m	
	x	= 2925 mm	and area = 4·83 m²
If desk		= 1500 × 750 mm	
	x	= 2775 mm	and area = 4·58 m²
If desk		= 1200 × 750 mm	
	x	= 2475 mm	and area = 4·10 m²

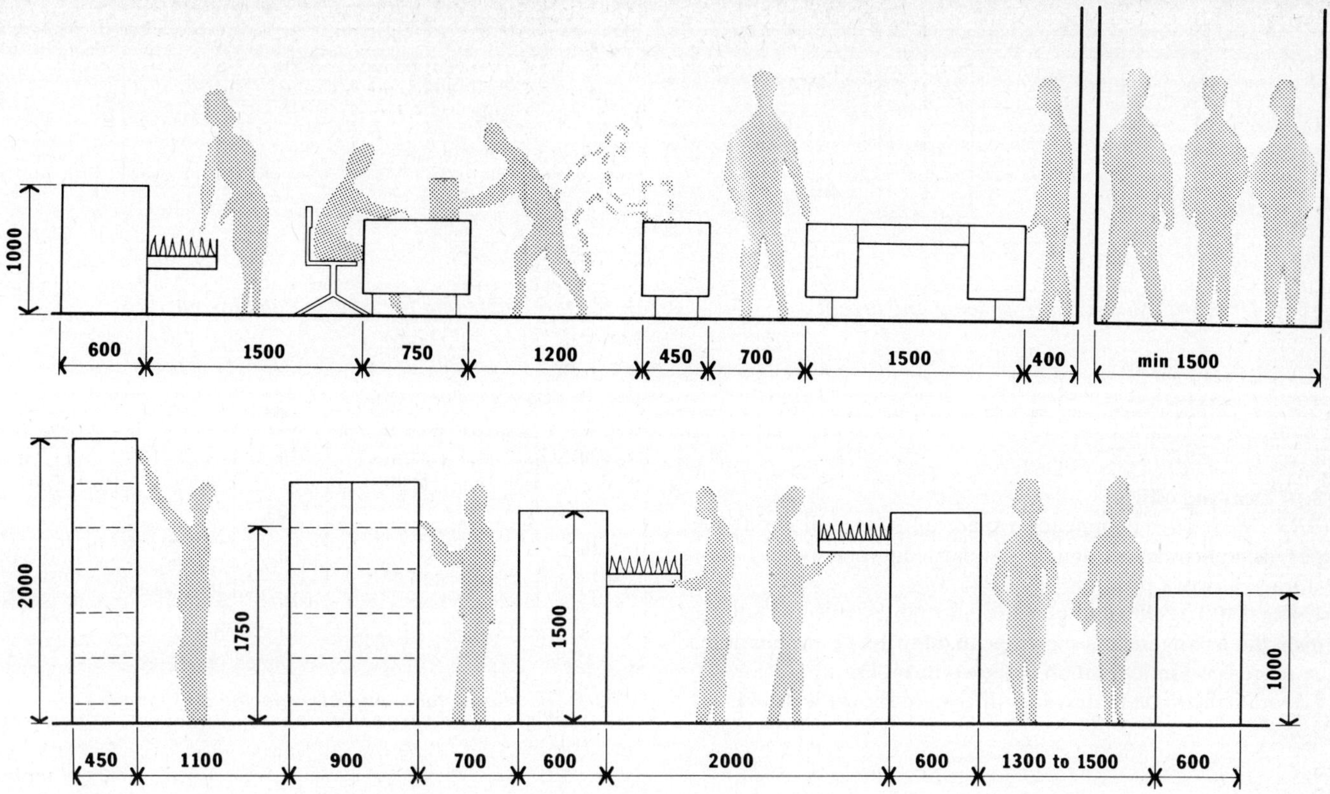

16.33 *Typical space and circulation requirements of filing and other office equipment*

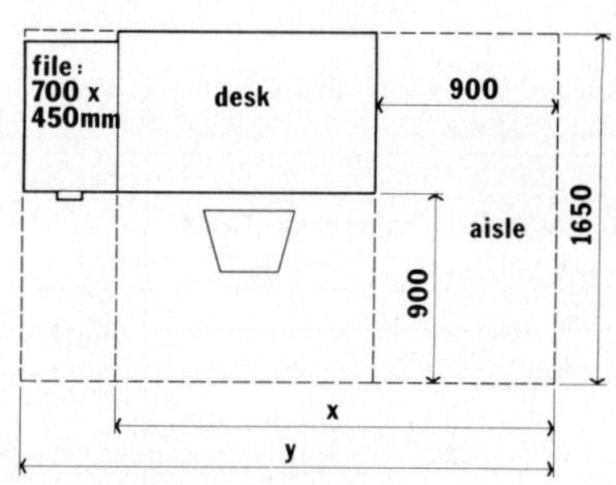

16.34 *Desk and file spacing and layout*

8.06 Desk spaces and layouts

Table XIV gives recommendations for spacing office furniture.

Table XIV Recommendations for minimum spacing of office furniture (mm)

a Distance from back to front of desks in a row (chair space)	
When each desk is on an aisle	900
When each desk is not on an aisle	900 to 1370
b Aisle widths	
Major aisle (large general office)	1500
Normal general office aisle	900
Minor general office aisle	750
c Distance required in front of filing cabinets	
Single row	900
Two rows facing	1220
When filing cabinets face an aisle, the width of the cabinets when open (1370 mm) should be added to the normal aisle width	
d Distance required in front of shelving	
Two-row facing	750 to 900
When shelving faces an aisle, the width of the shelving should be added to the normal aisle width	

Note Desks should not be placed tightly against and facing a solid wall or opaque glass partition.

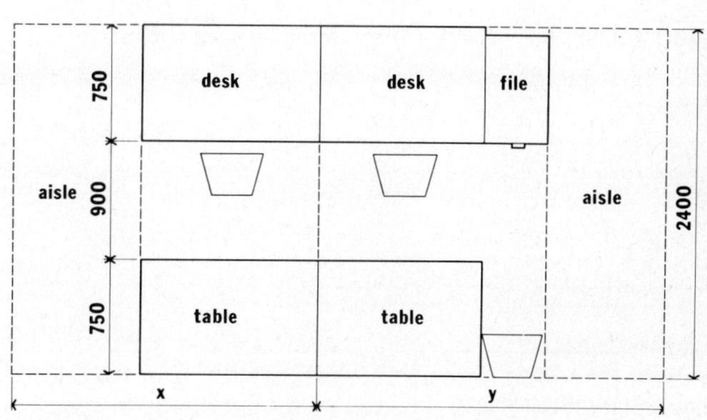

16.35 *Desk with tables, file and chair, spacing and layout*

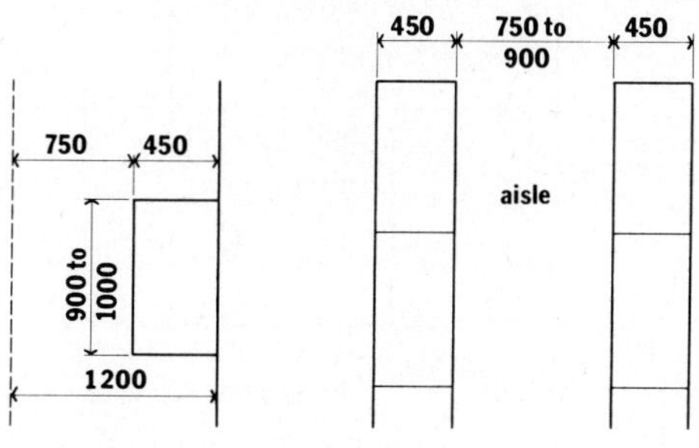

16.32 *Space requirements of lateral filing units*

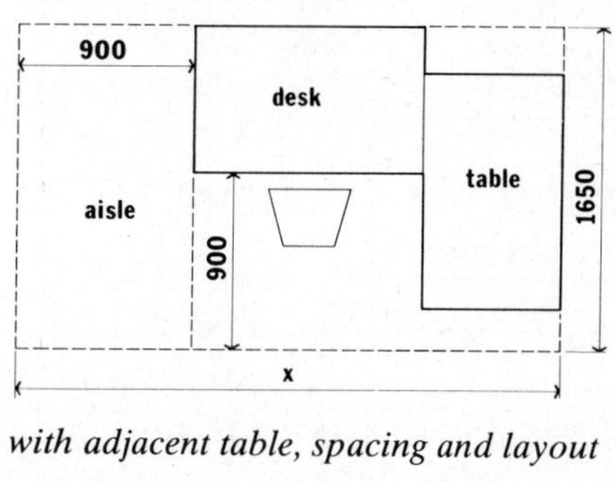

16.36 *Desk with adjacent table, spacing and layout*

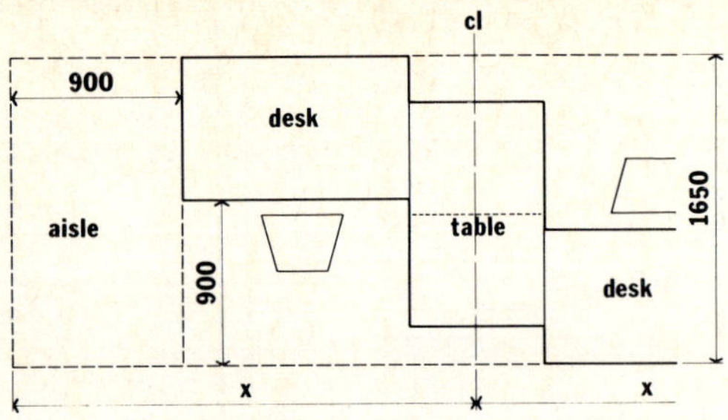

16.37 *Desk with shared table, spacing and layout*

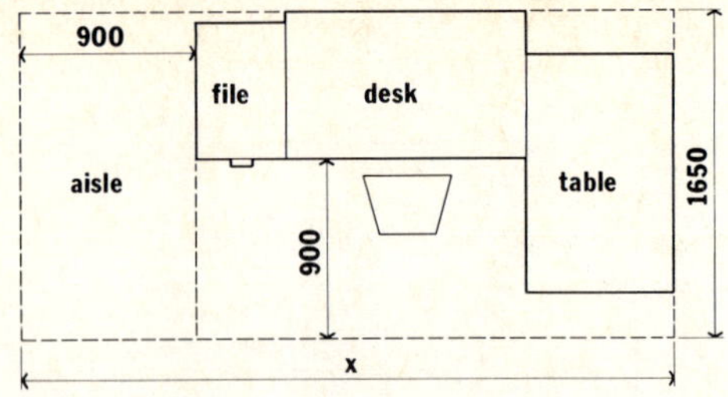

16.38 *Desk, table and file, spacing and layout*

8.07 Drawing offices

BRS Digest 20 recommends a space allowance of about 7 m² per draughtsman assuming standard equipment and some storage at work position.

Using 1500 × 900 mm benches and tables with a 1800 mm aisle the arrangements shown in **16.40** to **16.42** are possible.

A more economic solution is shown in **16.43.**

Layouts of complete drawing offices are shown in **16.44**.

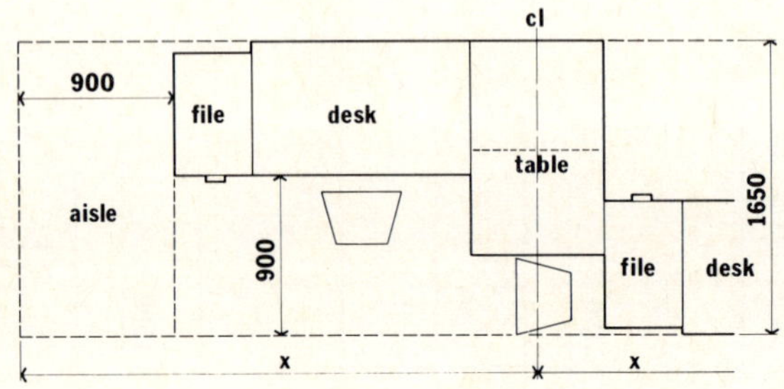

16.39 *Desk, shared table and file, spacing and layout*

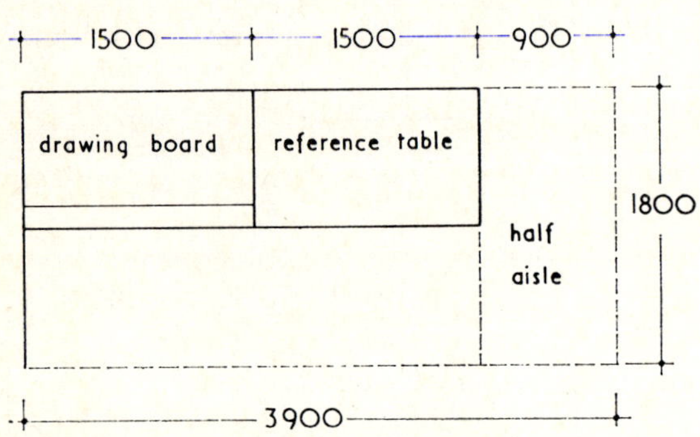

16.40 *Drawing board with front reference: area 7·0 m²*

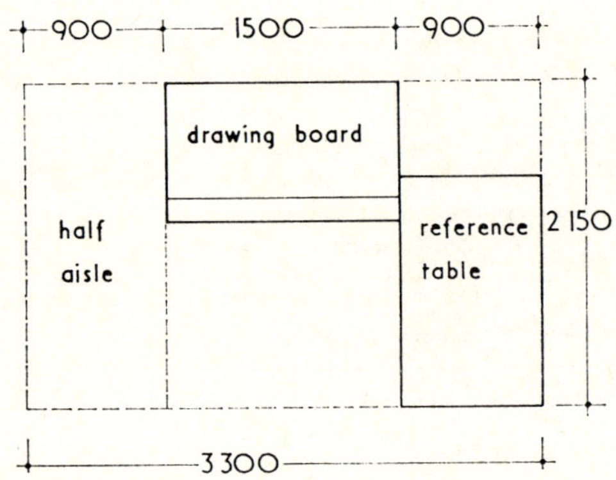

16.42 *Drawing board with side reference: area 7·1 m²*

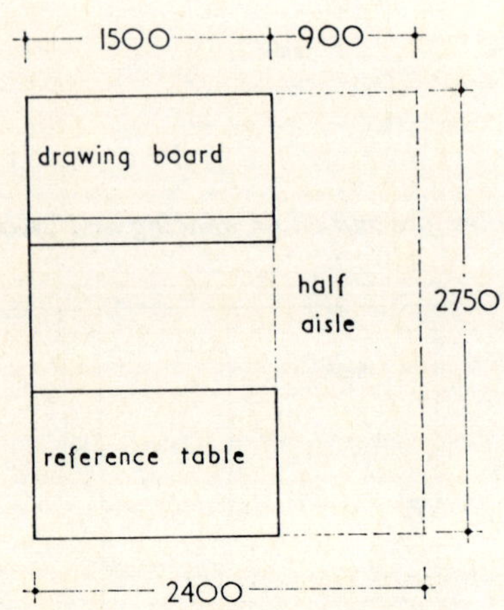

16.41 *Drawing board with back reference: area 6·6 m²*

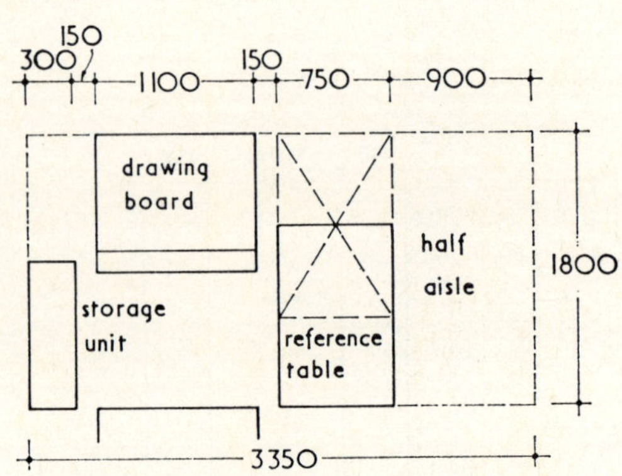

16.43 *Drawing board with mobile reference: area 6·0 m² (Building Design Partnership design)*

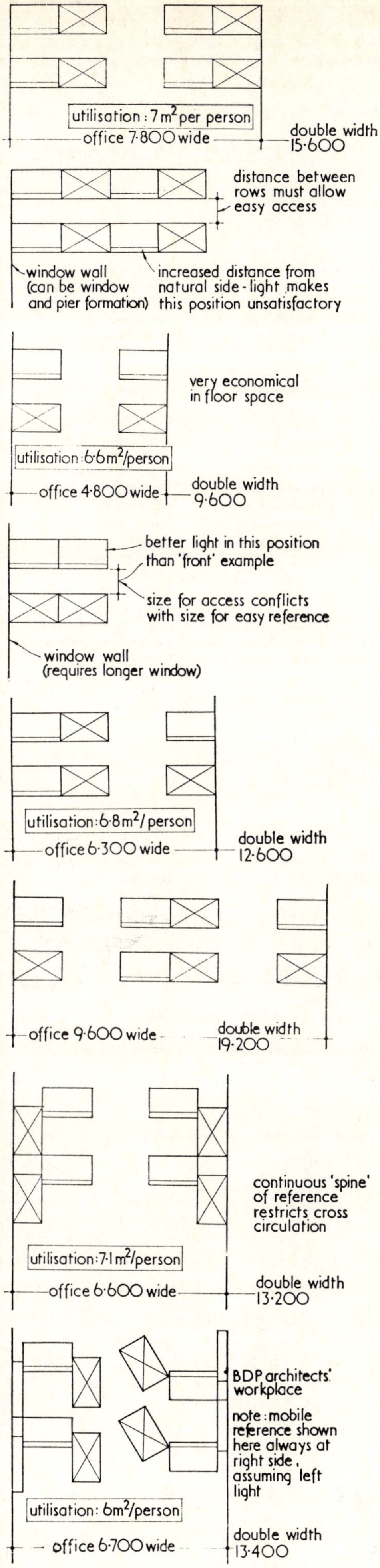

16.44 *Layout of drawing offices*

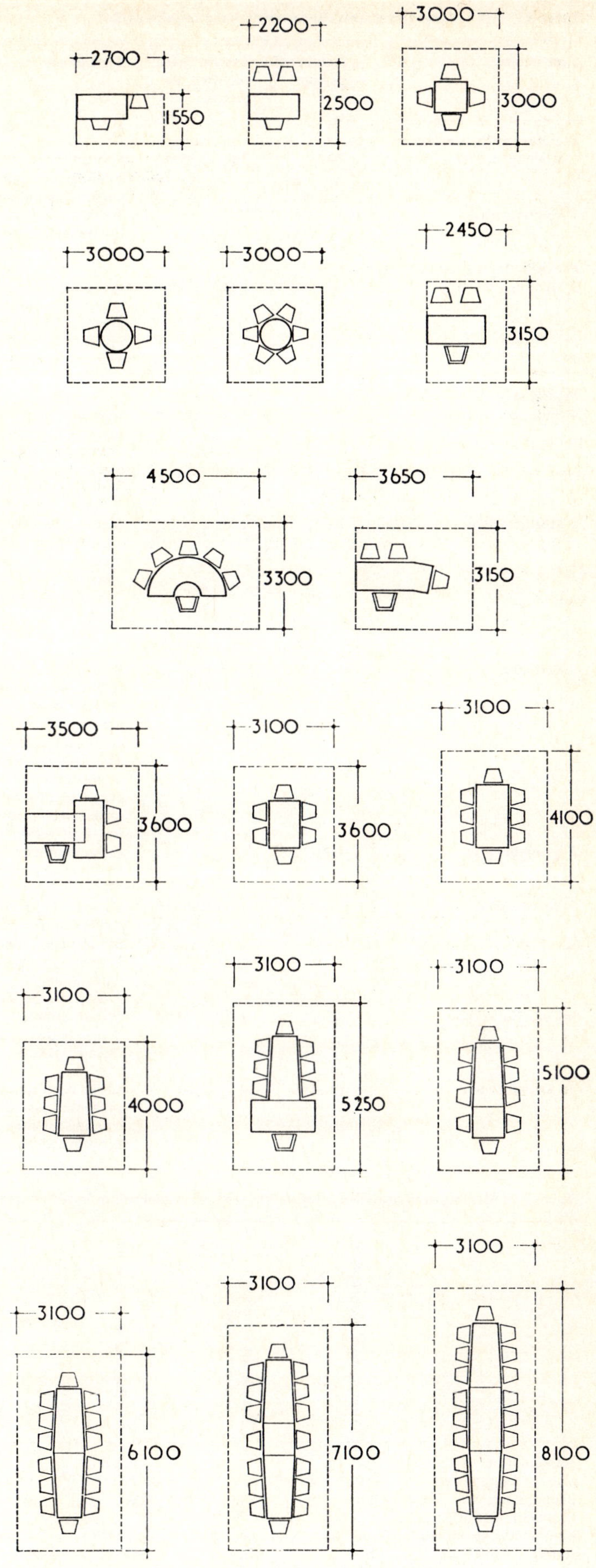

16.45a *Space requirements for informal meetings*

8.08 Meeting places and conference rooms

Typical requirements for meeting places are shown in tables XV and XVI. Conference room layouts are shown in **16.45**.

Table XV Types of meeting spaces, characteristics and requirements

Type of space	Number of persons	Typical space required per person	Type of use	Provision and equipment	Location
Provision at the workplace					
1 Meeting at desk	2–3	2·00–2·75 m²	Short discussions, briefing subordinates, personal interviews	1 or 2 visitor's chairs at work station	Located in screened area if in planned open office environment
2 Meeting area	4		Working discussion with the members of staff or visitors	Conference table and chairs Related equipment—pinboard, chalk board	
Provision for a group of workplaces					
3 Meeting area	6–8	1·50–2·25 m²	Working sessions between members within group or personnel from outside involved with same project. May last several hours	Conference table and chairs with some screening from surrounding work stations. Related equipment: flip charts, pin up space, chalk board. Provision of permanent notice board or chart board for use of group	Located in group area adjacent to primary circulation, to limit disturbance of individuals
Provision for all members of staff					
4 Interview room	2–3	1·50–2·00 m²	Interviewing personnel or sales representatives. Discussions with members of public. Short periods of use up to ³/₄ hour	Aural and visual privacy required	Adjacent to main entrance and departments with major usage. May require waiting area adjacent if used frequently
5 **Meeting room**	8–12	1·50–2·00 m²	Meetings with outside visitors or internal policy making and planning meetings. 2–3 hour meetings	Slides, overhead projector, flip charts, dimmer lights, good ventilation. Storage for drinks, audio-visual equipment	Ease of access to all departments. Easily serviced with refreshments. Access for outside visitors without going through work areas. Coats area adjacent
6 **Rest area**	12–15	2·25–4·00 m²	Primarily used for refreshment breaks, but may also be the area where general notices, scheduling charts etc can be displayed. This area may become an important point for the exchange of information and ideas. Used throughout day for short periods	Vending machines, stand-up counters, low tables and easy chairs. Display board. Screening from work areas	Adjacent to cloaks, wcs and rest room. Equally accessible to all personnel on each floor
7 **Assembly area**	100–150		Infrequent meetings. Involvement of all staff		May use cafeteria or recreation space
8 **Board room**	16–24	1·50–2·00 m²	Formal board meetings, signing of contracts. Management meetings. Business lunches and entertaining. 2–3 hour meetings	Formal layout. Audio-visual equipment. Good ventilation essential. Telephone extension. Space and facilities for stenographer	Anteroom (for refreshments and leaving coats) attached. Easy access for refreshments. Two visits
9 **Conference room**	15–20	1·50–2·00 m²	Presentations. Working discussions with outside visitors	Audio-visual equipment. Dimmer lights and black out. Storage for equipment and furniture. Allow sufficient space for alternative layouts	Easy access for visitors
10 **Lecture room**	50–100		Large conferences, presentations, lectures and training sessions furniture display systems	Closed circuit tv system. Control room for projector, lighting, curtains, tv and audio systems. Storage space for Several entrances	Adjacent area for audience to assemble before meeting.

Table XVI Meeting area provision for different types of organisations based on a selection of projects

	Meeting area at workplace	**Meeting area serving a group of workplaces**	**Meeting rooms 6–8 persons**	**Lounge/rest areas**	**Meeting room 12–16 persons**	**Meeting room 16–20 persons**	**Conference room 22–28 persons**	**Lecture room 100–150 persons**
Headquarters accommodation for an engineering organisation (population 1200)	1 per 15 office staff	1 per 10 office staff	1 per 80 office staff	1 per 280 office staff	1 per 120 office staff	None	1 per 1200 office staff	1 per 1200 office staff
Manufacturing administrative organisation (population 400)	1 per 40 office staff	1 per 12 office staff	1 per 45 office staff	1 per 80 office staff	None	1 per 60 office staff	1 per 200 office staff	None
Headquarters accommodation for a clerical organisation (population 1400)	1 per 18 office staff	1 per 26 office staff	1 per 55 office staff	1 per 400 office staff	1 per 280 office staff	1 per 230 office staff	1 per 1400 office staff	None
Consultancy organisation (population 80)	1 per 16 office staff	1 per 20 office staff	None	None	None	1 per 80 office staff	None	None

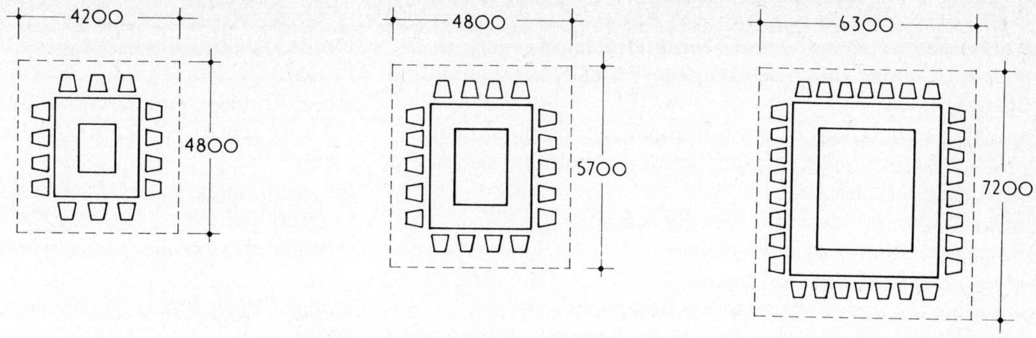

16.45b *Space requirements for formal meetings*

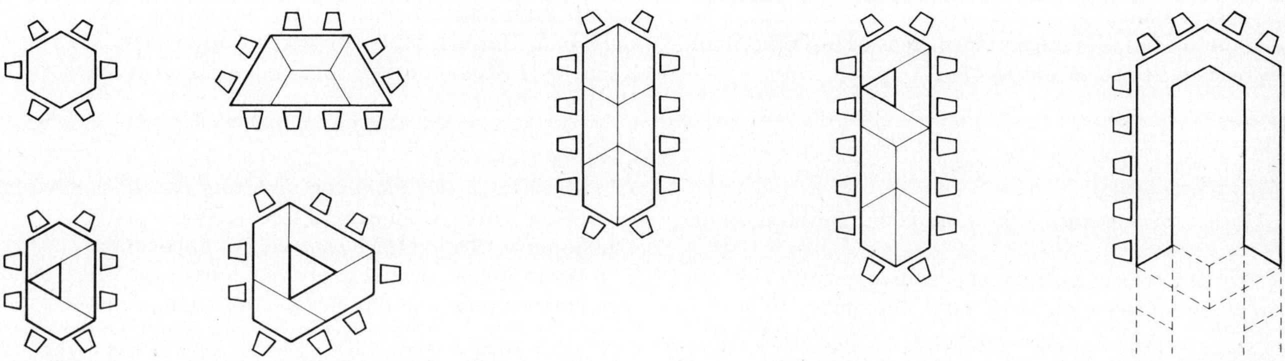

16.45c *System of unitary furniture to allow for alternative layouts*

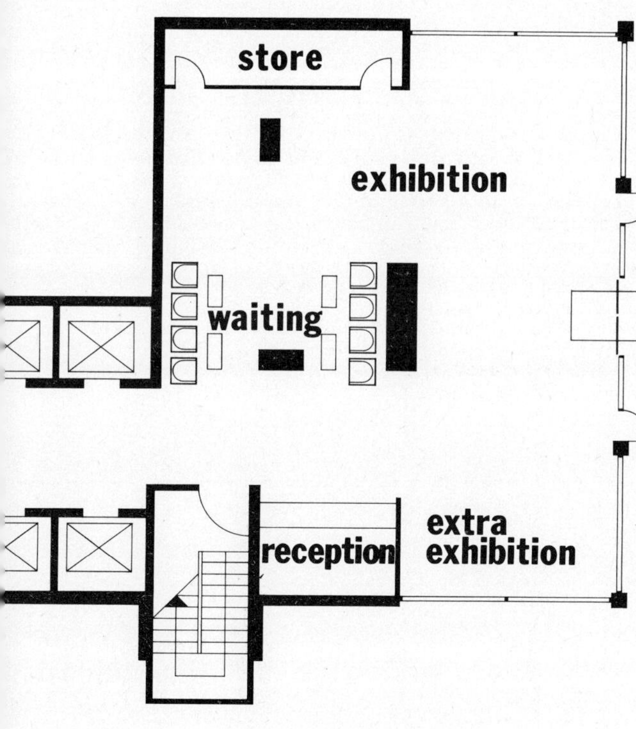

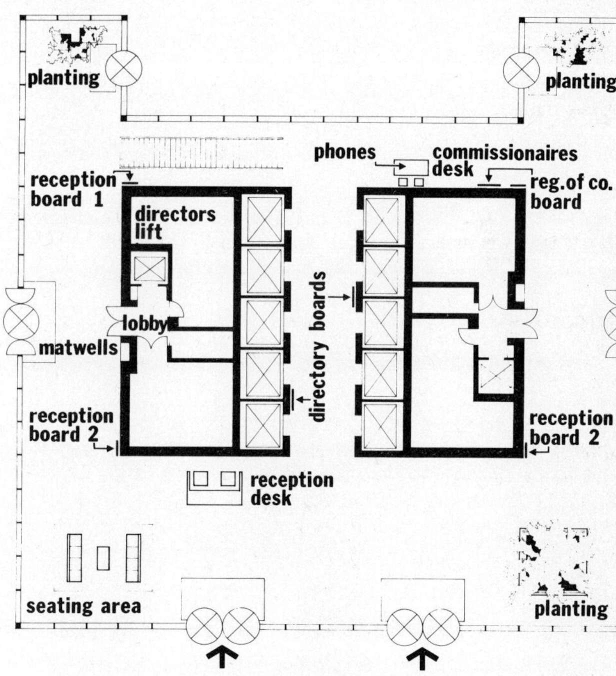

16.46 *Reception area in a government building provides good access for the disabled but little or no draught or dirt control. Extensive area provides a variety of arrangements of exhibition and waiting areas. Reception point, however, is partitioned off for security and to minimise the effect of draughts. This means that from the normal seated position the receptionist has no visual control of part of the exhibition area, the lifts and one entrance door. There is no easily accessible lavatory—because of the occupancy pattern of the building, the closest wc is on the tenth floor!*

16.47 *The expensively furnished reception area of an insurance company in London. Although at one side there is a commissionaire's desk and at the other a receptionist's, these two points cannot supervise at least two of the six entrances. At the main entrance the relationship between the receptionist, doors and waiting area is good. No exhibition space is provided. The floor finish is durable but inadequate matting was provided in the first instance, and had to be increased later*

9 ENTRANCE AND RECEPTION AREA

There are three main types of reception facility:

- *main* (buildings in single occupancy) usually prestigious and at street level
- *deferred* (buildings in multi-occupancy) commissionaire or just signboard at main entrance; reception points as tenancy or departmental basis
- *staff* and/or goods, ie for security reasons.

Reception areas normally involve:

- receiving staff, services, security
- waiting discussion space, toilets, telephone
- exhibition and display—what scale, permanent/temporary, who for?
- storage (goods and coats)
- lifts and/or stairs.

The major elements in the entrance areas of two office buildings are shown in **16.46** and **16.47**.

10 BIBLIOGRAPHY

Axel Boje *Open planned offices*, London, Business Books, 1971

P. R. Boyce User's assessment of a landscaped office, *Journal of Architectural Research,* vol 3, no 3, September 1974

A. Dickens *Land use and built form studies working paper 35,* Cambridge University School of Architecture, 1970

F. Duffy, C. Cave and J. Worthington (eds) *Planning office space*, London, Architectural Press, 1976

F. Duffy and A. Wankum *Office landscaping,* London, Anbar, 1969

Peter Manning (ed) *Office design: a study of environment*, Liverpool University, 1965

Office space: a primer for users and designers, London, HMSO, 1976

Planned open offices, a cost benefit analysis, London, DOE, 1971

John Pile *Interiors third book of offices,* Whitney Library of Design, New York, Watson Guptill, 1976

Robert Probst and Michael Wodka *The action office acoustic handbook,* Herman Miller Research Corporation, 1975

Kenneth H. Ripnen *Office space administration,* New York, McGraw Hill, 1974

M. F. Schmertz *Office building design,* 2nd edition, New York, McGraw Hill, 1973

Lila Shoshkes *Space planning, designing the office environment,* New York, Architectural Record Books, 1976

Christoper A. Sykes *Office planner,* A4 Publications, 1976

T. J. Wyatt A comparative analysis of four recent office buildings, *International Lighting Review,* vol 15, no 3, 1974

17 Retail trading

David Gosling

David Gosling is Professor of Architecture at Sheffield University

CI/SfB 34
CI/SfB (1976 revised) 34

Contents

1 INTRODUCTION

1.01 Type of facility

Shopping facilities fall in to one of six types:

- markets
- individual shops in shopping (or other) streets
- department stores
- supermarkets
- shopping centres
- hypermarkets.

1.02 Self or staff service

There are two contrasting philosophies: service by staff, and self-service. Design will depend on the particular method to be used, the decisions in each case relating to the degree of pre-packing of the product on sale. Virtually all items can now be obtained in the pre-packed form and this has extended the range of self-service facilities. However, the cost of packing in many cases exceeds the value of the product packed and it may be that future trends may return to staff service.

2 MARKETS

2.01 Open markets

Some markets are out of doors, in streets and open spaces. Layouts will depend on the size and shape of the space. The simplest solution is where the stalls are ranged in line along the kerbside, **17.1a**.

2.02 Covered markets

In covered markets the stalls around the perimeter are the more sought-after. Servicing of all stalls should cause minimal interference with the customer, **17.1b**.

3 SHOPS

3.01 Shops in new developments

Where shops are provided in the ground floor of new office and residential developments, the scale and grid are usually decided by the requirements of the accommodation above. However, for most types of shops the layouts can be varied to fit the size and shape of the space available.

3.02 Shop layout

Figures **17.2** to **17.5** indicate the nature of the layout for a particular trade, but individual shops can vary considerably from the norm. However, most shops will require some non-sales area for storage and must provide adequate lavatory and washing accommodation.

Rear access for loading and unloading is not always possible. If servicing has to be from the front, the arrangement of show-cases and counters must allow for this. The recommended minimum shop width is 5·4 m.

Various types of furniture and fittings for shops are shown in **17.6** to **17.7**.

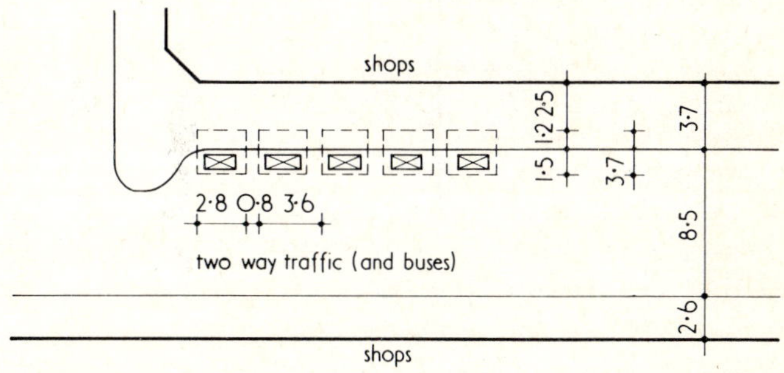

17.1 *Markets:*
a *Roadside: North End Road, Fulham*

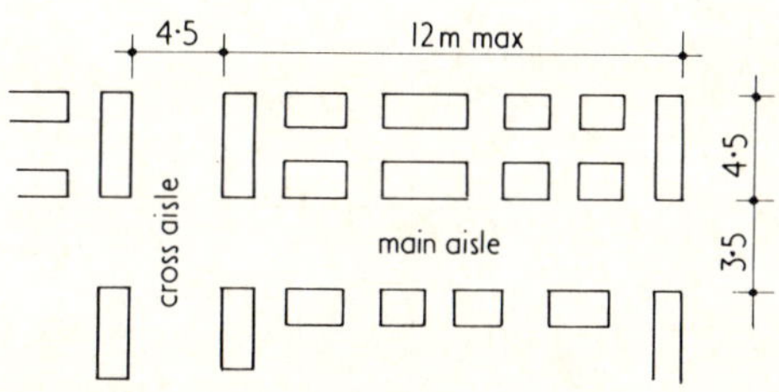

b *Part of typical market square*

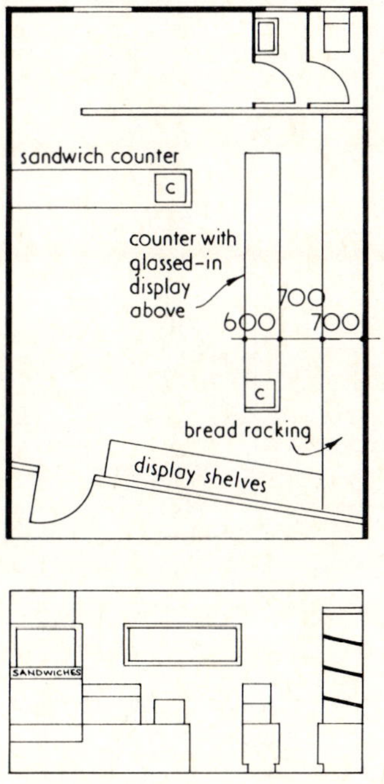

17.2 *Baker's shop, plan and section*

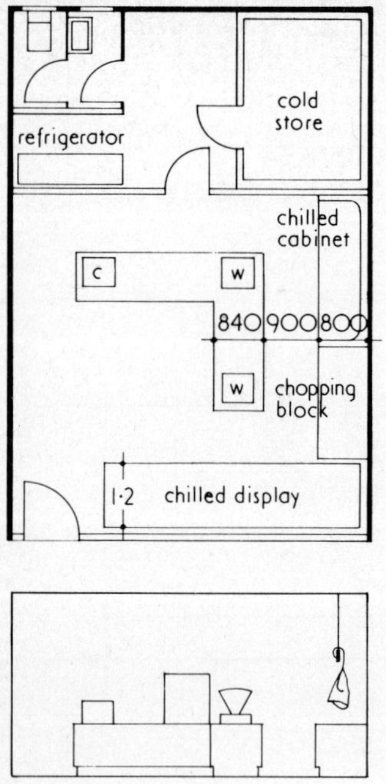

17.3 *Butcher's shop, plan and section*

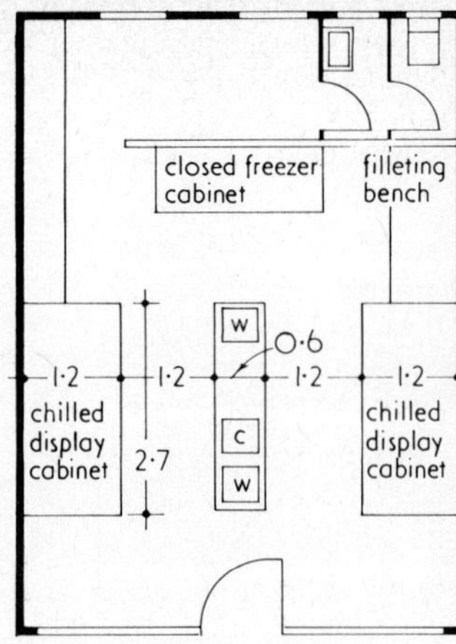

17.4 *Fishmonger's shop, plan*

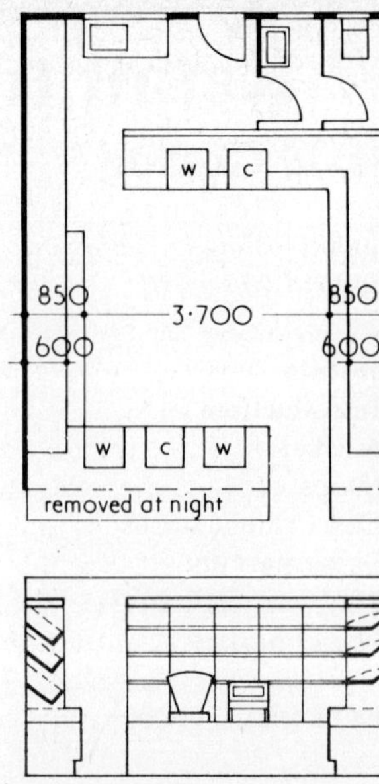

17.5 *Greengrocer's shop, plan and section*

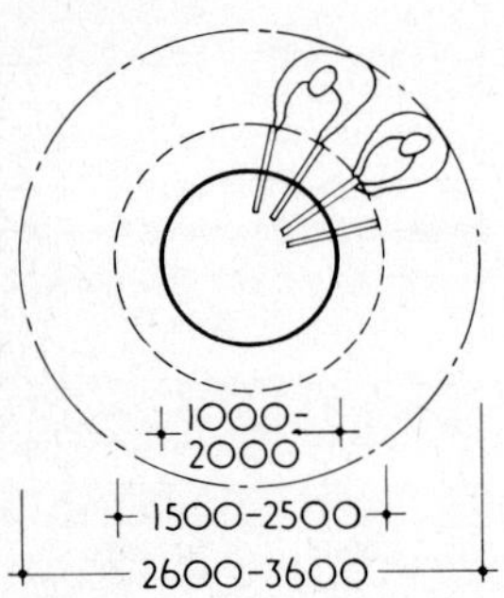

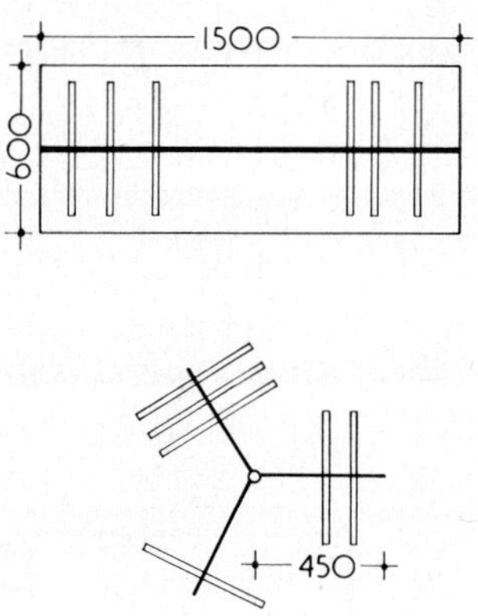

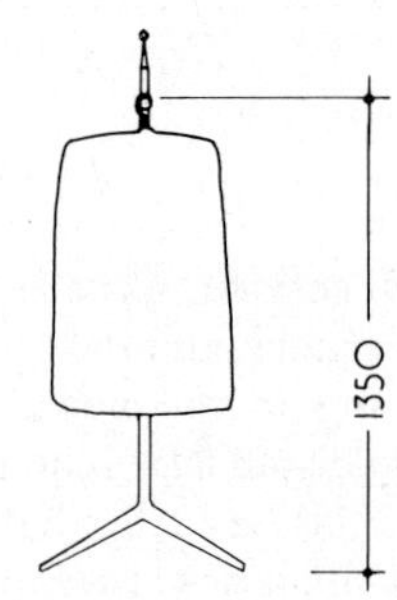

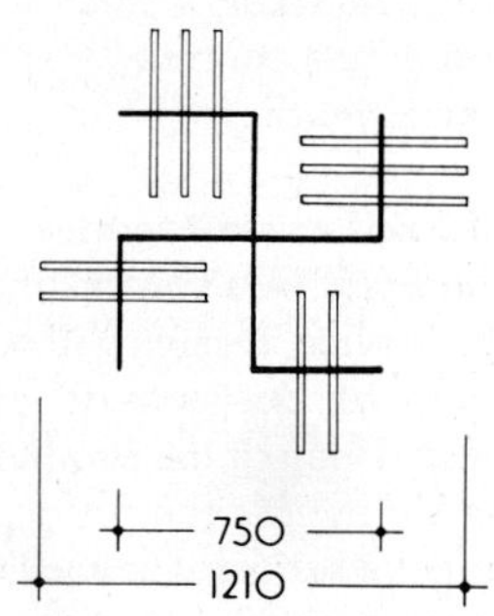

17.6 *Storage and display racks for clothing shops*

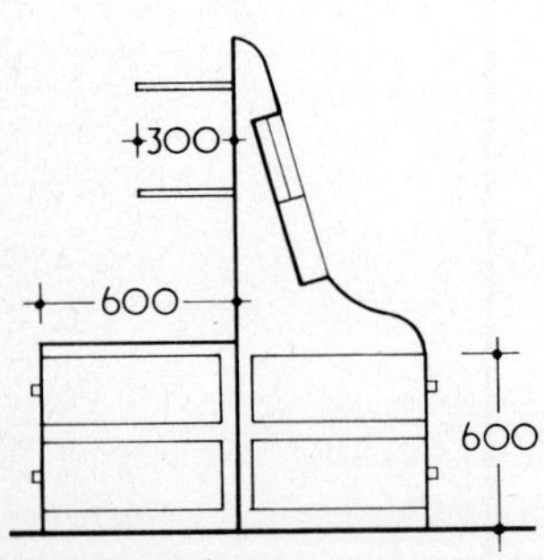

Greetings cards

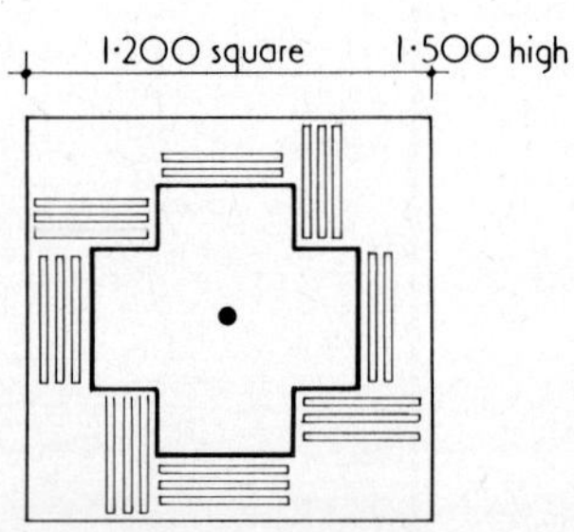

Paper-back book display unit

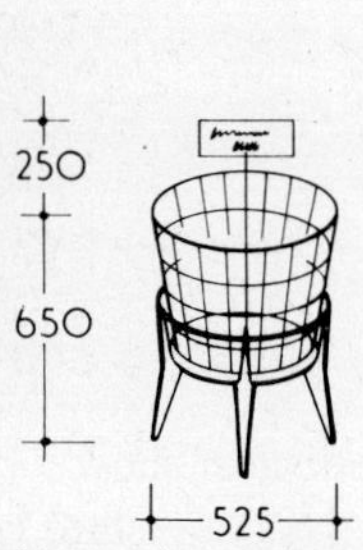

Bins used everywhere

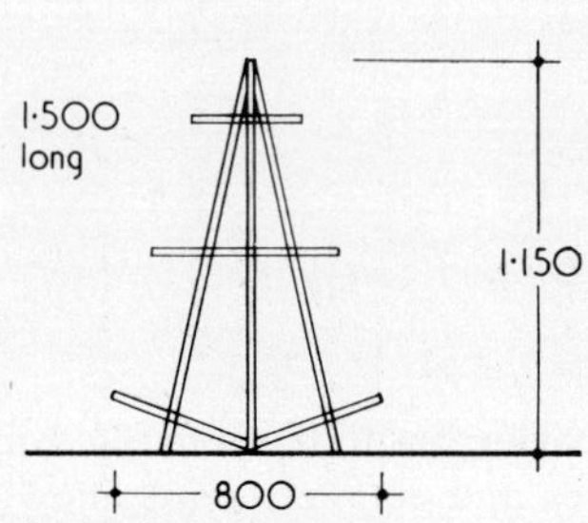

Books – island display

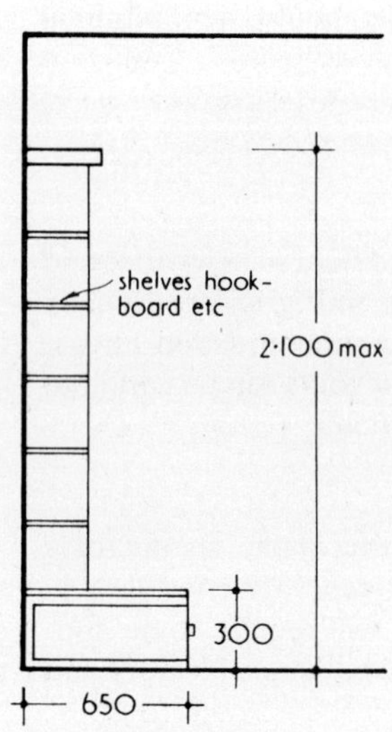

Stationery (or books)

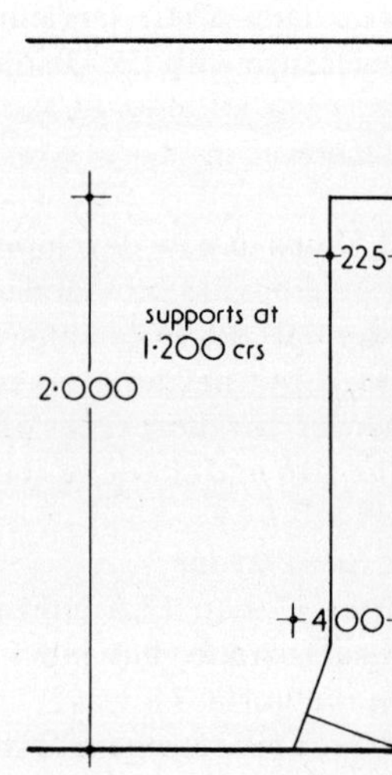

Books

17.7 *Storage and display fittings for stationery and bookshops*

3.03 Location

Shops such as these, can occur not only in traditional shopping streets, but also as separate areas in department stores, as tenants in an enclosed shopping centre, or as a speciality facility within a hypermarket complex.

4 DEPARTMENT STORES

4.01 Definition

These are very large shops, almost invariably on several floor levels, selling a wide variety of goods including, by definition, clothes. The goods are displayed in departments corresponding to traditional shop classifications, but these are flexible in size and position. Seasonal sales patterns dictate frequent adjustments, and the building should be designed for maximum freedom in this respect. Only the food department is purpose designed, with appropriate floor and wall finishes. Frequently it is organised on supermarket lines. Even in this case provision for expansion during the pre-Christmas period is made.

4.02 Flexibility

The structural grid may be determined by other, eg office, accommodation on floors above. Additionally, the developer may anticipate splitting the store into a number of separate tenancies at a later date. Consequently, the main structural grid is likely to be between 5·4 and 6 m. 5·4 m is considered the optimum.

4.03 Fire requirements

The need for wide open spaces conflicts with the requirement for fire compartmentation. Maximum compartment sizes vary, but in general 2000 m² and 7000 m³ is permitted, with twice these figures where an efficient automatic sprinkler system is installed. There are also requirements for fire resistance of the materials of construction, and for adequate means of escape to be provided and sign-posted.

5 SUPERMARKETS

5.01 Definition

Supermarkets sell food and similar domestic necessities, on self-service lines. While originally supermarkets were at least 200 m² and may be ten times that size, the term is now loosely used, even to refer to small grocers converted to self-service.

5.02 Display

The goods are arranged on a proprietary display system. Shelves can be against the walls of the shop **17.8**, or in the central area in islands known as gondolas **17.9**. These can be between 800 and 1300 mm wide, a minimum of 1400 mm apart. In order to accommodate the trolleys used in restocking the shelves, 'main aisles' width should be increased to 2 m.

5.03 Gondolas

The lengths of gondolas are broken at intervals by cross aisles, often considerably wider than the main aisles to accommodate baskets and displays of special offers. In the largest shops, gondolas can be over 30 m long. For such long aisles, the widths are usually increased to reduce the sensation of claustrophobia.

5.04 Checkout

A typical checkout is illustrated in **17.10**.

5.05 Ancillary accommodation, etc

Other items commonly illustrated are shown in **17.11** to **17.13**. Large supermarkets will need properly organised unloading

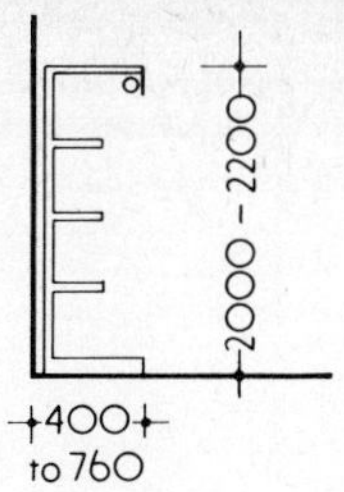

17.8 *Section through typical supermarket shelving against a wall*

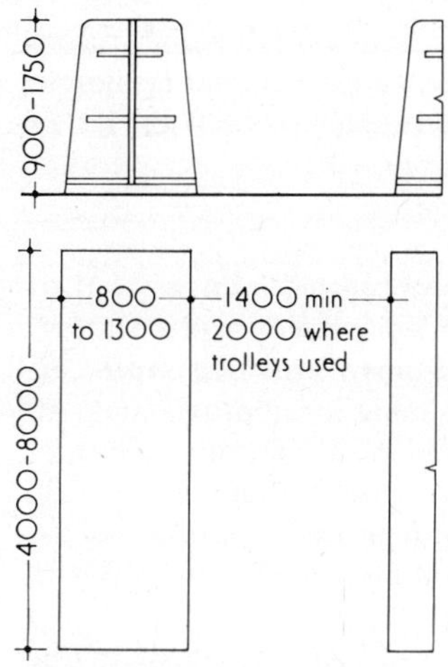

17.9 *Plan and section of island units*

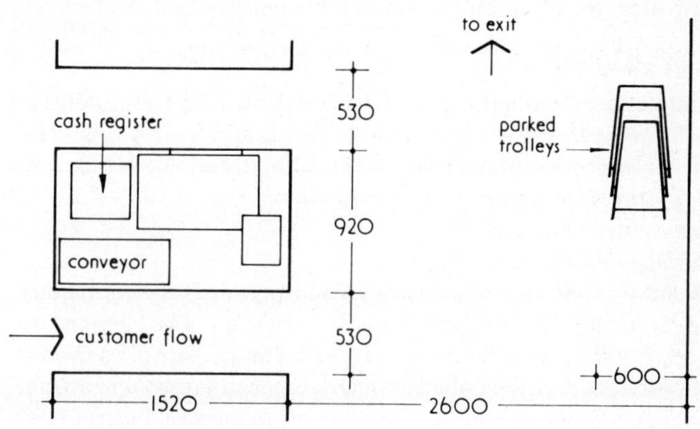

17.10 *Typical check-out*

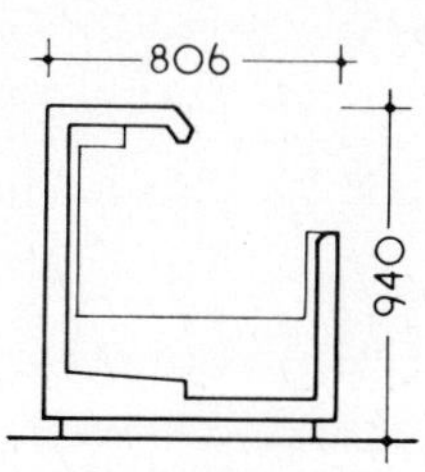

17.11 *Chilled food display cabinet*

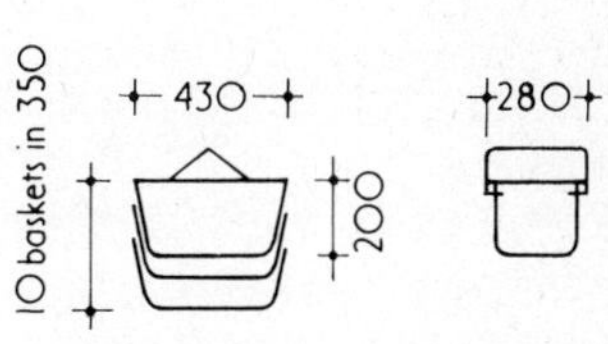

17.12 *Supermarket baskets*

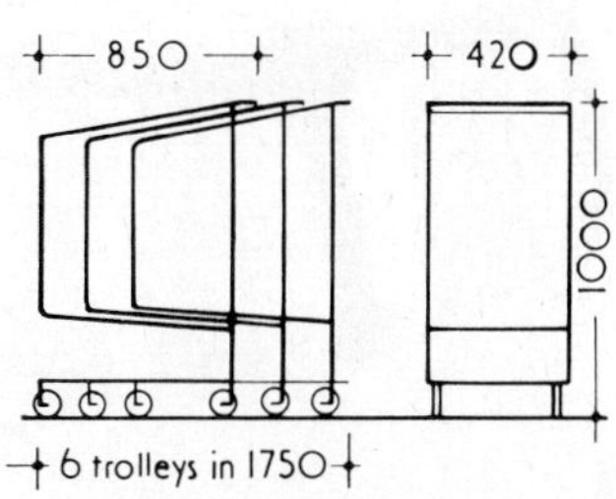

17.13 *Supermarket trolleys*

and storage facilities; cold rooms with a range of operating temperatures; and a meat cutting and packing facility, where meat is brought in as carcasses.

6 SHOPPING CENTRES

6.01 Definition

This somewhat loose term can apply to all the following:

- an area of the town where most of the ground floors are shops
- a pedestrianised street with shops
- a purpose-built covered area containing separate shops, department stores and supermarkets.

It does not apply to large department stores, although in recent years some of these have been arranged in a similar fashion to small shopping centres. This section applies solely to the last of the three definitions above.

6.02 Siting

Shopping centre sites are either town centre (or integrated), or out of town. In Britain we have tended to discourage the latter. Both sitings are normally generously provided with car-parking, the out-of-town centre aiming for approximately 80 per cent of custom car-borne, while the town centre development will be well served by public transport; but buses etc will be attracted to the out-of-town centre by the very success of the enterprise. Space should therefore be available for buses to reach as close as possible to the building.

When planning the car parking capacity, the needs of the centre's employees should not be forgotten, particularly those that may be working unsocial hours, and consequently travelling at times when public transport is not available.

6.03 Basic design

Simple centres will probably be single storey. The large centres will, of necessity, require two or more levels of shops. The larger multiples often prefer to spread between two levels, and department stores may need even more.

• *Malls and magnets*

A simple approach to the basic design envisages the centre as a covered street (mall) with shops on both sides. The elementary centre will consist of a single mall with a large store at each end. These stores are described as magnets because they will attract shoppers to their vicinity. The effective range of a magnet is 90 to 120 m, so the length of the dumbell mall described above would be no more than about 180 m. A third magnet near the centre of the run would enable the mall length to be doubled.

• *Central feature*

A focus of attention is usually provided within the mall area by a fountain, a large sculpture or simply a sitting-out area with some planting. Space is often provided adjacent to this for special exhibitions and displays, the mall, at this point, functioning as a public square.

• *Mall hierarchy*

From the simple dumbell mall enlarged to accommodate the focal feature, alternative plans of increasing complexity can be developed. Wide central malls form the axis of the centre and, in multi-storey centres, these may extend through two or more levels. Branch malls will link the squares or central malls, or may lead to the car parks, or ancillary facilities.

• *Alternative plan forms*

A regional centre of 100 000 m² of gross lettable area will have half of this area devoted to magnets. The depth of shop away from the mall is about 45 m, consequently 50 000 m² of smaller shops will require a frontage length of 1100 m. The dumbell with three magnets has a frontage of only about 700 m; and it is really not practical to extend the mall beyond a length of 350 m. Alternative plan forms to cope with this problem are:

- L-plan, **17.14**
- T-plan, **17.15**
- racetrack
- cruciform, **17.16**
- pinwheel, **17.17**
- figure of eight, **17.18**.

Some of these plans, although less profligate in the use of land than the dumbell, still necessitate excessive walking distances across the centre which must be added to the distance from a remote parking stall. For larger centres the multi-level solution becomes almost mandatory.

6.04 Detail planning

Much of the detail planning of the centre will depend on the choice of grid.

• *Mall width*

The mall width has increased over the years from the original 5·4 m. The typical British mall is 8 m wide, occasionally 12 m. The average French centre uses a 16 m mall, while in North America, the malls vary from 12 to 27 m in width.

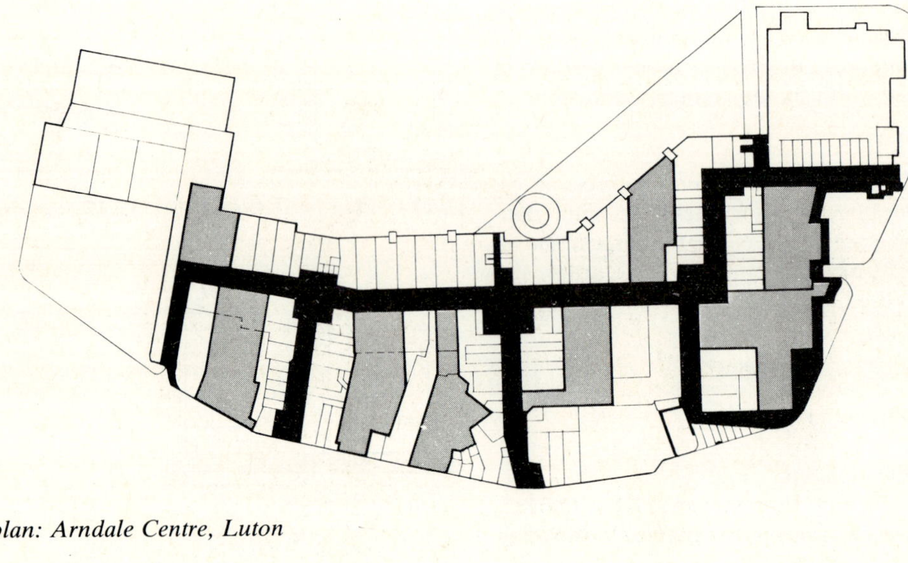

17.14 *L-shaped plan: Arndale Centre, Luton*

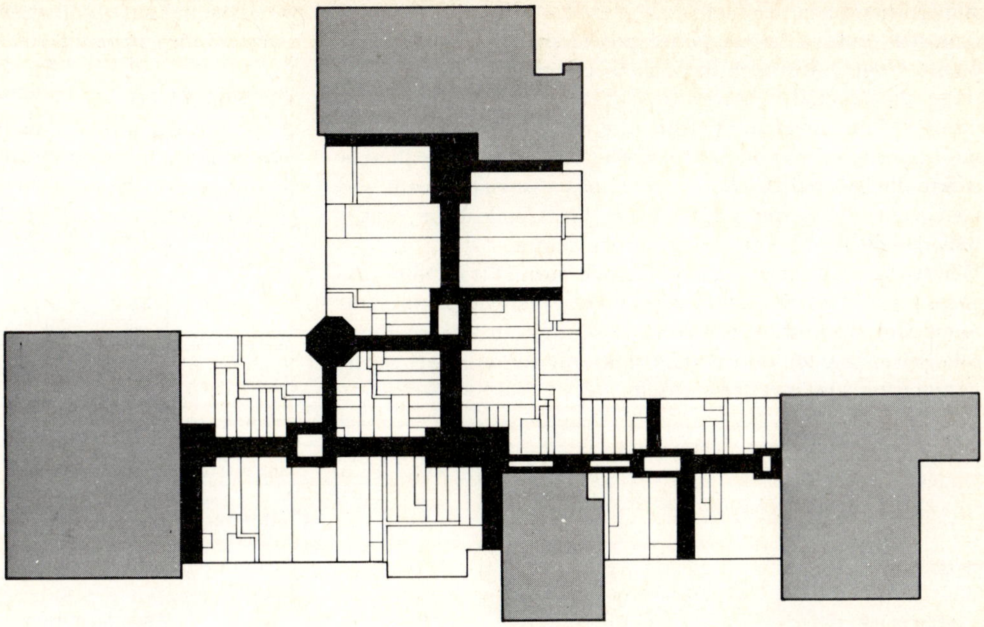

17.15 *T-shaped plan: Willowbrook Mall, New Jersey*

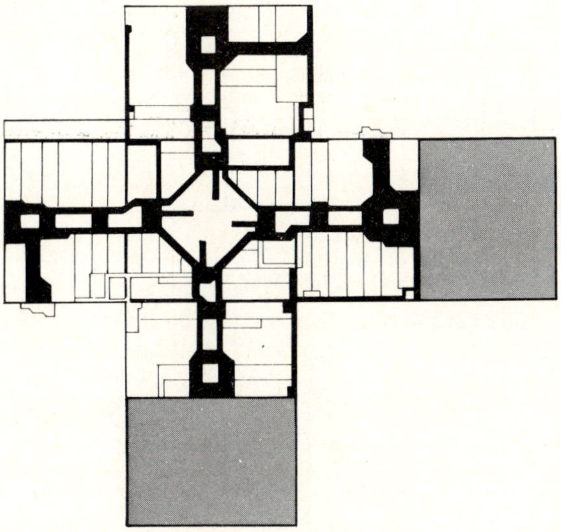

17.16 *Cruciform plan: La Puente, California*

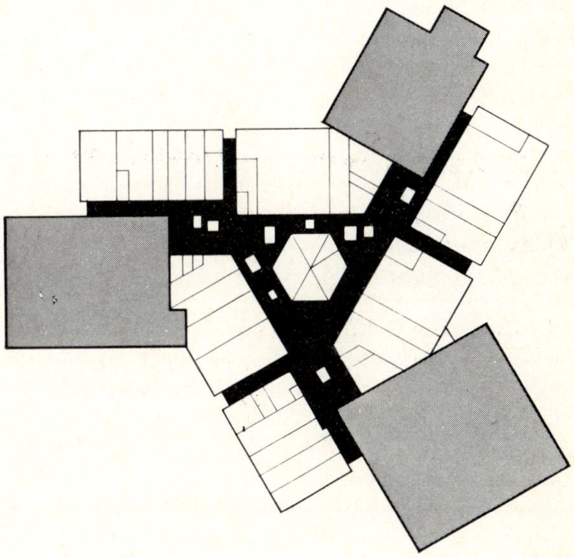

17.17 *Pinwheel plan: Randhurst, Illinois*

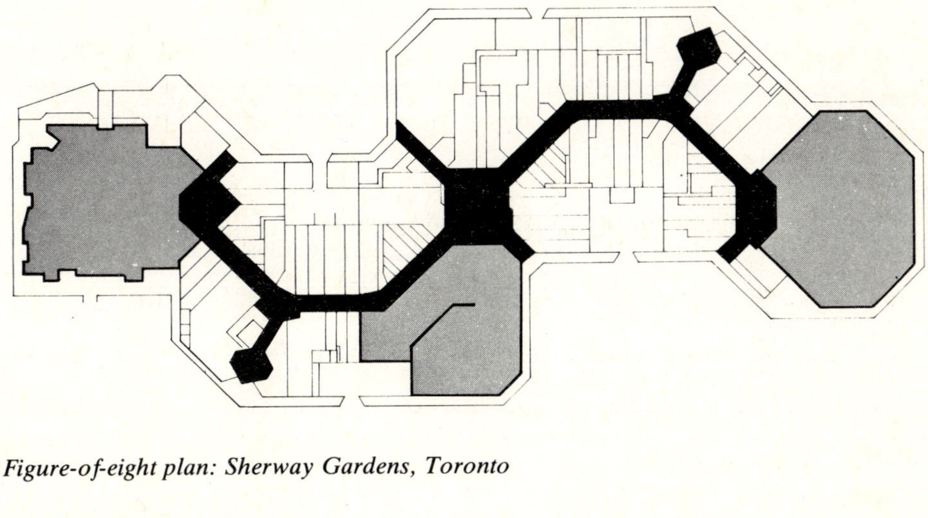

17.18 *Figure-of-eight plan: Sherway Gardens, Toronto*

• *Shop fronts*

In some centres the shops are provided with conventional glazed fronts. Although there is no environmental need for these, as the mall is heated to normal indoor comfort standards, this arrangement permits the malls to be open to the public at nights and on Sundays. This is particularly valuable where there are recreational facilities associated with the centre, or where the malls are on direct pedestrian routes between parts of the town outside the centre.

Where the malls are closed off outside normal trading hours, some form of barrier is still required between shop and mall. Some fire authorities insist on fire-shutters in this position, otherwise the open lattice-work shutter used to protect continental shop windows can be used.

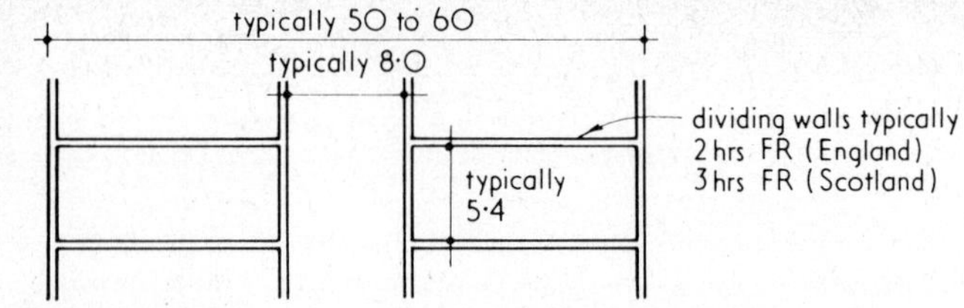

17.19a *Plan of typical mall showing fire separation between shop units*

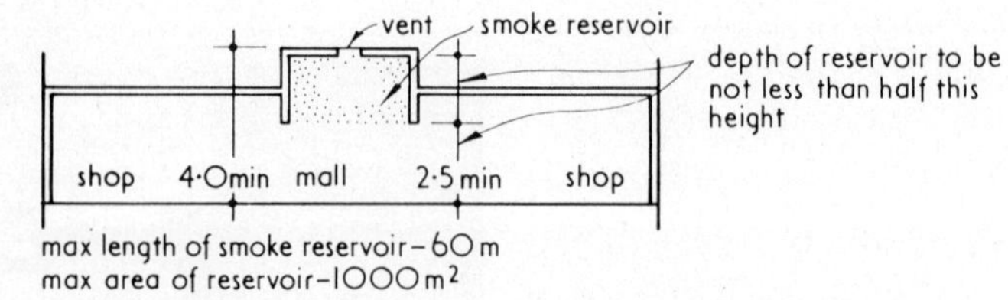

17.19b *Section through same mall showing smoke reservoir*

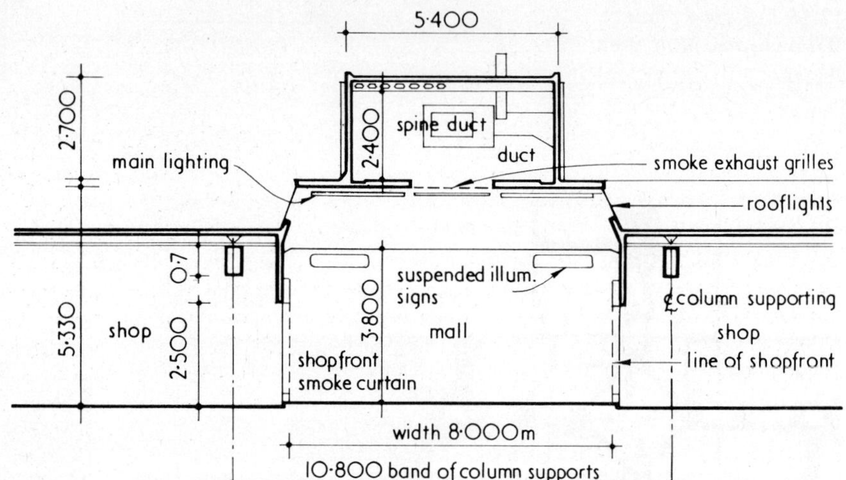

17.20 *Detail cross-section through shopping mall*

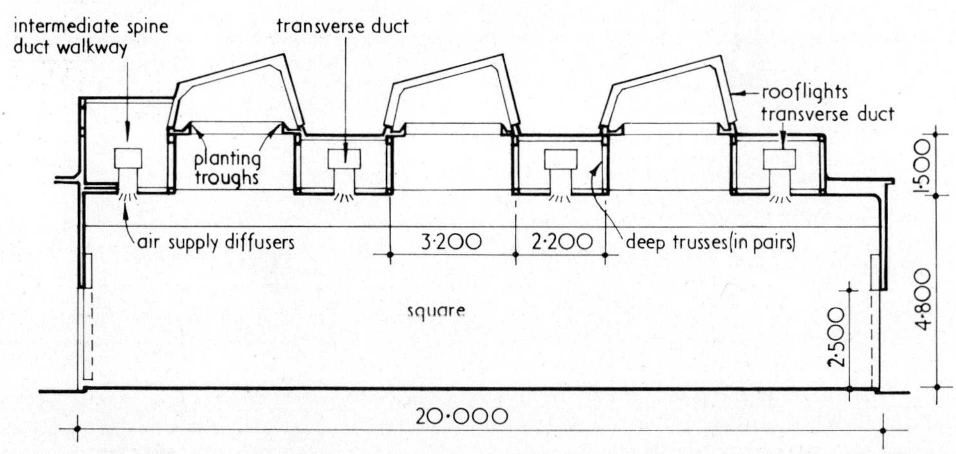

17.21 *Detail cross-section through central square*

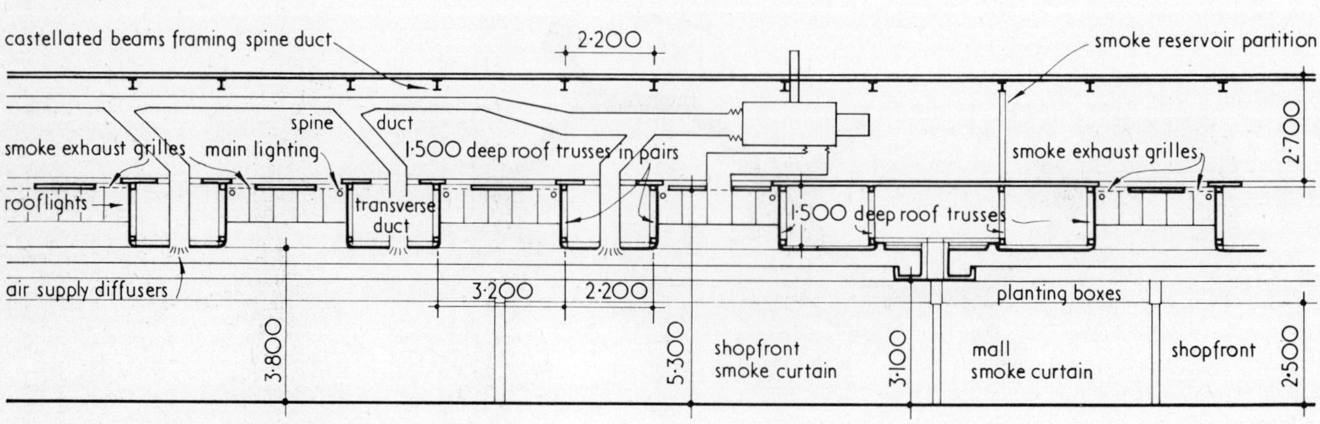

17.22 *Detail longitudinal section along mall*

• *Links between mall levels*
Where there is more than one level of mall, escalators are invariably used for transferring from level to level. For economy, the escalator bank may cope with only the up-traffic at busy periods, leaving the descending traffic to use the stairs. At least one lift should be installed for the benefit of the disabled. Where there are a number of medium-sized stores that spread between the levels, escalators provided for their own customers will also be used by others to change level.

6.05 Fire precautions
Design of shopping centres does not conform with conventional compartmentation techniques. In each case the authorities will stipulate appropriate standards.

• *Fire control*
To prevent spread of fire between shops, fire separation walls separate tenancies, **17.19a**. Fire can then only spread via the malls, but these are usually of non-combustible construction and the amount of incidental combustible material in them can also be controlled. Combined with the obligatory comprehensive automatic sprinkler system, it should be difficult for a fire to spread very far.

• *Smoke control*
This is the main danger to life. To control smoke, smoke reservoirs are created by downstand beams or fascias at the shop-front position, and at intervals across the malls, **17.19b**. Smoke detectors activate powerful extract fans in the reservoir, exhausting the smoke harmlessly to the open air at high level. At the same time, other fans can be used to supply fresh air at low level to prevent the further sucking of smoke into the escape routes.

• *Escape routes*
Most shopping centres have a considerable perimeter length and will have little difficulty in maintaining sufficient escape routes, both from the malls and from the separate occupancies directly (although this can lead to a considerable security problem).
More problems with escape routes are found with multi-storey centres. Where the site is itself sloping, it is practical to achieve level escape to the open air from at least two levels, possibly more. In town-centre developments, bridges are often provided across perimeter roads to stores on the other side. If these bridges are designed to ensure they do not themselves help to spread the fire or the smoke, they can be satisfactory escape routes from upper levels.
For the sizing of escape routes and staircases, a high occupancy should be assumed for the public areas. In addition, no more than 50 per cent of the people within shop areas should be assumed to escape from the rear, the rest must be assumed to use the mall.

• *Alarm systems*
Smoke detector systems will also need to activate an automatic fire brigade call via the post office telephone system. A central fire control panel will be installed at a continuously manned position, with repeater stations at each of the points easily accessible to fire brigade appliances.

• *Sprinkler systems*
Shop units, and to some extent the public malls, should be covered by sprinkler systems. The method of actuating these must be carefully considered, as more damage is often caused by the water from the sprinklers than by the fire itself, particularly when the latter is of a relatively trivial nature. Sprinklers are also vulnerable to vandalism.

• *Fire fighting access*
To ensure rapid response, the fire brigade can use small appliances based on a Land-Rover chassis, which can drive into the pedestrian mall, directly to the seat of any incident. The furnishing arrangement of the malls should anticipate this.
All mall entrances should be kept clear of parked cars and unloading vehicles.

6.06 Structure
The floors are nearly always of reinforced concrete construction. Where vehicular servicing is at basement level, it will be necessary to omit occasional columns to provide vehicle-turning circles etc. Sufficient height should be left to accommodate the increased depth of beams spanning these areas.
The wider malls and central squares will require long-span roofs. These are often economically constructed with some form of steel structure, provided the fire regulations will allow it. A certain amount of roof glazing is often provided. In one centre, the roof slides back to open the central square to the sky.
Where a lightweight truss is used with little glazing, the voids in the truss can be used for services.

6.07 Environmental services
A major problem is the integration of complex servicing, in particular, distinguishing landlord's services in respect of the malls and communal service areas, from the tenants' requirements. The latter is unpredictable, and may range from simple heating and ventilation by means of small fans and appliances, to full air-conditioning. The landlord will usually provide a boiler house, and sufficient plant to ventilate those areas for which he is responsible, together with main supplies of electricity and water. He will also have overall responsibility for the sprinkler and smoke control systems, fire hydrants, etc.
Sections through a typical centre showing the complexity of structure and servicing are shown in **17.20** to **17.22**.

6.08 Vehicular servicing
For details of unloading bays, and vehicular marshalling and circulation, see section 7 of this handbook

7 HYPERMARKETS

7.01 Definition
A hypermarket is a large store on supermarket lines, but at least 5000 m² in area serving a population of at least 50 000. In order to find space for such a large area, hypermarkets are normally out of town, and consequently orientated towards the car-driving customer. At least 20 per cent of the area is devoted to food sales, to attract the regular user. 40 per cent of the area will be taken up by storage and ancillary activities. The largest hypermarkets are about 25 000 m², none of which are in the United Kingdom.

7.02 Design
Hypermarket design is relatively simple. The main area will consist of a large basic 'box'. It is common to use rugged finishes, or none at all.
However, at the entrance to the hypermarket there is often a vestigial mall, with a number of speciality shops. The design of this area is similar to that of enclosed shopping centres.

• *Trolley size*
In the layout of the main hypermarket and the checkout area, the size of customers' trolleys must be considered. A large size of trolley is normally used, since the size and quantity of goods to be purchased is greater than at a supermarket. In addition, payment is less likely to be in cash. Facilities at each checkout

for the writing of cheques and for the processing of credit cards must be provided.

7.03 Example

The illustration shows a typical large hypermarket layout of some 20 000 m² of selling space, **17.23**. This net area generally represents only about half of the total building area, with the remainder occupied by warehousing, food preparation rooms, and office. In the largest stores the net area increases to about 65 per cent. For the large example illustrated, a total site area of 10 to 11 hectares would be required, including all the car parking.

7.04 Superstore

The superstore is a further development of the hypermarket. This one, **17.24**, has an unusual mix of retail and non-retail uses, occupying a site area of 20 hectares. The superstore itself has a net selling area of 9300 m² (a gross area of 12 000 m²) out of the total building area of 47 000 m². 19 000 m² is occupied by recreational facilities: swimming pool, sports hall, roller skating rink, golf, archery, squash courts, bars and restaurants. The site also contains a riding school, soccer pitches and athletics and cycle tracks.

The building has been designed as a pavilion with an umbrella structure 288 m long by 126 m wide by 15 m high. The large units of accommodation—hypermarket, sports hall and central plaza—are located in the central full-height band.

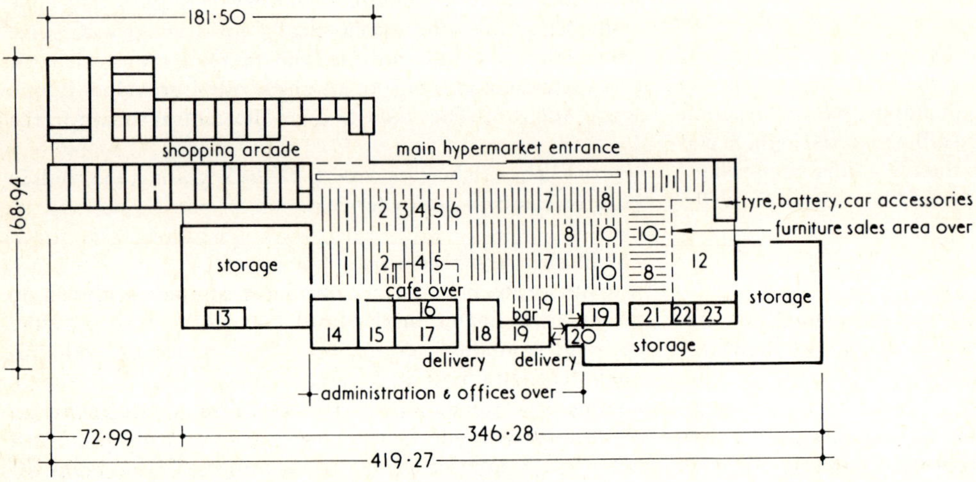

17.23 *General layout of a typical major hypermarket:*

1 soft drinks, wines, etc
2 groceries
3 pharmacy and cosmetics
4 dairy
5 fruit and vegetables
6 children's clothing
7 clothing
8 household goods
9 fancy goods
10 shoes
11 electrical goods
12 furniture
13 fire-fighting and sprinkler installation
14 electricity sub-station
15 butchery preparation
16 cooked meats counter
17 fruit and vegetable preparation
18 baker's and pastry shop
19 changing rooms
20 lavatories
21 boutique
22 after sales service
23 furniture repair workshop

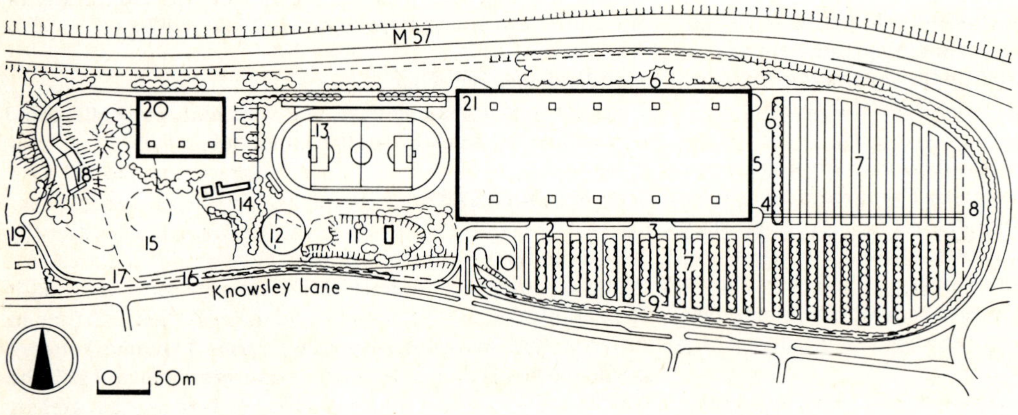

17.24 *Site plan of a superstore at Knowsley, Lancashire Architects: Foster Associates*

1 site entrance
2 bus
3 taxi
4 service road
5 service bay
6 staff car park
7 car parking
8 coach parking
9 perimeter exit road
10 petrol service station
11 adventure playground
12 open play space
13 sports pitches and athletic track
14 horticultural centre
15 riding school field
16 bridle path
17 cycle training and racing track
18 dry ski and toboggan slopes
19 site boundary
20 riding school
21 pavilion containing superstore, leisure pool, sports hall, restaurants, library, cinema, exhibition gallery, etc

18 Hospitals

Anthony Noakes
Superintending architect DHSS

Ergonomic material supplied by **Glyn Stanton**, *Consultant ergonomist*

CI/SfB 41
CI/SfB (1976 revised) 41

Contents

1 INTRODUCTION

1.01 The hospital and the health service

The term 'hospital' can mean a small building serving a remote rural area or a complex with over 1000 beds, including large teaching and research elements.

All hospitals provide 24-hour medical and nursing care of sick or disabled people. The boundary between primary (ie non-hospital) and hospital care depends largely on the form of the health service. In the USA, there are few GPs (general practitioners), so most people go directly to the office of a medical specialist, who will refer them to hospital only if in-patient care or special diagnostic or treatment services are warranted. In Britain, the patient only sees the specialist if his needs exceed the resources of his GP. The British specialist holds consultations in the general hospital, hence extensive OPDs (out-patients departments). By contrast, many American hospitals have little or no OPD, but their accident-emergency departments have to cope with poorer people who cannot afford private medical consultation, and who thus use the hospital for primary care. In Britain, as health centres become more widespread and as more specialists hold sessions at them, some work now carried out in the OPD will move off the hospital site.

1.02 General hospitals

The size of hospital has grown over the last 50 years, mainly through the development of medical technology, with its need for expensive equipment and staff with special skills. These resources become more economic if they are assured of a steady flow of work; this in turn demands a large catchment population. In Britain the term 'district general hospital' is used to describe hospitals which are large enough to provide for all but the more esoteric of specialist services. Their size, measured in terms of numbers of beds, ranges from about 200 to well over 1000. The majority of new general hospitals are planned with between 500 and 800 beds although the current nucleus hospitals are planned as viable first phases of 200 to 300 beds. The recent trend towards very large hospitals is now in reverse. This section deals primarily with general hospitals, but much of it is applicable also to teaching and community hospitals.

Teaching hospitals

Teaching hospitals provide for training of undergraduate medical students and invariably contain significant research facilities. All major general hospitals, however, contain teaching facilities for nurses, technicians and others. As more general hospitals are now providing experience for medical students the boundary between them and the traditional teaching hospitals is becoming less distinct.

Community hospitals

'Community hospital' is the term currently used in Britain for smaller hospitals (50 to 150 beds), intended to be staffed by GPs and requiring less intensity of medical, nursing and technical services. They provide for patients transferred from the DGH, for medical cases where home circumstances prevent domiciliary care, and for some longer-stay geriatric patients. A few also include maternity units. Community hospitals are increasingly associated with health centres.

1.03 Specialist hospitals

Specialist hospitals—dealing with one category of patient only—form a different class. A small group of these (the British ones being mainly in London) are among the leading centres of postgraduate teaching and research in their specialty. Much more numerous, though, are small maternity hospitals; there are a number for children only, and for other special groups. New general hospitals usually include all such specialties, and make it unlikely that many new specialist hospitals will be required.

By far the greatest volume of hospital work undertaken in existing specialist establishments is in the field of mental illness, mental retardation, and geriatrics: many of these 'patients' are primarily receiving sheltered communal living conditions rather than acute medical or nursing care. In these fields, too, the trend is away from large or specialised institutions, and more effort is being made to return them to community care. Many such hospitals are very large (over 2000 beds) and remote. The boundaries between medical, housing and social services are changing with regard to these groups of patients—most radically, in the case of mental retardation, which is increasingly regarded as an education and social service problem rather than one of medicine and nursing.

Hospices are a kind of specialist unit that are becoming more common. They provide a very good environment for terminal patients (and a few long-stay ones). They are run mainly by Catholic nursing orders. Patients may come in for periods of time, and spend as much time as possible at home.

1.04 Growth and change

The need for considerable and unpredictable growth of hospitals was justified before the National Health Service was set up, at a time of steady population growth, and rapid movement of population, especially to new suburbs. Now, with a coordinated health service, moderate changes of population can be dealt with by modification of catchment areas, and larger ones by building complete new hospitals; thus the many problems of designing a hospital of unguessable ultimate size may be avoided.

In addition to the growth arising from the hospital serving an increased catchment population, there is the growth resulting from the addition of facilities for new functions or new technical developments. The latter case poses a fundamental problem; remarkable advances in medical technology make it possible for 'progress' to continue. But the higher the technology, the higher the cost of each life saved. The demand for health services is limitless, but the funds available are not. Considerations of cost effectiveness are likely to become more significant.

The need for designing for change—to cope with changes in equipment and technique, and in methods of organising care—is however not in any doubt.

1.05 Size of hospitals

The size of a hospital is commonly indicated by numbers of beds. This can however be misleading, as the proportion of the hospital given over to the nursing units or bed-areas varies greatly from as little as 25 per cent of the whole in teaching hospitals, to over 60 per cent in psychiatric and geriatric hospitals. As the cost of nursing and of the 'hotel' element in medical care increases so do the range of procedures that are undertaken on an out-patient basis, eg in the development of day care units, which enable patients to undergo investigation or minor surgery without admission for overnight stay.

Bed numbers are nevertheless likely to continue to be used as size indicators, qualified by the kind of service given by the hospital. As a rough guide to correlate bed numbers to population served, the current NHS planning norms may be helpful (table I). These are general in nature, and are subject to variations according to local conditions (eg of housing) even in Britain, and elsewhere will be greatly affected by the nature of the health service.

Table I Provision of beds*

Service	No of beds per 1000 pop	Comments
Acute medical and surgical	2	Assumes increasing extent of domiciliary care in community
Maternity	0·3	Depending on birthrate and no. of beds for ante-natal care
Geriatrics	1 to 1·5	On basis of 10 beds per 1000 population over age of 65; areas with good housing and social services manage with less
Psychiatry	0·5	Numbers required for DGH-based service; actual number in specialist psychiatric hospitals is much higher, but is diminishing
Children	0·5	Provides for all children requiring acute or long-stay care

* This table excludes provision for mentally retarded adults.

2 SPECIAL PROBLEMS OF DESIGN AND CONSTRUCTION

The large size and technical complexity of hospitals make the management of the design process difficult. Methods adopted to reduce the timespan between briefing and completion, and thus to limit obsolescence, include:

- phasing—reducing size of contracts likely also to minimise overruns and extras)
- standardisation—more scope for smaller elements than for large ones (such as whole hospitals); standard designs need regular review and revision
- 'fast-track'—building in overlapping stages (eg siteworks, structure, cladding, fitting-out); the 130 000 m² McMaster (Ontario) teaching hospital took only 4 years to design and build.

Development of the brief and early planning studies are best undertaken in parallel; likely subsequent needs should be anticipated, as well as those of the first users.

3 THE FORM OF THE HOSPITAL

3.01 External factors

- *Air conditioning*

Air conditioning is one of the most significant influences on form. A few departments (eg operating theatres and intensive therapy units) need air conditioning for functional reasons. But the greater part of the hospital can function as well if not better by natural ventilation.

Total air conditioning is usually required if the climate has extremes of heat, cold, or humidity, if there is serious atmospheric pollution, or traffic or aircraft noise or if a small urban site requiring high planning density. If air conditioning proves necessary for much over half the accommodation it is usually more economic and efficient to air condition the whole building. (Use of interstitial floors for services—see para **3.07**—then becomes advisable.)

- *Location*

Rebuilding often takes place on the edges of towns, where land is cheaper and more plentiful. Such a site is often bought at the cost of impaired accessibility—especially where cross-town journeys are involved. A large and growing proportion of patients are elderly, and they and their visitors are the least likely to own cars. Also, the lower paid staff (eg porters and domestics) may prove hard to recruit if access by public transport is poor.

3.02 Whole hospital policies

Policy decisions affecting the hospital as a whole include staff changing (if centralised, cloakroom requirements in individual departments will be reduced, and circulation affected, influencing overall form), catering, and supplies distribution.

3.03 Relationships between departments

Analysis of traffic between hospital departments will suggest some preferred groupings, **18.1**. They will vary according to values placed on time, staff and patients and to the type of hospital and health service.

There is much difference of opinion as to the relative convenience of vertical and horizontal travel. Many tall hospitals have been built with a view to reducing the long walking distances characteristic of old pavilion type designs, **18.2**. Lift-waiting time is however either not allowed for, or over-optimistic. If, as often happens, lifts are specifically allocated (eg for visitors, beds, clean supplies, disposal, medical school), they become too dispersed for any one group to be adequate, and overlong waiting times result. The less a hospital depends on vertical transportation the better it will function in emergencies such as power cuts.

3.04 Fire safety and means of escape

In many high life risk areas of hospitals—namely those containing in-patients—it is impractical to move patients down staircases. Many of these patients are dependent on life-support systems, eg receiving intravenous transfusions. Horizontal evacuation, into fire compartments progressively further away from the seat of the fire, is the only realistic solution.

This implies a minimum of two fire compartments on each floor. The larger the continuous floor area on any storey, the greater the number of compartments and the better the scope for progressive evacuation. This applies particularly to high life risk areas. This principle has been incorporated into the 1976 revision of Hospital Design Note no 2 and the Building Regulations 1976.

High life risk and high fire risk departments (eg stores and kitchens) must be defined. If the former are located above the latter, special attention must be given to avoid the spread of fire.

3.05 Flexibility

The forms of flexibility most relevant to hospitals are those which facilitate patterns of growth and change. Both at whole hospital and departmental levels, radical changes of function must be expected within the lifetime of the hospital:

1 Sizes, shapes, and groupings of rooms should be such as to accommodate as many different operational policy options as possible.

2 Arrangement of services should be such that minor modifications (eg the installation of a lavatory basin) should be possible with a minimum of labour, materials, and functional disruption.

3 The design of structure, engineering services and fitting out elements (ceilings, partitions etc) should facilitate more major internal replanning with minimum disruption of adjoining areas, and without altering the structure or main service runs.

4 Departments likely to grow should be planned adjacent either to 'soft areas' containing functions that can easily be displaced elsewhere, or in a position where the building can be expanded. In the latter case, it is better that the department expected to grow should either be on the ground floor, or above other departments also likely to grow.

5 Provision must be made for the addition of completely new functions, caused either by new developments in medical technology, or new methods of organising care, such as the introduction of psychiatric departments in DGHs. The most common patterns of growth are by lateral expansion, addition of a floor or floors, infill of basement car-parks, and filling-in of courtyards or spaces between buildings. The last method is generally unsatisfactory both functionally and environmentally.

3.06 Phasing

It is usual for some hospital departments to be needed more urgently than others and for funds or land to become available progressively. Phasing can thus influence the location of departments within the hospital, and can lead to less than ideal inter-departmental relationships.

Existing site

In the phased redevelopment of an existing site, there is the further constraint that the number of possible locations for a first phase, especially if it is fairly large, will be limited. In order that the old hospital should continue to work, it is likely that several interim moves of displaced departments or parts of departments will be necessary. Maintenance of traffic routes both around and within the buildings need much attention.*

New site

A general hospital is not viable without a considerable amount of diagnostic, treatment and support services and in many cases an accident and emergency service will also be needed early on. This tends to make first phases disproportionately expensive. On an existing site it is usually possible to continue to use, say, the complete old X-ray department until the new one is in operation.

3.07 Long life, loose fit, low energy

Hospitals should have structural envelopes aimed at very long life within which equipment, engineering services, and internal layout can be renewed at appropriate intervals.

The most permanent elements, eg structure, can reasonably have a life span of well over 100 years. However in looking so far into the future many vital influences (political and economic, as well as in building technology and medical and social services) are hard to predict.

This approach leads to the 'loose-fit' solution rather than a building tailor-made to the particular needs of each individual department or function. Two significant and different interpretations of this principle have been developed, on the one hand by Llewelyn-Davies and Weeks, and on the other within the DHSS (Department of Health and Social Security—Chief

* DHSS Design Note 3.

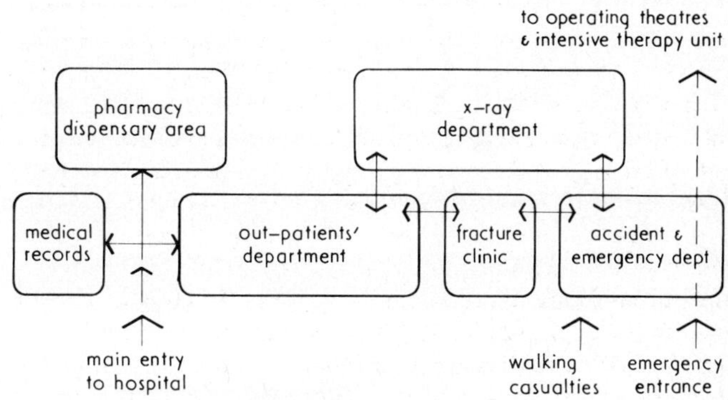

18.1a *Typical grouping of out-patient services*

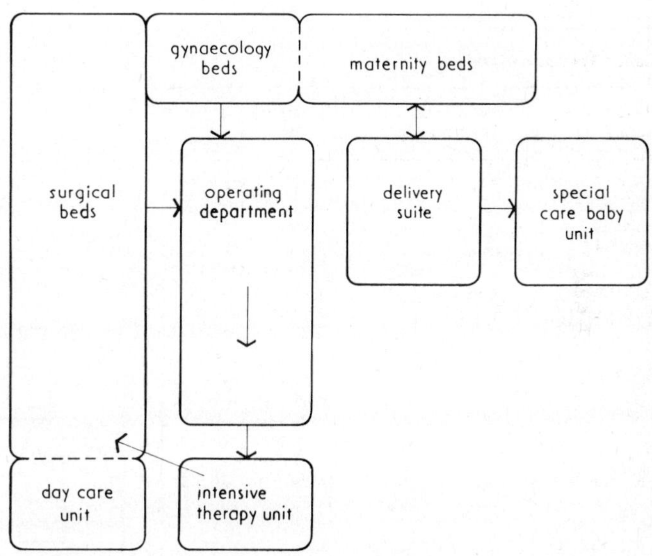

18.1b *Typical grouping of surgical and maternity services*

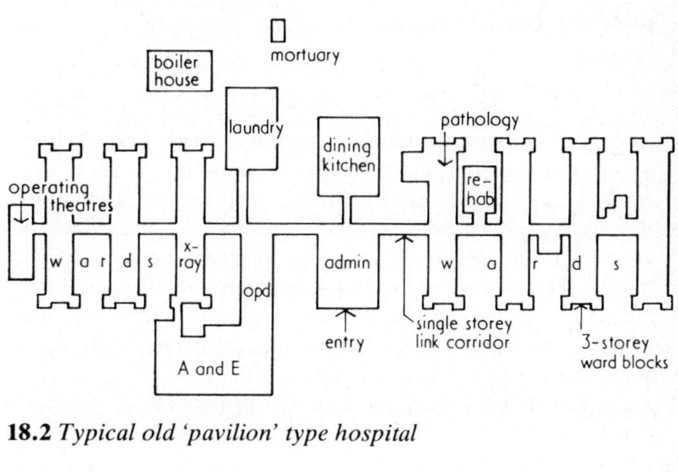

18.2 *Typical old 'pavilion' type hospital*

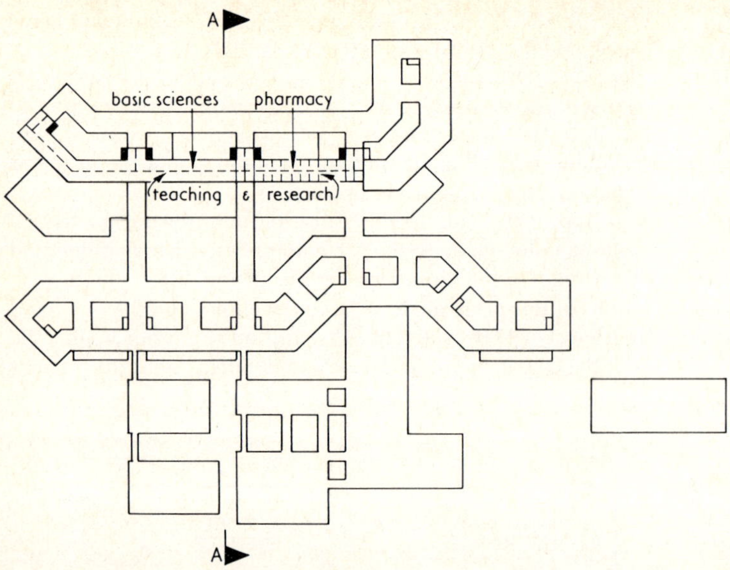

18.3 *Leuven Hospital (Belgium):* **a** *Plan*

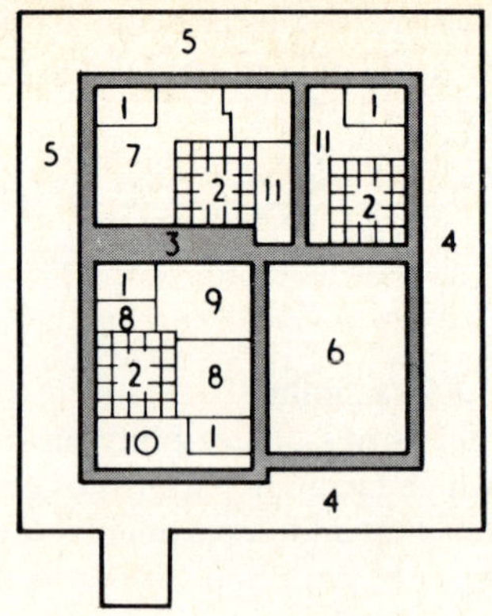

1 *vertical service shafts*
2 *courts*
3 *hospital street*
4 *surgical beds*
5 *maternity beds*
6 *operating department*
7 *delivery suite*
8 *dining*
9 *kitchen and supply core*
10 *intensive therapy unit*
11 *administrative department*

18.4 *Greenwich District Hospital:* **a** *Zoning plan of first floor*

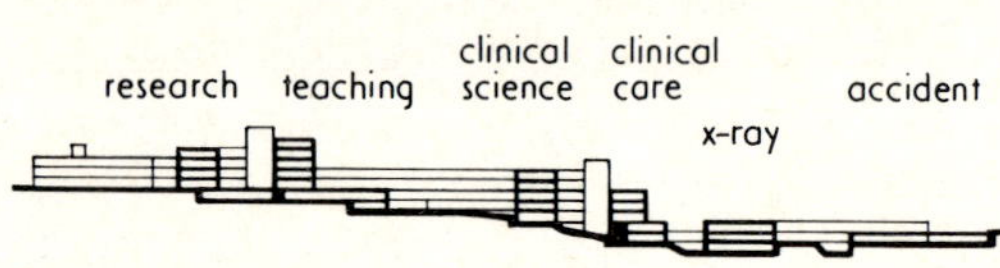

18.3b *Section A–A*

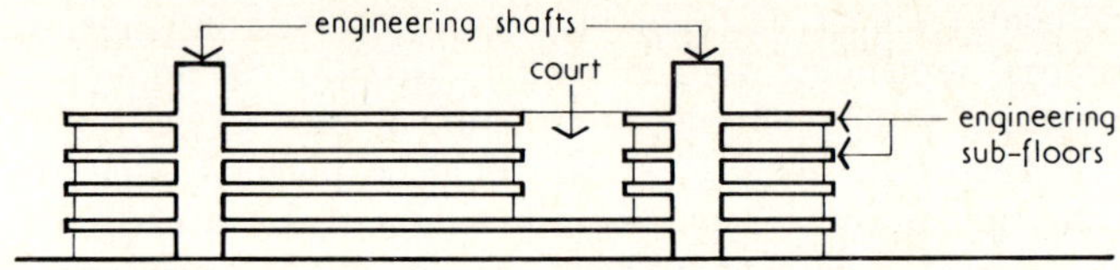

18.4b *Typical section (exaggerated vertical scale)*

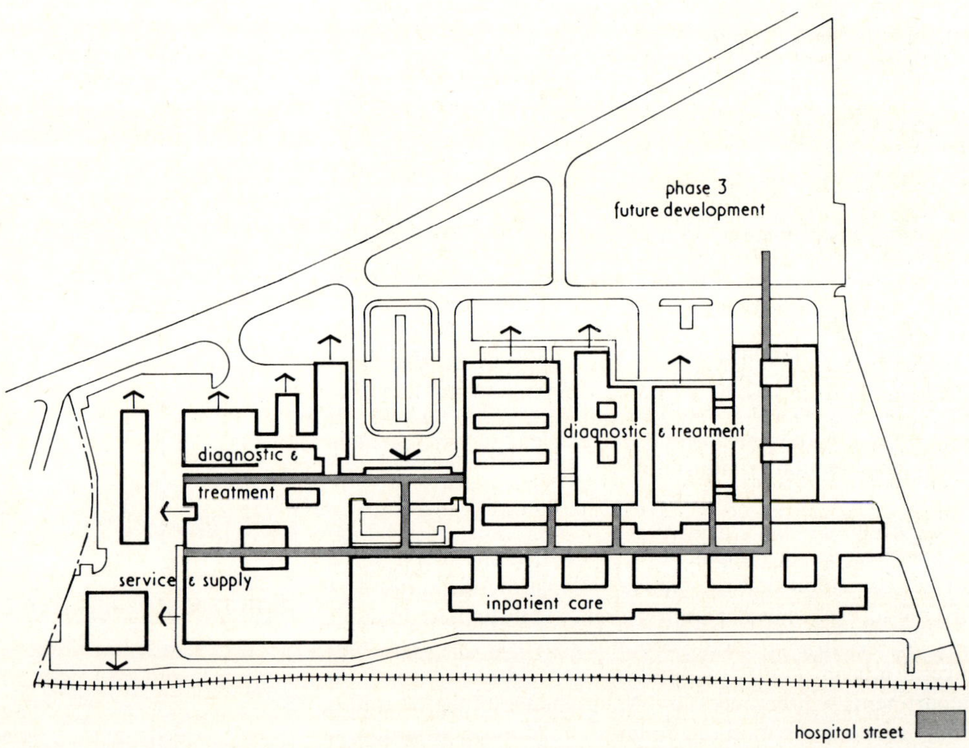

18.5 *York District Hospital, site plan*

Architect, Howard Goodman). A good example of the former is the teaching hospital at Leuven, in Belgium, **18.3**. This consists of linked strips of building, consisting of space modules of 7·2 × 7·2 m on either side of communicating corridors. At major intersections, there are vertical communications points; the whole thus forms a 3-dimensional lattice of 'indeterminate' space. This building, being on a large site, was low-rise enough and incorporated courtyards sufficiently large to minimise the use of mechanical ventilation.

At Greenwich, a DHSS development project, **18.4**, where an 800-bed hospital was built on a 3 hectare site, and the compact planning (as well as traffic noise and atmospheric pollution) necessitated air conditioning, the idea of 'universal floor space' has been taken even further. There are 3 such floors, each with 2 hectares of continuous space. The structure is wide span (19·5 m). Engineering services are collected into 4 vertical shafts and interstitial spaces 1·7 m deep between hospital floors. This gives the maximum freedom for initial planning, subsequent replanning and alterations to services. There is a cost penalty, usually between 1 and 2 per cent of the total building cost, in wide span structures (as compared with conventional spans of 7 to 10 m); some initial space savings can however usually be made, and there is a significant long-term advantage. This design minimises external surfaces, and thus heat loss.

Conventionally, a low energy hospital is considered one with maximum natural lighting and ventilation, with good insulation to walls and roofs, and windows less than (say) 30 per cent of external wall surface. It should minimise the use of lifts and other mechanical handling systems. The volume per floor area—that is, floor to floor heights—should also be kept low. Good examples of this are York's new DGH, **18.5** and the 'best buy' hospitals at Bury St Edmunds, **18.6**, Frimley, and King's Lynn.

Heat recovery

The use of heat recovery methods to transfer heat from the internal areas which tend to gain it, to the external ones which tend to lose it, can be 3 or 4 times as efficient as traditional forms of heating and cooling. Recirculation of air is now medically acceptable in most circumstances.

Mechanical transportation systems

The justification for conveyors etc might be made in terms of cost and availability of labour. Pneumatic tube systems are much used in the USA but not in Britain. This is mainly due to the paperwork generated by payments for drugs and service.

4 IN-PATIENT SERVICES*

4.01 Nursing sections

The term 'nursing section' describes the groups of patients under the care of a sister or charge nurse (or, in American usage, head nurse). The term 'ward' is avoided as being ambiguous in the context of modern hospitals. (Accommodation for the mentally ill and mentally handicapped are dealt with in **7**.)

4.02 General adult acute nursing sections

There are no radical differences between the needs of medical and surgical nursing sections, nor, with a few exceptions, between those of the different specialties (eg orthopaedics, gynaecology) of surgery.

The design of adjacent nursing sections with beds that can be under the care of either unit also makes for flexibility and increased utilisation, **18.4**, **18.6**. A grouping of 120 to 160 beds on one level is a large enough number to provide most of the flexibility benefits. (If 2 or more such groups can be placed on the same floor, then there are advantages in case of access to operating theatres, food service, etc.) These groups tend to correspond in Britain with the span of control of a Nursing Officer. They are also large enough to provide alternative forms of progressive patient care.

The most ill are cared for in ITUs (intensive therapy units), and others may be classified as having high, intermediate or low dependency on nursing care.

The number of beds in a nursing section can vary, according mainly on how dependent the patients are, but 25 to 30 is the most usual range. The trend is generally downwards, but not rapidly so, as smaller units cost more to staff. Peripheral banding, or other grouping of nursing sections can avoid the inflexibility inherent in single units of fixed size.

4.03 Nursing sections design

Nursing staff should be aware of patients' needs, and able to give the necessary care, with the minimum of effort and walking and patients should be as comfortable as possible. These principles sometimes conflict.

The 'nurses' station' is in most designs the organisational hub of the sections. It is not a place at which nurses are often stationary, except at night but is the most important place at which the nurse-call system registers, where all paperwork is done, and where staff report at changes of shift. The need for visual observation of patients from the nurses' station or from the corridors is controversial. Although true observation of a patient's condition calls for the nurse to be very close to him, nevertheless, a passing glance may detect someone's need for help. This is particularly true at night. Immediately adjacent to the nurses' station are the medical records and drug trolleys and storage (including refrigerator and controlled drugs cupboard). A treatment room is used for procedures such as dressings where there would be a risk of cross-infection if done at the patient's bedside. The Nuffield Report proposed an arrangement which has been widely used since, in which this room is placed between the clean and dirty utility rooms, with

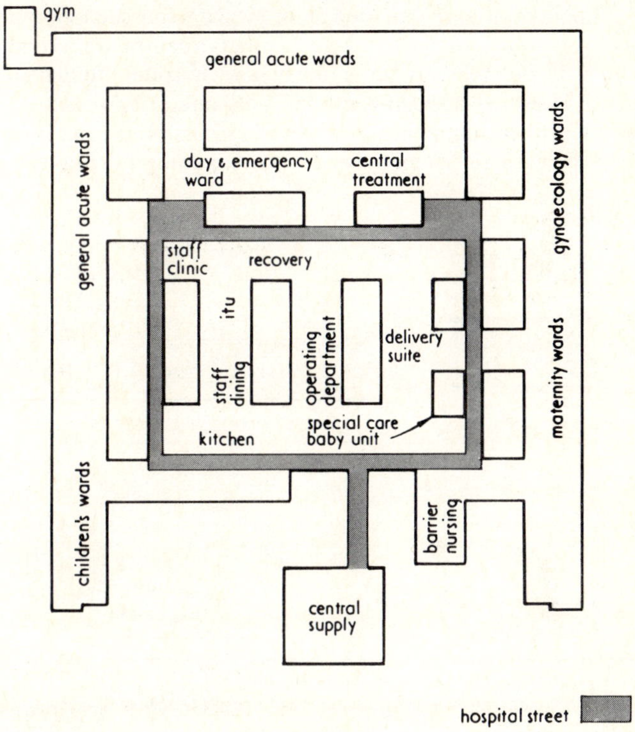

18.6 *Bury St Edmunds District Hospital:*
a *Arrangement of departments on first floor*

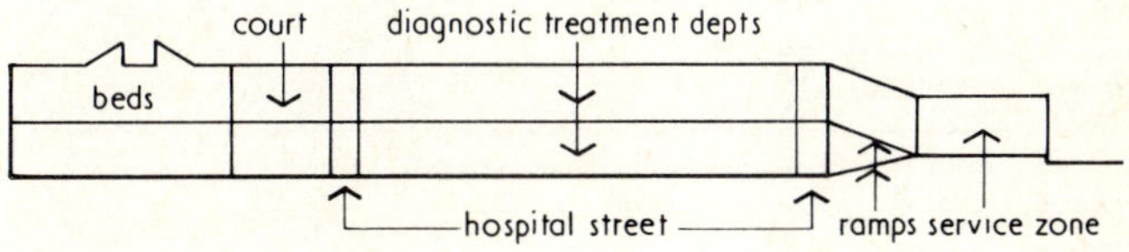

18.6b *Diagrammatic section*

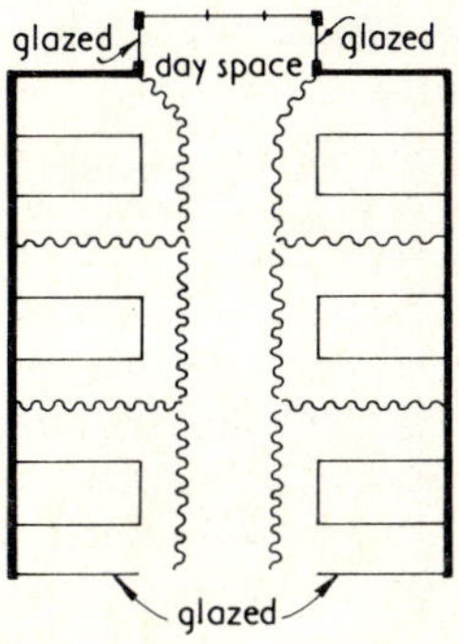

18.6c *Six bed ward with bay window as day space*

* The departments described in paras **4** to **6** are those specific to the general run of hospitals. Highly specialised functions such as renal dialysis units have been excluded, as have laundries, boiler houses, etc which are not peculiar to hospitals.

hatches to each, so that treatment trays can be laid out in advance and passed through, and all soiled items can be removed easily. The treatment room generally requires mechanical ventilation. The dirty utility room contains the equipment for dealing with bedpans; and urine testing (a hatch to a specimen toilet is an advantage) and space for dirty linen and items for disposal to await collection, **18.7**.

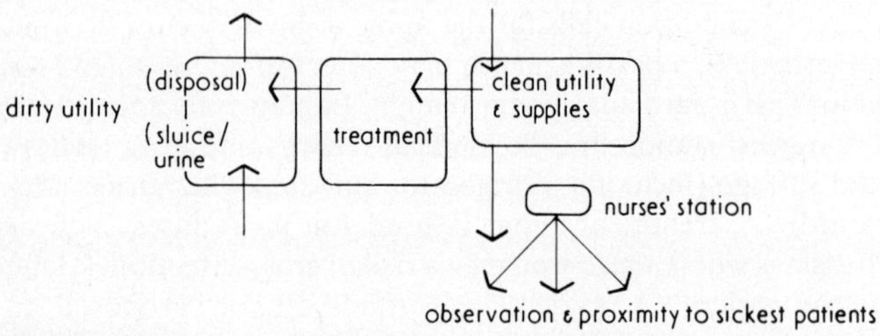

18.7 *Diagram of nursing unit*

Treatment rooms are well used on surgical nursing sections but less so on others. In the 'best buy' hospitals a central treatment suite was provided to serve the whole hospital, to which patients went by appointment, and only one or two sections had their own treatment room. Another possible saving is to share a treatment room between two adjoining nursing sec-

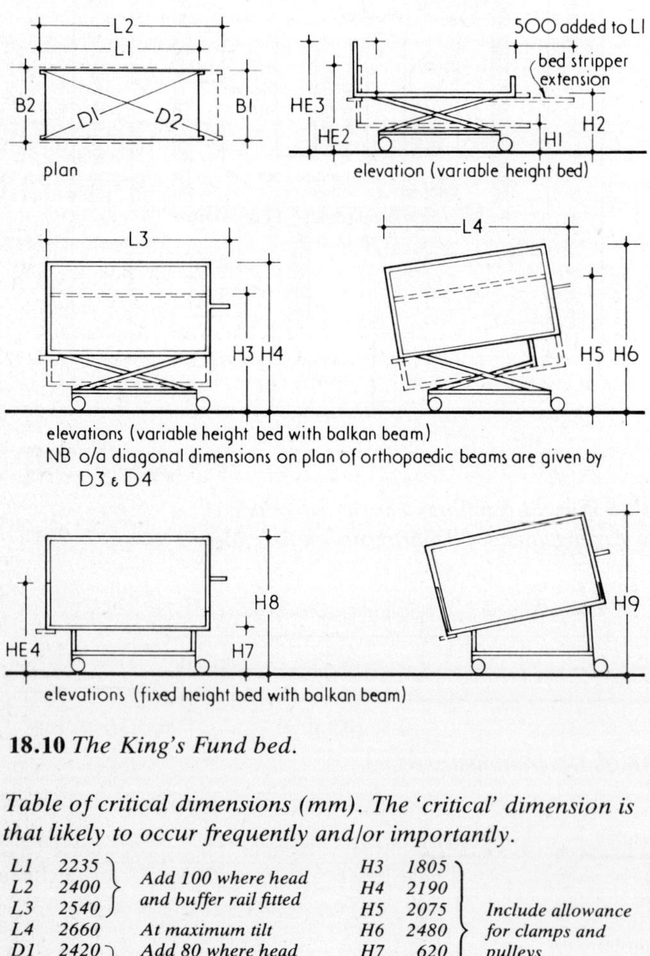

18.10 *The King's Fund bed.*

Table of critical dimensions (mm). The 'critical' dimension is that likely to occur frequently and/or importantly.

L1	*2235*	*Add 100 where head and buffer rail fitted*	*H3*	*1805*	*Include allowance for clamps and pulleys*
L2	*2400*		*H4*	*2190*	
L3	*2540*		*H5*	*2075*	
L4	*2660*	*At maximum tilt*	*H6*	*2480*	
D1	*2420*	*Add 80 where head and buffer rail fitted*	*H7*	*620*	
D2	*2500*		*H8*	*2010*	
D3	*2720*	*When horizontal*	*H9*	*2290*	
D4	*2900*	*When tilted 12°*	*HE1*	*810*	
B1	*960*	*Including corner buffers*	*HE2*	*1205*	*With bed in H1 position*
B2	*1060*	*With side rails*	*HE3*	*1590*	*With bed in H2 position*
H1	*410*				
H2	*840*	*N-E extends to 855 during 1st operation*	*HE4*	*1410*	

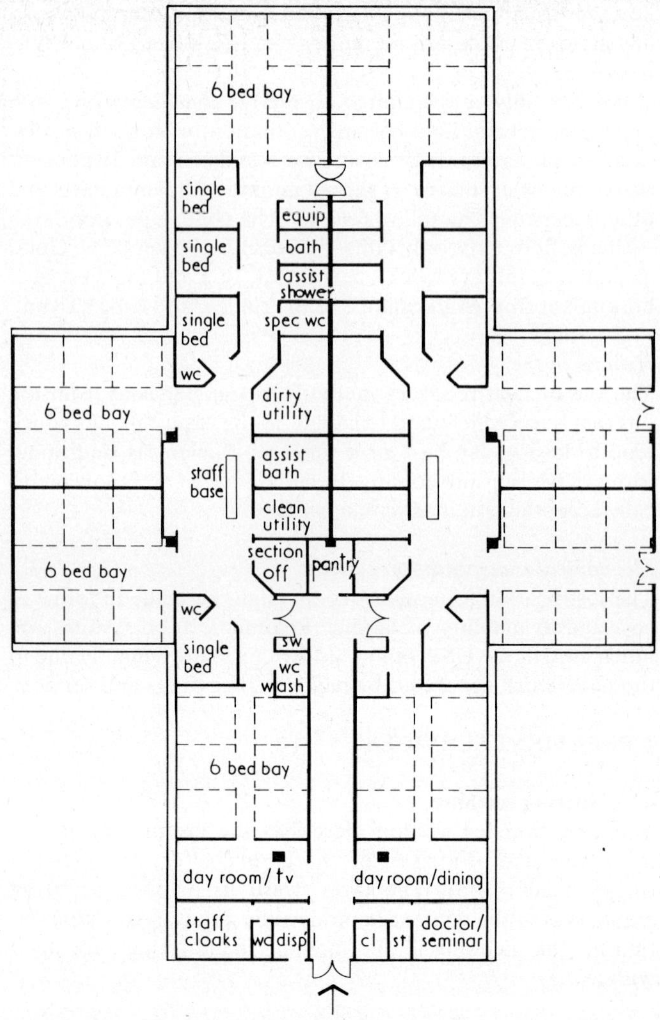

18.8 *Pair of 28-bed adult acute nursing sections for 'Nucleus' hospitals; designed for observation of 50 per cent of patients from staff base*

tions, but this may be a problem for busier wards. The large bathroom used for patients who need assistance in bathing (and in some cases have to be lowered by hoist into the bath) can in emergency be used as a secondary treatment room for non-sterile dressings and procedures such as enemas.

Other ancillary rooms needed are further bath and shower rooms, wcs and washing cubicles for patients if these are not provided off the bedrooms. A sister's room can well be designed and furnished more as a sitting room than an office, as its major function is confidential discussions with doctors, relatives and nurses. The size and grouping in relation to nursing sections of ward pantries and cleaners' rooms depend on whole hospital catering and cleaning policy. Flowers are sometimes arranged and vases kept in a special bay; otherwise this may be combined with a cleaners' room. Planning of corridors must allow for regular or occasional parking of linen, food and other trolleys, and of wheelchairs. Storage for bulky equipment is needed by some specialties (eg orthopaedic) more than others, but in all some provision should be made.

4.04 Patients' bed and day rooms

The conflict between patients' comfort and nursing convenience is illustrated by the main difference between current British and American designs. In the former, most patients are in multi-bed rooms or bays (generally of 5 or 6 beds), with relatively few (between 10 and 25 per cent) in single rooms, the latter being for those with special nursing or social needs

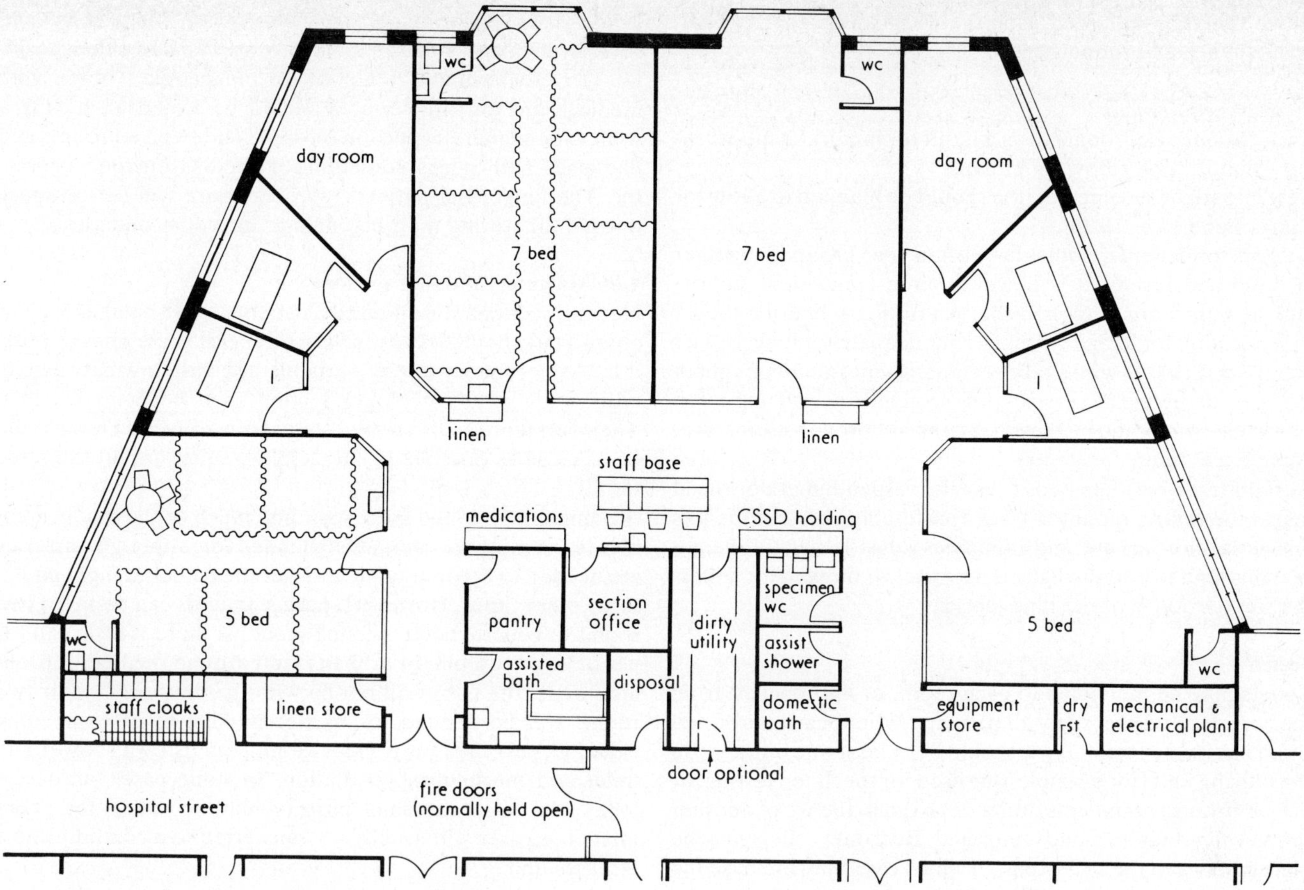

18.9 *28-bed nursing section designed for observation of 80 per cent of patients from staff base, and minimal walking distances*

(eg frequent observation and attention, infection, and for the noisy, the dying). The doors and partitions to the corridors are usually partly glazed for ease of observation, although these may be covered by bed curtains. The doors to multi-bed rooms have in some cases been omitted. Even if this option is taken, a frame to which doors can be fitted facilitates their addition later if required. **18.8** and **18.9** show two attempts at improving observation in nursing sections, each with four multi-bed and four single rooms. Many nurses still favour open Nightingale wards for good observation, especially at times of staff shortage, and they even have some advantages for patients (see *Ward Evaluation, St Thomas' Hospital*). It seems unlikely that new open wards will result, but upgrading is less likely to mean the subdivision of existing ones.

In North America semi-private insurance cover has led to widespread use of two-bed rooms. These provide the least privacy as the patient is aware of everything that his neighbour does. Compatibility is always a problem. The trend is now towards hospitals with entirely single rooms. A higher occupancy is possible with more (or all) singles, and risk of cross infection is reduced—but singles are much harder to observe, and require more staff. Two-way speech nurse call systems, and even closed-circuit television, have been used in mitigation. Bedroom space per patient is between 25 and 50 per cent

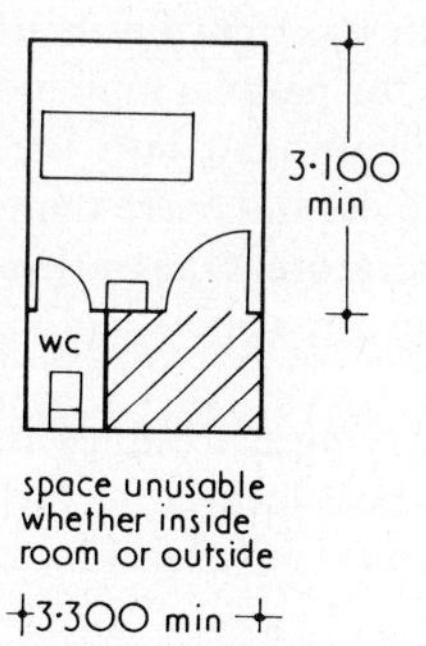

a *Conventional single room*

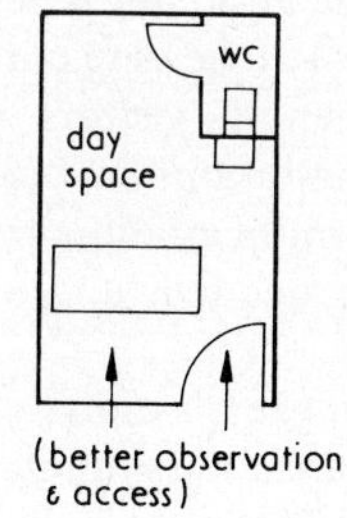

b *Improved single room*

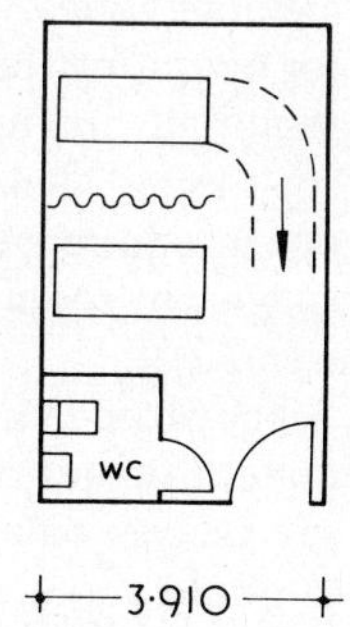

c *Double room. Note: must be wider to allow beds to pass*

18.11 *Bedroom details. Note: single-leaf 1200 mm doors are shown; in many ways preferable to more common 1½ leaf type*

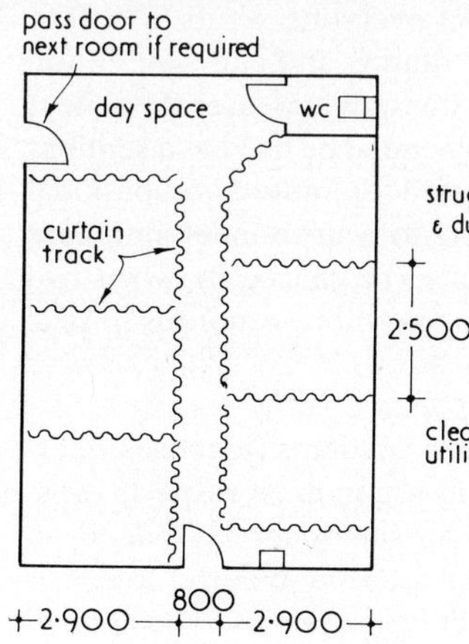

d *Six-bed room with own wc and day space*

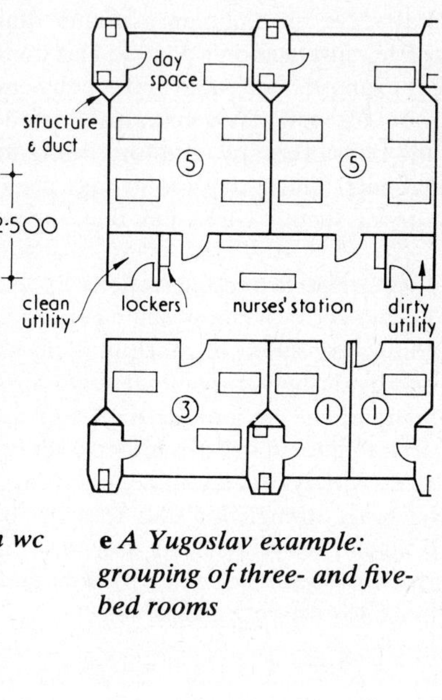

e *A Yugoslav example: grouping of three- and five-bed rooms*

above that for multi-bed rooms; this means higher capital, maintenance and running costs.

Bedroom equipment

Each patient's bedroom needs a basin for medical hand washing—that is, with elbow action taps.

New in-patient accommodation should be planned to allow for King's Fund Bed, **18.10**.

Lockers for patients' clothes have often been grouped together in multi-bed rooms, it is however more convenient for the patient if they are at his bedside. An ordinary bedside locker with a section for hanging clothes is inadequate for long dresses and coats. Lockers with a taller compartment sometimes get in the way of bedhead services. In Scotland, a wall mounted wardrobe combining bedhead services has proved useful, but, being fixed, limits flexibility.

Curtain track between beds is usually suspended at doorhead height. A ceiling mounted track (preferably recessed) is less unsightly, more secure, and eliminates a dust-ledge; full-height curtains tend to obscure light and restrict air movement, unless the top portion is of net-type fabric.

Toilets

Toilets may be individual to each room, or approached from the corridor and shared by all patients. Where each multi-bed room has its own toilet there should be at least one other wc in the nursing unit (for example, one used for specimen testing, or in a bathroom) as patients prefer not to go to the wc of another room when their own one is engaged. Bedroom toilets may be placed between the bedrooms, in which case they increase the units' length (and the walking distances), or on the corridor side, when they make the patients further from the staff, and impair visibility. A third option that is in most respects preferable is for the toilet to be on the perimeter wall; this has the further advantage of increasing the usable day space in the ward, and allowing the toilets to be naturally ventilated, **18.11**.

Day space

For day space on the nursing unit, Building Note 4 recommends a total of 1·7 m² per bed, but a somewhat smaller area than this is acceptable in view of the shortening stay of patients. If some of this space is within the bedroom and some outside, the patient can experience a progression as he becomes more ambulant. In many hospitals, a 5-bed room has a day space equivalent to that of a sixth bed. The bay window as day space, which is a feature of the best buy hospitals, **18.6c**, is one of the most pleasant solutions.

The day space away from the bed areas is better as two small rooms than one larger one, as not everyone wants TV. One may serve for ambulant patients' dining, and can well adjoin the nursing section's pantry. This division increases flexibility, for example, one room may when needed be used as a seminar room for staff. They may also be used for limited occupational and physiotherapy, and for visitors to wait to meet ambulant patients. These day rooms may also be shared between two nursing sections, and may thus appear more generous in size.

4.05 Geriatric nursing sections

The needs of the more acute geriatric patients (ie those undergoing assessment or rehabilitation) are in most respects satisfied by designs for general acute nursing sections. Indeed, so many of the patients in the latter are now elderly, that it is desirable for standards hitherto set for geriatrics to be applied more widely. For example, many, and perhaps all wcs should be large enough for staff to assist patients, and some to be designed for wheelchair use. Also, handrails should be provided on both sides of corridors and elsewhere where assistance for walking is desirable.

Longer-stay geriatric units, especially for the elderly and severely mentally infirm, should be more like home than hospital. An environment rather of an old people's home with medical and nursing services should be aimed at, having a somewhat higher proportion of single and two bed rooms and day space, with at least something of a bed-sitting room character. The longer the patients' stay the more will be the space needed for storing their belongings, including suitcases.

4.06 Maternity nursing sections

Nursing sections for maternity patients (post-natal as well as ante-natal) should be as much like general acute ones as possible. As occupancy is very variable, ebb and flow between an adjoining (eg gynaecology) section then becomes feasible. Their size tends to be smaller—eg about 25 beds. They should be as close as possible to the delivery suite and on the same floor.

'Rooming-in'—ie the baby spending much of the day in a cot beside the mother—should be planned for, although nurseries are needed to accommodate most of the babies at night and at some other times. However, space standards can be as for bed rooms in general acute nursing sections: there is little call for use of large equipment or many staff around the bed; if problems arise, the baby will not be with the mother. One or two rooms will be needed for barrier nursing of cases needing isolation: this calls for a single room with its own shower and toilet and mechanical ventilation. In some cases, an access lobby and separate small nursery may be called for. Very infectious cases will usually be transferred to a communicable diseases unit.

4.07 Children's nursing sections

The same principle as for maternity, of planning to a similar overall pattern as for general acute nursing sections, applies. However, the need for observation is greater, and of privacy less, so more partitions should be glazed. A number of bedrooms are needed in which a mother can stay with her baby or young child. Play space is required, and so is space for teaching. These may be shared with other functions. A quite different approach to interior design and furnishings is needed, to make a setting as homelike as possible, without compromising medical or nursing efficiency.

Special care baby units, although the responsibility of the paediatrician, functionally form part of the maternity department, and should adjoin or be very close to the delivery suite. They are for premature or other newborn babies who need continuous very close observation with facilities for control of infection and micro-climate. Many will be nursed in incubators. The extent of special gowning and changing precautions varies, but the discipline is generally less rigorous than for operating theatres. A viewing corridor for parents may be required, but increasingly, visitors are encouraged into the cot-rooms. These usually contain 4 to 6 babies, and more than one is required, to permit variation in temperature. One or two single cot-rooms are needed, and a room for sophisticated treatment.

A baby feed store and preparation room, serving also babies in maternity and paediatric nursing units, is best located in the special care baby unit; it should include a breast milk bank and mothers' room.

4.08 Intensive therapy units

This term is used in preference to 'intensive care', as indicating that patients are receiving treatment, involving much medical attention and complex equipment, as well as a very high level of nursing care. There is a major difference between British and American practice: the British ratio of 1 to 2 per cent of the total acute beds (excluding geriatric and psychiatric) tends to produce units of between 6 and 10 beds. In the USA, much larger units are common; this may be because the hotel type arrangement of general nursing units gives inadequate observation of patients with high nursing dependency who thus need to be in the ITU. The US Public Health Service has made recommendations for design of cardiac care units consisting of 4 or 5 single rooms with glazed partitions. In Britain, combined cardiac and general ITUs are acceptable if in two zones; a large hospital with a heavy cardiac service may provide a cardiac unit in association with the ITU. Alternatively, a cardiac unit may be planned in relation to medical nursing sections. Proximity to the operating department is highly desirable, and a link to the post-operative recovery unit is possible.

A large open room will provide for most patients, but two single rooms with glazed partitions are generally provided, especially where infection is involved. Generous space is essential, **18.12**; continuous rails are often provided at the bedhead, onto which monitoring and other equipment may be clipped (in the USA, ITUs with all single rooms are becoming more common). Windows are essential to prevent disorientation in patients. A staff room and a relatives' room, with windows (and a view if possible) and a totally non-clinical decor, are necessary for contrast with the highly clinical nature of the rest of the unit and its stressful nature for all concerned.

4.09 Communicable diseases units

These are sometimes referred to as isolation or infectious diseases units. In British practice, a unit of 20 beds (all in single rooms) is considered to be the minimum viable size for economy of nursing. It takes all serious cases of infection arising within the hospital, including dermatological ones. Units which are designed to receive cases of infection severe enough to make it hazardous for patients to arrive by any internal route through the hospital should have their own external entrance; this may lead to a ground floor location.

Occupancy is very variable. At Greenwich, an 18-bed unit was planned with a major section of 9 beds, and others of 4 and 5. The latter could be shut off when not needed for their main purpose, to form a single bedded staff sick bay.

Similar precautions concerning staff changing are required to those for special care baby units. Special consideration must be given to handling of disposal items, including food service, and to ventilation. Maximum use of disposables is desirable. In some units, but not all, each bedroom has a lobby, with a basin and provision for gowning. Each bedroom should have its own shower and WC.

4.10 Day care sections

A number of minor surgical and investigative procedures that have traditionally meant one or more night's stay as an in-patient are now dealt with on a day basis. The patient comes in early in the morning; if a minor operation is required, he goes to the theatre, and spends the rest of the morning recovering in bed. Such food as is appropriate is brought to him; as soon as he is ambulant he may go home with an escort or spend the afternoon in a day area. If he is not well enough to go home, he can be admitted, but this seldom proves necessary.

Some of these units have been provided with their own operating theatre and major treatment room (for cystoscopy etc). It avoids duplication if they are near the operating suite, and shares its facilities with in-patients. The size of the unit depends on local need; there are examples of between 12 and 25 beds. Due to the shortness of stay, larger multi-bed rooms may be acceptable, and one or two singles should be provided. A large and comfortable sitting up area and lockers are also needed.

5 HOSPITAL DEPARTMENTS: DIAGNOSTIC AND TREATMENT FUNCTIONS

5.01 Out-patient department

On arrival at the OP reception counter, the patient is directed to a sub-waiting space under the control of receptionist who looks after one or more clinics.

Design of suites

Some specialties (eg dentistry, ear, nose and throat, and ophthalmic) need purpose-designed suites of consulting and ancillary rooms, because of special equipment. But for the majority, the same spaces can be used in one session (a half day) for one specialty, and in another for a different one. Hospital Building Note no 12 illustrates two types of clinic layout, type B in which each consulting room has close to it one or more examining rooms, and type A consisting of combined consulting/examining (C/E) rooms **18.13**. In the former, those patients requiring

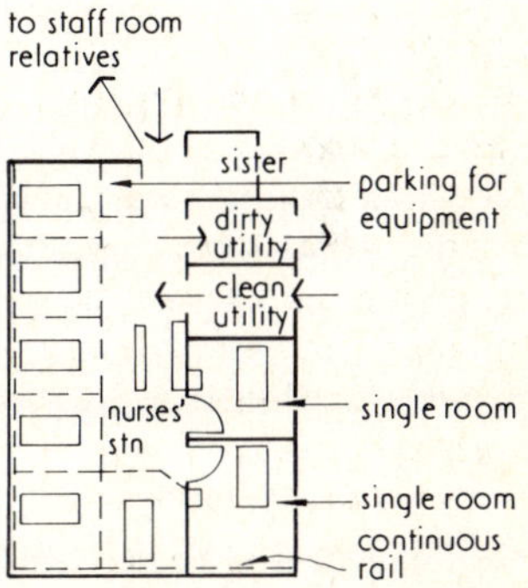

18.12 *ITU. Note: cubicle curtains not used, but movable screens may be. Location of bed within space variable with needs of patient, staff and equipment*

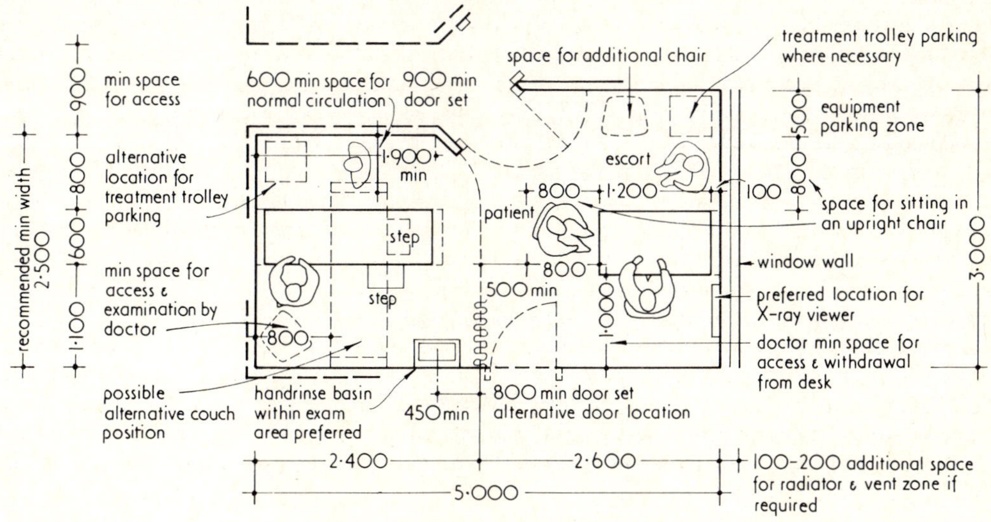

18.13 *Consulting/ examination room, minimum space requirements, preferred plan type A*

examination move to a separate room, where they undress, are examined, and dress again. (Separate changing cubicles are not desirable; patients in them can be overlooked, there are problems of security for clothes left in them, and they tend to be disagreeable environmentally.) In the latter, the patient remains in the one room: the doctor may, following consultation and while the patient is undressing, move to see a patient in an adjoining combined C/E room. For this reason, intercommunicating doors between C/E rooms are often provided; these need to have a high level of sound attenuation, or the essential confidentiality will be lost.

This latter arrangement provides maximum flexibility. In a clinic where most patients undress, and there is a rapid throughput, each doctor will need at least 2 C/E rooms; a typical 'group' consisting of consultant, registrar and house officer may occupy a 'string' of 6 or 7 C/E rooms. In a specialty like psychiatry, however, where each patient takes much longer, and physical examination is less general, each doctor will need one room only. Thus C/E rooms will be allocated in quite different groupings for each session. This facilitates a very high level of occupancy; at Greenwich, where this arrangement has been used for all general clinic purposes, a space reduction of the order of 30 per cent was achieved, **18.14**. Strings of at least 6 rooms, and preferably 12, optimise flexibility and thus occupancy, but the Scottish Building Note no 7 recommends a layout with strings of only 3 rooms, with careful attention given to outlook from and amenities in waiting spaces. A combination of the efficiency of Greenwich and the environmental concern shown by the Scottish guidance is likely to lead to the optimum solution.

Ante-natal clinics

Ante-natal clinics have traditionally been planned as separate departments. The rapid throughput, with much more time spent on examination than consultation, has led to a layout of open ended cubicles, with staff moving rapidly from one to the other. This is unsatisfactory. Although a combined C/E room type of layout takes more space, it enables this function to be carried out in part of the general OPD, with consequent improvement in overall economy and utilisation of space. (Hospital ante-natal work has been diminishing due to the declining birthrate and due to more work being done by GPs, especially in health centres.) A large space that can be enclosed—traditionally the ante-natal washing space—is however needed for parentcraft and other class work including showing of films.

Orthopaedic and fracture clinics

Orthopaedic and fracture clinics are often provided as part of the accident and emergency department, as many of their

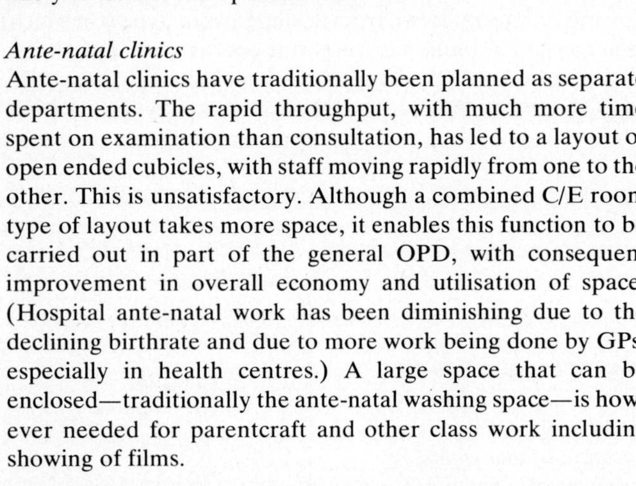

18.14 *Greenwich District Hospital out-patient department ground floor:*

1 *sister*
2 *clean supply*
3 *dirty utility*
4 *electrocardiographic technician*
5 *electrocardiography*
6 *toys*
7 *cleaner*
8 *clinical assistants*
9 *consulting/examination rooms*
10 *treatment*
11 *pathology out-station*
12 *audiometry*
13 *reception*
14 *porter*

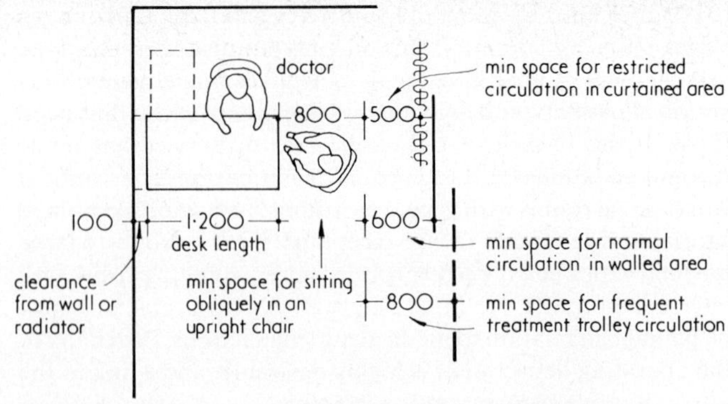

18.15 *Activity space requirements for room depth in consulting areas*

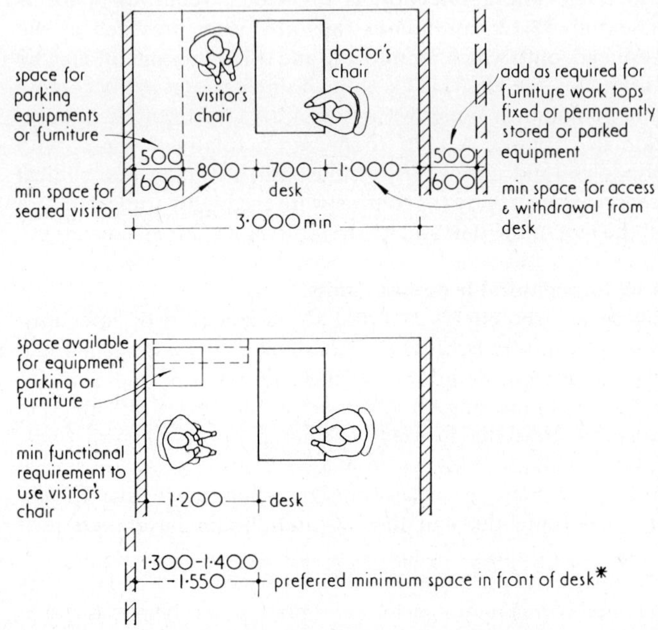

18.16 *Activity space requirements for room width in consulting areas*

** this functionally derived minimum dimension of 1200 mm tends to be psychologically unsatisfactory. The space in front of the desk should be significantly larger than that behind.*
1300 mm gives more flexibility in arrangement and use of space in front of the desk, and is more psychologically acceptable.
1400 mm is the minimum space permitting movement past a seated visitor.
1500 mm will allow passage behind a seated visitor

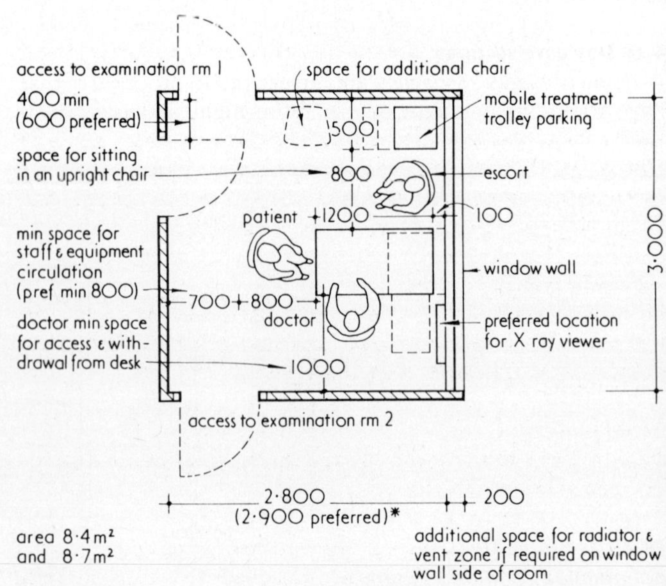

18.17 *Minimum space required for a separate consulting room, preferred layout*

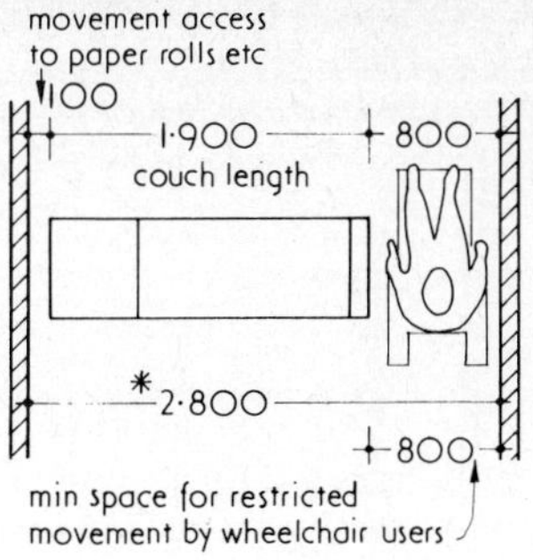

18.18 *Activity space requirements for room lengths in examination areas*

a *Access at foot end of couch for wheelchair movement*
** this dimension is also the preferred minimum room length where standing workspace at foot or head end of couch is required*

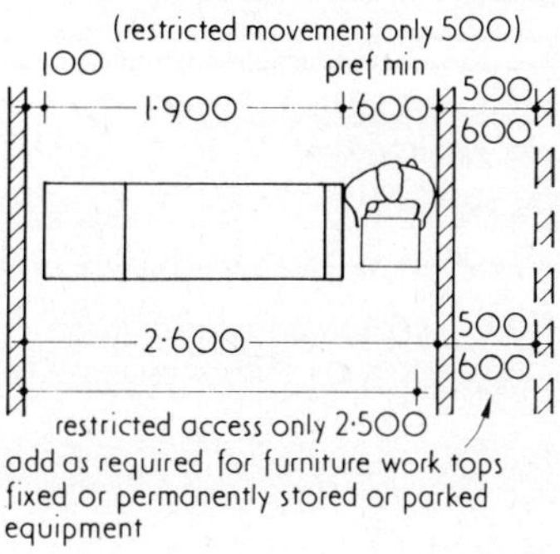

b *Where wheelchair movement at foot end not required*

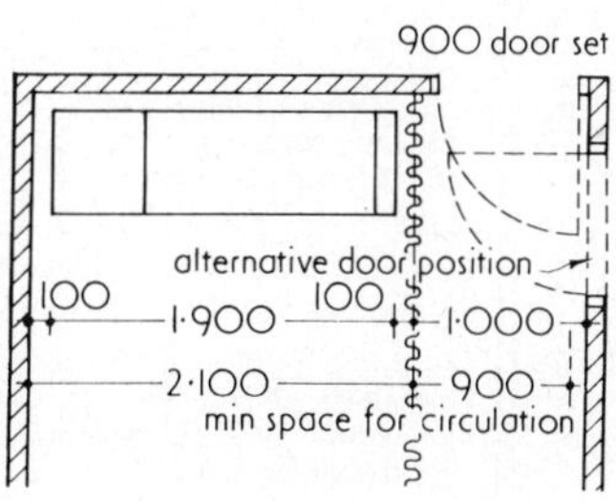

c *No access across foot end of couch*

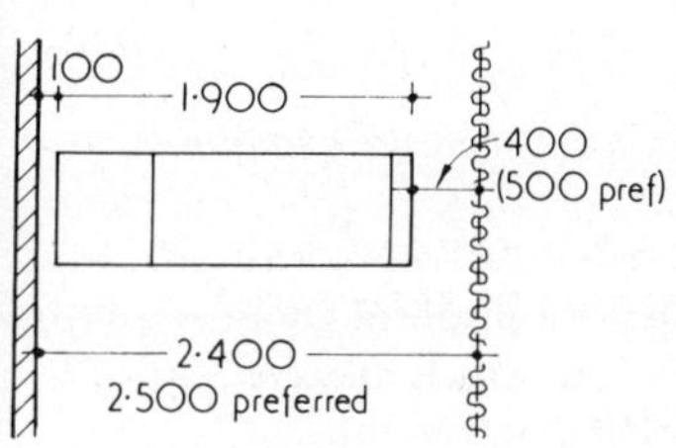

d *Minimum for restricted sideways access within curtained area*

18.19 *Activity space requirements for room widths in examination areas*

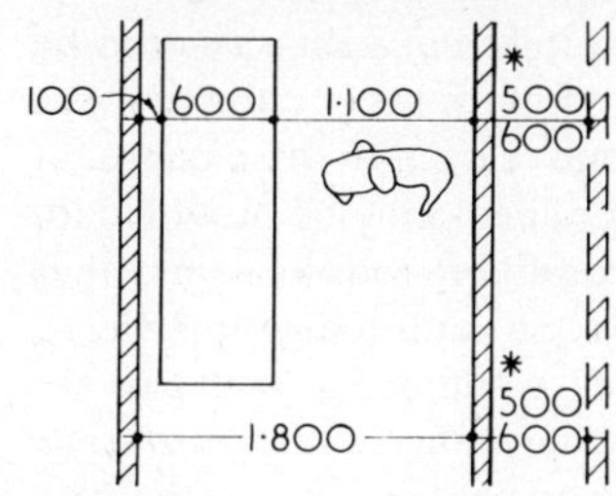

a *Access to one side of couch only*

1100 mm is the minimum space for an ambulant patient changing

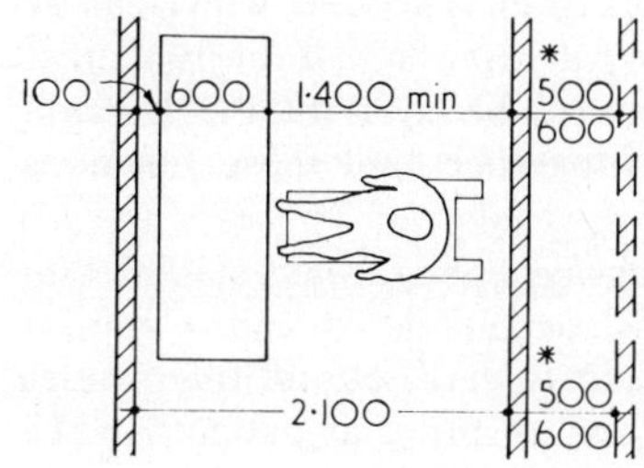

b *Access to one side of couch only*

1400 mm is the minimum space for a wheelchair patient changing

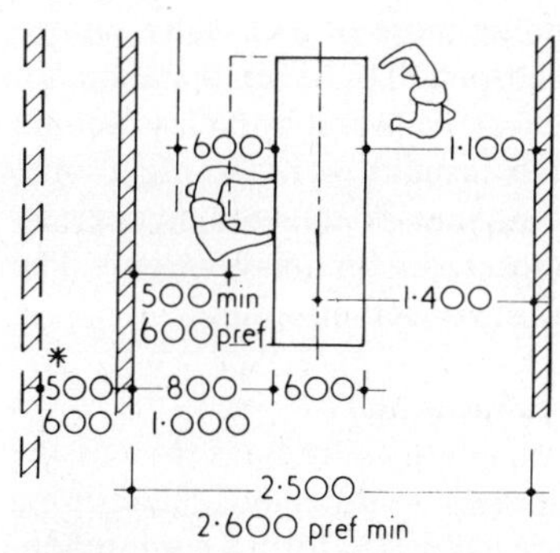

c *Access to both sides of couch*

600 mm is the essential unobstructed space for access and examination
1100 mm is the space at side of couch for changing
1400 mm is the space at side of couch for wheelchair access
800–1000 mm is the clear workspace at side of bed or couch for examination and treatment, preferred minimum 900 mm

** add as required for furniture worktop fixed, or permanently stored or parked equipment*

patients are receiving follow-up treatment resulting from injuries, and as it, like the AED, needs a room for application and removal of plasters (whose floor finish must cope with plaster dust, and whose sink needs a plaster trap). They too have a rapid throughput, and predominance of examination; they may be part of a general OPD with C/E rooms, or purpose designed, with separate examining rooms or cubicles.

Occupational health service
An occupational health service is usually now provided for hospital staff. In new hospitals this may take the form of a small adjunct to the OPD (see *DHSS Design Guide,* 1973).

Common areas
18.15 to **18.21** show typical standards for areas occurring for OP and other departments.

5.02 Accident and emergency department
The two main kinds of patients are those seriously ill or injured brought in by ambulance, and the 'casual' or walking sick. The different needs of these types requires two entrances (screened from each other). For the former, a short direct route to a fully equipped resuscitation room (or rooms) is essential; the patient may then need X-ray or surgery. Whether these facilities are provided as part of the accident and emergency department (AED) depends on the size of the department and proximity of X-ray and operating departments. They are however expensive to install and run and difficult to staff, so if the X-ray room that is used for emergency work has direct access from the AED, duplication can be avoided. Similar access to the OD (operating department) is more difficult, especially as the AED naturally tends to be on the ground floor, and the OD, if placed on the same floor as most surgical beds, is likely to be on an upper floor. (In this case, a lift should be reasonably close. A special lift should not be needed, if a general lift can be taken over in emergency, and the patients' route does not pass through corridors much used by the public.) A common compromise is to provide minor theatres or major treatment rooms in the AED, in which some surgical procedures can be carried

out. These do not generally require such rigorous disciplines of staff changing as the main theatres.
The walking patient will come to a reception desk and wait until he can be examined by a doctor; if treatment is needed, he will go to one of a number of treatment rooms or cubicles. If he proves to be seriously ill, he will move over to the emergency side of the AED. A small number of beds may be provided for observation of patients whose condition requires this; they may after a while be discharged, or else admitted as in-patients. This observation unit may be used for admission of all patients arriving at night, who go the next morning to an appropriate nursing unit. This avoids the disturbance that is otherwise caused to other patients by night admissions, and the unexpected workload on night staff. It is however becoming more common to admit observation patients directly to a day care or other nursing section. A plaster room is needed, which can be shared by the orthopaedic and fracture clinic, if this is nearby. In smaller hospitals, some space can be saved if OPD, fracture clinic and AED are planned together, and these functions overlap.
It is crucial that the AED is always open, always staffed with doctors and nurses, and with consultants on call. Compact planning should enable the staff to carry out all their duties with a minimum of walking and ensure that patients in the observation unit are not overlooked.
Because of the increasing emphasis on the needs of the most critical cases, the casual element which includes people not registered with any GP tends to be discouraged. In Britain, some health centres are providing more of a casualty service (especially those remote from hospitals) but even these are not open at night. In big cities, new arrivals and homeless people account for some of the hospitals casualty work. At night some people brought in are drunk, and may be infested, necessitating a bathroom for cleansing and a room for noisy patients. The latter will usually form part of the observation unit.

5.03 Comprehensive child assessment unit
Special units provide for assessment of children as soon as any variation from normal development is suspected. These bring together the many skills that are needed to form a comprehensive understanding of the physical and mental condition of each child, and to prescribe appropriate care and treatment. Medical specialties involved include ophthalmics, ear, nose and throat (ENT), and orthopaedics, and others include psychology and speech therapy. Rooms with observation panels for watching children at play are required. These units may not be provided in all DGHs; they should be physically associated with the OPD; and special provision (eg for audiometry) should be shared with those of the ENT part of the OPD. Facilities for case conferences and seminars can also be shared with adjoining departments.

5.04 Operating department
This brief introduction to the design of the most technically complex of hospital departments cannot be exhaustive. Much has been published on the subject, and designers should study the references quoted. *British operating theatres* contains several useful articles, for example, on wall finishes and sterile air, and illustrations of 12 recent designs. Its first chapter includes the following useful definitions, which have also been adopted in this chapter:
'An operating department is a unit consisting of one or more operating suites together with ancillary accommodation provided for the common use of these suites, such as changing rooms, rest rooms, reception, transfer and recovery areas and circulation space.
An operating suite comprises the operating theatre together with its own ancillary areas, namely: anaesthetic room; a room

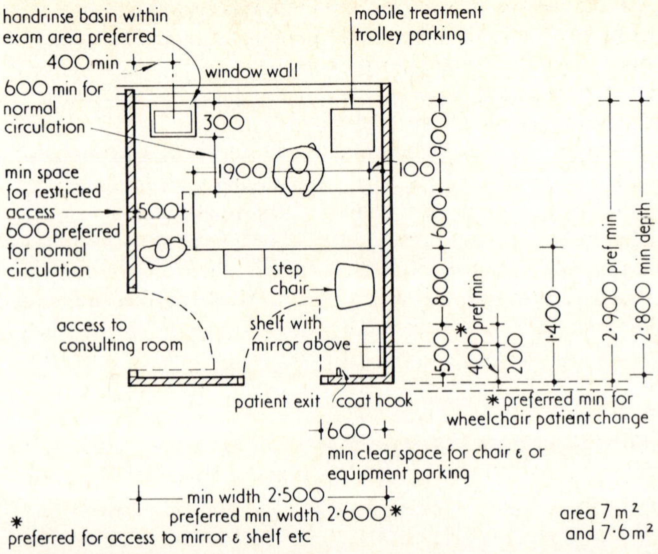

18.20 *Minimum space required for a separate examination room, preferred layout*

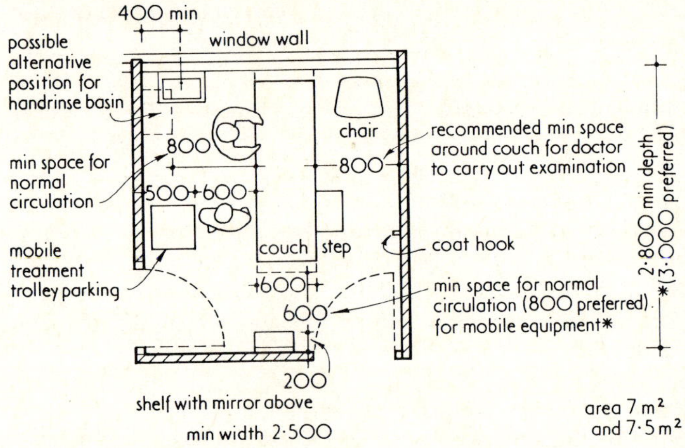

18.21 *Minimum space required for a separate examination room, alternative layout*

for preparing instrument trolleys; a disposal room; scrub-up and gowning area; and an exit area which may be part of the circulating space of the operating department.
An operating theatre is the room in which surgical operations and certain diagnostic procedures are carried out.'
The British Hospital Building Note No 26 contains valuable guidance on designing to minimise infection, and on ventilation and other engineering services. The Nuffield report has a long chapter on operating departments, which includes the results of thorough research; in spite of its date much of it is still relevant, as is chapter 12 of Gainsborough's *Principles of hospital design.* A joint working party chaired by Dr Lidwell, on *Ventilation in operating suites*, produced a report in 1972, published by the Medical Research Council and DHSS.

Size and location
Centralisation of all theatres in the hospital makes for flexibility and economy in deployment of staff, and thus reduces running costs. By improving utilisation it can lead to savings in the overall number of theatres, and avoids duplicating ancillary accommodation. Operating departments should be designed so that (with a very few exceptions) any theatre can serve for any purpose. Some specialties, such as ophthalmic surgery, call for some additional equipment, but one-off design should be minimised in the interests of long-term flexibility: once this department is in use, any major modifications prove extremely disruptive to the whole work of the hospital.

The number of theatres required is determined by the number of surgical beds, the length of in-patient stay and the number of operations performed per theatre. Current British practice is to provide one theatre for every 40 surgical beds. Because of the common configurations of operating suites, and the sharing of exit bays, it is usual for theatres to be designed in pairs. In spite of the advantages of pairing, at times of financial stringency an odd number of theatres may be acceptable.

The preferred location is on the same floor as all—or as many as possible—of the surgical beds, and in particular of the ITU. It should also be close to the AED. These requirements are generally incompatible. Since the most urgent work on accident cases is usually done in the resuscitation or major treatment room of the AED, a short journey, including use of lift, may be acceptable.

Although it is usual and preferable for the whole department to be on one floor, there are examples of two-storey layouts, especially where the floor space at each level is restricted. The elements that can most easily be on another floor are staff changing and sterilising services. (It is however now more common practice to provide an HSDU (hospital sterilizing and disinfecting unit) associated with the hospital's service zone rather than with the operating department.) The provision of an upper level viewing gallery is sometimes still required in teaching hospitals, but use of closed circuit television tends to obviate this need. In a hospital that is not generally air conditioned, the extensive plant required for this department is usually placed immediately above or below it. Elements like staff changing or viewing galleries can sometimes be placed in spare areas of a plant room floor.

This department location should permit subsequent extension; where an open-ended situation is impossible, an area into which it can grow can initially house functions whose position in the hospital is not critical and which can then be relocated.

Planning the operating department

Flow patterns for people, air and services that avoid causes of infection reaching clean areas are crucial. Air-conditioning systems are designed so that critical areas such as the theatres are pressurised to minimise entry of air-borne contamination. Separate venting of anaesthetic gases is now generally required.

When soiled items are bagged in the theatre, it is microbiologically acceptable for a single corridor to serve for all circulation of people and things. The first 'best buy' hospitals, at Bury St Edmunds and Frimley, were designed in this way, but the Mark 2 version at King's Lynn (which will be completed by the late 1970s) has a back disposal corridor, **18.22**, which has advantages of fire escape, access for infected patients, and servicing by engineering staff. Sterile supplies, however, are taken to the theatre by the main (or 'clean') corridor, which is the main internal circulation route for staff and patients.

Staff change on arrival into theatre clothes; toilets, basins and showers are needed with the changing rooms. One set of changing rooms for men and one for women is now usual rather than the former extravagant and anomalous practice of separate facilities for surgeons, nurses and orderlies. They emerge into the clean zone; the changing rooms thus form a barrier between this zone and the rest of the hospital. The staff enter the theatre via a scrub-up room. Another such barrier is the transfer area where the patient is moved from his bed to a theatre trolley. On the trolley, he goes to the anaesthetic room where he is anaesthetised. From there he is wheeled into the theatre. Sterile packs and trays are brought in from a supply or preparations room. After the operation, all items to be disposed of are bagged and taken out into the disposal corridor. The patient goes out by way of an 'exit bay' to a post-operative recovery area. Building Note no 26 recommends that this recovery area should contain 1½ beds per theatre. In North America there are generally no anaesthetic rooms, as the patient is anaesthetised in the theatre. This reflects a practice of performing far fewer operations per theatre per day, and thus needing more theatres per bed. American operating department designs are much simpler, but the average overall cost related to throughput is probably fairly similar on both sides of the Atlantic. A dark room, for X-ray processing, is usually needed; the X-ray machine will be mobile. Rooms or bays will be needed for resuscitation and monitoring equipment, ventilators and heart–lung machines.

The operating theatre

Theatres with an area of about 38 m^2 are adequate for all but a very small proportion of operations, which need extra space because of numbers of staff or amount of equipment involved. A square room is most usual, **18.23**, and tends to make for easier planning of the suite and the whole department. Polygonal or circular theatres have been justified as corresponding more closely with the space occupied during operations. Modular theatres are often octagonal. These may tend to be a little more expensive. Their advantage however is speed of installation, especially in the use for which they are most obviously suited, namely when an existing building is converted into an operating department. They also have an assured high quality of finish.

The ceiling height should be at least 3 m. Theatres as low as 2·7 m have proved adequate, but are difficult to design, especially with regard to the fixing of the suspended theatre light. Access to all services, for their maintenance and cleaning, should be from outside the theatre. For example, it is an advantage for the general lighting to have fixed lay-lights with access from above; a service floor above the theatres is thus beneficial.

One radical variation from the typical theatre layout that has been described and illustrated is the type of downward air-flow theatre, pioneered at Wrightington Hospital, Wigan, for hip-joint replacement, a procedure needing a very high degree of sterility. The operation takes place in a small enclosure with the minimum of staff inside—for example the patient's head and the anaesthetist are outside. Materials are passed in and out through hatches as required. Very highly filtered air is directed down onto the patient; this air finds its way out at the base of the enclosure. This technique has been adopted in the new Nuffield Orthopaedic Centre at Oxford; here there are two such enclosures in one double sized theatre. If this method comes to be used widely for other kinds of surgery, it is possible to envisage the main part of an operating department as a large open space, containing several of these small operating enclosures of 3 × 3 m or less. This could allow much greater freedom of planning, and probably some saving of space. A room within the operating department especially for endoscopic procedures (such as examinations of the urinary tract) is hard to justify unless the workload demands it; in this case it can be smaller than a full-scale theatre. Where there is a significant amount of orthopaedic work, a plaster room is usually provided.

Theatre windows are not needed—indeed, where they are provided, blackout facilities are required, and solar gain should be avoided, but some surgeons feel that an external view is a psychological benefit. It is hard to provide an outside wall, but a borrowed light across a corridor may more easily be achieved. If daylighting cannot be provided for the theatre, it is important that the theatre staff's rest room (or rooms) should have a pleasant view, and be designed and furnished in a manner to contrast with the inevitably clinical environment of the rest of this department.

5.05 Central delivery suite

Some delivery suites contain operating theatres for performing Caesarian sections. Such theatres are unused for much of the time, and it is more economical for procedures known in advance to need general anaesthesia to take place in the operating department; this does however imply proximity of the operating department and central delivery suite.

In cases where general anaesthesia is not anticipated the mother comes to the delivery suite reception point, where her notes will be brought without delay. She is then taken to the room where the delivery will, if normal, take place. This room, which must provide space for up to 6 people (including the husband) should be about 15 m²; natural daylight is required. In spite of a need for medical gases, and lighting adequate for suturing, the room should not look too clinical.

If it becomes evident that it will not be a normal delivery (eg if forceps are required), the mother is moved to an abnormal delivery room of about 24 m². One of these rooms will be used in emergency for unscheduled cases of Caesarian section.

The number of delivery rooms depends on the birthrate and age structure if related to population, and on length of stay if related to numbers of maternity beds. For a current British project, 9 normal and 3 abnormal rooms are being proposed for a population of about 200 000, but the need must be assessed individually to meet each local situation.

Ancillary rooms include toilets (1 per 2 delivery rooms for mothers), day room, clean and dirty utility rooms and staff bases for both normal and abnormal delivery rooms, bathrooms, showers and staff room.

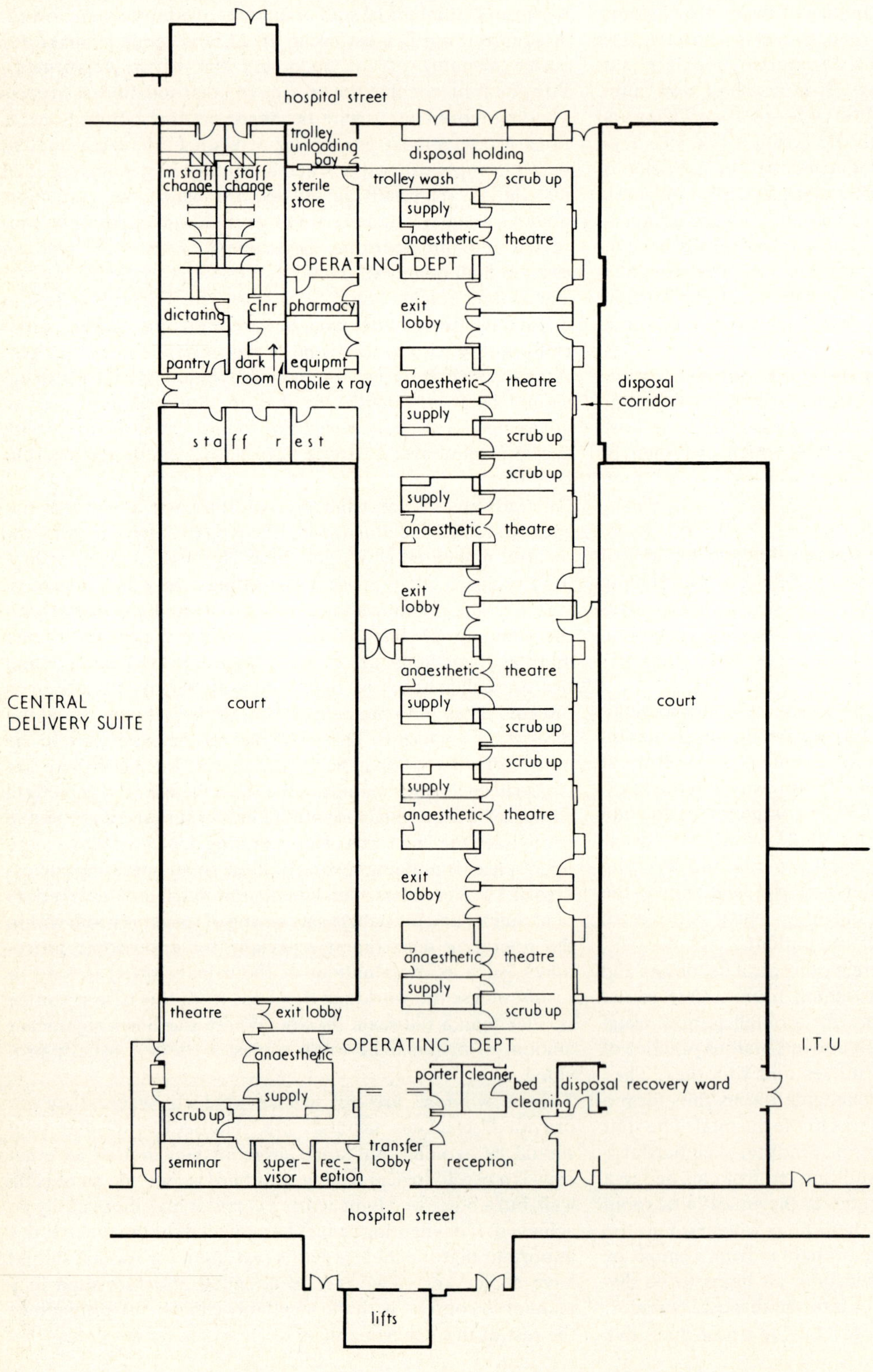

18.22 King's Lynn District Hospital, operations department with back disposal corridor

5.06 Diagnostic X-ray department

This title distinguishes the main use of X-rays for diagnostic purposes. Therapeutic use is referred to as radiotherapy, and is provided in certain hospitals as a sub-regional specialty.The use of radioactive materials for diagnostic and therapeutic purposes is also generally a sub-regional specialty, although a small unit may be provided at DGHs, associated with the pathology or the diagnostic X-ray department. More extensive and specialised facilities for radiotherapy and radio-isotopes may be provided on a regional basis, usually in teaching hospitals, and may be associated with a department of medical physics or nuclear medicine. Head or body scanners (for computerised axial tomography) will also be provided as regional or sub-regional specialties.

The use of X-rays for diagnostic purposes has been growing steadily, although some of the new techniques (such as scanning) may drastically alter the situation within the lifetime of the hospital. Provision of growth for the diagnostic X-ray department, or additions to related functions, should be possible, either by extension, or by conversion of adjoining 'soft areas'. It should be close to the OPD, but access for in-patients should also be as direct as possible.

At the entrance to the department, a reception and records area should have waiting space for ambulant patients to sit, and (in a screened area but within sight of reception) for patients in bed, trolley or wheelchair to wait before or after their X-ray. In the records area, clerical work takes place, and current records and film are stored. Longer-term storage of X-ray films and records can be in a separate archive. Film is very heavy to store. The most intensively used radiodiagnostic rooms should be nearest the entrance, and the room or rooms used most for accidents are nearest the AED. The more functions that each room can carry out, the greater the throughput the department can have. It may be a long-term saving to make all rooms a fairly generous size—say 34 to 38 m²—in the interests of long-term adaptability. Larger departments can justify a much smaller room for chest X-rays. The equipment in each room usually includes transformer and control gear, the X-ray tube in its stand or support (ceiling suspended, floor mounted, forming a floor to ceiling column running in a floor track, or part of the table), the X-ray table, and the control console behind the glazed screen with radiation protection from which the radiographer operates.

The design of X-ray machines is constantly developing, and each radiodiagnostic room should be able to accommodate more than one possible layout, both initially and subsequently. Maintenance is always under contract to the manufacturers, who can also advise on detailed arrangement of equipment. The danger of ionising radiation necessitates screening of partitions and doors. The Regional Health Authority's radiation protection officer will advise on this subject. Much equipment is suspended. A fixing direct to the soffit of the structure eliminates the need for complex strengthening of suspended ceilings. The room must have a clear height at least 3·1 m above floor level to allow the movement of equipment. Floor services will often require a greater depth of screed than usual. The nursing element in the X-ray department should be near the radiodiagnostic rooms used for special procedures and angiography.This includes a staff base, patient recovery area, lavage room with wc (for patients needing an enema before examination by the radiologist), another wc, shower, clean and dirty utility rooms.

Rapid automatic daylight processing is becoming increasingly common, one such machine generally serving a pair of radiodiagnostic rooms; even so a small darkroom is still needed as a standby. A viewing, photographic checking, and medical reporting area is needed close by. (In smaller departments, radiologists may prefer to make their reports on the films in their own office.) Other ancillary rooms include offices, storage (including unused films), seminar, and staff rooms.

Two newer diagnostic techniques which may be included are ultrasonics and thermography. Each needs a room of 17 to 20 m²; no screening is called for. The ultrasonics room should be close to the patients' wc and shower.

5.07 Pathology department

This department is called 'laboratories' in North America. It carries out tests on specimens received from in-patients, out-patients and those outside the hospital (eg at the request of a GP). The results of the tests form an aid to diagnosis or treatment for those who have requested them. It generally consists of four main sections:

- *Morbid anatomy* (sometimes called 'histology', or simply 'pathology') includes histopathology, the microscopic study of diseased tissue, and cytology, the study of individual cells. Tissue is often received from operating theatres or the post-mortem room. In the former case, an urgent test is frequently required in the course of an operation. The specimen is brought to this laboratory, a 'frozen section' carried out, and

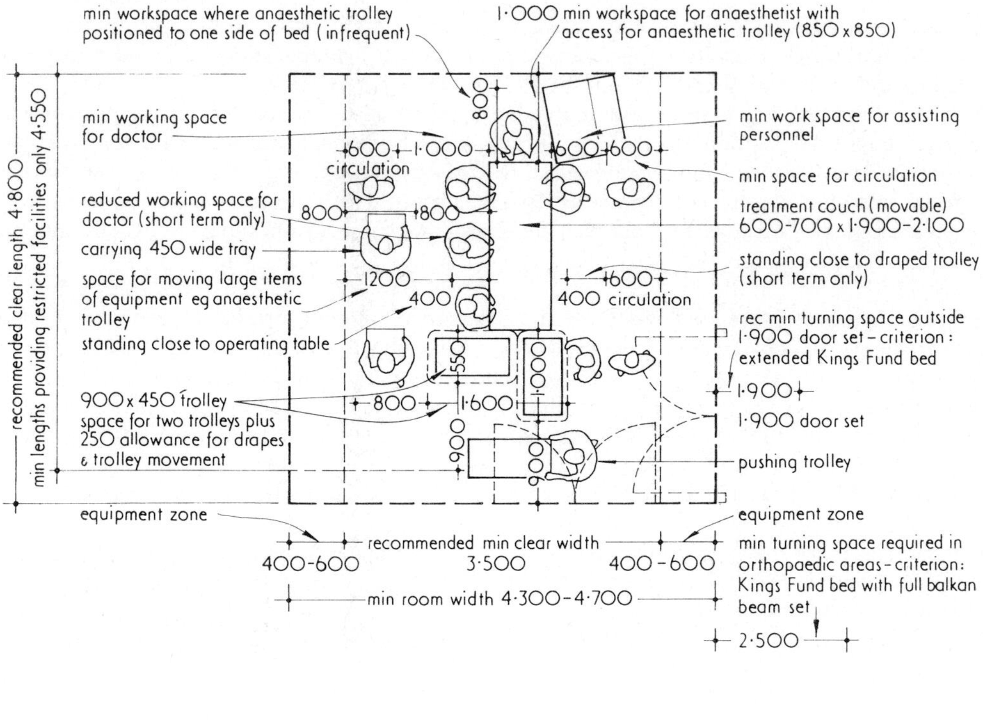

18.23 *Minimum space requirements for major treatment rooms*

the result telephoned to the operating department. This function does not call for immediate proximity between the two departments, or for the provision of a laboratory within the operating department. (The mortuary is the responsibility of the pathologist; it is convenient but not essential for it to adjoin this department. Its main components are a post-mortem room, a refrigerated body store, and a relatives' viewing room. Screened vehicular access for hearses is required.)

• *Haematology* is the study of the functions and disorders of blood. Part of this section deals with testing (for compatibility) of blood for transfusion received from blood donors. The blood-bank should be nearby, part of it being accessible in emergency at night.

• *Chemical pathology,* or biochemistry, is the study of the chemistry of the living tissues and fluids of the body. Much of the routine testing can be done on mechanised equipment; there is scope for associating this work with the similar routine testing of the haematology section, which can enable both to be automated by use of a computer.

• *Microbiology* (or bacteriology), includes parasitology. It is the study of the nature, life, and actions of micro-organisms. Some virology may be included, but this usually forms a separate section in the larger or more specialised laboratories—including those which, in England and Wales, serve also the Public Health Laboratory Services' needs. (The PHLS also performs microbiological tests on food samples etc.)

Shared facilities
Several functions such as offices and staff rooms, central wash-up, hot and cold rooms, and storage are shared between these sections. Patients' rooms for specimen taking are related particularly to the haematology and chemical pathology sections.

Expansion
The pathology department is one of the fastest-growing functions of the hospital, although automation has enabled many more tests to be undertaken without increase in size. Nevertheless, design for easy expansion is very important. If a location close to the OPD is impossible, then a possible alternative is to provide within the OPD an out-station where specimens can be taken, and some simple tests performed; this does have staffing consequences.

Planning and equipment criteria
The general planning and equipment criteria are not very different from those of other kinds of laboratories. Hospital Building Note 15 sets out the functions and the building and engineering design requirements in considerable detail, for example, on special ventilation for fume cupboards; although it will need to be updated to take into account the forthcoming Howie report on prevention of infection in clinical laboratories. It refers throughout to a single size of department, to provide an area service for a population of 200 000. Although this may be an optimum level for provision of service, it is of course possible to design pathology departments for both greater and smaller workloads; the size of department and degree of centralisation of pathology services (and whether satellite laboratories in smaller hospitals are required) will depend on the size and distribution of the catchment population.
Similar design principles apply for the teaching laboratories found in medical schools and teaching hospitals. Animal houses are usually associated with the larger, more specialised departments with an emphasis on teaching and research; they should be sited so that disagreeable smells do not inconvenience other hospital departments.

5.08 Rehabilitation
This is the term now used for a department (otherwise referred to as physical medicine or, in North America, physiatry) containing rehabilitative functions carried out under medical direction by physiotherapists (including remedial gymnasts) and occupation therapists (including art therapists and various craft instructors). Speech therapists may also be based on this department, which serves out-patients (who increasingly will spend half or whole days undergoing intensive rehabilitation, generally until they are ready to return to work) and in-patients.
Similar rehabilitation work also takes place in the geriatric and psychiatric day hospitals, where these are provided as separate entities, with rather more emphasis on occupational than physiotherapy. These two groups of patients have rather different needs from those recovering from injuries, and ideally three separate units are needed. However, when money is limited and skilled staff are hard to recruit, some degree of programmed sharing of spaces is reasonable, particularly in the more specialised and highly equipped elements. The type and form of the psychiatric department (see para 7) will affect this.
The entrance of the rehabilitation department must cope with the arrival of disabled out-patients. For efficiency of portering and transport arrangements and if the planning layout permits, it is convenient for this entrance to be shared with other users—eg the main entrance, OPD entrance if this is separate, walking patients' entrance to AED, or geriatric day hospital.
The design of the whole department must take into account the needs of disabled patients, for example, in provision of some assisted wcs, of handrails on most corridor walls, and inclusion of upright chairs in waiting areas, as these are easier to get in and out of. A main reception, records and waiting area should serve the whole department, and be close to the entrance point for in-patients and out-patients.

Physiotherapy
Individual therapy using electrical apparatus for various forms of heat treatment, with manipulative techniques, is tending to decline. For this purpose, an open area with curtained cubicles mostly 2·4 × 2·4 m, but with a few 2·4 × 3·7 m, and with passageways 1·8 m wide between curtained tracks, affords maximum flexibility. At least one enclosed treatment cubicle will be needed for children and noisy patients. This area needs ample storage for electro-medical equipment, walking aids etc.
Other rooms generally associated with this area are a wax treatment room, a splint room where plaster or plastic splints can be made (both of these need artifical ventilation and special fire precautions), a preparation bay, and a disposal room serving the whole department.
The individual open exercise area has special equipment such as mats, parallel bars, pulleys and steps. For flexibility, it can well adjoin the treatment area, and should be placed so that patients' changing rooms (with showers) and equipment storage can be shared with the gymnasium. Active treatment, usually for groups, takes place in the gymnasium, which should have one side for wall bars and other apparatus; there should if possible, be access to an outdoor terrace. A hardwood floor is recommended. Recommended dimensions are 9·75 × 19·5 × 7·6 m high. This permits ball games such as badminton. The gymnasium can then serve for staff recreation and (a need that is being recognised) for psychiatric patients. A full-scale gymnasium might be omitted in smaller hospitals if access to a suitable nearby hall (eg of a school) could be obtained.
Hydrotherapy pools provide for some patients the ideal form of treatment. They are however very expensive to install and run. Some have been built but not used very much; it is important to ensure that there are adequate staff to justify this provision, and that it will be fully used. A raised surround to

the pool (with water level above floor level) enables physiotherapists to operate from outside the pool as well as within it. Adequate space for manoeuvre is needed around the pool, also steps and hoists (which may be ceiling-mounted) for patients to get into it. Associated spaces are patients' and staff changing cubicles, lockers, WC, and shower, a utility area, an apparatus bay and a recovery room with couches for resting after treatment. Design of heating and ventilation system and choice of building materials and finishes must take into account the high humidity and temperature, and the corrosive effect of chlorine in the atmosphere. The pool water treatment plant should be separate from the calorifier or ventilating plant room to obviate chemical contamination.

Occupational therapy

A workshop for woodwork, metal and plastics work should allow 6·5 m² per place. A woodblock floor is appropriate. If spray painting is to be done, a partitioned booth is needed. Storage of materials and especially of long lengths of timber will be adjacent. Another room (sometimes called a light workshop) will provide light assembly work, typing and other office processes. As occupational therapy becomes more realistic, these functions will apply rather than the traditional art and craft activities (which are however still appropriate in geriatric day hospitals, and, with more emphasis on creative aspects, in psychiatric day hospitals). These rooms can be noisy, and should not disturb neighbouring users. Outdoor space adjoining will enable some work to be done there in summer; gardening can also be a rehabilitative activity.

A 'domestic unit' with a kitchen, bathroom, WC and bedroom enables patients to be retrained in everyday functions, using apparatus similar to that in typical homes. Adaptations found appropriate in the hospital can be fitted in patients' homes. The kitchen should allow for food cooked here to be eaten by the patients. (If no snack-bar is available in the hospital, meals may have to be brought in to a room in which day-patients can lunch.) This unit can well, if appropriately located, be used by patients in the geriatric (and possibly also the psychiatric) day hospital.

Consulting suite

If this department is close to the OPD it will be more efficient to allow for consultation and examination by the medical consultant in rehabilitation, and for speech therapy, to take place in the OPD. If not, a suite needs to be provided in the rehabilitation department, preferably near to the main waiting reception and records area.

6 SUPPORT SERVICES

6.01 Administration

The DHSS Building Note no 18, 'Administrative department', includes a wide range of elements in addition to offices and associated rooms for the hospital's management functions. Although some aspects of it are out of date, it has not yet been superseded, and the following notes cover the same range.

Management

Offices of senior administrative and nursing staff should be grouped together, and be served by typing and clerical facilities. Also in this area will be a committee room, interview and waiting spaces. This element should not be too remote from the main entrance, but can be on another floor. Offices and interview facilities for medical social workers will be needed, preferably near the OPD. Hospitals where patients pay for their care will require cashiers, billing, accounting and data processing facilities.

Staff changing

A central changing area should be reasonably near to the staff entrance. The conventional approach is to have one changing room for male and female staff (including part-timers), each containing the required number of full height lockers—preferably in echelon form, for best use of space. There are problems of ventilation and damage to lockers, which have led to the hanging basket arrangement, which separates the changing from the clothes storage elements; small multi-tier lockers are needed in addition for personal belongings. This saves space, but means employment of more counter staff. A more recent innovation is automated clothing storage, operated by electronically coded identity cards, which also deals with issue of clean uniforms, and saves both space and staff. In the last two systems, the small lockers needed for personal belongings, and the storage element, can be shared by both sexes. In all systems, showers and toilets should be provided with each changing room.

Medical records

Medical records staff deal with registration, appointments and transport at the OP main reception point. Current-use records should be adjacent. Records are kept for a statutory minimum of six years. Accommodation for statistical analysis of records is a growing need. An in-patient admission room where particulars are taken, and associated waiting space should be near the hospital main entrance. (The term 'patients' services' is often used now to include medical records, and some of the items listed in the next paragraph.)

Miscellaneous

The library is for:

- general reading (mainly for in-patients, to whom books are brought by trolley to the bedside, or who when ambulant may go to the library and read there)
- technical reference for professional and other staff, including study facilities.

Porters' room or rooms will be needed at a central point near the main or AE entrance. Also near the main entrance will be the main reception/enquiry desk, shop, public telephones and toilets, wheelchair and trolley store and postal sorting room. The telephone exchange and associated communication systems can be elsewhere.

A chapel is usually provided; at Wycombe General Hospital, a small chapel for private prayer becomes the sanctuary of a larger space when the sliding folding partitions to the adjoining committee room are opened. It should be possible for wheelchairs, and occasionally beds to be brought in. One or more chaplain's rooms are needed.

Amenities such as barbers' and hairdressers' shops for patients, bank or post office facilities and creches for children of OPs and visitors may also be needed.

A number of bedrooms, with associated bathrooms and toilets are needed for night duty staff (mainly medical) who are on call.

6.02 Education centre

Nurses and technicians

This department developed historically from the nurses' training schools, needing practical rooms and demonstration areas. These consist of key elements of a typical nursing section, including some beds, with dummy patient, lecture room or rooms, discussion rooms, and tutors' rooms and a library, if the hospital's technical library is not nearby. Many of these facilities are similar to those needed for training of technicians and other staff, and so the concept of an education centre evolved, ensuring that accommodation is sufficient for all the potential users, with careful timetabling enabling much fuller use to be made of space than if training is more widely dispersed.

Postgraduates

Postgraduate medical centres have developed in parallel; these provide a forum at which local general practitioners and hospital doctors can meet and pursue their in-service education. A common room and bar as well as office and lecture facilities are included.

Education centres are best planned as an integral part of the hospital. They may be linked with the medical photography and illustration department and the pathology museum.

6.03 Catering

The catering service can have a major effect on the design of the whole hospital. Important relationships are those between stores and kitchen, kitchen and staff dining room or rooms, and kitchen and nursing sections. If the kitchen is not on the same floor as the main store, then a hoist or lift giving easy access is required. The kitchen should be adjacent to the dining room, and, ideally, on the same floor as the largest possible number of beds.

Patients in modern hospitals select their menus every day; a card goes to the kitchen, and, as is now most common, is placed on a tray on a conveyor belt, and each food item requested is placed on the plate or tray as it passes. The plate is generally insulated to keep the food hot (or cool). This tray is then placed on an unheated trolley, which goes to the nursing unit; dirty items are returned on the same trolley to a central wash-up. This system gives a better service to patients, improves control by reducing wastage, and relieves nursing section staff of dishing up and washing up, which they had to do with a traditional heated bulk trolley system.

Variations include where the hot dish is placed in a heated compartment on the trolley and where hot meals are cooked in a micro-wave oven in the nursing section; this avoids food being spoiled if (for example) the doctor is examining a patient when the meal arrives. Other developments include frozen meals and accelerated freeze-drying.

A major problem arises with a hospital built in phases with long intervals between them. Kitchens are hard to extend while remaining in service.

If the old hierarchical and expensive pattern of different dining rooms for different grades and categories of staff is to be broken, the environmental standards must be levelled up. Sometimes part of a dining room may be waitress-served. A space of about 1·2 m² per diner is appropriate; 2½ to 3 settings at the midday meal (which will enable the size to be determined) are feasible, especially if there is an adjacent coffee room. This should if possible open on to a garden or roof terrace. Most hospitals require a snackbar for out-patients, visitors, and staff in a hurry. This, and the number of day-patient attendances, should also be taken into account when calculating the maximum—ie midday—number of meals to be provided.

6.04 Pharmacy

The pharmacy of a hospital is the department in which drugs are received, stored and dispensed to out-patients and issued to nursing units and other departments. In addition, in most larger hospitals, sterile and non-sterile preparations are manufactured in a 'sterile-products suite', as required by the medicines inspectorate (see *Guide to Pharmaceutical Manufacturing Practice*, HMSO 1977).

Building Note no 29 (DHSS 1973) gives guidance on the design of pharmacies, and specific guidance on subjects such as safety of flammable liquids and security of dangerous drugs. This department needs access to a service entrance for receipt of supplies. However, the receiving, checking and sorting function must be self-contained, and not shared with the other stores. The dispensing area should be as near as possible to the out-patient department (or OPD).

6.05 Supply and disposal services

The effective function of the clinical areas of the hospital depends on an efficient means of providing all the supply items (including linen and sterile supplies) that they need to carry out their work, and removing all items needing reprocessing (eg laundry) or disposal.

Laundries and CSSDs (central sterile supply departments) are now being provided on an area basis, so that for many new hospitals, clean linen and sterile supplies are received at the loading bay of the main stores. Area stores are also being introduced.

The means of transport depend on the form and size of the whole hospital and include:

- automatic conveyors that take containers or trolleys up to floors in a ward tower
- trolleys taken up a service lift by a porter
- use of a ramp.

Items for disposal include combustibles, which may go by chute or by trolley and lift to an incinerator room, and dirty linen which will travel similarly to the laundry (or pick-up point for laundry collection). Further categories are sterile supply and other items for reprocessing, and recyclable items such as cans.

However a high proportion of sterile supplies are brought in from an area CSSD, or purchased ready for use; it is most economical for some items, of which the most significant are operating theatre instruments, to be processed within the hospital. Also the current trend in Britain is to develop at the DGH and HSDU (hospital sterilising and disinfecting unit). The workflow is a steady progression without cross-traffic from dirty to clean to sterile. Used items and equipment are received, sorted, washed (and if necessary first decontaminated); then in a workroom the trays and packs for operating and other departments are checked and assembled; these may, in addition to instruments, contain utensils, linen, and other soft items. They are then steam-sterilised and go to the sterile goods store, or are despatched direct to their destination. Medical equipment used within the hospital, such as incubators and suction apparatus, is cleaned, and sterilised or disinfected following a similar but separate workline.

6.06 Central engineering plant and works department

The plant required to service a major hospital includes boilers, fuel storage, standby electrical generators, transformers, and, where air-conditioning is extensive, refrigeration. It is logical to locate the works department, and especially its mechanical and electrical workshops, close to the central plant installation. (See DHSS Building Note no 34, 1973.)

7 MENTAL HEALTH SERVICE BUILDINGS

7.01 Changes in policy

The changes taking place in these services were outlined in para **1.03**. There is no likelihood that architects will again be called on to plan large virtually self-contained communities either for the mentally ill or retarded* in Western countries.

* Mental illness may be caused by psychological or physiological processes, or both. The less severe forms are referred to as neuroses, the more severe as psychoses. Treatment can bring about complete or partial recovery.

Mental retardation (referred to now in Britain as mental handicap, or formerly, as mental subnormality) is a state, generally from birth, of low or very low intelligence. Some forms are preventible (eg by improved ante-natal and obstetric care), but it is not 'curable'. A similar condition may result from brain damage following an accident.

That such projects continue to be proposed in developing countries has caused some concern.

In Britain, three government publications (*Better services for the mentally handicapped,* 1971, *Hospital services for the mentally ill,* 1971, and *Better services for the mentally ill,* 1975) have set out the reasons for a change of policy that is intended to permit the closure of these large old institutions. Perhaps the strongest objection is that these large hospitals draw patients from a very wide catchment area making it hard for patients' friends and relations, GPs and local social workers to maintain effective contact with them: post-hospital rehabilitation becomes more difficult.

7.02 Mental illness

Many of the patients in the large hospitals for the mentally ill have been there for 20 years or more and are so institutionalised that it is more humane to leave them where they are than to attempt to discharge them. As the more recently admitted patients generally stay a shorter time, the numbers in most such hospitals have been reducing. This has permitted upgrading, with less overcrowding and better living conditions.

Vacancies have however often been filled by confused elderly people, for whom this is a most unsatisfactory situation. Confusion in the elderly may be caused by one of a number of physical diseases, so joint assessment by psychiatrists and geriatricians, with easy access to diagnostic facilities, is required. Those with mental illnesses such as depression will usually be treated with other mentally ill patients. Those whose condition is caused by physical illness will be under the care of the geriatrician. Those with senile dementia can probably be best cared for in small units more like old peoples' homes, but with trained nursing staff.

In Britain, in the 1950s and 1960s, as new drugs and other treatment methods made shorter stay possible for most new adult psychiatric patients, separate units were set up for this purpose, some attached to a DGH. Some psychiatrists found that they were able to treat virtually all patients in these units, and they stopped admitting new patients to the old psychiatric hospital serving their district. It is on this experience that *Hospital services for the mentally ill* is based, and the psychiatric department of the DGH which is proposed as the hub of the district service, is described in Building Note no 35 (1973). *Just an ordinary patient* by Winifred Raphael (King's Fund 1974), a survey of patients' reactions to DGH psychiatric departments, contains many valuable insights and indicates a marked preference by patients for these departments, as compared with psychiatric hospitals.

The psychiatric department consists of an in-patient element divided into nursing sections of about 30 beds, and a day hospital which includes consulting, treatment and social areas as well as occupational therapy. Each nursing section, and the therapeutic team responsible for it, covers (on the basis of the current norms of 0·5 beds per 1000 population) a sector of the catchment area with 60 000 population; there is progressive patient care within each section, from single and well observed multi-bed rooms, to a more hostel-like environment for the less dependent. Psychiatric out-patient clinics are, in most DGHs, held in the general OPD; many patients can be treated by consultation there, and never enter the psychiatric department. Still more attend between one and five days a week, but need not become in-patients. The present norm for day places is 0·65 per 1000 population; this includes the majority of in-patients who will spend much of their day in the day hospital. The balance between needs for beds and for day places is tending to change in favour of day places.

In each nursing section, the handover of duties when shifts change tends to take place in the sister's office, rather than an open nurses' station, due to the confidential nature of the discussion. As a result, the sister's office needs to be larger than in a general section; indeed it is often suggested that if the sister's office is adequate, and is well placed for observation, no nurses' station is needed. Building Note 35 recommends 5 single rooms, two 5-bed observation rooms with cubicle curtains, and two 5-bed rooms divided into more private cubicles by partitions or wardrobe and other fittings.

The day hospital includes provision for all day-patients and all but a few in-patients to take meals. At midday this generally involves two sittings. A wide range of occupational therapy activities, such as music, art, dressmaking and clerical work may be undertaken. A number of rooms designed with comfortable sitting space for 10 to 20 can function for group therapy, for some occupational therapy activities, and as general purpose sitting rooms. Electro-convulsive therapy (ECT) has in the past demanded a large and complex suite of treatment recovery and ancillary rooms (many in use for only two sessions, perhaps four hours, a week). But it is still used widely enough to need planning for. The aim should be that all spaces provided for it can serve for other purposes.

The main reason for planning this psychiatric service as a department of a DGH is to bring psychiatry into the mainstream of medicine. There are many interactions between physical and mental illness; psychiatric patients can easily receive X-ray and laboratory tests, and psychiatrists and psychiatric nurses can deal with the psychological problems of patients in hospital for physical illnesses. For example, attempted suicides (often by drug overdose) can be seen while still in AED, and may subsequently be admitted to the psychiatric department. Medical, nursing and other staff in training can more easily gain experience in mental health care. Such a department can also reduce the stigma for a patient seeking psychiatric help. This department should be integrated into the hospital enough to facilitate communications but should be independent enough to have its own (not too clinical) environment, and should be located so that, for example, noise from evening activities does not disturb nearby hospital functions. It is advantageous for it to have its own entrances.

Psychiatric departments sometimes occupy one floor of a multi-storey block of nursing sections; this clearly has many functional disadvantages, and the physical configuration tends to be limiting. They have however been much more successfully integrated if they have their own entrance, and clearly defined boundaries. If adjacent to the rehabilitation department, some spaces can be easily shared. The most popular solution is a linked satellite—a distinct block, but with an enclosed access route to the rest of the hospital. A satellite department can also be unconnected to the DGH (for example, across the road); this is less convenient for access to supplies, catering, etc, but is not in practice a severe drawback.

The appearance of the psychiatric department will, if integrated, be determined by that of the DGH; but there should be more freedom with satellite designs to provide a 'familiar' appearance, suited to the locale.

Internal planning should make for ease of movement. A two-storey building has advantages of scale, and minimises use of lifts. Orientation is helped by distinctive internal features, and views to outside. Sound attenuation is needed for all rooms used for confidential interviewing; noisy areas (eg music room, workshop) should be located to reduce nuisance to other building users. Demand for specially designed physical security measures is diminishing. Hospital standards applied indiscriminately lead to extravagance; for example beds can be closer together (2 to 2·2 m) than in general nursing units, and there is little or no need for piped medical gases.

Interior design should be part of the design process and not a cosmetic afterthought. Choice of carpets, curtains, furniture (eg divan beds) is crucial; soft floor coverings are appropriate for virtually all non-wet areas.

Other hospital buildings for the mentally ill include day hospitals without beds that may be needed if a centre of population is too far for daily access to the psychiatric department of the DGH; such day hospitals may form part of a 'community hospital' (see **1.02**). There are still a few new long-stay in-patients who need hospital run hostels (which would be near to, and staffed jointly with the DGH psychiatric department). These probably can be similar in design to hostels provided by local authority social service departments for people who have been mentally ill and still need sheltered housing. Such hostels may well be conversions of domestic buildings; so may specialist units based on living in small groups, eg for drug addicts, alcoholics, or mentally ill adolescents and children. The latter need some educational facilities.

7.03 Mental retardation

Early diagnosis is essential. In the past, many were thought to be much more retarded than was the case, because it was not realised that they were also deaf. The comprehensive child assessment units perform a valuable service for mentally retarded children.

Many of the most severely retarded, especially those who are physically disabled, incontinent, or both, will still need residential care. In Britain, this is the responsibility of the Health Service, although the medical and nursing elements are becoming less emphasized. However, the role of the psychiatrist is still significant with regard to behaviour disorders.

Units of fewer than 100 places are being developed. These should be integrated with the community as far as possible. In addition to comfortable living quarters with an increasing proportion of single bedrooms, a day centre is needed, for occupational pursuits and training. Some of these units have been associated with DGHs, but this is not necessary. Provision for children is generally separate from that for adults. Single storey buildings are best.

7.04 Secure psychiatric accommodation

Those accused of crimes (particularly of violence) but found 'unfit to plead' or 'guilty but insane', together with people in prisons or hospital facilities for the mentally ill or retarded who become too violent or disturbed to handle there, are provided for in Britain by the 'special hospitals' which are psychiatric hospitals with a high level of security. The trend in most countries is to much smaller secure establishments, generally of 100–200 places. A secure perimeter wall with discreetly guarded entrance, is usually required.

In England, the new 'regional security unit', has been proposed to take the most disturbed patients, both mentally ill, and mildly but not severely retarded from the Health Service. These are likely, also, to serve as halfway houses for 'special hospital' patients who no longer need maximum security. Within these units of between 50 and 100 beds, there should be provision for a number of small group living situations, eg for patients of both sexes, and for different levels of physical security provided by levels and quality of staffing that make for good observation of patients. The buildings must be designed to prevent absconding. As wide as possible a variety of experiences and activities will be needed.

8 ENVIRONMENTAL CONSIDERATIONS

The ultimate client of the hospital architect is the patient. While it is most important to give an appropriate environment to those who spend all their working life in the hospital, the designer has failed if he has not made a place which helps to reduce patients' anxiety, and which gives them aesthetic pleasure and material comfort.

8.01 Landscaping

External landscaping should help to relate the building to its site. Small ornamental trees will often be inappropriate in scale, and forest trees should be planted. These can effectively shade the building in summer. Landscaping of courtyards and grounds around the hospital should be designed for utility as well as appearance. Spaces where staff and patients can sit out should be planned and (where possible) orientated accordingly. For elderly patients and those undergoing rehabilitation raised flower beds will enable gardening to be a therapeutic activity. Landscaping of overlooked roofs should be considered especially if usable as terraces. The publications of the South Western Regional Health Authority and Bath University give valuable information on hospital landscaping and road layout, including that of ambulance access to the AED. Some car parking must be provided for staff for whom use of a car is essential. Whether car parking can be provided for other staff and for visitors or out-patients will depend on public transport services, local patterns of car ownership, and availability of land. In some cases, basement space may be used for this purpose. Large expanses of flat roofs of low rise buildings may also give scope for car parking, if a suitable location for an access ramp can be found. Outdoor parking should be subdivided, and broken up with planting. Access for service vehicles—and especially for hearses—should be screened.

8.02 Interior design

Hospitals, especially in Britain, have not been notable in this respect. There have been too many hard and shiny surfaces, and the colour grey has been too much in evidence. Whether this is due to lack of imagination, or to a mistaken interpretation of the clinical nature of the hospital is hard to say. McMaster Hospital in Ontario is a striking example of the opposite approach. There is hope now that, even where budgets are tight, standards may be improving. The greatest single factor is the increasing use of carpet and other soft floor coverings, as these have been shown to be hygienically acceptable in most areas, and economical, if reduced cleaning costs (compared with floor finishes such as PVC) are set against any extra initial costs.

Interior designers in regional health authorities and DHSS have collaborated with Supplies Divisions in DHSS and DOE to produce ranges of furniture designed for longer-stay patients. But apart from bedside lockers and a few other specialised items, choice of furniture—and of finishes and fabrics generally—should be made on similar criteria as for, say a hotel, or other public or residential buildings. Fabrics and upholstery should, however, be fire retardent wherever possible. The best use must be made of natural lighting. Internal rooms can be made more pleasant by use of windows in partitions to 'borrow' daylight from a corridor or other room. Internal bathrooms seem less claustrophobic with a high level window to an adjoining corridor. Artifical lighting should be an integral part of the interior scheme and not be designed in isolation by the engineer. Fluorescent lighting, although most economical, is not appropriate for some areas where a domestic atmosphere is specifically required.

The total design responsibility should include everything that is seen in the finished building even though the nature of health authorities can make this difficult to achieve.

8.03 Noise

Carpets are not only pleasant to the sight and touch, but also reduce noise, which has been one of the greatest nuisances of hospitals. Modern equipment, such as beds (eg the 'King's Fund Bed', now widely used in the NHS hospitals), tends to be quieter to move and operate than older models. Also the Nightingale wards and large OPD waiting halls were much noisier, due to the concentration of people and activity in a single space, than their present day counterparts; and noise begets noise, for example, people have to talk louder in order to be heard.

The design of doors and partitions is another important factor. Where confidentiality is important—eg in consulting rooms and interview rooms—at least 40 decibels sound attenuation is needed. Solid core doors, without grilles or gaps above or below should be used. Sound can travel by ducting to adjoining rooms, also over the partitions if these stop at the level of the ceiling rather than that of the structural soffit.

9 BIBLIOGRAPHY

G. Agron Application of the VA (Veterans Administration) hospital building system to the design of the Loma Linda VA hospital, *Architectural Record*, New York, 1973, September, vol 154, no 3, pp 161–4

G. Agron Building systems research for VA applied in both public and private hospitals, *Architectural Record*, 1972, June, vol 151, no 7, pp 115–22. Also in this issue, Joint venture studies for D.O.D. by architects and industry probe systems analysis for a new generation of military and civilian health care facilities, pp 123–5; and S. Clibbon, Industrial techniques and spaces make clinical spaces, and patient fostering spaces work in health care facilities, pp 126–32

R. W. Allen and I. V. Karolyi *Hospital planning handbook*, New York, Wiley, 1976

R. Aloi and C. Basso *Ospedali: hospitals*, Milan, Ulrico Hoepli Editore, 1973. Text and captions in both Italian and English

American Association of Hospital Consultants *Functional planning of general hospitals*, Ed B. Mills, New York and Maidenhead, McGraw-Hill, 1969

Architects' Journal and Ministry of Health *Hospital planning and design guide*, London, Architectural Press, 1967

Architectural Review, London, 1965, June, vol 137, no 820 Walton Hospital, Liverpool, p 425; Princess Margaret Hospital, Swindon, second stage, pp 426–9; Huddersfield General Hospital, pp 434–535; Wexham Park Hospital, Slough, pp 437–8

Best buy hospital: Bury St Edmunds *Architects' Journal*, 1974, June 19, vol 159, no 25, pp 1373–93

A. Borup *Hospital design today, hospital design tomorrow*, Copenhagen, National Health Service of Denmark, 1975

British hospitals: home and overseas 12th edn, April 1976, Maidstone, Whitehall Press, 1976

British ward design and equipment *Hospital Management, Planning, Building Equipment, Supplies*, Sevenoaks, 1971, May/June, pp 137–83

S. Clibbon Innovation in the design of health care facilities: some influences of systems building, *in* International systems building round table conference, Boston, 1971, Proceedings pp 121–7

C. Davies The Harness System, *Hospital Engineering*, 1974, January, vol 28, pp 3–11 paper given at one-day seminar on Harness)

Department of Health and Social Security *Better services for the mentally handicapped* (Cmnd 4683), London, HMSO, 1971

Department of Health and Social Security *Better services for the mentally ill (Cmnd 6233), London, HMSO, 1975*

Department of Health and Social Security *Hospital services for the mentally ill*, London, DHSS, 1971

Department of Health and Social Security *Planning, design and construction of hospital buildings for the National Health Service*, H. J. Cruickshank, London, DHSS, 1973

DHSS and Scottish Hospital Building and Equipment notes

District Hospital at Greenwich: London SE10 *The Architects' Journal*, London, 1969, November 26, vol 150, no 48, pp 1369–86

H. H. Field and others *Hospital design evaluation: an interdisciplinary approach*, Springfield, Va., National Technical Information Service, 1971

H. Gainsborough and J. Gainsborough *Principles of hospital design*. London, Architectural Press, 1964

R. H. Goodman Systems health building: a discussion on building technology in the 1980s, *Philosophical Transactions of the Royal Society of London*, series A, 1972, July 27, vol 272, no 1229 pp 611–19

J. R. B. Green *Health service facilities planning and design: a guide book*, Kensington NSW, School of Health Administration, University of New South Wales, 1974

B. Gruffydd *Landscape architecture for new hospitals*, London, KEHF, 1967

Hamilton, Ontario: McMaster University Health Science Centre *Architectural Forum*, 1971, June, vol 134, no 5, pp 30–5

Hospital By Oxford Method *The Architects' Journal*, 1972, July 12, vol 156, no 28, pp 83–6 (Battle Hospital, Reading)

Hospital Ward Block at High Wycombe: designed by Powell & Moya *The Architects' Journal*, London, 1968, April 17, vol 147, no 16, pp 821–36

S. Howells and B. Hitchcox Nucleus hospitals, *Hospital and Health Service Review*, 1976, June, vol 72, no 6, pp 206–11

King Edward's Hospital Fund for London *Hospital research and briefing problems*, by J. Green, R. Moss and C. Jackson, ed K. Baynes, London, KEHF, 1971

King Edward's Hospital Fund for London *Hospital traffic and supply problems: a report of studies undertaken by members of the Hospital Design Unit of the Ministry of Health within the context of the Greenwich District Hospital development project*, ed W. A. H. Holroyd, London, KEHF, 1968

Llewelyn-Davies Weeks Forester-Walker & Bor *Long-life loose-fit: a comparative study of change in hospital buildings*, London, Llewelyn-Davies, 1973, reprinted 1975

Manplan 6. Health and Welfare: Hospitals *The Architectural Review*, London, 1970, May, vol 147, no 879, pp 326–44

Mathers and Haldenby *Interstitial space study*, Toronto, 1979 and BDP Preston 1978 and 1979

New Lister Hospital *The Architects' Journal*, 1974, April 3, vol 159, no 14, pp 725–41 (Stevenage, Herts)

North Tees General Hospital, Harness unit for the elderly *Hospital Development*, London, 1975, May/June, vol 3, no 3, pp 10–12

Nuffield Provincial Hospital Trust *Studies in the functions and design of hospitals*, London, Oxford University Press, 1955

Oxford Regional Hospital Board *The evaluation of a deep ward plan*, Oxford, Oxford RHB, 1970

Research Bulletin, Medical Architecture Research Unit, Polytechnic of North London

I. Rosenfield and Z. Rosenfield *Hospital architecture and beyond*, New York, Van Nostrand Reinhold, 1969

I. Rosenfield and Z. Rosenfield *Hospital architecture: integrated components*, New York and London, Van Nostrand Reinhold, 1971

J. V. Sheoris The large scale planning module/unit theory design of hospitals, *World Hospitals*, Walsall, 1973, January, vol 9, no 1, pp 31–6

Two Hospitals *The Architects' Journal*, London, 1971, November 10, vol 154, no 45, pp 1061–78 (Eastburn Hospital near Keighley, Yorks; South Cheshire Hospital, Crewe)

Ward Evaluation: St Thomas' Hospital, Medical Architecture Research Unit, Polytechnic of North London, 1977

J. Weeks *Architectural Design*, London, 1973, vol 43, no 7, pp 436–43, 456–63. Reviews the current situation and possible future trends in the design and construction of hospitals

19 Health centres and group practices

Brian Brookes

CI/SfB 421
CI/SfB (1976 revised) 421

Brian Brookes was Assistant Director of the King's Fund Centre of the King Edward's Hospital Fund for London. He was an international expert in the field of medical architecture and was killed in a plane crash while assisting the World Health Organisation

Contents

1 HEALTH CENTRES

A health centre is a building provided, equipped, maintained and staffed (with the exception of family doctors and dentists who are tenants) by a regional or area health authority (England and Wales), the Secretary of State or a health board (Scotland) or an area board (Northern Ireland). Policies differ considerably between the countries and therefore the functional content of the building varies widely.

The purpose of a health centre is to draw together, and fuse into one organisation, any combination of (but rarely all) such traditionally separate health services as:

• *General practices*
general medical practice (always, varies between one and three GPs)
general dental practice (uncommon)
general pharmaceutical practice (rare, except in Northern Ireland)
general ophthalmic practice (very rare)

• *Community and school health and dental services*
This includes:
maternity and child welfare
ophthalmic
child guidance
speech therapy
physiotherapy
community nursing services
community health visiting services
chiropody (usually)
health education
social services (uncommon)

• *Hospital based services*
hospital out-patient services } (uncommon except for large
hospital diagnostic services } health centres)
X-ray services (uncommon except in Scotland)

• *Medical education*

2 GROUP PRACTICES

A group practice building is provided by a group (usually more than three) of general practitioners for their own use as a central shared surgery. It is not unusual to find other NHS services being offered at a group practice, but these are usually limited to community nursing and health visiting services; the range of services likely to be accommodated is far smaller than for a health centre.

(Because accommodation for a group or groups of GPs is always a constituent part of a health centre, this guide will concentrate on the latter.)

3 DESIGN PHILOSOPHY

No single pattern of health centre has evolved, and any health centre may contain a different mix of services. However, the philosophy of health centres is based on the improvement of general medical practice in an area, and this, together with a combination of community services, is the usual nucleus.

Because of the changing nature of general medical practice, and the changes that can be expected when general practitioners work with community services in a health centre, and because of the changing balance between hospital care and community care, a health centre should provide both for the flexible use of space, and for balanced growth.

• *Location*

Because of the community emphasis of a health centre, and to foster its role in health education, it should be sited with due regard to that section of its population most 'at risk'. This population, on the other hand, is determined not so much by the economics of scale, nor by any health planning criteria, but far more by the number of general medical practitioners who can be attracted to work in a building that they do not own. It is therefore not unusual to find that different services in the same health centres are catering for different sizes of population.

• *Size*

At present, the average health centre in the United Kingdom is accommodating six GPs and serving a population of 15 000 (a ratio of 1 GP to 2500). This average conceals wide variations between 1 GP and 30 GPs. Most literature divides health centre into those serving up to 10 000, 10 to 20 000, 20 to 30 000 and 30 to 40 000.

Obviously, size is one determinant to the range of services to be offered. It would be rare to find hospital X-ray services at a health centre serving less than 40 000 people, and unusual to find hospital out-patient sessions being held in centres serving less than 20 000, although no general rule can be applied.

• *Relationship to hospital*

In some circumstances, a health centre can be conveniently sited with a hospital; such an arrangement will encourage the joint use of facilities by GPs and hospital staff. However, this is only desirable when the accessibility to the centre by patients, especially the very young, the elderly and the handicapped, is not compromised.

• *Structure of health centre*

Because the principle of health centre practice is co-operation, it is not usually satisfactory to divide the building into separate units for GPs, community health etc, **19.1**. If the size of the

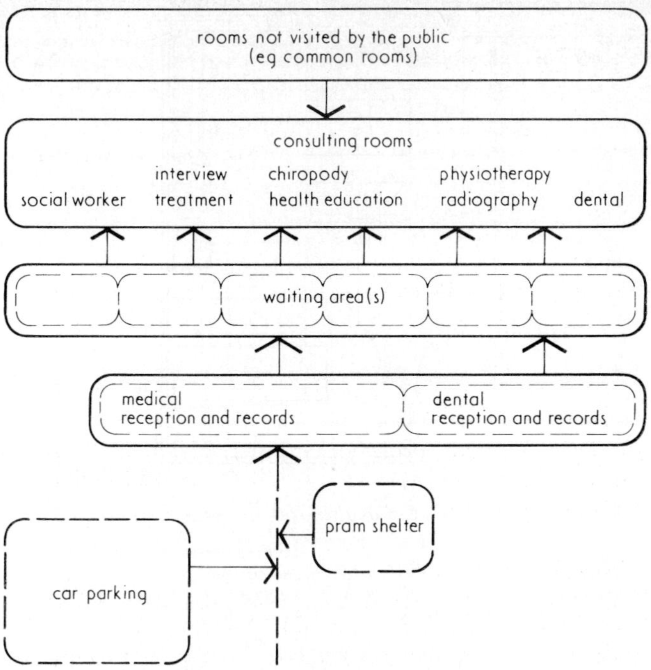

19.1 *Simplified sequence diagram*

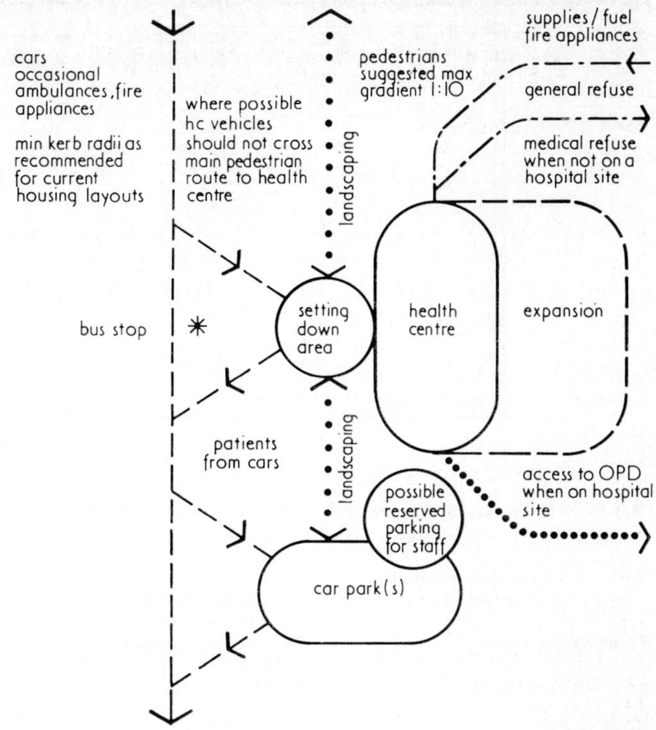

19.2 *Some requirements at a health centre site*

building is such (above 20 000 population) that the 'domestic' character traditionally associated with general practice is endangered, then the plan may be divided between different groups or teams (eg four GPs, two nurses, one health visitor etc) which have an identity of their own, but which can share central services such as reception, records etc. The degree of decentralisation can only be decided once the management framework is clear.

4 DESCRIPTION OF SPACES ETC

4.01 Health centres and group practices

Car parking

On-site car parking (25 m² per car) should be allowed for staff in the amount of 1·5 spaces per consulting room (qv), and for patients 2·5 spaces per consulting room, **19.2**. Space for manoeuvring ambulances should be allowed.

Main entrance

The main entrance should be clearly visible, identifiable and easily accessible to all patients and staff.

Reception

The reception staff has to cope with the direction of patients to waiting areas, the making of appointments etc. Allow 5·5 m² per receptionist on duty at peak hours, **19.3**.

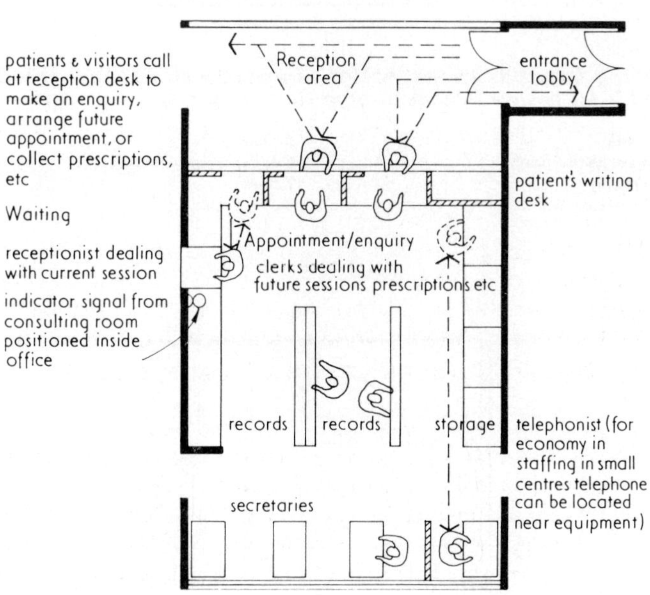

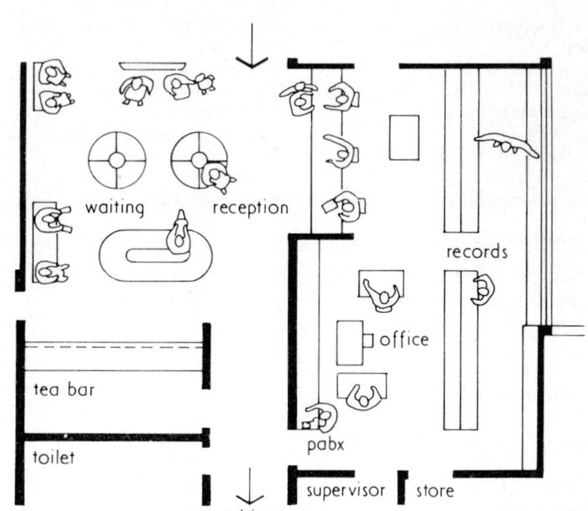

19.3 *Reception areas:*

a *For smaller centres. Note hatch overlooking waiting area for calling patients to their appointments. All administrative functions take place behind the enquiry desk, and visible (and audible) from it*

b *For larger centres. Secretarial work is done elsewhere, thus reducing noise from typewriters, etc. Most of area is screened from the enquiry counter*

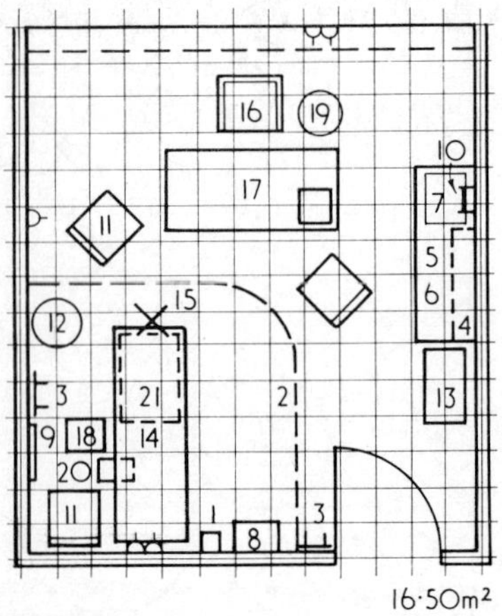

19.4 *Combined consulting/examination room as used in Scotland:*

1 *bracket for sphygmomanometer*
2 *ceiling mounted curtain track*
3 *coat hooks*
4 *high level storage*
5 *worktop*
6 *low level storage*
7 *washhand basin*
8 *writing shelf*
9 *mirror*
10 *paper towel dispenser*
11 *chair*
12 *disposal*
13 *equipment trolley*
14 *examination couch*
15 *mobile examination lamp*
16 *swivel chair*
17 *desk*
18 *couch steps*
19 *waste paper bin*
20 *scales*
21 *couch cover dispenser*

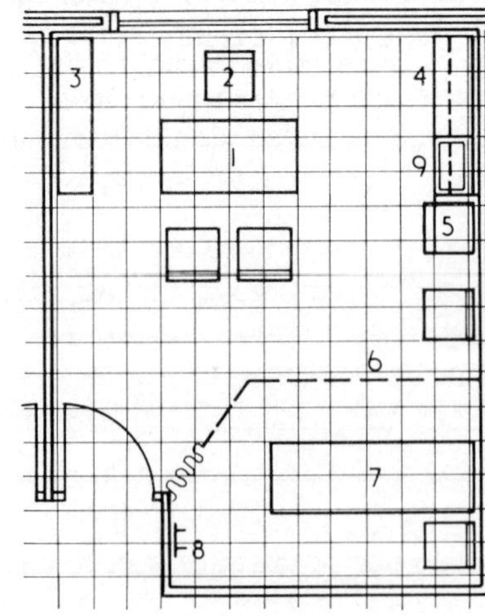

19.5 *Alternative plan of consulting/examination room:*

1 *desk*
2 *swivel chair*
3 *bookcase*
4 *worktop with cupboard above*
5 *trolley*
6 *ceiling mounted curtain track and curtain*
7 *couch*
8 *coat hooks*
9 *washhand basin*

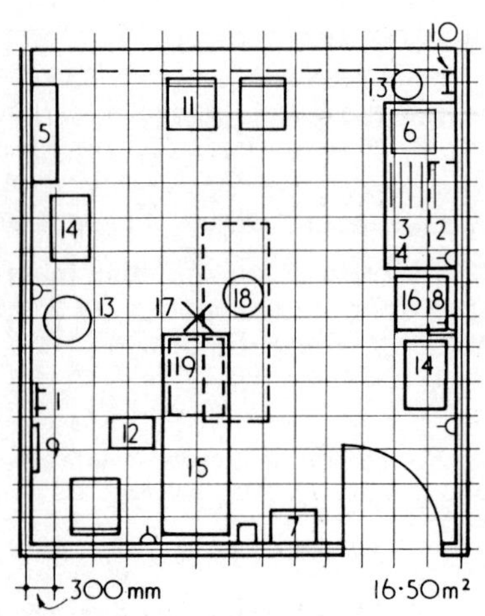

19.6 *Treatment room (Scottish):*

1 *coat hooks*
2 *high level storage*
3 *worktop*
4 *low level storage*
5 *shelving*
6 *sink and drainer*
7 *writing shelf*
8 *DDA cupboard*
9 *mirror*
10 *paper towel dispenser*
11 *chair*
12 *couch steps*
13 *disposal*
14 *equipment trolley*
15 *examination couch*
16 *refrigerator*
17 *mobile examination lamp*
18 *stool*
19 *couch cover dispenser*
20 *warning light*

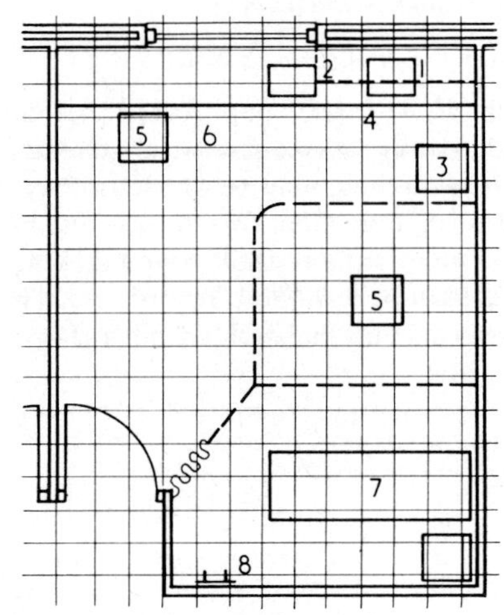

19.7 *Alternative treatment room (for one nurse)*

1 *sink*
2 *washhand basin*
3 *trolley*
4 *worktop with cupboard above*
5 *chair*
6 *administration area*
7 *couch*
8 *coat hooks*

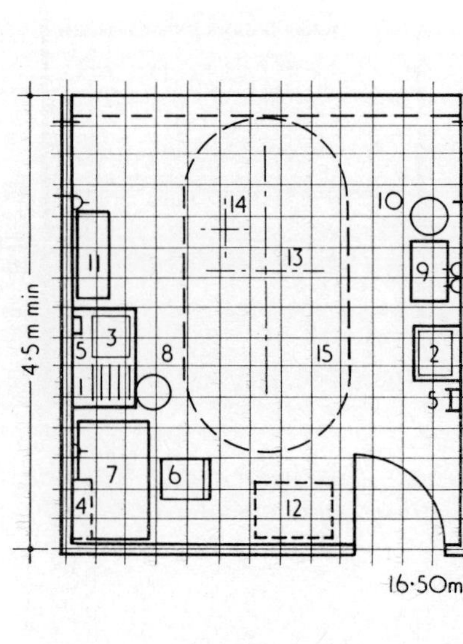

19.8 *Dental surgery (Scottish)*

1 *low level storage*
2 *washhand basin*
3 *sink and drainer*
4 *lockable cupboard for scheduled poisons*
5 *paper towel dispenser*
6 *chair*
7 *desk*
8 *disposal bin*
9 *instrument trolley*
10 *pedal waste bin*
11 *dental equipment cabinet*
12 *space for anaesthetic machine*
13 *dental chair*
14 *dental unit*
15 *dentist's seat or stool*

Record storage
Usually associated with reception, but ideally not part of it, is record storage. Assuming that all records (of A4 size) are centralised, allow 1·4 m² per 1000 patients. It is most important that this area is out of sight of waiting patients, and that it can grow.

Administration
Associated with the record storage is the clerical/administrative area, and the telephone switchboard. 5·5 m² should be allowed for each typist-secretary. In small centres, the switchboard may be manned by the receptionist; but if separate, allow 7·5 m² per telephonist.

Waiting areas
Waiting areas are usually located with the reception desk. This has a number of disadvantages in that private conversations at the desk can be overheard by waiting patients; but has the advantage that patients can be called to the doctor by the reception staff. In larger centres, waiting areas can be decentralised, but only when the policy for reception and control of patients is closely identified. It is vital that waiting areas are not part of the circulation pattern.
Six seats should be allowed for each consulting room and treatment room, 1·4 m² each.
Location of wcs and pram parks should be carefully considered in relation to waiting and reception areas.

Consulting rooms
Consulting rooms are usually provided for each doctor on a personal basis. Where this results in under-use, there might be advantages in scheduling the use of rooms for other purposes, eg hospital consultants, social workers etc. Patterns of consulting/examination practice vary, and can result in consulting rooms with separate examination rooms, combined consulting-examination rooms, **19.4**, **19.5**, etc. As a general rule allow 17 m² per combined room.

Treatment rooms
Treatment rooms are used primarily by nurses performing simple therapeutic techniques, **19.6**, **19.7**. For one nurse allow 17 m². If more than one nurse is to be employed at the same time, treatments may be performed on two patients in one room of 25 m²; although, on the grounds of privacy, it might be better to provide two rooms.

4.02 Health centres only
Other typical space requirements are given in table I.

Table I Space requirements for specialties in health centres

Space	m²
Dental surgery	16·5
Dental waiting room	7·5 one surgery D 1·4 for each additional surgery
Dental laboratory	28 for 1 or 2 technicians
Dental reception and records (if separate from main reception)	5·5 per receptionist D 0·9 per 3000 patients
Health education roomvphysiotherapy	28
Chiropody	11 per chair, 15 for two chairs
Staff room	0·9 per staff member, minimum 17
Health visitor's office	55·5 per desk
Nurse/ midwife's workroom	11
Speech therapy	14–23·5

5 DENTAL GROUP PRACTICES
A typical dental surgery is shown in **19.8**.

6 BIBLIOGRAPHY
Brian Brookes General practice surgery spaces: briefing guide. The Architects' Journal Information Library, vol 144, no 7, 17.8.66, pp 421–37; no 8, 24.8.66, pp 491–501; no 12, 21.9.66, pp 761–6
Ruth M. Cammock *Health centres: reception, waiting and patient call, some aspects of provision for reception, waiting and patient call, in a selected group of large health centres,* report of a study carried out for the DHSS, London, HMSO, 1973
Ruth M. Cammock *Health centres handbook*, London, Medical Architecture Research Unit, Polytechnic of North London, 1973
Department of Health and Social Security and Welsh Office *Health centres: a design guide*, London, HMSO, 1970
Ministry of Health *Buildings for general medical practice*, prepared by The General Practice Advisory Service Ltd, London, HMSO, 1967
Scottish Home and Health department *Design guide: health centres in Scotland*, Edinburgh, HMSO, 1973

20 Catering design

Fred Lawson

CI/SfB 51
CI/SfB (1976 revised) 51

Fred Lawson is a lecturer in the Department of Hotel, Catering and Tourism Management at the University of Surrey

Contents

1 RESTAURANTS

1.01 Space allowances

Table I shows space allowances for common types of seating and service. Space requirements per diner are shown in **20.1** and **20.2**

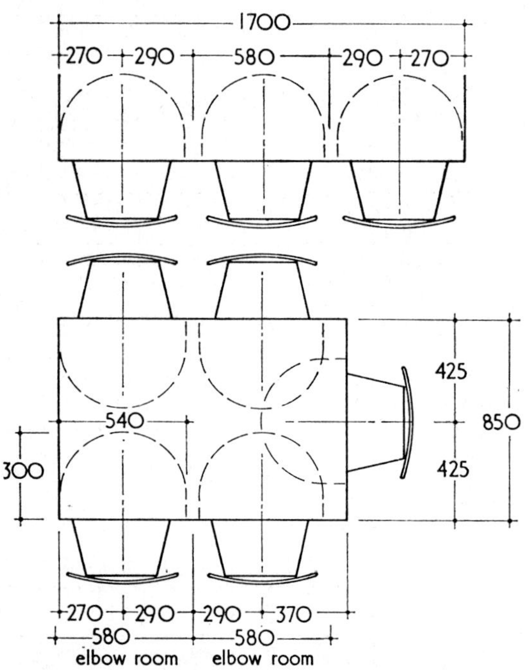

20.1 *Area required by individual diner*

Table I Space allowances

Type of seating and service	Area per diner (m²)
Commercial restaurants	
Table service	
● Square tables in rows	
Parallel seating 2	1·7 to 2·0
Parallel seating 4	1·3 to 1·7
Diagonal seating 4*	1·0 to 1·2
● Rectangular tables in rows	
Seating 4	1·3 to 1·5
Seating 6	1·0 to 1·3
● Circular tables in rows	
Seating 4	0·9 to 1·4
Fixed banquette seating	
● In booths seating 4	
Waitress service	0·7 to 1·0
Including counter for self service	0·9 to 1·4
Counter seating	
Tunnel counters	1·4 to 1·6
Single counters	1·7 to 2·0
Single counters used with wall units	1·2 to 1·5
Banquet groupings	
Multiple rows	1·0 to 1·2
Single row	1·1 to 1·4
Self-service (trolley clearance)	
Rectangular tables in rows	
● Dining area only	
Seating 4	1·5 to 1·7
Seating 6	1·2 to 1·4
Seating 8	1·1 to 1·3
● Including counter area	
Seating 4	1·7 to 2·0
Seating 6	1·3 to 1·8
Seating 8	1·2 to 1·6
Self-service (self-clearance)	
Rectangular tables in rows	1·3 to 1·5
● Dining area only	1·0 to 1·2
Seating 4	1·3 to 1·5
Seating 6	1·0 to 1·2
Seating 8	0·9 to 1·1
● Including counter service	
Seating 4	1·5 to 1·9
Seating 6	1·2 to 1·6
Seating 8	1·1 to 1·5
Canteens (industrial and office)	
Cafeteria service, tables for 4 to 6	1·1 to 1·4
Cafeteria service, tables for 8 or over	0·74 to 0·9
School dining rooms	
Primary schools	
Counter service	0·74
Family service	0·83
Secondary schools generally	0·9
Colleges of further education	1·1

* Economy in space is obtained with tables at 45°

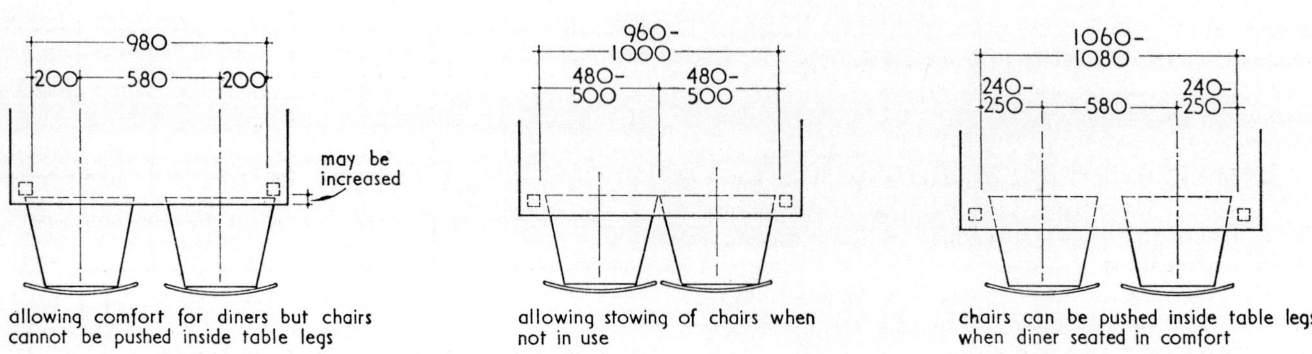

20.2 *Obstruction by table legs*

Table II Table sizes for one to twelve diners (to be read in conjunction with 20.3)

Table type		Length of table in mm for widths of table in 50 mm increments															Round
No of persons	Row	400	450	500	550	600	650	700	750	800	850	900	950	1000	1050	1100	Diam (mm)
1	D	950	900	850	750	700	650										
	H				750	700	650										
	Round																750
2	A								750								
	E				950	900	850	800	750								
	D	1350	1250	1150	1100	1100	1100	1100									
	G				950	900	850	800	750								
	I	1300	1250	1050	950	900	850	800									
	Round																850
3	B								950	850							
	D	1800	1700	1700	1700	1700	1700	1700									
	E				1350	1350	1350	1350	1300	1100	1100	1100	1100	1100	1100	1100	
	F				1250	1200	1100	1050	950	900	850						
	Round																950
4	A								1150	1100	1100	1100	1100	1100	1100	1100	
	C								1100	1050	1000	950	900				
	Round																1050
5	B								1350	1350	1300	1300	1250	1200	1100	1100	
	Round																1100
6	A								1700	1700	1700	1700	1700	1700	1700	1700	
	C								1550	1550	1550	1450	1400	1250	1250	1200	
	Round																1200
7	B								1850	1800	1750	1750	1700	1700	1700	1700	
	Round																1300
8	A								2300	2300	2300	2300	2300	2300	2300	2300	
	C								1950	1900	1850	1750	1700	1700	1700	1700	
	J															1600	
	Round																1500
9	B								2400	2400	2350	2300	2300	2300	2300	2300	
10	A								2850	2850	2850	2850	2850	2850	2850	2850	
	C								2550	2500	2400	2350	2300	2300	2300	2300	
11	B								3000	2950	2950	2900	2850	2850	2850	2850	
12	A								3450	3450	3450	3450	3450	3450	3450	3450	
	C								3100	3050	3000	2900	2850	2850	2850	2850	

1.02 Table sizes and arrangements

• *Table sizes*
A selection of table sizes (extracted from *Bord*) is given in table II and **20.3**. Tables for drinking only may be slightly smaller, **20.4**.
Banquette and wall booth spacing is given in **20.5**.
Aisle circulation widths are shown in **20.6**.
• *Heights of tables and chairs*
Recommendations by Dr B. V. Akerblom (in his study *Standing and sitting*) and by Eric Berglund (in *Bord*) are given in **20.7**. However, the present standard height of most tables is about 760 mm with chair seat heights of 455 mm to 460 mm.
• *Table arrangements*
Various arrangements with an indication of local densities are shown in **20.8**. No allowance is made for main circulation, entry, queing, payment and exit.
Typical banquet arrangements are given in **20.9** and **20.10**.

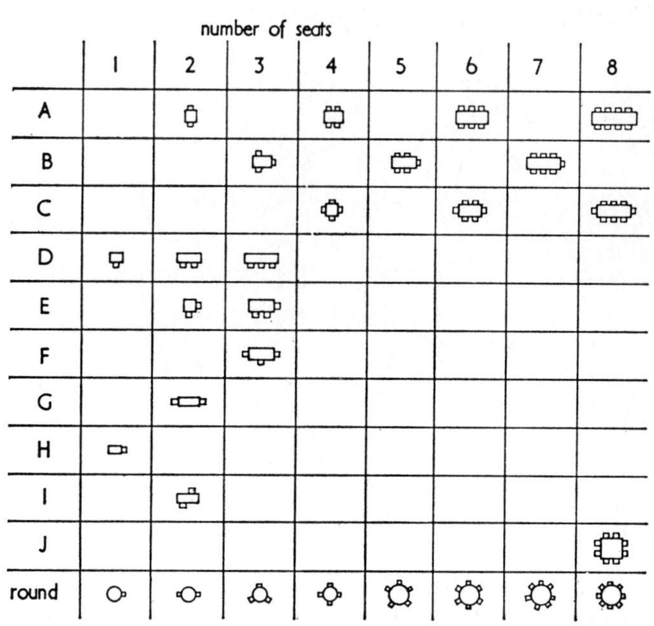

20.3 *Tables for varying numbers of people (to be read in conjunction with table II)*

number of seats	table size: drinking mm	table size: eating mm
1	450 to 600	600 to 700
2	600 square	750 square
4	750 square	900 x 950
	—	1500x 750
6	—	1400x 950
	—	1700x 750
8	—	1750 x 900
	—	2300 x 750

number of seats	table size: drinking mm	table size: eating mm
1	450 to 600	750
2	600	850
4	900	1050
6	1150	1200
8	1400	1500

20.4 *Selection of recommended table sizes related to number of persons*

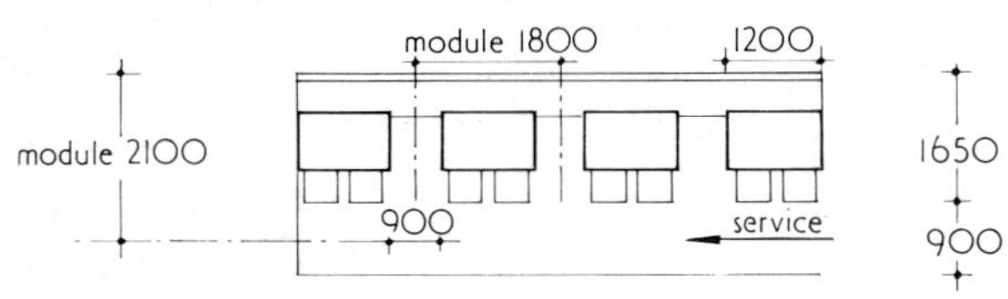

20.5 *Banquette seating; including space for access and service:*
a *Banquette along wall*

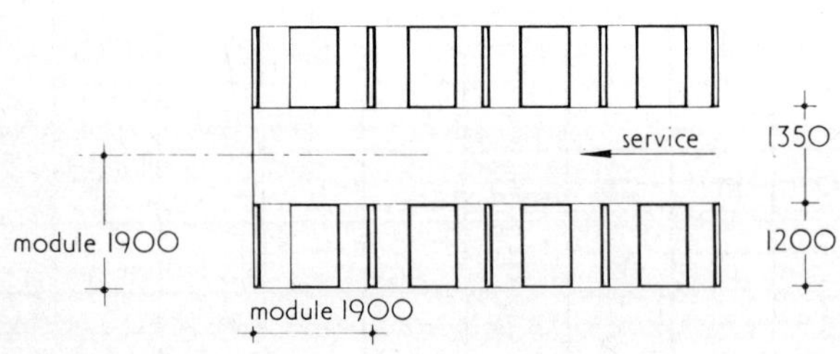

b *Booth seating*

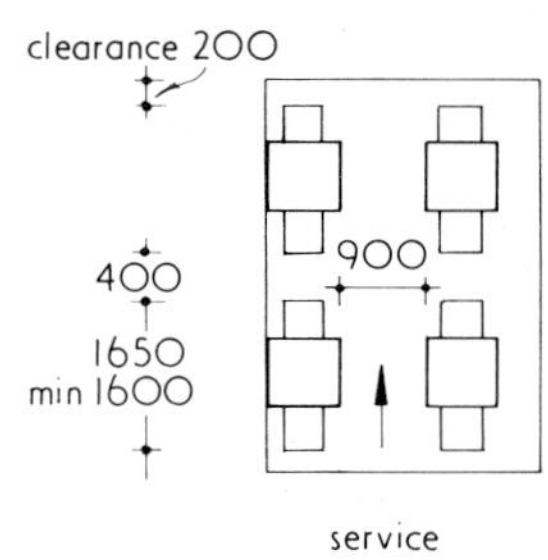

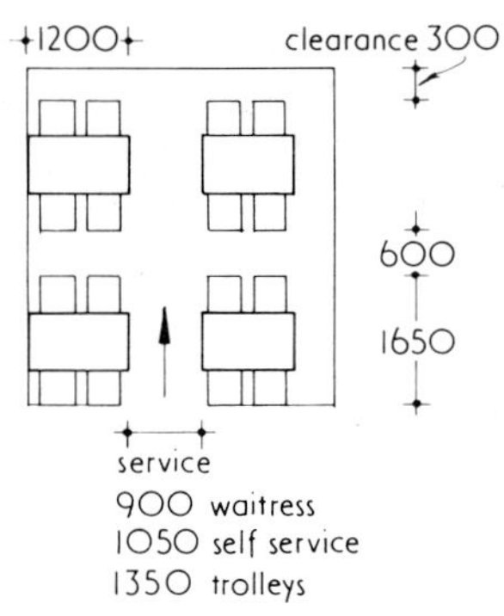

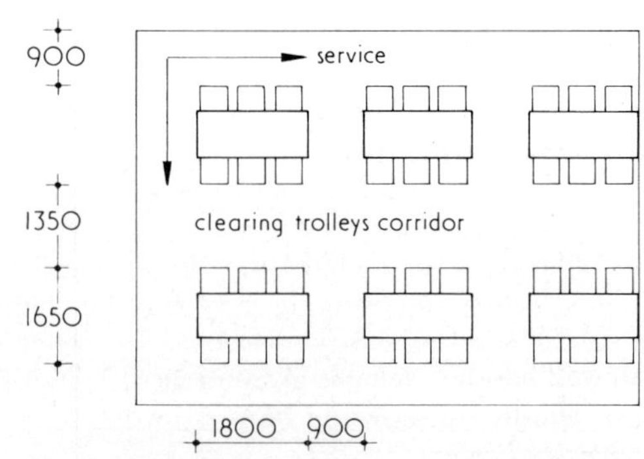

20.6 *Minimum spaces between tables to allow for seating, access and circulation*

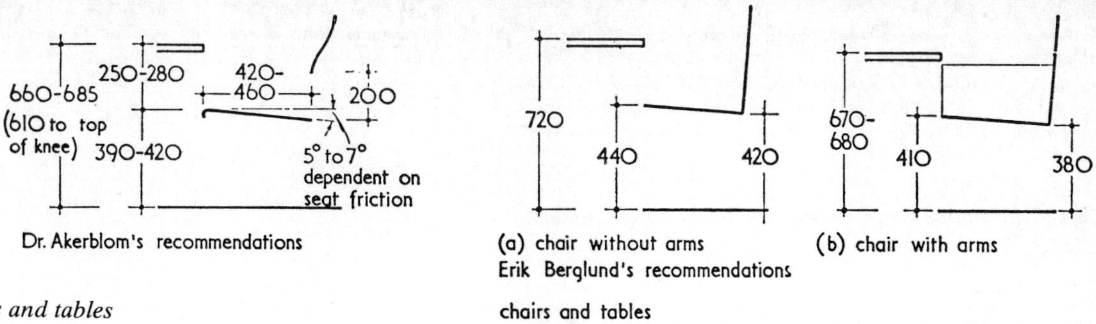

20.7 *Heights of chairs and tables*

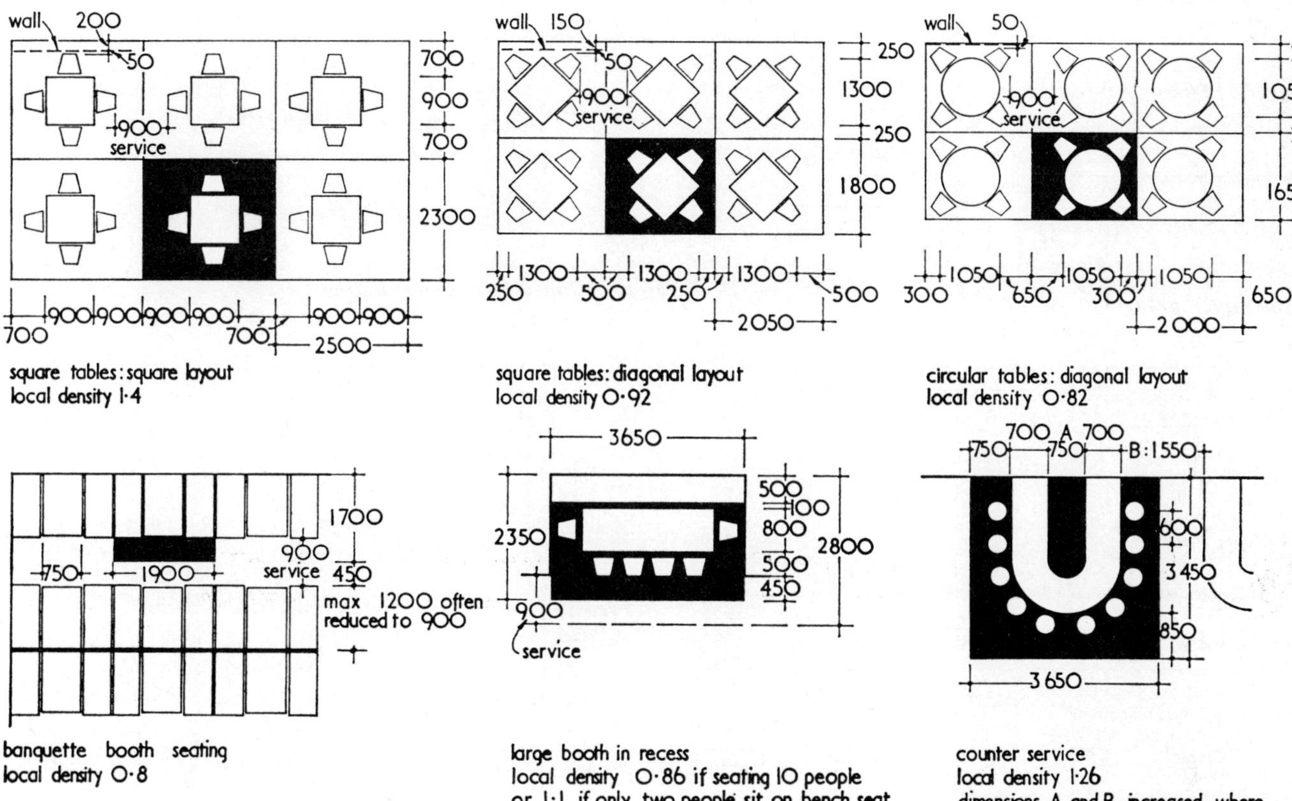

20.8 *Layout arrangements for restaurant tables*

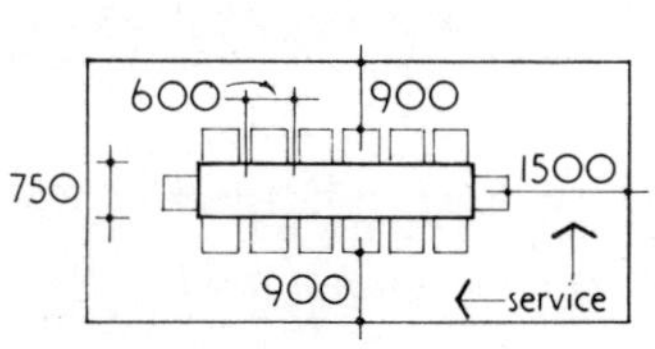

20.9 *Small formal dinner arrangement*

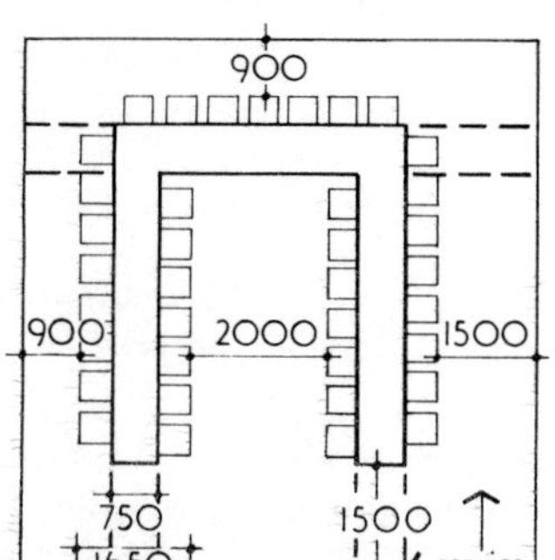

20.10 *Banquet layout. The U arrangement can be extended in both directions to the limits of the banqueting room*

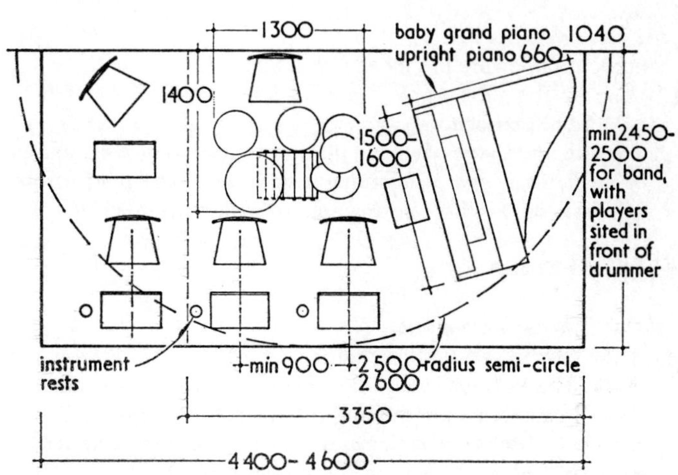

20.11 *Space required for band platforms*

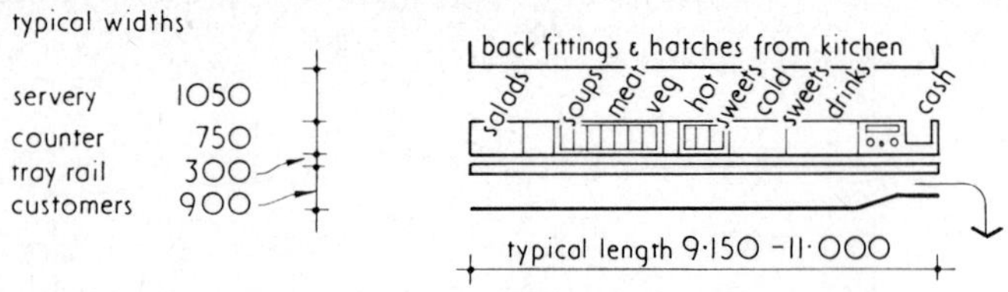

20.12 *Alternative arrangements for self-service counters:*
a *Single line counter, 60–90 customers per minute*

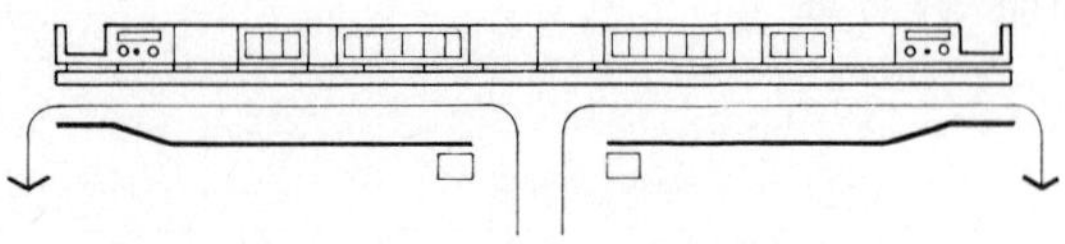

b *Divergent flow*

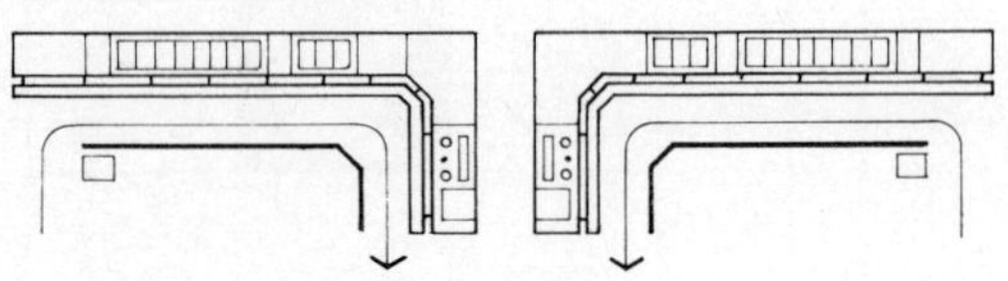

c *Convergent flow*

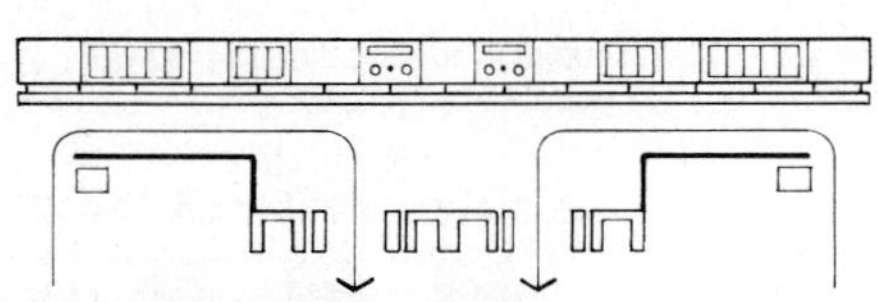

d *Multiple outlets*

Table III Range and sizes of tableware in general use

Type	Range	Size (rounded)
Pots	Tea	430, 570, 850, 1140 ml
(related to cups and pint sizes)		
Jugs	Coffee, hot milk/water	280, 570, 850, 1140 ml
	Cream	30, 40, 70 ml
	Milk	140, 280, 430 ml
Cups	Tea	170, 200, 230 ml
	Coffee (demi-tasse)	110 ml
Saucers	Size related to cups but should be interchangeable	
Plates	Side	165, 180 mm
	Dessert	190, 205 mm
	Fish/dessert	215, 230 mm
	Meat	240, 255 mm
	Oval meat	240, 255 mm*
Bowls	Cereal/fruit	155, 165 mm
	Sugar	90 mm
	Soup	215, 230 mm

* Usually maximum size for a dishwashing machine

- *Tableware*

Sizes of tableware are indicated in table III.

1.03 Band platforms

The size and shape of a band platform depends on the number of musicians, type of instrument and use of the platform by other artists. Suitable arrangements are shown in **20.11**.

2 CAFETERIAS

2.01 Self-service counter

Some typical self-service counter arrangements are shown in **20.12**. The average rate is 6 to 9 per minute. One cashier can serve 9 customers per minute, but the rate depends on:
- variety of menu (simpler menu gives faster throughput)
- whether beverage service is separate from the rest
- whether there are 'bypass' facilities

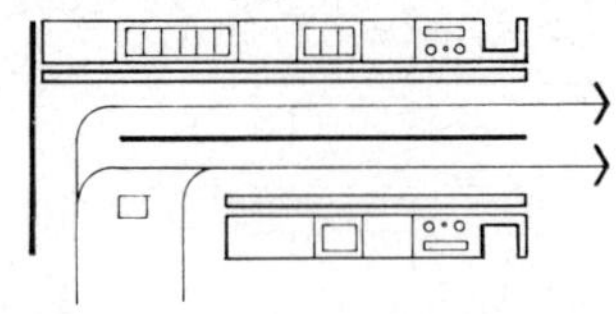

e *Parallel flow*

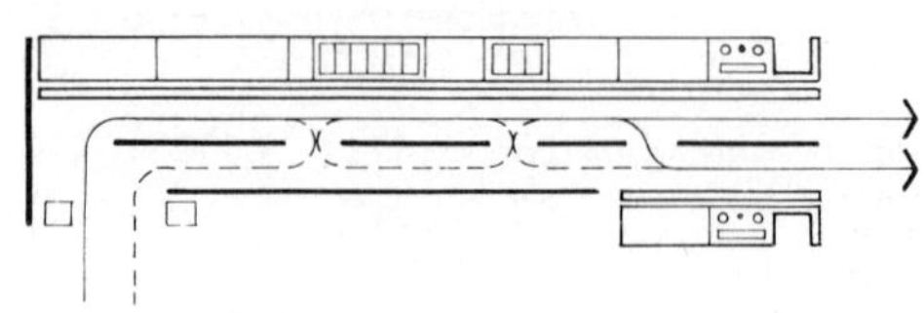

f *Bypassing*

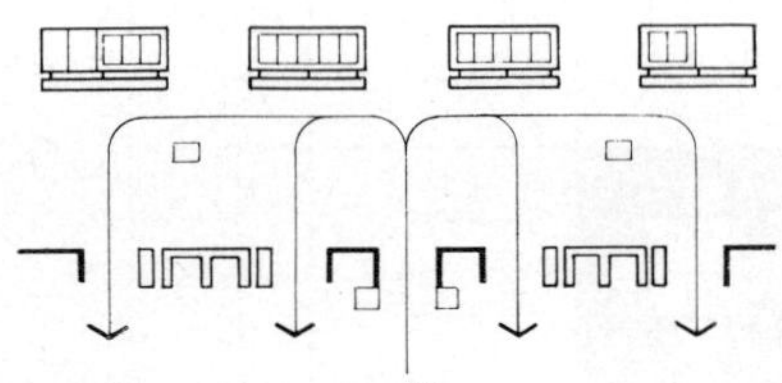

g *Free flow with counters in line*

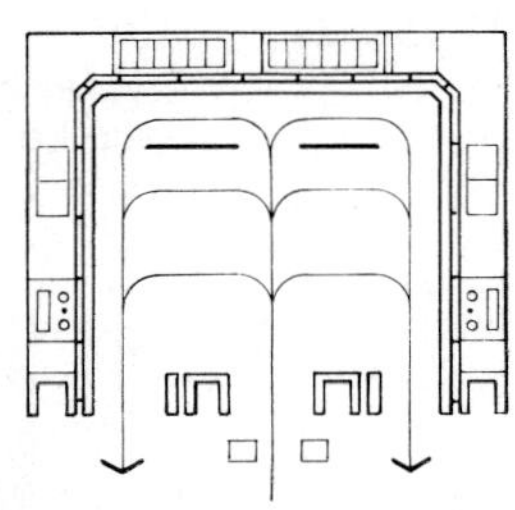

h *Free flow with counters in perimeter*

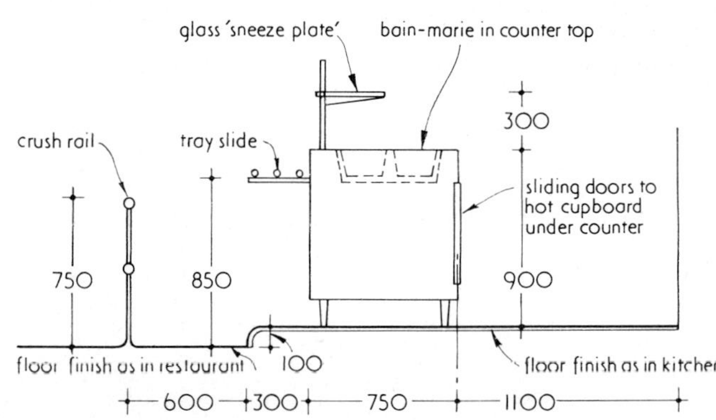

i *Section through counter for hot food*

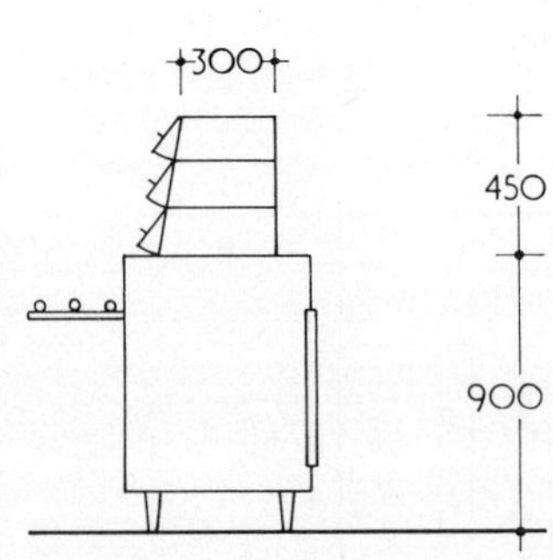

j *Alternative for cold servery*

Counter length depends on menu and rate of service:
- *School canteen:* between 9 and 11 m would be suitable for a menu offering 2 or 3 choices of main dish and serving about 80 to 90 customers in 10 minutes
- *Public cafeteria:* the same length would serve 60 to 70 in 10 minutes.

Table IV Equipment requirements guide: self-service cafeteria

Equipment for servery		**Meals served per day—based on main meal period**								
		50	100	200	400	600	800		1000	
							1 (1—single line 2—double line)	2	1	2
Trays										
Tray storage length	m	0·45	0·45	0·60	1·35	0·60	1·35	2 × 0·6	1·35	2 × 0·6
Bread, rolls, butter, etc*										
Unheated counter with self-service display above: length	m	0·45	0·45	0·75	1·20	1·65	1·80	2 × 1·2	2·30	2 × 1·5
Cold meats, salads, etc*										
Refrigerated counter	m	0·45	0·75	0·90	1·20	1·80	2·30	2 × 1·2	2·60	2 × 1·5
with dole plate and glass display above, refrigerator under of capacity:	m³	0·06	0·06	0·08	0·08	0·11	0·11	2 × 0·08	0·14	2 × 0·11
Hot foods										
Hot cupboard with sectioned bain-marie and heated service shelf: length	m	0·9	1·5	2·4	3·6	4·9	6·1	2 × 3·6	7·3	2 × 4·3
Beverages—hot drinks†										
Counter length	m	0·9	1·1	1·2	1·4	1·5	1·8	2 × 1·2	2·1	2 × 1·5
Comprising water boiler capacity	litres/hr	55	115	170	225	340	455	2 × 225	570	2 × 285
Tea/coffee urns										
No × capacity	litres	1 × 15	2 × 15	2 × 25	2 × 45	2 × 70	2 × 90	4 × 45	2 × 115	4 × 70
Storage racks under counter for cups/saucers:	capacity	50	100	150	200	250	350	2 × 200	450	2 × 250
Reserve cup and saucer storage behind counter:	capacity	—	—	50	200	350	450	2 × 200	550	2 × 250
Cold Drinks, etc†										
Counter length	m	0·6	0·9	1·2	1·8	2·1	2·4	2 × 1·8	2·7	2 × 2·1
Comprising refrigerator capacity	m³			0·06	0·08	0·08	0·11	2 × 0·08	0·11	2 × 0·08
Cold shelf length	m	0·45	0·6	0·6	0·9	1·2	1·2	2 × 0·9	1·5	2 × 1·2
Ice cream storage*	litres			4·5	9·0	13·5	18·0	2 × 9·0	22·5	2 × 13·5
Squash dispenser		1	1	1	1	1	1	2	1	2
Iced water point				1	1	1	1	2	1	2
Cutlery†										
Counter length	m	0·30	0·30	0·45	0·60	0·60	0·90	2 × 0·6	0·90	2 × 0·6
Cutlery boxes fitted in top-capacity pieces		250	300	400	600	900	1000	2 × 500	1700	2 × 650
Reserve cutlery under			200	600	1400	2200	3000	2 × 1400	3800	2 × 1850
Cashier counter-cut away for cash desk length	m	1·2	1·2	1·2	1·2	1·2	1·2	2 × 1·2	1·2	2 × 1·2
Standard cash desks		1	1	1	1	1	1	2	1	2
Automatic change machine						1	1		1	1

* Depends on type of meals and customer preferences
† May be located away from service counter
Based on equipment by Stotts of Oldham

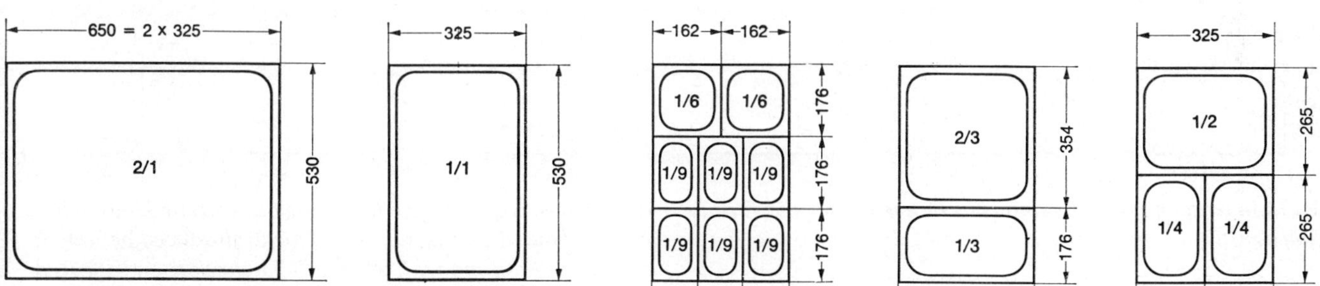

20.13 *Module sizes for 'Gastronorm' containers*

2.02 Equipment requirements and sizes

Servery equipment requirements are given in table IV. British Standards for equipment are given in the bibliography.

'Gastro-norm' containers

There are no universal standards for dimensions of equipment but the European Catering Equipment Manufacturers Association favours the Swiss 'Gastro-norm' system of dimensions, widely adopted on the continent, **20.13**.

3 KITCHENS

3.01 Size and type

Kitchen sizes depend on number of meals served at peak periods of the day: normally lunch, but in some types of restaurant the main meal is dinner.

Finishing kitchen

A finishing kitchen for the cooking or heating up of so called convenience foods will be smaller than the equivalent conventional kitchen, in which all the cooking and preparation processes are carried out. Typical requirements for a finishing kitchen are given in table V.

A kitchen for a residential or commercial restaurant serving a more varied menu over a longer period will be larger than for a school or factory canteen serving a more limited menu only at set times for the same number of people. As an approximate guide the ratio of dining to kitchen area will vary between:

- 3:1 where only one sitting
- 1:1 where two or three sittings per meal.

Table V Finishing kitchens for pre-cooked frozen meals

Area	Equipment		Meals served per day—based on main meal period						
			50	100	200	400	600	800	1000
Goods entry	Scales capacity	kg	12·5	12·5	25·5	25·5	25·5	25·5	25·5
	Trolleys for frozen meal trays and general use		2	2	3	4	5	6	7
Cold stores	Deep freeze room (based on 7 day stock)	m³	1·3	2·5	5·0	10·0	15·0	20·0	22·0
	Normal cold room	m³	0·3	0·5	0·9	1·2	1·5	1·9	2·2
	(based on daily delivery of dairy produce, otherwise 3 deliveries per week)								
Dry stores	Shelving—width 450 mm								
	length	m	6·1	9·1	15·2	21·3	27·4	33·5	39·6
	(for canned and dried items, 3 deliveries per week)								
Kitchen	Convection oven capacity	m³	0·09	0·18	0·37	1·02	1·10	1·47	1·84
	(based on reheating all frozen meals in 1 hour, ie 2 reheatings of average 30 minutes each)								
	Boiling tank or pressure steamer								
	(for 'boil-in-the-bag' food where this is to be used. Provision depends on type of equipment, eg rotating boiling tank 1000 mm (40") diameter produces a total of 120 bags/hour in 4 reheatings of approximately 15 minutes each)								
	Microwave oven								
	2 kilowatt units		1	1	1	1	2	2	3
	(Depends on number of snacks required and availability of alternative call-order equipment. Based on 30 second cycle of reheating with 4 snack items/loading)								
	Supplementary equipment								
	Griller	m²	0·2	0·2	0·3	0·4	0·5	0·7	0·9
	(for call order grills and toast. Based on surface area)								
	Griddle	m²	0·2	0·3	0·4	0·5	0·7	0·8	1·0
	(for snack catering, particularly at breakfast)								
	Fryer	kg/hr	10	25	45	90	135	180	225
	(for 'flashing off' blanched frozen meals. Based on 15 minutes use before and 15 minutes during meal period)								
	Boiling rings	No	2	2	3	4	5	6	7
	(for reheating of canned vegetables, soups, etc)								
Wash-up (crockery, cutlery, etc)	Wash-sterilizing unit								
	length for dishwashing	m	2·7	3·4	3·7				
	Capacity of machine	pieces/hr			1200	2400	3600	4800	6000
	Sink for serving dishes, etc (length)	mm	600	600	750	900	1050	1200	1200
	Burnishing machine		—	—	—	hand	hand	hand	hand
	Waste disposal units		1	1	1	1	2	2	2
	(based on 550 W machines assuming part refuse collection. For complete waste disposal 2230 W units employed)								
Source:	Paper by D. J. Cottam given to Catering Teachers Association Annual Conference Wolverhampton, October 1969. Based on equipment by Stotts of Oldham								

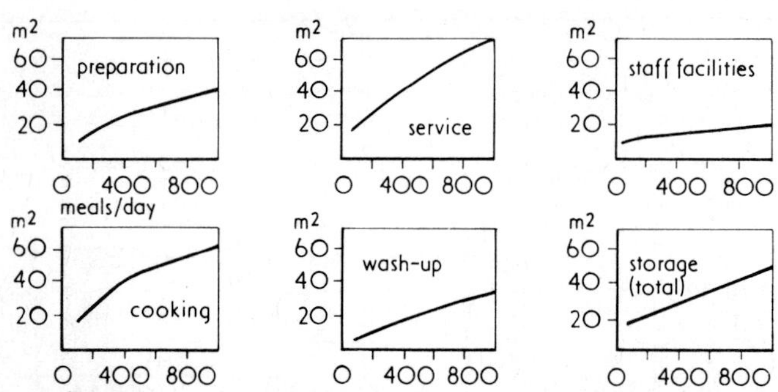

20.14 *Individual space requirements in kitchen for the various functions*

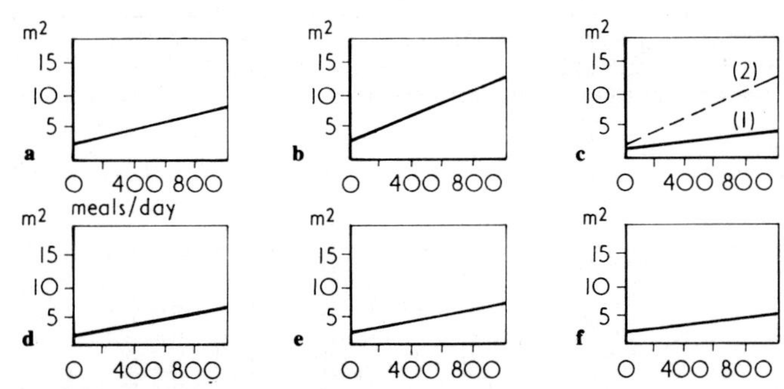

20.16 *Space requirements for storage of various kinds. These areas are based on the number of meals produced per day in a modern conventional kitchen using a proportion of frozen foods. Where mainly pre-cooked frozen meals are used, a separate vegetable store may not be needed, and more deep-freeze space should be provided*

***a** Vegetable store, three deliveries per week*
***b** Dry goods store, three days' supply*
***c** Deep freeze, 7 days' supply: 1 conventional use*
2 precooked frozen meals
***d** Cold room, daily delivery of perishable food*
***e** Goods entry area including weighing and checking*
***f** Refuse storage where bins are used*

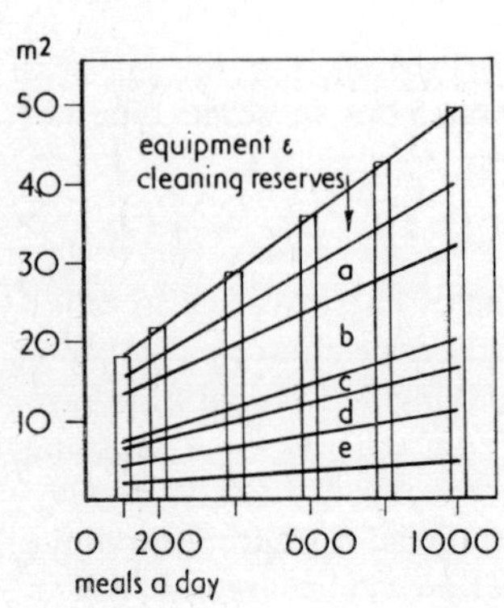

20.15 *Total storage requirements for conventional kitchens based on number of meals prepared per day. Segments are covered in detail in* **20.16**

Ratio of total kitchen area to area required for servery, stores, offices and staff facilities, **20.14**, will vary between:

- 2:1 for conventional kitchens
- 1.5:1 for finishing kitchens.

3.02 Storage area

Typical storage requirements, based on the floor areas of stores in relation to the number of main meals served per day are given in **20.15** and **20.16** for conventional kitchens where all normal cooking processes are carried out.

Storage of containers

Shelf and storage unit heights should approximate to heights of food containers, likewise widths. An interspace of 40 to 50 mm should be allowed between packages for easy access. The top

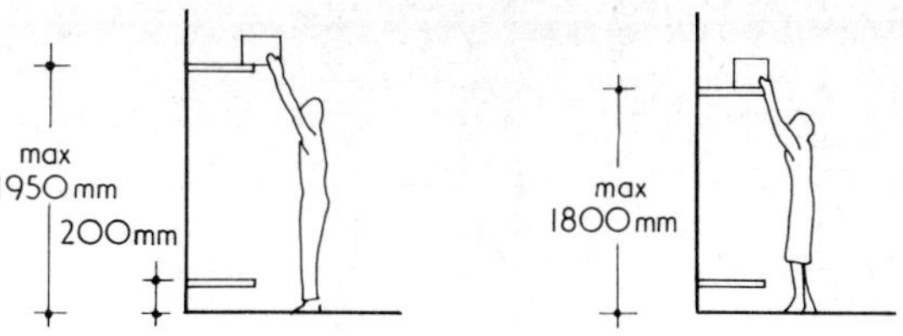

20.17 *Heights for storage shelving:*
a *Limits for maximum reach*

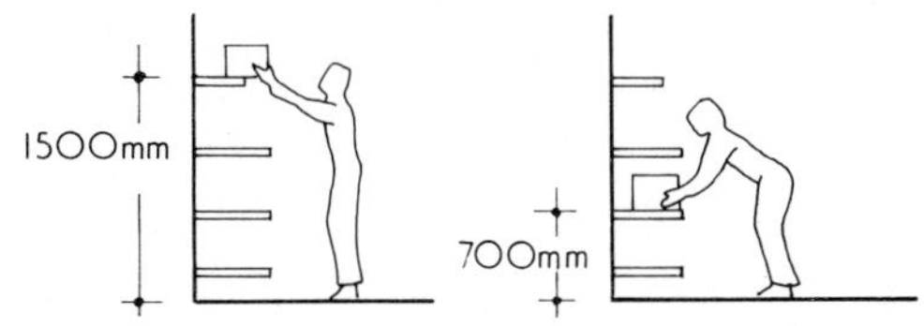

b *Within convenient reach for heavy or frequently used items*

shelf should be no higher than 1950 mm, **20.17**. Shelves and open bins must be kept at least 200 mm above the floor to allow a clear space for access and cleaning and to deter rodents. Shelves for frequently used or heavy items should be between 700 and 1500 mm high.

Expensive containers may have to be returned to suppliers and space should be allowed for the collection of these.

Goods access

For sizes of vehicles see section 7, External circulation. Typical delivery arrangements are:

- *dry goods*—weekly or fortnightly
- *vegetables*—once or twice weekly
- *perishable foods*—daily
- *refuse and waste removal*—usually once or twice weekly.

Bulk refuse containers may be used instead of bins. These are 0·57 to 0·85 m³ capacity. Refuse may be sited in a refrigerated area to reduce odour and fly nuisance, at a temperature between 2 and 5 °C.

Refrigerated stores

These may be either:

- *chilled stores:* at temperatures just above freezing point to preserve appearance and prolong life of perishable foods. Chilling is also used for cooling prepared foods and drinks
- *deep freeze stores:* at about −18 °C for prolonged storage of frozen food. They may be purpose built and can always be opened from within.

In large establishments, refrigerated stores may be provided in a number of areas, as in table VI.

3.03 Preparation areas

Kitchen layout is determined by:

- the sizes of equipment and fittings
- the spaces left for access and circulation, **20.18**.

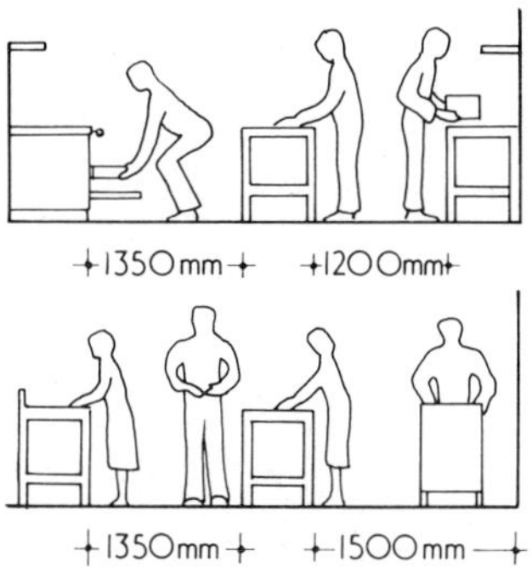

20.18 *Minimum space between equipment to allow for working and circulation*

Table VI Types of refrigeration storage

Types of storage, etc	Situation
Bulk cold stores with separate deep freeze storage	Near goods receiving area
Kitchen stores—refrigerated cabinets and cold rooms	In or adjacent to the preparation area
Refrigerated pass-through units, display cabinets and counters	In or adjacent to the servery
Ice-making machines	In or adjacent to the bar
Vending machines for drinks and meals	Near the servery or in an independent area
Beer cellars and wine stores	Accessible from goods entrance and to bars and restaurant counters
Cold stores for food waste	Near collecting area accessible from kitchen

Some typical dimensions are:
work top height: 865 mm (900 mm preferred)
sink top height: 900 mm (more common) and 865 mm
wall bench width: 600 to 750 mm
island bench and table width: 900 or 1050 mm
length of work area within convenient reach: 1200 to 1800 mm
length for two people working together: 2400 to 3000 mm
comfortable height (700 mm) when seated (430 mm).

There are usually four main areas of food preparation:

- vegetables
- meat and fish
- pastry
- general.

Preparation areas may be segregated:

- separate rooms adjoining the main kitchen
- low walls (approximately 1200 mm) between the areas which are otherwise open to the kitchen
- the arrangement of benches and equipment into specialist sections.

Alternative arrangements are shown in **20.19**.

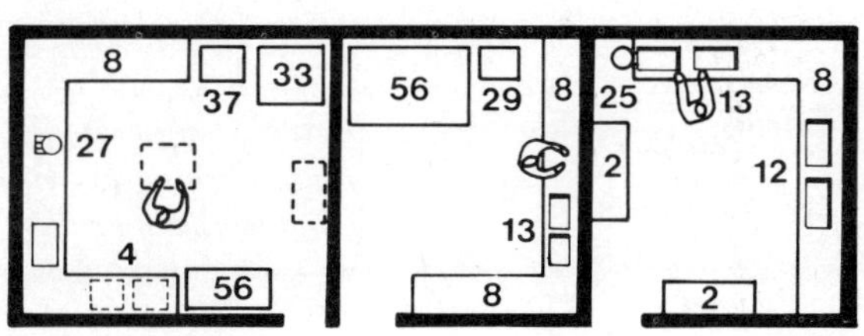

20.19 *Alternative preparation area arrangements (see* **20.20** *for key to numbers):*
a *Separate rooms*

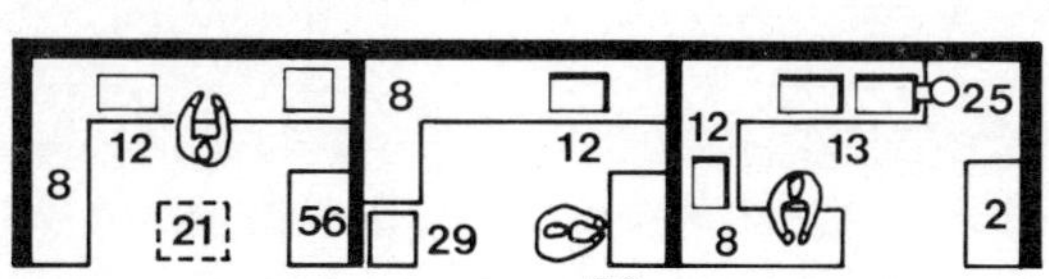

b *Bays*

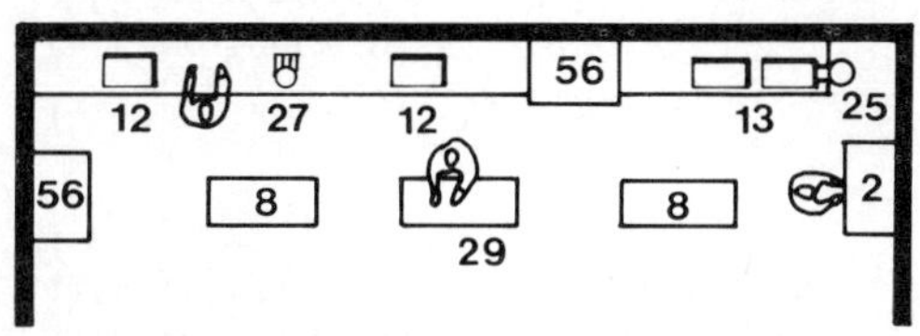

c *Open plan kitchen*

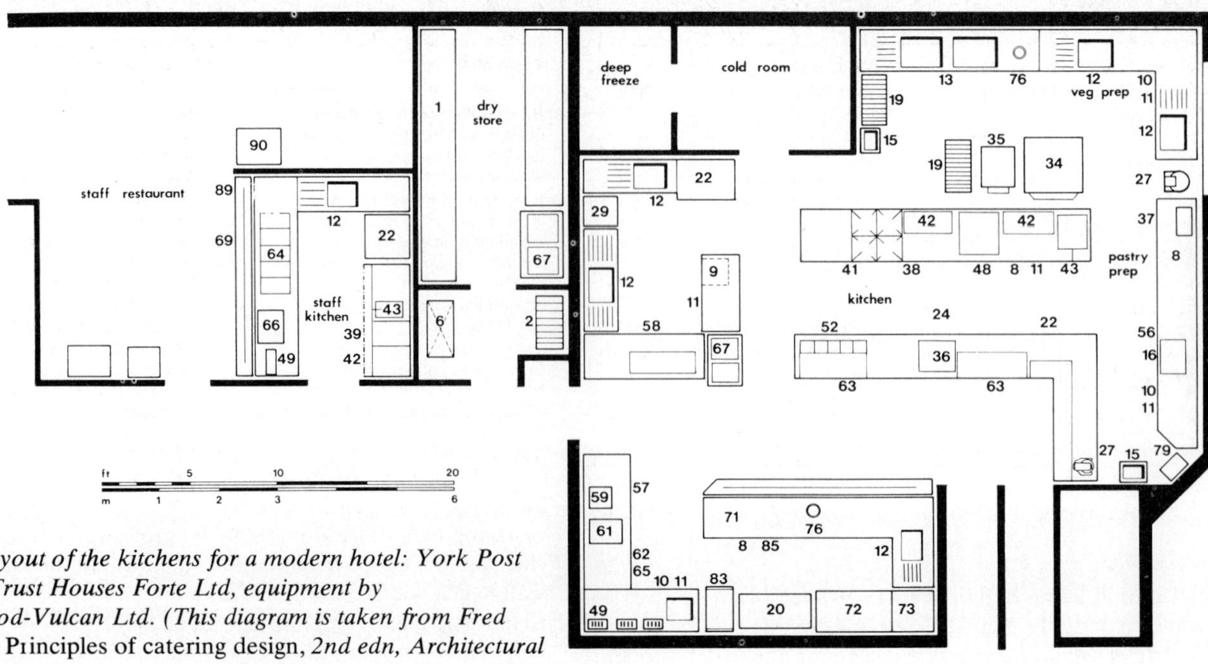

20.20 *Layout of the kitchens for a modern hotel: York Post House, Trust Houses Forte Ltd, equipment by Moorwood-Vulcan Ltd. (This diagram is taken from Fred Lawson,* Principles of catering design, *2nd edn, Architectural Press, 1978. Numbers are explained in the key below and are taken from a system further developed in the book.)*

1 *shelving*
2 *vegetable racks*
4 *storage bins*
6 *mobile racks*
8 *work table or bench for preparation*
10 *work bench with waste bin*
11 *work top with shelves*
12 *single sink and drainer*
13 *double sink unit*
15 *wash hand basin*
16 *marble-topped bench*
19 *pot rack*
20 *trolley*
21 *mobile tray racks*
22 *refrigerator*
24 *deep freeze cabinet*
25 *potato peeler*
27 *mixing machine*
29 *chopping block*
33 *pasty oven or pizza oven*
34 *forced-air convection oven*
35 *steaming oven*
36 *microwave oven*
37 *boiling top*
38 *boiling top with open top burners*
39 *boiling top with solid top*
41 *oven range with boiling top*
42 *grill or salamander*
43 *deep fat fryer*
48 *griddle plate*
49 *toaster*
52 *hot cupboard with bain-marie top*
56 *refrigerated under-cupboard*
57 *refrigerated cupboard with doleplate*
58 *refrigerated display cabinet*
59 *milk dispenser*
61 *coffee unit*
62 *unheated counter unit*
63 *counter with infra-red lamps over*
64 *counter display cabinet*
65 *compressor or boiler under counter*
66 *tray stand*
67 *ice cream conservator*
69 *tray rail*
71 *receiving table for dirty dishes*
72 *stacking table for clean dishes*
73 *semi-automatic dishwashing machine*
76 *waste disposal unit or scraping point*
79 *refuse bins*
83 *glass storage racking*
85 *glass washing machine*
89 *chilled water dispenser*
90 *beverage vending unit*

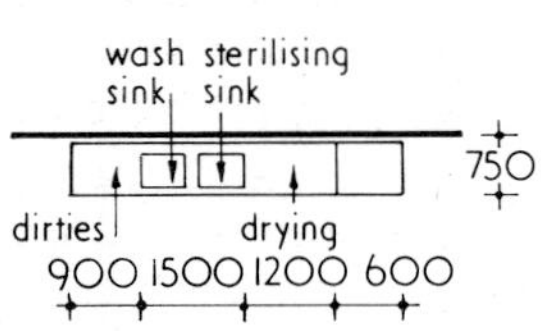

a *Manual washing-up with wash and sterilising sinks*

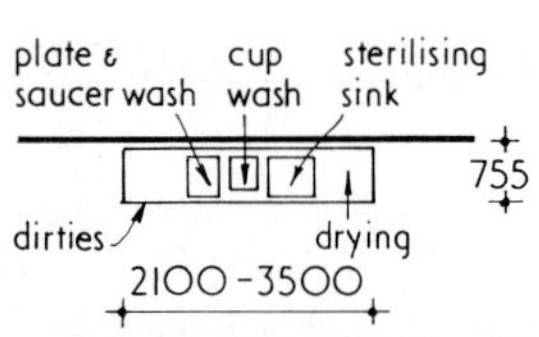

b *Brush type dishwashing machine*

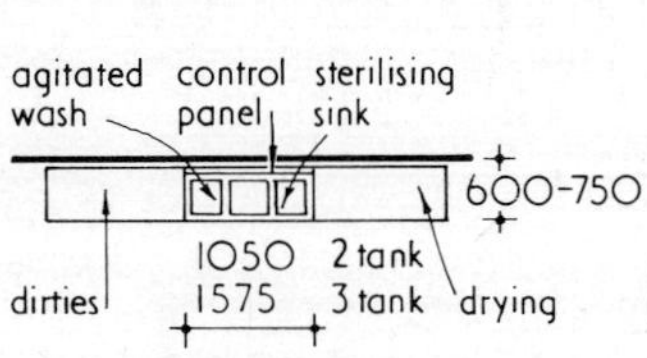

c *Agitated water type dishwashing machine*

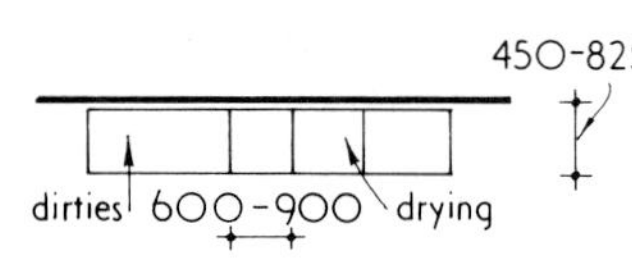

d *Small spray type dishwashing machine*

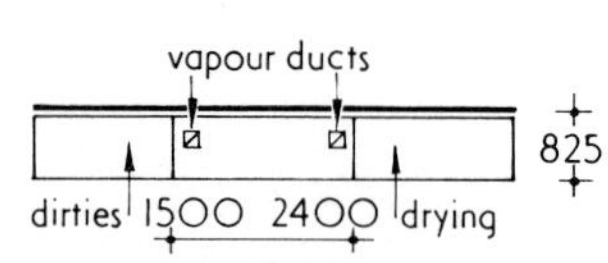

e *Medium size automatic conveyor type dishwashing machine*

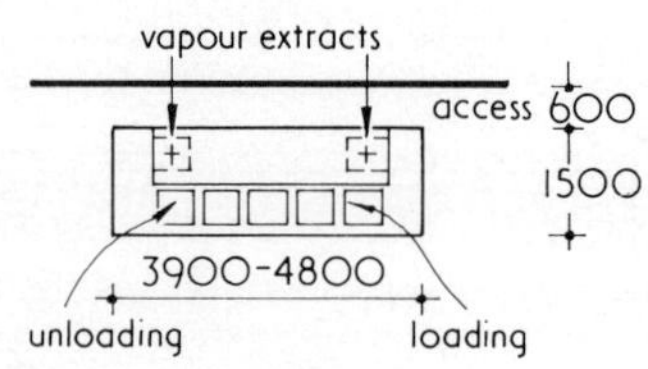

f *Rotary conveyor type dishwashing machine*

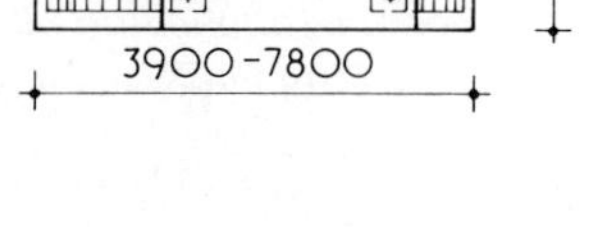

g *Large 'flight type' dishwashing machine with continuous escalator type conveyor*

20.21 *Alternative layouts for wash-up areas, with equipment dimensions:*

Table VII Oven and cooking equipment sizes and capacity

Oven type and capacity (m^3)	Internal* size (mm)	External* size (mm)	Loading (kW)
Oven ranges			
0·113	530 × 460 × 460	760 × 810 × 870	11–14
0·142	600 × 600 × 400	910 × 870 × 870	12–15
Forced-air convection ovens			
0·113		810 × 810 × 870	8
0·198		1020 × 910 × 1370	10–12
0·283		1120 × 1170 × 1580	13–15
Pastry ovens			
1 deck	760 × 610 × 230	960 × 1090 × 1380	14·4
2 deck	760 × 910 × 230	1270 × 1090 × 1680	14·4
Roasting cabinets			
0·48	610 × 660 × 1220	1170 × 910 × 1600	8·5
0·85	840 × 760 × 1320	1400 × 1020 × 1680	12
Atmospheric type steaming oven			
0·17	450 × 530	770 × 840 × 1520	6
Boiling tables			
(Consult individual manufacturers for number of plates etc)		1220 × 760 × 860	12·4
		1680 × 760 × 860	21·2
		2130 × 760 × 860	30·2
Gas fired boiling pans			
45 litres		910 × 990 × 760	7
90 litres		1040 × 1120 × 840	11
135 litres		1190 × 1240 × 860	14·2
180 litres		1190 × 1300 × 910	16·5
Tilting kettle			
40 litres		760 × 510 × 860	7·5
Deep fat fryers			
Single unit		910 × 810 × 880	15·8
Double unit		1830 × 810 × 880	31·5
Shallow tilting frypan		1180 × 820 × 930	12
Grillers			
Grill area			
410 × 380 × 230		610 × 510 × 510	4
560 × 380 × 230		760 × 510 × 510	5
760 × 380 × 230		840 × 560 × 510	7

* Width × depth × height

3.04 Cooking equipment

Dimensions, capacity and loading of typical equipment are given in table VII. A typical kitchen layout is shown in **20.20**.

3.05 Dishwashing

Typical space allowances for intermittent dishwashing are given in table VIII. Dishwashing equipment is shown in **20.21**.

Table VIII Typical allowances for dishwashing spaces (intermittent use)

Area/activity	Space (mm)	
Collection area for unsorted table ware prior to sorting and scraping	600 length per 10 meals* Minimum 900 Maximum 2400	
Stacking area for tableware sorted and stacked for manual washing	300 length per 10 meals† Minimum 900 Maximum 3600	
Loading onto racks for machine washing	Depends on rack/basket size Minimum 1000	
Draining and drying in racks or baskets after washing and sterilising	Manual process and brush type machines: Minimum 1200 Conveyor or spray type machines up to 3600	
Unloading baskets and racks for clean crockery awaiting removal	100 length per 10 meals Minimum 600 Maximum 2400	
Spray-type machines with mechanised conveyor systems	Space occupied by machine conveyor system	
	Width	Length
Rotary conveyor type (600–1000 meals/hour)	1500	3900 to 4800
Flight type escalator conveyor (over 1000 meals/hour)	750 to 1200	3900 to 7900

* Based on self clearance. Smaller areas suitable where part stacking is provided
† Assumes some accumulation of dishes before washing up. The lengths relate to tabling 750 mm wide

4 BIBLIOGRAPHY

Fred Lawson, *Principles of catering design*, London, Architectural Press, 1973; 2nd edn 1978

Advisory Committee on hotels and restaurants, *Reports on tableware, restaurant tables and chairs and on food storage requirements*, London, the Council of Industrial Design, 1965–70

BS 2512: 1963 *Gas heated catering equipment*

BS 4104:1967 *Catering equipment burning liquified petroleum gases*

Directory of products and members, 1972, Catering Equipment Manufacturers Association, London, 1972

Fred Lawson, *Principles of catering design*, London, Architectural Press, 1973

Fred Lawson, *Restaurant planning and design*, London, Architectural Press, 1973

S. Medlik, *Profile of the hotel and catering industry*, London, Heinemann, 1972

21 Public houses

Frank Bradbeer

CI/SfB 517
CI/SfB (1976 revised) 517

Frank Bradbeer is an architect whose passions, other than good architecture of all periods, include baroque organs, horses and pubs

Contents

1 INTRODUCTION

The greater proportion of building work involving licensed premises is in connection with the alteration of existing buildings. New licences, and therefore new buildings, form a small section of the total amount of work.

Public houses, **21.1**, are divided into two parts:

- *Public areas* (one or more bars, together with lavatories) where the public, or customers, are allowed.
- *Private areas* confined to staff use. These include serveries, cellars, kitchen and staff accommodation (see section 20, Catering).

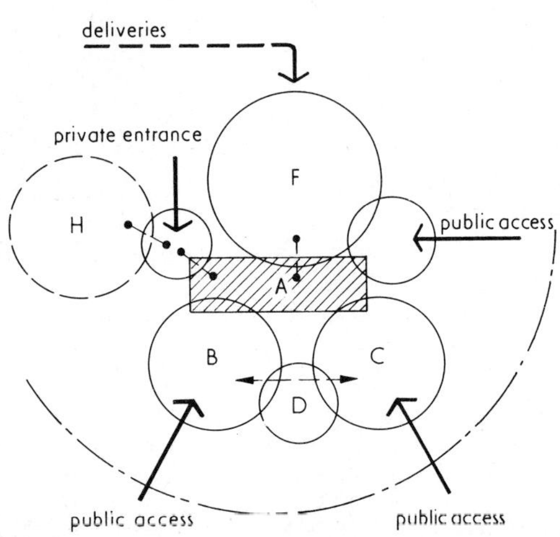

21.1 *Organisation of spaces:* **a** *Space relationship diagram*

A servery
B drinking room
C drinking room
D lavatories
E off-sales
F storage (cellars)
G licensee's entrance
H licensee's accommodation

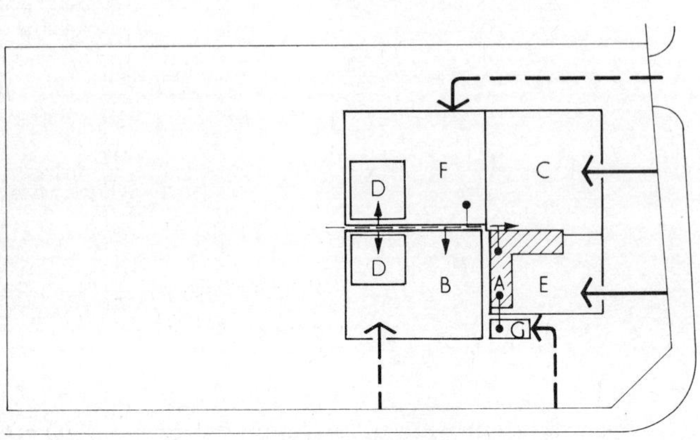

b *Corner site with building clear of all boundaries*

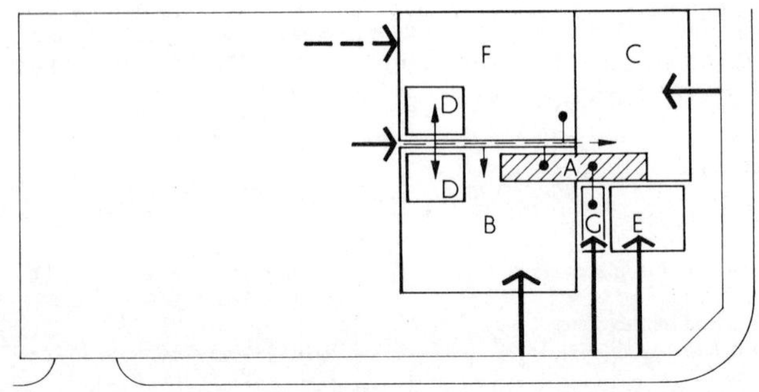

c *Corner site with building set against back boundary*

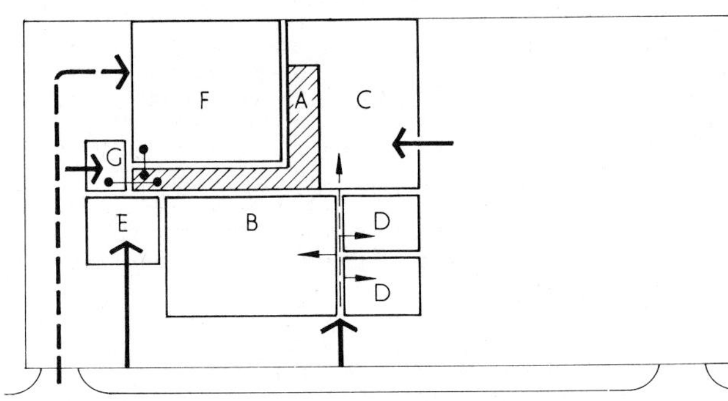

d *Straight site with delivery access at side*

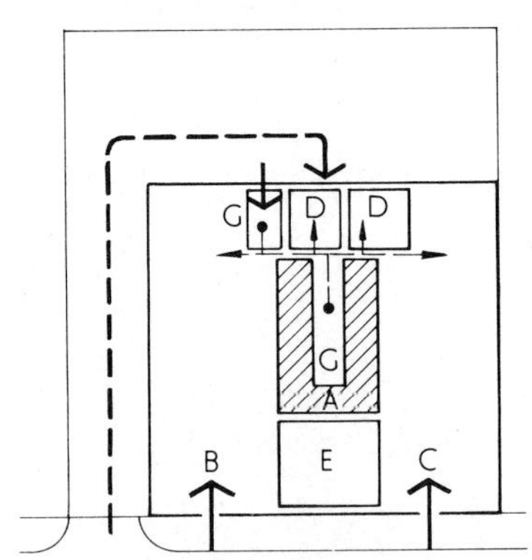

e *Straight site with delivery access at rear, and storage at lower level*

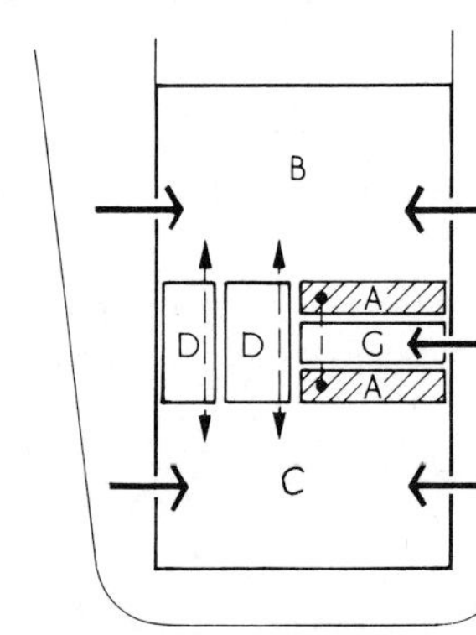

g *Island site with storage at lower level*

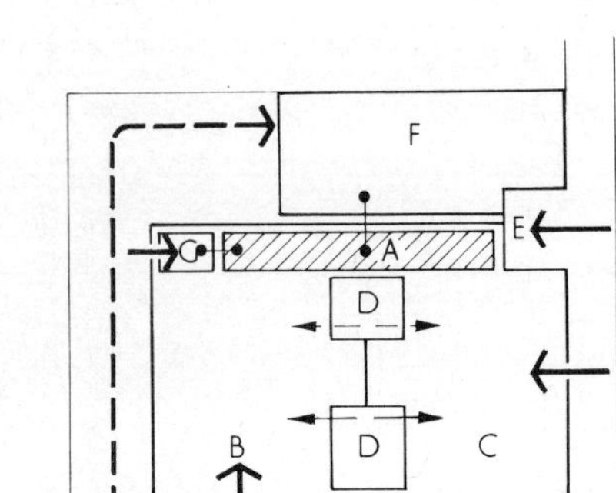

f *Restricted corner site*

Often, particularly in older properties, the charm or customer appeal can lie in the eccentricity, or even the inconvenience, of the layout. The planning of public areas depends on the flair and experience of the designer. Optimum seating may not be possible (or even desirable) and the clever designer will turn this to advantage. However, in the areas where the staff do their work, and where the liquor is stored, specific dimensional requirements assume great importance and closely affect the efficient running of the establishment.

Requirements peculiar to the pub are:

- drink delivery
- drink storage
- drink dispensing.

2 DRINK DELIVERY

Vehicles deliver drink in bulk (by tanker), in barrel (or keg), or by bottle.

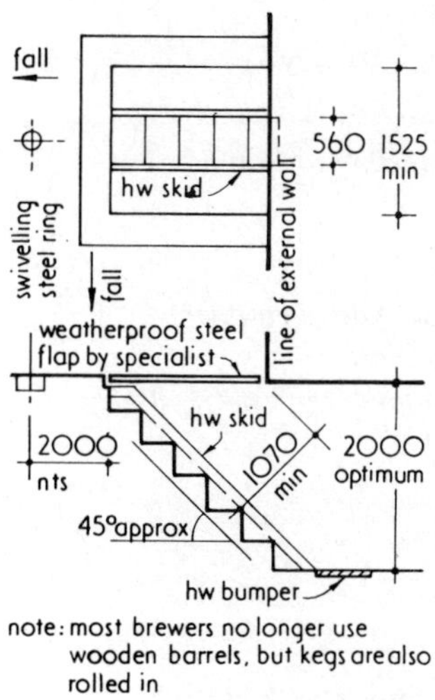

21.2 *Cellar flap and barrel chute for use with storage below ground*
Note: Most brewers no longer use wooden barrels, but kegs are also rolled in

21.3 *Cellarage for barrels and kegs:*
a *Keg with valve attachment*

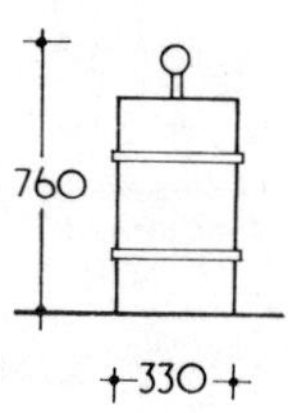

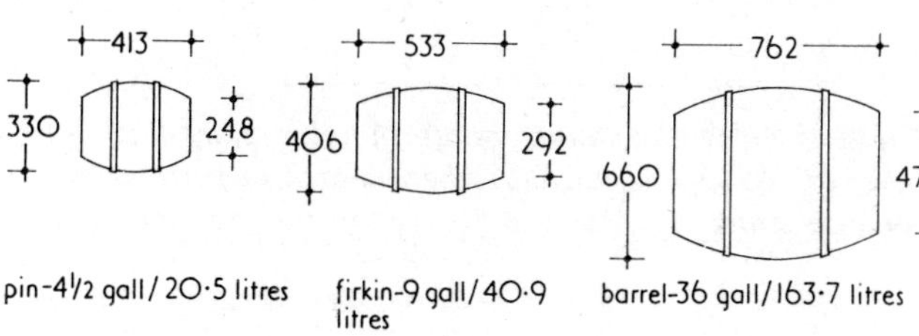

b *Metal barrels*

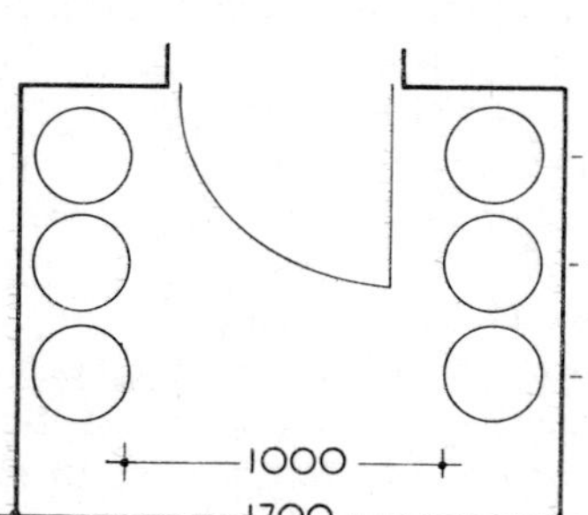

c *Bar store with kegs*

3 DRINK STORAGE

The vehicular access must be convenient for the cellars, which may be below ground, **21.2** at ground level, or in rare cases at an upper level.

The cellar must be planned to store the following in an economic manner:

- barrels, or, more likely, kegs, **21.3**
- bottles, usually in cases stacked on top of one another, **21.4**
- wine and spirits, usually stacked in a separate store on shelves
- bulk storage tanks, **21.5**.

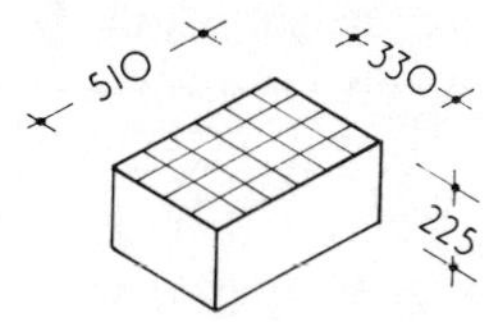

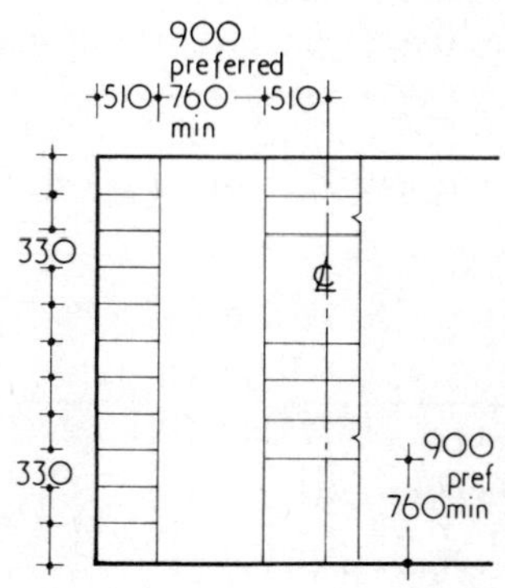

21.4 *Case storage. Each case contains 24 half-pint bottles. Non-standard cases are normally smaller*
a *Typical case*
b *Plan of case cellarage*

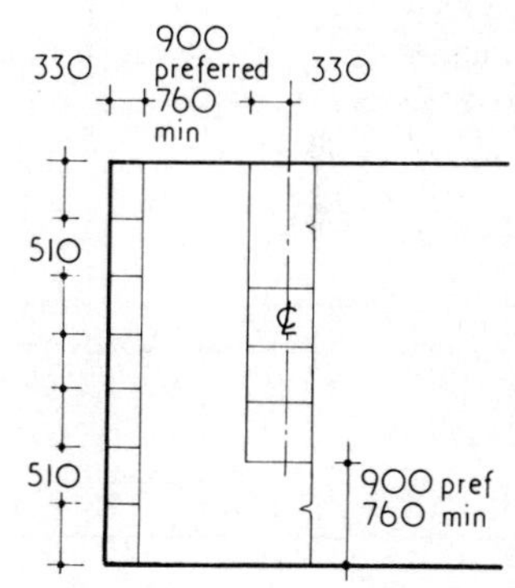

c *Alternative plan*

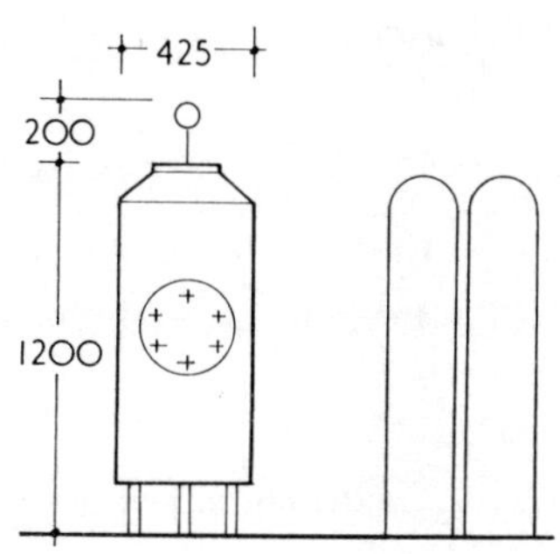

21.5 *Bulk storage with typical CO_2 canisters. Larger canisters exist*

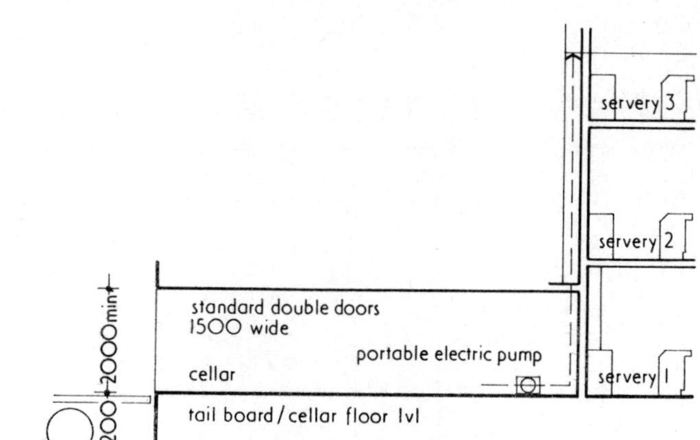

21.6 *Beer supplied through several storeys from ground level cellar. Standard electric cellar pump will raise beer up to 9 metres*

The wooden barrel accommodated on a tilting stillage is now extremely rare. The size of a case containing two dozen half-pint bottles is fairly standard, **21.4a**, as is the standard metal keg, **21.3**. These are the common modules of beer storage. Storage of empties is based on the same standards, except that it is easy to stack empty cases higher and therefore less area is required.
Any beer other than in bottles or cans is (for the purposes of design) referred to as 'draught'. Draught beer from below ground, or ground level cellarage is usually delivered under pressure. Beer travelling upwards one storey height is usually pumped electrically, **21.6**.
Bottled beer travelling upwards one storey height is usually delivered by a case hoist, **21.7**, which is installed as a unit and can be manually or electrically operated. Manual operation is simpler and considerably cheaper.

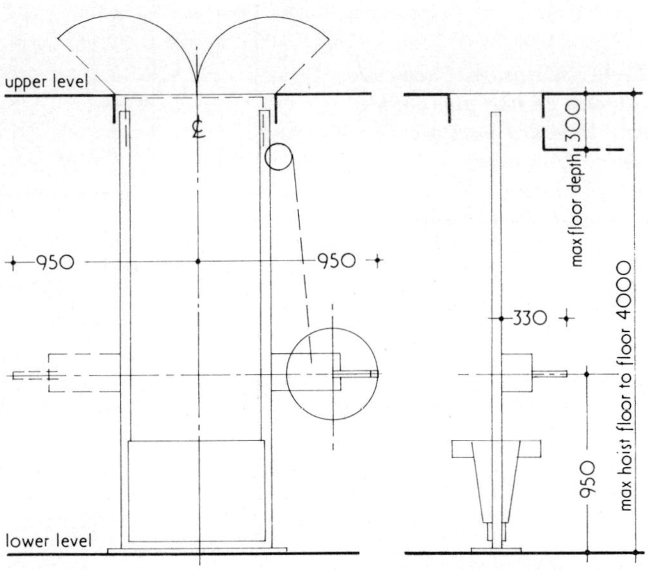

a *Front elevation* **b** *Side elevation*

21.7 *Small manual case hoist for two cases:*

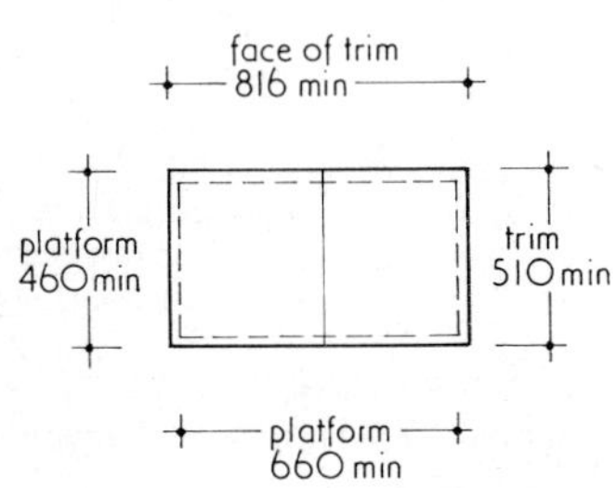

c *Plan at upper level*

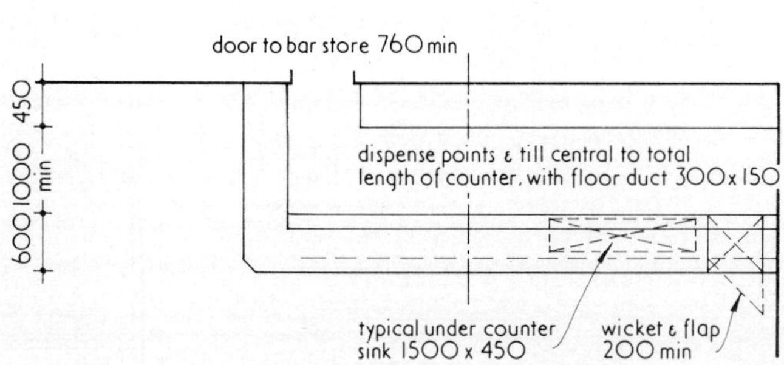

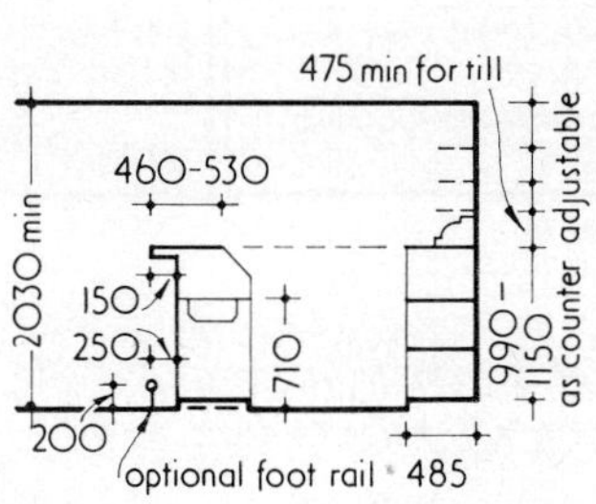

21.8 *Typical bar servery:*
a *Plan*

b *Section*

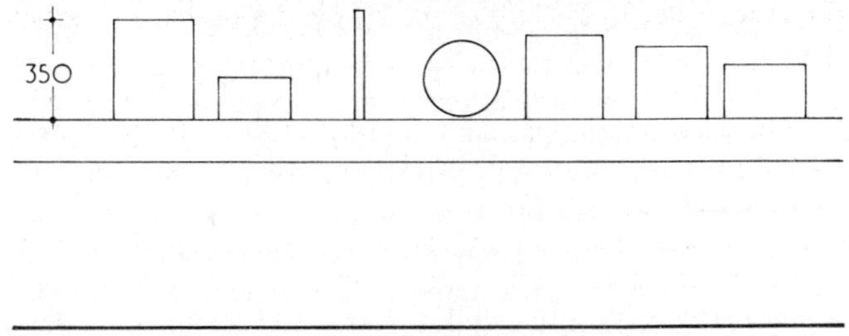

21.9 *Bar counter top:*
a *Plan, average centres of dispense points shown, sizes vary according to brand*

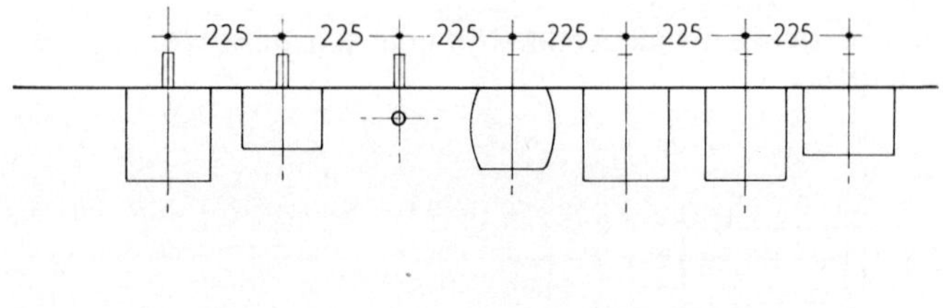

b *Elevation, heights also vary according to brand*

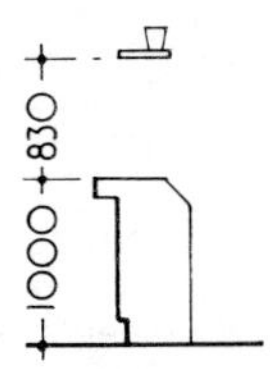

21.10 *Above counter canopies:*
a *Counter with glass shelf over*

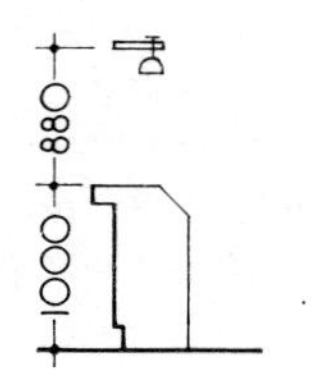

b *Counter with glasses suspended over*

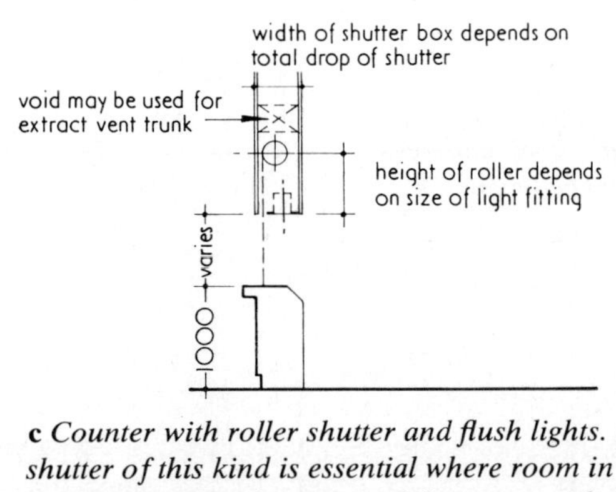

c *Counter with roller shutter and flush lights. Lockable roller shutter of this kind is essential where room in which the counter is situated is required for use outside licensed hours*

4 DRINK DISPENSING

There are three elements to the servery, **21.8**:

- the bar counter
- the back bar fitting
- the servery space between.

The length of the counter, **21.9**, will depend on circumstance, but the height and width can be regarded as standard within narrow limits. The width can be reduced in special circumstances, but the need to accommodate at least one sink will be an important factor. Glasses may be stored in racks above the counter, and standard heights are important, **21.10**. The servery space may be regarded as a standard width in most cases.

This is governed by the ease of working as opposed to economy of area.

The back bar fitting will comprise a sideboard top, often at the same height as the bar counter. Below this level will be stored bottled beer, sometimes on cooled shelves. These are fairly standard items, **21.11**, **21.12**.

Above the sideboard top will be a display fitting which may accommodate spirits (often with 'optics'), glasses and other products (including wine). Adjustable shelves are useful. The lowest shelf must be high enough to accommodate loose bottles, siphons, the till etc.

The sizes of special equipment such as coolers, glass washers etc should be determined from consultation with the client as to preferred manufacturers. The sizes of ducts for the beer pipes, which are usually lagged, should be confirmed with the client.

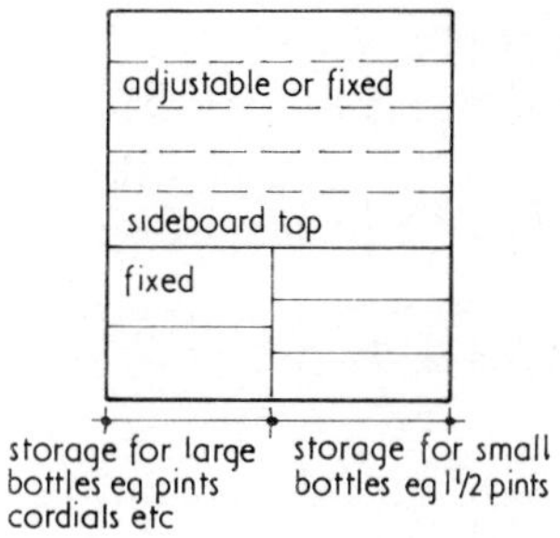

21.11 *Elevation to bar back fitting*

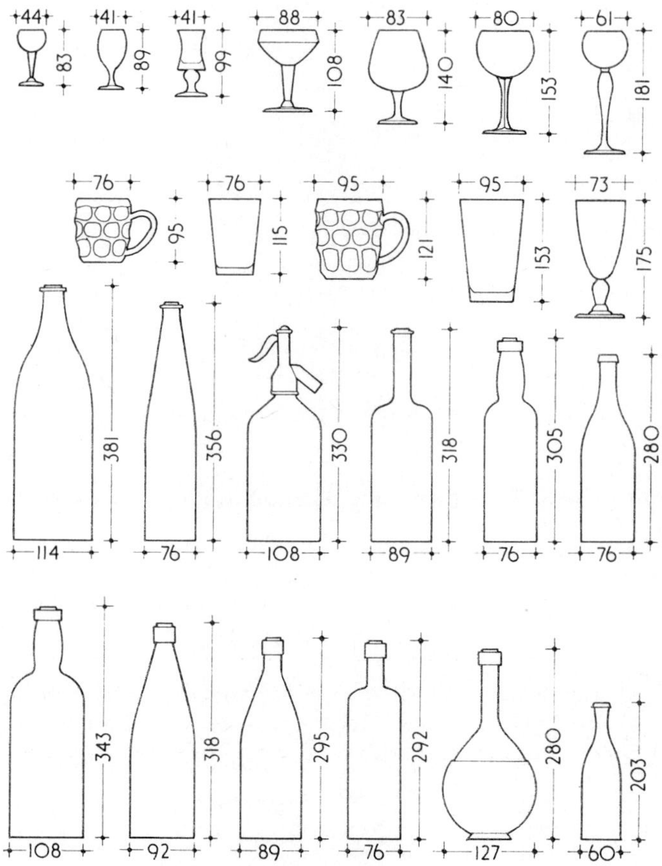

21.12 *Sizes of typical glasses and bottles*

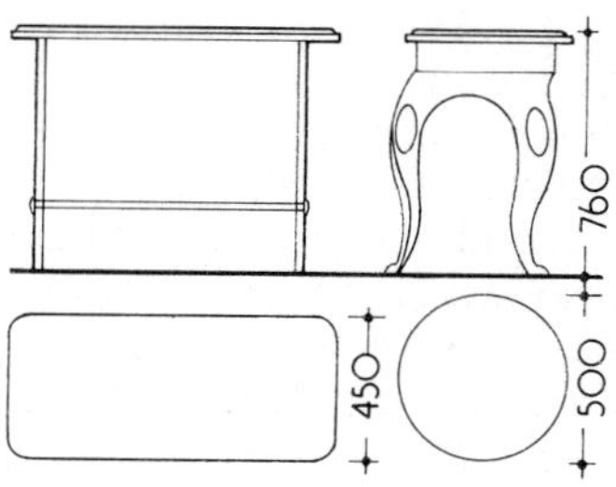

21.13 *Typical traditional cast iron pub tables, the round form is Britannia*

5 COOLING

The temperature at which beer is kept and served is an important factor. The two are not always the same. Recent fashion has tended towards colder dispensing; the exception to this is what has become known as 'real ale'.

Cooling may be either:

- in cellar
- on pipeline. This will probably take the form of an under-counter cooling unit.

Specific requirements should be checked with:

- the supplier
- the client.

6 BARS

Sizes of cast iron tables are fairly standard (eg Britannia tables), but other types are almost infinitely variable, **21.13**. Fixed seating can vary from practically minimum to opulently upholstered units, **21.14**. Stools are shown in **21.15**.

Cill heights should be arranged with due regard for the fixed seating.

Ceiling heights are infinitely variable and depend on the type of pub and the required effect.

Space may be required in the various bars for the following:

- dartboard, **21.16**
- bar billiards, **21.17**
- juke box, sizes vary considerably
- fruit machine, **21.18**
- piano or organ, **21.19**.

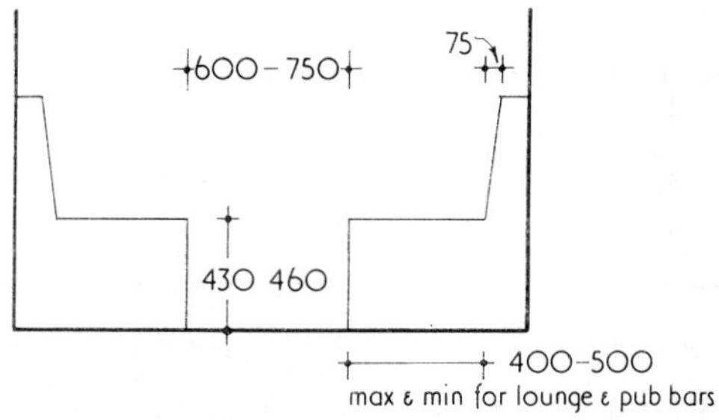

21.14 *Fixed seating of the bench type:*
a *Section through booth*

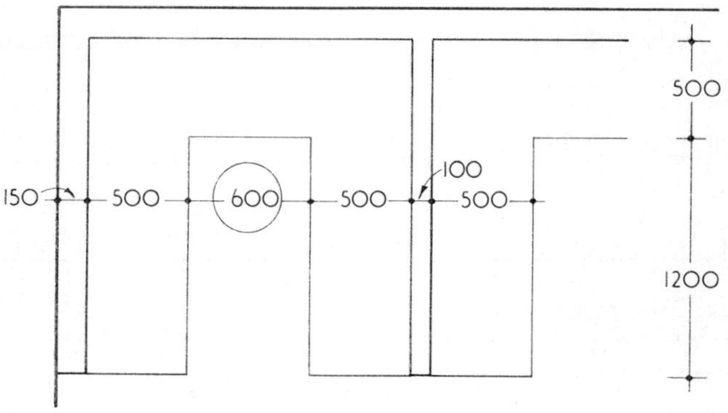

b *Typical plan of peninsular seating*

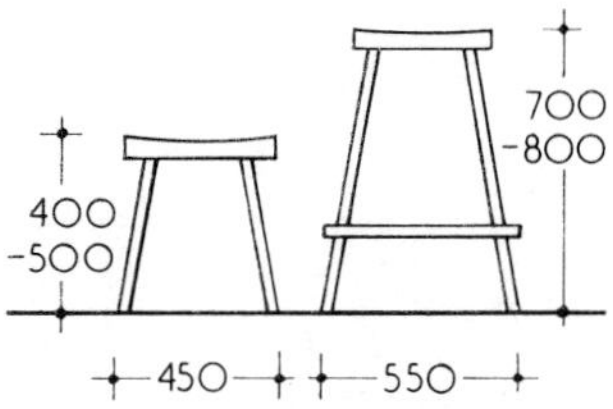

21.15 *Stools:*

a *Low stools for use at tables*
b *High stools for sitting at the bar*

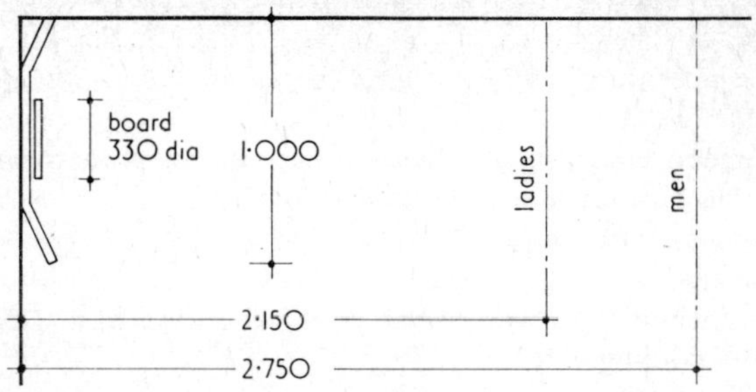

21.16 *Space required for darts*

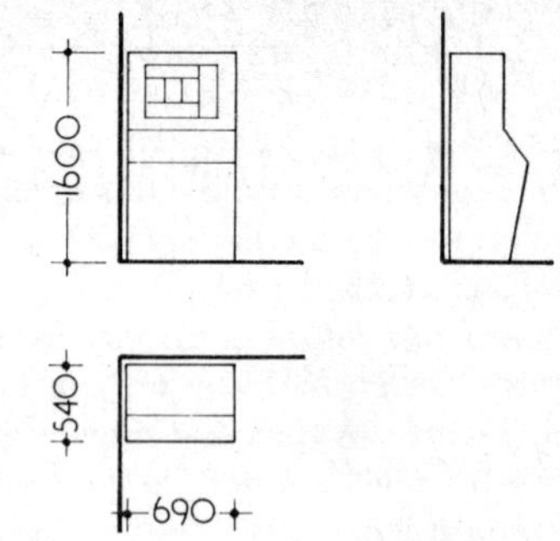

21.18 *Typical large fruit machine. Electricity supply required*

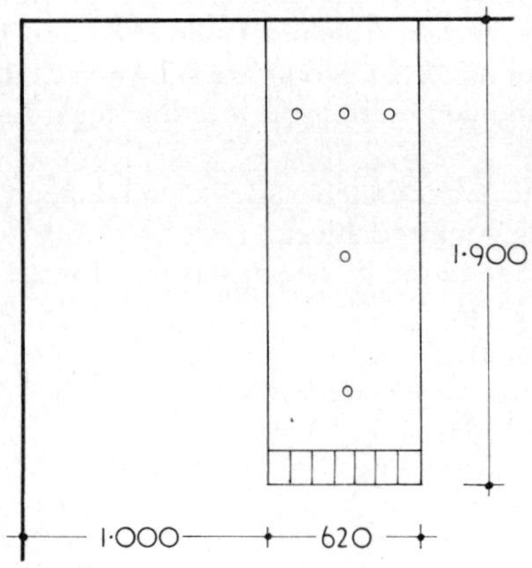

21.17 *Space required for bar billiards*

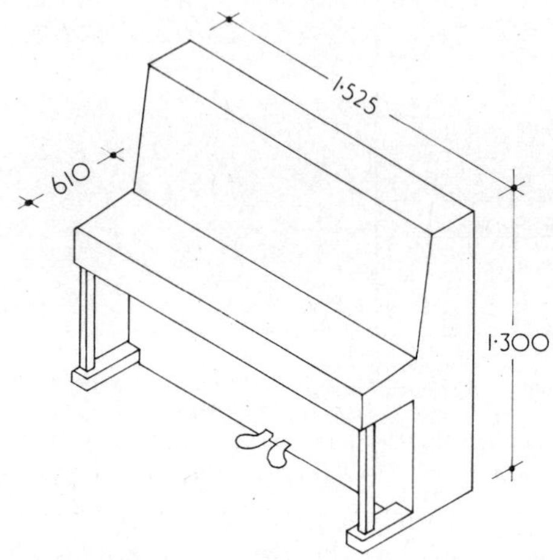

21.19 *Public house piano*

22 Television studios

David Binns

CI/SfB/784
CI/SfB (1976 revised) 528

David Binns is the senior partner of Sandy Brown Associates, architects and acoustic consultants

Contents

1 INTRODUCTION

1.01 Scope
A studio is an area in which activities are performed specifically for observation. (Television cameras are also used outside studios for surveillance in stores, banks and so on.)

1.02 Broadcasting studios
The greatest differences between TV studios will be in the ancillary areas rather than the studio per se. These differences reflect the nature and attitudes of the client: the BBC in the UK, for instance, is a public service organisation whereas the independent companies are not (although they must adhere to standards set by the Independent Broadcasting Authority).

1.03 Independent and educational studios
Increasingly, small independent studios are being developed, operating for private commercial use (like sound recording studios) or attached to higher education institutions.

2 STUDIO TYPES

2.01 Multi-purpose production studios
Previously, TV studios differentiated between music and drama. Now, all studios are multi-purpose (largely due to economic pressures) and have accepted acoustically 'dead' conditions, reverberation or presence being added electronically. Greater use of zoom lenses in preference to camera tracking means microphones are located further from performers, necessitating low reverberation times and background noise levels. Camera tracking requires a floor laid to very precise tolerances (currently ± 3 mm in 3 m). The floor is normally heavy duty linoleum sheet laid on an asphalt mastic screed and requires a specialist floor laying contractor to achieve the fine tolerances.
Studio length to breadth ratio should be in the region of 1:1·5. The minimum practical floor area for a small commercial studio would be 60 m² with static cameras, and typical broadcasting studios would range between 200 and 400 m². The studio height is determined by the clear space required below the lighting grid (a function of the longest camera angle). The minimum height for a small studio is 4 m and in the larger studios 13·5 m overall (11 m to the grid with a clear height above of 2·5 m). In these studios an access gallery is required at grid level approximately 4·5 m above studio floor level. This is normally to avoid obstruction of access doors and observation window. Access to the galleries from studio floor level is mandatory and direct access to lighting grid level is desirable.
A cyclorama or backdrop cloth is suspended below lighting grid level. It should be at least 1·25 m away from the walls to allow a walkway around the studio and is on a sliding track with radiused corners to enable it to be stored.

2.02 Interview and announcers' studios
These studios range in size from 30 to 60 m² with a height of 4 to 6 m. Static cameras and a simple form of lighting grid combined with floor lighting stands are used.

2.03 Audience participation studios
Some productions require audience participation and fixed theatre type seating on terraces is provided. In smaller studios this is demountable, so storage space has to be provided. Audiences place more stringent demands on the planning of a TV complex as segregated access and safe emergency escape routes have to be provided.

3 PLANNING

3.01 Layout
A typical layout for a TV broadcasting studio complex is shown in **22.1**. Larger installations will have workshop facilities adjacent to the scene dock and if flats and backdrops are made on site a paint frame will need to be the full height of the cyclorama curtain.

Equipment areas
Ancillary equipment areas will include separate areas for VTR (video tape recording) and telecine (transference of filmed material to video). The machine operator should be able to hear sound track and cues above the noise of other machines in the room which are usually enclosed in open fronted cubicles with heavily acoustically treated walls.

Master control room
Adjacent to these equipment areas will be the master control room, which is the last monitoring link in the video and audio chain before transmission. Here programme material, either recorded (VTR and telecine) or live from the studios, will be linked with continuity from the announcer's studio.

Dressing rooms
Artists' facilities adjoining the studio will include dressing rooms with associated wardrobe and laundry, rest and refreshment areas.

Rehearsal spaces
Separate rehearsal spaces are required as there is considerable pressure on studio floor time (much of which is used in setting and striking scenery, and setting up lighting and cameras for

productions). These need not be the full studio size as several sets will occupy the studio floor and scenes are rehearsed individually, often in remote assembly halls.

Service spaces
In addition to the areas detailed in **22.1**, space will be required for a sub-station, emergency generator and tape stores. The small commercial and education studios, which do not broadcast, will have simpler planning arrangement.

3.02 Control Suites

In recent designs control rooms do not overlook the studio they monitor, for the following reasons:

• The cyclorama track and studio scenery are likely to interfere with the producer's view. Production decisions should be made off a monitor screen.
• In colour TV the chroma of the glass in the observation windows must be adjusted to confirm colours reproduced by TV monitors. All current systems use an applied tinted finish and its life is limited.
• Windowless production suites do not need to be elevated, therefore production staff have direct access to the studio floor. A typical control suite layout of this type is shown in **22.2**.

Minimum clear height in the control room, including a false ceiling for services, should be 4 m.
Separate control areas for lighting, production and sound control are required with about 25 dB insulation between each. Visual contact between the areas is via observation windows (separate viewing area for production staff and visitors is desirable).
The disadvantages of such a layout is that the producer has no direct visual contact with his floor manager or performers.

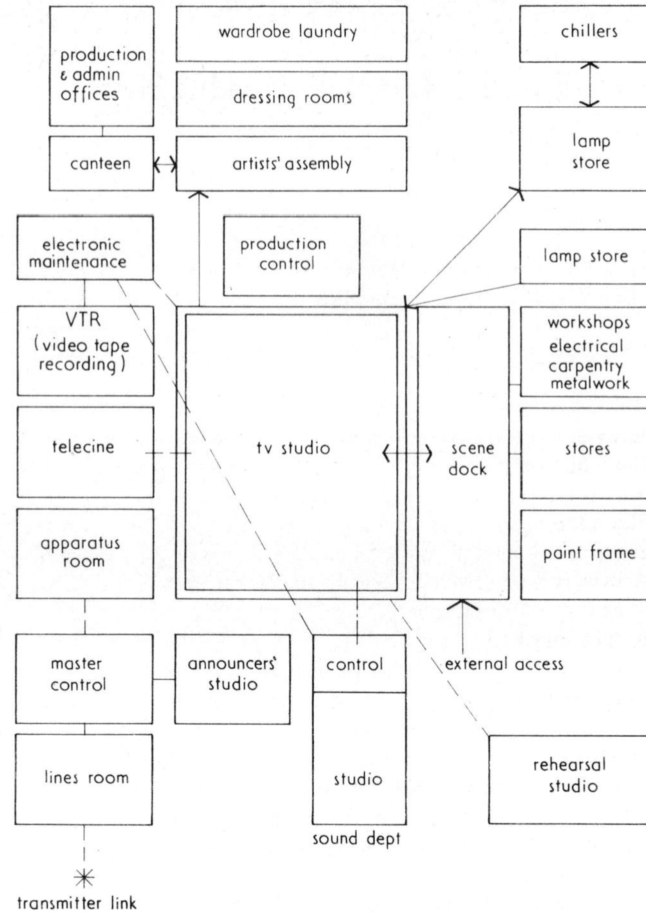

22.1 *TV studio complex: typical block planning diagram*

4 SERVICES

4.01 Lighting

Television studio lighting is highly specialised. The large production studio will have a remotely operated lighting grid, whereas the small studio will have a simple pre-set system.

Lamp support systems
There are three basic types of lamp support systems:

• The most elaborate is a grid of 'railway' tracks covering the whole studio. On these tracks run carriers which have a telescopic arm holding the lamp. The arm is motor driven (either electrically or hydraulically) and lowers the lamp to studio floor level for setting and adjustment. The lamp can be panned, tilted and dimmed remotely. An additional overhead rail at the perimeter of the grid will carry carts on to which the 'monopole telescopes' and lamps can be run off the grid to a lamp store. Where several studios exist this rail will interconnect them all via a central lamp store. The latest grids are equipped with an electronic memory to enable a whole production of lighting settings to be stored.
• A simpler form has lighting bars which can be raised or lowered electrically or manually fixed direct to the studio. The bars take several forms from the 'lazy scissors' principle to a simple bar on cables and pulleys.
• The third and simplest type is a fixed barrel grid. As in the second type no space is required above this grid for access as lamps are clamped direct to the bars and set from studio floor level.

Lighting to equipment and control areas needs to be carefully studied to avoid reflections and provide correct levels for viewing. Special fittings are often required.

4.02 Air conditioning

Air conditioning presents the designer with a number of unique problems: the large volume, high heat loads generated by lamps, low background noise levels and the need to provide comfortable conditions in parts of the studio obscured in all but one plane by scenery. Low air speeds have to be used to achieve the noise levels. The most successful system has been the 'dump' system where cooled air is fed from a large plenum chamber above grid level and returns via natural convection of heat from the lamps to a similar exhaust plenum at high level. Plant rooms, unless remote from the studios, require an iso-

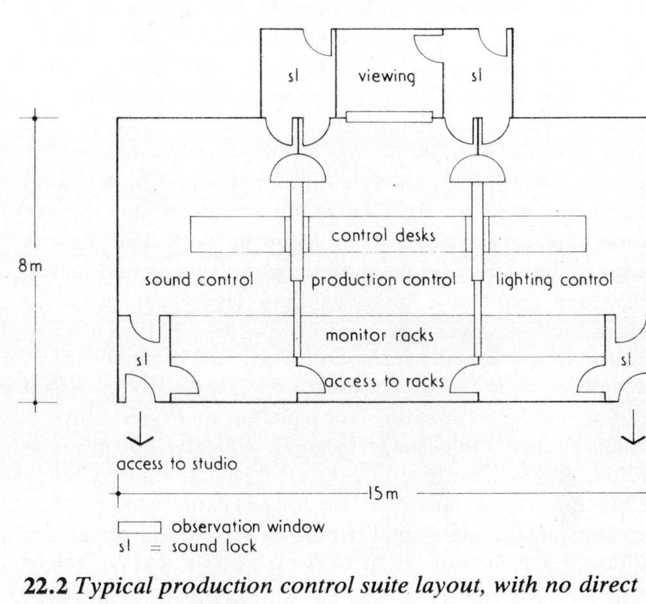

22.2 *Typical production control suite layout, with no direct visual access to studio*

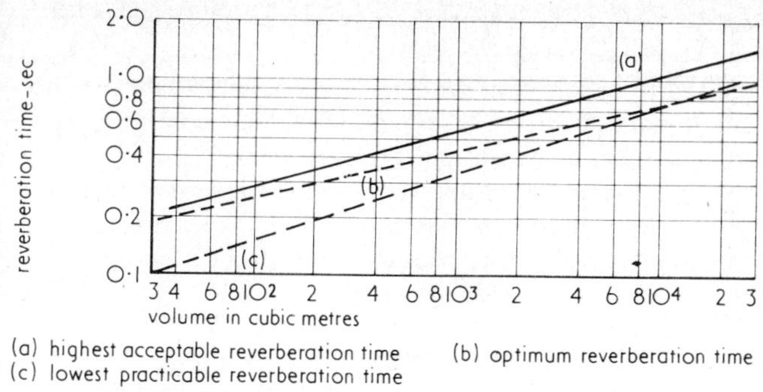

22.3 *Optimum reverberation times of television studios*

lated structure to control vibration transmission and adequate space must be allowed for attenuation. Mechanical engineers are familiar with duct-borne noise problems but do not normally investigate noise break-in through duct walls or the architectural acoustic problems. The architect should make certain that this forms part of the specialist consultant's brief.

4.03 Technical wiring

Extensive provision has to be made for power, audio and video wiring connecting the studio to control suites and equipment areas. Colour camera cables are approximately 50 mm in diameter and have a minimum bending radius of 0·5 m. Power wiring, which may include low voltage power, has to be run in separate trunking from audio wiring to avoid interference. Trunking is often concealed within the acoustic finishes and all perforations of the studio enclosure have to be sealed airtight to avoid sound transmission.

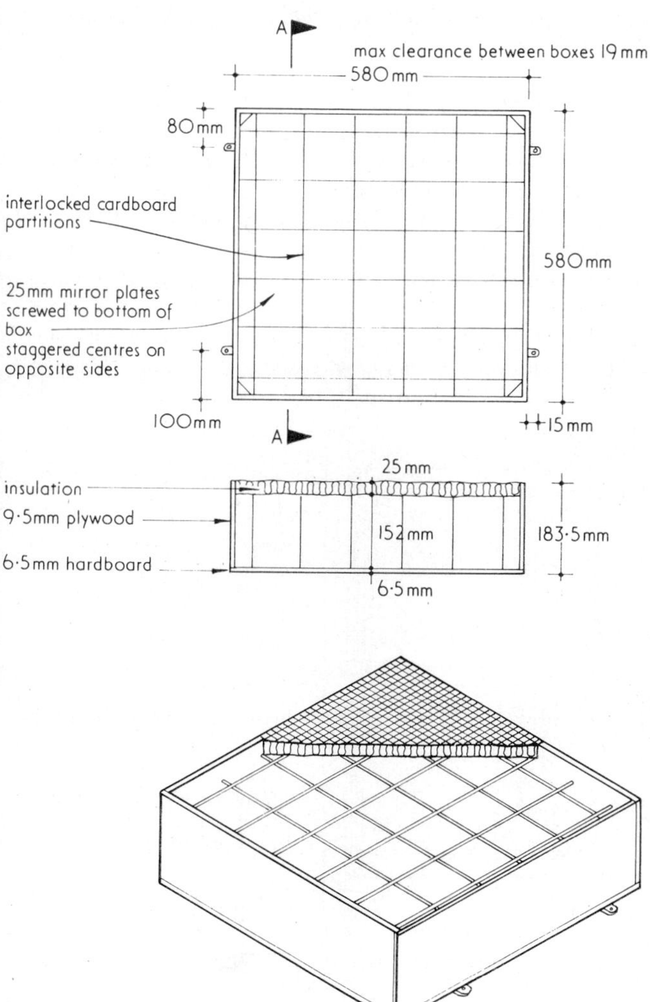

22.4 *Typical modular absorber*

4.04 Other services

Large production studios will require compressed air, gas, water (including drainage) and a smoke detection system, in addition to electrical services.

5 ACOUSTICS

5.01 Identify standards

The standards to be achieved should be identified by the specialist consultant and agreed with the client at the outset. The two main sources are the BBC and the ISO (International Standards Organisation).

5.02 Airborne sound insulation

For every location a full one-third octave band, site noise level survey must be carried out to determine the design of the enclosing structures.

Additionally, all internal transmission loss defined by frequency should be established and can be extended to provide the mechanical engineer with the requirements for in-duct cross-talk attenuation. For this it will be necessary to establish the maximum permissible noise levels from all sources in each room.

5.03 Vibration isolation

The noise and vibration levels of all mechanical plant should be studied and the architect must identify who should be responsible for defining maximum permissible levels and designing to achieve them. Structure-borne sound transmission, particularly on the upper levels of framed buildings, may necessitate the 'floating' of plant rooms and noise protected areas.

This involves isolating the walls, floor and roof from the surrounding structure. The walls are built off a secondary floor bearing on steel spring or rubber carpet mountings designed to a maximum natural frequency not exceeding 7 to 10 Hz. Footfall impact noise often requires floors to be carpeted with heavy underfelt or in extreme cases, the floating of studios.

5.04 Reverberation time

Figure **22.3** relates reverberation time to volume for television studios. Calculation will indicate the amount and type of absorption required. Details of a typical wide band modular

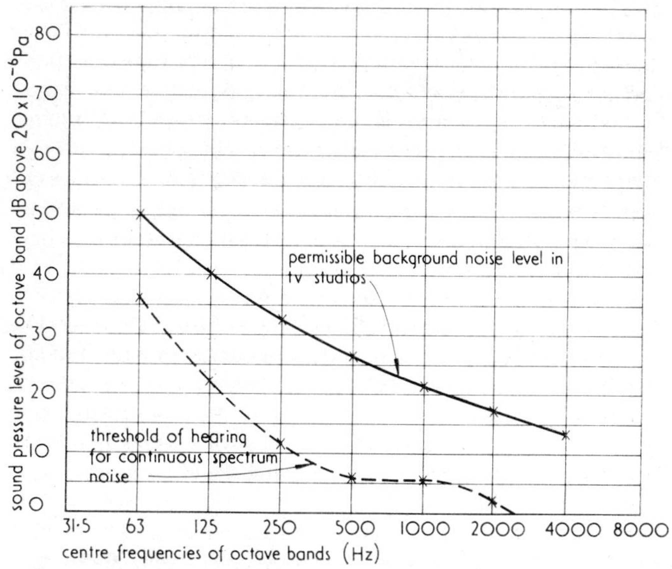

22.5 *Permissible background noise levels*

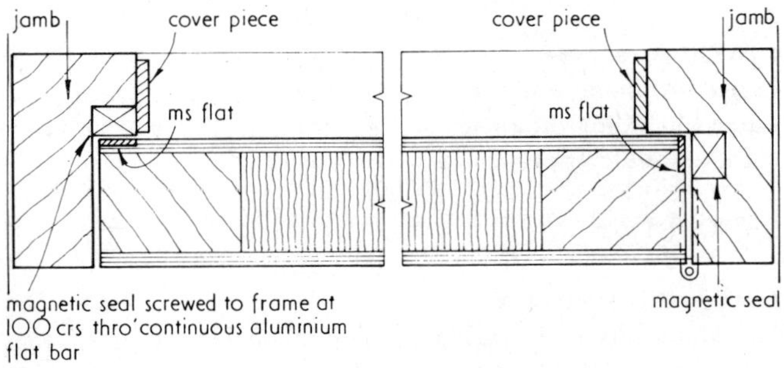

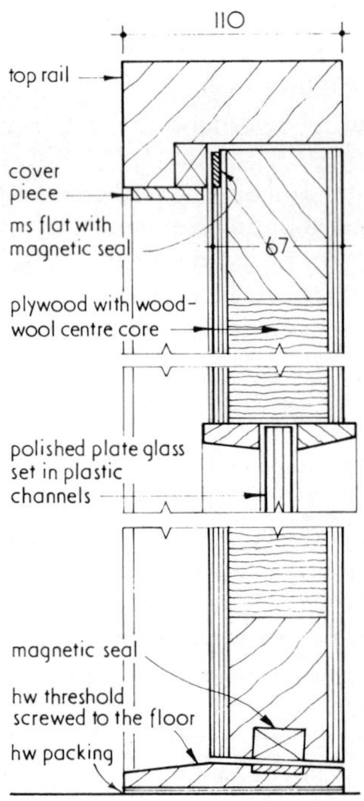

22.6 *Typical acoustic door*

absorber are shown in **22.4**. Approximately 200 mm should be added to the clear studio height and to each wall thickness to accommodate the acoustic treatment. Sound control rooms need to be similarly treated, with the other production control rooms and technical areas made as dead as possible.

5.05 Background noise levels

Maximum permissible background noise levels are shown in **22.5**. These should be related to the external ambient levels and to noise from air conditioning plant. In certain situations where plant rooms are adjacent to noise sensitive areas, maximum permissible noise levels at intake and extract louvres should be specified, to limit this noise breaking back in through the external skin, particularly at windows.

5.06 Special details

Acoustic doors and sound lock

Typical details for an acoustic door and an observation window are shown in **22.6** and **22.7**. All noise sensitive areas should be approached via a sound lock lobby consisting of acoustic doors, with either end of the lobby treated to be acoustically dead. The mean sound transmission loss of each door is 33 dB and sealing is affected by means of magnetic seals.

Scenery doors

The transfer of scenery into the studio requires an opening in the region of 5 m high by 5 m wide with a sound reduction index between 50–60 dB. This door will almost certainly be of steel construction. Hinged doors have been used but the forces required to ensure that the edge seals close airtight produce operational difficulties. A 'lift and slide' door is more satisfactory. An electric or hydraulic drive opens and closes the door while radius arms lower it inwards and downward to compress the edge seals all round. This type of door does not require an upstanding threshold as does the hinged door, and this is a considerable operational advantage.

6 STATUTORY REQUIREMENTS

Careful examination should be given at the planning stage to means of escape and fire resistance. Statutory requirements vary considerably in all parts of the world, but the most stringent are those operated in the UK, particularly the London area, where Class O flame spread may be required for all finishes and up to a four hour fire separation for the studio walls. This necessitates double steel roller shutters on all perforations through walls. Smoke vents are sometimes required and these must be designed to match the sound insulation of the roof.

7 BIBLIOGRAPHY

Data for the acoustic design of studios, BBC monograph no 64, Gilford, Burd and Spring

Christoper Gilford, *Acoustics for radio and television studios,* Stevenage, Peter Peregrinus Ltd, 1972

London Weekend television centre, buildings illustrated, *The Architects' Journal*, February 6, 1974

F. C. McLean, *The BBC television centre and its technical facilities*, Proceedings of the Institute of Electrical Engineers, vol 109, part 13, no 45, May 1962

V. S. Mankovsky, *Acoustics of studios and auditoria*, London, Focal Press, 1971

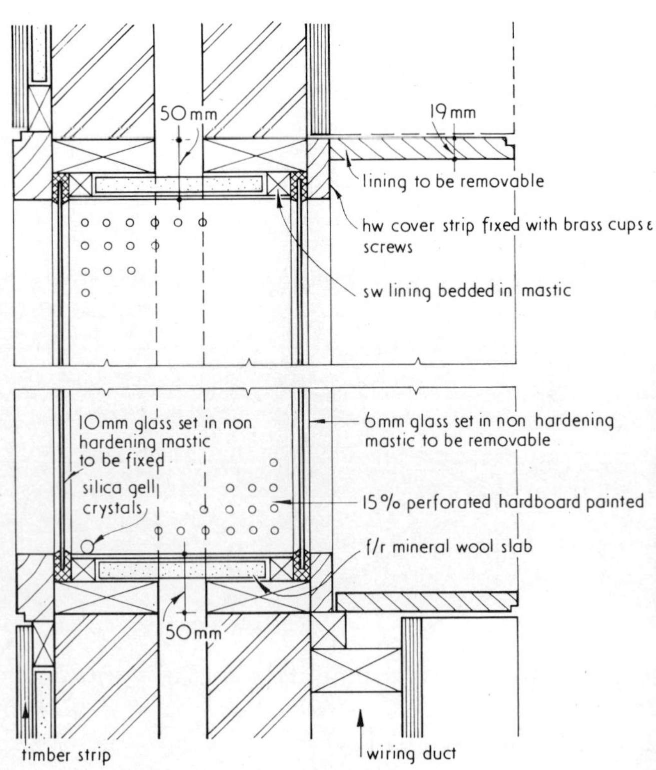

22.7 *Typical observation window*

23 Theatres

Peter Forbes

CI/SfB 524
CI/SfB (1976 revised) 524

Peter Forbes is an architect/planner who worked on the Haymarket Theatre, Leicester. He is an amateur actor.

Contents

1 RELATIONSHIP BETWEEN FUNCTIONS

There are basic procedures and inter-relations of functions common to all forms of theatre. Apart from the performance itself, the main activities in a theatre are:

- instigation and development of a production, using the spaces shown in **23.1**, including the preparation of scenery and costumes, **23.2**, and rehearsal
- organising performers, scenery and lighting and sound equipment before and during a performance, **23.3**
- performers' preparation for a performance, **23.3**
- access and egress by the public, and their welfare, **23.4**
- managerial function—policy making and administration, **23.5**.

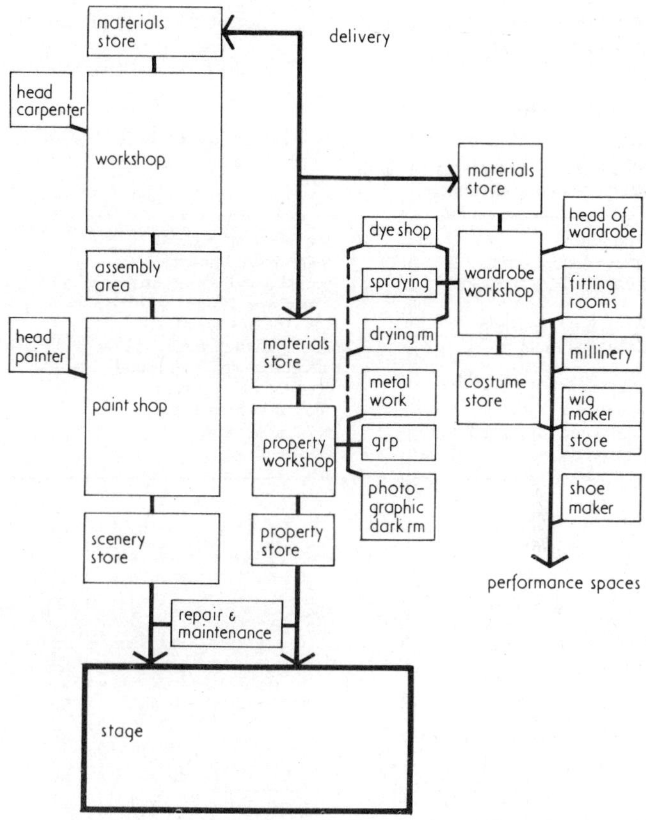

23.2 *Organisation of production spaces*

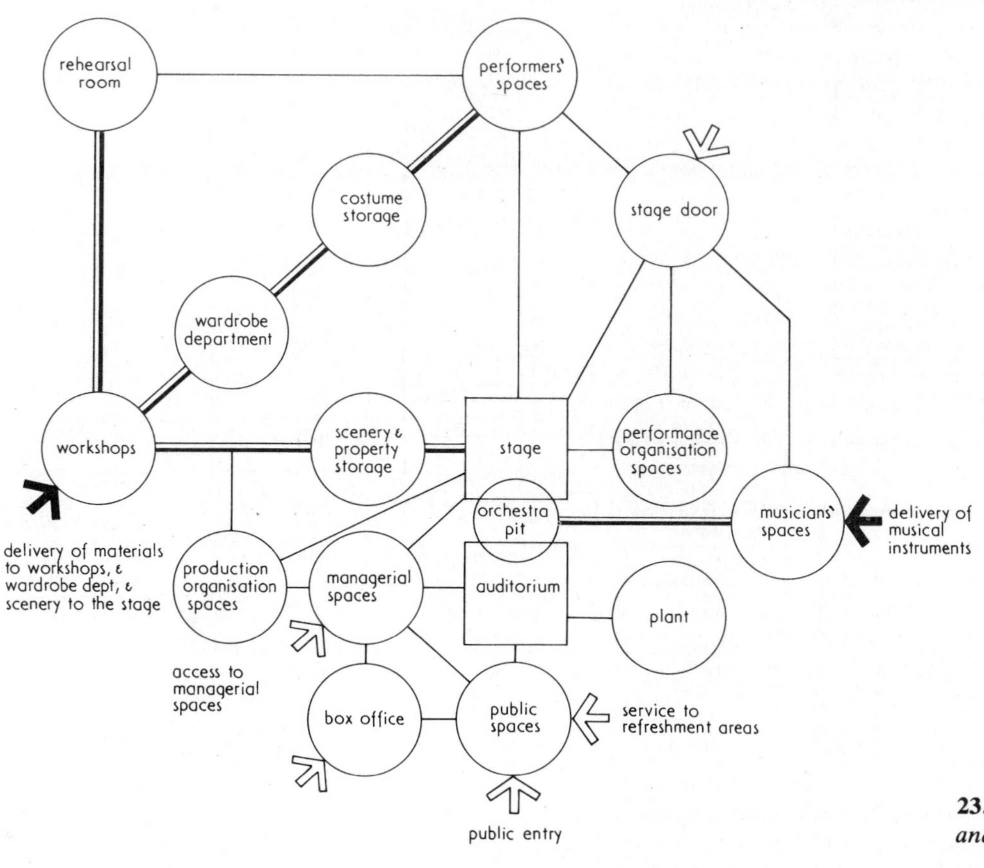

23.1 *Inter-relation of theatre and production facilities*

The extent of facilities required for each of these functions depends on the scale and category of the theatre.

2 LAYOUT OF THEATRE

The relationships between the spaces in a theatre with production facilities on site are shown in **23.1**. Where productions do not originate on site, but are brought in by touring companies, the workshops, wardrobe department and rehearsal room may be omitted. In this case, it is important that adequate access to the stage for delivery and removal of scenery is provided. An alternative means of access is necessary for when the stage is in use. This should be associated with a scene dock which can be shut off from the stage.

3 PLANNING

Some typical theatre forms are shown in **23.6–9**. Planning allowances are given in table I.

Table I Planning allowances

Auditorium	0·65–0·84 m² per seat
Stage: drama, variety, musical, opera and ballet	0·46–0·65 m² per seat reducing to 0·19 m² per seat with capacity + 1000
Workshop/paintshop	50–70% of stage area
Rehearsal room	Acting area of stage + circulation space all round: 35–50% of stage area
Foyers	0·65–0·84 m² per seat
Production rooms and offices	0·74 m² per seat
Control room	25–30 m²

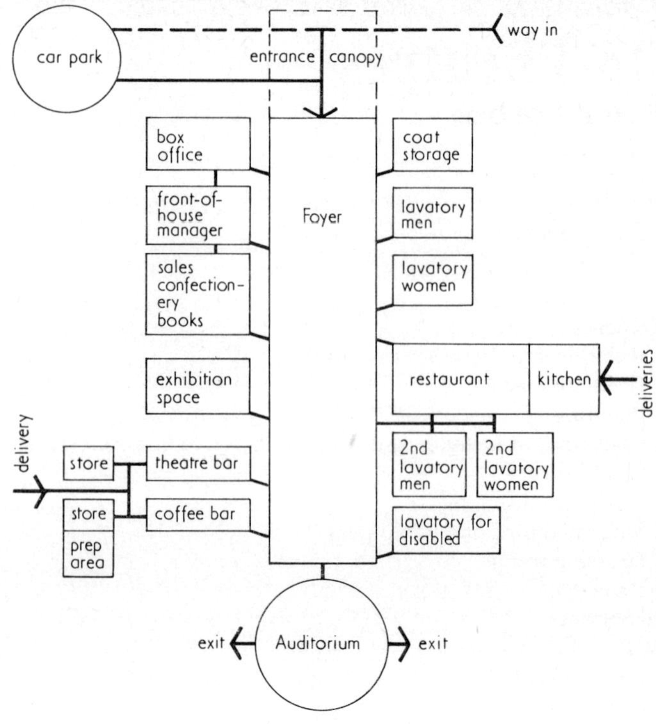

23.4 *Organisation of public spaces*

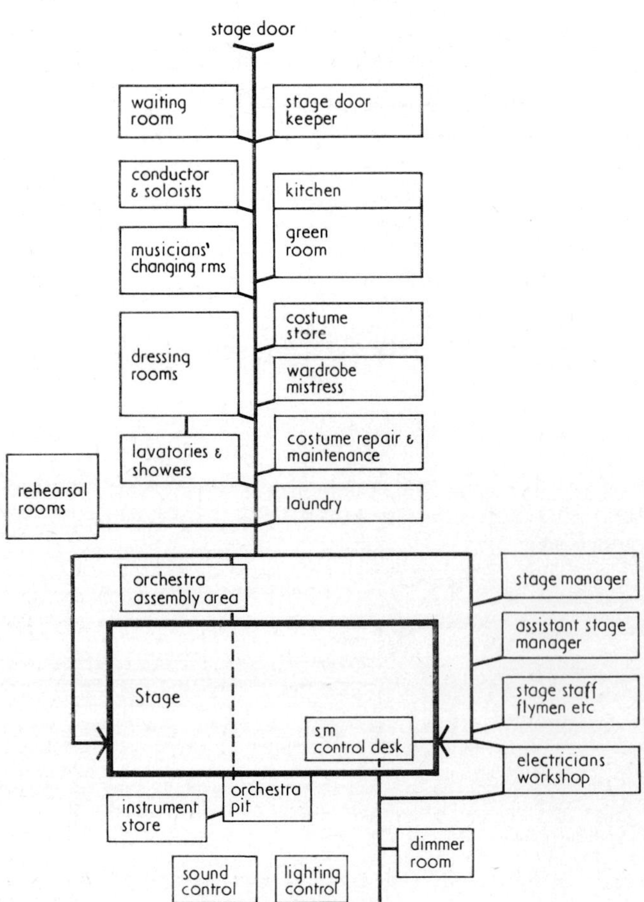

23.3 *Organisation of performance spaces*

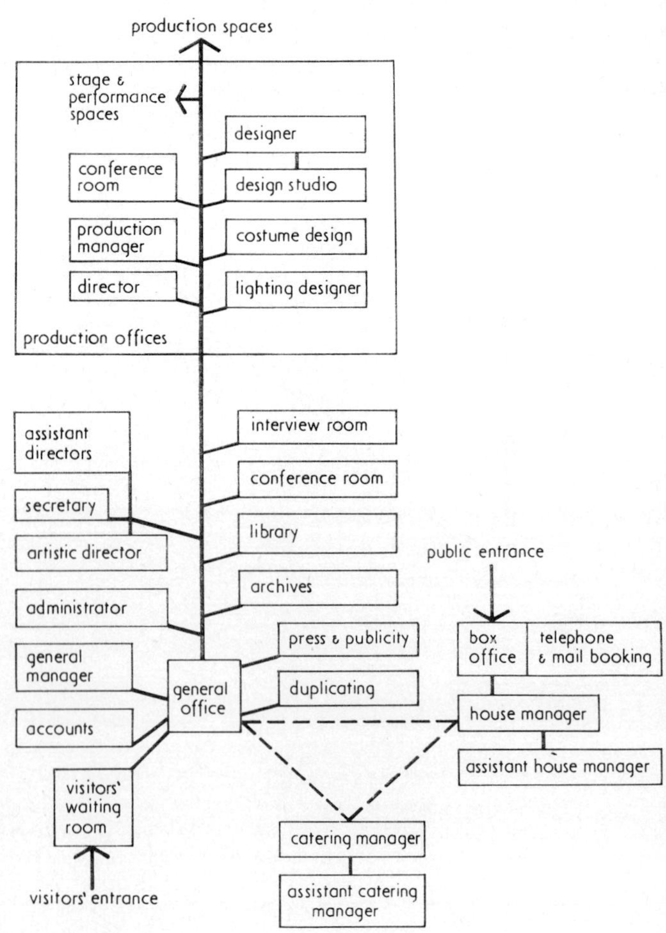

23.5 *Organisation of administrative spaces*

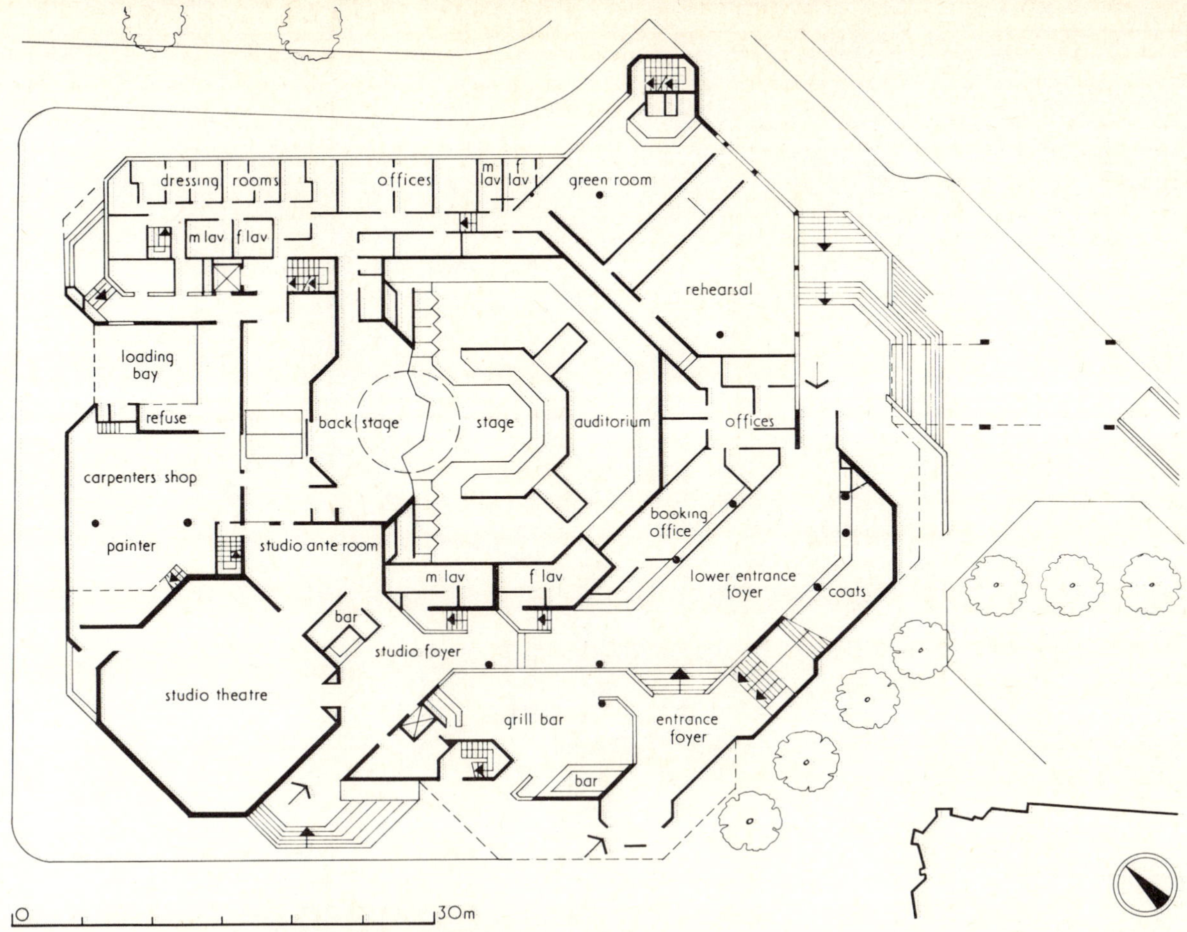

23.6 *Crucible Theatre, Sheffield. 1000 seats, open thrust stage*

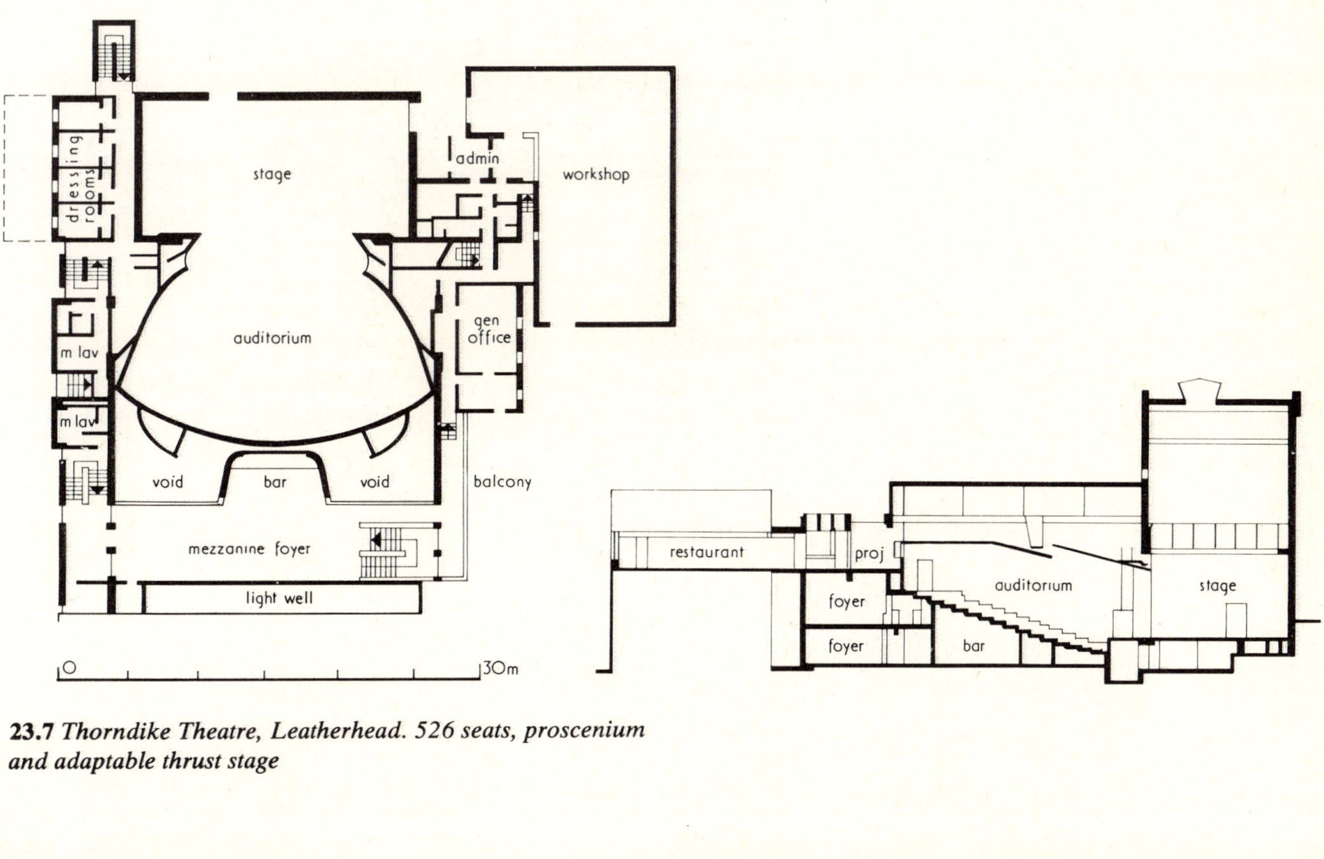

23.7 *Thorndike Theatre, Leatherhead. 526 seats, proscenium and adaptable thrust stage*

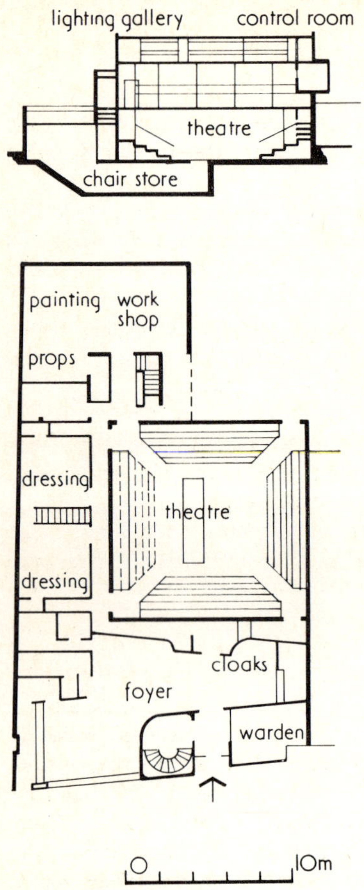

23.8 *Cockpit Theatre, Marylebone. 100 to 177 seats, adaptable from endistage to thrust stage or theatre in the round*

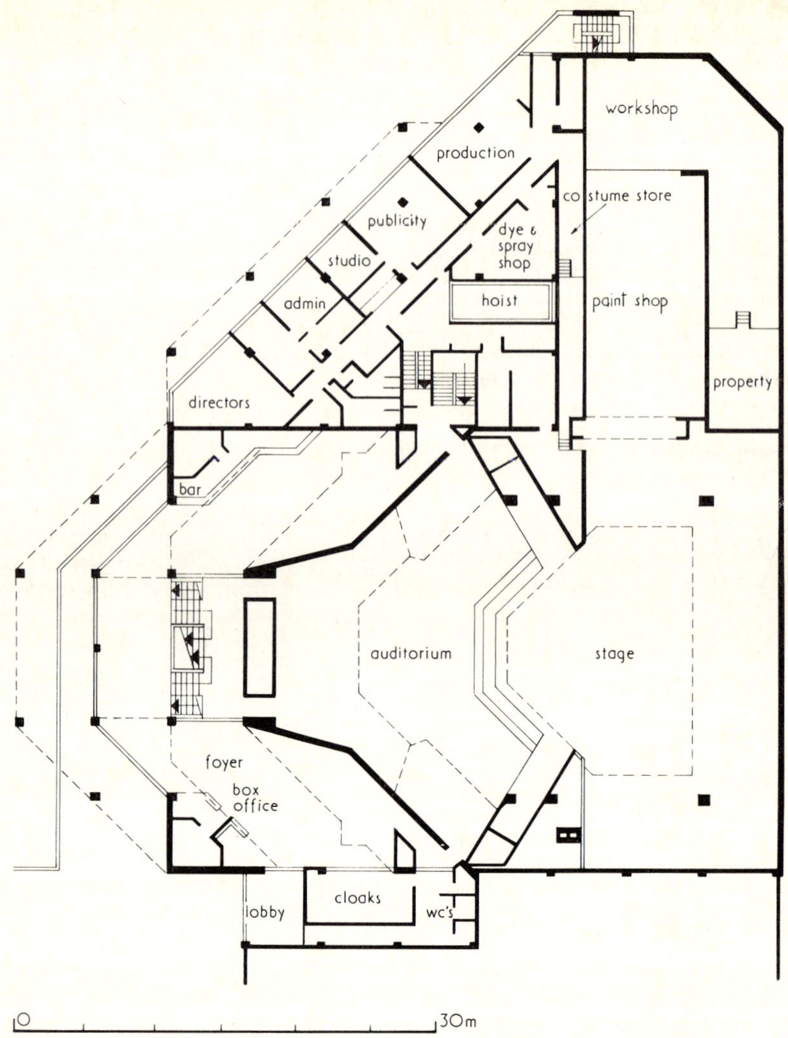

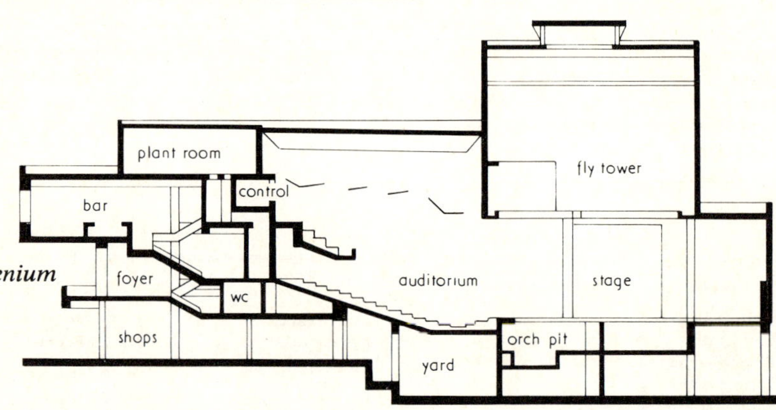

23.9 *Haymarket Theatre, Leicester. 705 seats, proscenium and fixed thrust stage*

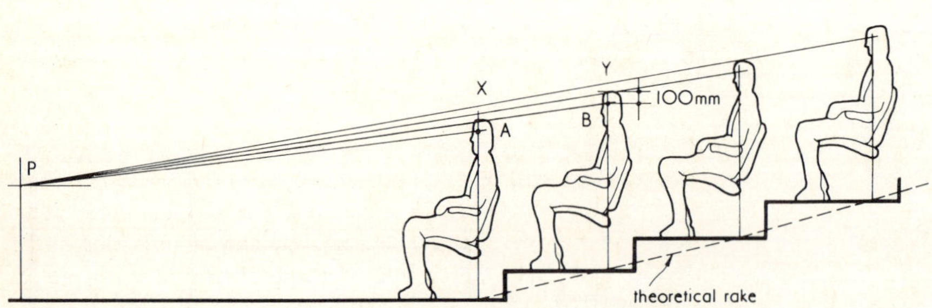

23.10 *Method of calculating vertical sightlines with low P. Maximum stepped rake 35°*

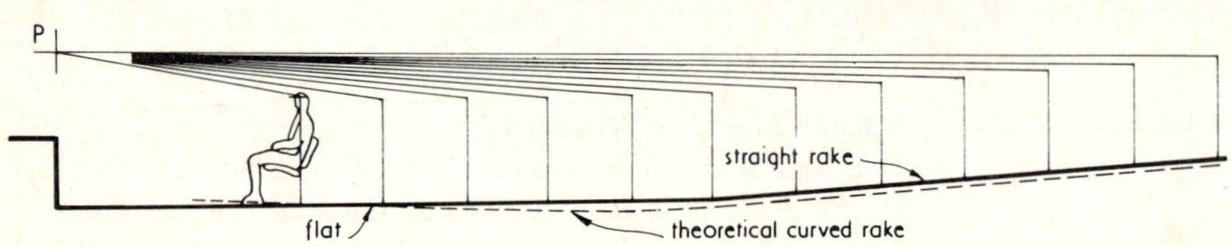

23.11 *Sightlines with high P: theoretical curved rake. Maximum gradient of aisles without steps 10%*

4 CHECKLISTS OF FACILITIES AND EQUIPMENT

Room areas given in table II are for a 500 to 700 seat theatre

Table II Facilities and equipment checklist

Foyers	
Box office	Exhibition area
Bookshop	Publicity display
Licensed bar(s)	Toilets
Coffee bar/refreshments	Cloakroom
Seating/tables	
Production wardrobe	**area required**
Cutting tables	12–14 m² space each
Sewing machines	1100 × 1100 mm space each
Dress racks	1800 × 600 mm each
Ironing boards	5–6 m² space each
Fitting room(s)	10 m² each
Supervisor's room	15 m²
Materials storage	1800 × 1500 mm
Costume storage	5–7 costumes per 1 m²
Sink and draining board	
Wig bench	750 × 1000 mm per person
Maintenance wardrobe	
Sewing table	5–6 m² space
Dress rack	1800 × 600 mm
Ironing board	5–6 m² space
Rack for shoes, hats etc	1800 × 600 mm
Laundry	
Washing machines	1 m² each
Tumbler driers	1 m² each
Dry cleaner room	7 m²
Sink and worktop	2400 × 600 mm
Drying cabinet	1 m²
Dye shop	
Dye vats	1300 × 750 mm each
Sinks	
Marble worktop	1800 × 750 mm
Drying cabinet	11 m²
Spray booth	10 m²
Workshop	
Circular saw	9000 × 15 000 mm space
Band saw	5000 × 5000 mm space
Lathe	2000 × 1500 mm space
Planer	2500 × 15 000 mm space
Workbenches	5400 × 1050 mm each
Welding area	10 m²
Timber racks	7000 × 900 mm each
Materials storage	6 m²
Storage for flats	0·5 m² per flat
Space for set assembly	15–20% of acting area
Paintshop	
Paint frame	15–20 m long
Paint storage	4 m²
Sink and worktop	
Paint mixing bench	3600 × 750 mm
Property shop	
Workbenches	1400 × 800 mm per person
Plaster sink	1200 × 900 mm
Drying cabinet	3 m²
Materials storage	4 m²
Control room	
Sound control equipment	
Lighting control equipment	
Stage manager's desk	

5 SIGHTLINES

5.01 Vertical

The following method can be used to set out the slope of the floor that will give satisfactory vertical sightlines for the main bank of seating, **23.10**, **23.11**.

- The lowest and nearest point the whole audience should be able to see clearly must be decided (P). If this is the edge of a horizontal stage, the nearest eye level should be above this horizontal plane.
- Verticals are then drawn through the eye positions for each row of seats. The point X, 100 mm above A on the first vertical, represents the top of the head of the person in the front row. When PX is extended beyond X, the point B at which it cuts the second vertical gives the eye position for the second row. 100 mm above B lies the point Y through which the next sightline is produced to cut the next vertical, and so on.
- When the eye positions for every row of seats have been determined, the floor line can be found by measuring 1120 mm below each of the verticals: this theoretical line is a shallow curve.

Reducing a steep rake

When this method is strictly applied the rake will usually be very steep, and where horizontal sightlines limit the width of the audience, it may be necessary to introduce another seating tier. The rake of the main bank of seating must then be reduced to make way for another level, and in this case the point P may be taken 600 or 900 mm above the edge of the stage on the assumption that the actual edge may be seen between heads only and an unrestricted view of the performers from the knees up is all that is necessary.

The situation is improved by staggering seats. The arbitrary choice of a position for point P and the variations in the dimensions of the individual members of an audience do not justify the refinement of a subtly raked curve. A varying rake may also introduce the complication of different heights of risers which are uncomfortable and sometimes dangerous for the public. The maximum gradient for an aisle without steps is 1:10, and the maximum slope for stepped seating is 35° (GLC), **23.12**.

Approximate method of determining the slope of a tier

This does not replace the more meticulous method described above which can be applied to tiers as well as main banks of seating.

- First fix the eye position for the front row of seats (A) and the depth (L) to the eye position of the back row, **23.13**.
- Vertically above A find point X so that AX = l/10.
- Draw a line from P (on stage) through X to cut the vertical through the eye position of the back row at O.
- The rake of the tier will then be parallel to AO, but 1120 mm below.

5.02 Horizontal sightlines

Horizontal sightlines are critical only in theatres with a proscenium stage **23.14, 23.15**. Given a desired acting area, sightlines will limit the width of seating that can be provided in the auditorium. Sight lines from side seats restrict the amount of the stage that can be used as acting area. The addition of a false proscenium will limit the acting area still further.)

6 AUDITORIUM SEATING

The minimum dimensions and gangway requirements which follow are based on those of the Greater London Council. Other licensing authorities have their own regulations which

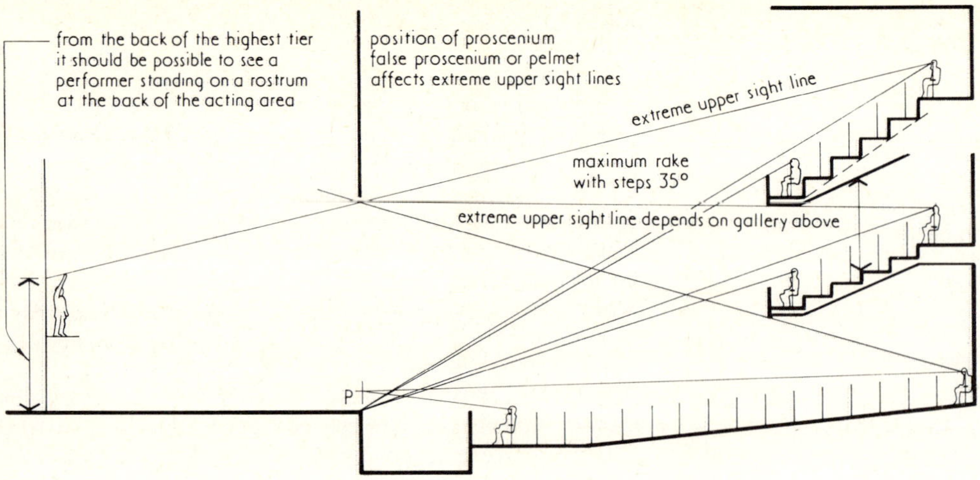

23.12 *Vertical sightlines with galleries*

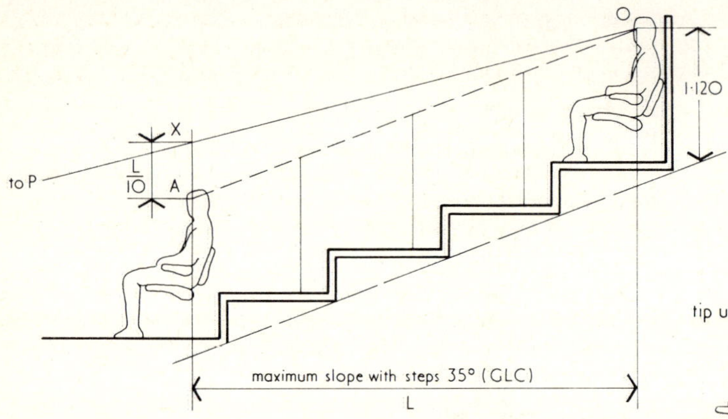

23.13 *Graphical method for finding balcony rake*

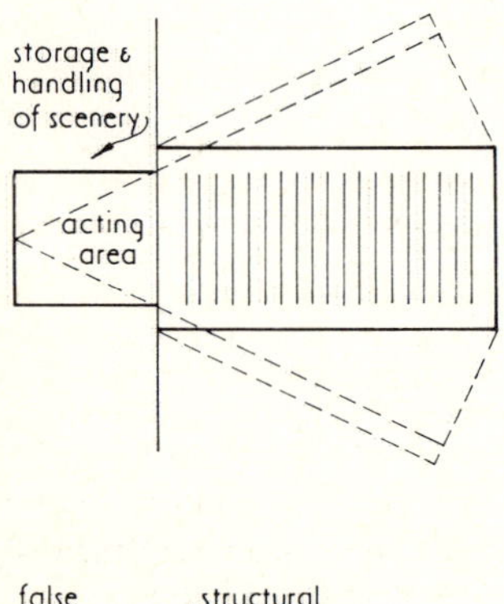

23.14 *Proscenium stage: horizontal sightlines limiting the seating area*

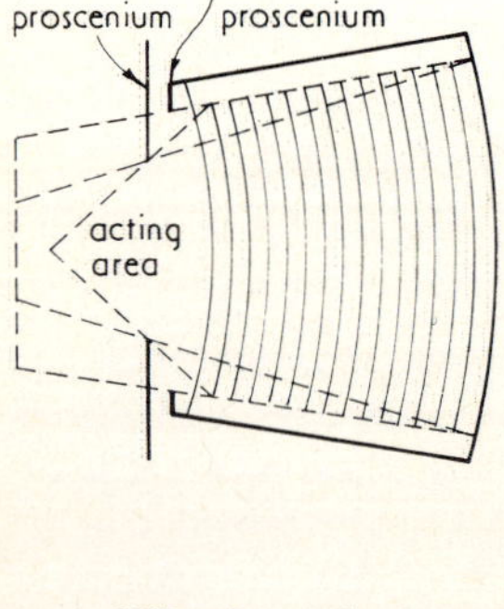

23.15 *False proscenium further limits the acting area*

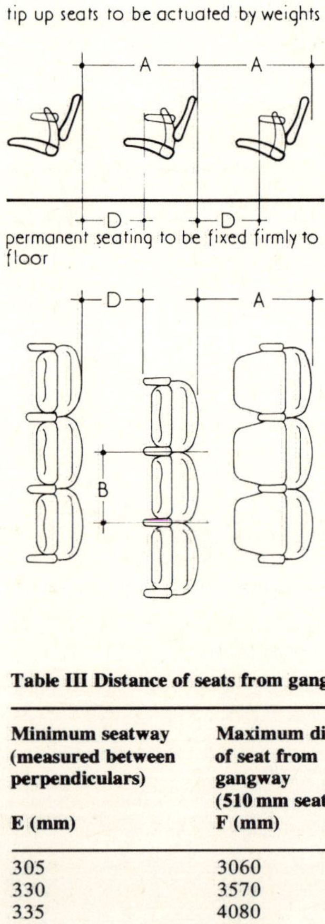

23.16 *Auditorium seating with arms*

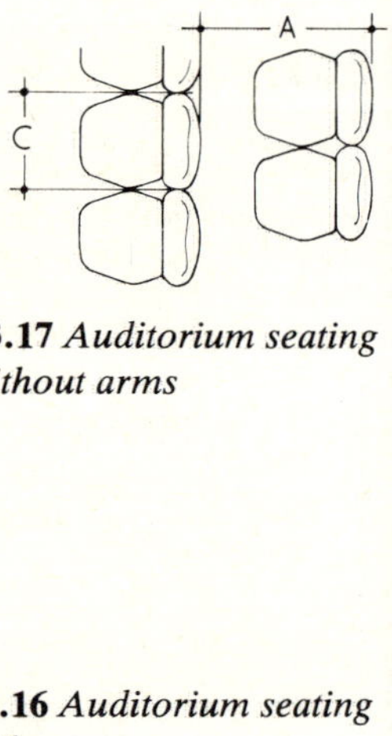

23.17 *Auditorium seating without arms*

may differ from these considerably. Minimum dimensions are given in table III and **23.16**, **23.17** and **23.18**.

Minimum dimensions

A Back-to-back distance between rows of seats: 760 mm (min)
B Width of seats with arms: 510 mm (min)
C Width of seat without arms: 460 mm (min)
D Unobstructed vertical space between rows (seatway): 305 mm (see table III)

Table III Distance of seats from gangways

Minimum seatway (measured between perpendiculars) E (mm)	Maximum distance of seat from gangway (510 mm seats*) F (mm)	Maximum number of 510 mm wide seats per row* Gangway both sides	Gangway one side
305	3060	14	7
330	3570	16	8
335	4080	18	9
380	4590	20	10
405	5100	22	11

* The standard dimension for seat is 508 mm (1 ft 8 in). Exact sizes must be checked with manufacturers

E For normal maximum distance of seat from gangway see table III, but rows with more than 22 seats could be possible, provided the audience was not imperilled
F Minimum width of gangway: 1070 mm.

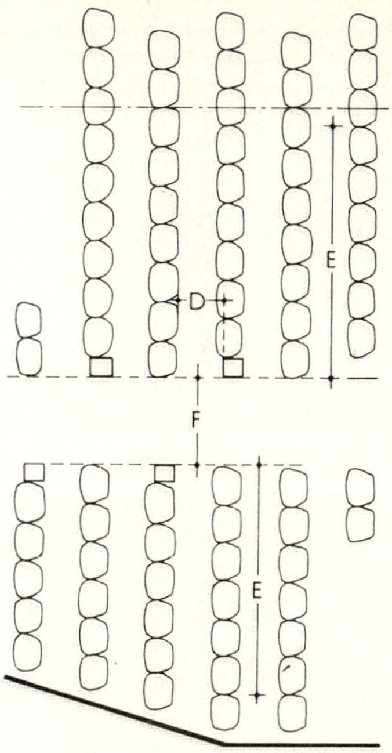

23.18 *Part plan of auditorium*

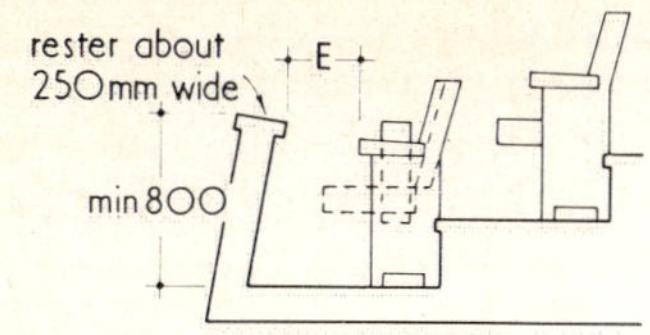

23.19 *Section through balcony front*

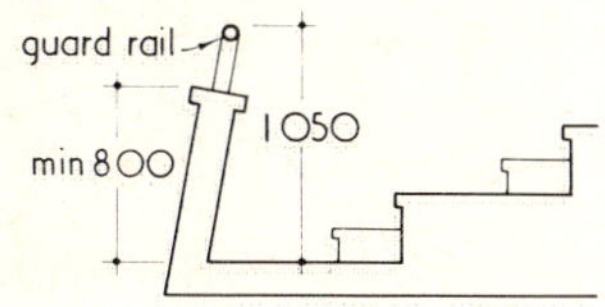

23.20 *Section through balcony front opposite gangway*

Balcony fronts

The top of the rester, **23.19**, should be designed to discourage it from being used as a resting place for small articles such as binoculars and handbags and to prevent them from falling on people below. If it is too narrow some patrons may feel giddy: about 250 mm is an acceptable width. A guard rail must be provided opposite the full width of the end of each gangway, **23.20**.

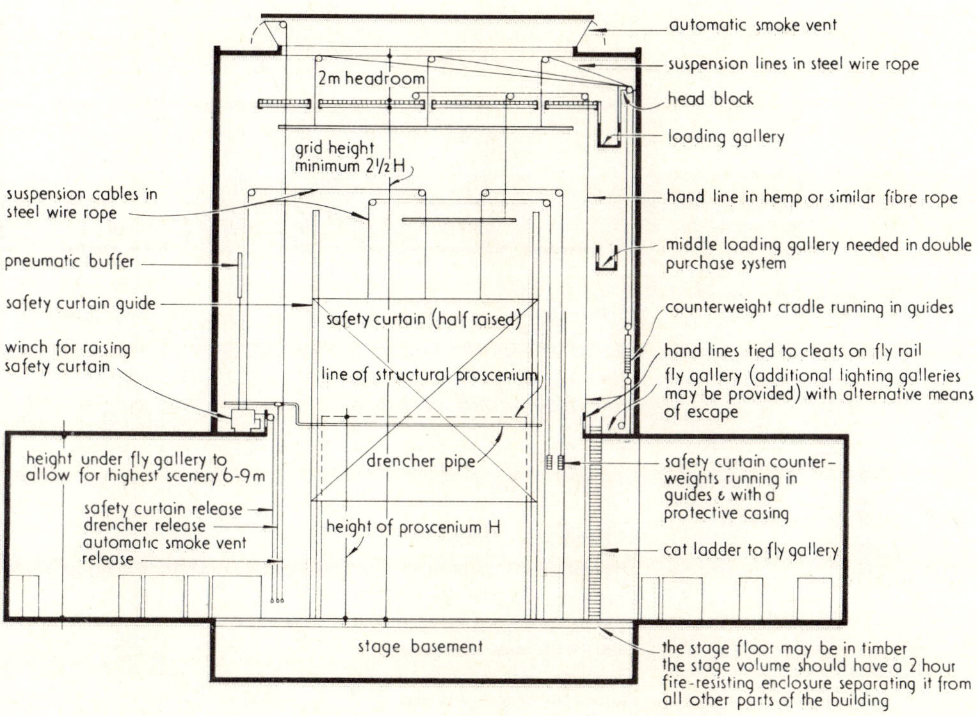

23.21 *Longitudinal section through stage and flytower equipped with double purchase flying system*

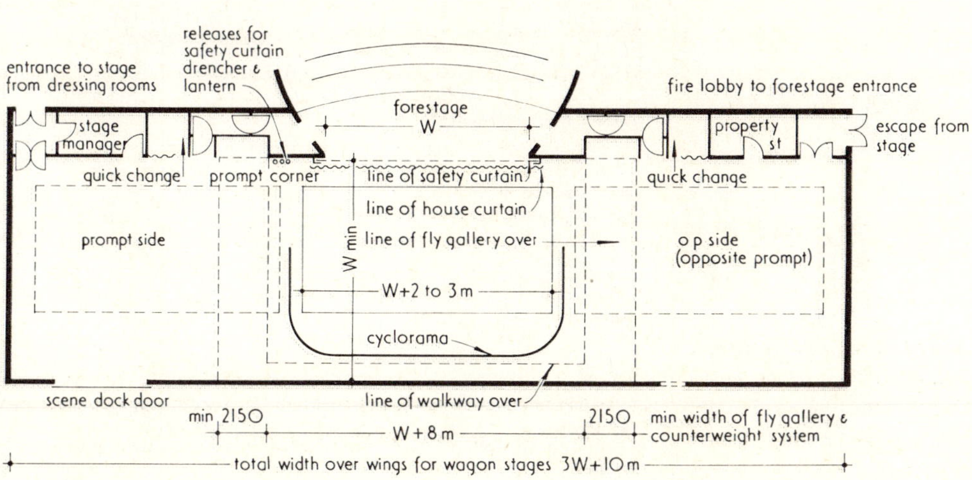

23.22 *Plan of stage showing wagon stages*

7 STAGE PLANNING AND EQUIPMENT

7.01 Stage plans

Some typical stage layouts are shown in **23.21**, **23.22** and **23.23**.

7.02 Orchestra pit

Level of orchestra pit should be adjustable between 1800 and 3000 mm below stage. Allow 1 m² per player, 5 m² for piano and 5 to 6 m² for tympani.

7.03 Stage basement

Maximum headroom 2500 mm.

7.04 Stage lifts

Lifts, operated by ropes and winches, screw jacks or hydraulic rams, can be used to vary the levels of all or part of the stage. The diagrams show a lift system used to provide, **23.24**, an orchestral platform, and, **23.25**, a multi-level stage and orchestra pit.

7.05 Stage revolves

Revolving platforms, set in the stage permanently, or fitted temporarily on the stage floor, permit rapid scene changes by having several sets built on them. They can either be single, **23.26**, or double, with one half of a set on each revolve, **23.27**. Their size will depend on the size and number of sets they are required to accommodate.

7.06 Stage wagons

Sets can be built on wheeled wagons which are rolled onto the stage from one or both side wings, or from the rear. Wagons must be wider than the proscenium opening and deeper than the acting area to allow for the propping of the set. Wagons can be used in conjunction with stage lifts, which can be adjusted to bring the wagons level with the rest of the stage floor **23.28**.

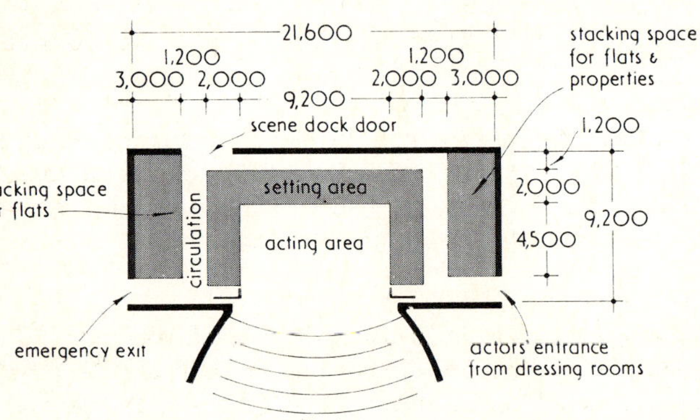

23.23 *Space requirements for a stage where sets are assembled and dismantled for each change. More width and more depth would be desirable*

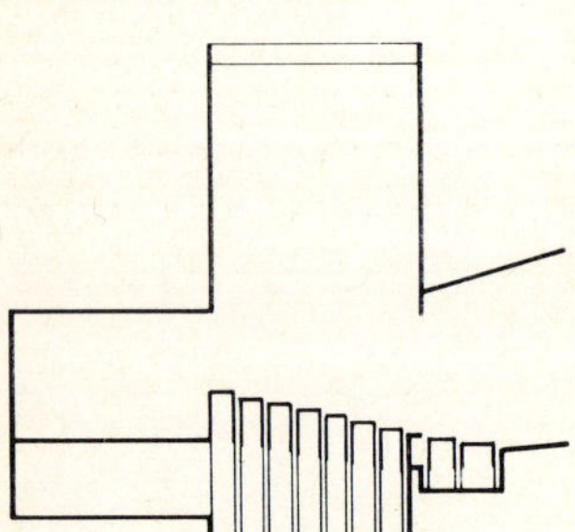

23.24 *Stage lift system arranged to give a concert orchestra platform*

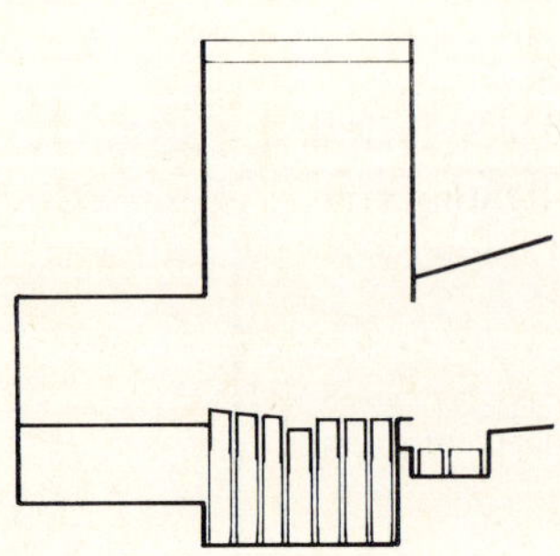

23.25 *The same system arranged for a multi-level stage, including a rake, plus an orchestra pit in lieu of forestage*

23.26 *Single revolve*

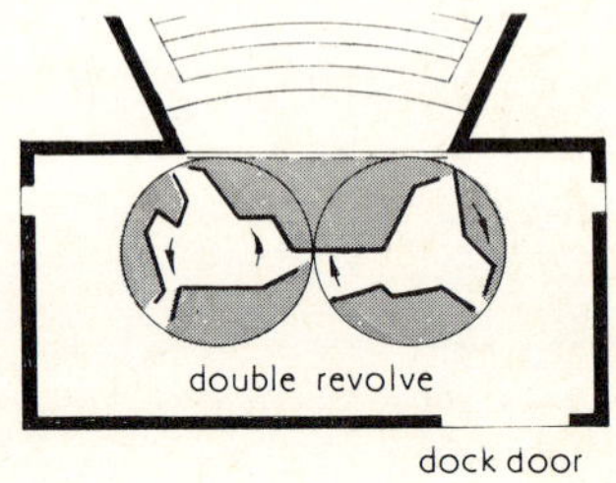

23.27 *Double revolve*

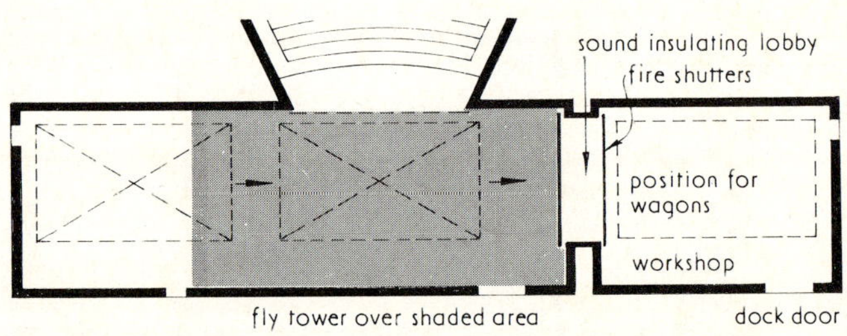

23.28 *Where one of the wagons is housed in the workshop, the change is slowed up by the opening and closing of the fire-resisting, soundproof and noiseless shutter necessary between stage and workshop*

8 DRESSING ROOMS

Dressing room arrangements are illustrated in **23.29** and **23.30**. Planning allowances are given in table IV.

Table IV Planning allowances

Clothes on hangers or hooks	600 mm centres
Length of hanging rail	750 mm (normal) 1250 mm for multi-change
Dressing room door width	850–900 mm
Wall space for washbasin	900 mm
Long mirror	1200 × 600 mm minimum height 750 mm above floor

Table V Exit requirements

Number of people accommodated on each tier or floor	Minimum number of exits*	Minimum width (mm)
200	2	1050†
300	2	1200
400	2	1350
500	2	1500
750	3	1500
1000	4	1500

* Plus one additional exit of not less than 1500 mm for each extra 250 persons or part thereof
† Would not normally apply to exit corridors or staircases serving auditorium of a theatre

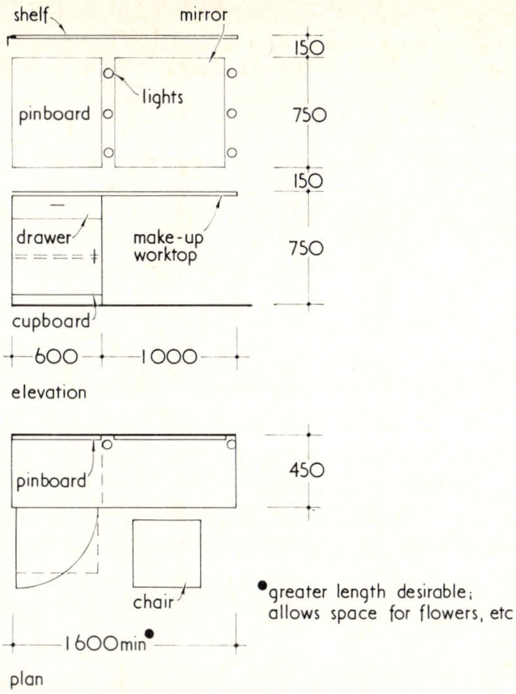

23.29 *Single dressing room*

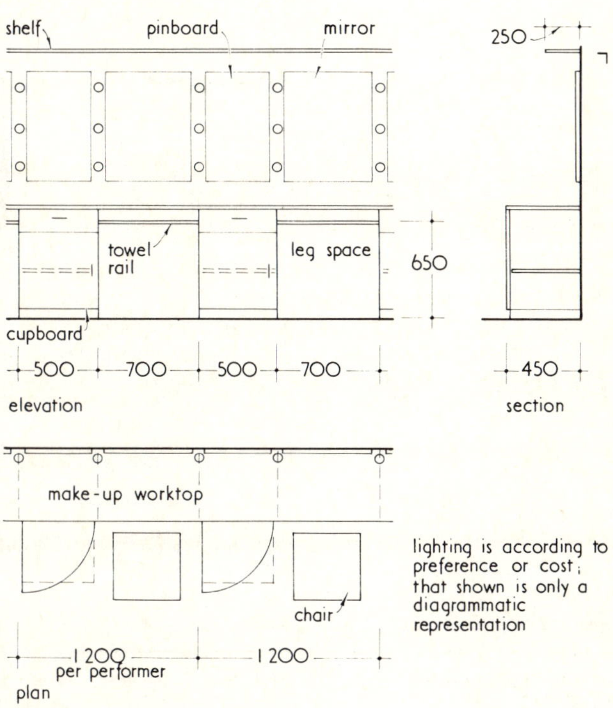

23.30 *Shared dressing room*

9 EXITS

Minimum requirements as to number and width of individual exits are given in table V, but more than the minimum number may be required, depending on such factors as maximum travel distances to an exit and layout of seating and gangways. Principal regulations applying to design of exit corridors and staircases are illustrated in **23.31** to **23.33**.

10 SERVICES

10.01 Heating systems

Auditorium

The most common air distribution system is one in which the air is delivered at high level and extracted below the seats. GLC regulations call for a supply of not less than 28 m^3 per person per hour, and 75 per cent of this must be extracted mechanically. With this system, heat from stage lanterns in the roof may be carried down to the audience. Additional extract from high level may therefore be needed to carry away hot air from round the lights, **23.34**.

Stage

Heating is more important than cooling, particularly to provide comfortable conditions for rehearsals. Warm air systems can cause movement in scenery and curtains. Recessed radiators are probably the most acceptable solution. During performances a certain amount of air from the auditorium will in any case flow onto the stage.

Backstage

Simple radiator systems are usually adequate, except where internal rooms require mechanical ventilation. Workshops and paintshops need space heaters or radiant panels.

Foyers

Mechanical ventilation is normally required, and air distribution may be by ceiling or wall grilles and/or convector units.

10.02 Lighting positions

Light sources directed at the actor should strike the actor's face at about 45° above horizontal. Since these sources are usually crossed they can be set at about 55° above horizontal to produce an angle of about 45° after crossing.

Method of locating theoretical positions

Lighting bridges above the auditorium are the best method of providing light from the direction of the audience **23.35**. Lanterns are hung from bars in front of a catwalk, and generally point through holes or slots in the ceiling. These must be large enough to permit the light beam to be adjusted as fully as is required in the vertical plane without causing shadows. Preferred sizes for lighting bridges are shown in **23.36** and **23.37**.

Wall slots

Wall slots in the auditorium are often used in conjunction with lighting bridges to provide the necessary coverage of the stage.

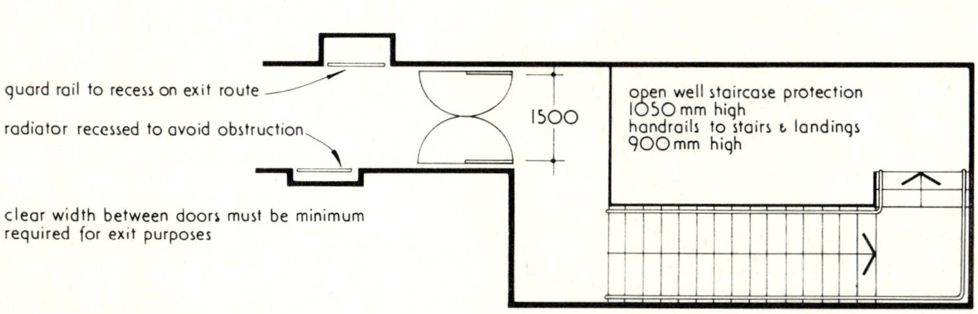

23.31 *Escape staircase*

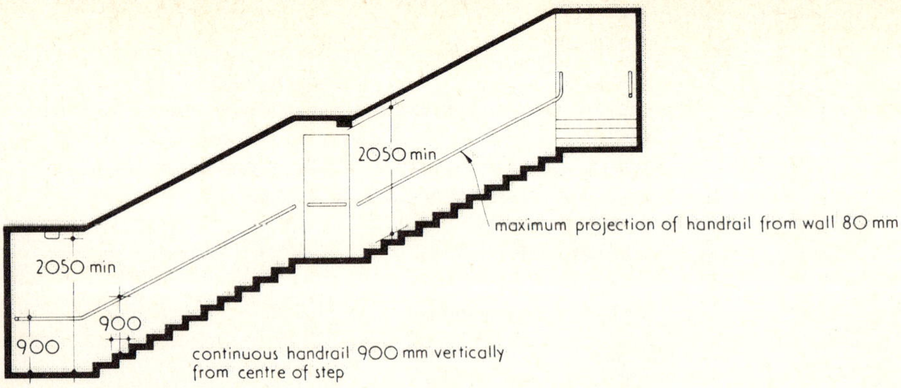

a *Plan*

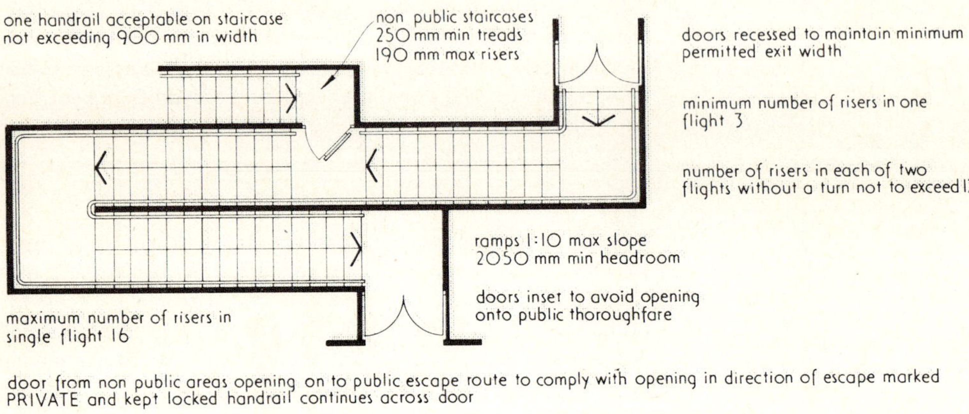

b *Section*

23.32 *Escape staircase with additional access from non-public area:*

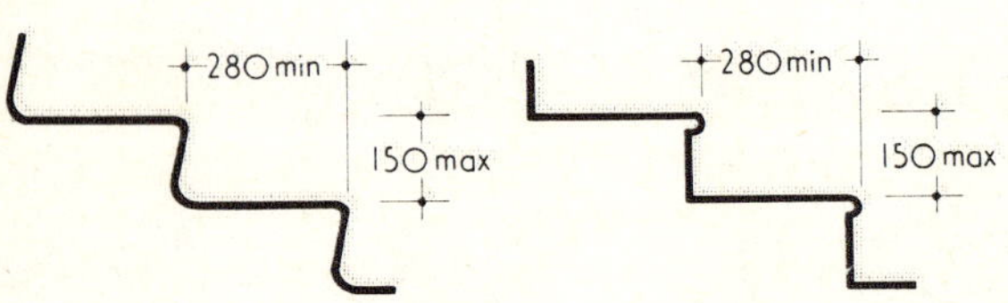

23.33 *Two acceptable details for escape stairs*

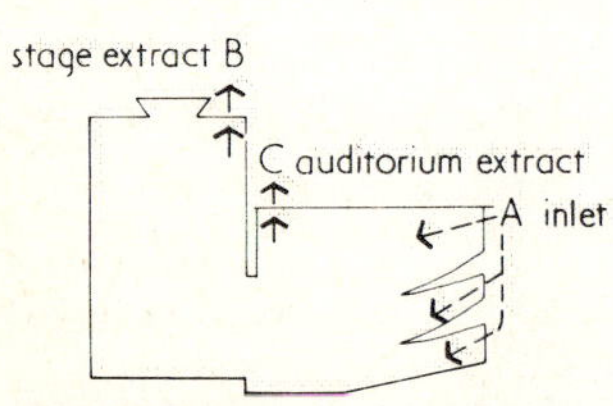

A inlet fan
B stage extract fan, capacity 40% of total
C auditorium extract fan, capacity 60% of total

23.34 *Displacement ventilation system (based on GLC requirements). A simple plenum system with one inlet fan and independent stage and auditorium extract fans*
Total capacity of extracts 75% of input
Usual order of starting: B, A, C
(Where convenient the stage extract for B and the auditorium extract for C can be combined into one extract fan common to both)

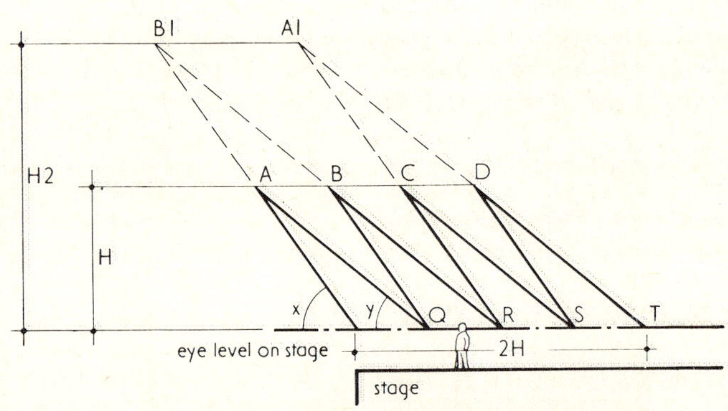

23.35 *Method of locating theoretical positions of spotlights. Spotlights at A will light an actor at the edge of the stage at 55° in section, about 45° to 50° after crossing, but as the actor moves in from the edge the angle will decrease. At Q it is only 40° in section, about 35° after crossing and this is the minimum. It is therefore necessary to provide another lighting position B, which will cover the area Q to R within the same range of angles; and then positions C and D, lighting areas R to S and S to T*

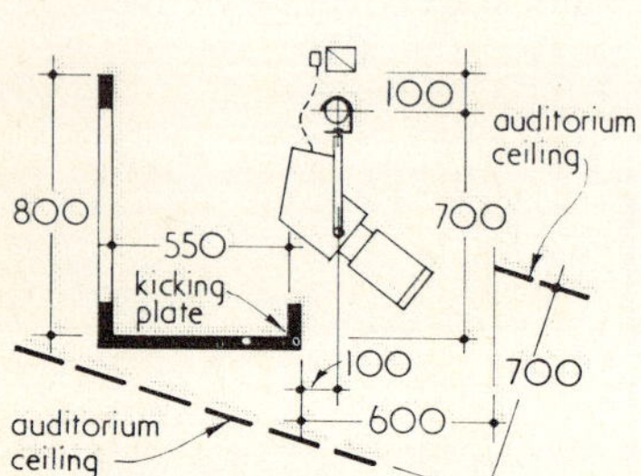

23.36 *Auditorium lighting bridge using standard lanterns*

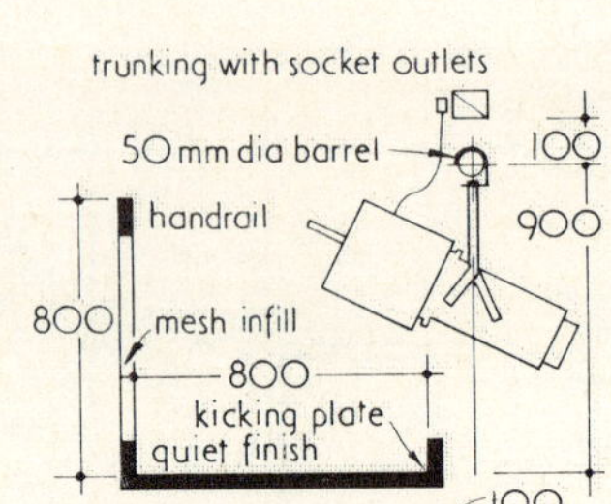

23.37 *Auditorium lighting bridge for large lanterns*

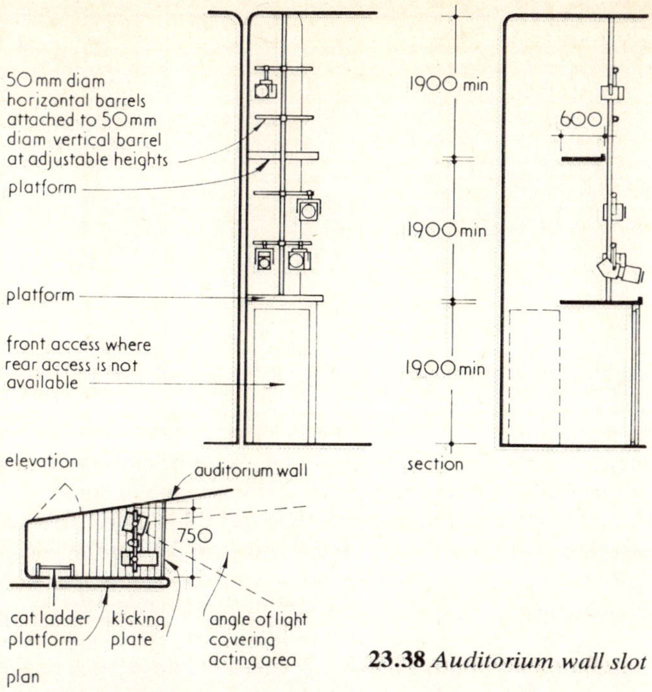

23.38 *Auditorium wall slot*

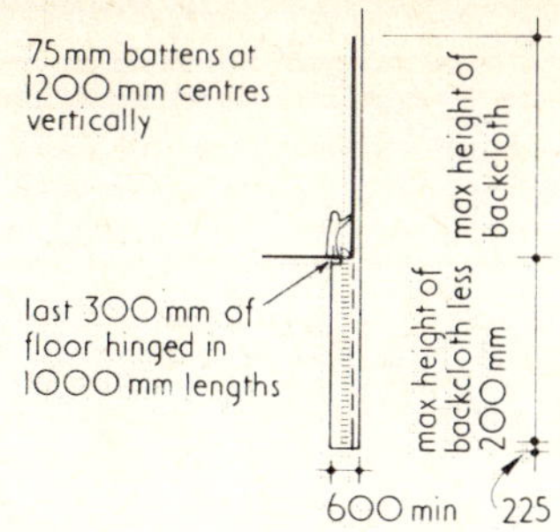

23.39 *Section through movable paint frame in slot*

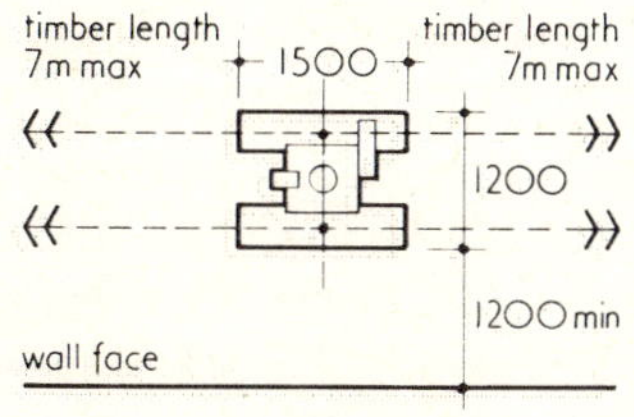

23.40 *Space requirements for woodworking machine*

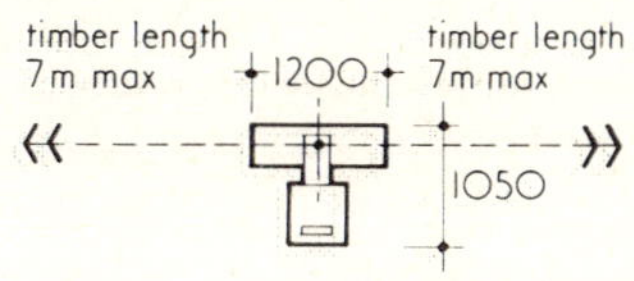

23.41 *Space requirements for mortiser*

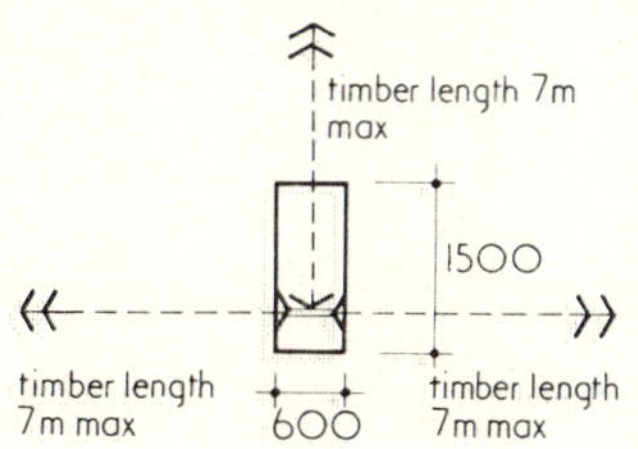

23.42 *Space requirements for circular saw*

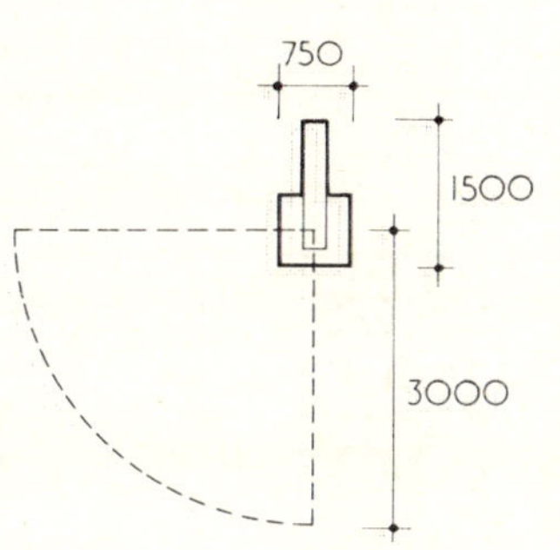

23.43 *Space requirements for bandsaw*

The diagram shows a common arrangement, **23.38**.

10.03 Acoustics

Ambient sound levels
Auditorium NC 25dBA up to 500 seats, 20dBA above 500
Rehearsal room NC 25dBA
Recording studios NC 25dBA
Workshop/paintshop NC 25dBA
Walls and roofs of auditorium and stage may be required to have a weight of 4 to 5 kg/m² to reduce externally generated noise to a level which will achieve the above results.

Lighting levels
Workshop/paintshop 200 to 250 lux
Wardrobe 150 to 200 lux
Auditorium house lighting 70 lux

11 PRODUCTION SPACES

11.01 Paint frame

A paint frame, **23.39**, consists of vertical timber battens on which backcloths are fixed for painting. Their maximum dimensions are based on the stage width and height, and they should be positioned in such a way that a rolled cloth can be taken straight onto the stage without having to be turned round. There are two types—a fixed frame with a travelling bridge in front of it, and a movable frame which can be raised and lowered within a slot in the floor deep enough to accommodate all but the top 2 m of the cloth. The second method allows the painter a full view of the cloth. The first does not require the special structural arrangements of the second, but it is more difficult for the painter to view his work.

11.02 Carpentry shop

Space requirements for other common production workshop equipment are given in **23.40** to **23.43**.

12 BIBLIOGRAPHY

Colin Amery, *The National Theatre: an architectural guide*, London, Architectural Press, 1977

The Architects' Journal, National Theatre building study, January 12, 1977, pp 59–86

Roderick Ham (ed), *Theatre planning*, London, Architectural Press, 1972.

24 Cinemas

CI/SfB 525
CI/SfB (1976 revised) 525

The editors wish to acknowledge the assistance of Dr Leslie Knopp, the cinema design consultant, in writing this chapter

Contents

1 INTRODUCTION

1.01 Developments in cinematographic techniques

The main objectives over the last two decades have been to improve the quality of both sound and image production (negated to some extent by the demand for smaller cinemas). Many large cinemas have been divided into two or more auditoria with smaller screens and reduced seating capacity.

1.02 Decline in attendance

As a result of the Second World War there was a period of eleven years or so during which no new cinemas could be built. By 1950 television had become the popular medium for mass entertainment and cinema audiences declined. During the 50s and 60s some 73 per cent of the cinemas in the United Kingdom closed down.

In an endeavour to arrest this decline, new techniques were devised. In 1953 Cinerama was introduced, requiring a large semicircular screen and the simultaneous use of three projectors, and the provision of stereophonic sound. The system called for considerable modifications and only a few existing auditoria were adequate.

1.03 Cinemascope and wide screen

In 1954 Cinemascope was introduced by Twentieth-Century Fox Ltd. This employed stereophonic sound and a special optical system (anamorphotic) that permitted the width of the screen picture to be doubled. This provided a picture of maximum size without impairing the sight lines from the rear stalls, which were invariably restricted by the height of the front part of the balcony soffit. Thus the picture had a height to width aspect ratio of 1:2·55.

Public acceptance of this aspect ratio led cinema owners to think that large screens were an answer to the competition of television, and they started projecting ordinary films at greater magnification, but reducing the height of the screen by fitting narrower apertures to the film gates of the projectors. These techniques were loosely termed 'wide screen'. It was usual to provide a screen as wide as the proscenium opening would permit, restricting the height to acceptable sight lines, often with unfortunate results in the picture presentation—principally cutting off close-ups at eyebrow level. Subsequently, a means of achieving a large picture with stereophonic sound was introduced by the late Michael Todd and the American Optical Corporation, known as Todd-AO. This system made use of film 70 mm wide, in place of the conventional 35 mm and thus permitted projection of a larger picture with less magnification of the image on the film, resulting in a much improved picture quality. The film is sufficiently wide to accommodate six magnetic tracks for stereophonic sound.

A similar system was subsequently used by Cinerama (using a single projector instead of three as hitherto) and for the D150 system the deeply curved screen which accepted an angle of 150 degrees.

Because of the extremely high production costs of 70 mm films and the decrease in the number of cinemas able to show them, none have been made recently. For similar reasons the number of Cinemascope films that have been produced is much reduced, and the use of stereophonic sound has almost been discontinued.

1.04 Television

Televised pictures in black and white have been projected on to screens 6 m in width using an external (xenon) light source. Recently a system has been evolved and installed in small cinemas by which colour pictures can be projected on a screen 2.25 m in width. The pictures, originating from broadcast material (and lines of video tape), are produced directly from three cathode ray tubes, each via a primary colour filter.

1.05 Main types of film

The main types of film in use today are shown in table I.

2 TYPES OF CINEMA

2.01 Cinemas for general public entertainment

These are the popular commercial cinemas operated by film companies, eg Rank Organisation and EMI (the major circuits) and smaller independent companies.

Requirements are practical seating arrangement allowing every member of the audience unobstructed view of the screen and maximum intelligibility of sound and dialogue; comfortable seats, warmth in winter and good ventilation.

2.02 Drive-in cinemas

Spectators drive their cars into a parking area from which they watch the film. Individual loud speakers (and, in winter, heater units) are provided at each parking bay. 35 mm film is invariably used, and since screens of up to 45 m width are not uncommon, screen luminance is low because the intensity of the light source is limited. The screens are usually a plastered surface, painted with titanium white and supported by a brick and steel structure. Drive-in cinemas have not found favour in Europe, due to objections from planning and transport authorities.

2.03 Exempted exhibitions

This term was introduced in the Cinematograph Act 1952 and covers certain cinematograph exhibitions for which licences are not required provided safety film only is used.

Table 1 Film type and application

Film type	Projection	Applications	Print quality	Light source/ screen size	Aspect ratio
8 mm					
Type 8R—a 'silent' film. There have been several attempts to add sound. Type 8S—more recent. Smaller perforations allow a larger picture area and sound track, magnetically or photographically recorded	Small, portable and (particularly when the film is contained within a cassette or cartridge) very simple to operate	In schools, in aircraft for entertainment and in small auditoria for instruction, demonstration and advertising	Satisfactory where prints made from the original negatives and sound recordings	Low powered incandescent source satisfactory for screens up to 900 × 655 mm. Tungsten-halogen lamp allows screens up to 1350 ×982 mm (35/40 person audience)	1:1·375
16 mm					
Silent type—perforations along both edges. Type with sound—perforations along one edge only leave room for either photographic or magnetic sound record	Either relatively light and readily portable, or heavy-duty for permanent installations	For educational, scientific, advertising, entertainment and TV purposes	Many prints made from inferior internegs, so resultant poor quality stopped shows of this gauge	Incandescent source permits screens up to 3 m × 2·2 m. Xenon or carbon arc* screens up to 6 m × 2·4 m. Power of source limited by amount of heat dissipated	1:1·135 (not readily changed)
35 mm					
	Portable with incandescent light source, or heavy duty for permanent installations	'Standard' type for commercial cinemas; also for high quality lecture halls and TV	Good	Incandescent—7 m × 5 m screen. Xenon or carbon arc* screen size proportional to power of light source; if very high, water cooled film gates provided	1:1·375
70 mm					
Space for six magnetic sound tracks	Specially designed, invariably will also accommodate 35 mm	Production costs very high—no films made in 70 mm recently	High quality photography and sound		

*In licensed premises, xenon and carbon arc projectors must be housed in separate projection rooms

Commercial club cinemas

These are cinemas which provide specialised entertainment for members only. In general these cinemas exhibit 35 mm films which have not received a certificate from the British Board of Film Censors. Club cinemas are usually small, seating up to about 200 persons, and when operated responsibly, their standards of presentation, seating and comfort are equal to the licensed cinema.

There are some specialised club cinemas, the members of which are immigrant races and Indian, Pakistani, etc, films, which, of course, do not carry a certificate of the British Board of Film Censors, are shown. Sometimes such films are shown on Saturday or Sunday mornings in licensed cinemas. The exhibitions remain 'exempted' and only those conditions of licence relating to safety apply.

'Charity' exhibitions

These may be given by film societies, charitable organisations, etc, to which the public may be admitted on payment or otherwise. Where payment is required, a certificate must be obtained from the Commissioners of Customs and Excise stating that the Commissioners are satisfied that the organisation is not conducted or organised for profit.

Trade or review cinemas

Some commercial and industrial organisations find it advantageous to have their own cinema in which to show technical, trade or advertising films to selected audiences. Either 16 mm or 35 mm films are shown, to audiences of up to 250 persons. A good standard of presentation and comfort is usually demanded. Many advertising agencies have their own cinemas, and within recent years have required the film presentation to simulate that of a 'commercial' on a domestic television receiver. This requires an appropriate screen luminance and colour quality of the projector light source. In some cases telecine equipment is installed in the projection room and the picture is provided by a large-screen TV projector, and in other cases, the film may be transmitted to one or more domestic type receivers.

Film companies incorporate small review cinemas in studios, processing laboratories or distributing offices. In some cases, such as laboratories, a high level and uniform distribution of screen luminance is a prime requirement to enable the picture quality to be assessed and defects more readily observed. In other cases, the screen size and luminance and the quality of the sound reproduction is required to simulate that of the average cinema.

3 AUDITORIUM DESIGN

3.01 General requirements

In commercial cinemas, every member of the audience should have an unobstructed view of the entire picture area without visual and physical discomfort and without picture distortion. Resolution (sharpness) and luminance of the picture must be uniform and satisfactory. The auditorium must be suitable for the sound reproduction used, which must be free from distortion and colouration arising from architectural deficiencies. In this, shape and nature of the surfaces play an important part: opposite surfaces of walls, and floor and ceiling should not be parallel. Design calls for a reasoned balance between screen size, viewing conditions, seating conditions and circulation requirements.

3.02 Viewing conditions

The size and position of the screen must relate to the size and shape of the auditorium, and upon the rake of the floor (if any). In lecture theatres and small cinemas exhibiting 16 mm films at an aspect ratio of 1:1·375, the limiting factors are shown in **24.1**.

In multi-purpose halls, the floor will invariably be flat, and the plan rectangular. The rear wall and the rearmost parts of the side walls will require acoustic treatment, and a suspended acoustic ceiling may be necessary. The screen must be raised higher when the floor is flat, than when it is raked.

Sight lines from the rearmost two or three rows should be drawn on a large scale to check that the bottom of the screen is visible from these seats. At most, the heads of people in front

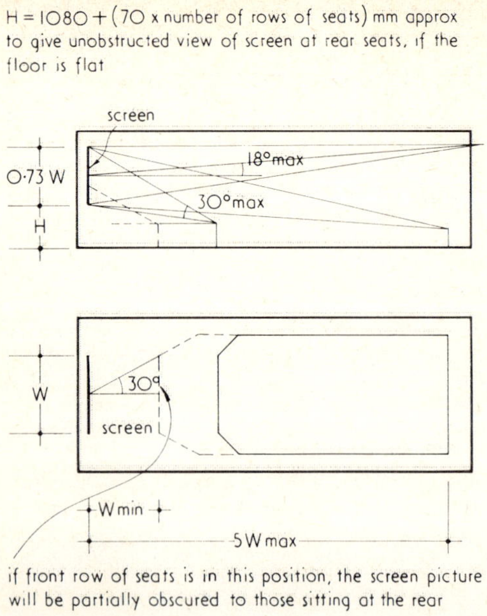

24.1 *Arrangement of a multi-purpose hall used for 16 mm film projection*

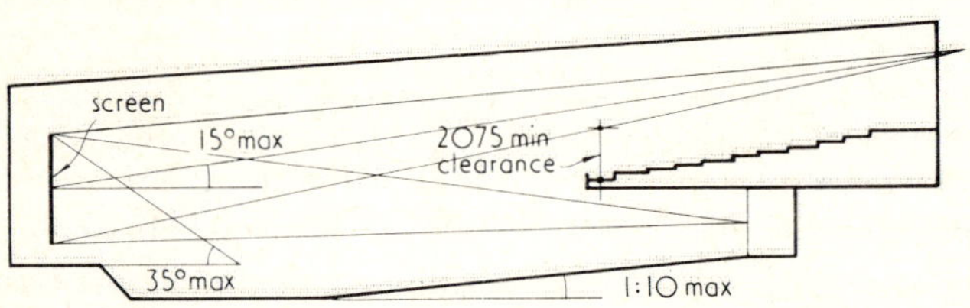

24.2 *Typical between the wars cinema with stalls and circle seating*

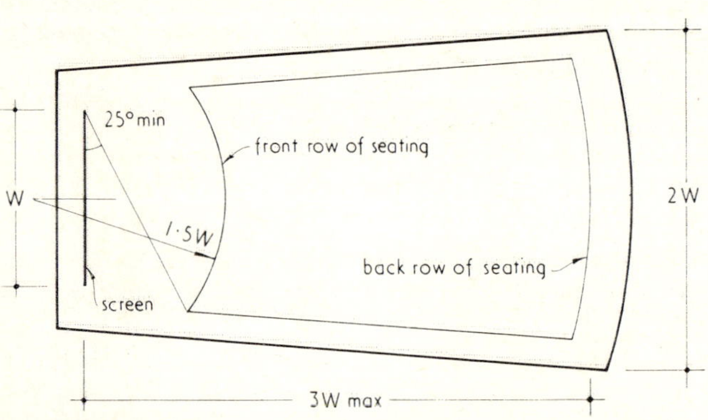

24.3 *Modern requirements for 35 mm projection*

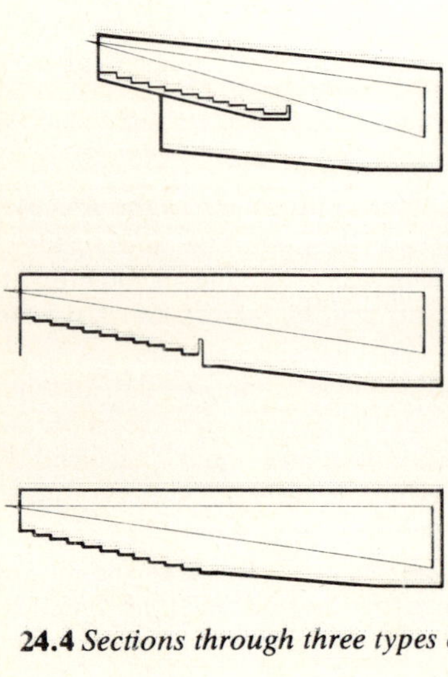

a *Stalls and balcony*

b *Stadium*

c *Semi-stadium*

24.4 *Sections through three types of 35 and 70 mm cinemas:*

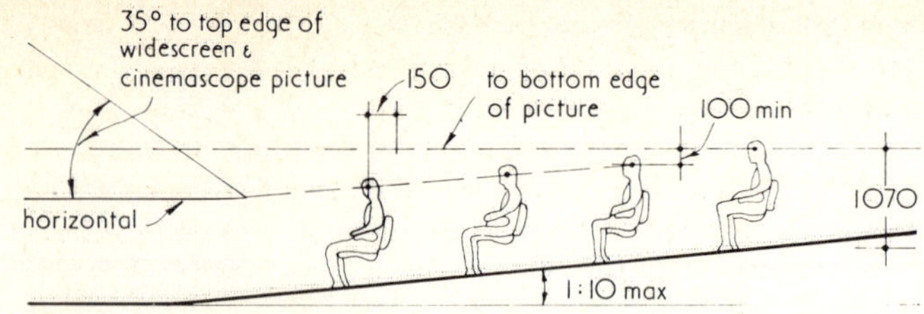

24.5 *Seating on raked floor*

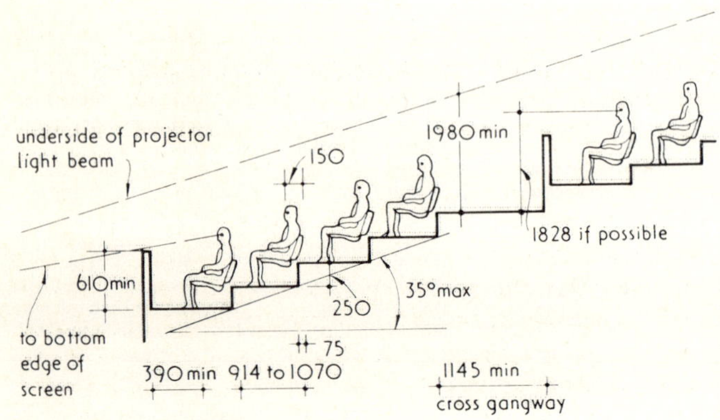

24.6 *Seating on stepped floor*

should not obscure more than 12½ per cent of the total height of the screen.

When the screen is set high, the front row of seats must be set back to preserve the 30° sightline to the top of the screen.

3.03 Types of cinema

Cinemas with stalls and circle seating, **24.2**, are no longer profitable to construct. The modern requirement for commercial showing of 35 mm film are more exacting, **24.3**.

35 and 70 mm cinemas are of three types, **24.4**:

- *stalls and balcony:* stalls have a raked floor, balcony above is stepped.
- *stadium:* front part of auditorium has a raked floor, and the rear elevated above it, is stepped. A barrier and the different levels make it necessary to have separate means of access.
- *semi-stadium:* similar to stadium, except that steps continue down to the raked floor, so separate entrances are unnecessary.

Seating capacity: in modern cinemas with seats spaced 900 mm back to back, the total area of the auditorium in m² divided by 1·05 will give seating capacity. With more closely spaced seats, the factor may be reduced to 0·85.

Location of projection room: optical axis of projectors should be as nearly as possible over the main axis (rear to front) of the auditorium. Projector level should be kept as low as possible to prevent distortion of the picture to a keystone shape. In 35 and 70 mm systems, distortion can be disguised by shaping the aperture in the film gate.

Rear projection: it is best to avoid placing a translucent screen between the projectors and the audience, or angling projection through a series of mirrors because:

- it is not possible to obtain maximum uniformity of screen luminance or directivity and seating area is therefore severely limited.
- mirror projection installation costs are very high, and 'float' glass is unsatisfactory (high quality polished, rolled plate is no longer manufactured).

3.04 Seating

Picture screens may be flat and matt white. Light will be reflected uniformly and the seating area can be rectangular in

plan; but, where directional screens are used (as in most 70 mm and wide screen cinemas) the forward reflectance is not uniform. The area nearest the observer would be the brightest (ie a person sitting on the left side of the auditorium would see a brighter image on the left-hand side of the screen only). To mitigate this lack of uniformity, the screen is curved, but this in turn restricts the seating area. Figures **24.5** and **24.6** show the seating areas for satisfactory viewing.
Space requirements for seating are given in **24.7** and **24.6–20**.

3.05 Gangways

Gangway provision depends on width of auditorium and number and width of seating blocks. The normal considerations are:

- small 200 person cinemas: often a single central gangway with both blocks of seats abutting side walls
- medium-sized cinemas: two gangways may suffice. Audiences prefer them at the two sides of the auditorium rather than dividing up seating
- three gangway situations not popular, as the central gangway is in the prime seating position; a four gangway arrangement is preferred in cinemas seating more than 500
- width of gangway usually 1067 mm clear, but continental seating with extra wide seating blocks is usually extended to 1372 mm
- central exit to 2 or 3 gangway auditorium necessitates rear cross gangway of 1067 mm
- 4 gangway auditoria usually have two exits in line with inner gangway and a linking cross-gangway (1067 mm) at the rear
- side exits should be avoided. They necessitate a cross-gangway through main bank of seating, and can be a visual nuisance
- for standing space, add 455 mm per row of persons to the width of gangways.

4 PROJECTION SUITE

Projection suites may include projection room, dimmer and switch room, workroom and store, projection staff rest room and lavatory.
Manual operation has been widely replaced by automatic control in commercial cinemas. Fresh developments require changes in the layout of equipment, if not a demand for more space, so a certain amount of latitude for change is desirable.

4.01 Projection rooms

Projection rooms are not required for 8 mm film. They are statutory requirements for 16, 35 and 70 mm film.

Activities
The projectionist controls the exhibition of films, ensuring correct focusing and direction onto the screen (critical when two or more projectors are in use at the same time), regulating volume and tone of sound reproduction, adjusting screen masking (by remote control), playing music during intervals and controlling the house and decorative lights and curtains. He also undertakes running repairs, replaces and rewinds film, and takes charge of all technical equipment in the building. He must be on duty in the projection room all the time the projectors are in use unless automatic control equipment is installed.

Minimum sizes
With minimal equipment—3·9 × 4 m; with effects lantern and spotlight—5·5 × 3·9 m; generously equipped—7·3 × 4·6 m; floor to ceiling height—not less than 2·6 m.

Essential equipment
Essential equipment for commercial and preview cinemas includes either two projectors, or one projector with long

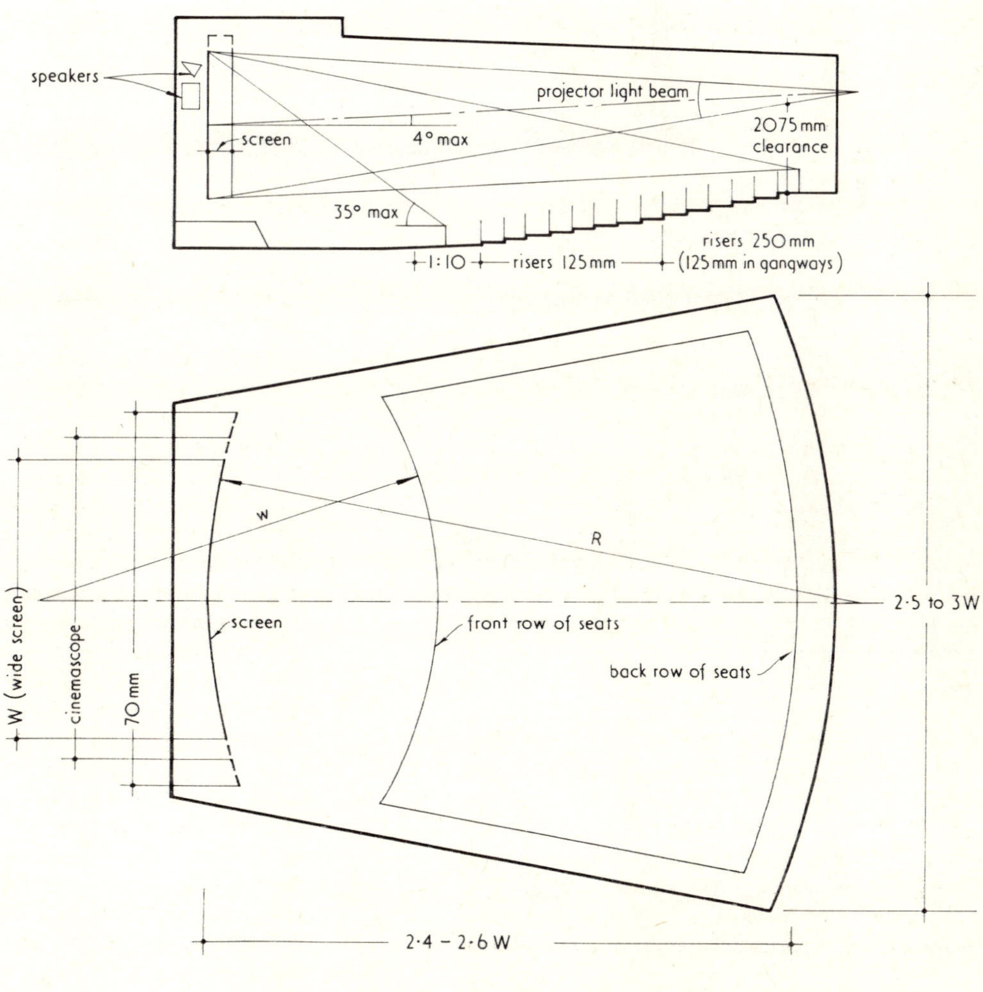

a *Section*

b *Plan*

24.7 *Criteria for semi-stadium auditorium for 35 and 70 mm projection:*

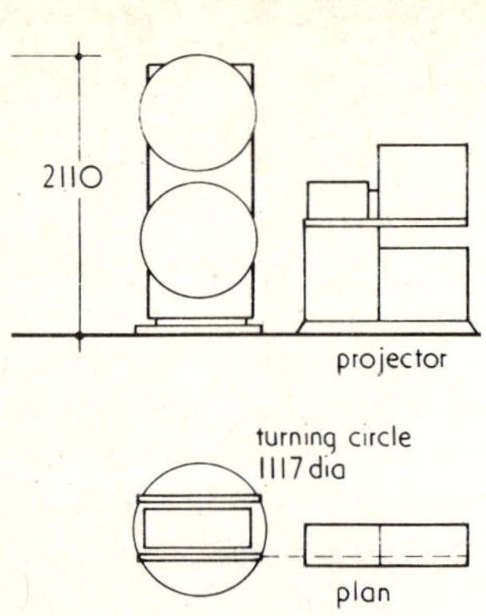

a *Westrex type 5035 film carrier*

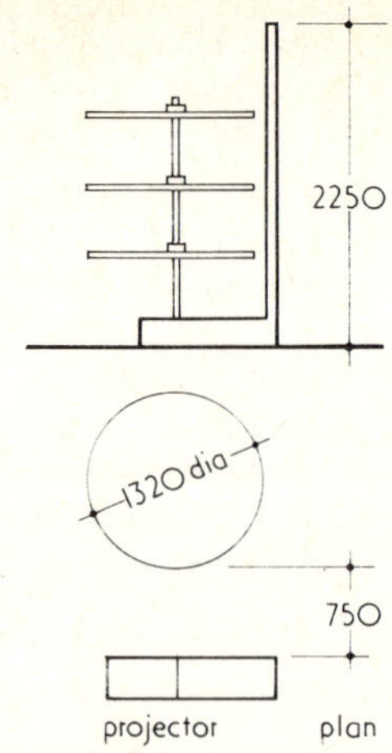

b *Phillips 'cakestand' long-run pedestal*

24.8 *Long-running film projection equipment*

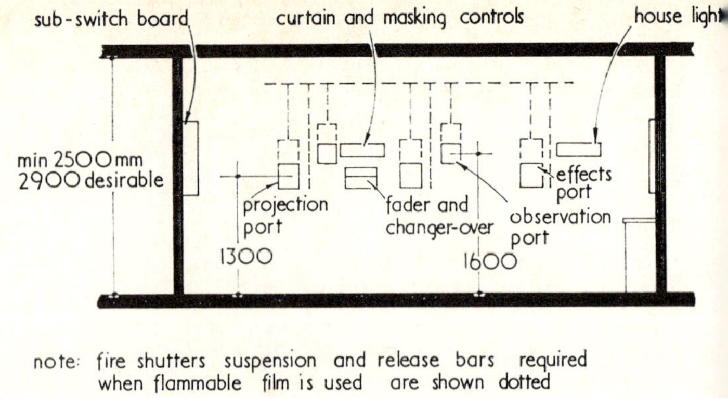

24.10 *Typical arrangement of ports in projection room wall. Broken lines show fire shutters, suspension and release bars required when flammable films are shown*

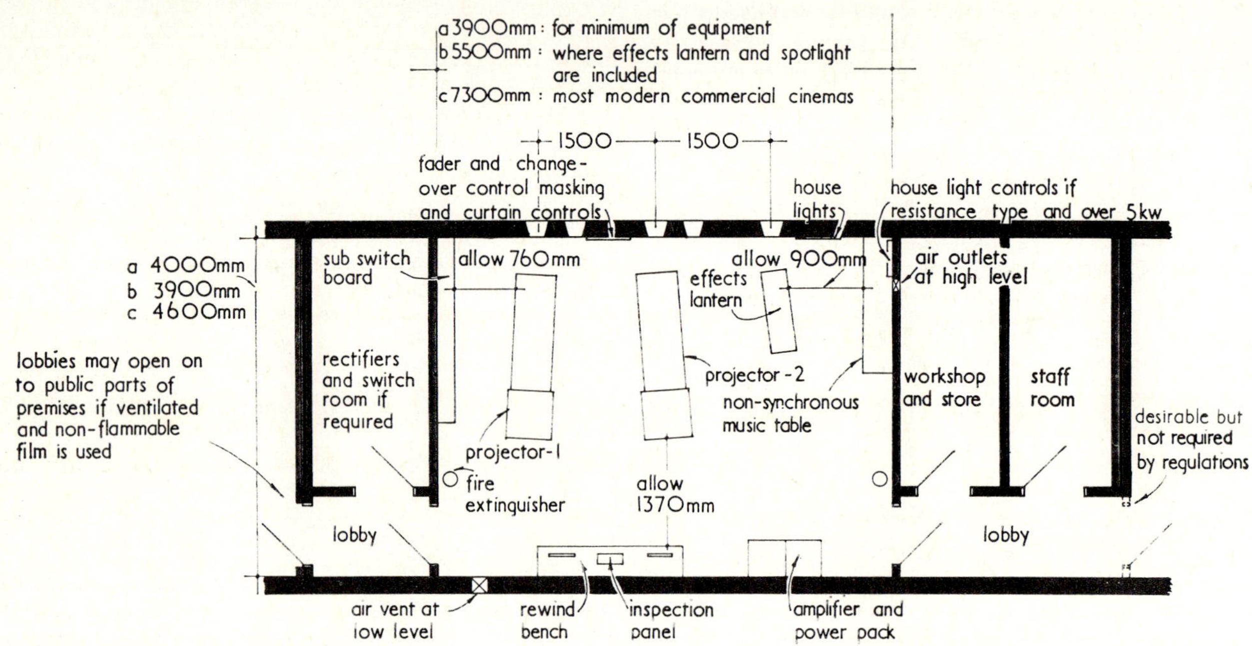

24.9 *Typical layout for projection suite*

running equipment, sound amplifiers, and controls and lighting dimmer controls. Long running equipment, **24.8** is common where one large cinema has been split into two or more smaller ones; the projectionist does not have to stay in the projection room, and so the one man can move around, attending to each cinema in turn.

Ancillary equipment
Includes spotlights, non-synchronous music desk, rewind bench rectifiers, curtain control, masking control, and switchboard. Some of these items may be in separate rooms.

Layout of equipment
A typical layout is shown in **24.9**. Twin projectors are usually placed at 1·5 m centres with a 1 m workspace between them (minimum 762 mm). The effects lantern, which may double as a spotlight, is normally to the right of the projectors, adjacent to the music and lighting controls. These will be used at the same time, and should have an observation port nearby.
Ports, **24.10**: ports are required for each projector, projectionist and effects lantern. Projection and lantern ports are glazed with optical quality glass while observation ports are fitted with plate glass. Hinged metal frames simplify cleaning.

Amplifiers: solid-state amplifiers can be mounted on the front wall of the projection room between the projectors. They usually contain change-over, tone and volume controls, which are linked to similar controls operable from beyond the other projector.

4.02 Services

Projector power supply
Requirements vary according to the type of projector in use:
8 mm 5 amp, three-pin switched socket;
16 mm 13 amp or 15 amp switched socket for portable equipment;
35 mm single phase 30 amp supply to rectifiers for arcs requiring up to 60 amps is usual but above this three-phase supply is desirable;
70 mm three-phase 60 amp supply is desirable.

Lighting
Lighting has to be arranged so that no light is accidentally directed onto the screen through the ports. Bracket fittings on the front wall of the projection room to the right of each projector, or narrow beam spotlights on adjustable arms suspended from the ceiling are usually considered suitable.

Water supply

35 and 70 mm projectors with high power carbon arcs may have to be water cooled (12 mm diameter supply pipe; 19 or 22 diameter drainage pipe). Where authorities do not permit direct connection to the water mains a closed circuit system can be used. This requires a tank of up to 680 litres capacity and a circulating pump. A stand-by pump should be fitted.

Heating

A minimum temperature of 10°C is necessary, eg tubular electric heaters on a frost-stat. Heating must be provided to maintain a suitable working temperature of 18°C.

Ventilation

A separate ventilation system (natural or mechanical) has to be provided for projection rooms and lobbies. When projector light sources are carbon arcs or those xenon arcs which give off ozone, ventilation must provide a uniform volume of air at low velocity to prevent uneven burning (and hence changes in light intensity during projection in the case of carbon arcs, and uneven cooling causing fracture of the quartz envelope in the case of xenon arcs). The ventilation system must include both inlets and outlets, quite apart from any skylights or windows. Any ductwork necessary can be worked out from the requirement that there shall be not less than 16 500 mm^2 effective (clear) area per projector installed.

Additional outlet ventilation is required for projectors and effects lanterns employing carbon or xenon arc light sources. This may connect directly with the open air or by ductwork, with an extract fan at the end if the ducting is at all complicated. In such cases, ductwork must be fitted with access panels so that it can be cleaned along its full length.

4.03 Rewinding

Most 16 mm film projectors incorporate a mechanical rewinding device but it is better to rewind on the bench so that films can be inspected.

Rewinding benches

These are bench tops preferably covered with white laminate material, 610 mm wide, at least 1620 mm long and 914 mm high, with a small frosted glass panel set centrally and illuminated from below. This is used to inspect film as it passes between rewind heads fitted at each end of the bench. These are made to accommodate various sizes of spool and have to be suitable for the type of projection system in use.

There are several standard spool sizes for 16 mm film. The largest holds 615 m of film, representing 55 minutes showing time. 35 mm film is kept in standard spools of 615 m (22 minutes showing time) but some projectors are made to accommodate 1846 mm spools (three standard lengths joined together).

70 mm film is kept in 1846 m spools and requires a power driven rewind head. When 1846 m spools are used, rewind benches must be at least 2 m long.

Rewind benches have metal lockers fitted with divisions and spring loaded self-closing flaps beneath or next to them, in which film is stored. Six divisions are a minimum.

Long run equipment is of two types. One, known as the 'cake-stand' comprises three metal discs, 1320 mm diameter, mounted horizontally on a vertical shaft which is rotated by an electric motor. The stand is positioned on the non-working side of the projector with the edge of the discs being 750 mm from it.

The other form comprises a pedestal, 1935 mm high, supporting two pairs of large spools of conventional type, but each capable of carrying 4155 m of film, sufficient for a programme of 2½ hour length. The pedestal is best positioned behind the projector and requires a turning circle 915 mm diameter.

4.04 Dimmer and switchroom

The simplest apparatus for dimming lights is a coil resistance and provided that the dissipated power does not exceed 5 kW it may be in the projection room. Above this loading, it must be placed elsewhere, and controlled remotely by quadrants or link gear passing through the dividing wall. The switchroom is often convenient.

A main cinema switchroom should be related to the position of the electricity mains entry to the building. However it is generally convenient and economically sound to have a switchboard for the projection suite. This is best placed on the left wall of the projection room, ie on its 'dead' side. Large installations call for a separate room, for which a size of 3·6 m × 2·5 m is usually adequate.

4.05 Automatic control

Equipment is available which will, in timed sequence:

1 switch on house lighting and non-synchronous music;

2 start the first projector, adjust screen masking and open the curtains;

3 stop the non-synchronous music and start the sound on the film;

4 change projectors at the appropriate time.

All the projectionist has to do is rewind the film, relace the machines (if no long running equipment is installed) and control picture focus and sound quality and level.

5 ANCILLARY ACCOMMODATION

In addition to the auditorium and projection suite, the principal accommodation will comprise:

5.01 Lobbies

Lobbies serve as insulating locks for both heat and sound. Air temperature should be about 2° higher than in the auditorium, to prevent draughts. Lighting should assist those entering the auditorium to adapt to darkness and not dazzle them when leaving.

5.02 Paybox

This is always in the entrance foyer, either against a side wall or centrally placed. A box 2 m × 1·5 m will accommodate two ticket issuing machines. Security should be provided with complete enclosure below working level and protection above (eg a glass screen). When an advance booking office is required, a working length of 2 m is necessary, with depth increased to accommodate up to 100 books of seat tickets, which are stored in pigeon holes.

5.03 Kiosk

This frequently adjoins the paybox so that in quiet periods the cashier can attend to both.

In addition to a service area and display cabinets for confectionery etc, space may be required for:

- drink coolers, infra-red grilles or ovens for hot dogs and similar snacks
- display shelving or glazed front cupboards, invariably arranged on the back wall
- security grilles, preferably of the roller type
- a wash basin with running cold water where 'protected' food only is sold, or running hot and cold, or warm water where 'open' food is sold
- extract ventilation where hot dogs, etc are sold.

5.04 Lavatories

Public lavatory accommodation is best arranged at the back of the auditorium and accessible from it. Public and staff toilet and sanitary equipment should comply with CP 305: part 1: 1974

5.05 Staffrooms
Staffrooms 6·5 m² should be provided for each sex, containing facilities for tea making, a sink, wash basin with hot and cold running water, table, chairs and clothes lockers.

5.06 Manager's office
This should be not less than 9·3 m², and located close to the entrance foyer.

5.07 Cleaners' store
The cleaners' store should have convenient access to the auditorium. It should have a sink with hot and cold water, shelves, space for vacuum cleaners, brooms, etc.

5.08 Refrigeration room
This should be easily accessible from the front of the auditorium and sufficiently large to accommodate two ice cream refrigerators and perhaps a cooler.
Shelving for sales-girls' trays, a loading table, sink with draining boards, wash-hand basin with hot and cold running water, hooks or cabinets for overalls should be provided.

5.09 Confectionery store
This store, about 9 m² in area, should be convenient to the sales kiosk. It contains shelving and cupboards and occasionally a refrigerator. It should be cool and well ventilated.

5.10 Dustbin store
This should preferably be in the open air and convenient to the auditorium. Where this is not possible, it should be large enough to accommodate at least six dustbins. The walls, ceiling and floor should be of impervious materials, and the room should be very thoroughly ventilated. Sometimes provision is made for hosing the floor and walls, in which case drainage is necessary.

5.11 Film store
A film store is desirable. It should be at ground level set into an external wall. Access to the store should be possible from outside, by a small fire resistant door, provided with a lock and key. The store should be about 1·5 m high, by 900 mm square, and provided with one shelf. This store enables film transporters to collect and deliver loaded film cans when the cinema is closed. No ventilation is necessary.

6 SOUND REPRODUCTION

6.01 Acoustic conditions
Acoustic conditions depend upon auditorium shape, acoustic absorption characteristics of surfaces (including people in the auditorium), and qualities of reverberation resulting from them.
Rectangular rooms with parallel floor and ceiling surfaces, especially if long and narrow, give the worst results. The ideal is a fan-shaped auditorium, which is also ideal as regards viewing conditions.

Reverberation time
Optimum reverberation times for cinema auditoria versus auditorium volumes are shown in **24.11** for monophonic and stereophonic sound. These curves are drawn for 500 Hertz, ie the approximate mean frequency of the human auditory range. These curves are the standard adopted by the British film industry.

Decay characteristics
The characteristics of the decay curve for a cinema should be such that each reflected sound wave is attenuated by not less than 15dB by comparison with direct sound or the preceding reflected sound wave. Reflection paths which exceed the direct path by more than 13·7 m (equivalent to a time interval of 40 milliseconds) should be avoided unless they are attenuated by not less than 15dB. These requirements apply to any cinema auditorium irrespective of size.

Speaker installation
A space of at least 1·35 m must be provided behind a cinema screen to allow for speakers. In large installations, the speakers should be mounted on brick piers or steel framing (the latter may be incorporated in the framing of the screen itself). Brick piers have the advantage, acoustically, of relatively high mass and hence freedom from sympathetic vibration at harmonic frequencies. Except for small installations, wooden supports are not to be recommended. In the case of multi-channel and stereophonic systems, additional speaker units are arranged throughout the auditorium, either in side or rear walls or in the ceiling.
Loudspeakers are usually housed behind the screen. Only one speaker unit is required for monophonic sound. For multi-

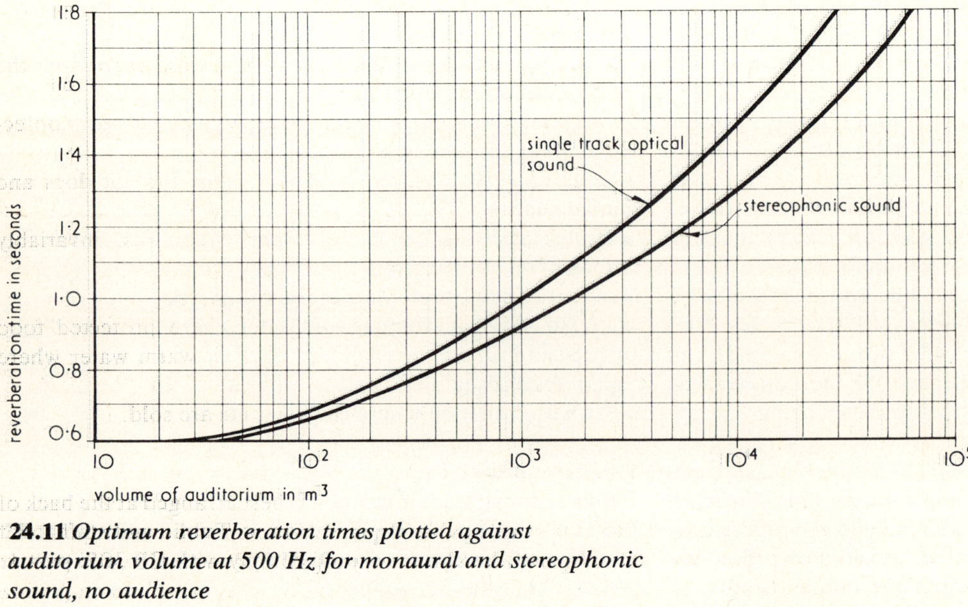

24.11 *Optimum reverberation times plotted against auditorium volume at 500 Hz for monaural and stereophonic sound, no audience*

channel and stereophonic sound reproduction from 35 mm film, three speaker units are necessary: one centrally placed, the others equidistant from it. 70 mm film sound reproduction requires five units, also symmetrically placed about a central unit. Each speaker unit comprises one or more cones in a single unit with bass response, and one or more responding to higher frequencies.

7 SCREEN AND CURTAIN DESIGN

7.01 Aspect ratio

8 and 16 mm film is projected with an aspect ratio of 1:1·375 **24.12**.

When large and wide screens are used (as with 35 mm) the picture frame has an aspect ratio of 1:1·375 but is invariably masked by the aperture in the projector gate to give an aspect ratio between 1:1·65 and 1:1·85. The international standard of 1:1·175 is to be preferred: although this ratio reduces the effective area of the film image by over 33 per cent, it gives a picture area 3 or 4 times greater than on the screens two decades ago. The aspect ratio of 70 mm films is 1:2·2.

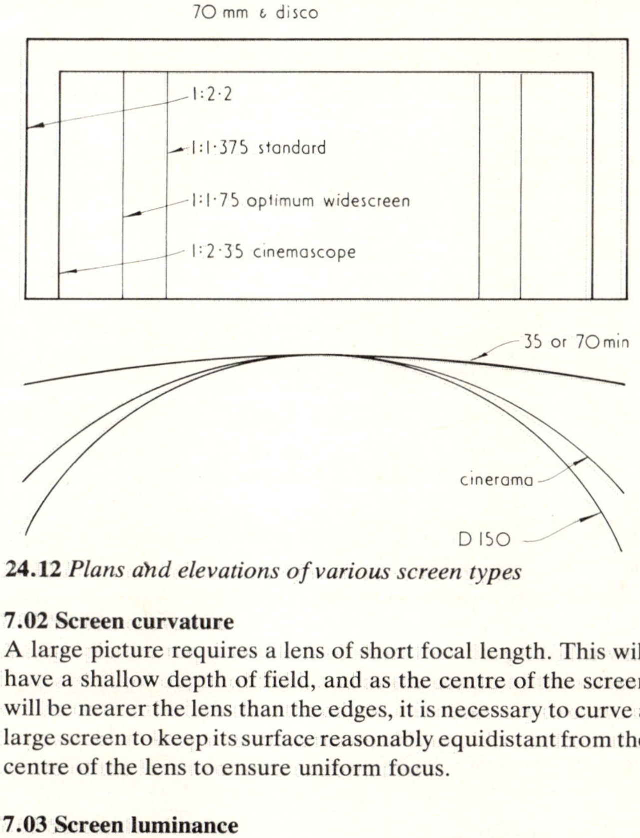

24.12 *Plans and elevations of various screen types*

7.02 Screen curvature

A large picture requires a lens of short focal length. This will have a shallow depth of field, and as the centre of the screen will be nearer the lens than the edges, it is necessary to curve a large screen to keep its surface reasonably equidistant from the centre of the lens to ensure uniform focus.

7.03 Screen luminance

Increases in area of cinema screens have not been accompanied by equivalent increases in light output from projection light sources. If the original type of matt white surface were used picture luminance would be inadequate. Instead screens surfaces have partial specular reflection characteristics which give an increased luminance within a limited forward sector. The screen must curve to provide a uniform distribution of the increased luminance but the required curvature depends upon the reflection characteristics. The actual curvature adopted is a compromise between the needs for picture focus and uniform distribution of luminance in the auditorium seating area. This requires the advice of a specialist consultant.

7.04 Screen position

Screens are almost invariably placed on the centre lines of auditoria and normal to them. In the case of curved screens centre lines are normal to the chord of the screen arc. Screens may be tilted from the vertical plane. The amount of tilt is a function of the location of projector, type of auditorium and system of projection. In some American cinemas the screen occupies the whole end wall of the auditorium. British and Continental (and most American) practice is to provide curtains which can be drawn across the screen and black masking to each edge of the screen. The masking is normally adjustable at the sides and sometimes at the top to contain the picture and obtain maximum apparent brightness. In commercial cinemas especially screens are set forward from the structural wall so as to provide space for loudspeakers and mechanical masking equipment. A clear space of 1·3 m is adequate. (Too much space behind a screen can harm sound quality.) Masking gear is best fixed at floor level for ease of maintenance. It can be fixed to brackets on the wall or on the screen framing.

7.05 Stage screens

Where auditoria are to be equipped for both stage performances and cinema exhibitions many compromises are necessary. If the picture screen is flat when in use and does not exceed about 6 m in length it can be fitted in the proscenium opening in front of the safety curtain or main curtain. It can then be housed in a roller in a wooden box set before the forestage and raised electrically or by hand. In such cases retractable davits are used to support the top of the screen. Alternatively, the housing box can be fitted under the proscenium arch and hidden from view by a pelmet or valance. The screen is then lowered into position when required. Variable masking is not practicable for these types of screen.

Curved screens or those more than 6 m wide have to be fitted on the stage behind the proscenium opening. Removal of such screens when not in use is difficult. Two alternatives are suspension ('flying') and movement on castors. Both methods are unsatisfactory as they tend to damage the screen surface. A more satisfactory method is shown in **24.13**.

Where auditoria are used primarily for stage performances, cinematograph screens are often fixed too high. They should be kept as low as possible and, if necessary, the front rows of the stalls left empty during cinematograph exhibitions.

7.06 Screen construction

Screens are generally made from pvc or metallised fabric stretched into position by cord lacing to hooks on a special frame (usually of steel or aluminium lattice construction). Lighter construction is possible when the frame is supported by brackets fixed to the rear wall, ceiling and floor. Size is determined by the largest type of picture that will be shown. Screen frames are 460 to 920 mm larger overall than the maximum size of picture to allow for fitting eyelets and localised rucking. Screen surfaces deteriorate and have to be replaced from time to time, therefore provision has to be made for delivering new rolled screens into the auditorium. These measure about 1 m in diameter by a length about 600 mm more than the height of the screen and must not be bent in manhandling. Front or side exits usually make the best way in. Masking is usually done with black wool serge fixed along the bottom edge of the screen but carried on rails attached to the top and sides of the screen and adjusted by cables over pulleys connected to hand or by electric 'controllers', controlled from the projection room.

7.07 Stage screens

The whole of the back of the screen frame (including the back and sides of loudspeakers) should be covered with heavy felt or asbestos cloth. Forestages should be carpeted with black carpet or heavy dark felt to prevent reflection of sound or light.

7.08 Curtains

Screen curtains are hooked into suspension bobbins which run on rails. The rails are supported by steel tubes which can be

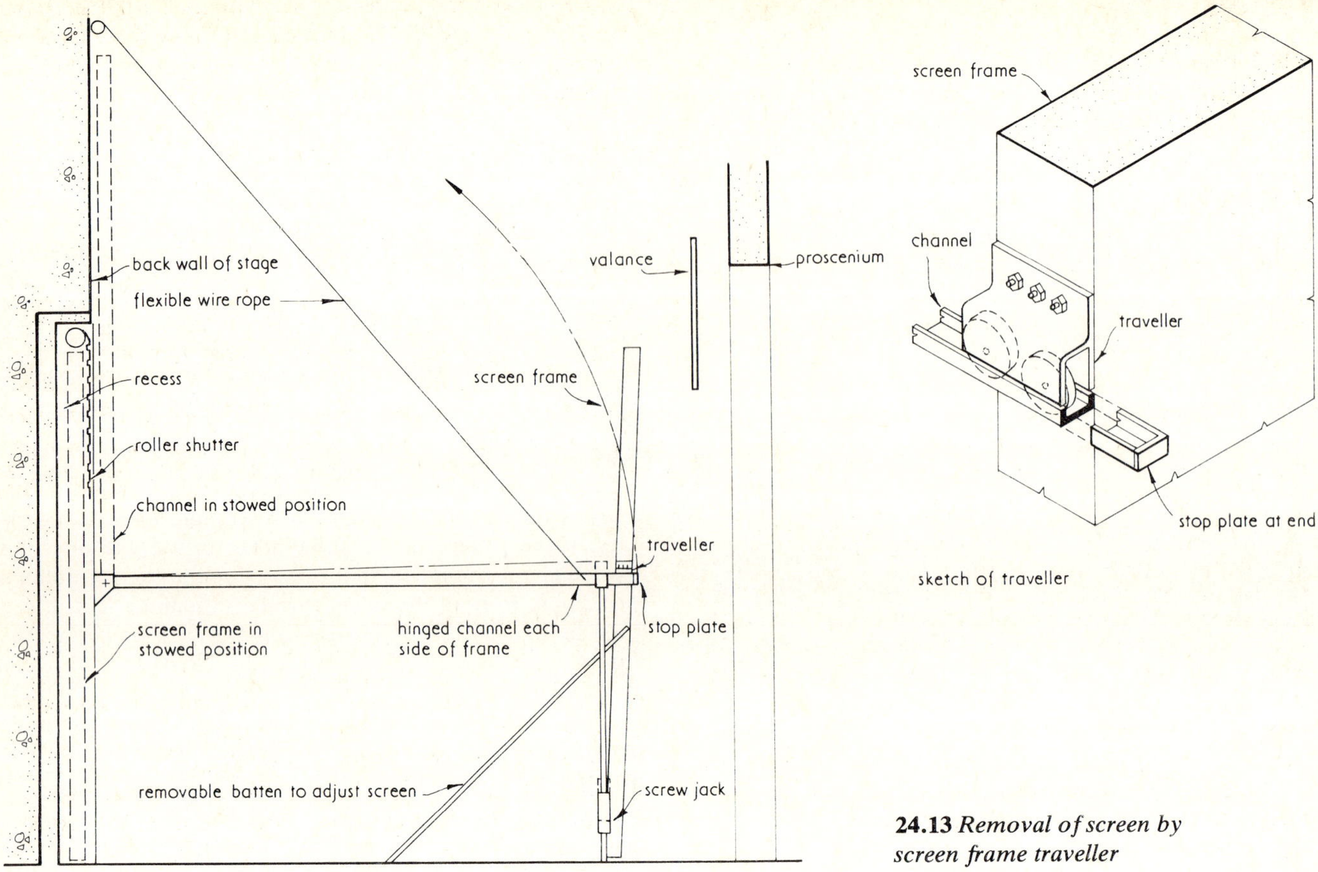

24.13 *Removal of screen by screen frame traveller*

bracketed from the screen frame, attached to overhead angle brackets, or fixed to battens on the ceiling.

Provision must be made for the curtains to overlap at the centre when closed, and to hang at each side when opened. The amount of storage space to be provided is shown in **24.14**, where stored curtains require a depth of 250 mm for satin, 305 mm for medium velour and 410 mm for heavy velour. Other dimensions are given in tables II to IV. When there is not enough space for storage at the sides, curtains may be raised vertically, into a storage space 610 mm deep for curtains of up to 8 m drop. Drop curtains are usually satin or very light velour, and because of their tendency to show storage creases, they are usually festooned.

It is customary to suspend a pelmet or valance curtain to hide curtain tracks. Curtains are usually illuminated for decorative effect by batten lights, floodlights or spotlights, or by a combination of these. Common practice is to house battens and spots in a recess at ceiling level. Spot and floodlights are sometimes arranged along the front of the balcony or on the side walls of the auditorium. Where a three-colour lighting system is provided, the number of spots, floods and batten compartments should be in multiples of three.

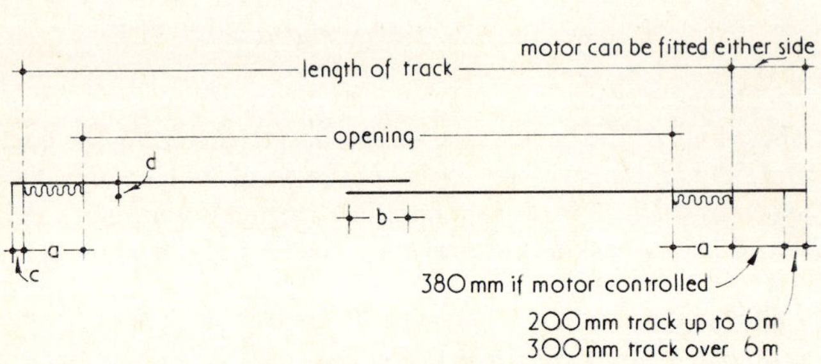

24.14 *Details of curtain storage (see Table II)*

Table II Space required for curtains and track 24.14

Length of track m	Space to stow curtains per side (a) mm	Centre overlap (b) mm	Allowance for pulleys (c) mm
6·2	300	460	75
up to 9·25	615	615	100
up to 12	925	615	100
up to 15	1230	925	125
up to 18	1460	925	125

Table III Approximate weight of curtains including fullness, weighting and bobbins

Drop m	Opening m	Heavy velour kg	Medium velour kg	Satin kg
8	15·40	122	107	92
7·38	13·85	102	90	78
6·15	10·75	68	60	52
4·95	7·70	41	36	31
4·30	6·15	28	25	22

Table IV Gauge of track

Gauge	Maximum length m	Maximum weight of curtain kg
Light	6·25	37
Medium	12·00	136
Heavy	Any length	364

8 LEGISLATION

8.01 Licences to exhibit films

The Cinematograph Act 1909 was introduced to limit the exhibition of nitro-cellulose film to licensed premises. With the introduction of non-flammable (safety) film, the Act was extended by the Cinematograph Act 1952 to cover safety film and large screen video and tv. Licences are granted by district councils and, in London, by the GLC. These control not only opening, censorship, and admission of children, but also safety, layout and structure. The GLC Places of Public Entertainment: Technical Regulations give the conditions under which licences are granted in London.

8.02 Dual function buildings

Separate licences are granted when places used for cinema are also used for theatre. These are granted by district councils under the Theatres Act 1968 and may impose additional conditions eg safety curtain, sprinkler and drencher system. Music and singing licences are not required if the music performance does not exceed one quarter of the total time of the cinematograph exhibition on any one day. Concerts given in cinemas may require a licence, depending on section 51 of the Public Health Acts Amendment Act 1890 and certain local acts.

8.03 Food and liquor licences

The Food and Drugs Act 1938 applies to most cinemas where ice-cream, confectionery and non-alcoholic drinks are sold. It is principally concerned with washing, cleaning and refrigeration facilities.

Liquor licences are granted by local justices under the Licensing Acts 1964 and 1967. Many cinemas now serve alcoholic drinks in their foyer. Licences are usually given if premises conform to 'approved' plans.

8.04 Legislation affecting detail planning

Cinemas have not yet been made 'designated premises' under the Fire Precautions Act 1971.

The Offices, Shops and Railway Premises Act 1963 controls some cinema staff, viz the manager and office staff, cashier and persons employed in the confectionery kiosk. The regulations relate to sanitary and washing facilities and clothes lockers. Section 5 of the Act (floor area per person) does not apply to a paybox or sales kiosk.

The Chronically Sick and Disabled Persons Act 1970 requires that, where practicable and reasonable, access should be provided to and within the building and in parking and sanitary facilities.

9 BIBLIOGRAPHY

BS CP 1007: 1955 Maintained lighting for cinemas

BS 5550: part 5; Section 5.2: Subsection 5.1.1: 1978 Definition testing and marking of motion-picture safety film

BS 2560: 1976 Exit signs (internally illuminated)

BS 4218: 1967 Self-luminous exit signs

BS 5382: 1976 Specification for cinematograph screens

L. Knopp, *The Cinematograph Regulations 1955*, London, Cinema Press, 1955

25 Water sports and recreation

Indoors: Gillinson Barnett and Partners (Marinas: Donald Adie)

CI/SfB 5
CI/SfB (1976 revised) 54

Gillinson Barnett and Partners specialise in leisure building design

Contents

Indoors

1 SWIMMING POOLS

1.01 Circulation
Certain principles underly the planning of an indoor pool. There must be a predetermined circulation pattern for bathers from the time they purchase a ticket, through the changing and pre-cleanse facilities to the pool. For hygiene purposes bathers must be segregated from spectators and visitors not using the pool. An ideal relationship between public entrance, changing and pre-cleanse facilities and ancillary accommodation is shown in **25.1**.

1.02 Olympic standard pools (regional centres only)
These pools are used mainly for competition and training, **25.2**.
Standard dimensions are:

length	50 m
constant water depth	1·8 m
lane width	2·5 m
pool width	21 m (8 lanes).

1.03 Pools of lower standard (district centres)
These pools are suitable for ASA, national and county championships, **25.3**.
Standard dimensions are:
• *Length* pools shorter than 50 m should be a factor of 100 m in length (eg 25 or 33·3 m). 25 m is adequate for all-purpose accommodation; 33·3 m is sometimes favoured for its greater water area, but it gives no advantage for swimming competitions where start and finish positions will be at opposite ends of the pool.
• *Depth* pool bottom can slope (minimum water depth 900 mm), **25.4**, **5**. Deck level filtration is acceptable provided the end walls are 1 m deep.
• *Swimming lanes* can be 2·1 m wide. 4 to 10, and even 12, lanes may be provided, depending on demand.
• *Diving* a 1 m removable springboard would be useful. It needs a minimum water depth of 3 m and necessitates a higher level of supervision.

1.04 Learner and training pools
These pools cater for three distinct activity groups:
• *Advanced training* length must be a factor of 100 m (eg 16·6, 20, 25 etc). Minimum water depth must be 750 mm and entry steps should not obstruct swimming lanes.

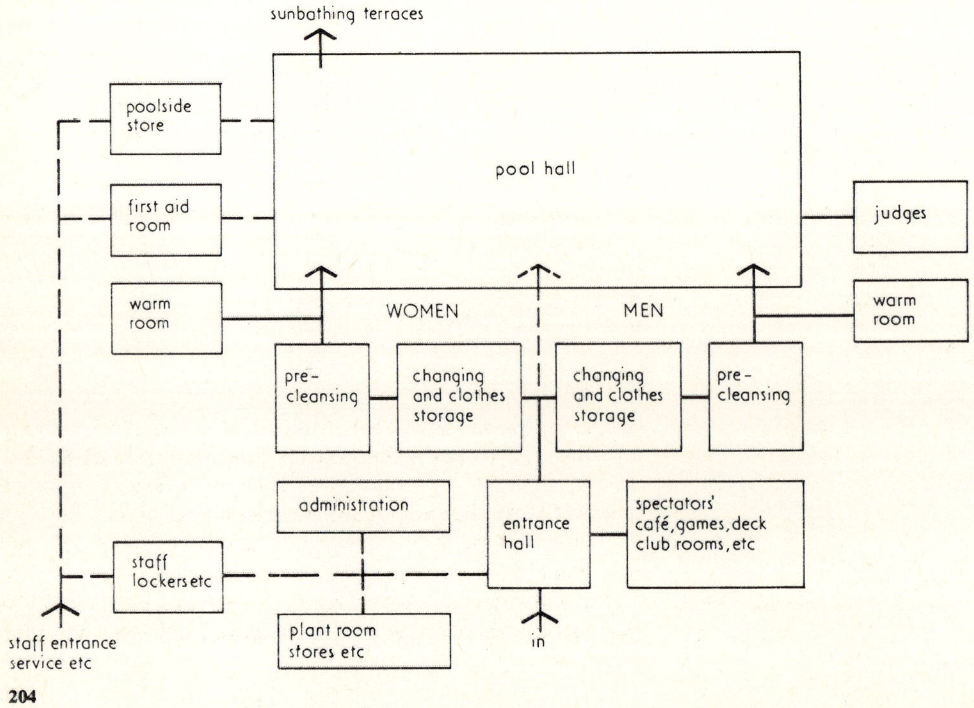

25.1 *Space relationships in swimming pool buildings*

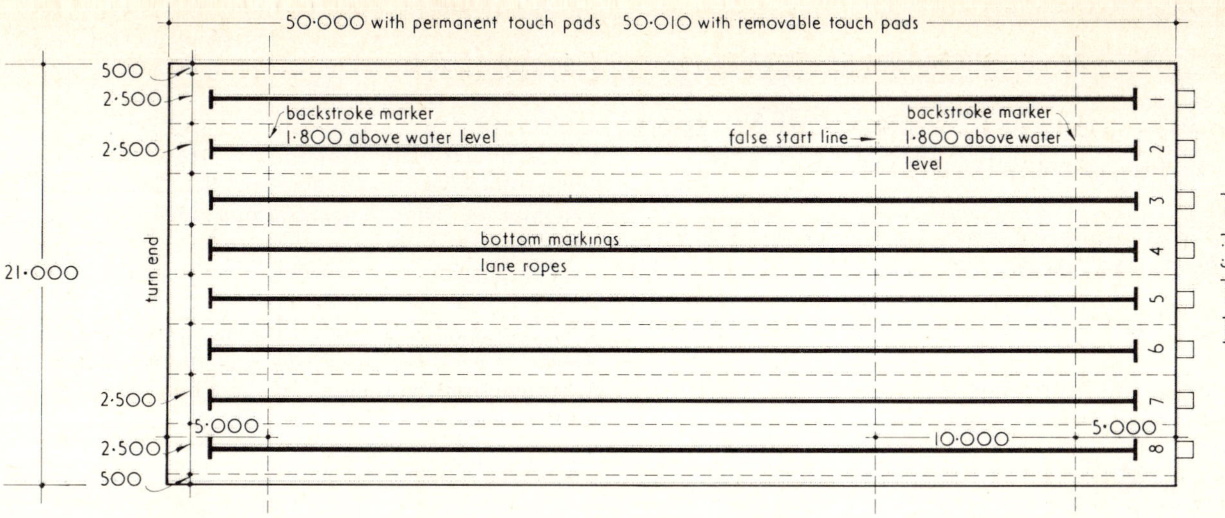

Against the nominal length of 50·000 a tolerance of plus 30 mm is allowed 300 above and 800 below the surface of the water

25.2 *The 50 m pool, to Olympic standard*

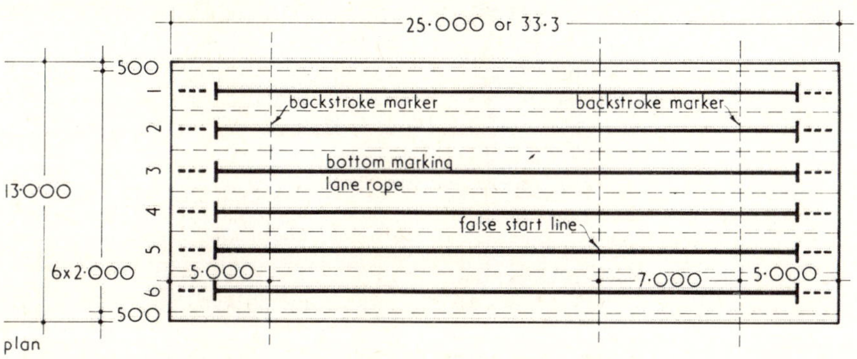

25.3 *The 25 m and 33⅓ m pools*

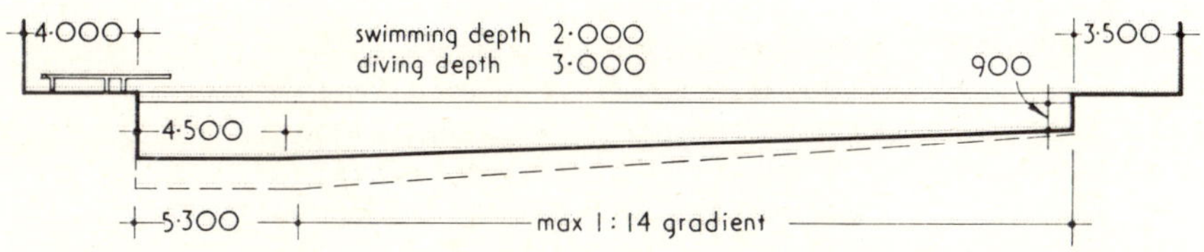

25.4 *Pool longitudinal section with deep and shallow ends (non-Olympic standard)*

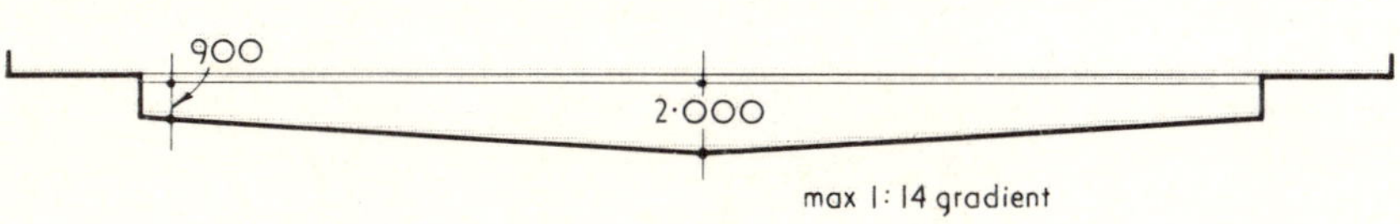

25.5 *Pool section with double shallow ends (also non-Olympic)*

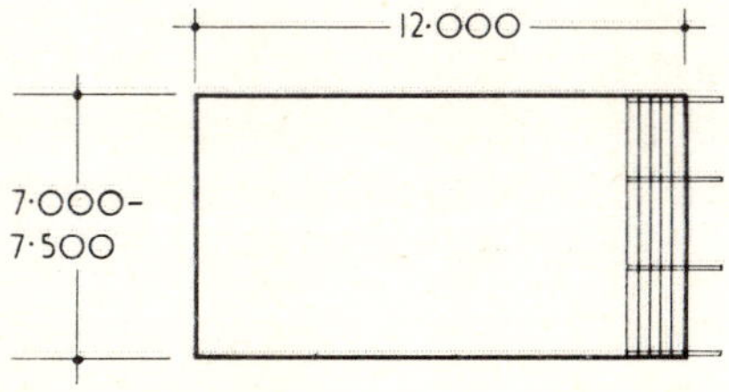

25.6 *The learner pool plan*

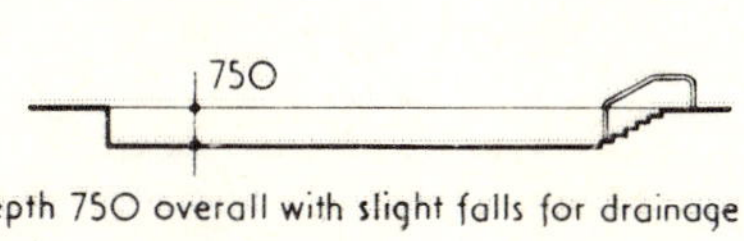

min depth 750 overall with slight falls for drainage

25.7 *Section through learner pool*

• *Learner pools* a width of 7 to 7·5 m allows the learner to take several strokes before reaching the other side **25.6, 25.7**. Length depends on the number of learners (say 12 m for 20 pupils); depth depends on instructor's choice—either constant at between 600 to 900 mm, or a slope between 600 and 900 mm (750 is recommended for teaching).

• *Non-swimmers and young children* splash pools should be 500 to 600 mm deep, with wide shallow entry steps and grab rails.

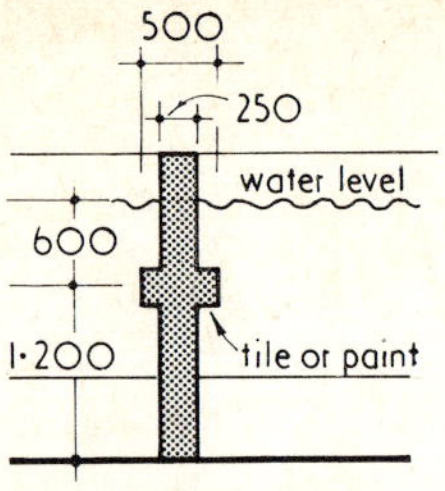

25.8 *Pool marking on end wall*

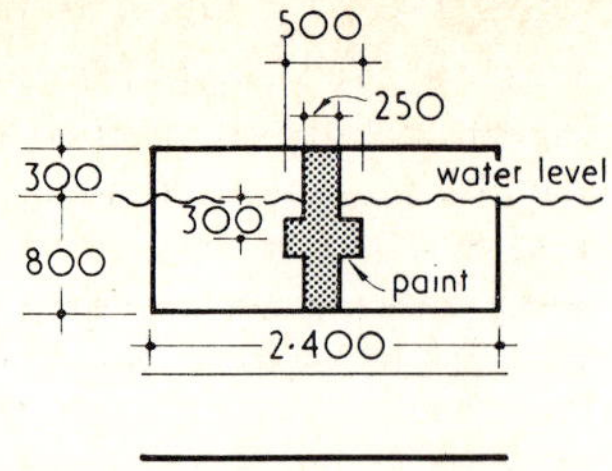

25.9 *Electronic touch pad on start and finish wall*

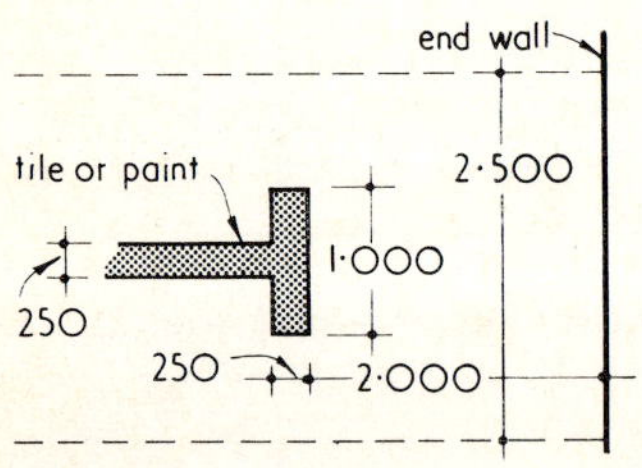

25.10 *Lane centre line on pool base*

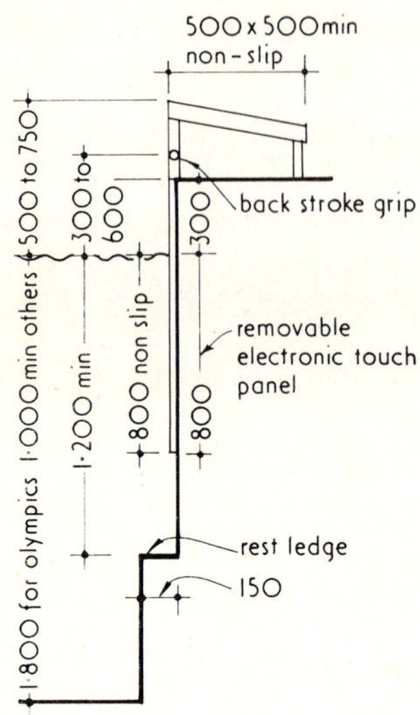

25.11 *Section showing end wall*

1.05 Markings

Competition wall markings are shown in **25.8** and **25.9**; on pool bottom in **25.10**.

1.06 Starting block

A section through an end wall shows starting block and position of touch panel, **25.11**.

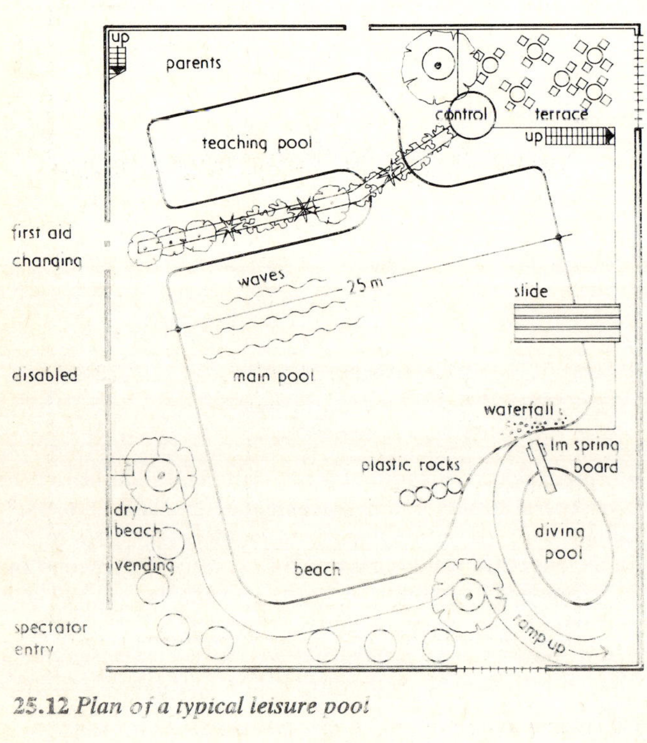

25.12 *Plan of a typical leisure pool*

2 LEISURE POOLS

The leisure pool is designed to accommodate canoeing, diving, aqua diving and water polo as well as swimming and refreshment services, **25.12**.

A leisure pool may incorporate:

- an increased amount of shallow water, beaching at zero depth from a deep end of approximately 2 m, **25.13**.
- an informal shape that still accommodates swimming lanes, water polo areas etc
- a wave machine set for waves to break on the 'beach'
- a dry beach area near the pool where swimmers and spectators can mix and buy refreshments
- areas of subtropical planting with species suitable for the pool environment, especially under a fully glazed roof
- a multi-lane water slide
- fountains, water-jets, plastic rocks and waterfalls
- underwater lighting and sound.

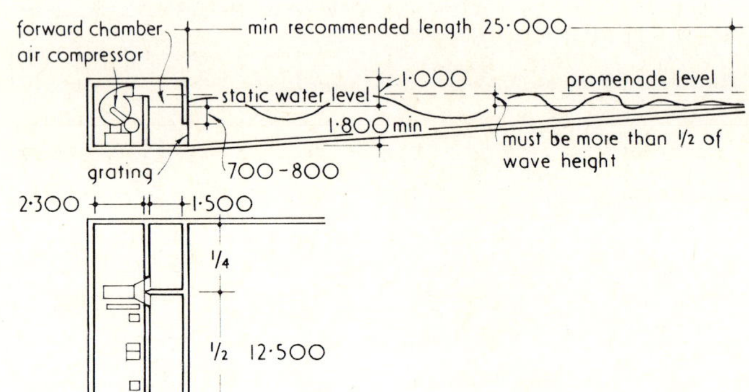

25.13 *Section through leisure pool showing wave making machine plant room and 'beaching' of pool*

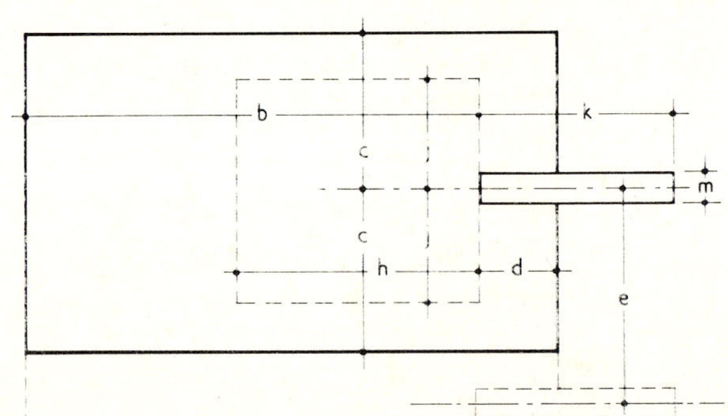

25.14 *Plan of diving pool (see Table I)*

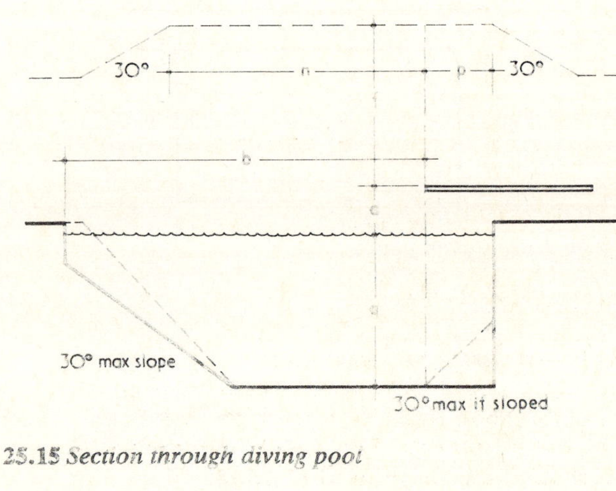

25.15 *Section through diving pool*

3 DIVING FACILITIES

The casual bather will use the 1 and 3 m springboards, but not the high, fixed boards which need only be provided on a regional basis. Diving facilities are expensive to provide and supervise: the keen diver sometimes suffers loss of the facility as a result of local experience of hooliganism.

Minimum dimensions for diving facilities are given in table I, **25.14** and **25.15**.

Table I Minimum dimensions for diving facilities

a Board height*	1·000	3·000	5·000	7·500	10·000
b Clearance forward	7·500	9·000	10·250	11·000	13·500
c Clearance to sides	2·500	3·500	3·800	4·500	4·500
d Clearance behind	1·500	1·500	1·250	1·500	1·500
e Centre of adjoining board	2·500	2·500	2·500	2·500	2·500
f Clearance overhead	4·600	4·600	3·000	3·200	3·400
g Depth of water	3·000	3·500	3·800	4·100	4·500
h Depth maintained forward	5·300	6·000	6·000	8·000	10·500
j Depth maintained to sides	2·200	2·700	3·000	3·000	3·000
k Board length	4·800	4·800	5·000	6·000	6·000
m Board width	0·500	0·500	2·000	2·000	2·000
n Clearance forwards overhead	5·000	5·000	5·000	5·000	6·000
p Clearance sides and behind overhead	2·750	2·750	2·750	2·750	2·750

* The 1 and 3 m boards are springboards. The 5, 7·5 and 10 m boards are fixed. The 7·5 m board is used mainly for training

A tolerance of ±0·100 is permissible on nominal board height. Relate all dimensions to front edge, centre, of each board

4 WATER POLO

The field of play is indicated by distinctive marks on the side of the pool above water level, **25.16**.

Dimensions of the field of play are:

- *Olympic standard:* 30 × 20 m, minimum depth nowhere less than 1·8 m
- *other competitions:* 20 × 8 m, depth 1 m.

Space must be provided for the referee to move freely from end to end of the field, and for the goal judges at the goal line.

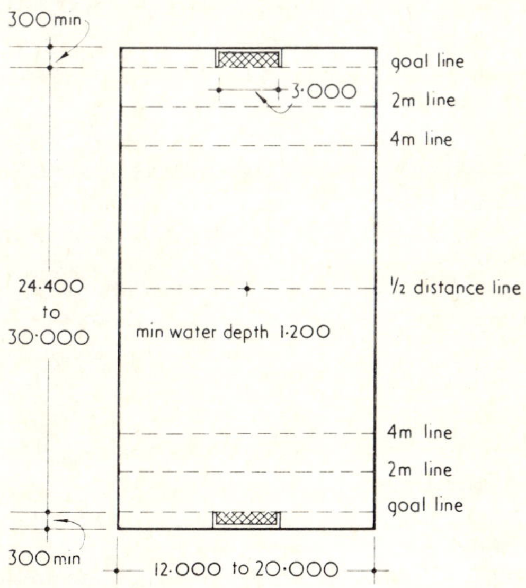

25.16 *Water polo field of play*

5 SUB-AQUA DIVING

5.01 Requirements for a small club

Normal training can take place in an ordinary pool.

The British Sub-Aqua Club suggests 5·0 × 3·6 m per 5 pupils with a water depth of 1·35 (min); 1·5 (preferred). Access to the pool should be as shown in **25.17**, **25.18**.

Other desirable requirements are:

- a compressor room of about 15 m²
- club room for up to 50 people, preferably with access to bar and restaurant
- separate storage space of about 15 m² with the floor draining to a gulley for the diving equipment.

5.02 Advanced training

Advanced training may need specialised installations and a full-time professional who would run a small equipment shop. Experience of pressure variations with depth calls for a minimum depth of 3·5 m and a maximum of 5·5 m.

Facilities should include:

- the equipment shop
- snorkelling and diving pools ranging from 1·5 to 5 m in depth with a diving pit up to 7 m deep
- lecture and meeting rooms
- social area with bar and catering facilities overlooking the pool
- changing rooms
- equipment store and compressor room, **25.19**
- administrative office and reception area.

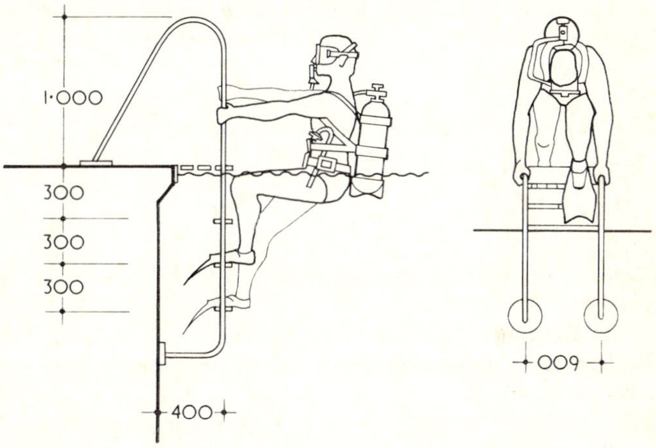

25.17, 18 *Access to pool for sub-aqua diving. A set of specially designed, removable steps can greatly assist the heavily laden diver*

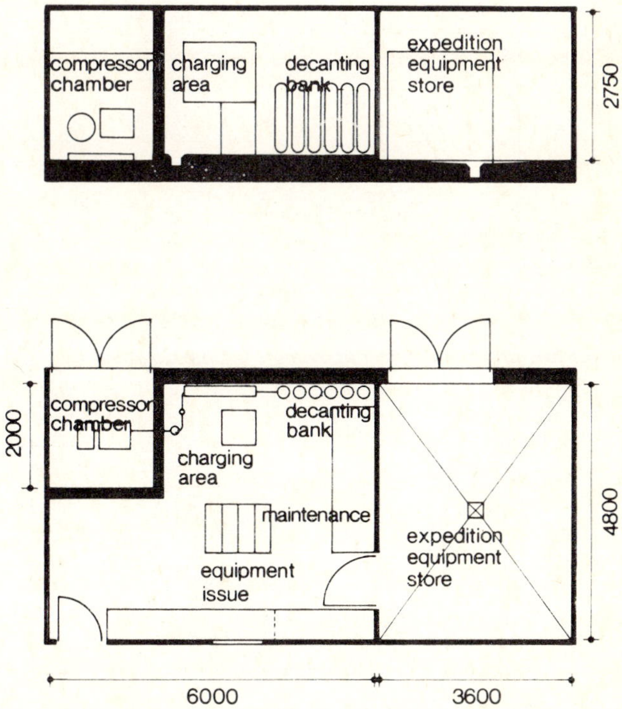

25.19 *Sub-aqua equipment store and compressor room*

6 CANOE TRAINING

The pool should have an area which is 600 to 1000 mm in depth so that the instructor can stand beside his student to help him should he become inverted and be unable to right himself.
The pool finish should not crack or chip easily if knocked by a canoe.
Access to the pool should be reasonably straight, with no tight corners. A canoe rarely exceeds 5200 mm in length, is usually not wider than 610 mm and is easily carried by one person, **25.20**.

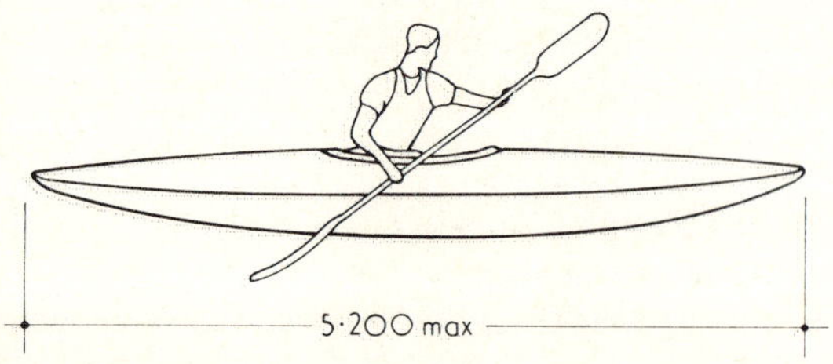

25.20 *Canoe size*

7 WAVE MACHINES

The most common type of wave machine used in the UK is the pneumatic type, **25.13**. The flow of compressed air to the forward chambers is controlled by hydraulic shutters. Number of chambers and size of the fans depend on the size of the pool and the spacing and height of waves required.
Pools 25 to 100 m long would accommodate a wave 0·5 to 1·5 m high. When wave height equals static water depth the wave automatically breaks.
Design of both the pool and the machine should be done with consultants' advice.

8 PHYSICALLY DISABLED FACILITIES

Provision should be made for access and use by disabled people wherever it is practicable and reasonable to do so, **25.21**, **25.22** (a requirement of the Chronically Sick and Disabled Persons Act 1970).
The main points to ensure are:

- ease of access, in the form of ramps at slopes of not more than 1 in 12

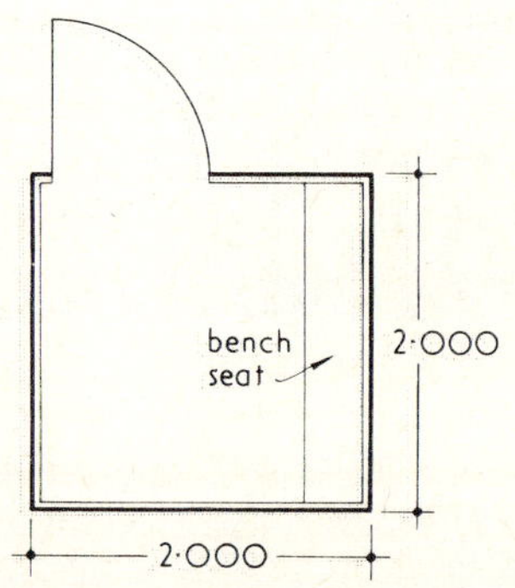

25.21 *Minimum changing provision for the disabled*

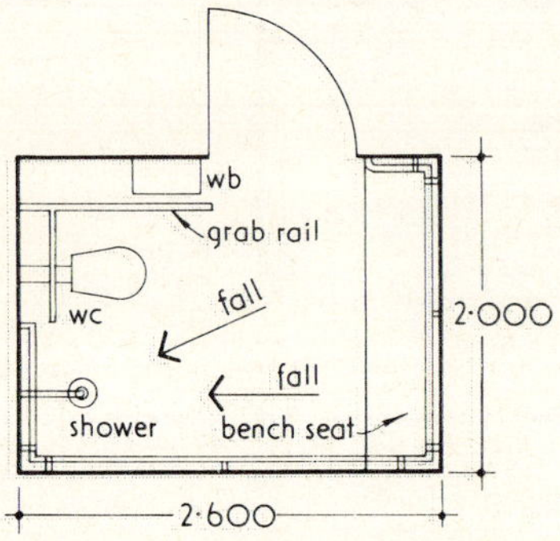

25.22 *Ideal provision for the disabled*

- doors and circulation routes should accommodate wheelchairs
- adequate means of escape for the disabled (ambulant or in chairs) in case of fire
- special changing and toilet facilities
- access to the pool via steps, chute or the slow natural slope of the pool floor (as in a leisure pool). Special equipment is usually only employed in therapeutic pools. The average disabled adult requires 1·2 m depth of water to keep the centre of gravity of the body below water level to maintain buoyancy.

9 CHANGING FACILITIES

9.01 Cubicles

Space requirements
Minimum dimensions for a changing cubicle are shown in **25.23** and **25.24**.

Method of storing clothes
There are two alternatives for storing clothes:

- in individual hanger baskets: the bather collects a basket on arrival, takes it into the changing cubicle, undresses, puts his clothes in the basket and leaves it at the central store on the way to the pool, collecting it after his swim, on his way back to the cubicle
- in individual lockers: the same routine is involved except that the bather has to carry his clothes in a bundle to the central locker store.

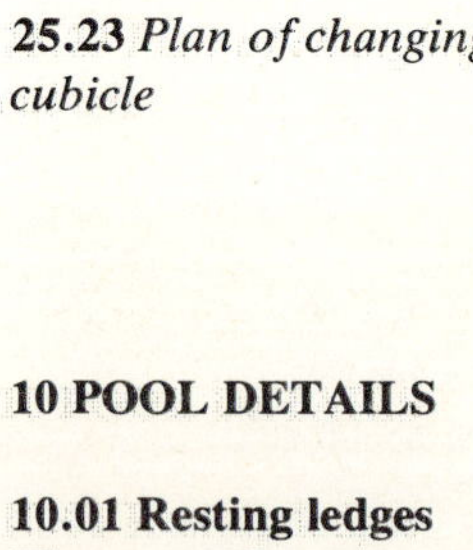

25.23 *Plan of changing cubicle*

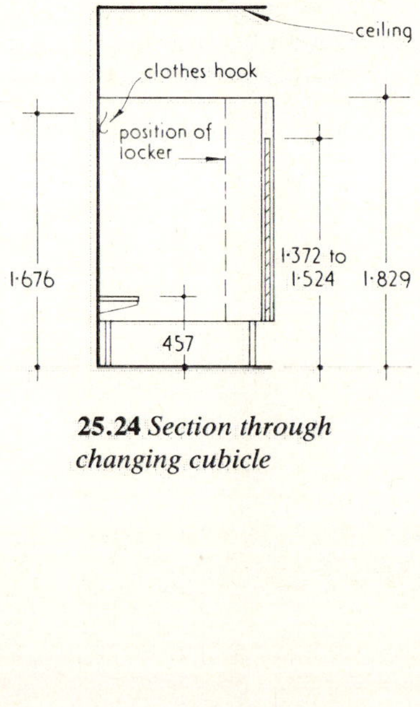

25.24 *Section through changing cubicle*

10 POOL DETAILS

10.01 Resting ledges

These small underwater projections enable the swimmer to support himself when out of his depth, **25.25**. They should never be placed in diving areas as they will encourage swimmers to congregate there.

10.02 Edge channels

These are usually part of the filtration system, serving as splash over or scum channels or surface water channels, **25.26**, **25.27**. The profile should minimise wave action and be large enough to prevent arms and legs getting trapped. The lip of the channel is usually moulded to provide an adequate hand grip.

Outdoors

11 LANDSCAPE AMENITY

11.01 Water in the urban landscape

Though perhaps the greatest magnet in the landscape scene and certainly the focal point of much recreation, water is only now beginning to be exploited properly. Use of rivers and canals as bases for linear park systems is growing, while water-based leisure complexes are being formed as in the Lee valley near London and by the Trent near Nottingham.

Inland water areas are subject to conflicting demands. On the one hand is the day tripper's desire for at least a view of the water, if not for complete access for participation in water sports, but at the same time there is an urgent need for more unpolluted drinking water supplies, coupled with an interest in wild life conservation. Multi-use is therefore inevitable and though some incompatibilities arise, they can be lessened by space zoning and time tabling (see tables II and III).

11.02 Inland water resources

Inland water resources include:

- *Natural lakes:* The more popular, accessible lakes are now seriously overcrowded during the holiday season.
- *Artificial lakes:* It is rare to create large lakes for recreational purposes, but some existing expanses of water such as old gravel or clay pits have been enlarged and reshaped, as at the National Water Sports centre at Holme Pierrepoint, Nottingham; **25.34**.
- *Enlarged gravel pits:* These appear rapidly along river valleys and have great potential in helping form new parks such as the Cotswold Nature Park.
- *Canal feeder reservoirs:* These maintain the water level in canals and are usually owned by the British Waterways Board, who may not, however, have riparian rights.
- *Supply reservoirs:* Recreational use can be made if the water has to be purified; otherwise, access will depend on the attitude of the local water board.

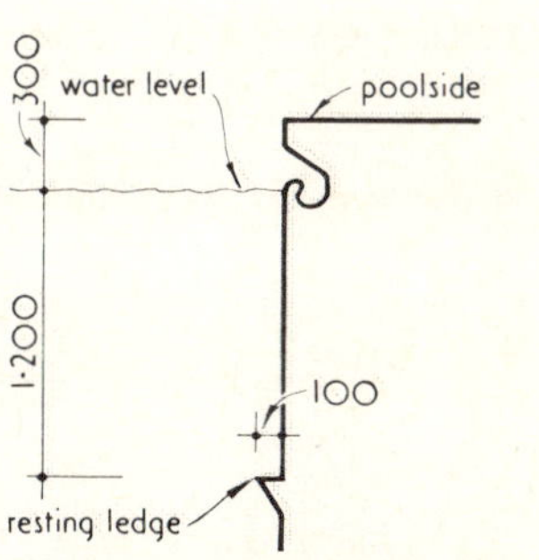

25.25 Section through pool wall showing resting ledge

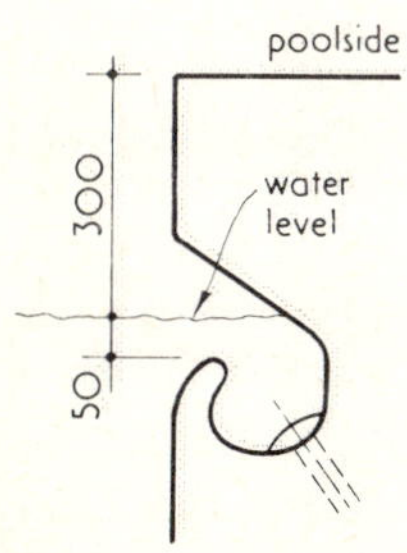

25.26 Splash over type of edge channel. Note that the normal water level will be 50 mm below lip of channel

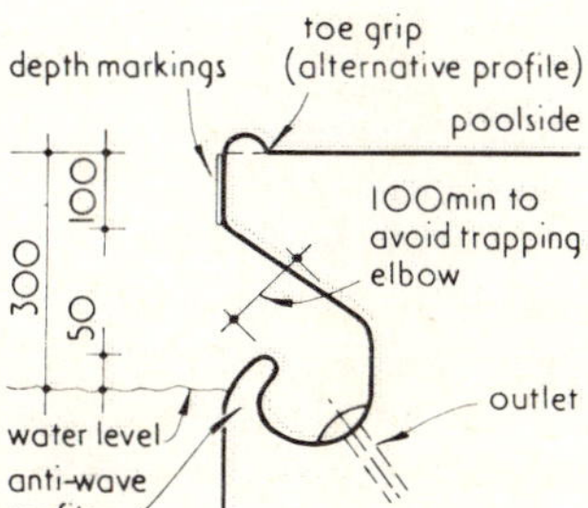

25.27 Surface water channel—note water level

Table II Compatibility of watersports

	Fishing	Swimming	Sub-aqua	Wildfowl	Canoeing	Rowing	Sailing	Water ski-ing	Hydroplaning	Power boats	Cruising
Fishing		X	X		PZ	PZ	PZ	X	X	X	PZ
Swimming	X			Z		Z	Z	Z	Z		Z
Sub-aqua	X					PZ	PZ	PZ	PZ	PZ	Z
Wildfowl		Z						X	X	X	
Canoeing	PZ					PZ	PZ	PZ	PZ	PZ	
Rowing	PZ	Z	PZ		PZ		PZ	P	P	P	PZ
Sailing	PZ	Z	PZ		PZ	PZ		PZ	PZ	PZ	Z
Water ski-ing	X	Z	PZ	X	PZ	P	PZ		PZ	PZ	N/A
Hydroplaning	X	Z	PZ	X	PZ	P	PZ	PZ			N/A
Power boats	X		PZ	X	PZ	P	PZ	PZ	PZ		N/A
Cruising	PZ	Z	Z			PZ	Z	N/A	N/A	N/A	

X incompatible; P programming; Z zoning; N/A not applicable

Table III Areas suitable for watersports

	Lakes	Canal feeders and compensation reservoirs	Water supply reservoirs	Rivers	Canals	Sea
Fishing	X	X	X	X	X	X
Swimming	X			X		X
Surfing	X					X
Sub-aqua	X			X		X
Diving	X	X	X	X	X	
Wildfowl	X	X	X	X		X
Canoeing	X	X	X	X	X	X
Sailing	X	X	X	X		X
Water ski-ing	X	X				X
Hydroplaning	X	X				
Power boats	X	X				X
Cruising	X	X		X	X	X

12 MARINAS

12.01 Siting

A successful marina must have convenient site access by way of existing or improved approach roads and reliable transport services.

12.02 Layout

The interrelation between the amenities provided, **25.28**, and the main activities, **25.29**, will control the basic layout. A marina may be off-shore, landlocked or anything in between, **25.30**. The nature of the site will largely dictate the solution which, almost invariably, will have equal amounts of space

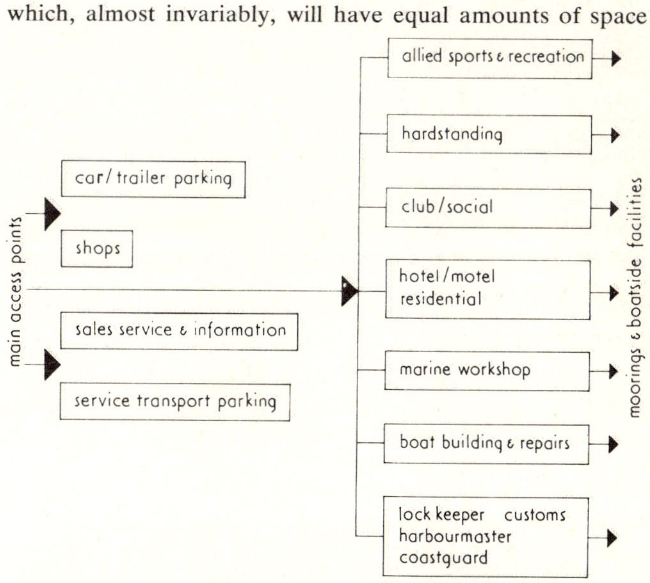

25.28 Amenity relationships for a marina for development into a basin layout

Table IV Spatial requirements and likely size ranges

	Min	Max
Land-to-water ratio	1:1	2:1
Density of boats/hectare (wet moorings)	62	162
Density of boats/hectare on hardstanding	25	75
Car-to-boat ratio	1:1	1·5:1
Density of cars/hectare (2·4 × 5·0 bays)	350	520
Ranges of boat length	4·8–13·7 m	4·3–21·3 m
Ranges of boat beam	1·8–4·3 m	1·5–6·0 m
Ranges of boat draught: inboard	0·64–1·27 m	0·48–1·65 m
outboard	0·30–0·56 m	0·20–0·64 m
sailing boats	1·14–1·77 m	1·00–2·16 m
Average boat length	5·5 m	9 m
Percentage total parking area to total water area	20%	50%
People-to-boats ratio	1·5:1	3:1
People-to-cars ratio	1:1	4·5:1
Cars-to-boats ratio	1:2	2:1

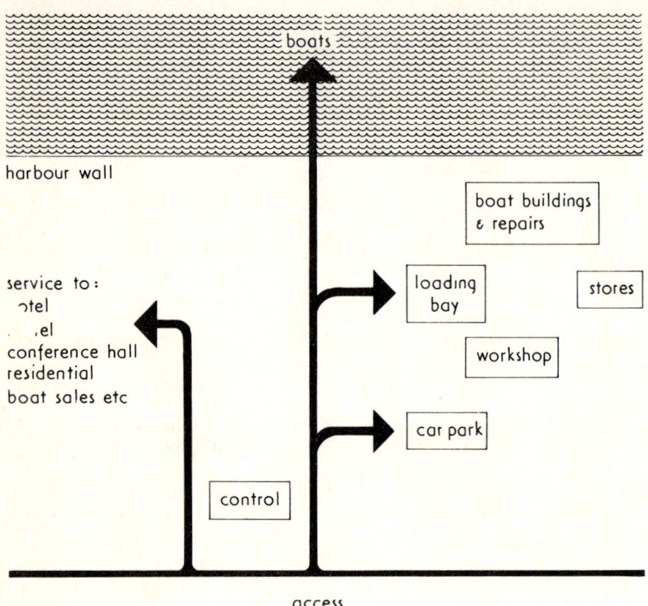

25.29 *Activities*

allocated to water and land areas, **25.31**, **25.32**. In **25.31**, a detailed breakdown of space requirements in a typical UK marina is given. The more general percentages given in **25.32** are based on averages from ten American marinas. The main difference between the two is a result of the differing parking bay sizes: 4·87 × 2·43 in the UK; 5·8 × 2·7 in the US. A detailed breakdown of spaces and ratios is given in table IV.

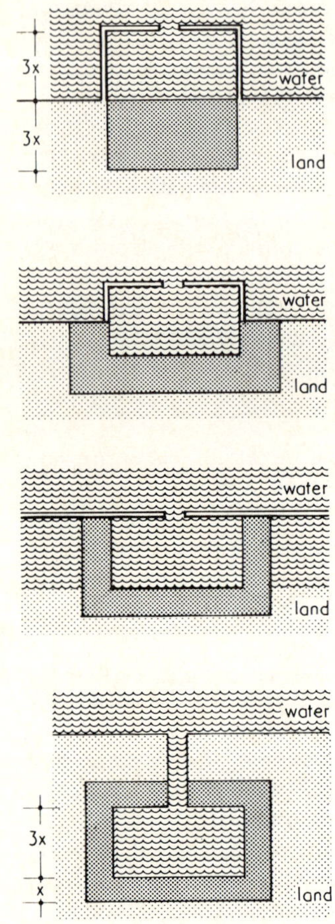

a OFFSHORE
Advantages:
minimum quay wall
minimum land take
minimum dredging
Disadvantages:
expensive in deep water
vulnerable to weather and currents
navigation hazards
minimum enclosure
silting by littoral drift

b SEMI-RECESSED
Advantage:
good for cut-and-fill economics
Disadvantage:
navigation hazard

c BUILT-IN
Advantages:
uninterrupted shoreline
large land/water interface
considerable enclosure
Disadvantages:
large land take
length of quay wall
amount of dredging

d LAND-LOCKED
Advantages:
maximum enclosure
minimum interruption of shoreline
Disadvantages:
maximum quay wall
distance from open water

25.30 *Types of land-to-water relationships, all with equal areas of land and water*

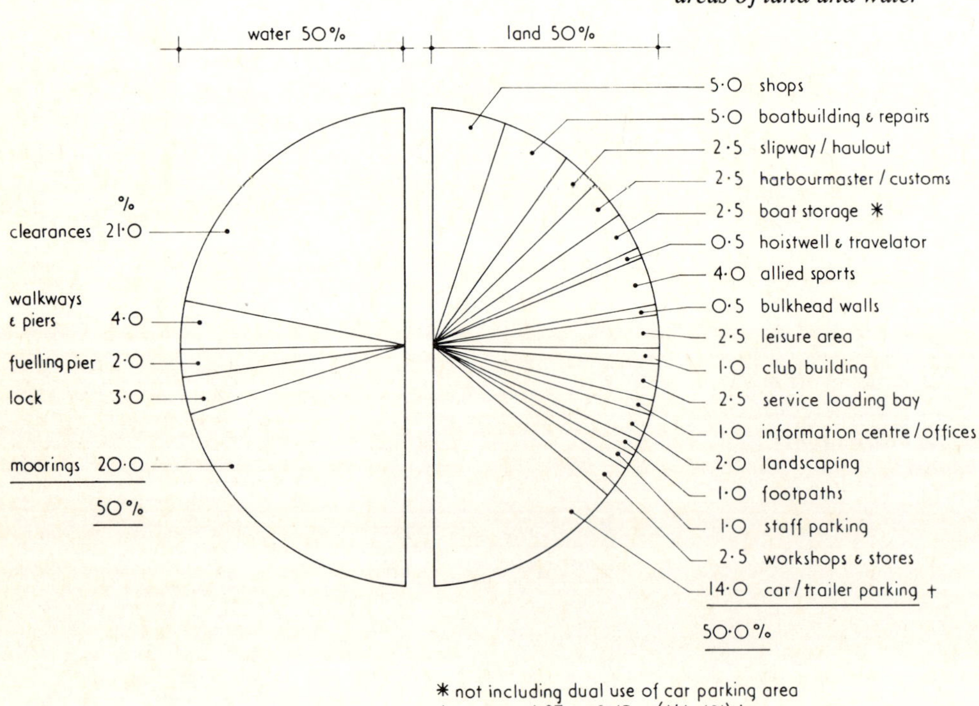

25.31 *A typical allocation of on and off shore space assuming a 50:50 land/water split. This is appropriate to European standards*

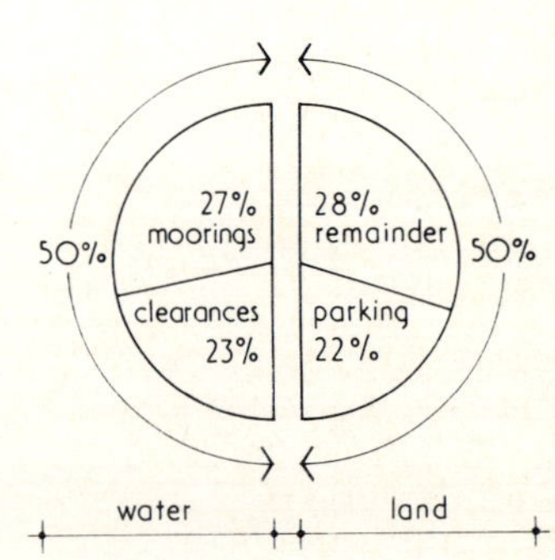

25.32 *Principal space allocations based on the average of ten American marinas. The difference between these figures and those in* **25.31** *are mainly due to the use of a 2·7 by 5·8 m parking bay*

Table V Moorings

References fig 25.33	Type of mooring	Examples	Advantages	Disadvantages	Remarks
A	Stern to quay, jetty or pontoon, bows to piles	Chichester La Grande Motte Rotterdam Kristiansund	Jetty economy	Not as convenient for embarking as alongside jetties or pontoons	
B	Ditto but bows moored to anchors or buoys	Deauville and the majority of Mediterranean marinas	Jetty economy	Not suitable with large tide range as excessive space required for head warps; danger of propellers being entangled in head warps	Particularly suitable for large yachts in basins with little tide range where gangways can be attached to sterns
C	Alongside finger piers or catwalks one yacht on each side of each finger	Cherbourg, Larnaca (Cyprus) and many American marinas	Convenient for embarking and disembarking		
D	Ditto but more than one yacht on each side of each finger	Port Hamble Swanwick Lymington	Ditto, also allows flexibility in accommodating yachts of different lengths	Finger piers must be spaced wider apart than in C though this may be compensated for by the larger number of craft between jetties	Fingers may be long enough for two or three vessels, if more than three then provision should be made for turning at the foot of the berths
E	Alongside quays, jetties or pontoons single banked	Granville	Ditto		
F	Alongside quays jetties or pontoons up to 3 or 4 abreast	St Malo Ouistreham St Rochelle	Economical in space and pontoons	Crew from outer yachts have to climb over inner berthed yachts	This is one of the most economical and therefore most frequently adopted types
G	Between piles	Yarmouth Hamble River Cowes	Cheapest system as no walkways, also high density	No dry access to land; difficulty in leaving mooring if outer yachts are not manned	Not recommended except for special situations such as exist in the examples quoted
H	Star finger berths	San Francisco			

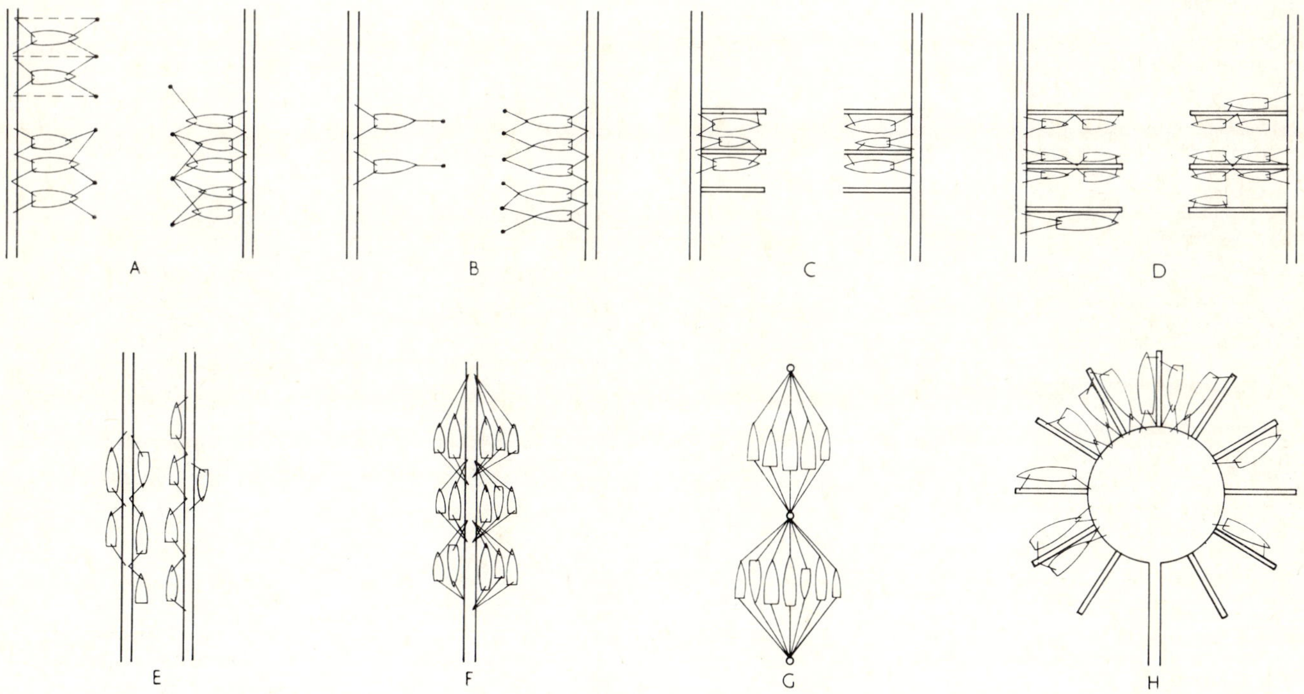

25.33 *Types of mooring (see table V)*

12.03 Moorings

A comparison between the main types of mooring is given in **25.33** and table V.

12.04 Checklist of marina accommodation and services

Social activities
Clubhouse, boat-owners' lounge, public house, bar, snack bar, restaurant, offices, committee rooms, starters' post, lookout, viewing terraces, sunbathing, reading room, navigational library, weather forecast board, chart room, television, children's play space, crèche, paddling pool

Shops
Food and general stores, tobacco, stationery, etc
bookshop
chandlery, clothes
hairdresser, beauty salon
barber's shop
sauna
masseur
chemist
laundry, launderette

Services and information
Marina office, information centre,
caretaker's maintenance workshop, storage and staffroom
banking
post office, Giro
visitors' information service (eg doctors, restaurants, entertainment)
flagpole, windsock,
weather and tides information
kennels

Allied activities
Customs house
harbourmaster's office
coastguard, weather station and information
radar, communications mast
Sea Scouts
lock-keeper's accommodation
police, security station

Boatside facilities
Storage lockers
lavatories (public and private)
showers, baths
drying rooms, cabinets
bottled gas service
electricity, lighting and power
plug-in telephone service
dockside laundry service
tannoy system
litter bins
mail service

General services
Gas, main, bottled or in bulk storage
electricity, lighting and power to piers and grounds (see safety equipment)
sewage and refuse disposal
water supply
telephones
centrally controlled security system

Boat services
Boat building, repair, maintenance yard, material store
new and second-hand boat and engine sales and hire
launching and hauling equipment (fixed and mobile)
hardstanding
launching ramps and slips
dry storage of boats
covered moorings (wet and dry)
information board of local services
brokerage, insurance, marine surveyors
divers' service
fuelling station or tender

Allied sporting activities: provision and instruction
Rowing
SCUBA, skin-diving equipment and instruction
water ski-ing, ski-kiting
swimming
fishing tackle (hire and sale of bait)
sail training
tennis, badminton and squash courts

Allied accommodation
Hotel, motel, holiday flats, public house, holiday inn

Transportation areas and services
Car parking and service (fuel, repairs and hire)
trailer bays and hire
bus bay
transport to and from local centres
carts for stores and baggage
motor cycle/bicycle sheds (open and covered)
boat trips and coach tours
marina staff electric runabout
marina workshops and transport areas

Safety equipment
First-aid post and observation platform
fire-fighting equipment, fireboat
life-saving equipment and instruction
warning or flood lights to breakwaters, lock and harbour entrance
general security system, fences and lighting
de-icing or aeration equipment
weather and tides information

Miscellaneous
Casual recreation area (eg picnic and kick-about areas)
swimming pool
vending machines, ice dispenser
paved and grassed areas
landscaping
gardener's stores and sheds etc.

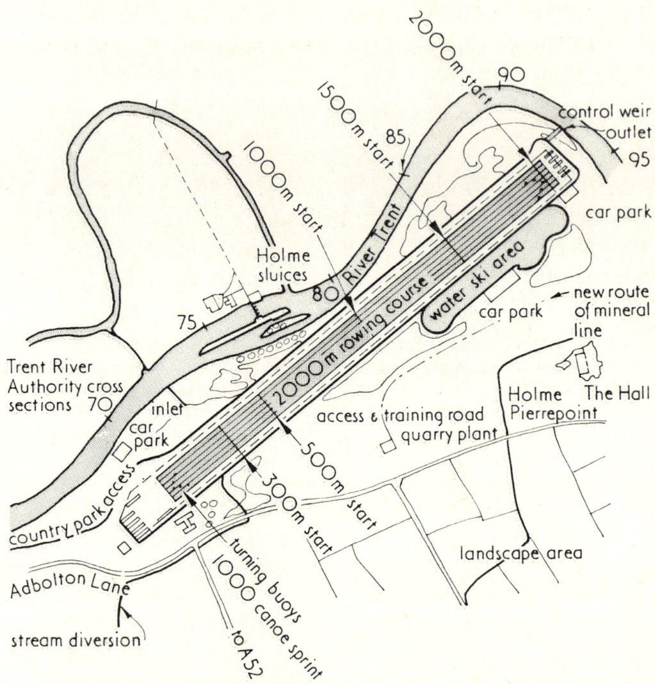

25.34 *Diagrammatic plan of the National Water Sports Centre at Holme Pierrepoint*

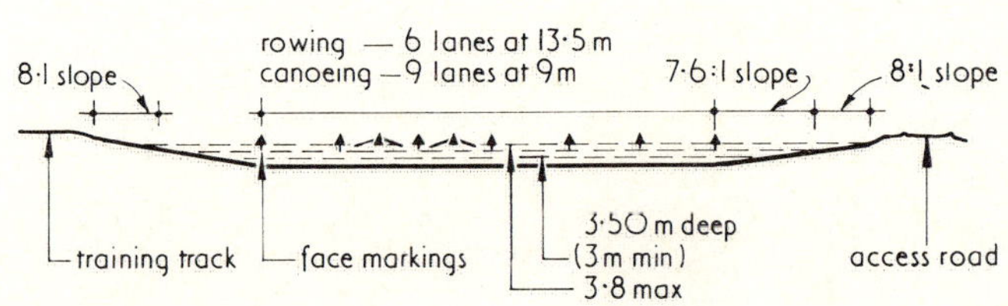

25.35 *Section through the 2000 m rowing course*

13 ROWING AND CANOEING SPRINTS

Regulations laid down by FISA (the International Rowing Federation) require six rowing lanes at 13·5 m in width (or 9 canoeing lanes at 9·0 m) with circulation width of 27 m outside the lanes on each side of the course. Optimum depth to reduce bias in certain lanes should be in excess of 3·8 m, but for economic reasons, 3·5 m is considered acceptable to all classes except perhaps the slower eights.

Courses

Rowing events: 2000, 1000 and 1500 m.

Canoeing sprint course: 500, 1000 m; juniors 500 and 300 m; also 10 000 m circuit around turning buoys.

The diagrammatic plan of the National Water Sports Centre at Holme Pierrepoint shows a typical layout, **25.34**, with a cross section through the 2000 m rowing courses, **25.35**. The course utilises existing lagoons which have been remodelled to meet the course requirements.

14 BIBLIOGRAPHY

Donald Adie *Marinas: a working guide to their development and design*, London, Architectural Press, 2nd edn, 1977

AJ Handbook of sports and recreation building design, Section 2, Swimming pools, *The Architects' Journal*, 10.8.77 to 15.5.77

Amateur Swimming Association, Swimming pools: notes for the guidance of designers

Elizabeth Beazley *Designed for recreation: a practical handbook for all concerned with providing leisure facilities in the countryside*, London, Faber, 1969

D. P. Bertlin Marina design and construction, paper presented at the Symposium on Marinas and Small Craft Harbours, Department of Civil Engineering, Southampton, 1972

R. T. Briggs *The environment and water sports: Holme Pierrepont Watersports Complex, Nottinghamshire*, IMunE Monograph no 6, 1972

British Sub-Aqua-Club *Pools for sub-aqua use*

Cement and Concrete Association *Launching ramps for boats*, 1971

Charles A. Chaney *Marinas: recommendations for design, construction and maintenance*, 2nd edn, New York, National Association of Engine and Boat Manufacturers Inc, 1961

H. L. Cohen, D. Hodges and F. L. Terret Brighton Marina: planning and design, paper presented at the Symposium on Marinas and Small Craft Harbours, Southampton, 1972

Fédération Internationale de Natation Amateur, Handbook 1972–6

B. Frazer If Marinas there must be . . ., *Boats and Sail*, June 1964

Charles R. Goldman, James McEvoy III, and Peter J. Richerson, eds *Environmental quality and water development,* San Francisco, W. H. Freeman, 1973

Gruen, Victor Associates *A development plan for Marina Del Rey small craft harbour*, 1960

Institute of Baths Management *Swimming pools design guide*

Sidney Kaye Marina architecture, paper presented at the Symposium on Marinas and Small Craft Harbours, Southampton, 1972

National Yacht Harbour Association *Yacht Harbour Guide*, London, Ship and Boat Builders' National Federation, 1963

G. Perrin Building for Recreation, *RIBA Journal*, October 1968

K. K. Sillitoe *Planning for Leisure*, Government Social Survey, London, HMSO, 1969

Thistle Foundation Edinburgh, *Sports centres and swimming pools*

R. Vian *General Layout of Pleasure Ports*, Service Maritime et de Navigation de Languedoc-Roussillon, 1968

R. Vian Water and Planning, *Town and Country Planning* (special issue), June 1966

26 Sports facilities: outdoors

Peter Ackroyd

CI/SfB 56
CI/SfB (1976 revised) 56

Peter Ackroyd is an architect in the Sports Council Technical Unit, prior to which he was a consultant for recreational and leisure building design

Contents

1 INTRODUCTION

A number of sports still quote critical dimensions in imperial units. These are not shown here except by the expression of dimensions to the second or third decimal point, which should not be rounded off.

Sports will be found in alphabetical order in the appropriate classification. Boundary lines are shown by a solid line, safety and other marginal areas by tone bounded by a broken line, the dimensions of which can vary and should be checked with governing bodies of sport. Court markings are usually indicated by fine lines.

For water sports such as rowing and canoeing see section 25, Water sports and recreation. For other information such as detail dimensions, other activities etc, see *Handbook of sport and recreational building design*. A bibliography will be found at the end of section 27.

2 ATHLETICS

2.01 Stadia

Facilities at a stadium capable of staging national and international meetings should include:

A 400 m, 8-lane floodlit track with one 10-lane straight of sufficient length to permit a 110 hurdles to be run with space for the athletes to pull up after passing the tape; a steeple-chase water jump, and full provision for all field events. A separate warming-up area is desirable. Changing and washing facilities for 200 athletes in the proportion two-thirds male to one-third female. Additional separate changing accommodation for boys and girls should be available.

Wherever possible, regional athletic stadia should be associated with other sports provision. An indoor sports centre or sports hall alongside the stadium would be a distinct advantage. Consult the AAA and NPFA regarding regional and local track gradings with specifications for minimum facilities. There is a need for a covered stand to seat least 2000 spectators with appropriate toilet and car parking facilities; an announcer's box and provision for the press, broadcasting and television; officials' room; equipment rooms and store. The perimeter of the track not covered by the stand should, if possible, be terraced to provide further spectator accommodation.

2.02 All weather surfacing

- Where surface is all-weather, 6 lanes are acceptable on circuit.
- Runways for long and triple jumps and pole vault should be all-weather. A reduced width of 0·90 m is acceptable. For the high jump an all-weather take-off strip 5 m wide is acceptable.

2.03 Layout

- The layout for the field events may be varied to suit local requirements.
- Where space allows, additional throwing facilities may be sited outside the track, provided there is proper control and safety.
- On cinder tracks the straight sprint and hurdle are run on the six outer lanes, thus avoiding the inner lane which is subject to heavy use during long distance events.
- If the central area is not required for winter games the distances from the shot circle to the inner edge of the track and the javelin runway should be increased to 10 m.
- The safety radius for the throws should be adjusted according to the standards expected to be attained by the competitors.
- Triple jump landing area should be increased to 3·35 m where space and funds permit.

2.04 Orientation

Particular requirements are:

Pole vault and all jump approaches should be sited so that the jumpers do not run towards the sun. The arc to be avoided for these events is approximately south-west to north-west (225° to 315°) in the UK. This also applies to grandstand siting.

2.05 Safety precautions

Detailed specifications and safety for field events are set out in the National Handbook. Extension wings should be provided to the safety throwing cage for the protection of the jumps and inner running lanes.

2.06 Discus and hammer circles

Hammer throwers prefer a smoother finish to the concrete than discus throwers. For this reason and also to allow training in both events simultaneously, separate circles are often provided, each protected by a cage.

2.07 Javelin runway

In order not to restrict the use of the running track, the runway should wherever possible be laid down clear of the track by extending it further into the arena. Ths would necessitate the reinstatement of the winter games pitch.

2.08 Tracks without a raised border

Where a track is marked out on grass or on a hard porous area without a raised or flagged border, the track length must be measured along a line 20 cm instead of 30 cm from the track side of the inner edge. This has the effect in the example shown of increasing the radius to the inner edge from 36·50 m to 36·60 m and of reducing the width of the first lane to 1·12 m.

2.09 Formula for other track proportions

Where a track of wider or narrower proportions or of different length is required, the appropriate dimension can be calculated from the following formula:

$$L = 2P + 2\,(\pi R + 300 \text{ mm})$$

where L = Length of track in metres
P = Length of parallels or distances apart of centres of curves in metres
R = Radius to track side of inner kerb in metres
$\pi = 3{\cdot}1416$ (not $\frac{22}{7}$)

The radius of the semicircles should not normally be less than 32 m or more than 42 m for a 400 metre circuit.

2.10 Alternative surfacing for D areas

If preferred the spaces at each end of the winter games pitch can be hard surfaced to the same specification as the track with the following advantages:

- maintenance is simplified
- runways do not have to be separately constructed and edged and their position can be varied as required
- portable landing areas for high jump practice and coaching can be placed where most convenient.

2.11 Running tracks

Layouts for running tracks scale 1:1000 are given in **26.1** to **26.3**. A typical district athletics centre is shown in **26.4**. Athletics centres for national and international level competitions vary with the money available—Olympic facilities now require Croesus' purse! Famous examples of less affluent centres are shown in the Architectural Press/Sports Council, *Handbook of sport and recreational building design*.

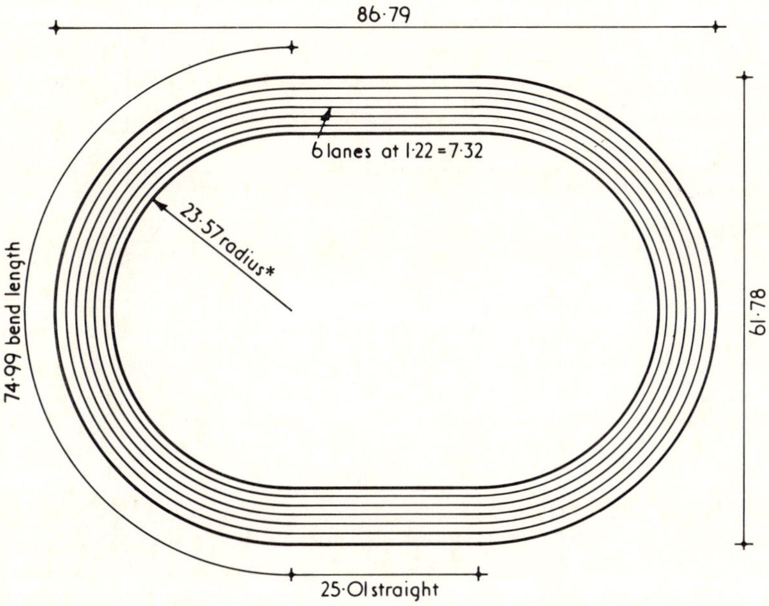

26.1 *A 200 m running track*
** radius to track side of raised or flagged edge. If chalk line only, radius is 23·67*

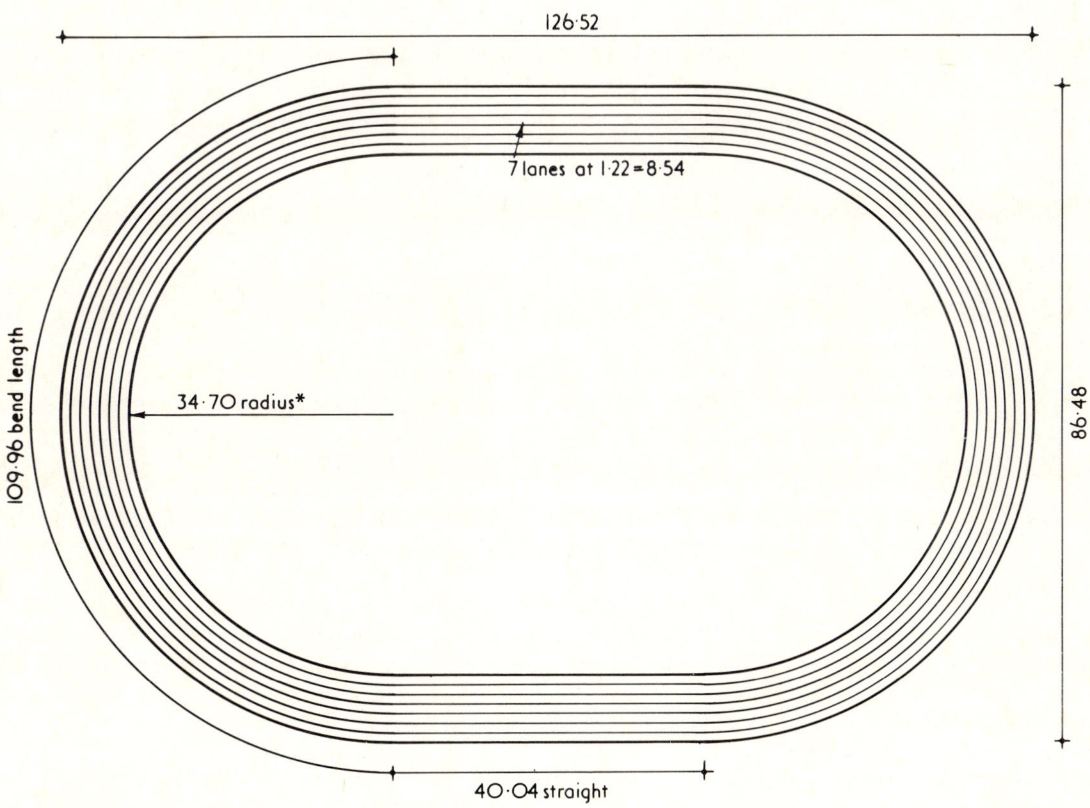

26.2 *A 300 m running track*
** radius to track side of raised or flagged edge. If chalk line only, radius is 34·80*

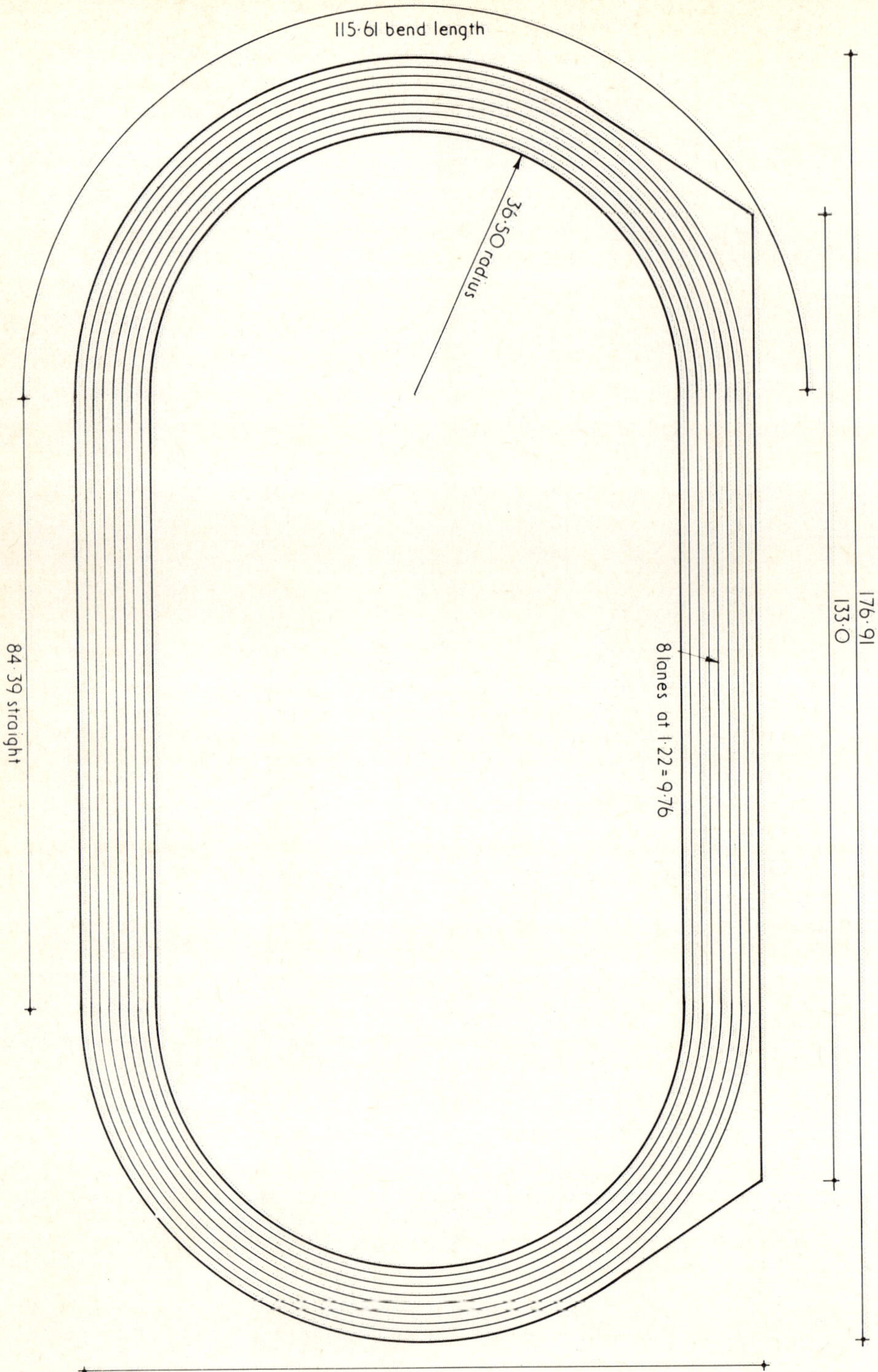

26.3 *400 m running track: the standard 7-lane club track. For 6-lane all-weather surfaces reduce overall dimensions by 2·44 m. For major competition tracks and regional facilities, 8 all weather lanes with a 10-lane sprint straight is required: increase overall dimensions by 2·44 m, and sprint straight as shown*

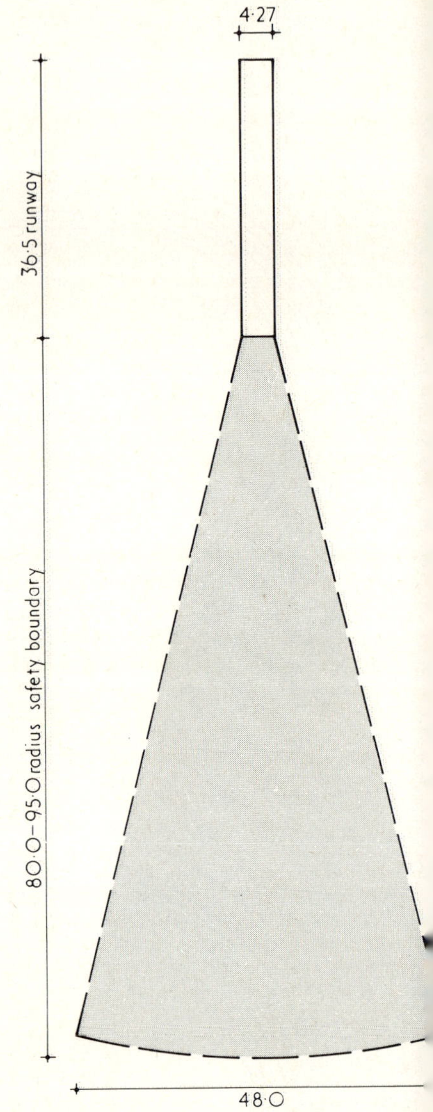

26.10 *Javelin*

2.12 Field events

The important dimensions for the main field events are shown in **26.5** to **26.12**, scale 1:1000.

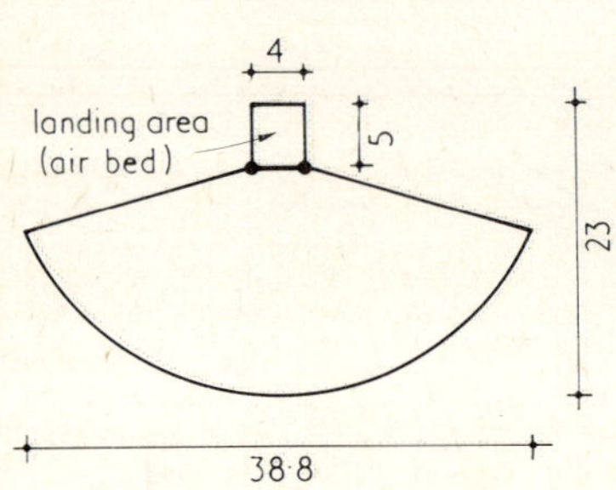

26.5 *High jump*

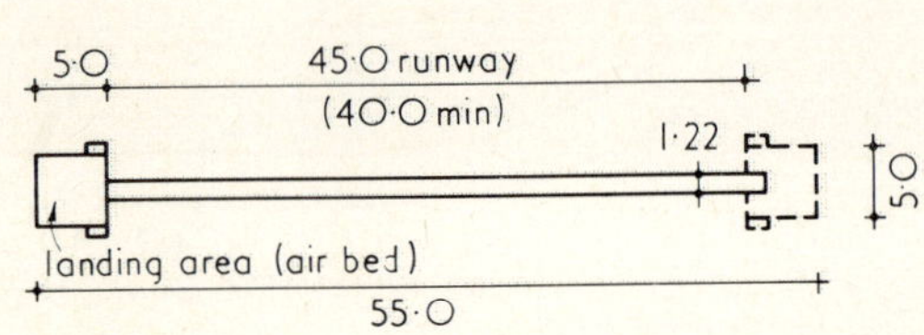

26.6 *Pole vault*

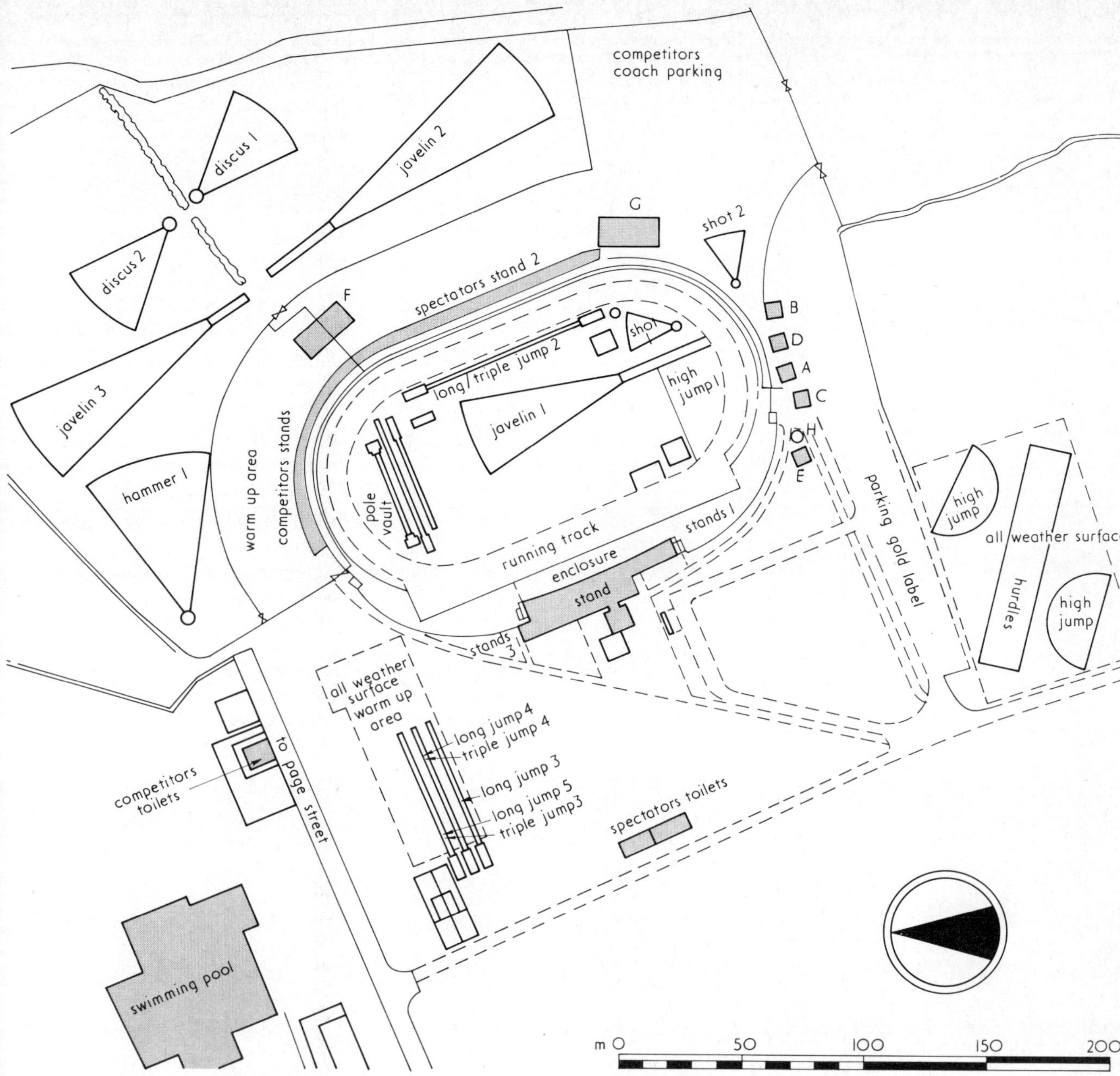

26.4 Plan of Copthall Sports Centre, courtesy of the London Borough of Barnet. This is an example of a good district athletics centre

A Transport office
B GPO telephones
C Police kiosk
D Billeting enquiries and lost property
E Box office
F Souvenir sales
G Public refreshments
H First aid

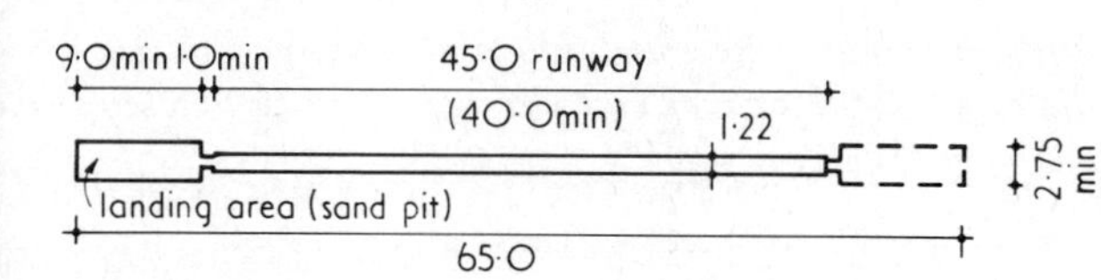

26.7 *Long jump: to avoid adverse wind conditions, landing areas are at both ends*

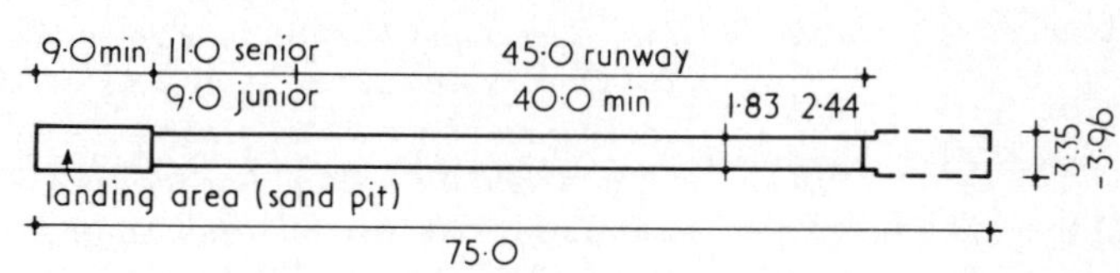

26.9 *Combined triple and long jump*

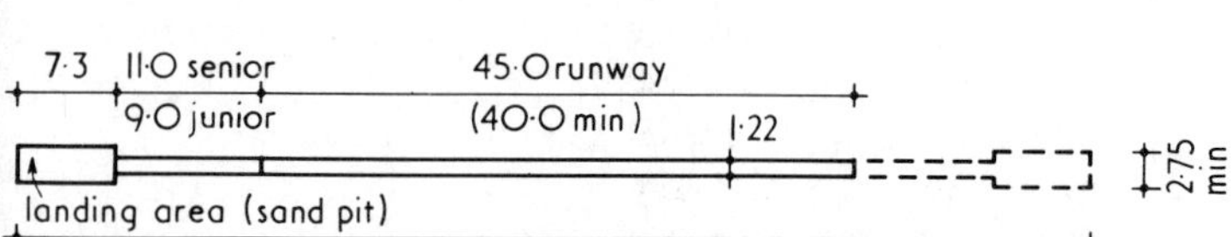

26.8 *Triple jump*

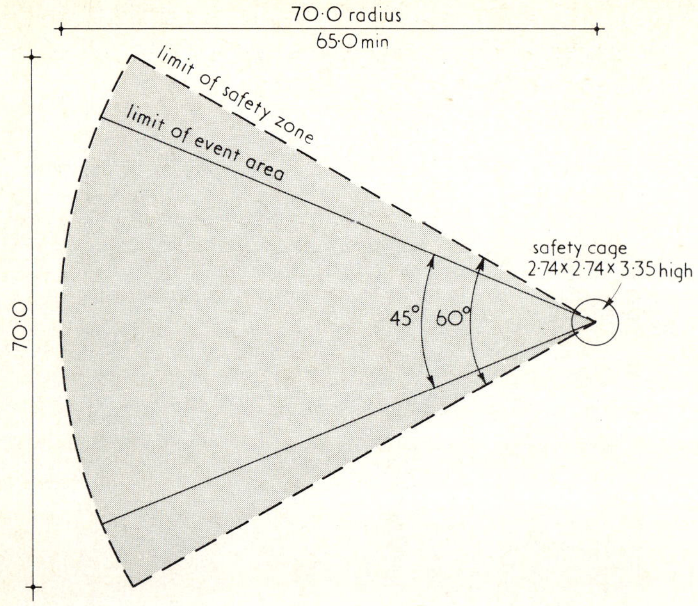

26.11 *Discus and hammer: discus base is 2·5 m dia, hammer base 2·135 m dia*

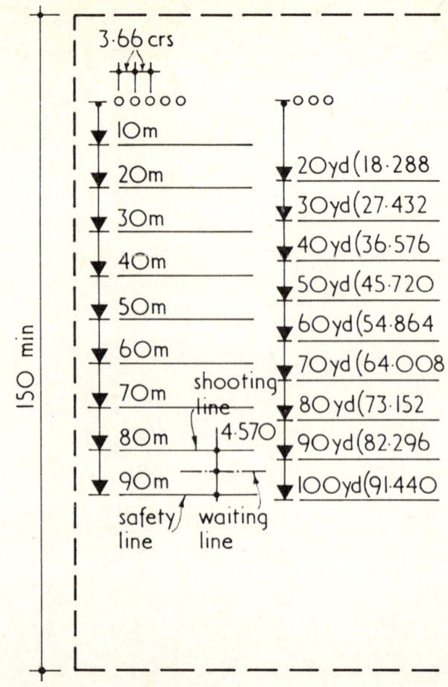

26.14 *Archery, target: club archery 100 m, championships over 150 m depending on the number of targets*
Some competitions are shot over metric distances and some are still and will always be shot over imperial lengths. Imperial and metric competitions can take place during the same meeting
The waiting and safety lines are moved to positions behind the correct shooting line for each competition. In this example, shooting is over a distance of 50 m

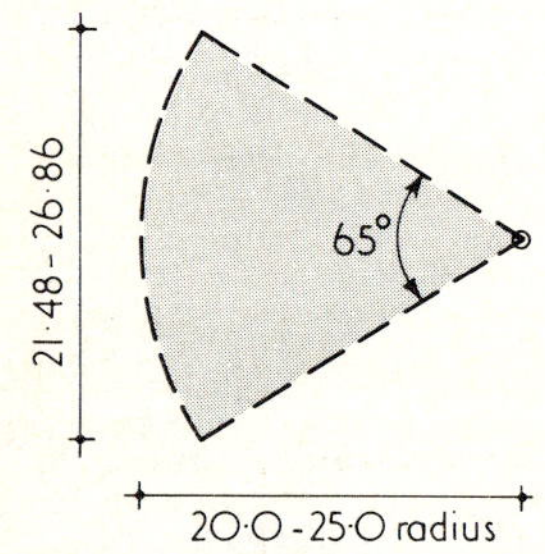

26.12 *Shot*

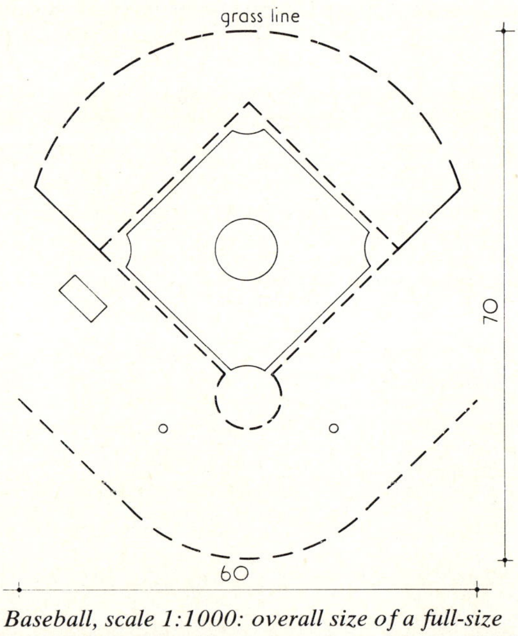

26.15 *Baseball, scale 1:1000: overall size of a full-size diamond. Little League (for young players) is two thirds the size*

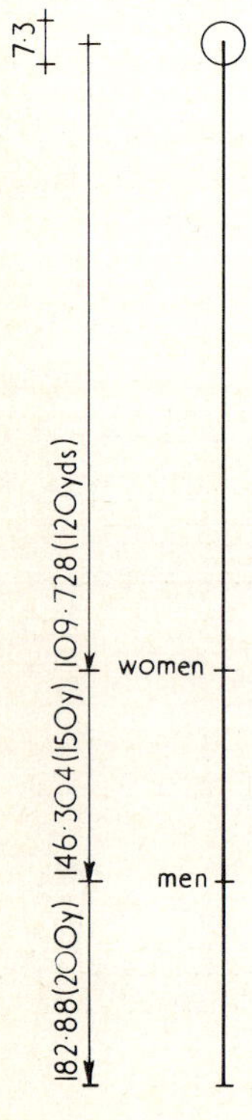

26.13 *Archery, clout. The arrows are shot high into the air to fall into circles marked on the ground, or a circular cloth pegged down, the centres of each being marked by a flag. The various shooting distances are clearly defined on the grass by white lines, tapes or spots and are always measured in yards. Archers move up and back to the distance position, and the waiting line moves accordingly. The overall distance for clout archery is about 230 m*

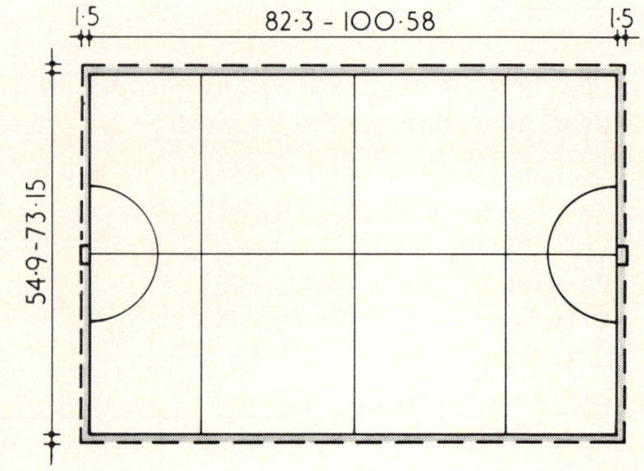

26.16 *Bicycle polo*

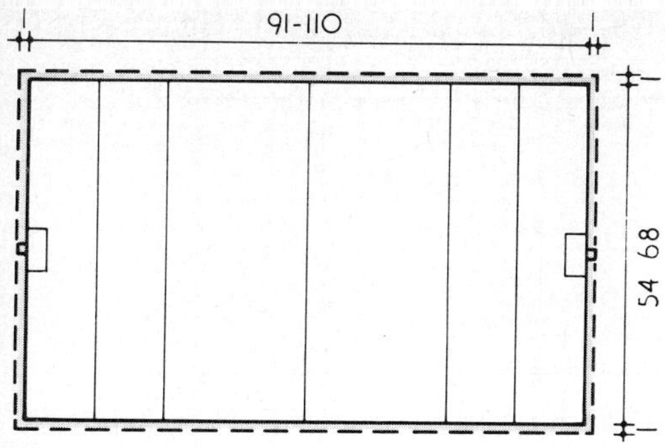

26.17 *Camogie*

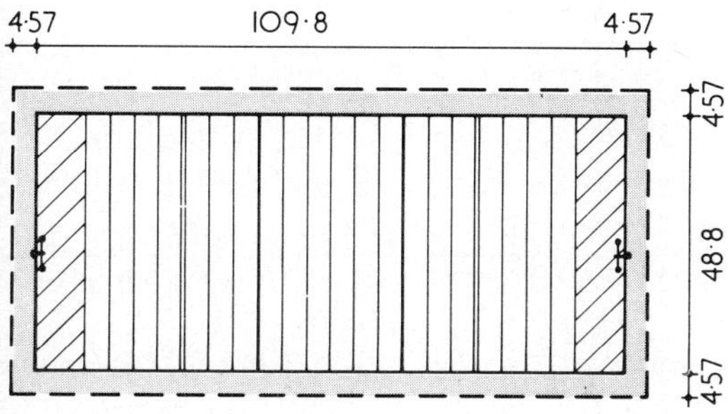

26.18 *Football, American*

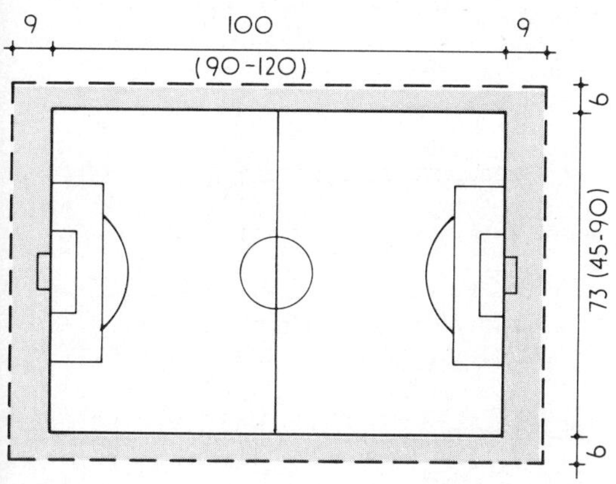

26.19 *Football, Association. The NPFA gives the following recommended sizes:*
senior pitches 96–100 × 60–64 m
junior pitches 90 × 46–55 m
international 100–110 × 64–75 m

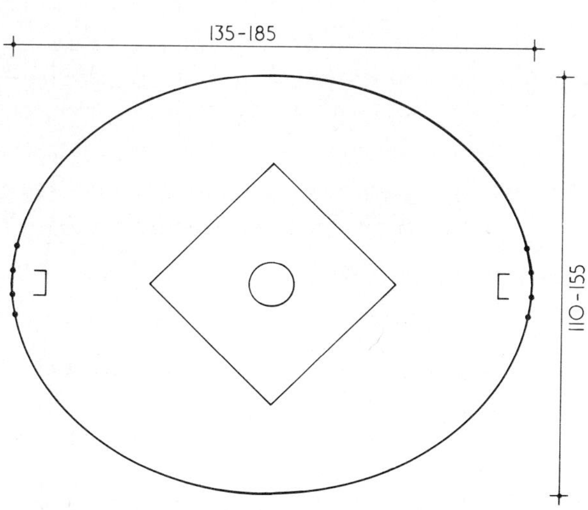

26.20 *Football, Australian*

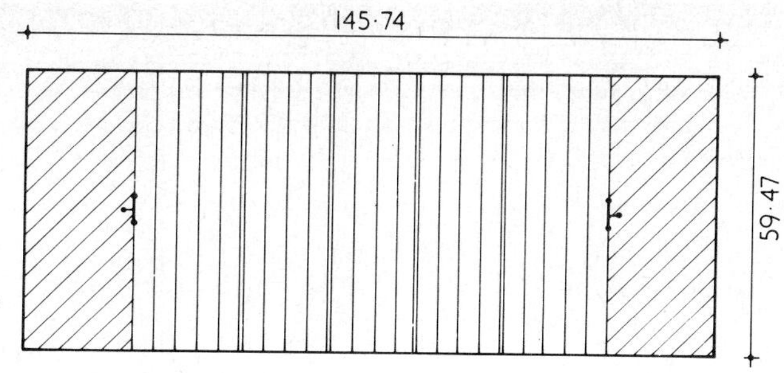

26.21 *Football, Canadian*

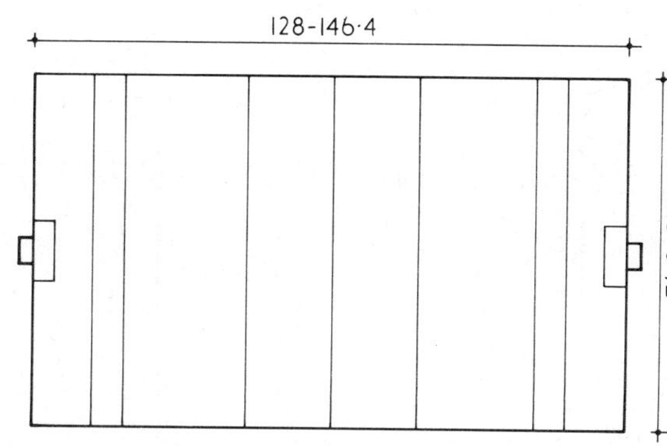

26.22 *Football, Gaelic*

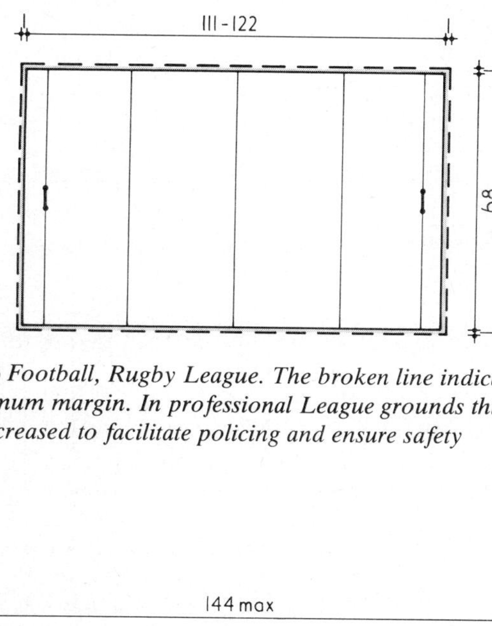

26.23 *Football, Rugby League. The broken line indicates 1 m minimum margin. In professional League grounds this may be increased to facilitate policing and ensure safety*

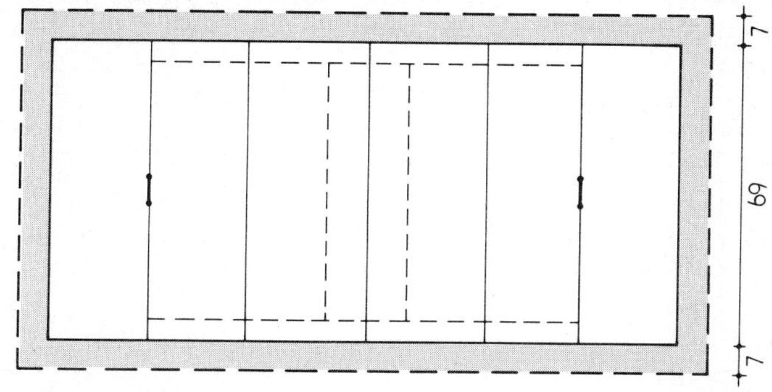

26.24 *Football, Rugby Union*

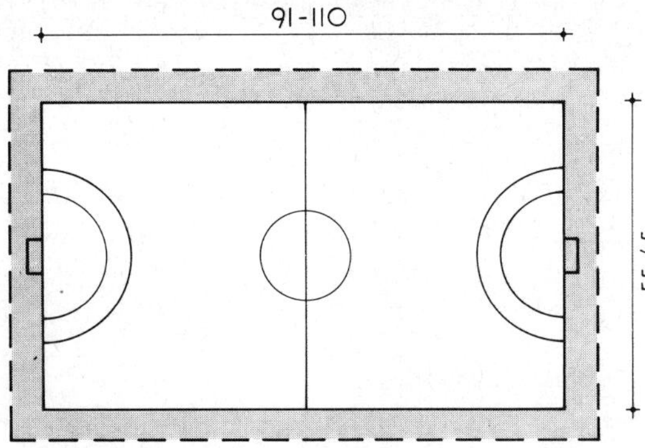

26.25 *Handball*

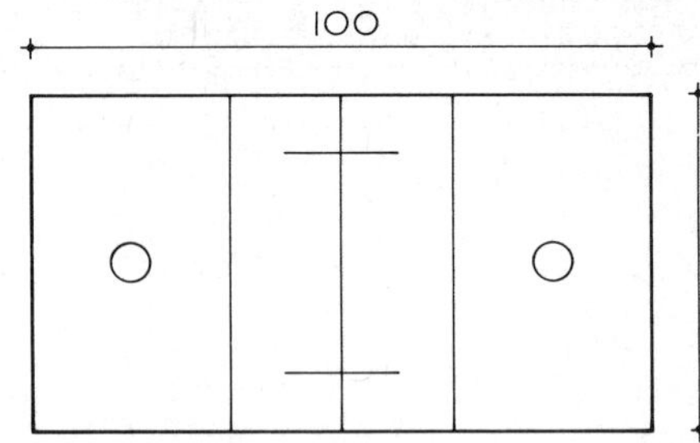

26.29 *Lacrosse, men's*

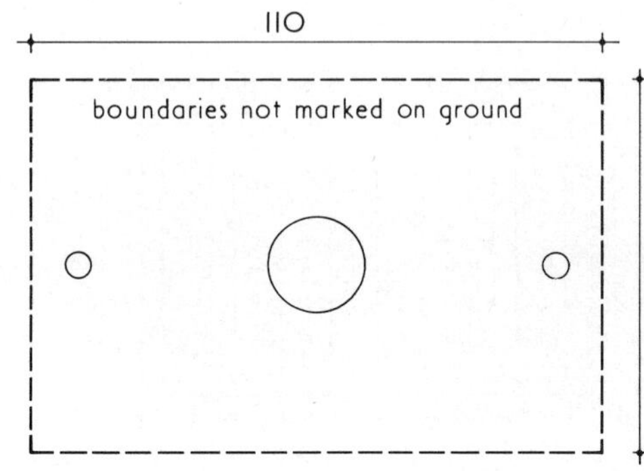

26.30 *Lacrosse, women's. The ground has no measured or marked out boundaries*

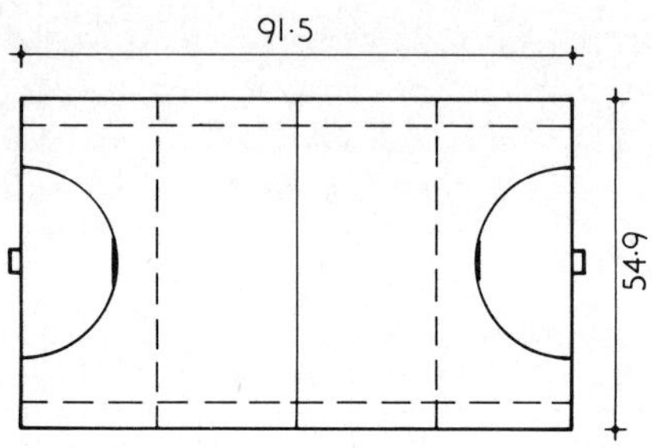

26.26 *Hockey. For county and club matches the NPFA gives a pitch size of 90 × 55 m in an overall space of 95 × 60.4 m, allowing for circulation about the pitch*

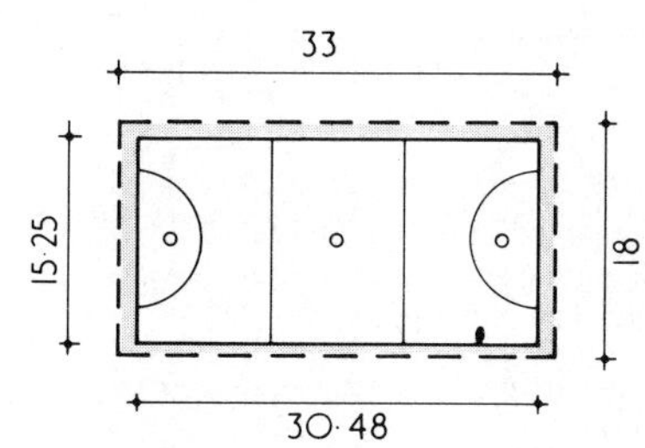

26.31 *Netball, scale 1:1000*

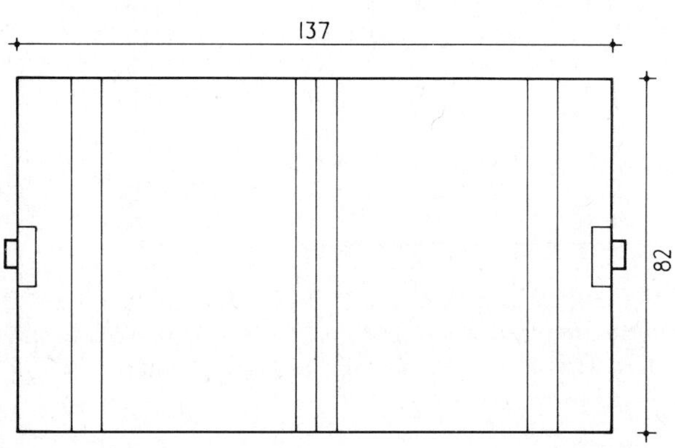

26.27 *Hurling*

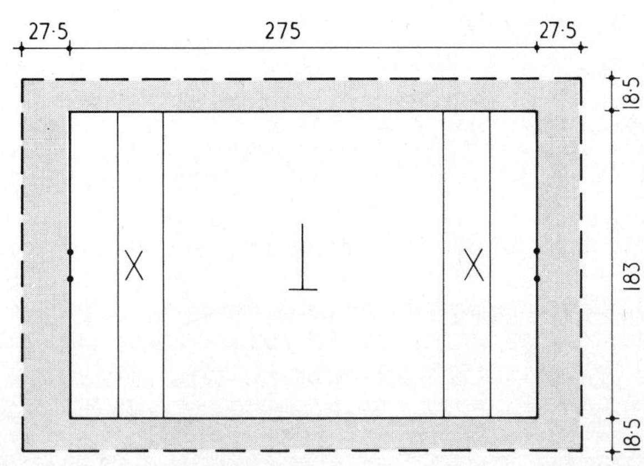

26.32 *Polo, scale 1:5000*

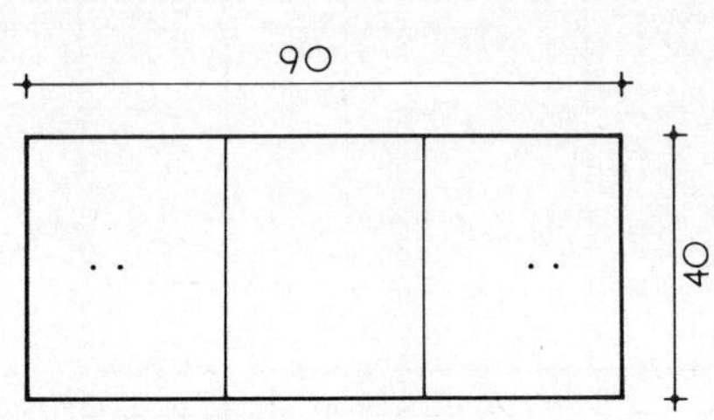

26.28 *Korfball*

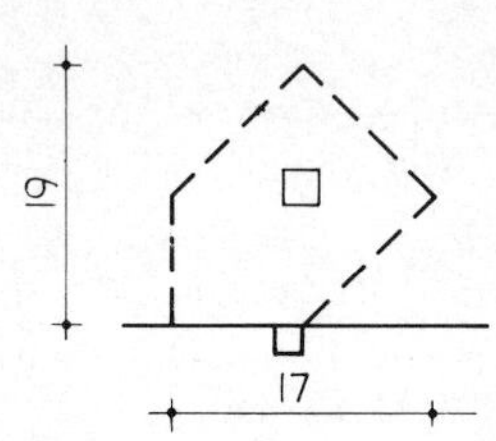

26.33 *Rounders, scale 1:1000*

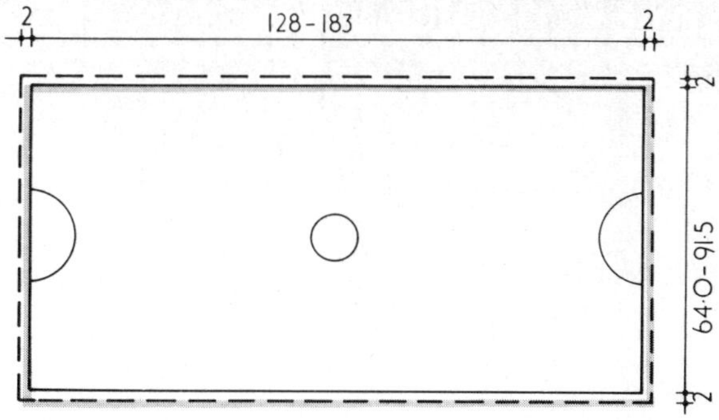

26.34 *Shinty*

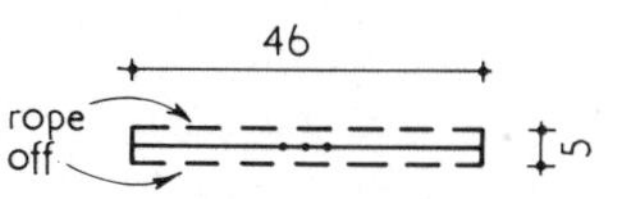

26.35 *Tug-of-war*

3 PLAYING FIELD SPORTS

3.01 Playing fields

Games and recreations that take place on ordinary playing fields are shown in **26.13** to **26.35**, scale 1:2000 except where shown otherwise

3.02 Sports grounds

For higher levels of competition in most sports, purpose-built facilities are usually provided. These incorporate special qualities of turf and its sub-grade, together with appropriate facilities for the players and for spectators.

In the simplest cases the spectators stand on the touchline or its equivalent. Where the standard is higher and the number to be accommodated greater, there are viewing slopes, terraces for standing and stands for sitting.

3.03 Viewing slopes

These are not suitable for large numbers, and should not be steeper than 17 per cent or 1:6.

3.04 Terraces

Plan and section of a terrace are shown in **26.36**. Barriers are provided at intervals as a protection against crowd surge; the spacings are given in table I. Gaps are provided in the barriers, but these should be staggered as shown on the plan. Gangways should be sunk 100 to 200 mm below the adjacent terrace to discourage standing in them, and radial gangways should be 'dog-legged' for the same reason. No point on a terrace should be more than 6 m from a gangway; the normal capacity is between 27 and 54 persons per 10 m^2. The front of the terrace should be no nearer the touchline than:

(1·75 ± H) cot A or 3 m, whichever is the greater, where
1·75 m is the height of an average male person
H m is the difference in level between the pitch and the bottom of the terrace, and
A is the angle the terrace makes to the horizontal.

3.05 Stands

The design of seating in stands is similar to that in theatres (section 23) and cinemas (section 24). The minimum area occupied by a seat is 460 mm wide × 610 mm deep, the preferred 550 × 760. There should be a minimum clearance of 305 mm between front and back of empty seats, although this is included in the above areas. The maximum run of seats with a gangway at each end is 28, half that if only at one end. No seat in a stand should be further than 30 m from an exit.

3.06 Exit from sports grounds

Large numbers of spectators in sports grounds are a source of danger to themselves, particularly from:

- tripping, slipping and falling
- crowd pressure on terraces and exits, and
- fire and hooliganism.

Generally, all spectators must be able to leave a sports ground within eight minutes. If there are combustible stands (such as constructed of timber) spectators must be able to be cleared from them within 2½ minutes.

The flow through an exit is about 40 persons per minute per unit of width of 550 mm. Where there are narrowings in the exit route there should be 'reservoir' areas to accommodate those that are waiting to pass. These should not be less than 15 m from an incombustible building, nor 45 m from one that is combustible, and should be designed to hold 54 persons per 10 m^2. Nowhere should an exit or escape route be less than

Table I Spacing of barriers on sports grounds (from Guide to Safety at Sports Grounds, Home Office 1973)

Angle of terrace			Peak areas of ground* A	Peak areas of ground* B	Other areas of ground A	Other areas of ground B
5°	8·8%	1:11.4	5·0	3·3	8·4	5·6
10°	17·5%	1:5.7	4·3	2·9	6·7	4·5
15°	27%	1:3.7	3·8	2·6	5·6	3·7
20°	37%	1:2.7	3·4	2·3	4·8	3·2
25°	47·6%	1:2.1	3·1	2·1	4·2	2·8
30°	58·8%	1:1.7	2·9	1·9	3·8	2·5

* Peak areas of ground are those where the crowd collects thickest, such as behind the goals in association football.
Type A barriers are tested for 6 kN/m loading, or designed for 5 kN/m.
Type B barriers are tested for 4·3 kN/m, or designed for 3·4 kN/m.
Barrier foundations are designed for a factor of safety against overturning of 2.

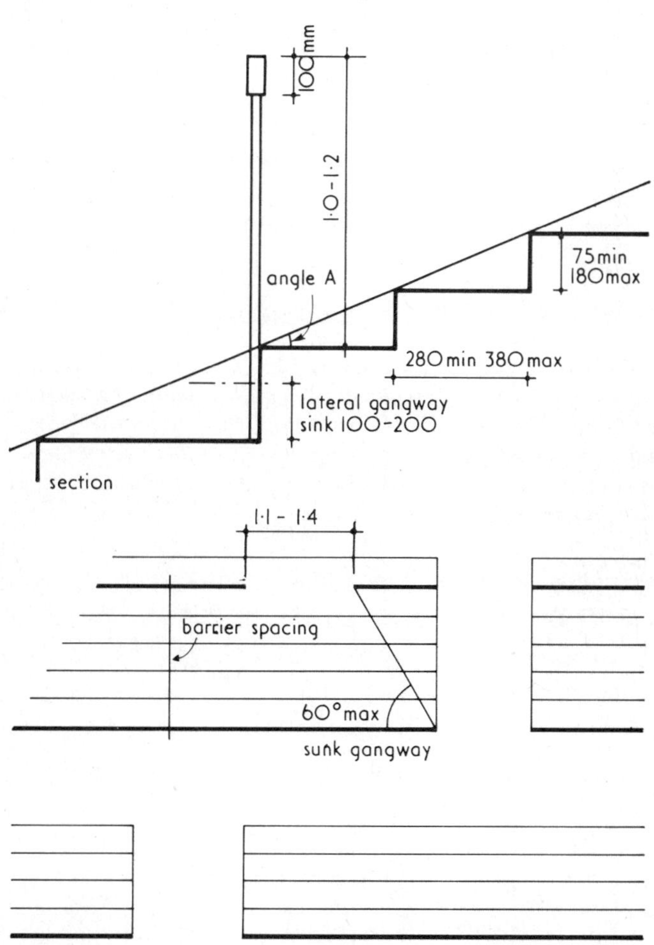

26.36 *Details of terraces for standing spectators*

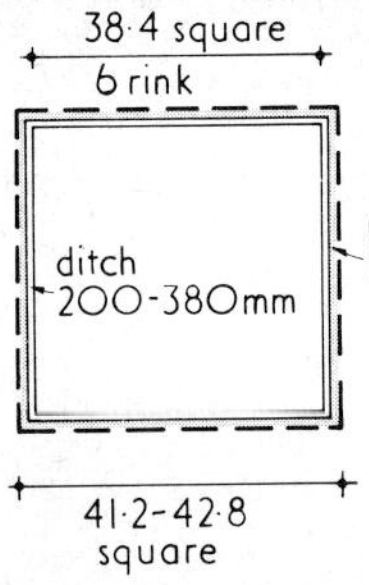

26.37 *Bowls. The green size may vary between 30·175 to 40·2 m. 38·4 m is recommended for six rinks. Rectangular greens are permissible with the short side minimum 30·175 m and the long side a maximum of 40·23 m*

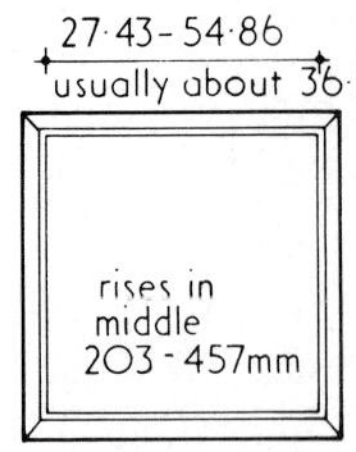

26.38 *Bowls, crown: a game played mainly in northern England and Wales, and in the Isle of Man. The centre of the green is from 0·2 to 0·46 m higher than the edges*

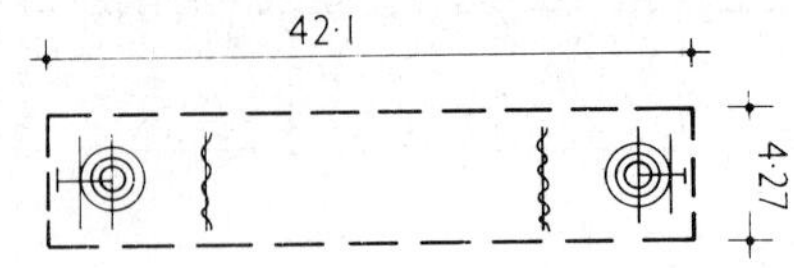

26.41 *Curling, scale 1:1000. This Scottish game is played on ice*

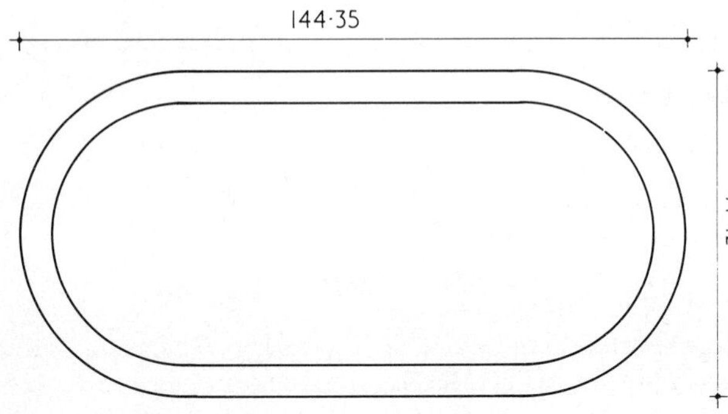

26.42 *Cycling 333⅓ m track*

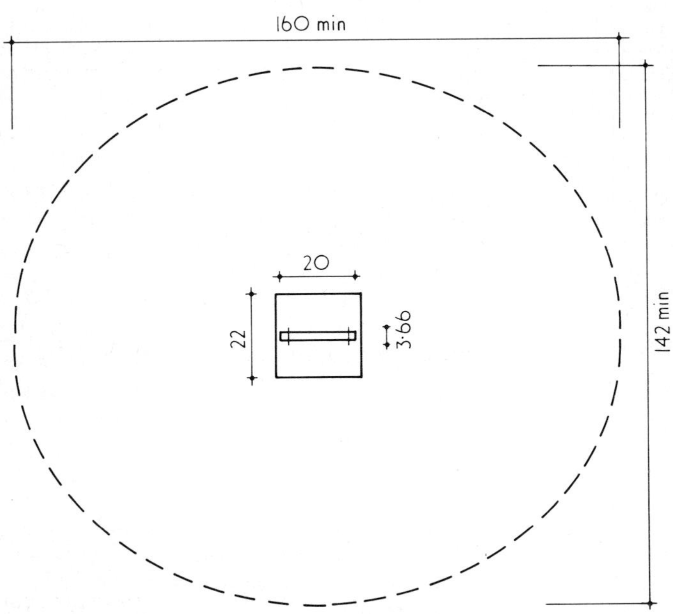

26.39 *Cricket. The centre square of about 22 m side is able to take the wickets in either direction. This is usually of special turf and grass species, and is roped off when not in use for playing. The outfield, on the other hand, is usually used for other field games (such as those in* **26.13** *to* **35**) *when cricket is not played*

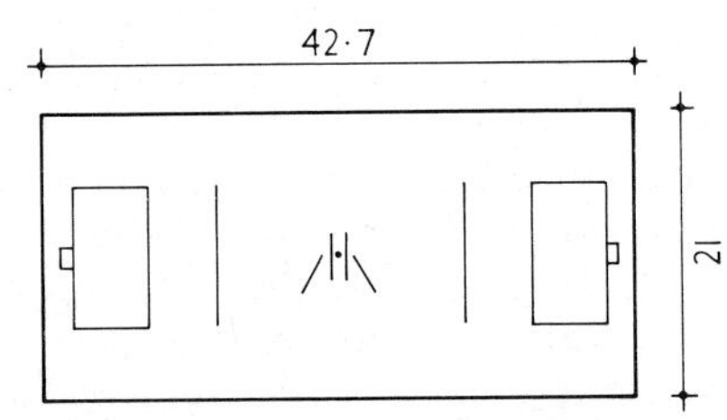

26.43 *Roller hockey, scale 1:1000. This demands a surface suitbale for roller skating: strip wood, terrazzo, smooth concrete or asphalt of the appropriate type*

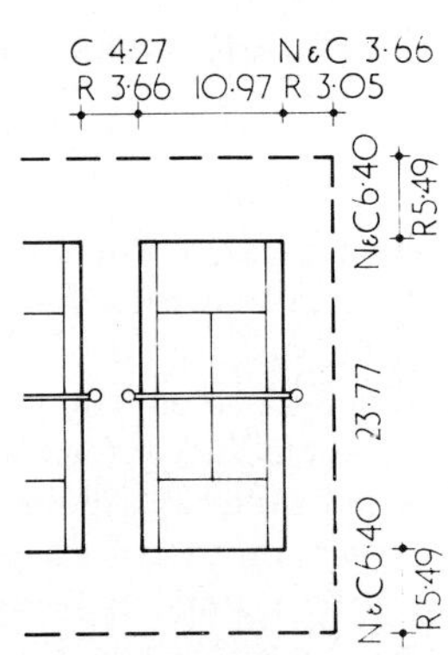

26.44 *Tennis, scale 1:1000. The surfaces of tennis courts may be grass, suitable asphalt or 'en-tout-cas'. The surrounds are of wire netting 3 to 4 m high*

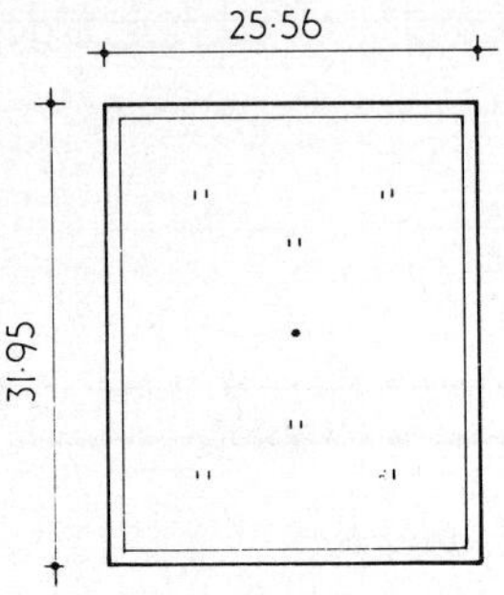

26.40 *Croquet. This can be played on an ordinary field, but the good game demands a turf similar to that of a bowls green, scale 1:1000*

1·1 m wide, minimum headroom 2·4 m. Steps should be a minimum of 280 mm going (305 preferred), and 190 rise (150 preferred). No flight should have less than 3 or more than 16 risers, and two flights with more than 12 risers should have a turn between. Ramps should not be steeper than 10 per cent (1:10).

4 SPORTS REQUIRING SPECIAL CONDITIONS OR CONSTRUCTION

A selection of special constructions are given in **26.37** to **26.44**, scale 1:2000 except where shown otherwise.

BIBLIOGRAPHICAL NOTE

Forthcoming revised edition of AAA/NPFA *Facilities for Athetics* containing technical requirements and specifications for all grades of outdoor athletics.

27 Sports facilities: indoors

Peter Ackroyd

CI/SfB 562
CI/SfB (1976 revised) 562

Contents

1 INTRODUCTION

Indoor sporting activity can be competitive, recreational or for training purposes. Facilities are designed to cater for all three, and are either general purpose spaces such as sports halls, or special to one activity or range of activities, such as a squash court or ice rink. The different sports and activities will be found in alphabetical order in paras **3**, **4**, **5** and **6**: whichever is appropriate. However, swimming and other water activities are covered in section 25, Water sports and recreation. For riding, see section 28, Stables.

In this section the information given about each activity will generally be confined to the required overall sizes at the various recognised levels:

N—international and national competition,
C—county and club competition, and
R—recreational

For further information such as detailed dimensions, equipment, environmental installations etc, refer to the *Handbook of sports and recreational building design*.

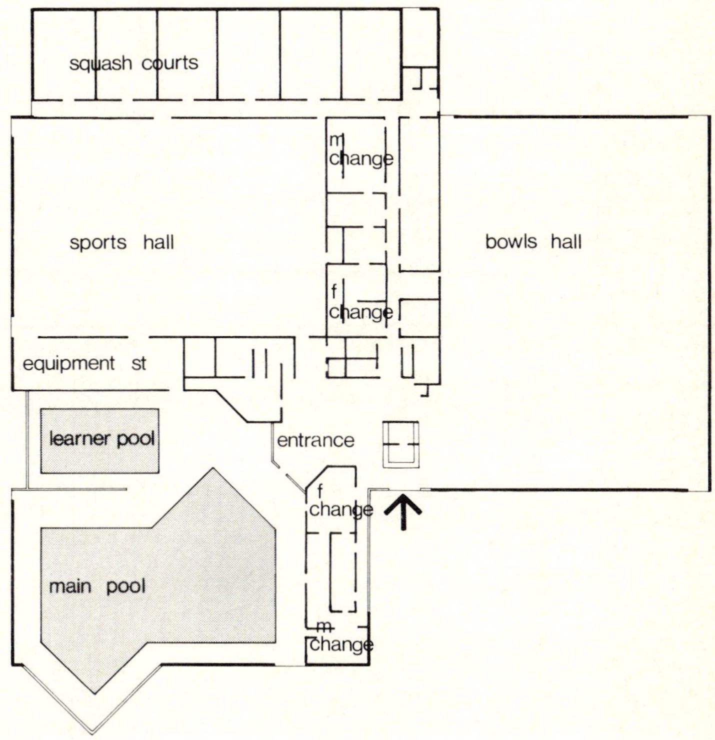

27.2 *Dunstable leisure centre: a leisure pool and dry facilities on a school site*

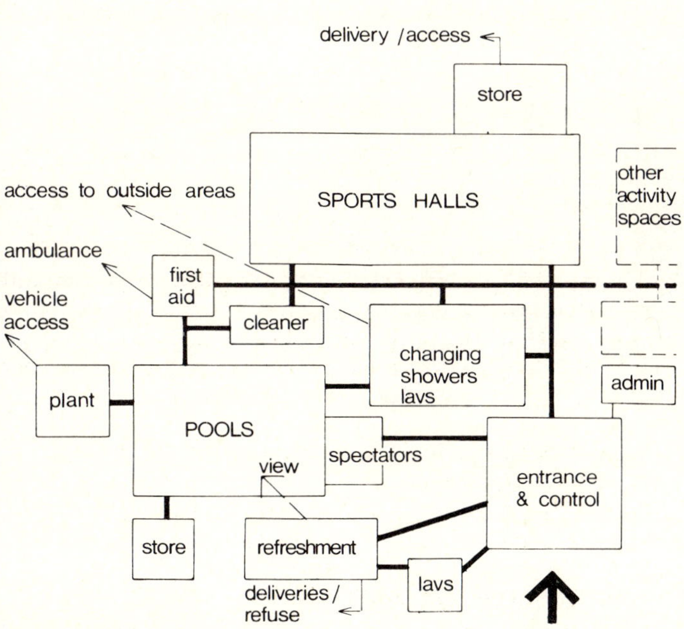

27.1 *Pattern of spaces and circulation in a large wet and dry sports centre*

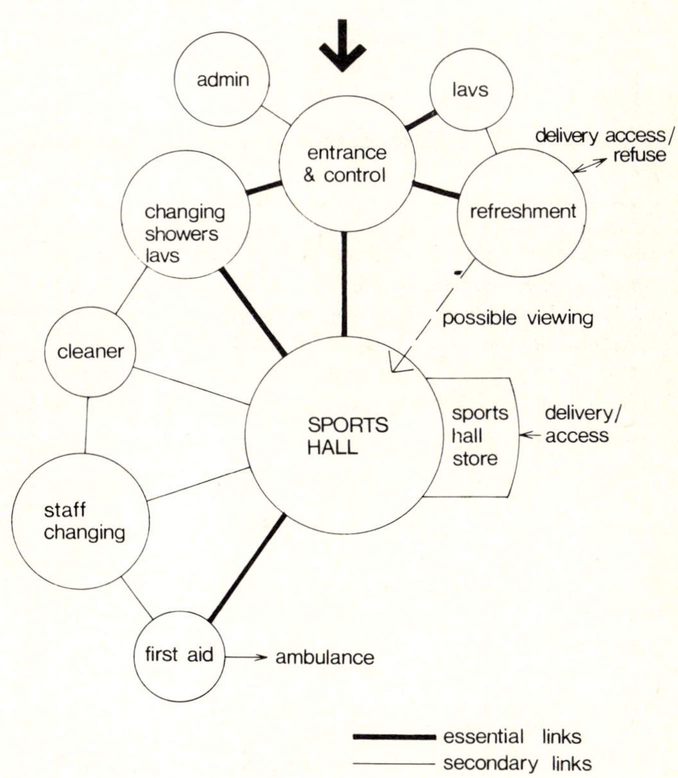

27.3 *The main elements of a dry sports centre*

2 SPORTS CENTRES

Some sports centres are large complexes encompassing *wet and dry sports*. Figure **27.1** shows the possible elements of such a complex, some of which are omitted in smaller centres.

Figure **27.2** is a plan of a large centre.

The essential elements of a small dry sports centre are shown in **27.3**, and a plan of a centre in **27.4**.

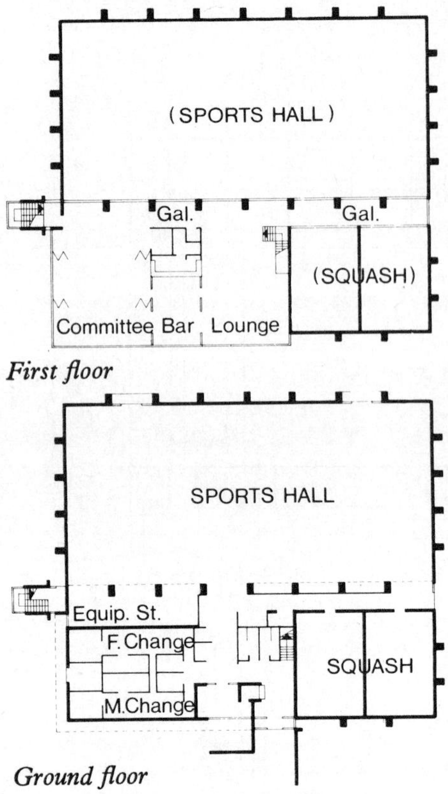

27.4 *Harpenden, a small compactly-designed centre. The social areas have been positioned to take advantage of the parkland site*

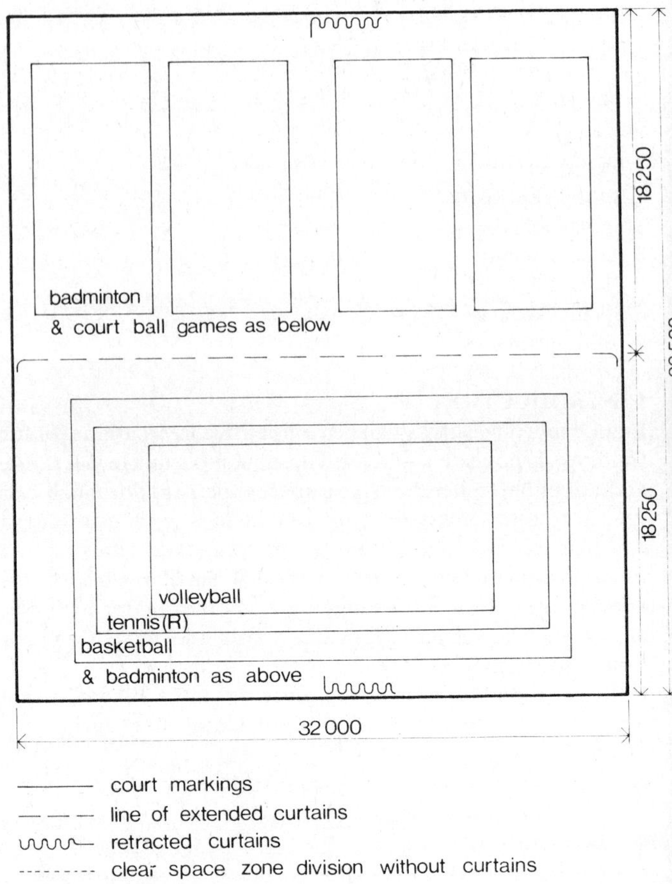

27.6

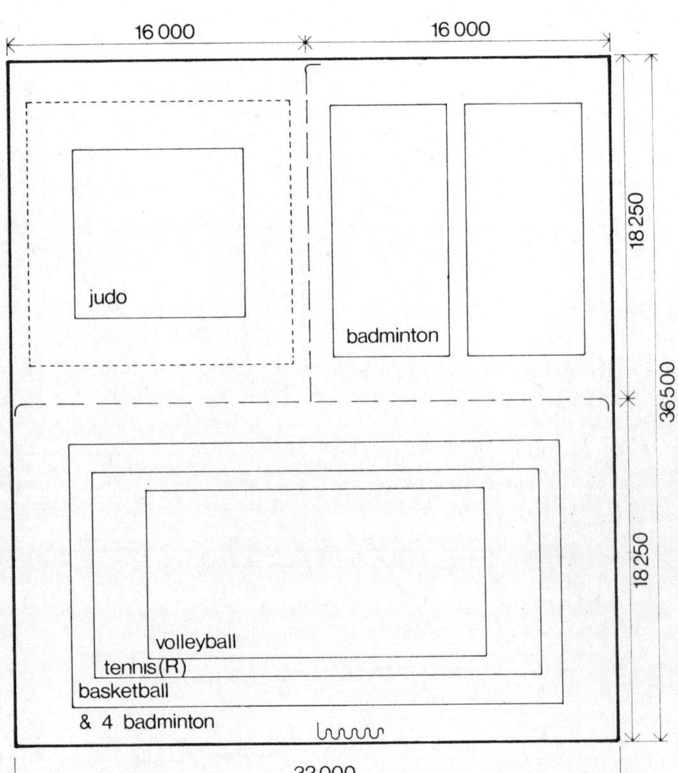

27.5–7 *Alternative court arrangements for large sports halls*

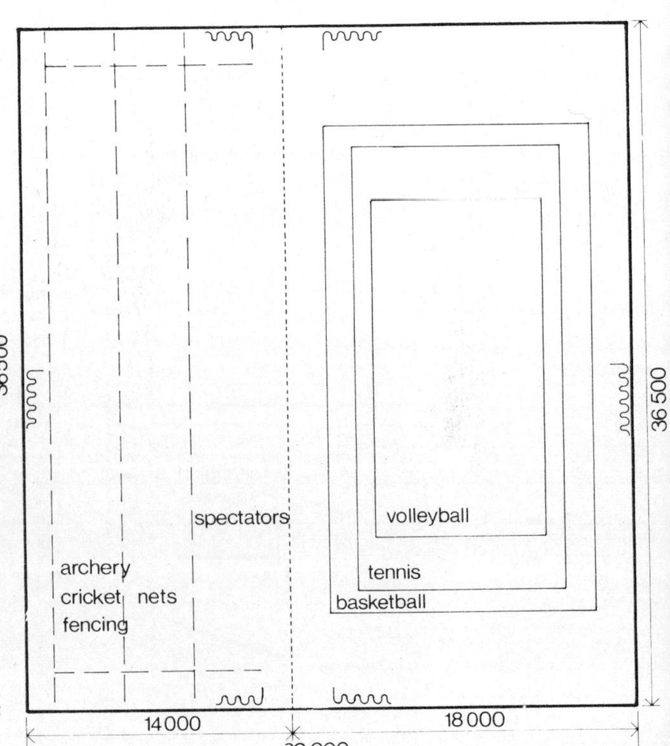

27.7

3 SPORTS HALLS

3.01 Use of facilities

Sports halls are general purpose spaces intended to cater for a great variety of activities. Some of these can take place simultaneously, but others need exclusive use for a time. In general all the activities in paras **3**, **4** and **5** and even some of those in para **6** can take place in a suitable sports hall. However, the demand for time in sports halls is so great that those activities that can be carried on in less expensive accommodation tend to be confined to projectile halls and ancillary halls. In this section information about the activities will be found under the most appropriate space.

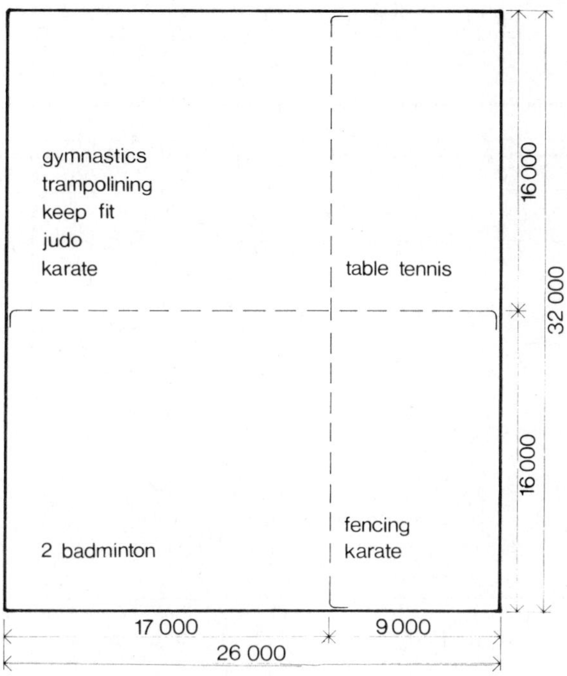

27.8–13 *Alternatives for medium-size halls*

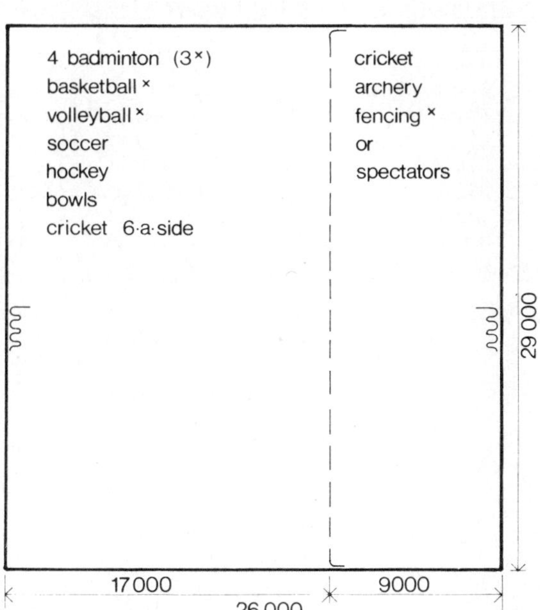

27.9

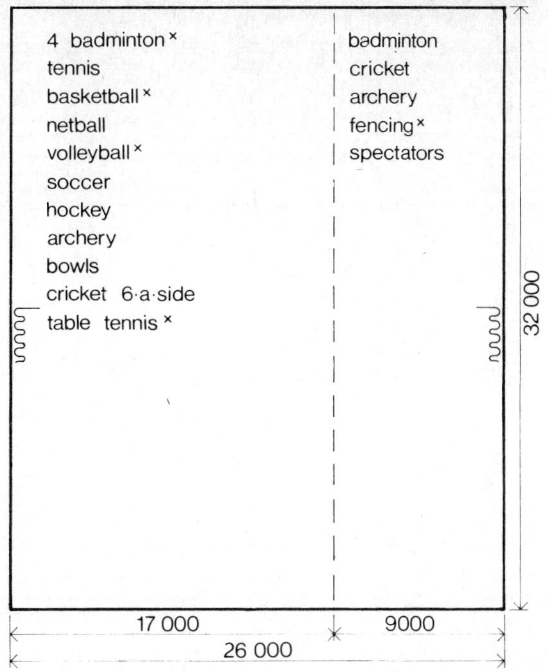

× to N & C standard in given location

27.10

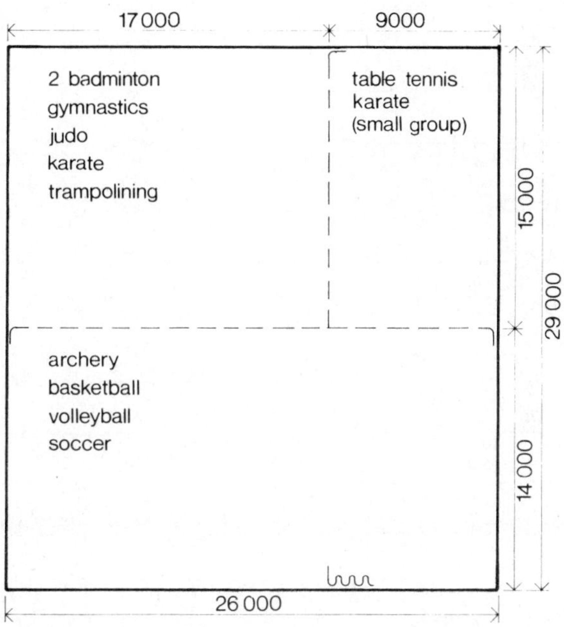

27.11

3.02 Sizes

Only the largest of halls will satisfy all required standards of play for all indoor sports, and therefore it will be necessary to decide on upon the range of sports and levels before determining the floor area. Table I shows what can be accommodated in the various standard sizes of hall.

The same floor area may provide for international standard in one or two sports and at the same time offer a wide variety of other activities at a lower standard. Typical arrangements are shown in **27.5** to **27.18**.

3.03 Height

The height of the underside of the roof structure, or the ceiling if there is one, above the floor is specified by each sport's governing body, and this is a critical design factor. Badminton, tennis and trampolining require an unrestricted height of

Table I Definition of sizes: maximum number of courts related to standards of play

	Large hall[fr]				**Medium halls[fr]**						**Small halls**								**Community halls**	
	36·5 × 32 × 9·1 m 1168 m^2		32 × 26 × 7·6–9·1 m 832 m^2		29 × 26 × 7·6–9·1 m 754 m^2		32 × 23 × 7·6–9·1 m 736 m^2		32 × 17 × 6·7–7·6 m 544 m^2		29·5 ×16·5 6·7–7·6 m 486·75 m^2		26× 16·5 × 6·7–7·6 m 429 m^2		22·5 × 16·5 6·7–7·6 m 371·25 m^2		17·0–20·0 × 15·6 × 6·7 m 265·2–321m^2		17·0 – 8·5 6·7 m 144·5 m^2	
	No.	Standard	No.	Standard	No.	Standard	No.	Standard	No.	Standard	No.	Standard	No.	Standard	No.	Standard	No.	Standard	No.	Standard
Aikido	4 6	N C	4	N	4	N	2 +3(*)	C R	2 3(1*)	N R	2	N	2	N	1 2	N R	1	N	–	–
Archery (length of shoot)		[S]30 m 25 m 18 m 20 yd		[S]25 m 18 m 20 yd		18 m 20 yd	[S]25 m	18 m 20 yd		[S]25 m 18 m 20 yd		18 m 20 yd		18 m 15 yd		18 m 15 yd	–	–	–	–
Badminton	8	N	5 6(2*)	N† R	3/4 4	NC R*	4 6	N[1] R	4	C**	3 4	C** R*	3	C**	3	R	2	R[1]	1	R[1]
Basketball	2	N	1 2	N C*/R	1 2	N R*	1	N	1	C**	1	C**	1	C**	1 1	R* Mini BB	– 1	– Mini BB	– –	– –
Bowls (portable non-competitive rinks)	7	R	5	R	5	R*	4	R	3	R	3	R*	–	–	–	–	–	–	–	–
Boxing (training rings)	9 12	N R	6 12	N R	4 9	N R	6 8	N R	3 6	C R	3 5	C R	2 5	C R	2 4	C R	2 4	C R	2 –	R –
Cricket six-a-side pitches ns	1 2	N C	1	C	–	–	1	C	1	R	–	–	–	–	–	–	–	–	–	–
Cricket nets	8	N	6	N	6	C	5	N	4	C	4	C	4	R	–	–	–	–	–	–
Fencing (pistes)	12 14	N C	8(3*) 9	N C	7 8	N C	6 8	N C	3/4 2/3	N/C R*	3/4* +2	N/C R*	3/4* +1	N/C R	3 4	N C*	3	C	2	R
Five-a-side football	1 2	N R*	1 2	C R*	1	R*	1	C	1	R*	1	R*	1	R*	1	R*	1	R*	–	–
Gymnastics (Olympic)	–	N	–	C	–	P	–	C	–	P	–	P	–	P	–	P	–	P	–	–
Handball. Mini handball	1	N*	1	C	1	R*	1	C	1	R	1 1	R* C	– 1	– C	– 1	– C	– 1	– R*	– 1	– R*
Hockey	1	C*	1	R	1	R	1	R	1	R	1	R	1	R	1	R	–	–	–	–
Judo	4 6	N R	2 4	N C	1 4	N C	2 4	N R	2 3	N R	1 2	N C	1 2	N R	1 2	N R	1/2 –	R –	– –	– –
Karate	4 12	N R	2/4 6	N/C R	2 4/6	N* C/R	2 6	N R	2 6	N R*	2 3	N*/C R	1/2 3	N/C R	1 2	N R	1 2	N R	2 –	R* –
Keep fit; Movement and dance; Yoga. ns		✓		✓		✓		✓		✓		✓		✓		✓		✓		✓
Kendo	4 6	N R*	2 4	N C	2 4	N* C	2 4	N R	2	N	2 2	N* C	1 2	N C	1 2	N R*	1 –	R –	– –	– –
Lacrosse F	1	N	1	C*	1	R	1	C*	1	C*	1	R	–	P	–	P	–	–	–	–
Lawn tennis	1 2	N* R	1	R*	–	–	1	R*	1	R*	–	–	–	–	–	–	–	–	–	–
Micro korfball	1	C	1	C	1	C	1	C	1	R*	–	–	–	–	–	–	–	–	–	–
Netball	1 2	N C*/R	1	R	–	–	1	R	1	R	–	–	–	–	–	–	–	–	–	–
Table tennisc/c	10 15/21	N C/C	6 10/15	N C/C	6 10/12	N C/C	6 10/12	N C/C	7/9 14	C/C R	7 12	C/C R	6/7 10	C/C R	4 8	C/C R	3–6 6–8	C/C R	4 –	R* –
Trampolining	12	N	8 12	N† R	8	N†	4 8	N† C*/R	4 6	C** R	4	C**	4	C**	4	R	2	R	1	R
Tug of war	–	N	–	C	–	R	–	C	–	C	–	R	–	–	–	–	–	–	–	–
Volleyball	2 3	N R	1 2 3	N† C R*	1 2	N† R	2* 2	N†/C R	1	C**	1	C**	1	C**	1	C**	1	R*	–	–
Weight lifting contests	–	N	–	N	–	N	–	N	–	C	–	C	–	C	–	C	–	C	–	–
Wrestling	4 12	N C	2 6	N C	6	C	2 6	N C	2 3	N* C	3 8	C R*	2 6	C R	2 6	C R	2 4	C R	2 –	R –

Key
N National/international standard
C County/club standard
R Recreational standard
P Practice area only

c/c For table tennis there are two grades of minimum space allowances for inter-county/inter-club standards of play
fr Fire regulations and maximum compartment volumes should be checked. Halls of 7000 m^3 or over need a DOE waiver. 'Volume' can include an unenclosed structural roof spaces
ns No standards have yet been laid down
S Area behind shooting line is below safety standard recommended. Acceptable space can be provided with a slight lengthening of the hall; or existing spaces may be used for practice purposes
* Below minimum space standard recommended by the governing body concerned, but capable of providing purposeful and enjoyable activity.
** Recreational standard where the hall is less than 7·6 m clear height for badminton and trampolining, or less than 7·0 m for basketball and volleyball 6·7 m height is suitable for mini basketball and mini volleyball
† County/club standard where the hall is less than 9·0 m clear height

9·1 m for international competition, while 7·6 is necessary at C level in all sports except those for which height is not critical. However, a height greater than justified by the intended use will increase running costs in heating, lighting and maintenance.

3.04 Construction

The construction and fabric of the hall should be such as to minimise damage, both accidental and from vandalism. Sports halls should only be naturally lit from above; any form of vertical glazing will produce some glare.

3.05 Activities

The sizes required for various activities in the sports hall are shown in **27.19** to **27.32**, scale 1:500.

4 ANCILLARY HALLS

To economise in the use of the large sports halls, larger centres have practice halls suitable for some smaller-scale activities. The two suggested sizes are:

- 15 × 12 × 3·5 to 4·5 m
- 21—24 × 12 × 4·5 m with a divider.

Sizes for various activities in this type of hall are given in **27.33** to **27.40**, scale 1:500. For yoga, each person will lie on the floor on a mat or blanket and will ideally need a clear area of 2·5 m diameter.

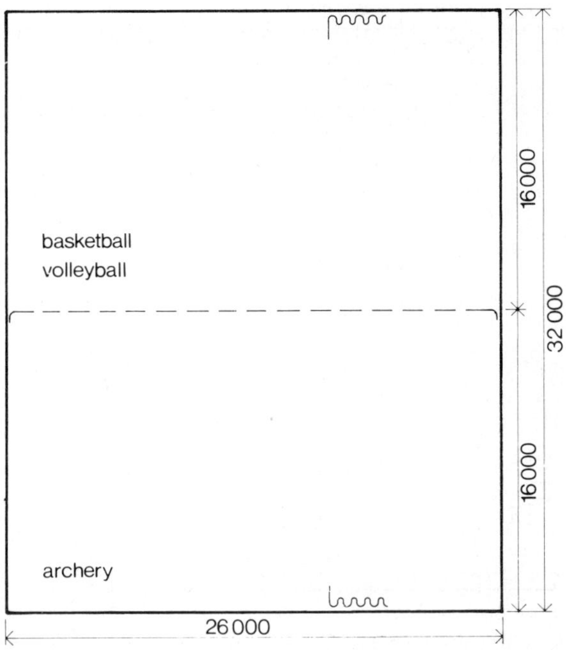

27.12

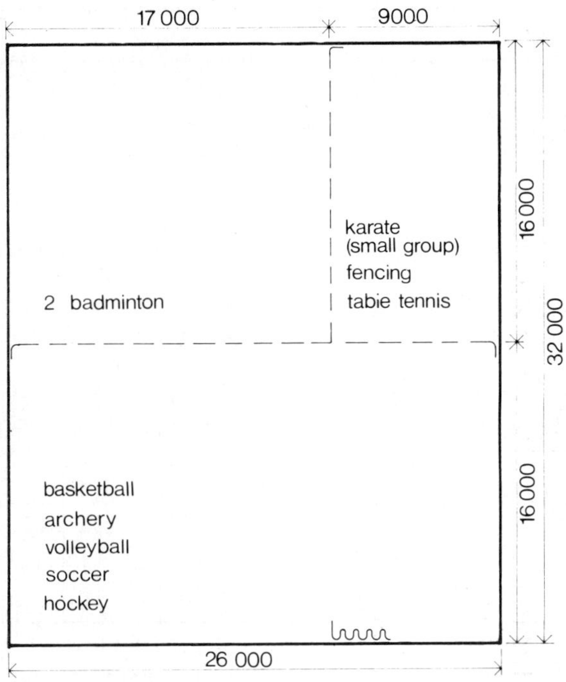

27.13

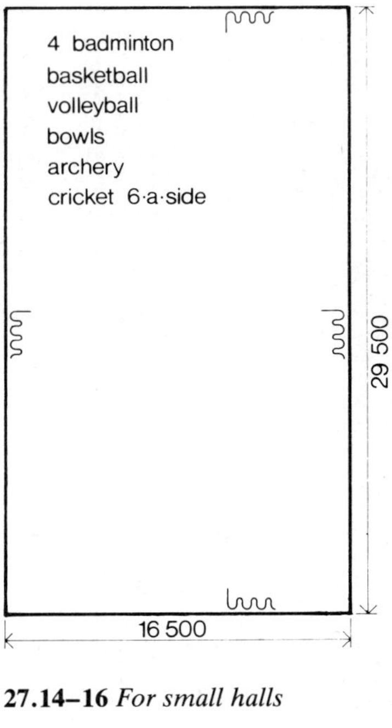

27.14–16 *For small halls*

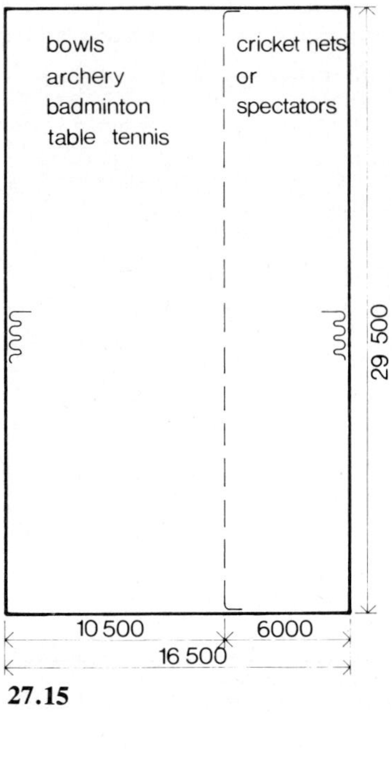

27.15

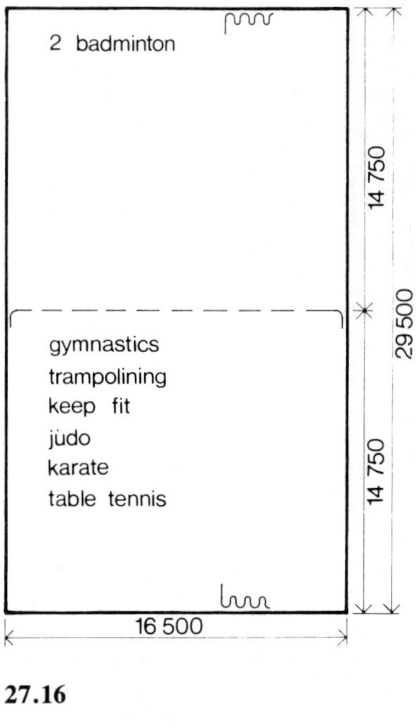

27.16

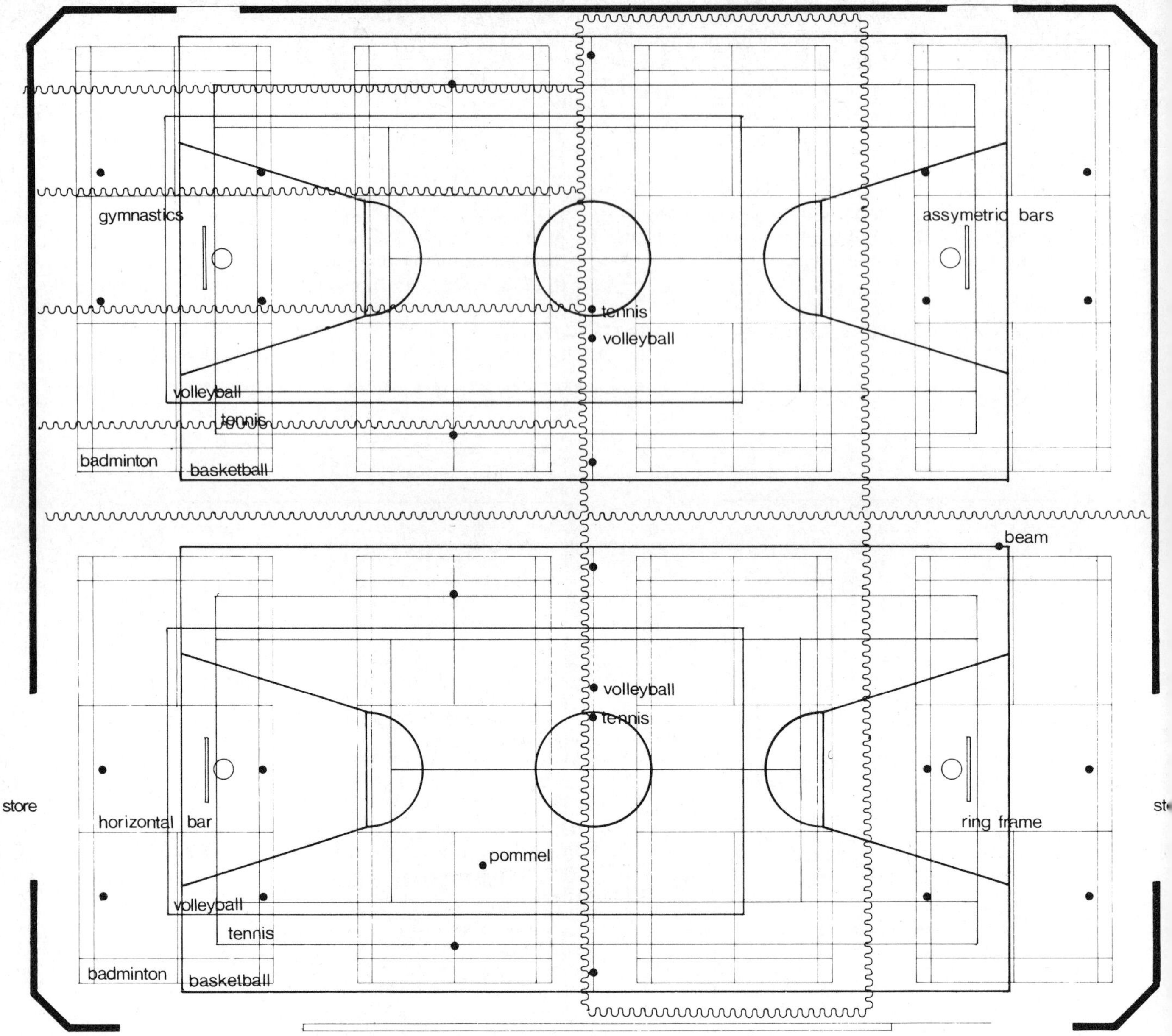

27.17 *Wycombe sports centre: plan of court markings and equipment fixings in sports hall*

27.18 *(below) Tamworth sports centre: plan of court markings and equipment fixings*

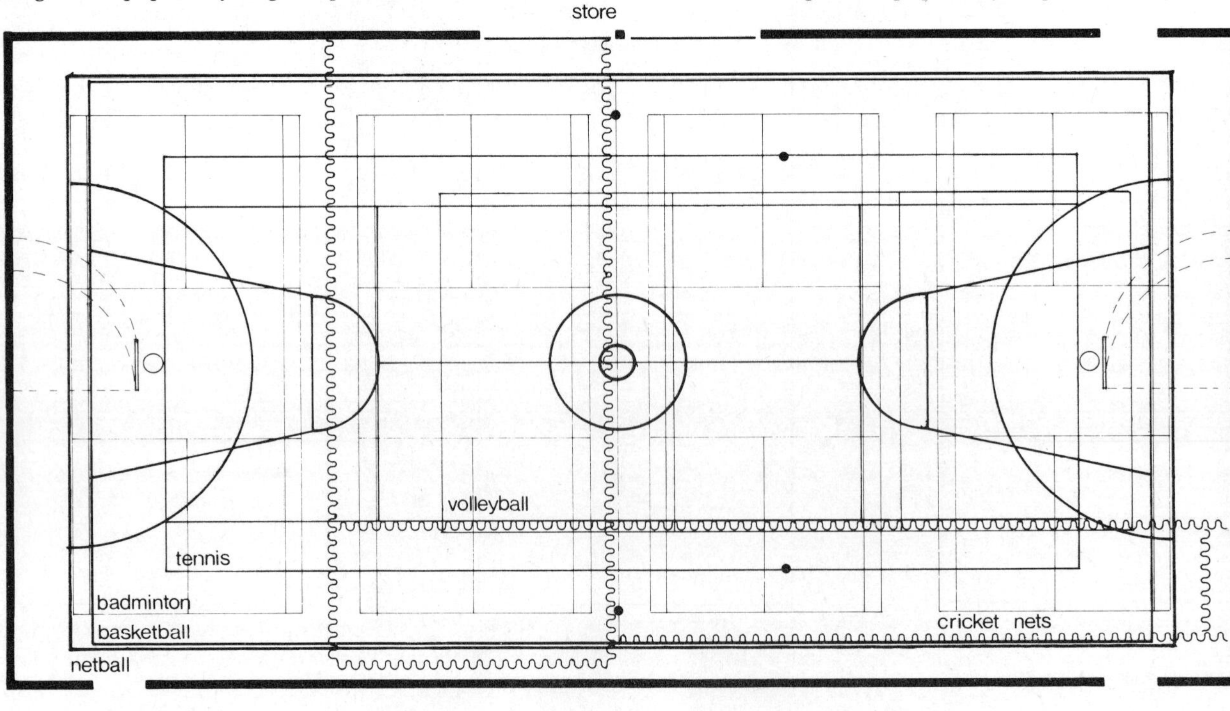

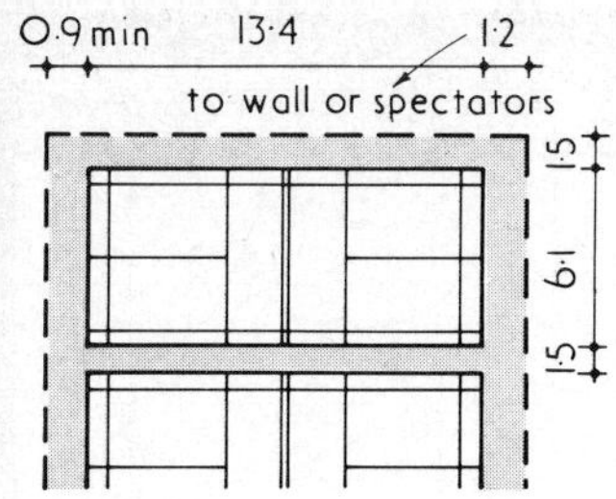

27.19 *Badminton, a doubles court for all standards of play. Where courts are placed side by side, tournaments are held with seating and play on alternate courts. Heights lower than 7·6 m are discouraged by the Badminton Association of England*

height
N 9·1
C 7·6
R 6·7–7·6

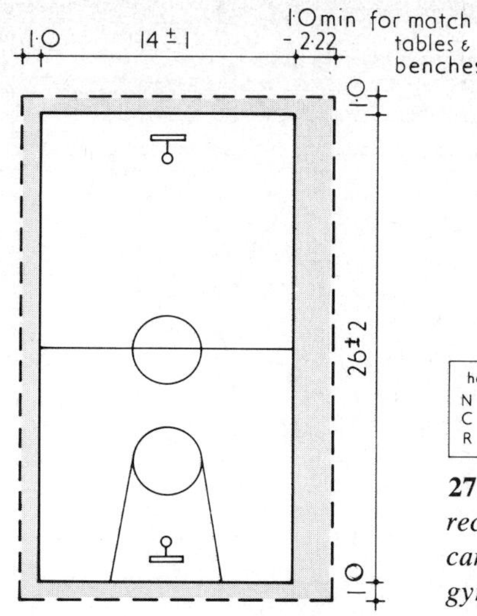

height
N 7·0
C 7·0
R 6·7

27.20 *Basket ball. At a recreational level this game can be played in a school gymnasium 21·3 × 12·2 m*

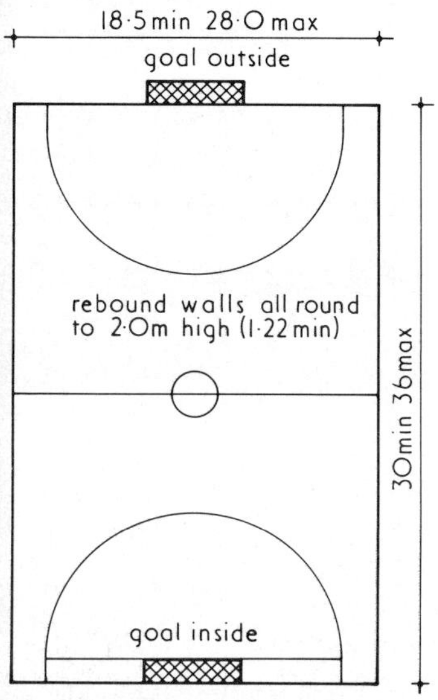

27.21 *Five-a-side football. This game needs rebound walls all round to about 2 m height, but can be adapted to the size of the available space. In a medium-size sports hall,* **27.10**, *the playing area is the floor of the hall. At a recreational level the game may be played in a small-size hall, about 30 × 15 m being regarded as a reasonable minimum. Depending on the age and sizes of players, their numbers on the pitch could be reduced as necessary for satisfaction*

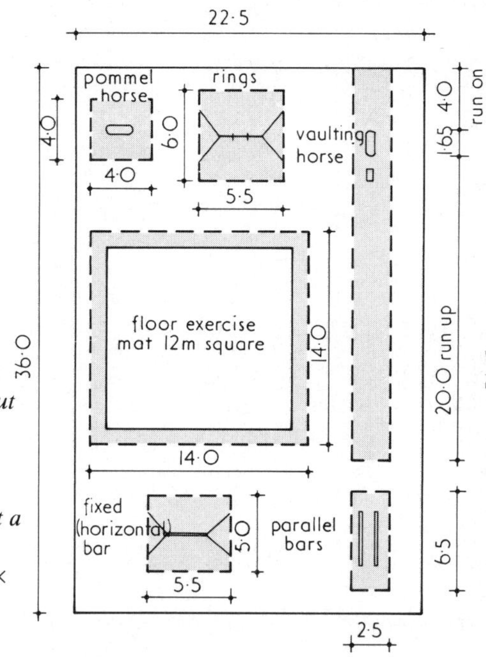

27.22 *Men's gymnastics. See* **27.55** *for special practice spaces*

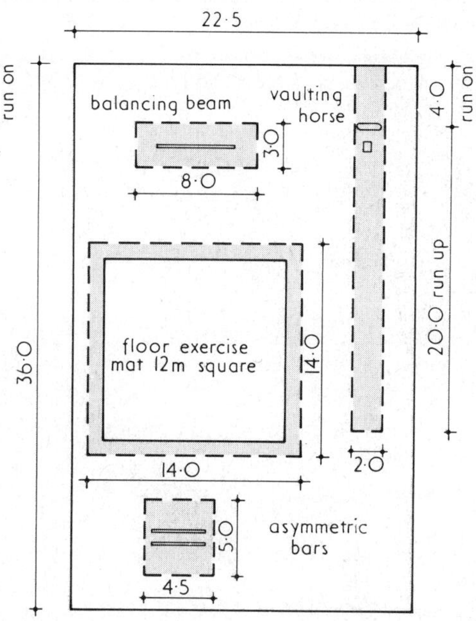

27.23 *Women's gymnastics. See* **27.55**

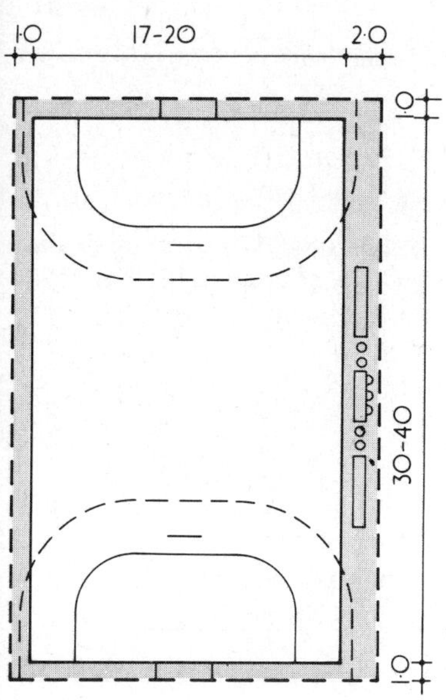

height
N 9·0
C 7·6–9·0
R 7·6

27.24 *Handball, seven-a-side*

height
7·6

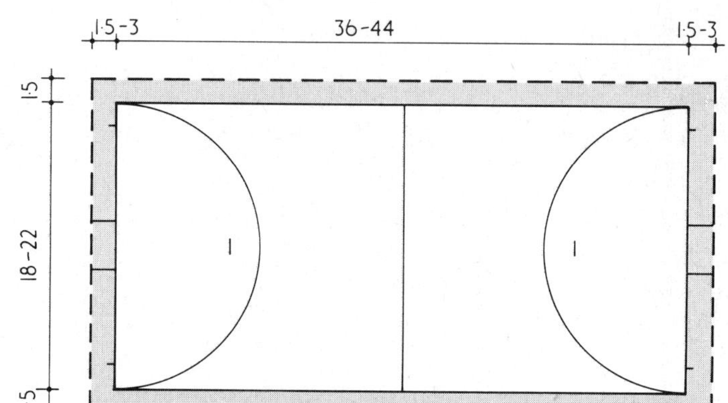

27.25 *Hockey. Team sizes are adjusted according to the size of the pitch available. Side boards should be provided 102 × 102 mm with a 20 mm inward tilt*

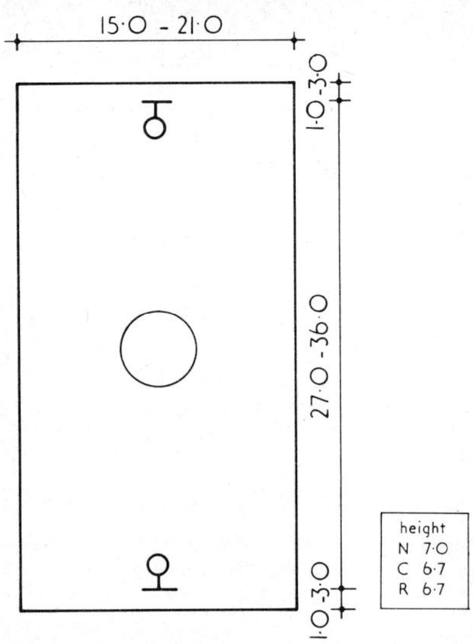

27.26 *Women's lacrosse. The men's game (not illustrated) requires a pitch 46–48 m long by 18–24 m wide, and thus an extra-large hall. At a recreational level it can be played within a 36·5 m length*

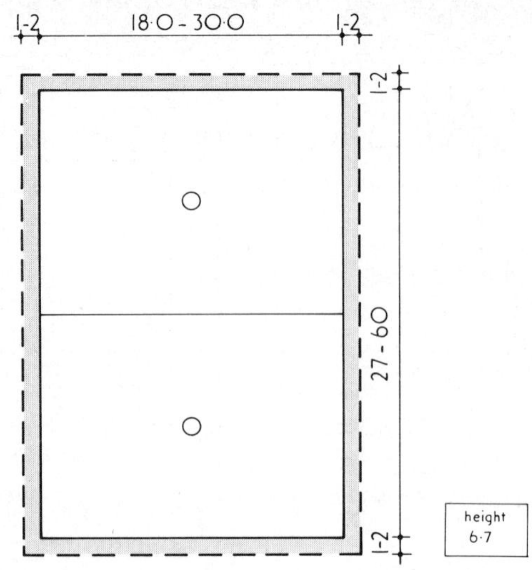

27.27 *Micro-Korfball*

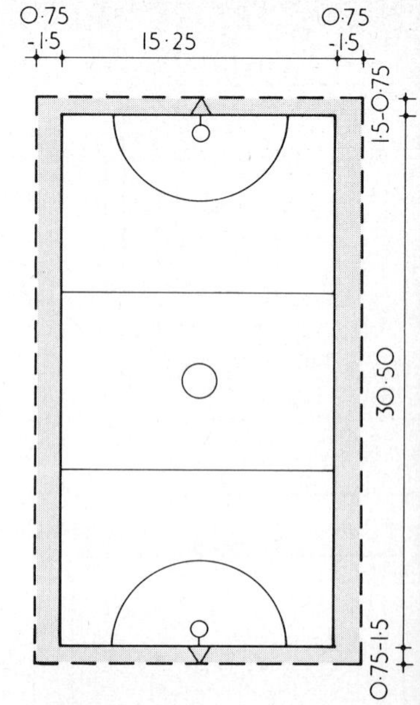

27.28 *Netball*

height
N 7·6
C 7·6-6·7
R 6·7

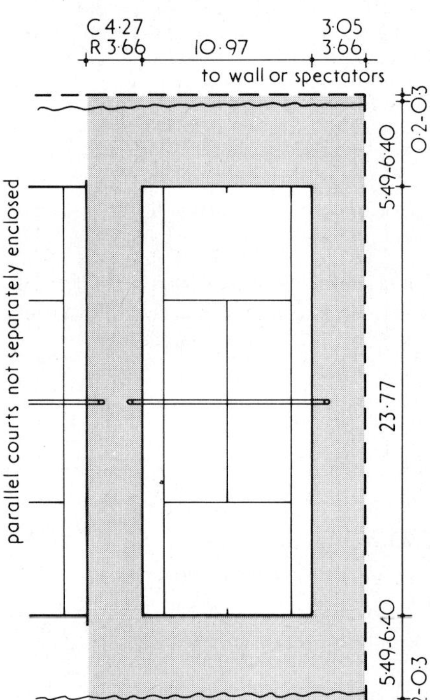

27.29 *Tennis*

height
N 10·67
C 9·0
R 8·0

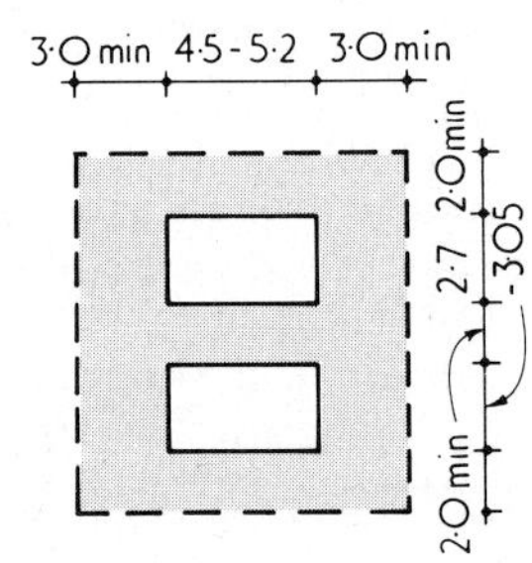

27.30 *Trampoline. The 'bed' is 0·95–1·05 m above the ground. Synchronised competitions on the trampolines must be parallel to each other and 2 m apart*

min ceiling
height
N 9·1
C 7·6
R 6·7

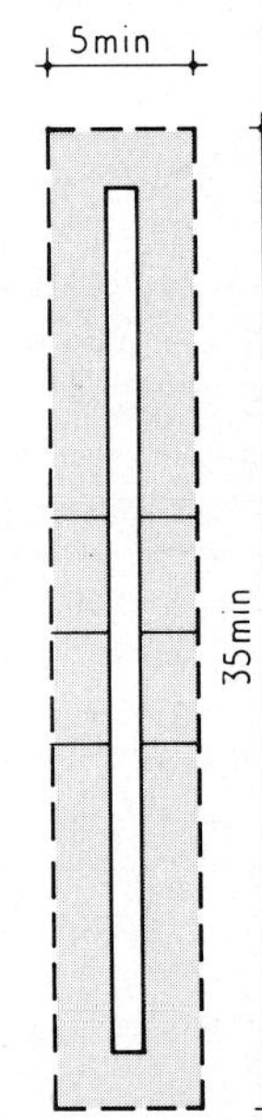

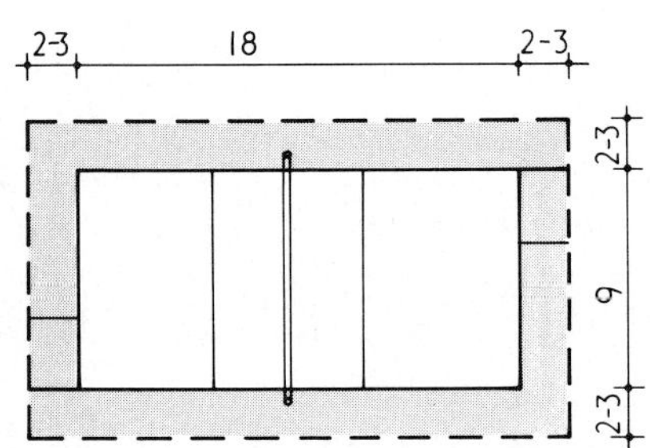

height
N 9·1
C 7·0
R 7·0
(6·7 min)

27.32 *Volleyball*

27.31 *Tug-of-War*

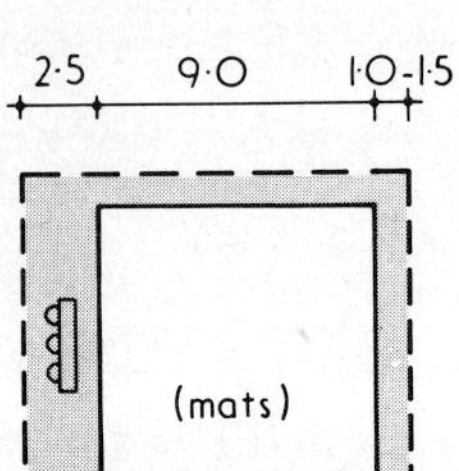

27.33 *Aikido*

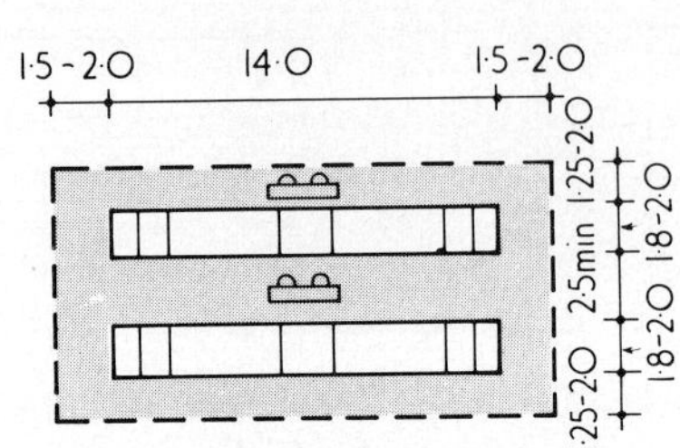

27.35 *Fencing pistes*

27.34 *Boxing. The ring for recreational purposes can be only 3·6 m square. For competitions, in addition to the ring and spectator accommodation, the following are needed:*

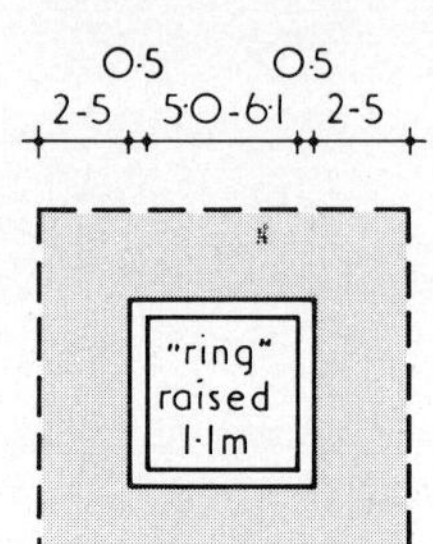

- *medical examination room*
- *weighing room*
- *gloving-up room*
- *administration facilities*
- *lighting above the ring*
- *water supply to each 'corner'*

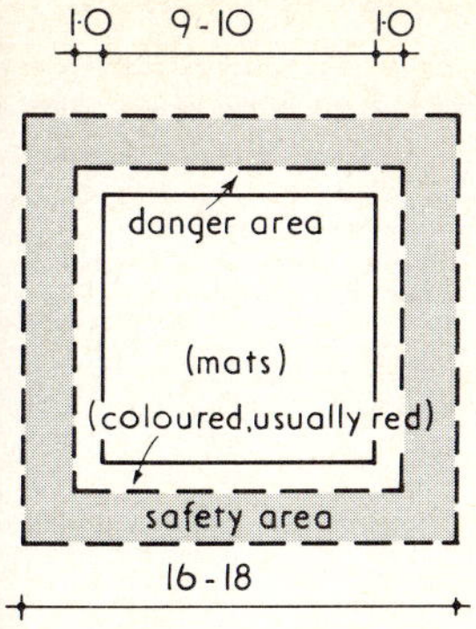

27.36 *Judo*

27.37 *Karate. Regional competitions require three combat areas of international size*

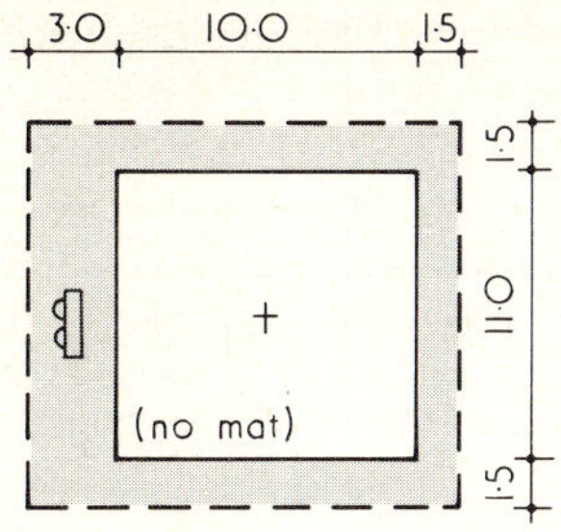

27.38 *Kendo*

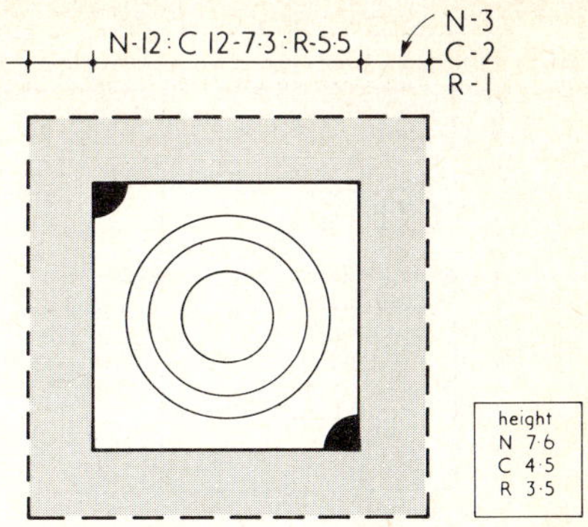

27.40 *Wrestling*

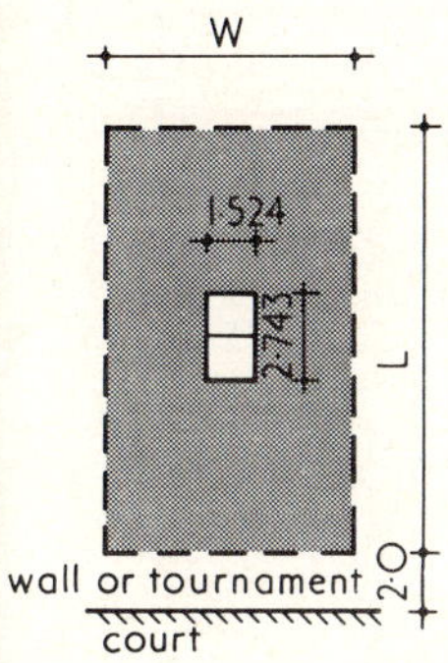

27.39 *Table tennis. See table III for overall dimensions. The table is 0·76 m high, and normally requires a space 1·4 × 1·6 ×0·5 for storage. When in use, each table requires individual lighting*

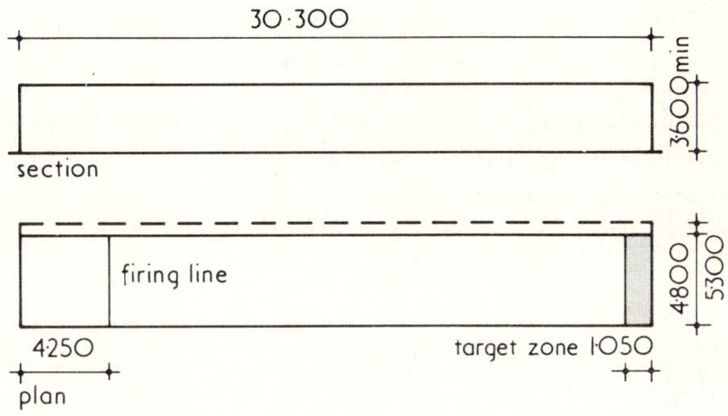

27.41 *Section and plan of a small projectile hall*

5 PROJECTILE HALLS

Figures **27.41** to **27.43** show plans and sections of a range of projectile rooms, and table II shows which sports can be covered by them. The spaces required are given in **27.44** to **27.48**, scale 1:500.

Where the projectile room is to be used for fire-arms shooting, the construction must be to safety standards and robust enough to withstand the use. It may be found that this use will severely restrict the projectile hall's use for other activities.

Table II Projectile halls

	Large 30·3 × 12·8 × 4·6	**Medium** 30·3 × 9·75 × 3·6–4·6	**Small** 30·3 × 5·3 × 3·6
Air rifle	12 firing points	8 firing points	4 firing points
Archery	3 details × 6 archers 3 targets	3 details of 4 archers range 18 m	2 details of 4 archers
Bowls	2/4·5 × 27 m roll-up rinks (if no shooting)	1 roll-up rink (if no shooting)	1 rink
Cricket	3 nets 6-a-side cricket	2 nets	1 net
Fencing	1 piste 4 practice pistes	1 piste	1 piste
Golf practice	4 ranges	3 ranges	1 range
Pistol shooting	7 firing points 10 with side screens	5 firing points	3 firing points
Rifle shooting	12 firing points ranges 25 m, 25 yd, 15 yd	9 firing points	4 firing points
Table squash	15 tables	8 tables	4 tables
Table tennis	15 tables	8 tables	4 tables

Table III Dimensions for table tennis playing space (m)

Standard of play	**L**	**W**	**Ceiling height**	**Clear height below lights**
International matches	14·0	7·0	4·20	4·05
Inter-league and inter-county matches	11.0 min –14·0	5.50 min –7·0	4·20	4·05
Practice and inter-club matches	10·0	5·0	4·20	4·05
Tournaments (more than one table)	8·0	5·0	4·20	4·05
Recreational play	7·6	4·6	–	2·7

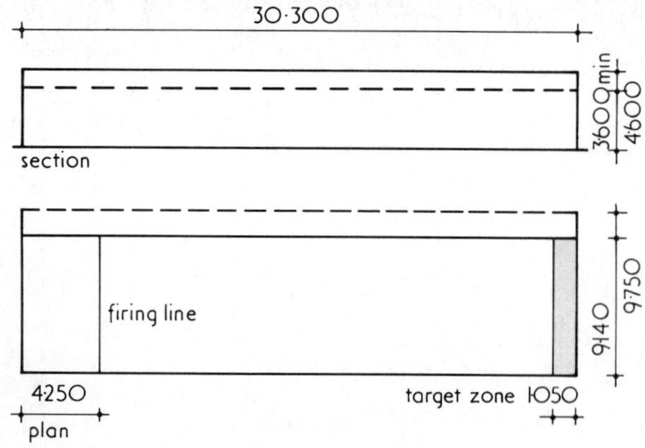

27.42 *Section and plan of a medium projectile hall*

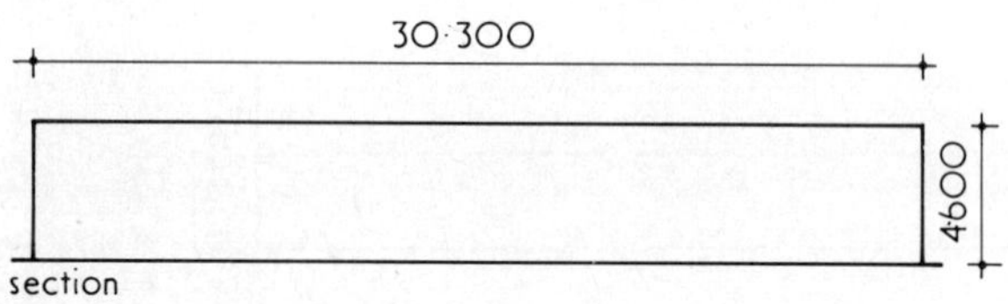

27.43 *Section of a large projectile hall*

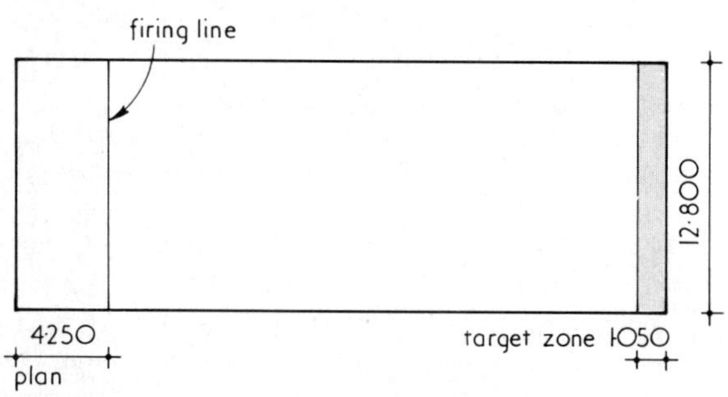

27.44 *Plan of a large projectile hall*

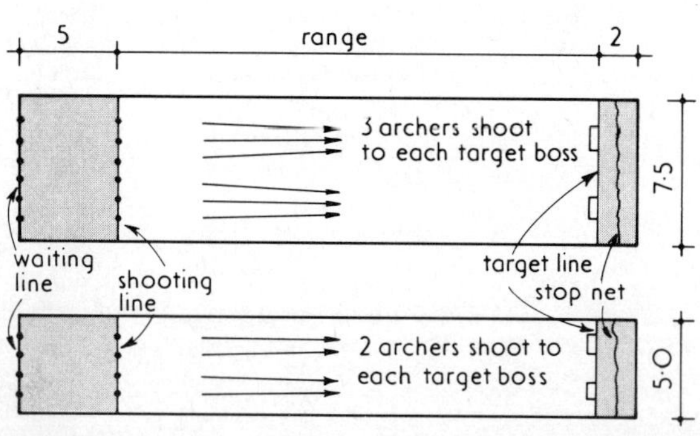

27.45 *Archery. International and national shoots require ranges of 30, 25 and 18 m and 20 yards (18·288 m). For club and recreational shoots 15 yards (13·716 m) will do, but 30 m is preferred for competition practice. Archers stand no closer together than 1·25 m when on the shooting line, with two or three to each target. The minimum ceiling height is 3 m. Where there is no public access the distance between the side wall and the first target should be at least 1·2 m. Where spectator accommodation is required, advice should be sought from the Grand National Archery Society. Storage is required for straw bosses and stands, preferably at the target end; and lockable storage for portable bow racks and tackle boxes*

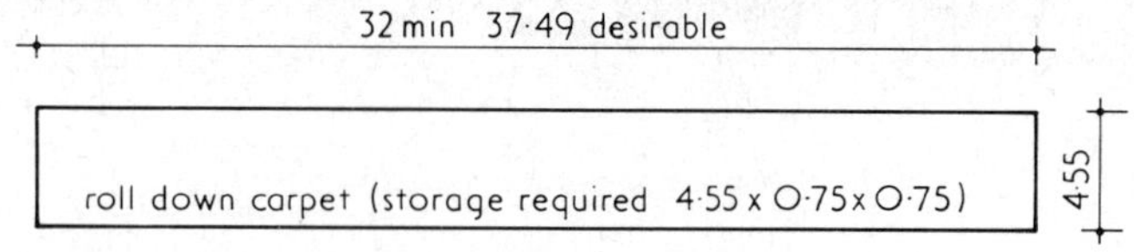

27.46 *Bowling. A single rink in a projectile hall. See also* **27.52**

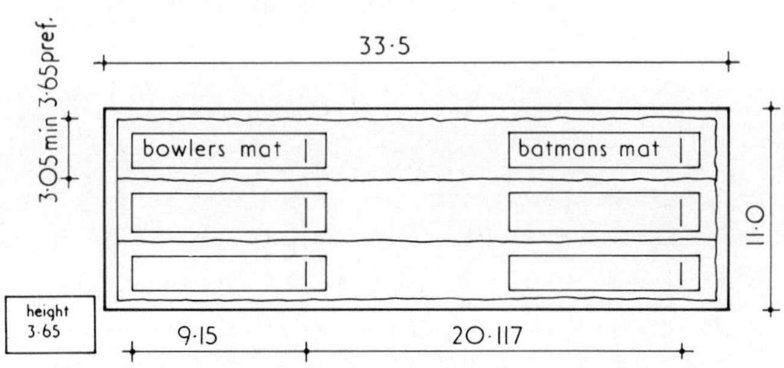

27.47 *Cricket practice nets. For the six-a-side game (not illustrated) the playing area is 30·4–36·5 × 18·9–30·4 × 6·1–7·6 m high*

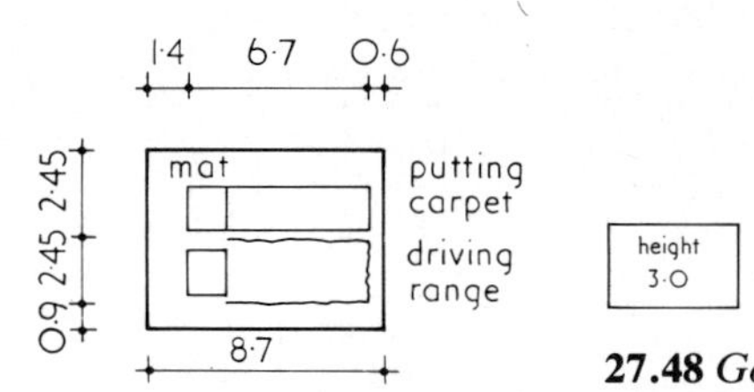

27.48 *Golf practice*

6 SPECIAL SPACES

There are a number of activities that need spaces permanently and exclusively reserved for them. This may be owing to the weight or size of the equipment, such as billiards/snooker, or because the playing area is closely defined, such as squash or real tennis. For some of these, semi-portable equipment is now being produced, but these are generally designed for special occasions such as national championships. The critical sizes for these special spaces are given in **27.49** to **27.59**, scale 1:500 except where shown otherwise.

7 BIBLIOGRAPHY

Geraint John and Helen Heard (eds) *Handbook of sports and recreational building design*, Architectural Press, to be published 1980

The Oxford Companion to Sports and Games, Oxford University Press, 1976, also as a Paladin paperback

Rules of the Game, Paddington Press, 1974, also as republished by Literary Guild and Bantam Books

Information published by the ruling bodies for each particular sport.

Table IV Dimensions for indoor athletics tracks (m)

Lap length	Length of straight (s)	Length of bend (B)*	Radius of bend (R)†	Overall length (L_1)		Overall width (W)		Space for sprint straight (L_2)
				6-track	4-track	6-track	4-track	
200	35‡	65	20·49	88	84	53	49	75·98
	50	50	15·716	93·44	89·44	43·44	39·44	81·44
	52·25	47·75	15·0	94·25	90·25	42	38	82·25
	65	35‡	10·94	98·88	94·88	33·88	29·88	86·88
160	35‡	45	14·124	75·25	71·25	40·25	36·25	63·25
	40	40	12·532	77·06	73·06	37·06	33·06	65·06
	45	35‡	10·941	78·88	74·88	33·88	29·88	66·88

* Measured 200 mm from inside of outer white line around flat-edged track, or 300 mm inside a raised border or edge framework.
† Nett radius allowing for 200 mm deduction. The smaller the radius, the greater the inclination of the banking, 10°–18° max.
‡ A European Athletic Association regulation minimum dimension.

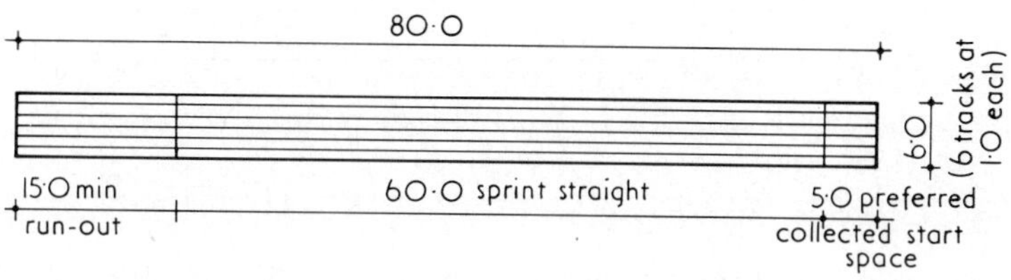

27.49 *Athletics: requirements for straight sprint, scale 1:1000*

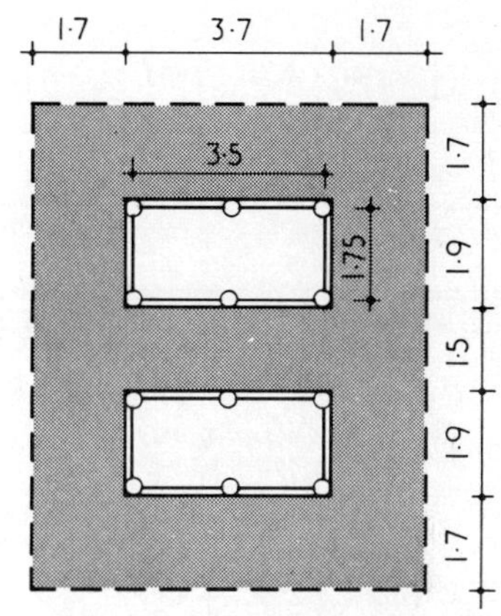

27.51 *Billiards and snooker, scale 1:200. The Billiards and Snooker Control Council has introduced, with worldwide agreement, the B&SCC 3·50 m standard table, which specifies for the first time the actual playing area within the cushion face. The present tables, specified over the frame, remain acceptable until the year 2025*

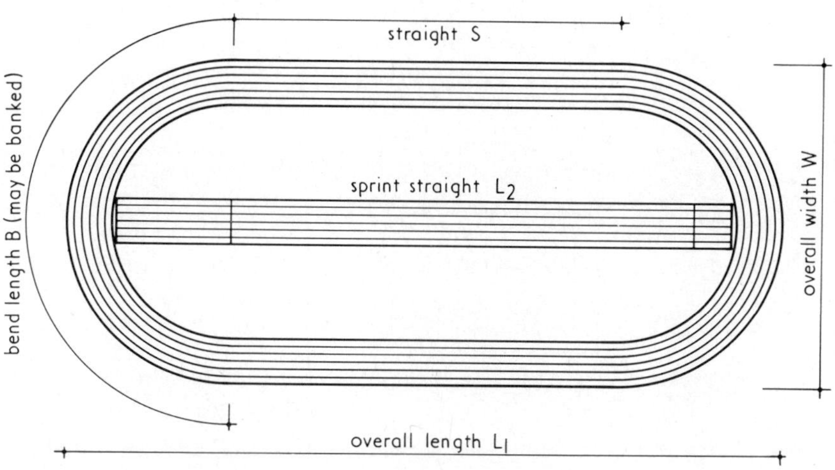

27.50 *Athletics: indoor tracks 200 and 160 m laps, with straight sprint in the centre, scale 1:1000. See table IV for the dimensions. It has been the practice to fit a suitable running track inside the cycling track in* **27.53**, *but this is no longer considered appropriate. When spectator accommodation is provided around the track it can be seen that indoor athletics requires a building of considerable clear span. Supports in the central area are not used*

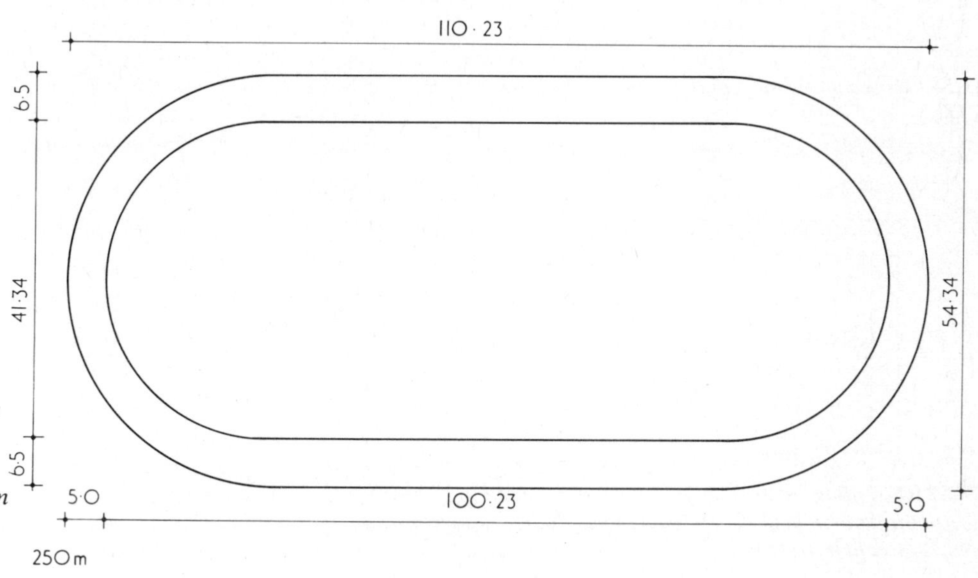

27.53 *Cycling, 250 m track, scale 1:1000. This is relatively steeply banked. See other sources for details of construction. The 33⅓ m track in* **26.42** *can be used internally*

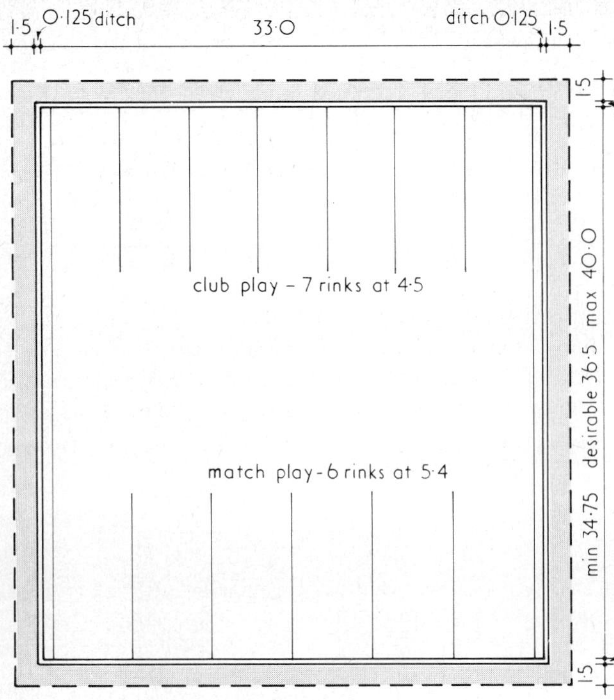

27.52 *Bowling. Four rinks are the minimum for recreation; six are required for tournaments*

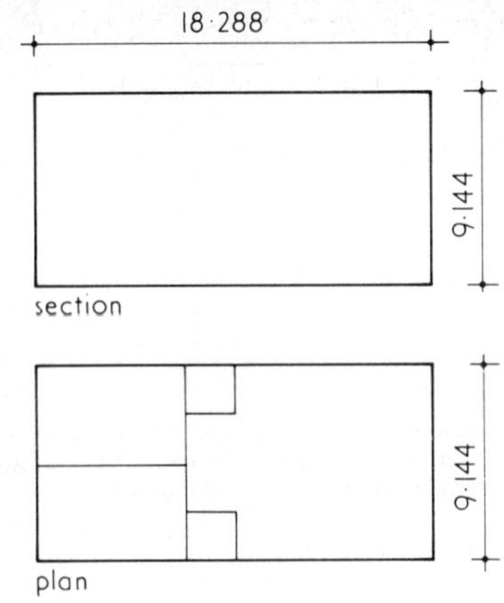

27.56 *Rackets (or racquets)*

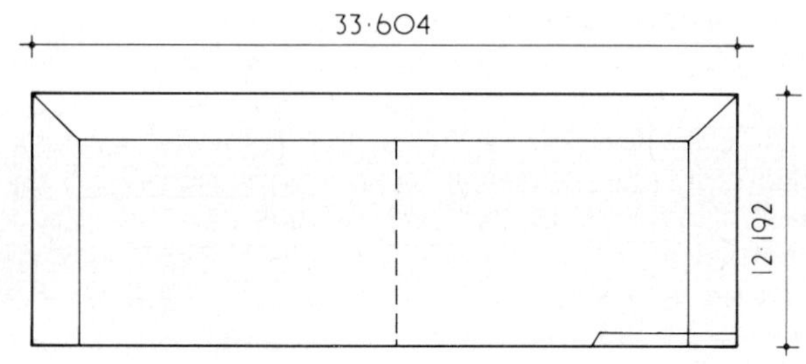

27.57 *Real tennis. The dimensions of the court at Hampton Court, which is reputed to be the widest and one of the longest*

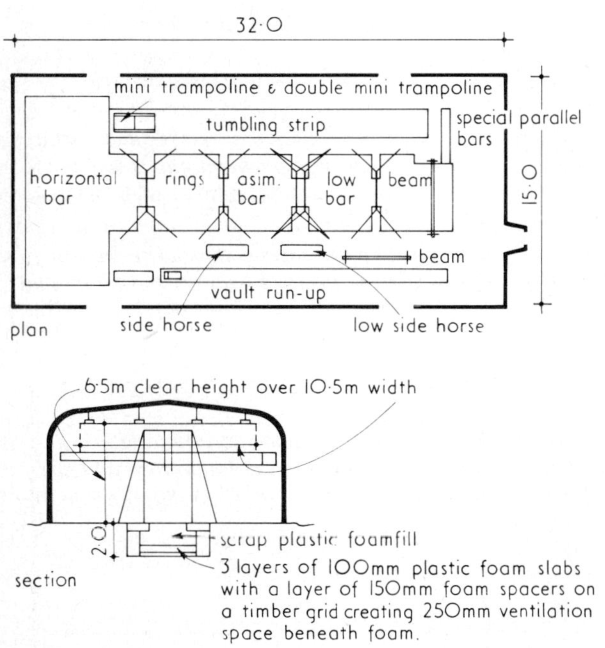

27.54 *Gymnastics practice: training hall at Lilleshall Hall NSC*

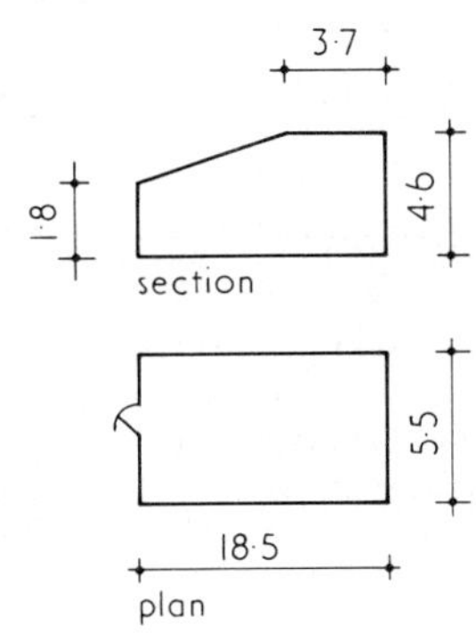

27.58 *Rugby fives*

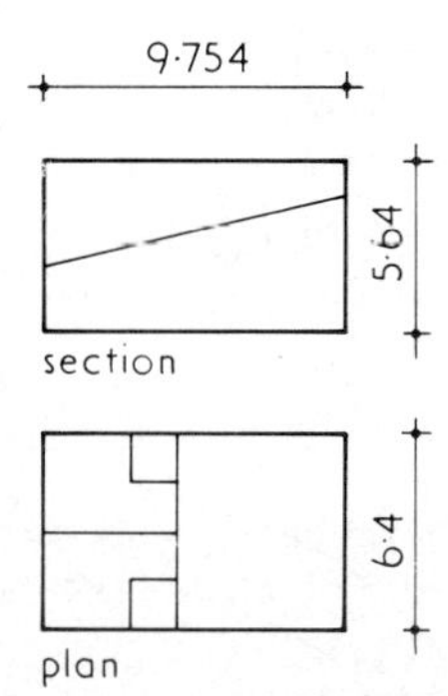

27.59 *Squash. All dimensions are highly critical and are to finished surfaces, which are normally plastered to a special specification*

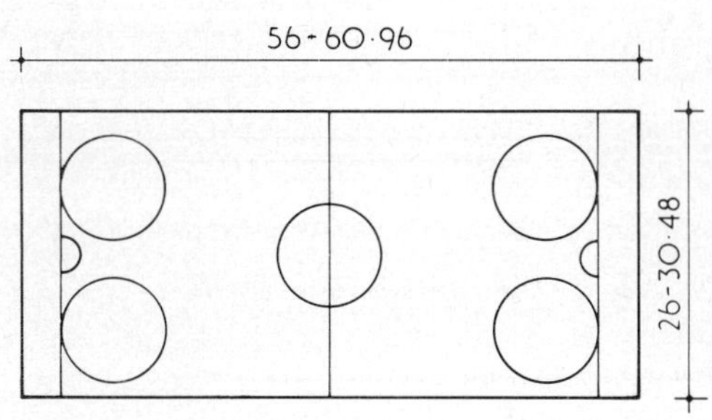

27.55 *Ice skating, scale 1:1000. Rinks are usually sized to accommodate an ice hockey 'pad'; this should be surrounded by a 1·2 m high barrier*

28 Stables

Frank Bradbeer

CI/SfB 565
CI/SfB (1976 revised) 565

Contents

1 GENERAL CONSIDERATIONS

1.01 Horse riding today

Horse riding is now mainly recreational, either as equestrian sport or in riding schools. Large commercial stables, such as the London Omnibus Company, the railways or Carter Pattersons are never likely to be built again, and the requirements for police or military stables (such as the new barracks at Knightsbridge) are of a special nature and a very complete brief would be available in the event of such a proposal.

1.02 Planning elements in private stables

Private stables for anything from a single horse to perhaps a string of polo ponies or race horses in training should not present too complex a planning exercise. The principal elements, **28.1**, are:

- loose boxes for fit or sick horses
- tack and cleaning room
- feed store—sacks and bins
- hay store—bales
- bedding store—for bales of straw, sacks of wood chips etc
- manure compound.

Additionally, space may be required for parking horse transport facilities (landrover, trailer, motorised box) and possibly, in a driving establishment, accommodation for carriages.

A typical plan is shown in **28.2**.

Table I Common breeds of horses and ponies, with heights in hands and equivalent metric measure, 23.3a
(1 hand = 4 inches, 12·2 hands = 12 hands + 2 inches)

Breed	Height in hands	Height in mm
Horses		
Cleveland bay	16	1625
Clydesdale	16	1625
Morgan	14–15	1420–1525
Percheron	16–16.3	1625–1700
Shire	17	1725
Suffolk	16	1625
Tennessee Walker	15.2	1575
Thoroughbred	16	1625
Ponies		
Connemara	14.2	1475
Dartmoor	12	1220
Exmoor	12.2	1270
Fell	13.1	1345
Highland	12.2–14.2	1270–1475
New Forest	14.2	1475
Shetland	39 to 42 inches*	990–1065
Welsh	12	1220

*** Shetland ponies are always described in inches.**

Table II Typical dimensions of horse or pony and rider, 28.9

Dimension	Thoroughbred	New Forest pony	Welsh pony
A	1600	1450	1200
B	550	500	415
C	900	815	675
D	1620	1470	1215
E*	2450	2225	1840
F	1625	1475	1220

* Assuming that the rider is in proportion to the horse or pony

1.03 Dimensional criteria

Dimensionally standardised criteria may be applied to planning for the horse in any one of four spheres:

- stabling and the care of the horse, **28.3** to **28.8**, table I
- schooling of horse and rider, **28.9** to **28.12**, table II
- equestrian sport, particularly in the dressage arena, **28.13**
- transportation, **28.14**, **28.15**.

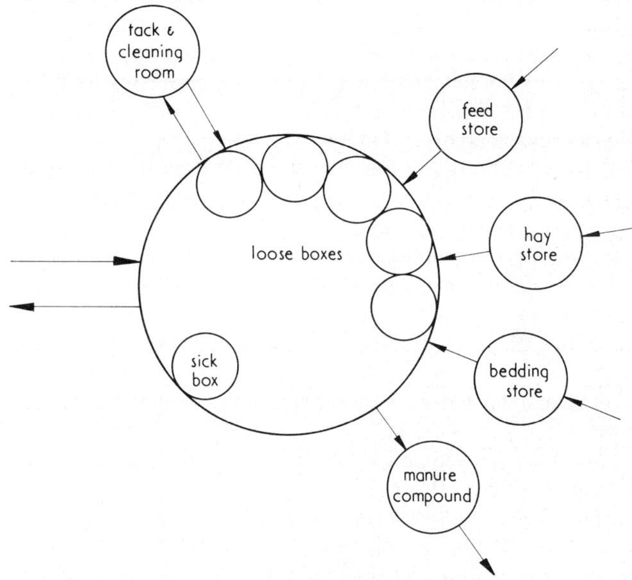

28.1 *Relationships between elements of the plan*

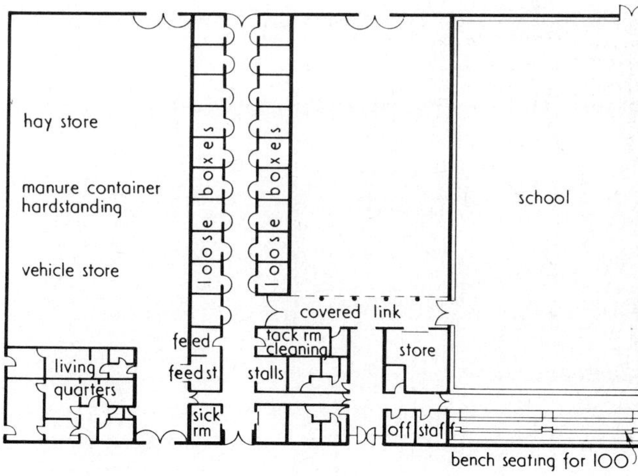

28.2 *Plan of Porter's Field Riding School, Leyton*

2 STABLING AND THE CARE OF THE HORSE

2.01 Environmental conditions

The following considerations should apply:

- *Aspect:* a southerly aspect is not desirable for stables; a northerly aspect is preferred.
- *Temperature:* contrary to popular belief, the temperature of

a stable should not be appreciably warmer than the temperature outside except in the most bitter weather. The stables should remain cool in hot summer weather. The stable, though adequately ventilated, should be free from draughts. In very cold weather, the top door may be closed.

- *Size:* ideally, the size of a loose box should vary with the size of the inmate, being as small as possible, compatible with comfort. This is partly on grounds of economy and partly through the difficulty of maintaining a 'genial' temperature in a large and lofty stable.
- *Noise:* daytime noise does not disturb horses, but noise interrupts their sleep at night. Victorian stable planners often avoided accommodating grooms and coachmen above the stables so that the animals would not be disturbed by their keepers returning from a local hostelry after it closed at night.

2.02 Detailing considerations

Details to consider for the safety and welfare of the horses include:

- *Floor:* ideally, a stable floor should be impervious, non-slip and laid to a fall not exceeding 1 in 80. Engineering brick laid on edge in herringbone pattern is easily kept clean and provides a good non-slip surface.
- *Mangers:* although horses like to see one another, they do not like to be disturbed when feeding. Mangers should therefore not be visible from box to box.
- *Drainage:* good drainage is important, but difficult to achieve in a stable, since fragments of bedding, feed and mud can penetrate gratings and clog drains. Internal drainage should be via surface channels leading to a gulley outside the building, which should be equipped with a perforated removable bucket to intercept the particles of rubbish which might be washed down.
- *Projections:* all projections should be avoided; doors should be hinged to swing through 180°.
- *Manure compound:* for reasons of hygiene, this should be as far from the stables as conveniently possible. The old way of mucking out was via wheelbarrow from stable to tip. Large modern establishments may use a small tractor and trailer, picking up at each box and driving to the tip which may then be conveniently further away.

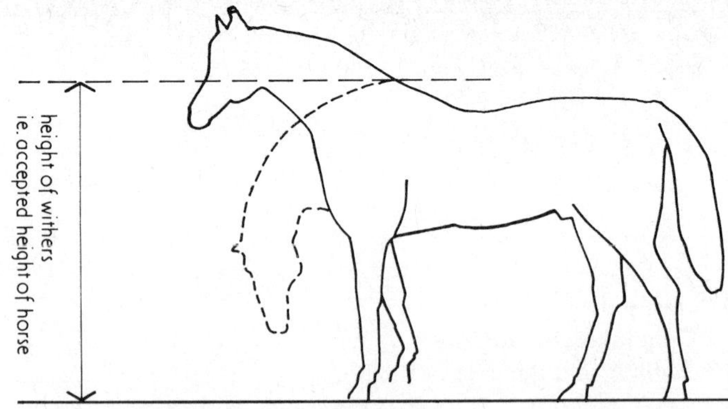

28.3a *Measurement of the height of a horse in hands at the withers. Table I gives the heights of a number of breeds of horses and ponies*

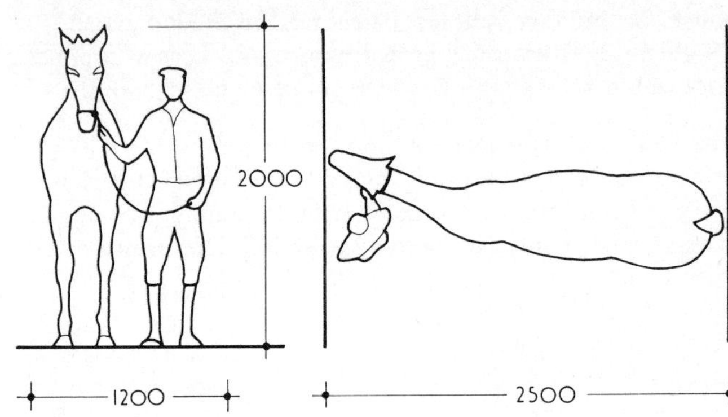

28.3b *The led horse*

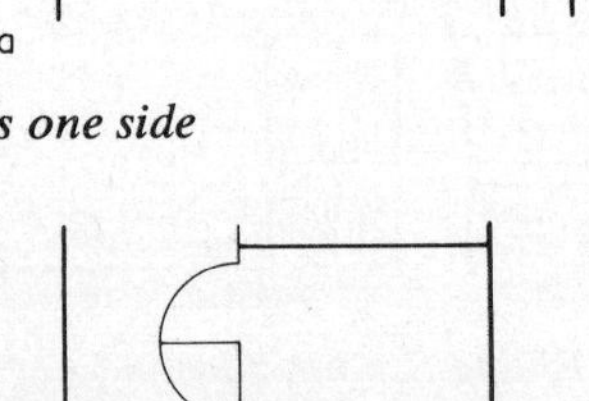

a *Stalls one side*

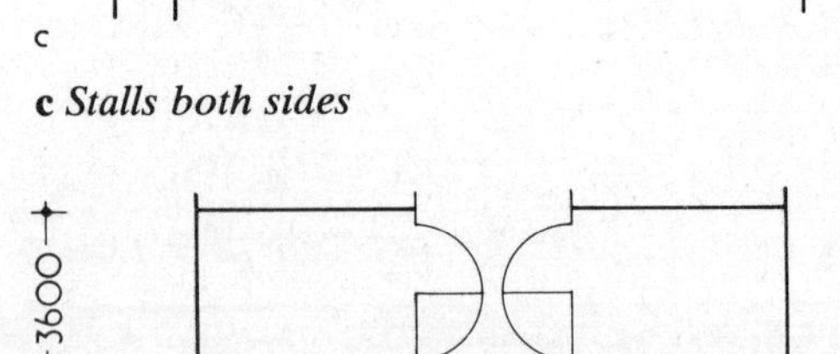

c *Stalls both sides*

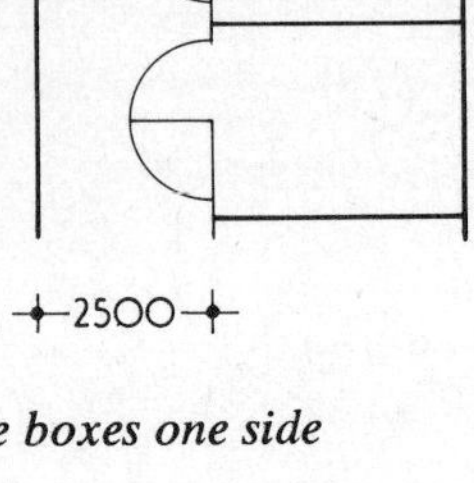

b *Loose boxes one side*

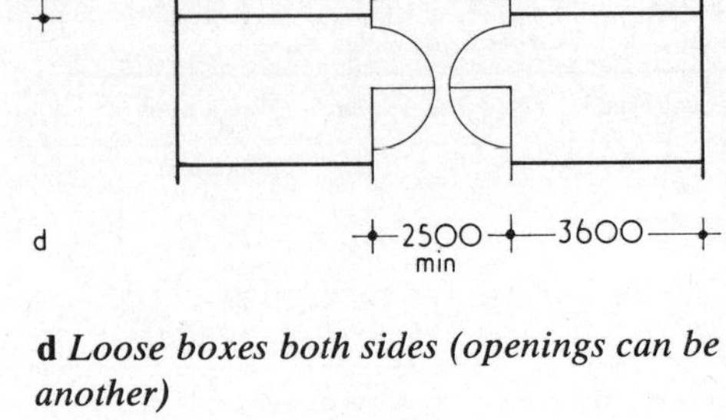

d *Loose boxes both sides (openings can be opposite one another)*

28.4 *Loose boxes and stalls*

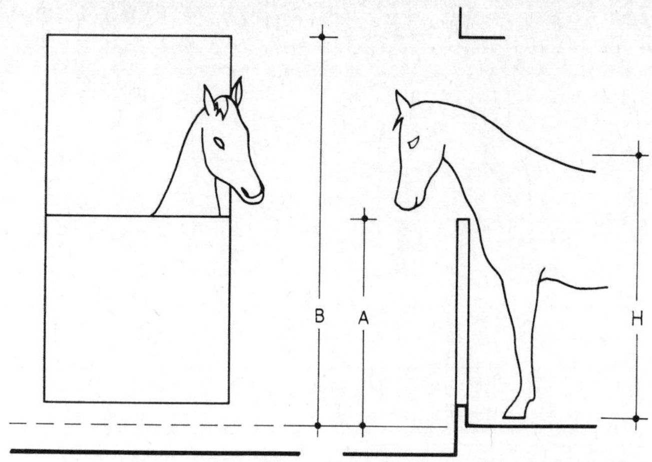

28.5 *The stable door. It is essential for its mental well-being that the horse can easily see out—horses are inclined to be very inquisitive. A = 0·8H (average 1300), B = 1·5H (average 2500), H = height at the withers* **28.3a**

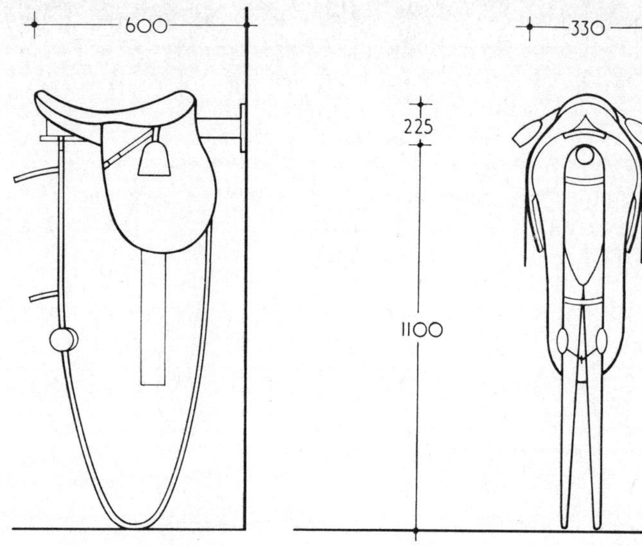

a *Saddles and bridles together*

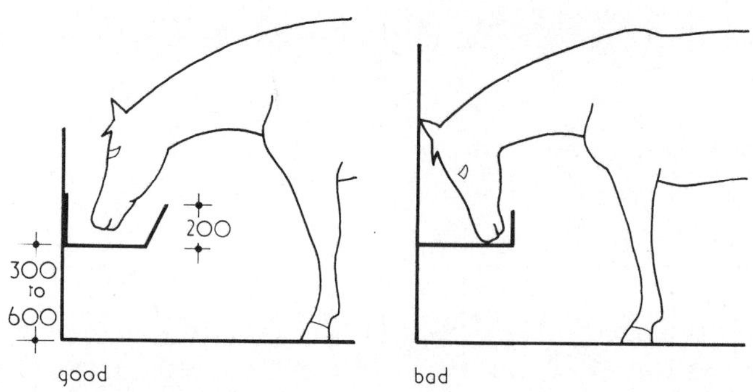

28.6 *Height of the manger*

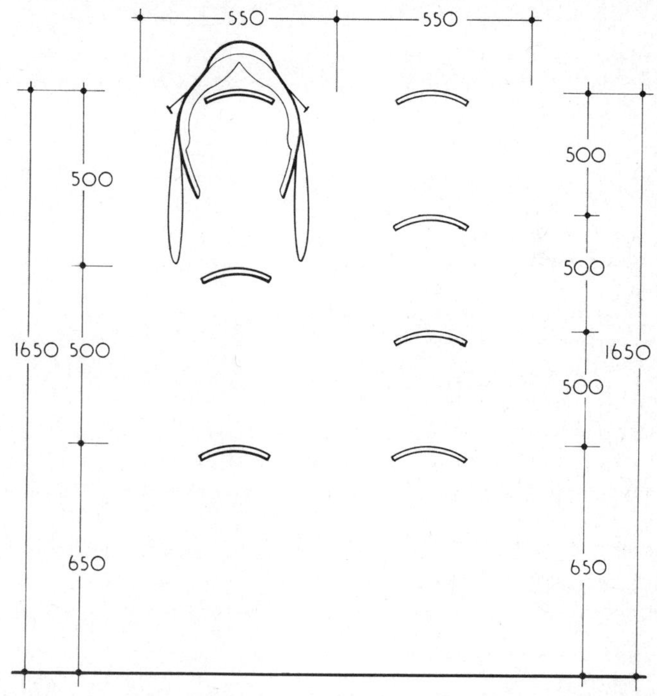

b *Saddles only*

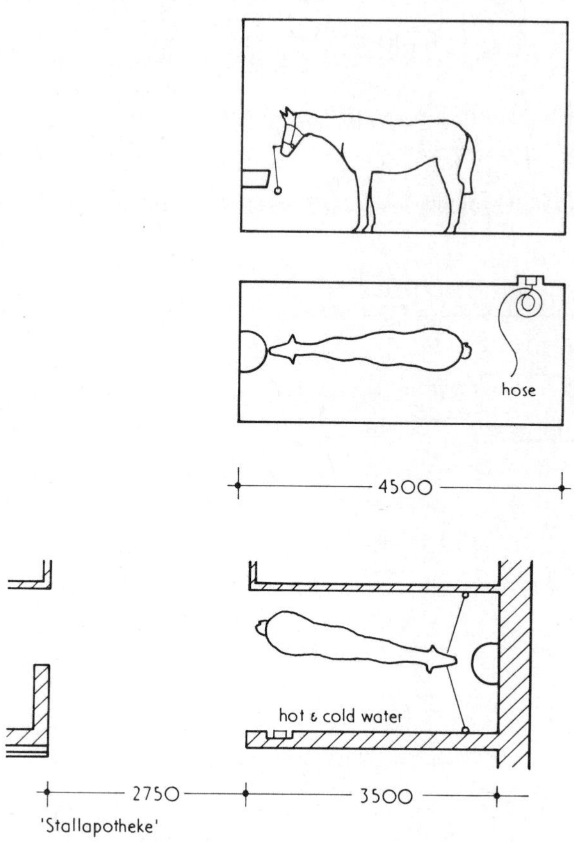

28.7 *Veterinary box*

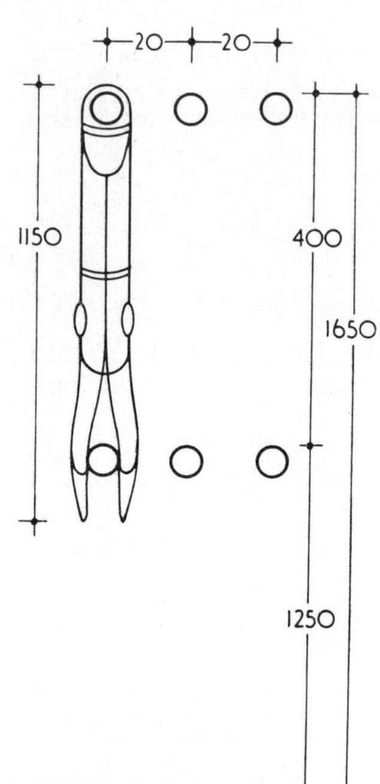

c *Bridles only, where kept separate*

28.8 *Tack rooms:*

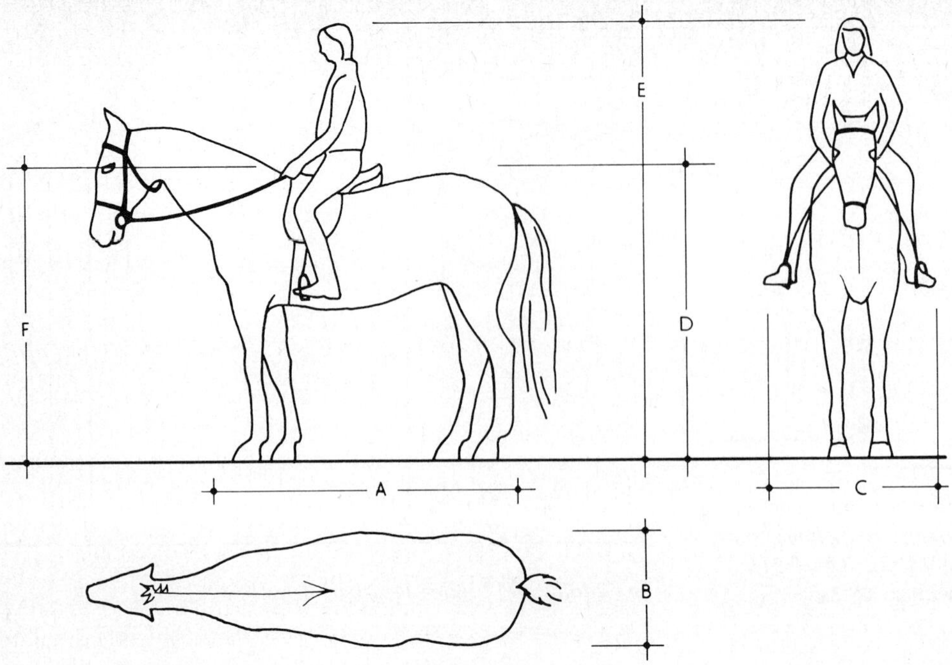

28.9 *Dimensions of horse and rider, see table II*

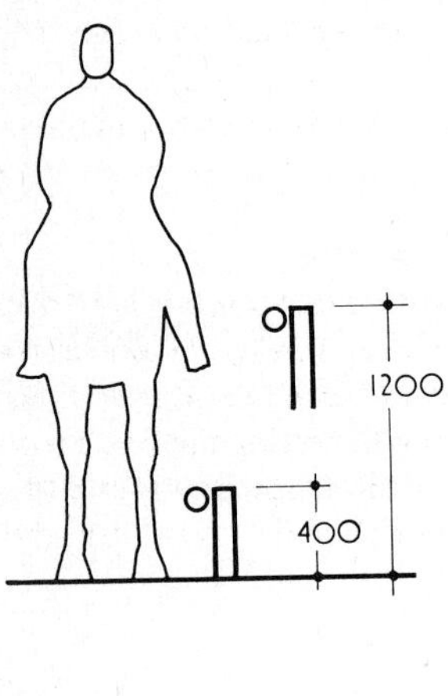

28.10 *Rails for the outdoor school*

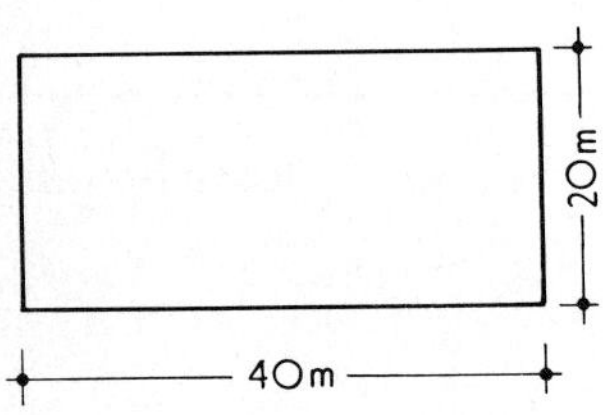

28.11 *Plan of indoor school*

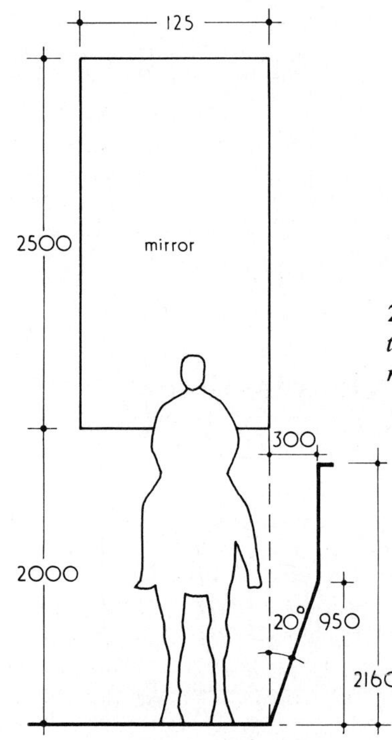

28.12 *Indoor school: batter to walls and arrangement of mirror tilted to give vision*

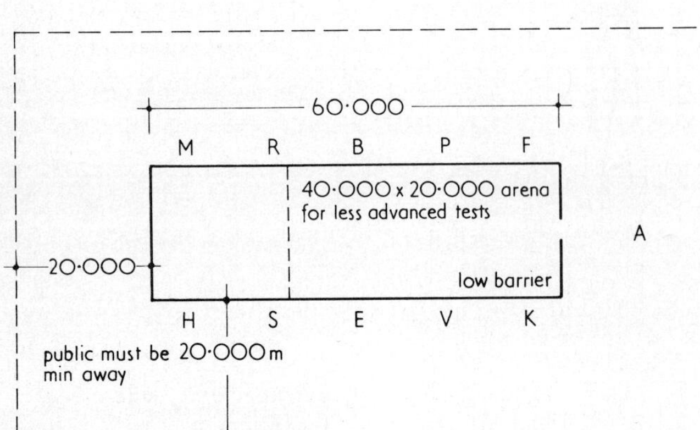

28.13 *Dressage arena*

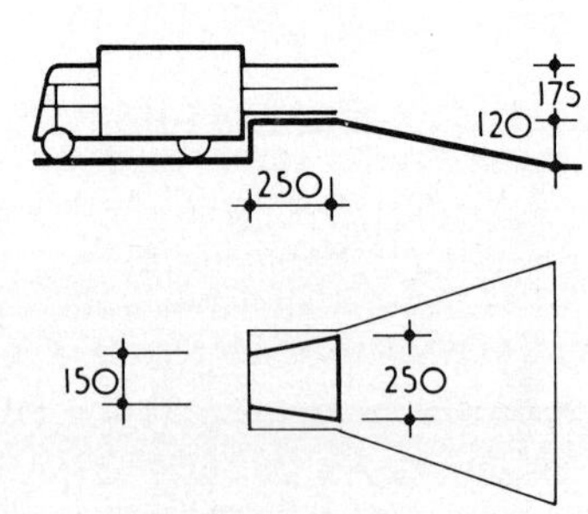

28.14 *Ramp for loading horses into horse-boxes or trailers*

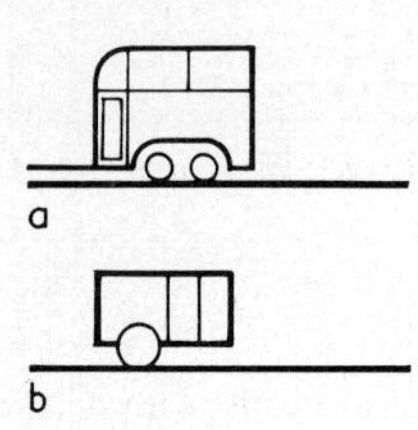

28.15 *Trailers*
a *Large*
b *Small*

29 Church buildings

Leslie Fairweather

CI/SfB 63
CI/SfB (1976 revised) 63

Leslie Fairweather is editor of The Architects' Journal

Contents

Part A Guide to denominations

1 INTRODUCTION

1.01 Definitions

A more exact title for this section would be 'places of Christian worship' because some Christians do not use the term 'church' at all. But 'church' is now becoming accepted as the generic word for any place where Christians assemble for worship. Places of local worship, only, are dealt with, and not larger buildings like cathedrals or those with social spaces, which are not significantly different from social spaces in secular buildings.

1.02 History and tradition

In church architecture the architect is bound to be more concerned with tradition than he is in other fields. Users of a church are more conscious of history (and more sensitive about it) than most other clients: they will not allow an architect to ignore established precedents and may even expect him to understand them. Current developments in the design of churches are very largely connected with the changes in our understanding of ecclesiastical history which have been brought about by the liturgical movement.

Exact details of the forms of worship and building procedures should be discussed with the local religious community, and the architectural implications thoroughly understood. A fairly detailed general guide to the history, procedures and forms of worship (with architectural implications) of the Church of England, the Roman Catholic Church, the Presbyterian Church, the Salvation Army, the Methodist Church and the Society of Friends, is given in *Church buildings,*[1] originally published as a series in *The Architects' Journal* and later in book form (now out of print). The remainder of Part A will be similarly sub-divided.

2 CHURCH OF ENGLAND

2.01 The buildings, and how they are used for worship

After the Reformation the Church of England inherited many medieval, mostly Gothic, buildings, strongly directional in their east-west orientation and with the main action remote from the congregation. The people therefore had lost the sense that they were engaged with the clergy in a common action and tended to become spectators with an individualistic rather than a corporate response to the liturgy. The church is now moving into a new period of experiment in which it must find expression of the corporate nature of worship and the equal importance of the Word (which must be heard) and of Sacrament within a building which must remain true to the Anglican sense of proportion.

2.02 Altar, priest and people

During the period of the Oxford Movement's influence* the 'north end position', **29.1**, was largely superseded by the 'eastward position' in which the priest would face the altar with his back to the people, **29.2**. This way of celebrating the Eucharist is contrary to the spirit of the Anglican liturgy, but is still the practice in most Anglican churches today. Now, in its turn, the

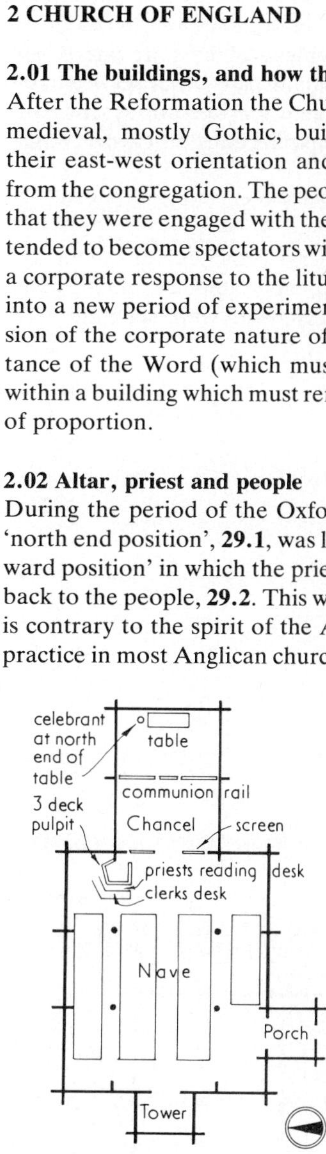

29.1 *Communion table is placed against east wall, priest stands at north end*

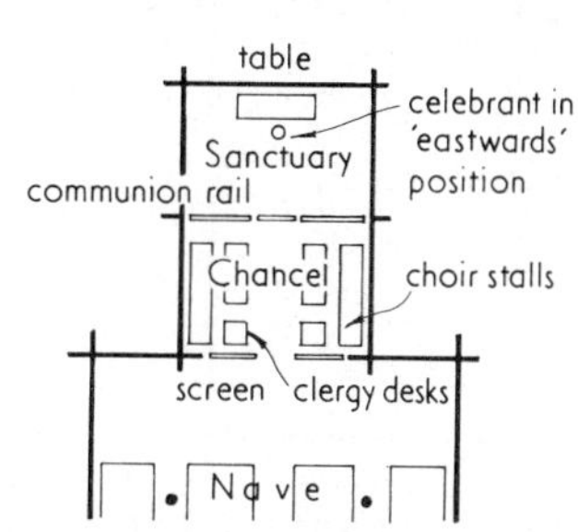

29.2 *Eastward position: priest faces altar with his back to the people*

* The Oxford Movement began in 1833 to revive Catholic doctrine and observance in the Church of England.

eastward position may well be in the process of being replaced by the one in which the priest faces the people across the altar, **29.3**. This change is by no means general yet, but if eastward position is still the fixed practice of a parish, the architect ought to consider whether provision should be made for a future change to 'facing the people'.

The change is in fact a return to a practice that was almost universal in the primitive church. The position of the priest is not a matter of churchmanship: 'facing the people' may be adopted by any parish. But even if most parishes do adopt the position of the celebrant 'facing the people', it is likely that some will not.

2.03 The main services

The six main services of the Church of England are: the Eucharist, the offices of matins and evensong, baptism, confirmation, the solemnisation of matrimony and burial of the dead. If the architectural requirements of these are catered for, so will the requirements be for almost any other service likely to take place in the church. These services are described in detail, with their architectural implications, in *Church buildings.*[1]

2.04 Church design

Apart from the altar, the general principles of layout and design are the same as for the Roman Catholic Church and the Methodist church. These are shown, with separate altar details, in para **9**.

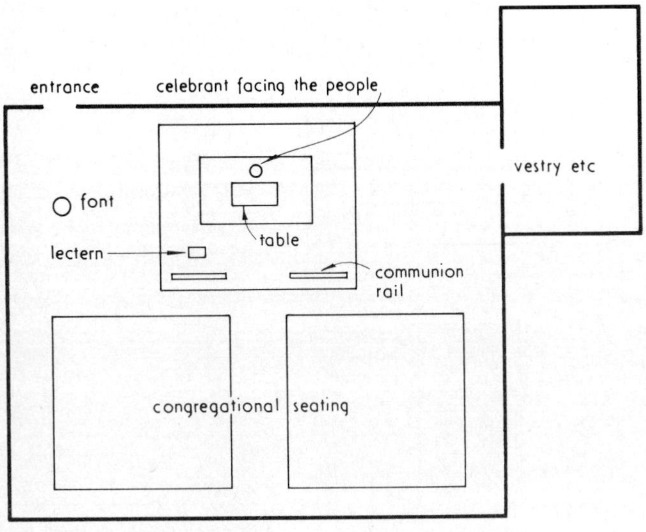

29.3 *One of the many variations of plan where the priest faces the people across the altar. The seating may extend around three sides of the altar*

3 ROMAN CATHOLIC CHURCH

3.01 Worship as a corporate act

The term 'Roman Catholic' (or to members simply 'Catholic') denotes the Christian community which has continuously accepted the authority of the Bishops of Rome; that is, of the Popes. Disregarding earlier divisions, the Roman Catholic community became distinguished from the Orthodox communities, which separated finally in the twelfth century; and from the Reformed or Protestant churches which separated in the sixteenth century.

The chief problem of a community which is building a Catholic church now is that, though its building must suit the liturgy as it now is, community and architect must also make some estimate of what the total ultimate change in form of services is likely to be.

The *corporate* nature of Catholic worship is again now being stressed. Catholic churchgoing has for centuries been highly individualistic. The congregation should participate in the liturgical action and not merely watch it. Baptism is being gradually restored as the corporate act of the local assembly: but it is still conducted as a private ceremony held at a time to suit the parents and which only they and their friends attend. The architect must make corporate baptism possible in his church, even though it may not be practised for some time.

The existence of societies within the parish is always important in its social life and the architect should find out which they are, what they do, and whether they are to be accommodated in any way.

3.02 The main services

The six main services of the Catholic church are: mass, the Easter liturgy (holy week ceremonies), baptism, marriage, burial of the dead, and devotions. Other liturgical activities include: blessing, dedication, consecration, confirmation, ordination. All of these activities are described in detail, with their architectural implications, in *Church buildings.*[1]

When the architect has read through these activities he will have a full picture of 'what goes on' in a Roman Catholic church, and will have a record of such data—dimensional and otherwise—as can be safely inferred from these activities. What he will not know is how exactly to arrange the different parts of his building: this is something that he and his priest client must work out for themselves.

3.03 Church design

Apart from the altar, the general principles of layout and design are the same as for the Church of England. These are shown, with separate altar details, in para **9**.

4 PRESBYTERIAN CHURCH OF SCOTLAND

4.01 Origins and buildings

The Church of Scotland claims to be a continuation of the Celtic church. This grew from the missionary activities of St Ninian, who first brought Christianity to Scotland in about AD 397; from St Columba who landed in Iona in AD 563; and from their followers and successors. When the Celtic church, which had become independent of Rome, met the church resulting from the Augustine mission in England, some divergence of practices caused controversy. The Synod of Whitby in AD 664 therefore finally decided in favour of the Augustinians so that there might be uniformity throughout. The present Church of Scotland is that continuing church reformed.

During the reformation one of the leaders, John Knox, was deported or exiled, ultimately finding his way to Switzerland where he came into contact with John Calvin, one of the great

Swiss reformers. Knox was greatly influenced by Calvin and it was basically his system of church government and structure, as well as much of his theology, which he brought back with him. This system of church government by courts is known as 'Presbyterian'.
The Reformation established the doctrine of 'the priesthood of all believers': it was not necessary for any human being to come between God and a worshipper, and the only mediator accepted was Jesus Christ. There is therefore no good theological reason for a chancel in its literal sense of a 'railed-off area' in a Presbyterian church.
All take part in the full worship and sacramental act. The sanctuary or chancel area is therefore now simply where the central act takes place, but the congregation are essentially participants in that act and the nearer they are to it the better. There is now a fairly general departure from the earlier rectangular form of church in favour of a more open form where the sense of gathering the people round the Word and sacrament, as represented by the pulpit and communion table, can be expressed.
Present thinking seeks to emphasise the close links between the worship room and the rooms for secular purposes, so that there may not be a complete divorce between the weekday and Sunday activities of the congregation. The economic situation may well give rise to consideration in future years of the possible use of multi-purpose buildings where only a portion of the worship room is kept entirely for worship, while a part or parts of it may be used for other purposes, with the use of partitions or screens.

4.02 The main services

The types of services normally held in a church are:
- normal morning public worship
- evening worship which increasingly is taking a variety of forms, or may be largely a repetition of the morning service
- as above with addition of one or other of the sacraments, holy communion or baptism
- as above, with ordination of elders or admission of new communicants
- marriages.

Funeral services in a church are very infrequent.
The ordinary conduct of worship is left almost entirely to the minister and he may choose to remain in the pulpit for the whole service or he may take part of the service from a prayer desk or a lectern, or from behind the table, using the pulpit only for preaching. He may move from the pulpit to the communion table to receive the offering. He will certainly do so to administer the sacrament. He will move to the font for administering baptism. He will move to the front of the sanctuary steps for admission of new communicants or for ordination of elders.
Full details are given in *Church buildings.*[1]

4.03 Church design

The Kirk session normally sit among the congregation. While there is great variety in the way services are conducted, most of the speaking is normally done by the Minister. The congregation *sit* for prayers and do not kneel in the pews. The pews are seen as an extension of the communion table, so the congregation is, in effect, sitting around the table. Worshippers remain in their seats throughout the services, the elements of communion being passed around by the elders.
Baptism must take place in the face of the congregation and the font should therefore be visible to all. It should be in advance of, and probably to one side of, the altar and at a slightly lower level. The font can be movable but it should have a permanent site in the sanctuary. It need *not* be at the entrance to the church. There are no special design requirements for Christmas or Easter services.
A central aisle is desirable for marriage ceremonies.
The choir should be visible to the congregation but should not be in the sanctuary.
There is little social activity connected with worship but ancillary accommodation for social and educational purposes may be needed outside the hours of service.
Many aspects of design are similar to those for the Church of England. The main differences are listed below:
- The communion table is a *table* and not an *altar*. It is usually rectangular but other shapes are not precluded. Basic sizes are given in para **9**.
- A large lectern is not essential but where provided should be to the sizes shown in para **10**. Sometimes pulpit and lectern are designed in one piece with upper and lower levels.
- Pulpit may be centrally in front or to one side of the table or, more rarely, behind the table. It should not be more elevated than is required for the congregation to see the preacher. General design is as shown in para **10**.
- A chair for the minister is provided centrally behind the altar with often at least one additional chair on each side for elders. Alternatively seats for elders may be provided against the back wall of the sanctuary.
- A session room should be provided in addition to a vestry. This is where the elders meet before services and where communion is prepared.

5 THE SALVATION ARMY

5.01 Origins

The Salvation Army (Salvationists) arose from evangelical meetings conducted in east London during 1865 by the Reverend William Booth, a minister of the Methodist New Connexion. Booth decided to take the church to the people. His services were held in the open air, in tents and in theatres: later he built barracks and citadels in which his converts could hold their meetings. His services were sensational: he used brass bands playing secular tunes to accompany hymns; converts (soldiers) wore uniform. He became a 'general' and directed the organisation in quasi-military style. Men and women had equal rights in office. Officers, full time, received a meagre allowance and his soldiers, part time, did their work without remuneration. Booth regarded social work, care of the poor and rehabilitation of the outcast as an essential part of his Christian mission.
The Christian mission, as it was first called, grew beyond all expectations and in 1878 became known as the Salvation Army. In its belief, it is orthodox, evangelical and prophetic. The corps assembles for worship in a hall—a multi-purpose building sometimes called a citadel, temple or barracks. Sometimes within the complex there are two halls, for senior and junior soldiers respectively. Religious services and social activities may be conducted in either or both halls.

5.02 Ceremonies

Ceremonies may be divided into two types:
- services held on Sunday—the holiness meeting on Sunday morning, the praise meeting or festival on Sunday afternoon, and the salvation meeting on Sunday evening (these may be termed, for convenience, ordinary meetings)
- ceremonies applicable to marriages, funerals, memorial services, covenants, swearing-in of soldiers and presentation of colours (these may be embodied within one of the ordinary meetings).

A detailed account of these ceremonies is given in *Church*

Buildings.[1] The architectural implications are shown in the series of diagrams **29.4** to **29.10**.

5.03 Design of the assembly hall

There will be no communion service or Lord's supper in the accepted sense, and visually there will be no elevation or placing of the sacred elements in a part of the building. The font, lectern, pulpit, communion table and altar are not essential to the Salvation Army form of worship. In every Salvation Army hall there is a place for the mercy seat. It is a simple wooden form, usually placed at the front of the platform in front of the congregation. It is a place where Christian and non-Christian penitents may kneel at any time.

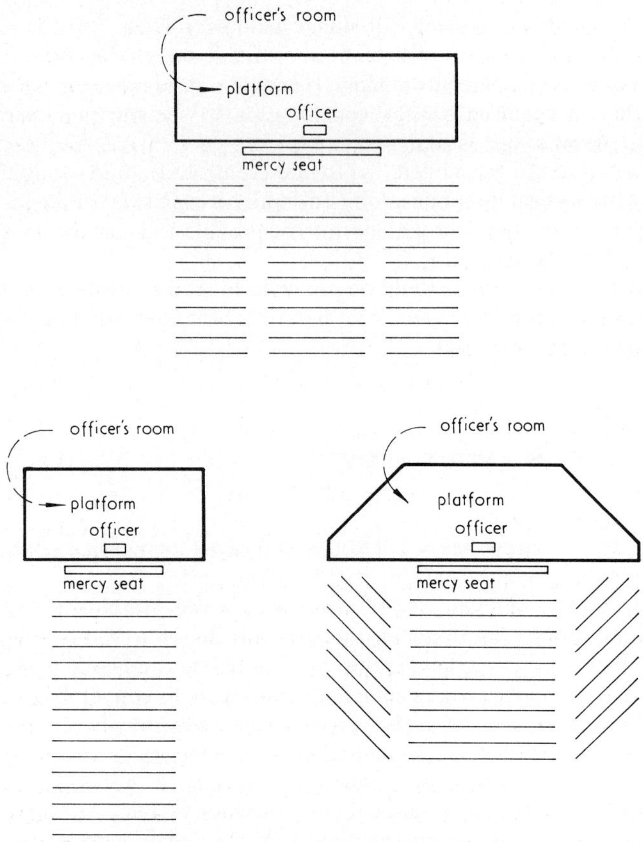

29.4 *Possible basic arrangements. (Note access from officers' room to platform may be on either side or in rear wall)*

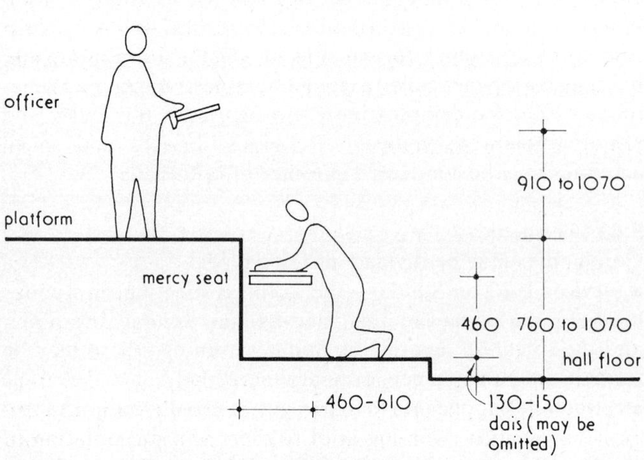

29.5 *Mercy seat*

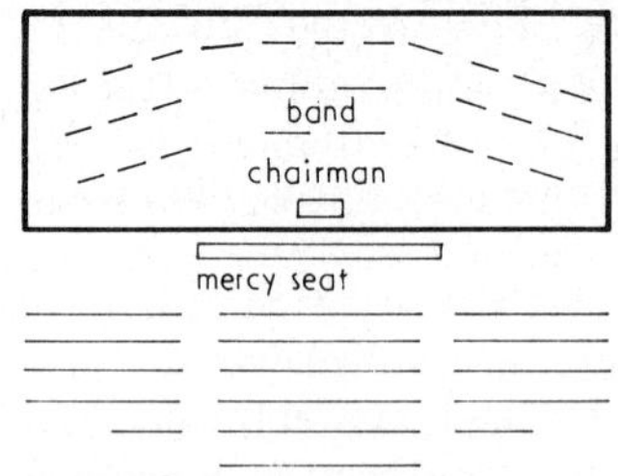

29.6 *Praise meeting or festival*

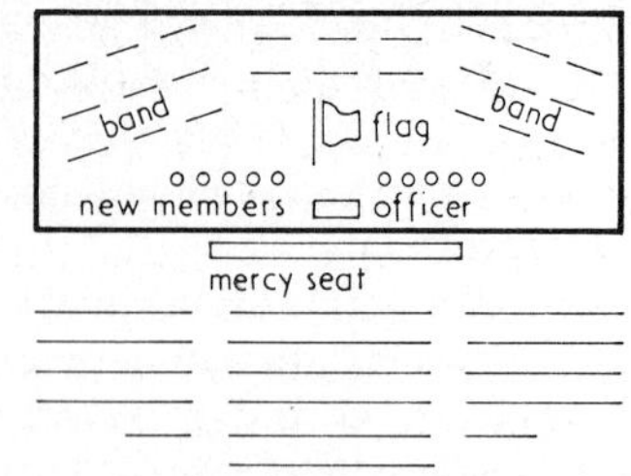

29.7 *Swearing-in ceremony*

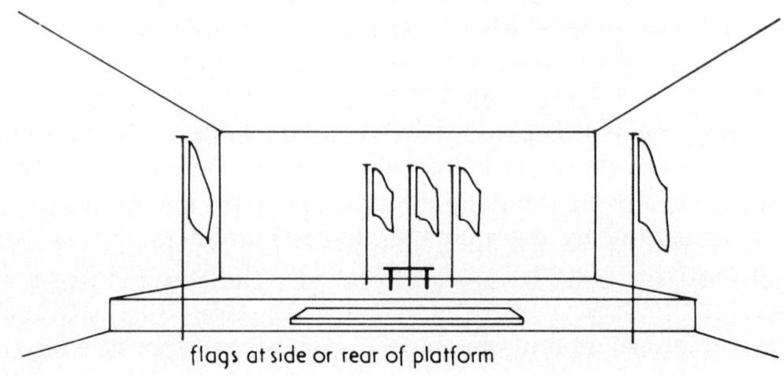

29.8 *Presentation of colours*

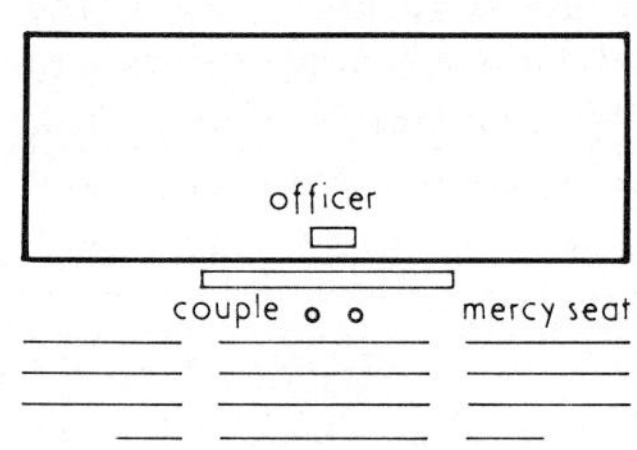

29.9 *Weddings*

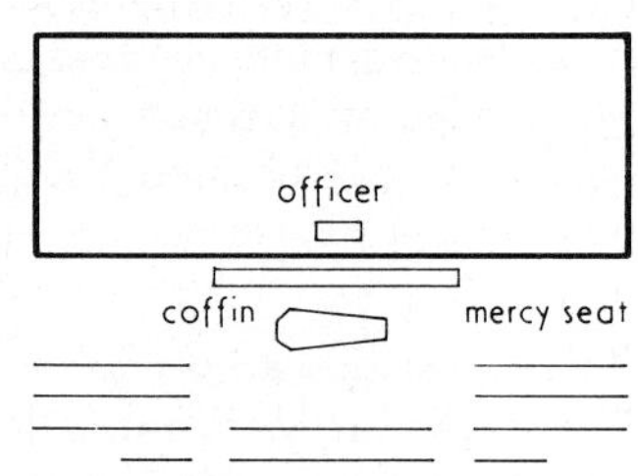

29.10 *Funerals*

Platform

The platform will be required to accommodate, at various times, officers taking a leading part in the services; the band (possibly a visiting one which may be larger than that designed for); the songster brigade or a visiting choir.

Size of the platform will depend on local conditions and requirements but approximately 0·5 to 0·6 m^2 per person (seated in rows) should be allowed. The platform should not be less than about 4·5 m depth, but will normally be more. Its height above the hall floor should not normally be less than 760 mm or more than 1100 mm but sightlines must be considered in relation to a level floor both on the platform and in the hall. If there is a gallery there must be a view of the mercy seat from every seat. Movable seats, not fixed pews, are always used both on the platform and in the hall: flexibility is important.

Hall

As a first estimate for the overall area of the hall, allow 0·56 m^2 for each person to be accommodated.

Aisles should not be less than 1·37 m wide. The space between the front edge of the mercy seat and the front face of the congregational seating should not be less than 1·5 m. A wide rectangle, polygon, or square will be a more satisfactory shape than a long narrow rectangle.

The requirements will be for a reasonable standard of acoustics for speech, choral singing and brass band playing. There should be a reasonable level of natural lighting, and a reasonable standard of artificial illumination, both on the platform and in the body of the hall. The ventilation in most cases should be natural.

Officers' room
The officers' room (vestry) is used as a meeting place by leaders or chairmen of meetings and officers of the Census Board who will assemble in the room before taking their place on the platform. The room should accommodate between two and ten persons.
Corps records will be kept in this room.
Lavatory accommodation will be required, normally one unit for each sex.

Cloakrooms
Lavatory accommodation should be provided for both sexes near or adjacent to the entrance vestibules of the senior or junior halls.
Lavatory accommodation will be required for women in the songster room, and for men in the band room.

Storage
Storage will be required in a band room for brass band instruments, music stands and music. The room will also be used for assembly, briefing and cloakroom. Minimum area usually 23 m^2 to 28 m^2. Instruments are stored in individual lockers to suit sizes of instruments with small store for reserve instruments and cupboard with shelves for music.
Storage will be required in a songster room for a wind type portable organ or electric portable organ with amplifier, and music. The room will also be used for assembly, briefing and cloakroom for the women members who would number about 20 to 30 in an average sized brigade: 18·5 m^2 would be a minimum area.
Where there are junior and senior corps activities, there may be a junior band and singing company (junior choir). Storage will be required for brass band instruments for this junior band and for music for the singing company.
Where there is a separate junior hall, storage compartments will be contained in that hall.
Other social and club activities may also require storage.

6 METHODIST CHURCH

6.01 Origins
The Methodist Church (Methodists) grew out of the Church of England during the eighteenth century and was founded by John and Charles Wesley (born in 1703 and 1707) who were brought up in the atmosphere of the established church tradition. Although John Wesley maintained until the end of his life that he had lived and died a faithful member of the Church of England, and although he regarded the societies and congregations he established as being within the Church of England, a break was inevitable and by the time of his death it was clear that the new movement could no longer be contained within the established church.
A distinct feature of the new movement was the introduction of lay preachers or, as they came to be known, local preachers. In the nineteenth century there were schisms in the Methodist church, the earliest being between the Wesleyan Methodists and the Primitive Methodists. Other schismatic groups continued to emerge during the century until in 1907 a United Methodist church was established by uniting the smaller groups. A further act of union took place in 1932 when the Wesleyan, Primitive and United Methodists joined to form the Methodist church, as it is now known.
The strong emphasis on evangelical preaching is rooted in the Methodist tradition. 'Methodism is nothing if it is not evangelical.' More details are given in *Church buildings.*[1]
The church or chapel is normally reserved for worship alone, although meetings, lectures and musical recitals of a specifically religious character may also take place there. A multi-purpose hall, with a sanctuary that can be screened or curtained off, may also be used for public worship during the early stages of the development of a local congregation. There may be within the complex other units such as classrooms, club rooms and assembly halls, both for religious and secular purposes.

6.02 Ceremonies and buildings
The kind of church required today will seat between one hundred and three hundred people and will in addition have accommodation for catering, adult activities and young people's organisations.
The influence of the liturgical movement shows in certain specific ways, and particularly in a new emphasis on the place of the sacraments in the life of the church, and in the increasing use of set prayers and liturgical forms of service. The altar, therefore, is seen again as the Lord's table, used for the family. The font is no longer viewed as making possible a semi-magical act. It is the place at which and the means by which the person baptised, whether child or adult, is sacramentally incorporated into the body of Christ: this rediscovery of the laity, the Lord's table and the font has to be considered in designing a church, for in the end it is the function of the church which is of supreme importance.
These theological rediscoveries will affect the building of a Methodist church:
- in relation to the shape of the church
- in the arrangement of the sanctuary
- in the siting of the choir.

There are normally two services on Sunday, one in the morning and one in the evening. They are normally conducted by a minister or local preacher from the pulpit, though in some cases (and they may well increase during the next few years) the earlier part of the service may be conducted from a lectern, the pulpit being used only for the sermon.
Holy communion is normally celebrated once a month. In order to emphasise the theological statement that the ministry of the Word is of equal importance to the ministry of the Sacrament, pulpit and communion table may be placed close to each other. The minister dispenses the bread and the wine (the latter in small individual glasses) to each of those who have come forward from the congregation to kneel at the communion rail. A small trough or perforated shelf should be fitted to the inside of the communion rail, about 50 mm or 65 mm deep and slightly lower than the top of the rail, where communicants may place the empty wine glasses. See also para **9**, for dimensions of communion rails.
All other main services are similar to Church of England practice and the architectural requirements are the same, although a credence table is not required near the communion table. Specific guidance is contained in 'The Methodist Church builders' decalogue' (from the Methodist Property Division, Central Buildings, Oldham Street, Manchester M1 1JQ).

7 SOCIETY OF FRIENDS

7.01 Origins
The Society of Friends (Quakers) originated through the experiences and preaching of George Fox (1624–91). His early preaching, which began during the troubled period of the Cromwellian revolution, was violently iconoclastic, denouncing all parish churches as 'steeple houses'.
Disillusionment drove him in search of help. He therefore abandoned his job, left his relatives and travelled the country questioning and reasoning with any who had a local reputation for wisdom in those first Commonwealth days of ferment. Fox gathered other seekers to join him in silent meetings in the Midlands and north-west counties, and the movement which

gained momentum in the north in 1652, spread to the south of England in 1654. Persecution rapidly followed.
When Fox was imprisoned at Derby in 1650 for heresy, he bid the magistrates tremble at 'the name of the Lord', and thus received the nickname 'Quaker'. Early Quakers reacted strongly against the current religious practices and liturgy of the established church. All through their history, Quakers have been concerned with a sense of duty to the community. Because Friends believe that God can communicate with man direct, they do not partake of the outward sacraments.
They have no separated priesthood, they do not require their members to subscribe to a creed, and their worship does not make use of a liturgy. They are however in broad agreement with the main emphases of Christian belief and would claim to be (using the words in their wider meaning) both orthodox and evangelical.

7.02 Ceremonies and buildings

In a meeting for worship, Friends gather in silence as a congregation of seeking souls, and 'wait upon the Spirit'. No one directs the worship.
There may be extempore speaking but there is no music or hymn singing. There is no ordained or appointed priest or minister. Out of the 'gathered' silence of united worship one or another may be led to engage in vocal ministry or in prayer or to read from the scriptures. The building must be designed to help the quiet worship and the vocal ministry of the participants.
There will be no need to concentrate on the things that ceremonies symbolise, for there is no observance of the Lord's supper or orthodox communion service; no baptism or initiation ceremonies involving ritual. There will therefore be no need for font, lectern, pulpit or altar.
Sizes of meeting halls vary, but they will mostly be designed to accommodate 50 Friends or less. A square, rectangular or polygonal room may meet the requirements. It is likely that seating will be required in a square or a circle. Meetings may prefer fixed or movable, tiered or level seating.
There is no need for a table in the meeting room, though there normally is one to act as a focal point. The position is not usually literally at the centre of the room; the seating will be arranged in the way felt by the group concerned to be most conducive to a good meeting for worship, and the table, if placed anywhere, will be fitted in with the general arrangement of the room. Its main use is for business meetings.
Other accommodation might include a multi-purpose hall, library, small kitchen foyer and cloakrooms and a caretaker's flat.

8 PASTORAL CENTRES

8.01 Form and function

Pastoral centres are alternatives to conventional church buildings and are increasingly being considered where there is to be a large new population with non-existent or inadequate ministry and buildings.
A pastoral centre could take many forms—basically it would be a small building or suite of rooms with facilities for consultation, meetings, refreshments and occasional worship. It could be a modified house, a transportable structure or a specifically designed community building: it could even be a converted shop or a caravan. Its purposes would be to shelter a Christian 'presence', comparable to a doctor's surgery or a citizen's advice bureau: ministers would be present for counselling and office work at advertised hours. Small meetings could be held and modest hospitality offered. It could be used for acts of worship though it would be linked to a major worship centre elsewhere. It should be sited in the local centre and could be integrated into a complex of amenity buildings or a shopping centre. If resources were available, and if the population distribution were suitable, two pastoral centres might be provided in a neighbourhood of 6000 to 10 000 people.

Part B Design data

9 ALTAR OR COMMUNION TABLE

9.01 Symbolism

The altar symbolises three things:
- the body of Christ
- the altar of sacrifice
- the table of the last supper.

9.02 Canon law

A Church of England altar may be of wood, stone or other suitable material and may be movable. It should be covered with silk or other 'decent stuff' during divine service, and with a fair white linen cloth at holy communion. A Roman Catholic altar must be of natural (not reconstructed) stone with the top member (the mensa) in one piece and containing the relics of two canonised martyrs or saints. However, this requirement may be fulfilled by a small portable altar which is in effect an altar stone often about 300 mm square (or less) and 50 mm deep containing a sealed cavity (the sepulchre) for the relics. The altar may be fixed (stone cemented to the structure) or unfixed (timber with inset portable stone altar as described above). Only a fixed altar may be consecrated. Alternatively, relics may be sunk into floor below altar.

9.03 Position

Altar must be related to congregational seating in such a way that what is done at it is, and is seen to be, a corporate action of the whole assembly. Altars are being brought further forward in the sanctuary and with congregations grouped around, there is less need for the altar to stand so high as in the past, especially when the priest faces the congregation across the altar.

9.04 Size

For liturgical reasons, altars can be less long but possibly slightly deeper than in the past. Average sizes are shown in **29.11**, and anthropometric data in **29.12**.

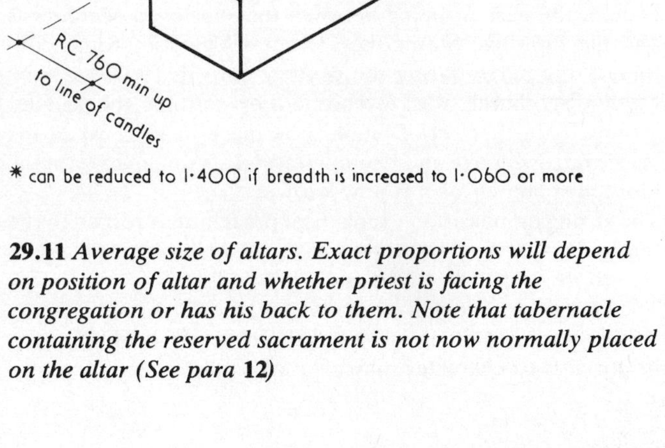

29.11 *Average size of altars. Exact proportions will depend on position of altar and whether priest is facing the congregation or has his back to them. Note that tabernacle containing the reserved sacrament is not now normally placed on the altar (See para* **12***)*

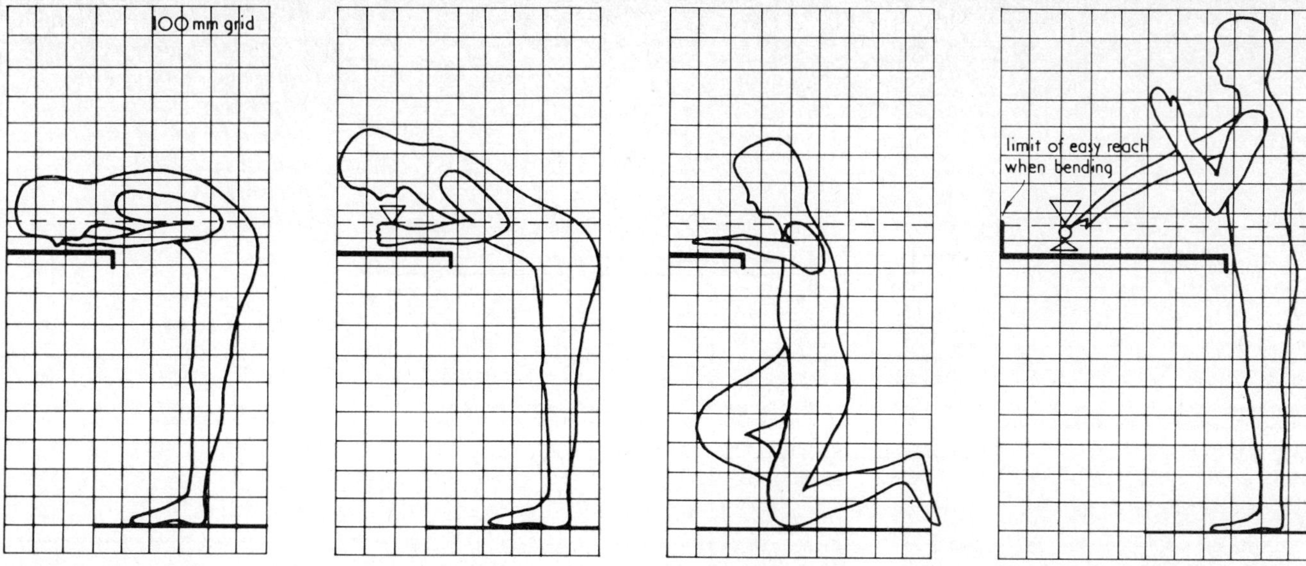

a *Kissing the surface of the altar*

b *Saying the words of consecration*

c *Genuflecting*

d *Reaching forward when standing upright*

29.12 *Critical actions of priest when standing at the altar: The horizontal dotted line represents the eye level of the average adult member of the congregation when kneeling. This should, if possible, be not less than 75 mm above the altar top, which emphasises the importance of keeping the altar top low. This represents some sacrifice to the priest, since actions* **a** *and* **b** *would be easier if the top were at the traditional height of just over 1 m. The genuflecting position* **c** *emphasises the value of recessing the supports of the altar, at least on the priest's side, to prevent him from bumping his knee:* **d** *illustrates the extent of his reach when standing upright*

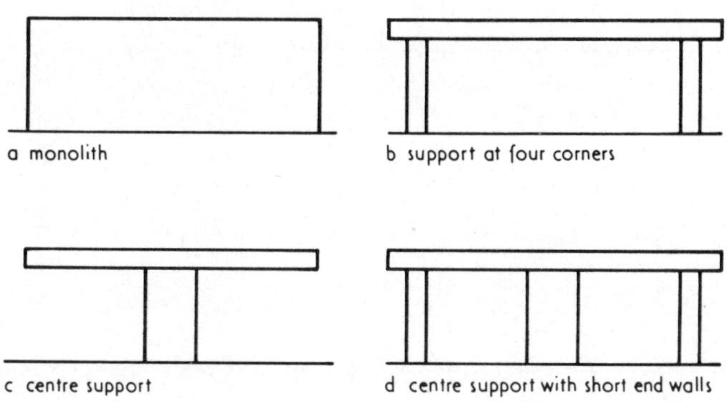

29.13 *Support of mensa (altar top). Note that RC altar should not oversail its supports by more than about 150 mm so that the Bishop may pass his thumb over the joint at the Consecration*

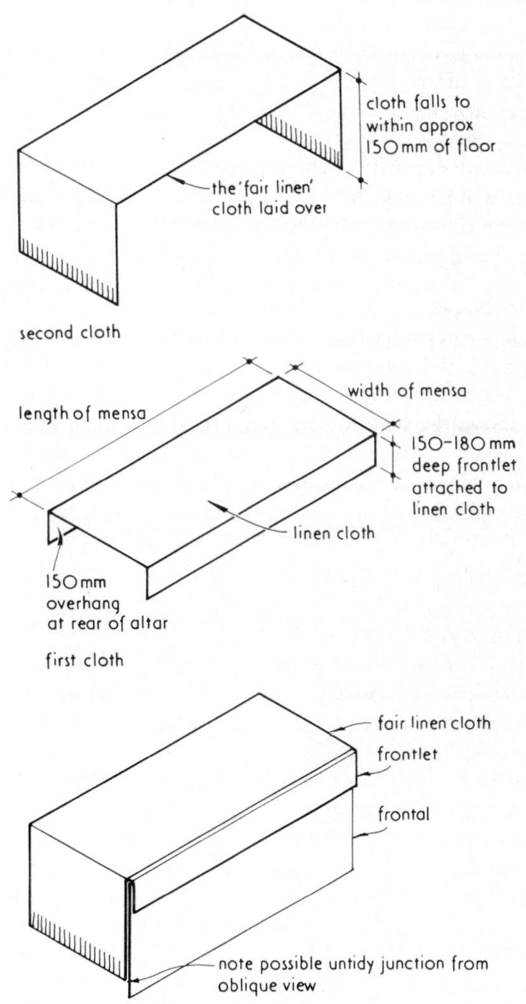

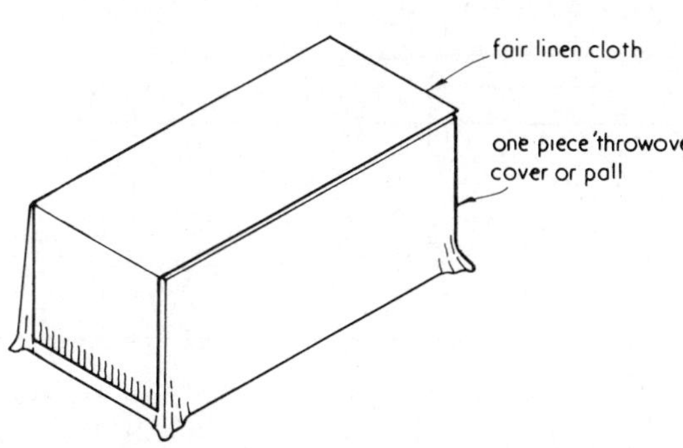

29.15 *One-piece throwover cover. This may have loose draped corners as shown or tight fitted corners*

29.14 *Clothing the altar*

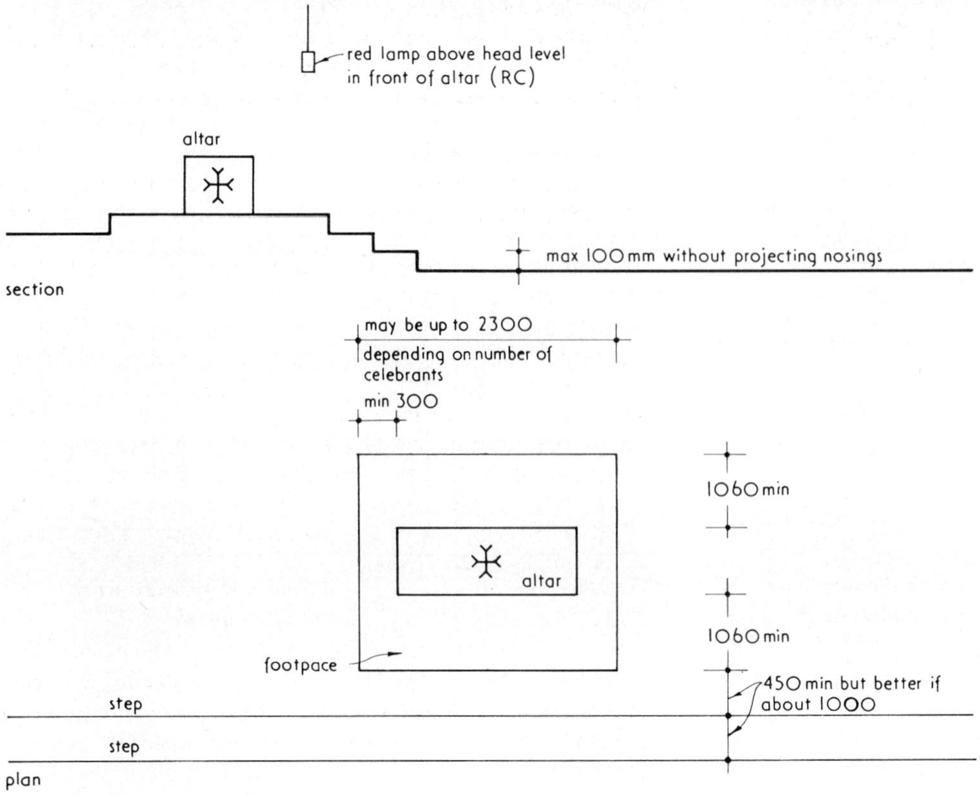

29.16 *Space around the altar. Keep altar as low as possible, usual maximum 3 steps. Position of altar can also be defined by use of canopy, structure, lighting, floor patterns, etc*

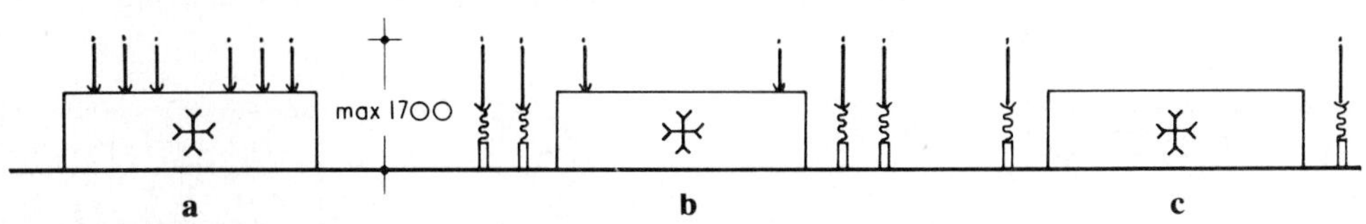

29.17a, b *Arrangements of six candles*
c *Two candles only. Candles may also be placed on lower step than altar to reduce height and possible obstruction*

9.05 Supports and coverings
Methods of supporting the mensa are shown in **29.13**. With the altar now normally nearer the congregation, the question of clothing the altar may need to be rethought. A traditional C of E altar covering is shown in **29.14** and **29.15**.

9.06 Footspace
The platform or base on which the altar rests must be large enough for the number of priests expected to stand around while celebrating the Eucharist. It may be raised one or two steps above the level of the congregation, but in general the altar should be kept as low as possible to avoid any sense of separation between it and the people. Possible dimensions are shown in **29.16**.

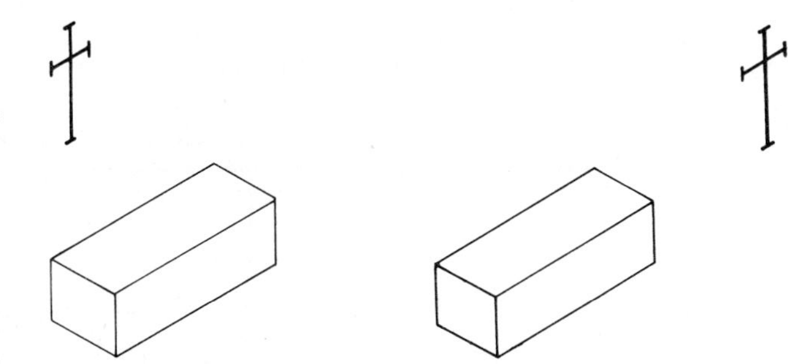

29.18a *Cross may be placed forwards of altar at one side or suspended above*
b *Cross may be behind altar at a height where it will not be obscured by the priest. Crosses are not usually placed on the altar for 'facing the people' churches*

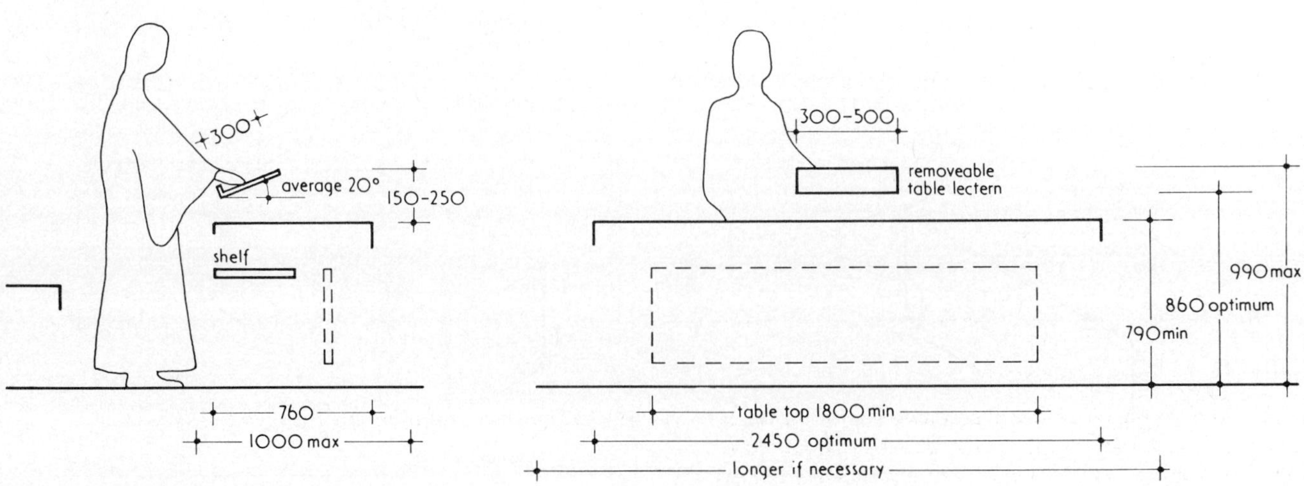

29.19 *Communion table with table lectern for use in Presbyterian churches instead of the altars described above*

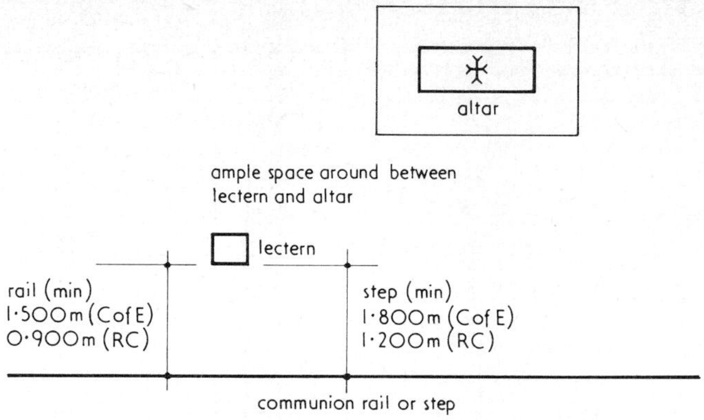

29.20 *Minimum dimensions around lectern depending upon whether communion rail or only a communion step is provided*

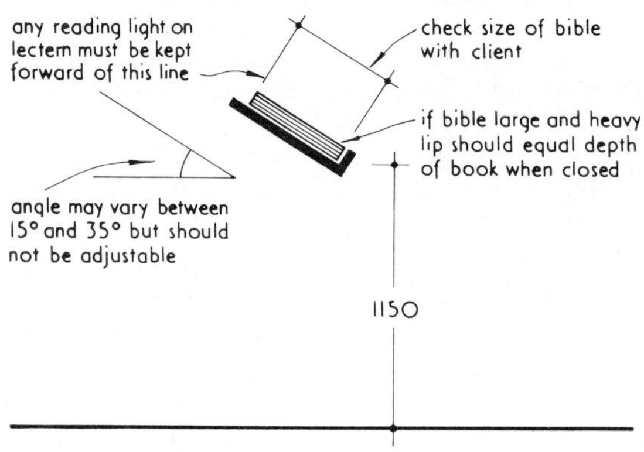

29.21 *Minimum requirements for lectern*

9.07 Cross and candles

The exact requirements must be discussed with the priest. Candles and cross should not form a barrier between priest and people. Some possibilities are shown in **29.17** and **29.18**.

9.08 Communion table

A communion table with table lectern as used in Presbyterian churches is shown in **29.19**.

10 SANCTUARY FURNITURE AND PULPIT

10.01 Lectern

This should be in or near the sanctuary but sufficiently apart from the altar to constitute a separate focus, and conveniently sited to be accessible to the congregation to read from, **29.20**. Occasionally two lecterns may be asked for, one each side. The lectern must be in a position where everybody can see the reader's face including those within the sanctuary. It may be fixed or movable, and it may also serve as a pulpit. Dimensions are shown in **29.21**.

10.02 Altar rail

A rail may be dispensed with in some churches, but most will still provide lengths of rail for kneeling to receive communion. It need not be continuous, but must be rigid and firm. Minimum dimensions are given in **29.22** and **29.23**.

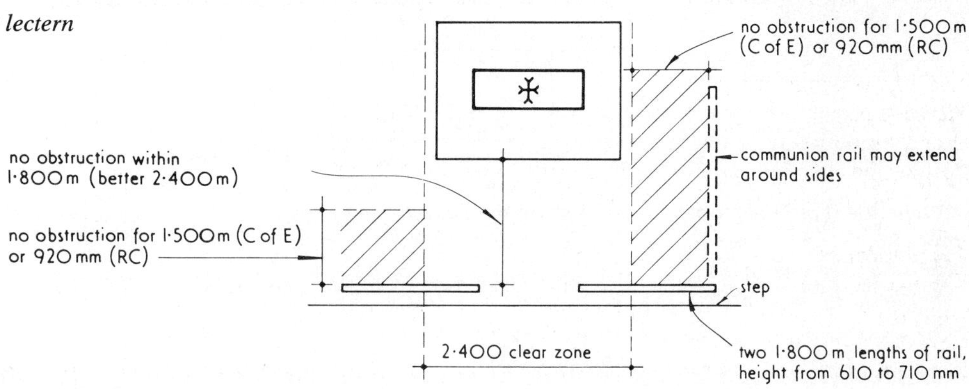

29.22 *Clear space requirements around communion rail*

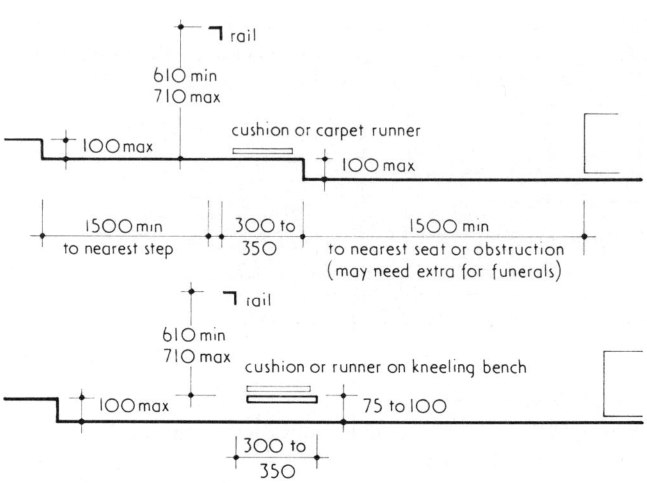

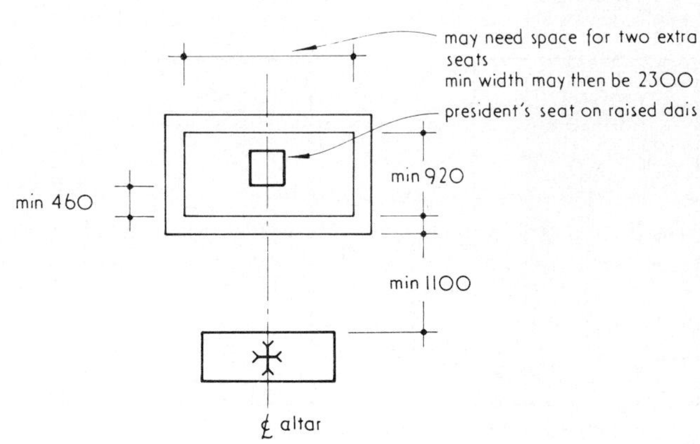

29.23 *Dimensions for altar rail (ie outside central zone of sanctuary): top, with stepped floor; above, with flat floor. In a new Roman Catholic church there is unlikely to be a second stap in the sanctuary and the minimum unobstructed distance on the sanctuary side of the rail is 920 mm. It is also normal for the whole of the sanctuary to be one step above the nave, but too many steps should be avoided*

29.24 *Dimensions around the President's seat*

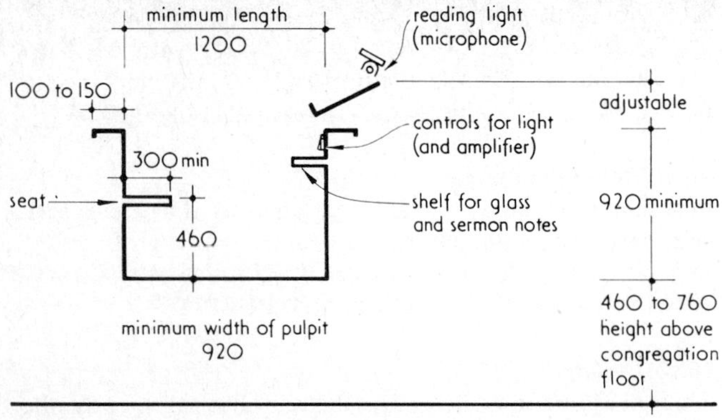

29.25 *Pulpit dimensions*

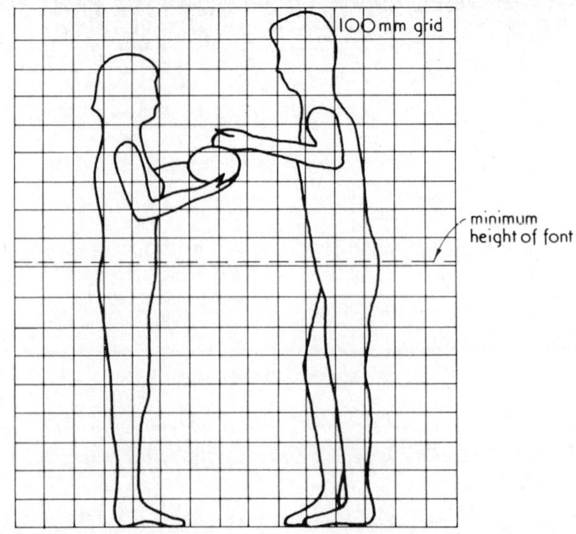

29.26 *Anthropometric diagram showing the critical action at the font*

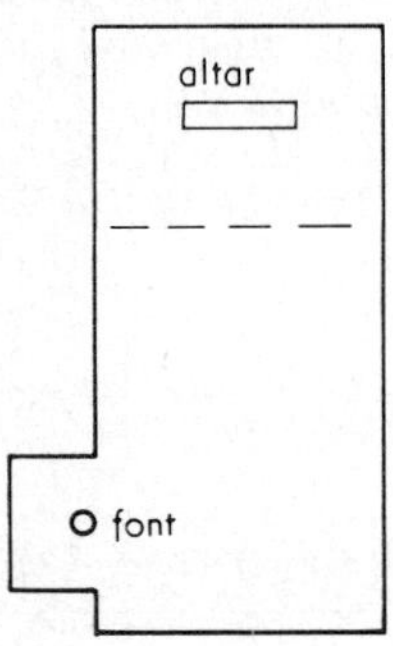

29.27 *Possible variations on position of font*
a *In separate baptistery. Acceptable if only baptism proper is performed there with rest of service in the main body of the church*

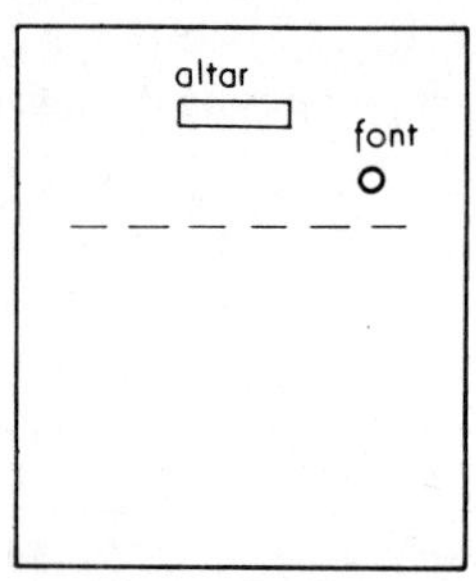

b *Font in sanctuary. May compete with altar or be made insignificant by it. But in Presbyterian church, font is often in this position although lower than the table and possibly on a side extension of the lowest sanctuary step*

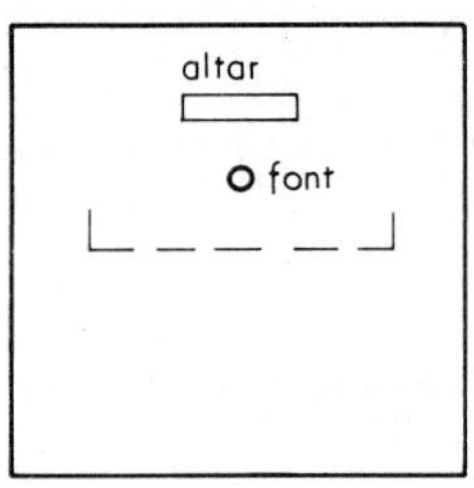

c *As for* **b** *but the font is also an obstruction*

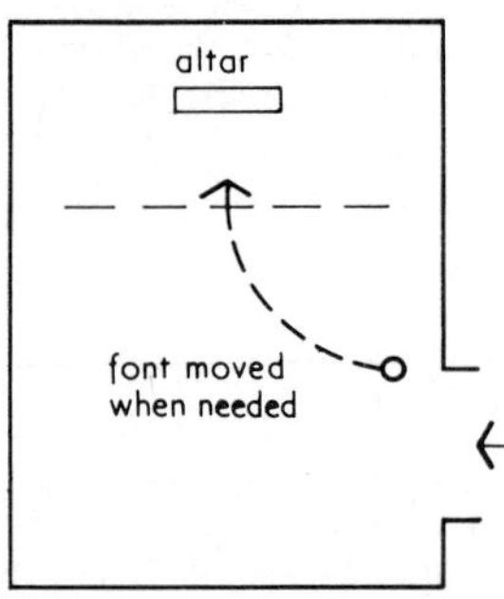

d *Care is needed with a movable font to retain the dignity of one of the church's great sacraments*

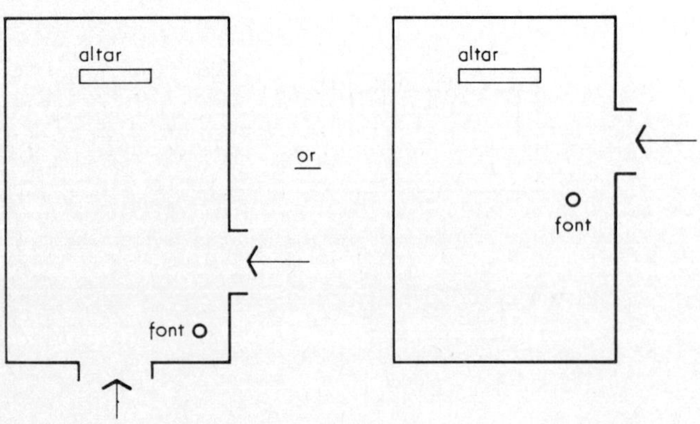

e *When font is placed near entrance, it is important to have the entrance near the font and not the font near the entrance. This may imply that main entrance is not at the back of people's seats, but in front*

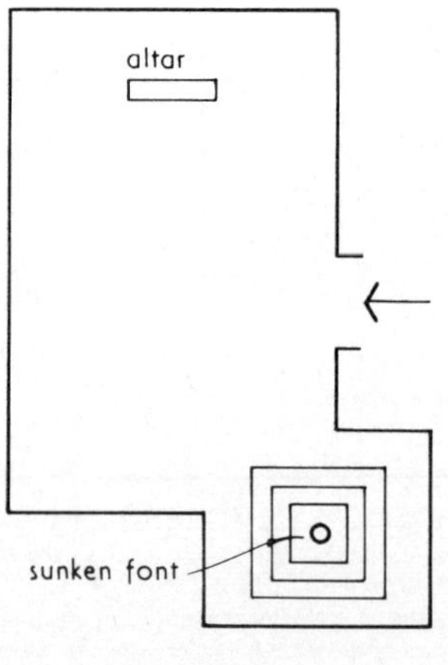

f *Font is sunk into a 'dry pool' suggesting a 'drowning' symbolism*

10.03 Credence table

This serves as a sort of sideboard for water and wine, but check if anything else is to be placed on it (eg offertory money, service books etc) in which case a shelf below the table may be needed. It should be to the right of the celebrant within the sanctuary but not where it might cause a visual obstruction. It could be a shelf instead of a freestanding table. Dimensions will vary from about 610 × 760 mm to 1200 × 460 mm, with a minimum height of 820 mm.

10.04 Seating for officiants

These will vary considerably and requirements must be established with the local community.

Ceremonial seats for priests. In RC churches the 'president's seat' will be on the centre line behind the altar, raised up on one or two steps, **29.24**. The seat which is used by the priest must not look like a Bishop's throne and will usually not have a back to it. The Bishop's throne is portable, and placed in front of the altar whenever he visits the church. In C of E churches it is often better to have the ceremonial seat (which is used by the Bishop) at the side of the sanctuary facing the altar.

Incidental seating should be kept to a minimum. It will be needed for additional priests, servers, and lay people on special occasions. It will usually be in the form of benches at the side of the sanctuary.

Stalls will be needed (especially in C of E churches) for priests at matins and evensong at the side of the sanctuary. Dimensions are similar to those for congregational seating (see para **14**) except that as priests are vested they will need more room to move in and out. Surface for resting a book should be deeper (say 300 mm) and almost horizontal— not sloping—with a shelf below.

10.05 Pulpit

Sited outside but close to the sanctuary. No particular location is now insisted on apart from functional reasons of good sight and sound. Minimum internal area is about $1{\cdot}2\,m^2$. Access should desirably be from the side (if from back, fit a door). Other details are shown in **29.25**. A Presbyterian pulpit may be behind the altar or at either side.

11 FONT

11.01 Description

In RC churches movable fonts are not permitted and they are not generally approved of in the Anglican communion. They can be used in Presbyterian and Methodist churches. In C of E churches fresh water is used for each baptism drained to a separate soakaway. In RC churches water is blessed once a year and the baptismal water is stored in the font which has two compartments.

11.02 Dimensions and shape

Anthropometric requirements are shown in **29.26**. Shape is governed by the needs of the priest (space for service book and other small objects); comfort of priest in holding the baby; and the safety of the baby (priest should not have to lean too far over).

11.03 Position of font

Various alternative positions are shown in **29.27**. The font must be approached from the church (congregation) side but be divided from it in some way (eg by difference in floor or ceiling levels). It must be in a prominent position and be seen by the congregation when seated, or space provided for most of the congregation to stand around. It should be a separate focus from the altar and therefore possibly not in the sanctuary.

12 RESERVATION OF THE SACRAMENT

12.01 Siting

The reserved sacrament in RC churches (the consecrated bread of the Eucharist) is not now normally kept in a tabernacle on the altar. It is normally kept:

- in a separate chapel outside the sanctuary
- in a position which is architecturally important
- where its location will be easily recognised by the congregation.

The requirements are complex and the practice of the local church community must be established and agreed with the Diocesan Advisory Committee. **29.28** shows height limitations. The reserved sacrament is occasionally found also in the Church of England, but is more rare and many methods of reservation are practised.

For fuller explanation see AJ information sheet 1529 contained in *Church buildings.*[1]

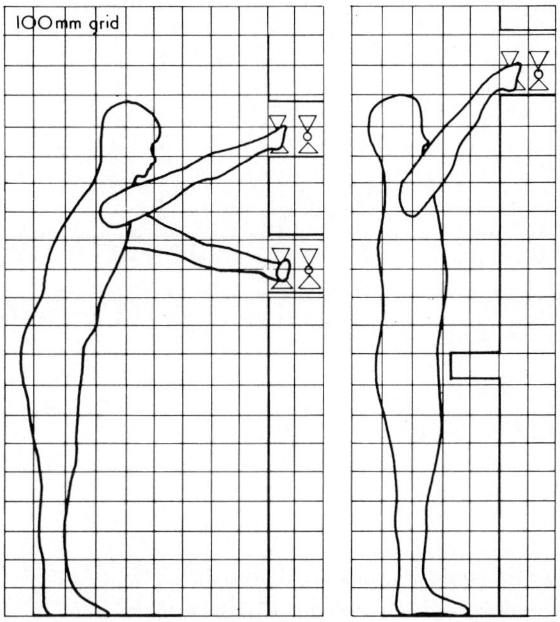

29.28 *Left, removing ciborium from tabernacle and putting on lower shelf. Right, maximum height to give view of back of tabernacle*

13 ENTRANCE AREAS

13.01 Design requirements

- provide wind lobbies outside entrance doors
- minimum clear door width (for processions and funerals) should be 1·7 m (1·1 m in small church); minimum clear height (for processional cross) 2·3 m, otherwise 2·05 m is sufficient
- provide proper access for the disabled (see *Designing for the disabled*)[2]
- provide secondary exit door, especially for weddings
- provide gathering spaces inside and outside the church as part of the normal exit route where people can naturally gather to talk
- where needed (especially in RC churches) place a holy water stoup on the entrance side of each doorway leading into the church. Rims should be 710 to 760 mm above the floor.
- provide considerable space for notices of all types in a conspicuous position
- allow space, where required, for one or two small credence tables just inside the entrance
- provide facilities for selling of publications etc
- where possible provide access from the entrance area to the wc and to the sacristy or vestry
- as a general rule try to keep the purely *secular* activities more in view when the congregation is *leaving* than when entering for a service.

14 CONGREGATIONAL AND CHOIR SEATING

14.01 Arrangement of seating

The congregation should be *continuous* with the minister with no strong dividing line between them. Equally the congregation must be united with one another and all must have good access to circulation and to the sanctuary. Seats facing each other around an altar should not be closer than about 6 m, and nobody should look sideways on to anybody else closer than 1·5 to 3·0 m.

14.02 Spacing and dimensioning of seats

Anthropometric details are given in **29.29**. Note also:

- minimum dimension of seat plus kneeling space front and back is 920 mm. Where congregations do not kneel (eg Methodists, Presbyterians), the dimension can be reduced to 760 mm
- leave a space of about 280 mm between front edge of seat and back edge of kneeler
- allow minimum width of 510 mm per person
- maximum length of row is 10 persons (5·1 m) with access from both ends, or 6 persons (3·06 m) with access from one end
- the ledge for hymn books should be about 150 mm wide (300 mm for the choir).

14.03 Circulation

Basic dimensions are shown in **29.30**. Space must be allowed for invalid chairs[2] during the service where the occupant can participate but not block circulation.

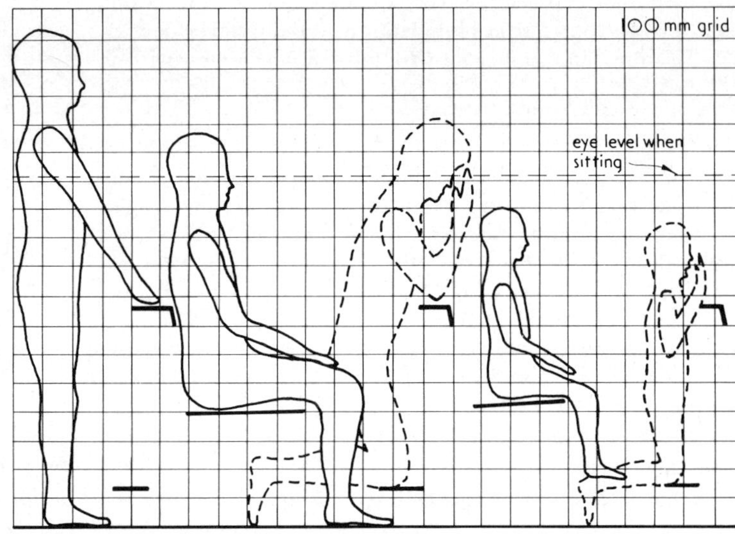

29.29 *Critical dimensions for seating. Black lines indicate one particular solution which has been found very satisfactory in practice. The dotted line indicates 'eye' level when sitting and concerns the relationship with the altar top*

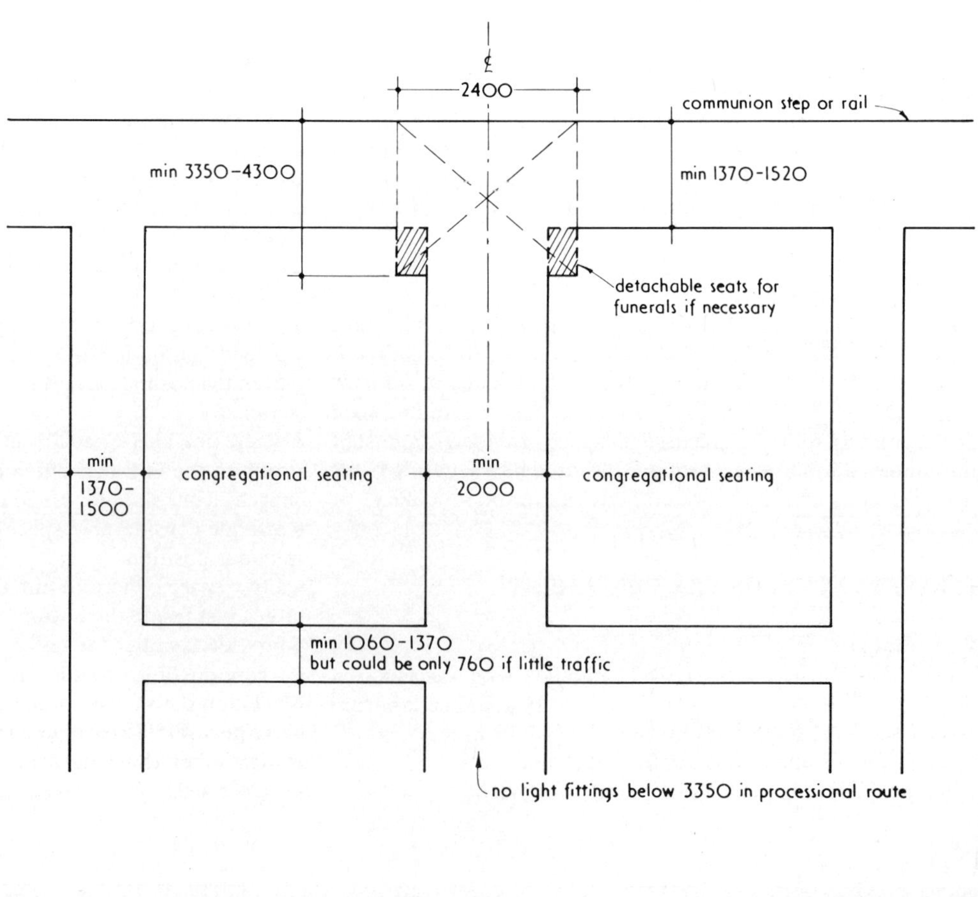

29.30 *Aisle widths. Size and pattern will depend on overall plan and liturgical considerations. Seating may be fan shaped or in blocks around the sanctuary*

15 VESTRIES AND SACRISTIES

15.01 Accommodation

Accommodation requirements will vary considerably. The most lavish could include:

- priest's sacristy (sometimes called priest's vestry)
- server's vestry
- choir vestry and practice/committee room
- women's choir vestry
- churchwardens'/interview room
- cleaners' room } sometimes placed together
- flower arrangers' room } and called working sacristy
- priest's wc
- men's and women's wcs
- general storage
- small kitchen.

15.02 Planning relationships

The traditional position of the vestry/sacristy complex opening directly on to the sanctuary has some disadvantages: it emphasises the separation of priest from congregation and makes sharing of toilet facilities difficult. The vestry could be near the entrance (but ensure security) so that priest and procession go through the congregation to their place in the sanctuary.

Since children often participate in adult worship for only part of its duration, movement between the main worship space and ancillary rooms should be considered.

There should be a lobby with double doors between the choir vestry and the worship room, and a space out of sight of the congregation for processions to form up. A door direct to the outside is essential as priests and choir should not have to go through the worship room to reach their vestries.

15.03 Detailed design

There is a very wide range of objects and vestments to be stored and the precise requirements must be ascertained. A vestment storage cupboard is shown in **29.31**. Doorways should be 1700 mm wide × 2300 mm high.

More detailed lists are given in *Church buildings.*[1]

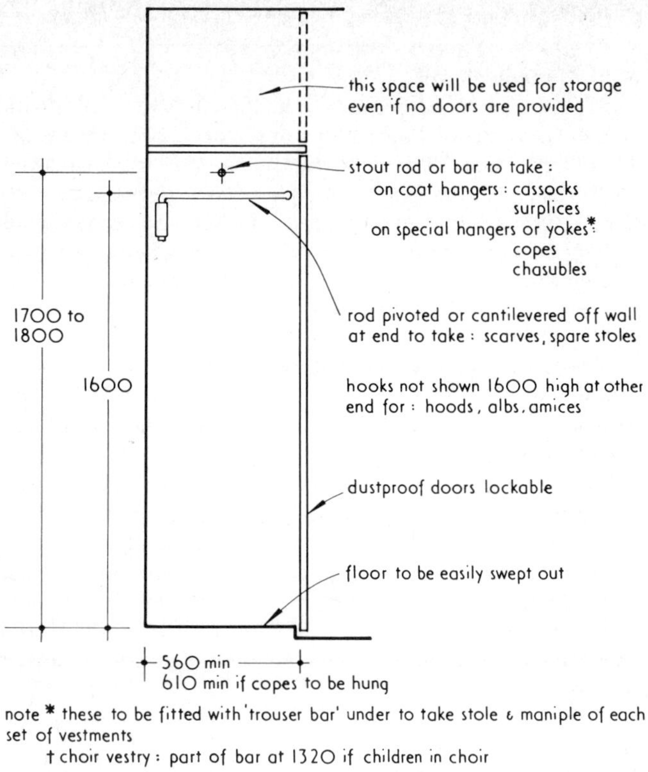

29.31 *Section through vestment cupboard*

16 CONFESSIONAL

16.01 Design requirements

Priest and penitent must be able to hear what each other has to say without being overheard by others. The Bishop's requirements must be established, and the psychological expectations of the parish.

Either priest and penitent may be enclosed in an acoustically isolated box or, more desirably, they may be visible but placed far enough away from other people to make a physical acoustic barrier unnecessary.

Basic dimensions are shown in **29.32**.

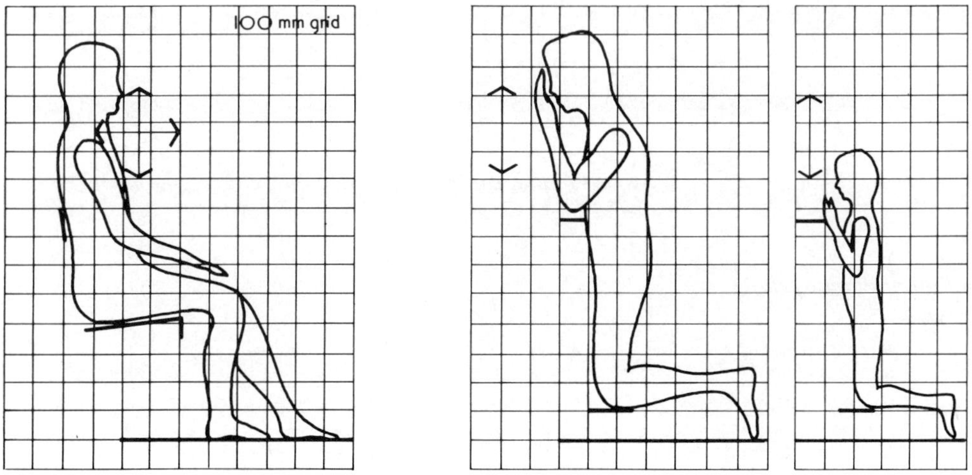

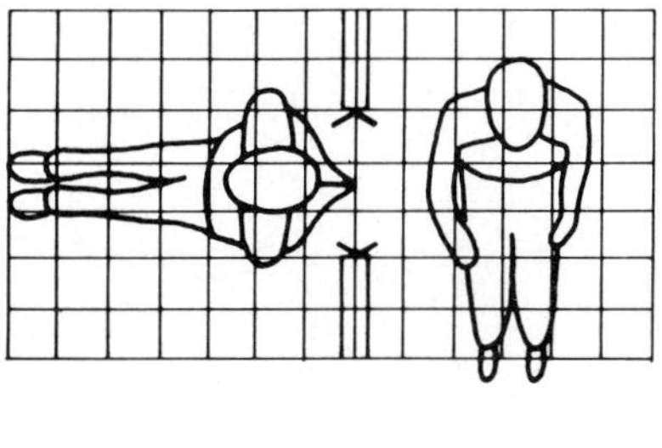

29.32 *Anthropometric study of confessional. Left, relationship of grille to seated priest, and plan view of priest and penitent. Above, kneeling penitents showing positioning of grille where it will serve both adult and child*

17 ORGAN AND CHOIR

17.01 Music in church

Use of music in worship can be a highly emotive subject with the type of organ and the placing of the choir two of the most difficult problems. For new churches organs based on the 'werk prinzip' (sometimes called 'neo-classic') should be considered rather than the more traditional Victorian organs with high wind pressures and powerful but muffled tone (see *Church Buildings*).[1] The organ builder should be brought in at an early stage of the discussions.

17.02 Relationship of musical elements

A few basic principles must be observed:

- the player must be with his instrument (ie the organ console must be near the pipes), **29.33a**
- the choir should be as near as possible to the organ, **29.33b**
- ministers should not be separated from either people or organ, **29.33c**
- people should not be placed between the choir and ranks of pipes, **29.33b, c**
- the organ must be in the main volume of the building and raised above the floor so that the pipes are above the heads of the listeners.

Different possible locations of choir, organ and congregation are shown in **29.34**.

18 BIBLIOGRAPHY

[1] Architects' Journal, *Church buildings*, London, Architectural Press, 1967

[2] Selwyn Goldsmith, *Designing for the disabled*, London, RIBA Publications, 3rd ed, 1976

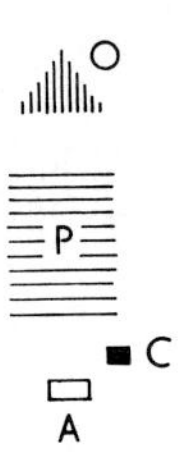

O organ
C organ console
P people
CH choir
M minister
A altar

29.33a *Bad organ arrangement*

29.33b *Bad position of choir and people*

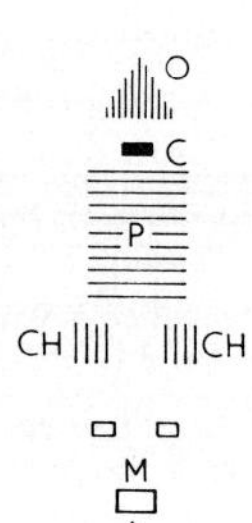

29.33c *Bad position of choir, ministers and people*

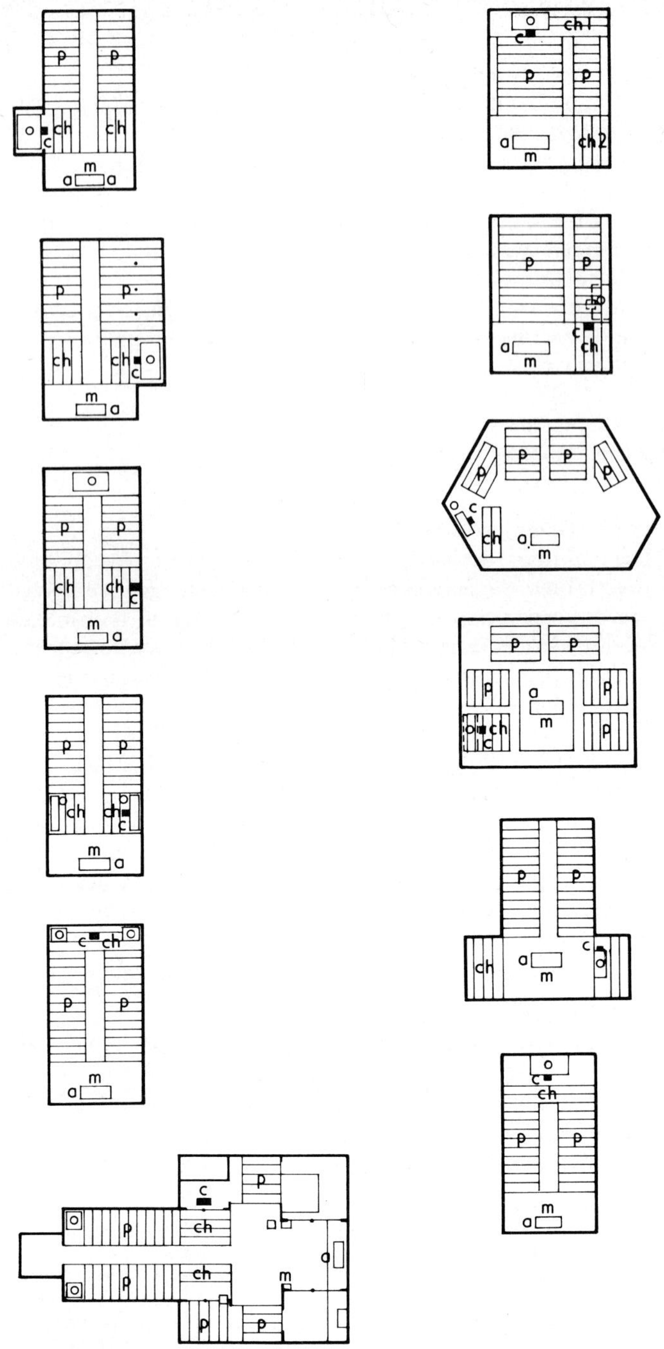

29.34 *Relationships of organ, choir, congregation and minister*

30 Schools

Jeremy Wilson

CI/SfB 71
CI/SfB (1976 revised) 71

Jeremy Wilson is a Principal Architect in the Architects' and Building Branch of the Department of Education and Science

Contents

1 TYPES OF SCHOOLS

The main types of schools are shown in **30.1**.

2 THE DESIGN OF PRIMARY SCHOOLS

2.01 General principles

Primary education today demands a building that is responsive and flexible in the way it can be used. This is most likely to be achieved by the judicious choice and disposition of spaces rather than by reliance on sophisticated hardware or an entirely 'open' solution.

30.1 *The structure of school education in the United Kingdom*

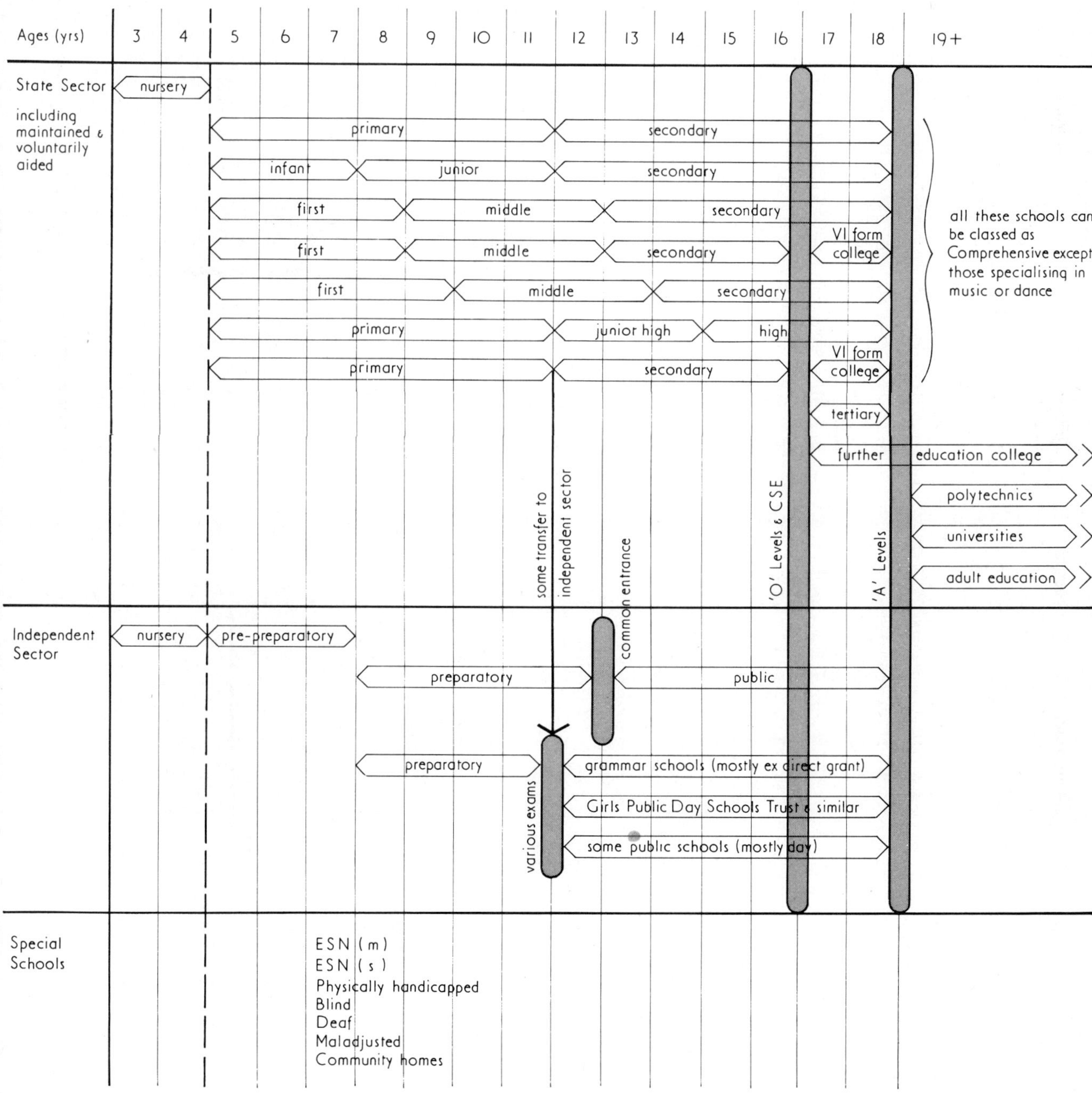

'Flexibility in use' means that a number of diverse activities (noisy and quiet, messy and clean) must co-exist in varying proportions throughout the school day. If the building is not to constrain its occupants, some enclosed (and acoustically isolated) spaces are required. Teaching methods too will vary over the life of the school. Some teachers find they prefer to work as individuals with their own group of children; others may wish to co-operate with their neighbours. Additional adults (specialist teachers, parents, students) may need to be accommodated so as to make the most of their contribution.

In practical terms flexibility in use can be considered on several levels:

Long term This is inherent in the original design. It is dependent on the basic structure, the relationship of open and closed spaces together with specialised facilities, the pattern of circulation and shared facilities and the environmental servicing.

Medium and short term The response to the changing pattern of teaching from week to week and month to month related to changes in school roll, group size and organisation, staffing and general educational needs. Given a particular design, such changes can often be achieved by modifications in the layout of furniture such as divider units and display screens, and in the use of curtains. Certain basic provisions in the design will encourage (or hinder) the development of a wide range of day-to-day activities. For example, the choice of floor finish will welcome some activities and discourage others. The use of teaching aids often involves special storage provision, together with suitable services at their potential points of use.

2.02 Community involvement

Community involvement during school hours

Adults, mainly parents, may support the teaching staff by listening to reading, helping with cooking, etc. The possibility of this involvement should be recognised from the outset. If a separate financial contribution is available, it may be appropriate to provide a social base to these activities.

Community involvement out of school hours

Some of the major facilities of the school may be made available to the community for evening and weekend use. Commonly these will be the hall (or PE space), drama and craft facilities and the playing fields. The school should be planned so as to allow access to these facilities, while enabling other areas to be kept private to the school, **30.2**. Services, particularly heating and lighting, should be zoned to allow economy in running.

Parallel school and community use

Some schools (eg Abraham Moss) have been integrated with community amenity and social service provisions.

2.03 Outside spaces

Statutory requirements determine the minimum areas of sites, hardplay, playing fields and car parking. Within this framework educational requirements will vary with the age range being considered. Nursery gardens should offer a wide

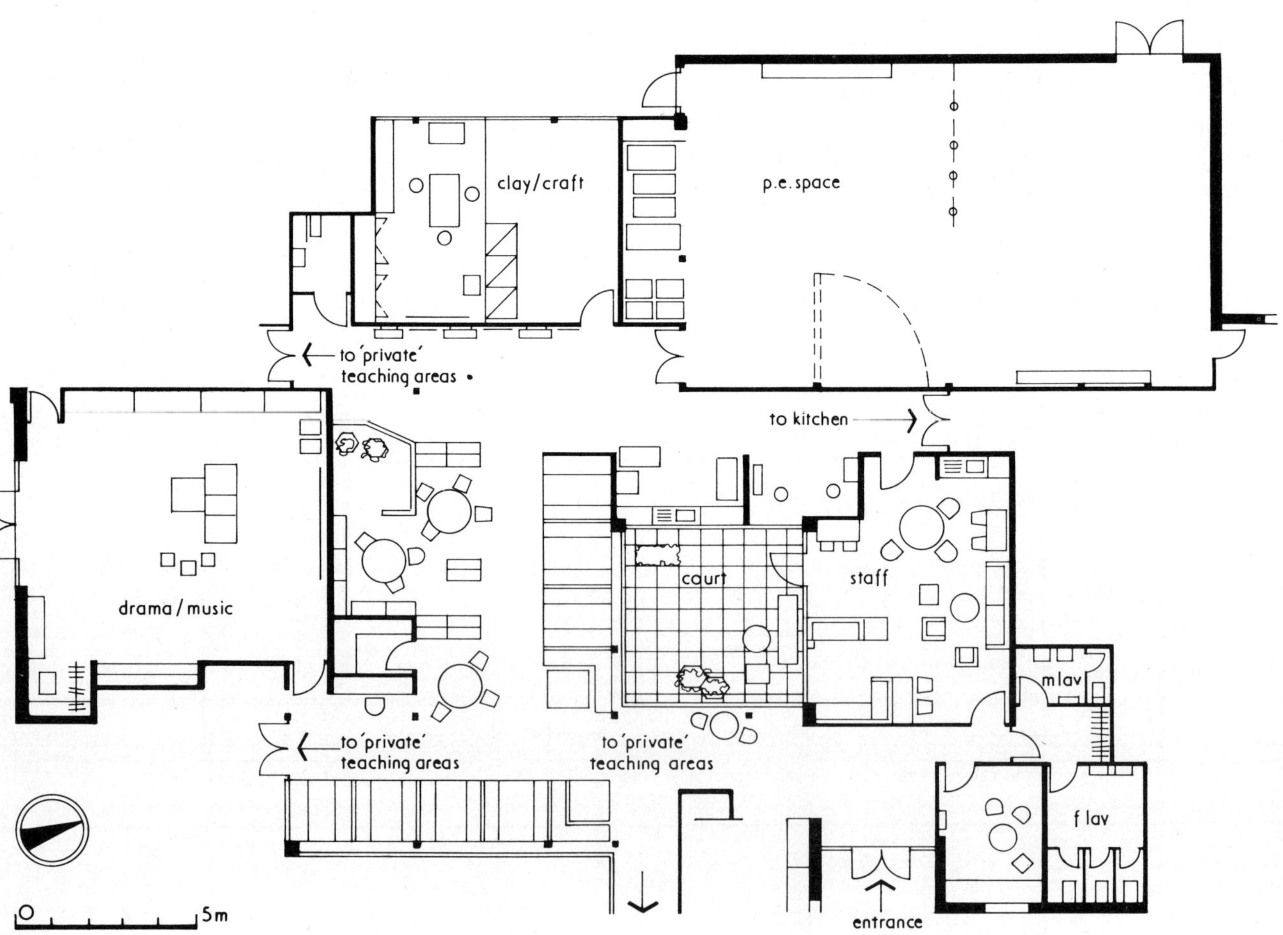

30.2 *Guillemont Junior School, Farnborough, Hants, showing areas available for general community use*

range of opportunities for exploring, games and other activities. In addition to some fixed climbing equipment, materials for building, sand and water play will be needed. The gardens need to provide seclusion and a safe place to play from which children cannot easily stray. Infant hardplay areas can be informal in shape so as to cater for (and possibly separate) quiet play and ball games. Informal games may take place on grassed areas. At the junior age range some of the hardplay should be rectangular for formal games (ie netball) and the playing fields should allow one or two small (75 × 40 m) football pitches to be marked out. A nature area with a pond is an advantage, together with facilities for imaginative play. Provision for outside work, quiet play and social gatherings is also required. In general the problem of vandalism must be recognised. Internal courtyards will offer a degree of protection in this respect.

2.04 Nursery provision

Age range 3 to 5 years (mainly part-time)
Normal roll size 30 to 60 full-time equivalent (2 part-time = 1 full-time)
Gross area approximately 4·18 to 4·64 m^2 per cost place
Playroom area (minimum) 2·3 m^2 per pupil
Nursery garden (minimum) 9·3 m^2 per pupil (minimum 3·7 m^2 paved).

General

Nursery units and classes will normally be associated with infant schools and depend on them for some facilities (including meals).

Example: planning ingredients for a nursery group of 40

Playroom area: 92 m^2 minimum. Activities will include large-scale floor play, painting, clay modelling, sand and water play, music and movement in small groups. Activities will often spill outside and a covered verandah is desirable.
Quiet room 10 m^2 (this will also be used by the staff)
Reception/coats 16 to 17 m^2
Office 8 to 9 m^2
Store 9 to 10 m^2 (also accessible from outside)
In addition to pupils' and staff lavatories, a kitchen/utility room is desirable.

2.05 Infant schools

Age range 5 to 7 years
Normal roll size 240 to 320
Gross area approximately 3·7 m^2 per cost place
Usable teaching area 2·2 m^2 per cost place (statutory minimum approximately 2·0 m^2).

General

The design should satisfy the child's need for security at the same time allowing the opportunity for developing wider relationships by contact with neighbouring teachers and children.

Example

Planning ingredients of teaching accommodation for a group of 90 to 105 children with three teachers, **30.3**.

Open areas

(*a*) General activity space: children mainly working at tables (reading, writing, drawing, number work).
(*b*) Practical areas: messy tasks (water play, painting, modelling, collage). Sinks, working surfaces and storage are required.
(*c*) Covered outdoor work area: relates to (*b*) and (*d*).

Closed or semi-closed areas

(*d*) Play activity studio (29 m^2): floor play, large-scale craft projects, various spontaneous (and possibly noisy) group activities.
(*e*) Three group rooms (13 m^2) or group bays. Group rooms may be used by small groups for story telling, listening to (or making) music, watching TV. Group bays are more likely to be used as a withdrawal area for 2 to 3 children or for perhaps a parent with an individual child.

Facilities shared with remainder of school

(*f*) Large group room (20 m^2). Large group activities (including audio visual aids).
(*g*) Dining: This activity is likely to be accommodated by dual use of some of the teaching space such as individual study area (ideally not the hall).
(*h*) Staff: a room for relaxing, facilities for informal work and preparation of materials, head teacher's room (may be associated with teaching areas), secretary's room (secretary may act as receptionist).

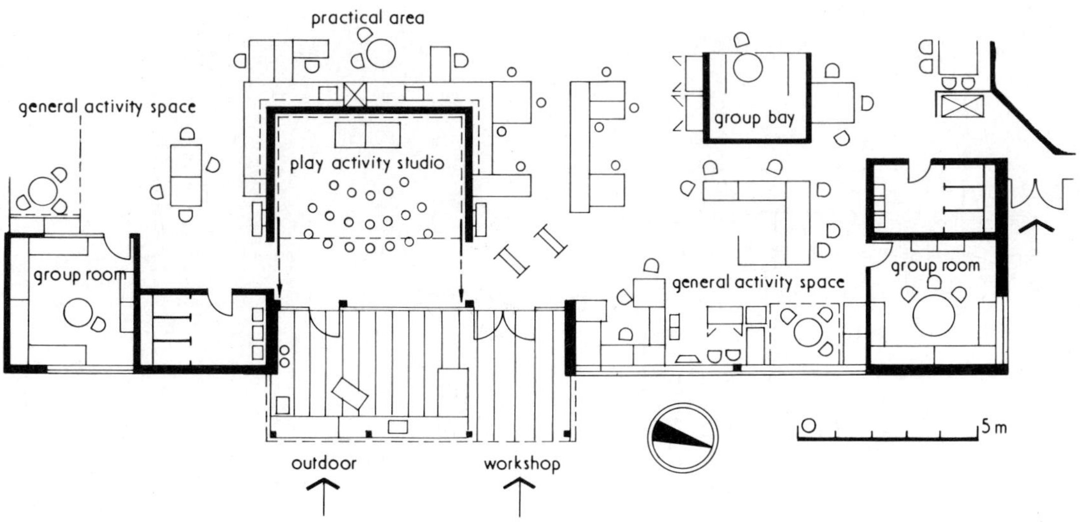

30.3 *Chaucer Infant and Nursery School, Ilkeston, Derbyshire, showing the ingredients of infant teaching accommodation*

2.06 Junior schools

Age range 7 to 11 years
Normal roll size 320 to 480
Gross area approximately 3·7 m^2 per cost place
Usable teaching area approx 2·2 m^2 per cost place (statutory minimum approximately 2·0 m^2).

General

The design should allow for variety in both organisation and teaching methods; it should provide a recognisable 'home base' area for each child while providing opportunities to move away and to use more sophisticated equipment.

Example

Planning ingredients of teaching accommodation for a group of 90 to 105 children with three teachers, **30.4**.

Open areas

(*a*) General activity space: from this might be formed a pair of home bases. By use of display and furniture each teacher can establish a separate territory. Children will be working mainly at tables individually or in groups of 2 to 6.
(*b*) Practical areas: if these are regarded as nuclei, activities will overflow back into (*a*). They need to provide the basic services and storage for work in science, cookery, painting, modelling. Access to the outside is desirable.

Closed areas

(*c*) Small group room (17 m^2): this provides the essential balance to the open areas allowing quiet (or noisy) work by 10 to 12 children. In addition up to 35 children can be accommodated for short periods (for a story, TV programme, etc).
(*d*) Large group room (50 m^2): this room (if equipped with a sink) is capable of use as a traditional primary classroom. However if the three teachers choose to treat the 90 to 105 as a single unit, this room might develop as a science, mathematics or languages base.

Facilities shared with remainder of school

These suggestions are appropriate to a school of 480 children. Not all can be provided in smaller establishments.
(*e*) Physical education space (128 m^2): in addition to PE and movement, if robust in construction will allow ball games and lunch-break play on wet days. Minimum clear ceiling height of 4·2 m is required.
(*f*) Drama/music room (60 m^2) used by groups of up to 35. In addition will accommodate larger groups for audio-visual activities and half-school assemblies.
(*g*) Clay/craft room (40 m^2): for preparation of basic materials (clay, wood, metal) before their dispersal around the school. Also small group craft work.
(*h*) Covered work area: (should relate to (*g*)).
(*j*) Library/central resource area: for books and display, reading and individual work. Can also be seen as a social nucleus.
(*k*) Tutorial (8 to 9 m^2): where one adult and one or a small group of children can work privately.
(*l*) Dining: this activity is likely to be accommodated by dual use of some of the teaching space such as individual study area (ideally not the hall or PE space)
(*m*) Staff: a room for relaxing, facilities for preparation and reproduction of teaching material (may be dispersed), head teacher's room, school office (adjacent to reception area), medical inspection room (may also be used as tutorial).

2.07 Middle schools

Age range 8 to 12, 9 to 13 years
Normal roll size 280 to 560
Gross area approximately 5·2 m^2 per cost place
Usable teaching area approximately 3·5 m^2 per cost place.

General

An extension of the junior school approach is appropriate for the 8 to 12 school, allowing more scope for practical investigation work and some increase in the facilities shared by the whole school. The 9 to 13 school may need more emphasis on specialised areas including storage and preparation areas.

Example (8 to 12 school)

Planning ingredients of teaching accommodation for a group of 90 to 105 children with three (or four) teachers. Read in conjunction with **2.06**.

Open areas

(*a*) General activity areas as in junior example; practical areas and facilities expanded from the nucleus concept.

Closed areas

(*b*) Small group room (17 m^2). Functions as in Junior example.
(*c*) Large group room (50 m^2). Can function as in Junior example, but it is likely that around the school some will be used mainly for languages.

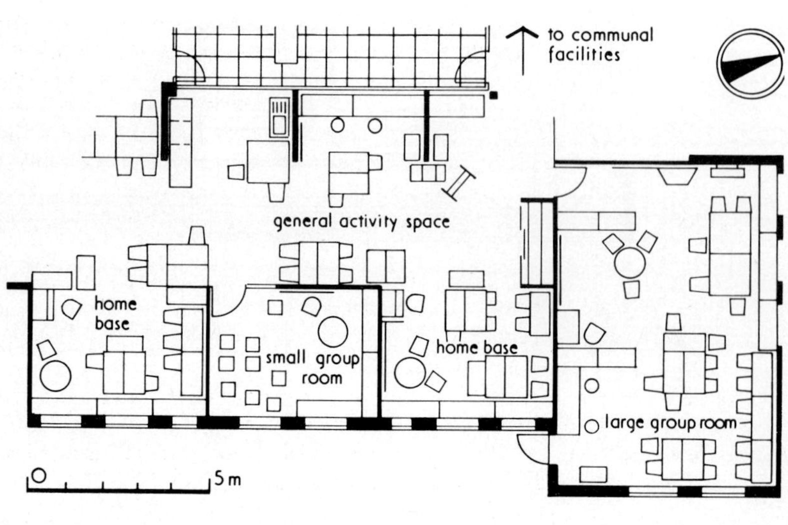

30.4 *Chaucer School showing the ingredients of juniors' teaching accommodation*

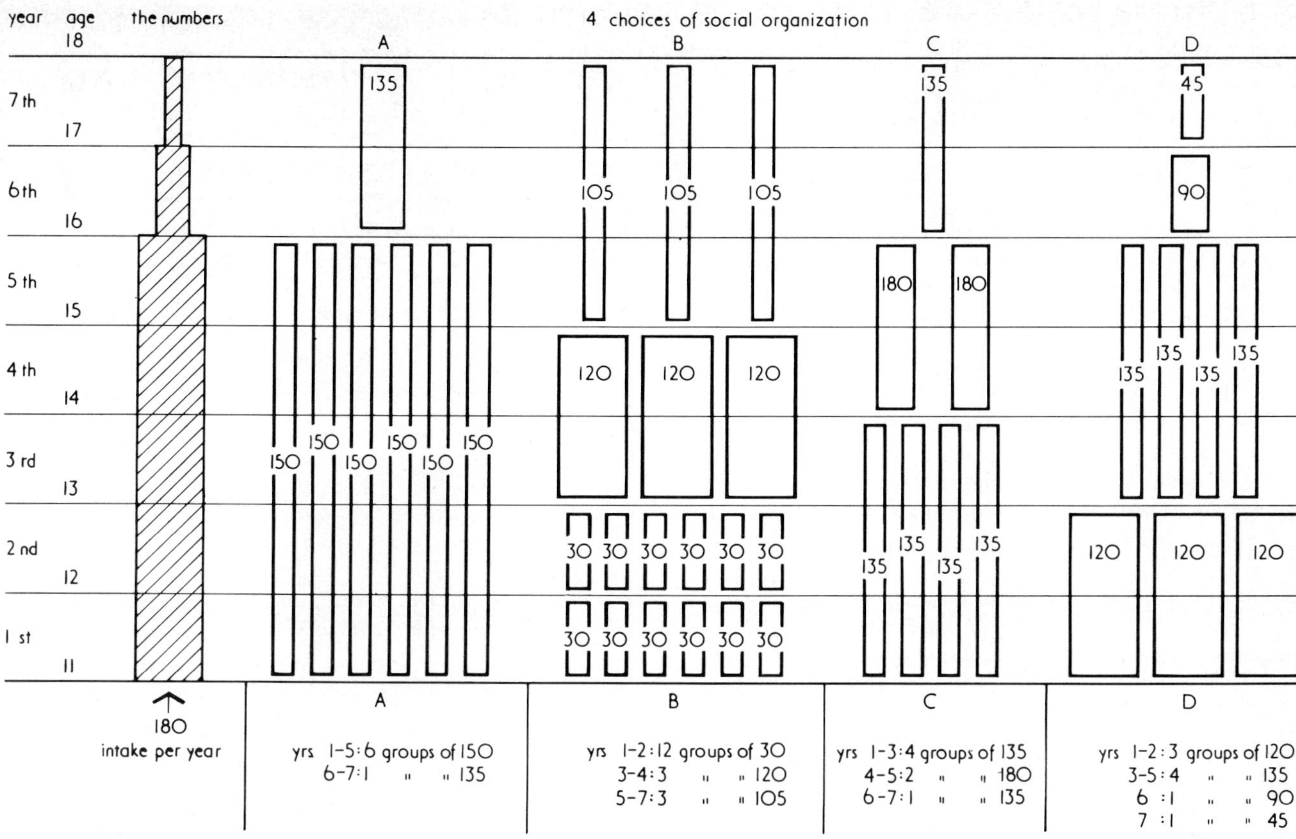

30.5 *The individual in the school, choices in social grouping. A school for 11 to 18 year olds with an annual entry of 180, and 135 students over 16. This diagram suggests four possible sub-divisions of the enrolment into smaller communities in which the individual could feel a sense of identity. This grouping may be quite different to the pattern of working groups*

Facilities shared with remainder of school

This again assumes a large school.

(*d*) Physical education space	(135 m^2)
(*e*) Drama room	(60 m^2)
(*f*) Music room	(36 m^2)
(*g*) Music practice room	(7–8 m^2)
(*h*) Clay/craft area	(70 m^2)
(*j*) Library/social area	(50 m^2) (may include some or all dining)
(*k*) Tutorial	(8 to 9 m^2)
(*l*) Staff: see junior example	

Note: at this age range changing rooms will definitely be required.

3 THE DESIGN OF SECONDARY SCHOOLS

3.01 Organisation

In any discussion of organisation patterns, the question of promoting and sustaining the unity of the school cannot be overlooked. The school as a whole creates its own ethos, reputation and traditions. The individual and all the groups to which he belongs contribute to this, as can the design of the building and the relationships of its different parts.

Comprehensive reorganisation will usually lead to the creation of larger schools. This in turn can lead to the widening of the range of activities available to each pupil, together with an increase in the number of specialist teachers. This makes it more difficult for the traditional form master to maintain close links with one group of pupils; some structure for pastoral care is necessary. It is generally agreed that there is a need for carefully organised divisions of the whole school into smaller communities. For most schools these either depend on age groups or on a vertical house system, **30.5**.

In either case, recognisable bases are required where the group can meet and from where it can be organised. Careful design may enable the dual use of space but an overriding consideration should be the sense of security essential to the well-being of the younger children. The design should allow for a variety of approaches to curriculum and teaching method to be followed. Accommodation which too exactly entrenches any particular pattern may well restrict options both now and in the future. Immediate needs must be met but buildings should also be designed with as many options for change of use left open as possible. Loose fit principles should apply to location and access no less than to selection of structure and to internal design and equipment **30.6**.

3.02 Community involvement

The less formal disciplines developing within the curriculum can benefit from contacts with agencies within the community and from the largely untapped resources of specialised knowledge and skills among individuals in the community. Conversely, it is obviously of benefit to the community to have access to the school's facilities. The concept of joint school and community planning is widely accepted and forms part of the policy of the majority of education authorities. Nevertheless the realisation of this concept has not been widespread. This may, in part, be due to various practical problems (none of which are incapable of solution), for example:

Finance Integrated joint provision schemes require a co-ordinated flow of capital funds from a number of different sources. (See AJ 26.5.76 for discussion by DES—Further problems of growth and change.)

Location The sites reserved for new schools are customarily on the outskirts of new urban or suburban developments, on the far side of a housing estate and well away from the natural social and commercial foci of the community.

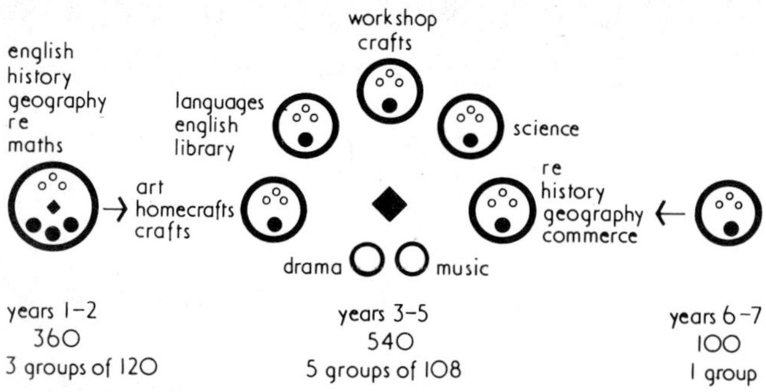

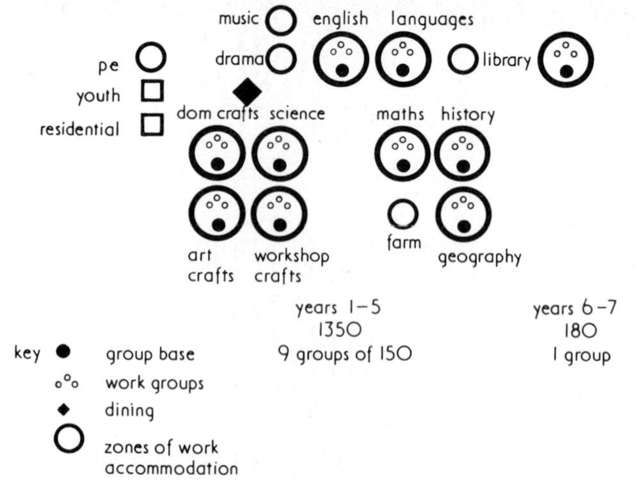

30.6 *Relationships between accommodation for social groups and work groups.*
a *A school for 11 to 18 year olds with an annual entry of 180 and 100 over 16. Years 1 and 2 are divided into three groups of 120, based in a separate building where they work about 60 per cent of their time, and where they can also dine and assemble. Years 3, 4 and 5 are in five groups, each with a common room used partly for work but not in this case for dining, and associated with a particular work zone. Years 6 and 7 have a common room and snack bar with additional provision for seminars and individual study*

b *A school for 11 to 18 year olds with an annual entry of 270 and 180 over 16. Years 1 to 5 are divided into nine groups of 150, each with a social base incorporated in one of the departments as shown. Years 6 and 7 have their own base in close association with the library*

Management To keep facilities open for up to 14 hours a day, 7 days a week throughout the year imposes considerable strain on those responsible for their management and maintenance.

3.03 Outdoor spaces

Statutory requirements determine the minimum area of the site, paved or hard porous area, playing fields and car parking. The balance of provision for formal games varies from school to school, but will be derived in parallel with the general curriculum content and timetable. Some of the following activities should be provided for: association and rugby football, hockey, athletics, cricket (including practice nets), tennis, netball, lacrosse (see section 26, Sports facilities: outdoor).

3.04 Time allocation to individual subjects

The normal school day follows, with some local variations, the following pattern: 9 to 4 each day, with a lunch break of up to $1^1/_2$ hours; a five-day week; three terms of roughly 13 weeks. Many areas operate a 40 period week, some 35 and a few are even organised on a 20 period basis. Most schools run a weekly timetable.

A general pattern is emerging of a compulsory collection of subjects for the first two years, some choice in the third year, and options leading to examinations in the fourth and fifth years. The curriculum content can be generally summarised as shown at the foot of this page.

Within a school larger groupings may be formed, for example:

Communications (English, languages, humanities)
Technology (Science, maths, crafts)
Movement (Art, music, PE)

Alternatively departments might be smaller and the groupings different. Ideally the building should be able to respond to these various requirements.

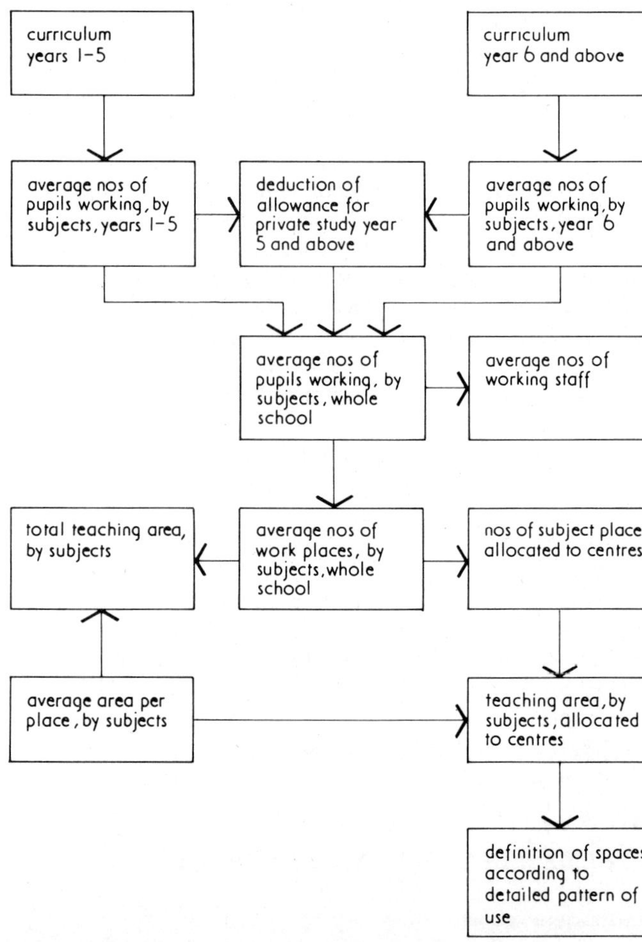

30.7 *Network showing derivation of teaching area accommodation*

English Mathematics Science Foreign languages	History Geography RE	Humanities Social sciences	Art Music Drama Crafts (metalwork, technical drawing, woodwork)	Home economics (cookery, needlework) Commerce PE and games

3.05 Space allocation to individual subjects

Whatever the age range of pupils or organisation of the curriculum, it is not possible to use every space in a school to its maximum capacity all the time—a utilisation factor of 70 to 80 per cent is most likely to be achieved. If a specific schedule of accommodation is not available the distribution of teaching area can be derived along the lines, shown in **30.7**.

1 A curricular framework is proposed
2 From this is derived the average percentage of time spent on each subject and thence
3 The average number of pupils working in that subject at any one time
4 Estimate the proportions of time in a subject that may be spent in different kinds of activity, together with the areas required for each activity
5 Derive the amount of teaching area for each subject (or group of subjects).

3.06 Examples

General
Age range 11 to 18
Normal roll size 600–2000
Gross area approximately 7·0 m² per cost place
Usable teaching area 4·2 m².
Typical distribution of teaching area (% of usable teaching area). Teaching groups generally of 30 children.
Large spaces (17 per cent) Drama, PE, movement. Minimum of 1 space per 16 teaching groups.
General purpose teaching area (26·5 per cent) English, maths, modern languages, humanities. May include for some general work on science, music and drama.
Serviced accommodation with some flexibility in use (26·5 per cent) Art and design, science (including rural studies), music, home economics. May include some maths (and commerce) and humanities.
Heavily serviced accommodation (9 to 10 per cent) Craft and technology, home economics (cookery).
'Balance' teaching area (20 to 21 per cent) Library/resource, private study areas, tutorial space, social areas, dining.

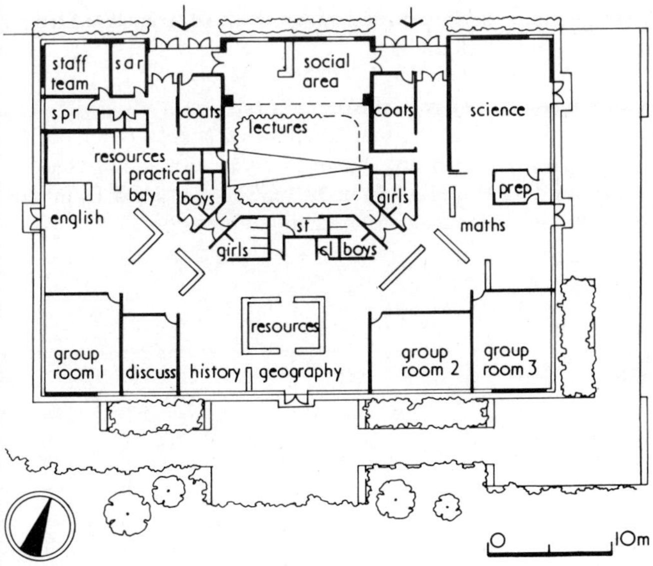

30.8 *Maiden Erlegh Secondary School, Berkshire, showing lower school*

Example of lower school
This provides the home base for 480 pupils in years one and two, **30.8**. They will spend about half their time here; subjects such as home economics, workshop, crafts, languages, music, PE and some science are taken in the main school. A total of 275 work places are provided (average of 240 pupils).

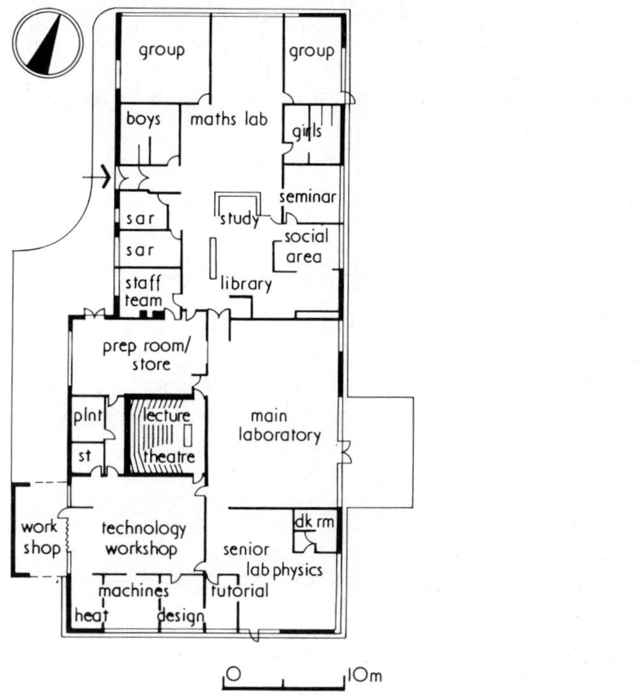

30.9 *Maiden Erlegh School: centre for physical science, mathematics and technology*

Example of centre for physical science, mathematics and technology
This centre, **30.9**, houses facilities for all pupils from years 3 to 7 to work in mathematics, physical science (physics and some chemistry), engineering, crafts and allied subjects. This combination is designed to allow interplay between 'practical' and academic work and the breaking down of subject boundaries. A total of 245 work places are provided (average of 200 pupils).

Example of sixth form centre
This example, **30.10**, shows a sixth form centre built as an addition to an existing girls' school, to accommodate approximately 300 girls. At this age range emphasis is on individual study, seminar and tutorial work. A large proportion of a student's time is likely to be spent inside the centre. The design aim is for each space to cater for the widest possible range of activities and not become too closely associated with particular departments.

4 FURNITURE AND EQUIPMENT

BS 3030:1972 gives metric dimensions for pupils' chairs and tables. However, the Department of Education and Science has been carrying out research on the heights of school furniture, and has published its results in Building Bulletin 52.

Heights
The comfortable heights of chairs and tables for an individual child are related to his height, or *stature*. **30.11** shows the comfort zones for these heights, and also for the difference in height between them. However, children are divided into classes by age rather than by stature, so **30.12** gives the relationship between age and height for various percentiles of the British school population. The following example illustrates the use of these diagrams:

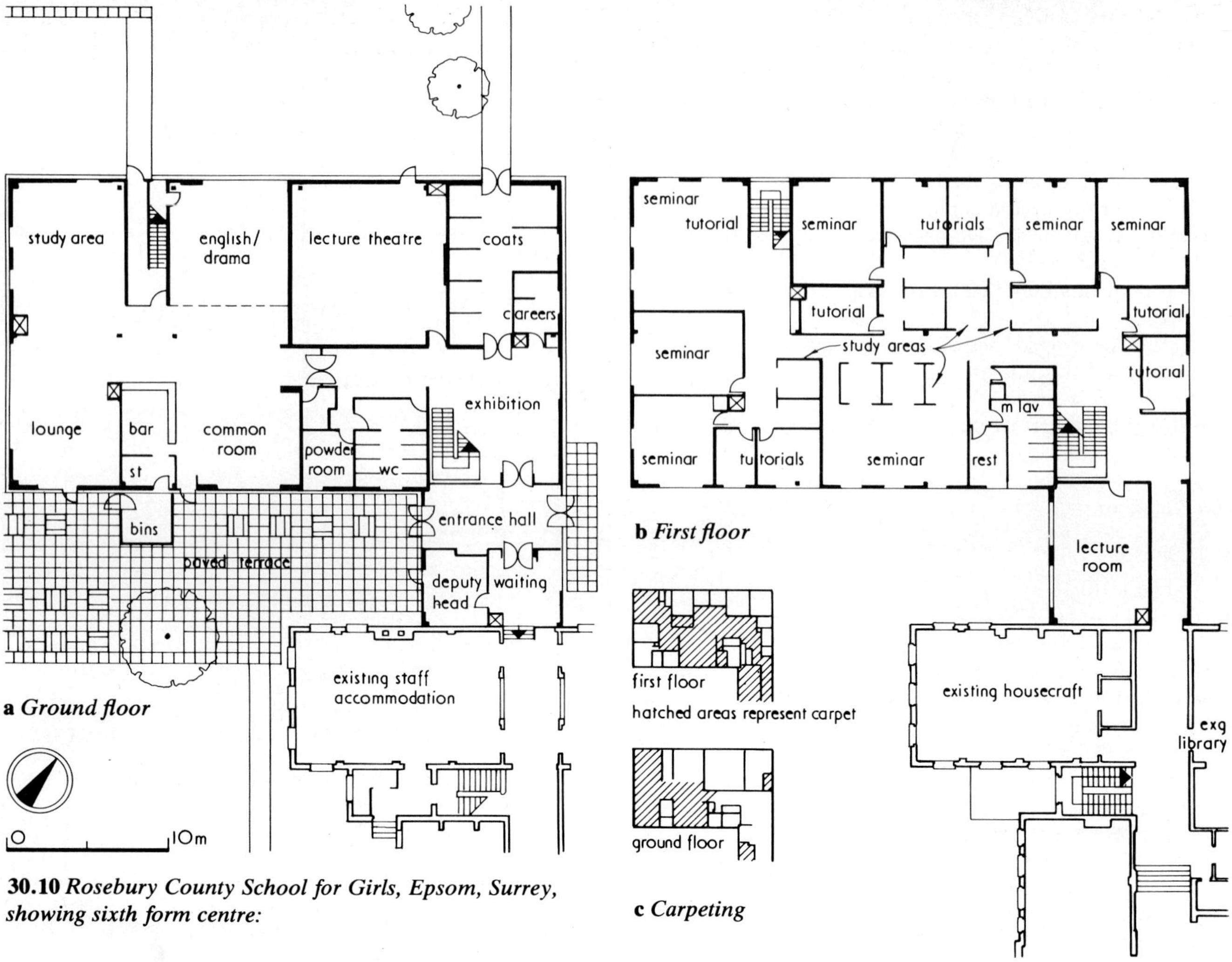

a *Ground floor*

b *First floor*

c *Carpeting*

30.10 *Rosebury County School for Girls, Epsom, Surrey, showing sixth form centre:*

What height table and chair would suit 75% of a class containing 9 and 10 year olds?

From **30.12**,
25% stature of a 9 year old = 1310 } 75% of class has stature
75% stature of a 10 year old = 1460 } between these limits

From **30.11**,
seat height 340 to 345 }
table height 550 to 570 } would suit children between above limits
difference 210 to 230 }
Hence a seat 340 and table 550 would be satisfactory.

BB 52 also suggests new standard furniture heights to supersede those given in BS 3030:1972. The latter is due to be replaced by a new International Standard in the near future. Table I gives the recommended heights from BB 52, together with some appropriate values for other dimensions, **30.13**. Average height and reach are shown in **30.14** (translated from DES Building Bulletin 41).

Work spaces for senior pupils

Guiding dimensions for individual work spaces (from DES Building Bulletin 25) are shown in **30.15** and a study carrel (from Building Bulletin 41) in **30.16**. See also DES Building Bulletin 38 for standing and reaching dimensions.

Table I Typical dimensions of school furniture (see **30.13**)

		Chairs				Tables or desks								
Size code from BB 52	Approximate age range	Key height H1	Min depth B1	Min width L1	Backrest height H2	Key height H3	Min depth B2	Min length: 1 person	Min length: 2 persons	Min height for thighs H4	Min height for tibia H5	Min depth for knees B3	Min depth for tibia B4	Min width for 1 person L3
A	3–5	260	260	250	470–510	455	450	600	1200	350	250	300	400	450
B	5–8	300	290	270	550–580	510	500	600	1200	410	250	300	400	470
C	8–11	335	330	290	615–645	565	500	700	1300	470	300	300	400	470
D	11–15	385	360	320	695–715	640	500	700	1300	530	300	350	400	470
E	15+	420	380	340	750–780	700	500	700	1300	590	350	400	450	470

Primary school furniture

A selection of useful sizes of primary school furniture (from Building Bulletin 36) is illustrated in **30.17** to **30.31**. Letters in the captions refer to the classification in Building Bulletin 52. Other sizes than those shown are also available.

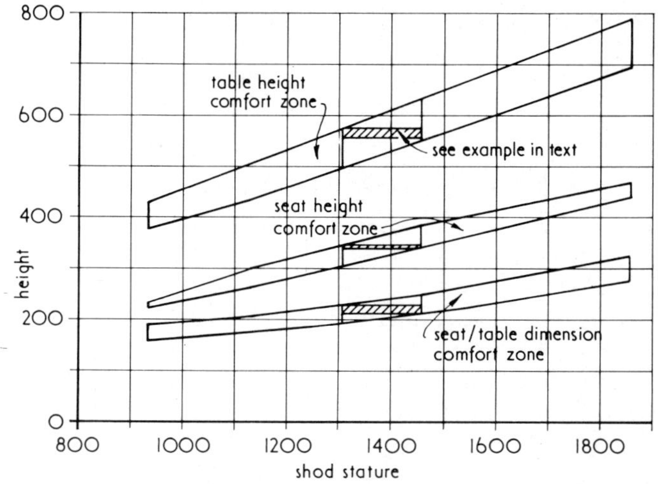

30.11 *Seat and table height comfort zones related to pupil's stature (from DES Building Bulletin 52 figures)*

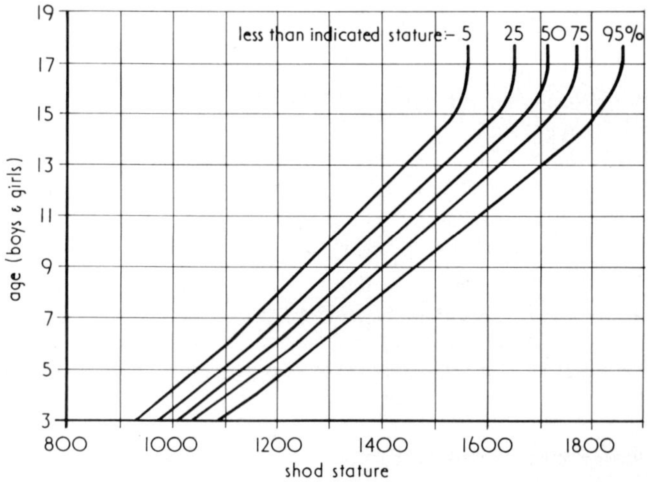

30.12 *Shod stature related to age for girls and boys combined (originating from DES Building Bulletin 46)*

5 MAIN STATUTORY REQUIREMENTS

The Education Act 1944

Statutory Instrument 1972 No 2051, The Standards for School Premises Regulations 1972, HMSO.

Department of Education and Science Building Bulletin 7: *Fire and the design of schools*, London, HMSO, 5th edn, 1975 (this relates to the above Regulations, para 51). Parts I, II, IV of the Health and Safety at Work etc Act 1974 apply.

Notes on procedures for the approval of school building projects in England, London, HMSO, 1972. With subsequent amendments for insertion. Known as 'the blue book'.

Department of Education and Science Administrative Memoranda 16/75 (Welsh Office 3/75) 18 December 1975.

Constructional standards for maintained and direct grant educational buildings in England and Wales (revision), DES, 1975.

At present compliance with National Building Regulations is technically on a voluntary basis.

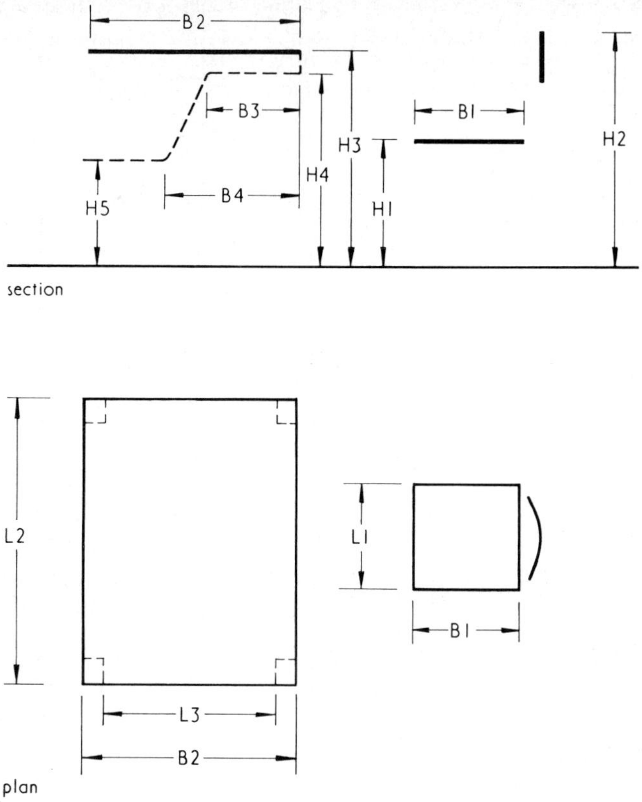

30.13 *Critical dimensions for school furniture, see table I*

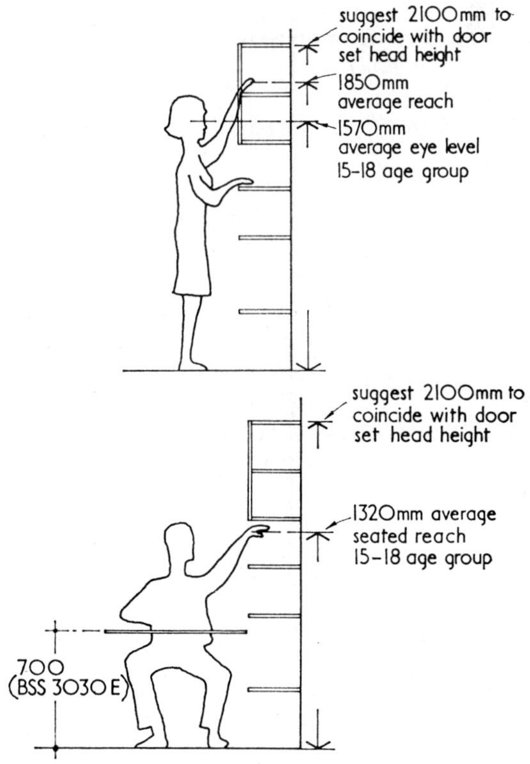

30.14 *Average height and reach for a sixth former*

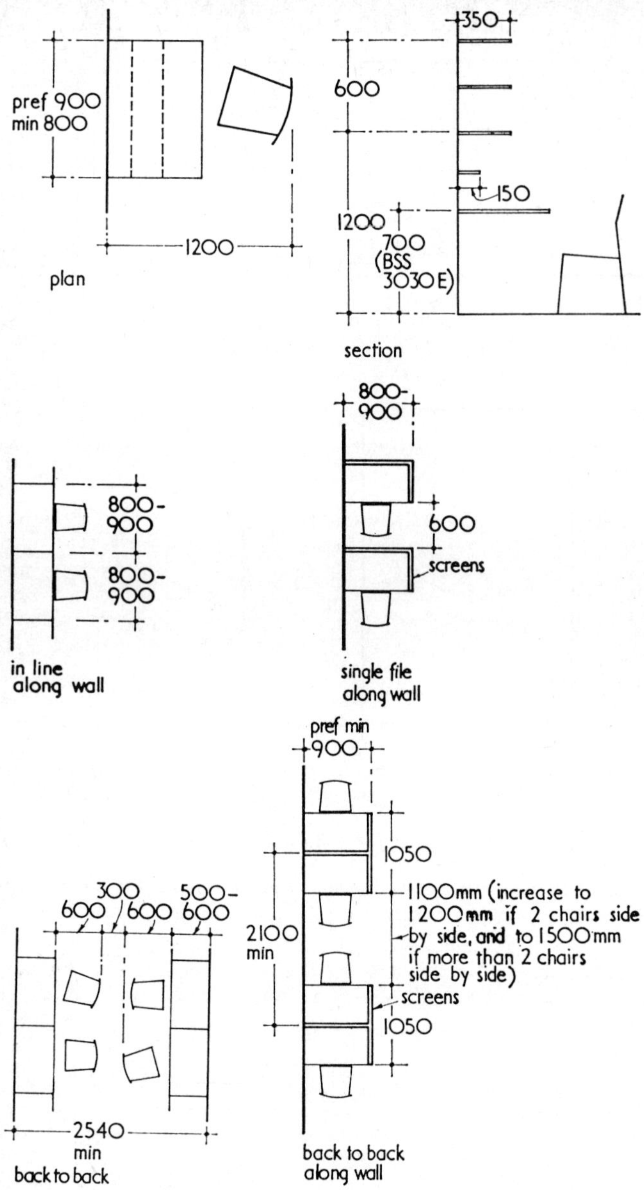

30.15 *Study bays: guiding dimensions for individual work spaces for senior pupils*

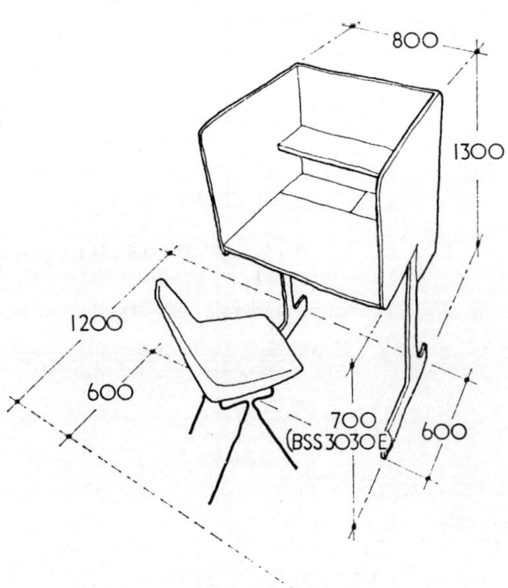

30.16 *Study carrel*

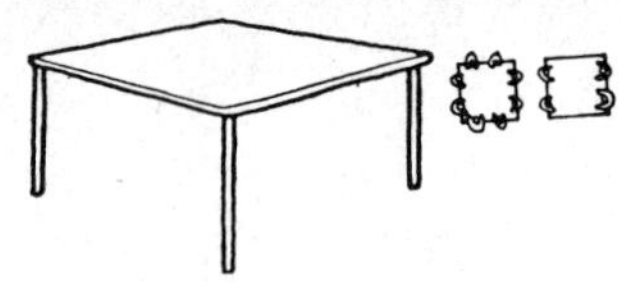

30.17 *Square table for primary schools:*
A 900 × 900 × 455
A 1200 × 1200 × 455
B 1200 × 1200 × 510

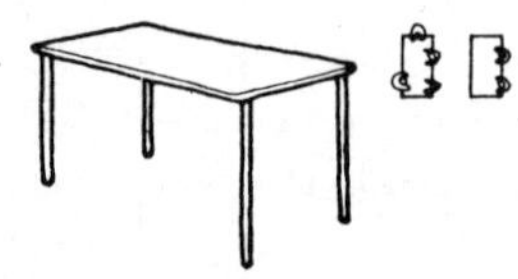

30.18 *Oblong table for primary schools:*
A 900 × 450 × 455
A 1200 × 600 × 455
B 1200 × 600 × 510
C 1350 × 700 × 565

30.19 *Round table for primary schools:*
A 850 dia × 455
B 850 dia × 510
C 1270 dia × 565

30.20 *Square pedestal table:*
B 1200 × 1200 × 510
C 1350 × 1350 × 565

30.21 *Teacher's table:*
1200 × 750 × 700

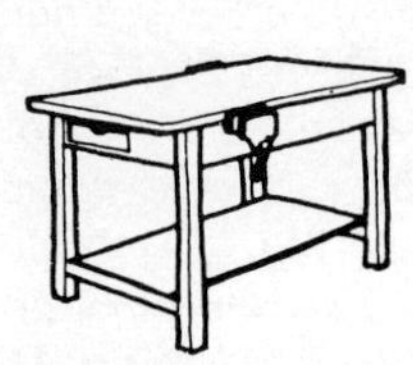

30.22 *Workbench:*
1200 × 600 × 550, 600, 650 or 700

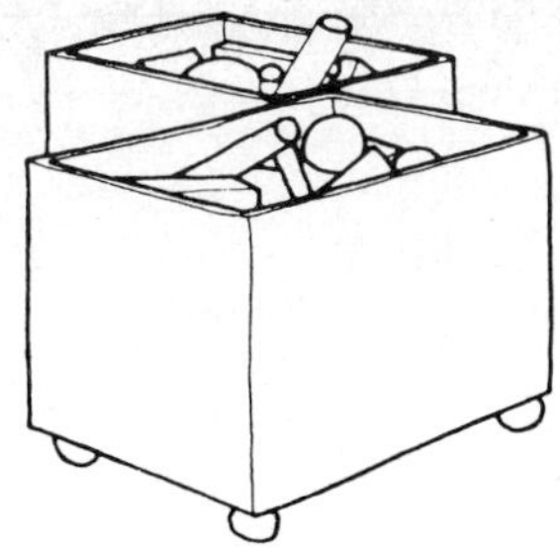

30.23 *Mobile storage bins: 600 × 450 × 500*

30.24 *Trolley locker: 900 × 450 × 600, 650 or 700*

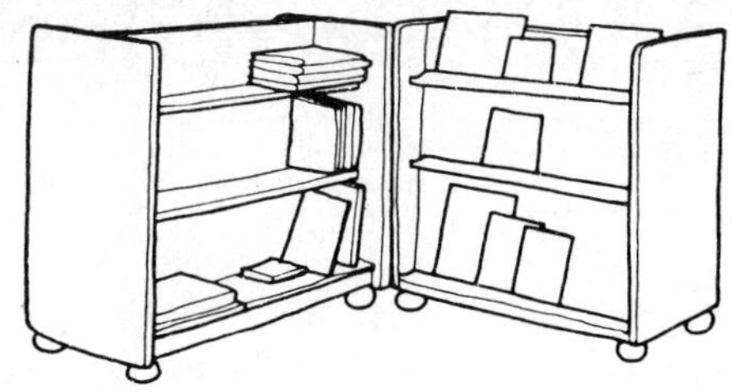

30.25 *Book storage and display trolley: 900 × 450 × 900*

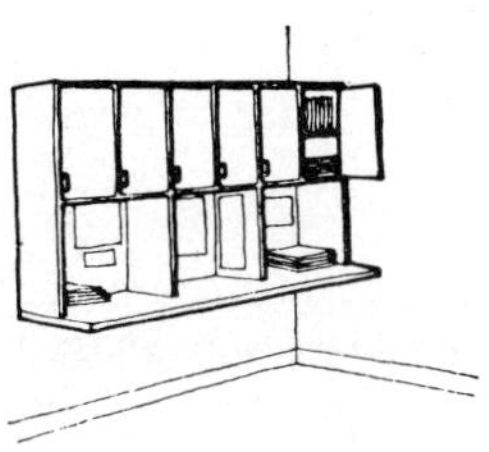

30.26 *Staff locker and writing unit: 2400 × 350, writing flap overall width 500 mm, height 1200 mm*

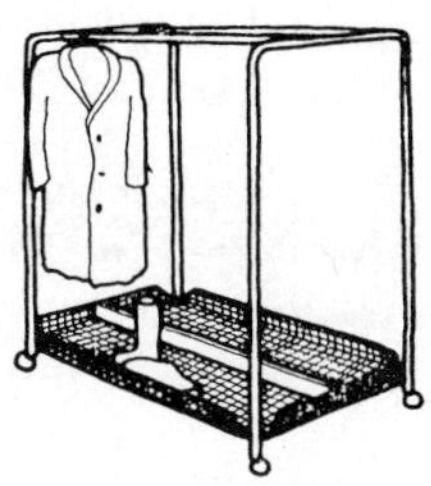

30.27 *Coat trolley: 1400 × 1000 × 650*

30.28 *Easel: 800 × 1200 high*

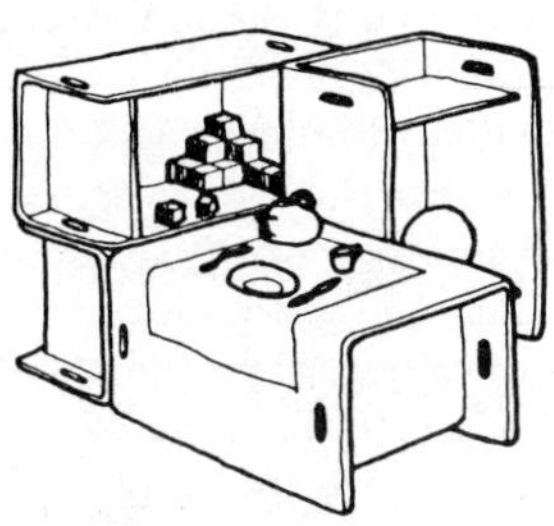

30.29 *Oblong and trapezoidal rostra: 900 × 450 × 225*

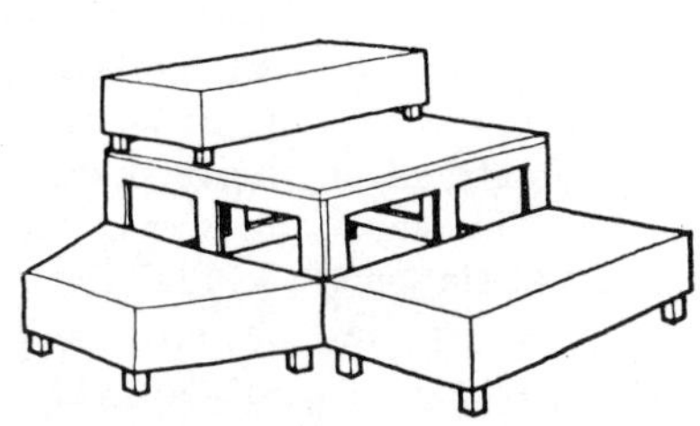

30.30 *Rostrum boxes: 550 × 400 × 280*

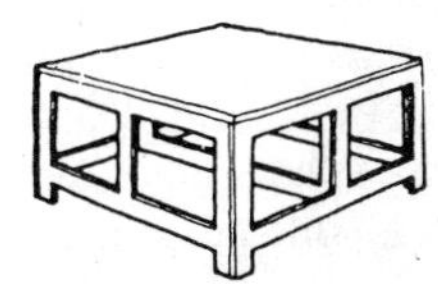

30.31 *Folding rostrum: 900 × 900 × 450*

6 BIBLIOGRAPHY

(See also para **5**, Main statutory requirements)

British Standards Institution BS 3030 School furniture, part 1: 1959 Materials, workmanship and finish; part 2: 1959 Performance tests; part 3: 1972 Pupils' chairs and tables; part 4: 1959 Chalkboards

British Standards Institution BS 4330: 1968 Recommendations for the co-ordination of dimensions in building: controlling dimensions (see also section 4, Dimensional co-ordination)

Building Bulletin 9 *Colour in school buildings*, HMSO, 4th edn, 1969

Building Bulletin 28 *Playing fields and hard surface areas*, London, HMSO, 1966

Building Bulletin 38 *School furniture dimensions, standing and reading*

Building Bulletin 52 *School furniture, standing and sitting postures*, London, HMSO, 1976

DES Building Bulletin 11 *Kitchens*, London, HMSO

DES Design note 11 *Chaucer Infant and Nursery School, Ilkeston, Derbyshire*, 1973 .

DES Design note 14 *School and Community 2*, 1976

DES *Guidelines on environmental design in educational building*, Department of Education and Science, 1972

Scottish Education Department Educational Building Note 14 *School kitchens*, London, HMSO, 1976

Peter Smith *The design of learning spaces*, London Council for Educational Technology, 1974

31 Higher Education

John Jordan

CI/SfB 72
CI/SfB (1976 revised) 72

John Jordan is the AJ consultant on education and an architect director of Architects' Co-Partnership Inc. He is a regular consultant to UNESCO.

Contents

1 INTRODUCTION

Higher education is taken to mean all post-secondary education. Table I gives the main types of institution covered in this section, although the Open University will not be specifically detailed.

No particular institution is without its peculiarities of one sort or another: siting, constituents or functions. What follows, therefore, is a series of generalisations which may or may not apply in another time or place.

Many of the building types found in higher education have their counterparts elsewhere. Factors controlling their design will therefore be found in other sections of this Handbook, and will not be repeated here.

2 UNIVERSITIES AND POLYTECHNICS

2.01 The exemplar

Other types of higher education facility tend to be modelled on universities, or to be miniature versions of them. The major part of this section will therefore deal with universities in some depth. Polytechnics, as a version of university with special aims, are also covered here.

2.02 Non-UK considerations

The criteria developed and published by the UK, US and other western government agencies for the design and management of institutions of higher education can be used as a basis for other parts of the world. However, local considerations may necessitate modifications:

- climatic
- socio-religious, as the segregation of the sexes in Moslem countries
- standard of living.

Caution is therefore needed in transposing western source data to projects elsewhere. It is recommended where doubt exists to re-synthesise space planning data from detailed net workstation areas, in consultation with the future users or other experienced local equivalents.

2.03 Types of university

There are three basic types of university:

- Oxbridge, consisting of a number of semi-autonomous colleges providing residential and catering facilities for students and staff together with some small-scale teaching space, with an amount of central shared facilities jointly administered. This type is unique to Oxford and Cambridge.
- London, consisting of a number of almost independent colleges, many of a specialist nature, each virtually self-contained universities. There are some central services, nearly all duplicating college facilities. This type is unique to London.
- Provincial, consisting of a number of subject departments or faculties, and various central facilities including usually an element of residential accommodation. This is the archetype,

Table I Categories of higher educational institutions

UK designation	Features	Designations elsewhere for institutions with similar requirements	UK mandatory requirements
University	Full-time courses to first and succeeding degrees Research	University University college	*Notes on procedure*, 1974, revised 1975, published by University Grants Committee (UGC) *Planning norms for University buildings* (UGC)
Polytechnic	Full-time courses to diploma, first and succeeding degrees, usually with a vocational bias, often scientific and technical Research, mainly scientific and technical	Technical college Technical university Vocational training college (scientific/ technical bias) Specialised technical training institute such as: maritime academy agricultural college	*Notes on procedure for the approval of polytechnic projects* (DES)
College of Further Education (CFE)	Full and part-time courses to diploma level for vocational and recreational subjects	Technical college Technical high school Sixth form college Vocational training college (arts and crafts bias) Non-advanced further education centre Adult education centre	*Notes on procedure for the approval of Further Education Projects (other than Polytechnics)* June 1972 (DES)
College of Education (Teacher training college)	Full-time course for non-graduates for Bachelor of Education or equivalent Full-time course for graduates for Certificate of Education	Teacher training college	*Notes on the procedures for the approval of College of Education building projects*, revised 1969 (DES)
Open University	Courses by correspondence, also using radio and television. Summer schools and evening tutorials at other educational establishments borrowed for the purpose	Correspondence colleges	Staff accommodation as for universities No student accommodation

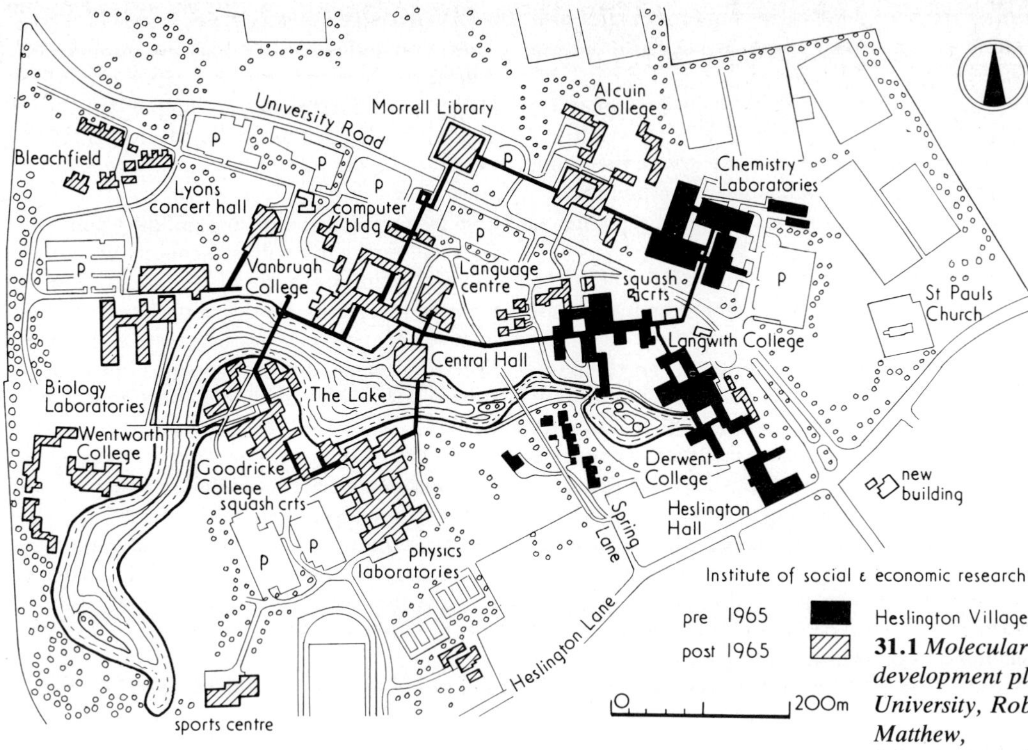

31.1 *Molecular type of development plan: York University, Robert Matthew, Johnson-Marshall and Partners*

and most of what follows applies to this type of university.

A provincial type of university or polytechnic can be built in one of two ways, or a combination of them:

- integrated and dispersed, where separate buildings and facilities are found among the local community as and where sites become available. Often facilities are fitted into converted existing buildings, when space standards as described later may have to be modified. Otherwise, the design principles are not different from:
- campus, where the buildings, or most of them, are arranged on one large site.

2.04 Types of campus

When a new university or polytechnic is to be built, a development (or master) plan is drawn up, showing how it is intended for the institution to cope with the expected expansion over the years to come.

Expansion usually occurs by the increasing size of the existing

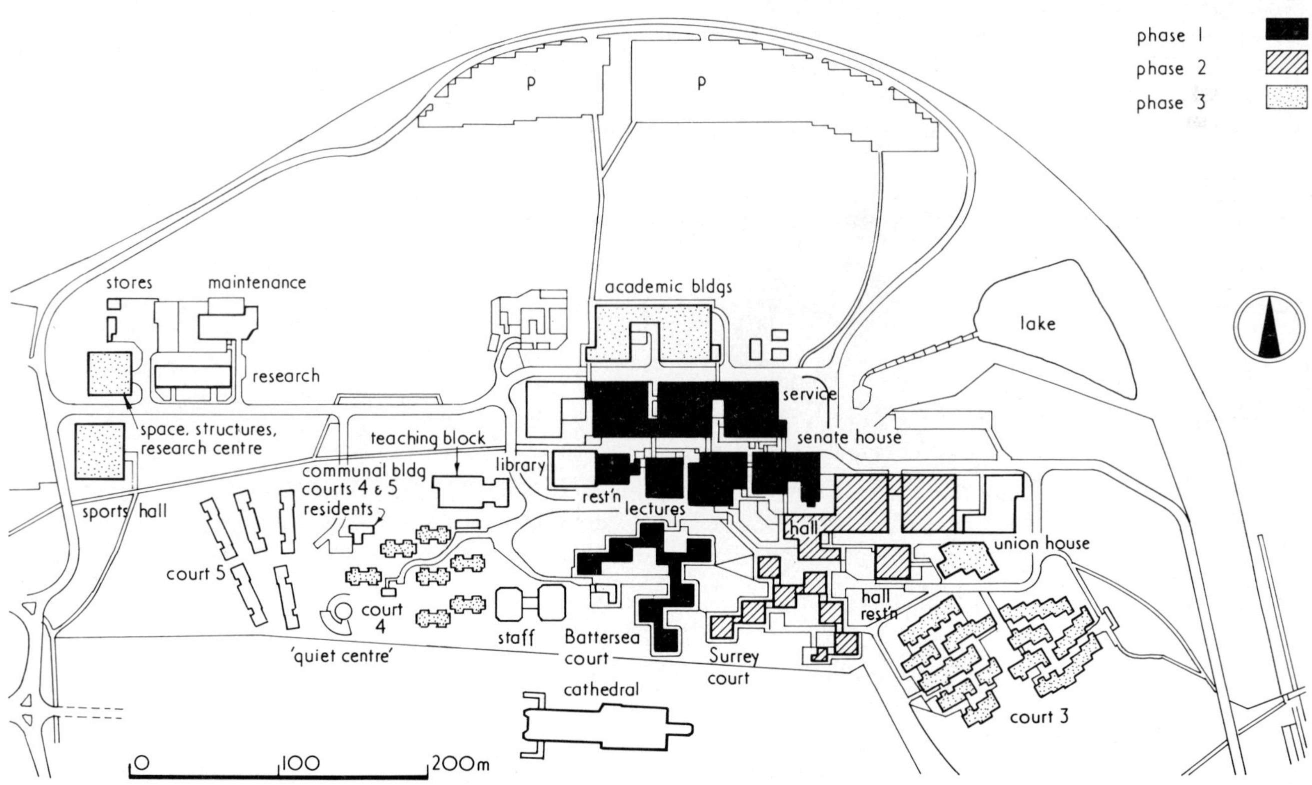

31.2 *Linear development: Surrey University at Guildford, Building Design Partnership (original development plan and Phases 1 and 2)*

departments rather than by the establishment of many new departments. There are three ways in which a department can expand:

- extension to its existing buildings externally, for which space must be available
- displacement of adjoining departments, for which the buildings must have been designed with flexibility in mind, and
- fragmentation over a series of separated buildings, which is normally deplored.

The form of the initial development of the campus will reflect the decisions on methods of expansion. The common forms are:

- molecular, as at York, **31.1**, where departments and facilities are in widely separated buildings, leaving ample space for expansion. The disadvantage of this scheme is that there are long distances to be covered between facilities, and some minor functions such as parking, lavatories and refreshments have to be repeated at each 'nucleus'
- linear, as at Surrey, **31.2**, which is designed with three strips containing residential, general and academic accommodation respectively. These strips can be extended at either end, and the academic accommodation is designed for easy conversion, enabling displacement to be facilitated
- radial, such as Essex, **31.3**, where expansion takes place all round.

2.05 University building types

The main types of buildings are shown in **31.4**, which also indicates where information can be found elsewhere in the Handbook. The form of the campus will also be determined by a number of important policy decisions regarding these buildings.

2.06 Non-specialist teaching building policy

Most departments will have their own seminar and tutorial rooms, and may even use academic staff offices for such functions. A policy on whether departments should have their own lecture theatres, classrooms or even libraries must be established. In most new universities such facilities are usually shared between some or all departments for more economy of usage.

2.07 Residential accommodation policy

Students may live:

- in accommodation provided by the university on campus
- in accommodation provided by the university off campus
- in lodgings, with or without meals
- in privately rented accommodation, usually shared between a number
- at home (in their parents' house)

Before constructing students' accommodation it is usual to

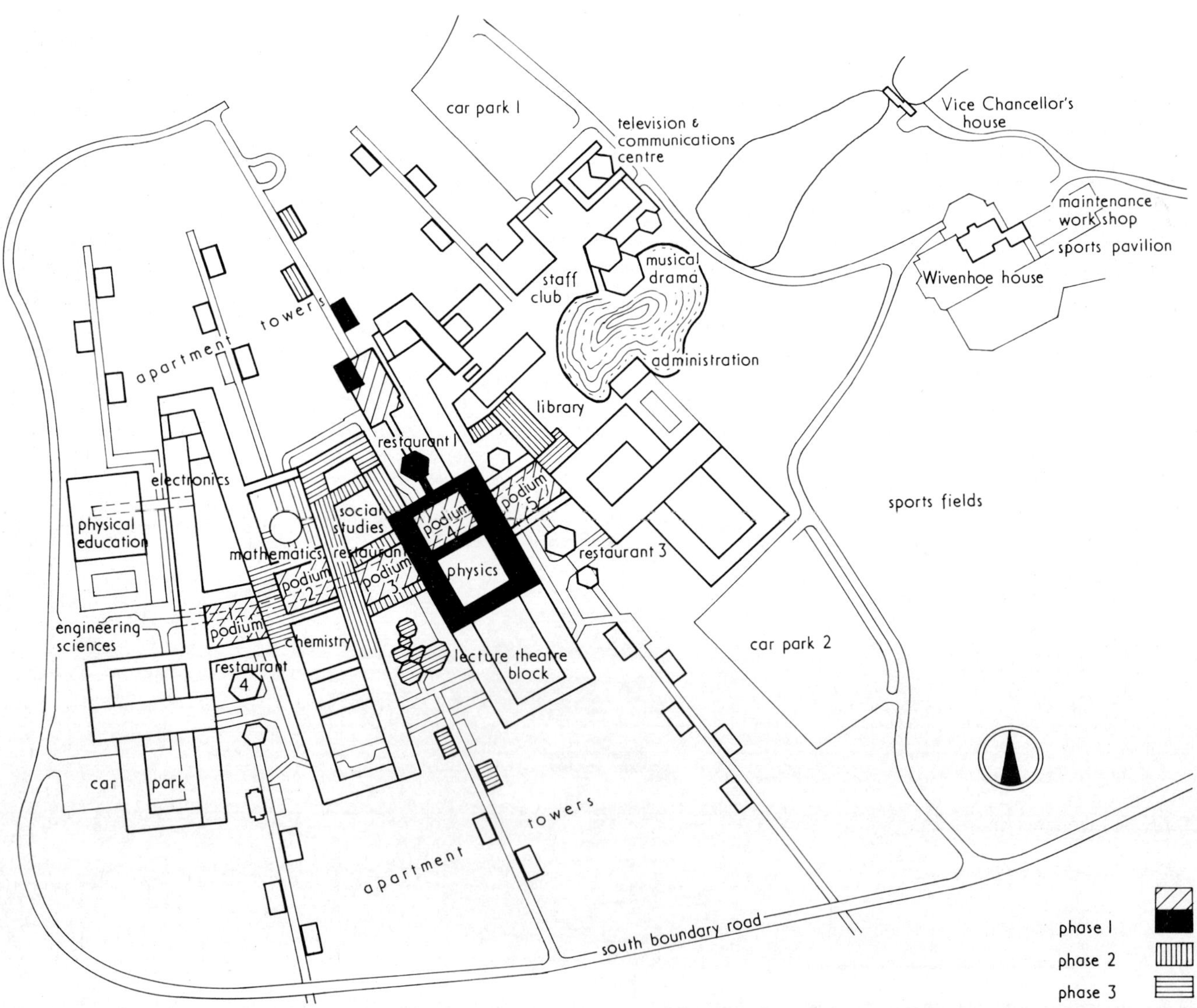

31.3 *Radial development: Essex University at Colchester, Architects' Co-Partnership*

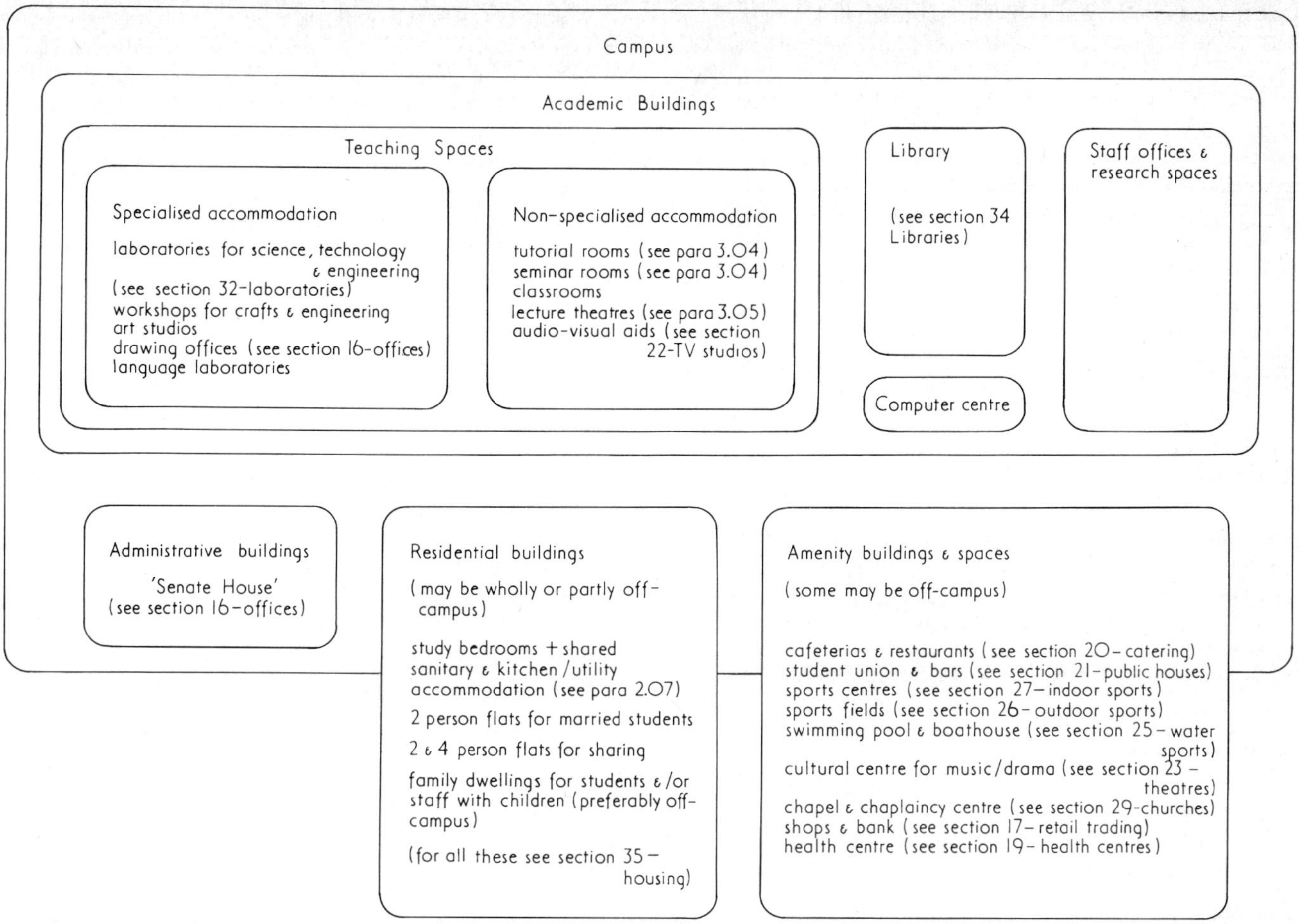

31.4 *Schematic diagram of a university campus*

conduct a survey of lodgings and rentable accommodation in the locality. When doing this it is important to estimate other demands on such resources: other higher educational establishments, specialist industrial enterprises, etc. From such surveys, it can be determined what number of students will need to have accommodation provided directly or indirectly by the university. Of this number, some may be situated on campus, although there are arguments for and against such accommodation:

Residence on campus

Advantages

Savings of time and money in travel.
Ability to prepare all meals oneself.
Reduction in private study facilities in other university buildings.
Propinquity to library, etc over weekends.

Disadvantages

Mutual disturbance by noise, etc.
Lack of contact with locality.
Need for parking facilities for students' vehicles on campus.

The types of accommodation that might be provided are given in **31.4**. Children have been found to be undesirable on campus, and family accommodation is not often provided. However, it is found that flats shared between four or more students are more popular than individual study bedrooms. This scheme also simplifies the problem of cleaning and maintenance of sanitary accommodation. In all cases some provision must be made for simple food preparation and storage. With study bedrooms, a kitchen/breakfast room shared between about 14 students may be the solution, **31.5**. Other communal accommodation that should be provided includes a TV/sitting room shared between about 100 students. Space standards will be found in table II.

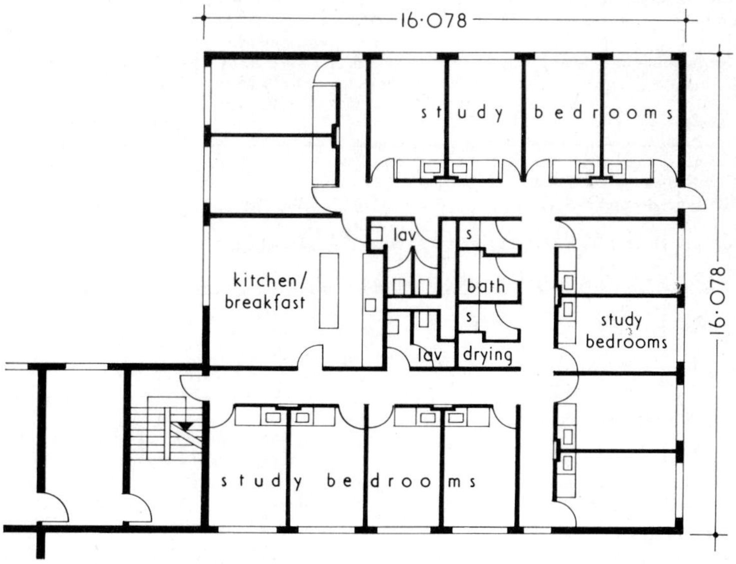

31.5 *Battersea Court, University of Surrey, showing 14 study bedrooms with associated kitchen/breakfast and sanitary accommodation (BDP)*

Table II Space standards for universities and polytechnics

	Staff offices and research spaces	Administrative, technical and secretaries	Classrooms, seminar rooms, etc
TEACHING AREAS			
1 Arts, social sciences, mathematics, architecture	1.55 m²/student	0.5 m²/student	0.65 m²/course student
2 Science, engineering science, electronics	4.35	0.45	0.35
3 Engineering	4.50	0.45	2.4
4 Preclinical medicine	3.80	0.45	0.35
5 Clinical medicine	6.15	1.0	0.35
6 Clinical dentistry	5.05	1.0	0.35

Additional areas for specialised accommodation:		
1 Languages and social psychology	0.8 m²/course student	0.8 m²/research student
Mathematics	1.1	1.1
Education, traditional geography, archaeology	2.7	2.7
Scientific geography	5.5	5.5
Experimental psychology	6.9	6.9
Architecture	6.55	6.55
Music (departments of 50 students only)	7.5	7.5
2 Laboratories and ancillary accommodation for:		
Biology	5.0 m²/course student	15.2 m²/research student
Physics, engineering science, electronics	4.9	13.8
Chemistry	5.0	14.25
3 Engineering laboratories, workshops, preparation, storage	3.95 m²/course student	14.3 m²/research student
4 Preclinical medicine teaching laboratories: multi-disciplinary	7.50 ,,	
anatomy	1.88 ,,	
5 Clinical medicine reasearch space including ancillaries		16 m²/research student
6 Clinical dentistry: teaching laboratories including ancillaries	5 m²/course student	
research and writing-up space including ancillaries		11 m²/research student

Additional space in association with lecture theatres for audio-visual facilities: allow for TV studio accommodation and ancillaries (see section 22)	
for between 3000 and 6000 students	450 m²
over 6000 students (provides second studio)	40 m²

Balance areas	for general teaching areas excluding workshops	40 %
	for workshops	25 %
	for academic staff workrooms	50 %
	for non-academic staff workrooms	40 %

LIBRARIES		
Basic provision:	1 reader space for 6 students	0.40 m²/student
	books: 3.8 m run of shelving/student	0.62 ,,
	administrative and support facilities	0.2 ,,
	Total (say)	1.25 m²/student
Additional area in law schools to provide	1 reader space for 2 students	0.80 ,,
	Additional area for book stacks to accommodate excess of accessions over withdrawals for ten years	0.20 m²/student
	Additional area for special collections of books, manuscripts or pamphlets	as required
	Addition for reserve store, separate from main library	50 m² plus 3.5 m²/1000 volumes
	Balance area	25 %

ADMINISTRATION		
for central administration, including Senate House, conference room, committee rooms		
	up to 3000 students	450 m²
	additional students	0.35 m²/student
for maintenance depot, including central stores and workshops, but excluding furniture stores		
	up to 3000 students	0.25 m²/student
	additional students	0.15 ,,
balance area		50%

AMENITY BUILDINGS		
Restaurants and cafeterias	Dining areas (based on 60% usage)	0.2 m²/student
	Kitchens, etc	0.17 ,,
	or can be calculated:	
	Kitchen area:	
	for 3 main meals including breakfast	0.45 m²/meal/sitting
	1 meal per day	0.4 m²/meal
	cooked snacks	0.3 m²/snack
	coffee and sandwiches	0.1 m²/snack
	balance area for catering spaces	25%
Communal and social areas	students	0.7 m²/student
	academic, senior administrative and research staff (excluding medical schools)	0.19 m²/student
	ditto in medical schools	(0.30)
	non-academic staff	0.16
	Total	1.05 m²/student
	Total in medical schools	1.16 ,,
	large hall or space for use in conjunction with social space between 3000 and 6000 students	450 m²
	balance area for communal spaces	30%

Table II Space standards for universities and polytechnics (*continued*)

Students' Union offices and administration	up to 3000 students	0.15 m²/student
	additional students	0.02 ,,
Sports facilities	Indoor sports (see section 27)	
	up to 3000 students	0.47 m²/student
	additional students up to 6000	0.13 ,,
	additional students	0.02 ,,
	Outdoor sports (see section 28)	
	Grass pitches, playing fields	
	up to 3000 students	28 m²/student
	additional students	14.5 m²/student
	Pavilion and groundsman's store	
	up to 3000 students	0.18 m²/student
	additional students	0.10 ,,
Health services (see section 19)	Simple consultancy suite for doctor and nurse treatment based on NHS provision for a group practice to service an equivalent number of patients	
	up to 3000 students	0.03 m²/student
	additional students	0.015 ,,
	Dental services are only provided if unavailable locally	
	A central sickbay may be provided unless located within residential accommodation	2 beds/1000 students
	Complete health service, including dentistry	
	up to 3000 students	0.10 m²/student
	additional students	0.03 ,,
RESIDENTIAL ACCOMMODATION (for students)		
Medium rise buildings with no lifts		420 students/hectare
High rise buildings		600 ,,
Allocations of space:		
study bedrooms		8.4–13 m²/place
ablutions		1.21 ,,
storage		0.54 ,,
amenities		1.0 ,,
utilities		0.5 ,,
communal space		0.65 ,,
balance area for circulation		25%
additional area for self-catering dining and kitchens		1.2–1.7 m²/place
Where a warden is in residence, allow for warden's residence		107–120 m²
offices for: warden		9.3 m²
domestic bursar		9.3 ,,
secretary		7.0 ,,
porter		6.5 ,,
records		5.6 ,,
In independent housing with self-catering		
study bedroom		9.3 m²
ancillary		1.5
amenity		2.2
balance		3.0
	Total	16.0 m²

Table III Part-time students

Type of student/Description	Full-time equivalent (FTE) for planning purposes
Full-time student Has no other occupation. Probably attends minimum 20 hours a week. May live in	1
Thick sandwich student Attends full-time for three academic years in rota but works in industry for at least a year during the period	1
Thin sandwich student Attends full-time for six months, works in industry the other six months including the long vacation. Repeats as long as necessary	1
Block release student While being trained in industry (eg an apprentice) attends full-time for a block of three or four months	1/3
Part-time day student Attends one day a week plus two or more evenings	2/9
Evening student Only attends in evening	no allowance

2.08 Catering policy

The third policy decision affecting campus shape is concerned with the communal catering service. This can be:

• completely centralised preparation and consumption (one large kitchen and dining room)

• centralised preparation, dispersed consumption (one large kitchen, separated dining accommodation)

• dispersed preparation and consumption (separate dining rooms, often specialising in different kinds of food and catering, each with its own kitchen)

Dispersed facilities can be centred on residential buildings to resemble Oxbridge colleges, as at York; or can be distributed at random as at Surrey.

2.09 Space standards

Allocations of space for different functions cannot be made to rigid rules, as each circumstance will be specific. However, the figures in table II can be used as an initial design guide.

2.10 Part-time students

Not all students, even in universities and polytechnics, will be full time. Various forms of higher education are intended to keep the student from becoming completely divorced from the real world of industry and commerce to which he will return at the end of his course. Table III gives the forms of part-time involvement common in the UK, and the equivalent full-time student (FTE) factor to be taken in connection with the space standards in tables II, VII and VIII.

2.11 Balance area

The areas given in table II are mainly *net usable areas*. To these have to be added *balance areas*, given as a percentage of the net usable area:

net usable area + balance area = gross area.

Balance area includes allowance for corridors and stairs, entrance foyer, enquiry counter, cloakrooms, locker spaces, lavatories, cleaners' stores, maintenance workshops, gardeners' stores, boiler rooms, electricity sub-stations and meter rooms, delivery bays, porters' rooms, plant rooms, service ducts.

The percentage allowances for balance area are given in the appropriate places in table II.

3 TEACHING SPACES

3.01 Density of academic development

The numbers of students that can be accommodated on a campus are given in table IV.

Table IV Density of facilities for academic areas

	Number of students per hectare	
Plot ratio	**Art based**	**Science and technology**
0.5:1.0	395	200
1.0:1.0	790	400
1.5:1.0	1185	600
2.0:1.0	1580	800
2.5:1.0	1975	1000

3.02 Teaching places

The numbers of teaching places that will be required for any type of institution can be calculated from the following formula:

$$N_t = N_s \times \frac{H_s}{H_w} \times \frac{100}{F}$$

where N_t = number of teaching places required
N_s = number of students
H_s = hours per week per student in the accommodation
H_w = total number of available hours a week for the accommodation
F = net utilisation factor

Example

200 students require an average of 10 hours a week of lectures in a working week of 40 hours and assuming a net utilisation factor of 80 per cent. What number of teaching spaces should be provided?

$N_s = 200$, $H_s = 10$, $H_w = 40$ and $F = 80$

hence $N_t = 200 \times \frac{10}{40} \times \frac{100}{80}$

$= 63$ spaces.

Table V Usable area per working space for teaching accommodation
(for balance areas see table VII under 'Teaching Spaces')

Non-specialised	
Tutorial rooms	
Rooms with informal seating	1.85 m²/space
Rooms with tables or desks	2.30 ,,
Rooms with demonstration area	2.50 ,,
Lecture theatres	
Rooms with close seating	1.00 ,,
Drawing offices: A1 and smaller	3.70 ,,
A0 and bigger	4.60 ,,
Laboratories	
Advanced science and engineering	5.60 m²/space
Non-advanced science and engineering	4.60 ,,
Management and Business Studies	
Work study	4.60 m²/space
Typewriting	3.20 ,,
Accounting	2.80 ,,
Workshops	
Crafts involving large-scale machines and equipment, eg welding, motor vehicles, machine tools	8.40 m²/space
Crafts requiring workbenches and smaller scale machines and equipment, eg carpentry, plumbing, electrical	5.60 ,,
Craft rooms, eg dressmaking, cookery	5.60 ,,

3.03 Areas of teaching spaces

The areas required for various forms of teaching accommodation, related to teaching spaces rather than to total student population, are given in table V.

3.04 Tutorial and seminar rooms

Tutorials often take place in academic staff offices. Some prefer special rooms for the purpose, **31.6**. Seminar rooms are shown in **31.7**.

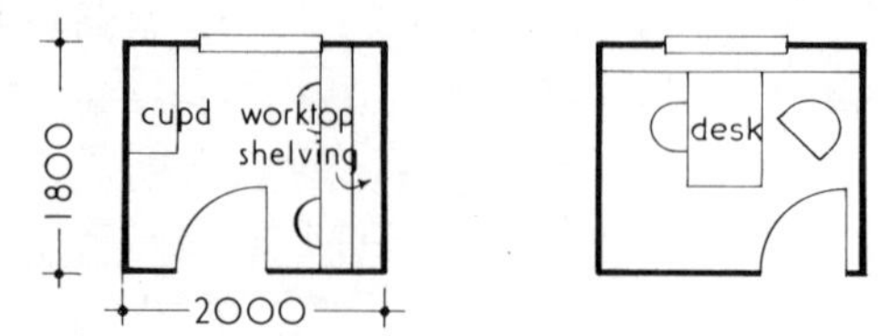

31.6 *Two tutorial rooms*

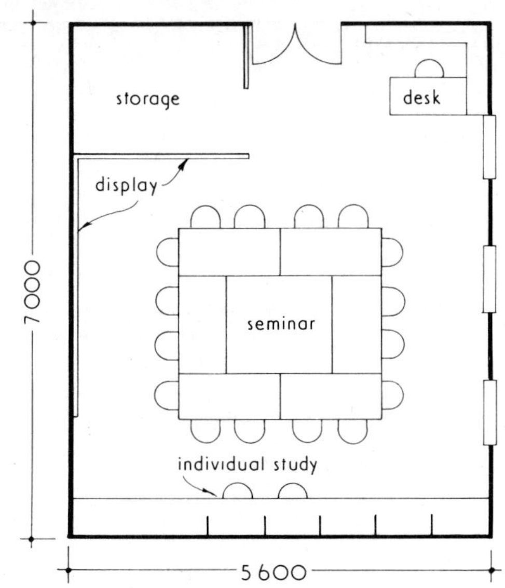

31.7 *A seminar room*

3.05 Lecture theatres

These are expensive facilities which are often under-used. They range in size from large classrooms accommodating 50 students, to large theatres for 500.

However, investigation has shown that the common lecture group is between 30 and 60, so the larger sizes are appropriate only when conference facilities are needed, or where use as an assembly hall or cinema is also envisaged. A typical lecture theatre is shown in **31.8**.

3.06 Lecture theatre seating

Generally the layout of seating in a lecture theatre is similar to that in an ordinary theatre such as is covered in section 23. However, there is usually some provision for writing notes, which will affect the seating layout. The common forms of provision are:

- fixed writing shelf, **31.9**, involving a greater spacing between rows
- folding writing shelf, **31.10**
- fixed tablet arm, **31.11**, involving a greater spacing between seats in the row
- hinged tablet arm, **31.12**.

While the folding facilities are more economical in space, they are more subject to misuse and damage.

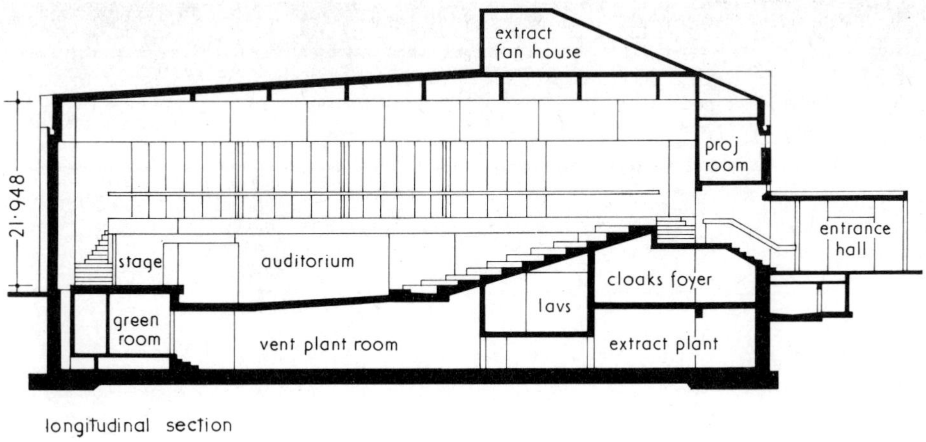

a *Long section*

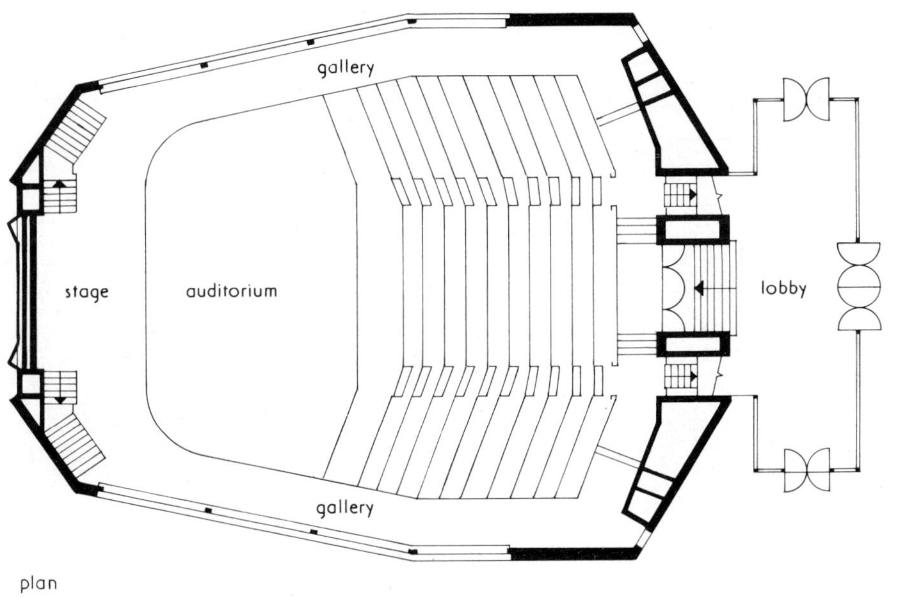

b *Auditorium plan*

31.8 *Lady Mitchell Hall, Cambridge. A large lecture theatre for 450 students. Casson, Conder and Partners*

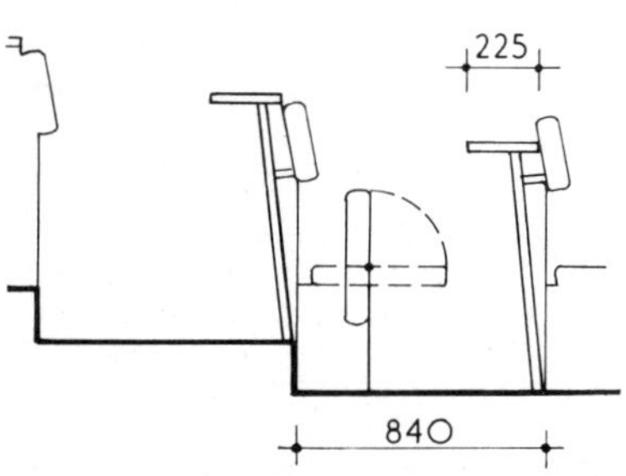

31.9 *Lecture theatre seating with fixed shelf and tip-up seat*

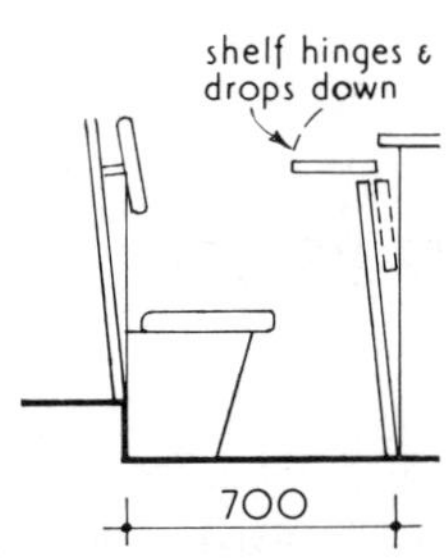

31.10 *Lecture theatre seating with fixed seat and lift-up shelf*

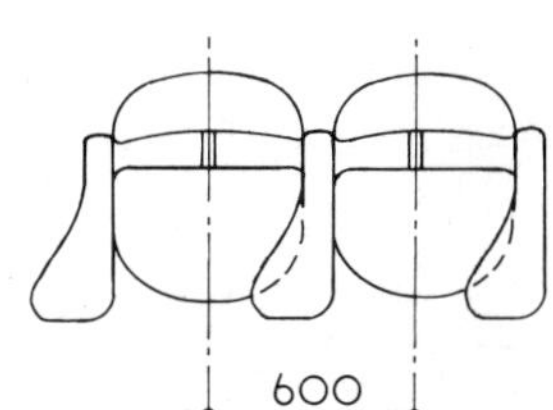

31.11 *Lecture theatre seating with fixed tablet arm*

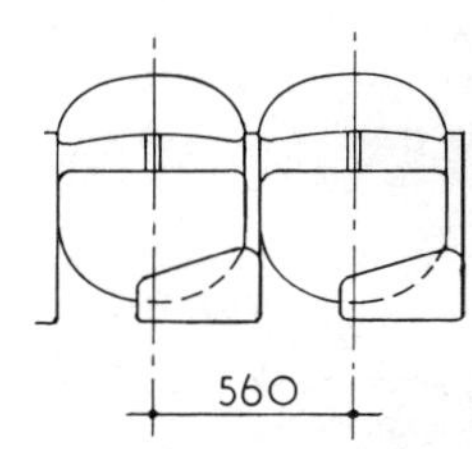

31.12 *Lecture theatre seating with hinged tablet arm*

3.07 Display

Chalkboards, both fixed and movable, are still employed in many cases. Viewing angles are important, particularly in relation to glare and reflections from lighting. Visibility is restricted to about twelve rows.

Overhead projectors are often used instead. Ideally, a tilted screen should be provided to reduce 'keystone' distortion.

3.08 Audio-visual facilities

Much material is now videotaped. While projection TV is available, it is still uncommon and expensive. It is more usual to suspend a number of conventional TV sets at high level. Minimum viewing distance and coverage are given in table VI and **31.13.**

3.09 Demonstration

Unless no scientific work is envisaged, some demonstration facilities will be needed. Since these require considerable preparation in many cases, a degree of portability is desirable. In elaborately equipped theatres, demonstration benches have been designed to rise from the depths like a cinema organ! However, it is normally sufficient to anticipate trolleys wheeled from an adjacent preparation room.

4 COLLEGES OF FURTHER EDUCATION

These have a higher proportion of part-time and evening students. Areas for teaching spaces must therefore be calculated

Table VI TV viewing criteria (see **31.13**)

	Seating row spacing							
	900		**1300**		**1550**			
Size of screen	D_{min}	**H**	D_{min}	**H**	D_{min}	**H**	D_{max}	**area covered** m^2
425 (17 in)	1650	650	1250	425	1125	350	4425	14.4
465 (19 in)	1700	675	1300	450	1150	375	4550	14.4
520 (21 in)	2125	850	1625	562	1450	462	5700	24.2
570 (23 in)	2150	875	1650	575	1475	475	5800	24.2
595 (24 in)	2400	962	1825	620	1625	520	6425	30.2

by the method in para **3.02**. Areas for other facilities will be found in table VII.

5 COLLEGES OF EDUCATION

Space standards for institutions training teachers are given in table VIII.

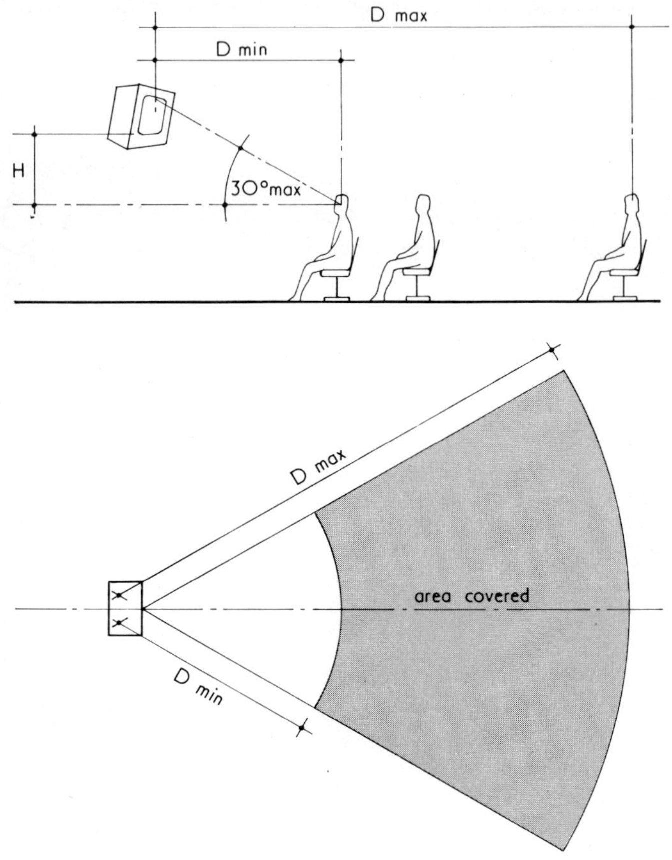

31.13 *Criteria for viewing on TV sets in lecture theatres, read with table VI*

Table VII Space standards for colleges of further education

Teaching spaces		
see para **3**		
Libraries		
	colleges with 30% advanced work	colleges with less than 30% advanced work
first 500 FTE students	390 m^2	300 m^2
additional FTE students	0.44 m^2/student	0.38 m^2/student
balance area	25%	
Non-teaching areas		
for the following:		
principal's and vice-principal's rooms		
registrar's and departmental heads' rooms		
main offices		
rooms for principal's and departmental heads' secretaries		
offices for welfare and advisory services		
building maintenance officer's room		
interview room		
enquiry kiosk		
porter's room		
bookshop		
medical room		
storage for the above at 15%		
*up to 500 FTE students	255 m^2	
500 to 2500 FTE students	0.128 m^2/student	
additional students	0.05 ,,	
academic staff rooms (other than departmental heads)	0.36 m^2/student	
non-academic staff allocated to departments	0.20 ,,	
Balance areas: administrative	50%	
academic staff workroom	50%	
non-academic staff workroom	40%	
communal	30%	
Communal accommodation		
for the following:		
physical recreation including changing rooms		
student and staff common rooms		
students' union/staff association		
music/indoor sports		
storage for the above		
*up to 500 FTE students	590 m^2	
500 to 2000 FTE students	0.42 m^2/student	
additional FTE students	0.14 ,,	
additional area for full-time and sandwich course students who make fuller use of the facilities	0.5 m^2/full-time/ sandwich student	
dining rooms, allow for quarter to half of student body	1.12 m^2/space	
cooking and service areas, see universities		
Student common rooms	0.75 m^2/student	
Staff common rooms	1.85 m^2/member	
Lockers, baths, showers, laundry/drying space for day/lodging students	0.9 m^2/student	
(A single-sex changing room is about 74 m^2)		
Balance areas: communal	30%	
catering	25%	
Residential accommodation		
see universities and polytechnics		

*approximation to complex formula

Table VIII Space standards for colleges of education

Teaching space	
Total space provided	4.65 m^2/student
tutorial/seminar rooms	13.5 m^2/staff member excluding principal and vice-principal
lecture rooms: first 100 places	1.1 m^2/student
additional spaces	0.9 ,,
general teaching rooms	1.85 ,,
additional area for storage	10%
balance areas: general teaching spaces	40%
academic staff workrooms	50%
non-academic staff workrooms	40%
Libraries	
including private study areas	
*first 200 students	1.1 m^2/student
additional students	0.95 ,,
balance area	25%
Non-teaching areas	
see colleges of further education	
Communal areas / **students' residences** } see universities and polytechnics	

*approximation to complex formula

6 BIBLIOGRAPHY

Academic building system studies: 1 Environmental study: science and engineering buildings; 2 Cost performance study; 3 Information manual: Indiana and California universities, US Department of Health, Education and Welfare, EFL, 3 vols, 1971

AJ Information Library: AJ Design Guide, College and university buildings, AJ 10.1.68 pp 103–116

Bullock, Dickens, Steadman *A theoretical basis for university planning land use and Built form studies* Report 1, University of Cambridge School of Architecture, 1968

P. Cowan (ed) *The university in an urban environment,* London, UCL Research Group, 1974

DES Design Note 8 *Polytechnics: planning for development,* London, HMSO, 1972

DES *Furniture and equipment dimensions: further and higher education* Building Bulletin 44, London, HMSO, 1970

DES Report on Education No 90 *The management of non-university higher education,* 1977

DES Welsh Education Office *Notes on procedure for the approval of further education projects (not including polytechnics)*, 1972

DES Welsh Education Office *Notes on procedure for the approval of polytechnic projects*, 1972

C. J. Duncan *Modern lecture theatres,* Newcastle, Oriel Press, 1962

E. Faure *Learning to be—the world of education today and tomorrow,* UNESCO, 1972

UNESCO *Planning buildings and facilities for higher education,* London, Paris, Architectural Press and the UNESCO Press, 1975

University Grants Committee *Planning norms for university buildings*, London, HMSO, revised February 1974

32 Laboratories

Tony Branton

CI/SfB 732
CI/SfB (1976 revised) 732

Tony Branton is an architect with the Laboratories Investigation Unit, a research and development group sponsored by the Department of Education and Science and the University Grants Committee. The opinions expressed in this section are those of the author and do not necessarily represent those of the DES or UGC.

Contents

1 INTRODUCTION

1.01 Type of laboratory—changing requirements

All laboratories must be designed to cope with growth and change irrespective of the scale of the work, or the scientific discipline involved.
The three major types of laboratory are:

- research
- teaching
- routine.

Of these, the need to cater for rapidly changing requirements is most evident in research work.
With teaching, an increasing diversity of educational methods coupled with multi-disciplinary use of laboratory space has produced similar needs for adaptable layouts.
Routine work is slower to change, and therefore changes are more predictable than in research. Nevertheless, requirements will change during the life of the building and this must be anticipated in the initial design. The designer must encourage users to think beyond immediate needs when preparing the brief.

1.02 Scope

The information contained in this section is related primarily to bench scale laboratories. Standards common to all scientific disciplines are stressed to highlight areas where interchange of use is viable. While special requirements must be met, they should, wherever possible, be catered for within an adaptable building framework based on common requirements. The designer must develop a basic structure and service distribution system within which a variety of alternative layouts is possible. In the case of a laboratory building, the dimensional requirements of laboratory spaces and their associated services should primarily determine this system, other spaces, such as lecture rooms, offices and stores, being co-ordinated into the system as necessary. Where the laboratory is only a small part of a larger scheme, as in a hospital or school, other constraints will predominate. Whatever the context, design decisions on the size of laboratory spaces, frequency of services, etc must be based on a general analysis of detailed requirements for all present and future likely activities in the laboratory.

2 AREAS

2.01 Areas per workplace

The figures given in tables I to IV are expressed as areas per workplace. In most cases this includes an allowance for shared facilities within the laboratory, eg instrumentation, wash-up sinks, fume cupboards, which are additional to the basic area required for the student or scientist's work station. If some of these facilities are centralised in order to allow sharing between laboratories, an appropriate portion of calculated area must be transferred to the centralised space and the laboratory area reduced accordingly. Areas per workplace are given as 'usable areas', ie the area needed exclusively for research, teaching or routine laboratory purposes. The percentage addition for 'balance areas' is accounted for by ducts, lavatories, cloakrooms, boiler houses, plant rooms, lifts and general circulation space.
Tables I to IV give area guidelines for schools, polytechnics and colleges of further education, universities and research. In addition, there are:

Table I Schools
These areas, derived from recent DES bulletins, distinguish levels and scale of work, but not subjects

Level of work		Scale of work*	
General science	$2{\cdot}8\,m^2$	Bench scale work	$3{\cdot}2\,m^2$
Individual projects	$3{\cdot}6\,m^2$	Workshop scale	$4{\cdot}6\,m^2$

* For calculating overall laboratory areas, including storage and preparation

Table II Polytechnics and colleges of further education
Individual subjects are not distinguished

Level of work	Area per workplace	% addition for service rooms	Balance area addition*
Advanced science and engineering	$5{\cdot}6\,m^2$	25	40%
Non advanced science	$4{\cdot}6\,m^2$	15	40%

* The balance area is a percentage addition to the area per workplace and the percentage addition for service rooms

Table IV Research (government and industrial)
*An individual scientist's or research assistant's area requirements can vary considerably depending on the sizes and amounts of equipment used for experiments. The figures give a likely range**

Chemistry	$8–12\,m^2$
Physics	$6–8\,m^2$
Biology	$6–8\,m^2$

* University figures for storage and preparation, and ancillary laboratory areas may be used as guidelines for research accommodation but ad hoc requirements are more likely to occur in ancillary laboratory areas and must be allowed for on an individual job basis

Table III Universities

	Area per workplace m²	Addition for storage and preparation %	Addition for other teaching and research ancillaries %	Balance area addition† %
Pure sciences*				
Teaching laboratories				
Biological sciences (general purpose labs)	4·0	15	Ad hoc in accordance with needs (say 15%)	30*
Biological sciences (other than gen purpose)	5·0	15	,,	30*
Physics	5·0	15	,,	30*
Chemistry	5·0	15	,,	30*
Research laboratories				
Individual or advanced research	11·0	15	,,	30*
MSC courses	7·5	15	,,	30*
Other technological and scientific subjects				
Teaching laboratories				
Elementary or intermediate	3·7	15	15	45
First and second year honours and general	4·2–4·6	15	15	45
Final year honours	5·6–6·5	15	15	45
Research laboratories				
Research students in groups of 4 or more	7·4	15	15	45
Individual or advanced research	11·0	15	15	45

(Other technological and scientific subjects: balance area % includes allowance for plant rooms, etc)

* Additional balance area allowances will be needed for plant rooms, ducts, boiler houses and entrance halls: physics—up to 12½% of workplace, storage/prep and ancillary areas; chemistry and biological sciences—up to 20% of workplace, storage/prep and ancillary areas
† The balance area is a % addition to the workplace, storage/prep and ancillary/areas

• *Routine (industrial or hospital) laboratories*
Private industrial firms may have their own standards. Simple routine work can have requirements similar to teaching, eg 5·6 m² per workplace. More complex routines may be better related to the research standards given in table IV.
Laboratory areas for hospital pathology routines are given in DHSS Hospital Building Note 15.
• *Machine or equipment orientated laboratories*
In this type of laboratory (or workshop), whether teaching, research or routine, the sizes of spaces will be determined by the shape, size and number of machines, rigs or pieces of equipment rather than by the number of workplaces. In general these areas must be assessed on an ad hoc basis. The following standards are given for further education colleges: workshops, machine laboratories—8·4 m² per workplace plus 25 per cent addition for storage plus 25 per cent addition for balance area.

3 FURNITURE AND EQUIPMENT DIMENSIONS

3.01 Worktop height

Most laboratory work takes place in a vertical zone above the worktop. The lowest convenient worktop height should be specified as this increases the volume with easy reach, **32.1**. In addition to catering for prolonged periods of written work in the laboratory, a seated worktop height, **32.2**, can also be specified for small-scale work involving fine manipulations, eg fine electronics assembly or work at the microscope. In some cases a further lowering of the supporting surface can usefully increase the vertical zone within easy reach. An extreme example of this is the low chemistry bench for tall glassware rigs, **32.3**.
Worktop heights for school science must be considered in relation to other non-science subjects. The heights given in **32.4** are for general use and thus allow for an interchange of use in practical work areas.
The relationship of seat height to worktop height is critical and these should be related as shown in **32.4**.

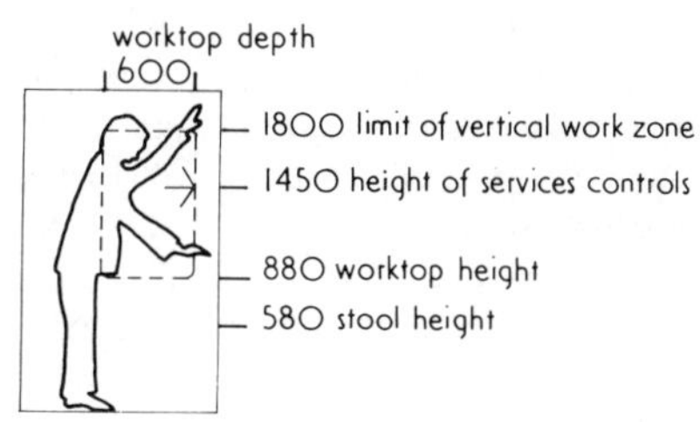

32.1 *Limits of reach when standing, or sitting on a stool*

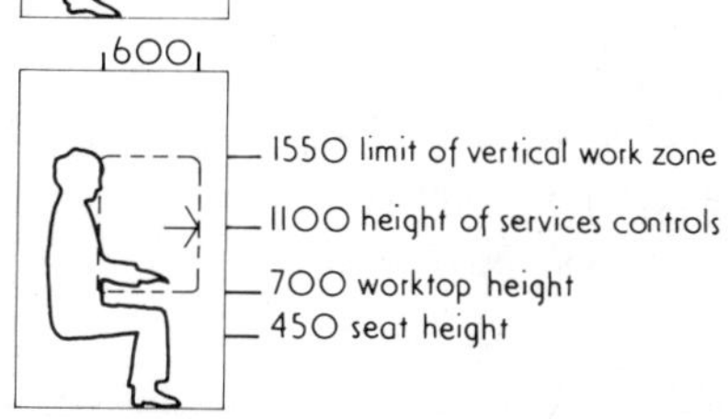

32.2 *Limits of reach when seated in a chair*

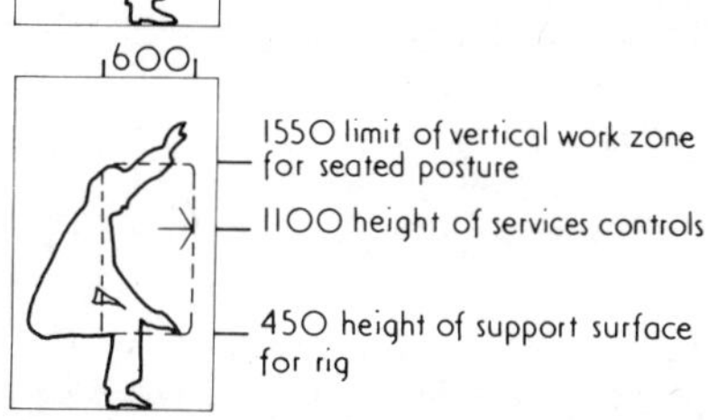

32.3 *Limits of reach when seated working on tall rigs*
Height of service controls specified in **32.1**, **32.2** *and* **32.3** *may have to be reduced in practice, but the reduction should be minimised*

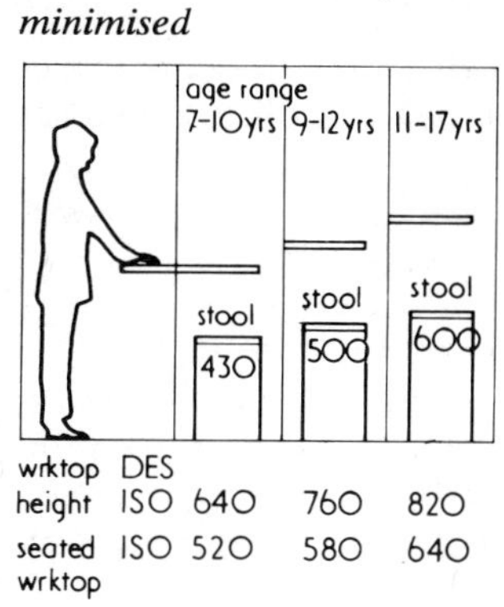

32.4 *Standing and sitting heights for schools*

3.02 Worktop depth

If the worktop is too deep the back of it will be used for dead storage, making access to services controls difficult. The depth should be based on maximum convenient reach. A 600 mm depth will meet most requirements but there may be cases where 700 or 750 mm will be needed for large bench mounted instruments.

3.03 Height of services controls

Services controls should be as near to shoulder height as is possible. This is the shortest distance in terms of reach and controls are less likely to be concealed by apparatus.

3.04 Sinks

For comfortable working the rim of wash-up sinks should be slightly higher than the general worktop, **32.5a**. The same applies to any associated draining top. As an alternative, a shallow tray with rim at worktop height can be specified and a deep bowl used with it when a greater depth of water is required for soaking, washing or cooling larger apparatus, **32.5b**. Smaller sinks, drip cups or continuous troughs used for the disposal of liquids are not critical and can be set at worktop height. As an alternative to the small sink a shallow tray can be placed on the worktop to drain into a drip cup or trough. The tray can be stored when not in use. For use with large free-standing rigs, or for demonstration purposes, a movable sink **32.5c**, can be employed.

3.05 Storage

The maximum height and lowest level of storage which is frequently used should be based on convenient reach. Extreme high and low zones tend to be used for dead storage, **32.6.** Underbench storage should not exceed 50 per cent of the underbench space and must be movable. At least 600 mm width of knee-hole space must be provided under the bench at each work-station. In schools with a small bench length requirement, virtually all underbench space will be required for knee-hole space.

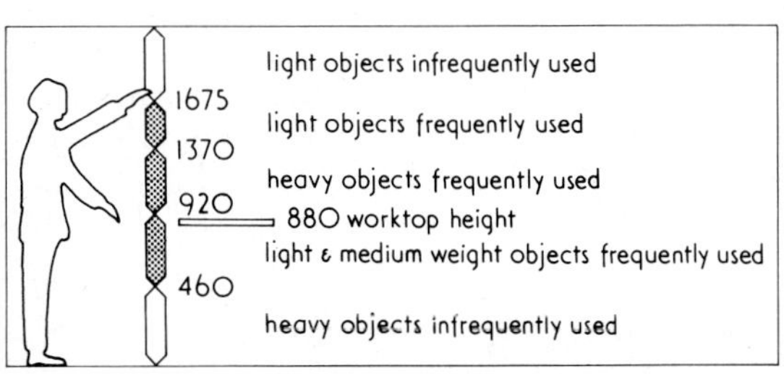

32.6 *Vertical storage zones based on common patterns of use*

The depth of storage cupboards should not normally exceed 500 mm. Access to the back of shelves deeper than 500 mm is difficult particularly in the case of underbench units. The 500 mm depth will fit neatly under a 600 mm deep worktop (allowing for a back rail to the table).

Shelves and trays within storage units should be adjustable in height. Drawers are not recommended as they cannot be adjusted.

Comfortable reach into a 300 mm deep cupboard above floor level is as follows:

7 years—1100 mm
9 years—1170 mm
10 years—1260 mm
11 years—1300 mm
12 years—1375 mm
17 years—1640 mm
18 years—1675 mm

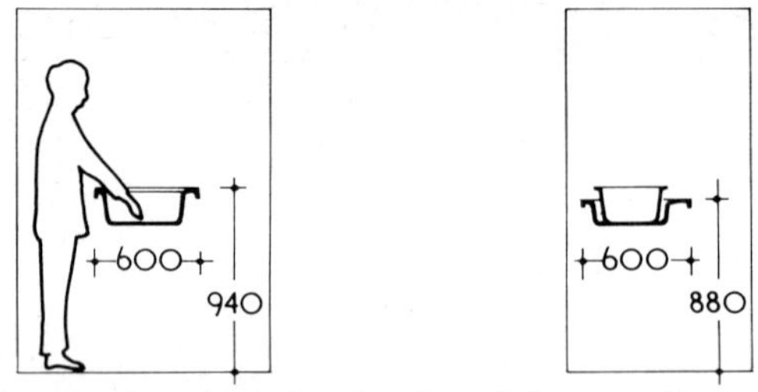

32.5 *Wash-up sink heights for adults over 18 years old:* **a** *sink raised above normal 880 mm worktop height for comfortable working* **b** *shallow tray at worktop height with a loose, taller washing-up bowl*

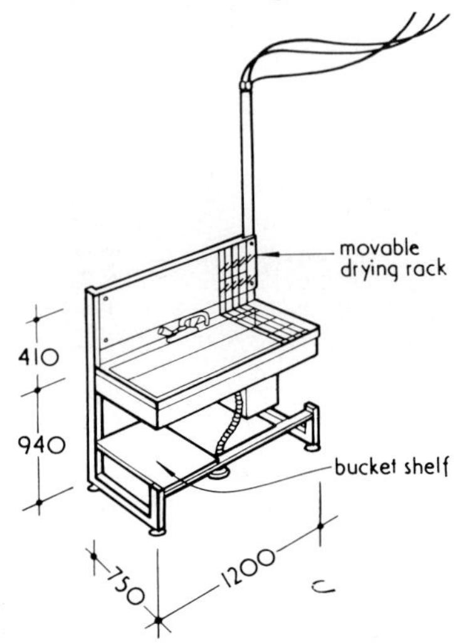

c *general purpose movable wash-up sink with heater*

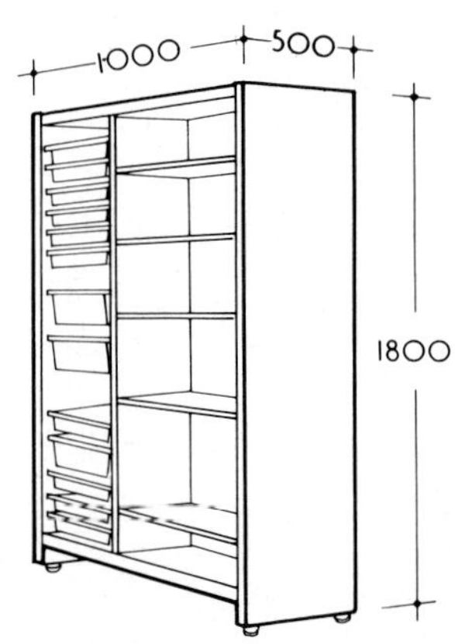

32.7 *Tall storage cupboards, preferred to underbench storage*

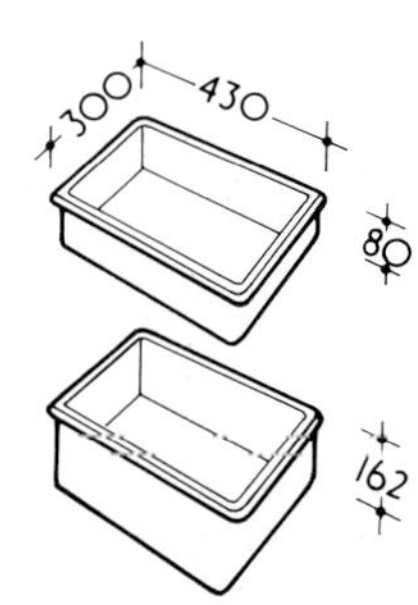

32.8 *Standard storage tray sizes*

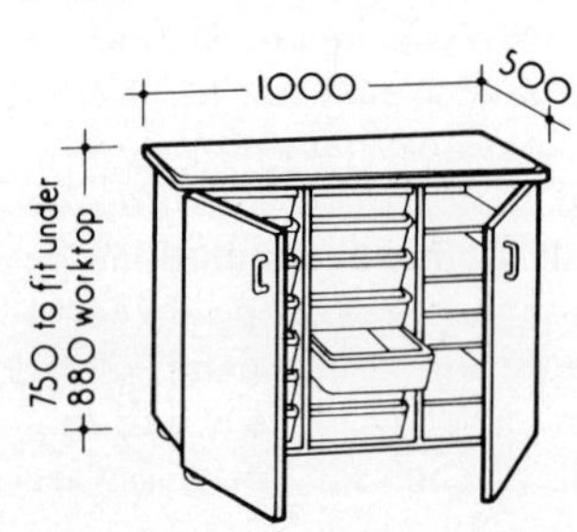

32.9 *Mobile underbench unit*

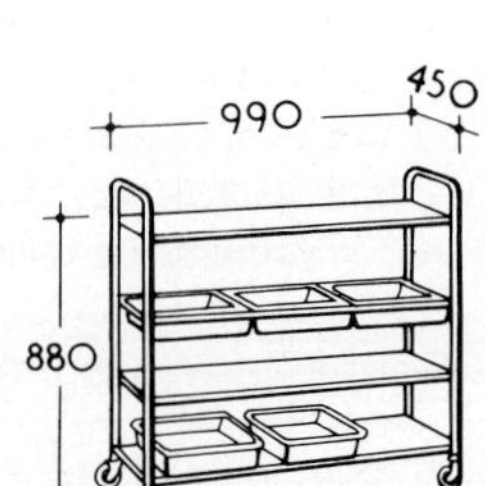

32.10 *Service trolley*

3.06 Storage units

A number of science storage systems now available, based on a standard tray size, **32.8**, are intended for the storage of small

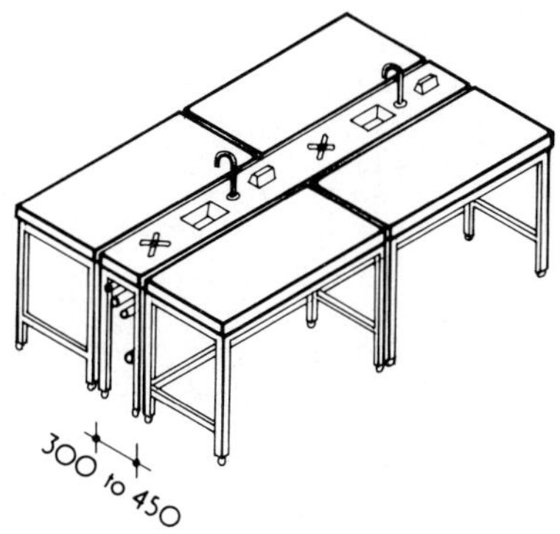

32.11 *Workbench services between and below the benches, taking the most space*

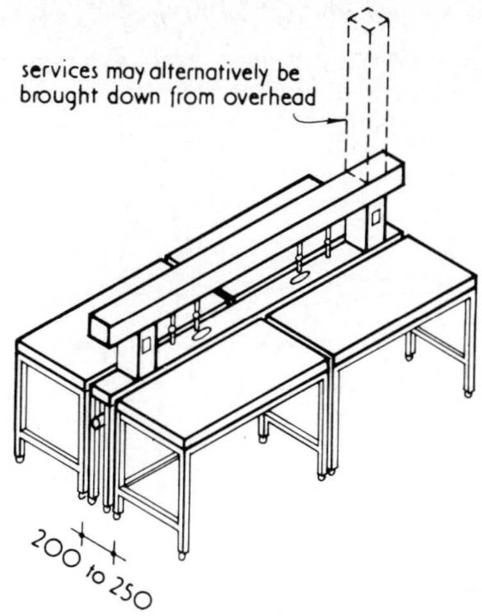

32.12 *Workbench services above the worktop suspended at optimum control height (see* **32.1***)*

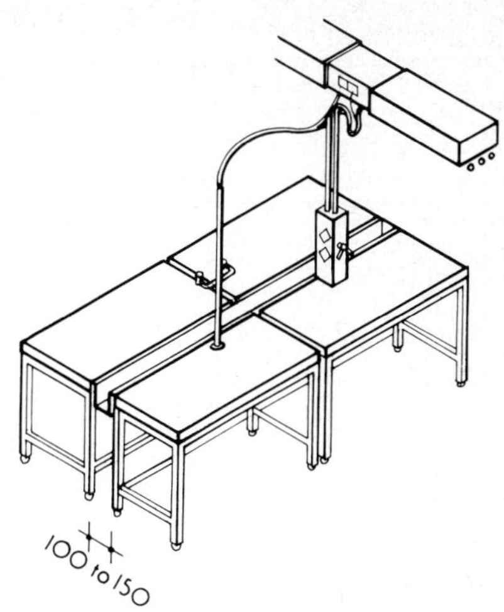

32.13 *Individual bollards linked flexibly from trunking just below ceiling level (most economical in space). Note drainage trough between benches*

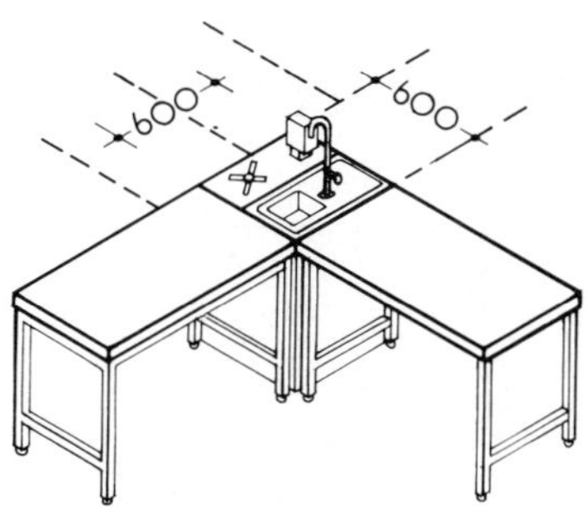

32.14 *Services brought up from floor in bollards*

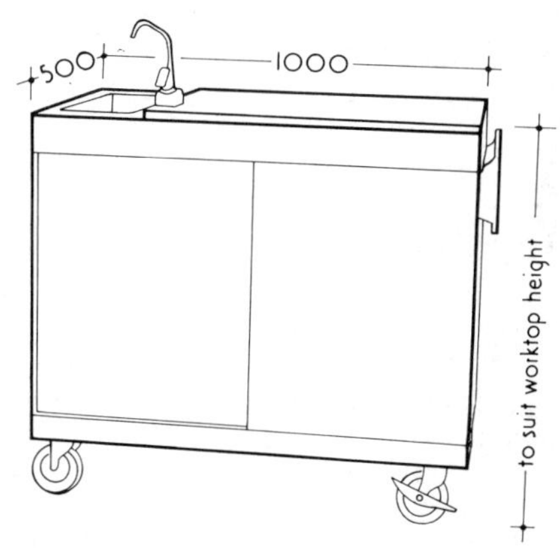

32.15 *Mobile demonstration trolleys, developed for use in schools. It contains its own water supply and drainage in tanks under the sink. The unit is plugged into the nearest 13 amp socket outlet to enable demonstrations and simple tests to be carried out in unserviced areas*

equipment and experimental kits. Tall cupboard units, **32.7** should be used in preference to underbench storage as these give the range of choice as to storage level referred to in **32.6**. If underbench units are required they must be related to bench unit sizes. The unit shown in **32.9** is dimensioned to suit a 600 × 1200 bench unit. Trolleys are available for serving laboratories from storage and preparation areas, **32.10**. All units should be based on standard tray and shelf sizes and the interior of units should be capable of taking any combination required with full vertical adjustment.

3.07 Services space related to worktop

When assessing floor area for worktop and equipment, the area taken up by associated services must be included. This area should be kept to a minimum in order to give as much floor space as possible for laboratory work. By raising the service runs, apart from drainage, above the worktop (see recommendations for the height of controls), **32.12**, the width of services spine can be usefully reduced from that shown in **32.11**. Further savings can be made by raising the local service runs into a boom below the ceiling with flexible connections to bollards on the worktop. The space between the worktops having to accommodate a drainage trough only, **32.13**. The dimensions of floor-mounted service bollards, **32.14**, are commonly related to the width of benching, allowing worktop to abut any side as requirements dictate.

3.08 Mobile services units

Where the requirement for laboratory services is infrequent or where centralised piped services are unable to supply the quality of service required, localised mobile sources provide a highly adaptable solution to the problem, **32.15** and **32.16**. If they are used floor space must be allocated.

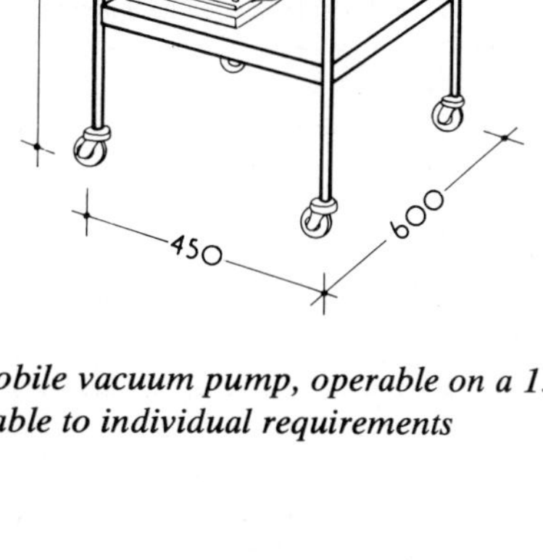

32.16 *A mobile vacuum pump, operable on a 13 amp supply, and adjustable to individual requirements*

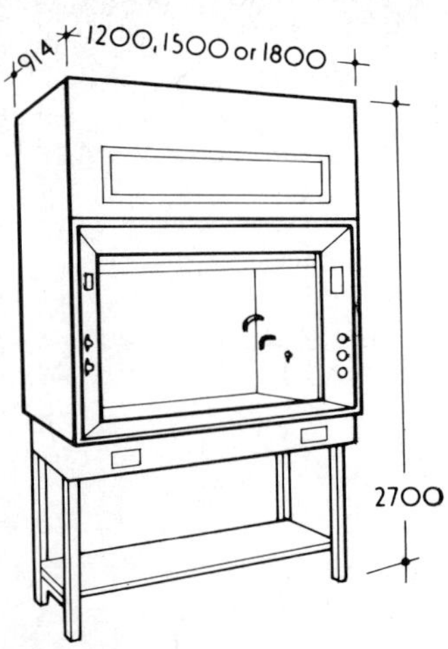

32.17 *Semi-movable fume cupboard with aerofoil front and bypass air grille over sash*

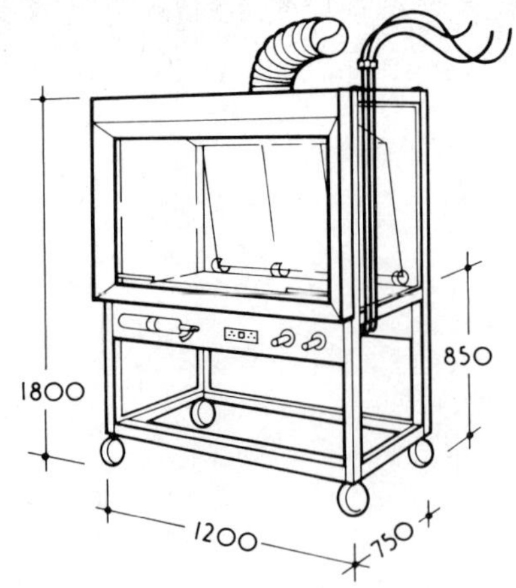

32.18 *A mobile fume cupboard which has glazed baffle, back and sides for demonstration purposes*

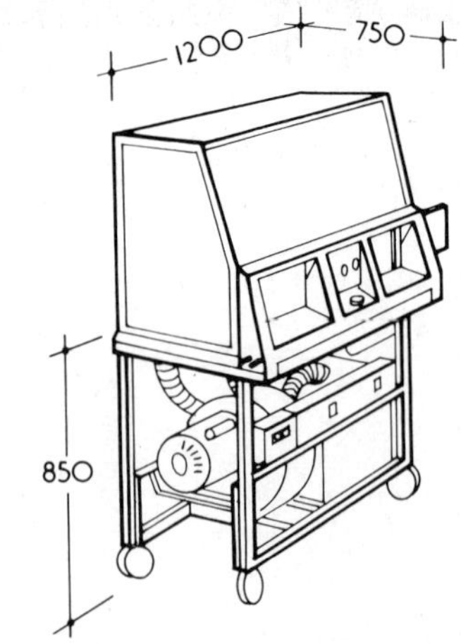

32.19 *A mobile recirculation fume cupboard incorporating filters*

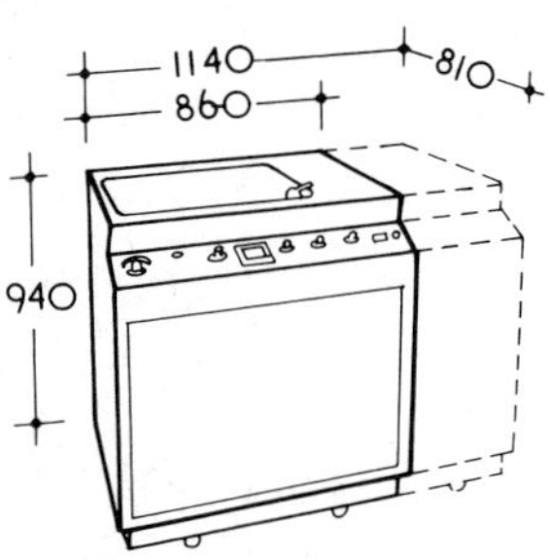

32.20 *Floor-mounted centrifuge*

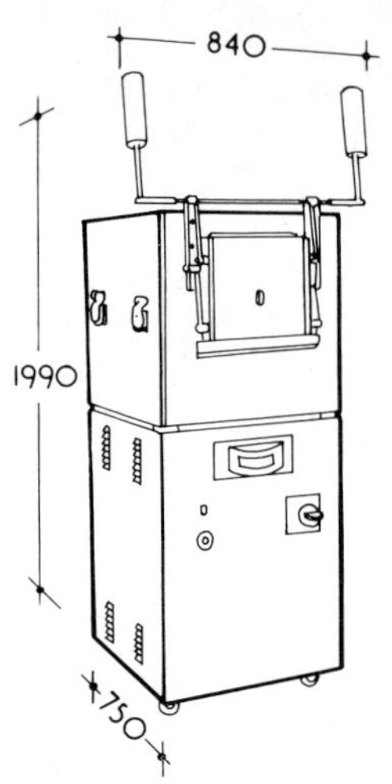

32.21 *Box furnace*

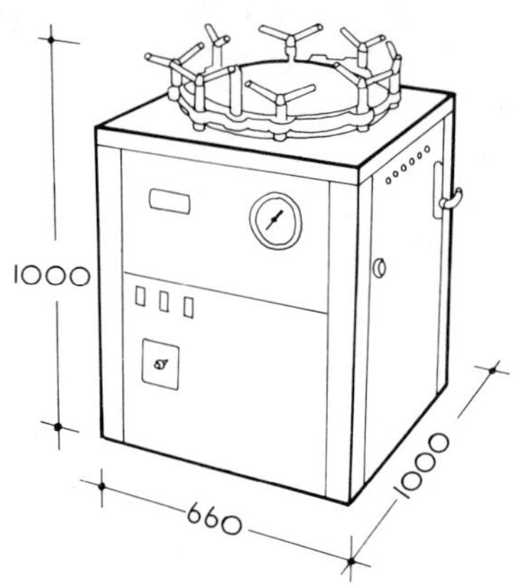

32.22 *Autoclave*

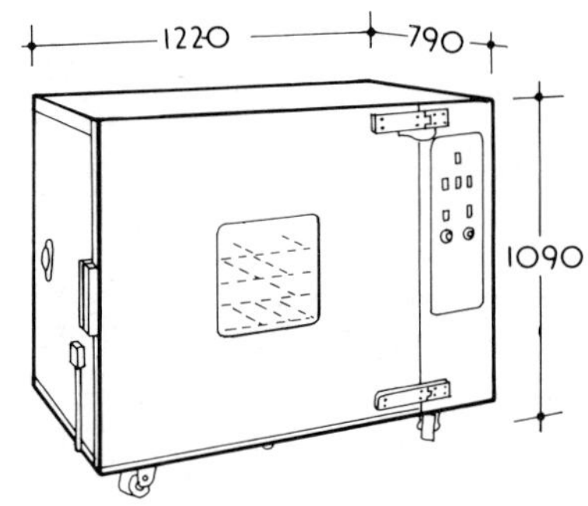

32.23 *Humidity oven*

3.09 Fume cupboards

Fume cupboards, **32.17** to **32.19**, are costly items and demands for them vary between scientific disciplines (table V). The facility to reposition them and build up concentrations in different parts of a building will help to ensure their full use when changes in the use of laboratory space occur. Critical dimensions are given in table VI; linear air velocities in table VII.

Table V Numbers of cupboards required (rough guide)

FE, university teaching	
Physics	1 to 40 students
Organic chemistry	1 to 5 students
Inorganic chemistry	1 to 5 students
Physical chemistry	1 to 15 students
Biology	1 to 13 students
Biochemistry	1 to 9 students
Research work	
Chemistry	1 to 1 or 2 workers
Biochemistry	1 to 2 to 4 workers
Biology	small demand ad hoc to individual needs
Physics	small demand ad hoc to individual needs

Table VI Critical dimensions of fume cupboards

	Schools	FE, university, industrial
Worktop height	as **32.4**	as **32.1**
Clear width of front opening	750–1200	900 +
Height of front opening	750	840–900
Worktop to top of cupboard	1050 +	1050–1800
Depth of clear working space	600–750	600–900

Table VII Linear air velocities required through open fronts

Schools	0·5 m/sec
FE, university, industrial	0·5 m/sec
Special toxic hazards	Higher rates may be recommended

3.10 Floor mounted equipment

In many cases space must be allocated for floor-mounted equipment. Typical examples are shown in **32.20** to **32.23**, but sizes will vary with different performance requirements and manufacturers' models. Space allocated on plans should never be precise and laboratory furniture should be movable to accommodate changes in the quantity and sizes of such equipment during the life of the laboratory.

3.11 Layout at the workstation

Where more than one worktop unit is required, the opportunity to create workbays should be allowed for. This more readily defines an individual's territory and puts a greater area of worktop or equipment space within convenient reach from a single position, **32.24**. Worktop lengths are given in table VIII.

3.12 Space requirements for personnel

The application of the dimensions given in **32.25** will depend on the form of layout adopted in a given situation. The lower figures are suitable for short peninsula or island layouts (up to 2400 mm in length). Longer runs of worktop will require the

Table VIII Space requirements at the workstation for worktop/equipment*

Level of work	Worktop length per person	Activity
Teaching (school) all subjects	600 900 1200	Working in pairs Individual working Advanced work
Teaching (college, polytechnic, university)		
Physics	900 to 1800	Elementary to advanced 1800 includes a 50% contribution of space for a small sink between 2 persons
Chemistry	1200 to 1800	
Biochemistry	1200 to 1800	
Biology	1200	
Research (low requirement)		
Plant physiology, Botany, Zoology	1800 to 2400	
Research (medium requirement)		
Chemistry, Physics, Microbiology, Pathology, Animal physiology	2400 to 3600	For overall calculation a mean length should be taken
Research (high requirement)		
Biochemistry	3600 to 4800	

* The requirements above do not include space for shared equipment, fume cupboards, wash-up sinks, etc

maximum shown (4800+lengths). These situations shown in **32.25** are related primarily to linear worktop layouts and necessarily indicate (as in **b** and **e**) workspace and circulation intermixed. Where possible this should be avoided. This can be done of short peninsula or island layouts, or workbays, **32.24**, are used. Separation of workspace and circulation will reduce the likelihood of accidents and ease the problem of escape from hazards (see para **4**).

4 LABORATORY LAYOUTS

4.01 Planning for variety

Furniture will normally have a minimum useful life of 15 years and over such a period changes in requirements will demand adjustments or even total rearrangements. All laboratory furniture should be movable.

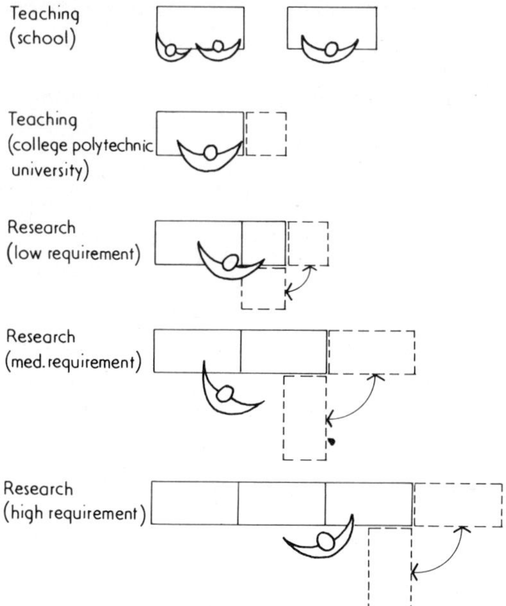

32.24 *Worktop/equipment areas based on 1200 × 600 mm units*

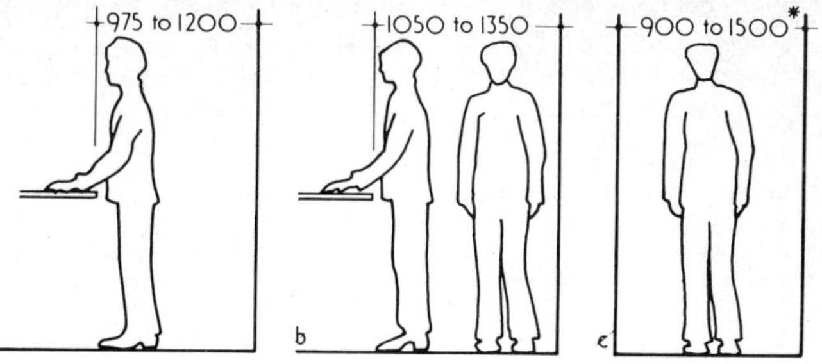

32.25 *Space required between worktops or equipment:*
a *one worker, no through traffic*
b *one worker plus passageway*
c *passageway only*
* *1500 mm allows two to pass each other*

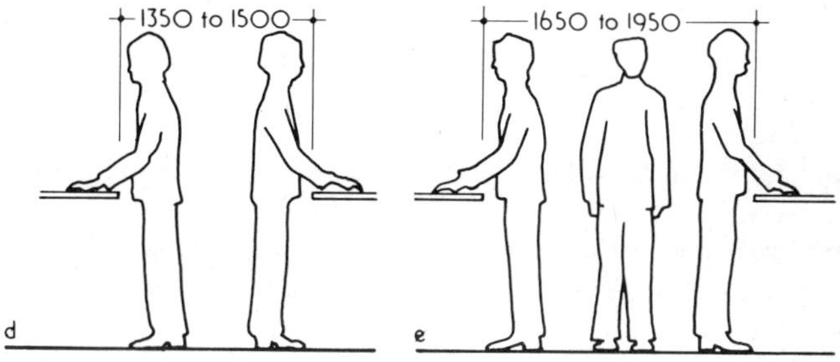

d *two workers back to back, no through traffic*
e *two workers back to back plus passageway*

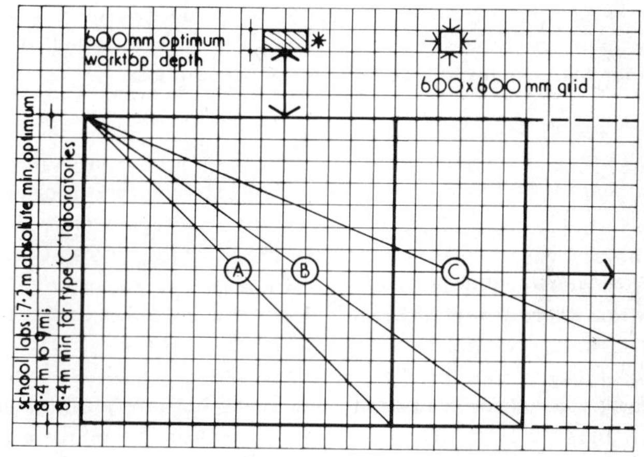

32.26 *Laboratory shape for teaching. Where constant voice and visual contact between teacher and all students is important, the laboratory space should be designed within a plan proportion of 1 × 1 to 1× 1½ (A to B). In universities and further education large classes of up to 100 students with intermittent lecturer contact may have the proportion extended beyond 1 × 1½ (C)*
* *Furniture units should be sized to suit the amount of area available. If furniture is increased in size, circulation space will have to be reduced and standards of safety may suffer*

There are no ideal layouts which suit a wide range of needs. Requirements will vary according to the educational research or routine needs of the moment. The shape of the laboratory should not therefore be based on a single layout of furniture and equipment but should rather be assessed on its ability to accommodate a variety of layouts. Long and narrow, or irregular shapes are less satisfactory than the squarer shapes in **32.26**.

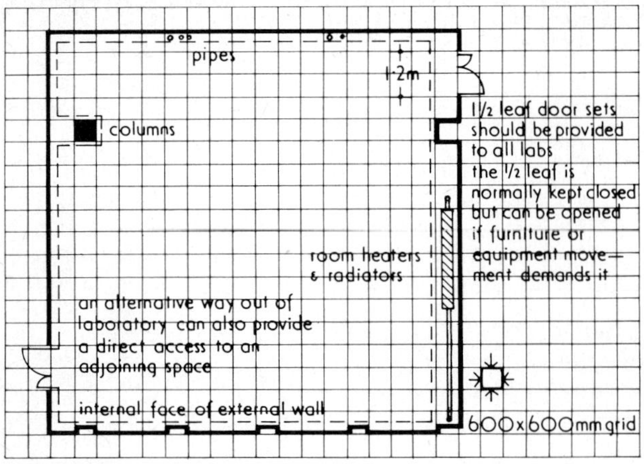

32.27 *Effective space and access*

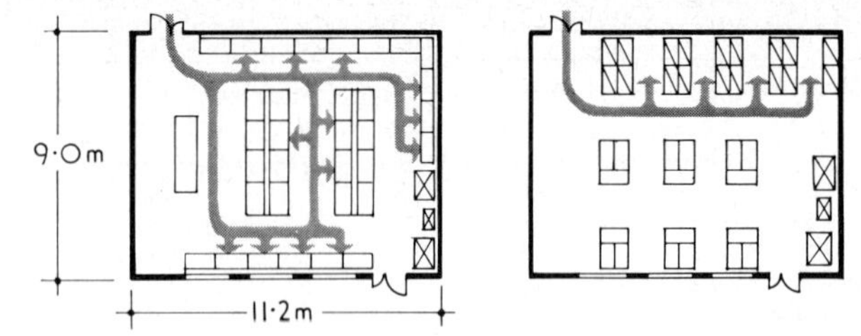

a *traditionally dispersed*

b *centralised for more efficient management and access*

32.28 *Distribution of storage within a teaching laboratory:*

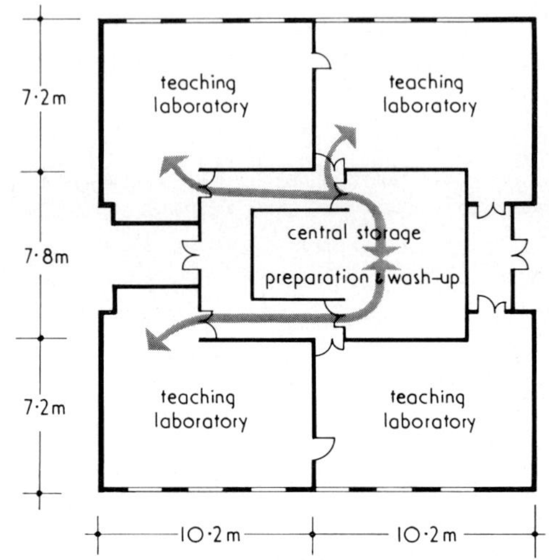

32.29 *Centralised storage for a group of laboratories*

4.02 Effective space and access

In addition to providing a suitable shape of space, it is vital that full use is made of the floor area for laboratory purposes. Irregularities on the perimeter of a space will reduce the effective area, and environmental as well as laboratory services elements require careful co-ordination with the building structure, **32.27**. If such irregularities are unavoidable, as is often the case when converting existing buildings, the perimeter can be smoothed out with a continuous rail at bench top height. This then provides a straight abutment for benching and defines the effective area.

An alternative way out of the laboratory should be provided for escape purposes. Doors to laboratories should be 1½ leaf minimum to cater for some items of equipment which are wider than the normal single leaf door opening.

4.03 The location of storage

The efficient storage and distribution of apparatus and equipment is critical to the effective use of teaching laboratories. The traditional practice of distributing storage round the laboratory in underbench units, **32.28a**, is no longer appropriate in school laboratories. The wide range of experimental work undertaken by different classes in the same laboratory necessitates centralising storage within the laboratory, **32.28b**. This approach enables apparatus and experimental kits to be checked by the teacher or technician without interfering with the class. Also the class can visit the storage in a controlled manner. Organising storage within the laboratory in this way is also appropriate in higher and further education where the multidisciplinary use of laboratories by various groups of students is proposed. Centralised storage/preparation is possible for groups of laboratories if there is adequate technical/management assistance available, **32.29**. Experimental kits can be trollied from the central area to any teaching laboratory to meet individual class needs. This leads to more intense use of equipment, reduces the need for duplication and allows teachers and lecturers to devote more time to personal contact with the students.

4.04 School laboratories

The layout of the laboratory should provide:

- good pupil/teacher contact
- safe movement
- convenient access to equipment and materials.

32.30 and **32.31** show alternative layouts for general science work using the same furniture units. Both layouts will accommodate up to 36 pupils for general work (or 22 for advanced studies). In both cases storage is organised as shown in **28b**. The long linear arrangement of benching requires wide gang-

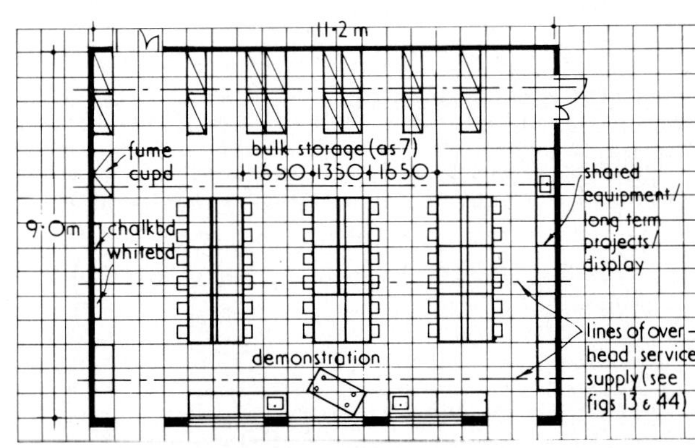

32.30 *School laboratory layout with linear bench arrangement accommodating 36 pupils at 2·8 m² each, or 28 pupils at 3·6 m² each*

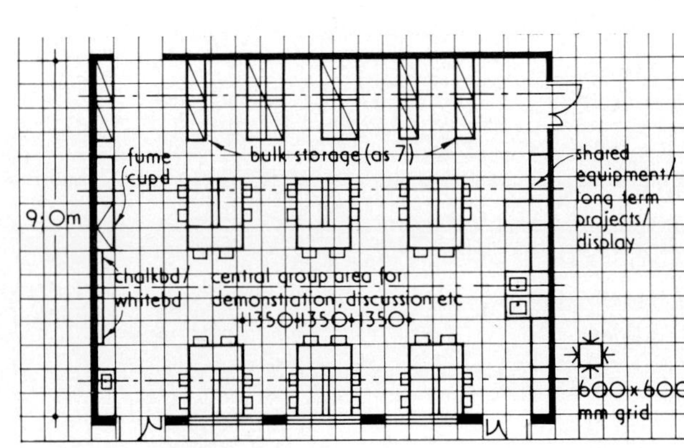

32.31 *The same numbers as in* **32.30** *are accommodated at short peninsular and island benching. Both schemes have overhead servicing*

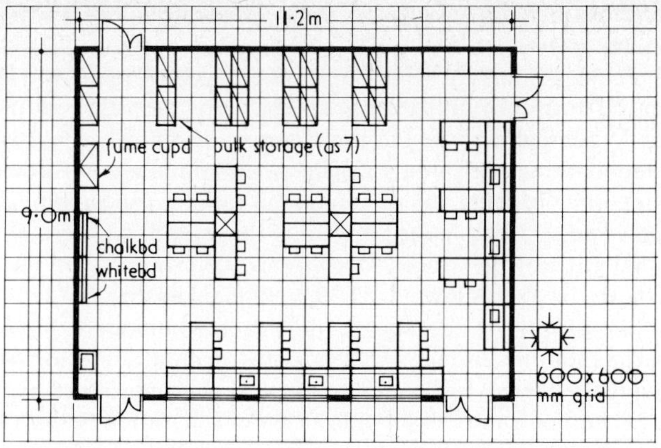

32.32 *The same capacity with less adaptable perimeter and underfloor servicing*

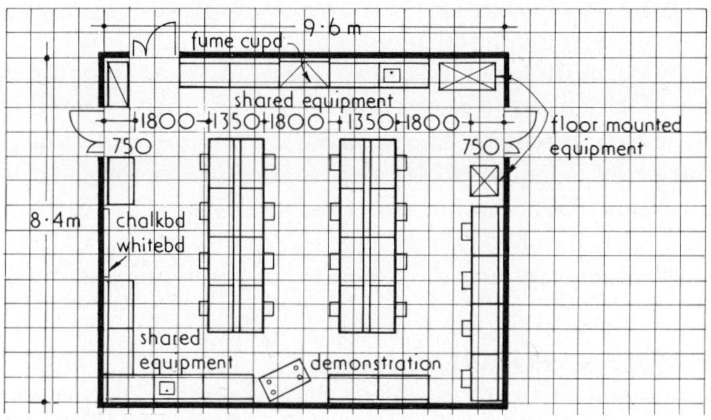

32.33 *University laboratory layout accommodating 20 students at 4·0 m² each, or 16 students at 5·0 m² each*

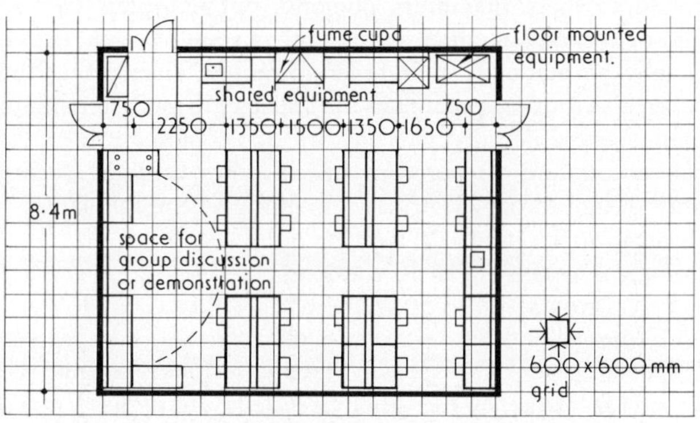

32.34 *The same capacity as* **32.33** *with shorter bench arrangements and a discussion/demonstration area*

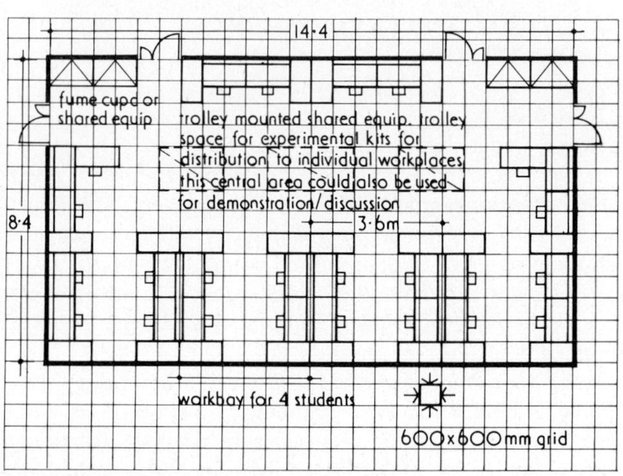

32.35 *University layout for 24 students at 5·0 m² each with shared equipment centrally located*

ways to ensure safe back-to-back working and allow the teacher to circulate safely while work is in progress. The short peninsula and island benching, **32.31**, provides a clearer separation between individual workspace and general circulation. Gangways between benching can be reduced and space for a central group area gained. Both **32.30** and **32.31** are serviced from overhead supplies as shown in **32.13** (refer to later notes on services distribution for further information **32.43**).

4.05 School laboratories

32.32 shows an alternative method of servicing from under the floor using fixed bollards (see **32.14**) with wall-mounted spines. Although this layout meets the requirement of separating workspace and general circulation, it is less adaptable than overhead booms due to the fixed nature of the bollard, which while allowing alternative arrangements of benching nevertheless predetermines their basic spacing.

No fixed position for demonstration is indicated on any of the layouts. To fix the use of floor area with demonstration facilities which will have only intermittent use is wasteful. It is more efficient to use any suitably located bench or a trolley (see **32.15**) of a height to suit the age range using the laboratory, ie normal worktop height as specified for general laboratory work.

4.06 University, polytechnic and college teaching laboratories

Planning for safe movement and convenient access to shared equipment is vital (as for school laboratories). The need to plan for direct contact with the lecturer will vary depending on the level of work undertaken. Large laboratories make constant direct contact impossible but this is educationally acceptable with the use of project-orientated teaching methods. **32.33** and **32.34** show typical layouts suitable for a wide range of subjects with some variations, eg fume cupboard provision would increase for some chemistry subjects (see para **3.09**). The extended linear layout of **32.33** and the short peninsula and island arrangements of **32.34** are generically related to **32.30** and **32.31** respectively and the same observations apply. Discussion and demonstration should be catered for, but continuous lecturing should be provided for in an appropriate space. If storage and shared equipment are centrally located (as in **32.29**) multipurpose use of the laboratories shown is feasible, leading to a possible higher utilisation of their high cost facilities.

For larger groups the layouts illustrated can be repetitively extended based on the areas per student given.

Multipurpose use of teaching laboratories requires convenient access to and rapid changes of special equipment and apparatus. **32.35** shows a reversal of the traditional layout, ie space for special apparatus which is to be shared is concentrated in a central area as opposed to being distributed around the perimeter. If mounted on trollies, apparatus and experimental kits for a whole variety of subjects can be conveniently brought into the laboratory from an adjoining preparation/storage area.

4.07 Research laboratories

Research is commonly carried out by small teams of workers and lends itself to accommodation within repetitive units of space. The range of area requirements (para **2.01**) indicates a useful unit size of 24 m². This will accommodate four workers at 6 m² each, three at 8 m² and two at 12 m², in the science disciplines listed.

The degree of adaptability within the unit is dependent on the shape adopted. The rectangular unit, **32.36**, provides for straight runs of worktop and equipment. The square unit, **32.37**, provides for straight runs of worktop and equipment

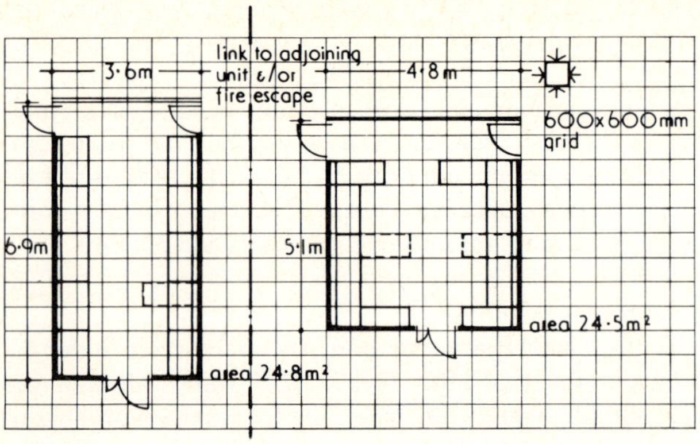

32.36 *(left) Research laboratory based on a rectangular unit of space*

32.37 *(right) Research laboratory based on a square unit of space*

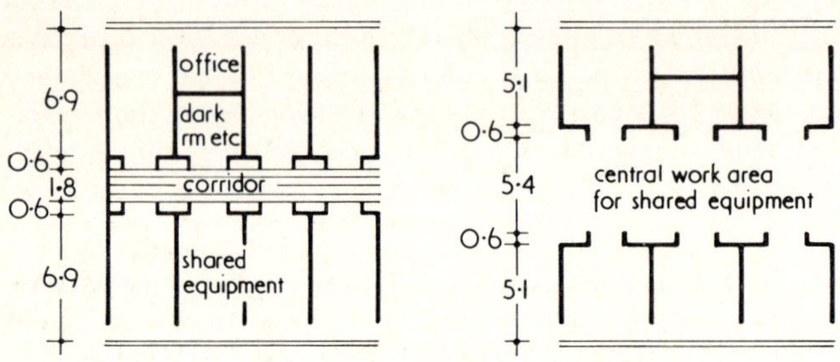

32.38 *Plans incorporating the two laboratory types:*

a *rectangular units with central corridor on a 3·6 m module*

b *square units with a shared central work area on a 4·8 m module*

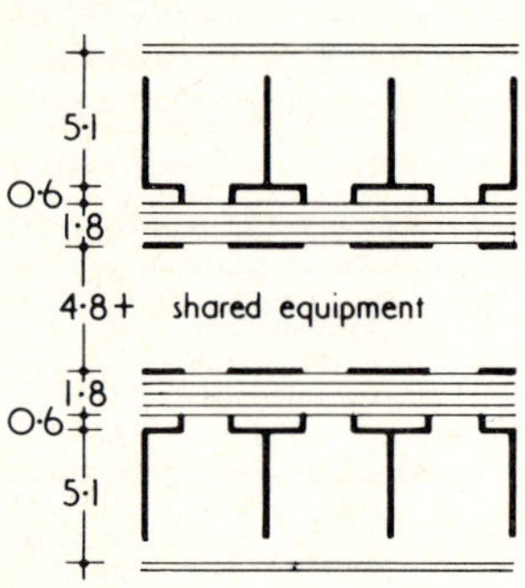

c *square units with a double corridor, the core holding shared equipment, also on a 4·8 m module*

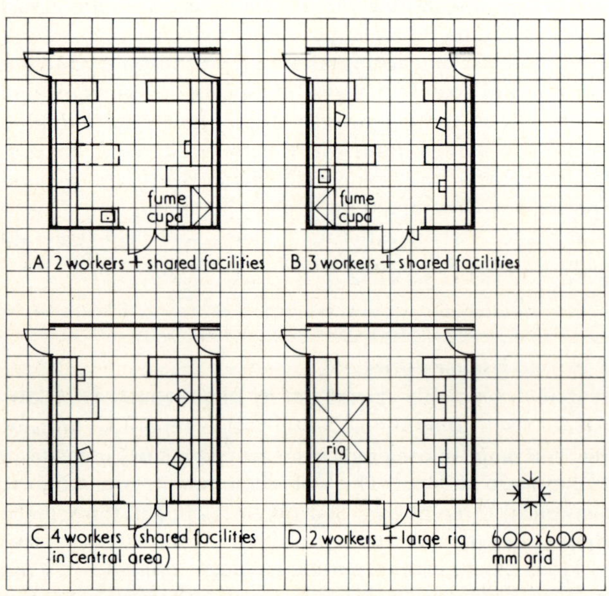

32.39 *Alternative layouts within the square laboratory unit shown in* **32.37**

layout including the creation of individual workbays as previously recommended (para **3.11**).

The rectangular unit provides the most economical plan form if a central corridor only is used, **32.38a**. Growth in the use of shared equipment has led to the planning of central areas for accommodating such items which, when combined with the square unit, provide highly adaptable laboratory buildings, **32.38b** and **c**. Both laboratory units illustrated can be subdivided to provide offices, darkrooms, coldrooms etc. The alternative layouts shown in **32.39** are not exhaustive but indicate the ability of the square unit to accommodate large experimental rigs and more importantly to allow definition of individual workers' territories at different densities of occupation. The opportunity to provide a clearer separation between personal circulation through the laboratory should be noted (cf recommendations in para **3.12**).

4.08 Routine laboratories

The layout of laboratories for routine experimental work may follow recommendations for research if closely related to it, thereby permitting change of use. Large routine laboratories with a high density of occupation are closer to those found in high density teaching, **32.33**, **32.34**, and **32.35**, and similar layouts can be applied. Growth in the use of large-scale electro-analytical equipment (as in hospital pathology laboratories) has made the ability to exchange the use of floor space between furniture or equipment of greater importance.

5 SERVICES REQUIREMENTS

The demand for services outlets depends on the nature and level of work. Table IX gives only general assessments of need; requirements in individual situations should be checked. If the number provided is fixed, a large number will have to be installed to meet maximum demand, but if outlets can be added or subtracted easily at any time, a smaller initial provision may be made. It is more important to ensure that the services distribution system has the capacity to meet maximum demands. Accepted diversity factors are 20 per cent for electrical outlets and 40 per cent for all other outlets. Recent studies of laboratory electrical demands give the following:

- physical work—up to 20 watts/m^2
- biological work—up to 30 watts/m^2
- chemical work—up to 50 watts/m^2.

Requirements for other services—steam, vacuum, compressed air, special gases and special electric supplies—should be assessed for frequency and quality needed. Where the demand is intermittent or a particular quality is required, localised sources may provide the best answer, **32.16**.

6 STRUCTURAL AND ENVIRONMENTAL REQUIREMENTS

6.01 Structure

- *Loadings (superimposed):* For bench scale laboratories design for 5·0kN/m^2 + point loads of 3·6kN to 5·0kN. Average loads from suspended services (including ventilation ductwork) 0·25kN/m^2. The point load capacity will deal with most bench scale items of floor-mounted equipment (up to, say, 1 tonne in weight, assuming the load is spread). Loads in excess of this, ie heavy engineering equipment and rigs, are most economically located on ground floors.
- *Spans:* These must be related to preferred laboratory sizes, see **32.26** and **32.38**, the basic planning grid adopted (600 mm preferred) and the spacing of services distribution. Rectangular column grids may be utilised for linear building development but square grids are preferred if two-way development is envisaged, ie if the building form turns through 90°.

Table IX Average numbers of outlets per person

Laboratory	Electric* (13 amp)	Cold water	Natural or town gas	Additional notes
Teaching				
School				
Elementary level	1†	1	1	If a laboratory is set aside for physics use only, cold water and gas outlets should be reduced in number
Advanced level	2	2	1–2	
Teaching				
University				
Polytechnic				
College				
Physics	2	1 per 6 people	–	Gas requirement minimal, bottled gas can be used
Chemistry	2	2–3	2	
Biology	2	2–3	1	
Research				
University	2–6	2–10	Up to 4	Chemistry subjects have a high requirement for water and gas (if used as a heating medium). Physics subjects have a high requirement for electrics
Polytechnic				
College				

* Socket outlets—all scientific work is using increasing numbers of electrical instruments. This results in high demands for socket outlets. In most cases the loads involved are small and these demands can be met by using plug-in splitter boards fused at 13 amps

† For schools the use of low voltage (110v AC) is recommended by some authorities for safety reasons. However, not all electrical instrumentation is readily available in low voltage ratings. An alternative is a 240v supply with earth leakage protection rated at 10 milliamps

6.02 Environment

• *Heat:* 20 to 21°C required for comfort conditions. Special requirements for experimental work, eg coldrooms, temperature controlled rooms should be dealt with on a local basis (complete package rooms are available).

• *Ventilation:* Rate of air change: up to 6 air changes per hour for general work. Rates above 2 per hour will require mechanical ventilation or full air conditioning. Extract must be balanced if fume cupboards are provided in laboratories. Certain activities may require higher air change rates, eg glassblowing 20 per hour; or special filtration, eg aseptic area.

• *Lighting:* General laboratories 350 to 400 lux; glare index 19 (schools 16). In routine laboratories constant reading of instruments may demand 600 lux; glare index 19. Where such activity is intermittent local additional lighting provides a more economical solution.

• *Sound:* Background noise levels in individual spaces, eg office, research laboratory or teaching laboratory (up to 36 places) NC 35. For larger multipurpose laboratories (teaching, research or routine) NC 45.

7 LABORATORY SERVICES DISTRIBUTION RELATED TO THE BUILDING SHELL

Routes for the distribution of services must be co-ordinated with the structural design of the laboratory building shell. Sub-main distribution is commonly based on one of two variants, vertical sub-mains, **32.40**, or horizontal sub-mains, **32.41**. Vertical sub-mains incorporated in a regular provision of builders' work ducts tend to be used in conjunction with repetitive laboratory units as shown in **32.36** and **32.37** or where standardisation of bench spacing is imposed throughout the building. While readily accommodating fume extract ductwork, the floor area taken up by the vertical ducts (around 4 per cent of the gross area) is higher than for the horizontal method (between 1 and 2 per cent depending on fume extract requirements). The horizontal method with its limited number of vertical mains risers allows a high degree of planning flexibility. Horizontal sub-mains can be designed on the ring main principle which is better able to deal with high demands for piped supplies in any part of the building. Both methods of distribution enable branches to be isolated on the floor that is being served (ie without interference to floors above and below), but the horizontal method allows the isolation of large areas on one floor by isolating the sub-mains from the mains riser at the same floor.

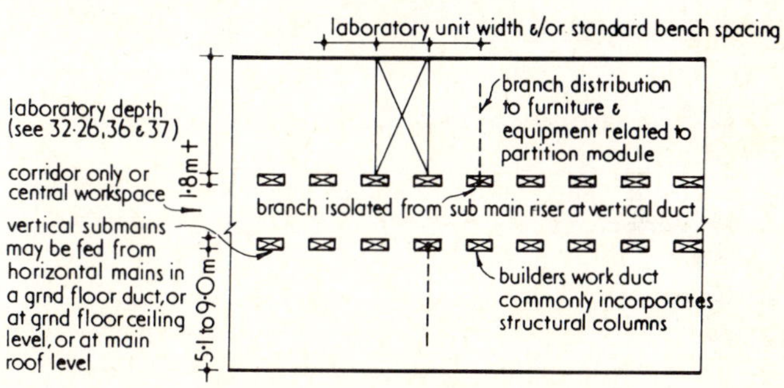

32.40 *Vertical sub-main distribution (relate to* **32.38***)*

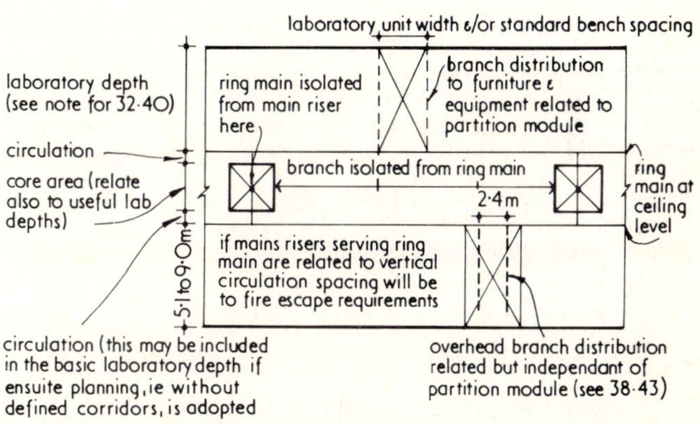

32.41 *Horizontal sub-main distribution incorporating the ring-main principle*

8 SERVICES DISTRIBUTION IN THE LABORATORY

The method of distributing services from the sub-mains to furniture and equipment has a major influence on the flexibility of layout possible during the laboratory's total life. Where fixed service runs are integrated into the furniture, **32.42**, the spacing and configuration of furniture must necessarily be standardised if the initial economic advantages of repetitive service runs are to be gained. **32.42** shows a selection of typical configurations, each of which requires some variation in the routing of services to and from them. Where layouts abut vertical ducting as in A, the route is simple and direct. Island layouts as in B involve more indirect routes, the introduction of doors as in D adding similar complications. Peninsula layouts from the external wall as in C require distribution along the

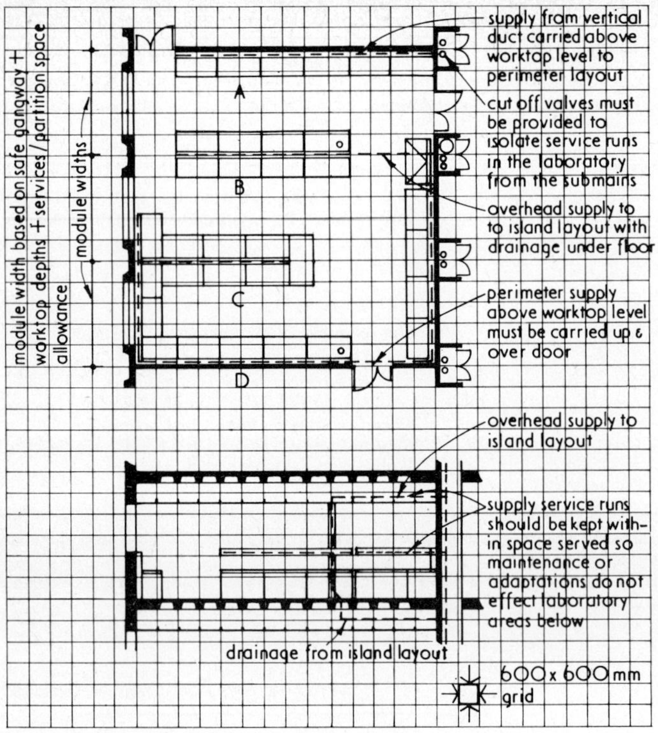

32.42 *Rigid services distribution integrated with furniture layout:*
a *plan. Although vertical sub-mains are shown, service runs could alternatively be supplied from horizontal sub-mains as in* **32.43**
b *section*

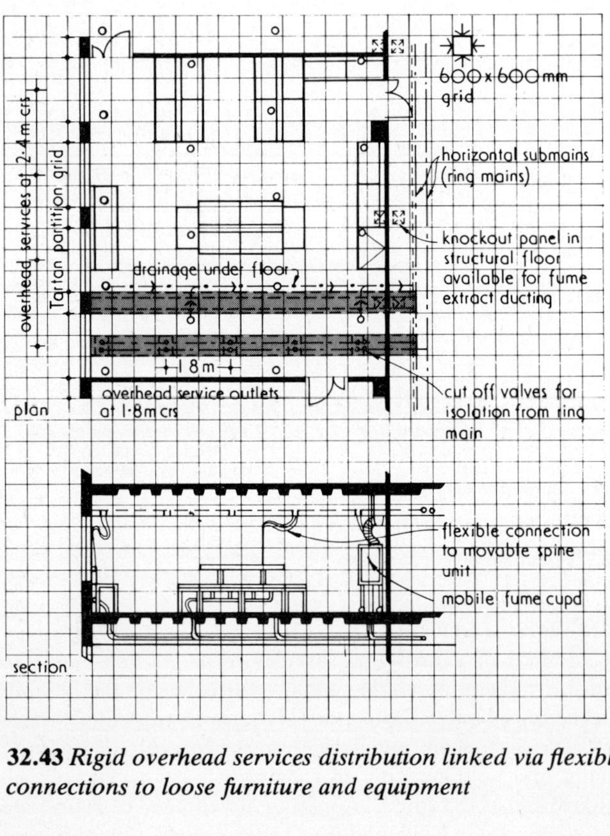

32.43 *Rigid overhead services distribution linked via flexible connections to loose furniture and equipment*

external wall. If this is extended along the whole building it becomes a horizontal sub-main (a variation on **32.41**). This sub-main location takes up space below windows and for ease of maintenance and adaptation, perimeter heating must be independently located at ceiling level. Service runs in furniture may be based on **32.11** or **32.12** or **32.13**.

Although loose worktop and storage units (or equipment) can be associated with rigid service runs, **32.42**, any major change in layout will involve adaptation or extension of the service runs. If, however, the distribution of rigid service runs within the laboratory can be separated from furniture, equipment and partitions, and flexible connections made to loose services units, **32.43**, layouts can be adjusted directly by users to meet new requirements. Standardisation of furniture layouts is no longer critical.

Island, perimeter and peninsula arrangements can be accommodated and gangways adjusted to meet safe requirements for individual situations (see **32.25**). While a wide variety of layouts can be provided, the rigid services distribution in the laboratory is completely standardised, ie both supply services and drainage are provided on a regular grid. Loose services units associated with furniture can be based on **32.12** or **32.13**.

9 VERTICAL DIMENSIONS AND SPACE FOR SERVICES

Vertical dimensions

The clear height in bench scale laboratories should not be less than 2·7 m. If a height approaching 4·8 m is required for tall equipment or rigs, this may be increased as indicated in **32.44** to allow the possibility of introducing a mezzanine floor for work which does not require the double height, thereby achieving a more effective use of volume.

Horizontal space for services

Variations are shown in **32.45**, **32.46** and **32.47**. The space shown in **32.47** is expensive to provide and can be justified only if intensive use of the laboratory is predicted 24 hours a day, everyday; and services require constant adaptation and maintenance. Suspended ceilings are not always essential but may be necessary for cleanliness, acoustic or aesthetic reasons

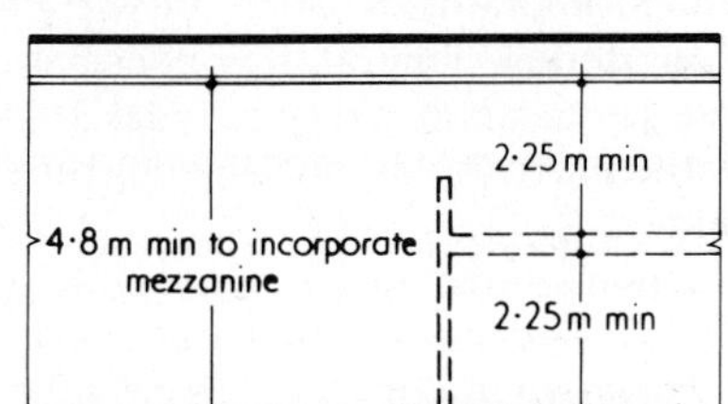

32.44 *Double height laboratory incorporating a temporary mezzanine floor*

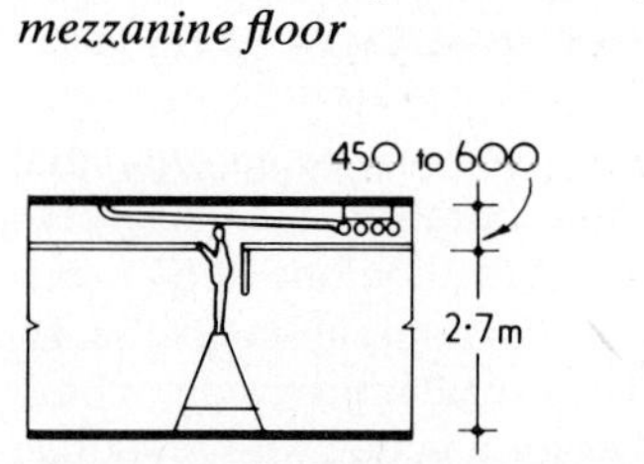

32.45 *Space for pipework, electrics and drainage*

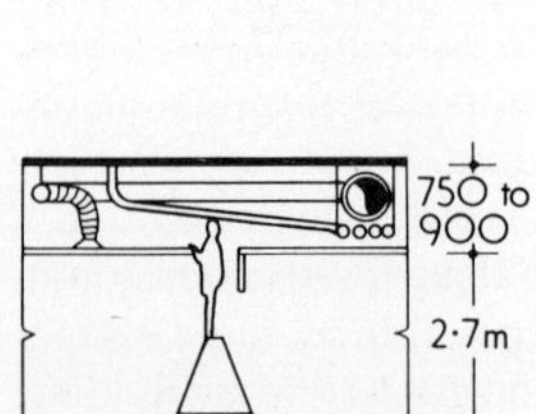

32.46 *Space for pipework, electrics drainage and ventilation ductwork*

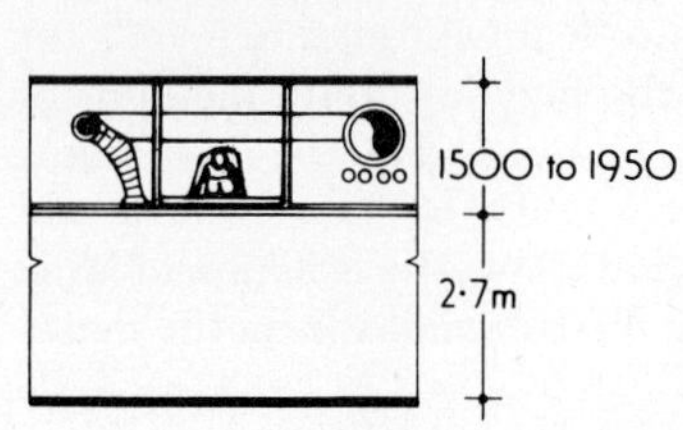

32.47 *All services, plus space for a crawlway or walkway*

(or to support services); 50 to 75 mm should be allowed for their construction.
Suspended ceilings which allow direct access to services above should not form the fire protection for the structural floor above—the access requirement will compromise the fire resistance function.

Vertical space for services
If vertical sub-main ducting is used, horizontal space requirements will be minimal. Minimum size to accommodate fume ducting plus pipe-work would be 600 × 1200 mm, **32.38**.

10 APPENDIX

10.01 Legislation
Most relevant legislation has now been brought under the umbrella of the Health and Safety at Work Etc Act 1974. Other relevant statutes are listed below.
Offices, Shops and Railway Premises Act 1963 This Act will almost always apply because most laboratories require more than twenty-one hours of secretarial work a week to be done.
Control of Office and Industrial Development Act 1965 This will affect siting of laboratory buildings with office spaces in excess of 3000 ft^2.
Factories Act 1961 This Act may affect laboratory buildings or spaces. Its relevance should be checked when workshops for repair or maintenance are provided, especially as regards means of escape in case of fire (sections 40 to 45).
Petroleum (Consolidation) Act 1928 May apply wherever petroleum spirit in excess of three gallons, or compressed gases, calcium carbide and calcium disulphide are stored.
Radioactive Substances Act 1960 An explanatory memorandum is published to explain the stringent controls that surround the use and disposal of radioactive materials. The DoE should be consulted about current legislation when new projects are undertaken. See basic safety for radiation practice.
Dangerous Drugs Acts 1965, 1967, Pharmacy and Poisons Act 1933, Therapeutic Substances Act 1956 These Acts all make provisions to ensure security in storage of drugs.
Clean Air Act 1956, Public Health Acts 1925, 1936 and 1937, Public Health (London) Act 1936 These Acts may affect disposal of laboratory waste products either by incineration or as effluents.
Useful information related to statutory requirements:
DES Building Bulletin No 7 Fire and the Design of Schools (1975). Although this deals with school requirements, the general principles have wider application.
DES Safety Series No 2 Safety in Science Laboratories, London, HMSO, 1976.

10.02 The layout of ancillary laboratory spaces
Spaces containing special equipment and/or routine processes necessary to support general laboratory activities are too numerous in type to be dealt with here. The following references provide detailed information.
Note that many of the design solutions for fitting out spaces illustrated in the information listed above are of the 'built-in' type. 'Building-in' should be avoided wherever possible and preference given to loose adaptable design approach, eg loose laboratory furniture and services fittings are more readily adjustable by users when new equipment has to be accommodated; coldrooms and temperature controlled rooms can be purchased as prefabricated movable enclosures as an alternative to building them in.
a W. Schramm *Chemistry and Biology Laboratories: design, construction, equipment*, Oxford, Pergamon Press, 1965. Gives layout information on a wide range of special laboratory equipment.
b AJ Information Sheets:
1546 (December 13, 1967) Fume cupboards—design and layout
1547 (December 13, 1967) Chromatography rooms
1548 (December 13, 1967) Controlled temperature rooms
1549 (December 13, 1967) Electron microscope rooms
1550 (November 29, 1967) Glass washing facilities
1551 (November 29, 1967) Solvent stores
1552 (November 29, 1967) Balance rooms
1597 (May 29, 1968) Animal house space—general design
1598 (May 29, 1968) Animal house space—rabbits and small rodents
1599 (May 29, 1968) Animal house space—dogs
1600 (June 12, 1968) Animal house space—cats and primates
1601 (June 12, 1968) Animal house space—specific pathogen free and gnotobiotic units.
c K. Everett and D. A. Hughes *Guide to Laboratory Design*, London, Butterworths, 1975. Contains recommendations on the layout and specification of laboratories for radioactive and biohazard work.

10.03 Sources

ABS *Information manual*, Universities of Indiana and California, EFL and US Department of Health, Education and Welfare, 1971
AJ Information sheet 1312 (No 2), *The Architects' Journal*, 3.2.65
Architecture Research Unit *An extension of the Department of Zoology laboratories*, University of Edinburgh, 1968
BS 3202, Laboratory furniture and fittings, London, BSI, 1959
DES Building Bulletin No 45—JDPCLASP *System building for higher education*, London, HMSO, 1970
DES *Guidelines on environmental design in educational buildings*, DES, 1974
Electrics 72/73 *Educational and scientific laboratories*, Electricity Council, 1973
Laboratories Investigation Unit paper no 3 *Growth and change in laboratory activity*, LIU, 1971
Laboratories Investigation Unit paper 6 *Adaptable furniture and services for education and science*, LIU, 1972
Laboratories Investigation Unit paper 9, *The Charles Darwin Building, Bristol Polytechnic*, London, HMSO, 1977
Nuffield Foundation *The design of research laboratories*, London, Oxford University Press, 1961
University Grants Committee *Notes on procedure*, London, HMSO, 1971

Acknowledgement
Particular thanks are due to Frank Drake, my colleague in the LIU, for contributions to the section on school laboratory layouts, and other helpful comments.

33 Museums and art galleries

Gemma Hunter

Robin Wade Design Associates, specialist museum designers

CI/SfB 75
CI/SfB (1976 revised) 75

Contents

1 INTRODUCTION

It is almost impossible to generalise or to particularise about museums. Almost the only thing they have in common is that all the objects on display were never designed to be there! They can range in size from large general collections like the British Museum to a small local authority museum with a few hundred square metres, an entrance lobby and a postcard rack. They can contain artefacts, or natural objects such as minerals, fossils and stuffed animals; specialise in steam engines, blast furnaces, coins and fabrics.

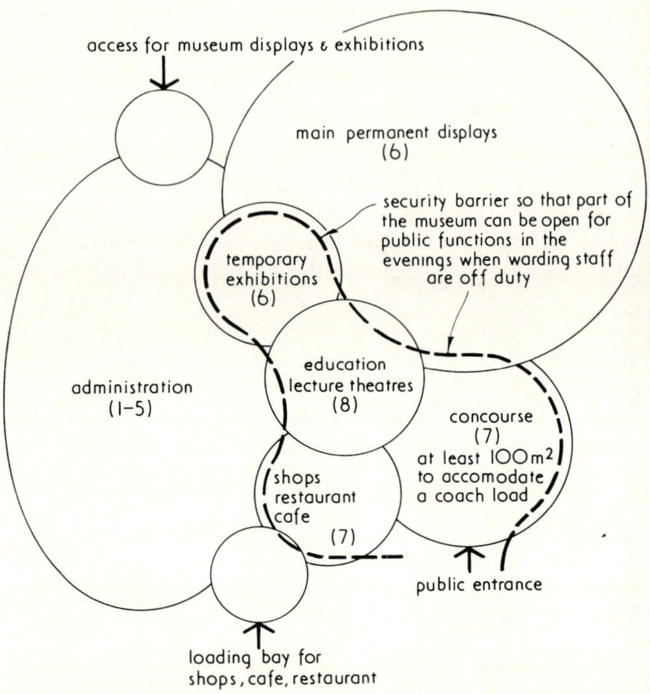

33.2 *Relationships between the spaces in a museum (numbers refer to functions in* **33.1***)*

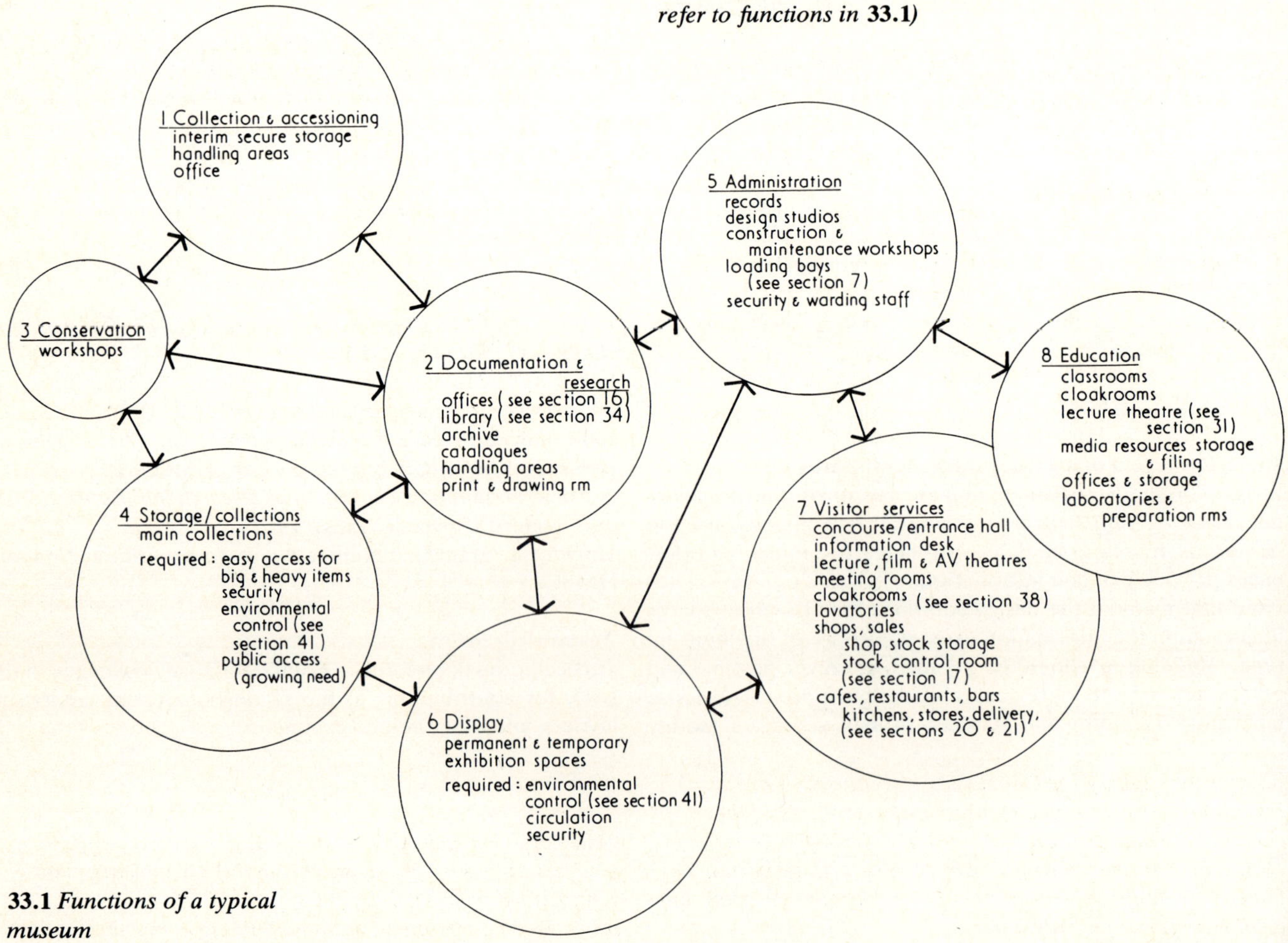

33.1 *Functions of a typical museum*

2 GENERAL PLANNING

When looking for common factors, they emerge more as functions and their spatial relationships rather than prescriptions for standardised dimensions. These functions (not the spaces) are shown in **33.1**. Many parts, such as shops, workshops, lecture rooms, etc are similar to spaces in other building types, and reference should be made to the relevant sections of this handbook.

Not all museums will encompass all the functions shown in **33.1**. Sensible relationships between the corresponding spaces are shown in **33.2**. The Fox Talbot Museum, **33.3**, is an example of a small museum that works well. The Art Gallery at Christ Church, Oxford, **33.4** exemplifies a satisfactory small art gallery.

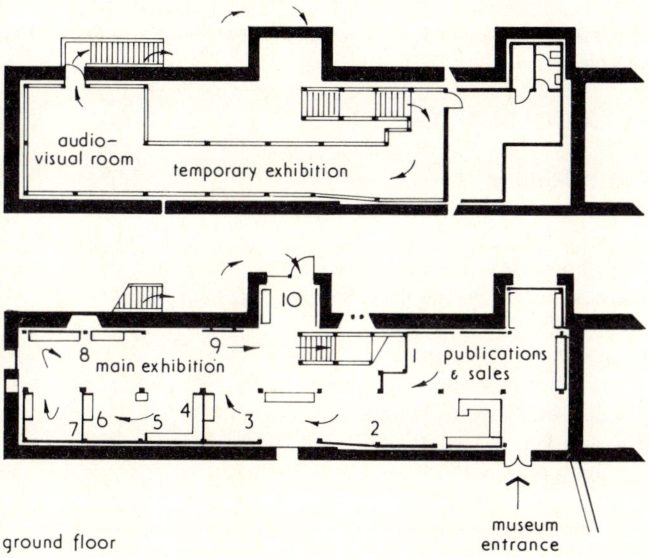

33.3 *The Fox Talbot Museum, Lacock Abbey, Wiltshire. A museum to commemorate the invention of photography, fitted into a 17th century barn.*

Exhibition sequence:
1 introduction
2 the village, the Abbey and the family
3 Fox Talbot—the man
4 camera obscura
5 the invention
6 other inventors of photography
7 the pencil of nature and the reading establishment
8 Fox Talbot's other scientific works
9 further developments
10 browsing area—scrap books

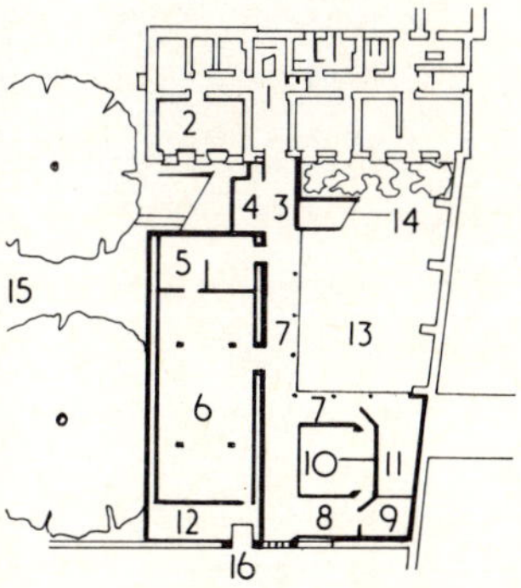

1 original building
2 picture storage
3 entrance
4 custodian
5 gallery 1—Italian primitives
6 large gallery
7 cloister
8 smaller pictures
9 modern exhibits
10 print room
11 study and picture store
12 air-conditioning plant room
13 garden
14 grass ramp up
15 Dean's garden
16 exit for tall pictures

33.4 *Picture gallery for Christ Church, Oxford. Architects: Powell and Moya. A small gallery in the grounds of the college. The paintings are mostly Italian from the 14th to the 17th centuries, the drawings from 15th to 17th centuries. The building is designed with storage for the whole collection, but exhibition space for only a part.*

3 LAYOUT OF DISPLAY AREAS

3.01 Random layout

Display is an important function of a museum, and can be designed in a variety of ways. In a traditional museum, display is often in a series of galleries of rooms, arranged in a random sequence, **33.5**. The objects and pictures may have been acquired in a piecemeal way and accommodated wherever space was available. In particular, large exhibits had to go

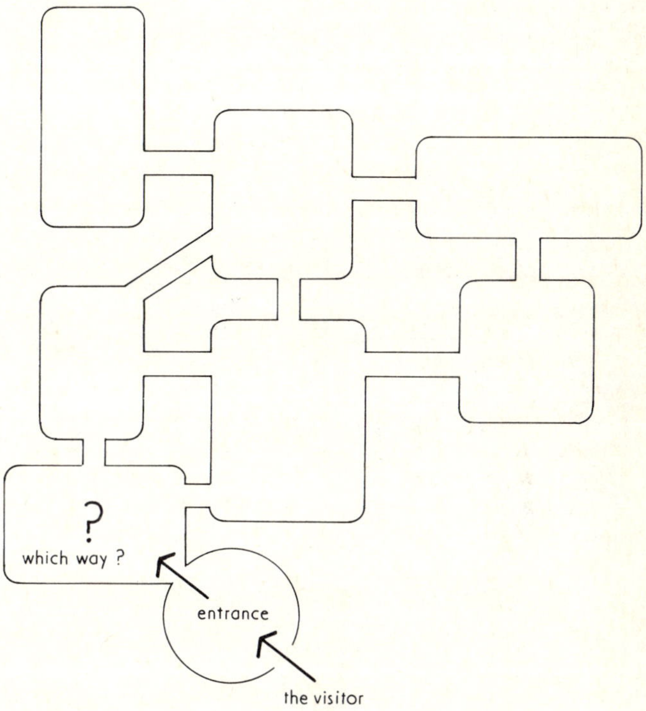

33.5 *Typical large traditional museum with random arrangement. There are many of this sort, catering for the casual browser or for people who already know the museum. The visitor has problems of orientation and circulation which must be solved by a comprehensive system of signposting*

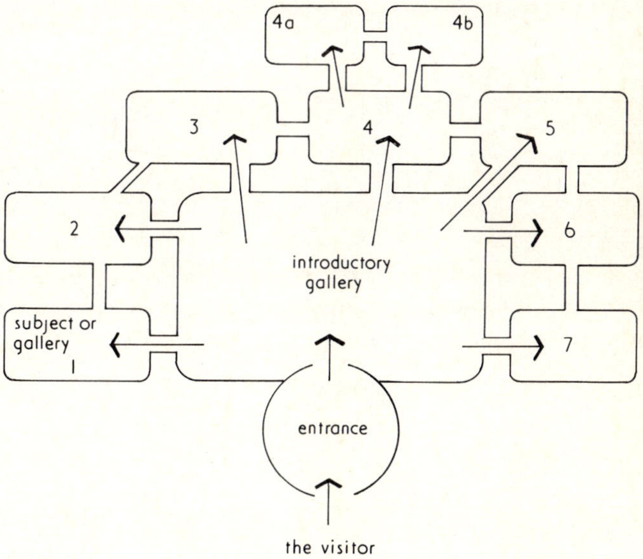

33.6 *Large museum with an introductory gallery, different sections and galleries leading off. In this arrangement the visitor is led from the introductory to galleries which give a more detailed treatment, thus maintaining theme links*

where they would fit. The display ends up showing what the museum possesses, rather than what is suitable or needed. In addition, the visitor can become confused on entering if there is a plethora of signposts and notices which have been put up at different times. A unified system of guidance should be considered at the planning stage.

3.02 Layout with introductory gallery
An alternative arrangement is to lead the visitor from the entrance to an introductory gallery, **33.6**, which may contain a display summarising the themes and subjects of the museum. If his eye is taken by one particular subject, he is immediately directed to a side gallery which treats this subject in greater depth. The same principle can be extended further in large museums.

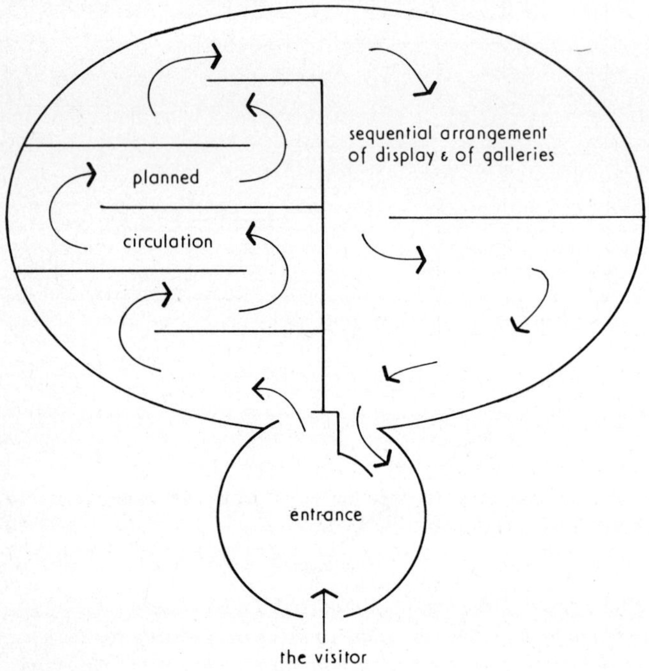

33.7 *Sequential circulation*

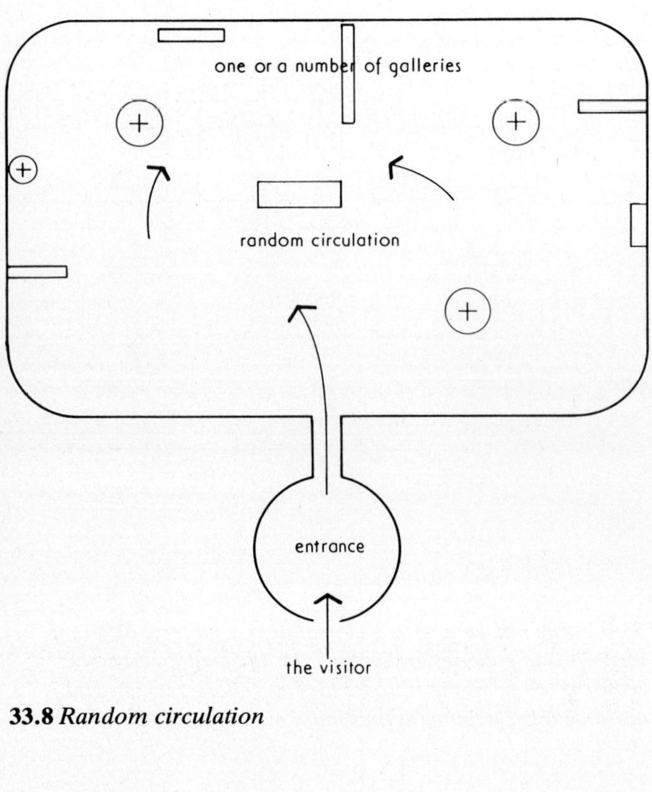

33.8 *Random circulation*

3.03 Specific accommodation
Where a new building is being constructed to house an existing collection, the galleries can be designed around the exhibits. For example, if ships' masts are to be accommodated, the necessary clear height is provided.

3.04 Arrangement within galleries
Of the two main circulation patterns, sequential circulation, **33.7**, is appropriate for a chronological arrangement or where there is a storyline that needs to be followed from start to finish. The other, random, pattern, **33.8**, is suitable for subjects where browsing at random is called for.

3.05 Disabled access
Where the display route involves changes of level by steps, or other difficulties for wheelchair users, lifts or ramps should be provided as in other public buildings. Provided these are adequately signposted, they need not exactly parallel the ambulant route. In general, one wheelchair access to each level is sufficient.

4 DISPLAY DETAILS

4.01 Exhibition
In museums, objects can be displayed

- in showcases
- freestanding on the floor, on plinths or supports
- on walls
- on panels.

In art galleries, pictures are displayed on walls and panels, but can also be shown on easels.

The two major aspects of display are the ease of viewing, and the lighting.

4.02 Vision and viewing
The normal limit of vision without moving the head is a cone of 40°, **33.9**. A picture, therefore, can only be comfortably viewed as a whole from a distance of about double the

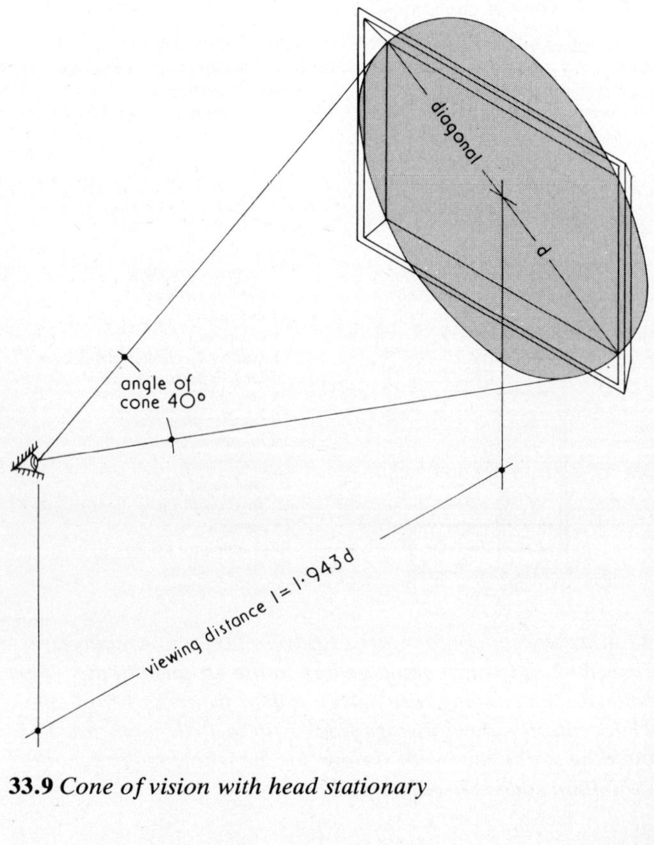

33.9 *Cone of vision with head stationary*

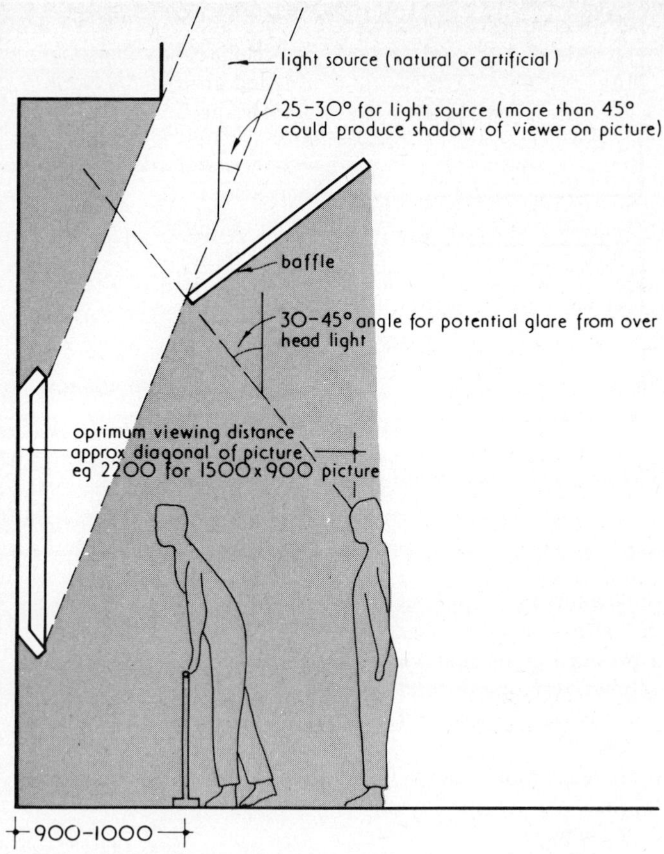

33.10 *Factors for satisfactory viewing, distance and lighting. With suitable design of top light, baffle may not be needed*

diagonal. It is generally accepted, though, that a distance equal to the diagonal will enable the viewer to appreciate the details of the picture, but he will need to move his head to compass it all, **33.10**.

4.03 Lighting
Lighting for pictures should not come from an angle less than 45°, **33.10**, but the source should be screened against glare. Normal windows, **33.11**, tend to leave adjoining walls, and any displays on them, in shadow. Where top light is not used, side lights, **33.12**, can be substituted.

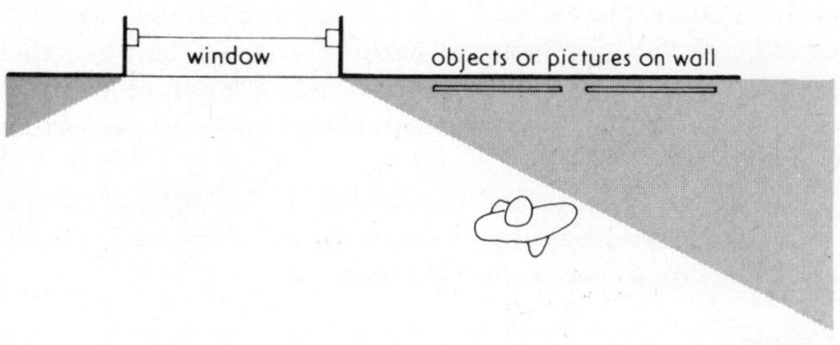

33.11 *Wall in shadow adjacent to normal window*

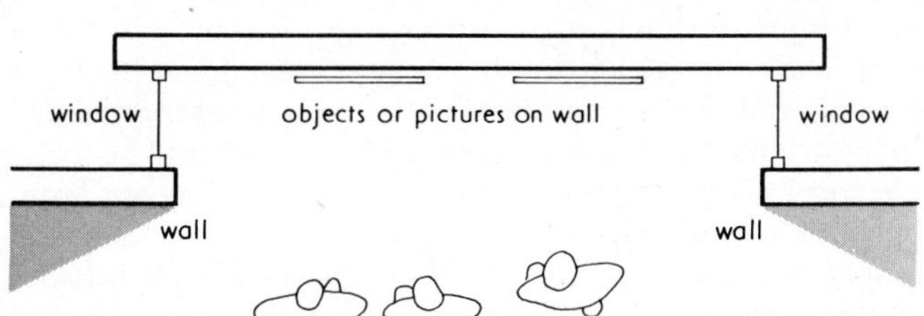

33.12 *Side lighting as an alternative to top light*

4.04 Showcases
Showcases are normally viewed from close up, **33.13**. Here the lighting is usually artificial, but should be screened from direct view. The source should be isolated from the exhibit so that maintenance can be carried out without breaching security. The exhibits should also be protected against the heat of the lighting, and from the danger of damage during maintenance.

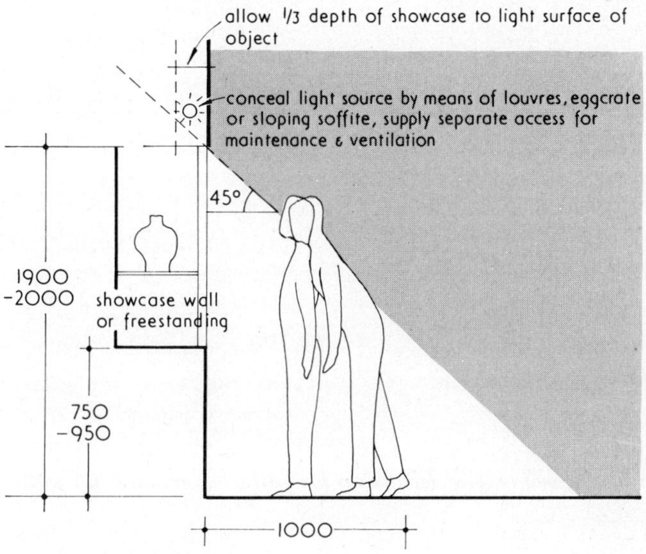

33.13 *Viewing and lighting a showcase*

4.05 Labels and descriptive panels
Text and captions should be of a type size relative to the distance from the viewer, **33.14**.

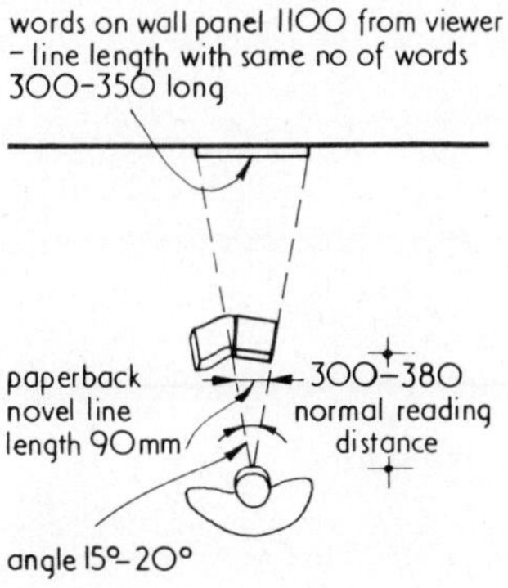

33.14 *Method of sizing type for descriptive material, labels, etc*

4.06 Circulation
Adequate space should be provided for people to view the exhibits, and also to pass between groups of viewers, **33.15** and **33.16**. If objects are placed too near corners, **33.17**, congestion will tend to occur. Where there is a designed sequence, there may be queueing at peak periods for the more popular items, and space must be allowed for this. Star exhibits should have extra viewing space, and should not be placed too near to one another.

5 SECURITY

5.01 Methods
Museums and art galleries contain objects of value. Some have priceless articles, and must maintain the highest level of security. Even those with relatively mundane contents, however,

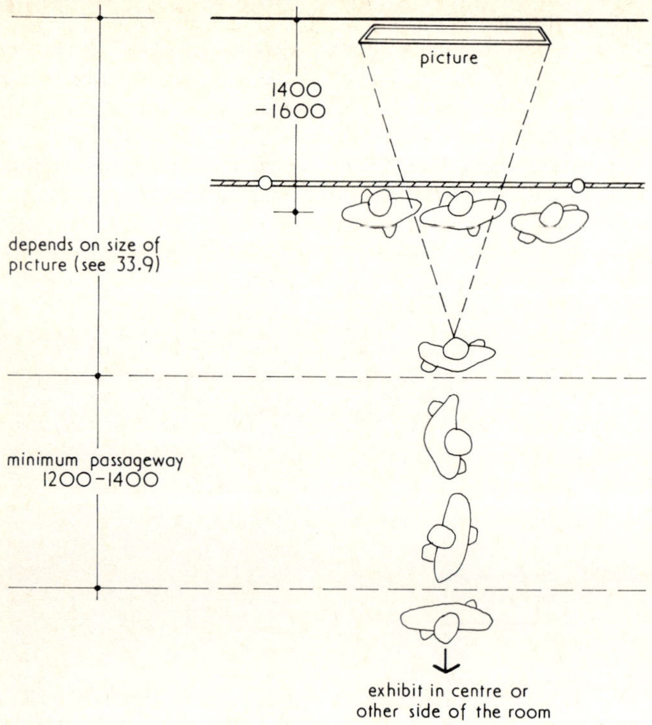

33.15 *Viewing and circulation for objects or pictures on walls*

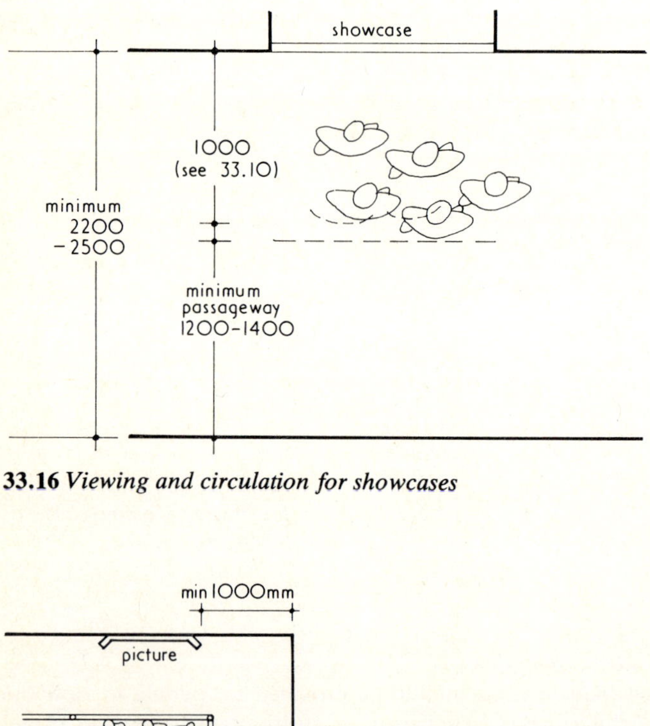

33.16 *Viewing and circulation for showcases*

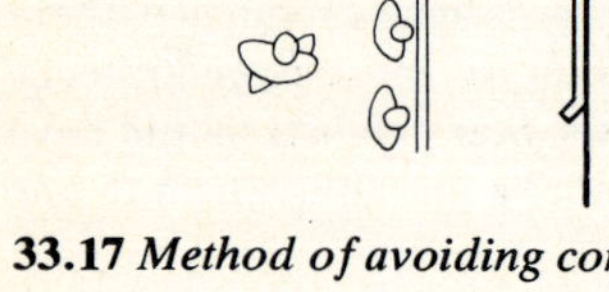

33.17 *Method of avoiding congestion in corners*

should be provided with good security systems. Traditionally, reliance has been mainly on attendants or wardens; in these cases it may be important to design each gallery for maximum visual coverage from the warding position, **33.18**.

Modern practice still considers the human element to be

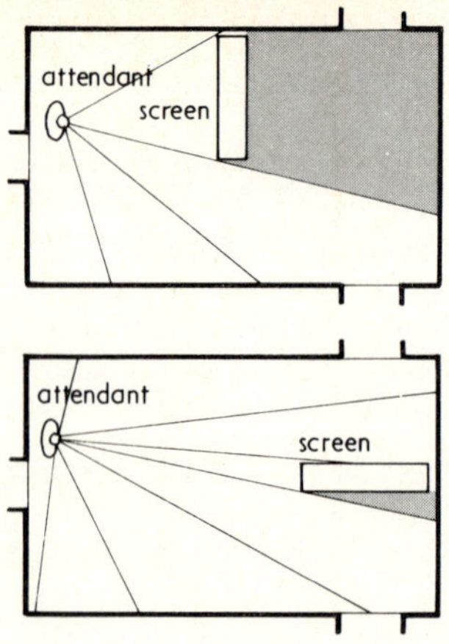

a *Poor arrangement with large area not under surveillance*

b *Good arrangement, most of area visible*

33.18 *Ensuring maximum vision for attendants*

important, but supplements it by mechanical and electronic measures. These are usually multi-level:

1 prevention of removal of object by placing in a secure cabinet, or fixing to solid structure, then

2 detection of successful removal by alarm, visual or audial, followed by

3 prevention of removal from building.

5.02 Damage

Occasionally the aim of the criminal is not theft but iconoclasm. The Portland Vase and Rembrandt's *The Night Watch* are two examples of such vandalism. Only protecting the object behind glass, perspex or polycarbonate is fully effective, and this detracts from the enjoyment of the innocent majority. It is normal to prohibit the carrying of obviously harmful objects such as umbrellas into museums and art galleries, so cloakrooms need to be provided when planning these buildings. In the display areas there should be no ledges, nooks or crannies where parcel bombs could be hidden.

5.03 Entry and exit

The checking of entry into a museum or art gallery by turnstiles or electronic detection helps to maintain security even if no admission charge is made. If similar arrangements are made at the exit, the clearing of the building at closing time can be assured. This should also be facilitated by ensuring that there are no hiding places, such as cleaners' cupboards, opening directly off the public areas, and the accesses between the public areas and the administrative section are fully secure. All external doors and windows should of course be protected from illegal entry.

It is particularly important to ensure control of egress during a fire alarm, as a false alarm, or even a small real fire, can be used as a diversionary tactic during a theft.

6 FIRE

Damage to objects in museums and art galleries is caused not only by the fire itself but even more by the use of water to fight fire. Emphasis should therefore be on prevention rather than cure. The structure and finishes of new buildings should be as incombustible as practicable. Since smoking is never permitted in these buildings, the chief cause of fire starting will be faulty electrical wiring or accessories. Leakage detectors are available that will almost completely eliminate this possibility.

Smoke and heat detectors should be installed, but sprinkler systems only rarely (such as for some industrial archaeological museums). Fire extinguishers should be of types that minimise damage.

7 ENVIRONMENTAL CONTROLS AND CONSERVATION

7.01 Decay
Everything in the world tends to decay. Museums try to exclude the agencies of decay, and to slow up the inherent processes. The most vulnerable categories of exhibits are fabrics, metal, wood, water colour paintings and photographs. The Museums Association publishes useful information sheets on aspects of conservation, including fuller details of the following.

7.02 Temperature and humidity
Too damp or too dry an atmosphere can be very harmful. Humidity is closely linked to temperature (see section 41, Thermal comfort, para **5**). People working in or visiting the museum also need comfortable conditions. Comfort zones for these, and safety zones for different uses are plotted on the psychrometric chart, **33.19**; and these must be used to control conditions not only throughout the display areas but also in all areas used for storage and conservation work. Some exhibits may also need protection from the heat produced by the lighting.

7.03 Deleterious elements in the air
There are various harmful chemicals that can be present in the atmosphere (see section 41, para **3.01**). Crop spraying, for example, does more damage than the carbon monoxide from cars. There are also grit and dust. When eliminating these, electrostatic filters should not be used. Malfunctioning of this type of equipment can generate ozone, which is very harmful.

7.04 Light
Both natural and artificial light fades fabrics, and deteriorates water colour paintings and photographs. It also can have a harmful effect on natural history specimens such as stuffed birds and animals, which are a whole field of conservation on their own! While ultra-violet filters can be beneficial, expert advice should be sought for most objects of value.

7.05 Insects etc
Woodworm have more often than not invaded old agricultural implements, furniture, etc, all of which should be treated. Obviously, fabrics should be protected from the ravages of moths. Natural history specimens and their infestation are another particular problem.

7.06 Material used in display
Care must always be taken when selecting the modern materials used in display. Certain felts that could be used for lining showcases contain acid. Some plastics, such as thermoplastic tiles and their adhesive, can affect photographic material. This should always be mounted on an acid-free base, such a rag board, not a mechanical wood-pulp board; and the special adhesive used.

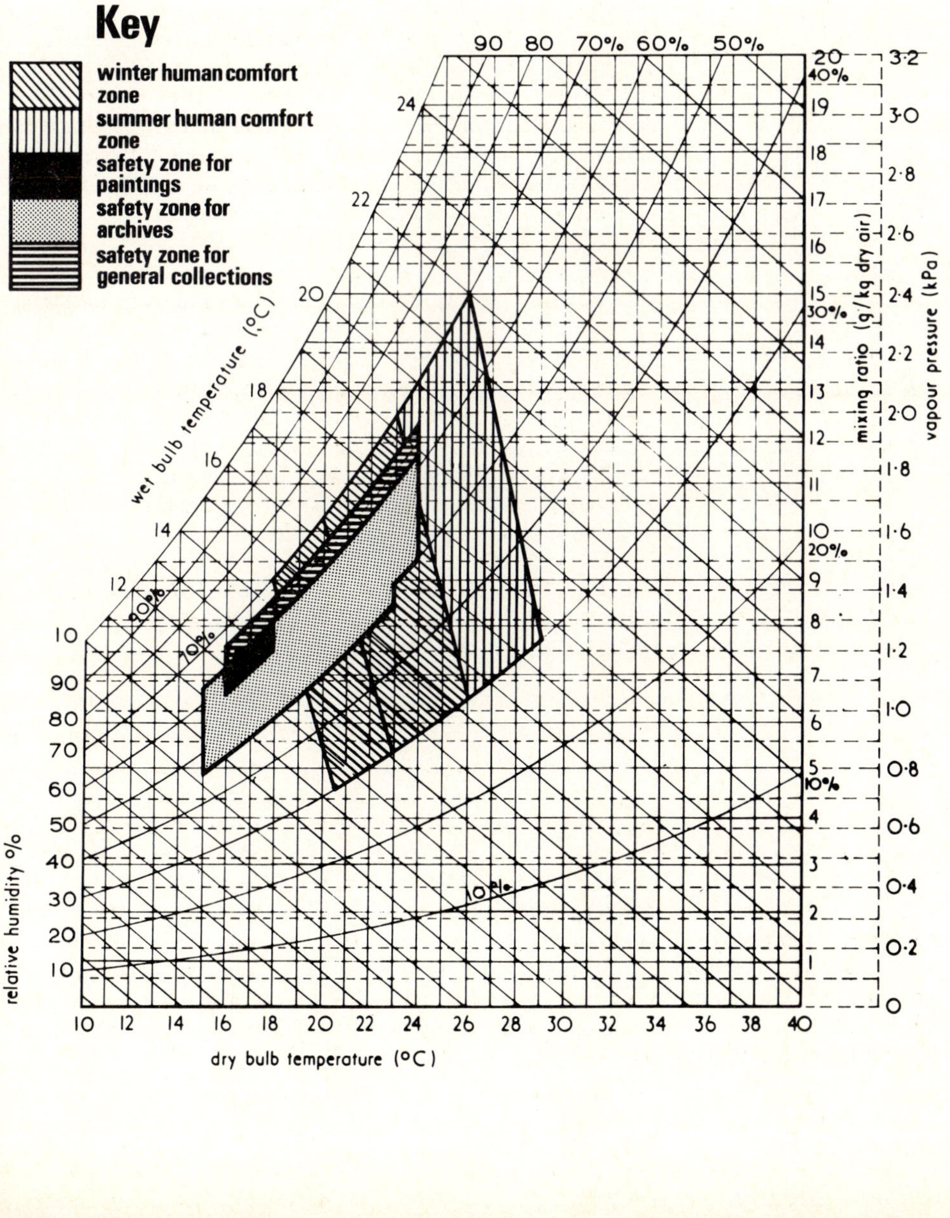

33.19 *Psychrometric chart (see section 41, Thermal comfort) showing safety and comfort zones for museums and art galleries*

8 BIBLIOGRAPHY

Michael Brawne, *The new museum*, London, Architectural Press, 1965

Michael Brawne, The picture wall, *Architectural Review*, May 1959

Michael Brawne, Object on view, *Architectural Review*, November 1959

A. F. Clapp *Curatorial care of works of art of paper*, Intermuseum Laboratory, Ohio

L. V. Coleman *The museum in America*, Museum Publications, Washington, reprinted 1970

Countryside Commission *Countrylife museums*, Countryside Commission, Scotland

M. Crook *British Museum*, Harmondsworth, Pelican

DHSS *Museums in education: education survey No 12,* 21 September 1971

F. K. Fall *Art objects—their care and preservation*, L. McGulvey

P. E. Guildbeck *Care of historical collections*, American Association for State and Local History

R. O. Harrison *The technical requirements of small museums*, Canadian Museums Association

HMSO Command 4676: Future policies for museums and galleries, London, HMSO, 19 May 1971

H. Hudson and A. Nicholls *Directory of museums,* London, Macmillan

Illuminating Engineering Society *Lighting of art galleries and museums: technical report of the Engineering Society No 14* Illuminating Engineering Society, 1970

International Institute for Conservation *Conservation in museums and galleries*

C. K. Keck *Handbook of the care of paintings*, New York, Watson Guptill TABS

Leicester Museum *Museums and the Handicapped*, Leicester Museums and Art Galleries Records Service

34 Libraries

Godfrey Thompson

The author is Guildhall Librarian and Director of Art Gallery, City of London

CI/SfB 76
CI/SfB (1976 revised) 76

Contents

1 INTRODUCTION

In 1955, 20 000 books were published in Great Britain. In 1970 there were 34 000. Libraries have to absorb this growth. Academic and research libraries grow by about 5 per cent per year. Public libraries, because they weed out as well as replenish and extend stocks, do not grow so quickly. Much depends on the size and style of readership in the catchment area. Libraries deal with films, gramophone records, cassettes, tape recordings and microfilms of information as well as books, but, until the late 1980s at least, books will continue to form the majority of the library stock.

2 DESIGN CHECKLIST

The following checklist covers the kinds of information that may be needed to draw up a design brief. It is relevant mainly, but not only, to public libraries.

User services
Opening hours
Peak use times
hours
days of the week
times of the year
(particularly for educational libraries)

Numbers of readers
(preferably separate figures for each part of the library)
bibliographical
general reference
adult lending
children's lending
periodicals, newspapers
music
special reference (eg commercial, technical)
local history
arts
other departments
typing room
exhibition area

Associated activities
meeting rooms
lecture rooms

Reader facilities
document copying
microform viewing
video cassette viewing
audio reproduction
poster display

Refreshments
Storage of readers' belongings
coats
bags
Lavatories
Telephones
Book shop
Vending machines

Staff services to users
Number of staff on duty at the following points
security points
book issue and return
reader enquiries

External activity for which the library is the headquarters
branch library supply and services
school library supply and services
welfare libraries (handicapped readers, prisons etc)
privileged readers (school teachers to select books from a display range)
mobile libraries (garaging, servicing)

Technical services
Number and types of staff

Offices
administrative
executive

Areas for
accessioning
cataloguing
processing
receipt and despatch
post and packing
printing
photography
binding
poster drawing

Staff rooms
lounge
tea room and kitchen
lavatories

Storage
strong room
stationery
furniture
Cleaning materials
Car parking

3 AREA ALLOWANCES

3.01 Public libraries

Space requirements vary considerably from job to job, but the International Federation of Library Associations has worked out averages based on actual libraries. These are given in table I.

Table I Public library serving population of 100 000

Function*	Floor area m^2	Comment
Adult lending	750	Plus 10% if exhibition area needed
Adult reference		
Book stock	200	Plus 20% overall for staff workrooms and offices
Seating	375	
Periodicals	100	
Children's library	350	Plus 20% overall for circulation area
Stack	100	
Staff rooms	180	
Total	2945	

*These areas exclude music library, audio-visual material and administrative offices if the library is headquarters for branches of mobile libraries, etc.

3.02 University and college libraries

Published space recommendations vary for universities but an approximate guide would be:

- one seat for each of 30 per cent of students
- between 2·3 m and 3·5 m floor area per seat—overall area
- between 50 and 65 volumes per m^2 of overall floor area.

For libraries in educational establishments under the control of the Department of Education and Science allowances should be:

colleges with at least 30 per cent of advanced work
390 m^2 for the first 500 full-time students and then
0·44 m^2 for each additional student.

colleges with less than 30 per cent advanced work
300 m^2 for the first 500 students and then
0·38 m^2 for each additional student.

Spaces taken by library readers are given in table II.

Table II Reader space requirements

User	Floor area m^2
Student or general reader	2·3
Research worker	3·25
Carrel user	3·70
Actual floor area occupied by reader at table	0·93 to 1·20

4 BOOKSHELF CAPACITY

The capacity of standard 900 mm bookshelves to hold books, periodicals and reports is indicated in **34.1**, **34.2** and **34.3**. These shelves are assumed to be only three-quarters full to allow for expansion and book movement.

The average space requirements of each type of book are given in table III.

Table III Books per 300 mm run of shelf

Type	Number	Recommended shelf depth
Children's books	10–12	200–300
Loan and fiction stocks in public libraries	8	200
Literature and history, politics and economics	7	200
Scientific and technical	6	250
Medical	5	250
Law	4	200

NB Various space-saving devices, most of them proprietary, are available, eg hinged shelf, sliding drawer, parallel rolling and right-angled rolling stacks

Table IV Book capacity at various sizes of structural grid

Grid size m	Spacing of stacks	No of double sided stacks	Books per structural bay
5·6	1·4	3	5012
6·0	1·5	3	5460
6·0	1·2	4	6860
6·5	1·1	4	6160
7·0	1·55	4	8310
7·25	1·45	4	8610
7·2	1·2	5	10276
7·5	1·25	5	10780
7·7	1·1	6	12992
7·8	1·56	4	9380
7·8	1·30	5	11290
8·4	1·68	4	10220
8·4	1·2	6	14364
8·4	1·4	5	12292

Structural grid

Spacing of book stacks, and therefore the capacity of the library will be affected by the structural grid chosen. A layout for stacks within a 6900 mm structural grid is given, **34.4**, and

Table V Shelf depth and spacing

Type of book	% of total	Spacing mm	Depth mm
Popular (light novels)	50	225	230
General	97	280	230
Bound periodicals	–	300	230
Oversize books	3	500	300–400

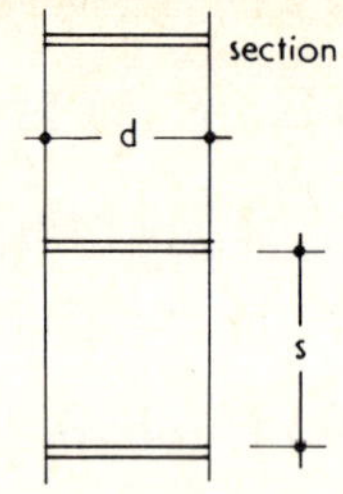

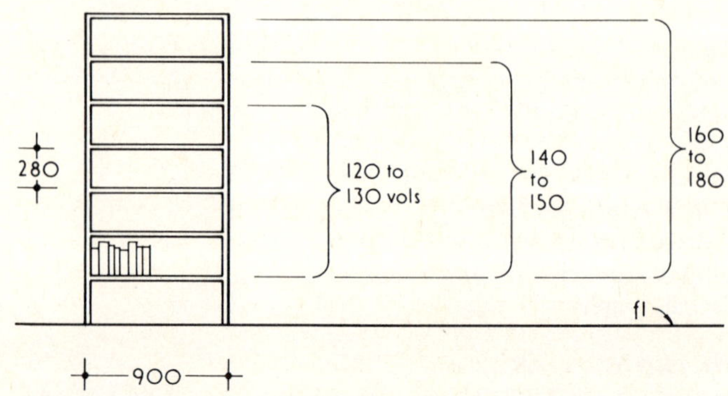

34.1 *Capacity of shelves to hold books, ³/₄ full to allow for expansion and movement*

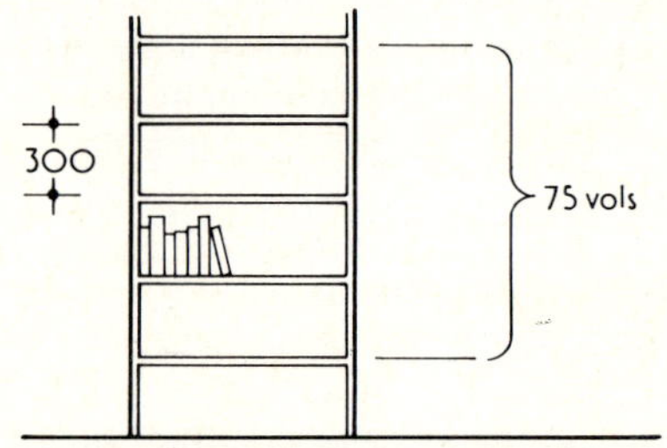

34.2 *Capacity of shelves to hold periodicals in bound volumes*

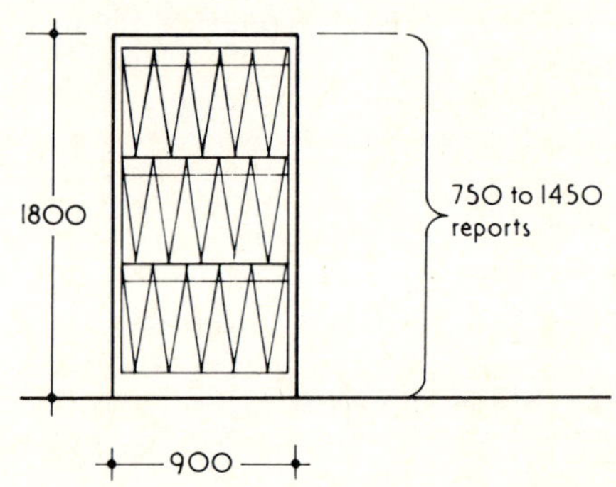

34.3 *Capacity of lateral filing cabinets to hold reports*

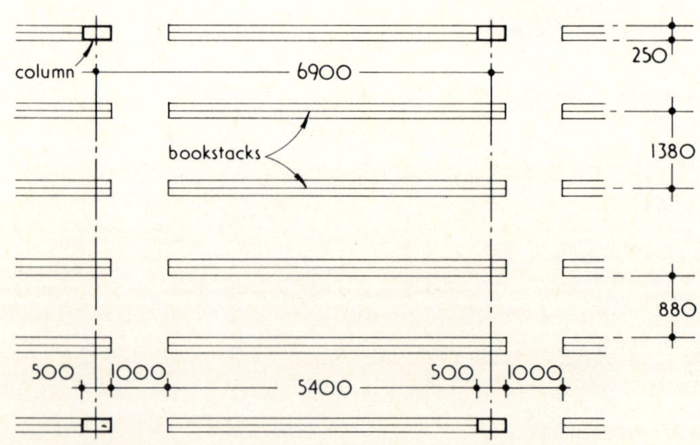

34.4 *Capacity of bookstacks within a 6900 mm structural grid:*

2 stacks on grid line	*= 10800*
8 stacks in bay	*= 47200*
	≃ 58 m run of shelves
58 m of shelves 7 high	*= 406 m run of books*
406 m at 20 books/m	*= 8120 books per bay*
or	*169 books/m² floor area*

table IV lists the book capacity of structural grids ranging from 5·6 to 8·4 m.

Shelf depth and spacing
Table V gives the recommended shelf depth and spacing for the main types of book.

5 BOOK ISSUE COUNTERS

5.01 Shape

- In the smallest library a single control, **34.5** covers both issue and return. A desk may be used, but a counter has the advantage that it is at standing height and can have cupboards fitted. This layout has the disadvantage that the inevitable crossing of traffic routes can be troublesome at peak times.
- The layout in **34.6** offers more security control and fewer traffic problems, but two members of staff are necessary at all times to operate it.
- This shape, **34.7**, with slight variation, is the commonest of all. It shields the staff and allows one person to serve at quiet periods. In larger libraries, a variation can be this shape doubled so that a reserved book and enquiry area can be manned at quiet times with minimal staffing, **34.8**.
- In the largest libraries, particularly in universities, the

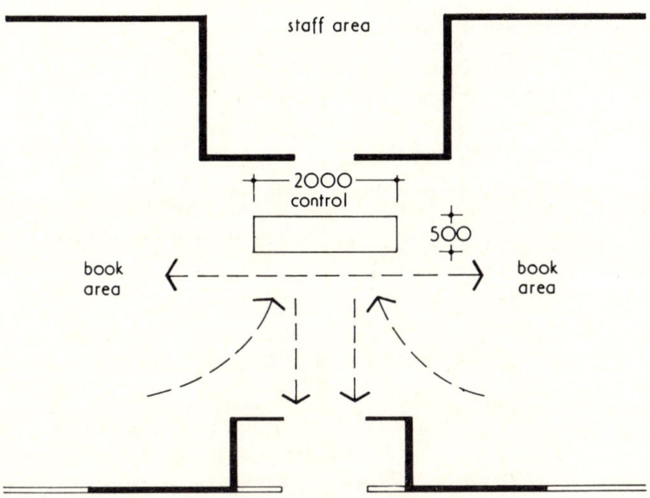

34.5 *Control counter for a small library, suffers from congestion at busy times*

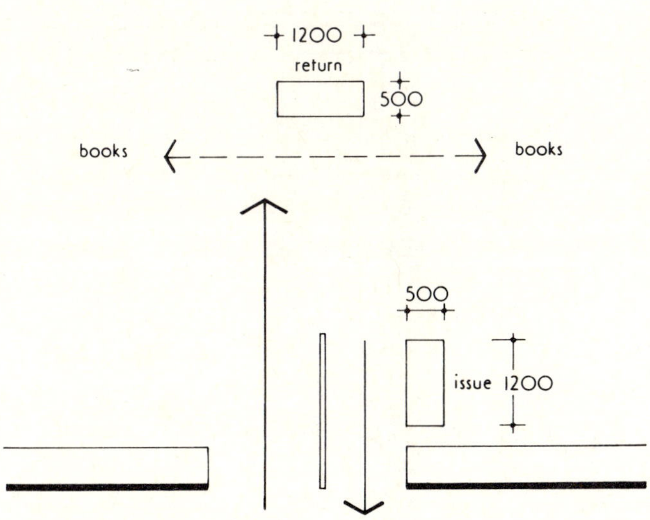

34.6 *Control arrangements for a slightly larger library: more security and less congestion, but requires at least two staff at all times*

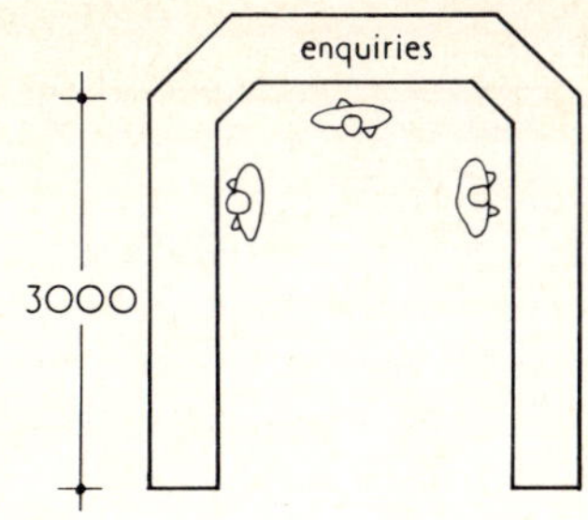

34.7 *Most common type of control counter: can be operated by one during quiet periods*

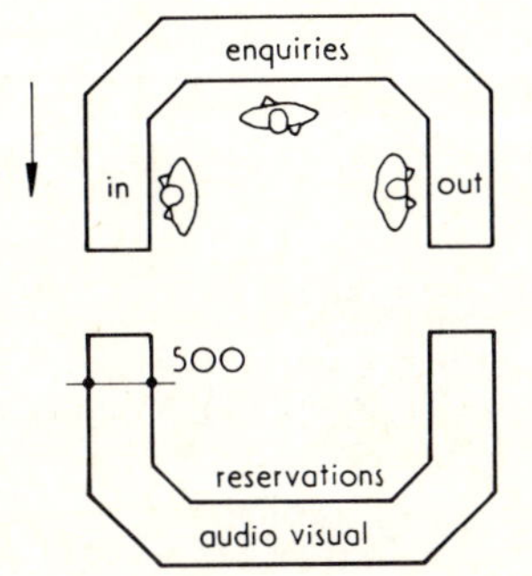

34.8 *Variation of* **34.7** *for larger libraries—can still be operated by one*

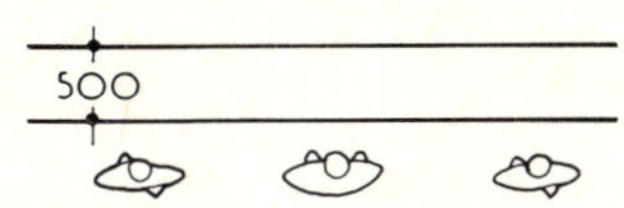

34.9 *Type of control counter preferred in very large libraries, universities, etc*

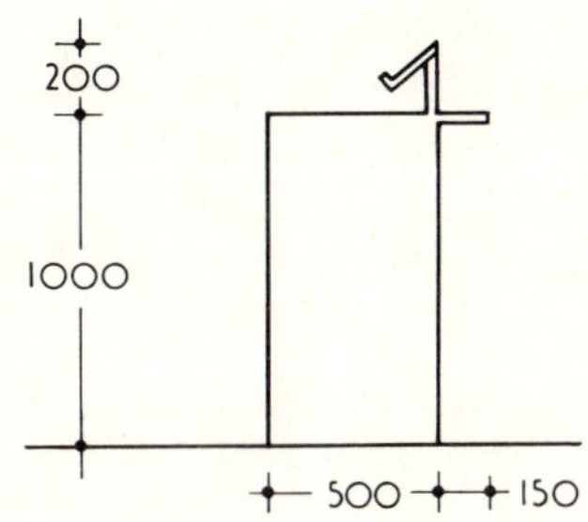

34.10 *Typical section through control counter*

counter acts as a barrier between readers and staff working areas. Here the shape can fit in with that of the building, but it is usually straight, **34.9**. In some university libraries such a counter can be more than 20 m long.

5.02 Width

Counters are usually 500 mm wide and sometimes also have a 150 mm bag rail on the outside of the in counter.

5.03 Height

Both readers and staff will stand, so height should be 1·20 m. A popular variation has a slope on the in counter where books can be placed while being checked in, **34.10**. Alternatively, the same design can be used, but with a counter 750 mm high, with an overall height of 950 mm. Staff can then sit to receive books.

6 MICROFORM STORAGE

Microforms are:

- *microfilm,* stored in cabinets, **34.11**. The cabinets will hold 675 reels of 35 mm film, or 125 reels of 16 mm film.
- *microfiche,* in a number of sizes varying from 75 × 125 to 100 × 150 mm but a new British Standard recommended for international use specifies one size: 105 × 148 mm.
- *micro-opaques,* the usual sizes of which are 125 × 75, 225 × 150 and 215 × 165 mm.

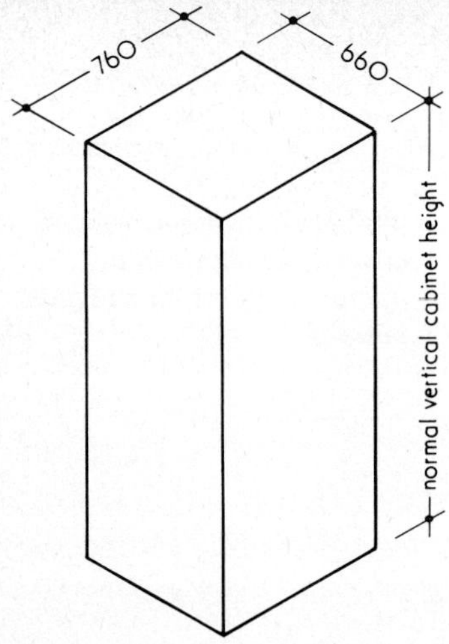

34.11 *Storage cabinet for microfilms*

7 LAYOUT AND CRITICAL DIMENSIONS

Typical layout requirements are shown in **34.12** to **34.29**. Critical reach dimensions are shown in **34.30** to **34.33**; typical furniture in **34.34** to **34.36**.

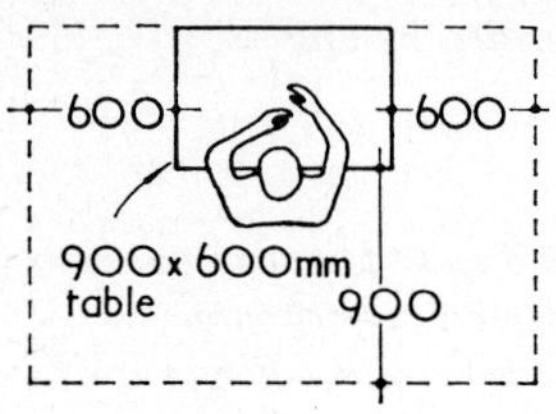

34.12 *Recommended minima for one-person reading tables*

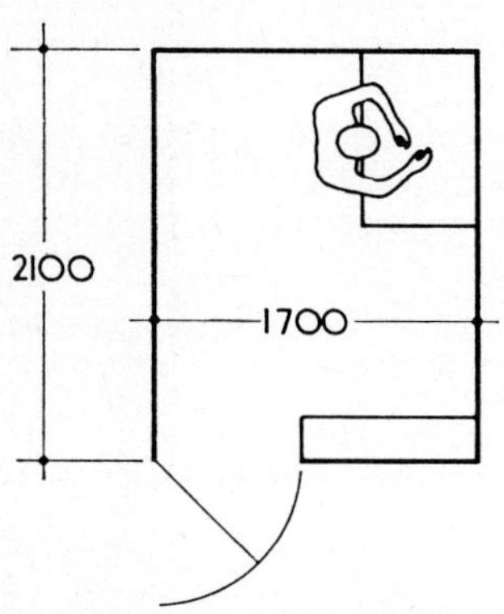

34.13 *Recommended single person enclosed carrel*

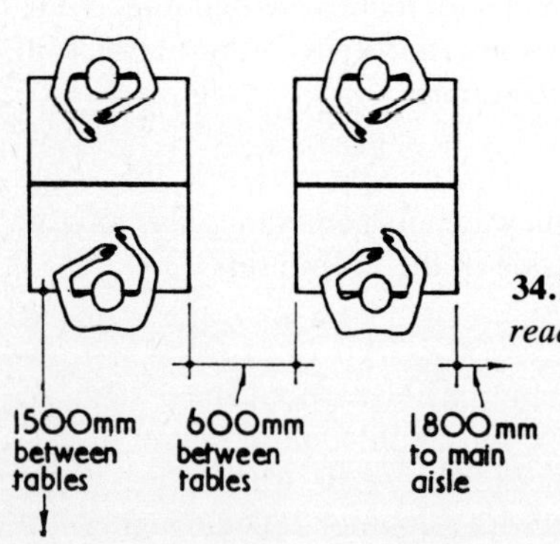

34.14 *Minima for dual reading tables*

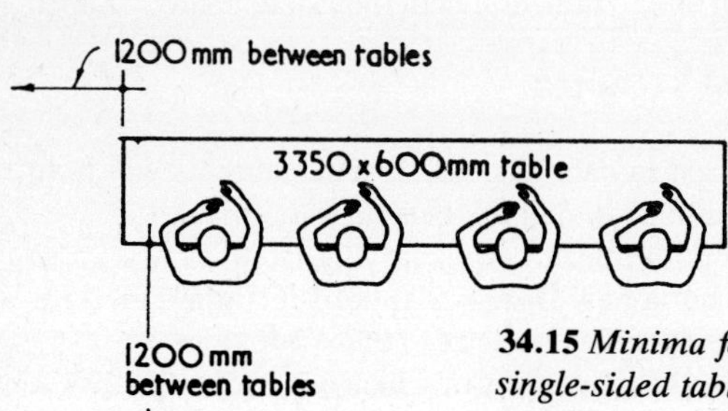

34.15 *Minima for single-sided tables for four people*

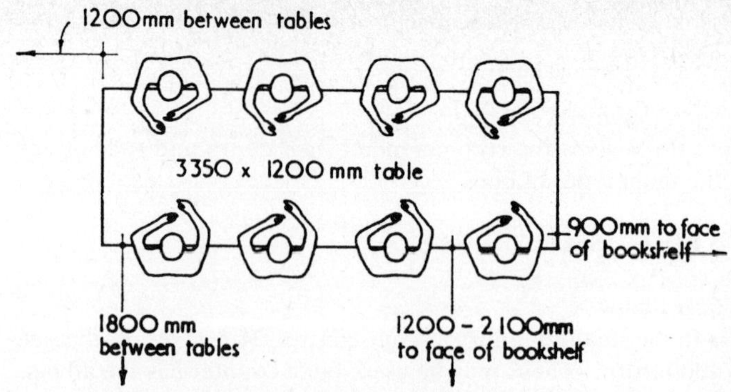

34.16 *Minima for eight-person reading tables*

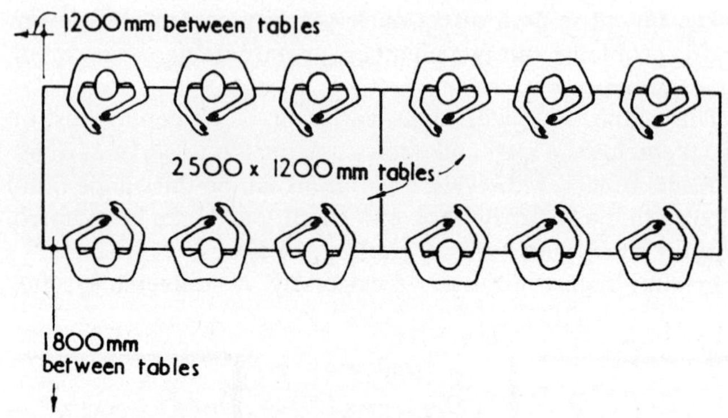

34.17 *Minima for six-person reading tables*

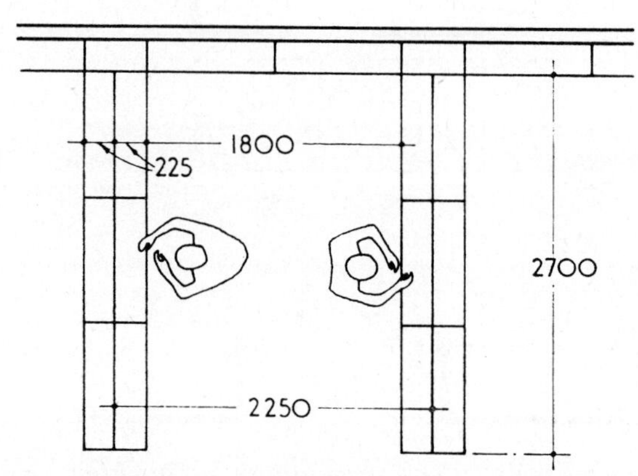

34.18 *Recommended minima for open access bookshelf areas arranged as alcoves*

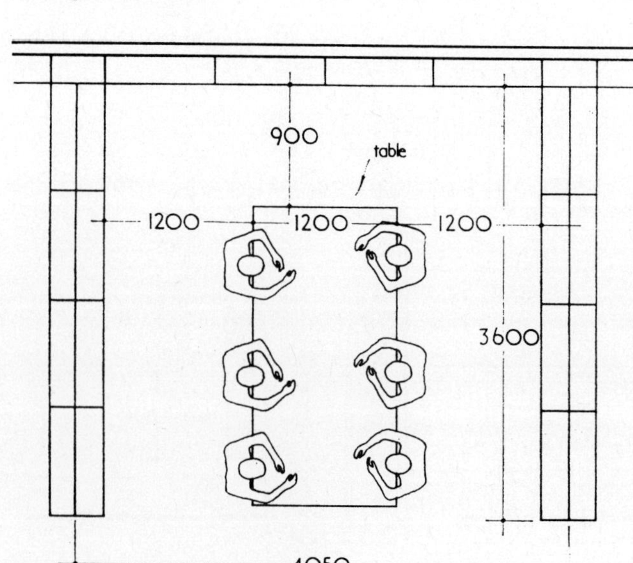

34.19 *Recommended minima for open access bookshelf areas arranged as alcoves containing reading tables*

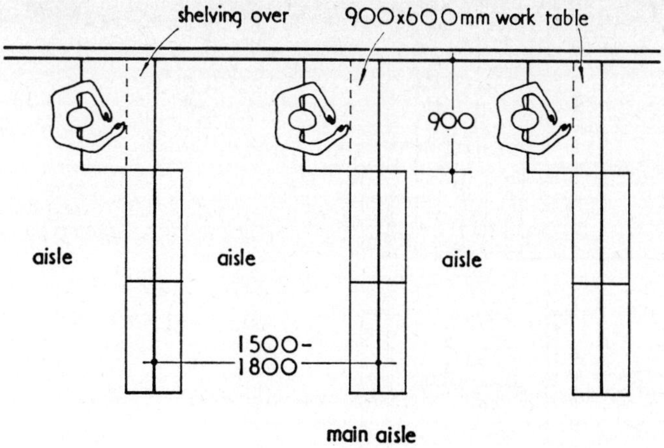

34.20 *Suggested arrangement for open carrels in bookshelf area*

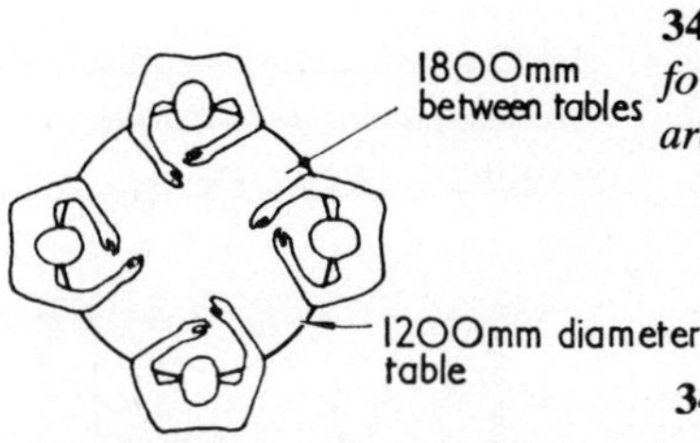

34.21 *Round reading tables*

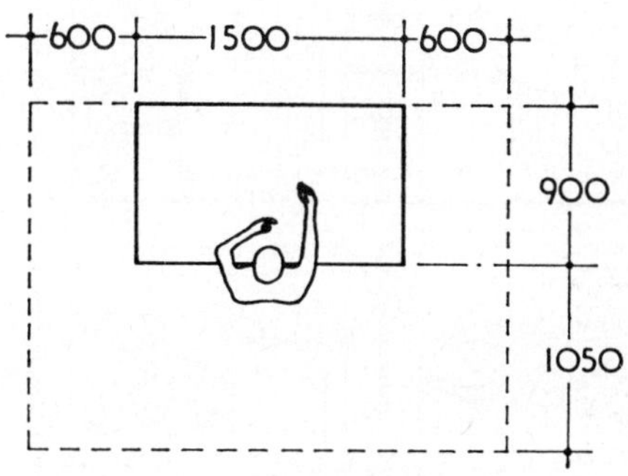

34.22 *Recommended minima for library staff*

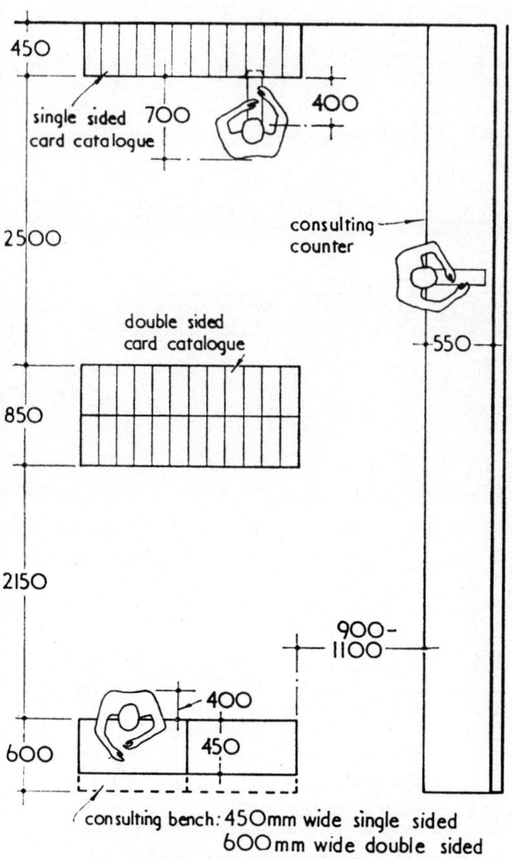

34.23 *Recommended minima in card catalogue area*

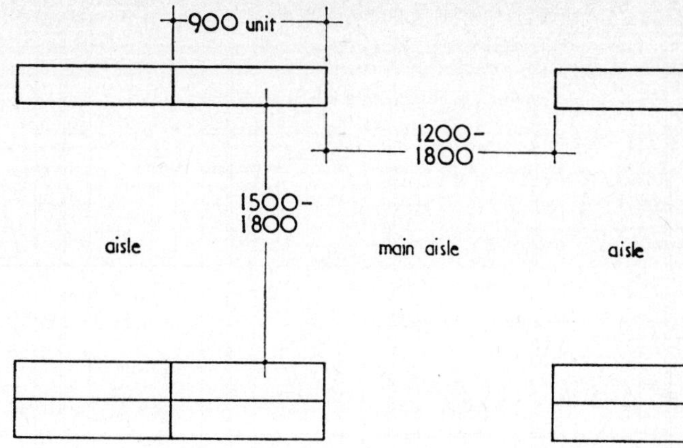

34.24 *Recommended minimum aisle widths in open access bookstacks*

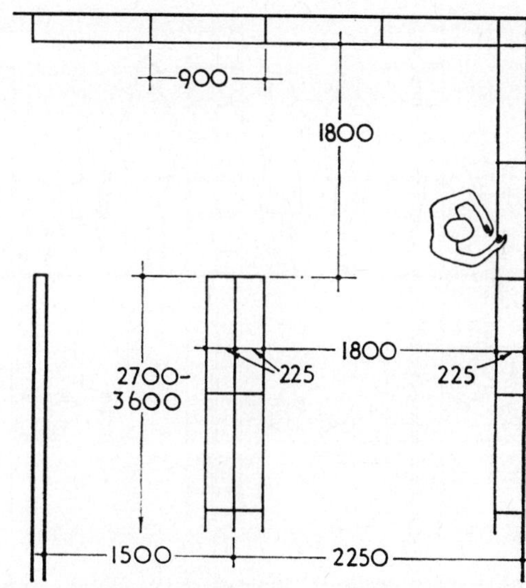

34.25 *Recommended minima in open access bookshelf areas*

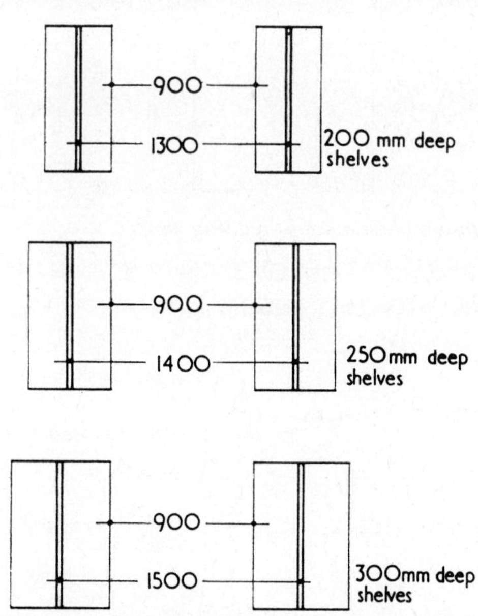

34.26 *Recommended aisle widths in closed access bookstacks for various depths of shelf*

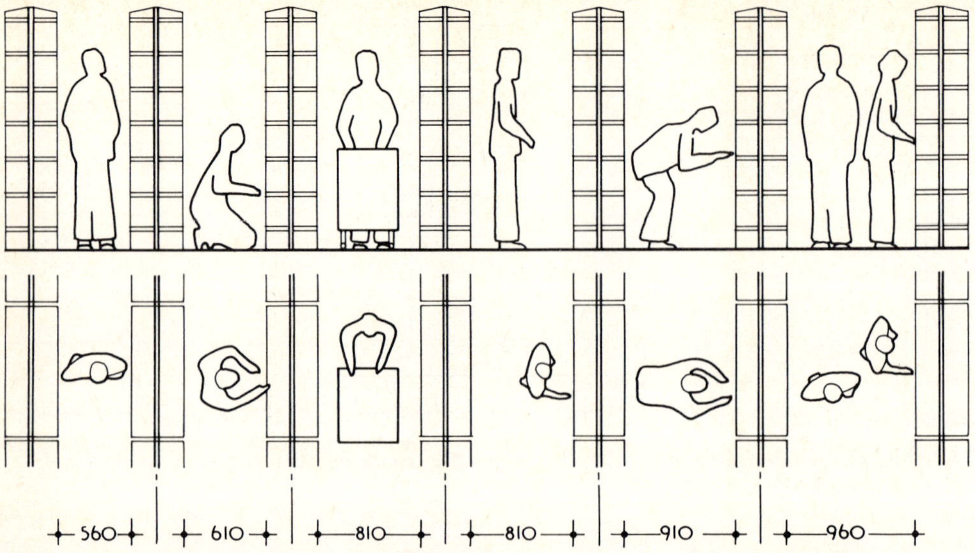

34.27 *Minimum clearances in shelving areas for various attitudes: narrow aisles*

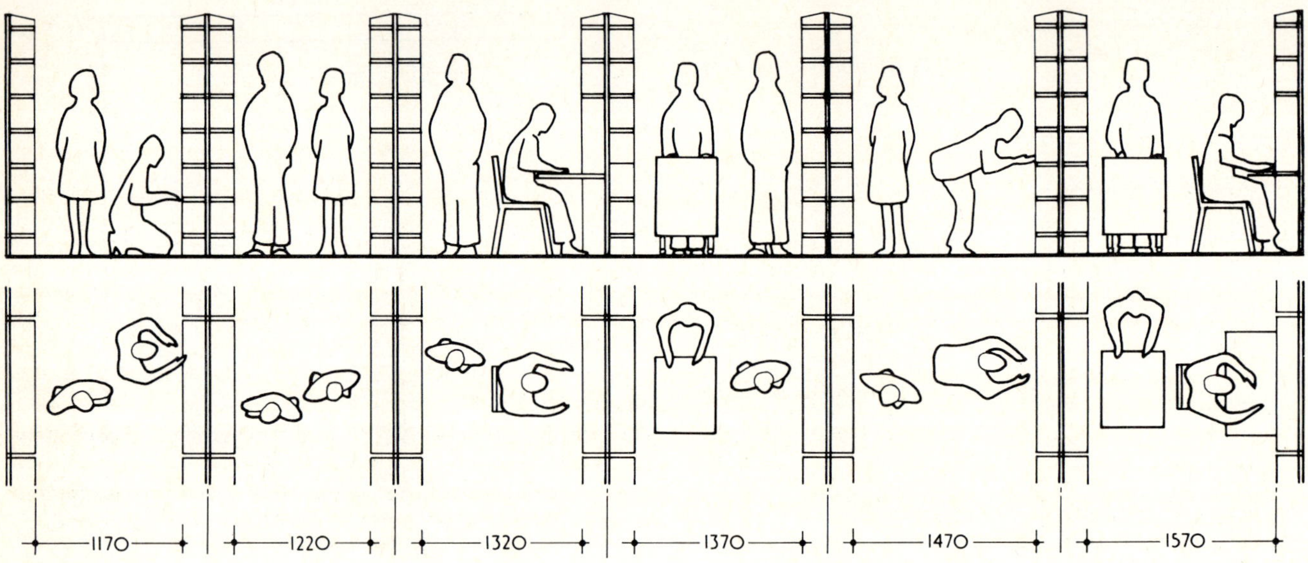

34.28 *Minimum clearances in shelving areas for various attitudes: wide aisles*

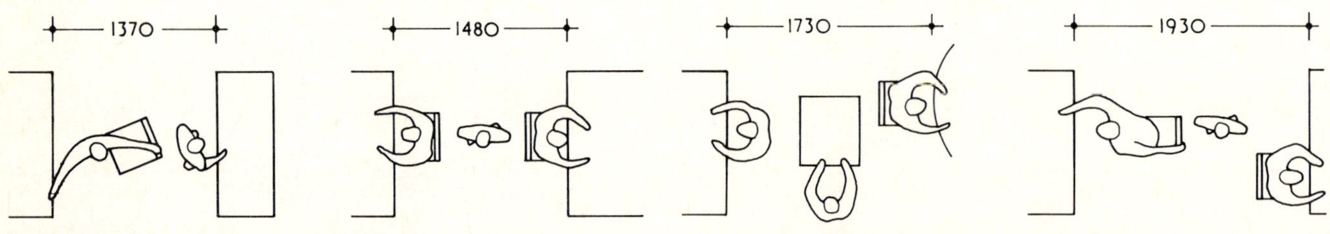

34.29 *Minimum clearances in reading areas*

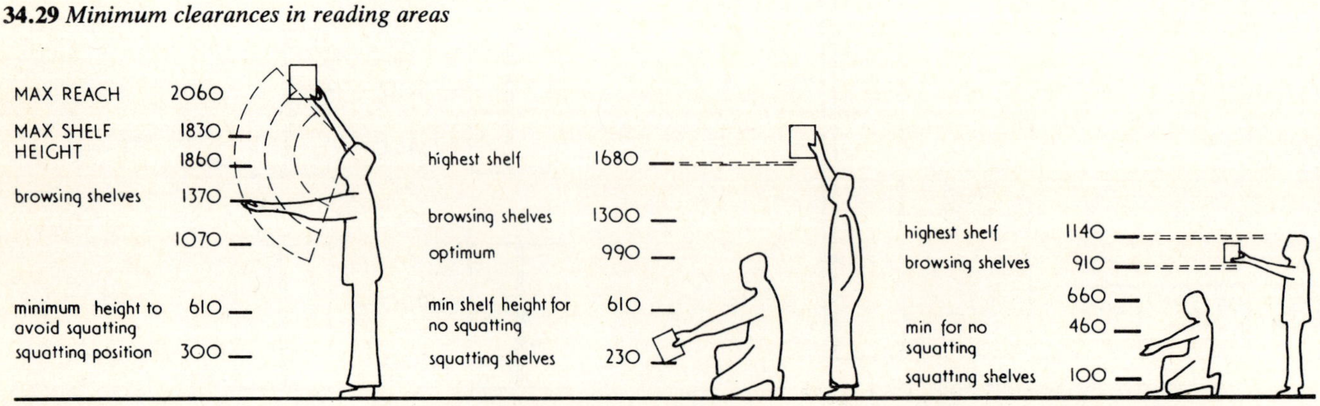

34.30 *Optimum shelf heights for adults*

34.31 *Optimum shelf heights for teenagers*

34.32 *Optimum shelf heights for children*

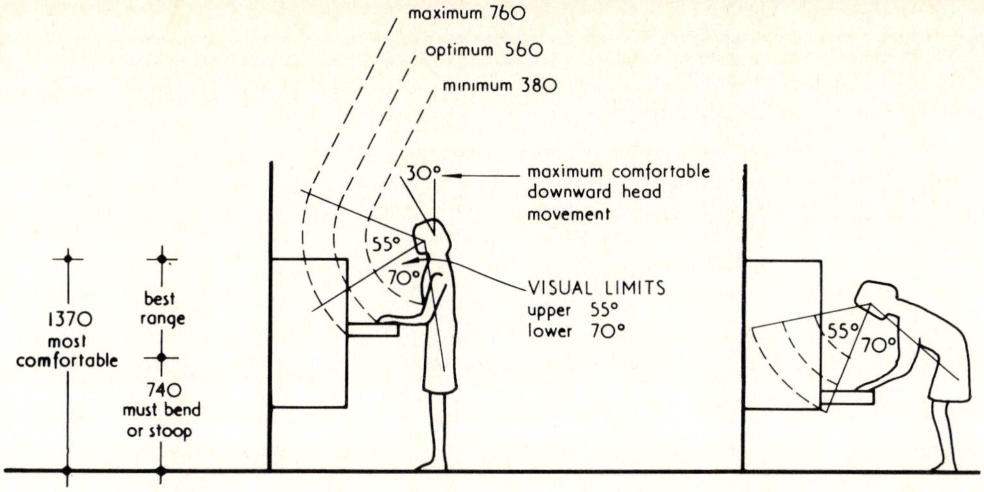

34.33 *Recommended drawer heights in card catalogue areas*

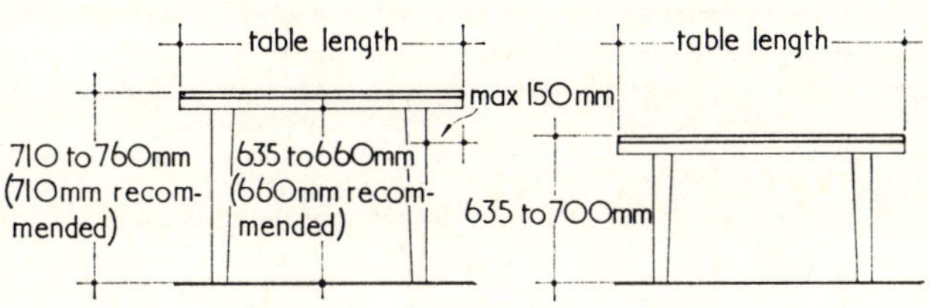

34.34 *Reading table heights for adults and children*

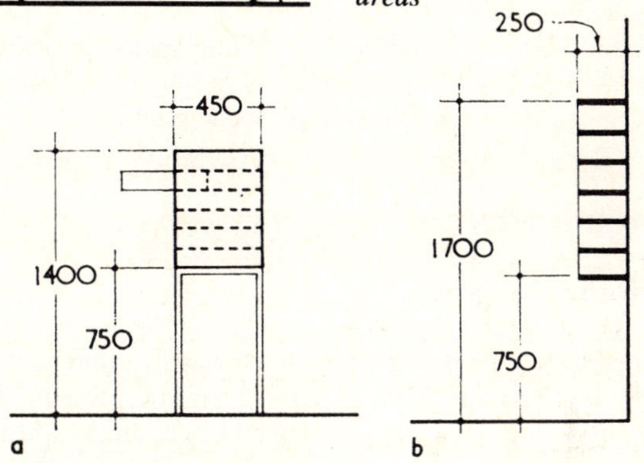

34.36 *Recommended heights for:*
a *card catalogue cabinet*
b *sheaf catalogue binder shelves*

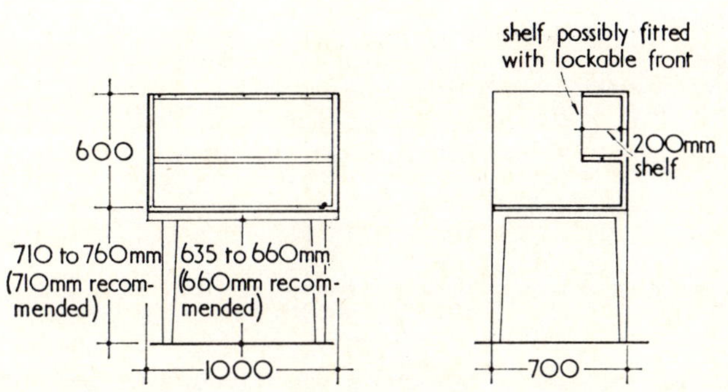

34.35 *Typical open carrel*

8 BIBLIOGRAPHY

AJ Information sheet 1318 Library planning 2: Space standards, *The Architects' Journal*, 24.2.65

AJ Information sheets 1319 and 1320 Library furniture and equipment 1: Book storage, 2: General, *The Architects' Journal*, 3.3.65

AJ Information sheet 1593 Library planning: structural modules, *The Architects' Journal*, 28.2.68

Godfrey Thompson *Planning and design of library buildings*, London, Architectural Press, 1973; 2nd edn 1977

35 Housing

CI/SfB 81
CI/SfB (1976 revised) 81

Contents

1 INTRODUCTION

1.01 Housing standards

Historically, government has only felt it necessary to set minimum standards for housing which was in some part paid for out of public funds. While the prime consideration has been to ensure value for money, the social objective has been to provide the council tenant with something better than low cost private rentals. The notion that certain sizes of household require certain sizes of living space formed the basis for most standards systems culminating in Parker Morris standards for various household sizes which have applied to all post 1967 local authority housing. It is generally accepted that Parker Morris (PM) standards of space and amenity meet the requirements of most local authority tenants. The Housing Development Directorate (formerly the R & D Group at MHLG) Sociological Research Division has conducted many surveys of local authority estates which suggest that dissatisfaction arises now from factors outside the physical confines of the house. The more important problems appear to be in the spaces around the house, in social support systems, education, job opportunities and the like, rather than in the physical dimensions of the home itself. Private sector housing has not been subject to the standards applicable to local authority housing, despite the recommendation of the Parker Morris Committee that its standards should apply to all housing, and a private developer is free to build to whatever space standards he can find a market for, provided the statutory requirements of planning, public health, safety, etc, are complied with in the development. In recent years this has led to an imbalance between the standards which the market (ie the private sector) will bear, and the standards developed in the public sector based upon equating social aims with value for money. The standards of space provided by the PM Committee derived from a disinterested attempt to relate floor areas to the activities thought to be necessary for normal family life, and as such may be thought to be more credible than standards based upon market forces alone. The standards applied to the public and private sectors are summarised in para **4**.

2 THE DESIGN OF MASS HOUSING

2.01 Policy development

The established methods of providing low cost housing for owner-occupiers and local authority tenants have been severely questioned, possibly because of an enhanced public awareness of housing problems, but more likely because costs are increasing rapidly as a result of inflation and the energy crisis. Traditionally, housing policy has been dictated by different fiscal, legislative, and subsidy arrangements rather than physical solutions; it is often said that the problems lie in land and money rather than built form. The role of housing associations and co-operatives perhaps offers some indication that the

Table I Sources for further information on standards applied to house building

Subject	DOE publications	Circular (para nos in brackets)
Generally		31/67 (2–5) 36/67 (4–6) and App II (4) 27/70 (MHLG) App IV
Car accommodation	DB10 DB12	24/75 (32) and Annex IV
Children's playspace	DB27 HDD OP 2/76	27/72 (except App II) 61/75 App 5
Density		36/67 (13) 24/75 (19)
Dimensional co-operation	HDN Part 1 (1–9 generally) DB16	31/67 (2–5)
District heating		82/71
Drainage services	DB30	
Floor space		24/75 (22)
Handicapped people's dwellings	HDD OP 2/74 HDD OP 2/75	74/74 92/75
Heating		121/73
Hostel accommodation	HAN 1/75 HAN 3/75 HAN 4/75	170/74 App F
Insulation (sound)	Building Regulations 1972 Guidance Note—Sound insulation (DOE/HMSO 1975)	110/75
Insulation (thermal)	HDN IV 1–2	105/75
Landscaping	HDN II 1–4	
Metrication	DB 16	27/70 (MHLG) App IV
Mobility housing	HDD OP 2/74	74/74, 92/75
Noise	DB 26	10/73
Old people's dwellings	DB1 DB2 DB11 DB13 DB31 HDD OP 1/76	82/69 61/75 App 3
Parker Morris	*Homes for today and tomorrow* (HMSO, 1961, reprinted 1975)	36/67 (1–7) and App 1 27/70 (MGLG) App IV
Private builders' schemes		24/75 (14) and Annex II
Residents' attitudes to estate layout	DB25	
Roads and footpaths	DB32	79/74 24/75 Annex IV
Safety	DB3 part 7 DB13	
Sanitary plumbing	DB30	
Single people	DB23 DB29	12/76
Space in the home	DB6 DB14 DB24 1 and 2	
Ventilation		121/73
Vertical dimensional		31/67 (2–5)
Wheelchair housing	HDD OP 2/75	92/75

DB—Design Bulletin
HDN—Housing Development Note
HAN—Housing Association Note
HDD OP Housing Development Directorate Occasional Paper

distinctions between subsidised housing and private development are becoming less sharp, while the ideas put forward by John F. C. Turner, Martin Pawley and others provide rather more radical solutions to housing problems. The idea of subsidising people rather than buildings and allowing housing standards to be decided by individuals is attractive to some, but others maintain that this might destroy the careful system of checks and balances embodied in existing regulatory controls by pre-empting all possibility of ensuring that new buildings are safe in terms of structure, construction, and environmental health. Inevitably, any changes in the present system are likely to be evolutionary rather than revolutionary, resulting in a compromise solution which it is hoped will reflect some of the more forward thinking proposals. For the present, housing remains tightly controlled by a series of procedures, standards, and financing arrangements which must be considered for all new developments. Table I provides a reference list of DOE publications which have direct bearing upon housing design in Britain.

2.02 Estate or neighbourhood layout

The shape of a housing development depends very much upon its planning density and the structure of the intended residential population. Broad issues like the relationship to schools, shops, workplaces, transport, etc are generally beyond the control of individual designers but the extent to which these facilities are already available will determine the amount of non-housing accommodation to be provided in new development.

The manner in which necessarily simple house types are grouped together, the hierarchy of vehicular and pedestrian access, the extent of landscaping, play provision, and many other considerations in the detail design of the layout are just as important to the final quality of a residential environment as the design of the houses themselves. Simplistic attitudes to such a complex design problem can only produce poor results which are not only unpopular with the residents and the community at large but can be wasteful of resources.

2.03 Built-form choice and costs

The choice between different types of building will also depend upon such factors as the optimum cost per unit in each category and the density of the development (usually decided prior to the appointment of the designers in Britain). Whether or not the scheme will be financially viable depends very much upon the quality of the decisions made at this early stage in design. It is well known that the construction cost of units in high-rise buildings is higher than low-rise houses, but it is possible that the additional cost of locating suitable units in high blocks may free more land for low-rise houses, or provide space for facilities and amenities which would have to be omitted if the development simply concentrated upon low-rise buildings. It is equally well known that the costs of low-rise buildings increase with the complexity of the design of the house or block. Where it is not possible to develop a site using simple houses, the design contortions required to maintain even relatively low densities using three to four storey development lead to considerable additional costs in site development and complicated construction, in many cases without immediately apparent advantages in appearance, amenity, or even user satisfaction. Clearly, the degree of ingenuity and skill required to design housing layouts at the densities usually required to off-set high land costs within low overall cost limits is considerably greater than for housing which is not strictly circumscribed by the limits of cost and density of development: standard semi-detached houses at densities of 10 to 12 to the acre may be laid out easily on cheap green field sites, but this kind of development is unlikely to be economically viable on expensive land which dictates higher densities and considerably more design expertise.

The nature of the buildings themselves will become more complex as densities rise but several basic needs for successful schemes may be thought to remain constant. The following discussion of factors affecting user satisfaction may provide some insight into the aspects of estate layout which residents think important. See Design Bulletin 25, Housing Development Directorate, DOE.

2.04 Amenity and appearance

Very little hard evidence exists about the characteristics which constitute good amenity and appearance in the eyes of users. The design professions will express a clear-cut aesthetic, but their opinions rarely appear to agree with those of the lay public. The opinions of the users must be realistically considered in the design of housing which, while essentially cheap, must accommodate the lifestyles of its occupiers over considerable periods of time and offer sufficient facilities to enable residents to have a proper pride in their home. This is perhaps the most difficult aspect of mass housing since the designers very seldom have the opportunity, or possibly the desire, to discuss intimate details with the wide range of possible users. However, research into residents' attitudes to appearance begins to suggest a preference for bright, light, modern buildings with big windows over back-to-back brick houses with roads and garages, suggesting that some residents at least are not hostile to non-traditional appearance. Amenities considered favourably include 'a sense of spaciousness', plentiful greenery, good maintenance, pleasant entrances, convenient access, and car parking which can be easily supervised by passing residents. The institutional appearance of large, massive, dull grey or drab buildings, large areas of exposed concrete and grey paving, grey or dirty looking entrances and exposed concrete finishes are strongly disliked.

How far these broad generalisations can be applied to private housing is problematic. The ubiquitous three up, two down semi is quoted by most private developers as the overall best seller, while adventurous firms who have produced housing which attempts to improve upon ribbon development suburban environment tend to find their market restricted to higher income groups, possibly because of difficulties in obtaining loans on unconventional property. Other questions related to the overall status of the neighbourhood, its location in relation to schools, shopping centres, job opportunities and other broader issues are clearly important factors which can also influence the reaction of residents to their home in both private and public sector housing.

2.05 Maintenance

The quality of maintenance on all residential estates affects long-term appearance. In the public sector complaints about litter, dirty lifts and stairs, and the length of time taken to get repairs done are frequent sources of dissatisfaction. More staff and better supervision may be cures for vandalism while specifications of materials which retain their freshness might reduce the unkempt appearance of many existing estates. Maintenance is considered the owner's province in private sector housing where occupiers exert rather more control over the quality of their surroundings.

2.06 Children's play

Children's play can be a problem on all types of estates, irrespective of built form and tenure. Difficulties tend to be more pronounced at higher densities, but detail design and the extent of provision appear to have little effect. While social rather than physical aspects would seem to be more important, well-designed facilities can contribute to children's development by offering a wide variety of play opportunities and experiences. Although supervised play areas may alleviate some emotional/social problems, most spaces and features on an estate can offer some opportunity for play. Children's safety and avoiding disturbance of other residents, particularly old people, are also important.

2.07 Car accommodation

The most popular form of car accommodation is probably a garage within the curtilage of the house but, where this cannot be provided because of costs or density, uncovered low cost spaces within sight of the owner's dwelling are the best alternative (140 m is a reasonable walking distance). Water stand pipes and electric power points are useful extras. The extent to which car accommodation should be provided merits careful consideration in the light of rapidly increasing costs of fuel and cars.

2.08 Roads

Design Bulletin 32 *Residential roads and footpaths: layout consideration* discusses the issues connected with the design of layouts for pedestrian and vehicular access in residential areas in considerable detail. A corporate approach is recommended which strikes a balance between housing, planning and highway objectives and allows for a wide range of options in design. Much material from this Bulletin will be found in Section 7, External Circulation.

The main text sets out the objectives which should normally be pursued at the start of each section, followed by supporting commentaries which draw on empirical evidence as well as current practice. The basic themes considered are:

- relating the housing scheme to its setting
- minimising danger and nuisance from non-access vehicular traffic
- reducing vehicle flow and speed
- providing for pedestrian movement
- making effective provision for off-street parking
- providing for vehicular movement
- making provision for statutory and other services.

Appendices give information on: sizes of vehicles for use when determining the geometric characteristics of residential access roads; geometric characteristics of vehicles when turning; a method of constructing forward visibility curves on bends; a summary of studies into the use of single-track roads with passing places; considerations for the provision of forward visibility between passing places on narrowed carriageways; and data on vertical curves. There is a further appendix which gives an example of provision for statutory undertakers' services when shared vehicular/pedestrian surfaces are used.

2.09 Private open space

It seems important that open space directly connected to dwellings should be demonstrably private, no matter what tenure arrangements apply, and that the territorial rights of the occupiers should be clearly marked.

Gardens and balconies are popular but only if they are a reasonable size. A flat located near the centre of cities may be preferred to a house with a garden away from the centre, even for a proportion of families with older children, especially if a usable balcony is available.

2.10 Influences of energy conservation on design

Few physical design factors have an individual effect upon the thermal behaviour of buildings which is sufficient to make a significant reduction in domestic energy consumption, but combinations of different treatments can begin to achieve substantial savings. Well-informed patterns of household activity can maximise these savings through economical control of heating and ventilating systems, but it is important that residents are provided with clear and simple instructions about the heating system in their new home if this potential saving is to be realised.

Code of Practice 3 Chap 2: 1970 describes the thermal behaviour of buildings in great detail and carries a broad requirement that the organisation of the natural relationship between outdoor climate, building structure and patterns of living should produce an acceptable indoor environment with the least assistance from mechanical plant.

Apart from increasing insulation values in walls and roofs, the internal plan arrangement of dwellings, the size and location of doors and windows, the provision of draught lobbies and space for flued heating appliances (or heat pumps), with improved methods of heating and ventilation controls, and better sealing within the structure in general can all contribute to energy savings both individually and in different combinations. However, the interaction between these factors remains more or less uncertain so that innovations in construction should be considered very carefully with respect to long-term consequences for building integrity.

The orientation of dwellings to take the best advantage of natural features on the site can also provide opportunities for maximising solar gain and minimising exposure to prevailing winds and rain. Similarly, different configurations of assemblies of dwellings can provide a more sheltered environment by the careful design of their relationships to natural features which are more exposed to severe weather conditions.

Most of the factors mentioned above may be thought to increase housing costs, but it is possible that these extra physical provisions may reduce the cost of heating installations as well as minimise operating costs and energy consumption in future.

3 BUILT FORM

3.01 Choice and family type

The relationship of family characteristics to built form may be considered in many ways and is open to individual interpretation, depending upon attitudes to political and social objectives as much as the geometric problems of providing the most accommodation on the smallest ground area consistent with acceptable amenity standards. Clearly, 'acceptable amenity' varies according to the status, power, and income of residents in this imperfect world, but generalisations might be made about the sorts of houses which are more likely to suit different sizes and ages of family. Table II shows notional relationships between families (age and size) and houses, flats, or maisonettes which may be thought more suited to their activities. It can be seen that one, two or three storey houses can meet the needs of most family types, but that flats or maisonettes may be as suitable for small adult families, single people and the elderly. The categories included are, of necessity, broadly based but the table illustrates that the choice of built form for any development may be sensibly based upon an appreciation of its proposed population structure.

3.02 House plan-form

The suitability of any plan-form for mass housing units can be assessed only in common-sense terms and the wide variety of possible users makes attempts to standardise desirable spaces here seem unprofitable, not to say naive. A simple checklist may be of use to inexperienced designers new to housing!

- Does the plan allow changes to be made to accommodate altered lifestyles over long periods of time, eg young family to adult family in 10 years?
- Can both noisy and quiet activities take place within the house without creating annoyance to the occupants?

Table II Family characteristics related to notionally suitable built-forms

Household size	Age of family: Young	Adult	Old	Mixed adult and old	Notionally suitable built-form considering building economics and sociological information: Houses 1 storey	2 storey	3 storey	Flats 2–3 storey	Over 3 storey	Maisonettes 4 storey	Over 4 storey	Blocks of any height providing communal facilities and services
1	☆				★			★	★			★
		☆			★			★	★			★
			☆		★			★	★			★
2	☆				★							★
		☆			★			★	★			
			☆		★			★	★			★
				☆	★	☆		★	★	☆	☆	★
3	☆				★	☆						
		☆			★	☆		★	★	☆	☆	
			☆		★			★	★			★
				☆	★	☆		★	★	☆	☆	★
4	☆				★	★						★
		☆			★	★		★	★	★	★	
				☆	★	☆		★	★	☆	☆	☆
5	☆				★	★	★					
		☆			★	★	★	★	☆	★	☆	
				☆	☆	☆	☆	★	★	★	★	
6	☆				★	★	★					
		☆			★	★	★					
				☆	★	★	★					
7	☆				★	★	★					
		☆			★	★	★					
				☆	★	★	★					

★ Preferred accommodation
☆ Acceptable accommodation

- Are the basic spaces conveniently arranged, eg kitchen adjacent to dining area?
- Can young children be easily supervised from the kitchen in family size homes?
- Is it possible to reach the garden or balcony directly from living areas?
- Is there space to deal with visitors at the front door without necessarily inviting them into the living areas?
- Is the circulation space as efficient as possible?
- Is there opportunity for personalising the main entrance, ie can a recognisable threshold be easily created by the occupants?
- Does the layout of the scheme allow a reasonable amount of privacy, avoid overlooking, etc for each home?
- Does the layout of the scheme allow good views out, exploit sunlight and daylight, and provide easy identification of the occupant's home? The provision of a wide variety of plan-forms within individual schemes, or the use of easily altered partitions, etc may assist in increasing user satisfaction.

3.03 Component co-ordination and the metric dimensional framework

Component co-ordination has been adopted in Britain as a policy which aims to increase value for money in house-building by encouraging the mass production of components dimensioned to fit a range of preferred sizes as shown in Section 39, Materials, in the belief that mass production techniques can lower price and improve quality. The financial control exercised by central government has made it possible to apply this discipline to local authority housing, and the use of some components has spread to the private sector; but the substantial results which had been hoped for have not materialised, partly because of the fragmentation of the building industry in general, and partly because of the resistance of some designers who regard the dimensional framework as a straight-jacket which inhibits design freedom. On the other hand, the use of standard designs for metric house shells (produced by the NBA) and the recent rationalisation of GLC house types for low-rise developments into a preferred range suggests that a substantial number of client authorities regard rationalisation of this nature as a sensible activity. It is also widely accepted that the British Standard Specifications could be even more useful if they were:

- dimensionally compatible
- performance based.

Design Bulletin 16, available from HMSO, sets out the objectives of the co-ordination of components in housing and describes the dimensional framework summarised in tables III and IV and **35.1**.

4 LOCAL AUTHORITY HOUSING

4.01 Design standards and guides

The standards which apply to local authority housing may be a useful starting point. However, reference should be made to the many relevant DOE circulars at a reasonably early stage in order to avoid abortive work (see table I for general sources). Many authorities have their own specification and construction preferences as well as standardised design guides for both housing design and estate layout which must also be considered before even broad design decisions can be made. In addition the complexities of the Building Regulations and the changes implicit in the Health and Safety at Work Etc Act 1974 do not make the production of satisfactory residential

Table III Dimensional co-ordination in the horizontal plane for housing

Dimensional basis		
The dimensional basis for housing design is a 300 × 300 mm grid	a	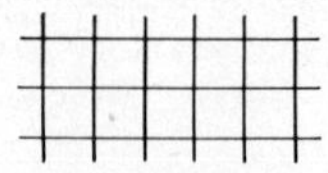
To allow for second preference conditions this grid is subdivided to form a 100 × 100 mm grid	b	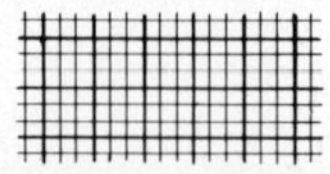
Zones for external walls, columns and beams are 100, 200, 300 or 400 mm thick		
and are placed with the inner boundary on a 300 mm grid line	c	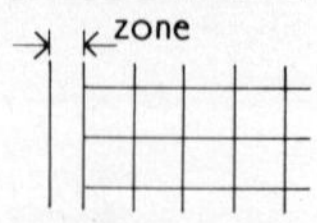
Second preference: zones for non-floor-supporting external walls may be placed with the inner boundary on a 100 mm grid line	d	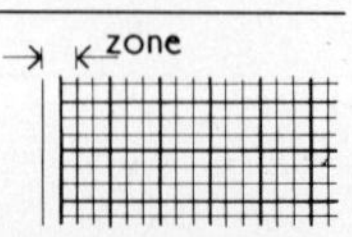
Zones for internal floor-supporting walls, columns and beams are 100*, 200, 300 or 400 mm thick		
100 mm (non-separating) walls are placed with one boundary on a 300 mm grid line	e	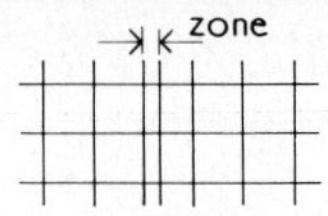
if 200, 300 or 400 mm thick, with both boundaries on 300 mm grid lines *NB With the 200 and 400 mm zones the 300 mm grid is interrupted*	f	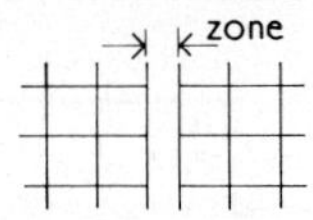

Table III (continued)

Zones for non-floor-supporting internal walls are 50* or 100* mm thick		
and are placed with one boundary on a 300 mm grid	g	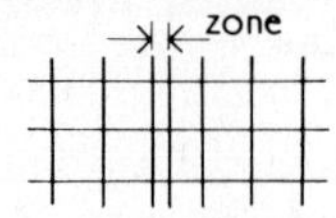
Second preference: one boundary on a 100 mm grid	h	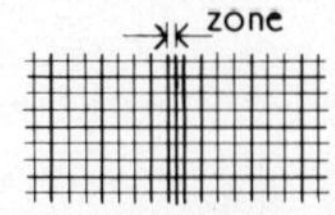
Exception: In flats non-floor-supporting separating walls may be 200, 300 or 400 mm thick and placed with one boundary on a 300 mm grid	i	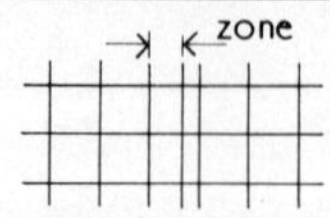

* These thicknesses may not be used for separating (party) walls between occupancies.

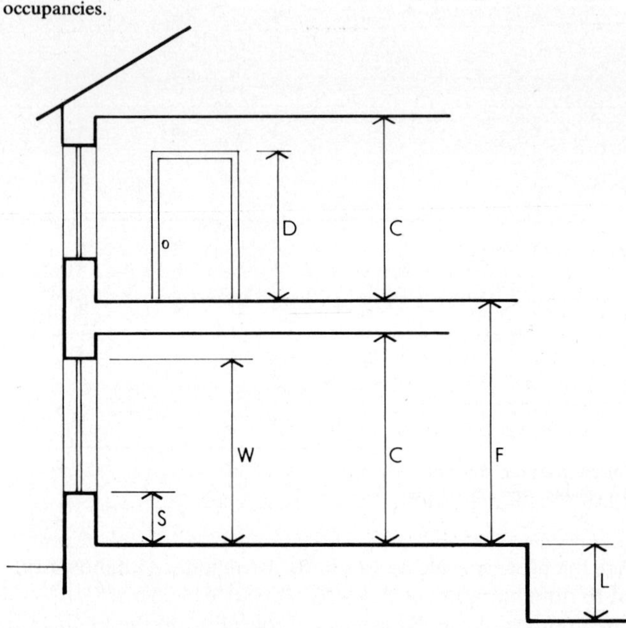

35.1 *Vertical dimensional co-ordination in housing, see table IV*

Table IV Dimensional co-ordination in the vertical plane for housing
To be read in conjunction with **35.1**

	F **floor-to-floor height**	**C** **floor-to-ceiling height**	**T** **floor zone thickness**	**L** **change of level**	**D** **door head height**	**W** **window head height**	**S** **window sill height**
Publicly financed housing							
Mandatory	2600						
First preference		2350	250	600 1300*	2100	2300	
Second preference		2400 2300	200 300			2100	
Other permissible		2100†					0 200 600 800 1000 1400 1800
Other housing	2600 2700 and upwards in increments of 300	2100† 2300 2350§ 2400 and upwards in increments of 100	200 250§ 300 and upwards in increments of 100	300 600 900 1200 1300§ 1400§ 1500 1700§ 1800 and upwards in increments of 300 to 2400, then 600	2100 and upwards in increments of 100	2100 2300 and upwards in increments of 100	0 200 300 600 700 800 900 1000 1100 1200 1400 1800 2100

* Only between dwellings   † Only for garages   § Only where F = 2600

Table V Aggregate space standards mandatory in public sector housing

Homes built for occupation by:	7 people	6 people	5 people	4 people	3 people	2 people	2* people (old)	1 person
• Should be designed with a *net floor area* of: (exclusive of general storage and stores for dustbins and fuel, garage and balcony)	m^2	m^2	m^2	m^2	m^2	m^2	m^2	m^2
Built form								
3 storey house, 3 storey house and garage	112	98	98	—	—	—	—	—
2 storey centre terrace	108	92·5	85	74·5	—	—	—	—
Semi or end terrace	108	92·5	82	72	—	—	—	—
1 storey centre terrace	—	84	75·5	67	57	44·5	44·5	30
Maisonette	108	—	82	72	—	—	—	—
Flat	—	86·5	79	70	57	44·5	44·5	30
Flat (balcony access)	—	—	—	67	—	—	—	—
Shared accom*								
Flatlets, Bedsit with shared bath	—	—	—	—	—	—	—	27
1 bed private bath*	—	—	—	—	—	—	39	27

• and should include *general storage space* (exclusive of stores for dustbin, pram, fuel or through access space (700 mm wide) in terraced houses)

	7 people	6 people	5 people	4 people	3 people	2 people	2* people (old)	1 person
Houses	6·5	4·5	4·5	4·5	4	4	4 self contained	3
Flats and maisonettes		3·5	3·5	3·5	3	3	3	2·5 (2·6 for old people)
Shared accom*								
Flatlets: 1 bed private bath								2·5 (2·6 for old people)
Bedsit and shared bath*								1·9 (2·6 for old people)

• as well as fuel storage at the following rates:
1·5 m^2 for 1 appliance only
2 m^2 for 2 appliances or in rural areas
1 m^2 in flats if there is no auxiliary storage

* Standards for old people and single accommodation were not stated in the Parker Morris report but were provided in later government circulars

development in the public sector easy, but it would be ingenuous to expect that this should be so, particularly when public money is a large consideration. However, when the briefing elements are simplified, as in the majority of private housing developments, it can be seen that the overall environmental quality achieved is far from satisfactory, despite the relative simplicity of designing low density developments. The emergence of design guides issued by local planning authorities indicates the concern felt in many areas about the poor quality of recent residential environments.

4.02 Mandatory standards in public sector housing

Table V summarises the aggregate space standards recommended by the Parker Morris Committee in *Homes for today and tomorrow*, and subsequent government circulars. These are mandatory for housing subsidised directly by the Exchequer. In the following, mandatory requirements are printed in *italics*. Private sector housing is subject only to those space standards required by the Public Health Act.

4.03 Sanitary provision

• *In one, two and three person dwellings, one WC is required, and may be in the bathroom,* **35.2**

• *In four person, two storey or three storey houses and two-level maisonettes, and in four person and five person flats and single storey houses, one WC is required in a separate compartment,* **35.3a**

• *In two or three storey houses and two-level maisonettes at or above the minimum floor area for five persons, and in flats and single storey houses at or above the minimum floor area for six persons, two WCs are required, one of which may be in the bathroom* **35.3b**

• *Where a separate WC does not adjoin a bathroom, it must contain a washbasin,* **35.4**

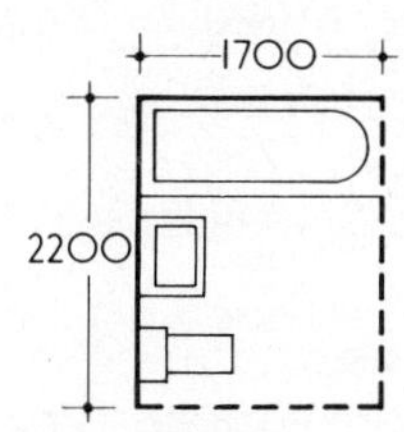

35.2 *Minimum bathroom including* WC

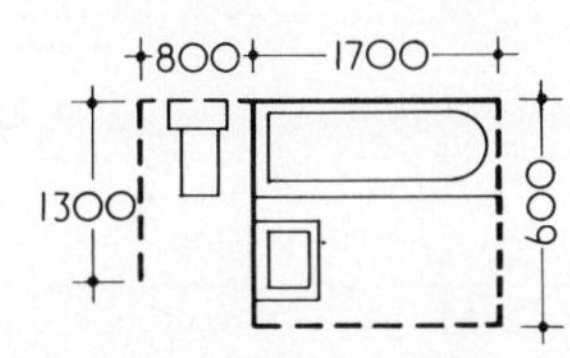

35.3a WC *in separate compartment adjacent to bathroom*

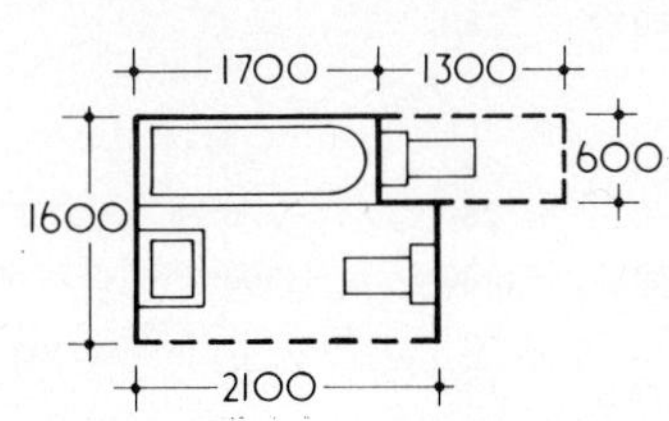

35.3b WC *in separate compartment adjacent to bathroom also containing* WC

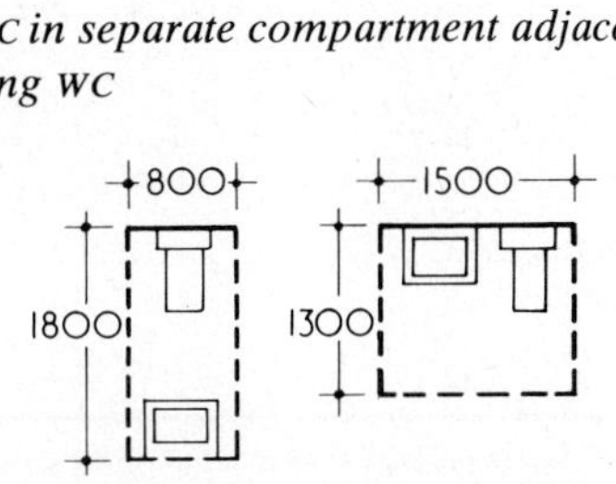

35.4 *Alternative* WCs *with washbasins*

4.04 Kitchens

Worktops shall be provided on both sides of the sink and cooker positions. Fitments shall be arranged to provide a work sequence comprising worktop/cooker/worktop/sink/worktop (or the same in reverse order) unbroken by a door or other traffic way, **35.5**.

Kitchen fitments, **35.6**, *comprising enclosed storage space in connection with*

- *preparation and serving food and washing-up*
- *cleaning and laundry operations*
- *food*

shall be provided as follows:

Three person and larger dwellings $2{\cdot}3\,m^3$
One and two person dwellings $1{\cdot}7\,m^3$

This provision must include a ventilated cool cupboard and a broom cupboard that need not be located in the kitchen.

Where standard fitments are used, the cubic capacity is measured overall for the depth and the width, and from the underside of the worktop to the top of the plinth for the height.

Tables VI and VII give the storage requirements of the average family, based on an investigation carried out by the Council of Scientific Management in the Home in 1963/4, and on a subsequent study of the space requirements of stored items made by Queen Elizabeth College, University of London.

Worktop heights

The preferred sink and worktop height adopted in British Standards is 900 mm because of problems in achieving adjustability. However, research has shown that sink heights of 975 mm and worktop heights of 950 mm suit 75 per cent of women, **35.7**. It may be noted that even the highest sink is 75 mm too low for the average man; with the increasing involvement of men in kitchen activities this is a problem that will have to be solved.

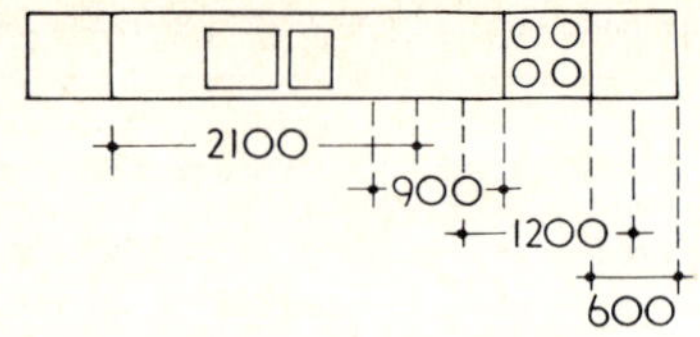

35.5 *Typical kitchen layout*

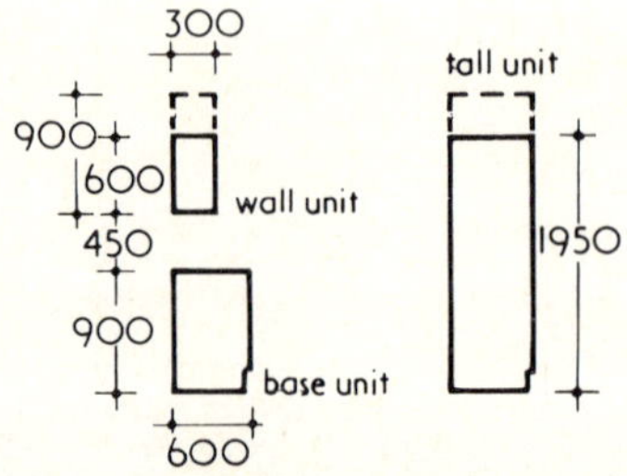

35.6 *Standard kitchen storage units*

Table VI Space requirements: Cupboards and shelves

Item	Shelf length (mm)	Shelf depth (mm)	Vertical clearance (mm)	Storage (m²)
Dry goods*	2800 / 250	150 / 230	200 / 330	0·5
Tinned and bottled goods	1100 / 300	150 / 150	150 / 300	0·2
Drinks	Length immaterial	150 minimum	350	0·3
Bread, cake, biscuits + bread bin base	900 / 300×250 mm	300	150	0·3
Pet foods	300	150	200	0·05
Dairy goods, meat, fish, poultry, frozen foods				0·9
Total food (including refrigerator)				2·25
Table china and glass	3250 / 450	280 / 280	150 / 200	1·6
Cooking china and glass	1000 / 1600	450 / 280	150 and a small amount of 250	0·45
Saucepans, frying pans, and so on	760 / 1700	280 / 280	150 / 300	0·7
Miscellaneous china and glass	900	280	300	0·25
Cookery books	300	150	230	0·05
Empty jars and bottles				0·2 in any form
Broom cupboard				0·7
Bucket and soaps cupboard				0·5
Total non-food				4·45
Total shelf area				6·70

* Figures given are for items stored in packets as purchased. If they are stored in canisters the area required could be as much as 0·8 m² with 3900 mm of 150 mm deep shelving and 1100 mm of 230 mm.

Vertical zoning of storage related to accessibility, **35.8**

The principles generally employed for determining suitable storage heights are:

- frequently needed articles should be placed in a zone which extends from arms outstretched at shoulder height to the tips of fingers when arms are down at attention (for the average housewife 700 to 1300 mm height range), **35.9**
- lighter articles can be placed in a zone extending higher to the full reach of arms and lower to the hand height associated with half trunk bending (500 to 1900 mm), **35.10**, **35.11**.
- the zones above and below these should be set aside for the dead storage of seldom used articles
- the need to be able to hold the articles safely when placing and removing and when reaching to the back of shelves dictates a shelf location about 100 mm shorter than the comfortable heights determined by these rule of thumb methods.

Eating area in kitchen

It is useful to have a small table in the kitchen for eating, unless one is built-in, such as a breakfast bar, **35.12**. Otherwise a rectangular shape is best for a small kitchen, with a minimum width of 750 mm. If the eating area is bounded by walls or worktops on both sides of the table, total width should not be

Table VII Space requirements: Baskets, drawers

Item	Method of storage	Number required	Dimensions (mm)	Storage area (m²)
Fruit and vegetables	Baskets	2	480 × 380 × 80	0·4
Table cutlery	Drawer	1	450 × 450 × 80	0·8
Kitchen cutlery and equipment*	Drawers	1½ / ¾	450 × 450 × 100 / 450 × 450 × 150	
Baking tins	Drawers	1¼	450 × 450 × 150	
Linen	Drawers	2	450 × 450 × 120	0·4
Total				1·6

* Including bread board, chopping board, rolling pin

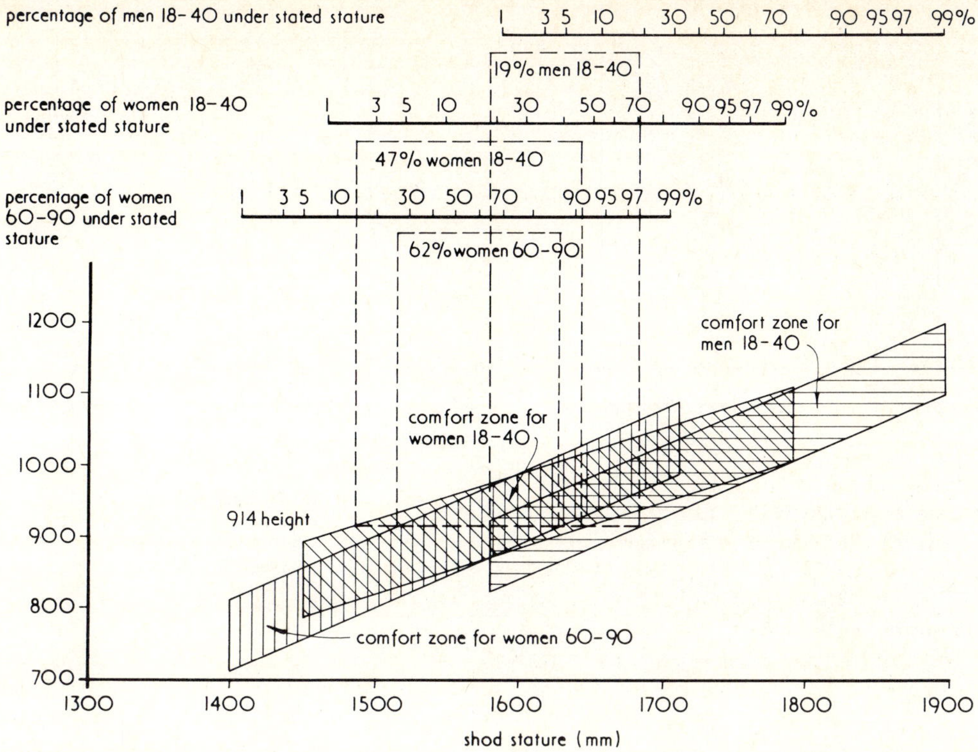

35.7a *Comfortable heights of worktops. This diagram shows the 'comfort zones' for worktops for their heights plotted against the stature of the user wearing shoes. The assumption here is that the worktop should be no higher than the user's elbow, nor lower than 100 mm below elbow height. (Some authorities recommend that the lower limit is 75 mm below elbow height.) The scales above the main diagram indicate the percentiles of the population of the appropriate stature. The dotted lines demonstrate that the standard height of 914 mm suits only 47% of women 18 to 40, and 19% of men in that age range, although it does suit 62% of elderly women*

less than 2300 mm; or 2600 mm if there needs to be free passage behind the chair on one side.

Allow 700 mm between the oven and a chair in use, and 800 mm between the most-used worktop and a pushed-in chair.

Recommended table sizes:
Two people: 750 × 750
Three people: 750 × 1000
Four people: 750 × 1200
Five people: 750 × 1350

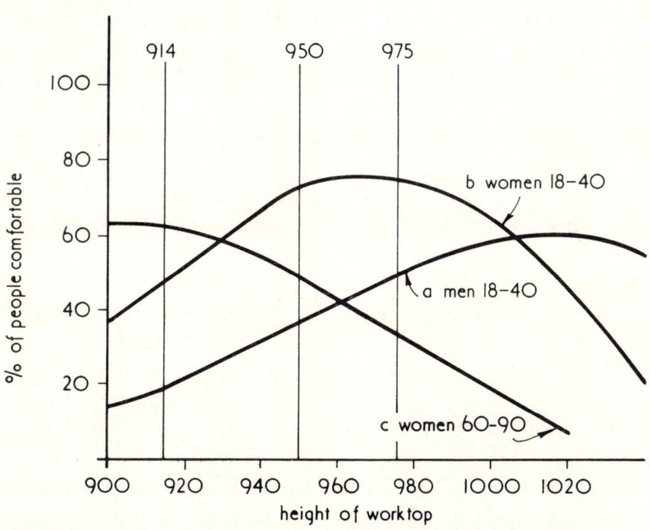

35.7b *Graphs of percentages comfortable at each worktop height. These curves are derived from* **35.7a**. *They clearly show that for younger people of both sexes a worktop height of about 975 mm will be more suitable than either of the standards of 914 and 950 mm. Ideally, different kitchen units should be used depending on the expected ages of the occupants. Too low is in many ways much worse than too high, as the user can often stand on a loose platform to work at a high worktop. Constant stooping at a low worktop leads inevitably to 'lower back pains'. When using a sink the effective worktop is the bottom of the sink. The sink surround should therefore be at least 75 mm higher than the appropriate worktop*

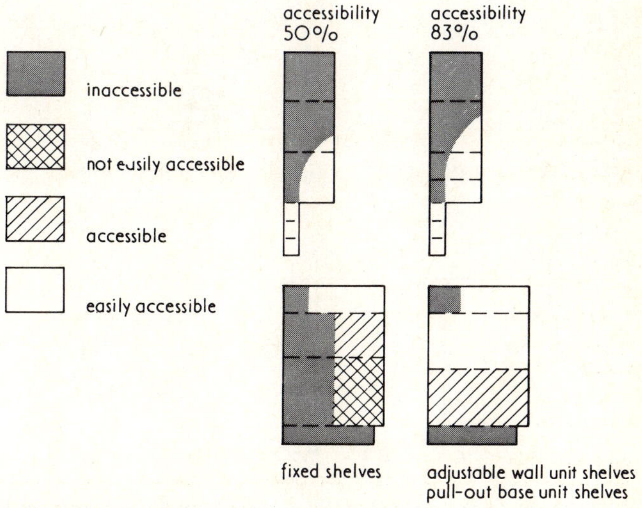

35.8 *Accessibility of storage*

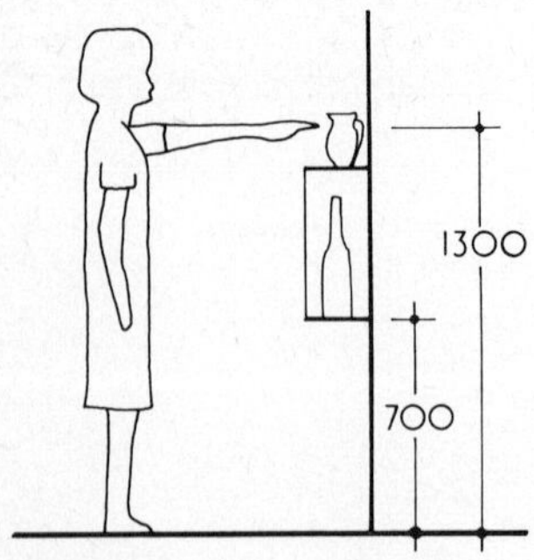

35.9 *Frequently needed articles*

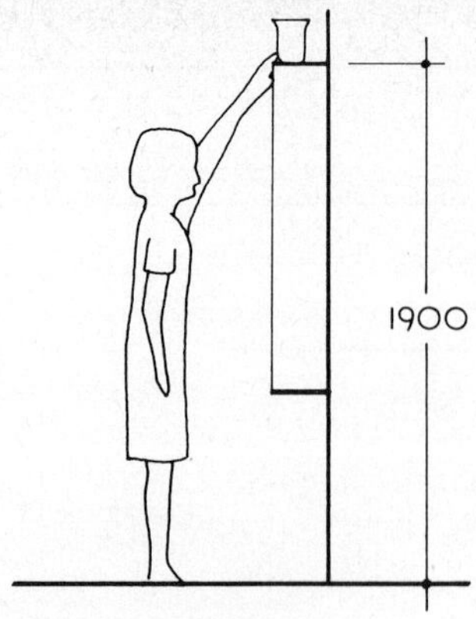

35.10 *Less often needed articles put higher*

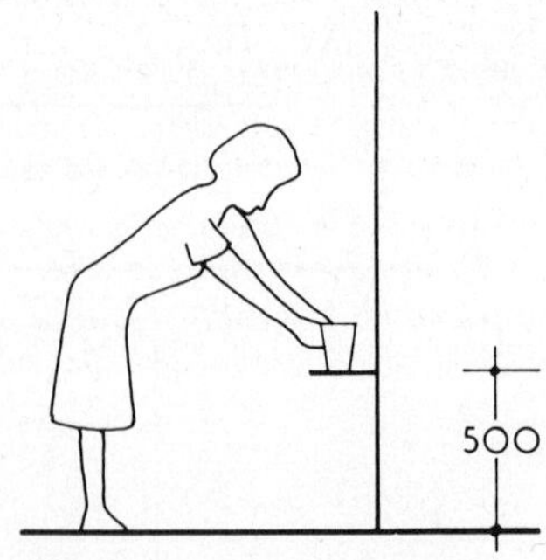

35.11 *Less often needed articles put low down*

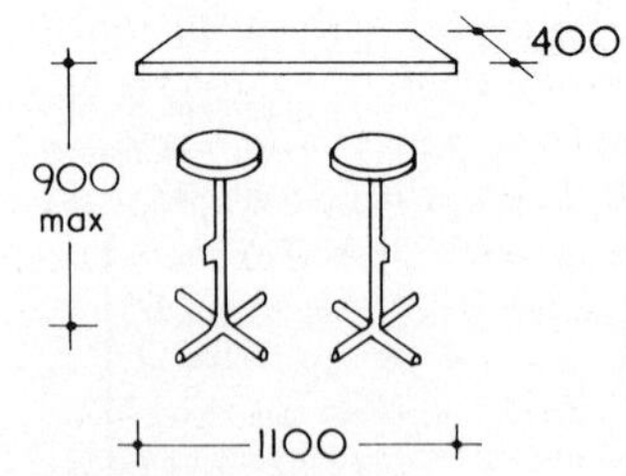

35.12 *Breakfast bar*

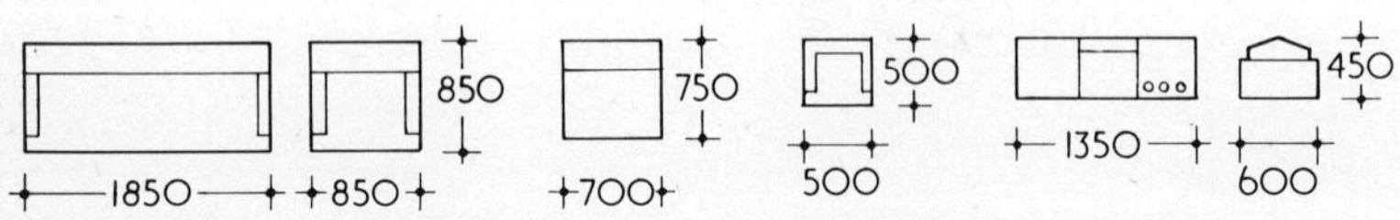

a *Three-seat settee*
b *Easy chair with arms*
c *Easy chair without arms*
d *Occasional chair*
e *Radiogram*
f *Television*

35.13 *Living room furniture*

4.05 Living areas

All dwelling plans must show the furniture drawn on.

Living rooms should be designed to accommodate the following furniture, 35.13, according to the number of people provided for:

- 2 or 3 easy chairs
- settee
- TV set
- small tables
- reasonable allocation of other possessions, such as a stereo deck, bookcase.

Eating areas in living rooms or dining rooms should be provided with a table, preferably extendable, at least 850 mm wide. Recommended sizes (before extension) are:

Three to four people: 850 × 1050
Four people: 850 × 1200
Five people: 850 × 1350
Six people: 850 × 1500
Seven people: 850 × 1800

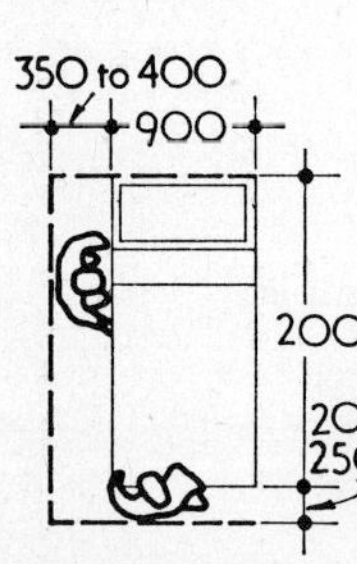

35.14 *Making a single bed*

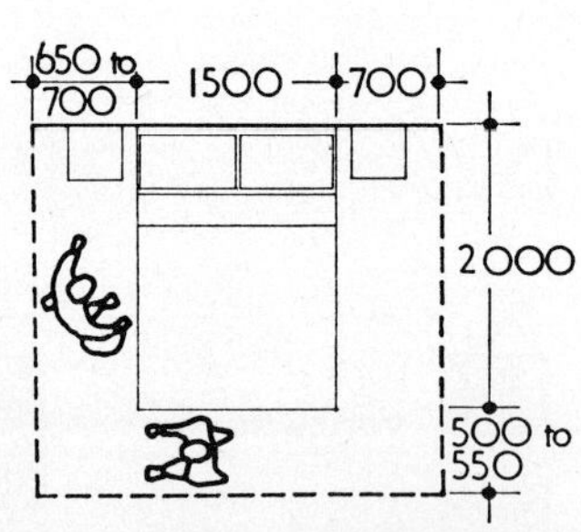

35.15 *Making a double bed*

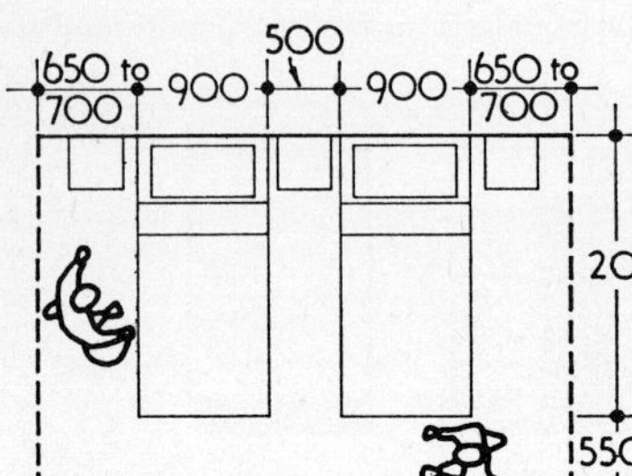

35.16 *Circulation around twin beds*

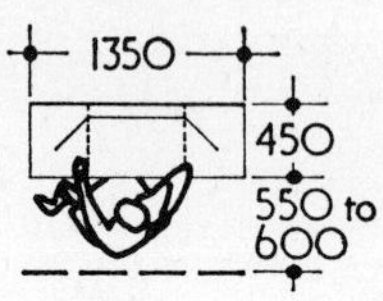

35.17 *Sitting at a dressing table*

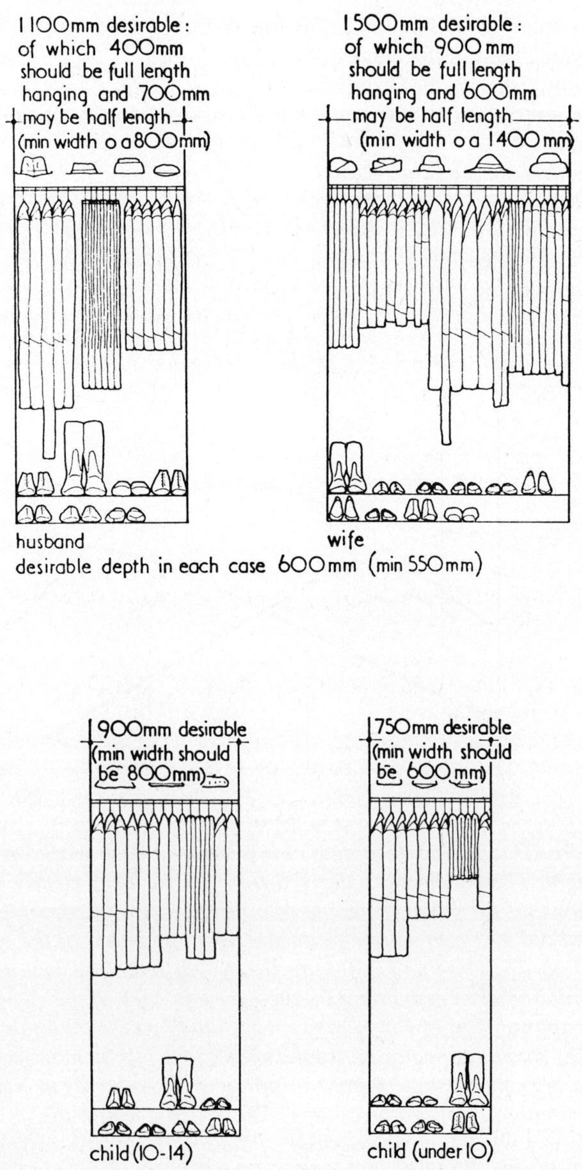

35.18 *Optimum hanging space for a family of four*

4.06 Bedrooms

Beds require space around them for making, **35.14** to **35.16** **35.17** shows the space needed for using a dressing table. **35.18** to **35.21** show clothing storage normally found in bedrooms. *Bedrooms should have sufficient space for a single bed and a bedside table for each occupant, although double bedrooms should allow for a double bed.* There must also be room for a wardrobe or built-in cupboard not less than 550 mm deep with 600 mm run of hanging space per person. In addition there should be chest of drawers in each room and a dressing table in a double bedroom or with easy access.*

* Double bedrooms capable of accommodating a double bed but not two single beds as an alternative may be permissible.

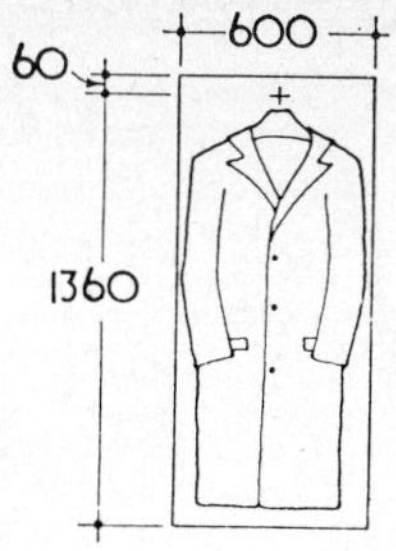

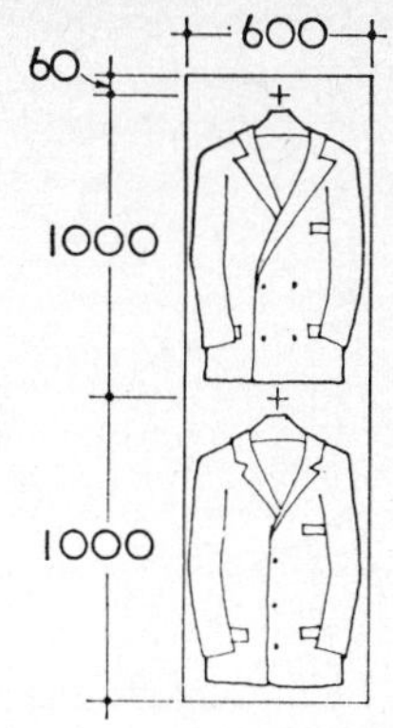

35.19 *Dimensions of wardrobes for full and half length hanging*

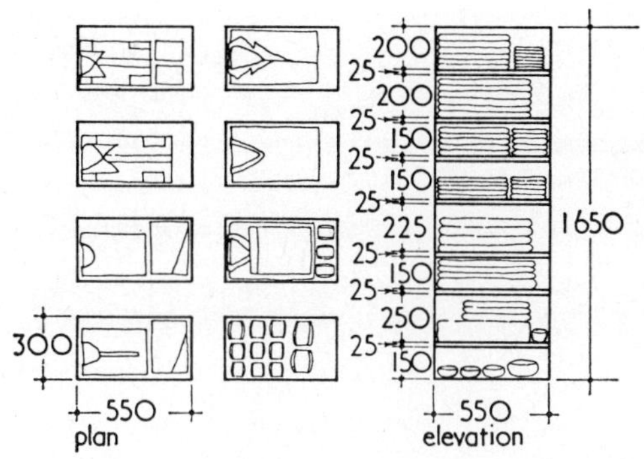

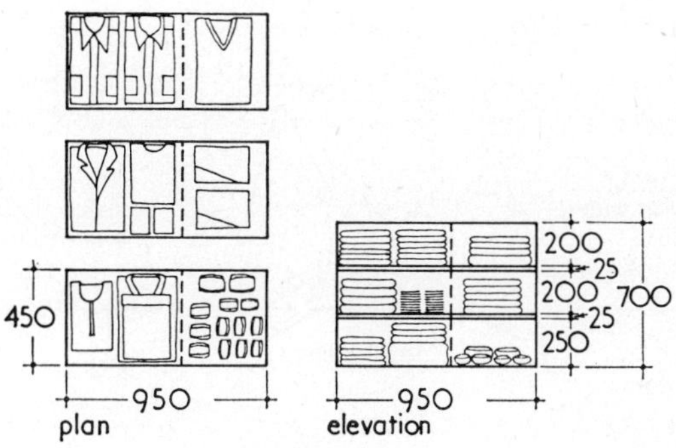

35.20 *Shelf or drawer storage for men's clothes*

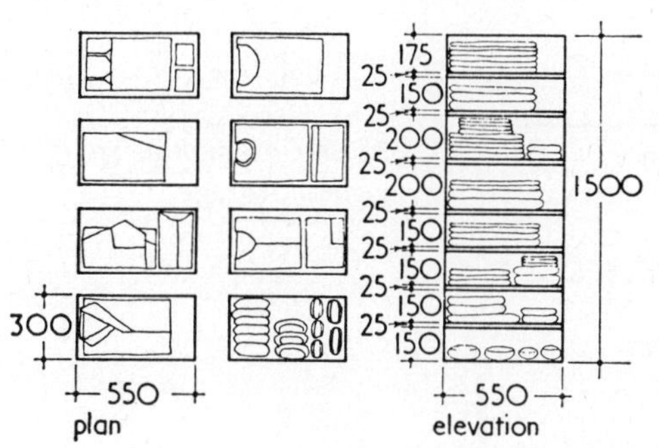

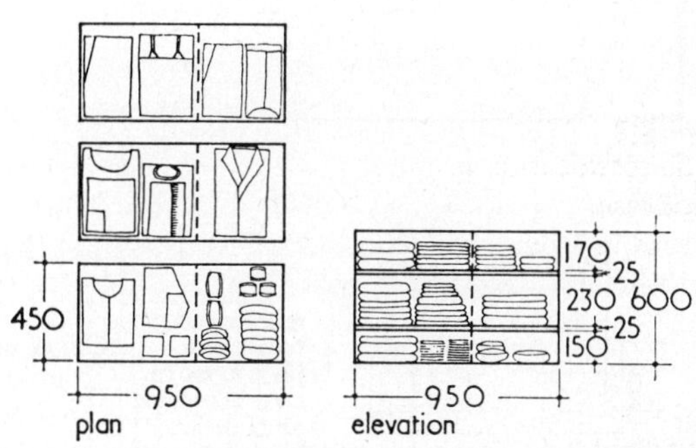

35.21 *Shelf or drawer storage for women's clothes*

4.07 Storage

An airing cupboard shall be provided giving 0·6 m² of clear storage space in four person dwellings and larger; or 0·4 m² in smaller dwellings, **35.22**.

The average five person household will need further storage for general household linen, **35.23**. This will normally be situated in a bedroom.

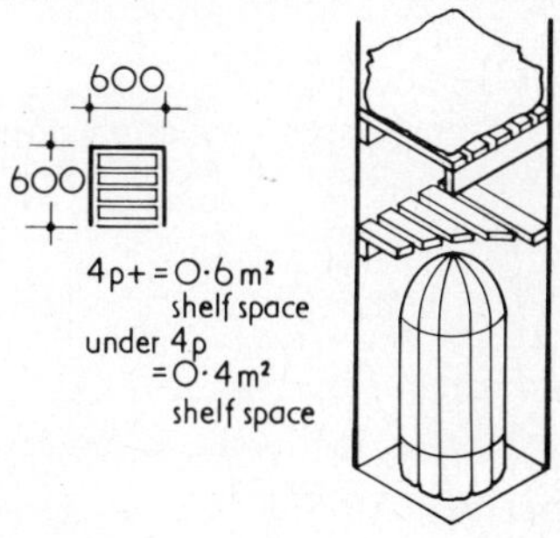

35.22 *Airing cupboard for linen including hot water storage cylinder*

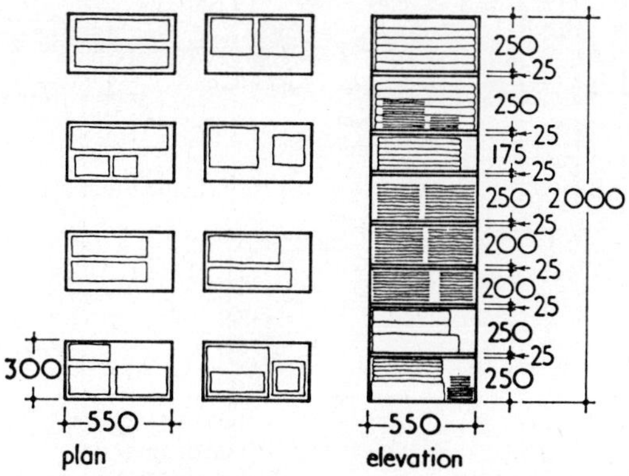

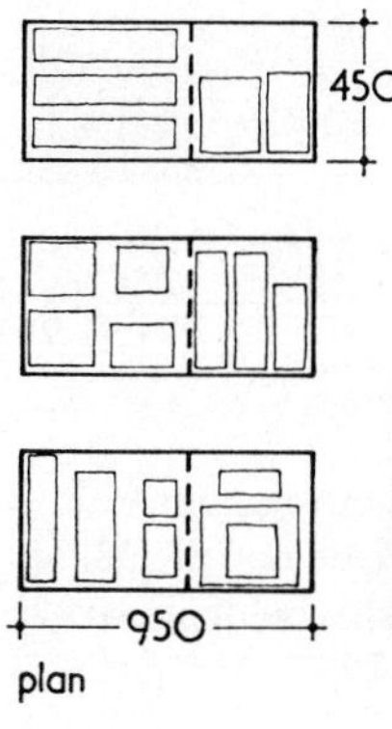

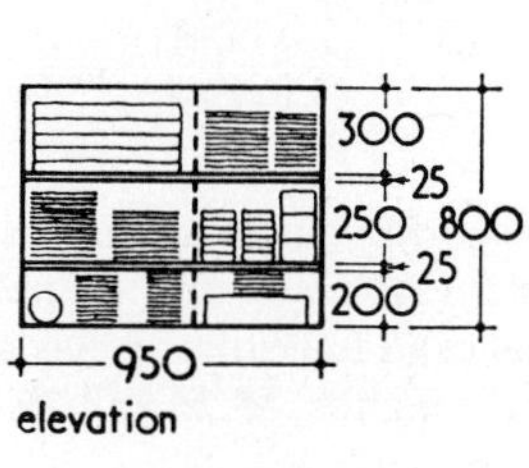

35.23 *Alternative storage arrangements for linen for a family of five*

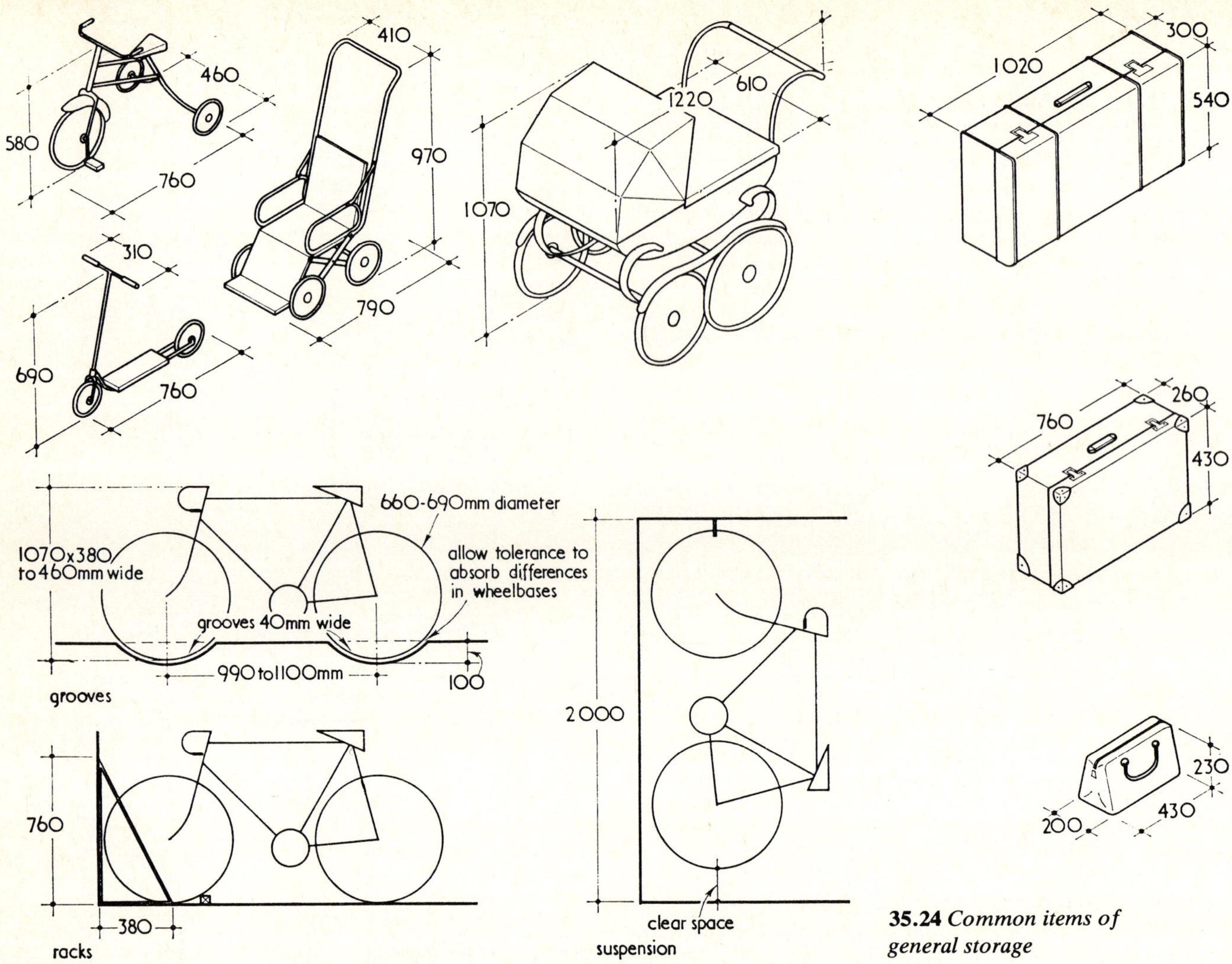

35.24 *Common items of general storage*

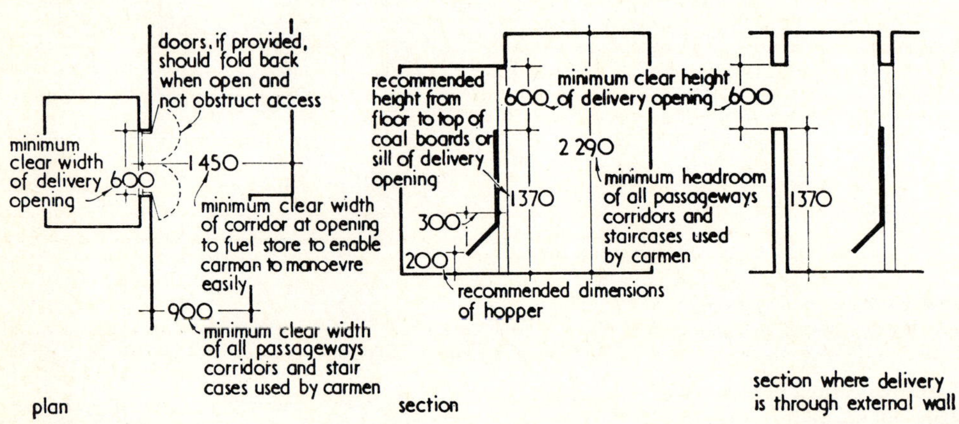

35.25 *Minimum dimensions of solid fuel storage areas*

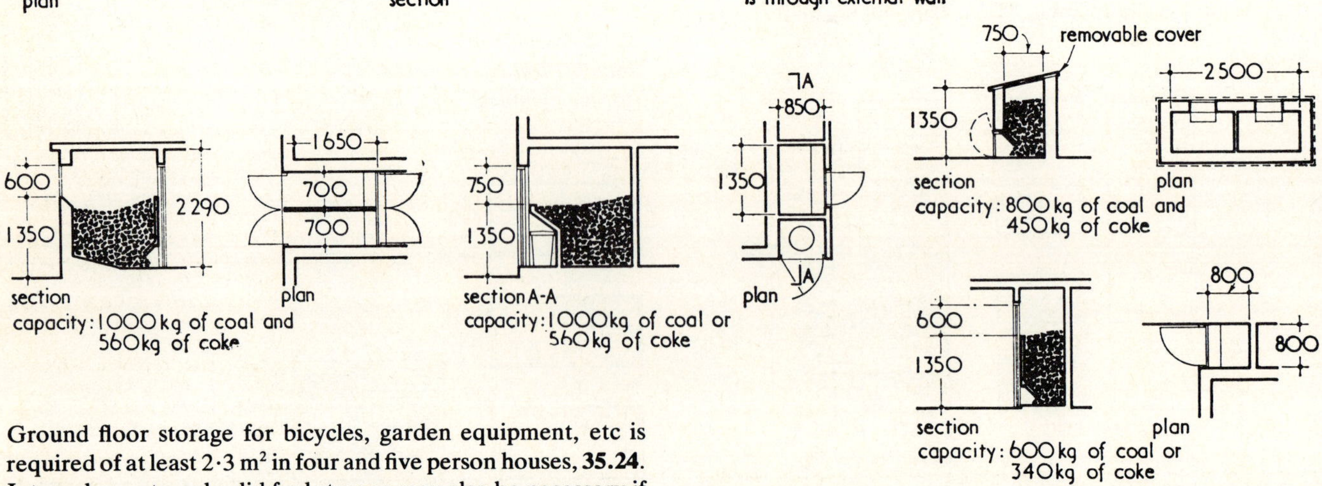

35.26 *Types of domestic solid fuel stores with capacities*

Ground floor storage for bicycles, garden equipment, etc is required of at least 2·3 m² in four and five person houses, **35.24**. Internal or external solid fuel storage may also be necessary if the heating system demands, **35.25** and **35.26**.

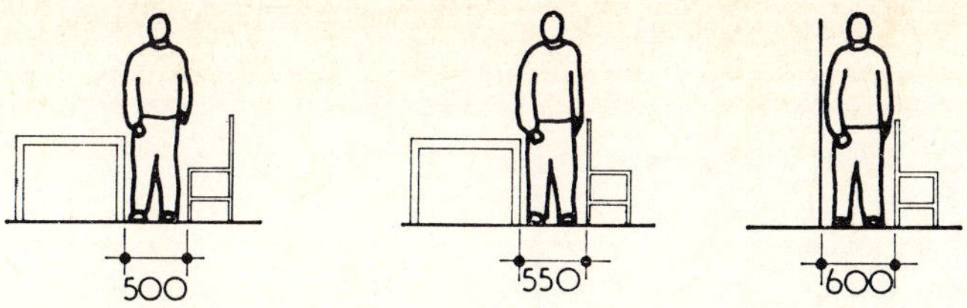

35.27 *Passing between two items of furniture, each table height or lower*
35.28 *Passing between two items of furniture, one table height or lower, the other higher or the wall*
35.29 *Passing between tall furniture and the wall*

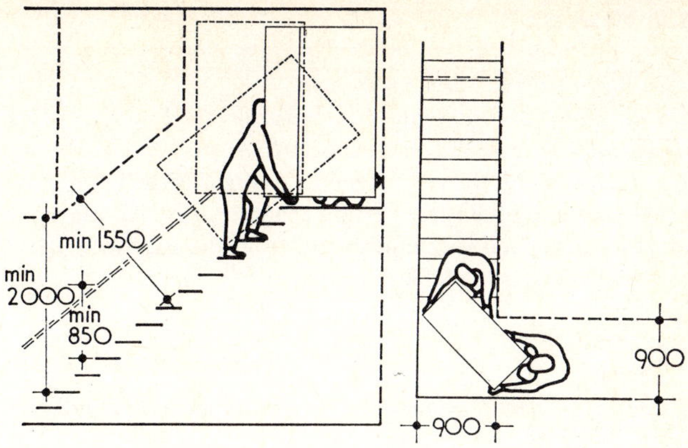

35.30 *Moving a double wardrobe up a staircase, showing minimum headroom, clearance, handrail height. Going 215 mm, rise 190 mm*

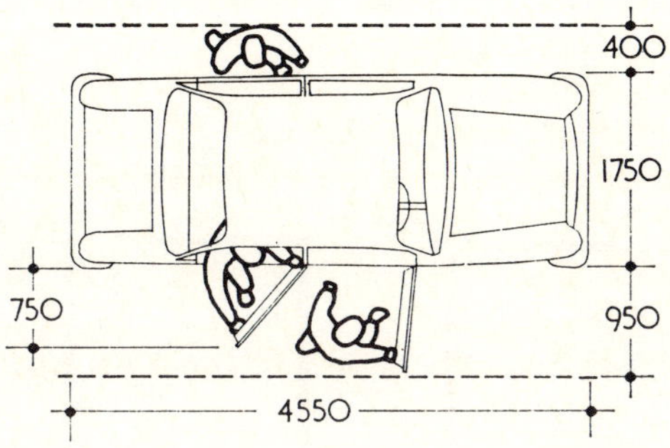

35.31 *Minimum required for getting in and out of a car*

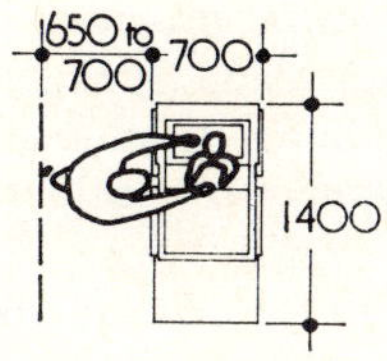

35.32 *Getting a pram ready*

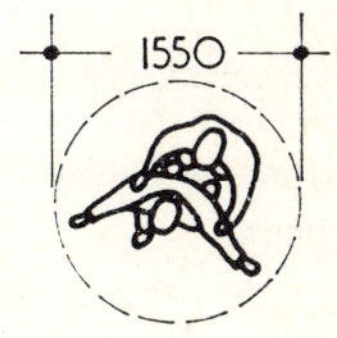

35.33 *Helping on with a coat*

4.08 Circulation

Spaces for general circulation should be adequate for their purposes, **35.27** to **35.33**.

4.09 Electric socket outlets

Table VIII gives the minimum and desirable scales of provision of socket outlets.

4.10 Space heating

The minimum standard shall be an installation with appliances capable of maintaining the kitchen and circulation spaces at 13 °C, and the living and dining spaces at 18 °C, when the outside temperature is −1 °C.

An installation also capable of heating the bedrooms to 18 °C is desirable.

Table VIII Provision of electric socket outlets

Part of dwelling	Desirable provision	Minimum provision
Working area of kitchen	4	4
Dining area	2	1
Living area	5	3
First or only double bedroom	3	2
Other double bedrooms	2	2
Single bedrooms	2	2
Hall or landing	1	1
Store/workshop/garage	1	–
Total for average family	20	15
Single study-bedrooms	2	2
Single bed-sitting rooms in family dwellings	3	3
Single bed-sitting rooms in self-contained bed-sitting room dwellings	5	5

Table IX Allocations of children's playground equipment*

Size of scheme	10–19 child bedspaces	20–49 child bedspaces	50–99 child bedspaces	100 or more child bedspaces
Space	At least 1 play area of 30 m² minimum	At least 1 play area of 50 m² minimum	At least 1 play area of 100 m² minimum	At least 1 play area of 150 m² minimum
Equipment* see list in para **4.11**	At least 1 piece of equipment from the list	1 At least 1 piece of equipment from the list for every complete 50 m² of provision 2 Each separate play area must have at least 1 piece of equipment from the list	1 At least 2 pieces of equipment from the list on the area of 100 m² and in addition a minimum of 1 piece of equipment for next complete 50 m² and every following complete 100 m² of provision 2 Each separate play area must have at least 1 piece of equipment from the list	1 At least 3 pieces of equipment from the list on the area of 150 m² and in addition a minimum of 1 piece of equipment for next complete 50 m² and every following complete 100 m² of provision 2 Each separate play area must have at least 1 piece of equipment from the list unless 1 area is a rectangular ball games area with surfacing and goal posts in a high enclosure 1 piece of equipment may be deducted from the required number for the first 50 m² and every following 100 m² of ball games area

* The requirements as regards minimum numbers of pieces of equipment may be fulfilled by the provision of more than one of the same item or items listed in para **4.11** (items 8 and 9 may be provided as long as one of items 1–7 is also included in the scheme)

4.11 Playspace

Allocation of space and equipment
*On all schemes containing 10 or more child bedspaces, playspace must be provided on the basis of 3 m² per child bedspace.**
Table IX sets out the requirements as regards the allocation of space and equipment for certain scheme sizes. All playspaces must be furnished with equipment from the following list:
1 swing (but not a plank swing)
2 slide
3 climbing frame
4 seesaw
5 merry-go-round or similar (but not an 'ocean wave')
6 rocking horse or similar
7 pendulum seesaw or similar
8 sandpit
9 paddling pool
Typical sizes for such equipment, and safety areas for them are shown in Section 40, Landscape.
The number of pieces of equipment to be provided varies according to the size of the play area provision (see table IX) but this does not preclude a local authority from providing additional different pieces of equipment if it considers this appropriate.
All play equipment must conform with the relevant British Standard.
All play areas must contain at least one seat or bench suitable for adults.

4.12 Family dwellings

35.34 to **35.36** show plans of typical family dwellings in the public housing field.

5 HOUSING FOR SINGLE PEOPLE

The following information has been extracted from DB 29.

5.01 Recommended areas and volumes

Dwelling areas are shown in table X; storage and fitting recommendations in table XI.

Table X Minimum area of dwelling (including storage) in square metres

Dwelling type	1 person		2 sharing		3 sharing	4 sharing
	Categories* a	b	a	b	a and b	a and b
Houses 1 storey	33	25	48·5	45	65	85
over 1 storey	—	—	—	—	—	90
Flats	32·5	25	47·5	45	65	85 and pro rata for larger shared flats
Flatlets (with bed-sitting room and shared bathroom)	28·9					
Flatlets (with one bedroom and private bathroom)			41·5			
Maisonettes	—	—	—	—	—	90

* Category a is: 'middle-aged permanent'. Category b is: 'young mobile'

* The number of child bedspaces in a scheme is to be calculated by subtracting all the bedspaces in old people's dwellings, all bedspaces in one and two person dwellings, and two bedspaces in family dwellings from the total number of bedspaces in the scheme.

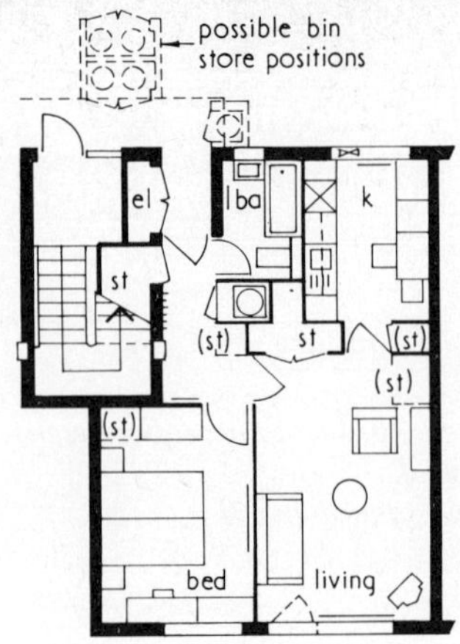

35.34 *Plan of a two person flat in three storey block*

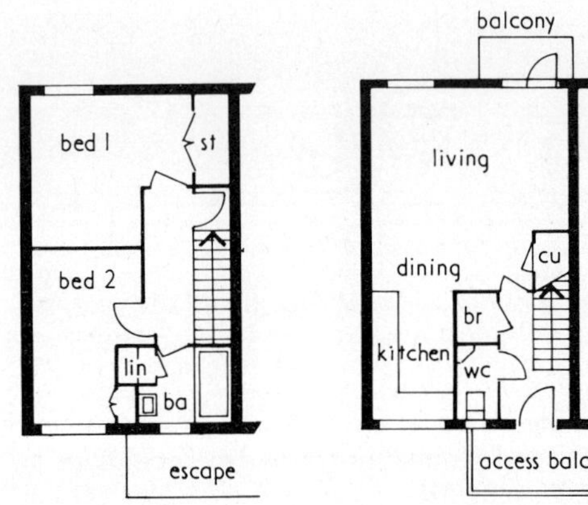

35.35 *Plan of a four person maisonette, with 2 two person bedrooms*

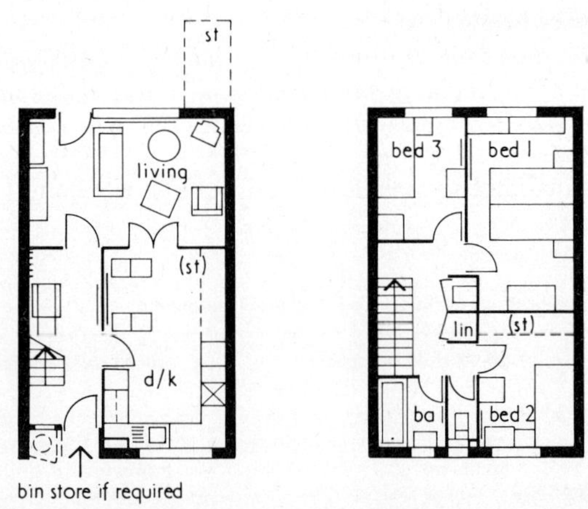

35.36 *Plan of a four person, two storey house, 4·8 m frontage*

5.02 Common room areas

Where common rooms are provided the areas should be at least:

- if up to 25 young mobile people are in the scheme: 20·0 m²
- for each additional young mobile person above 25 persons add: 0·4 m² per person
- for each older (middle-aged permanent) person in the scheme (provided there are at least 25 young persons in the scheme) add: 0·2 m² per person.

Table XI Minimum storage and other recommendations

Type of storage	Persons in the dwelling 1	2	3	4	For each person over 4 add to previous column
Personal storage (m^3)	3	6	9	12	3
Including shelves or drawers with an area not less than (m^2)	2	4	6	8	2
Dwelling storage (m^3)	0·5	0·5	0·5	0·5	—
Including shelves or drawers with an area not less than (m^2)	0·8	0·8	0·8	0·8	—
Kitchen storage (m^3)	1·4	2·1	2·8	3·5	0·7
Including shelves or drawers of an area not less than (m^2)	5	7	9	11	2
13 amp sockets	9	12	15	18	3
Bath or shower	1	1	1	1	
Washbasin	1	1	1	2	
wc (separate)	1*	1*	1	2*	

* One may be in bathroom
Areas of shelving or drawers are given in addition to volumes so as to avoid vast empty cupboards. Generally speaking the more shelves you can get into a storage unit the more you can store.

5.03 One person flats
35.37 and **35.38** show flats to Parker Morris standard for middle-aged tenants wishing to set up a permanent home, **35.39** and **35.40** show the smaller flat of 25 m^2 suitable for young mobile workers.

5.04 Shared two person flats
Alternative forms of two person flats are shown in **35.41**.

5.05 Four person flats
The greater choice of layout in four person flats is shown in **35.42**.

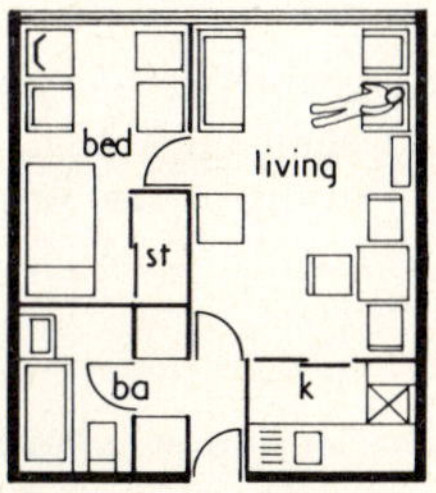

35.37 *Plan of single person flat, suitable for someone of middle age*

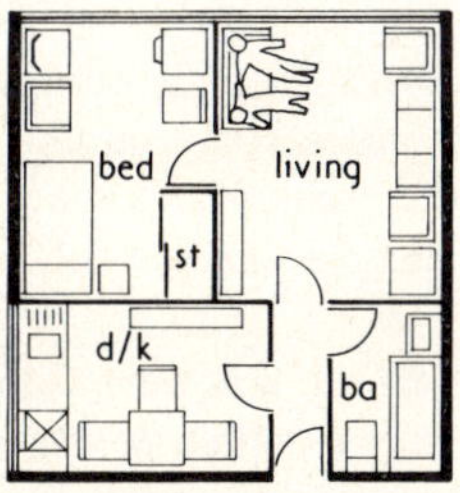

35.38 *Variation with kitchen/diner*

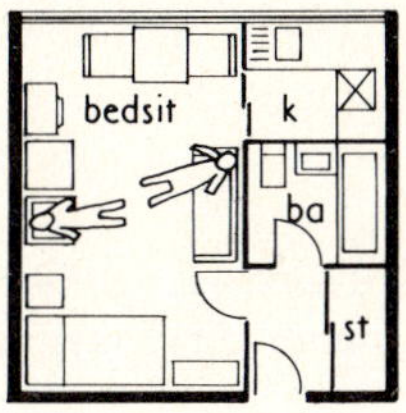

35.39 *Bedsitter flat, more suitable for a young person*

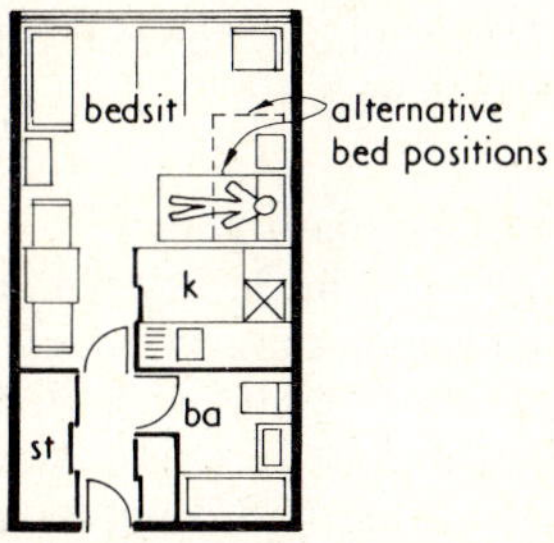

35.40 *Variation with narrower frontage*

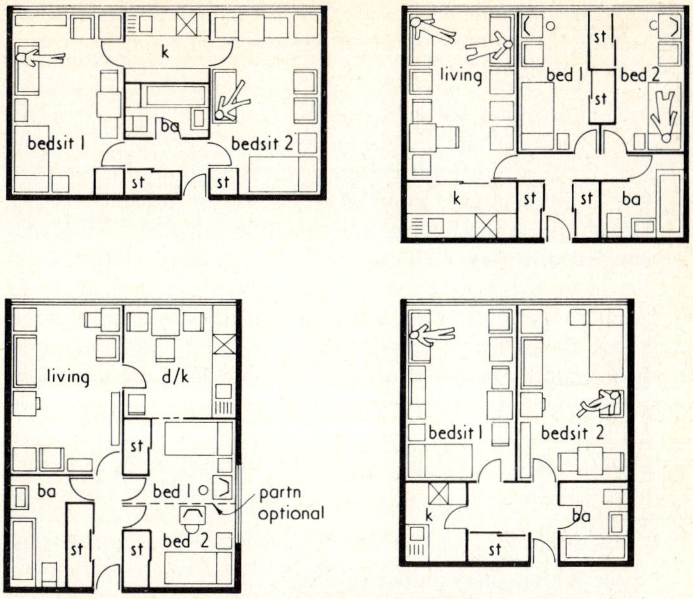

35.41 *Two person flats not intended for married couples. These are usually more popular than single person units for students and young people generally*

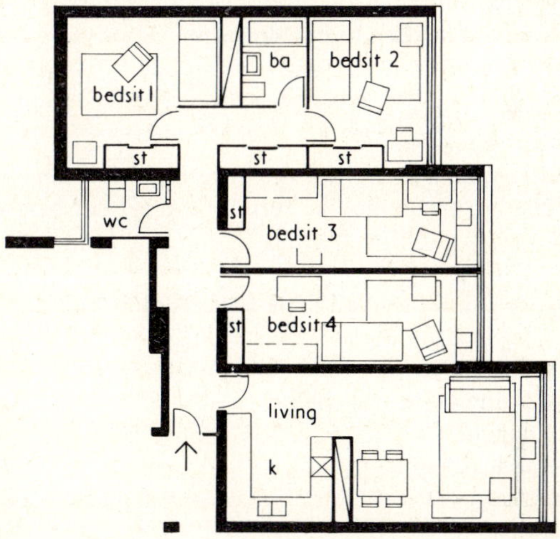

35.42 *Four person flats for sharing. These are likely to be the shape of housing for single people as they are more economical than other forms*

6 STANDARDS FOR OLD PEOPLE'S DWELLINGS (BASED ON APPENDIX 1 OF CIRCULAR 82/69)

6.01 Minimum standards
Mandatory minimum standards relate to all local authority housing for old people with special provisions within the dwelling taking the form of a high standard of heating, certain safety measures and other aids.

6.02 Dwelling categories
Dwellings for old people are of two types, known as Category 1 and Category 2:
(1) Self-contained dwellings for one or two of the more active old people designed to the standards set out in para **6.03**. Schemes involving this category of dwelling may include the optional extras set out in section J, paras 1, 2 and 4 of circular 82/69.
(2) Accommodation in grouped flatlets to meet the needs of less active elderly people and designed to the standards set out in para **6.03**. Such dwellings must as a minimum be provided

with the communal facilities set out in section J, para 3 and may include the optional extras set out in section J, para 4.

6.03 Plan arrangement

(*a*) All access stairs shall be enclosed.

(*b*) Where access to dwellings involves a climb of more than one storey from the point of pedestrian or vehicular access (whichever is the lower), a lift shall be provided. All access above four storeys high shall be enclosed and two lifts shall be provided. In blocks of five and six storeys a minimum of 12 old people's dwellings on the 5th and 6th floors is considered necessary to justify a second lift. Old people should not, therefore, be accommodated on the 5th and 6th floors of such blocks unless this condition is met.

(*c*) The dustbin or refuse disposal point and the fuel store (where one is provided) shall be conveniently accessible from every dwelling and shall be under cover and shall be capable of being lit at night.

(*d*) All dwellings shall have an entrance lobby or hall with space for hanging outdoor clothes.

(*e*) Kitchen working surfaces shall be provided on both sides of the sink and the cooker position. Kitchen working surfaces shall be provided and arranged in the following sequence: worktop/cooker/worktop/sink/draining board (or the same in reverse order).

(*f*) In two and three person dwellings sufficient space shall be provided in the kitchen to enable casual meals to be taken by two people.

6.04 Furniture

All dwelling plans must show the furniture drawn on and shall be designed to accommodate furniture as set out below:

Kitchen

In category 1 cases a small table unless one is built in.

Living space

A small dining table and chairs
2 easy chairs or
1 settee and 1 chair
A TV set
Small table
A reasonable quantity of other possessions such as a bookcase.

Bedrooms (single)

1 single bed
1 bedside table
1 small dressing table and chair
A built-in cupboard 600 mm (2 ft) wide or a space for a single wardrobe.

Bedrooms (double)

2 single beds (a double bed should be shown dotted)
2 bedside tables
A small chest of drawers
A small dressing table and chair
A built-in cupboard 1200 mm (4 ft) wide, or a space for a double wardrobe.

Bed recess

As for single rooms.

6.05 Kitchen fitments

General situations

Kitchen fitments shall comprise a minimum storage capacity of 1·7 m^3 in connection with:

- preparation and serving of food and washing up
- materials for cleaning and laundry operations
- the storage of food, including refrigerator with a minimum capacity of ·07 m^3 (2·5 ft^3) or a ventilated cool cupboard and space for a refrigerator.

NB

- Part of this provision shall comprise a broom cupboard which may be provided elsewhere than in the kitchen.
- The cubic capacity of kitchen fitments shall be measured overall for the depth and width and from the underside of working top to the top of the plinth for height.
- The maximum height for kitchen working surfaces shall be 850 mm. The maximum height for shelves shall be 1520 mm.
- In grouped flatlet schemes (category 2), gas or electric cookers large enough to cook a full meal for three people shall be provided of a design specially adapted for safety in use by old people.

6.06 Linen cupboard

A cupboard shall be provided with a minimum of two shelves neither lower than 300 mm nor higher than 1520 mm. The capacity of the storage area shall not be less than 0·4 m^3.

6.07 Electric socket outlets

The minimum provision shall be:

Kitchen	4
Living area	3
Bedroom	2
Hall or lobby	1
Bedsitting room	5

6.08 Space heating

The minimum standard shall be an installation with appliances capable of maintaining:

- the living area, bathroom, hall or lobby, bedroom,* kitchen, communal rooms (if any)—at 21 °C.
- circulation areas in grouped flatlet schemes at 15·6 °C when the outside temperature is −1 °C.

6.09 Space standards

Self-contained dwellings (category 1)

Standards: Flats or bungalows for one or for two old people which are self-contained (category 1) shall be designed to provide areas of net space and general storage space not less than those set out in the table V.

6.10 The WC and washbasin

(*a*) All WCs shall be provided with at least one hand-hold conveniently set at the side of the pedestal. (Example: inclined grip handle as illustrated in MOHLG Design Bulletin no 1.)

(*b*) Doors to WC compartments shall open outwards and shall be fitted with special locks openable from the outside.

(*c*) In grouped flatlets schemes (category 2), in one person flatlets a WC compartment with handbasin shall be provided. (NB In all other dwellings for old people the WC may be in the bathroom.)

6.11 The bathroom

(*a*) All baths shall be flat bottomed and of such a length that an old person cannot become completely immersed (a maximum standard length of 1550 mm).

(*b*) At least one hand-hold to assist an old person into or out of the bath shall be provided. (Examples: grip handles

* Except the second bedroom in a three-person dwelling. The temperature shall be controllable by the tenant.

incorporated in the design of the bath, or other aids as described in MOHLG Design Bulletin no 1).
(*c*) Doors to bathrooms, shall open outwards and shall be fitted with special locks openable from the outside.
In grouped flatlet schemes (category 2):
(*d*) Bathrooms for single person flatlets shall be provided on a shared basis of not less than one bathroom to four flatlets—some of the baths in these bathrooms may be sitz-baths.
(*e*) Each two person flatlet shall be provided with a private bathroom within the dwelling.

Showers
In any proposed group of not less than four bathrooms, one shower compartment may be provided in lieu of one of the proposed bathrooms. Where a shower compartment is provided the following shall apply:
(*a*) The floor of the shower compartment shall be non-slip and safe from hazards.
(*b*) The hot water output to feed a shower shall be thermostatically controlled to give a maximum output temperature of 49°C.
(*c*) The shower compartment shall contain a secure hand-hold and wall-mounted seat. Spray outlets shall be adjustable to varying heights.

6.12 Communal facilities

Self-contained dwellings with common rooms (category 1)
Where self-contained dwellings to Parker Morris standards are provided with common rooms, ie lounge, TV room, workshop or hobbies room, the following shall apply:
(*a*) The common rooms shall have a minimum floor space allowance of 0·95 m^2 per person. The route from the dwellings need not be covered but shall be kept as short as possible.
(*b*) At least one wc and handbasin shall be provided, located conveniently for the common room.
(*c*) A small pantry or tea kitchen with a sink and a means of heating water shall be provided adjacent to the common room.
(*d*) A space for hats and coats shall be provided.
(*e*) A small cupboard for cleaning materials shall be provided.
(*f*) A store adjacent to the common room with an area not less than 2 m^2 shall be provided.

Self contained dwellings with emergency alarm systems (category 1)
Where self-contained dwellings to Parker Morris standards are provided with emergency alarm systems, the systems shall communicate with a nominated person or be easily noticeable by those passing the dwelling.

Grouped flatlets (category 2)
Grouped flatlet schemes planned to areas set out in section G shall have the following provisions:
(*a*) A warden's dwelling, self-contained to the standards set out in Appendix I to circular 36/67 (Welsh Office Circular 28/67), except that where a scheme of fewer than 15 flatlets is associated with a residential home provided by the welfare authority, this requirement is waived if warden service is to be provided by the matron of the home and her staff.
(*b*) An emergency alarm system connecting each dwelling with the warden's quarters whereby tenants can communicate with the warden in cases of emergency. (In the case of the linked schemes referred to above, the alarm system will communicate with the residential home.)
(*c*) i A common room or rooms with a minimum floor space allowance of 1·9 m^2 per person
ii A wc and handbasin convenient to the common room
iii A small pantry or tea kitchen with a means of heating water close to common room
iv A space for hats and coats
v A store adjacent to the common room with an area of not less than 2 m^2
(*d*) A laundry room with not less than 1 sink, 1 automatic washing machine, 1 tumbler drier and a table or bench for folding clothes.
(*e*) A cupboard with a minimum capacity of 1 m^3 to hold cleaning materials for the communal areas.
(*f*) A telephone with seat adjacent for use by tenants.
(*g*) All accommodation shall be accessible by enclosed and heated circulation areas.
(*h*) Goods delivery shall be made possible from door to door preferably direct or by means of delivery hatches or alternatively to grouped lockers.

6.13 Guest accommodation and warden's office
In schemes of self-contained dwellings (category 1) a guestroom may be provided, and in grouped flatlet schemes (category 2) a guestroom and a warden's office may be provided. The guestroom should be situated near a communal toilet.

7 BUILDING FOR SALE

7.01 NHBC standards
A private developer need comply only with statutory requirements applicable to all building and is not obliged to conform to any codified set of space and amenity standards for residential development. But in order to provide house purchasers (and building societies) with a recognisable seal of quality backed by guarantee, the National House Building Council (NHBC) has developed a set of standards for private housing. The NHBC maintains a register of all housebuilders who meet its standards and offer all purchasers the NHBC House Purchasers' Agreement. The Council's standards are described in the *Registered House-Builders Handbook*, but they differ considerably in content and intention from Parker Morris standards. There are no floor space minima, but heating, kitchen layout, kitchen and linen storage, wc provision, and the number of electric socket outlets are included. Basic NHBC requirements are summarised in para **8**. General requirements and specifications relate to construction and finish and are revised to take into account the more common failures reported to the Council. Virtually all new houses offered for sale in the private sector are covered by the NHBC certificate. Dwellings built by local authorities for sale to owner-occupiers should at least comply with NHBC Standards and be built by registered builders under the national scheme.

7.02 Design guides
It is important also that design guides issued by local authorities for private developments should be used as guides only, and not as expedient checklists by designers or development control officers.

8 STANDARDS GENERALLY APPLIED IN PRIVATE SECTOR HOUSING FOR SALE
NHBC Standards are observed by most developers building for sale. The following details are based upon the *Registered House-Builders Handbook*, Part II, revised 1974.
Figures also appropriate to dimensional requirements in the private sector will be found in para **4**.

8.01 Internal planning

Measurement of floor areas
The area of one or more floors enclosed by the walls of the dwelling should be measured to unfinished surfaces and

include the space on plan taken up on each floor by any staircase, partitions and chimney breast, flue and heating appliance, and the area of any external wc. It should exclude the floor area of dustbin store, fuel store, garage or balcony, and any area in rooms with sloping ceilings where the height of the ceiling does not exceed 1·5 m and any porch, lobby or covered way open to the air.

8.02 Kitchens

Kitchens should be planned and equipped to include the following **35.43**:
(*a*) A sink and at least one drainer, not less than 1·0 m long overall with a minimum drainer area of 0·28 m^2.
(*b*) The drainer should not overlap a return.
(*c*) A cooker energy outlet.
(*d*) A cooker space of 510 mm for dwellings of up to 50 m^2 in area or 600 m for dwellings over 80 m^2, which should not be located beneath a window.
(*e*) A clear space of at least 100 mm between the cooker space and any return.
(*f*) The sink and the cooker space should be separated by at least 600 mm at a return.
(*g*) Work surface spaces not less than 500 mm in depth should be provided on each side of the sink and on each side of the cooker (the drainer may count as a work surface space).
(*h*) The total work surface space should not be less than 1·5 m long, including the drainer measured at the outside edge.

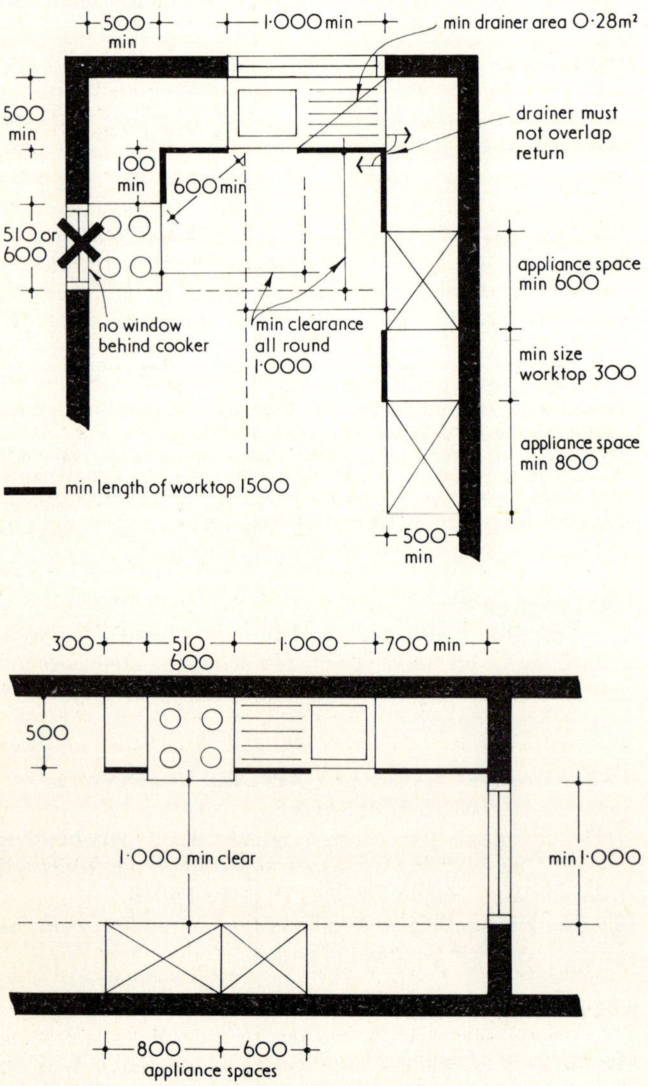

35.43 *Alternative forms of kitchen to NHBC standards*

(*i*) No work surface space should be less than 300 mm long.
(*j*) At least two appliance spaces should be provided, one being not less than 800 mm long, and the other not less than 600 mm long. (Appliance spaces may be provided in alternative suitable locations, eg space for a washing machine in a utility room).
(*k*) A clear space of at least 1·0 m in front of all fittings and spaces required under (*a*) to (*j*) above.

8.03 Storage accommodation

(*a*) Enclosed domestic storage accommodation should be provided as follows:

Area of dwelling (m^2)	Minimum volume of storage (m^3)
Less than 60	1·3
60–80	1·7
Over 80	2·3

(*b*) At least half the above volume should be in the kitchen with the remainder easily accessible to the kitchen.
(*c*) Part of the storage accommodation should be suitable for brooms and similar equipment.

8.04 Bedrooms

(*a*) Any bedroom should have an adequate bed space, not less than 2·0 m long, and be sufficiently wide to accommodate a single bed conveniently.
(*b*) The bed space should not overlap or otherwise restrict the use of any wardrobe cupboard provided by the builder.

8.05 Wardrobe cupboards

Should be of sufficient size (measured) either from front to back, or from side to side to accommodate clothing on hangers and contain an adequately supported hanging rail.

8.06 Airing cupboards

Every dwelling should be provided with an airing cupboard, containing the following:
(*a*) Not less than 0·5 m^2 of shelving.
(*b*) Shelf spaces not less than 500 mm high.
(*c*) The hot water storage cylinder or an equivalent source of heat.

8.07 Services

Sanitary fittings
(*a*) Every dwelling should be equipped with at least one sink, one bath, one washbasin and one wc.
(*b*) Any compartment, including an external compartment, containing a wc should also contain a washbasin, unless the compartment immediately adjoins a bathroom containing a washbasin.
(*c*) Dwellings equipped with only one wc should provide separate wc compartment unless the floor area of the dwelling is less than the following:

No of storeys in dwellings	Floor area (m^2)
2 or more	80
1	75

Water services
(*a*) Every dwelling should provide a cold and a hot water service.
(*b*) A drinking point must be provided off the incoming main in the kitchen.
(*c*) The cold water system may be either a storage or a mains pressure type.

(*d*) In the case of a conventional storage system, the actual volume should, unless prohibited by the local water undertaking, be not less than 115 litres if for cold storage only, or 230 litres if for cold storage and hot feed.
(*e*) The hot water system may be of either a storage or an instantaneous type.
(*f*) In the case of a storage system, the temperature at the outlet of the storage cylinder should be not less than 60°C under normal operating conditions.
(*g*) Hot water storage capacity in a conventional system should be not less than 115 litres.
(*h*) The full volume of the cylinder should recover the temperature of 60°C as follows:

Fuel	Recovery time (hrs)
Continuous burning solid fuel	3
All others	2

(*i*) Boilers should have an output of at least 3 kW available for domestic hot water.
(*j*) Outlets should be of not less than the following sizes:

Fitting	Outlet diameter (mm) Cold	Hot
Sink	13	13
Bath	20	20
Washbasin in internal compartment	13	13
Washbasin in external compartment	13	Not mandatory
WC	13	Not mandatory
Shower	Shall have a water supply of adequate pressure	

(*k*) Insulation should be provided to cisterns located in roof voids, hot water cylinders; pipes located in roofs and hollow ground floor voids, and underground pipes where cover is less than 750 mm.

Heating to main living room
In non-centrally heated dwellings, the main living room should be equipped with a fixed appliance capable of producing heat at a rate of at least 42 W/m^3 with a minimum capacity of 2 kW.

Central heating
The provision of heating in addition to that required above is at the option of the builder. Where central heating is to be provided, the purchaser shall be advised which of the following standards is to be provided. In no case shall the heating to the main living room be below the standard required above.

Whole house heating—Grade 1

The system should achieve not less than the internal temperatures given in the table below when the external temperature is −1°C and where two air changes per hour are assumed.

Room	Air temperature (°C)
Living room	21
Dining room	21
Kitchen	18
Hall and landing	16
Bedrooms	16
Bathroom	21
WC	16

Whole house heating—Grade 2

The system should achieve not less than the internal temperatures given in the table below when the external temperature is −1°C and where two air changes per hour are assumed.

Room	Air temperature (°C)
Living room	17
Dining room	17
Kitchen	13
Hall and landing	10
Bedrooms	10
Bathroom*	10
WC	

* Where Grade 2 applies, these compartments may be deemed to be heated by infiltration.

Background heating—Grade 3
Where the builder is to provide heating standards lower than either of the above standards for whole heating, the purchaser should be advised which rooms are to be heated and to what standard. The minimum standard should not be lower than the temperatures below when the external temperature is −1°C.

Room	Air temperature (°C)
Living room	13
Dining room	13
All others	10

Gas service
The provision of a gas service is at the option of the builder.

Electricity service
The following outlets must be provided:
(*a*) Every room should have at least one lighting outlet.
(*b*) 13A socket outlets should be provided as indicated below.

Room	Outlets*†	Notes
Kitchen	4	Two may be in dual unit. One may be in cooker control unit.‡ A maximum of one outlet may be conceded if the builder provides a wired-in appliance
Dining room	2	May be in a dual outlet
Living room	3	Two may be in a dual unit. One shall be near the television outlet.
Living/dining room with distinctive areas	5	Four may be a dual unit
Each bedroom	2	May be in dual unit in third and subsequent bedrooms if floor area is less than 7·5 m^2
Spare or box room	1	
Nursery	1	
Study	1	
Landing**	1	
Hall**	1	'Hall' shall mean 'any space giving access to two or more rooms' and, except in single-storey dwelling the staircase

* All socket outlets should be well separated.
† Socket outlets on walls should be at least 150 mm above floor or work surface level as appropriate (measured to bottom of socket plate).
‡ Where a cooker control panel is provided, it should be located to the side of the cooker position. Final sub-circuits for cookers must be not less than 30A rating.
** There should be adequate provision for lighting to halls, landings and staircases with two-way switching provided to staircases.

(*c*) Unless the area of the dwelling is less than 60 m^2, at least two final sub-circuits for lighting should be provided.
(*d*) Immersion heaters, where provided, should have a switch with the on/off position clearly indicated.
(*e*) Provision should be made for the installation of television in the main living room. A conduit and draw wire should be provided from the roof to the outlet or alternative provision should be made.

8.08 Insulation
Sound insulation should be at least to the mandatory standards in Building Regulations and in addition provide:
(*a*) An average sound reduction index of not less than 35 dB over the frequency range 100 to 3150 Hz when tested in accordance with BS 2750 for any partition separating a WC compartment from a living room or bedroom.
(*b*) Any soil pipe passing through a living room or bedroom should be encased and insulated so as to minimise sound nuisance.
Thermal insulation should meet Building Regulations.

8.09 Access

Access to lofts
(*a*) Every roof void should be provided with an access door.
(*b*) Access openings should be not less than 550 mm wide in any direction.
(*c*) Access openings should not be located directly over stairs or in other hazardous locations.
(*d*) Gangway boarding should be provided from the access opening to each cistern.
(*e*) At least 1 m^2 of boarding should be provided around each cistern.

8.10 External facilities

External access
(*a*) A drive should be provided to connect any garage, carport or paved standing area within the cutilage of the dwelling to the highway.
(*b*) The gradient of any drive should permit an average family saloon car unimpeded access.
(*c*) Paths should connect the highway to the main and kitchen entrances of every dwelling except mid-terrace dwellings where the kitchen is at the back of the house.
(*d*) A path should connect the dwelling to any garage, carport or paved standing area within the curtilage of any dwelling.
(*e*) The minimum width of any path required under (*c*) or (*d*) should be 600 mm, or, if it adjoins a building, the minimum width should be 700 mm.
(*f*) Steps should be incorporated in any path so that no part slopes at a gradient of more than 1:6.
(*g*) Steps or other suitable means of access should be provided to any inaccessible part of the garden.
(*h*) Where the total rise in any flight of steps required under (*f*) or (*g*) above exceeds 600 mm, or where any path required under (*c*) or (*d*) above adjoins a vertical difference in levels of more than 600 mm, a handrail or balustrade should be provided.

Area immediately surrounding the dwelling
(*a*) Precautions should be taken to prevent water logging.
(*b*) Field drainage systems should connect either to surface water drains or drain away in a suitable manner.
(*c*) A flat area not less than 1·0 m wide shall be provided around the dwelling to give free access on sloping sites only.
(*d*) Ground or path levels should be at least 150 mm below dpc.
(*e*) Ground or paths adjoining the dwelling should slope away to a slight fall.
(*f*) Soil should be kept clear of airbricks and the like.

Garden area
(*a*) Old foundations, concrete bases and similar obstructions occurring within 500 mm of the ground surface must be removed.
(*b*) Disturbed ground should be reshaped to conform with the general shape of adjacent ground.
(*c*) Sub-soil must not be left on top of vegetable soil.
(*d*) Vegetable soil disturbed must be reinstated or replaced. (The provision of further vegetable soil is not required.)
(*e*) Rubbish and debris must be removed.
35.44 to **35.46** show typical private sector house and bungalow plans.

9 TRANSPORTATION OF PREFABRICATED HOUSING UNITS

Proposals to use volumetrically prefabricated units in a housing scheme should recognise that the transport by road of wide loads is subject to control under the Road Traffic Act 1972.

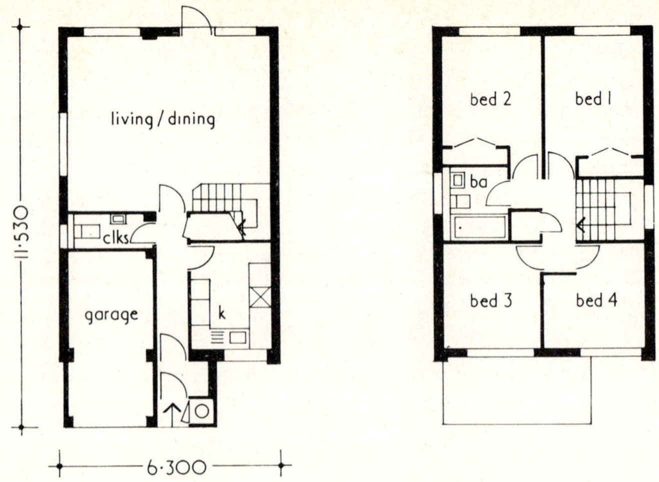

35.44 *Plan of four bedroom two storey private sector house: Costain Homes Ltd*

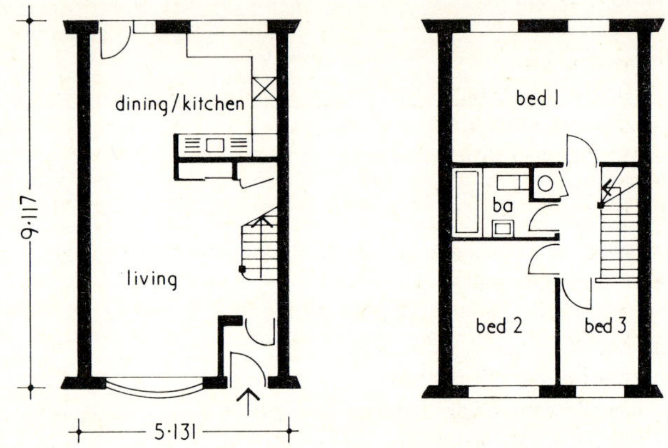

35.45 *Plan of three bedroom terrace two storey house: Peter Warwick, Architect*

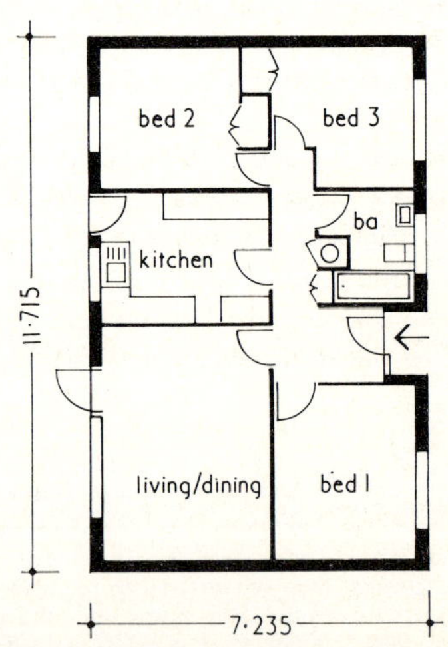

35.46 *Plan of three bedroom bungalow: Costain Homes Ltd*

Control is imposed by Statutory Instrument No 1101, 1973 and is related to the width of the load:
- for loads up to 4·3 m wide, provided certain conditions and instructions specified in the SI are met, no reference to the Department of Transport is required.
- the movement of loads between 4·3 and 6·1 m wide requires the written authority of the Secretary of State but no particular route is specified.
- loads more than 6·1 m wide require a Special Order from the Secretary of State and a mandatory, agreed route is given.

Department of Transport policy regarding loads over 4·3 m in width is to authorise moves only where the manufacturer is unable to design a unit within the width limitation or in sections for assembly on site and to confine authorisation to those cases where no alternative means of transport are available, or the number of journeys is small, or the journey is short.

Some manufacturers of factory-built houses have been advised that the Department is not prepared to authorise the transport of loads of this nature and they have been recommended to redesign their unit so that its width does not exceed 4·3 m.

The onus for obtaining authorisation for the transport of wide units is upon the manufacturer. When housing proposals involve this type of unit written evidence should be obtained from the manufacturer that he has obtained agreement in principle to the movement by road.

10 BIBLIOGRAPHY

Many of these publications are referred to in table I.

A very helpful *Housing circulars and publications: index to current departmental circulars and publications* (as at 30.6.77) is available from HMSO

New construction

82/69 Housing standards and costs: accommodation specially designed for old people
27/70 Metrication of housebuilding: progress
79/72 Children's playspace
10/73 Planning and noise
74/74 Housing for people who are physically handicapped
92/75 Wheelchair and mobility housing: standards and costs
61/75 The housing cost yardstick
24/75 Housing needs and action
8/76 Housing co-operatives
12/76 Housing for single working people
18/77 Housing capital expenditure

Improvement/renewal

160/74 Housing Act 1974: improvement of older housing
13/75 Housing Act 1974: renewal strategies
14/75 Housing Act 1974 Parts IV, V, VI: Housing action areas, Priority neighbourhoods and General improvement areas
14/76 Housing Act 1974: housing action areas, priority neighbourhoods and general improvement areas
38/77 Housing Act 1974: house renovation grants

Housing associations

170/74 Housing Act 1974: Housing corporation and housing associations

Procedures

Code of procedures for local authority housebuilding, London, HMSO, 1975

DOE bulletins and papers

More detailed advice on design is contained in Housing Development Directorate bulletins, notes and papers. These generally refer to new local authority housing. Bulletins are available from HMSO. Notes and papers are available, free from DOE.

DB13 *Safety in the home*, metric ed 1976
DB14 *House planning: a guide to user needs with a checklist*, 1968
DB24 *Spaces in the home*
Part 1 *Bathrooms and WCs*, 1972
Part 2 *Kitchens and laundering spaces*, 1972
DB25 *The estate outside the dwelling: reactions of residents to aspects of housing layout*, 1972
DB26 *New housing and road traffic noise: a design guide for architects*, 1972 reprinted 1974
DB27 *Children at play*, 1973
DB29 *Housing single people 2: a design guide with a description of a scheme at Leicester,* 1975
DB30 *Services for housing: sanitary plumbing and drainage*, 1974
DB31 *Housing for the elderly: the size of grouped schemes*, 1975
DB32 *Residential roads and footpaths: layout considerations*, 1977

Housing Development Notes

Housing on a dimensional framework:
Part 1 *Objectives*
Part 2 *Spaces for components*
Part 3 *Accuracy*
Part 4 *Dimensions for joints*
Part 5 *Drawings*
Part 6 *References*
Part 7 *Choice of brick size*
Part 8 *Brickwork details (65 mm height bricks in section)*
Part 9 *Brickwork design* (90 *mm wide bricks), with reference to the building regulations*
Landscape of new housing:
Part 1 *Background*
Part 2 *Trees in housing*
Part 3 *Shrubs in housing*
Part 4 *Grass and other small plants*
Thermal insulation in housing:
Part 1 *The case for better insulation*
Part 2 *Relationship between construction and heating costs*

Water services for housing

Part 1 *Performance requirements for cold water services*
Part 2 *Cold water installations*
Part 3 *Performance requirements for domestic hot water systems*
Part 4 *Domestic hot water installations*

House shells

Part 1 *Feedback studies—four-, five- and six-person houses*
Part 2 *Dwellings for small households: three-person two-storey houses*

Parking in new housing schemes

Parts 1 and 2

HDD Occasional Papers

2/73 *Local authority housing: a comparative study of the land use and built form of* 110 *schemes*
3/73 *Vandalism: a constructive approach*
1/74 *The quality of local authority housing schemes*
2/74 *Mobility housing*
1/75 *The social effects of living off the ground*
2/75 *Wheelchair housing*
1/76 *Housing the elderly: how successful are granny annexes?*
2/76 *Children's playgrounds*
3/76 *A survey of standards of design and specification in new local authority housing*
1/77 *Local authorities and building for sale*

Area Improvement Notes

Published by HMSO
Area Improvement Note 1
Sample house condition survey, 1971
Area Improvement Note 2
House condition survey, 1971

Area Improvement Note 3
Improving the environment, 1971
Area Improvement Note 4
House improvement and conversion, 1972
Area Improvement Note 5
Environmental design in four GIAs, 1972
Area Improvement Note 6
The design of streets and other spaces, 1973
Area Improvement Note 7
Parking or garaging in GIAs, 1972
Area Improvement Note 8
Public participation in GIAs, 1973
Area Improvement Note 9
Traffic in GIAs, 1974
Area Improvement Note 10
The use of indicators for area action
Area Improvement Note 11
Networks in area improvement

Area Improvement Occasional Papers
Occasional Paper 1/75 *Aspects of housing strategy, with reference to the use of census indicators*
Occasional Paper 2/75 *Gradual renewal*
Occasional Paper 3/75 *Research related to the renewal of older housing areas*

Housing Corporation Papers
HCP/22R *Rehabilitation of existing property* Technical brief for housing associations and their consultants.

Private sector housing
Certain design standards are conditions of the National House Building Council guarantee arrangements. They are contained in the NHBC handbook, and referred to in NHBC Advisory Note 1 *Guide to internal planning.* These are available from the National House Building Council, 58 Portland Place, London, W1N 4BU.
Further design recommendations for private sector housing are contained in:
Domestic interior planning, and *Houses of today and tomorrow*, published by the National Council of Women; *New housing in SE England: purchasers' likes and dislikes,* and *New housing in the Midlands: purchasers' likes and dislikes*, published by the Housing Research Foundation.

Planning standards
Architects should consult individual planning authorities to discover what design requirements are published, eg *Design guide for residential areas* Essex CC.

36 Hotels

CI/SfB 85
CI/SfB (1976 revised) 852

Contents

1 BASIC LAYOUTS

1.01 Bedrooms and public areas

Various relationships between the two sorts of accommodation—bedrooms and public areas—are shown diagrammatically in **36.1** to **36.4**.

1.02 Height

Choice between a high or low building depends on site value and any limitations imposed on the development. In a rural setting of relatively low land value, planning and construction costs usually favour a low building. There are substantial savings in foundation and structural work; lifts are not obligatory up to three storeys and there may be economies in the use of pumping equipment. Maintenance is usually easier and cheaper. But long corridors with more staircases are needed and there is an appreciable increase in engineering service costs—heating, insulation, plumbing and drainage.

In an urban location an expensive site may be justified only by a tall building. Generally, a compromise must be reached taking into account planning limitations on the average numbers of floors which may be constructed over the plot, rights of light and other legal restrictions protecting nearby property. The most common arrangement is a tall bedroom block over a much larger area of low public rooms, **36.4**.

High costs of site purchase in a city centre may be recompensed by letting high value frontage space at ground floor level, and above, as shops and other commercial activities unconnected with the hotel. It may be advantageous to provide hotel accommodation above office or shop development, on the upper floors of a tall building. In a provincial town having limited demand for hotel accommodation, a combined residential development including self-contained flats might be commercially attractive.

1.03 Apartments

The bedroom areas are formed from relatively small cell-like

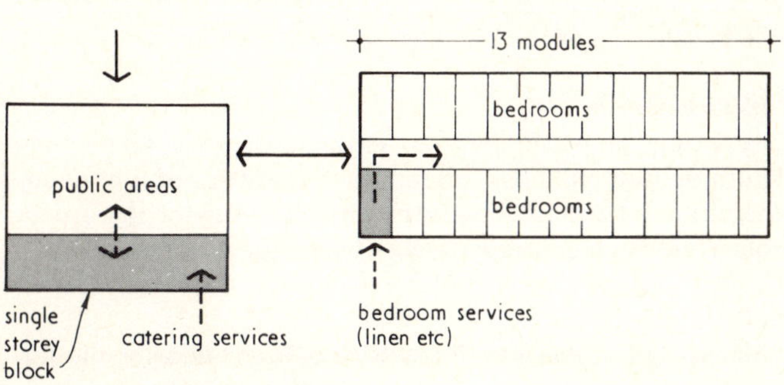

36.1 *Simple bedroom block with detached single storey catering building as in primitive motels. No cohesion or rationalisation of circulation*

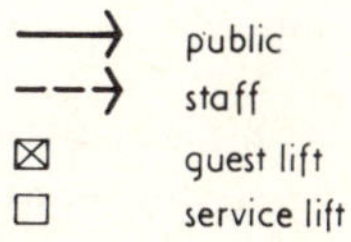

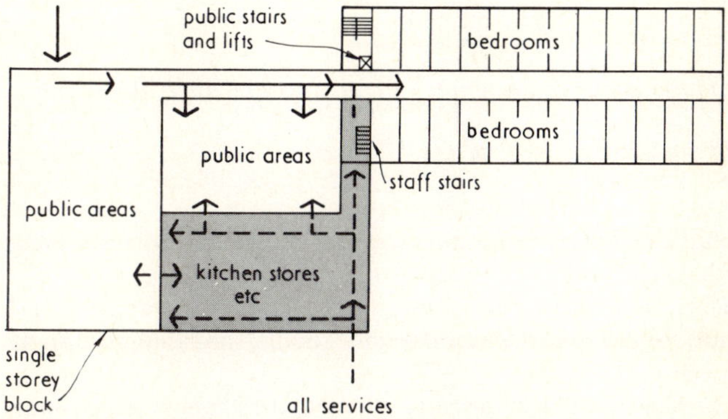

36.3 *Enlargement of* **36.2**. *More public and functions rooms (still single storey) which may be appropriately subdivided—but guest and staff circulations should not mix*

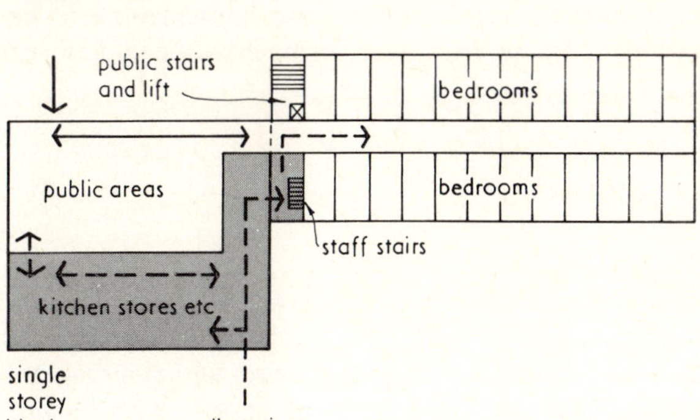

36.2 *Simple rationalisation of circulation. All bedroom and catering services collected in controllable zones and routes. Public areas are single storey*

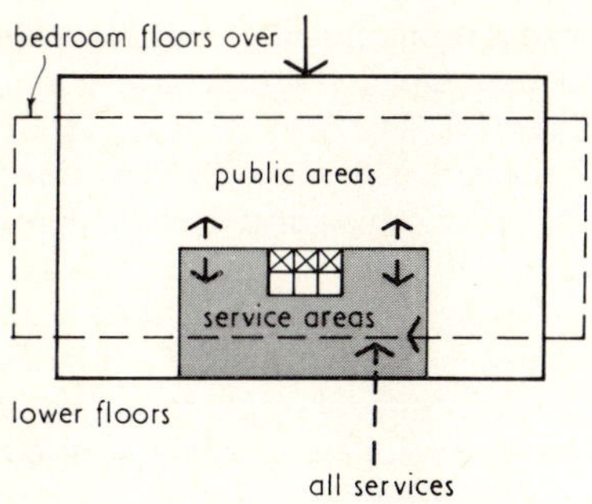

36.4 *Same circulation principles apply where bedrooms are built over public areas. Much simpler to plan on vertical than horizontal circulation. Note grouping of lifts to serve guests on one side and staff on the other*

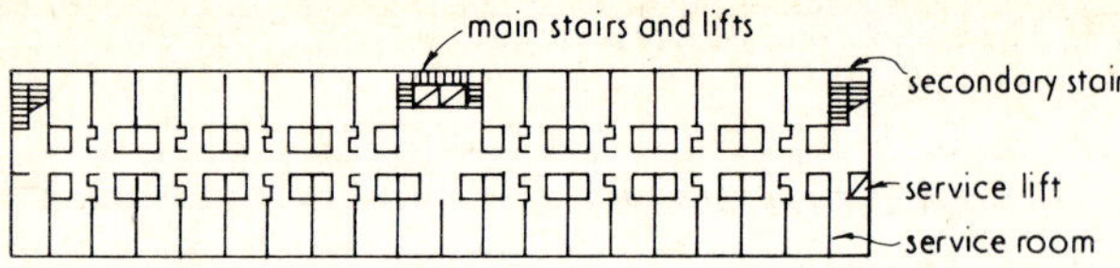

a *Linear arrangement of rooms*

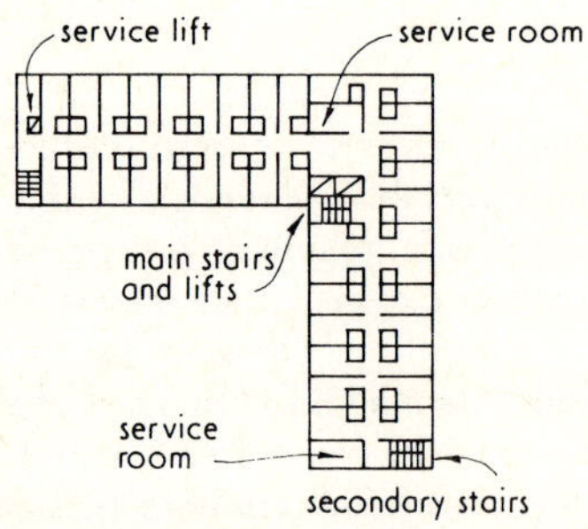

b *L-shaped room arrangement*

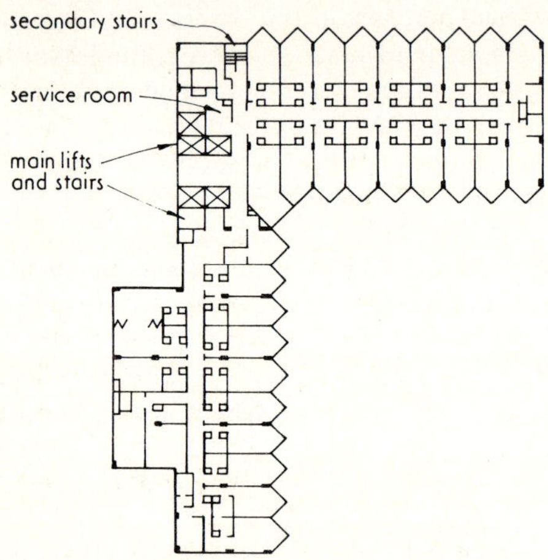

c *Good American example of L-shape: Chicopee Motor Inn*

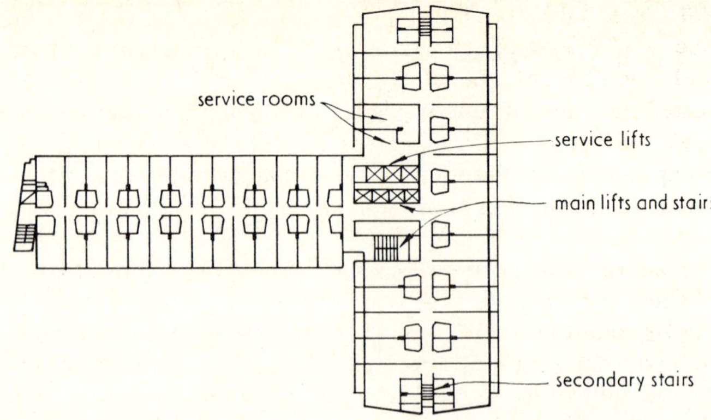

d *T-shaped arrangement: Royal Garden Hotel, London*

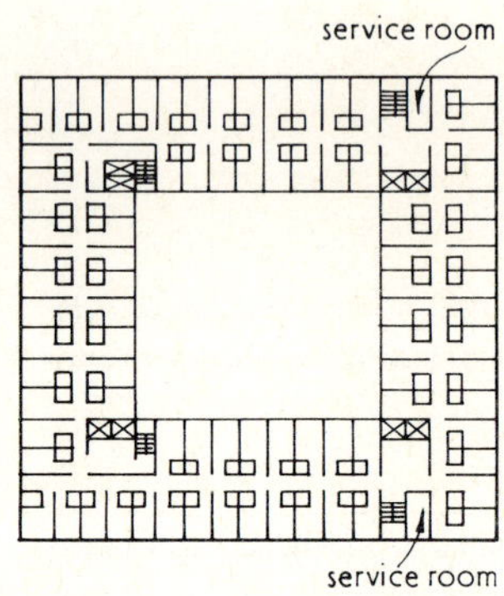

e *Rooms around a square court*

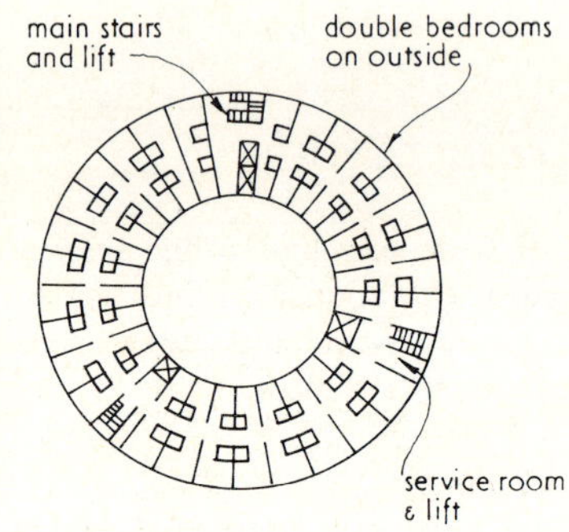

f *Rooms around a circular court, as at Ariel Hotel, Heathrow*

36.5 *Typical block plan forms*

units divided by separating walls, and interspaced with ducts which convey soil, waste and other services.

To minimise noise, the bedrooms usually form the upper structure rising above the base of public rooms, shops and reception areas.

On plan the block forms an elongated rectangle which may have straight or curved walls, be joined at each end to form a rectangle or circle or be linked at one end to give an L, T or Y shape, **36.5**.

Initially the arrangement of the bedroom block will depend on size and shape of site and nature of surroundings but, in preparing subsequent details, structural framework and vertical services (eg soil, waste, mains and lifts) must be related to layout of public areas underneath—which have a quite different structural requirement. The structural module for the bedroom block is determined by sizes of rooms concerned. Similarly, positions of vertical ducts also depend on room sizes and their arrangement.

The length of each wing of the bedroom block is usually limited by the maximum distance the occupants must travel to reach a staircase in the event of fire. For practical and identification purposes lifts and staircases are normally placed together.

1.04 Public rooms

These areas should be as free from restriction as possible and the structural framework is essentially based on large span units. Public rooms generally need to be sited at or near ground level for convenience, although restaurants may also be provided at the top of the building to command a view.

1.05 Orientation

Restrictions imposed by site, particularly in a town, may determine the buildings' orientation regardless of other considerations. Account, however, must be taken of the aspect from various rooms and effects of sunlight. Bedroom blocks with the long axis nearer north–south than east–west will normally be preferable.

Kitchens and rooms used for storage of food and wine should, if possible, be sited on the north or north-east side to facilitate temperature control.

1.06 Environment

The relationships of different parts of the hotel and the effects of noise and pollution must be considered in the preliminary layout. Bedrooms must be positioned to minimise noise from traffic, machinery and kitchens. Care must also be taken in positioning room service areas as much work is done while guests are still asleep.

2 CIRCULATION

2.01 Movement of guests and staff

Routes taken by resident guests, non-resident diners, and staff follow distinct patterns and these establish clear relationships between the hotel's various parts. Layout and planning of the hotel must facilitate movements of people and, as far as possible, provide for the separation of guests, staff and maintenance personnel. This is important not only to avoid disturbance and annoyance of the guest but also to enable service facilities to be designed for efficient use.

2.02 Secondary circulation

Secondary circulation is often desirable to separate resident and non-resident guests—eg by providing direct access to restaurants and banqueting halls. This has a dual purpose in that residents' needs can be given exclusive attention at the reception area, without unnecessary congestion, while people entering and leaving both areas can be better controlled and supervised.

Having determined the sequence of movement, the relative positions of rooms and their ancillary services can be located. A diagram of the main circulations is given in **36.6**. This is *not* necessarily a suitable layout for any hotel. It is included merely to give some idea of possible circulation patterns and interrelationships.

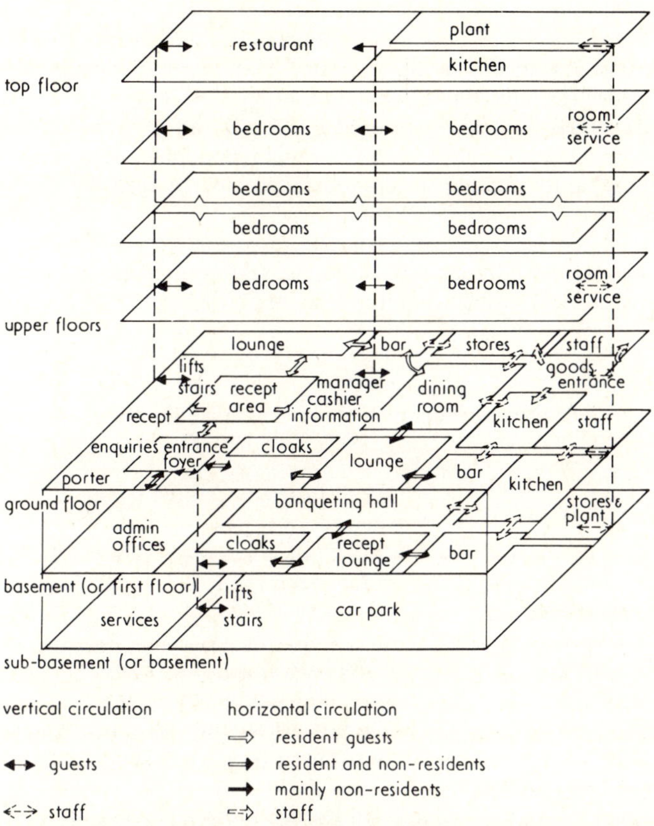

36.6 *Typical circulation pattern diagram showing desirable relationships. This is not intended to imply a particular layout*

3 SPACE STANDARDS

3.01 Figures for preliminary sketches

Precise space standards may be established by the brief, otherwise the following figures may be used for very preliminary sketches and estimates *only*; they may well be revised during detailed design stages. This is particularly true of the figures given for cooking, which must be treated with caution.

Table I Service areas required in m² according to number of guest rooms

	100 rooms	250 rooms	500 rooms	1000 rooms
Housekeeping and general storage	1·40	1·11	0·93	0·74
Administration department	0·46	0·46	0·37	0·28

3.02 Food and beverage department

Food service areas	
Dining rooms (luxury)	1·7 m² to 1·9 m² per seat
Coffee shops and reasonable restaurant standard	1·3 m² per seat
Lounge and bar	1·1 m² to 1·4 m² per seat
Banquet	0·9 m² to 1·3 m² per seat
Staff canteens	0·7 m² to 0·9 m² per seat
Service facilities	
Kitchen for dining room and coffee shop (exclusive of stores)	60 per cent of dining room and coffee shop or 0·9 m² to 1·0 m² per seat
Kitchen for coffee shop only	45 per cent of coffee shop or 0·6 m² per seat
Food and liquor and china storage	50 per cent of kitchen or 0·5 m² per seat in dining room and coffee shop; or 0·3 m² per seat where coffee shop only
Kitchen or pantry to banquet rooms	20 per cent of banquet facility or 0·24 m² per seat
Banquet storage	8 per cent of banquet area or 0·05 m² per seat

3.03 Rooms and service departments

Guests	
Bedrooms	Sizes vary enormously. Double room plus bathroom and lobby could be about 20 m² to 24 m². Minimum standards for bedroom only (British Travel Association Memorandum) are: Single: 5·57 m² Double: 8·36 m² Twin-bedded: 10·22 m²
Bathrooms	Allow about 3·8 m² to 4·2 m² if calculated separately from bedroom, but see also above.

3.04 Offices normally required

Areas will vary between about 7·5 m² and 20 m² or more depending on number of guest bedrooms and other facilities. Exact standards are difficult to define: the areas given below are therefore roughly grouped to give an indication only.

Group A: 7·5 m² to 9·5 m²
Group B: 9·5 m² to 11·5 m²
Group C: 11·5 m² to 14·0 m²
Group D: 14·0 m² to 18·5 m²
Group E: 18·5 m² to 23·0 m²

Administrative	*100 bedrooms*	*250 bedrooms*
Manager	B	C
Executive assistant or assistant manager	A	B
Budget and control director or analyst	*	A
Secretaries' offices (possibly two, each)	A	C
Sales and catering	*	C
Purchasing	*	C
Personnel manager, hiring room, auditing	*	A
General office	D	E

Food and beverage		
Food and beverage manager	A	B
Chef, steward	*	B
Banquet manager and banquet head waiter	*	A
Room service	*	A
Housekeeping		
Housekeeper	A	A
Receiving clerk, timekeeper	A	A
Engineer	A	A

* Indicates room may not be necessary

3.05 Other spaces

Cloakrooms	
Fixed rows of hooks	Allow $0{\cdot}1\,m^2$ per user (includes allowance for staff circulation and for public space around counter)
Hooks plus seating or lockers	Allow $0{\cdot}2\,m^2$ to $0{\cdot}3\,m^2$ per user
Public assemblies	
Conferences	Allow $0{\cdot}5\,m^2$ to $0{\cdot}6\,m^2$ per person plus allowance for stage, translation booth. etc; or overall average of $0{\cdot}7\,m^2$ to $0{\cdot}9\,m^2$
Dances	Allow $0{\cdot}6\,m^2$ to $0{\cdot}9\,m^2$ per person exclusive of band space which may be up to $12\,m^2$ for six-piece band
Swimming pool (internal or external)	Can be any size or shape. For serious swimming smallest pool size is 25 m × 12·5 m (depth 0·9 m to 2 m, but up to 3 m or more if for diving). Plus surround (min 2 m), changing rooms
Sports and recreations	Depends on facilities offered; eg badminton needs space 16·5 m × 8·5 m × up to 9·15 m high; table tennis 9 m × 3 m per table
Circulation and reception areas	
Overall allowance for general circulation areas	$2{\cdot}3\,m^2$ to $2{\cdot}8\,m^2$
Waiting areas	$0{\cdot}5\,m^2$ to $1{\cdot}4\,m^2$ depending on number of seated people and cross flows. The lower figure is for 100 per cent standing with no cross flows

4 CAR PARKING AND ACCESS

4.01 Cark parking

Planning requirements vary according to the size and type of hotel and its situation. Required provision must be sought from planning authority. Typical provision is given in section 7: External circulation.

But apart from any regulations, the hotel itself may determine the number of parking spaces—eg a motor hotel on the outskirts of a large town or city may need one car space for every bedroom. Inside London fewer car spaces are needed per bedroom but the requirement for conference parking may be greater. As a rough guide, between thirty-two and thirty-six cars can be manoeuvred and parked in a space 27·5 m × 30·5 m.

While the car park will almost always be reached through a separate entrance and is often some distance from the reception area, provision must be made for passengers to alight safely and conveniently at the main hotel entrance whether from taxi or private car. In the latter case it should also be made easy for the driver to return from the car park direct to the reception area to rejoin his passengers.

In an urban area, the need for car parking space may pre-empt the use of basement or roof area of low rise buildings and, in turn, the siting of boiler plant and mechanical services tends to be displaced to the sub-basement or upper storeys of the building.

4.02 Access

An important consideration of planning control is the potential risk of traffic congestion and obstruction. Pedestrian and vehicular access to the hotel needs to be determined and agreed at an early stage. Commercially, the position of the hotel entrance is usually a critical feature which then determines the location of the main frontage.

Secondary access is required for goods and service vehicles with adequate provision for turning, unloading and loading.

A separate staff entrance is usually provided with direct access to changing and service rooms.

It is necessary to make separate provision for receiving and handling different types of goods, taking into account their nature and storage requirements. Main divisions are:

Goods	*Stores*
Beers, wines, spirits	Beer cellars, wine and spirit stores, crate storage
Food	Cold stores, vegetable stores, dry goods stores
Laundry and soft furnishings	Linen stores
General	Crockery and cutlery stores. Cleaning equipment stores. Storage for maintenance plant, furniture and general goods
Fuel	Oil storage tanks and/or solid fuel enclosures

4.03 Refuse

Provision must be made for refuse storage and collection, including the temporary accumulation of empty containers, which often tends to be under-estimated. This is a potential source of nuisance due to decomposing residues of food and rubbish, rats and flies. Properly covered bins and containers are essential, and these should be sited in an impervious enclosure equipped with means of hosing down and drainage.

5 BEDROOM LAYOUTS

5.01 Situation

Rooms can be situated on one side of the corridor or on both. The former allows natural daylight into the corridor, making it pleasanter. It also allows a short duct connecting internal bathrooms with the external face of the corridor, cutting out mechanical ventilation. However, the latter arrangement is considerably cheaper, saving up to 20 per cent in area and between 10 per cent and 15 per cent over the total cost of the bedroom wing.

Structurally, bedroom wings are generally of two types: post and beam construction used mainly on high rise buildings only, and cross walls, used mainly in low and medium rise schemes. Cross walls have the advantage that the structure itself provides adequate sound insulation between suites, but post and beam might save a very small amount of space and a considerable weight, thus reducing foundation costs. All standards today are rising more rapidly than ever before. This tendency will probably grow. It is therefore most important to design easily renewable fixed services with sufficient spaces, ducts, and so on for further requirements.

5.02 Numbers of bedrooms per floor
Owing to restricted sites it is sometimes difficult to control this factor, but their number should be related to the staff's capacity, **36.7**. One chambermaid can cope with about six bedrooms in luxury hotels and up to twenty or even more in lower grade premises. To design multiples of her capacity per floor will save staff expenditure. Similar considerations apply to room service.

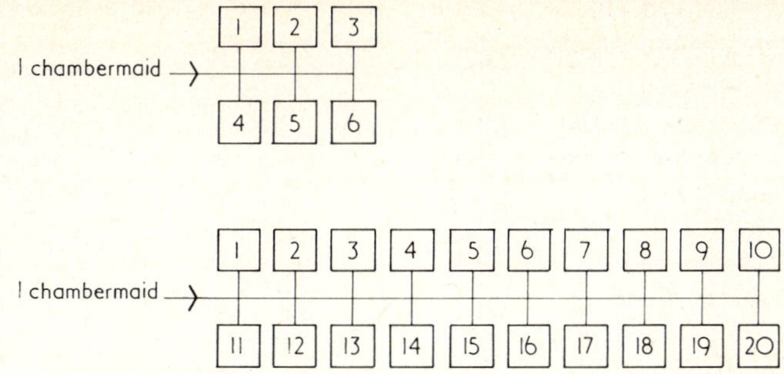

36.7 *Chambermaid 'modules'. Degree of luxury and standard of service will determine how many rooms are to be serviced by one maid. This in turn decides the number of modules and hence the number of bedrooms per floor*

5.03 Mechanical services
Although elevationally a 'handed' terrace appearance may result, it is wise to plan bedroom and sanitary accommodation units in pairs. This cuts out noise from adjoining bathrooms and avoids separate ducts for each bathroom. Special care must be taken with ventilation ducts as these may conduct airborne noise from one bathroom unit to the other. Small details like recessed soap dishes can also cause trouble unless staggered to avoid a direct noise path.

5.04 Corridors
To avoid an institutional appearance corridors should be modulated or not be too long. Modulation can be achieved by recessing certain elements of the bedroom (generally the lobby) and using different heights of false ceilings with variation of the light intensity in the corridor.

5.05 Disabled people
All guests' bedrooms should be accessible without having to negotiate steps. At least 50 per cent of guests' bedrooms must be accessible to chairbound disabled people. Selected rooms having private bathrooms must be planned so that there is room inside the bathroom to enter in a wheelchair and to close the door, and also to transfer from a lateral position on to the wc. (See *Designing for the disabled*, section 77.)

5.06 Escape route geometry
This part deals mainly with the acceptable travel distance dimensions. The official guide to The Fire Precautions Act 1971 about hotels describes most of these dimensions and a detailed analysis of the requirements was made by Alan Parnell *et al* in *The Fire Precautions Act in Practice*. Diagrams **36.8** to **12** may help to interpret the following description of the three stages of escape route provision.

5.07 Stage 1
In this context Stage 1 is the distance from any point in a room, to a doorway from that room. The maximum allowable travel distance from a doorway to the most remote corner across bedrooms, and two-room suites, is 9 m, **36.8**. For multiroom suites, no single cross room dimension should exceed 9 m, **36.9**, and any associated private corridor should not exceed 7·5 m, **36.10**. For rooms of high fire risk (eg kitchens) the acceptable maximum distance is 6 m. These distances are expected to be small enough to allow sufficient time for any occupant of a room containing a fire to reach an adjacent corridor (or other space) before the fire in the room produces untenable conditions.

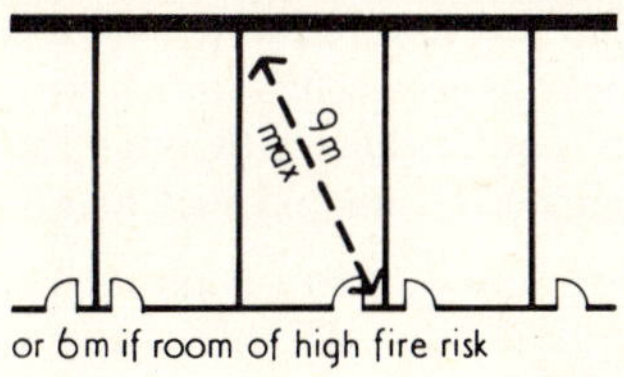

36.8 *Maximum allowable travel distance from the most remote corner of room to doorway*

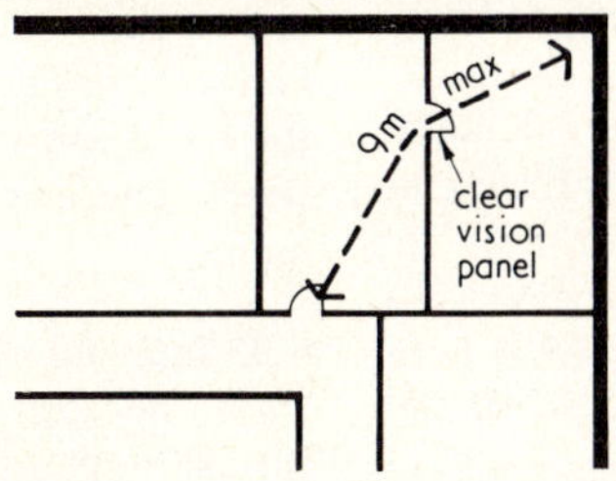

36.9 *In multi-room suites no single cross room dimension should exceed 9 m*

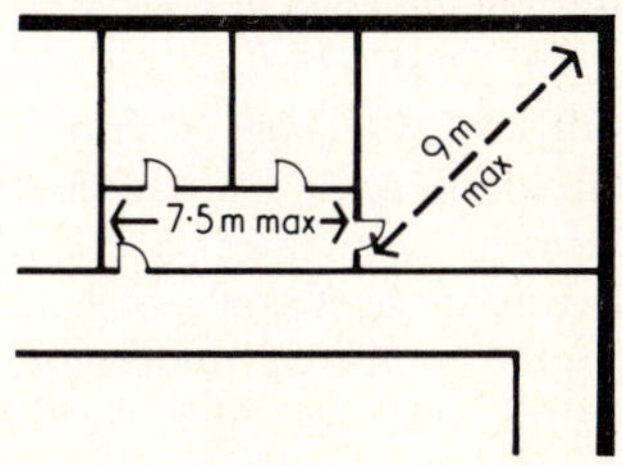

36.10 *In multi-room suites any associated private corridor should not exceed 7·5 m in length*

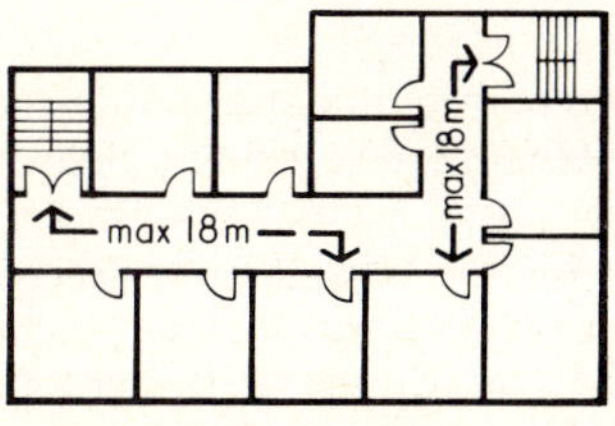

36.11 *Stage 2 escape; no further than 18 m is allowed from exit door of room to entrance of protected escape route*

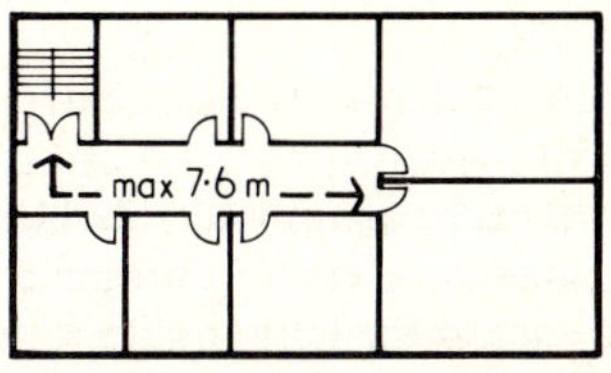

36.12 *The dead end of any escape corridor should not exceed 7·6 m*

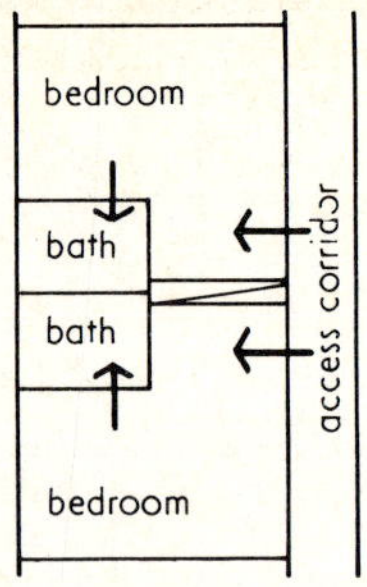

36.13 *Plan arrangement with bathrooms on external walls*

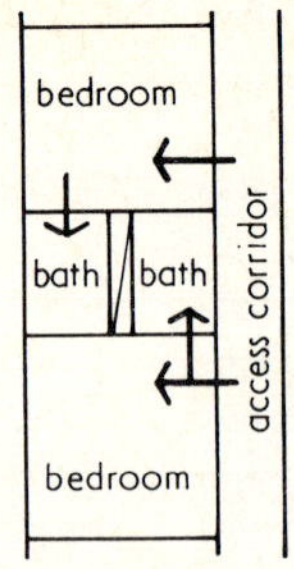

36.14 *Bathrooms between bedrooms: one external and one internal*

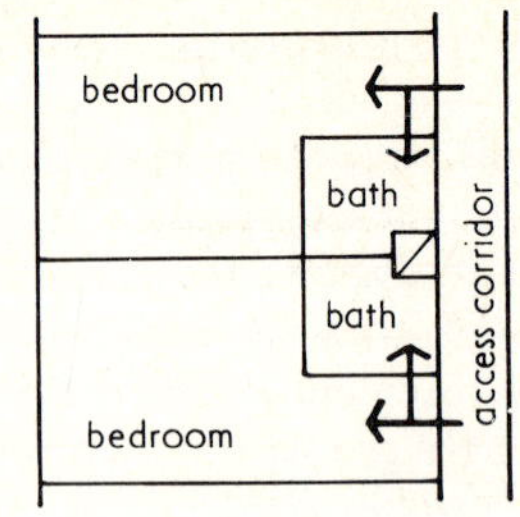

36.15 *Plan arrangement with all internal bathrooms*

5.08 Stage 2
Once people have reached the corridor, they are protected from smoke and combustion gases by the room doors opening on to the corridor, which should have the same fire resistance as the enclosing wall—up to 30 minutes. In addition, the corridor should be interrupted by self-closing smoke-stop doors, at most 30 m apart. In most cases no further than 18 m is allowed from the exit door of a room to the entrance of a protected route, **36.11**, and all protected horizontal routes to protected staircases or open air exits should be separated from the bedrooms by doors with a 30 minute fire resistance. For rooms from which there is escape in one direction only (dead ends), this portion of the route should not exceed 7·6 m, **36.12**. Where alternative routes join they must be separated by fire-resisting doors.

5.09 Stage 3
Stage 3 of the route is within a protected staircase, which descends to a suitable level from which the escapees would continue horizontally (still in a protected corridor) to the open air.
In small buildings, not more than four storeys high, a single staircase may be acceptable; but in larger establishments at least two staircases will be required to give an alternative vertical route. Where an alternative stair is required, an external staircase will be acceptable to most local authorities—except in Scotland, where a staircase protected from the weather is required. Spiral staircases are also acceptable, provided that no more than 50 people are likely to be involved. The final exits of staircases should not be close to each other and the outside space must not be an enclosed courtyard but a space from which continuing escape can be made.
The widths of doorways, corridors and stairs are calculated from the widths required to discharge all occupants from chosen spaces in 2½ minutes (or a more appropriate time). The discharge route commonly accepted through doorways is 40 persons per minute per unit width (500 mm). Normal hotel door widths will usually be sufficient for escape, although multi-purpose function suites need careful consideration. Similarly, corridors wide enough for luggage movement will be generally adequate for the number of people on bedroom floors.

6 BEDROOMS

6.01 Dimensions
Exact dimensions of each room should not be settled until the room has been planned in detail: quite small variations may make all the difference to the proper placing of bed and other furniture.
Essential minimum standards are laid down by the British Travel Association:

Room	*Area (m^2)*
Single bedroom	5·57
Double bedroom	8·36
Twin-bedded room	10·22

More space will normally be required to achieve a satisfactory layout with correctly sized furniture.

6.02 Basic units
The shape and to some extent bedroom size will be governed by positioning of bathroom. Most new hotels will provide individual bathrooms for each bedroom. There are broadly three relationships, as follows.

6.03 Bedrooms with bathrooms on external wall, 36.13
There is only one important advantage in this arrangement, that of easy natural ventilation. In some cases a lobby omitted may reduce costs, though reduction in privacy is inevitable. Its greatest failing is that the service duct can be inspected only from the bathroom or bedroom and workmen pass through the room to reach it. In most cases where rooms are on both sides of the corridor, two separate drainage systems are necessary. Since the external envelope of the bedroom units is very much greater than with an internal bathroom arrangement, their building cost is higher and as the bedroom window in most cases is recessed, a certain amount of light and view from the bedroom is lost.

6.04 Bathrooms between two adjacent bedrooms, 36.14
The worst feature of this arrangement is the undue elongation of the corridor. This is costly in capital expenditure and slows down servicing. Since, if there are two bathrooms adjacent, one of them must be internal, the ventilation problem is only half solved. The common duct can be reached only from one of the bathrooms, which again requires access through the bedroom. There are few arguments to commend such an arrangement.

6.05 Internal bathrooms, 36.15
These necessitate a lobby, but normally this can be used for furnishings and thus at least part of the area can be regained by reducing the bedroom area. They also require artificial ventilation and lighting. However, the common duct can be, if well planned, serviced from the corridor, and widths of bedrooms along the outer walls kept to a minimum, thus shortening the length of the corridor. The lobby can be open to the bedroom or separated by a door. If open it can be used as dressing area between room and bathroom but sound insulation of the corridor must be greater than if a lobby-bedroom door is also provided.
This type of layout is commonest and examples are given in **36.16** and **36.17**.

6.06 Single or double rooms and suites
The optimum ratio of single to double rooms will vary from hotel to hotel and will be decided by the client at briefing stage,

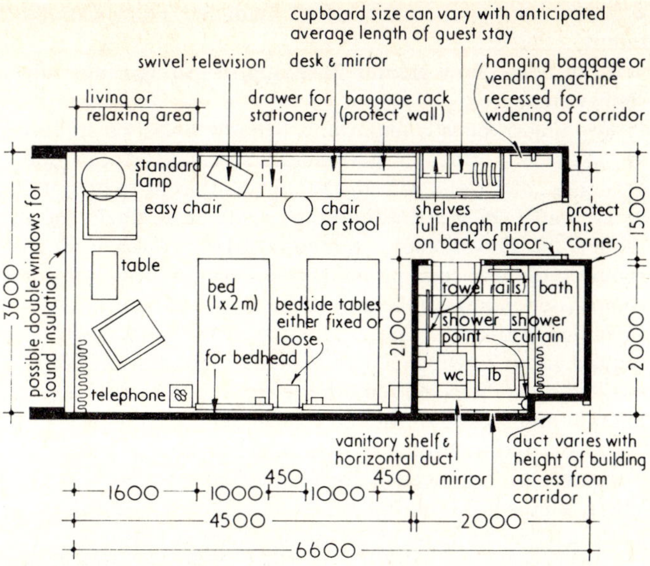

36.16 *Typical twin-bedded room with clothes storage and dressing table along party wall. Sizes vary with luxury standard and site conditions*

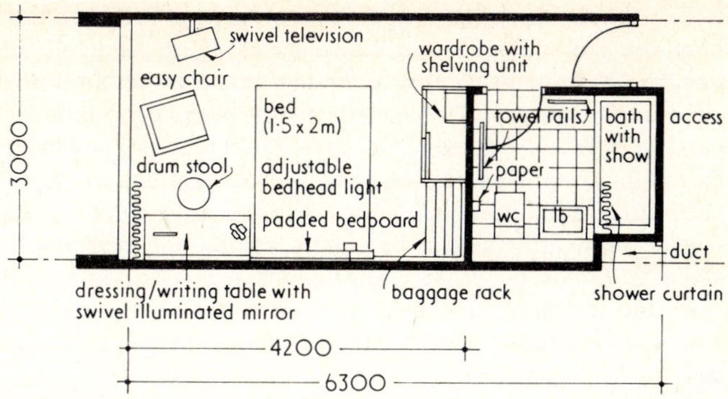

36.17 *Layout for single bedroom. Note double bed for use as double room if required*

arising out of the market feasibility studies. A city hotel catering mainly for businessmen may require 80 per cent single rooms or more. In other situations where there is a greater flexibility of use there may be advantage in providing rooms of standard size fitted with twin beds for either single or double use. Combination rooms facilitate this interchange by using divan beds which can be readily converted to couches.

Intercommunicating doors between adjoining rooms will further extend bedroom flexibility. The rooms linked in this way can, if required, form a suite. Difficulties arise from the need to soundproof the connecting doors when the rooms are separated but this can usually be met by fitting two doors in tandem. The doors must also be capable of being locked on each side. Suites could be used for business, in which case beds would have to be stored and replaced with other suitable furniture.

Suites of this sort should have a common lobby, leading to both room lobbies. This common lobby can often be designed relatively easily at points where corridors turn. To obtain a clean solution at these turning points is extremely difficult and the inclusion of a private lobby to two bedrooms may ease the problem.

6.07 Terraces and balconies

While such amenities are pleasant, particularly for the long-stay guest, they are also costly. Not only because they add to the volume of building, but because they lead to problems of security, wind and waterproofing, necessitating special equipment and more expensive joinery. In addition a safety problem arises, in that a raised threshold is virtually always essential and clients may slip on the polished or metal surfaces and claims for damages may result. It is wise therefore to restrict such features to rooms with a prominent view, and to those in the higher price range only.

7 BEDROOM FURNITURE AND FITTINGS

7.01 Different requirements

A decision must be taken as to whether the furniture is to be:

- freestanding or built-in,
- primarily manufactured for the domestic market or of specially commissioned design.

The requirements for different hotels will vary according to whether the guest stays for a short or long period, and to the standard of accommodation, which is broadly related to the prices charged.

7.02 Built-in and fitted furniture

Selection depends on the following factors:

In a hotel bedroom it is important to make the best use of space—which is expensive. Fitted furniture helps.

Furniture of this type is regarded as fixtures and hence capital investment.

Units can be designed to facilitate cleaning—by hanging them from the wall clear of the floor and extending them to the ceiling.

Added fitted furniture to partition walls can improve sound insulation between rooms. Against this, however, fixing units against a wall is likely to increase sound transmission when doors, drawers and hangers are moved.

7.03 Freestanding furniture

Advantages of free-standing furniture, particularly if of standard design and manufacture, are its:

- cheapness
- flexibility
- easy maintenance
- availability in a wide range.

7.04 Construction and maintenance

All hotel furnishings and fixtures must be robust and those attached to walls, firm. Mobile fitments should be few, particularly in motels, where client control is more difficult than in conventional premises with a main entrance, through which everybody must pass. It has been known for everything movable to be stripped from a room, including television set and beds.

In his account, the hotelier generally writes down the value of his furnishings and décor within five to eight years. While furniture on the whole does last for this time if well made, soft furnishings and décor generally deteriorate more rapidly.

If any room needs to be refurnished, recarpeted, redecorated, it is likely to be out of commission. Walls and ceilings should therefore be of materials easily and rapidly cleaned. Carpets, which wear greatly in certain areas and not at all in others, should be of modular squares, so that their life could be extended by replacing only worn ones.

Some newer materials help the designer in this respect, though they are generally more expensive. However, such a small increase in cost, probably amounting to no more than $\frac{1}{4}$ per cent to 1 per cent over the whole scheme may be well warranted if it avoids redecorations, refurnishing and gaps in room lettings.

There are two major sources of damage to furniture and fittings, apart from obvious misuse—hot receptacles, liquids and cigarettes, and heavy weights of persons sitting on fixtures. The first is almost insoluble, a cigarette stub ground into a carpet damages irreparably. Horizontal surfaces should be made as heat and burn resisting as possible and carpets fitted in interchangeable squares. The remedy to the second problem is extremely firm anchorages to walls (the best place for any fixed furnishings is on load bearing cross-walls). Should there be no such stout building components, fixtures must be on legs. Catches and fittings must be simple, strong and reasonably silent. Considerable damage often results from doors swinging open too far. Stops and stays must be fitted when required. Loose keys should not be used, and doors and drawers must be easy to open and close.

A large proportion of any hotel's income is spent on overheads—staff, mechanical equipment, management, maintenance, and so on. If the designer can reduce these costs the hotel will be that much more viable.

7.05 Typical furniture and equipment

A useful source of advice on the design of bedroom furniture is a report by a COID advisory committee on which the recommendations given in **36.18** are based.

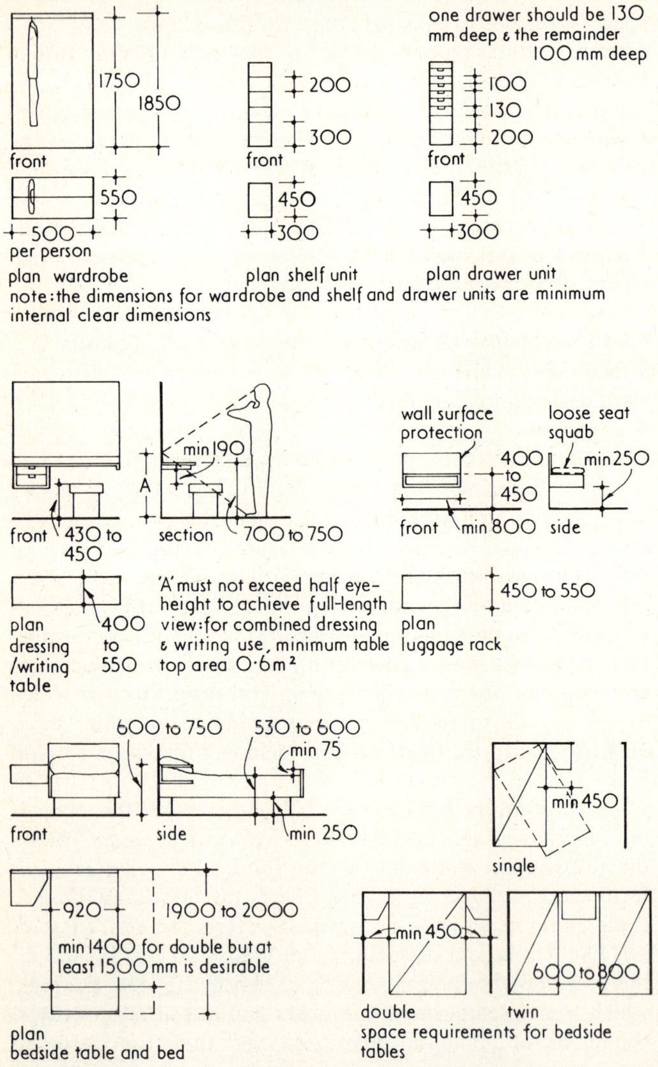

36.18 *Space requirements for hotel bedroom furniture*

7.06 Beds

Hotel beds must satisfy a number of requirements:

- comfort, height, length and width
- durability and resistance to edge damage
- ease of movement for making up
- ease of dismantling for removal and storage
- absence of creaking
- appearance—particularly of the bedhead which may be a fixture
- divan beds allow greater flexibility of use and must meet similar standards.

Positioning of beds is important. Broadly only two solutions are feasible: either a peripheral, so-called studio room arrangement, or one with the bedhead adjacent to one of the party walls, **36.17**. Both arrangements have their merits, though the first is likely to require a larger area in double rooms, and the second probably requires more space in single rooms (depending on the size of the bathroom unit).

Including bedheads, where adjacent to the party wall, beds require about 2·10 m in length. Adding a minimum of 800 mm passage space between the other party wall and the bed end, the minimum width of the room must be 2·90 m. However, if the associated bathroom has a full length bath, basin and wc the bathroom also needs to be about 2·10 m wide.

There is thus no room in the entrance for any furnishings, so wasting space completely. Allowing a further 600 mm for furniture placed in the lobby, a wall to wall clear width of 3·50 m is required. This is a fairly common module, in modern medium priced bedroom wings. It successfully balances the need for shortening corridors and the provision of reasonable space for individual suites.

Beds vary greatly in width and height. The most commonly used single bed is 900 mm to 920 mm wide and the double 1400 mm. Measurements for new metric beds are to be: double 2 × 1·5 m; single 2 × 1 m. These should always be used in future. There is a strong case for using 2 × 1·5 m beds in single rooms so that they could be used as doubles if needed. For appearance a height of 350 to 400 mm (including mattress) is best. For ease of stripping and making, however, (such low beds cause fatigue and backaches) the preferred height should be from 530 mm to 600 mm or even up to 700 mm. This allows for lying down without a strenuous climb. Many folding or convertible beds have been designed. Since these cause additional trouble to the staff, they should be used in special circumstances only.

7.07 Fitted furniture

A long hanging wardrobe space of between 300 mm and 500 mm is needed for the first person, plus 300 mm for each extra person—and preferably a short hanging space (somewhat smaller). At least two shelves or drawers for the first and one for each additional person are required, with a width of 300 mm and a minimum depth of 450 mm.

A luggage rack, which could have space for a trolley underneath, 800 mm to 900 mm long, is required. There must be guards against abrasion by metallic objects, such as studs, and it should be high enough to avoid fatigue in those packing.

A dressing-table at least 900 mm long which can also double up as a writing desk is required. It should contain at least one drawer. There ought to be a mirror above it, preferably with two swivel side mirrors. It must have a good light, preferably day and night type, with separate switches and a long full length mirror with a suitable light source.

Bedside tables large enough for books, glasses, water, and a telephone are needed.

There should be separate reading lights over the beds and as well as a general light the room requires a light in the entrance lobby.

It is now usual to build in a bedhead. This holds bedside tables, often mounted as a mobile fitment, to allow the position of the bed to be changed. All lights, radio and television should be included in bedhead controls. Telephone, intercom and other control equipment can be fixed to it—eg mechanical curtain control, heating and ventilation control, tea making machine.

In the highest range of hotels a drink dispenser (ice cube

dispenser along a corridor) a refrigerator and a safe are often installed.

7.08 Loose furniture

Each bedroom should have one or two occasional chairs, at least one easy chair per person, a swivel dressing-table chair, an occasional table, standard or table lamps, ashtrays, maybe a trouser press, very often a television set.

8 PUBLIC AREAS

8.01 Approach and entrance

The impression created by the main entrance is important since it tends to typify the type of hotel. It must always be clearly defined and provide direct access to the hotel reception. If this is on an upper floor, stairs and lifts *must be exclusive* to the hotel.

8.02 Port-cochère

Something more protective than a canopy is desirable to provide shelter at the entrance from wind and rain. A port-cochère should be wide enough to allow two cars to pass and possibly high enough to accommodate buses or coaches. Special lighting may be needed to accentuate the entrance.

8.03 Entrance doors and lobby

Doors should have sufficient clear width to allow for porter plus bags (minimum 900 mm clear opening) or trolley. With revolving doors, side hung escape doors will also be required. A draught lobby should normally be provided. Consider automatic opening of doors.

8.04 Disabled guests

All public entrances must be accessible to ambulant disabled people and at least one must be available for the chairbound. At least one entrance served from the hotel garage must be accessible to chairbound disabled people. (See *Designing for the disabled*, section 77.)

8.05 Flooring

An area of transition is needed at the entrance before using fine floor finishes. Allow for dirt and patches of wear from foot traffic.

8.06 Baggage handling

Method of handling will be established by client. Guest/baggage routes should be carefully analysed and appropriate handling systems designed at each stage. A special baggage entrance may be necessary, especially if there are mass arrivals of baggage from tours. Control is normally by a porter with easy access to baggage lift(s). Where baggage is taken by hand or conveyor belt directly from the street to a baggage room and thence by goods lift to the bedrooms, guests must feel confidence in the system of identification as they are nervous of being separated from their baggage.

8.07 Reception

Reception desk should be located so that it is easily identified by the guest immediately on entry and is also on his route to lifts and stairs. Separate counter facilities are usually provided for inquiries and for guests checking in and out of hotel. The top of the desk should be suitable for writing and there should be a bag shelf.

Arising out of the brief the following accommodation and equipment may be required in the reception area:

Porters' station Visual control over lifts and entrance doors. Easy access to baggage room. Space for trolleys.

Head porters' position Incorporated in inquiry counter or separate unit. If separate unit, must have easy communication with and control of porters and messenger boys. Telephone or microphone communication with garage, luggage room, cashier, reception. Fire alarm and service bells near head porter.

Must be in strategic position for supervising entrance, guests coming and going and in control of external 'taxi' sign or rank if any, call boxes if any.

Requirements are: counter; storage for timetables, brochures, small pieces of luggage left by guests for short periods; cash drawer for minor transactions with guests, and so on.

Cashier and accounting machines Apart from space needed for receptionist and clerical staff, space and equipment may be needed for cashier staff. Machines may be simple adding machines or complicated electronic accounting equipment. A computer terminal or Telex may be needed. Foreign currency service will affect requirements for drawers and storage.

Key racks and key depository Often associated with letter racks at rear of counter.

Postbox, letter racks, stamp machine Posting and collection box for mail may be required, either (*a*) for internal operation by hotel staff, or (*b*) as PO official box. Can usefully be located near or in inquiry counter. Slot size recommended 200 × 50 mm. If PO box, must conform to official requirements.

Pneumatic tubes

Electro-writers

Telephone meters For metering guest calls from their rooms.

Stationery and record store

Strongroom or safe

Parcel or baggage storage

TV monitors, loudspeakers

News teleprinter Special news service as a possible amenity for guests. Should be in position where noise is not a nuisance.

Room call system

8.08 Offices

A list of possible offices and approximate sizes is given in para **3.04**. To facilitate supervision and attention the manager's office should be sited near the reception area. Other offices, accounting rooms and record files need not necessarily be in the vicinity but there must be good communication between reception desk and offices, including means of rapid transportation for documents. In a large hotel much of the accounting and paperwork is mechanised in the interests of speed of attention and staff economy.

8.09 Telephones

If a telephone is provided in each bedroom the number of public phones can be relatively small—say one to three according to size of hotel, assuming others are provided for spaces such as banquet hall, bars.

Equipment will be to PO requirements. See AJ Handbook Building services and circulation, section 6. Space is needed for switchboard and adjacent switchroom.

Public telephones may be in kiosks or acoustic heads. House telephones, for visitors in foyer who wish to speak to guests in rooms, should be adjacent to inquiry desk. Each phoning position should have a certain sense of privacy yet be fully in view of inquiry desk for supervision.

8.10 Reception lobby

Establish priority between publicity value of being able to see into the reception lobby or entrance foyer from the street and the comfort and privacy of the guests inside. In some locations there may be interest for the guest in looking into the street or entrance court. This will affect not only the position of the lobby but also the glazing and type of curtaining.

8.11 Shops

Number and type will be decided by the brief. Note especially access, display requirements, position in regard to main circulation, delivery of goods and storage.

8.12 Furniture and fittings

Furniture should be of contract quality and able to withstand hard wear. Consider the following:

- easy chairs and settees
- desk chairs and writing-desks
- low tables
- ashtrays—table and pedestal
- carpets
- curtains—heavy (dress only or complete) and net
- electrically controlled curtains
- anti-glare glazing to windows
- staff seating behind counters and in foyer for messenger boys.

8.13 Clocks, calenders

Local system (clockwork or electrical) or master system. Twelve- or twenty-four-hour dial; easily seen from staff counters and by guests. Date clock and time clock of the world's capitals may be wanted for general interest.

8.14 Lounge

The traditional image of a lounge as a distinct room is changing and lounge space may now often be limited to the irregular areas joining the reception to other public rooms or, more profitably, it may be associated with the bar. Isolated lounges earn no revenue. The lounge must be associated with liquor service or have a special function such as a tea lounge. Resort hotels will have lounges for entertaining guests, and these lounges are intended to foster a feeling of community. Lounges should be informal and relaxing.

No particular space standards apply as much depends on class of hotel, availability of other public rooms and spaces for waiting and meeting, and the amount of use which might be anticipated: this should emerge from the brief. A rough planning figure is $1{\cdot}1\,m^2$ to $1{\cdot}4\,m^2$ per seat. Furniture is normally easy chairs and low coffee tables.

Telephone jack points may be required, especially in high class hotels.

8.15 Bar lounge or bar

Design of bar will be largely influenced by the number of areas it has to serve, eg lounge, restaurant, coffee shop, banqueting rooms, room service; and the degree to which waiter service is employed. When the bar is closed, if the lounge is to be used, the area must not appear 'dead'.

8.16 Check list

Bar design has been dealt with in detail in section 21, Public houses. The following check list therefore merely notes the main items:

- seating
- cash system
- wash up facilities
- stewards service
- lock-up for bar: security generally
- motif
- bar store supply, and disposal of empties
- bar counter:

storage for glasses	security
shelves, cold shelves	hoists
ice making	draught beer system
wine racks	back bar flooring
cigarette storage	food buffet
bottle display	drip outlet or barrier

- telephone jack points
- acoustics
- ventilation (check cigarette smoke staining)
- lighting level
- wine cooling (if no separate bar for restaurant).

Lounges and bars must be accessible to disabled people (see *Designing for the disabled*, section 77).

8.17 Dining room, restaurant, coffee shops

The dining-room is usually open to non-residents so there should be a convenient entrance from outside the hotel in addition to direct access for resident guests. Most larger hotels have several dining-rooms which may be on different floor levels. The range of dining accommodation may include general restaurants, speciality restaurants and buffet bars. Within each of these descriptions there are innumerable variations in both scale and type.

In every case, the dining-room requires to be positioned adjacent to a kitchen or servery from which there is direct and convenient access. Where several kitchens are involved these may be operated as a collective unit with the bulk of the food preparation and precooking being carried out at a central point.

8.18 Check list

Restaurant design has been dealt with in detail in section 20, Catering.

The following check list notes the main items:

- number and type of seating
- ventilation and heating system
- type of service
- theme and motif
- division of room
- dance floor
- orchestra facilities
- dumb waiters
- head waiter's desk
- cash system, cash desk
- liquor service system
- system of lighting, zoning, dimming
- curtain
- telephone jack points
- acoustic treatment
- possible separate external public access and cloaks
- colour and lighting related to food
- hard wearing floor at service entrances and exits
- speakers
- tables—mats, cloths, modular.

If catering is subcontracted, separate kitchen, changing-rooms, etc, will be required within hotel.

Restaurants must be accessible to disabled people (see *Designing for the disabled*, section 77).

8.19 Function rooms

Large rooms are usually designed to be multi-functional since costs involved in providing this space are high and may be justified only by frequent use.

A room of this type may be adapted as a ballroom, banquet hall, conference or exhibition hall by substituting different furniture and floor covering, or by other means. In more elaborate arrangements, moving screens may be fitted to divide the space into smaller areas when required, although difficulties in providing effective soundproofing are usually experienced.

Considerable storage space is needed for furniture, carpets and other equipment which must be sited conveniently near.

To facilitate public use of the hall an entrance should be

provided direct from the hotel foyer so that the public need not pass through the main reception area. Emergency escapes are also necessary and there must be at least two independent exits.

Particular attention is required to the acoustic properties of the hall and to sound insulation of noise liable to enter or be emitted to bedrooms and other areas. A high standard of environmental services is necessary.

Space allowances for a variety of functions are given in para **3.05**.

8.20 Check list

Restaurant design is dealt with in section 20. The following check list should be used in designing for other functions:

- ante-rooms, liquor service
- system of dividing room
- public and service access
- degree of soundproofing required
- storage of movable walls
- furniture store, tables, stacking chairs
- crockery and equipment store (banqueting rooms have own crockery store)
- dance floor (removable carpet)
- wall protection against damage by backs of chairs
- air-conditioning and zoning (if rooms are divided it may not be necessary to condition total area)
- food and liquor service system to ante-rooms and main rooms
- toilet and cloak facilities
- cine projection, 8 mm or 16 mm (separate projection room for 16 mm)
- external access for heavy equipment and exhibition material (may include cars or boats)
- exhibition services, eg gas, water, drainage
- electrical dimming: busbars (*a*) three-phase supply, (*b*) spotlights
- electronic equipment—TV (closed circuit and other amplification); may be linked to bedrooms TV
- changing-rooms for (temporary) staff—may have to accommodate men or women, sometimes both
- bandstand or dais—temporary or permanent
- theme motif or names
- telephone jack points
- acoustic treatment, floor/floor
- press box
- multi-lingual broadcast or relay
- performers' changing-rooms
- fire regulations: exits, lighting.

8.21 Special accommodation

Need for other facilities will arise out of the brief:

- staffroom for chauffeurs or guests' other staff
- day nurseries (especially in resort hotels)
- house doctor or nurse—sick bay or surgery (see section 19; Health centres)
- businessmen's secretarial facilities or offices (see section 16, Offices)
- gymnasium (see section 27, Indoor sports)
- Turkish or sauna baths
- cinema (see section 24, Cinemas)
- sample and meeting rooms
- press, interview rooms, lecture rooms
- casino and card rooms
- changing-rooms and cabanas for swimming pools (public?)
- night clubs,
- kosher kitchens and dining-rooms
- manager's flat,
- service flats or suites,
- VIP rooms,
- chapels or other religious facilities (see section 29, Churches).

8.22 Cloakrooms and lavatories

Cloakrooms and lavatories should be along main circulation routes near public rooms. They must be conspicuous but entrances should be separated and discreet. It must not be possible to see the inside of any lavatory from the public area—even if the doors are open, nor must they communicate directly with a room used for food.

Women's lavatories should include a powder room of adequate size. Note that women prefer to collect their coats from the cloakroom within the lavatory area, men prefer the cloakroom entrance or counter to be outside the lavatory area.

The number of fitments are based on the maximum number of people likely to use the area served.

Detailed information on cloakrooms and lavatories is contained in section 38, Sanitary installations.

9 REFERENCES

AJ Editors, *Principles of hotel design*, London, Architectural Press, 1970

Selwyn Goldsmith, *Designing for the disabled*, London, RIBA, 1977 (3rd edn)

Fred Lawson, *Hotels, motels and condominiums: planning and design*, London, Architectural Press, 1976

Jane Taylor and Gordon Cooke, *The Fire Precautions Act in Practice*, London, Architectural Press, 1970

37 Homes for old people

Ian Smith

Hubbard Ford & Partners

CI/SfB 447
CI/SfB (1976 revised) 447

Contents

1 MAIN ELEMENTS OF THE PLAN

The design of homes for old people should help to create a homely, comfortable and friendly atmosphere, and the importance of avoiding an institutional character is stressed in most design guides and instructions to architects. This section deals with the design of homes in which the residents are in need of special care and attention. Sheltered housing and grouped flatlets for old people are described in the housing section, **35**. The special facilities provided may vary, depending on the degree of infirmity and mobility of the residents, but the basic relationship between the main elements of the plan are common to all homes for old people. The main areas of the home are:

- the residents' rooms and closely linked bathrooms and lavatories
- dining room, lounges and rooms for other communal activities
- kitchen, storage, washing up and service areas
- the administrative rooms, matron's office and medical room
- staff accommodation for resident staff and day staff.

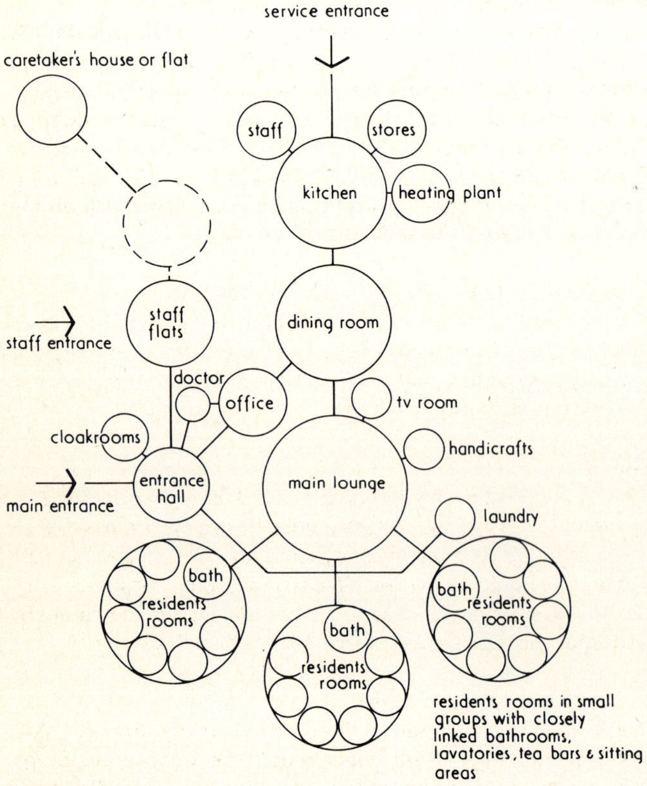

37.1 *Relationships between elements of the plan*

2 RELATIONSHIP BETWEEN ELEMENTS OF THE PLAN

The diagram **37.1** shows how the main areas of the building are interrelated. The aim should be to encourage social contact but at the same time to preserve individual privacy. The residents' rooms are often grouped round a small sitting room and services area containing a bathroom and lavatories. Circulation routes to the communal lounges and dining room should be as short as possible, although routes through the residents' groups should be avoided. Communal areas may either be centralised or divided between the residential groups, but most homes have a main dining room, which should be close to a sitting area. The administrative offices should be close to the entrance hall, and, if possible, within easy reach of the kitchen. Staff accommodation should be provided in self-contained flats with separate outside entrances.

Table I Planning allowances

Accommodation and facilities		
Residents	Single bedsitting rooms	9·6 m²—12 m²
	including private wc	15·3 m²
	Double bedsitting rooms	14·8 m²—16 m²
	Bathrooms and lavatories	
	Sitting areas and tea bars	8·8 m²
	Stores	
Communal rooms	Entrance hall and visitors' cloakroom	
	Lounges	2·3 m² per person
	Dining room	1·5 m² per person
	Handicrafts or sewing room	15 m²
Kitchen	Larder and dry store	12·15 m²
	Food preparation and cooking	42·50 m²
	Washing up	15 m²
	Cloakroom and non-resident staff room	12 m²
Administration	Matron's office	11 m²
	Doctor's room	10 m²
	Visitors' room	10 m²
Ancillary rooms	Sluice rooms	6 m²
	Laundries	20 m²
	Linen storage	8 m²
	Cleaners' stores	4 m²
	Box rooms	8 m²
	Boiler and plant room	25·30 m²
	Garden store and wc	10 m²
Staff accommodation		
Self-contained flat for matron		70 m²
Self-contained flat for assistant matron		60 m²
2 staff bedsitting rooms		12 m²
Staff bathroom		
Staff kitchen		6 m²
2 staff garages		
Staff lounge		12 m²

NB Room areas based on typical 40 person home

3 PLANNING ALLOWANCES

Typical accommodation allowances are given in table I.

4 PLANNING EXAMPLES

The plans of two typical homes are shown in **37.2** and **37.3**.

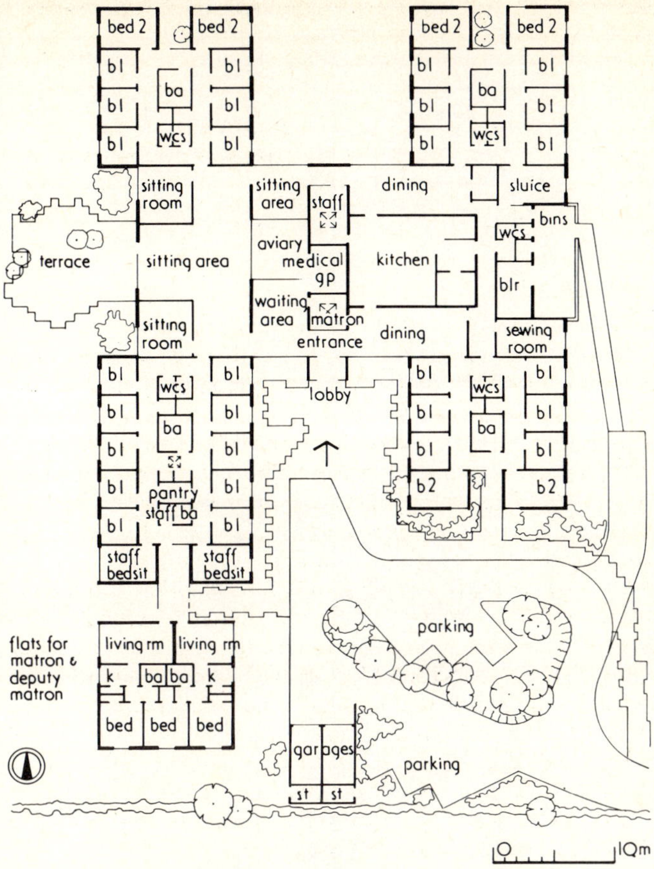

37.2 *Plan of Glebe House, Southbourne*

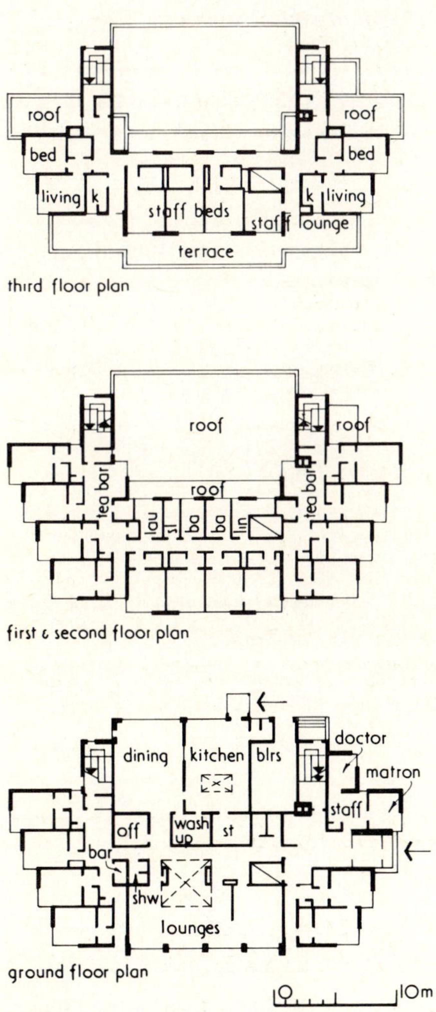

37.3 *Plans of Maidment Court, Dorset*

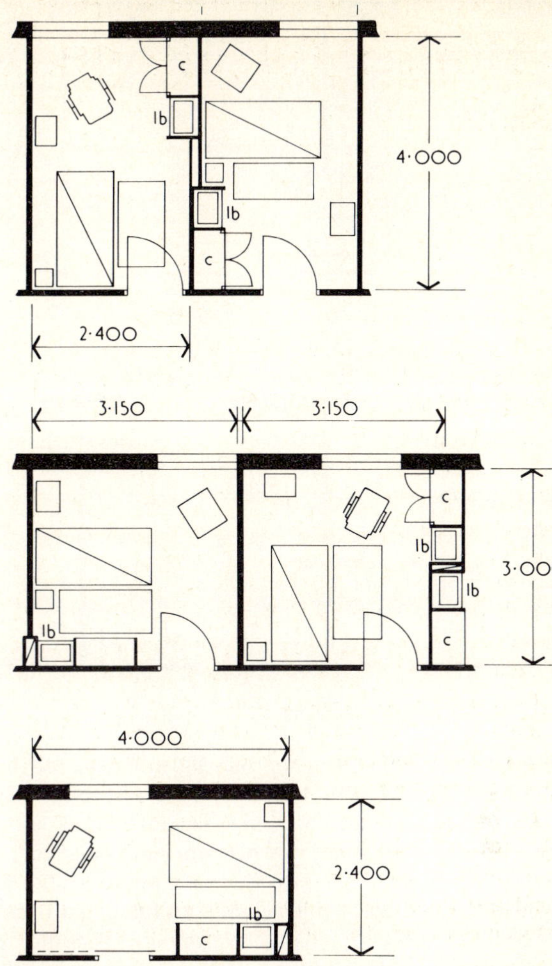

37.4 *Room data and space requirements for single rooms*

5 ROOM DATA AND SPACE REQUIREMENTS

Typical layouts are given for single rooms, **37.4**, and double rooms, **37.5**. These layouts, from DHSS Building Note 2, within rooms of varying proportion show possible ways of providing a flexible arrangement within clearly defined sleeping/sitting areas.

With narrow rooms, corridor circulation is reduced to a minimum, but other types may well be suitable where a different overall plan-form is chosen.

6 BUILDING EQUIPMENT AND FITTINGS

It is desirable that old people should be encouraged to do as much as possible for themselves. To facilitate this, the design of the accommodation and appliances should take into account the limitations imposed by age.

Taps

Choose taps that can be manipulated by arthritic fingers. Surgeon's taps are not recommended, however, as in extreme cases ordinary taps can be modified to provide similar facility. Within one building, it is sensible to maintain consistency as to the location of hot and cold, eg hot *always* on the right. In addition, the tops should always be boldly colour-coded.

Wash-hand basins and baths

Wash-hand basins should be fitted with their rims between 800 and 850 mm high. Bathrooms should be large enough for undressing and dressing, and for someone else to lend a hand. Low-sided baths are available, as the rim, which should be easy

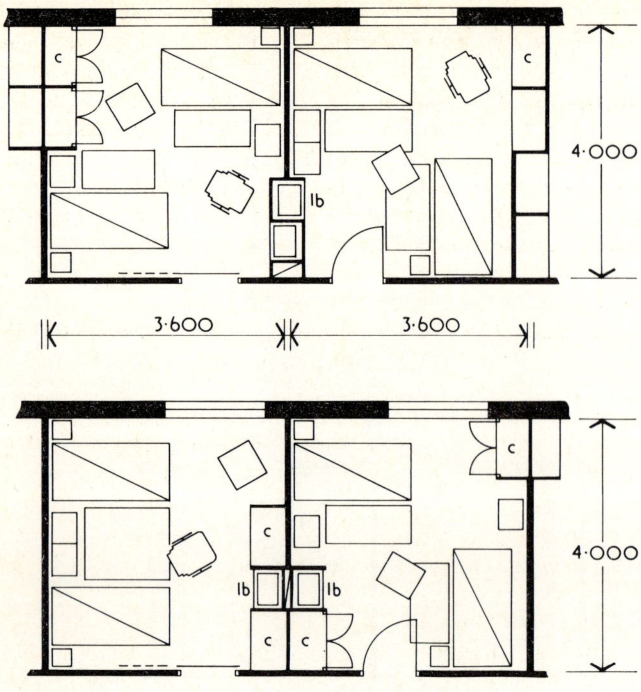

37.5 *Double room requirements*

to grip, **37.6**, should not be higher than 380 mm from the floor. Alternatively, the bath may be set with the trap below floor level. It should have as flat a bottom as possible and should not be longer than 1·5 m; lying down is not encouraged. Grab handles and poles should be provided as in figure **37.7** to help getting in and out. A seat at rim height is useful for sitting on to wash legs and feet. Bathroom and lavatory doors should open out, with locks operable from the outside in emergencies, **37.8**.

Showers

Some old people find showers more convenient to use than baths, **37.9**. If the floor of the compartment is of smooth non-slip material with a fall to a drain of 1:40, there is no need for a tray with an upstand to be stepped over. The compartment should be well-heated, with pegs for clothes on the dry side, divided from the wet with a shower curtain. The water supply should be automatically controlled to supply only between 35° and 49°C. The shower head should be on the end of a flexible hose, with a variety of positions available for clipping it on.

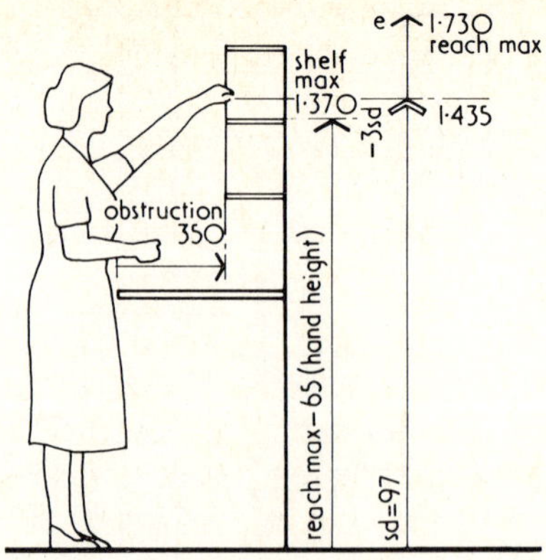

37.11 *Shelf and cupboard design, giving maximum reach over worktop*

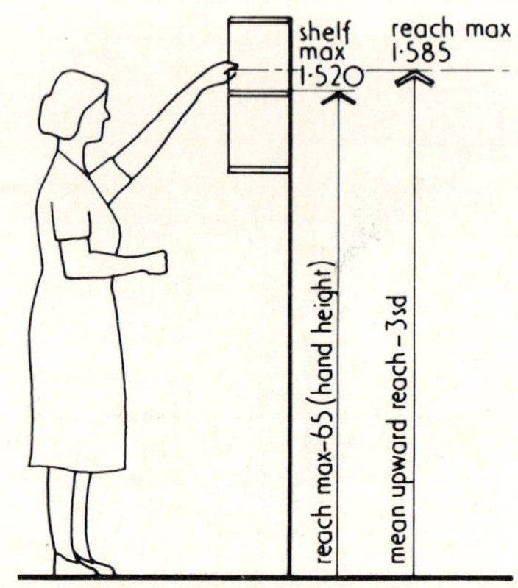

37.12 *Shelf and cupboard design, giving maximum reach to unobstructed wall-mounted cupboard*

WCS

WCS should have a seat height of 380 mm, and handles provided as in **37.10**.

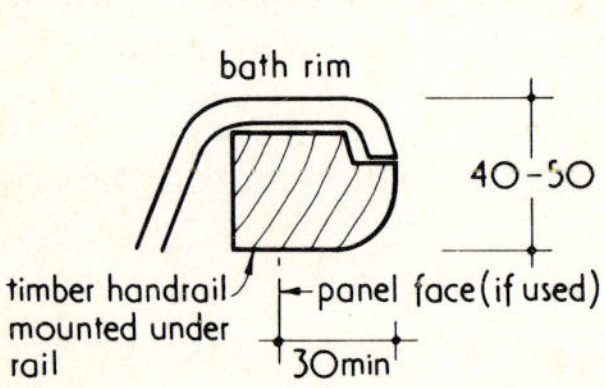

37.6 *Bath rim adapted for easy gripping*

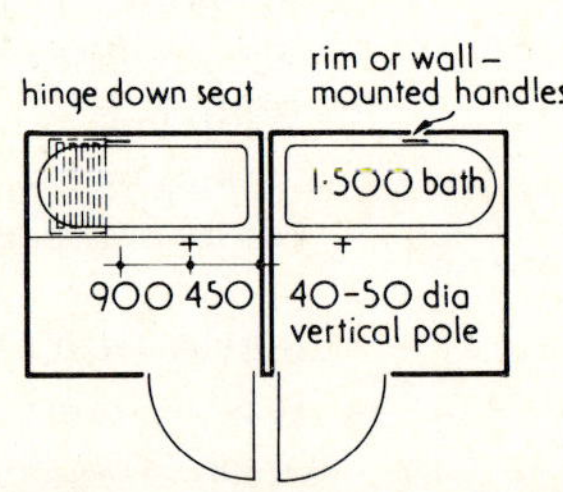

37.8 *Plan of bathrooms showing handing to suit people with disability of either right or left leg, and position of pole aid*

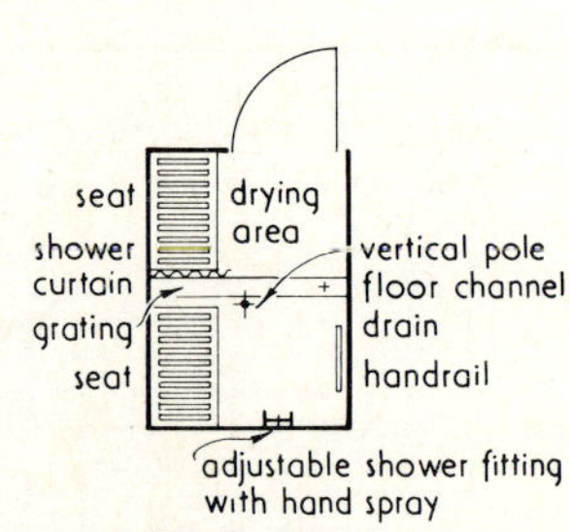

37.9 *Plan of shower room showing seats and aids*

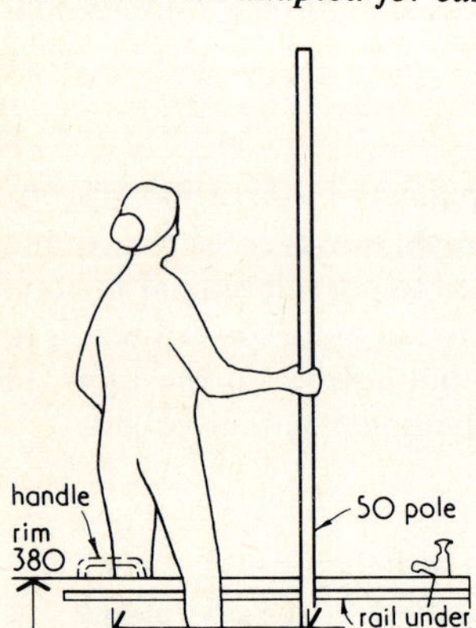

37.7 *Aids for getting in and out of the bath: pole, handle and rim. Maximum height of rim from floor*

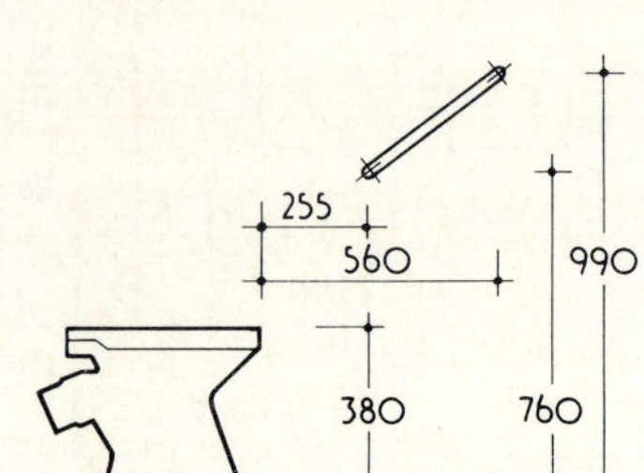

37.10 *Inclined rails mounted on walls of* WC

Cupboards

Shelves and cupboards should acknowledge the limitations of old people. The clothes cupboard rail should be mounted 1·5 m from the floor, and the cupboards should be at least 550 mm deep, **37.11**, **37.12**.

7 FURNITURE

Easy chairs

A variety of chair types should be provided in sitting and common rooms, to ensure maximum comfort for all the old people. Seats should not be too low, as this makes the chair difficult to get out of; but if too high, the feet may end up off the floor. A height between 400 and 430 mm is about right, with footstools available for those with extra-short legs. A seat depth between 410 and 470 mm is ideal: any more and cushions become necessary. The back should be angled at 28° to the vertical, and high enough to support the head, for which an adjustable pad is useful. Armrests 230 mm above the seat at the front facilitate getting up, but if lower at the back, make sewing and knitting easier. There should be a gap under the seat to allow the heels to be drawn right back when rising. Generally, the padding should not be too soft and generous, as this can put strain on the tissues rather than allowing the bone structure to support the body.

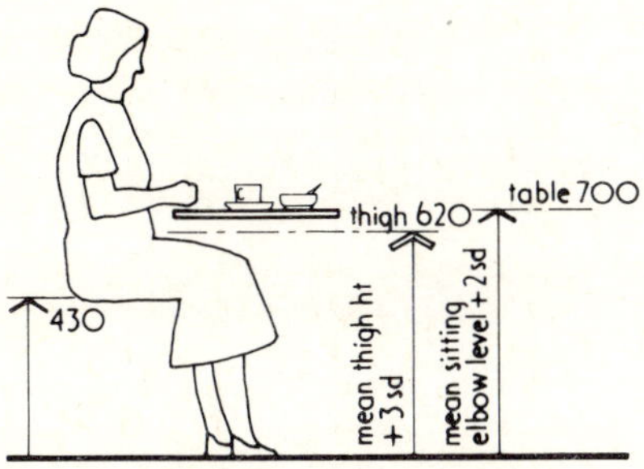

37.13 *Table and sitting-worktop design, giving height and thigh clearance*

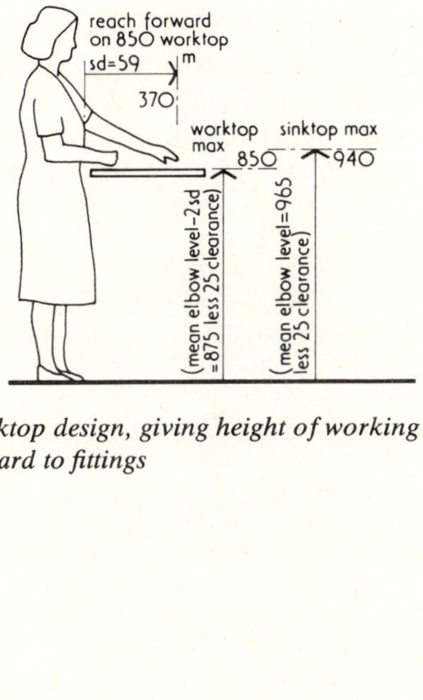

37.14 *Worktop design, giving height of working surface and reach forward to fittings*

Tables and dining chairs

Occasional tables in common rooms should not be lower than chair seat height. Dining tables should be 700 mm high, and used with chairs having a seat height of 430 mm and a depth of 380 mm. There should be a gap for the thigh between the chair seat and the underside of the table top of at least 190 mm, **37.13**.

Worktops

Comfortable reach to worktops are shown in **37.14**.

8 BIBLIOGRAPHY

David M. Boswell and Janet M. Wingrove *The handicapped person in the community*, London, Tavistock Publications and Open University Press, 1974

BS CP 96 *Access for the disabled to buildings*, Part 1 General recommendations, London, BSI, 1967

Cheshire County Council, Department of Architecture *Made to measure—domestic extensions and adaptation for handicapped persons*, Chester 1974

Chronically Sick and Disabled Persons Act 1970

DHSS Joint Circular 12/70 Chronically Sick and Disabled Persons Act 1970, London, HMSO, 1970 (explanatory circular)

DOE Joint Circular 92/75 Wheelchair and mobility housing, standards and cost, London, HMSO, 1975

Julia Farrant and Alice Subiotto *Planning for disabled people in the urban environment* (research study by the Planning Research Unit, Department of Urban Design and Regional Planning, University of Edinburgh) London, Central Council for the Disabled, 1969

Selwyn Goldsmith *Designing for the disabled*, 3rd edn, London, RIBA Publications, 1976

Selwyn Goldsmith, DOE, HDD Occasional Papers 2/74 and 2/75 *Mobility housing* and *Wheelchair housing* (*The Architects' Journal* 3.7.74 and 25.6.75)

38 Sanitary installations and cloakrooms

Alan Tye Design Ltd

CI/SfB 94
CI/SfB (1976 revised) 94

Contents

1 INTRODUCTION

1.01 Installation standards

Most sanitary installations are unsatisfactory. Professor Kira has emphasised two key factors:

- at the production end, there are no installation manufacturers, only material producers. Companies make dissimilar products such as dinner plates and wcs because they are ceramic, others make taps because they are metal and no one can conceive of an integrated entirety.
- at the user end, the public are unaware of what to demand. The manufacturer does not know what the user needs as he sells through builders' merchants who somewhat arbitrarily decide and control what is to be sold.

Strong words, but in essence undeniably correct and one of the reasons why sanitary installations are so inadequate. Architects need to be much more critical and demanding about what manufacturers supply.

Sanitary installations are places where one is obliged to perform the most private functions in public with strangers of the same sex.

Quite different gangways are needed on a tube train where to brush closely against Mr X is acceptable, as opposed to between two urinal rows where to brush against the same Mr X is almost criminal. Apart from the football club type of situation where camaraderie permits closeness or the traditional factory where closeness is born of economic necessity and lack of care, the fundamental point of planning spacing in public installations is that psychological not just physical clearances and spacing are required.

The purely physical and unacceptably tight spacing of proprietary sanitary wall systems naturally encourage the user not to care about the installation which is not really designed for him.

Certainly more design work is needed on this subject for at the moment we generally both design and accept degrading and crude installations.

2 EARLY PLANNING

2.01 Guide to planning areas

In the early planning stages, if you merely wish to establish an overall sanitary installation area you can obtain the approximate size of an installation for any number of persons from the graph, **38.1**.

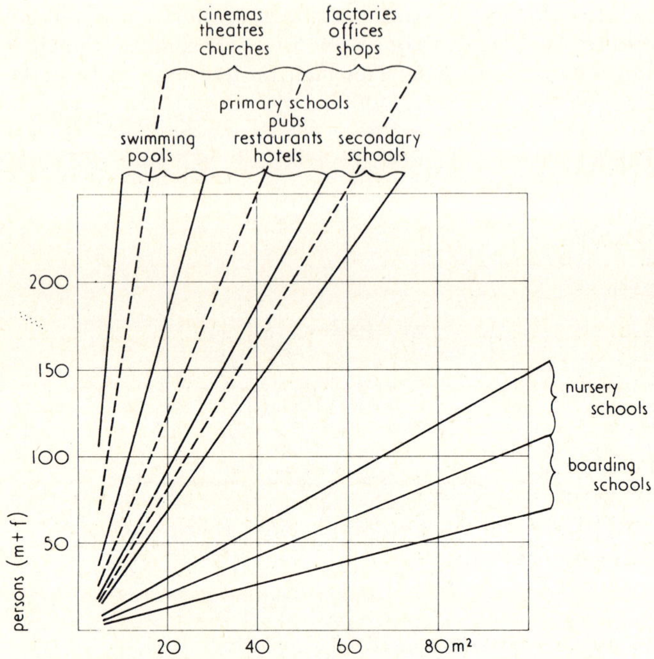

38.1 *Approximate guide to areas of sanitary installations for use in early planning stage. Based on CP 305, and equal male and female numbers are assumed. Obviously, no installation can be less than about 3 m² (*WC *and* WB *each for male and female)*

3 NUMBERS OF APPLIANCES REQUIRED

The recommendations given in table I to XI are derived from CP 305 Part 1: 1974. The specialist section authors' alternative recommendations are given where the CP was thought to be inadequate. They give the minimum requirements.

Designers should consult CP 305 for more detailed information and current guidance documents produced for each building type. In all situations, attention is drawn to the necessity to provide facilities for the disabled and also for the disposal of sanitary towels.

4 PLANNING THE SPACE

It is desirable that circulation of people through the sanitary area space is essentially one way, **38.2**. Single entry/exit plans can however work satisfactorily provided that the paths of users do not cross each other and the entry is wide enough. Placing the appliances in order of use simplifies circulation and reduces the distance walked. Hygiene should be encouraged by placing washing and drying facilities between the WC and/or urinal and the exit.

Actual doors on the entrance should be avoided wherever possible, see para **5.03**, Privacy.

There are significant differences between the sanitary accommodation of different building functions for the same number of people, **38.3**. It is therefore vital (1) that a proper calculation is made of numbers required for the particular building function, and (2) that correct appliance and activity spaces are adopted. In each case the example shown is for 100 people.

Table I Swimming pools

Appliances	For spectators Males	Females	For bathers* Males	Females
wcs	1 for 1–200 2 for 201–500 3 for 501–1000 Over 1000, 3 plus 1 for every additional 500 or part thereof	1 for 1–100 2 for 101–250 3 for 251–500 Over 500, 3 plus 1 for every additional 400 or part thereof	1 for each 60 persons	1 for each 40 persons
Urinals	1 per 50 persons		1 for each 60 persons	
Washbasins	1 per 60 persons	1 per 60 persons	1 for each 60 persons	1 for each 60 persons
Showers			1 for each 40 persons	1 for each 40 persons

* Number of persons based on pool capacity calculated in accordance with MHLG Design Bulletin No 45 Swimming Pools, 1962

Table II Hospitals*

Appliances	For medical staff quarters Male	Female	For nurses' homes
wcs	1 per 4 persons	1 per 4 persons	1 per 6 persons
Washbasins	1 in each bedroom, plus at least one with each group of wcs	1 in each bedroom, plus at least one with each group of wcs	1 in each bedroom, plus at least one with each group of wcs
Baths	1 per 4 persons	1 per 4 persons	1 per 6 persons
Hair shampoo basins		1 per 18 persons	1 per 18 persons
Cleaners' sinks	At least one per floor		

* *Administrative buildings*—staff may be subject to the Regulations under the Offices, Shops and Railway Premises Act 1963
Wards Sanitary provision for wards must also be considered; the type and number required will vary with the type of hospital. The norm is 1 wc per 6 patients. Sanitary accommodation should also be provided for both male and female visitors, preferably associated with a waiting room

Table III Office buildings and shops

Appliances	Accommodation other than principals, etc For male staff	For female staff
wcs* (no urinals provided)	1 for 1–15 persons 2 for 16–30 persons 3 for 31–50 persons 4 for 51–75 persons 5 for 76–100 persons For over 100 plus 1 for every additional 25 persons or part thereof	1 for 1–15 persons 2 for 16–30 persons 3 for 31–50 persons 4 for 51–75 persons 5 for 76–100 persons For over 100 plus 1 for every additional 25 persons or part thereof
wcs* (urinals provided)	1 for 1–20 persons 2 for 21–45 persons 3 for 46–75 persons 4 for 76–100 persons For over 100 plus 1 for every additional 25 persons or part thereof, but 1 in 4 of the additional fitments may be a urinal	
Urinals* 1 stall or 600 mm‡ (2 ft) of space	0 for 1–15 persons 1 for 16–30 persons 2 for 31–60 persons 3 for 61–90 persons 4 for 91–100 persons For over 100 additional provision determined by the number of wcs, (see previous item)	
Washbasins, trough or washing fountain†	1 for 1–15 persons 2 for 16–30 persons 3 for 31–50 persons 4 for 51–75 persons 5 for 76–100 persons For over 100 plus 1 for every additional 25 persons or part thereof	1 for 1–15 persons 2 for 16–30 persons 3 for 31–50 persons 4 for 51–75 persons 5 for 76–100 persons For over 100 plus 1 for every additional 25 persons or part thereof
Cleaners' sinks	At least 1 per floor, preferably in or adjacent to a sanitary apartment	
Bins, incinerators or macerator units for sanitary dressings disposal		1 in every sanitary accommodation regularly used by females

Based on the requirements of the Offices, Shop and Railway Premises Act 1963
* Sanitary Conveniences Regulations 1964 SI No 966
† Washing Facilities Regulations 1964 SI No 965
‡ 625 mm min recommended in appliance space diagrams

Table IV Dwellings

Type of dwelling	Appliances	Dwellings suitable for up to: 2 persons	3 persons	4 persons	5 persons	6 persons and above	All types and sizes of dwelling
On one level, eg bungalows and flats	wc	1*	1*	1†	1†	2‡	
	Bath	1	1	1	1	1	
	Washbasin	1	1	1	1	1	and in addition 1 in every separate wc compartment which does not adjoin a bathroom
	Sink and drainer	1	1	1	1	1	
On two or more levels, eg houses or maisonettes	wc	1*	1*	1†	2‡	2‡	
	Bath	1	1	1	1	1	
	Washbasin	1	1	1	1	1	and in addition 1 in every separate wc compartment which does not adjoin a bathroom
	Sink and drainer	1	1	1	1	1	

* May be in the bathroom
† Recommended to be in a separate compartment
‡ Of which, one may be in a bathroom
NOTE 1 In the private sector, a shower is equal to a bath (Public Health Act 1961, section 33)
NOTE 2 In Scotland, a washbasin is required in every wc compartment (Regulation Q7 (6))
NOTE 3 In Scotland, a shower may be provided instead of a bath in all dwellings both private and public (Regulation Q7)
This table is based on *Homes for today and tomorrow* (the Parker Morris Report) and Circular 36/37 of the Ministry of Housing and Local Government, and the new Scottish Housing Handbook Bulletin No 1, Metric space standards. Attention is directed to the Building Standards (Scotland) (Consolidation) Regulations, 1971 where these apply.

Table V Accommodation for elderly people

Type of accommodation	Appliances		Remarks
Self-contained for 1 or 2 old people, or grouped flatlet for 2 less active old people	wc	1 }	wc may be in bathroom
	Bath	1 }	Bathroom within dwelling
	Washbasin		
	Sink and drainer	1	
Grouped flatlet for single less active old person	wc	1	
	Washbasin	1	in wc compartment
	Sink and drainer	1	
	Bath—not less than 1 to four flatlets		Some may be sitting baths
Where communal facilities are provided, the following additional provisions must be made:			
Common room for self-contained or grouped flatlet dwellings	wc	1	Minimum number required
	Washbasin	1	In or adjoining the wc compartment
Tea pantry or kitchen	Sink and drainer	1	Adjacent to common room
Laundry room for grouped flatlet schemes	Sink	1	
	Washing machine	1	
	Spin drier	1	

This table is based on Circular 82/69 of the Ministry of Housing and Local Government and Circular 84/69 of the Welsh Office, Housing standards and costs—accommodation specially designed for old people. In Scotland the requirements are different; reference should be made to the New Scottish Housing Handbook, Bulletin No 3, Housing for old people with design standards for the disabled

Table VI Residential homes

Appliances*	Basic bed-sitting accommodation with other communal rooms including dining room
Generally	
Baths	1 to 15 persons
Hair washing cubicles	1 per floor
Footbaths	1 per floor
wcs (each with washbasin en suite)	1 to 6 men/women
Utility room	1 per floor
Sluice room	1 per floor
Cleaner's room	1 per floor
Washbasin	1 to each bed-sitting room
Mental homes	
wcs (each with washbasin)	1 to 5 people
Baths and showers	1 to 10 people (min 2)
Old people's homes	
Bathroom (incl wcs and washbasin)	1 to 15 people
Additional separate wcs (with washbasin)	1 to 4 people

Table VI Residential homes (*continued*)

Appliances*	Basic bed-sitting accommodation with other communal rooms including dining room
Community homes	
Washbasins	1 per child in single bedrooms. 1 to 3 children in shared bedrooms
Baths	1 to 8 children
Showers	1 for 5 to 12 children, 2 for more than 12
wcs	1 to 5 children (if upstairs provide 1 to 20 children on the ground floor)

* Separate washing and wc facilities for visitors should be considered

Table VII Factories*

Appliances	For male personnel	For female personnel
wcs†	1 per 25, or part thereof. For over 100, add at rate of 1 per 40 —if 'sufficient' urinals also provided	1 per 25, or part thereof
Urinals†	No specific interpretation of 'sufficient'; for guidance see table III	
Wash basins, troughs or washing fountains†	1 per 20—clean processes 1 per 10—dirty processes 1 per 5—injurious processes	1 per 20—clean processes 1 per 10—dirty processes 1 per 5—injurious processes
Baths or showers	As required for particular trades or occupations	

*For offices in factories, see table III
† Attention is called to the Factories Act 1963, Sanitary Accommodation Regulations, S R and O 1938 No 611 and other Regulations applying to specific trades or occupations which may require more extensive provision

Table VIII Schools and higher educational establishments

Appliances	Special schools	Primary schools	Secondary schools
All 'fittings' ie wcs and urinals	1 per 10 pupils	Less than 50 pupils—2 fittings per 40, or part thereof, plus 2 additional fittings More than 50 pupils— 2 fittings per 40, or part thereof, plus 4 additional fittings	Less than 1000 pupils—2 fittings per 30 More than 1000 pupils— as approved in each case
wcs only	Girls—all fittings Boys under 8 years—optional all fittings	Girls—all fittings Boys under 8 years—optional all fittings	Girls—all fittings
Urinals	Boys over 8 years—$^1/_3$ of fittings to be wcs, balance to be urinals Boys under 8 years— up to $^2/_3$ of fittings may be urinals	Boys and mixed over 8 years—$^1/_3$ of fittings to be wcs, balance to be urinals Boys under 8 years— up to $^2/_3$ of fittings may be urinals	Boys and mixed $^1/_3$ of fittings to be wcs, balance to be urinals
Washbasins	1 per 10 pupils	Not less than the number of fittings	Not less than the number of fittings
Showers			Sufficient for physical education
Washbasin (medical inspection)	At least 1, accessible to wc	At least 1, accessible to wc	At least 1, accessible to wc
Cleaners'/slop sink	At least 1 per floor	At least 1 per floor	At least 1 per floor
Nursery schools and play schools			
wcs	1 per 10 children under 5 years old		
Washbasins	1 per 5 children under 5 years old, where more than 50% stay for a midday meal or 1 per 10 children		
Sinks	1 per 40 children under 5 years old		
*Boarding schools**			
wcs	1 per 5 pupils		
Washbasins	1 per 3 pupils for first 20 pupils 1 per 4 pupils for up to 100 pupils 1 per 5 pupils for every additional 5 pupils		
Baths	1 per 10 pupils		
Shower baths	May be provided as alternative to not more than half the number of baths		

* Where sanitary accommodation for day pupils is accessible to, and suitable for the needs of boarders, these requirements may be reduced to such extent as may be approved in each case.
Teaching and other staff Some may be subject to Regulations under the Offices, Shops and Railway Premises Act 1963.
Note Based on Standards for school premises 1972, SI 2051, DES.

Table IX Cinemas, concert halls, theatres and similar buildings used for public entertainment, churches, etc

Appliances	For male public*	For female public*
wcs	In theatres, concert halls and similar premises: Minimum 1 up to 250 males plus 1 for every additional 500 males or part thereof In cinemas: Minimum 1 up to 250 males plus 1 for every additional 500 males or part thereof (GLC recommendations are: 1 for up to 200; 2 for 201–500; 3 for 501–1000; over 1000 3+1/500.) Accommodation for churches and other places of religious worship: 1†	In theatres, concert halls and similar premises: Minimum 2 up to 75 females plus 1 for every additional 50 females or part thereof In cinemas: Minimum 2 up to 100 females plus 1 for every additional 100 females or part thereof Accommodation for churches and other places of religious worship: 1†
Urinals	In theatres, concert halls and similar premises: Minimum 2 up to 100 males plus 1 for every additional 50 males or part thereof	

table continues on p. 340

Table IX Cinemas, concert halls, theatres and similar buildings used for public entertainment, churches, etc (*continued*)

Appliances	For male public*	For female public*
Urinals	In cinemas: Minimum 2 up to 200 males plus 1 for every additional 100 males or part thereof (GLC recommendations are: theatres 1 for every 25 males; concert halls etc, 1 for every 50 males)	
Washbasins	1 in respect of each wc and, in addition, 1 in respect of each 5 urinals Accommodation for churches and other places of religious worship: 1†	1 in respect of each wc Accommodation for churches and other places of religious worship: 1†
Cleaners' sinks	Adequate provision should be made for cleaning facilities including at least one cleaners' sink	

* It may be assumed that the total audience will consist of equal numbers of males and females
† In cases of buildings or places (including churches, places of historic interest etc) where a number of people congregate, sanitary accommodation should be provided having regard to the anticipated maximum number of people congregating there
Staff Some members may be subject to the Regulations made under the Offices, Shops and Railway Premises Act; for other members, the Regulations should be used as a guide.
Performers The sanitary accommodation provided will depend upon individual circumstances but guidance may be obtained from the Regulations made under the Offices, Shops and Railway Premises Act.

Table X Hotels

Appliances	For residential public and staff	For public rooms males	females
wcs	1 per 9 persons omitting occupants of rooms with wcs en suite	1 per 100 up to 400. For over 400, add at the rate of 1 per 250 or part thereof	2 per 100 up to 200. For over 200, add at the rate of 1 per 100 or part thereof
Urinals		1 per 50 persons	
Washbasins	1 per bedroom and at least 1 per bathroom	1 per wc	1 per wc
Bathrooms	1 per 9 persons omitting occupants of rooms with baths en suite		
Cleaners' sinks	1 per 30 bedrooms, minimum 1 per floor		

Non-residential staff Some may be subject to the Regulations under the Offices, Shops and Railway Premises Act

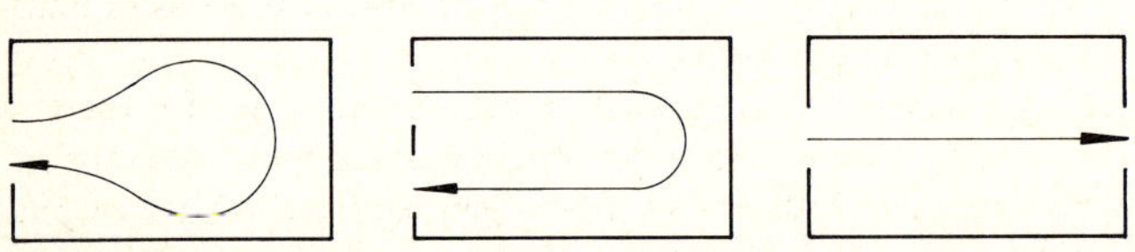

38.2 *Circulation through sanitary installation*

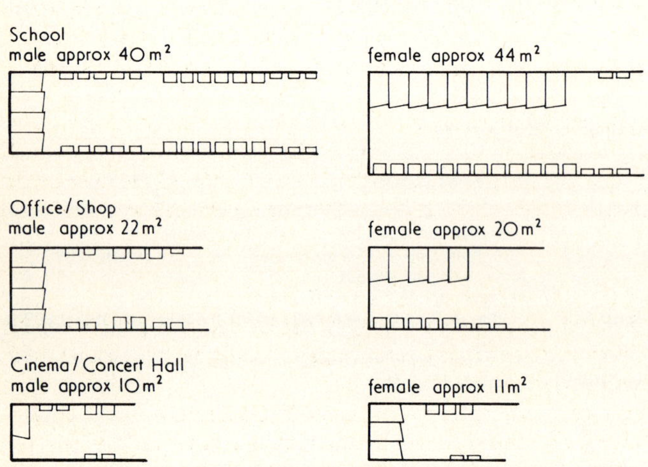

38.3 *Scales of provision for 100 people in different building types*

Table XI Restaurants, public houses and canteens

Appliances	For male public*	For female public*†
wcs	1 per 100 up to 400. For over 400, add at the rate of 1 per 250 or part thereof	2 per 100 up to 200. For over 200, add at the rate of 1 per 100 or part thereof
Urinals	1 per 25 persons	
Washbasins	1 in respect of each wc and, in addition, 1 in respect of each 5 urinals	1 in respect of each 2 wcs

* It may be assumed that there will be equal numbers of males and females
† A more satisfactory provision is: 1 for 1–15, 2 for 16–35, 3 for 36–65, 4 for 66–100, plus 3% for more than 100
This table is based on the Technical Regulations for Places of Public Entertainment in Greater London, 1965. In the case of licensed buildings which are conversions, the licensing authority may permit less than these requirements, where the total number of persons does not exceed 200.
Staff Some may be subject to the Regulations under the Offices, Shops and Railway Premises Act.
In canteens, the need for sanitary accommodation for persons using the canteen should be determined by the accessibility of normal workplace facilities, but at least one wc with washbasin should be provided for each sex.

5 OTHER CONSIDERATIONS

5.01 Ducts

As it is equally unacceptable to have pipes inside a room or outside it, ducts are an inevitable detail of sanitary installations.
Although pipes are frequently buried into wall structures, notably in Germany where appropriate pipework fittings exist, the UK situation is that we consider access is needed to critical points such as traps and cisterns. To achieve this access, ducts may be walk-in or have access from one or other sides.
For hygiene, cleaning maintenance, structural soundness and planning flexibility, plan **38.4a** is superior to **38.4b**.

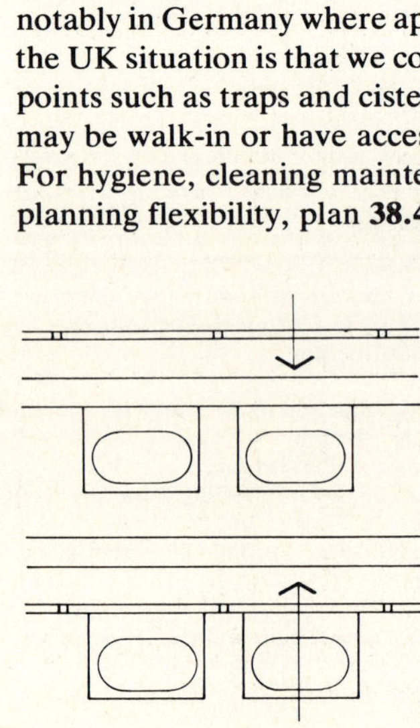

38.4 *Service ducts:*
a *Access from rear*
b *Access from front*

Although if the appliance could be part of the wall, plan **38.4b** would be very useful, in practice this plan usually results in the adoption of a standard ceramic appliance on a sheet plastic panel with its attendant impractical problems, details and module (which, in addition, seldom correctly relate to required activity spaces or correct operational heights).

On the other hand, **38.4a** demands more detailed work from the architect and it is not always possible to utilise adjacent room areas for access.

Once the duct is provided it is logical and more hygienic to bring the water supplies directly through the duct wall to wall mounted valves, rather than the pipework passing through the wall, sanitaryware, then on to deck mounted valves.

It is similarly illogical to have traps hanging from appliances in the room if there is space in the duct for them although some sanitaryware makes this unavoidable. Ducts should avoid maintenance problems and not have ferrous metal in their construction.

5.02 Tiles and modules

As UK manufacturers have never accepted the 100 mm modular tile (except as a special) and as UK manufacturers will not make sanitaryware modular, it is difficult to find a reason for choosing one module or another for the sanitary installation. It is, however, obvious, in view of the quite differing requirements of the various appliances and the individual situation, that any module should be as small as practicable. The best module is probably that used in Alvar Aalto's office ± 1 mm.

5.03 Privacy and screening

Vision is traditionally seriously considered in the planning of lavatories, although sound and odour are sources of considerable concern for many people and should also be considered, particularly in larger installations.

Vision

In larger installations, vision should be obstructed by the configuration of the entrance and in principle entrance doors should be avoided **38.5**. In smaller installations doors should open inwards and be hung so as to screen the appliances and the user as far as possible when opened. The doors to adjacent male and female rooms should not be next to each other as this is psychologically disturbing and aggravates vision problems. Consideration should also be given to positioning of mirrors and to the gap created by the hinges. Doors should be self-closing wherever possible.

Noise

It is difficult and costly to satisfactorily insulate lavatories acoustically and this problem should be tackled by planning isolation if possible.

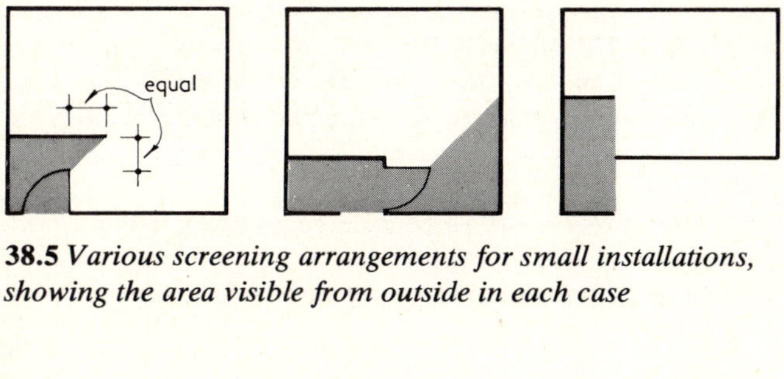

38.5 *Various screening arrangements for small installations, showing the area visible from outside in each case*

Odour

Except in extremely well naturally ventilated installations, some form of forced ventilation or air conditioning is desirable, particularly so in confined areas. Manually switched fans which continue to run for a set period after being switched off are useful in domestic situations.

5.04 Vandalism

No unsupervised installation can resist vandals. Even with the most vandal-resistant equipment (which would have to exclude all ceramics) an unsupervised facility will inevitably become substandard. In such situations the use of an attendant will result in a high standard being maintained, possibly with reduced costs.

A well-designed installation, easily kept clean, with an open layout, a high level of general lighting and robust equipment securely fixed will reduce the problem. Where vandal-resistant appliances are thought necessary, stainless steel is considerably less prone to damage than ceramics, but all designs should allow for individual items to be replaced.

Pipework, traps, cisterns, electrical supplies, etc should all be fully concealed and this is of course highly desirable from hygiene and appearance viewpoints anyway. The modular plastic panel is not desirable in areas likely to be vandalised.

5.05 The disabled

Disabled people are remarkably adaptable and often of necessity extremely determined to manage for themselves, albeit with considerable discomfort, in buildings designed primarily for able-bodied people. For many ambulant disabled people, the difficulties are surmountable, but for wheelchair users the problems are more serious for if an area is not negotiable by a wheelchair, then the user is forbidden entry and this is intolerable in any new buildings.

It is therefore vital that proper consideration should be given to the provision of wc and washing facilities for the disabled. Selwyn Goldsmith's book *Designing for the disabled* (3rd edition) is the most comprehensive study available. There is hardly any ergonomic evidence on this subject and the standard plans should be regarded as principles rather than unalterable working drawings.

Selwyn Goldsmith suggests that wc compartments for the disabled can often usefully be unisex; this has several advantages:

- husbands and wives can assist each other which is not possible in single sex compartments
- they avoid the need for and cost of duplicated facilities for each sex; one decent unisex facility can be considerably more economic than two inadequate single sex facilities
- they simplify signposting and access to disabled facilities.

WC compartments for ambulant disabled

Two alternatives from Selwyn Goldsmith's book are shown, **38.6**. The narrow width is desirable as it allows the user to use support rails on both sides of the compartment to raise themselves.

WC compartment for wheelchair users

A wc compartment for general use by disabled people should allow for frontal or lateral transfer from the wheelchair, with space for an attendant to assist.

In wcs for wheelchair users in public buildings or special buildings for disabled people a hand rinse basin should be installed where it can be conveniently reached by a person seated on the wc. However it is desirable that the basin is also usable from the wheelchair. These opposing criteria together with the

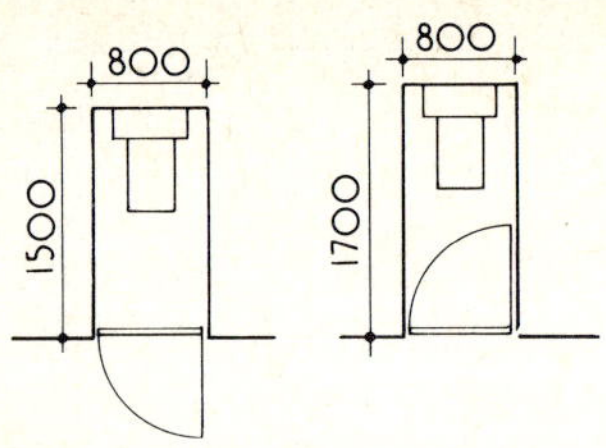

38.6 *Alternative* WC *cubicles for the ambulant disabled (from* Designing for the Disabled, *by Selwyn Goldsmith)*

38.9 *Family bathroom incorporating second washbasin for children's use*

requirements for handrails and supports present a difficult problem usually resulting in a poor or even unworkable compromise.

Selwyn Goldsmith's recommended WC compartment for chair-bound users, **38.7**, may be compared to an alternative plan by Alan Tye Design, **38.8**, embodying the principles of Selwyn Goldsmith's recommendations in a neat and pleasant facility.

Support rails

Alexander Kira points out that grab rails **37.10** for normal people, the elderly and children would virtually obviate the need for special provision for the disabled. Every bath needs a vertical pole, **37.7** particularly if used for showering.

The preferred diameter for support rails is 35 mm diameter with 50 mm clearance between the rail and the wall. Most proprietary rails are incorrectly sized. Rails must be securely fixed to the structural surface, and horizontal or inclined rails should be capable of carrying a static load of 150 kg. In practice, rails often need to be tailored to particular situations.

5.06 Children

As sanitary installations become less primitive and better thought out, it is correctly becoming standard practice to make provision in public buildings for the disabled, but strangely, other than in schools, children are usually neglected in public conveniences where appliances are adult size and fixed at adult height. This is quite unsatisfactory.

Even in fairly small public installations, at least one washbasin and urinal pod should be installed at reduced heights for children.

Children also deserve and appreciate consideration in the placement of dryers, towels, coat hooks and mirrors, etc and in domestic family situations a second lower basin should be considered, **38.9**.

5.07 Cleaning manual/mechanical

Cleanliness can either be visual or bacteriologically sterile or both. Most people are happy with visual cleanliness and would be unhappy with a dirty looking though sterile installation. Good design and detailing therefore plays a vital part in sanitary accommodation cleanliness.

Cleaning is rarely considered very seriously in sanitary areas, for most proprietary appliances are badly designed and uncleanable, producing dark shadows around down-to-floor appliances, cubicle feet, etc and too often one sees pipes and cistern exposed.

It is important to carefully consider the cleaning method to be employed and be critical of what manufacturers supply, for besides hygienic and aesthetic factors, dirty installations encourage sloppy use and vandalism. Generally one should ensure that all surfaces are capable of thorough cleansing, floor details should be coved wherever possible and all possible interruptions of the floor, such as legs, pedestals and pipes, avoided.

Wall mounted appliances are preferred, although many wall mounted WCs leave so little floor clearance as to actually aggravate the cleaning problem. Appliances designed to eliminate uncleanable areas should be chosen, they should be cleanable not only in the appliance itself but also in its junction with the structural surfaces as well. Generally the appliance is frequently cleaned but the wall is not, so all appliances need upstands. Appliances should not be placed so close together that cleaning between them is hampered. Wall-mounted valves over washbasins promote better cleaning than deck-mounted ones.

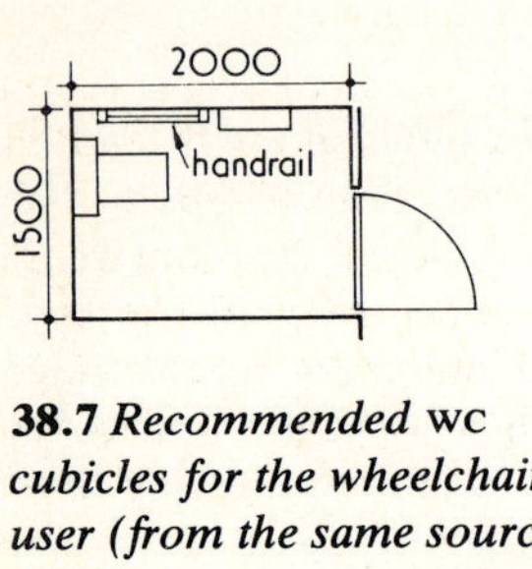

38.7 *Recommended* WC *cubicles for the wheelchair user (from the same source)*

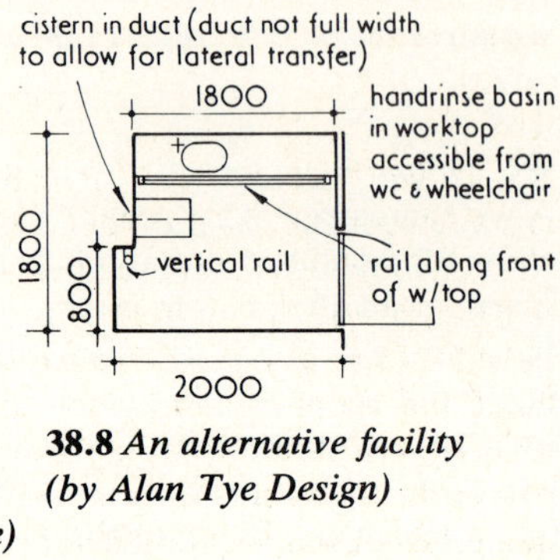

38.8 *An alternative facility (by Alan Tye Design)*

6 DETAIL PLANNING

For planning convenience we have combined appliance and activity spaces into one area. Two figures are given for each dimension—minimum (not absolute minimum) and desirable (not maximum) dimension. Where the dimensions given are larger than those normally used, the traditional figures are considered to be inadequate, **38.10** to **38.16**.

A degree of overlap of appliance/activity spaces is acceptable, depending on the likelihood of simultaneous use of all appliances. However in no case should appliances be mounted so close as to inhibit cleaning between them.

With the exception of the WC cubicles (and possibly urinal slabs without divisions where the position of the user is flexible), the figures given assume that the appliance/activity areas are not bounded at the sides by walls, but are adjoining other similar areas. Where the side of an appliance/activity space is bounded by a wall, **38.17**, it is recommended that about 400 mm should be allowed from the centre of the appliance to the wall (this is only of significance where the appliance/activity space is less than 800 mm).

6.01 Circulation

When an area is used by only one person at a time circulation space can be minimised, however it is vital in planning public lavatories that ample circulation space is provided, **38.18** to **38.21**.

Appliances should be arranged so that this space is concentrated into larger areas as it is psychologically and practically preferable to be able to see the whole of the room on entering. Narrow dead ends and corridors should be avoided and the circulation pattern planned to ensure that washing facilities are provided between wc/urinals and the exit to encourage hand washing.

Circulation areas must be considered as being around both appliance and activity spaces rather than merely around appliances, although some encroachment of the circulation

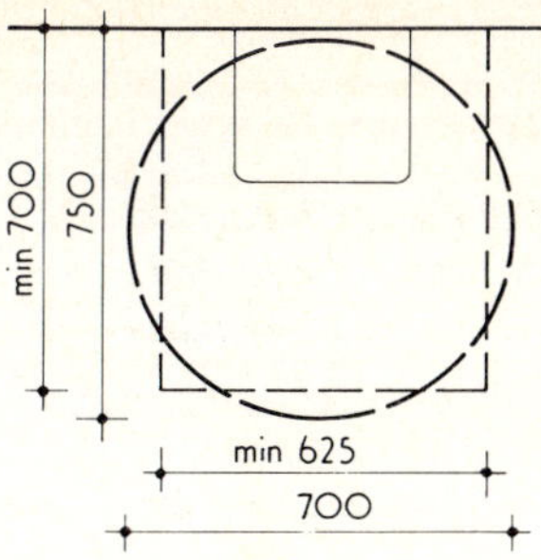

38.10 *Pod urinal*

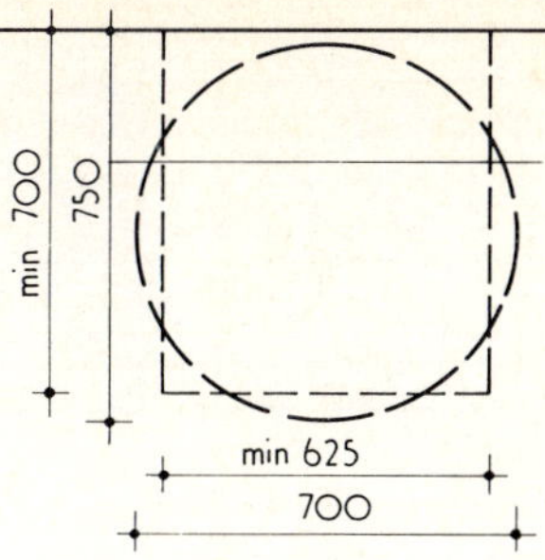

38.11 *Slab urinal*

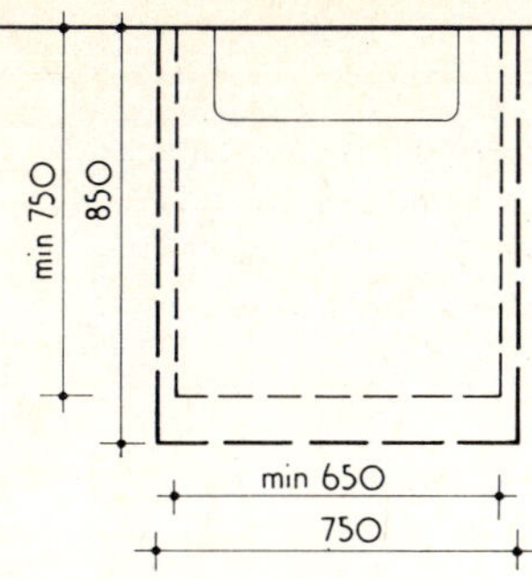

38.12 *Towel or drier*

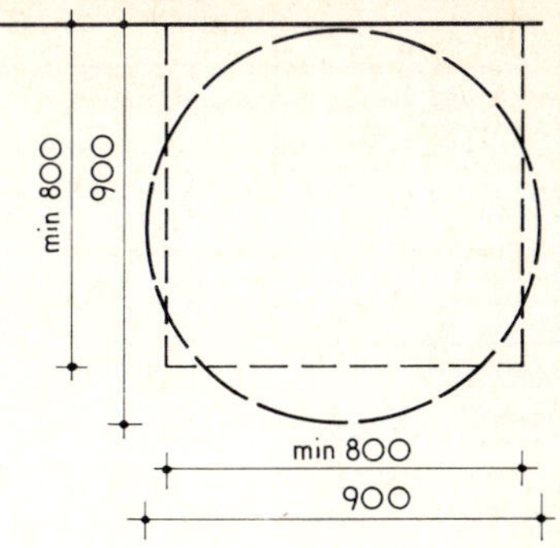

38.13 *Shower*

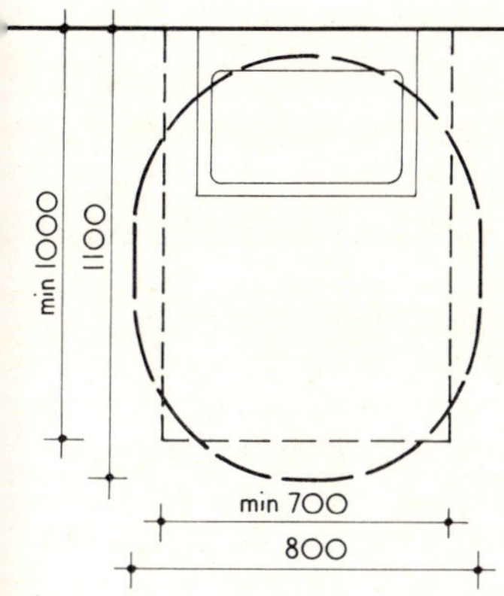

38.14 *Washbasin*

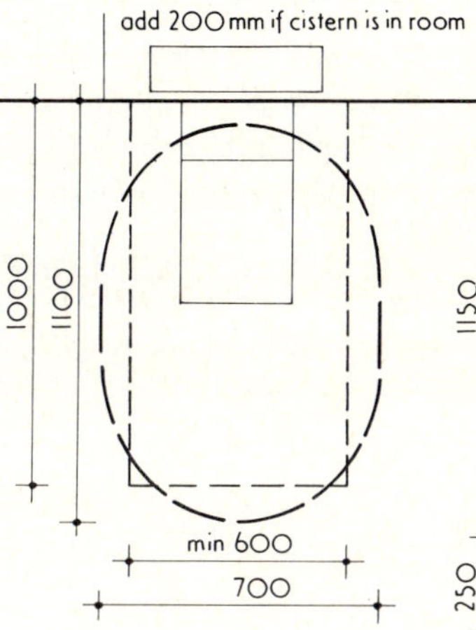

38.15 WC *and bidet*

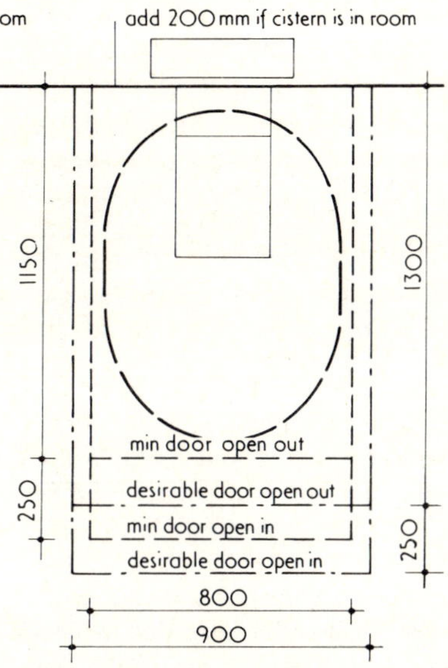

38.16 WC *cubicle*

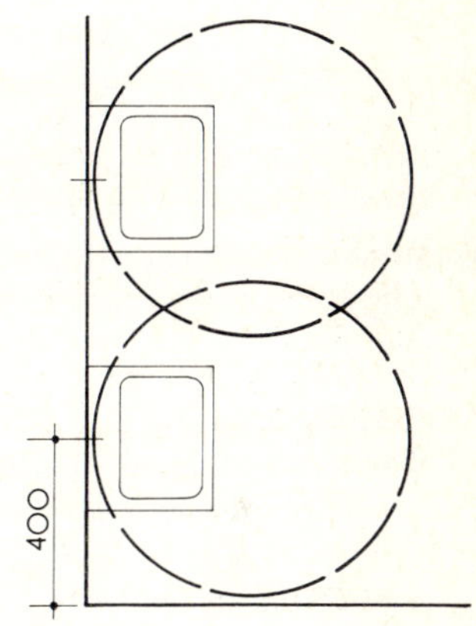

38.17 *Appliances against walls*

area into the appliance/activity space will normally be acceptable depending on the likelihood of full use of appliances. Allowances have been made in **38.18** to **38.21** for a degree of overlap, so the dimensions given should be thought of as straight additions onto the appliance/activity spaces given in **38.9** to **38.15**.

The architect should decide which diagram is applicable, depending on the likely density of traffic. The two figures given in each case are minimum (not absolute minimum) and desirable (not maximum) dimension respectively.

On larger installations, gangways must be carefully considered relative to the length of the room and number of appliances in a row. Circulation spaces generally have to be larger in sanitary installations than in other public spaces.

6.02 Fixing heights of appliances

Washbasins

The traditional fixing height of basins is 785 or 800 mm to the rim. For adults this height requires considerable bending as one is actually washing one's hands below the rim height. Alexander Kira suggests a height of 915 to 965 mm, but for normal use by a wide range of users 850 mm is preferred. However, the needs of children must be remembered and if they are likely to use the facility at least one basin should be mounted at a lower height of about 700 mm.

WCs

There has been a great deal of controversy regarding WC heights. The traditionally accepted height (for adult use) is 400 mm. Various lower heights are available for children, and a height in excess of 450 mm can be necessary for elderly and disabled persons. Wall hung WCs have the advantage that they may be positioned at a variety of heights providing there is adequate clearance below for cleaning.

These traditional heights are far higher than the physiologically desirable heights which are as low as 250 mm. However there seems to be little move towards the 'health closet' at present, as it is hampered by western tradition, the difficulty of the elderly or infirm to use a low level WC and the uselessness of a low WC as a seat or urinal which are common functions in domestic bathrooms.

Urinal pods

The preferred rim height for adults is 600 mm and again in any installation which young boys may use, at least one pod should be mounted at a lower level of about 500 mm to the rim. Children needing a lower height than this are normally helped by parents at a WC.

7 SAUNAS

7.01 Origin of the sauna

The sauna, **38.22**, is essentially Finnish and in its original form is a one-room hut built of logs, with a rudimentary furnace or stove, over which rocks are piled, in one corner. Steps lead up to a slatted wooden platform along one side of the room where naked bathers sit or lie in the hot air under the roof. The stove heats the room by convection and the rocks reach a high temperature. After sitting in the dry heat for some minutes, the bather produces steam from time to time by throwing small quantities of water onto the hot rocks.

38.18 *Minimum width in domestic bathroom*

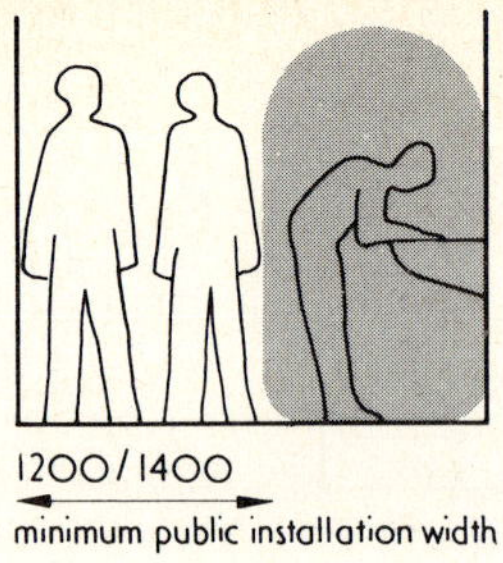

38.19 *Minimum width in public installation: appliances one side*

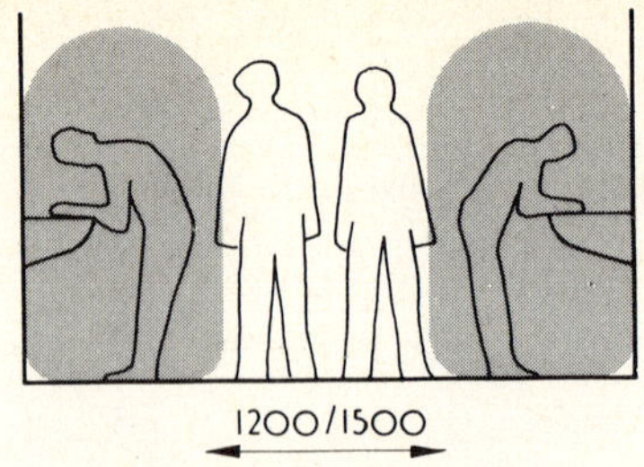

38.20 *Minimum width in public installation with appliances both sides*

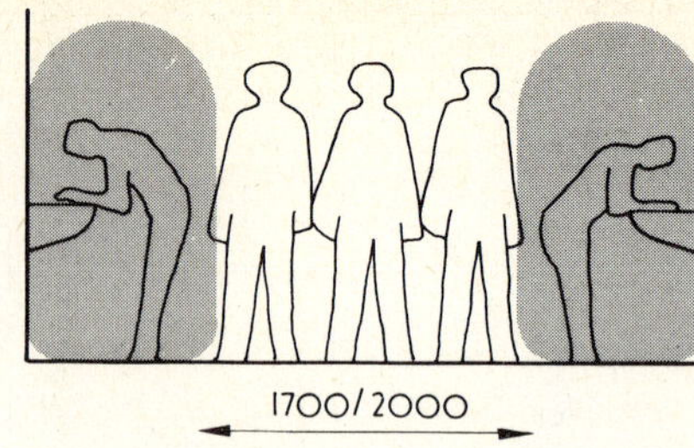

38.21 *Width of larger public installation*

The temperature varies from 88°C to 110°C and, provided that the moisture is properly absorbed by the wooden walls of the room, the air will not become saturated. Because the human body can stand a higher degree of dry heat than wet heat, the temperature is higher in a sauna than in a Turkish bath. After perspiration, bathers beat themselves with leafy birch twigs, wash and plunge into a nearby lake or take a cold shower. The cycle is repeated a few times until finally there is a period of rest while the body cools down completely. The time taken for the complete operation varies from 90 to 120 minutes.

7.02 Dimensions

The sauna room should be between 2·300 and 2·600 m high and have a minimum floor area of 1·800 × 2·100 m. Benches should be 600 to 760 mm wide and the platform at least 460 mm wide. The stove will take up 0·560 to 0·650 m² of floor area and will stand about 1·070 m high. Outside the sauna room, showers are required and if possible a cold 4 to 10°C plunge bath. Space for dressing and resting should be provided. Cubicles will strictly limit the maximum number of bathers, and an open layout is more flexible. Provision should be made for clothes lockers and a few dressing cubicles, and the rest of the area is occupied by resting couches/chairs and small tables.

Of the total number of bathers in an establishment at any time, 20 to 25 per cent are likely to be in the sauna room, an equal number in the shower/washing room and the remainder in the dressing/resting areas.

8 HYDRO-THERAPY SPA BATHS

This healthy new development in bathroom equipment is an alternative to the sauna. Spa baths or 'swirlpools' are small hydro-therapy pools that provide turbulent hot water as massage for the relief of aches, tensions and fatigue or simply for pleasure.

These pools are usually of one-piece glass-fibre construction, available in a variety of sizes and shapes which are relatively easy and low-cost to install and are used indoors or out.

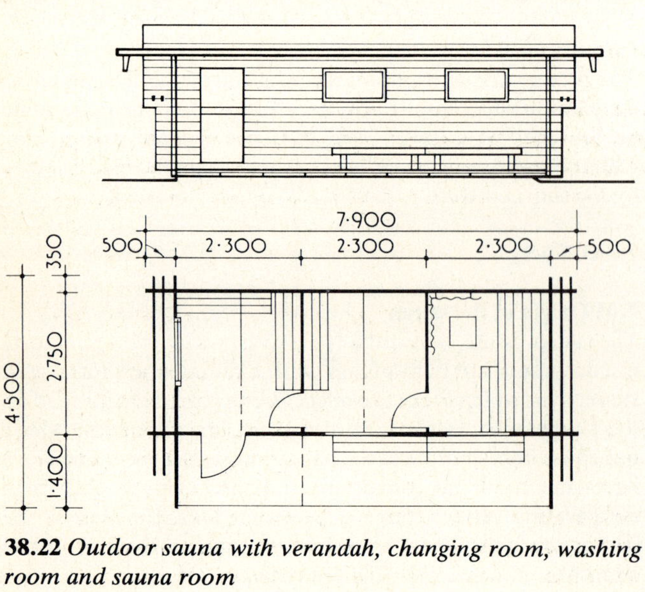

38.22 *Outdoor sauna with verandah, changing room, washing room and sauna room*

9 PUBLIC CLOAKROOMS

9.01 Items commonly stored

Typical sizes of items stored are given in **38.23**.

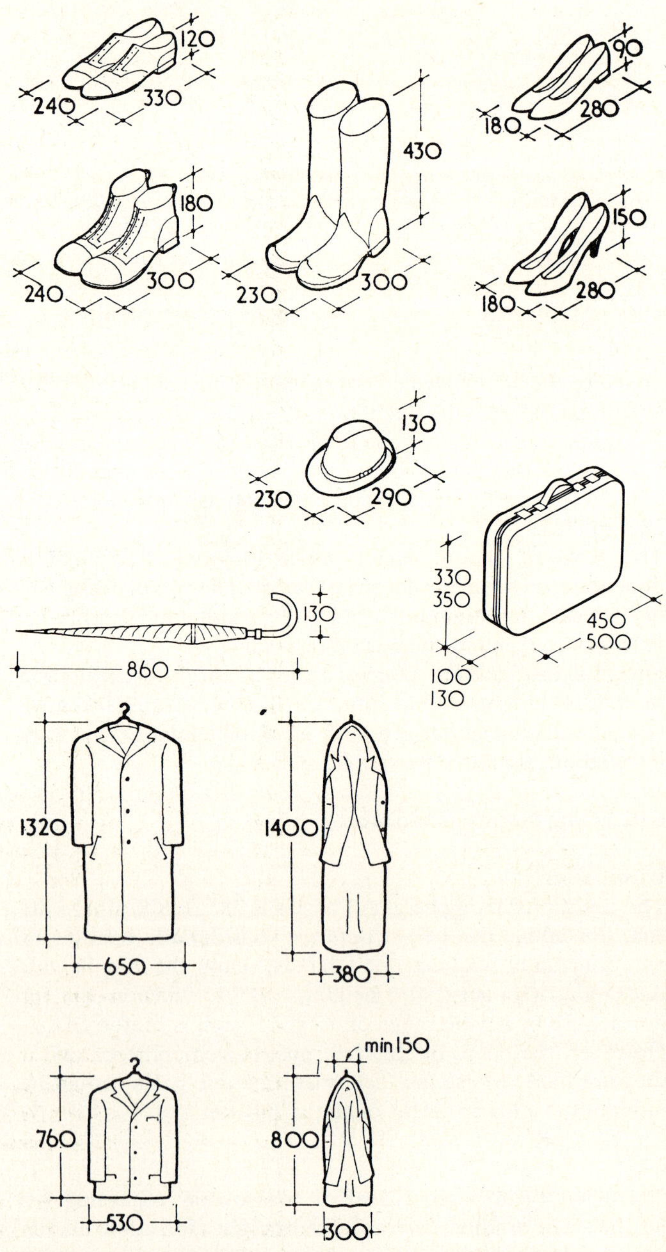

38.23 *Sizes of items commonly stored in cloakrooms*

9.02 Attended storage

Typical arrangements and space requirements per user are shown in **38.24** and **38.25**.

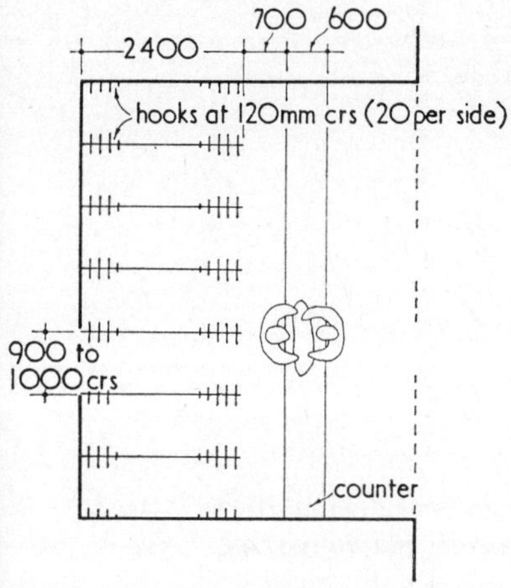

38.24 *Fixed rows of hooks. 0·08 m² per user including counter, 0·1 m² including 1200 mm on public side*

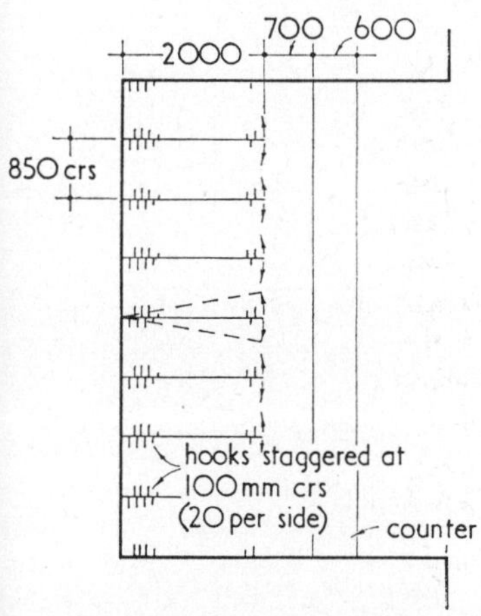

38.25 *Hinged rows of hooks. 0·07 m² per user including counter, 0·09 m² including 1200 mm on public side*

9.03 Unattended storage

The space allowances per user in **38.26** to **38.30** are based upon hangers or hooks at 150 mm in rows 3600 mm long with 1050 mm clear circulation space at ends of rows.

The proprietary system given in **38.31** provides unattended locked storage for coats and umbrellas.

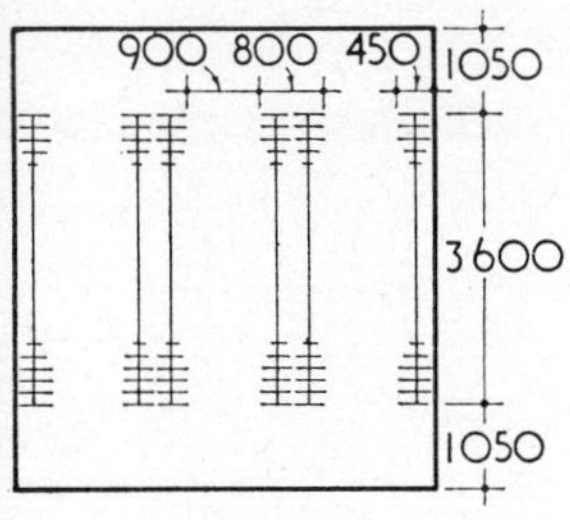

38.26 *Method of calculating space required by each user:*
6 rows each 3600 mm long of double-tier hangers at 150 mm centres = 300 hangers
area of room = 5·7 × 5·2 = 29·6 m²
Hence, space allowance 0·098 m² per user

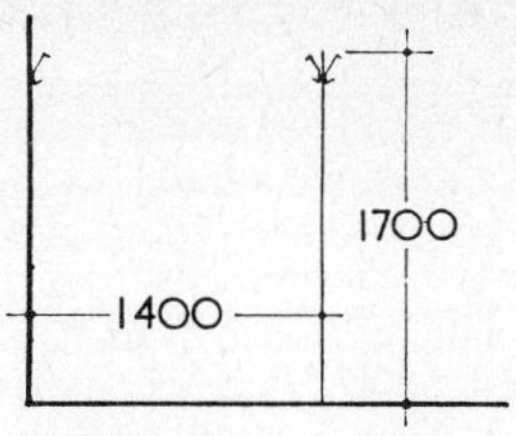

38.27 *Hooks in line: 0·16 m² per user including circulation*

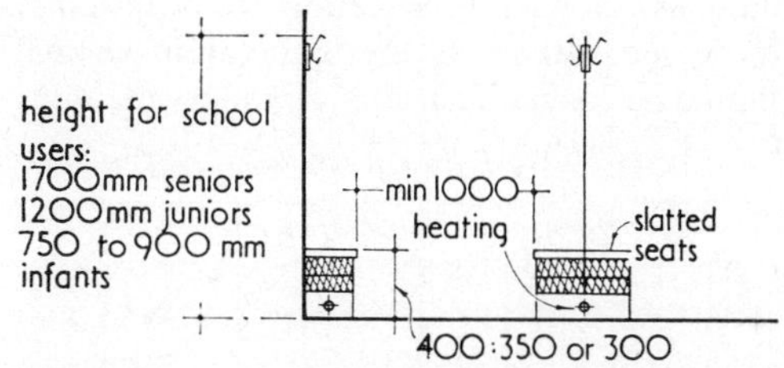

38.28 *Hooks with seating: 0·2 m² per user including circulation*

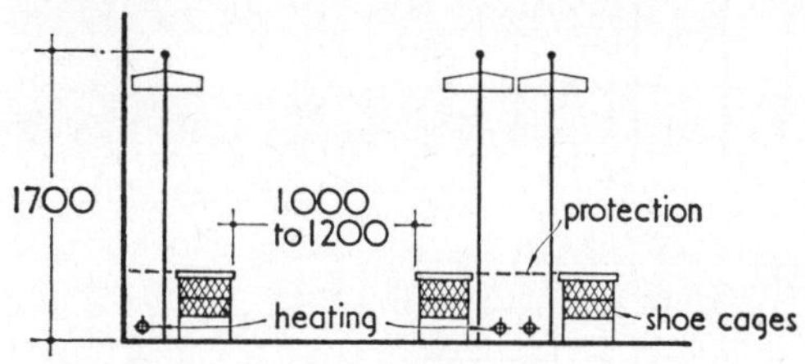

38.29 *Hangers with seating: 0·26 m² per user including circulation*

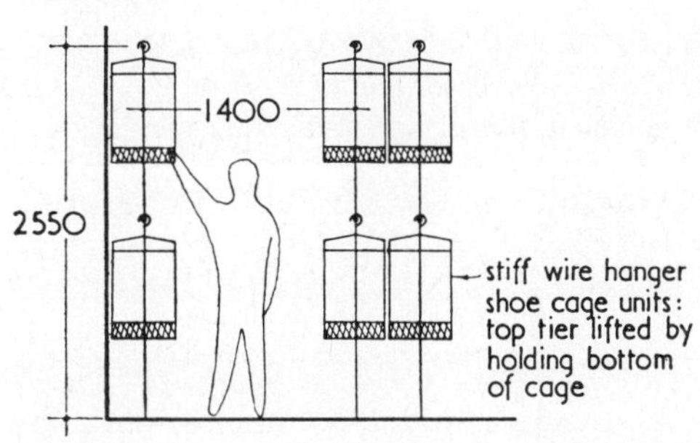

38.30 *Double tier hangers: 0·13 m² per user including circulation*

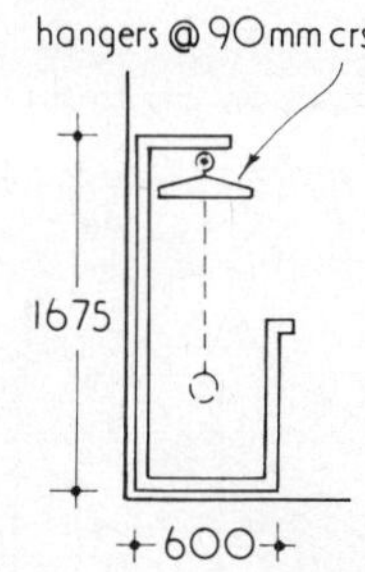

38.31 *Proprietary system affording security: 0·16 m² per user including circulation*

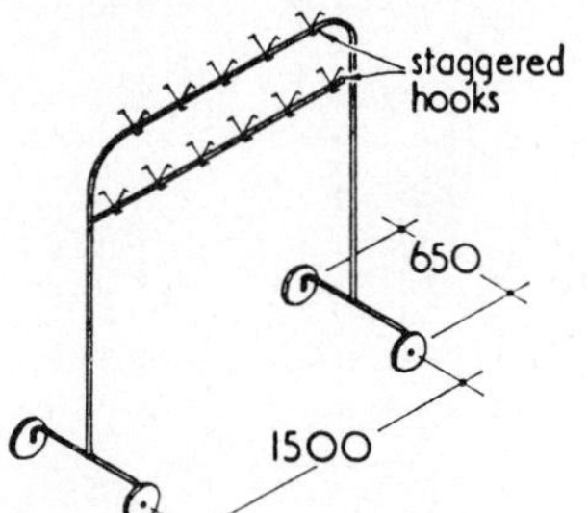

38.32 *Mobile coat rack*

9.04 Mobile storage

These are proprietary units and the measurements shown in **38.32** are approximate.

9.05 Lockers

Lockers may be full height with a hat shelf and space to hang a coat and store shoes or parcels; or half height to take a jacket; or quarter height to take either parcels or folded clothes, **38.33** and **38.34**. Combination units such as **38.35** are also available. *Note:* Many of the units shown in this section are proprietary systems and metric measurements are only approximate. Manufacturers should be consulted after preliminary planning stages.

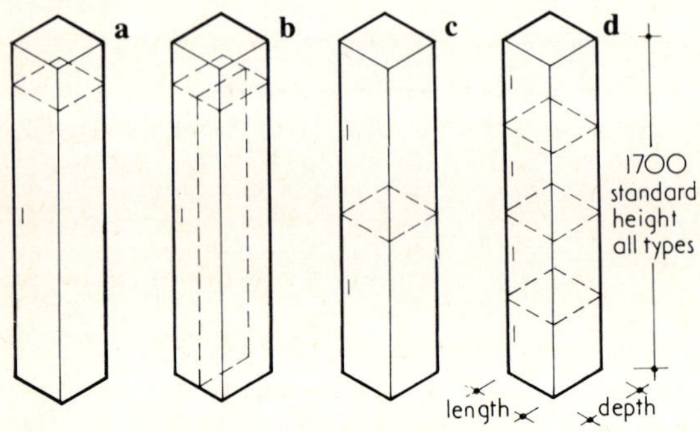

	reference size (mm)					
	A	B	C	D	E	F
length	300	300	300	400	400	500
depth	300	400	500	400	500	500

38.33 *Lockers:* **a** *type 1a, with hat shelf;* **b** *type 1b, with hat shelf and vertical divider. Not available in A, B or C;* **c** *type 2, two tier;* **d** *type 3, three, four, five or six tier*

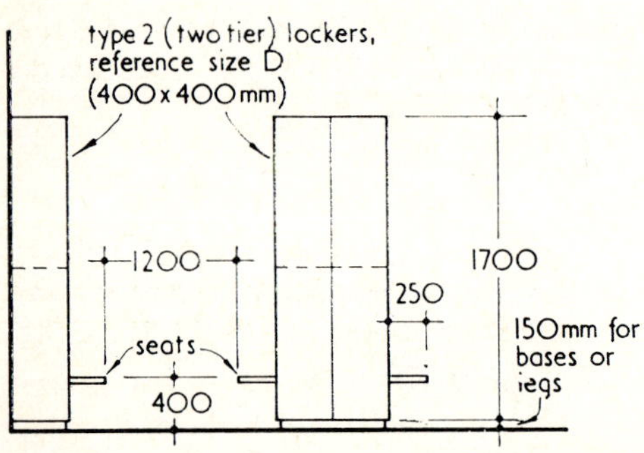

38.34 *Typical cross-section of lockers with seats*

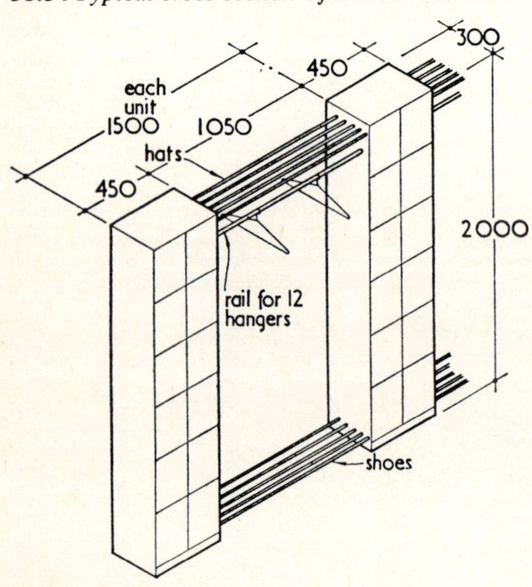

38.35 *Lockers with coat rail, hat and shoe racks*

9.06 Calculating cloakroom areas

In the early planning stages, if you merely wish to establish an overall cloakroom area, you can obtain the size of a cloakroom to suit any number of coats from the graph, **38.36**.

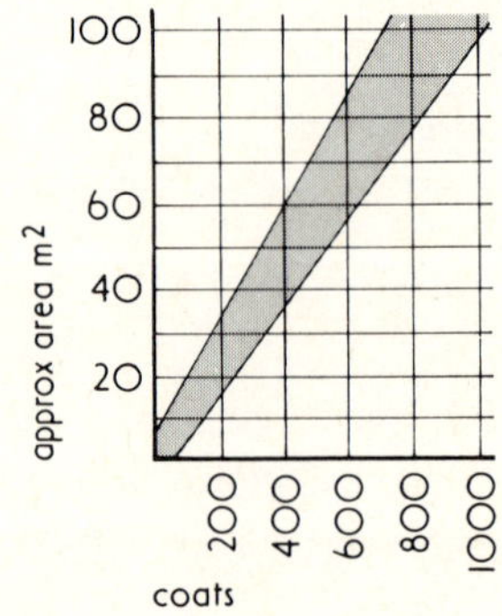

38.36 *Approximate guide to areas of cloakroom accommodation for use in early planning stage*

39 Materials

David Adler

CI/SfB Yy
CI/SfB (1976 revised) Ya

Contents

1 STEEL

Most steel in the UK originates with the British Steel Corporation. From time to time the BSC issues lists of sizes and sections currently in normal production.

All steel products are sized in metric dimensions and billed in kilograms or tonnes. However, only simple geometric sections such as flats and rounds, and angle sections are made in metric co-ordinated sizes. I-beams, channels, etc are still made to the traditional 'inch' sizes, although in many cases these sizes were in fact nominal. It is not, apparently, intended to change this situation in the immediate future.

I-beams, etc are available in metric co-ordinated sizes from the continent, as they have been ever since steel construction started at the end of the nineteenth century.

Tables I to XVI show the common sizes of the main steel profiles.

2 TIMBER

Timber used in building is either softwood or hardwood, depending on species; each may be supplied sawn or finished. Sizes are usually quoted 'ex', meaning the sawn size.

2.01 Softwood

The normally available sizes of sawn softwood are given in table XVII. The various finishing processes result in reduction to the finished size, which is given in table XVIII. This reduction takes into account:

- occasional under-measure in sawn size
- kiln-drying from 20 per cent to required moisture content resulting in shrinkage
- conversion from larger sizes with loss of saw kerf
- straightening material with slight bend or twist
- surface and finish machining
- sanding one or two faces.

To the figures given in the table a tolerance of ±0·5 mm should be allowed. Joinery standard allows for high degree of straightness, and both back and front are finished. Trim standard allows for the back to be rough as concealed, and a lower requirement for straightness.

Standard softwood profiles are available in a number of varieties, of which tongue-and-groove floorboarding is the commonest. Table XIX gives the standard dimensions of these.

Softwood is normally supplied in lengths varying from 1·8 m in increments of 0·3 m up to a maximum of 6·3 m. The maxima vary for different species and size of section.

2.02 Hardwood

Hardwood is normally supplied in planks of specified thickness but arbitrary width and length, depending on species and thickness. The standard thicknesses are: 19, 25, 32, 36, 38, 44, 50, 63, 75, 100, 150, 200, 250 and 300 mm.

Table I Mild steel plate standard sizes

THICKNESS	2000 × 1000	2500 × 1250	3000 × 1500	4000 × 1500	4000 × 1750	4000 × 2000	4000 × 2500	5000 × 1500	5000 × 2000	5000 × 2500	6000 × 1500	6000 × 2000	6000 × 2500	8000 × 1500	8000 × 2000	8000 × 2500	10000 × 2000	10000 × 2500
6		×	×	×	×			×			×			×				
8		×	×	×	×			×			×			×				
10		×	×	×	×		×	×	×	×	×	×	×	×	×	×		
12·5		×	×	×	×	×	×	×	×	×	×	×	×	×	×	×	×	×
15		×	×	×		×	×	×	×	×	×	×	×	×	×	×	×	×
20		×	×	×		×	×	×	×	×	×	×	×	×	×	×	×	×
25		×	×	×		×	×	×	×	×	×	×	×	×	×	×	×	×
30		×	×	×		×	×	×	×	×	×	×	×	×	×	×	×	×
35		×	×	×		×	×	×	×	×	×	×	×	×	×	×	×	×
40		×	×	×		×	×	×	×	×	×	×	×	×	×	×	×	×
45		×	×	×		×	×	×										
50	×	×	×	×		×	×	×										
60	×	×	×	×		×	×											
65	×	×	×	×		×												
70	×	×	×	×		×												
75	×	×	×	×		×												
80	×	×	×	×		×												
90	×	×	×	×		×												
100	×	×	×	×		×												
110	×	×	×	×		×												
130	×	×	×	×		×												
150	×	×	×	×		×												

Table II Hot rolled flats

Width (mm)	Standard thickness (mm) 3	5	6	8	10	12	15	18	20	25	30	32	35	40	45	50	60	65
25	×	×	×	×	×	×			×									
30	×	×	×	×	×	×	×		×	×								
35	×		×	×	×	×	×		×	×								
40	×	×	×	×	×	×	×		×	×								
45	×		×	×	×	×	×		×	×								
50	×	×	×	×	×	×	×	×	×	×	×		×	×				
55			×	×	×	×			×	×	×							
60	×		×	×	×	×	×		×	×	×		×	×	×	×		
65	×	×	×	×	×	×	×	×	×	×	×		×	×	×	×		
70		×	×	×	×	×	×		×	×	×			×				
75			×	×	×	×	×	×	×	×	×	×	×	×		×		×
80		×	×	×	×	×	×		×	×	×		×	×	×	×		
90			×	×	×	×	×		×	×	×		×	×	×	×	×	×
100		×	×	×	×	×	×		×	×	×		×	×	×	×	×	×
110					×	×	×		×	×	×			×				
120						×	×		×	×				×				
130		×	×	×	×	×	×		×	×	×			×	×	×	×	
140						×	×											
150			×	×	×	×	×		×	×	×			×		×		
160					×	×	×		×	×	×		×	×				
180					×	×	×		×	×	×		×	×				
200						×	×		×	×	×		×	×	×	×	×	
220						×	×		×	×	×		×	×	×	×	×	
250						×	×		×	×	×		×	×	×	×	×	
275						×	×		×	×	×		×	×	×	×	×	
300						×	×		×	×	×		×	×	×	×	×	
325							×		×	×	×		×	×	×	×	×	
350							×		×	×	×		×	×	×	×	×	
375									×	×	×		×	×	×	×	×	
400									×	×	×		×	×	×	×	×	
425									×	×	×		×	×	×	×	×	
450									×	×	×		×	×	×	×	×	
475										×	×		×	×	×	×		
500										×	×		×	×	×	×		
525										×	×		×	×	×	×		
550										×	×		×	×	×	×		
575										×	×		×	×	×	×		
600										×	×		×	×	×	×		

Table III Channels to BS4: part 1:1972

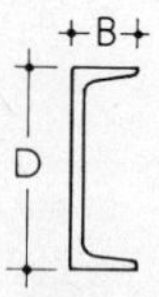

Designation: Nominal size mm	Mass per metre kg	Depth of section D mm	Width of section B mm
432 × 102	65·54	431·8	101·6
381 × 102	55·10	381·0	101·6
305 × 102	46·18	304·8	101·6
305 × 89	41·69	304·8	88·9
254 × 89	35·74	254·0	88·9
254 × 76	28·29	254·0	76·2
229 × 89	32·76	228·6	88·9
229 × 76	26·06	228·6	76·2
203 × 89	29·78	203·2	88·9
203 × 76	23·82	203·2	76·2
178 × 89	26·81	177·8	88·9
178 × 76	20·84	177·8	76·2
152 × 89	23·84	152·4	88·9
152 × 76	17·88	152·4	76·2
127 × 64	14·90	127·0	63·5
102 × 51	10·42	101·6	50·8
76 × 38	6·70	76·2	38·1

Table IV Equal angles to BS4848: part 4:1972

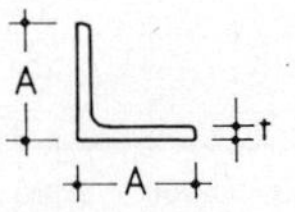

Designation: Size A mm	Thickness t mm	Mass per metre kg
200 × 200	24	71·1
	20	59·9
	18	54·2
	16	48·5
150 × 150	18	40·1
	15	33·8
	12	27·3
	10	23·0
120 × 120	15	26·6
	12	21·6
	10	18·2
	8	14·7
100 × 100	15	21·9
	12	17·8
	8	12·2
90 × 90	12	15·9
	10	13·4
	8	10·9
	6	8·30

table continues on p. 349

Table IV Equal angles to BS4848: part 4:1972 (*continued*)

Designation: Size A mm	Thickness t mm	Mass per metre kg
80 × 80	10	11·9
	8	9·63
	6	7·34
70 × 70	10	10·3
	8	8·36
	6	6·38
60 × 60	10	8·69
	8	7·09
	6	5·42
	5	4·57
50 × 50	8	5·82
	6	4·47
	5	3·77
45 × 45	6	4·00
	5	3·38
	4	2·74
40 × 40	6	3·52
	5	2·97
	4	2·42
25 × 25	5	1·77
	4	1·45
	3	1·11

Table V Unequal angles to BS4848: part 4:1972

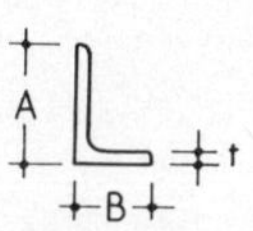

Designation: Size A × B mm	Thickness t mm	Mass per metre kg
200 × 150	18	47·1
	15	39·6
	12	32·0
200 × 100	15	33·7
	12	27·3
	10	23·0
150 × 90	15	26·6
	12	21·6
	10	18·2
150 × 75	15	24·8
	12	20·2
	10	17·0
125 × 75	12	17·8
	10	15·0
	8	12·2
100 × 75	12	15·4
	10	13·0
	8	10·6
100 × 65	10	12·3
	8	9·94
	7	8·77
80 × 60	8	8·34
	7	7·36
	6	6·37
75 × 50	8	7·39
	6	5·65
65 × 50	8	6·75
	6	5·16
	5	4·35

Table VI Universal beams to BS4: part 1:1972

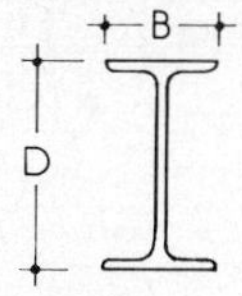

Designation: Serial size mm	Mass per metre kg	Depth of section D mm	Width of section B mm
914 × 419	388	920·5	420·5
	343	911·4	418·5
914 × 305	289	926·6	307·8
	253	918·5	305·5
	224	910·3	304·1
	201	903·0	303·4
838 × 292	226	850·9	293·8
	194	840·7	292·4
	176	834·9	291·6
762 × 267	197	769·6	268·0
	173	762·0	266·7
	147	753·9	265·3
686 × 254	170	692·9	255·8
	152	687·6	254·5
	140	683·5	253·7
	125	677·9	253·0
610 × 305	238	633·0	311·5
	179	617·5	307·0
	149	609·6	304·8
610 × 229	140	617·0	230·1
	125	611·9	229·0
	113	607·3	228·2
	101	602·2	227·6
533 × 210	122	544·6	211·9
	109	539·5	210·7
	101	536·7	210·1
	92	533·1	209·3
	82	528·3	208·7
457 × 191	98	467·4	192·8
	89	463·6	192·0
	82	460·2	191·3
	74	457·2	190·5
	67	453·6	189·9
457 × 152	82	465·1	153·5
	74	461·3	152·7
	67	457·2	151·9
	60	454·7	152·9
	52	449·8	152·4
406 × 178	74	412·8	179·7
	67	409·4	178·8
	60	406·4	177·8
	54	402·6	177·6
406 × 140	46	402·3	142·4
	39	397·3	141·8
356 × 171	67	364·0	173·2
	57	358·6	172·1
	51	355·6	171·5
	45	352·0	171·0
356 × 127	39	352·8	126·0
	33	348·5	125·4
305 × 165	54	310·9	166·8
	46	307·1	165·7
	40	303·8	165·1
305 × 127	48	310·4	125·2
	42	306·6	124·3
	37	303·8	123·5
305 × 102	33	312·7	102·4
	28	308·9	101·9
	25	304·8	101·6
254 × 146	43	259·6	147·3
	37	256·0	146·4
	31	251·5	146·1
254 × 102	28	260·4	102·1
	25	257·0	101·9
	22	254·0	101·6
203 × 133	30	206·8	133·8
	25	203·2	133·4

Table VII Universal columns to BS4: part 1:1972

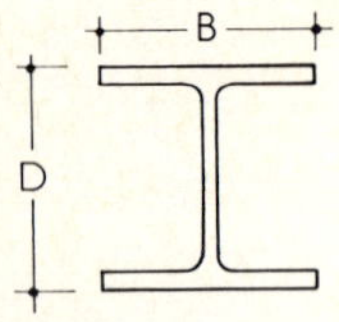

Designation		Depth of section	Width of section
Serial size mm	Mass per metre kg	D mm	B mm
356 × 406	634	474·7	424·1
	551	455·7	418·5
	467	436·6	412·4
	393	419·1	407·0
	340	406·4	403·0
	287	393·7	399·0
Column	235	381·0	395·0
core	477	427·0	424·4
356 × 368	202	374·7	374·4
	177	368·3	372·1
	153	362·0	370·2
	129	355·6	368·3
305× 305	283	365·3	321·8
	240	352·6	317·9
	198	339·9	314·1
	158	327·2	310·6
	137	320·5	308·7
	118	314·5	306·8
	97	307·8	304·8
254 × 254	167	289·1	264·5
	132	276·4	261·0
	107	266·7	258·3
	89	260·4	255·9
	73	254·0	254·0
203 × 203	86	222·3	208·8
	71	215·9	206·2
	60	209·6	205·2
	52	206·2	203·9
	46	203·2	203·2
152 × 152	37	161·8	154·4
	30	157·5	152·9
	23	152·4	152·4

Table VIII Joists to BS4: part 1:1972 where marked*

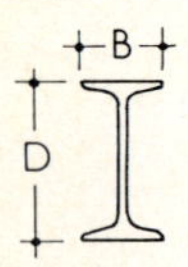

Designation		Depth of section	Width of section
Nominal size mm	Mass per metre kg	D mm	B mm
254 × 203	81·85	254·0	203·2
254 × 114	37·20	254·0	114·3
203 × 152	52·09	203·2	152·4
203 × 102	25·33*	203·2	101·6
178 × 102	21·54*	177·8	101·6
152 × 127	37·20	152·4	127·0
152 × 89	17·09*	152·4	88·9
152 × 76	17·86	152·4	76·2
127 × 114	29·76	127·0	114·3
127 × 114	26·79	127·0	114·3
127 × 76	16·37	127·0	76·2
127 × 76	13·36*	127·0	76·2
114 × 114	26·79	114·3	114·3
102 × 102	23·07	101·6	101·6
102 × 64	9·65*	101·6	63·5
102 × 44	7·44	101·6	44·4
89 × 89	19·35	88·9	88·9
76 × 76	14·67	76·2	76·2
76 × 76	12·65	76·2	76·2

Table IX Circular hollow sections to BS4848: part 2: 1975

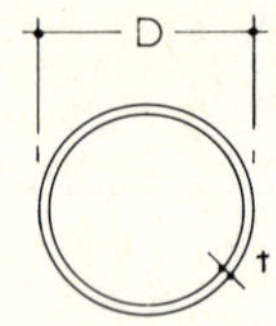

Designation		
Outside diameter D mm	Thickness t mm	Mass per metre kg
21·3	3·2	1·43
26·9	3·2	1·87
33·7	2·6	1·99
	3·2	2·41
	4·0	2·93
42·4	2·6	2·55
	3·2	3·09
	4·0	3·79
48·3	3·2	3·56
	4·0	4·37
	5·0	5·34
60·3	3·2	4·51
	4·0	5·55
	5·0	6·82
76·1	3·2	5·75
	4·0	7·11
	5·0	8·77
88·9	3·2	6·76
	4·0	8·38
	5·0	10·3
114·3	3·6	9·83
	5·0	13·5
	6·3	16·8
139·7	5·0	16·6
	6·3	20·7
	8·0	26·0
	10·0	32·0
168·3	5·0	20·1
	6·3	25·2
	8·0	31·6
	10·0	39·0
193·7	5·4	25·1
	6·3	29·1
	8·0	36·6
	10·0	45·3
	12·5	55·9
	16·0	70·1
219·1	6·3	33·1
	8·0	41·6
	10·0	51·6
	12·5	63·7
	16·0	80·1
	20·0	98·2
244·5	6·3	37·0
	8·0	46·7
	10·0	57·8
	12·5	71·5
	16·0	90·2
	20·0	111
273	6·3	41·4
	8·0	52·3
	10·0	64·9
	12·5	80·3
	16·0	101
	20·0	125
	25·0	153
323·9	8·0	62·3
	10·0	77·4
	12·5	96·0
	16·0	121
	20·0	150
	25·0	184
355·6	8·0	68·6
	10·0	85·2
	12·5	106
	16·0	134
	20·0	166
	25·0	204
406·4	10·0	97·8
	12·5	121
	16·0	154
	20·0	191
	25·0	235
	32·0	295

table continues on p. 351

Table IX Circular hollow sections to BS4828: part 2: 1975 (*continued*)

Designation: Outside diameter D mm	Thickness t mm	Mass per metre kg
457	10·0	110
	12·5	137
	16·0	174
	20·0	216
	25·0	266
	32·0	335
	40·0	411

Table XI Rectangular hollow sections to BS 4848: Part 2: 1975

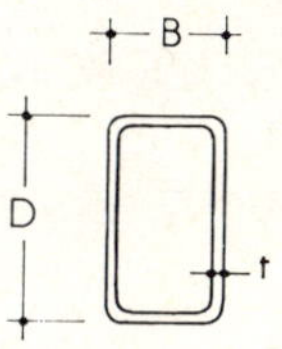

Designation: Size D × B mm	Thickness t mm	Mass per metre kg
50 × 30	2·6	3·03
	3·2	3·66
60 × 40	3·2	4·66
	4·0	5·72
80 × 40	3·2	5·67
	4·0	6·97
90 × 50	3·6	7·46
	5·0	10·1
100 × 50	3·2	7·18
	4·0	8·86
	5·0	10·9
100 × 60	3·6	8·59
	5·0	11·7
	6·3	14·4
120 × 60	3·6	9·72
	5·0	13·3
	6·3	16·4
120 × 80	5·0	14·8
	6·3	18·4
	8·0	22·9
	10·0	27·9
150 × 100	5·0	18·7
	6·3	23·3
	8·0	29·1
	10·0	35·7
160 × 80	5·0	18·0
	6·3	22·3
	8·0	27·9
	10·0	34·2
200 × 100	5·0	22·7
	6·3	28·3
	8·0	35·4
	10·0	43·6
	12·5	53·4
	16·0	66·4
250 × 150	6·3	38·2
	8·0	48·0
	10·0	59·3
	12·5	73·0
	16·0	91·5
300 × 200	6·3	48·1
	8·0	60·5
	10·0	75·0
	12·5	92·6
	16·0	117
400 × 200	10·0	90·7
	12·5	112
	16·0	142
450 × 250	10·0	106
	12·5	132
	16·0	167

Table X Rectangular hollow sections (square) to BS 4848: part 2: 1975

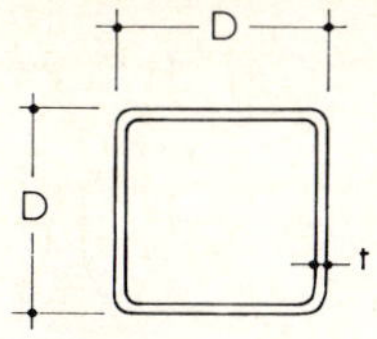

Designation: Size D × D mm	Thickness t mm	Mass per metre kg
20 × 20	2·0	1·12
	2·6	1·39
30 × 30	2·6	2·21
	3·2	2·65
40 × 40	2·6	3·03
	3·2	3·66
	4·0	4·46
50 × 50	3·2	4·66
	4·0	5·72
	5·0	6·97
60 × 60	3·2	5·67
	4·0	6·97
	5·0	8·54
70 × 70	3·6	7·46
	5·0	10·1
80 × 80	3·6	8·59
	5·0	11·7
	6·3	14·4
90 × 90	3·6	9·72
	5·0	13·3
	6·3	16·4
100 × 100	4·0	12·0
	5·0	14·8
	6·3	18·4
	8·0	22·9
	10·0	27·9
120 × 120	5·0	18·0
	6·3	22·3
	8·0	27·9
	10·0	34·2
150 × 150	5·0	22·7
	6·3	28·3
	8·0	35·4
	10·0	43·6
	12·5	53·4
	16·0	66·4
180 × 180	6·3	34·2
	8·0	43·0
	10·0	53·0
	12·5	65·2
	16·0	81·4
200 × 200	6·3	38·2
	8·0	48·0
	10·0	59·3
	12·5	73·0
	16·0	91·5
250 × 250	6·3	48·1
	8·0	60·5
	10·0	75·0
	12·5	92·6
	16·0	117
300 × 300	10·0	90·7
	12·5	112
	16·0	142
350 × 350	10·0	106
	12·5	132
	16·0	167
400 × 400	10·0	122
	12·5	152

Table XII Rounds for all purposes

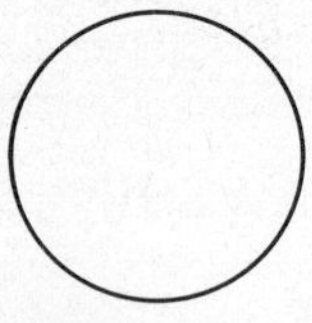

Diameter Standard mm	Mass/length kg/m	Diameter Standard mm	Mass/length kg/m	Diameter Standard mm	Mass/length kg/m	Diameter Standard mm	Mass/length kg/m
5·5	0·187	20·5	2·59	58·5	21·1	170·0	178
6·0	0·222	21·0	2·72	60·0	22·2	175·0	189
6·5	0·260	21·5	2·85	62·0	23·7	180·0	200
7·0	0·302	22·0	2·98	65·0	26·0	185·0	211
7·5	0·347	23·0	3·26	67·0	27·7	190·0	223
8·0	0·395	23·5	3·40	68·0	28·5	195·0	229
8·5	0·445	24·0	3·55	70·0	30·2	200·0	247
9·0	0·499	25·0	3·85	71·5	31·5	210·0	272
9·5	0·556	25·5	4·01	75·0	34·7	220·0	298
10·0	0·617	26·0	4·17	78·0	37·5	230·0	326
10·5	0·680	26·5	4·33	80·0	39·5	240·0	356
11·0	0·746	27·0	4·49	85·0	44·5	250·0	385
11·5	0·815	28·5	5·01	90·0	49·9	260·0	417
12·0	0·888	32·0	6·31	95·0	55·6	270·0	449
12·5	0·963	35·0	7·55	100·0	61·7	280·0	483
13·0	1·04	38·0	8·90	105·0	68·0	290·0	518
13·5	1·12	40·0	9·86	110·0	74·6	300·0	555
14·0	1·21	42·0	10·9	115·0	81·5	320·0	631
14·5	1·30	43·0	11·4	120·0	88·0	330·0	672
15·0	1·39	45·0	12·5	125·0	96·3		
16·0	1·58	46·0	13·1	130·0	104		
16·5	1·68	50·0	15·4	135·0	112		
17·0	1·78	51·0	16·0	140·0	121		
17·5	1·88	52·0	16·7	145·0	130		
18·0	2·00	52·5	17·0	150·0	139		
18·5	2·11	55·0	18·7	155·0	148		
19·0	2·23	55·5	19·0	160·0	158		
20·0	2·47	57·0	20·0	165·0	168		

Table XIII Squares

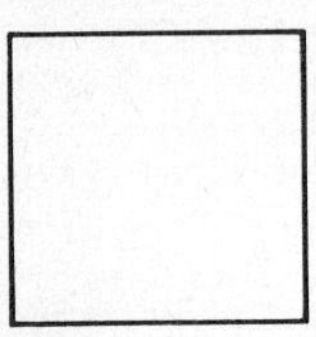

Side Standard mm	Mass/length kg/m
7·0	0·385
8·0	0·502
10·0	0·785
12·0	1·13
13·0	1·33
15·0	1·77
20·0	3·14
25·0	4·91
30·0	7·07
35·0	9·62
40·0	12·6

Table XIV Hexagons

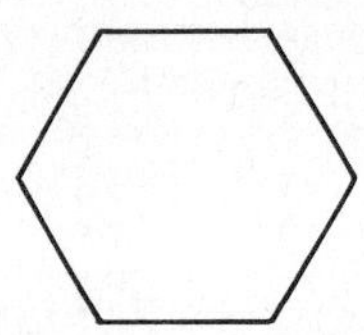

A/F Standard mm	Mass/length kg/m
9·5	0·61
11·0	0·82
12·5	1·06
14·5	1·43
15·0	1·53
16·	1·85
20·5	2·86

STEEL REINFORCEMENT FOR CONCRETE

Table XV Reinforcing bars

Size mm	Cross sectional area mm^2	Mass kg/m
6*	28·3	0·222
8	50·3	0·395
10	78·5	0·617
12	113·1	0·888
16	201·1	1·58
20	314·2	2·47
25	490·9	3·85
32	804·2	6·31
40	1256·6	9·86
50	1963·5	15·4

Sizes/qualities are available as follows:
Mild steel to BS 4449:1969 – all sizes
Bent reinf to BS 4466:1969 – all sizes and qualities
Torbar to BS 4461:1969 – 8 mm to 40 mm
Unisteel 410 to BS 4449:1969 – 16 mm to 50 mm

* 6 mm Torbar has been replaced by 6 mm indented wire to BS4482: 1969 'hard drawn mild steel wire for the reinforcement of concrete'

Table XVI Welded steel fabric reinforcement to BS4483:1969
Standard metric meshes

Mesh	British standard reference	Mesh sizes nominal pitch of wires		Size of wires		Cross sectional area				Nominal mass	
		Main mm	Cross mm	Main mm	Cross mm	Main mm^2/m	in^2/ft	Cross mm^2/m	in^2/ft	kg/m^2	lb/yd^2
Square mesh fabric	A393	200	200	10	10	393	·186	393	·186	6·16	11·35
	A252	200	200	8	8	252	·119	252	·119	3·95	7·28
	A193	200	200	7	7	193	·091	193	·091	3·02	5·57
	A142	200	200	6	6	142	·067	142	·067	2·22	4·09
	A98	200	200	5	5	98	·046	98	·046	1·54	2·84

table continues on p. 353

Table XVI Welded steel fabric reinforcement to BS4483: 1969 (*continued*)
Standard metric meshes

Mesh	British standard reference	Mesh sizes nominal pitch of wires		Size of wires		Cross sectional area				Nominal mass	
		Main mm	Cross mm	Main mm	Cross mm	Main mm^2/m	in^2/ft	Cross mm^2/m	in^2/ft	kg/m^2	lb/yd^2
Structural mesh fabric	B1131	100	200	12	8	1131	·534	252	·119	10·90	20·09
	B785	100	200	10	8	785	·371	252	·119	8·14	15·00
	B503	100	200	8	8	503	·238	252	·119	5·93	10·93
	B385	100	200	7	7	385	·182	193	·091	4·53	8·35
	B283	100	200	6	7	283	·134	193	·091	3·73	6·88
	B196	100	200	5	7	196	·093	193	·091	3·05	5·62
Long mesh fabric	C785	100	400	10	6	785	·371	70·8	·033	6·72	12·39
	C636	100	400	9	6	636	·300	70·8	·033	5·55	10·23
	C503	100	400	8	5	503	·238	49·0	·023	4·34	8·00
	C385	100	400	7	5	385	·182	49·0	·023	3·41	6·29
	C283	100	400	6	5	283	·134	49·0	·023	2·61	4·81
Wrapping fabric	D98	200	200	5	5	98	·046	98	·046	1·54	2·84
	D49	100	100	2·5	2·5	49·1	·023	49·1	·023	0·77	1·42

Preferred sheet and roll sizes

Flat sheets	4·8 m × 2·4 m (11·52 m^2)	All BS fabrics
Rolls	48 m × 2·4 m (115·2 m^2)	A142, A193, B283, B385, C385
	72 m × 2·4 m (172·8 m^2)	A98, B196

Table XVII Basic sizes of sawn softwoods

Thickness	Width 25	38	50	75	100	125	150	175	200	225	250	300
12	*	*	*	*	*		*					
16			*	†	†	†	†					
19	*	*	*	†	‡	‡	‡					
22				†	†	†	†					
25	*	*	*	‡	‡	†	‡	‡	†	‡	†	†
32				†	†	†	‡	‡	†	‡	†	†
36				†	†	†	†					
38		*	*	‡	‡	†	‡	†	†	‡		
40				†	†	†	†	†	†	†		
44				†	†	†	†	†	†	†	†	†
50			*	‡	‡	‡	‡	‡	‡	‡	†	†
63					†	†	‡	†	†	†		
75				*	‡	†	‡	†	†	‡	†	†
100					‡		†		†		†	†
150							†		†			†
200									†			
250											†	
300												†

KEY
† Sizes laid down in BS 4471:1969 Specification for dimensions for softwood (metric units). These sizes may be available from stock.
‡ Sizes from BS 4471:1969 that will probably be available from stock.
* Sizes not included in BS 4471:1969 that may be available from stock, or resawn from larger standard sizes.

Table XVIII Reduction from basic size to finished size of sawn softwoods

Purpose	For sawn sizes of width or thickness (mm) 15 to and inc 22	Over 22 to and inc 35	Over 35 to and inc 100	Over 100 to and inc 150	Over 150
Trim	5	5	7	7	9
Joinery and cabinet work	7	7	9	11	13

Table XIX Floor boards in mm

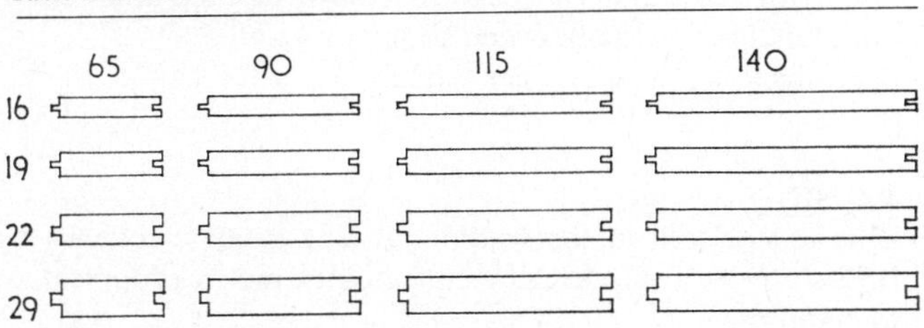

3 BRICKS AND BLOCKS

3.01 Bricks

The work size of the standard brick is 215 × 102·5 × 65, and this brick is supplied in an enormous variety of face colours and textures, strengths and other properties.

Other brick sizes are made, but the selection in each size is more limited. Near-standard sizes are:

215 × 102·5 × 50
215 × 170 × 66

Metric modular sizes are slowly increasing in popularity. These include:

190 × 90 × 90
190 × 90 × 65
290 × 90 × 90
290 × 90 × 65

3.02 Blocks

The sizes and types of blocks are laid down in BS 2028: 1364:1968 amended 1970 and 1978. Unfortunately, manufacturers do not seem to conform to all this standard's requirements. Table XX shows these requirements, and indicates the number of major manufacturers' standard products. From this it can be seen that:

- no major manufacturer appears to make type C blocks
- only one makes type A blocks
- the most easily available block sizes are 440 × 215 × 60, 75, 90, 100, 140, 150, 190, 200 and 215.

Other sizes are obtainable from most manufacturers by special order if a substantial number is required. Price and delivery problems are often experienced in these cases.

4 PRECAST CONCRETE

4.01 Paving flags

These are manufactured to the requirements of BS 368:1971 in two thicknesses: 50 and 63 mm, in the dimensions given in Table XXI. Some manufacturers make other sizes eg 500 × 500 × 50, 500 × 250 × 50 and 250 × 225 × 50 (co-ordinating sizes).

4.02 Concrete kerbs and edgings

These have not yet been metricated. Table XXII shows the current sizes expressed in millimetres*.

*BS 340: 1979 has recently superseded BS 340: 1963

Table XX Standard block sizes and availability

Work thickness	590 × 215	590 × 190	440 × 290	440 × 215*	440 × 190	440 × 140	390 × 190	390 × 90
	Work face size of blocks							
50				B – 3		B – 1		
60			B – 1	B – 4	B – 1	B – 1	B – 1	
	C	C	C	C	C		C	
				A – 1			A	A
75	B – 1	B	B – 1	B – 5	B – 1	B – 1	B – 1	B
	C	C	C	C	C		C	
90				A			A – 1	A
	B	B	B – 1	B – 3	B – 1	B – 1	B – 2	B
				A – 1			A – 1†	A – 1†
100	B – 1	B	B – 1	B – 5	B – 1	B – 1	B – 2	B
125	B – 1							
				A – 1			A – 1	A
140	B	B	B – 1	B – 4	B – 1 B – 1	B – 1	B – 2	B
150	B – 1		B – 1	B – 3	B – 1	B – 1		
							A – 1	
175	B – 1							
				A			A – 1	A
190	B	B	B – 1	B – 4	B – 1	B – 1	B – 2	B
200	B – 1		B – 1	B – 3	B – 1	B – 1		
				A				
215	B	B	B – 1	B – 4	B – 1			
				A – 1				
220				B – 1				
225	B – 1			B – 1				
230				B – 2				
255				B – 2‡		B – 1		
305				B – 1		B – 1		

* One manufacturer makes 448 × 215
† This block type has a special finish
‡ One of these blocks is actually 250 mm thick
KEY
In heavy type are the dimensions specified in the 1970 amendment to BS 2028, 1364:1968:
Type A: a block of density not less than 1500 kg/m^3, strength 3·5 to 35 N/mm^2, shrinkage not more than 0·06%; for general use in building, including below ground level DPC. If solid of dense concrete or a denser lightweight aggregate.
Type B: Any density, minimum strength 2·8 N/mm^2, shrinkage 0·09% maximum; for general use including below ground level DPC in internal walls, inner leaf of cavity walls.
Type C: Any density and strength, shrinkage 0·09% maximum; primarily for internal non-loadbearing walls, partitions, frame panel infill.
In light type are the numbers of manufacturers supplying the block in question as a standard product out of seven major block manufacturers. Where a block type and size is not preferred in BS 2028, the grading type A, B or C is shown in light type also.

Table XXI Flag dimensions

Flag type	Co-ordinating size mm	Work size mm	Maximum limit of manufacturing size mm	Minimum limit of manufacturing size mm
A	600 × 450	598 × 448	600 × 450	596 × 446
B	600 × 600	598 × 598	600 × 600	596 × 596
C	600 × 750	598 × 748	600 × 750	596 × 746
D	600 × 900	598 × 898	600 × 900	596 × 896

NOTE. When ordering, it will be necessary only to specify the type followed by the thickness, eg 'A 50'.

4.03 Flooring units
Precast flooring units are made for a wide variety of spans and loadings, both reinforced and prestressed. Some are intended to be used in conjunction with an in situ structural topping. Note that the quoted dimensions usually refer to the work sizes, not co-ordinating dimensions, despite being rounded figures.
In general, structural floor thicknesses will follow the following series, the thickness being greater for the greater spans and heavier loadings:
100 mm, 130, 150, 175, 180, 200, 225, 230, 240, 250, 290, 300, 350, 375, 400, 450, 500, 550 mm.

5 ALUMINIUM

5.01 Aluminium bars
Aluminium bars are made to the requirements of BS 4229: part 1:1967, and to BS 1474:1972. Tables XXIII to XXX indicate the main dimensions.

5.02 Aluminium sheet for roofing
Aluminium sheets for roofing come to site in coils 457 mm wide, and are passed through a machine to run up the edges to form seams when applied in situ. Because of the supply in coils it is not usually necessary to form joints transverse to the standing seams.
This roofing method has not yet been dimensionally co-ordinated. The material is available in thicknesses of 0·71 and 0·91 mm, and the seams will be at 365 mm centres.

6 COPPER ROOFING
Two methods of copper roofing are available: traditional and long-strip. Table XXXI indicates the dimensions of the traditional method; long-strip roofing is in accordance with table XXXII.

7 LEAD
Sizes of milled lead sheet and strip are laid down in BS 1178:1969. The BS code numbers given in table XXXIII correspond to the traditional weights expressed in lb per square foot.
Intermediate thicknesses can be supplied for special purposes. The colour marking for these is yellow, together with the colour for the standard thickness below, eg blue and yellow would indicate a size between code nos 4 and 5.
Lead sheet is priced in £ per tonne.

8 ZINC
Zinc is still sold in the traditional zinc gauges, covered by BS 849:1939. Table XXXIV indicates the metric properties of the material.

Table XXII Concrete kerbs and edgings to BS 340: 1963

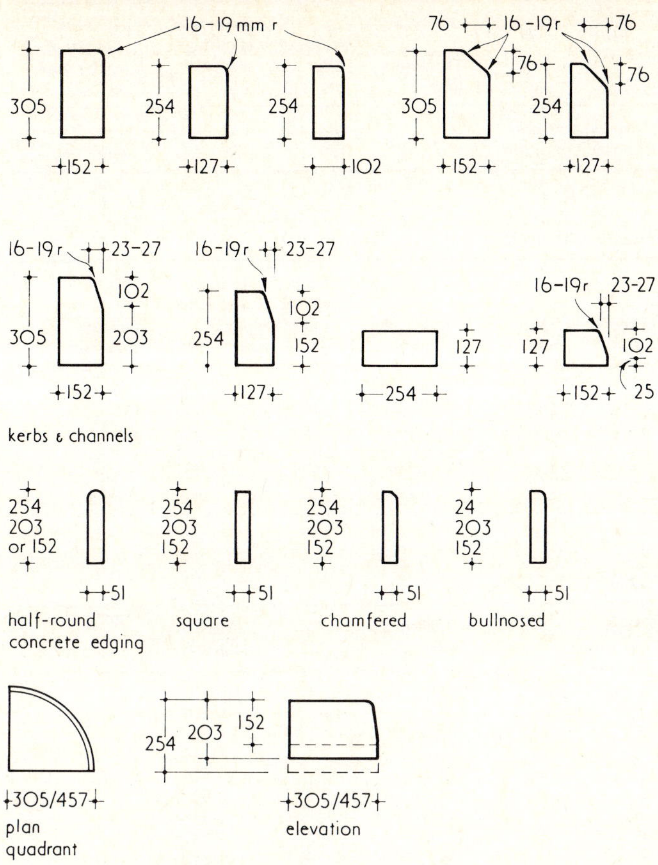

Table XXIII Recommended metric sizes for aluminium and aluminium alloy round bars

Diameter (mm)				
3·0	12·0	30·0	65·0	130·0
4·0	14·0	32·0	70·0	140·0
5·0	16·0	35·0	75·0	160·0
6·0	18·0	40·0	80·0	180·0
7·0	20·0	45·0	90·0	200·0
8·0	22·0	50·0	100·0	
9·0	25·0	55·0	110·0	
10·0	28·0	60·0	120·0	

Table XXIV Recommended metric sizes for aluminium and aluminium alloy square bars

Side (mm)				
3·0	8·0	20·0	50·0	120·0
4·0	10·0	25·0	60·0	160·0
5·0	12·0	30·0	80·0	200·0
6·0	16·0	40·0	100·0	

Table XXV Flat bars

	Thickness (mm)								
Width mm	1·6	2·5	3	4	6	10	12	16	25
10	—	—	X	—	X	—	—	—	—
12	—	X	X	X	X	X	—	—	—
16	—	X	X	X	X	X	—	—	—
20	—	X	X	X	X	—	—	—	—
25	X	X	X	X	X	X	X	X	—
30	—	X	X	X	X	X	X	—	—
40	—	X	X	X	X	X	X	—	—
50	—	—	X	X	X	X	X	—	X
60	—	—	X	—	X	X	X	—	—
80	—	—	X	—	X	X	X	—	X

Table XXV Flat bars *(continued)*

	Thickness (mm)								
Width mm	1·6	2·5	3	4	6	10	12	16	25
100	—	—	X	—	X	X	X	X	X
120	—	—	—	—	X	X	—	X	—
160	—	—	—	—	X	X	—	X	—
200	—	—	—	—	—	X	—	X	—
250	—	—	—	—	—	X	—	X	—

Table XXVI Equal angles

	Thickness (mm)			
Leg length mm	1·6	3	6	10
10	—	X	—	—
12	X	X	—	—
16	—	X	—	—
20	X	X	—	—
25	X	X	X	—
30	X	X	—	—
40	—	X	X	—
50	—	X	X	X
60	—	X	X	—
80	—	X	X	X
100	—	—	X	X

Table XXVII Unequal angles

	Thickness (mm)		
Leg length mm	1·6	3	6
12 × 10	X	—	—
16 × 12	X	—	—
20 × 12	X	X	—
25 × 12	X	X	—
× 16	X	X	—
× 20	X	X	—
30 × 16	—	X	—
× 20	—	X	—
× 25	—	X	—
40 × 12	—	X	—
× 20	—	X	—
× 25	—	X	—
× 30	—	X	—
50 × 20	—	X	—
× 25	X	X	X
× 40	—	X	X
60 × 25	—	X	—
× 40	—	—	X
80 × 25	—	X	—
× 40	—	X	—
× 50	—	X	X
100 × 25	—	X	—
× 50	—	—	X
× 80	—	—	X

Table XXVIII Channels

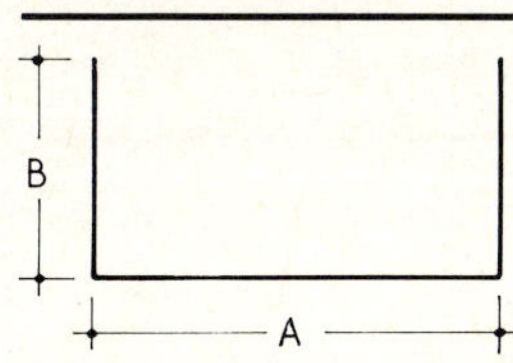

	Thickness (mm)		
Web mm Flange mm	1·6	3	6
A B			
10 × 10	X	—	—
12 × 12	X	X	—
20 × 10	X	—	—
× 12	—	X	—
× 20	—	X	—

table continues on p. 356

Table XXVIII Channels *(continued)*

Web mm	Flange mm	Thickness (mm) 1·6	3	6
25	× 12	—	X	—
	× 20	—	X	—
	× 25	X	X	—
30	× 20	—	X	—
	× 30	—	X	—
40	× 25	—	X	—
	× 40	—	X	—
50	× 25	—	X	X
	× 40	—	—	X
	× 50	—	X	X
60	× 30	—	—	X
80	× 25	—	X	—
	× 40	—	—	X
	× 50	—	X	X
100	× 25	—	—	X
	× 50	—	—	X

Table XXIX T Sections

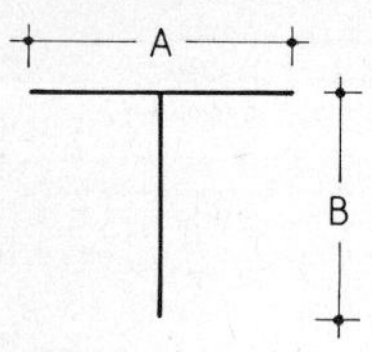

Size mm		Thickness (mm) 3	6
A	*B*		
12	12	X	—
20	20	X	—
25	25	X	—
40	40	X	—
50	50	—	X
80	80	—	X
100	100	—	X

Table XXX Z Sections

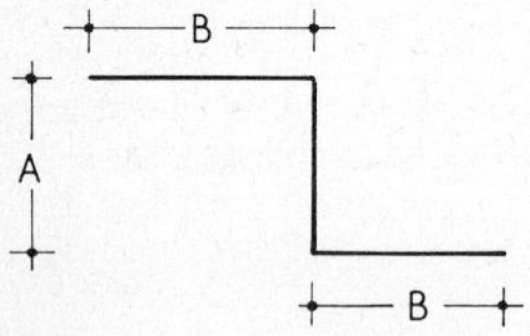

Web mm	Flange mm	Thickness (mm) 1·6	3
A	*B*		
25	25	—	X
30	20	X	—
40	25	—	X
50	25	—	X

Table XXXI Recommended thicknesses and sizes of copper sheet (for traditional roofing)

Thickness mm	Bay width Standing seam mm	Roll mm	Standard width of sheet to form bay mm	Length of each sheet m
0·45	525	500	600	1·8
0·60	525	500	600	1·8
0·70	675	650	750	1·8

Table XXXII Recommended maximum widths and lengths of copper strip for the long strip system

	Thickness mm	Width of strip mm	Centres of standing seams mm	Length of each panel m
Normal conditions	0·6	600	525	8·5
Exposed conditions	0·6	450	375	8·5

Table XXXIII Milled lead sheet and strip sizes

BS code no	Thickness mm	Weight kg/m²	Total weight in kg of sheet 2·4 m wide and of lengths: 3 m	6 m	9 m	Colour marking
3	1·25	14·18	102	204	306	green
4	1·80	20·41	147	294	441	blue
5	2·24	25·40	183	366	548	red
6	2·50	28·36	204	408	612	black
7	3·15	35·72	257	514	772	white
8	3·55	40·26	290	580	870	orange

Table XXXIV Zinc gauges

Gauge	Thickness mm	Mass kg/m²
1	0·102	0·73
2	0·152	1·10
3	0·178	1·28
4	0·203	1·16
5	0·254	1·83
6	0·279	2·01
7	0·330	2·38
8	0·381	2·74
9	0·432	3·11
10	0·483	3·48
11	0·559	4·02
12	0·635	4·57
13	0·711	5·12
14*	0·787	5·67
15*	0·914	6·58
16*	1·041	7·50
17	1·168	8·41
18	1·295	9·33
19	1·448	10·43
20	1·600	11·52
21	1·778	12·80

* Normal gauges for external building work nos 14, 15 and 16

9 PLASTERBOARD

The major manufacturer of plasterboard changed over to metric measurement on April 1, 1970. From that date the standard stock sizes have been as follows:

- *Gypsum wallboard*

Widths: 600, 900 and 1200 mm
Lengths: 1800, 1829, 2286, 2350, 2400, 2438, 2700, 3000 and 3048 mm
Thicknesses: 9·5 and 12·7 m

- *Gypsum plank*

Width: 600 mm
Lengths: 2350, 2400, 2700, 3000 and 3200 mm
Thickness: 19 mm

- *Industrial plastic-faced board*

Width: 600 mm
Lengths: 1200, 1219, 1524, 1800, 1829, 2134, 2286, 2400, 2438, 2700, 2743, 3000, 3048 and 3353 mm

- *Sarking boards*

Width: 600 mm
Lengths: 1800 and 1829 mm
Thickness: 9·5 mm

- *Gypsum lath*

Width: 406 mm
Lengths: 1200, 1219 and 1372 mm
Thicknesses: 9·5 and 12·7 mm

• *Baseboard*
Width: 914 mm
Lengths: 1200, 1219 and 1372 mm
Thickness: 9·5 mm
• *Dry partitions*
Widths: 600, 900 and 1200 mm
Lengths: 1800, 2286, 2350, 2400, 2438, 2700, 2743, 3000, 3048, 3300, 3600 and 3700 mm
Thicknesses: 38·1, 50, 57·2 and 63·5 mm
All plasterboards and dry partitioning is invoiced in square metres.

10 BAGGED PLASTERS

Building and industrial plasters, jointing compounds and gypsum material are packed in 50 kg bags, with the exception of Thistle projection plaster which is supplied in 40 kg paper sacks. Invoicing is in kilogrammes or metric tonnes.

11 BUILDING BOARDS

Co-ordinating sizes for rigid flat sheet materials used in building are specified in BS 4606:1970. They are as follows:
Widths: 600, 900 and 1200 mm
Lengths: 1800, 2400, 2700 and 3000 mm
Thicknesses: no recommendation for thicknesses under 25 mm. For thicker boards the recommendations of BS 4011 are advocated.
The suppliers of such boards do catalogue materials to these dimensions, but they are by no means the larger amounts stocked. The following are the normal available sizes:
• *Hardboard*
Widths: 610, 915, 1220, 1525 and 1600 mm
Lengths: 1220, 1525, 1830, 2440, 2745, 3050 and 3660 mm
Thicknesses: 2, 3·2, 4·8, 6·4, 7·9, 9·6 and 12·7 mm
• *Softboard*
Width: 1220 mm
Lengths: 2440, 3050 and 3660
Thicknesses: 12·7, 19 and 25 mm
• *Wood chipboard*
Widths: 600, 1220 and 1830 mm
Lengths: 2400, 2440, 2750 and 3050 mm
Thicknesses: 2·5, 3·2, 4, 6, 9, 12, 15, 18, 22 and 25 mm
• *Plywood*
Widths: 1220 and 1525 mm
Lengths: 1525, 2440, 2745, 3050 and 3660 mm
Thicknesses: 4, 6·5, 9, 12, 12·5 and 18 mm
• *Blockboard*
Widths: 1220 and 1525 mm
Lengths: 1830, 2440, 3050 and 3660 mm
Thicknesses: 12, 13, 16, 18, 22 and 25 mm

12 GLASS

Glass used in building falls into four categories: plain, patterned, wired and safety.

12.01 Plain glass

Plain glass is either sheet or float, the latter having almost completely replaced plate glass.
Sheet glass is supplied in horticultural, ordinary glazing (OG), special glazing (SG) and extra-special glazing (SSG) qualities. OG is most generally used in normal building.
Thickness is as follows:

Thickness mm	Maximum size* mm	Weight kg/m^2	Decibel reduction dB
3	2000 × 1200	7·5	23

Float glass is available in the following thicknesses:

Thickness mm	Maximum size* mm	Weight kg/m^2	Decibel reduction dB
4	2760 × 1220	10	25
5	3180 × 2100	12·5	26
6	4600 × 3180	15	27
10	6000 × 3300	25	28
12	6000 × 3300	30	29
15	3500 × 3000	37·5	33
19	3000 × 2900	47·5	35
25	3000 × 2900	63·5	36

* Because glass is supplied in certain maximum sizes, it should not be assumed that panes of these sizes would necessarily be safe. A procedure for determining the safe thickness of glass for a specific pane size is given in CP 152:1972 Glazing and fixing of glass for buildings.

12.02 Patterned glass

This is available in a wide range of different patterns, thicknesses, colours and degrees of obscurity. Refer to manufacturers' catalogues.

12.03 Wired glass

This is obligatory to comply with fire regulations in a number of situations. It should be noted that the clear version of wired polished plate is about three times as expensive as the obscure variety. All are nominally 6 mm thick.

Type	Maximum size mm	Weight kg/m^2	Light transmission
Diamond polished wired (DPW) } Georgian polished wired (GPW) }	3300 × 1830	15·9	80%
Georgian wired cast (GWC)	3700 × 1840	17·1	75%

12.04 Safety glass

There are two types of true glass for use where safety is important:
• tempered glass, the manufacturing process of which makes it five times as strong as normal glass and which, when broken, shatters into thousands of small, relatively harmless pieces
• laminated glass, which is a sandwich of plastic between sheets of normal glass. When broken, the plastic prevents the glass fragments from coming apart.
In addition, there is:
• polycarbonate sheet, not a glass but an extremely tough transparent plastic which is 250 times stronger than safety glass. This is used where security is required or vandalism is a problem. Polycarbonate is, however, vulnerable to fire.
All these materials are supplied in the finished size required, and cannot be worked on site. Refer to manufacturers' literature for limits to sizes.

13 WINDOWS

Windows are generally available in three materials:
• wood
• steel
• aluminium.
A major problem in co-ordinating sizes has been the mandatory storey height of 2·6 m for publicly-financed housing in the UK. As this is a non-co-ordinated dimension, it has led to the necessity of window heights in non-preferred increments. In addition, the joinery industry continues to manufacture frames to the old imperial dimensions translated to metric figuring.

13.01 Wood windows

The British Woodwork Manufacturers' Association run the EJMA Certification Scheme for wood windows. The standard (unco-ordinated) range contains:
heights: 769, 921, 1073, 1225, 1377 and 1529 mm
lengths: 438, 641, 933, 1225, 1809 and 2393 mm.
Various combinations of height, length and design are available. See **39.1** for co-ordinated range.

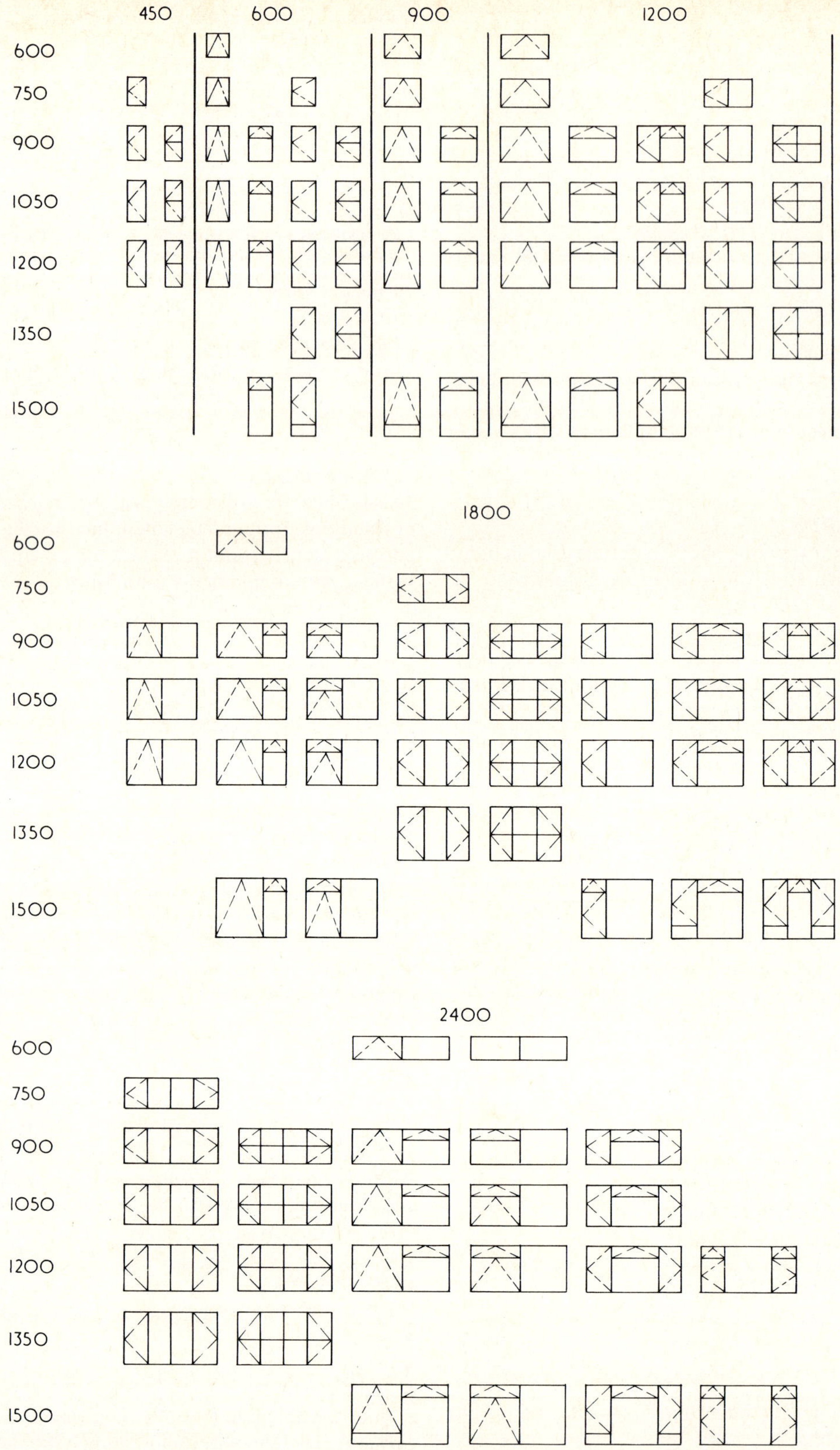

39.1 *Co-ordinating sizes of timber windows, showing available types of glazing sub-division and opening lights*

13.02 Steel windows

The steel window industry now classifies the old imperial size windows as specials. All the standard frames are in co-ordinated sizes, but are in two ranges, **39.2**:

- BS 990—Module 100 designed for light duties
- metric W20 range, for general and heavy duty use.

13.03 Aluminium windows

Only co-ordinated sizes are now covered in the standard aluminium range, **39.3**. Unlike other materials, there are standards for sliding types as well as casements.

14 DOORS

Although doors of steel and of aluminium construction are manufactured for mainly external use, these are usually classified as windows, and will be found in that section.

Doors of basically timber construction are available in two ranges:

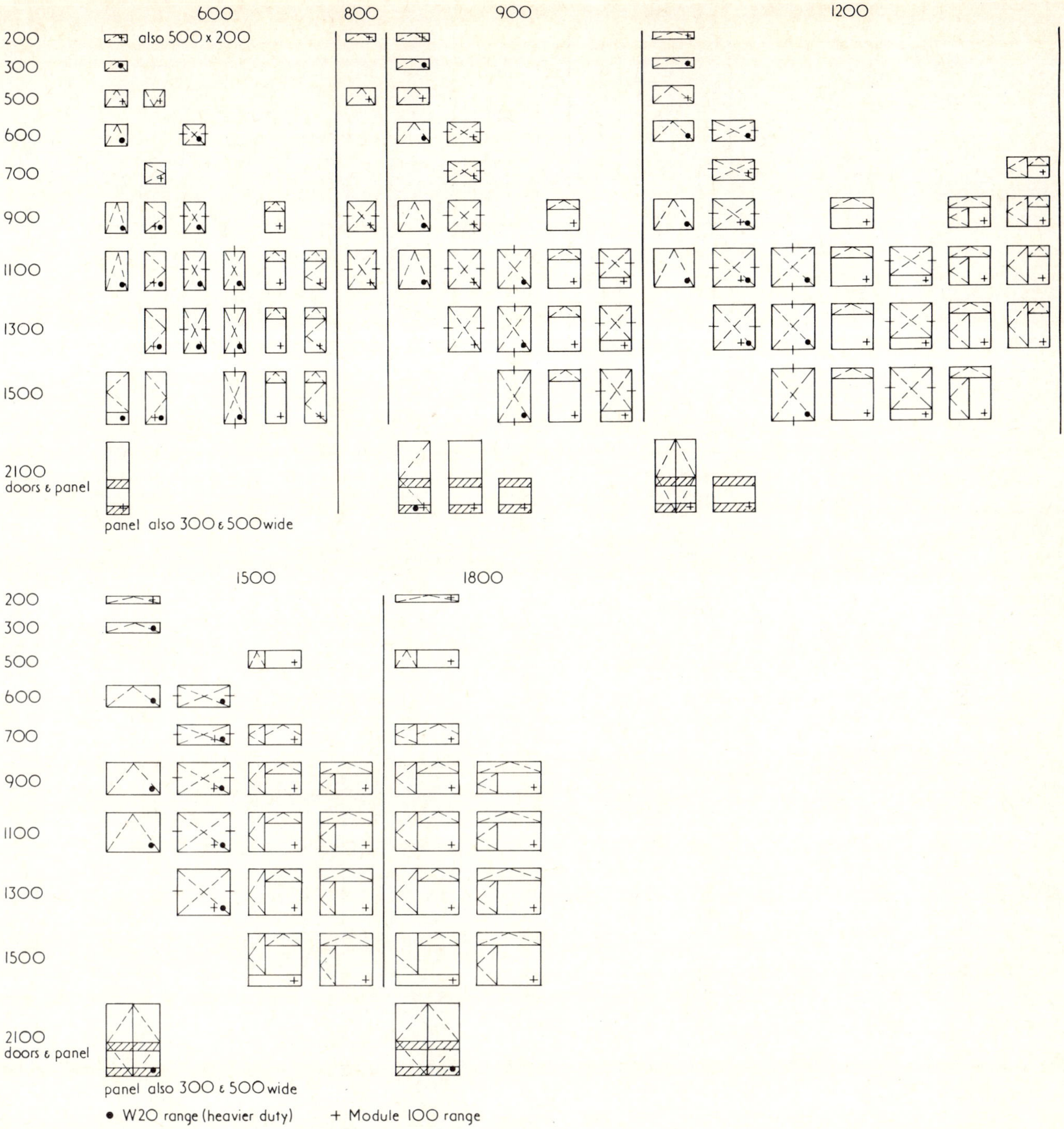

39.2 *Co-ordinating sizes of steel windows, showing available patterns*

- metric sizes
- dimensionally co-ordinated metric sizes.

14.01 Metric sizes

Despite the name, these are straight translations of the old standard doors based on imperial increments. They are currently sold in the following sizes:

- *internal doors*

Thickness: 35 mm
Lengths: (also described as widths) 610, 686, 762 and 838 mm

- *firecheck doors*

Thicknesses: 44 mm (½ hour) and 54 mm (1 hr)
Lengths: 838 and 914 mm

- *external doors*

Thickness: 44 mm
Lengths: 762 and 838 mm

Frames, linings and complete door-sets are available to fit these sizes.

14.02 Dimensionally co-ordinated metric sizes

These will eventually supersede the above. All standard doors are to the provisions of BS 4787:part I:1972. The old system of standard-sized leaves led to variations in structural openings as the frames and linings differed to fulfil functional requirements. On the other hand, dimensionally co-ordinated door sizes are based on standard structural openings. Since external door frames are more substantial than internal frames, the leaves will be correspondingly smaller for the same co-ordinated size.

Standard doors are available in two basic constructions, **39.4**:

- traditional panelled type, with panels either solid or glazed
- flush type, either solid or with glazed openings.

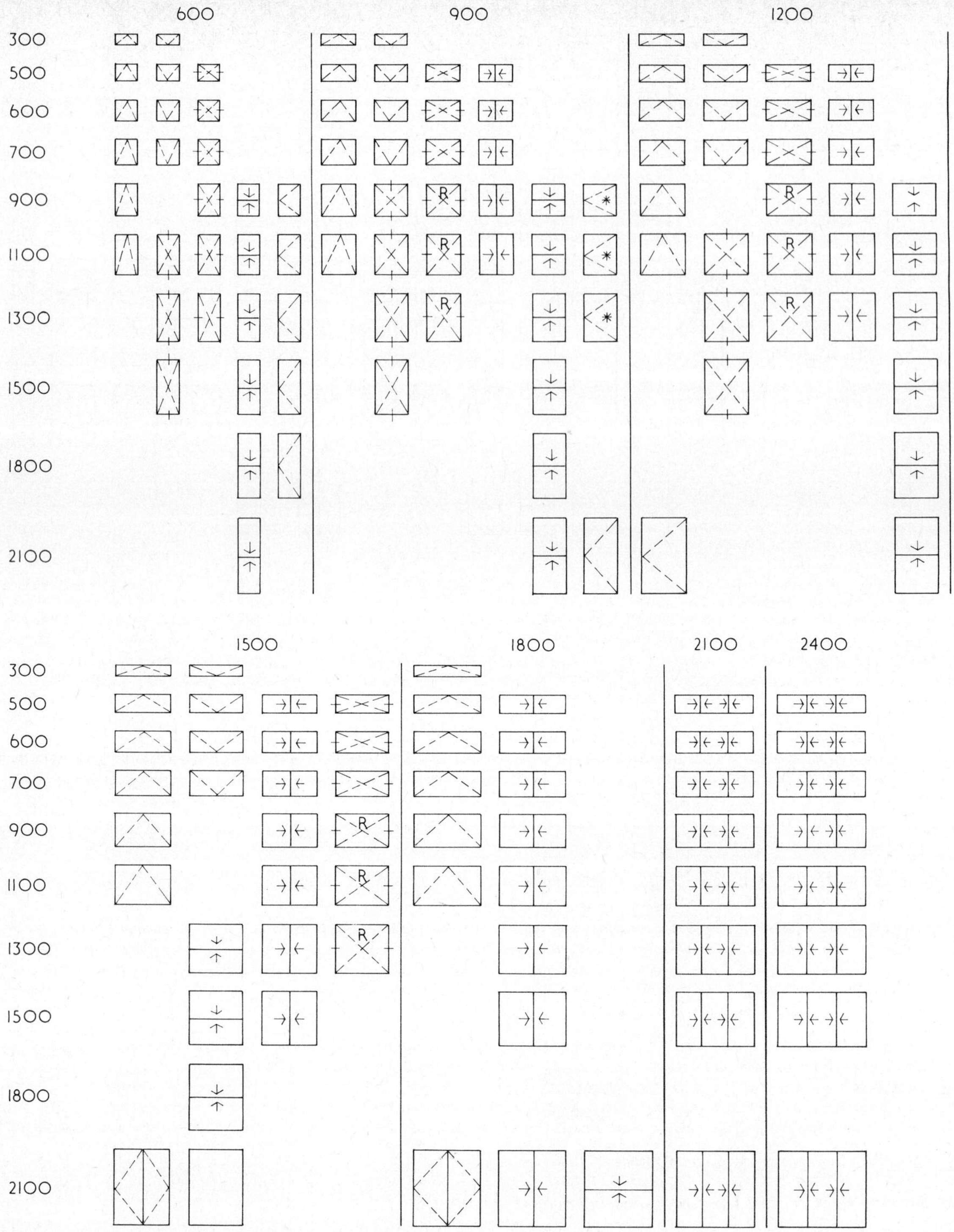

39.3 *Co-ordinating sizes of aluminium windows, showing available patterns*

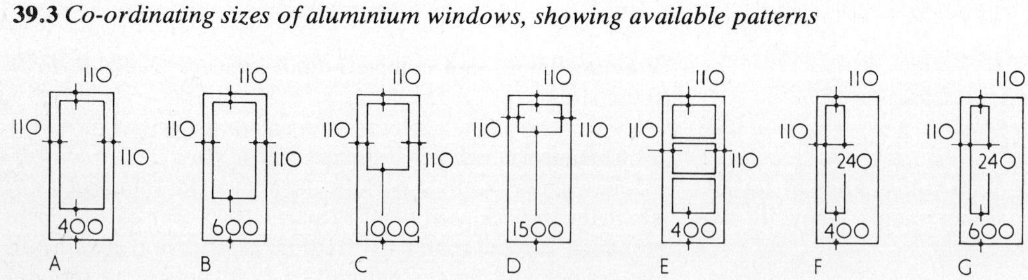

39.4 *Standard glazing in doors to BS 4787:*
a *Flush doors*

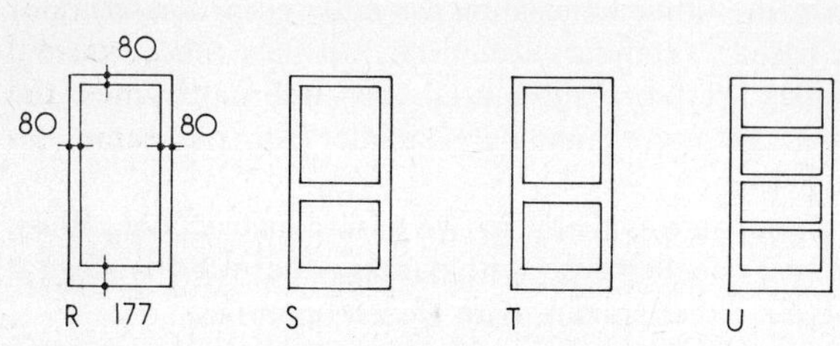

b *Panelled doors*

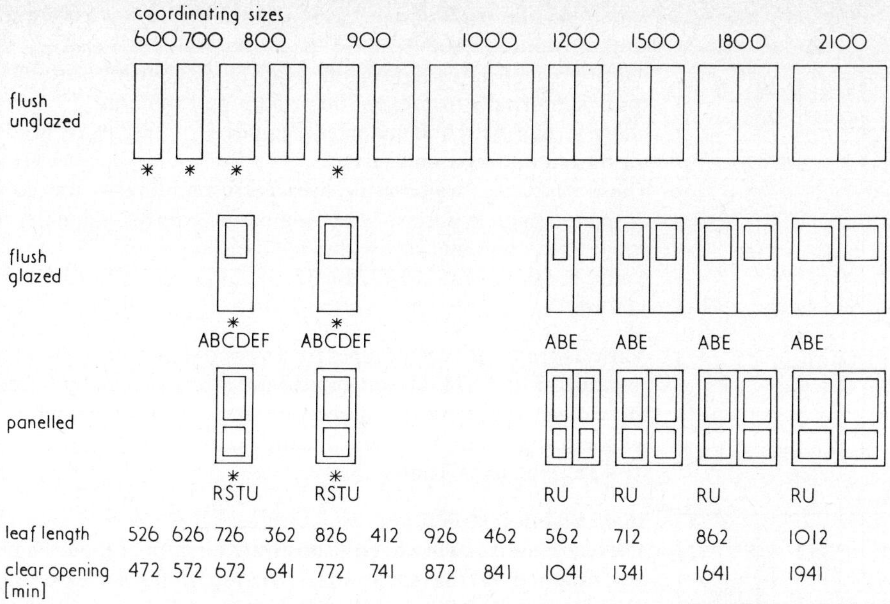

39.5 *Types of internal door leaves.*
All internal door leaves:
thickness 40 mm
height 2040 mm
** ex stock*

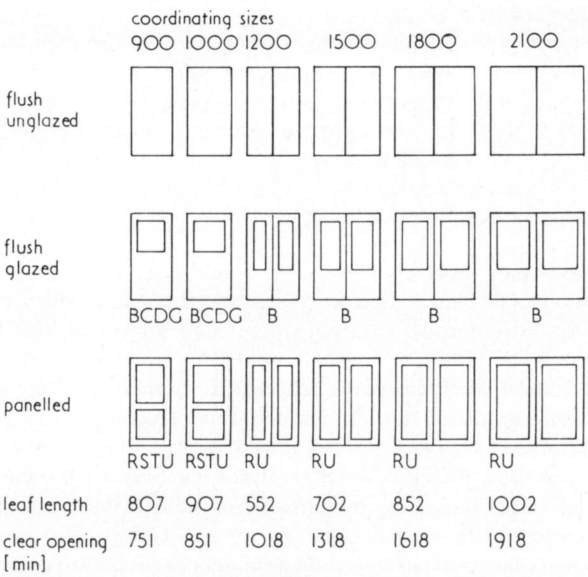

39.6 *Types of external door leaves.*
All external door leaves: thicknesses 40 and 44 mm
height 2000 mm

Figures **39.5** and **39.6** show the standard sizes for internal and external door leaves with the available types. Stock items are indicated by *.
Figures **39.7** and **39.8** show standard door sets for internal and external openings.

15 OTHER JOINERY ITEMS

The British Woodwork Manufacturers' Association sponsor standard ranges of joinery under the EJMA trade mark. Apart from windows and doors, these include kitchen units and stairs for private dwellings.

15.01 Kitchen units

The basic units for these, **39.9**, correspond to the standard modules generally adopted for all kitchens. Corner units are not shown, and worktops are supplied to lengths required, allowing for gaps in the base units to accommodate towel or tray storage, washing machines, etc.

15.02 Staircases

Five standard designs are available, **39.10**, with variations for handed versions and for extra bullnosed bottom treads. They conform to the provisions of BS 5395:1977 Code of Practice for Stairs, and also to the Building Regulations 1976.

16 PIPES FOR PLUMBING AND DRAINAGE

16.01 Types of pipe

Pipes for the conveyance of liquids and gas are made of the following materials:

- steel
- copper
- stainless steel
- cast iron
- plastics
- glass (for specialist laboratories, etc)
- vitrified clay.

Steel and copper are used in thick and thin-walled versions, depending on the system of connection. The methods in general use are:

- screwed joints—steel, plastic and copper (rarely nowadays)
- welding—steel
- spigot and socket—cast iron, plastics
- compression fittings—copper, stainless steel, light-gauge steel, plastics
- capillary soldering—copper.

Pipes carrying hot or chilled liquids, or in exposed conditions, will be insulated. The thickness of insulation will be between 25 and 75 mm depending on material used and the size of the pipe.
After allowing for any such insulation, the space allowed for any pipe should be between two to three times the actual diameter of the barrel. This will allow for sockets, joints, bends and clearances.

16.02 Steel pipes for screwed joints

These are available in light, medium and heavy qualities. The sizes given in table XXXV are to international standard ISO/65, and are based on nominal inch sizes.

16.03 Copper pipes

Copper pipes are specified in accordance with BS 2871:1971. Part 1 of this standard deals with tubes for water, gas and sanitation. Joints in these pipes are made with compression fittings or capillary soldered fittings. There are three quality grades, in increasing wall thicknesses:

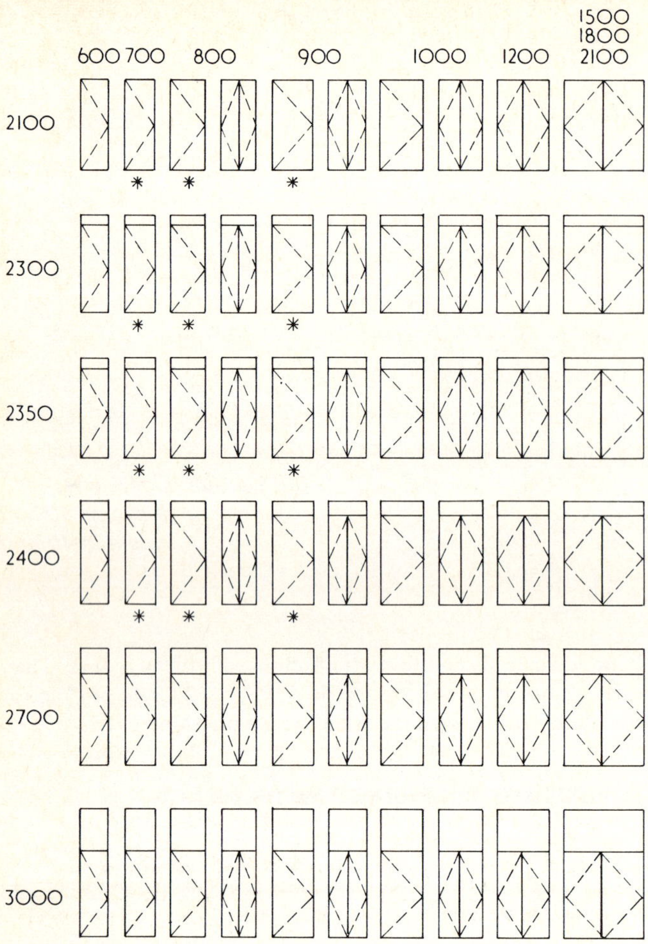

39.7 *Types of internal door sets*
** ex stock*

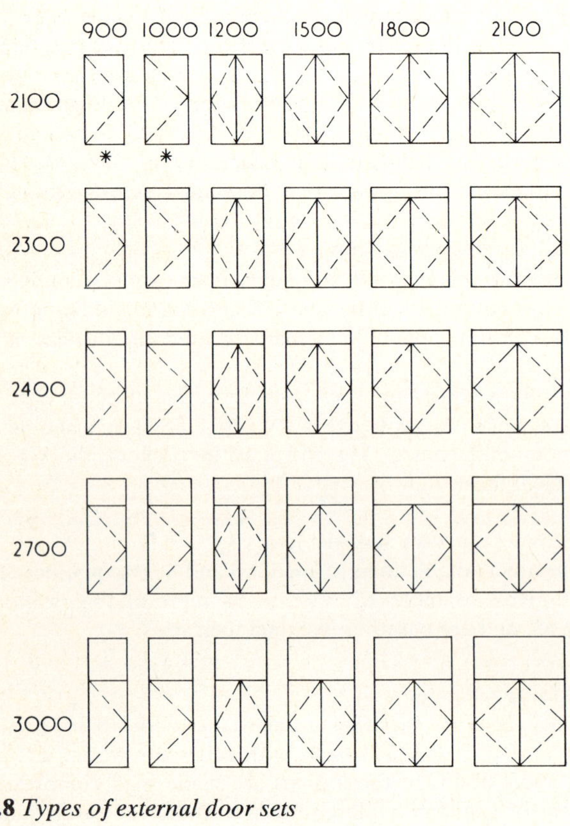

39.8 *Types of external door sets*
** ex stock*

Table Z: thin-walled and hard-drawn. Used for straight lengths, cannot be bent.
Table X: half-hard, light-gauge, can be bent using suitable springs internally to prevent deformation of the bore.
Table Y: half-hard and annealed. Supplied in coils, and is used for buried underground services.
These pipes are supplied in sizes corresponding to the co-ordinated metric series, with minimum external diameters equal to the nominal size within a close tolerance:
6 mm, 8, 10, 12, 15, 18, 22, 28, 35, 42, 54, 76·1, 108, 133 and 159 mm

Various grades of copper pipes for screwed joints are available to part 2 of BS 2871, although these are rarely used in building these days. These are specified by nominal bore size in inches, with external diameters almost identical to those of light quality steel pipe as in Table XXXV.

16.04 Stainless steel pipes
These are used as a lower cost substitute for copper pipe. They are supplied in the same sizes, corresponding to external diameter, as above.

16.05 Cast iron
Cast iron pipes are made to the following three specifications:

BS 460:	Light grade rainwater pipes
BS 416:1968	for soil waste and vent pipes above ground
BS 437:1970	for underground pipework, whether buried or in ducts.

They are made in nominal inch sizes, the external diameters of which are given in table XXXVI.

16.06 Plastic pipes
The types of plastic used for pipes, and the uses of the pipes, are both numerous. Consequently, there are a considerable number of British Standards governing this material. BS 3867:1969 lays down some general information on sizing and pressure gradings, not all of which has been followed in later standards.
BS 3867 specifies two series of pipe sizes. Series 1 is a metricated series based on nominal bore sizes in inches, and is consistent with international standard ISO/161. Types and sizes of plastics pipes based on nominal bore sizes are given in Table XXXVII
Series 2 in BS 3867 is a rationalised metric series. The nominal sizes correspond to the *minimum* external diameter of the pipe. The full series is as follows (dimensions in mm): 2·5, 3, 4, 5, 6, 8, 10, 12, 16, 20, 25, 32, 40, 50, 63, 75, 90, 110, 125, 140, 160, 180, 200, 225, 250, 280, 315, 355, 400, 450, 500, 560, 630, 710, 800, 900, 1000.
Very few of these sizes are in fact available. The principal ones are as follows:
UPVC rainwater pipes to BS 4576:1970: 63 and 75 mm sizes (a square section rainwater pipe 56 mm side is also available).
UPVC underground drain pipes to BS 4660:1973: 110 and 160 mm sizes.

16.07 Pipes of vitrified clay
These traditional drainage pipes are metricated. Since it is not possible to define a thickness of material (which varies as the nature and quality of the clay from which it is made) the sizing is based on a nominal bore. The actual bore can vary on either side of the nominal by about −2 per cent, +5 per cent for British Standard pipes and by about −4 per cent, +7 per cent for British Standard surface water pipes. The series of sizes is as follows, with non-preferred sizes marked * and the commonly available sizes in italics:
75, *100*, *125, *150*, *175, *200, *225*, *250, 300, 375, 450, 525, 600, 675, 750, 825 and 900 mm.

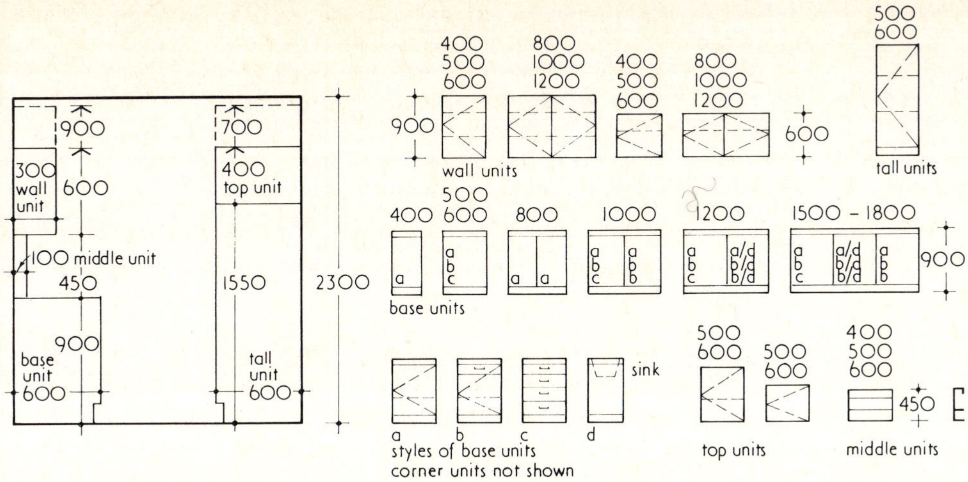

39.9 *Standard kitchen units. Corner and other special units not shown*

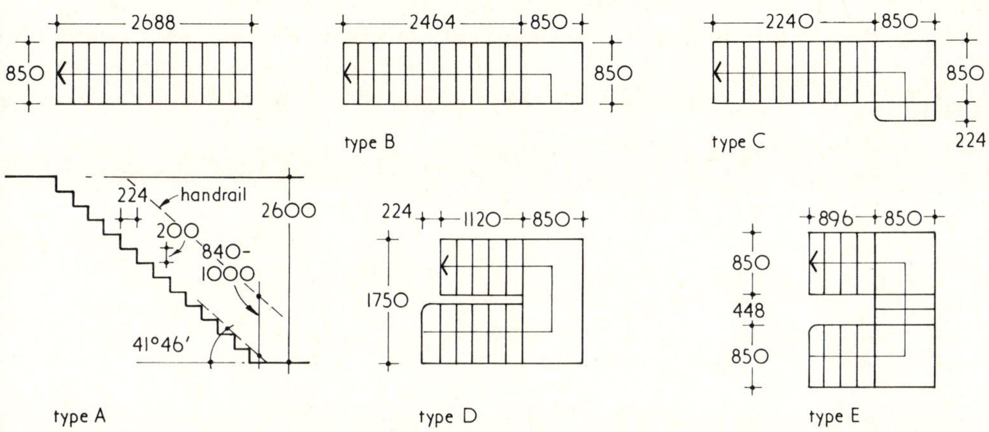

39.10 *Standard wood stairs for private dwellings*

Table XXXV Mean outside diameters of steel pipes in mm

Nominal bore in	mm	Light	Medium and heavy
1/8	6	9·9	10·1
1/4	8	13·4	13·6
3/8	10	16·9	17·1
1/2	15	21·2	21·4
3/4	20	26·6	26·9
1	25	33·5	33·8
1 1/4	32	42·2	42·5
1 1/2	40	48·1	48·4
2	50	59·9	60·3
2 1/2	65	75·6	76·0
3	80	88·3	88·8
4	100	113·4	114·1
5	125	–	139·6

Table XXXVI Maximum external diameters of cast iron pipe

Nominal bore		BS 460 rainwater pipes		BS 416 above ground s, w and v		BS 437 below ground	
in	mm	pipe	socket	pipe	socket	pipe	socket
2	50	54	69	64	89	65	117
2 1/2	65	67	89	76	103	–	–
3	75	79	94	89	116	92	150
3 1/2	90	–	–	102	130	–	–
4	100	105	120	114	143	119	185
5	125	130	146	140	171	–	–
6	150	156	175	165	197	173	239
9	225	–	–	–	–	256	337

17 MATERIALS FOR ELECTRICITY SUPPLY AND DISTRIBUTION

Electricity supply and distribution cables are mainly:

- armoured cable for intake (not covered here, see technical literature)
- PVC insulated, in conduits of steel or plastic
- PVC insulated, PVC sheathed
- mineral insulated copper (or aluminium) conductors
- wiring harnesses (manufactured for specific locations, hence non-standard).

PVC double-insulated cables are normally used in electrical distribution for the smaller building types. Table XXXVIII gives the dimensions of these, which are often accommodated in small ducts or voids in the construction. Conductor cross-sectional areas of 1 and 1·5 mm² are used for lighting circuits, while ring mains are composed of cables with conductors of 2·5 mm². Since earth-continuity conductors are now used in all circuits, those cable types with integral earth conductors are becoming the norm. Cables with three insulated cores plus earth are used for circuits with two-way switching of lights. Flexible electrical cords for the connection of mobile and portable equipment are detailed in table XXXIX. Mineral insulated cables (MICC or 'pyro') are used mainly when space is at a premium, or in external situations. Table XL gives the dimensions of these.

Steel conduits are covered in BS 4568:1970, and conduits of plastics in BS 4607:1970. These British Standards also give details of the round connection boxes used for junctions, looping and for the attachment of ceiling roses (known colloquially as 'beezer' boxes). Tables XLI and XLII summarise the dimensional information in both the standards. As far as steel conduits are concerned, it is worth noting that these are obtainable in four classes of protection:

Class 1: Light protection inside and out (priming paint)
Class 2: Medium protection inside and out (stove enamel, air-drying paint)
Class 3: Medium heavy protection, inside as class 2, outside as class 4
Class 4: Heavy protection inside and out (hot-dip zinc coating, sherardised).

Table XXXVII Mean outside diameters of plastics pipes to various standards (metricated inch series)

British Standard: Material	Use	Nominal size (in): 1/8	1/4	3/8	1/2	3/4	1	1 1/4	1 1/2	2	2 1/2	3	4	6
BS 3867:1969 Any plastics	Circular pipes for the conveyance of fluids, not covered by any other BS	10·2	13·5	17·2	21·3	26·9	33·7	42·4	48·3	60·3	75·3	88·9	114·3	168·3
BS 1972:1967 Polythene (type 32)	For cold water services (classes B, C and D)			17·2	21·4	26·8	33·6	42·3	48·3	60·4	—	89·0	114·3	—
BS 3284:1967 Polythene (type 50)	For cold water services (classes C and D) (thinner walls)			17·2	21·4	26·8	33·6	42·3	48·3	60·4	—	89·0	114·3	168·2
BS 3505:1968 Unplasticised polyvinyl chloride	For cold water services (classes B, C, D, E and 7)			17·2	21·4	26·8	33·6	42·2	48·2	60·4	75·4	88·9	114·3	168·2
BS 4514:1969 Unplasticised polyvinyl chloride	For soil and ventilating pipes in above-ground drainage systems to convey normal domestic effluents											82·6	110·2	160·3
BS 4576:1970 Unplasticised polyvinyl chloride	For rainwater pipes for external use										68·15			
BS 5254:1976 Polypropylene	Waste pipes for the conveyance of normal domestic effluents							34·6	41·0	54·1				
BS 5255:1976 ABS* MUPVC Polypropylene	Waste pipes for the conveyance of normal domestic effluents							36·3	42·9	55·9				
ditto Polyethylene (polythene)	ditto							38·1	44·45	57·15				

* Abbreviations: ABS = acrylonitrile-butadiene-styrene, MUPVC = modified unplasticised polyvinyl chloride, PP = polypropylene, PE = Polyethylene (polythene)

Rectangular boxes of the type used to accommodate wiring accessories such as switches, socket outlets, cooker points etc, are covered by BS 4662:1970. This is summarised in table XLIII.

Table XXXVIII Dimensions of PVC insulated, PVC sheathed electric cables (mm)

Cross-sectional area of single conductor (mm^2)	Approximate normal rating (amps)	Single (dia)	Flat twin	Flat three	Flat twin + earth	Flat three + earth
1·0	6	4·2	6·7 × 4·4	9·0 × 4·4	7·8 × 4·4	10·2 × 4·4
1·5	8	4·4	7·2 × 4·6	9·8 × 4·6	8·3 × 4·6	11·0 × 4·7
2·5	11	5·0	8·6 × 5·4	11·9 × 5·5	9·7 × 5·4	13·0 × 5·5
4·0	15	6·2	10·7 × 6·5	15·0 × 6·7	12·0 × 6·5	
6·0	19	6·8	12·0 × 7·3	16·7 × 7·3	13·8 × 7·3	
10·0	26	8·1	14·9 × 8·8	21·0 × 8·9	17·4 × 8·8	

Table XXXIX Dimensions of flexible electric cords (for connection of mobile appliances to outlets)

Cross-sectional area of single conductor (mm^2)	Suspension capacity (kg)	Approximate normal rating (amps)	Tough rubber sheathed			circular cotton braided		PVC sheathed			
			twin	three-core	four-core	twin	three-core	twin flat	twin round	three-core	four-core
0·5	2	3	6·2	6·6	7·3	5·1	5·5	4·0 × 6·1	6·1	6·5	7·0
0·75	3	6	6·6	7·2	7·8	6·3	6·8	4·2 × 6·5	6·5	6·9	7·5
1·0	5	10	7·2	7·8	8·3	6·7	7·2		6·9	7·3	8·2
1·5	5	15	8·8	9·3	10·3	7·3	7·8		7·5	8·1	9·0
2·5	5	20	10·2	10·9	12·1					9·8	10·8
4·0	5	25	12·1	12·8	14·2					11·6	12·7

Table XL Sizes of mineral insulated cables

		Copper sheathed			Aluminium sheathed		
No of conductors	Cross-sectional area of single conductor (mm^2)	Approximate normal rating (amps) single-phase	three-phase	Cable diameter over copper sheath (mm)	Approximate normal rating (amps) single-phase	three-phase	Cable diameter over aluminium sheath (mm)
1	1·0	22	18	3·1			
	1·5	27	23	3·4			
	2·5	36	31	3·8			
	4·0	46	41	4·4			
	6	63	56	6·4			
	10	85	75	7·3			
	16	112	99	8·3	100*	88*	8·6
	25	146	128	9·6	130*	114*	9·9
2	1·0	17		5·1	19		5·4
	1·5	22		5·7	24		5·9
	2·5	29		6·6	33	6·8	
	4·0	38		7·7	44		7·9
	6	53		10·9	55		9·0
	10	71		12·7			
	16	94		14·7	84*		15·0
	25	124		17·1	113*		18·8
3	1·0	14	14	5·8	16	16	6·0
	1·5	18	18	6·4	21	21	6·6
	2·5	26	26	9·3			
	4·0	34	34	10·4			
	6	44	44	11·5			
	10	59	59	13·6			
	16	78	78	15·6	71*	71*	16·0
	25	103	103	18·2	94*	94*	20·0
4	1·0	15·	15·	6·3	16	16	6·0
	1·5	19	19	7·0	21	21	7·3
	2·5	27	27	10·1			
	4·0	35	35	11·4			

* Aluminium conductors

Table XLI Electrical conduits

	Minimum internal diameter d_2						
	Steel			Pliable plastic self-extinguishing		Non-circular conduits	
Nominal size corresponding to maximum external diameter d_1	Light gauge plain ends	Heavy gauge screwed ends	Rigid PVC	Plain	Corrugated	Maximum outside dims	Minimum inside dims
13	–	–	–	–	–	13·0 × 8·1	11·0 × 6·1
16	13·5	12·7	13·0	10·7	11·7	16·3 × 9·9	14·3 × 7·9
20	17·5	16·2	16·9	14·1	15·5	22·6 × 11·4	20·6 × 9·4
25	21·9	21·1	21·4	18·4	19·8	28·7 × 11·4	26·5 × 9·2
32	28·9	28·1	27·8	24·4	26·4	32·5 × 11·4	30·3 × 9·2
40	–	–	–	31·2	34·0	–	–
50	–	–	–	39·7	43·5	–	–
63	–	–	–	49·6	56·0	–	–

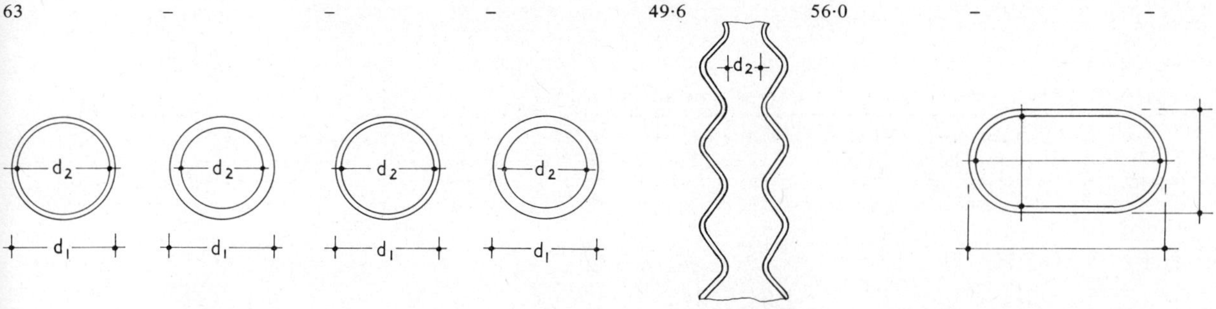

Table XLII Circular boxes for electrical conduit systems: minimum outside dimensions (diameter × depth)

Box type	Nominal conduit size	Rigid PVC	Cast iron	Steel
Small circular box	16	64·4 × 28·6	64·3 × 27·0	63·2 × 26·5
	20	64·4 × 28·6	64·3 × 27·0	63·2 × 26·5
	25	64·4 × 31·8	64·3 × 30·0	63·2 × 29·5
Cover thickness		–	2·4	1·15
Circular looping box	16	64·6 × 32 (nom)	67·5 × 34·0	66·4 × 33·5
	20	64·6 × 32 (nom)	67·5 × 34·0	66·4 × 33·5
	25	–	67·5 × 34·0	66·4 × 33·5
Cover thickness		–	2·0	1·47
Large circular box	20		86·6 × 37·8	85·7 × 36·5
	25		86·6 × 37·8	85·7 × 36·5
	32		86·6 × 40·8	85·7 × 39·5
Cover thickness			3·2	1·47
Extension rings	20	nominal depths of 13, 20, 25 and 32 mm	–	–

Table XLIII Rectangular boxes for the accommodation of electrical wiring accessories

Box type	Nominal conduit sizes	External face dimensions	Nominal external depth Insulating material	Cast iron	Steel
UA1	16, 20 and 25	75 × 75	17·5	18·0	17·0
			27·0	27·3	26·2
			37·0	37·3	36·2
			43·0	43·3	42·2
			49·0	49·3	48·2
UA2	16, 20 and 25	135 × 75	17·5	18·0	17·0
			27·0	27·3	26·2
			37·0	37·3	36·2
			43·0	43·3	42·2
			49·0	49·3	48·2
UA3	20 and 25	195 × 75	43·0	43·3	42·2
UA4	20 and 25	135 × 135	43·0	43·3	42·2
UA6	20 and 25	195 × 135	43·0	43·3	42·2

40 Landscape design

Steve Scrivens
Colin Moore

CI/SfB 90
CI/SfB (1976 revised) 998

Steve Scrivens is a horticulturalist landscape designer who lectures on the subject at Pershore College of Horticulture, Worcestershire

Colin Moore, who was responsible for preparing the tables of plants, is a graduate member of the Landscape Institute and works in private practice

Contents

1 INTRODUCTION

Landscape design is the design of all the space outside buildings: hard and soft, urban and rural, public and private. It ranges from large urban parks **40.1**, to small gardens for housing **40.2** and roof gardens **40.3**. Most of this chapter deals with soft landscape (ie planted areas), but a few of the more important hard landscape matters are covered in para **11**, and in section **7**, External Circulation.

2 BASIC HUMAN DIMENSIONS

The space requirements of people outside buildings are shown in **40.4** to **10**. A family group of six people on a lawn or terrace occupy a rough circle 4 m diameter; for ten people (the largest convenient single group) the dimension becomes 6 m. This constitutes the minimum useful lawn size for domestic use.

3 GROUND FORMING

Changes of level in the ground can be made with retaining walls (see **40.32**) or by slopes. Unreinforced slopes must not be steeper than the safe angle of repose of the material of which they are formed (table I).

Such ground materials are called soils (which in this context does not refer to the vegetable humus in which plants grow). They fall into three basic categories:

- Rocks, mostly strong and homogenous and can be cut to slopes approaching the vertical. They can only be deposited in fragmented form, when they are similar to gravels.
- Sands and gravels, can be cut or deposited up to the safe angle of repose. Beyond this, dry particles slide down the slope until equilibrium is reached.
- Clays, of which the significant property is cohesiveness or their tendency to stick together. The magnitude of this depends inversely on the moisture content, and clays in their natural state are more cohesive than when they have been disturbed or 'remoulded'. The problem with clay slopes is that they can easily be formed to steep angles, but with time they become unstable and slip with disastrous effect, **40.11** to **13**. This can also happen with a previously safe slope if it becomes waterlogged through some alteration in the drainage pattern.

Slopes steeper than the recommended safe angles are possible using various forms of reinforcement. The simplest method is using paving to reduce the effects of wind and water. Reinforcement should not be used before taking advice from an expert in soil mechanics.

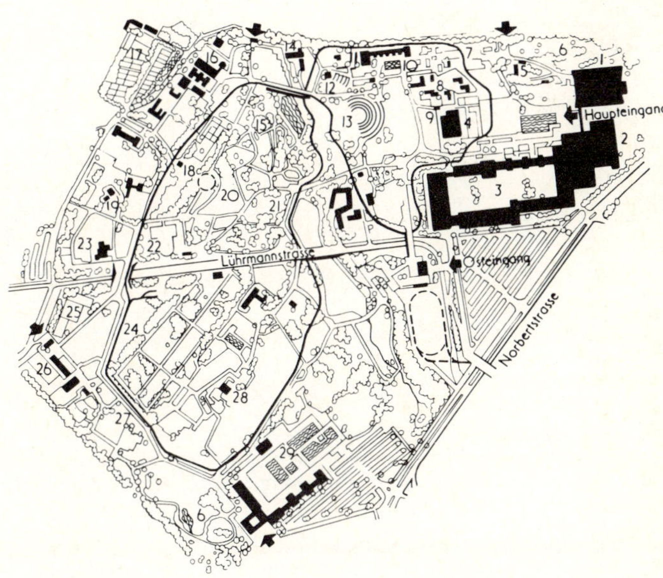

40.1 *A large urban park: Gruga Park, Essen*

1 Gruga hall
2 school of building
3 exhibition hall
4 flower garden restaurant
5 milk bars
6 playground
8 tower
11 terrarium
12 concert pavilion
13 rotunda
15 shrub garden
16 hot house
17 small sample gardens
20 lake
21 shrub garden
23 school gardens
24 cemetery and memorial
26 roller skating rink
27 playgrounds
28 leisure gardens
29 baths

Table I Angles of repose for various soils

Very wet clay and silt	15°
Wet clay and silt	25°
Dry clay, sands, chalk	35°
Moist sand, gravel	40°
Crushed rock, hardcore	45°
Dry sand and gravel	50°
Some types of rock	up to 90°

The angles of slope also depend on the use of the ground. Table II gives maximum angles for uses and ground cover.

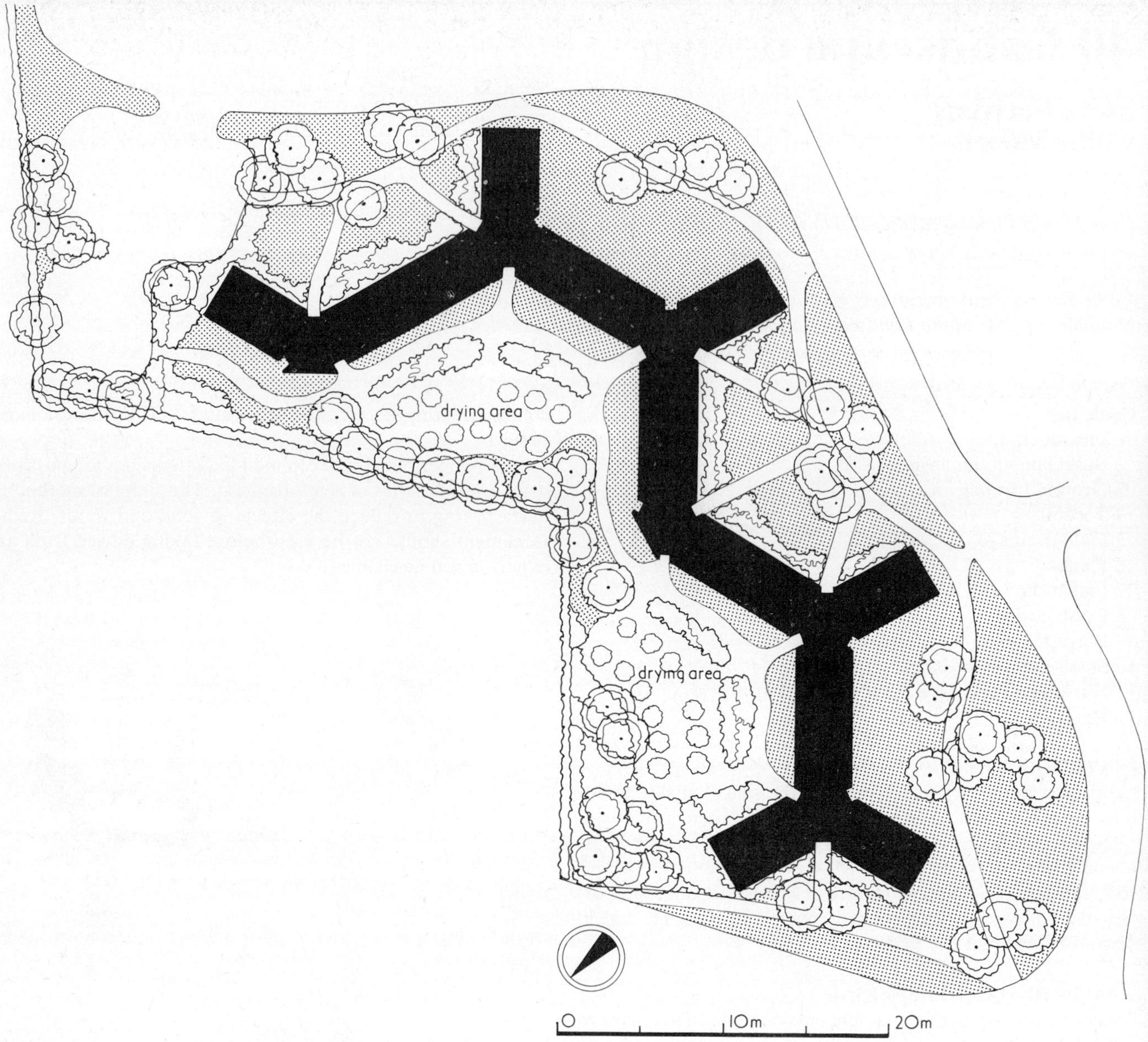

40.2 *Landscape around public housing: Gypsy Castle Lane, Hay-on-Wye (Brian Clouston and Partners)*

Table II Slope limits for various land uses
(Slopes steeper than 30 per cent, 17°, are expressed in degrees, others in grade per cent)

Running tracks cross-slope (tracks should be effectively level along the length)	1%
Tennis, netball, badminton	$1^1/_4$%
Winter games and cricket outfields	2%
Other sports (steepest gradients should be transverse to direction of play)	$2^1/_2$%
Grass areas to be mown with tractor drawn machines	18°
Grass to be mown with small machines; Banks planted with shrubs etc	25°
Grass to be mown with special bank machines	35°
Grass to be left unmown or maintained by hand scythe	45°

Various features are often installed or left below the surface of the ground. Table III indicates the minimum cover required.

Table III Covers required for subterranean items

Demolished buildings below grass, ground cover and perennial planting areas; gas mains, electricity mains, branch drains	450 mm
Demolished buildings below shrub planting areas	600 mm
Water branches, land drains	750 mm
Demolished buildings within 2 m of trees, water mains	900 mm

Note: Gas, electricity and telephone cable branches may be laid at depths less than the general minimum of 450 mm.

4 TOPSOIL

Topsoils can be:

- light sandy loams, good if water and nutrients are ample. Usually only suitable for heath type plants
- medium loams, good for supporting a wide range of plants
- heavy loams, usually good but can be difficult to work when wet
- clay soils, usually require drainage if subject to intensive use. They benefit from the inclusion of organic matter.

Topsoil should be evenly spread to the depths in table IV, **40.14**.

Table IV Required depth of topsoil after firming

Banks	75 to 100 mm
Grass areas	100 mm
Sportsfields	150 mm
Shrub areas	400 mm
Tree pits	600 mm

5 FIELD DRAINAGE

Poor drainage is either due to a high permanent water table that needs to be lowered by a drainage system, or to a perched water table that can be removed by sub-soiling, spiking or sand slitting.

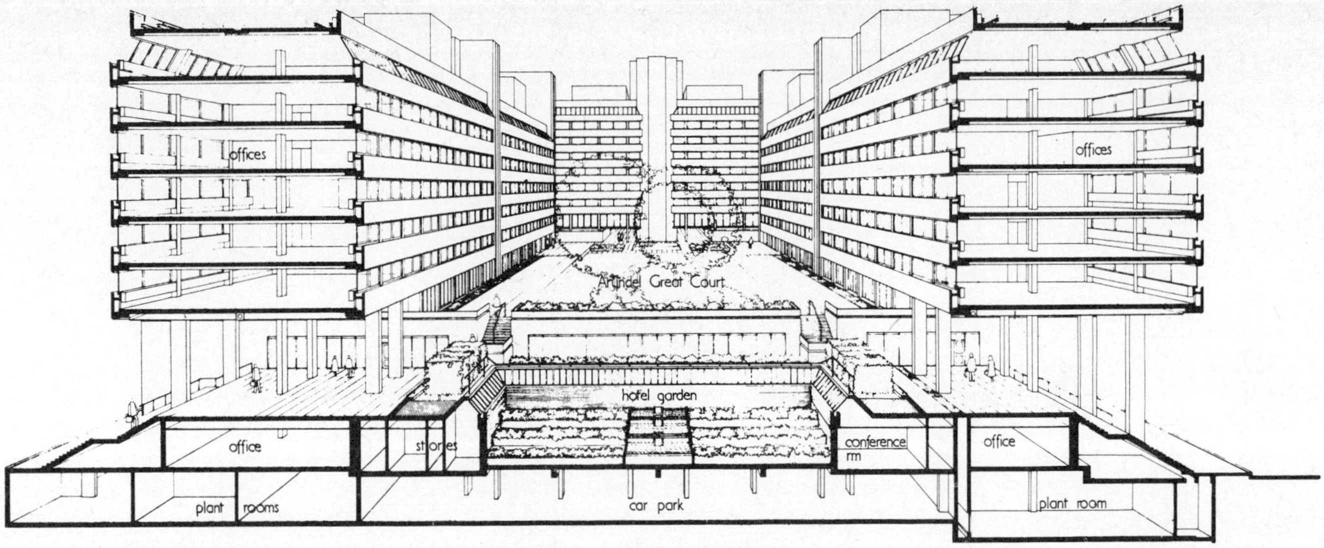

40.3 *Sectional perspective of Arundel Great Court, London WC2, showing garden on roof of loading bay (Frederick Gibberd and Partners)*

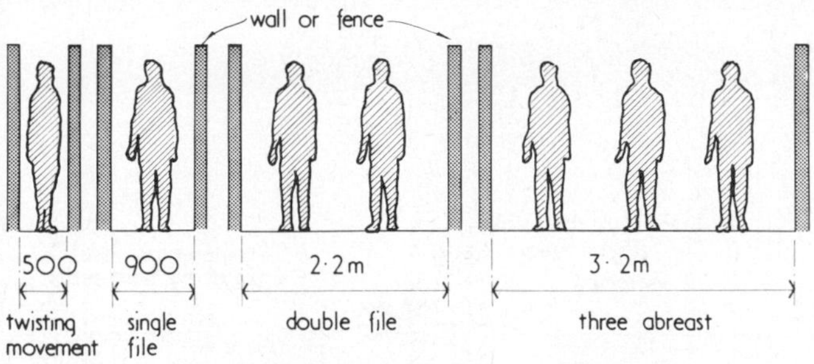

40.4 *Pedestrians between walls or fences, minimum dimensions. Add 25 per cent for freer movement, prams, wheelchairs and bicycles*

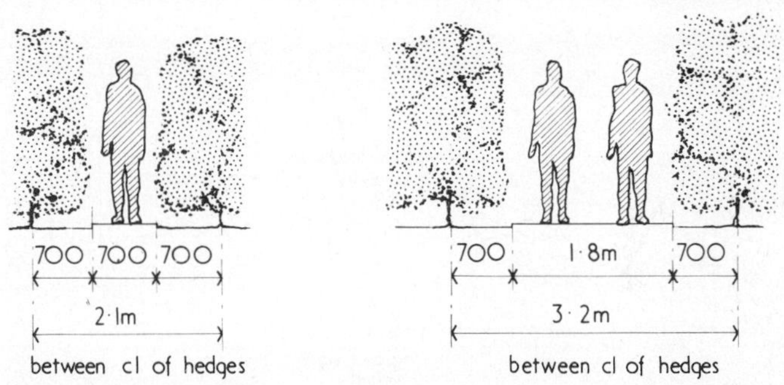

40.7 *Walking between clipped hedges with careful movement*

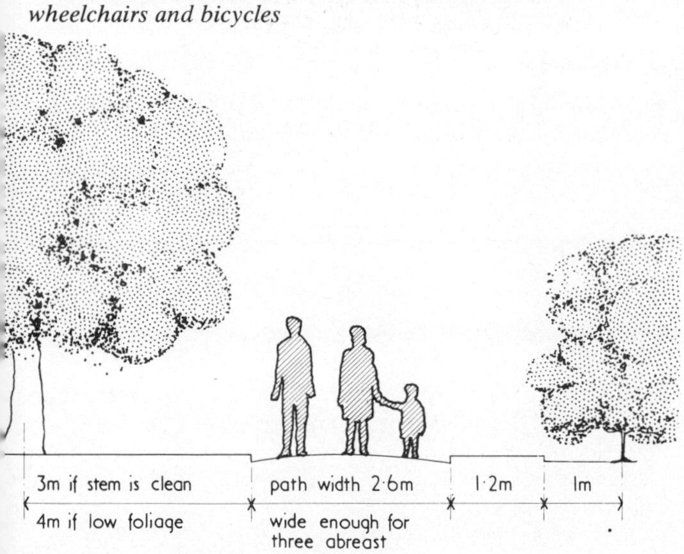

40.5 *Narrow path across open space*

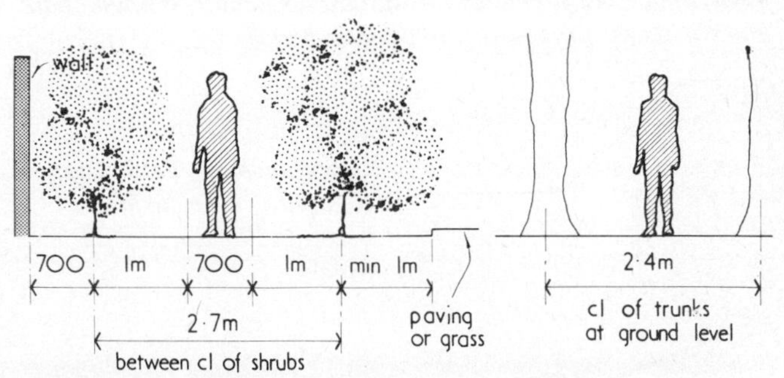

40.8 *Minimum dimensions for pedestrians between free growing shrubs. Where prams are used allow 3 m between centres of planting*

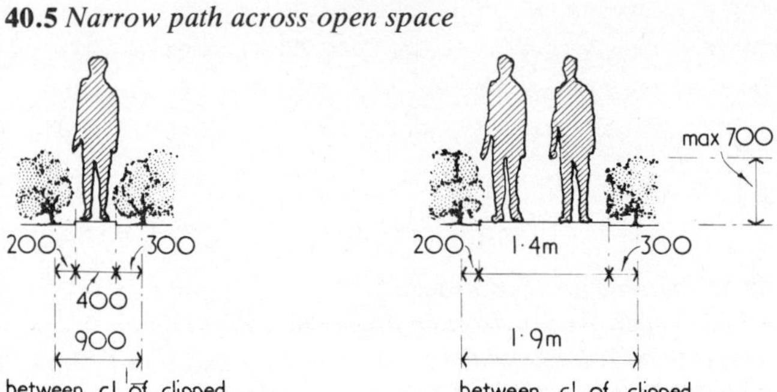

40.6 *Minimum path widths between low planting, impassible for prams. Planting beds should be 400 mm wide for clipped hedges, 600 mm wide for herbaceous plants*

Land drains are laid in patterns as in **40.15**. They are formed as:

- tile drains, lengths of clayware, porous concrete, perforated plastic or pitchfibre pipes laid butt-jointed in a trench
- french drain, **40.16**, the excavated trench is filled with selected coarse rubble at the bottom, and overlaid with fine rubble
- mole drain, unsupported drainage channels formed in clay soils by a mole plough.

Pipes are usually 75 mm diameter clayware or 50 mm diameter perforated plastic pipes for the collectors. Main drains can be as much as 300 mm diameter. As a guide, an outfall from a 2·5 ha area should be 150 mm, from 8 ha 225 mm. **40.17** shows the spacing required for the collector drains.

6 PLANTS

6.01 Grass

New grass can be provided as seed or by turfing.

Seeding is best done in late summer so that the grass can

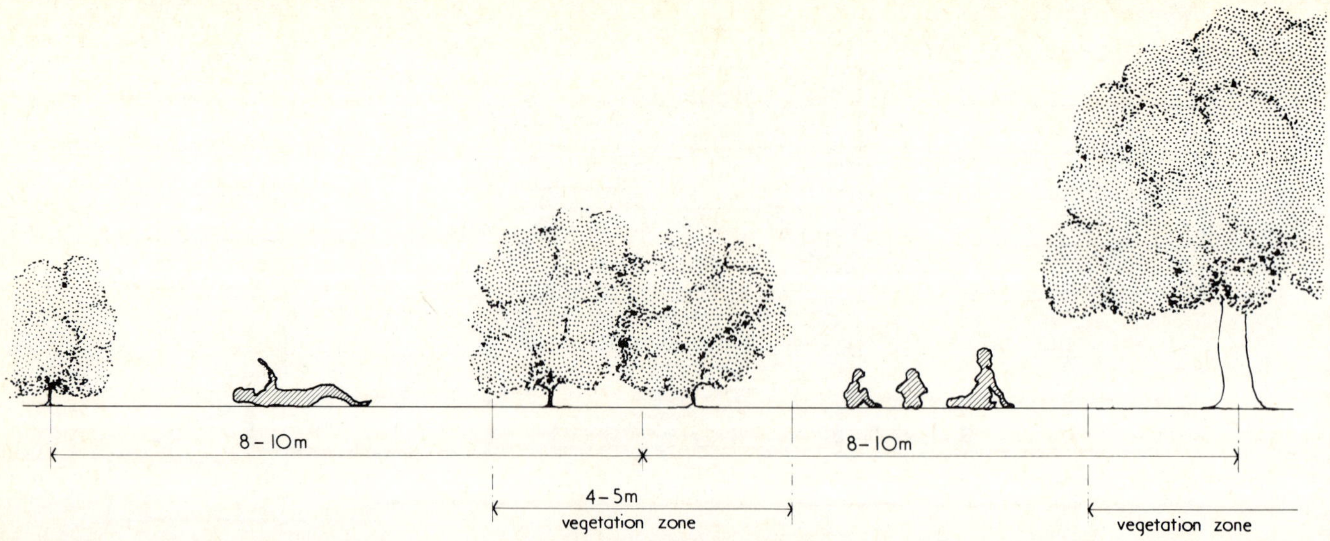

40.9 *Distances between individuals and casual groups in heavily used urban areas*

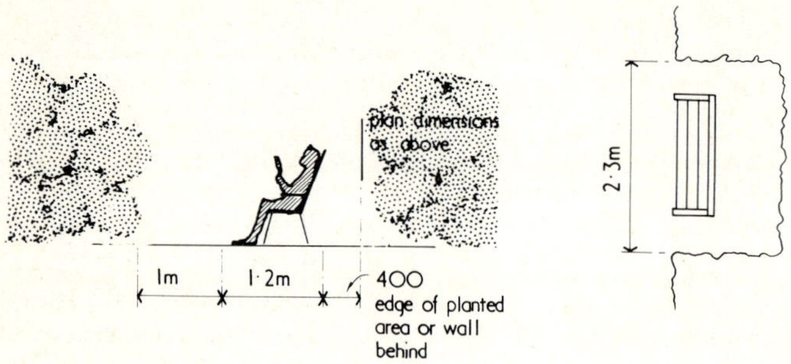

40.10 *Space requirements for upright deckchair, garden chair, and for recess for two-or three-seater bench*

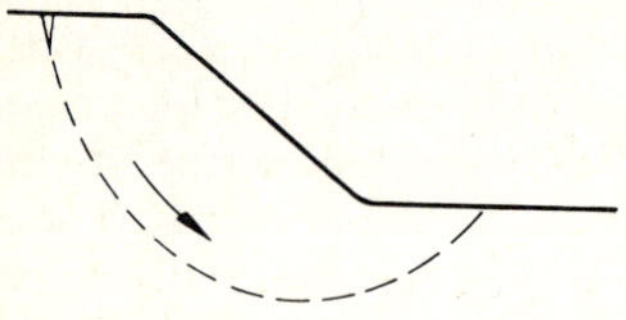

40.11 *Unstable clay slope. Crack develops behind top edge, slip circle surface becomes lubricated with water*

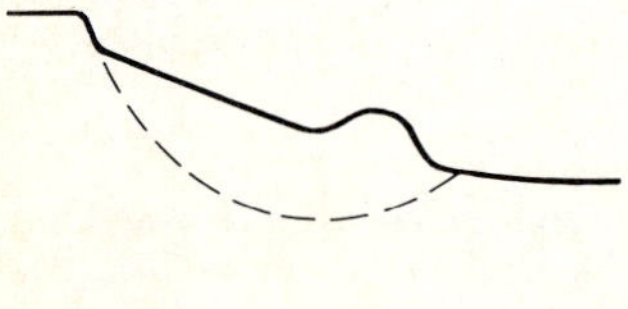

40.12 *The same slope after failure. Note 'hump' at bottom of slip*

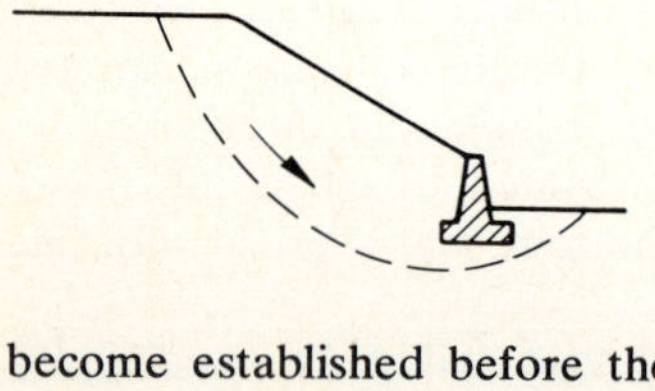

40.13 *The use of a retaining wall at the toe of a slope does not prevent a slip circle failure*

become established before the following summer although spring sowing is possible. The grass types found in established areas will reflect the influence of the environment rather than the seed sown. However, mixes of seed types for various situations will be found in AJ Handbook of Urban Landscape.

Turfing is more expensive than seeding but provides a quicker effect. Turf is sometimes used in single rows to provide an edge to seeded areas. The standard turf sizes are 300 × 900 mm in southern England and 300 mm square in northern England and Scotland. Turfing is best done in autumn, but can be carried out all year round provided it is not dry or frosty weather.

Grass should be cut frequently to the heights in table V.

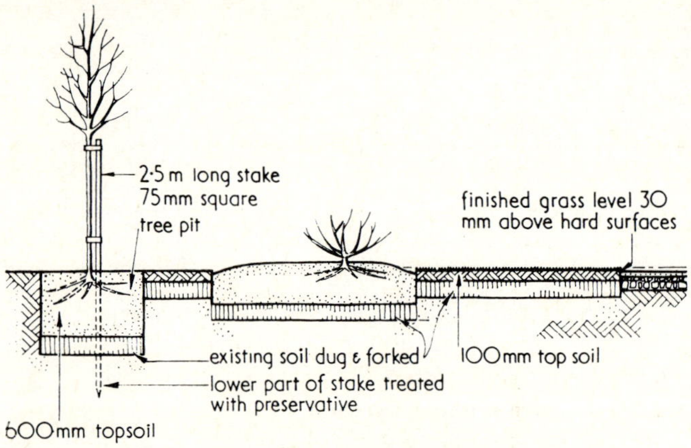

40.14 *Minimum recommended topsoil depths after light consolidation for tree pits, shrub beds and lawns*

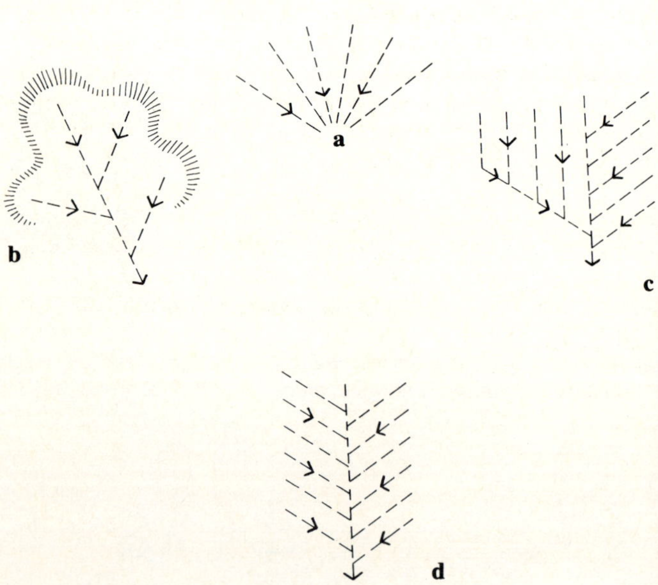

40.15 *Patterns for land drains:*
a *Fan shaped, no single main drain, all drains laid to converge on a single outlet*
b *Natural pattern adapted to suit the land contours*
c *Gridiron, lateral drains discharge into sub-mains which connect to the main drain*
d *Herringbone, lateral drains not more than 60 m long discharge into one main drain*

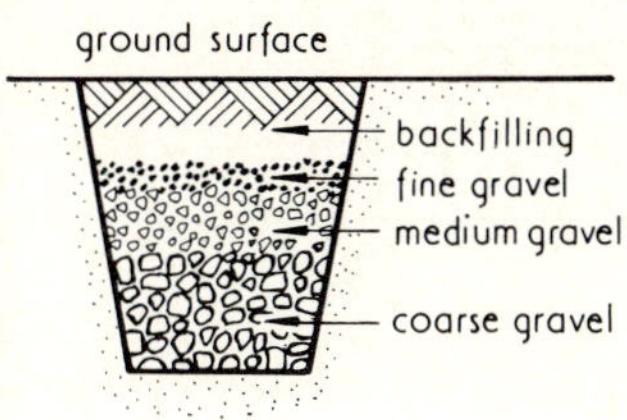

40.16 *Cross section of french drain*

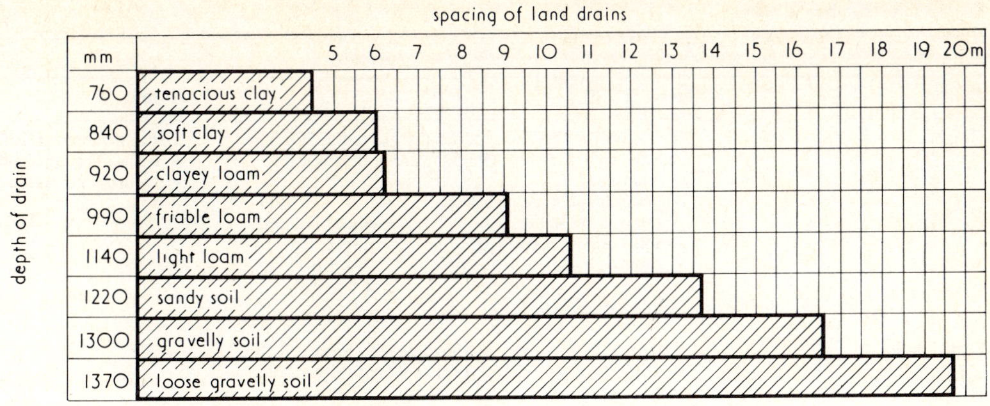

40.17 *Spacing and depth of land drains*

Table V Heights of cut grass

Bowling greens	4– 5mm
Fine lawns	5– 6mm
Ornamental lawns	10– 15mm
Golf fairways	12– 20mm
Amenity landscape	20– 30mm
Soccer pitches	25– 50mm
Pasture and rough landscape areas	100mm

Quality areas of grass also require high maintenance in the form of aeration, fertilising and weed control but amenity areas require less attention. Details in the AJ Handbook of Urban Landscape.

6.02 Trees, shrubs and climbers

In rural areas use native species suited to the location, soil, geology, microclimate, water etc. In general, the same species as grow locally in similar conditions is advisable.

On the urban/rural fringe and in villages use predominately native species, but some naturalised species will be acceptable. Ornamental varieties should be restricted to private gardens. The use of naturalised species is as important as for rural areas.

In urban areas use native species wherever possible, especially for trees. The choice is often restricted by space and vandalism. Naturalised and introduced species are acceptable, particularly in parks and gardens, although species need to be suited to conditions.

6.03 Trees

When specifying or ordering trees quote the required factors from table VI in addition to the species. The terms used tend to vary with the nursery.

Trees over 2 m high must be supported for some years after planting by staking.

The area covered by a root system can be very variable. In general the roots of single trees can be expected to extend as far as their heights. For poplars, willows and elms the roots tend to reach half as far again. Where trees are planted in rows or groups, the roots may extend up to $1\frac{1}{2}$ times the height of any single tree.

Table VII covers some of the common trees.

Where the ground level adjacent to trees has been altered in the course of building, the trees can be helped to survive by the methods shown in **40.18** and **19**. 150mm is the maximum the ground can be raised by simply depositing topsoil, and soil or water should never be allowed to accumulate against the trunks. Isolation on an area of raised ground can render a tree susceptible to drought.

Where trees are near buildings it is usually necessary to prune them to avoid damage and obstruction to light. **40.20** to **22** show correct pruning methods.

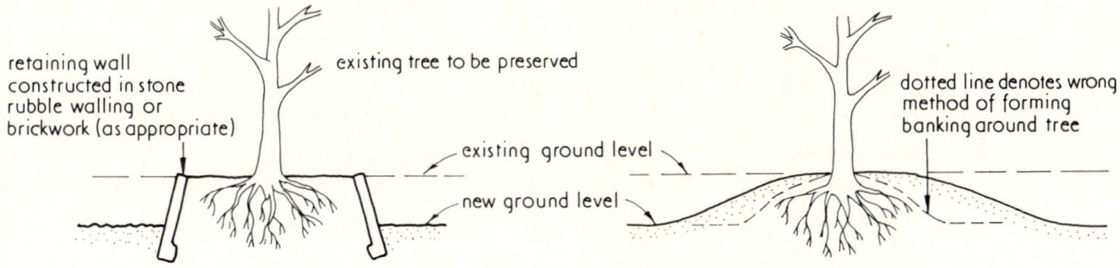

40.18 *Lowering the ground level in vicinity of an existing tree*

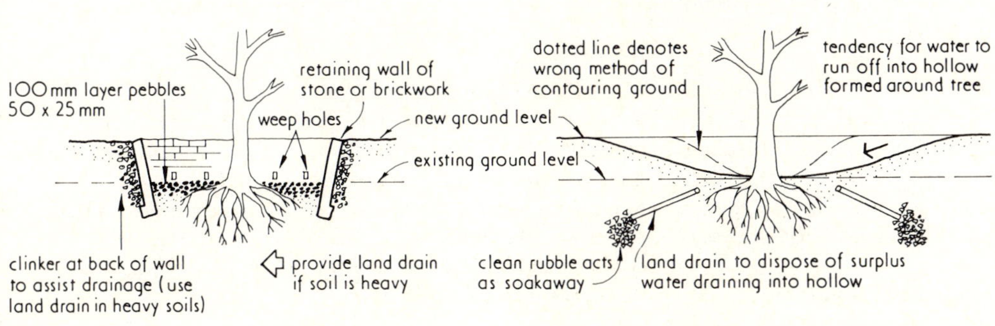

40.19 *Raising the ground level around an existing tree*

Table VI Specification of plants

When ordering or specifying try to avoid using terms like 'extra heavy standard', 'advanced nursery stock', 'semi-mature tree' etc since they tend to mean different things at different nurseries. The information below gives commonly found interpretations.

Factor	Trees	Shrubs
Height	This means the total height, usually quoted in 500 mm intervals, eg 4·5–5·0 m. Heights vary with species, see catalogues Generally heights are as follows: forestry transplants — under 300 mm whips (usually feathered) — under 3 m standards — 3–4·5 m extra heavy standards — 4·5–6 m semi-mature trees — over 6 m	Usually quoted in 150 mm intervals, eg 450–600 mm. For shrubs that spread rather than grow upwards, quote spread instead
Girth	Stem circumference at 1 m above ground level. Sometimes the stem diameter is specified instead, this is normally 50 mm for each 4 m in height. Always specify a minimum. The girths of trees in the form of a standard are: heavy standard — 12–14 cm extra heavy standard — 14–16 cm and — 16–20 cm The general practice within the EEC is for trees to be designated by stem circumference only in the series of even numbered centimetres, eg 18–20 cm	
Spread	For trees over 4 m high should be at least one third of total height. Not applicable to smaller trees. Some species do not spread until older. Specify range of 500 mm, eg 1·5–2 m	For shrubs that spread rather than grow upwards such as ground cover plants, specify spread instead of height in 150 mm intervals
Form (when purchased, not necessarily the form of the mature plant)	Forms are given in BS 3936: part 1:1965. Common forms are: feathered — branches from ground level standard — lowest branch 1·7–1·8 m from ground three quarter standard — lowest branch 1·5–1·6 m from ground half standard — lowest branch 1·1–1·4 m from ground single leader — one main vertical stem	
Root preparation	Depends on species, but generally for deciduous trees: up to 4·5 m — bare root, protected for transit with hessian 4·5–6 m — root wrapped, the assembly of roots and packing material securely wrapped in hessian or plastics sheeting over 6 m — root balled	Depends entirely on species. For out-of-season planting and for some species pot-grown plants are essential Plants grown in open ground are supplied bare roots or root balled. Species which have to be lifted with ball are now normally pot grown

Key and notes to tables VII, VIII and IX

Growth rates:
- F fast
- S slow

Soils:
- A clay, deep and damp
- B clay, shallower and drier
- C dry over sand/gravel
- D shallow over chalk
- E shallow over limestone
- F wet

Rural/Urban:
- R rural
- (R) useful for 'urban' or rural/urban fringe
- U urban
- (U) urban use limited to parks and open space

Uses:
- G street and/or paved areas (mainly applicable to trees)
- H park and open space
- I specimen
- J group (all shrubs usually grouped or massed)
- K industrial sites
- L highly polluted areas
- M against buildings and walls (shrubs and climbers only)
- N cold exposed sites
- O maritime sites
- P shelter belts
- Q shade tolerant
- R formal hedge (for trees, planting distance given in brackets)
- S informal hedge
- T screen
- U close to buildings (trees only)
- V river/lakeside

Planting distances: distances in brackets refer to spacing in hedgerows.

Notes:
- W do not plant close to buildings, especially on clay subsoils
- X naturalised – reproduce on their own in the wild
- Y support required (climbers only)
- Z 'vandal resistant' (shrubs only)
- * there are many more species and cultivars than are listed, and these are used predominantly in urban areas.

Table VII Trees

Common name	Botanical name	Height (m)	Spread (m)	Growth Rate	Evergreen/ Conifer	Soils	Rural/ Urban	Use	Notes
Native trees									
Alder	Alnus glutinosa	15+	5+	F		ABEF	R(U)	KV	Dominant on wet peaty soils
Ash	Fraxinus excelsior	25+	20+			ABDE	R(U)	HKNOP	W, dominant on soils type E
Aspen	Populus tremula	15+	10+	F		ABE	R(U)	V	W
Beech	Fagus sylvatica	25+	30+			CD	R(U)	HINP R(300)	Dominant in south on soils type D
Birch, hairy	Betula pubescens	20+	10+	F		ABCE	RU	GHJKNU	Codominant on poor light acid soils, mainly north and east
Birch, silver	Betula pendula	20+	10+	F		ABCE	RU	GHJKNU	Codominant on poor light acid soils, mainly south
Cherry, bird	Prunus avium	8+	4+			BE	RU	GHJ	* native mainly in north and west

Table VII Trees *(continued)*

Common name	Botanical name	Height (m)	Spread (m)	Growth Rate	Evergreen/ Conifer	Soils	Rural/ Urban	Use	Notes
Cherry, wild	Prunus padus	15+	12+			ABD	RU	GHIJK	*
Crabapple	Pyrus malus	8+	6+			AB	R(U)	HK	* native mainly in south
Elm, English	Ulmus procera	25+	15+			AB	R(U)	HJ	W } not planted at present due
Elm, wych	Ulmus glabra	25+	25+	F		ABE	R(U)	HI	W } to Dutch elm disease
Hawthorn	Crataegus monogyna	8+	6+	S		AB	R(U)	GHKNO R(250)	* often large shrub
Holly	Ilex aquifolium	10+	5+	S	E	ABCD	R(U)	JKLNQ R(450)	
Hornbeam	Carpinus betulus	15+	10+	S		AB	R(U)	GHNP R(300)	Native mainly in south-east
Lime, broad-leaved	Tilia platyphyllos	25+	15+			E	RU	GHIJK	
Lime, small leaved	Tilia cordata	25+	15+			E	RU	GHIJK	
Maple, field	Acer campestre	10+	10+	S		AE	R(U)	HK	
Oak, English	Quercus robur	15+	20+	S		A	R(U)	HIJNP	* W, dominant in south and midlands on soils type A
Oak, sessile	Quercus petraea	15+	20+	S		B	R(U)	HIJNP	W, dominant in north on soils type B
Pine, Scots	Pinus sylvestris	20+	12+		EC	BC	R(U)	N	Codominant on poor light acid soils
Rowan	Sorbus aucuparia	6+	4+			ABD	RU	GHIJKNO	* native mainly in south
Whitebeam	Sorbus aria	10+	10+			DE	RU	GHJKNOU	*
Willow, crack	Salix fragilis	15+	10+	F		F	R(U)	KNOV	W *
Willow, white	Salix alba	20+	20+	F		F	R(U)	KNOV	W *
Yew	Taxus baccata	15+	15+	S	EC	ACDE	R(U)	KNQ R(450)	Poisonous to cattle and horses
Introduced trees									
Cedar of Lebanon	Cedrus libani	30+	25+	S	EC		U	HI	*
False acacia	Robinia pseudoacacia	20+	15+	F			U	GHKL	
Horse chestnut	Aesculus hippocastanum	25+	25+				(U) (R)	HIJK	* X
Indian bean tree	Catalpa bignonioides	12+	15+				U	GHIL	
Larch, European	Larix decidua	30+	15+	F	C		U(R)	HJP	
Lime, common	Tilia x euchlora	20+	15+				U(R)	GHIJK	
Maple, Norway	Acer platanoides	20+	15+				U(R)	GHK	*
Oak, holm	Quercus ilex	20+	20+	S	E		U(R)	HJOP	
Pine, Corsican	Pinus nigra maritima	30+	10+	S	EC		U(R)	HIJOP	Drought tolerant
Plane, London	Platanus x hispanica	25+	25+				U	GHKLN	
Poplar, black Italian	Populus 'Serotina'	25+	15+	F			(U) (R)	KLT	W*
Spruce, Serbian	Picea omorika	25+	10+	F	EC		U	HIJL	
Sweet chestnut	Castanea sativa	20+	15+	S			(U) (R)	H	X
Sycamore	Acer pseudoplatanus	20+	15+	F			(U) R	KNOP	X
Tree of Heaven	Ailanthus altissima	20+	20+	F			U	GL	
Tulip tree	Liriodendron tulipifera	30+	15+	F			U (R)	GHIJL	
Willow, weeping	Salix x chrysocoma	15+	15+	F			(U)	HV	W*
Walnut, common	Juglans regia	20+	15+	S			U (R)	HIJ	

Table VIII Shrubs

Common name	Botanical name	Height (m)	Spread (m)	Growth Rate	Evergreen	Soils	Uses	Planting Distances (m)	Notes
Native shrubs									
Blackthorn	Prunus spinosa	4+	4+			ABCDE	HNOPS	1·5 (0·3)	Z
Box	Buxus sempervirens	4+	4+	S	E	D	KQRT	1 (0·5)	
Bramble	Rubus fruticosus	1+	2+	F		AB		1·0	* Z
Dogwood	Cornus sanguinea	2+	1+			ABDE	H	1·0	* Z
Elder	Sambucus nigra	4+	4+	F		ABCDE	ILNOPQT	1·5	Z
Gorse	Ulex europaeus	1+	1·5+		E	ABC	NO	0·9	
Guelder rose	Viburnum opulus	3+	2+			A	HNV	1·0	*
Hazel	Corylus avellana	4·5+	4·5+			ABDE	HQRST	1·5 (0·5)	
Heath	Erica carnea	0·3+	0·5+		E	C	N	0·3	* Ground cover
Heather	Calluna vulgaris	0·4+	0·5+		E	C	N	0·3	* Ground cover
Ivy	Hedera helix	0·2+	1·0+	F	E	AB	HKOQ	0·9	* Ground cover (also climber)
Privet	Ligustrum vulgare	2+	3+		semi E	ABDE	HIKNOPR	1·5 (0·3)	*
Rose, dog	Rosa canina	2+	1·5+			AB	H	1·0	* Z
Sallow, common	Salix atrocinerea	5+	5+	F		ABF	HV	1·5	* Z
Sea buckthorn	Hippophae rhamnoides	3·5+	3+			C	HNOST	1·5 (0·6)	Z
Spindle	Euonymus europaeus	3+	1·5+			ABDE	H	1·0	*
Wayfairing tree	Viburnum lantana	3·5+	4·5+			ABDE	HT	1·5	*
Willow, goat	Salix caprea	5+	5·5+	F		ABF	HNOTV	1·5	* Z
Introduced shrubs									
Bamboo	Arundaria japonica	4+	2+		E		HIOT	1·0	* Z
Barberry	Berberis darwinii	3+	3+		E		HKRST	1·0 (0·6)	* Z
Broom, Warminster	Cytisus x praecox	1·5+	1·5+	F			HKO	0·6	*
Butterfly bush	Buddleia davidii	3+	2·5+	F			HKO	1·5	* Z
Ceanothus	Ceanothus dentatus	3+	3+		E		EM	1·5	* Hardy only in south
Cotoneaster	Cotoneaster dammeri	0·05+	1+	F	E		HQ	0·6	* Ground cover
Cotoneaster	Cotoneaster horizontalis	0·5+	2+				HKM	1·0	*
Cotoneaster	Cotoneaster x watererii	4+	4+		semi E		HIOT	1·5	*
Elaegnus	Elaegnus pungens	3+	3+	F	E		HKLNOQRT	1·0 (0·6)	*
Fatsia	Fatsia japonica	3+	3+		E		HIKOQ	1·0	Hardy only in south
Fig	Ficus carica	5+	3+				HIMO	1·5	
Firethorn	Pyracantha coccinea	4+	4+		E		HMOQRST	1·5 (0·6)	* Z
Forsythia	Forsythia x intermedia	2+	1·5+	F			HIKR	1·0 (0·5)	*

Table VIII Shrubs (*continued*)

Common name	Botanical name	Height (m)	Spread (m)	Growth Rate	Evergreen	Soils	Uses	Planting Distances	Notes
Japonica (quince)	Chaenomeles speciosa	1·5+	1·5+				HKM	0·6	* Z
June berry	Amelanchier canadensis	3+	3+				GHIKL	1·0	* Z
Laurel	Prunus laurocerasus	6+	7+		E		HQRST	1·5 (0·9)	*
Laurustinus	Viburnum tinus	2·5+	2+		E		HLQST	1·0 (0·5)	*
Lavender, old English	Lavandula spica	1+	1+		E		ORS	0·5 (0·3)	*
Lilac, common	Syringa vulgaris	3+	2+				HI	1·5	*
Lonicera	Lonicera nitida	1·5+	1·5+		E		HLR	0·6 (0·3)	*
Oregon grape	Mahonia aquifolium	1·5+	2+		E		HKNOQ	1·0	* Z Ground cover
Periwinkle, lesser	Vinca minor	0·1+	0·5+		E		HOQ	0·3	* X Ground cover
Potentilla	Potentilla fruticosa	1·5+	1·5+				HOS	0·6 (0·3)	*
Rose, ramanas	Rosa rugosa	2+	1·5+				HORS	1·0 (0·6)	* Z
Rose of Sharon	Hypericum calycinum	0·4+	0·5+		E		HKQ	0·3	* Ground cover
Rosemary	Rosmarinus officinalis	1·5+	1+		E		ORS	0·5 (0·3)	
Stags horn sumach	Rhus typhina	4+	4+				GHI	1·5	* Z

Table IX Climbers

Common name	Botanical name	Height (m)	Growth Rate	Evergreen	Soils	Uses	Notes
Native climbers							
Honeysuckle	Lonicera periclymenum	8+	F		ABC	TM	* Y
Ivy	Hedera helix	15+		E	AB	M	*
Travellers' joy	Clematis vitalba	8+			D	M	Y Native mainly in the south
Introduced climbers							
Clematis	Clematis montana	8+	F			M	* Y
Hydrangea, climbing	Hydrangea petiolaris	10+	F			QM	
Japanese crimson glory vine	Vitis coignetiae	15+	F			M	Y
Jasmine, common white	Jasminum officinale	10+				M	* Y
Russian vine	Polygonum baldschuanicum	10+	F			TM	Y
Virginia creeper	Parthenocissus quinquefolia	15+	F			QM	*
Wisteria	Wisteria sinensis	20+				M	* Y

Key and notes to tables VII, VIII and IX

Growth rates:
- F fast
- S slow

Soils:
- A clay, deep and damp
- B clay, shallower and drier
- C dry over sand/gravel
- D shallow over chalk
- E shallow over limestone
- F wet

Rural/Urban:
- R rural
- (R) useful for 'urban' or rural/urban fringe
- U urban
- (U) urban use limited to parks and open space

Uses:
- G street and/or paved areas (mainly applicable to trees)
- H park and open space
- I specimen
- J group (all shrubs usually grouped or massed)
- K industrial sites
- L highly polluted areas
- M against buildings and walls (shrubs and climbers only)
- N cold exposed sites
- O maritime sites
- P shelter belts
- Q shade tolerant
- R formal hedge (for trees, planting distance given in brackets)
- S informal hedge
- T screen
- U close to buildings (trees only)
- V river/lakeside

Planting distances: distances in brackets refer to spacing in hedgerows.

Notes:
- W do not plant close to buildings, especially on clay subsoils
- X naturalised – reproduce on their own in the wild
- Y support required (climbers only)
- Z 'vandal resistant' (shrubs only)
- * there are many more species and cultivars than are listed, and these are used predominantly in urban areas.

Tree belts can be used to give a measure of sound shielding. However, it requires dense planting 100 m thick to achieve a reduction of 80 dB. To provide shelter from wind, a belt composed of both trees and shrubs is planted to produce a screen of uniform thickness and 50 per cent permeability. **40.23** shows the area sheltered by such a belt.

6.04 Shrubs

Table VIII gives details of common plants for the various situations in which shrubs are useful, and table VI should be followed when specifying and ordering. Before planting the ground should be well dug to a depth of at least 300 mm.

The spacing of the plants depends on the effect required, and on the time available to achieve it. Hedges should be planted in two rows with the plants staggered.

Shrubs useful for ground cover are indicated in the Notes column of table VIII. Natural ground cover tends to be herbs and grasses rather than shrubs. These are rarely planted if required but just allowed to invade.

6.05 Climbers

Table IX shows a few plants suitable for climbing over buildings, pergolas, etc.

7 IRRIGATION

Extra water supply to establish new planting is usually necessary even in temperate zones, such as Great Britain. In warmer climates the losses of water from the soil in evaporation and from the plants themselves in transpiration are nearly always such as to require continuing artificial replenishment of the soil water and in some cases there is danger of salinity causing salts poisoning. If this soil water is reduced beyond the 'permanent wilting point' the planting will never recover.

Table X gives the common types of irrigation systems and their advantages. **40.24** shows a system where there is insufficient precipitation for normal horticulture.

8 ROOF GARDENS

Roof gardens are of two types:

40.20 *Before and after pruning a tree by crown thinning, permitting more light to penetrate the canopy*

40.21 *Before and after pruning by crown lowering, reducing the overall dimensions*

40.22 *Pruning by crown raising, removing the lower branches to permit light to pass under the canopy.*

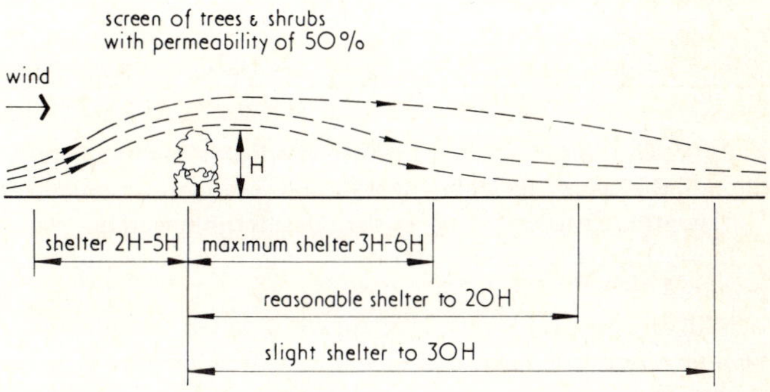

40.23 *Wind shelter provided by a screen of trees and shrubs*

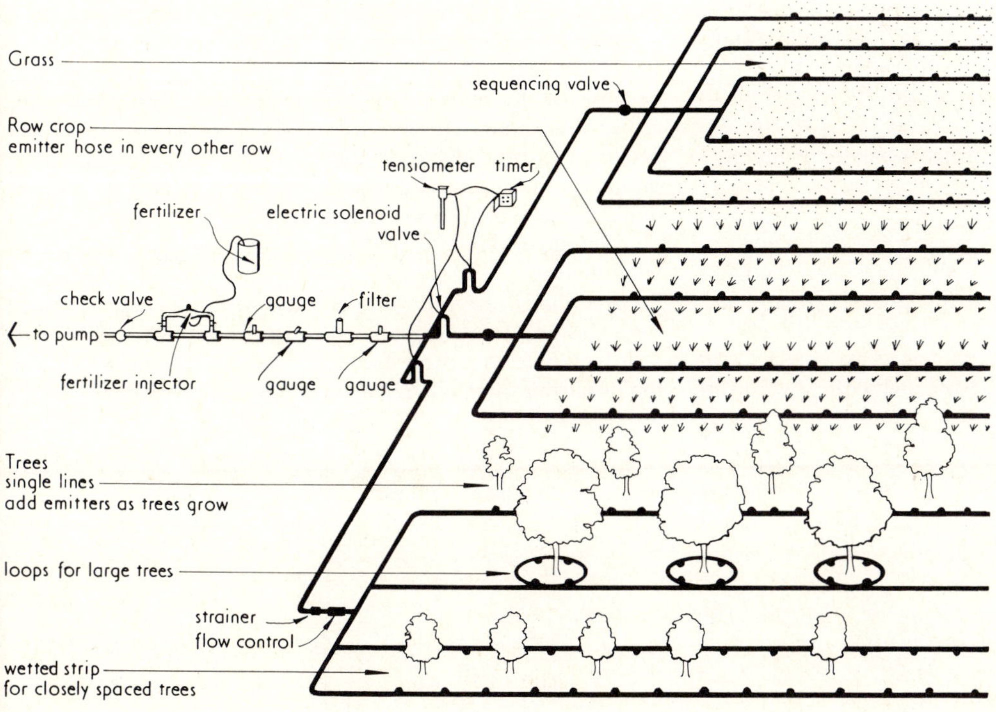

40.24 *Example of an irrigation system. This type would be appropriate for a nursery, but illustrates the various elements that could be used in a landscape context*

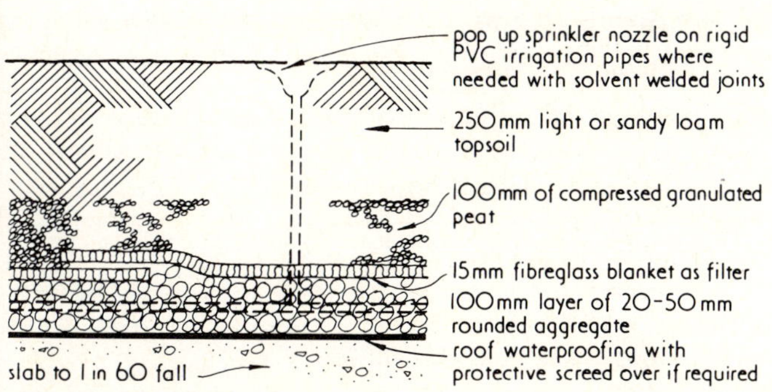

40.25 *Basic soil profile for a roof garden, the thicknesses of the layers can be varied*

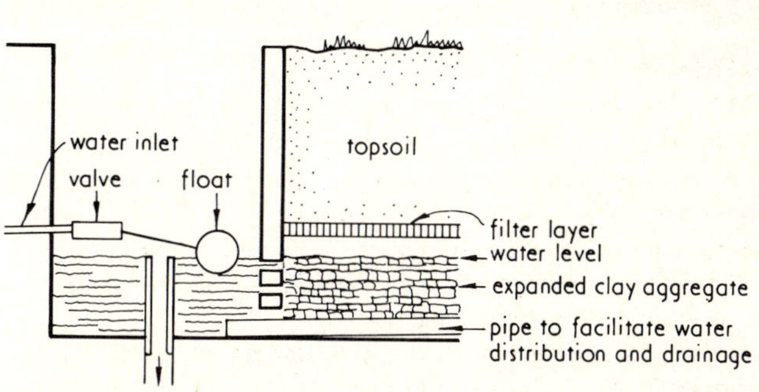

40.26 *Semi-automatic irrigation for roof gardens by capillary action*

Table X Comparison of irrigation systems

Criteria	Surface (flood)	Sprinklers	Trickle	Sub-surface trickle	Managed water table
Efficiency of use	Poor	Medium	Good	Good	Good
Control of soil moisture tensions	Poor	Poor to reasonable	Good	Good	Good
Losses during water transport	High	Low	Low	Low	Variable
Danger of over-irrigation	Application point	Related to uniform plant cover	Low	Low	Possible
Percolation losses	High	Medium–Low	Low	Low	Variable
Salinity build-up	High	Medium–Low	Low	Medium/High	High
Requirement for flat land	High	High	Low	Low	Possible
Ability to use saline water	Poor	Poor	Good	Poor/Medium	Poor
Surface evaporation	High	High	Low	Low	Low
Volume of water applied	High	Medium–Low	Low	Low	Low
Maintenance requirements	High	Medium	Low	High	High
Frequency of blockages	Occasional	Rare	Occasional	Frequent	Rare

Order of effectiveness
Trickle—Sub-surface trickle—Managed water table—Sprinkler systems—Surface systems

- those with sufficient soil to ensure survival of plants through all but the worst drought, **46.25**. This requires a structure sufficiently reinforced to carry the considerable weight.
- those using irrigation methods to reduce the depth of soil required.

In both cases the roof is usually built with a fall of 17 per cent to facilitate free drainage, and is waterproofed with a layer of asphalt, bitumastic or a butyl rubber liner. Below the soil is a drainage layer of gravel and a glass fibre mat or synthetic capillary matting.
Irrigation is achieved by manual watering, pop-up sprinklers at 12 m centres, or by capillary action, **40.26**. Electronic systems are available to run a fully automatic operation.

9 BARRIERS

Table XI compares the different forms of barrier available.

9.01 Fences

Timber fences are shown in **40.27**. They can be of sweet chestnut, larch, oak or western red cedar, which need no treatment. Other timbers can be used if they are properly treated with preservatives.
Chain link fencing, **40.28** is frequently used for security. The galvanised mild steel wire core is protected with a pvc sheath available in two shades of green, black, brown and white. Table XII gives the heights used.
Steel rod vertical bar fencing, **40.29** is often called unclimbable! For playgrounds the type with hairpin tops should be used. Spaces between the verticals should not be less than 100 mm nor more than 120 mm.

9.02 Walls

Masonry walls may be built of brick, concrete blocks or stone. In landscaping they may be simple barriers, or they may retain ground at a higher level on one side.
Brick walls should always be built with a high quality brick because of the severe exposure. The mortar should not be richer than grade III (see section 45, Structure) and a dpc provided 150 mm above ground level. This DPC should not be bituminous but consist of two courses of engineering brick, or a double course of slate in grade I mortar. An appropriate capping should also be provided to prevent moisture penetration from the top.
Freestanding or garden walls (ie those with soil at equal levels on each side) should be built with a height to thickness ratio of 7 or less in normal exposures and 4 in severe conditions. 2 m high walls may be built a half-brick thick in staggered form, as

Table XI Function related to choice of barrier

Form of enclosure	Physical barrier (security)	Visual barrier (privacy)	Noise barrier	Windbreak	To define space	Durability	Climbable	Permanence	Remarks
Trees	×	√	×	√	√	High	×	High	
Walls: brick, stone, concrete	√	√	√ (i)	√	√	High	×	High	(i) If properly placed and sized
Fences: timber	√	√	√ (i)	√	√	Low	× (ii)	Low	(i) If properly placed and sized
Fences: precast concrete with timber panels	√	√	√	√	√	Med	× (i)	Med	(i) Depending on design
Fences: precast concrete with wires	√	×	×	×	√	High	√	Med	
Metal: wrought iron and mild steel	√	×	×	×	√	High	× (i)	High	(i) Depending on design
Chain link and woven wire fence	√	× (i)	×	×	√	Med	√ (ii)	Med	(i) Woven wire can be a directional visual barrier eg glare fences on motorways (ii) Chain link if large mesh
Strained wire fence	√	×	×	×	√	Med	√	Med	
Guard rails	√ (i)	×	×	×	×	Med	√	Low	(i) Only for the law abiding
Hedge bank	√	× (i)	×	√ (ii)	√	Med	√	Med	(i) Unless very high (ii) If high enough
Ha-ha	√	×	×	×	×	High	√	Low	
Cattle-grid	√ (i)	×	×	×	×	High	×	High	(i) For animals
Hedges, shrubs	√ (i)	√	×	√	√	Med	×	Med	(if) If spiky eg hawthorn, blackthorn
Bollards	×	×	×	×	√	High	√	Med	

Table XII Heights of fencing (mostly chain link)

House garden fronts and divisions	0·9 m
Minimum for children's playgrounds; general agricultural	1·2
House gardens; playing fields; recreation grounds; highways; railways	1·5–1·8
Commercial property	1·8
Industrial security fencing	2·1

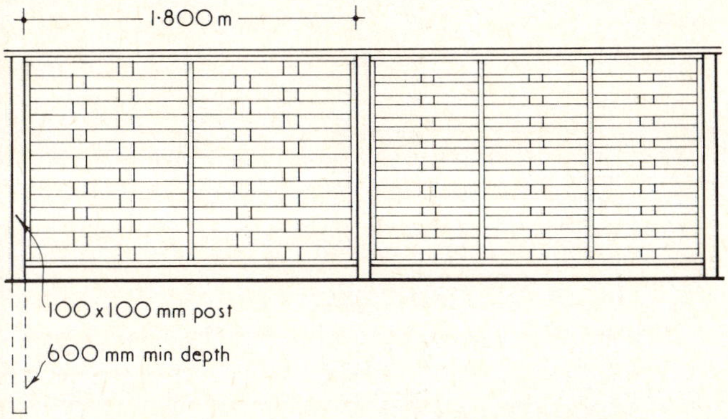

40.27a *Woven wood fence*

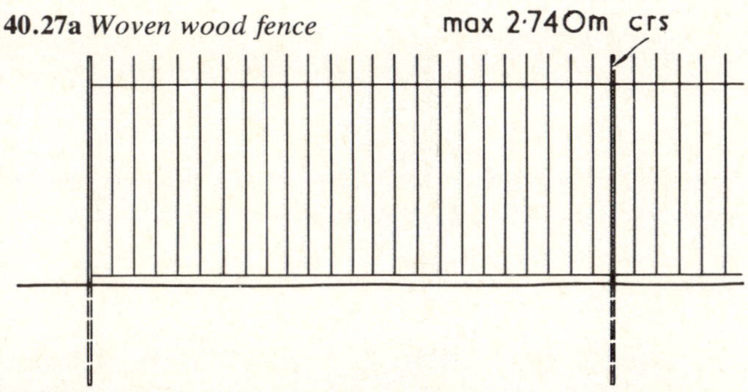

40.29 *Vertical bar steel fencing, claimed to be unclimbable! Where used in children's playground the type with hairpin tops is preferred*

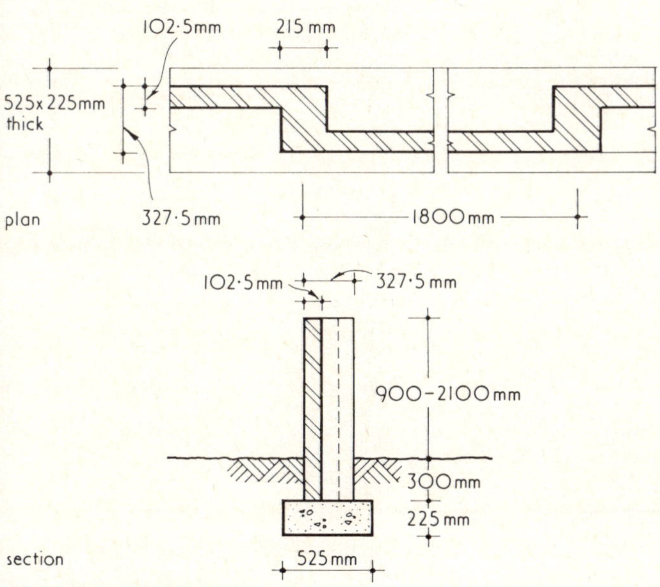

40.30 *Staggered brick wall*

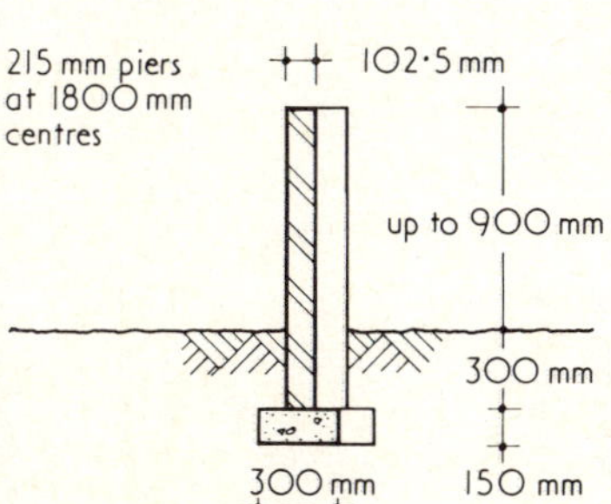

40.31 *Wall with piers at intervals on one side*

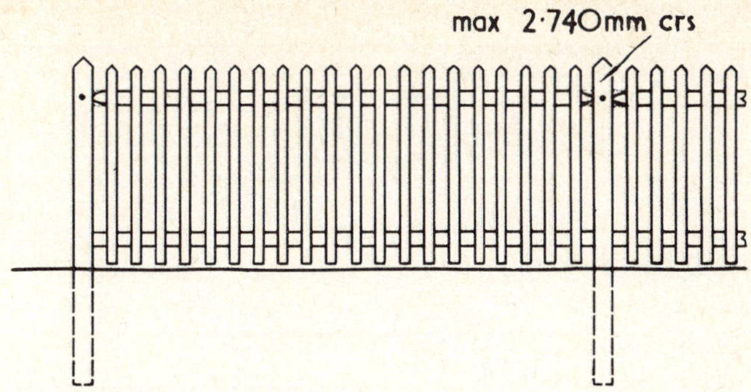

40.27b *Timber palisade fence*

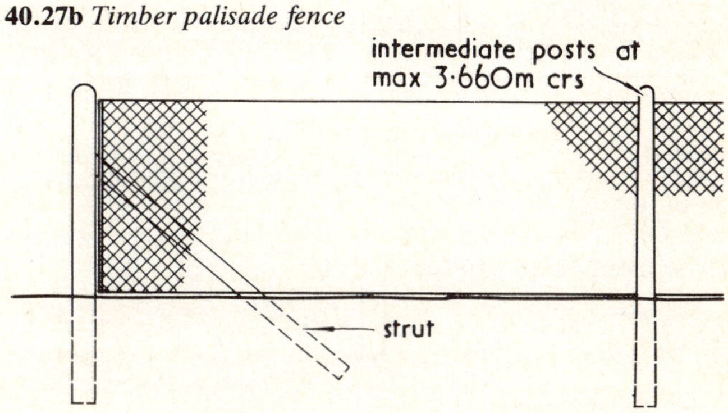

40.28 *Diagonal mesh chain link fencing on precast concrete posts*

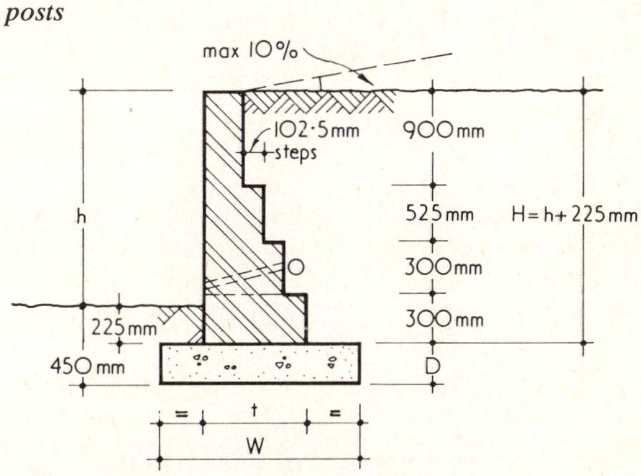

40.32 *Section through brick retaining wall*

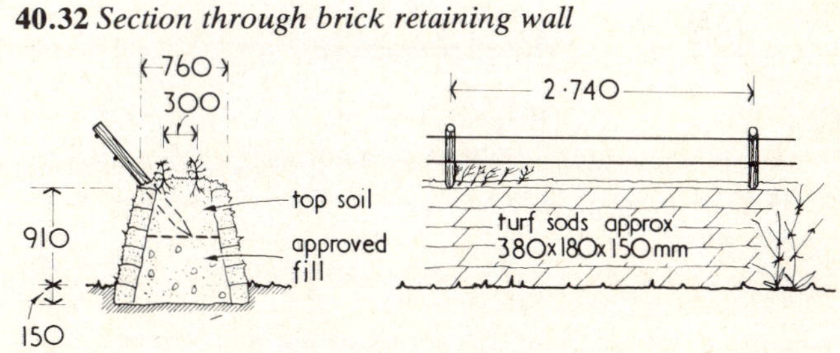

40.33 *A hedge bank appropriate in a rural context*

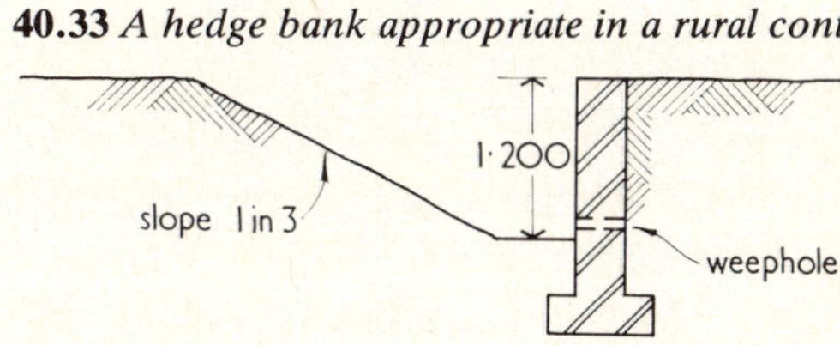

40.34 *Section through a ha-ha. This is a useful device for separating formal gardens from lifestock without a visual barrier*

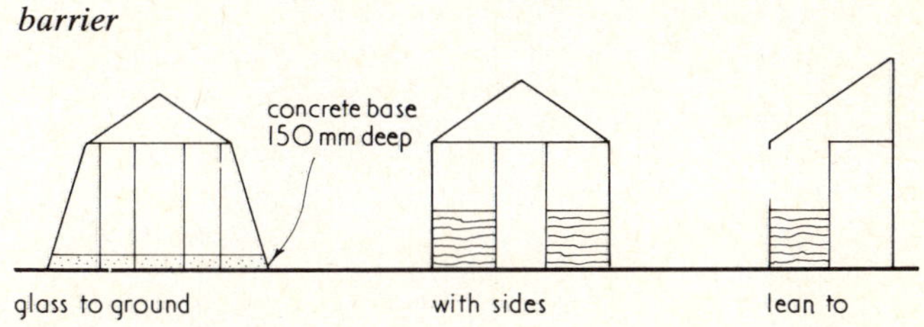

40.35 *Various forms of glasshouse*

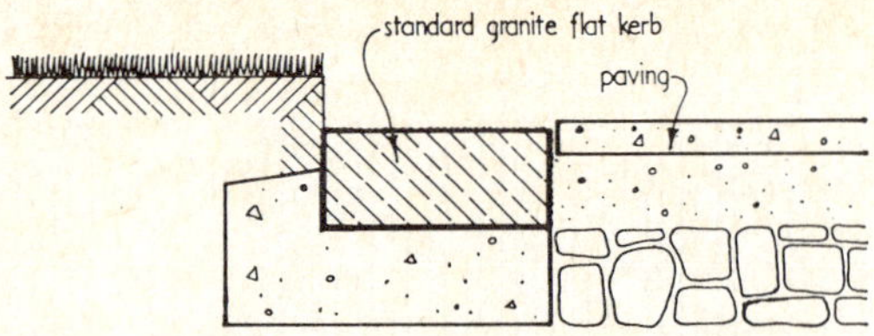

40.36 *Kerb set flush prevents spread of base of paving*

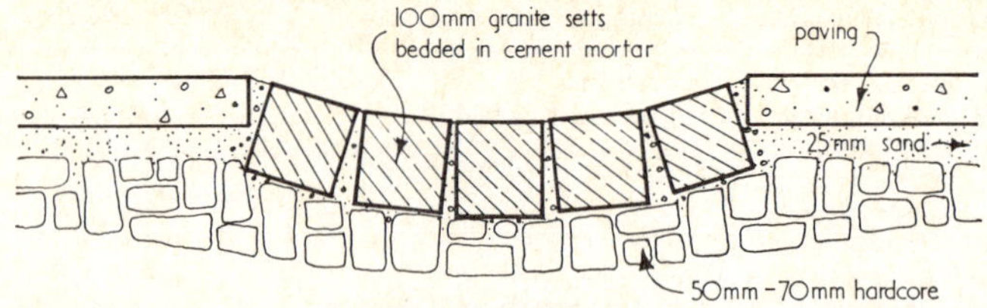

40.40 *Granite setts used as drainage channel in area of precast paving*

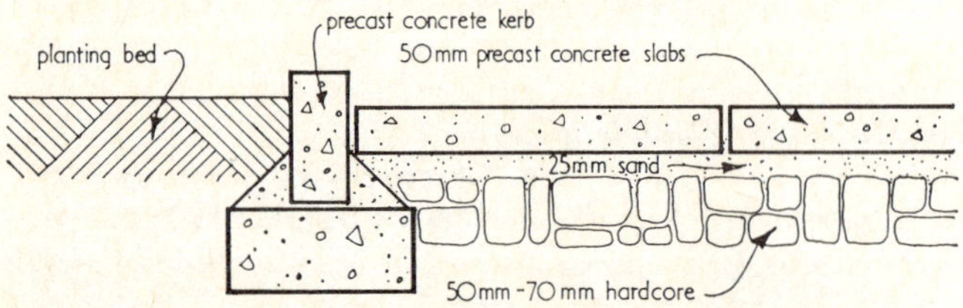

40.37 *Use of standard precast concrete kerb to separate path from planting bed*

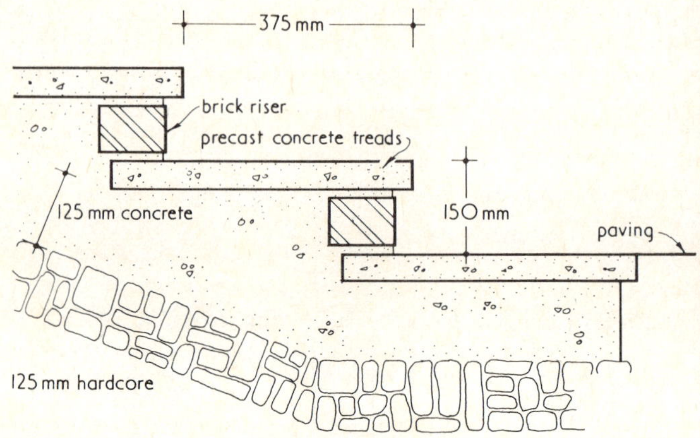

40.41 *Precast concrete or natural stone steps, brick or granite sett risers on a concrete base. Minimum tread thickness 50 mm*

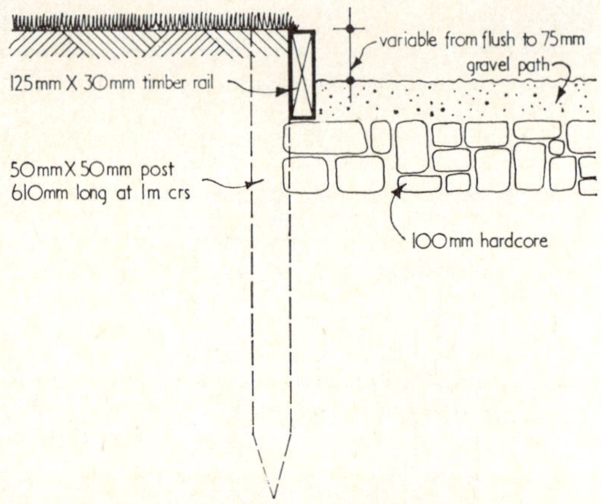

40.38 *Use of continuous timber rail to separate gravel path from rough grass in less formal context*

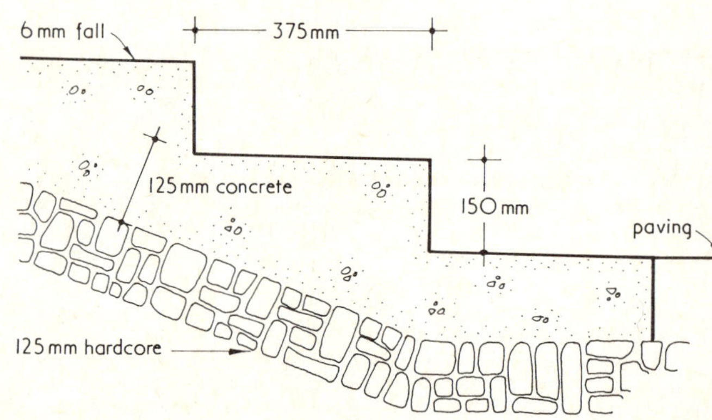

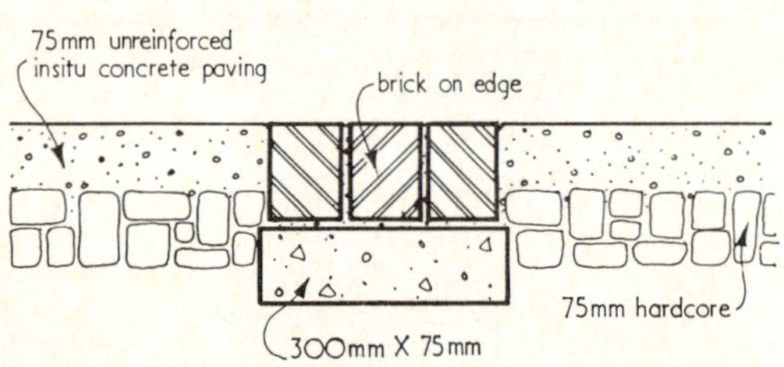

40.39 *Brick on edge used to break up areas of concrete hardstanding*

40.42 *In situ concrete steps. A non-slip finish can be produced by brushing with a stiff broom before the concrete hardens*

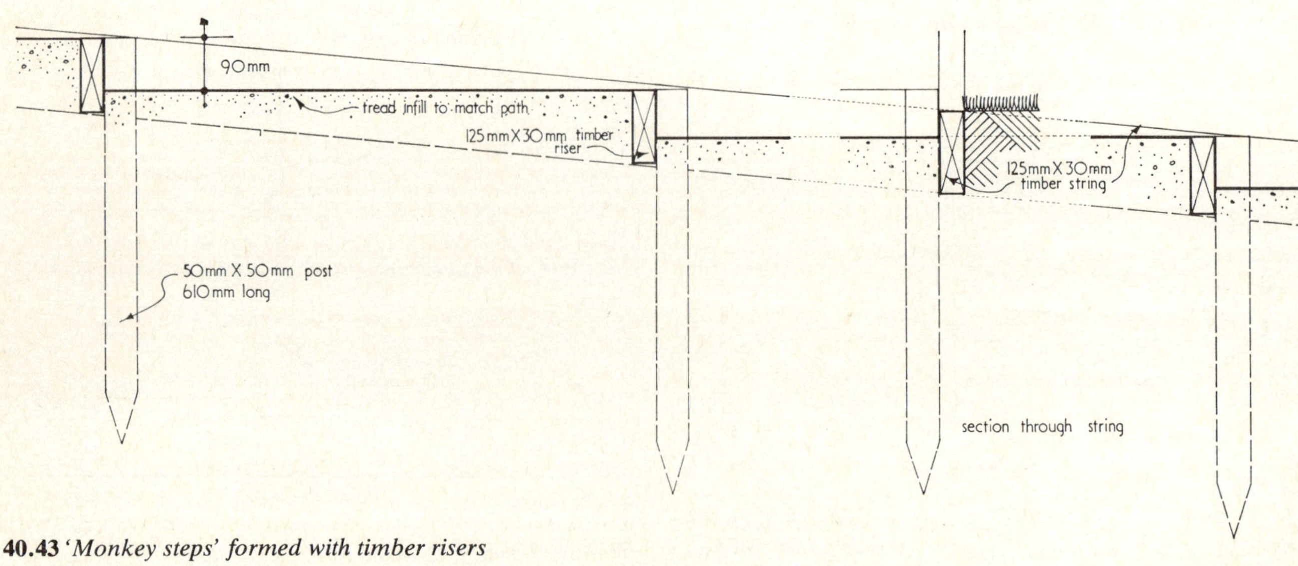

40.43 *'Monkey steps' formed with timber risers*

40.30. A 900 mm high wall may be constructed in half-brick thickness with piers, **40.31**.

High retaining walls are usually built in reinforced concrete, but masonry retaining walls may be built 215 mm thick for 900 mm high, and one additional half-brick thickness of 112·5 mm for each 300 mm of height retained up to a maximum of 1·8 m, **49.32**. Since the strength of this design of wall depends to some extent on the tensile value of the mortar, the bricks must be sufficiently porous to bond firmly with the mortar which should be grade II and of sulphate-resisting cement. The ground surface from the top of the wall should not slope upwards away from it more than 10 per cent. There should be a moisture-resistant membrane to protect the back of the wall, hardcore or gravel fill behind the wall and weepholes to relieve the water pressure.

Walls built of stone are usually constructed to the local traditional method. In many areas this is dry-stone: that is without mortar in the joints, and requires a craftsman. In general, mortared masonry will follow the dimensional rules used for brickwork.

9.03 Other forms of barrier

Hedges are covered in para **6.04**

40.33 illustrates a hedge bank used in some country districts, and **40.34** shows a ha-ha which provides a barrier invisible from one side. Water may also be used as discreet protection, such as in zoos (refer to the AJ Handbook of Urban Landscape).

10 SMALL GARDEN BUILDINGS

In this category comes glasshouses, toolsheds and summer houses.

Glasshouses are obtainable from 2·3 × 1·8 × 2·1 m up to 5·1 × 3·8 × 2·7 m (or larger) in forms as shown in **40.35**. Garden toolsheds come from 1·9 × 1·8 × 2·1 m to 3·9 × 2·6 × 2·2 m. Summer houses can be made by adapting some of these, or can be specially designed in a multitude of forms.

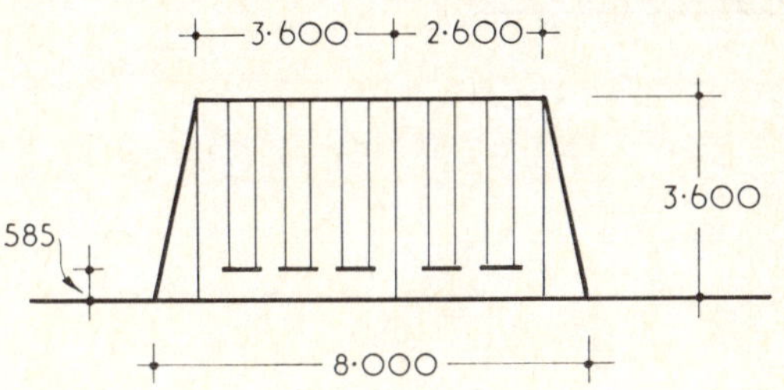

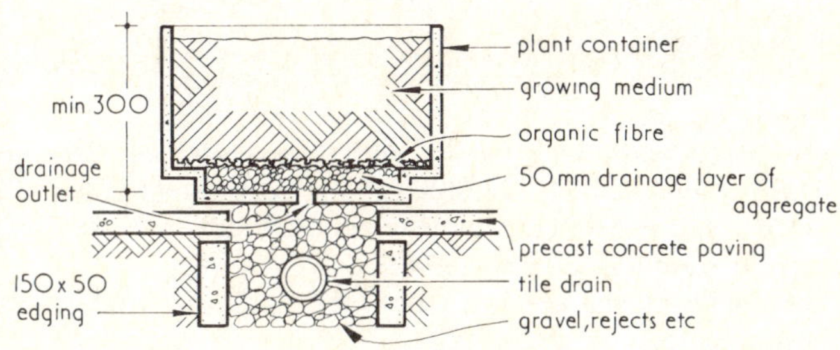

40.44 Section through plant container in paved areas, showing recommended method of drainage

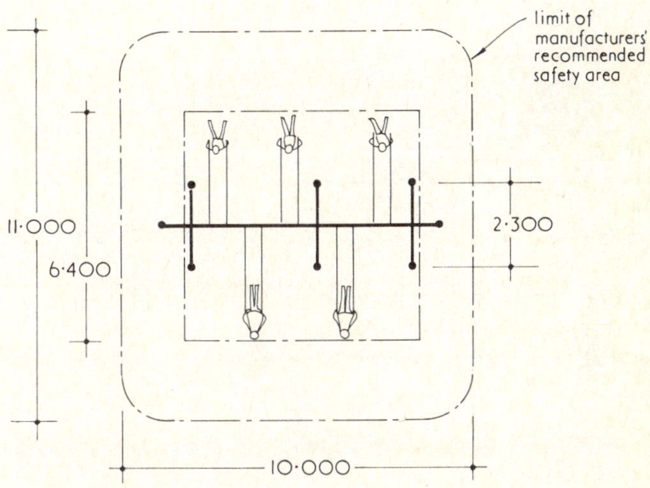

40.47 *Typical arrangement of full-size swing in a park. Smaller sizes are common*

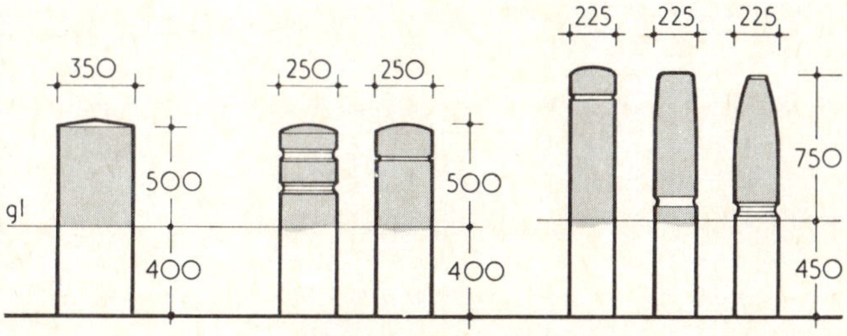

40.45 *Various styles of precast concrete bollards. Some of these are supplied in alternative version with built-in lighting*

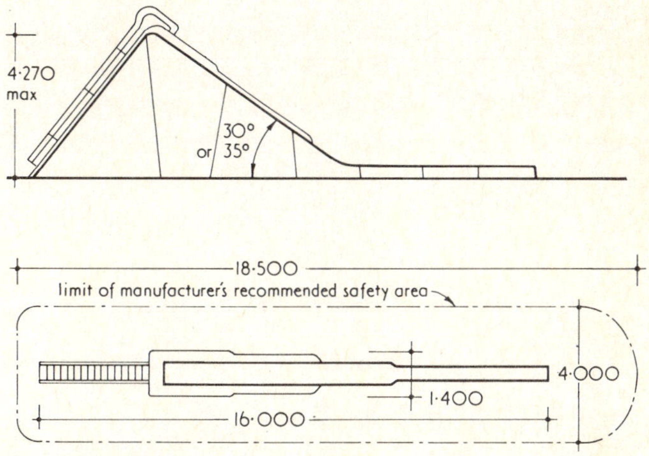

40.48 *Large slide for park. These are now normally installed on an earth mound to reduce the risk of falls*

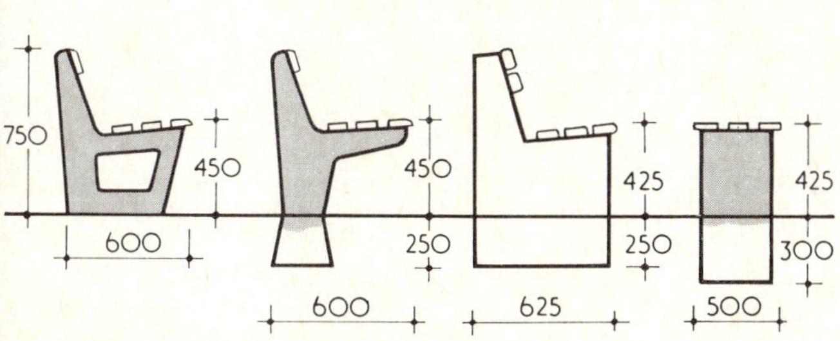

40.46 *Precast concrete framed seating. Less elegant but more robust than metal and wooden types*

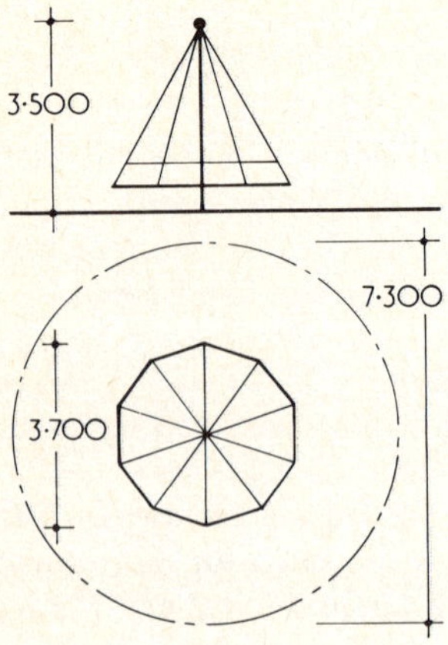

40.49 *Ocean wave (or 'witch's hat'). This device is now considered somewhat unsafe*

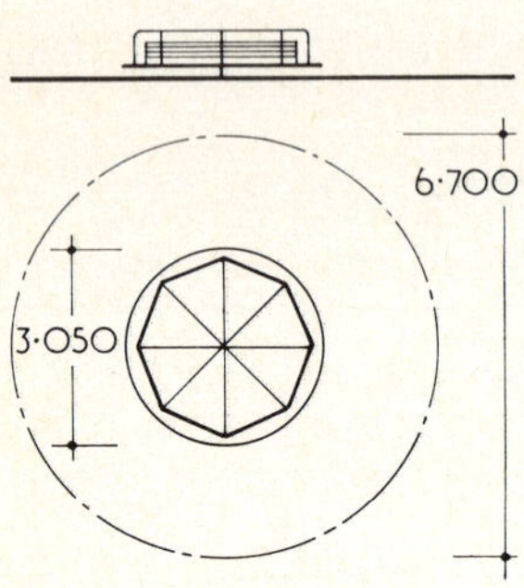

40.50 *Roundabout*

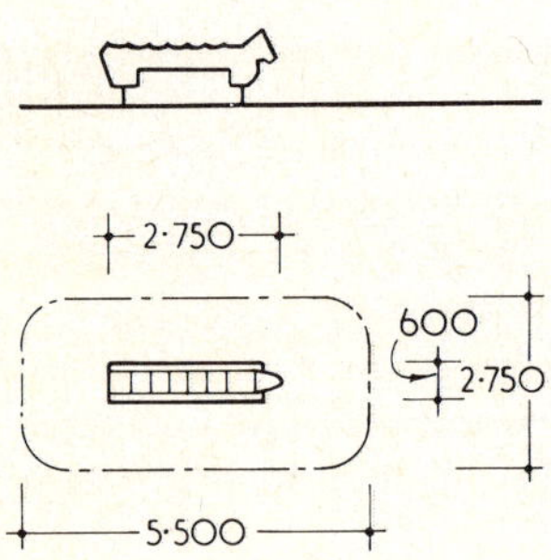

40.51 *Rocking horse. Only safe types of this device should be installed*

11 HARD LANDSCAPE DETAILS AND EXTERNAL FURNITURE

40.36 to **40** show some common details to form edgings to paths etc. Standard precast concrete kerbs are covered in section **39** Materials, table XXII. Two types of external steps are shown in **40.41** and **42**. The goings and rises for outside work should not be proportioned as for interior staircases: an angle of 20° is the maximum usually adopted. Alternative provision should be made for prams and wheelchairs by ramps of 8 to 10% wherever possible, but ambulant people usually prefer steps. **40.43** shows 'monkey steps' used for light traffic. These can be cheaply built to almost any steepness.

40.44 illustrates a concrete planter in a paved area such as a pedestrianised street. **40.45** shows typical bollards used to prevent access by vehicles into such areas, and **40.46** shows seating for parks, streets etc. Typical equipment for children's play areas are covered in **40.47** to **51**.

12 REFERENCES

Elizabeth Beazley *Design and detail of the space between buildings,* London, Architectural Press, 1960

Brian Clouston (ed) *Landscape design with plants,* London, Heinemann

B. Colvin *Trees for town and country*, London, Lund Humphries, 4th ed 1972

Olwen C Marlowe *Outdoor design: a handbook for the architect and planner,* London, Crosby Lockwood Staples

A. Mitchell *A field guide to the trees of Britain and Northern Europe*, London, Collins, 1974

National Building Agency *External Works. Detail Sheets*, London, Architectural Press, 1977

A. G. Tansley *The British Islands and their vegetation*, Cambridge University Press, 1939

Readers' Digest encyclopedia of garden plants and flowers, London, 1971

BS 5236:1975 Cultivation and planting of trees in the advanced nursery stock category

Cliff Tandy (ed) *AJ Handbook of urban landscape*, London, Architectural Press, 1972

41 Thermal comfort

David Adler

CI/SfB (J)
CI/SfB (1976 revised) (M)

Contents

1 INTRODUCTION

A pleasant thermal environment will result within a building once the following factors are within acceptable limits:

- air temperature (table XIII)
- temperature of surrounding surfaces (para **4.02**)
- humidity of atmosphere (para **5.07**)
- air movement (para **3.04**).

2 PRINCIPLES OF THERMODYNAMICS

2.01 Principles

An elementary knowledge of the basic principles of thermodynamics (or the science of heat) is essential to the comprehension of what follows. This exposition is a highly simplified version of a complex science.

Heat is a form of energy; in fact, it can be considered the lowest form of energy because all other forms eventually revert to heat.

2.02 First law

The first law of thermodynamics can be stated as follows:

Energy cannot be created or destroyed.

This law nowadays is said not to be completely true, as energy has been obtained from matter by atomic fission. However, it can be said that matter itself is a form of energy. In the field of thermal environment such considerations are hardly relevant.

2.03 Energy transformations

While energy cannot be created or destroyed, it can be transformed from one form to another. Heat can be produced by transformations that are:

- mechanical (such as friction)
- chemical (combustion)
- electrical or
- physical.

2.04 Efficiency

Above it was said that all other forms of energy eventually revert to the lowest form which is heat. The result of this is that the efficiency of processes of transformation into heat is 100 per cent, although not all the heat produced may be useful. The stated efficiency of these processes relates to the useful heat produced.

Heat energy can be used to produce higher forms such as mechanical energy. However, it is impossible to do this without expending more energy in the transformation than is received at the end. The lost energy is not destroyed, but must be eliminated from the process area. The percentage of transformed energy from the input is called the *efficiency* of the process, and is always considerably less than 100 per cent.

2.05 Heat quantities

Heat energy cannot exist apart from matter. Heat can be considered as related to the molecular activity of the matter with which it is associated. The total quantity of heat energy contained within a given mass of matter is not measurable, but changes in the contained energy are highly significant.

The two quantities we deal with in connection with heat energy are:

- amount of energy, measured in joules
- temperature, measured in kelvins or Celsius degrees.

The concept of temperature is best visualised as analogous to potential energy given by height, or to electrical voltage. While the results of temperature are well known, what it actually is remains a mystery.

2.06 Second law

The second law of thermodynamics can be stated as:

Heat will only flow from a higher temperature to a lower, **41.1**.

Again, this law may appear to be broken by such devices as the refrigerator, but such heat pumps still work in strict conformity to the laws.

2.07 Heat content

The total heat energy content of a given mass of matter (or a *body*) depends on three factors:

- the nature of the material of which it is composed (iron, stone, water)
- its *state*
- its temperature.

2.08 State

The term *state* requires some clarification. There are three basic states: solid, liquid and gas. In the case of a gas, the energy content varies with the pressure which can be said to be part of its state.

2.09 Specific heat

For a given material and state, the quantity of heat change per kilogram producing 1 kelvin of temperature change is called the *specific heat*. Table I gives the specific heats of a number of common materials. Strictly, the specific heat varies slightly with the level of the temperature, but this can be ignored for our purposes.

2.10 Latent heat

Since materials at different states contain different quantities of heat energy, a change of state is always accompanied by a change in energy content. For example, when ice melts heat is required to be absorbed by the water, **41.2**. Such heat is known as the *latent heat* of the change of state. When the water is again frozen into ice, the heat is given up.

There are two extreme ways in which a change of state can occur, most real changes being partly one and partly the other.

An *isothermal* change, **41.3**, is one that takes place with no change in temperature. Since the total energy contents before and after the change are different, energy must be supplied or

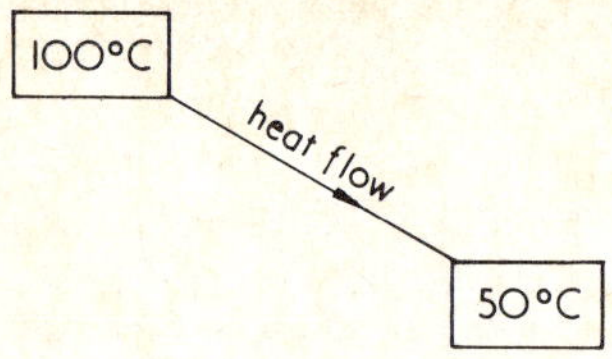

41.1 *Heat passing from a higher temperature to a lower, as water flowing downhill, illustrating the second law of thermodynamics*

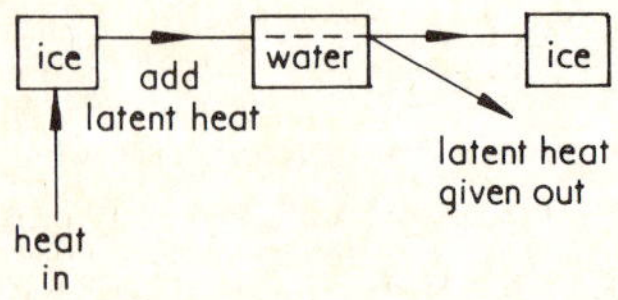

41.2 *Changes of state, with absorption and emission of heat*

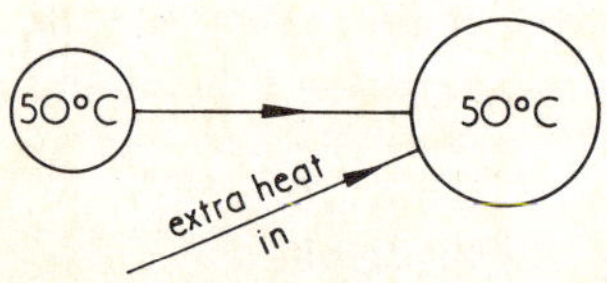

41.3 *Isothermal change of state: no temperature change, but change of energy content*

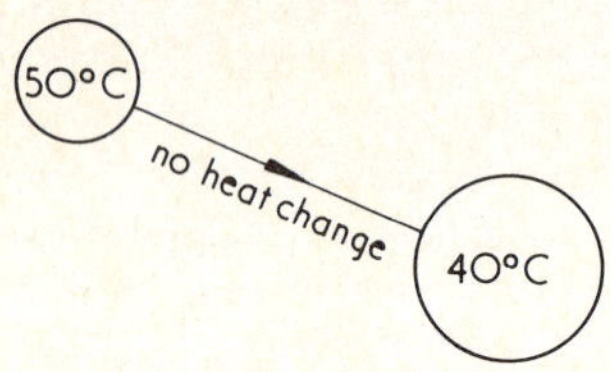

41.4 *Adiabatic change of state: no emission or absorption of energy but change of temperature*

Table I Specific heats of various materials

Substance	Specific Heat kJ/(kg.K)	Specific Heat cal/gm/°C
Air	1·013	0·242
Aluminium	0·883	0·211
Asbestos	0·84	0·20
Brass	0·38	0·09
Brick	0·84	0·20
Carbon dioxide	0·842	0·201
Clay	0·92	0·22
Coal	1·00	0·24
Concrete	0·84	0·20
Copper	0·39	0·093
Glass, crown	0·67	0·16
fibre	0·80	0·19
flint	0·50	0·12
Hydrogen	14·3	3·42
Ice	0·21	0·05
Iron, cast	0·523	0·125
Lead	0·13	0·03
Mineral fibre	0·75	0·18
Plywood	1·21	0·29
PVC	1·00	0·24
Rubber	2·01	0·48
Sand	0·80	0·19
Steel, mild	0·448	0·107
Timber	1·76	0·42
Water	4·187	1·00
Wax	2·89	0·69
Zinc	0·39	0·094

removed from the system during the change.

An *adiabatic* change, **41.4**, is one that takes place without change of energy content. The energy required or released by the change of state is allowed to cause a change in the temperature of the body, the degree of temperature change being related to the specific heat of the new state.

2.11 Volumetric changes

It is well known that temperature changes are always associated with volumetric changes. For solids and liquids the expansion is small for normal temperature rise but the force required to restrain it would be considerable. On the other hand, gases expand generously for a small rise in temperature, but little restraint effort (called *pressure*) is needed to maintain constant volume.

2.12 Gas equation

The process is covered by the *gas equation*:

$$PV = RT$$

where P = pressure of gas in Pa
V = volume of gas in m^3
T = thermodynamic temperature of gas in kelvins (K). 0 K is the absolute zero temperature equivalent to −273·15°C
and R = a constant. Where V is the volume of 1 *mole* (mol) of gas, R is known as the *gas constant* and has the same value of approximately 8313 J/(kmol. K) for any gas.

2.13 Critical temperature

Some gaseous materials when subjected to increase in pressure, change to the liquid state. The temperature above which this is not possible is called the *critical temperature*, and below this the material should be referred to as a *vapour.* A true gas is defined as a substance in the gaseous state at a temperature above its critical temperature.

2.14 Heat pump

We can illustrate these elementary principles of thermodynamics by returning to para **2.06**, and seeing how a heat pump, **41.5**, such as a refrigerator, works.

A gaseous medium (in this case a *refrigerant*) is made to expand by being pumped into a chamber of greater capacity. At first the expansion can be considered adiabatic, no heat energy change taking place. Since the new state of the gas would have greater total heat content than the original for the same temperature, equilibrium can only be maintained by a reduction in the temperature of the gas. This puts it at a lower temperature than its surroundings, and by the second law heat can be transferred from the surroundings into the refrigerant which then rises in temperature.

If the gas is then pumped back into its original chamber it must be compressed. For no temperature change the heat content is reduced. If the compression is adiabatic, the temperature of the medium will rise putting it above the temperature of the surroundings. Heat will therefore flow out of the gas, and can be used, for example, to raise the temperature of domestic hot water.

This heat energy originated from matter at a much lower temperature, and has been induced, or pumped, into matter at a higher temperature. The cost of this operation is the energy input into pumping the refrigerant around the circuit.

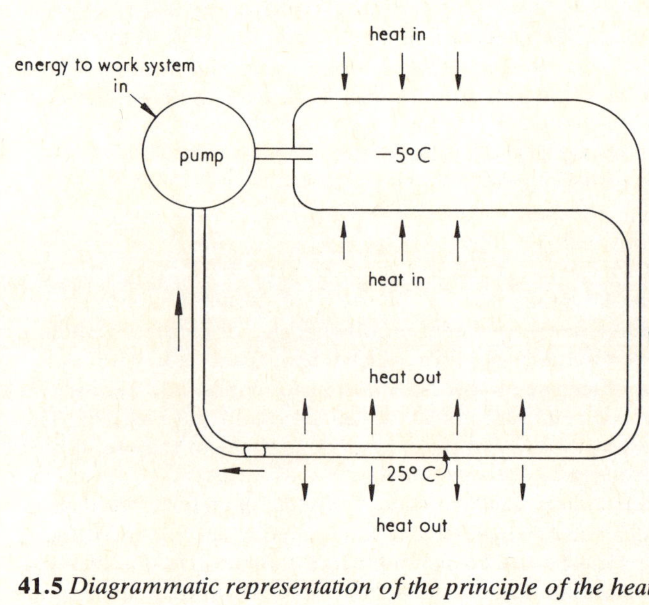

41.5 *Diagrammatic representation of the principle of the heat pump*

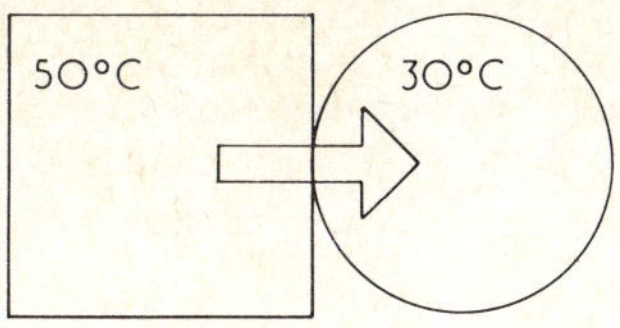

41.6 *Heat transfer by conduction*

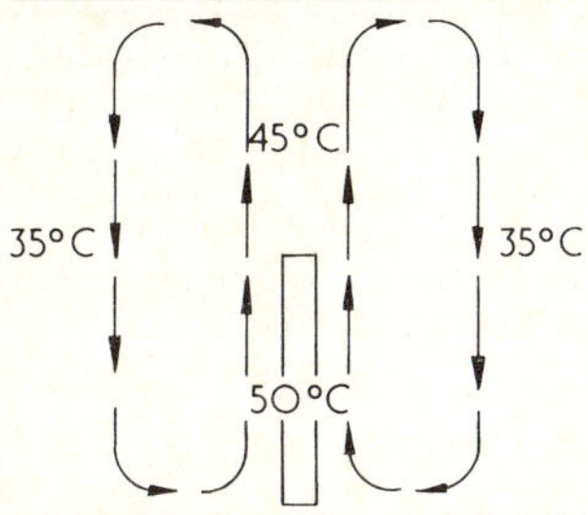

41.7 *Heat transfer by conduction*

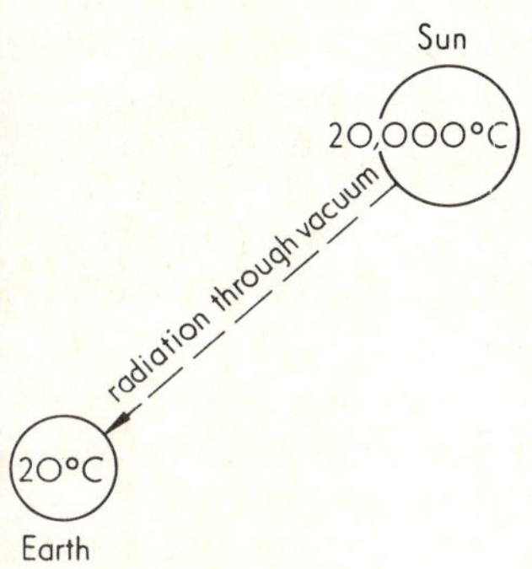

41.8 *Heat transfer by radiation: how the sun's energy reaches the earth*

2.15 Heat transfer

The second law of thermodynamics speaks of the transfer of heat energy from one body to another. The mechanisms through which this occurs are:

- *Conduction*—where a hot body is in direct physical contact with a cooler body, **41.6**, the heat flows through the molecules. Conduction also occurs within a mass of material when one part is at a higher temperature than another.
- *Convection*—this is really a special case of conduction, where the heat is transferred to a liquid or gas in contact with the hotter body. **41.7**. Since hot liquids and gases are generally lighter than cool ones, a circulation of the medium occurs which enables the hot liquid or gas to reach the cool body and transfer the heat energy to it.
- *Radiation*—this a form of heat transfer that is more difficult to comprehend. It is a fact that the energy contained within a body somehow causes a form of electromagnetic wave to be emitted. This energy radiation requires no medium to transmit it (as does convection). The energy from the sun comes to us through the vacuum of space, **41.8**. Sometimes radiant heat energy is combined with emissions in the visible spectrum: heated steel glows. It is perhaps less obvious that radiation occurs from sources at a comparatively low temperature.

2.16 Other aspects

Other aspects of the fundamental laws of thermodynamics will emerge as the various factors of environmental control are dealt with.

3 AIR

3.01 Constituents of air

Normal air consists of about 20 per cent oxygen, 80 per cent nitrogen and significant quantities of carbon dioxide and water vapour. It may also be contaminated with particles of solid dirt, with trace amounts of substances emanating from human and animal bodies, from the by-products of tobacco smoking, cooking or industrial processes often referred to as 'smells'!

For an acceptable environment none of the minority constituents must exceed a certain limit. This limit is a compromise between the minimum that is noticeable and the value that is injurious, and is known as the *threshold limit value* or *TLV*. In general, this value can be exceeded for a proportion of the time provided there is a corresponding reduction below the TLV at other times to maintain the average. However, in a few cases there is a value that must never be exceeded and this is called the *ceiling limit*. Tables II, III and IV give values for a selected number of substances.

Table II Threshold limit values for gases and vapours from the American Conference of Governmental Industrial Hygienists (Committee on Threshold Limits 1971)

Substance	TLVs ppm	TLVs mg/m³
Acetone	1000	2400·0
Ammonia	25	18·0
Benzene*†	25	80·0
Camphor (synthetic)	2	12·0
Carbon dioxide	5000	9000·0
Carbon monoxide	50	55·0
Carbon tetrachloride†	10	65
Chlorine	1	3
Chloroform	25	120
Ethyl alcohol	1000	1900
Formaldehyde*	2	3
Hydrogen chloride*	5	7
Hydrogen cyanide†	10	11
Hydrogen peroxide 90 per cent	1	1·4
Hydrogen sulphide	10	15
Liquified petroleum gas (LPG)	1000	1800
Methyl alcohol	200	260
Naphtha (coal tar)	100	400
Naphtha (petroleum)	500	2000
Ozone	0·1	0·2
Phenol†	5	19
Phosgene	0·1	0·4
Propane	1000	1800
Sulphur dioxide	5	13
Turpentine	100	560

* Ceiling limit
† Skin

Table III Threshold limit values for dusts, fumes and mists from the American Conference of Governmental Industrial Hygienists (Committee on Threshold Limits 1971)

Substance	TLVs mg/m³
Ammonium chloride fume	10
Antimony and compounds	0·5
Arsenic and compounds	0·5
Asphalt (petroleum) fume	5
Cadmium (metal dust and soluble salts)	0·2
Calcium oxide	5
Carbon black	3·5
Coal tar pitch volatiles	0·2
Cobalt, metal fume and dust	0·1
Copper dusts and mists	1
Cotton dust (raw)	1
Cyanide (as CN)	5
DDT	1
Iron oxide fume	10
Lead (inorganic compounds, fumes and dusts)	0·15
Mercury (except alkyl compounds)	0·05
Nickel	1
Nicotine	0·5
Oil mist (mineral)	5
Paraquat	0·5
Platinum (soluble salts)	0·002
Sulphuric acid	1
Warfarin	0·1
Wood dust	5

Table IV Threshold limit values for mineral dusts

Substance	TLVs million particles/m³	TLVs mg/m³
General value for inert dusts*	1060	10
Asbestos (fibres longer than 5μm)	5	
Coal dust	–	2

Table IV Threshold limit values for mineral dusts *(continued)*

Substance	TLVs million particles/m³	mg/m³
Graphite (natural)	530	
Mica	700	
Portland cement	1750	
Silica, amorphous	700	

* The general value for inert dusts can be used for the following, among others:
Calcium carbonate
Cellulose (paper fibre)
Emery
Glass (fibre or dust less than 5–7 μm in dia)
Graphite (synthetic)
Gypsum
Kaolin
Limestone
Marble
Plaster of Paris
Portland cement
Starch
Vegetable oil mists (with irritant exceptions)

3.02 Ventilation

Ventilation is used to exchange the air inside a room with fresh air from outside. This ensures:

- sufficient oxygen supply for breathing and other purposes
- removal of carbon dioxide, which is unsatisfactory beyond a concentration of 0·5 per cent
- removal of unwanted moisture (preventing condensation)
- reduction in 'smells', and
- removal of unwanted heat.

The amount of ventilation can be expressed in three ways:

- m^3/s
- litres/s (1000 litres/s = 1 m^3/s)
- air changes/hour, which depends on the volume of the room.

Table V gives the minimum required amount of ventilation where the density of occupation is known. This may not be sufficient to maintain comfort conditions.

Example 1

A room 25 × 18 × 2·5 m high contains 125 people, and smoking is not prohibited. Give the minimum required amount of ventilation in each of the three forms above.

Volume of room = 1125 m³ = 9 m³/person.
Fresh air supply from Table V = 10·4 litres/s × 125
= 1300 litres/s
= 1·3 m³/s
= 4680 m³/h
= 4680 ÷ 1125 = 4·16 air changes/h

Table V Minimum ventilation rates where density of occupation is known

Air space per person (m³)	Fresh air supply per person (litre/s) Recommended (smoking permitted)	Recommended (no smoking)	Minimum
3	22·6	17·0	11·3
6	14·2	10·7	7·1
9	10·4	7·8	5·2
12	8·0	6·0	4·0

In factories and offices the statutory minimum volume per person is 11·5 m³. The corresponding minimum fresh air supply is 4·72 litre/s/person.

Ventilation is the only method of removing unwanted heat. This subject will be covered fully later, but one of the chief sources of such heat is the human body. Table VI gives figures for this factor.

Example 2

If the people in room of example 1 are carrying out light work, how much fresh air at 12 °C is needed to maintain a temperature of 21 °C, ignoring all other gains or losses?

Total heat from occupants (table VI) = 140 × 125 = 17·5 kW
= 17·5 kJ/s
The specific heat of air (from table I) = 1·013 kJ/(kg.K)
To raise the temperature of 1 kg of air by 9 K or 9 deg C will require 9 × 1·013 = 9·117 kJ
Consequently 17·5 kJ will raise $\frac{17.5}{9.117}$ kg of air by 9 deg C = 1·92 kg
The density of air under normal conditions is about 1·2 kg/m³
Hence the 17·5 kJ will raise $\frac{1.92}{1.2}$ = 1·6 m³ by 9 deg C
Hence 1·6 m³ is required every second, the required ventilation is 1·6 m³/s = 5·12 air changes per h

Table VI Heat production by people

Degree of activity	Total adult male heat production (W)
Seated at rest	115
Light work	140
Walking slowly	160
Light bench work	235
Medium work, dancing	265
Heavy work	440

3.03 Infiltration

Ventilation can be *natural* or *forced*. Natural ventilation takes place through windows, doors, ventilators and flues. It also occurs through *infiltration*, which is the fortuitous leakage of air through imperfections in the structure of the building. Table VII gives figures for infiltration through wall construction. In practice this is negligible in comparison with infiltration through the joints around windows, etc. This can be calculated from the following formula:

$$Q_b = f\,C\,p^{0.63}$$

where: f is the internal arrangement coefficient (table VIII)
C is the window type coefficient (table IX)
p is the pressure difference from one side of the wall in question to the other in Pa
and Q_b is the basic filtration rate in litre/(m.s)

This pressure difference arises from two effects: wind, and a *stack effect* caused by temperature difference between inside and outside.

Table X gives the pressure difference across a complete building for three wind speeds corresponding to common situations in the United Kingdom. The pressure drop between the windward face of the building and the interior will be about 0·5 to 0·8 of this figure, and between the interior and the leeward face about 0·3. It is normal to assume a wind pressure difference of half the tabulated figure.

The stack effect is covered by using a factor from table XI or XII.

Example 3

A suburban building 65 m high is fitted with weather-stripped horizontally pivoted windows covering 28 per cent of the facades. Internally it is divided by partitions into a number of small rooms. What will the basic rate of infiltration be?

From table VIII, f = 0·8
From table IX, C = 0·05
From table X, pressure difference across building = 57 Pa
difference due to wind = 0·5 × 57 = 28·5 Pa
Hence $Q_b = f\,C\,p^{0.63}$
$= 0.8 \times 0.05 \times 28.5^{0.63}$ = 0·33 litre/(m.s)

Table VII Infiltration through walls in litre/(m²s)

Type		Pressure difference (Oa) 12·5	25	50	75	100
One brick	plain	0·425	0·765	1·36	2·04	2·38
	plastered	0·004	0·007	0·012	0·017	0·023
1½ brick	plain	0·425	0·680	1·19	1·70	2·04
	plastered	0·001	0·003	0·004	0·008	0·009
Lath and plaster		0·008	0·013	0·019	0·025	0·027

Table VIII Internal arrangement coefficients (f)

Window type	Internal structure	f
All types	Open plan	1·0
Short length of well-fitting window opening joint (say 20% of facade openable)	Rooms opening off corridor with few inter-connections, many partitions	1·0

Table VIII Internal arrangement coefficients (f) *(continued)*

Window type	Internal structure	f
Long length of well-fitting window or short length of poor fitting window joint (say 20 to 40% of facade openable)	Corridor plan	1·0
	Liberal partitioning	0·8
Long length of poor fitting window joint (say 40 to 50% of facade openable)	Corridor plan	0·8
	Liberal partitioning	0·65
Very long length of poor fitting window joint (say more than half of facade openable)	Corridor plan	0·65
	Liberal partitioning	0·4

Table IX Window infiltration coefficients (C) in litre/(m.s.Pa)

Window type	C
Horizontally or vertically pivoted, weather-stripped	0·05
Horizontally or vertically sliding, weather-stripped	0·125
Horizontally or vertically pivoted or sliding, no weather-stripping	0·25

Table X Pressure difference across a building from wind

Building height m	Total difference in Pa Open country wind speed 9 m/s	Suburban 5·5 m/s	City centre 3 m/s
10	58	21	6
20	70	31	11
30	78	38	15
40	85	44	21
50	90	49	23
60	95	55	26
70	100	59	31
80	104	63	34
90	107	67	37
100	111	71	40

Table XI Winter height factor (x) for lowest floor

No of storeys	x
5	1·03
10	1·1
20	1·2

Table XII Summer height factor (y) for highest floor

No of storeys	y
5	1·03
10	1·06
20	1·08

To find the infiltration rate for a room with one window wall, use the formula:

$$Q_r = x\,B_b\,L$$

where x is the height factor for the particular floor derived from table XI by interpolation from 1 for the midheight floor to the value for the lowest. In summer the factor y is used instead, and this is interpolated between 1 for the midheight to the value from table XII for the highest floor.

L is the total crack length in the room in m

Q_r is the total room infiltration in litre/s

Where a corner room has two adjacent walls with openable windows, the value of Q_r is increased by half as much again.

Example 4

The crack length for the building in example 3 is 1·4 m run per m² of facade area. There are 24 storeys. What is the winter infiltration rate for an office on the fifth floor 5 m square?

From table XI, x for the lowest floor is 1·2.

The midheight is 12 storeys, so for the 5th floor

$x = \frac{5}{12} \times 0{\cdot}2 + 1 = 1{\cdot}083$

The area of wall if the office has one window wall is

$5 \times 2{\cdot}67 = 13{\cdot}35\,m^2$

The crack length is therefore $1{\cdot}4 \times 13{\cdot}35 = 18{\cdot}7\,m$

Therefore $Q_r = xQ_bL$

$= 1{\cdot}083 \times 0{\cdot}33 \times 18{\cdot}7$

$= \underline{6{\cdot}1 \text{ litre/s}}$

If the office is a corner one, there will be 26·7 m² of wall and 37·4 m of crack length.

$Q_r = 1{\cdot}083 \times 0{\cdot}33 \times 37{\cdot}4 \times 1{\cdot}5$

$= \underline{20{\cdot}0 \text{ litre/s}}$

To illustrate the use of the foregoing, consider the following example.

Example 5

How much heating will be required in the corner office in example 4 to maintain the temperature at 20 deg C above outside temperature, assuming only infiltration ventilation?

To raise 1 kg of air by 1 deg C requires 1·013 kJ (specific heat from table 1).

Therefore to raise 1 kg by 20 deg C requires $20 \times 1{\cdot}013 = 20{\cdot}3$ kJ

to raise 1 m³ by the same $20{\cdot}3 \times 1{\cdot}2 = 24{\cdot}36$ kJ

to raise 1 litre by the same needs 24·36 kJ.

Now the infiltration is 20·0 litre/s.

to raise this 20 deg C requires $24{\cdot}36 \times 20{\cdot}0$ J/s (or W)

$= \underline{481\,W}$

The amount of energy required to heat the infiltration to the temperature of the inside of the building is called the *ventilation allowance*. Table XIII gives empirical values of infiltration and ventilation allowance for a range of building types, together with recommended internal temperatures.

3.04 Natural ventilation

Apart from infiltration, there will be additional natural ventilation due to the occupants opening windows. It is not usually necessary to calculate this, as the system is largely self-governing! In warm climates it is usual to design the building so that *through ventilation* is facilitated, **41.9**, although elsewhere this may be considered a draught. Figure **41.10** shows the maximum velocity for air movement that is acceptable; less than 0·1 m/s will lead to stuffiness.

Table XIII Recommended design values for internal environmental temperatures and empirical values for air infiltration and ventilation allowance (for normal sites and winter heating)

Type of building	t_{ei} (°C)	Air infiltration rate (h^{-1})	Ventilation allowance (W/m² °C)
Art galleries and museums	20	1	0·33
Assembly halls, lecture halls	18	½	0·17
Banking halls:			
Large (height > 4 m)	20	1	0·33
Small (height < 4 m)	20	1½	0·50
Bars	18	1	0·33
Canteens and dining rooms	20	1	0·33
Churches and chapels:			
Up to 7000 m³	18	½	0·17
> 7000 m³	18	¼	0·08
Vestries	20	1	0·33
Dining and banqueting halls	21	½	0·17
Exhibition halls:			
Large (height > 4 m)	18	¼	0·08
Small (height < 4 m)	18	½	0·17
Factories:			
Sedentary work	19		
Light work	16		
Heavy work	13		
Fire stations; ambulance stations			
Appliance rooms	15	½	0·17
Watch rooms	20	½	0·17
Recreation rooms	18	1	0·33
Flats, residences, and hostels:			
Living rooms	21	1	0·33
Bedrooms	18	½	0·17
Bed-sitting rooms	21	1	0·33
Bathrooms	22	2	0·67
Lavatories and cloakrooms	18	1½	0·50
Service rooms	16	½	0·17
Staircase and corridors	16	1½	0·50
Entrance halls and foyers	16	1½	0·50
Public rooms	21	1	0·33
Gymnasia	16	¾	0·25

Table XIII Recommended design values for internal environmental temperatures and empirical values for air infiltration and ventilation allowance (for normal sites and winter heating) *(continued)*

Type of building	t_{ei} (°C)	Air infiltration rate (h^{-1})	Ventilation allowance (W/m^3 °C)
Hospitals:			
Corridors	16	1	0·33
Offices	20	1	0·33
Operating theatre suite	18–21	½	0·17
Stores	15	½	0·17
Wards and patient areas	18	2	0·67
Waiting rooms	18	1	0·33
(See also DHSS Building Notes)			
Hotels:			
Bedrooms (standard)	22	1	0·33
Bedrooms (luxury)	24	1	0·33
Public rooms	21	1	0·33
Corridors	18	1½	0·50
Foyers	18	1½	0·50
Laboratories	20	1	0·33
Law courts	20	1	0·33
Libraries:			
Reading rooms (height > 4 m)	20	½	0·17
(height < 4 m)	20	¾	0·25
Stack rooms	18	½	0·17
Store rooms	15	¼	0·08
Offices:			
General	20	1	0·33
Private	20	1	0·33
Stores	15	½	0·17
Police stations:			
Cells	18	5	1·65
Restaurants and tea shops	18	1	0·33
Schools and colleges:			
Classrooms	18	2	0·67
Lecture rooms	18	1	0·33
Studios	18	1	0·33
(See also DES Bulletins)			
Shops and showrooms:			
Small	18	1	0·33
Large	18	½	0·17
Department store	18	¼	0·08
Fitting rooms	21	1½	0·50
Store rooms	15	½	0·17
Sports pavilions:			
Dressing rooms	21	1	0·33
Swimming baths:			
Changing rooms	22	½	0·17
Bath hall	26	½	0·17
(See also MOHLG Design Bulletin 4)			
Warehouses:			
Working and packing spaces	16	½	0·17
Storage space	13	¼	0·08

The values quoted for rates of air infiltration in this table should not be used for the design of mechanical ventilation, air conditioning or warm air heating systems.

3.05 Forced ventilation

Forced ventilation is used in situations where natural ventilation cannot operate satisfactorily:

- in deep plan buildings where accommodation is far from an external wall
- in internal rooms, especially bathrooms and wcs
- in underground areas
- in very high buildings where natural ventilation would cause problems.

The recommended amount of mechanical ventilation for various building types is given in table XIV.

The air exchange required can be accomplished by:

- forcing air into the room, and allowing the stale air to find its own way out, **41.11**, known as *fresh air supply*
- sucking the stale air out of the room and allowing its replacement to find its own way in, **41.12**, known as *exhaust*
- a combination of the two.

3.06 Air supply

Where fresh air is supplied to a room, it is uncommon for this to be introduced directly from the open air. It is usually heated or cooled to avoid discomfort, and it is often filtered to reduce unwanted dirt, or *conditioned* to adjust the content of water vapour. Fresh air from outside is often mixed with air already circulated to save energy. This is called *recirculation.*

3.07 Exhaust

This is the simplest method of mechanical ventilation. In many cases the air that takes the place of that exhausted comes from other parts of the building and is therefore at approximately the correct temperature and humidity for comfort.

This method is very commonly used for internal bathrooms and wcs in domestic combination. The system is designed for 3 air changes/h, or 7 litre/s per wc or bathroom without wc, or 14 litre/s for a bathroom containing a wc, whichever is greater. In public lavatories, the air changes are usually from 5 to 10 per hour.

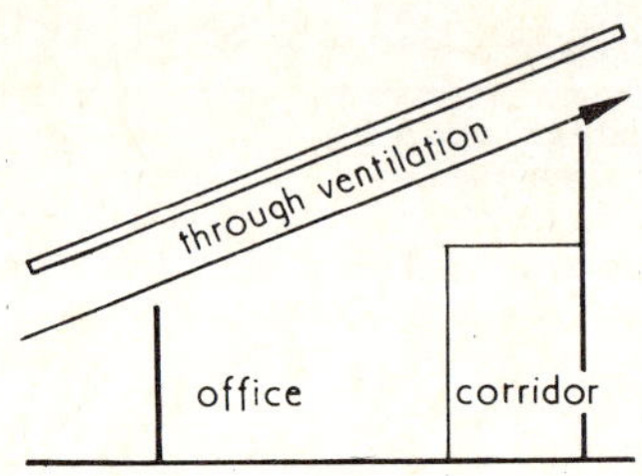

41.9 *Section through a building for a warm climate designed to facilitate natural through ventilation*

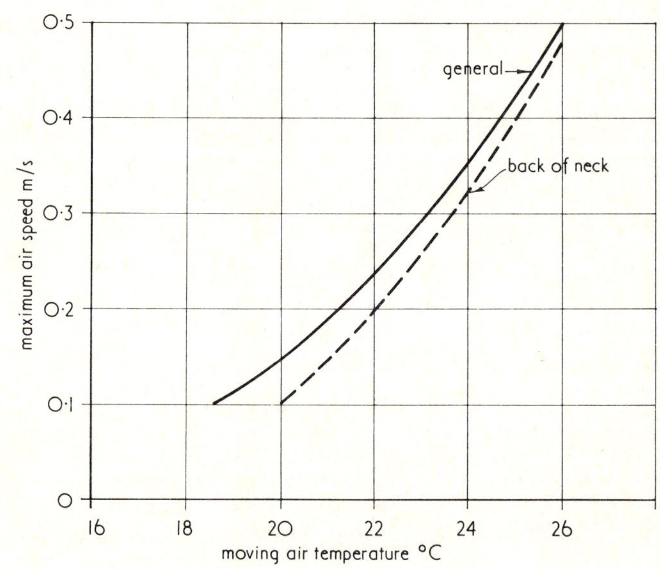

41.10 *Comfortable air speeds: less than 0·1 m/s feels stuffy, greater than the curve is a draught! (From IHVE Guide, 1970)*

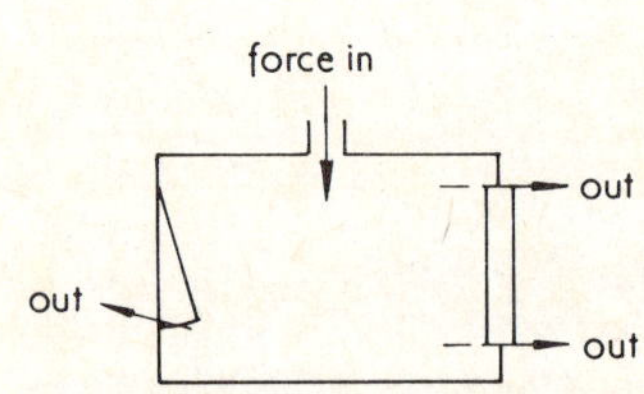

41.11 *Ventilation by fresh air supply*

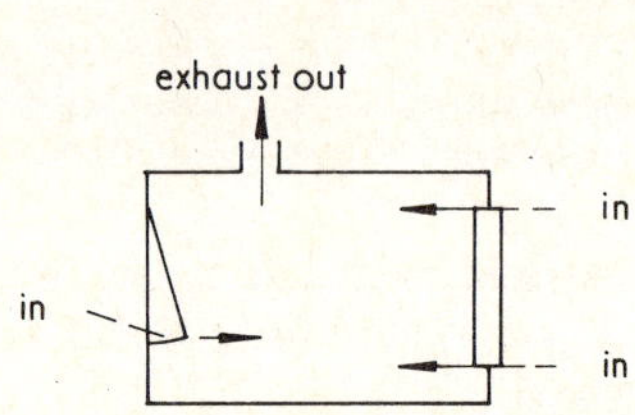

41.12 *Ventilation by exhaust*

Table XIV Mechanical ventilation rates for various types of building

Room or building	Recommended air change rates* (h^{-1})	Notes
Boilerhouses and engine rooms	15–30	The recommended air change is intended to include the air for combustion. The actual requirements should be checked in all cases in relation to the particular equipment installed
Banking halls	6	
Bathrooms, internal	6†	
Battery charging rooms	up to 5†	
Canteens	8–12‡	
Cinemas	6–10‡	Under GLC and other regulations, a minimum outdoor air rate of 8 litre/s per person is required
Dance halls	10–12‡	See note after Cinemas
Dining and banqueting halls, restaurants	10–15‡	
Drying rooms	up to 5	
Garages, public (parking)	6† minimum	
repair shops	10† minimum	
Hospitals, treatment rooms	6	
operating theatre	15–17	(470–570 litre/s input)
post-mortem room	5	
(for further details see Hospital Building notes)		
Kitchens, hotel and industrial	20–60†	The stated air change rates may be used for preliminary design, but should be checked against requirements related to the installed equipment
local authority	up to 10†	
Laboratories	4–6	Inlet air change rate must be checked against extraction rate via fume cupboards
Laundries	10–15	See Factories Act
Lavatories and toilets, internal	6–8†	The recommended rate of air change applies to normal public toilet areas incorporating both wcs and wash basins, etc. For small or congested toilet areas, comprising only wcs and urinals, a better basis is 10 litre/s per m² of floor area. In any case, the extract points should be situated over the wcs and urinals and the extraction rate will correspond to 19–24 litre/s per point. An inlet supply will be required to the lobby with a louvred connecting door
Libraries, public	3–4‡	
book stacks	1–2	Criterion is the air circulation around books to prevent mould formation. Air volume required depends on disposition of air inlets in shelving to maintain positive air movement. Humidity and SO_2 control may be required
Offices, internal	4–6‡	The London Building Act requires not less than 6 litre/s per occupant or per 4·7 m² of floor area, whichever provides the greater ventilation rate
Sculleries and wash-ups, large scale	10–15†	
Smoking rooms	10–15	
Swimming baths:		
bath hall		Ventilation rates must be related to the control of condensation. Criterion is the water area and recommended basis is 20 litre/s per m² of water surface, plus a margin (say 20%) to allow for effect of wet surrounds. See also MOHLG Design Bulletin No. 4
changing areas	10	Extract from this space should be via clothes storage baskets
Theatres	6–10‡	See note after Cinemas

Notes:
* The recommended air change rates do not apply in cases of warm-air heating, when the rate may be dictated by the heat requirements of the building or room.
† Refers to extract ventilation.
‡ The supply air at the recommended rate will not necessarily be all outdoor air; the required quantity of outdoor air must be checked against the number of occupants at a desirable rate per person.

In public and semi-public buildings such as offices, a mechanical ventilation system for lavatories, **41.13**, would comprise a common extract duct with a continuously running exhaust fan at or near the outlet. The various areas would be protected with fire dampers within the ductwork. Fans are required by local authority regulations to have standby units in case of failure.

In multi-storey flats it is more usual to have intermittent running. There is an extract point over each appliance incorporating small individual dual fan units. These are automatically switched on together with the electric light in the room (as the room is internal, the light will be switched on for each use). However, the circuitry is arranged so that the fans continue to run for some time (preferably 20 minutes at least) after the light is extinguished.

Ideally, each extract should be self-contained right up to the point of discharge into the open air. In practice, for tall buildings a shunt system, **41.14**, is sufficient.

3.08 Air conditioning

The term *air conditioning* strictly applies to cases where mechanical ventilation includes arrangements for adjusting the humidity of the supply air. Since this usually involves cooling somewhere in the process (see para **5.08**) the term is loosely used wherever refrigeration is applied to air.

An air-conditioning system, **41.15**, will supply almost all the air required, as little as possible being allowed to infiltrate. In a full system, the air will be filtered to remove dirt, heated or cooled as required by the season, and conditioned. The sizes of the expected particles as given in table XV will determine the filter system. There will also be an exhaust system, and in most cases to further economy, a proportion of the exhausted air will be *recirculated*. Recommended supply rates are given in table XVI.

The two types of installation are:

- unit air conditioners—small units incorporated in the external wall of a room providing a service for that room alone.

They usually require only a power point, but there needs to be a drain for the disposal of a small amount of excess moisture, and filters will have to be changed at intervals.

• central air-conditioning plant. These tend to be large and require substantial cooling towers to eliminate the heat produced. They also need a lot of ductwork around the building for supply, exhaust and recirculation. Buildings with this type of installation have to be planned with this in mind from the start.

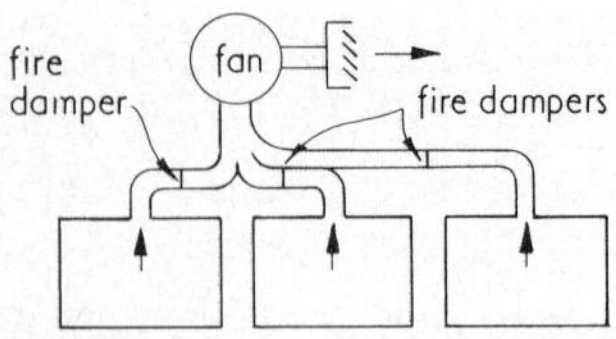

41.13 *Typical mechanical ventilation for lavatories, etc*

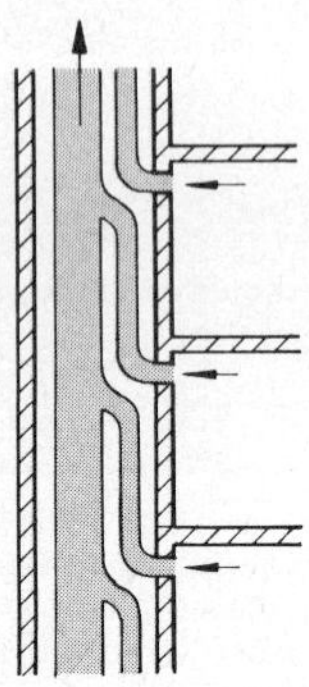

41.14 *Shunt duct system for lavatory ventilation in tall blocks*

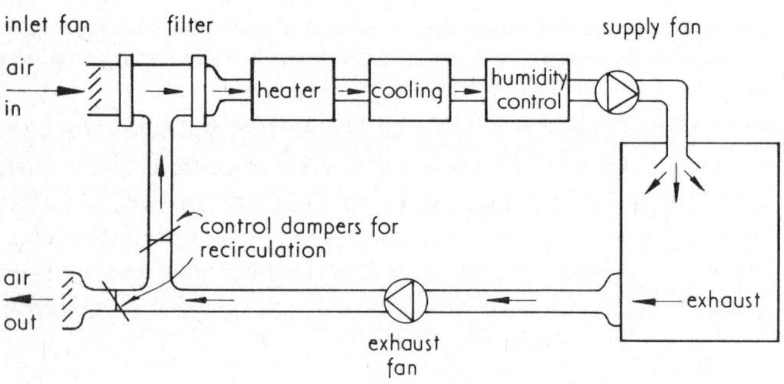

41.15 *Diagrammatic representation of an air conditioning system*

Table XV Particle sizes

Substance	Size μm (micrometers or microns)
Dusts	1–100
Pollen	20–50
Fogs, mists	2·5–40
Bacteria length	0·5–26
width	0·5–2
Smallest visible particle	10
Fumes	less than 1
Tobacco smoke	0·01–1
Viruses psittacosis	0·3
poliomyelitis	0·012

Table XVI Recommended outdoor air supply rates for air conditioned spaces

		Outdoor air supply (litre/s)		
		Recommended	Minimum (Take greater of two)	
Type of space	Smoking	Per person	Per person	Per m^2 floor area
Factories*†	None			0·8
Offices (open plan)	Some			1·3
Shops, department stores and supermarkets	Some	8	5	3·0
Theatres*	Some			–
Dance halls*	Some			–
Hotel bedrooms†	Heavy			1·7
Laboratories†	Some	12	8	–
Offices (private)	Heavy			1·3
Residences (average)	Heavy			–
Restaurants (cafeteria)†‡	Some			–
Cocktail bars	Heavy			–
Conference rooms (average)	Some			–
Residences (luxury)	Heavy	18	12	–
Restaurants (dining rooms)†	Heavy			–
Board rooms, executive offices and conference rooms	Very heavy	25	18	6·0
Corridors	A *per capita* basis is not appropriate to these			1·3
Kitchens (domestic)†				10·0
Kitchens (restaurant)†				20·0
Toilets*				10·0

Notes:
* See statutory requirements and local bye-laws.
† Rate of extract may be over-riding factor.
‡ Where queueing occurs in the space, the seating capacity may not be the appropriate total occupancy.
1 For hospital wards, operating theatres see Department of Health and Social Security Building Notes.
2 The outdoor air supply rates given take account of the likely density of occupation and the type and amount of smoking.

4 TEMPERATURE

4.01 Effects of temperature

There is no ideal internal temperature: requirements vary for:
• inhabitants (age, sex, race, acclimatisation)
• activities
• season
• humidity.

While the human body can adjust to great extremes, comfort demands a more limited range of acceptable temperatures.

4.02 Environmental temperature

There are two basic temperatures controlling comfort:
• air temperature, from which the human body loses or gains heat by conduction and convection, **41.16**
• the temperature of the surrounding surfaces, from which the effect of radiant heat is significant.

The temperature of the surrounding surfaces will depend on:
• the temperature on the other side of the construction
• the *thermal response* of the construction. This is the time taken for the fabric of the building to approach the air temperature both within and without, or to absorb the effect of radiant heat, say, from the sun. Buildings of massive construction usually have a slow thermal response. This is utilised in hot climates; the traditional Arab house is heated up by the sun only by the time it is setting, and the heat is useful to stave off the chill of the night. Light constructions incorporating a high degree of insulation often have a fast response, so that they feel cold as soon as the heating is turned off.

Because of the importance of both these factors, various formulae have been devised to produce an effective temperature as an indicator of comfort. Of these the *environmental temperature*, t_{ei} is the simplest, but not significantly less useful than the others:

$t_{ei} = \frac{2}{3} t_r + \frac{1}{3} t_a$ where t_r = mean radiant temperature of surroundings
t_a = air temperature

4.03 Resultant temperature

Another formula gives the *resultant temperature* which is important only because it has been used to define the *comfort*

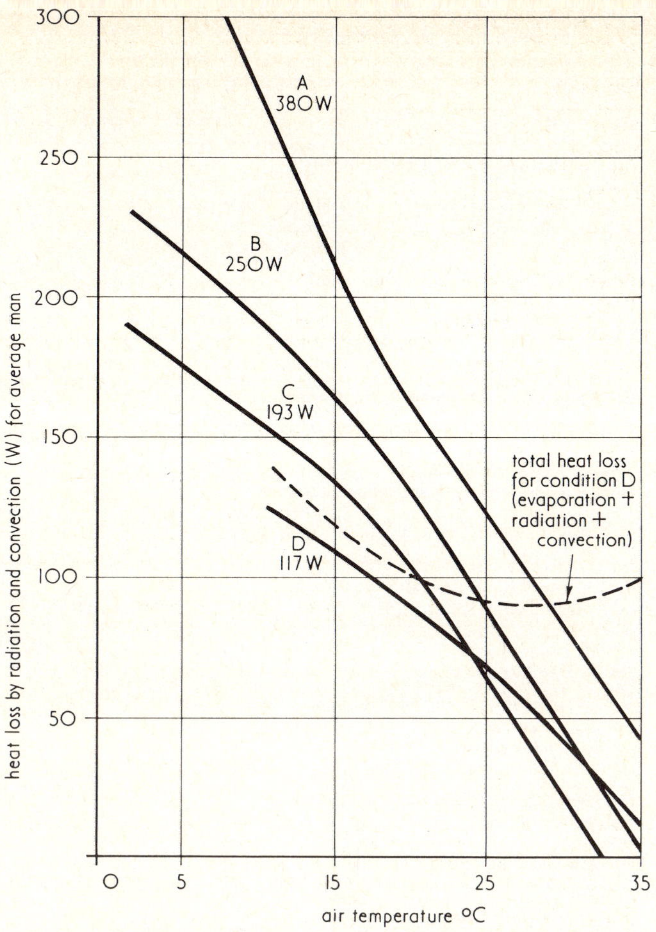

41.16 *Relationship between radiation and convection heat loss from the human body and the dry-bulb temperature for still air. Curves A to C are for people working at the stated metabolic rates, curve D is for people seated at rest with a metabolic rate of 117 W. Variation from humidity is small, curves follow the average. (From ASHRAE Guide)*

zone as shown in **41.17**, and gives temperature figures related to activity and clothing, **41.18**.

4.04 Heat losses

The achievement of a satisfactory internal temperature within a building depends on the following:

- the heat provided by the heating and ventilation system
- heat emanating from other sources, such as lighting, machinery and the bodies of the occupants
- heat losses from the building.

Recommended temperatures have already been given in table XIII. To determine the required size of installation it will be necessary to calculate the magnitude of the other two factors, particularly the heat losses. In fact, the total heat input to the building must equal the heat losses whatever the temperature which does not directly appear in the equation. It will be found, however, that the magnitude of the losses is almost directly proportional to the internal temperature, or to be precise, to the temperature difference between the inside and the outside. There are two main heat losses in a building:

- losses caused by the air changes (ventilation allowance)
- losses through the fabric of the building, eg walls (including windows), floors and roof.

Example 6

What are the heat losses due to air change in a building of volume $150 m^3$, where the internal temperature is 20°C, outside temperature 0°C, and 3 air changes per hour are required?

Total air exhausted from building = $3 \times 150 m^3/h$
= $0{\cdot}125 m^3/s = 0{\cdot}15 kg/s$

To raise this volume from 0 to 20°C, ie 20K, needs specific heat × temperature rise × mass (para **2.09**)
= $1{\cdot}013 \times 20 \times 0{\cdot}15 = 3{\cdot}04 kJ/s = \underline{3{\cdot}04 kW}$

4.05 U values

The loss of heat through a wall, floor or roof depends on the U value of the construction. Table XVII gives U values for a number of common types of construction. Where the U value of a construction is unknown, it can be calculated from the *thermal resistivities* of the materials it comprises. Table XVIII gives resistivities and conductivities for a number of materials. The *conductivity* of a material is the amount of heat that will pass through unit volume for unit difference in temperature:

$$Q = \frac{k A (t_2 - t_1)}{L}$$

where Q is the amount of heat flow in W
k is the conductivity in W/(m.K)
A is the area of the material in m^2
t_1, t_2 are the temperatures on either side of the thickness of the material concerned, $t_2 - t_1$ expressed in K
L is the thickness of the material under consideration in m.

The resistivity, r, is the reciprocal of the conductivity

$r = \frac{1}{k}$, expressed in m.K/W

Example 7

How much heat will pass through a 20 mm thick plasterboard, 2 × 1·5 m, if the temperature difference is 6 K? (Take k = 0·16)

Use the formula above $Q = \frac{k A (t_2 - t_1)}{L}$

$= \frac{0{\cdot}16 \times (2 \times 1{\cdot}5) \times 6}{0{\cdot}02}$

$= \underline{144 W}$

4.06 Surface resistance

In fact, the resistivity of the bulk of the material is not the only factor affecting the flow of heat through something like a wall. At each change of material, or surface, within and on either side of the wall, there is a local effect called the *surface resistance*. In practice, resistance where solid meets solid can be ignored. Table XIX gives values of surface resistance for internal and external surfaces.

4.07 Cavities

Where there are cavities within a wall, the effects of the cavity air and the surface resistances are combined. Values for unventilated cavities are given in table XX, for ventilated cavities in table XXI.

4.08 Solar gain

In this section we have dealt up to this point only with heat *loss* through the construction. There is also, even in the United Kingdom, a problem with excessive heat *gain* through the structure, particularly the glazing. This particularly affects air-conditioned buildings, and the extra cooling load has to be allowed for in the design of the plant. To conserve energy, and particularly in hot climates, measures to avoid solar gain should be taken such as:

- external sunbreakers
- overhanging shading (not effective against morning and evening sun)
- solar control glass
- internal venetian blinds, etc.

Table XXII gives figures for some window types in the United Kingdom.

4.09 Examples

Example 8

Compute the U value for the wall construction in table XVII, item A6, **41.19** *for a normal exposure wall.*

Add the thermal resistances for each element and surface in turn:

Inside surface (table XIX)	0·123
16 mm plaster (table XVIII) 2·0 × 0·016	= 0·032
100 mm lightwt concrete block (table XVIII) 5·3 × 0·1	= 0·53
unventilated cavity (table XX)	0·18
105 mm brick (table XVIII) 1·19 × 0·105	= 0·125
Outside surface (table XIX) normal exposure	0·055
Total R	= 1·045 m²K/W

Therefore U value = $\frac{1}{R}$ = 0·96 W/(m²K)

Example 9

In the last example, what would be the effect of filling the 50 mm cavity with foamed polyurethane?

The resistance of 50 mm of foamed polyurethane is
0·05 × 38·5 = 1·925
deduct the cavity 0·18

1·745

Total resistance of construction is
1·045 + 1·745 = 2·79

therefore the new U value is $\frac{1}{2.79}$ = 0·36 W/(m²K)

The cavity fill therefore reduces the heat loss to ⅜ of the original.

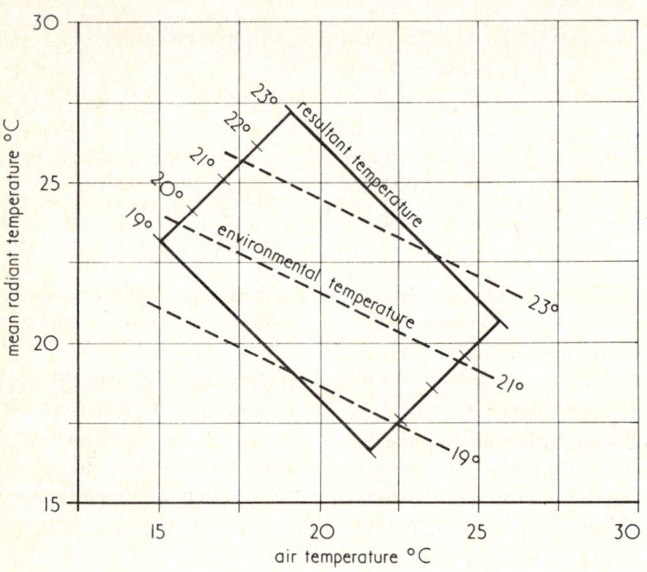

41.17 *Comfort zone for sedentary occupation with air velocity 0·1 m/s*

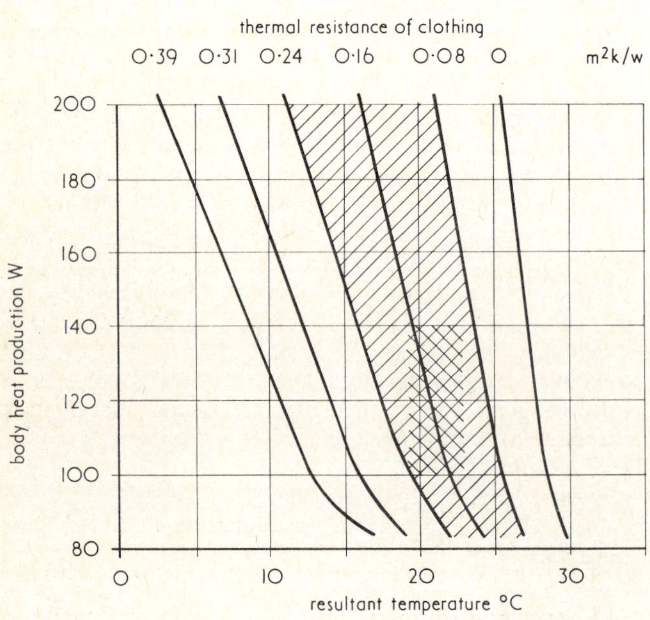

41.18 *Room temperature in relation to activity and clothing: normal range of clothing shown single hatched comfort zone shown cross-hatched (From IHVE Guide, 1970)*

Table XVII U values for various constructions

Construction		U Value (W/m² °C) Sheltered	Normal (Standard)	Severe
A External walls, masonry construction				
Brickwork				
1. Solid wall, unplastered	105 mm	3·0	3·3	3·6
	220 mm	2·2	2·3	2·4
	335 mm	1·6	1·7	1·8
2. Solid wall, with 16 mm plaster on inside face				
(a) With dense plaster	105 mm	2·8	3·0	3·2
	220 mm	2·0	2·1	2·2
	335 mm	1·6	1·7	1·8
(b) With lightweight plaster	105 mm	2·3	2·5	2·7
	220 mm	1·8	1·9	2·0
	335 mm	1·4	1·5	1·6
3. Solid wall, with 10 mm plasterboard lining fixed to brickwork with plaster dabs	105 mm	2·6	2·8	3·0
	220 mm	1·9	2·0	2·1
	335 mm	1·5	1·6	1·7
4. Cavity wall (unventilated) with 105 mm outer and inner leaves with 16 mm plaster on inside face	260 m			
(a) With dense plaster		1·4	1·5	1·6
(b) With lightweight plaster		1·3	1·3	1·3
5. As 4, but with 230 mm outer leaf and 105 mm inner leaf	375 mm			
(a) With dense plaster		1·2	1·2	1·2
(b) With lightweight plaster		1·1	1·1	1·1
Brickwork/lightweight concrete block				
6. Cavity wall (unventilated), with 105 mm brick outer leaf 100 mm lightweight concrete block inner leaf and with 16 mm dense plaster on inside face	260 mm	0·93	0·96	0·98
7. As 6, but with 13 mm expanded polystyrene board in cavity		0·69	0·70	0·71
Lightweight concrete block				
8. Solid wall, 150 mm aerated concrete block, with tile hanging externally and with 16 mm plaster on inside face		0·95	0·97	1·0
9. Cavity wall (unventilated) with 75 mm aerated concrete block outer leaf, rendered externally, 100 mm aerated concrete block inner leaf and with 16 mm plaster on inside face 50 mm cavity		0·82	0·84	0·86
Concrete				
10. Cast	150 mm	3·2	3·5	3·9
	200 mm	2·9	3·1	3·4
11. Cast, 150 mm thick, with 50 mm woodwool slab permanent shuttering on inside face finished with 16 mm dense plaster		1·1	1·1	1·1
12. As 11, but 200 mm thick		1·1	1·1	1·1
13. Pre-cast panels, 75 mm thick		3·9	4·3	4·8
14. As 13, but with 50 mm cavity and sandwich lining panels, composed of 5 mm asbestos-cement sheet, 25 mm expanded polystyrene and 10 mm plasterboard		0·79	0·80	0·82
15. Pre-cast sandwich panels comprising 75 mm dense concrete, 25 mm expanded polystyrene and 150 mm lightweight concrete		0·71	0·72	0·73
16. Pre-cast panels 38 mm on timber battens and framing with 10 mm plasterboard lining and 50 mm glass-fibre insulation in cavity (Assumed 10% area of glass fibre bridged by timber)		0·61	0·62	0·63
B External walls, framed construction				
Tile hanging				
1. On timber battens and framing with 10 mm plasterboard lining, 50 mm glass-fibre insulation in the cavity and building paper behind the battens (Assumed 10% area of glass fibre bridged by timber)		0·64	0·65	0·66

Table XVII U values for various constructions *(continued)*

Construction	U Value (W/m² °C) Sheltered	Normal (Standard)	Severe
Weatherboarding			
2. On timber framing with 10 mm plasterboard lining, 50 mm glass-fibre insulation in the cavity and building paper behind the boarding (Assumed 10% area of glass fibre bridged by timber)	0·61	0·62	0·63
Corrugated sheeting			
3. 5 mm thick asbestos-cement (No allowance has been made for effect of corrugations on heat loss)	4·7	5·3	6·1
4. As 3, but with cavity and aluminium foil-backed plasterboard lining	1·7	1·8	1·9
5. Double-skin asbestos-cement with 25 mm glass-fibre insulation in between	1·1	1·1	1·1
6. As 5, but with cavity and aluminium foil-backed plasterboard lining	0·76	0·78	0·79
7. Aluminium:			
(a) Bright surface outside and inside	2·4	2·6	2·9
(b) Dull surface outside bright surface inside	2·6	2·8	3·0
8. As 7, but with cavity and aluminium foil-backed plasterboard lining:			
(a) Bright surface outside	1·7	1·8	1·9
(b) Dull ,, ,,	1·8	1·9	2·0
9. Plastic-covered steel	5·0	5·7	6·6
10. As 9, but with cavity and aluminium foil-backed plasterboard lining	1·8	1·9	2·0
C External walls, curtain wall construction			
Composite cladding panels			
1. Comprising 25 mm expanded polystyrene between 5 mm asbestos-cement sheets set in metal framing, 50 mm cavity, 100 mm lightweight concrete block inner wall, finished with 16 mm plaster rendering on inside face (Assumed 5% area of expanded polystyrene bridged by metal framing)	0·79	0·81	0·83
2. Obscured glass, 38 mm expanded polystyrene cavity 100 mm lightweight concrete back-up wall, dense plaster	0·51	0·51	0·52
3. Stove-enamelled steel sheet, 10 mm asbestos board, cavity, 100 mm lightweight concrete back-up wall, dense plaster	1·1	1·1	1·1
Curtain walling panelling with 5% bridging by metal mullions, 150 mm × 50 mm wide			
4. With mullion projecting outside, flush inside:			
Panel construction 2	0·8	0·9	0·9
,, ,, 3	1·4	1·4	1·5
5. With mullion projecting inside and outside:			
Panel construction 2	1·3	1·5	1·8
,, ,, 3	1·9	2·1	2·4
Curtain walling panelling with 10% bridging by metal mullions 150 mm × 50 m wide			
6. With mullion projecting outside, flush inside:			
Panel construction 2	1·2	1·2	1·3
,, ,, 3	1·7	1·7	1·8
7. With mullion projecting inside and outside:			
Panel construction 3	2·2	2·5	3·0
,, ,, 3	2·7	3·1	3·6
D Glazing, without frames			
Single window glazing	5·0	5·6	6·7
Double window glazing with air space			
20 mm or more (¾in)	2·8	2·9	3·2
12 mm (½in)	2·8	3·0	3·3
6 mm (¼in)	3·2	3·4	3·8
3 mm (⅛in)	3·6	4·0	4·4
Triple window glazing with air space			
20 mm or more (¾in)	1·9	2·0	2·1
12 mm (½in)	2·0	2·1	2·2
6 mm (¼in)	2·3	2·5	2·6
3 mm (⅛in)	2·8	3·0	3·3

Table XVII U values for various constructions *(continued)*

Construction	U Value (W/m² °C) Sheltered	Normal (Standard)	Severe
Roof glazing skylight	5·7	6·6	7·9
Horizontal laylight with skylight or lantern light over			
Ventilated	3·5	3·8	4·2
Unventilated	2·8	3·0	3·3

E Typical windows

Window type	Fraction of area occupied by frame	U values for stated exposure (W/m² °C) Sheltered	Normal	Severe
Single glazing:				
Wood frame	30%	3·8	4·3	5·0
Metal frame	20%	5·0	5·6	6·7
Double glazing:				
Wood frame	30%	2·3	2·5	2·7
Metal frame with thermal break	20%	3·0	3·2	3·5

Construction	U Value (W/m² °C) Sheltered	Normal (Standard)	Severe
F Pitched roofs (35° slope)			
1. Tiles on battens, roofing felt and rafters, with roof space and aluminium foil-backed 10 mm plasterboard ceiling on joists	1·4	1·5	1·6
2. As 1, but with boarding on rafters	1·3	1·3	1·3
3. As 2, but with 50 mm glass-fibre insulation between joists	0-b6m49	0·50	0·51
4. Corrugated asbestos-cement sheeting	5·3	6·1	7·2
5. As 4, but with cavity and aluminium foil-backed 10 mm plasterboard lining	1·8	1·9	2·0
6. Corrugated double-skin asbestos-cement sheeting with 25 mm glass-fibre insulation between (No allowance has been made for effect of corrugations on heat loss)	1·1	1 1	1·1
7. As 6, but with cavity and aluminium foil-backed 10 mm plasterboard lining; ventilated air space	0·79	0·80	0·82
8. Corrugated aluminium sheeting	3·3	3·8	4·3
9. As 8, but with cavity and aluminium foil-backed 10 mm plasterboard lining	1·8	1·9	2·0
10. Corrugated plastic-covered steel sheeting	5·7	6·7	8·1
11. As 10, but with cavity and aluminium foil-backed 10 mm plasterboard lining; ventilated air space	1·9	2·0	2·1
G Roofs, flat or pitched			
1. Asphalt 19 mm thick or felt/bitumen layers* on solid concrete 150 mm thick (treated as exposed)	3·1	3·4	3·7
2. As 1, but with 50 mm lightweight concrete screed and 16 mm plaster ceiling	2·1	2·2	2·3
3. As 2, but with screed laid to falls, average 100 mm thick	1·7	1·8	1·9
4. Asphalt 19 mm thick or felt/bitumen layers* on 150 mm thick autoclaved aerated concrete roof-slabs	0·87	0·88	0·89
5. Asphalt 19 mm thick or felt/bitumen layers* on hollow tiles 150 mm thick	2·1	2·2	2·3
6. As 5, but with 50 mm lightweight concrete screed and 16 mm plaster ceiling	1·5	1·6	1·7
7. As 6, but with screed laid to falls, average 100 mm thick	1·4	1·4	1·5
8. Asphalt 19 mm thick or felt/bitumen layers* on 13 mm cement and sand screed, 50 mm woodwool slabs on timber joists and aluminium foil-backed 10 mm plasterboard ceiling, sealed to prevent moisture penetration	0·88	0·90	0·92

Table XVII U values for various constructions *(continued)*

Construction	U Value (W/m² °C) Sheltered	Normal (Standard)	Severe
9. As 8, but with 25 mm glass-fibre insulation laid between joists	0·59	0·60	0·61
10. Asphalt 19 mm thick or felt/bitumen layers* on 13 mm cement and sand screed on 50 mm metal edge reinforced woodwool slabs on steel framing, with vapour barrier at inside	1·4	1·4	1·5
11. As 10, but with cavity and aluminium foil-backed 10 mm plasterboard ceiling below steel framing (Bridging effect of framing neglected. Assumed that aluminium foil acts as vapour barrier)	0·88	0·90	0·92
12. Asphalt 19 mm thick or felt/bitumen layers* on 13 mm fibre insulation board on hollow or cavity asbestos-cement decking, with vapour barrier at inside	1·4	1·5	1·6
13. As 12, but with 25 mm glass-fibre insulation in cavity, with vapour barrier	0·72	0·73	0·74
14. Felt/bitumen layers* on 25 mm expanded polystyrene on hollow or cavity asbestos decking, with vapour barrier	0·85	0·87	0·89
15. Asphalt 19 mm thick or felt/bitumen layers* on 13 mm fibre insulation board on metal decking, with vapour barrier	2·1	2·2	2·3
16. Felt/bitumen layers* on 25 mm expanded polystyrene on metal decking, with vapour barrier	1·1	1·1	1·1

* The difference between the thermal resistance values of 19 mm of asphalt and three layers of roofing felt set in bitumen is sufficiently small to be ignored.

Table XVII U values for various constructions *(continued)*

Construction	U Value Heat flow downwards	Heat flow upwards
H Intermediate floors		
Wood		
20 mm wood floor on 100 mm × 50 mm	1·5	1·7
joists, 10 mm plasterboard ceiling, allowing for 10% bridging by joists	1·4	1·6
Concrete		
150 mm concrete with 50 mm screed	2·2	2·7
with 20 mm wood flooring	1·7	2·0
Hollow tile floors		
with 50 mm dense concrete over and between tiles		
tile thickness 150 mm	1·7	2·0
200 mm	1·6	1·9
250 mm	1·5	1·8
as above with 20 mm wood flooring		
tile thickness 150 mm	1·4	1·6
200 mm	1·3	1·5
250 mm	1·2	1·4

J Ground floors	
Concrete on ground or hardcore fill	1·13
with grano, terrazzo or tile finish	1·13
with wood block finish	0·85
Timber boards on joists, space ventilated one side	1·70
with parquet, lino or rubber cover	1·42
Timber boards on joists, space ventilated on more sides	2·27
with parquet, lino or rubber cover	1·98
with 25 mm fibreboard under boarding	1·08
with 25 mm corkboard under boarding	0·95
with 25 mm corkboard under joists or forming cavity	0·79
with 50 mm strawboard forming cavity (between joists)	0·85
with double sided aluminium foil draped over joists	1·42

Table XVIII Thermal properties of building materials (from various sources)

Material	Density kg/m³	Thermal resistivity m.K/W	Thermal conductivity W/(m.K)	Vapour resistivity MN.s/(g.m.)
Brickwork		0·69–1·38	0·72–1·45	25–100
outer leaf	1700	1·19	0·84	
inner leaf	1700	1·61	0·62	
Bricks, common				35–52
sandlime				45–370
Clay tiles	1900	1·2	0·84	
Concrete blocks	600	5·3	0·19	30–160
aerated, outer leaf	750	4·2	0·24	
inner leaf	750	4·5	0·22	
Clinker blocks				2000–2400
Concrete, gravel	2100	0·69	1·45	200
lightweight	1200	2·4	0·42	30–160
aerated	500	6·25	0·16	30–60
Rendering, cement/sand	2100	0·83	1·2	100
Lime mortar				45–50
Plaster	1300	2·08	0·48	60
lightweight	600	6·24	0·16	
Plasterboard	950	6·24	0·16	35–55
Timber	650	6·93	0·144	45–75
weatherboarding	650	7·1	0·14	
Plywood		6·93	0·144	1500–6000
Fibreboard	240–400	15·2–18·7	0·053–0·066	15–60
Hardboard	1010	6·93	0·144	450–750
Woodwool	560	8·66	0·115	15–40
Compressed strawboard	260–350	9·7–11·8	0·085–0·103	45–75
Asbestos board	700	9·1	0·11	
Asbestos cement sheeting	1500	2·8	0·36	
Asphalt	1700	2	0·50	
Roofing felt	960	5·3	0·19	
Mineral wool	50	25·6	0·039	1·1
Glass fibre	25	28·6	0·035	
Expanded polystyrene	25	27·72	0·036	100–600
Foamed urea formaldehyde	12	27·72	0·036	20–30

Table XVIII Thermal properties of building materials (from various sources) *(continued)*

Material	Density kg/m³	Thermal resistivity m.K/W	Thermal conductivity W/(m.K)	Vapour resistivity MN.s/(g.m.)
Foamed polyurethane (open and closed cell)	30	38·5	0·026	30–1000
Expanded ebonite	64	27·72	0·036	11 000–60 000
Glass, window	2500	0·95	1·05	
Stone: granite	2600	0·34	2·92	
limestone	2100	0·65	1·53	
sandstone	2200	0·77	1·30	

Table XIX Surface resistances R_s m²K/W

Surface	High emissivity* E = 0·9	Low emissivity E = 0·05
Internal surfaces, R_{si}		
Walls, heat flow horizontal	0·123	0·304
Ceilings or roofs, flat or pitched Floors, heat flow upwards	0·106	0·218
Ceilings and floors, heat flow downward	0·150	0·562
Outside surfaces, R_{so}		
Walls, sheltered†	0·08	0·11
normal exposure	0·055	0·067
severe exposure	0·03	0·03
Roofs, sheltered	0·07	0·09
normal exposure	0·045	0·053
severe exposure	0·02	0·02
Floor, underside exposed to air	0·09	

* Emissivities: ordinary building materials 0·9
dull aluminium 0·2
polished aluminium 0·05

† Exposures: Sheltered—up to third floor in city centres wind speed 1 m/s
normal—most suburban and country buildings, up to eighth floor in city centres wind speed 3 m/s
severe—buildings on the coast, exposed on hill sites, above five floors in suburbs or country, above eight floors in city centres wind speed 9 m/s

Table XX Thermal resistance of unventilated air spaces

Type of air space Width	Surface emissivity	Thermal resistance* Heat flow horizontal or upwards (m² °C/W)	Downwards (m² °C/W)
5 mm	High	0·11	0·11
	Low	0·18	0·18
20 mm or maximum	High	0·18	0·2
	Low	0·35	1·06
High emissivity plane and corrugated sheets in contact		0·09	0·10
Low emissivity multiple foil insulation		0·62	1·76

* Includes internal boundary surfaces.

Table XXI Thermal resistance of ventilated air spaces

Width 20 mm minimum	Thermal resistance* (m² °C/W)
Space between asbestos cement or black metal cladding with unsealed joints and high emissivity lining	0·16
Do with low emissivity lining	0·3
Space between flat ceiling and pitched roof with covering of asbestos or black metal	0·14
Do with bright aluminium covering	0·25
Space between flat ceiling and unsealed tiling	0·1
Do with felt sarking and beamfilling	0·18
Space between the tiles and sarking felt	0·12
Space behind tiles on tile-hung wall	0·12
Airspace in cavity wall	0·18

* Includes internal boundary surfaces.

Table XXII Solar gain (as measured by additional cooling loads)

Window type	Solar gain (W/m²) for orientations: east	south	west
Single glazing unshaded	510	371	478
Single glazing with internal venetian blinds	328	305	305
6 mm clear float double glazing with internal venetian blinds	265	302	301
6 mm 'Spectrafloat' and 6 mm clear float with internal venetian blinds	220	250	250
6 mm 'Antisun' and 6 mm clear float with internal venetian blinds	195	221	224
Suncool 26/22 (gold)	111	132	135

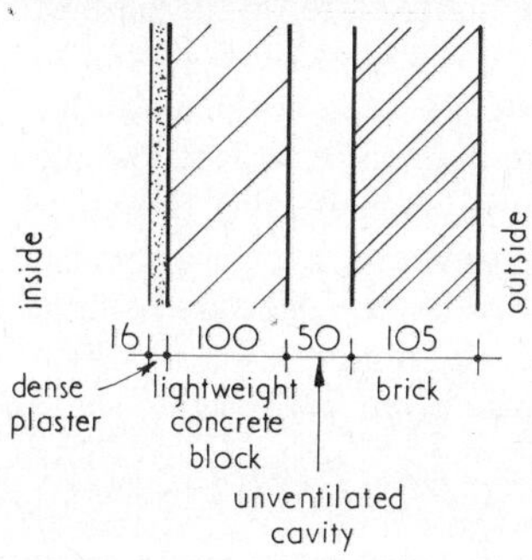

41.19 *Wall construction for example 8, table XVII item A6. A common form of construction in the UK*

5 HUMIDITY AND CONDENSATION

5.01 General gas equation

In para **2.12** the gas equation PV = RT was given. In this form the gas constant R is 8313 J/(kmol). K is for 1 mol of any gas. If the amount of gas is given in kg, the general form of the equation becomes:

$$P V = m R_1 T$$

where P is the pressure in Pa
V is the volume in m³
m is the mass of the gas in kg
T is the temperature in kelvins (= deg C + 273·15)
and R_1 is the gas constant for that gas as in table XXIII

Table XXIII Gas constants for some gases

Gas	Constant R_1 J/(kg.K)
Air	287
Carbon dioxide	189
Hydrogen	4150
Nitrogen	296
Oxygen	259
Sulphur dioxide	130
Water vapour	461

Example 10

What is the density of air at atmospheric pressure (101·325 kPa) and a temperature of 20°C?

Consider 1 m³,

$P V = m R_1 T$

$P = 101{\cdot}325 \times 10^3$

$V = 1$

$R_1 = 287$ (from table XXIII)

$T = 20 + 273{\cdot}15 = 293{\cdot}15$

Therefore

$101{\cdot}325 \times 10^3 \times 1 = m \times 287 \times 293{\cdot}15$

$m = \underline{1{\cdot}204\,\text{kg/m}^3}$

Because the molecules in a gas are so far apart, two different gases such as air and water vapour can occupy the same volume as though they were alone in it. If two such gases occupy the same space, the volume and temperature is the same for each.

for the first gas $P_1 V = m_1 R_1 T$

for the second $P_2 V = m_2 R_2 T$

by dividing $\frac{P_1}{P_2} = \frac{m_1}{m_2} \times \frac{R_1}{R_2}$

5.02 Mixing ratio

All air contains some amount of water vapour. The amount can be expressed in a number of different ways. The simplest is the *mixing ratio* expressed in g/kg dry air.

Example 11

What is the vapour pressure when the mixing ratio is 10 g/kg? The pressure of the dry air may be assumed to be 101·325 kPa.

In the formula $\frac{P_1}{P_2} = \frac{m_1}{m_2} \times \frac{R_1}{R_2}$

$P_2 = 101{\cdot}325 \times 10^3$

$\frac{m_1}{m_2} = 10 \times 10^{-3}$

$\frac{R_1}{R_2} = \frac{461 \text{ (table XVI)}}{287} = 1{\cdot}606$

Hence $P_1 = 101{\cdot}325 \times 10^3 \times 10 \times 10^{-3} \times 1{\cdot}606$

$= 1{\cdot}627$ k Pa

5.03 Dew point

In para **2.13** the *critical temperature* was defined as that above which the gas cannot revert to the liquid state on increase of pressure, that is, the level at which it changes from a vapour to a true gas. Water vapour in the atmosphere never reaches this point. For any given pressure there is a temperature at which the vapour will *condense* into a liquid (such as rain). Similarly there is a pressure for a given temperature at which condensation will occur. This is known as the *saturation pressure* for that temperature. For a given pressure, the temperature at which condensation occurs is called the *dew point.*

5.04 Psychrometric chart

Figure **41.20** is known as the *psychrometric chart.* On the right-hand side, the vertical axis represents the amount of water vapour in the atmosphere, expressed as both mixing ratio and vapour pressure. On the horizontal axis at the bottom there is a scale of temperature from 0 to 25°C. (This is shown as dry bulb temperature. This simply means that it indicates the actual temperature of the atmosphere.) Rising from the left-hand side to the right is a line called the *saturation line*, which marks the top edge of the chart and 100 per cent saturation.

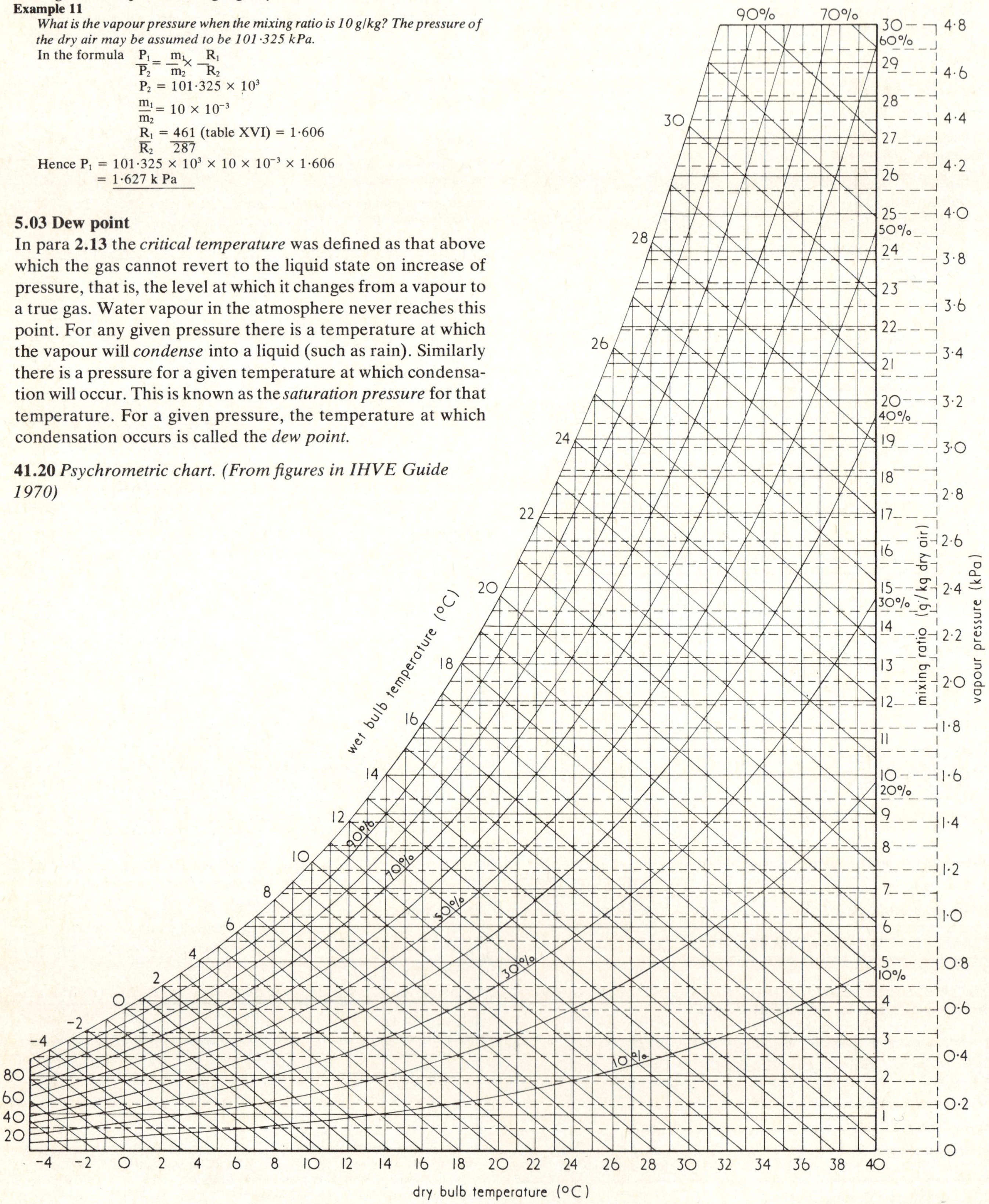

41.20 *Psychrometric chart. (From figures in IHVE Guide 1970)*

Example 12

What is the temperature at which the atmosphere in example 7 would become saturated?

With a mixing ratio of 10, the horizontal line representing this value is followed to the left until it reaches the saturation line **41.21**. This corresponds to a temperature of 14 °C on the bottom axis.

5.05 Relative humidity

The normal method of indicating the water vapour content of air is not vapour pressure or mixing ratio, but is the *relative humidity*. This is the mass of the vapour expressed as a percentage of the mass required to cause saturation. At the temperature of 14 °C in the example 8 above, a mass of 10 g/kg is needed for saturation. If there is 3 g/kg then the relative humidity will be 30 per cent. The relative humidity scale on the psychrometric chart is on the vertical axis on the left-hand side, and refers to the *curved* lines on the chart that run roughly parallel to the saturation line. If the 30 per cent line is followed, **41.22**, it is seen that it cuts the mixing ratio line for 3 g/kg at the temperature of 14 °C expected.

Example 13

What is the relative humidity of an atmosphere with a vapour pressure of 0·8 kPa at a temperature of 22 °C? What is the effect of reducing the temperature to 14 °C?

Following the line representing 0·8 kPa and 5 g/kg to the left, **41.23**, until it cuts the temperature line for 22 °C, it is seen that this coincides with the curved line for a relative humidity of 30 per cent. Continuing to reach the 14 °C line, this corresponds with a relative humidity of 50 per cent.

Hence, reduction in the temperature increases the relative humidity of the atmosphere. If the temperature in the example above was reduced as far as 4 °C, the atmosphere would become saturated.

5.06 Wet bulb

On the other hand, if the atmosphere was allowed to become locally saturated, its temperature would fall. This would be because the change from the water to the vapour would be adiabatic, as described in para **2.10**.

On the chart as shown in **41.24** the point A represents the vapour pressure of 0·8 kPa at a temperature of 14 °C as discussed in example 9. An adiabatic increase of water vapour content would move along the diagonal line AB to meet the saturation line at B, corresponding to a temperature of 9 °C. An isothermal move up the vertical 14 °C line would require

41.21 *Psychrometric chart for example 12: evaluation of saturation temperature*

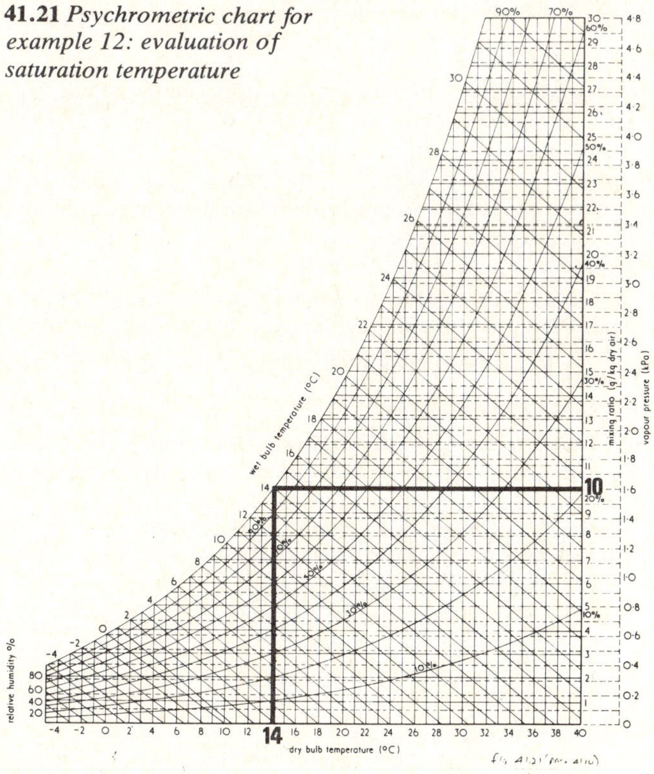

41.22 *Relative humidity at a mixing ration of 3 g/kg*

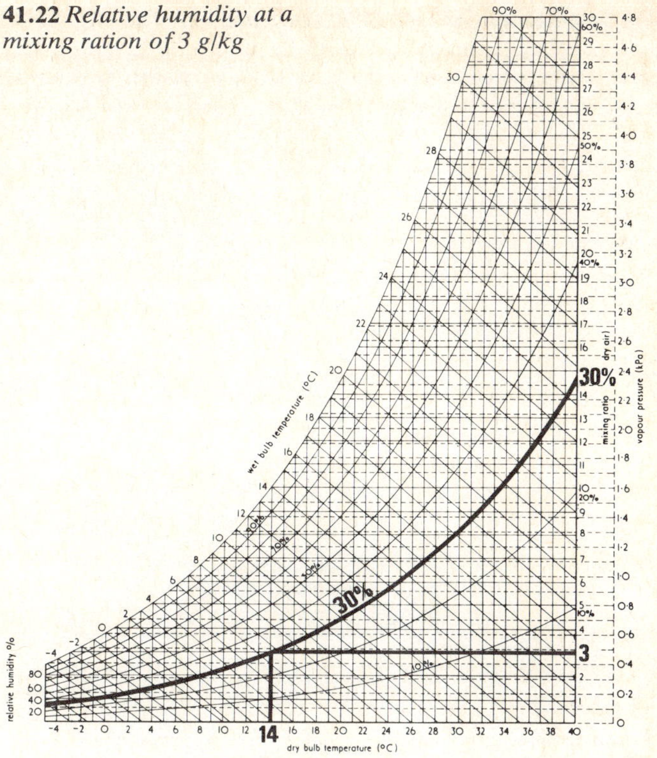

41.23 *Psychrometric chart for example 13: relative humidity and effect of reducing temperature*

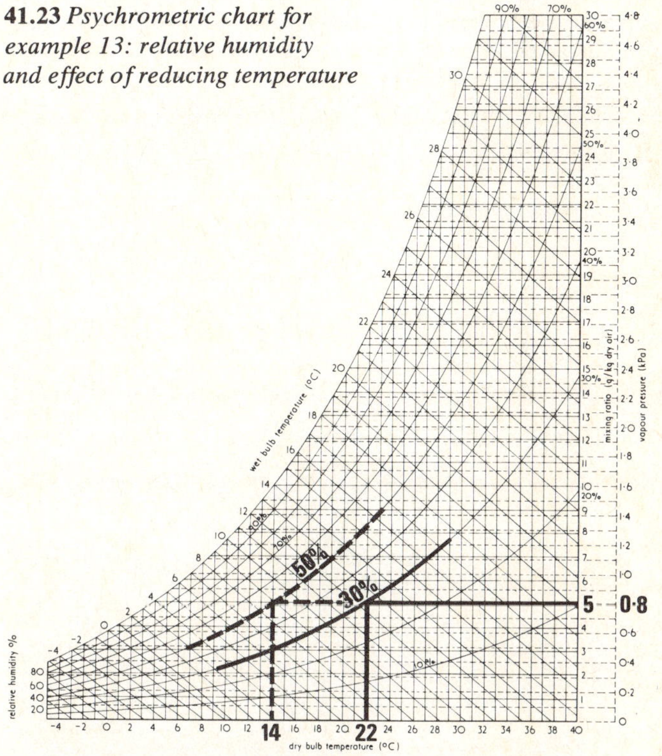

energy to be added to the system, by heating or otherwise.

It will be seen that there is a temperature scale along the saturation line described as *wet bulb temperature*. The wet bulb is simply an ordinary thermometer with muslin around the bulb, the end of which is dipped in a water reservoir. The air around the wet bulb is therefore always saturated. The difference between the reading on the wet bulb and that on the dry gives the relative humidity directly.

Example 14

What is the relative humidity when the dry bulb reads 19 °C and the wet 13·1 °C?

On the chart, **41.25**, locate point C on the saturation line corresponding to wet bulb temperature of 13·1 °C. Follow the diagonal line to D on the vertical line corresponding to dry bulb reading of 19 °C. This coincides with the relative humidity line for 50 per cent.

41.24 *Adiabatic change due to saturation*

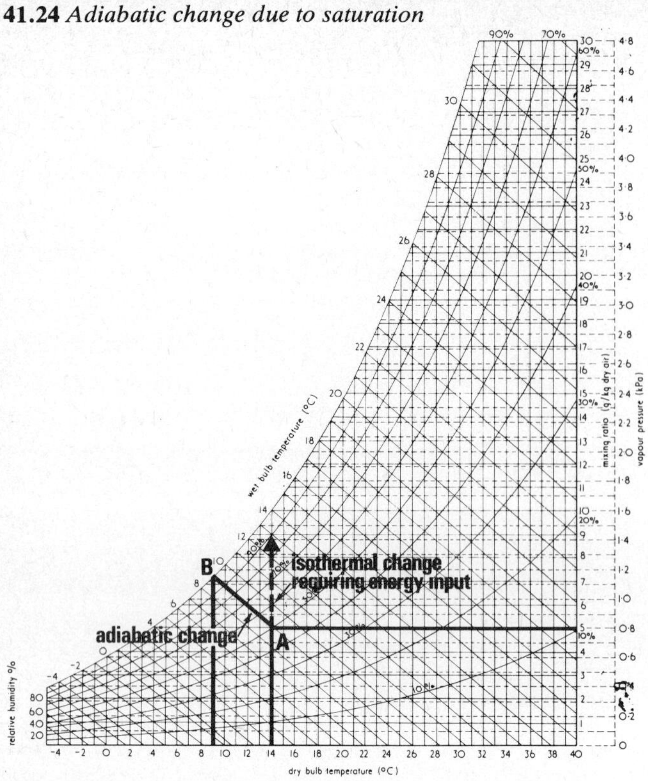

41.26 *Psychrometric chart for example 15: reduction of humidity by cooling*

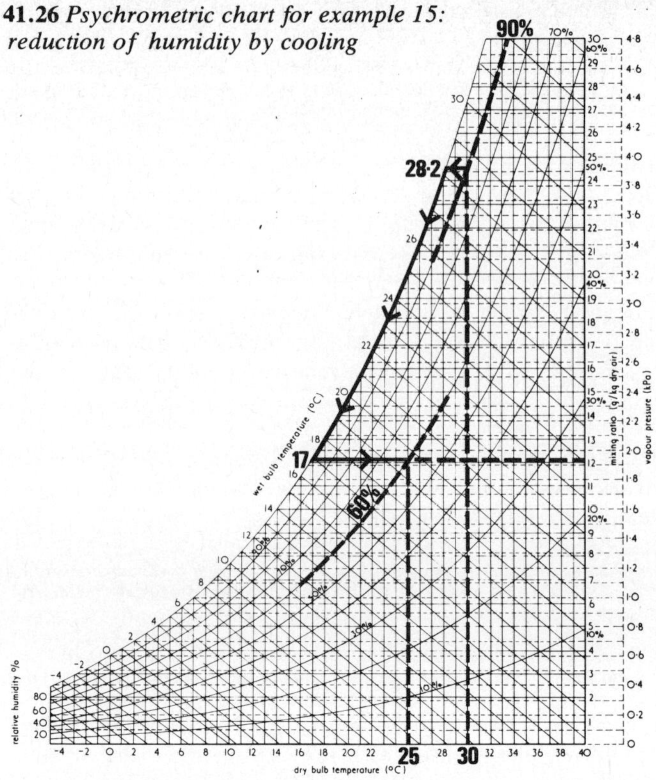

41.25 *Psychrometric chart for example 14: wet and dry bulbs*

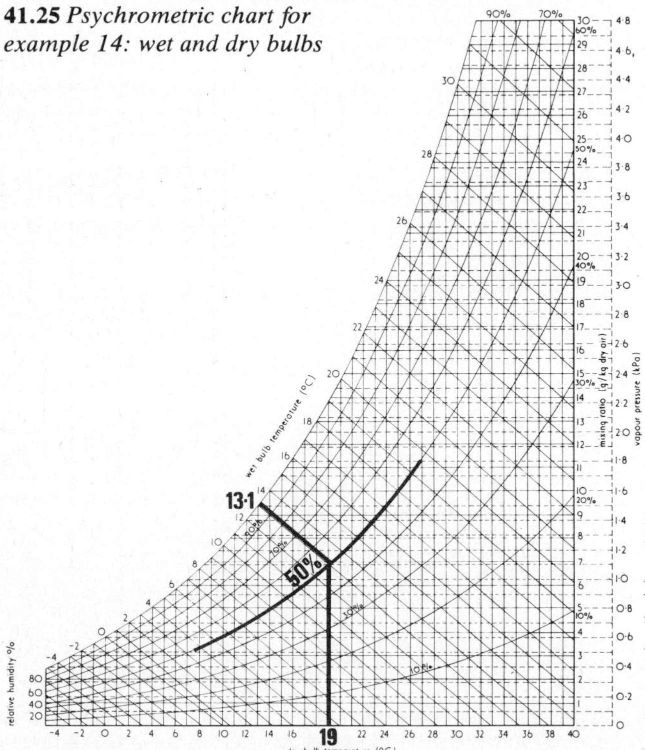

5.07 Humidity and comfort

In hot climates it is usually the humidity or lack of it that causes the discomfort rather than the temperature. In fact the range of acceptable relative humidities is quite large: for sedentary occupations it lies between 40 and 70 per cent. Very high and very low humidities both have a deleterious effect on building fabric and furniture. The build-up of static electricity is encouraged by low humidity. Many industrial processes such as pharmaceutical production and printing demand rigorous limits.

5.08 Humidity control

Air conditioning, as described in para **3.08**, means adjusting the humidity of the air. Air that is too dry can be moistened by passing it through washers. Very humid air is dried by cooling.

Example 15

An internal environment with relative humidity of 60 per cent and temperature of 25°C is required where the outside RH is 90 per cent and temperature is 30°C. How could this be achieved?

The point on the psychrometric chart, **41.26**, corresponding to RH 90 per cent and 30°C, corresponds to a mixing ratio of 24·6 g/kg. If the air is cooled to 28·2°C, it can be seen that the air will become saturated. Further cooling will cause water to condense out of it, and the saturation line on the chart will be followed down. At a temperature of 17°C, the mixing ratio will be 12·2 g/kg, which is the same as for the required combination of RH 60 per cent and temperature 25°C. The air, now conditioned, can be allowed to reach the required temperature by passing it over the ducts carrying the inlet air at 30°C, without, of course, allowing it to mix with it.

This example, which is greatly simplified, illustrates the principle by which air conditioning works.

5.09 Condensation

Condensation is one of the most serious problems in building. It occurs whenever the saturation line on the psychrometric chart is reached, for either of the reasons:

- the water vapour content of the atmosphere is increased from cooking, laundering or from the products of human respiration and perspiration
- the temperature of the atmosphere is lowered.

It is rare for condensation within buildings to occur other than on surfaces or within construction. Only in very large buildings (such as the Houston Astrodome) can 'rain' occur internally.

5.10 Surface condensation

Condensation problems, therefore, often appear to derive from the surfaces on which it occurs, rather than from the factors causing it. To prevent condensation:

- reduce the water vapour content of the atmosphere. This can be achieved by extract fans over sources of excess moisture, such as cookers and bathroom appliances
- increase the temperature of all surfaces, either by more heating or improved insulation. It is particularly important to avoid cold bridging of insulation.

5.11 Interstitial condensation

Sometimes condensation occurs within the material of the

construction rather than on the surface. This is because the dewpoint is reached within its thickness. The two factors affecting this are the thermal properties of the construction and also the vapour resistivity of the materials. The vapour resistivities of some materials are given in table XVIII.

Some materials are highly resistant to vapour penetration and can be used in a very thin form, such as a membrane. Table XXIV gives vapour resistances for some of these membranes.

Table XXIV Vapour resistance of membranes

Material	Vapour resistance MN.s/g
Bitumen	300
Bitumen/polymer (0·06 mm)	300
Average gloss paint film	7·5–40
Polythene sheet (0·06 mm)	250
Aluminium foil	4000
Building paper	2–40
Foil-backed building paper	10 000
Roofing felt	100

5.12 Method of calculating interstitial condensation

In order to demonstrate a method of determining the likelihood of interstitial condensation, consider the following example:

Example 16

A wall is constructed as **41.27a**. *The interior is at a temperature of 18 °C, RH of 50 per cent. The outside is at 0 °C, RH 80 per cent.*

First construct the following table for the given construction:

	Inner face	paint	acoustic tile	plaster	brick	cavity	brick	outer face
Thickness (m)			0·025	0·015	0·1025	0·05	0·1025	
Vapour Properties resistivities ($m^2K/g.m$)				60	25		25	
resistances (m^2K/g)		7·5*		0·9	2·56		2·56	
Thermal properties resistivities (m.K/W)			27	2·1	0·69		0·69	
resistances (m^2K/W)	0·12		0·68	0·03	0·07	0·18	0·07	0·0

* See example 18

Now, draw a diagram of the construction with the thickness of each element proportional to the vapour resistance to some suitable scale such as 10 mm to 1 m^2K/g, **41.27b**.

Construct a vertical scale of vapour pressure to scale of, say, 10 mm to 0·1 Pa. From the psychrometric chart **41.20** establish the internal and external pressures and corresponding saturation temperatures:

Temperature	Relative humidity	Vapour pressure	Saturation temperature
18°	50%	1·04 Pa	7·5°
0°	80%	0·49 Pa	−2·5°

Plot these points on the appropriate positions on the construction opposite the points on the vapour pressure scale. This will enable the vapour pressures at the other interfaces to be determined:

Position	Vapour pressure	Saturation temperature
plaster/brick	0·96	6·4°
brick/cavity/brick	0·72	2·3°

A third diagram is now constructed, this time with the thicknesses of each element in proportion to the thermal resistance, to a scale, say, of 10 mm to 0·1 m^2K/W. A vertical temperature scale is set up, say 10 mm to 1 °C, and the actual temperatures on each face are joined by a line giving the temperature gradient within the construction, **41.27c**. An additional line is drawn for the saturation temperature corresponding with the vapour pressure at that point in the construction. If the saturation temperature at any point is above the actual, at that point interstitial condensation will occur. In this case, this will happen in the acoustic tiling where it is fixed to the plaster, and also in the plaster itself.

Example 17

(a) With the same internal conditions and the external vapour pressure the same, at what external temperature will the interstitial condensation cease?

(b) What internal temperature will be needed to prevent the interstitial condensation, all other factors remaining the same?

By drawing the appropriate lines on **41.27c**, it can be seen that the answers to this question are:

(a) 2·5 °C and (b) 23 °C.

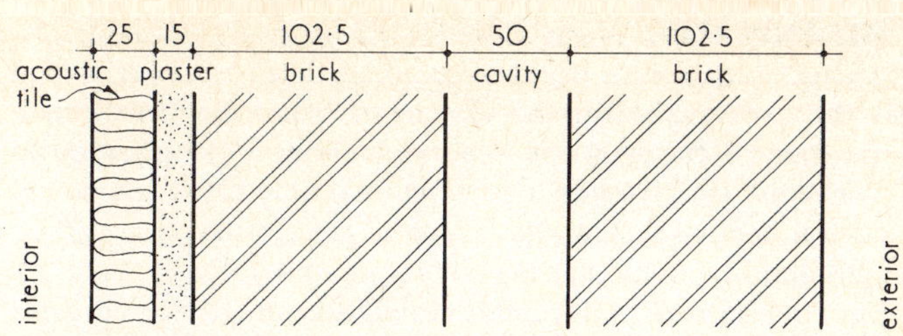

41.27 *Diagrams for examples 16 and 17:*
a *Cross-section of construction of wall*

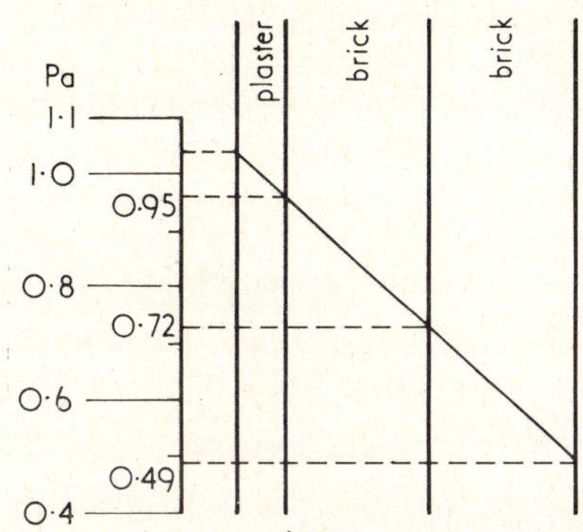

b *Cross-section drawn with horizontal scale proportional to vapour resistance of materials, line shows gradient of vapour pressure*

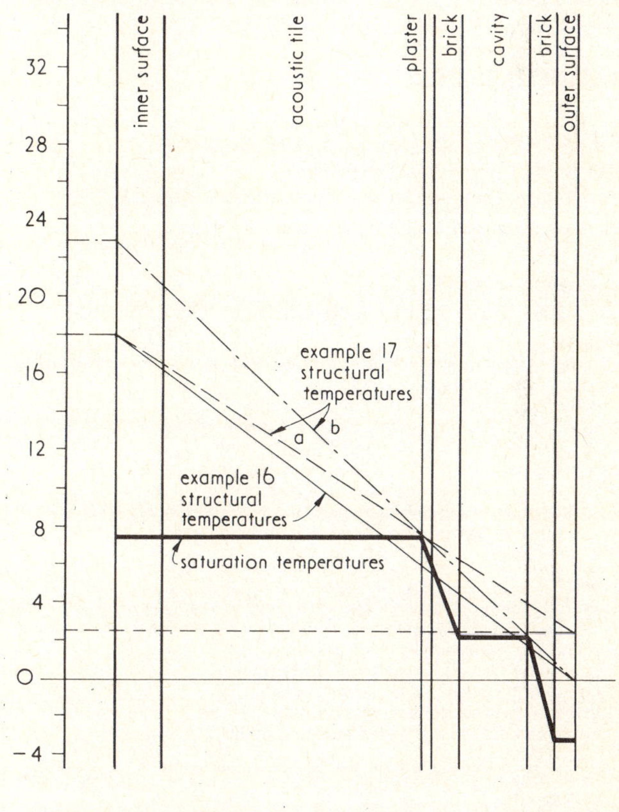

c *Cross-section drawn with horizontal scale proportional to thermal resistance of materials, lines show temperature gradient across the wall*

These examples show that:

- low outside temperatures encourage interstitial condensation
- increasing the inside temperature considerably will inhibit condensation
- in cases similar to that shown, the inside temperature would have to be raised by 2 °C for each 1 °C the outside temperature went down, if condensation was to be controlled by this method.

5.13 Thermal lag

An additional factor in condensation is the *thermal lag*, which is the time taken for the system to reach the *steady state* illustrated in the diagrams. Where a building is being heated up from cold, such as with a conventional time-clock controlled central heating system, the inside of the construction will be considerably cooler than the air temperature for some time. As can be seen from **41.27c**, this would encourage even greater condensation.

5.14 Problems of condensation

Apart from the unsightliness of condensation stains, this defect also has the following effects:

- since water is a poor heat insulator, the value of the insulating material in the construction once it is saturated will be seriously reduced. This will increase the heat losses, and also cause further condensation.
- moisture in the fabric of the building can lead to fungal attack, rot and decay.

5.15 Alternative solutions

Example 18

Reconsider the conditions of example 16 with the addition of one coat of gloss paint on the inside face of the acoustic tiling, **41.28a**.
The table to example 16 already incorporated the item for a gloss paint film, from table XXIV. **41.28b** and **c** show the effects of this addition, and it is seen that interstitial condensation is much less likely.

The general rule is that there should be *a vapour-proof membrane on the warm side of any insulation.* This is by far the most effective method of preventing this defect.

6 WATER HEATING

6.01 Purposes

Water is heated for two purposes:

- for washing (referred to as 'domestic hot water')
- for transporting heat in a convenient form.

6.02 Energy required

In para **2.09** the principle of specific heat was explained.

Example 19

What heat output would be required to produce a bathful of hot water (50 litres at 65 °C) in half an hour?
Specific heat of water is 4·187 kJ/(kg.K) (table I)
One kilogram of water raised 1 kelvin requires 4·187 kJ
there are approximately 1000 kg of water in 1 m^3
hence 1 kg per litre
If the water is supplied at 4°C, it must be raised by 65–4 = 61 K
One litre of water raised 61 K requires 4·187 × 61 kJ = 255·4 kJ
This amount of energy in half an hour represents
255·4 ÷ (30 × 60) = 0·142 kJ/s = 142 W

Example 20

A central heating system circulates water which is at 85 °C when it leaves the boiler, and 80 °C by the time it returns. How much water needs to be circulated per hour to provide an output of 3 kW?
1 kg of water raised 1 K requires 4·187 kJ (table I)
1 litre of water raised 1 K requires 4·187 kJ
1 litre of water raised 5 K requires 20·9 kJ
3 kW represents 3 kJ/s = 180 kJ/min = 10 800 kJ/h
hence, $\frac{10\,000}{20.9}$ litres raised 5 K will be needed per hour,
= 517 litre/h

6.03 Domestic hot water

Table XXV gives the hot water storage requirements for a number of common building types.

Domestic hot water can be provided in one of two alternative ways:

- instantaneous heating (usually using gas as the fuel)
- normal heating followed by storage.

Where hot water is stored until required for use, almost any fuel may be used. *Boilers*, **41.29**, are devices in which water is circulated in water passages (usually formed in cast iron). Hot gases from the combustion of the fuel pass on the other side of the passage wall, and the water is heated by conduction and

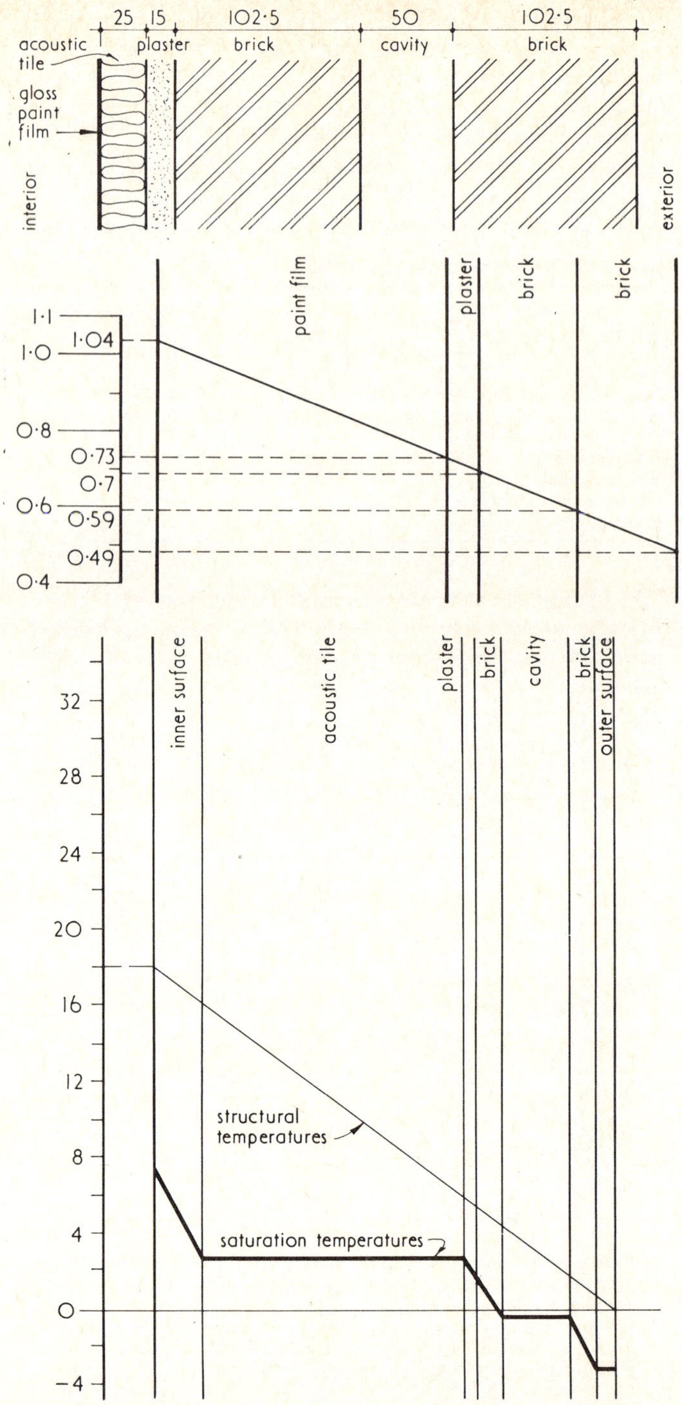

41.28 *Diagrams for example 18, construction as* **41.27** *but with addition of internal paint film*

convection. The fuels commonly used are solid, such as coal, coke, anthracite, wood; or fuel oil or gas.

Where electricity is the fuel the method of heating is by *immersion*, **41.30**. An electrical heating element is immersed in the hot water storage tank. In buildings where water-borne central heating is used, **41.31**, the heating water is usually employed to heat the domestic supply in a *calorifier.* This is a storage tank with a coil of water pipe immersed in it. The water from the heating circuit is passed through this coil. The advantage of this method over direct heating in a boiler is that the circuit water is used over and over again. This means that scale is only produced from the initial heating and the small amount of new water required to replace losses. Where domestic water is heated directly in a boiler, scale is continually produced, and the boiler has to be frequently de-scaled. Scale is solid material held in solution in hard water, which comes out of solution when it is heated. Soft water is rare in the United Kingdom,

Table XXV Water storage in various building types

Building type	Cold water storage (24 hours supply) litre/person	Hot water storage (at 65°C) litre/person
Dwellings, houses and flats:		
high rental	90	45
medium rental	90	30
low rental	90	45
Hotels:		
first class	135	45
average	135	35
Hostels	90	30
Nurses' homes	115	45
Colleges and schools:		
boarding	90	25
day	30	5
Sports pavilions	90	35
Restaurants (per meal)	7	3* (average)
Offices:		
with canteen	45	5
without canteen	35	5
Factories	depends on process	5
Hospitals:		
general and maternity	—†	30
infectious	—	45
mental	—	25

* Restaurants vary between 450 litre storage for 50 meals per day, to 1100 litres for 400 meals per day, and to 3400 for 1500 meals per day.
† Cold water storage for hospitals can vary between 180 litres/bed/day to 1800. Refer to Hospital Engineering Research Unit publications for figures for particular cases.

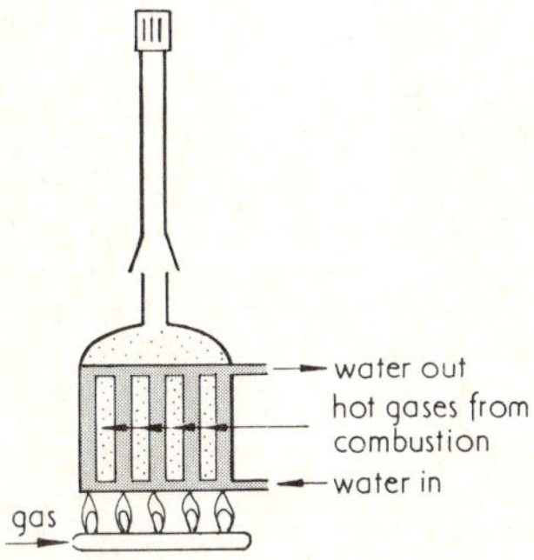

41.29 *Diagrammatic representation of a water boiler, in this case using gas*

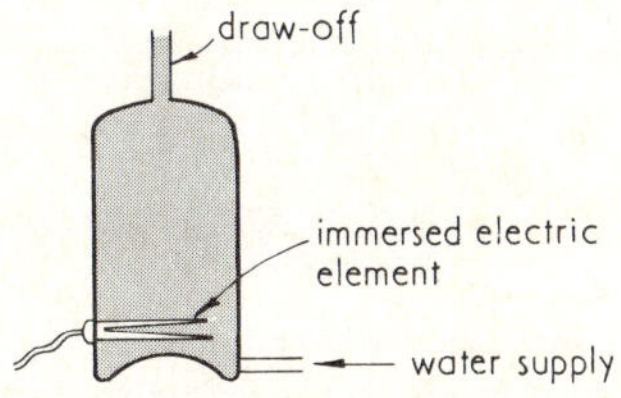

41.30 *Diagram of electric immersion heater in hot water cylinder*

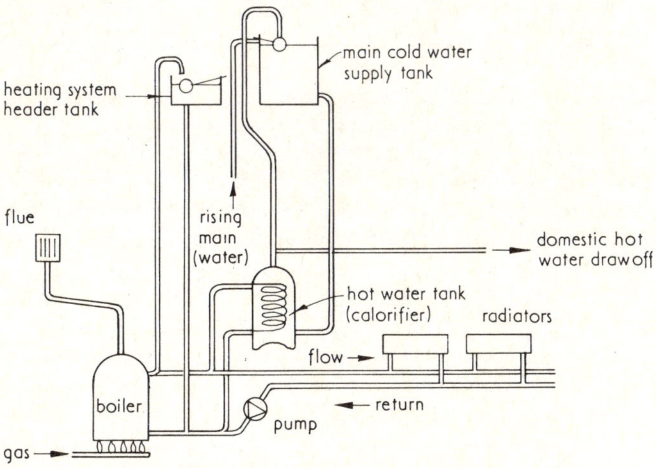

41.31 *Diagram of typical domestic central heating system using gas boiler, small bore pumped supply to radiators on two-pipe system and gravity circulation to heat domestic hot water*

and has its own disadvantages when used in a public water supply.

7 FUELS

7.01 Types of fuels

A fuel is a substance that contains energy that can be released and used for various purposes. The term can be extended, as it is here, to cover any source of useful energy.

The categories of fuels are:

- elemental—wind, flowing water, tides, solar energy, nuclear energy
- fossil—coal, oil and natural gas
- organic—wood, dung
- animal—human, horse, ox, etc
- secondary—electricity, coal gas, water power from pumped storage systems.

7.02 Choice of fuel

The choice of a fuel in any particular case will depend on four factors:

- availability
- cost of fuel
- capital cost of equipment to utilise the fuel
- convenience in use.

The last factor will include such considerations as whether the desired fuel has undesirable pollutant problems, or requires tall flues that are unsightly.

Table XXVI Calorific values of fuels

Fuel	Calorific value MJ/kg	Sulphur content per cent
Solid		
Wood (dry)	20	
Peat (dried)	22	
General purpose coals (701*–902)	23·8–26·7	1·7
High volatile coking coals (401–602)	27·6–29·5	1·7
Heat altered coals (200H–303H)	29·7–30·3	1·2
Medium volatile coking coal (301a & b)	30·4–30·7	1·2
Coking steam coals (202 and 204)	30·7	1·0
Dry steam coal (201)	30·5	1·0
Anthracite (101 and 102)	29·6–30·3	1·0
Coke	28·6	—
Low temperature coke	29·1	—
Liquid		
Heavy fuel oil, class G	42·5	3·8
Medium fuel oil, class F	42·9	3·5
Light fuel oil, class E	43·4	3·2
Gas oil, class D	45·5	1·0
Methylated spirits	25·6	0·1
Alcohol, commercial	29·3	0·1
Coal tar creosote	39·1	0·7
Benzole	42·1	0·9
Paraffin (kerosone)	44·0	0·1
Gaseous		
Natural gas, North Sea (A, B and C)	51·3–53·1	0·02–0·06
Natural gas, Algerian	55·0	
Town gas	32·8	0·003
Butane	49·5	up to 0·02
Propane	50	up to 0·02
Acetylene	50·2	

* National Coal Board Rank Code Number

Table XXVII Animal power

Animal	Power W	Daily energy MJ/d
Man:		
on treadmill	100	2·8
hauling up weight on rope	40	3·5
shovelling earth	11	0·5
wheeling barrow up ramp	13	0·4
turning capstan	72	2·0
Draught horse:		
turning mill	400	11·5
pulling cart or boat	600	16·5
Ox	400	11·5
Ass	240	7·0
Mule	475	13·5

7.03 Economic factors

Costs of fuels vary considerably, both with time and space. The one unvarying factor is the *calorific value*, or equivalent, which gives the amount of useful energy that can be derived from a given mass or quantity of the fuel. Table XXVI gives calorific values for a number of fossil and organic fuels. This table also includes the sulphur contents, as high sulphur content may render a particular fuel unusable in certain areas.

Table XXVII gives the power output that can be derived from various animals at any moment. However, with animals (including man) the greater the power the shorter the time for which it is available. Normal daily energy outputs are also given in the table.

No figures are quoted for the various forms of water power, as these vary considerably with circumstances. Table XXVIII gives figures for available wind power in the United Kingdom.

Table XXVIII Wind power

Wheel diameter m	Power (wind velocity 6·7 m/s) W
1	52
2	127
3	410
4	940
5	1976

Formula for general use:

$$\text{Power in watts} = \frac{A\,V^3}{3{\cdot}855}$$

where A = area of sails in m^2

V = wind velocity in m/s

7.04 Solar power

While solar power itself is free, the capital costs of the installations is often so high as to render them uneconomic. However, even in the United Kingdom limited use of this energy source is now a fact. The power of the sun reaching the British Isles, **41.32**, is about 0·9 kW/m^2 at sea level midsummer midday. Table XXIX gives monthly and annual figures for Kew, for a number of different aspects of collector. Table XXX gives figures for different parts of the world.

7.05 Uses of solar power

By far the commonest use is to heat water. Figure **41.33** shows that the efficiencies of collectors decrease as the temperature rises, so that heating to a low temperature is most efficient. It has therefore been found an excellent method of heating the water for swimming pools, which do not need to exceed 30°C. Domestic water can be heated to 65°C, but will usually require some 'topping-up'.

8 CONTROLS

8.01 Types of controls

Controls are necessary to ensure efficiency and economy in environmental systems.

There are four categories of control depending on the factor that initiates further action:

- manual
- time operated
- internal condition operated
- external condition operated

8.02 Types of operation

The system of control usually operates to control the flow of fuel or heat transfer medium. There are two main types of such control:

- two position—the control is either on or off
- proportional—the position is relative to the actuating factor value.

8.03 Domestic heating controls

Domestic heating systems usually comprise control by both time and temperature. Time clocks are fitted to ensure that energy is not wasted during periods when the house is not occupied, or when the occupants are asleep. In addition, a room thermostat maintains a comfortable temperature, turning off the circulation pump or motorised valve when the temperature exceeds the set value. There are additional controls on the boiler itself which will turn it off as soon as the circulating water reaches the desired temperature, and these will be actuated when the circulation ceases.

The position of the room thermostat is highly critical. In a circulation area, it may cause the temperature in the living

Table XXX Solar radiation on horizontal surface for various places

Country	Annual total GJ/m^2	Daily average over the year MJ/m^2
Northern Europe	3·1–3·8	8·6–10·1
Mediterranean	5·6–8·0	15·5–21·6
Sahara desert	8·3	22·9
Central Africa	5·8–8·3	15·8–22·7
South Africa: Desert	9·1	25·2
Central high veldt	7·5	20·5
Coastal regions	6·1	16·7
India and Pakistan	6·8–8·3	18·7–22·7
Australia: Darwin	7·4	20·3
Alice Springs	8·0	22·0
Melbourne	5·4	14·8
Japan	4·2–5·6	11·7–15·1
United States: Northern states	5·0	13·9
Southern states	7·6	20·9
Canada: North	3·1	8·3
South	4·7	12·8

Table XXIX Monthly and annual totals of solar radiation (MJ/m^2) at Kew, England

Month		Facing south				Facing south-east or south-west				Facing east or west			
Inclination	horizontal	30°	45°	60°	vertical	30°	45°	60°	vertical	30°	45°	60°	vertical
January	65	90	104	104	101	79	86	90	79	61	61	54	47
February	104	137	144	140	126	126	130	122	101	97	97	86	72
March	241	295	310	295	241	277	277	266	216	230	223	202	155
April	338	360	346	324	238	349	338	313	248	317	299	274	205
May	482	479	454	414	288	479	454	414	302	454	425	382	292
June	529	515	475	425	281	508	479	439	306	493	457	414	313
July	490	475	450	400	266	479	450	407	292	457	428	389	288
August	425	436	421	378	277	428	414	378	277	400	374	338	256
September	306	349	349	335	266	331	328	313	248	288	266	248	198
October	180	234	248	245	212	216	223	216	176	173	162	151	122
November	86	119	130	137	126	104	112	115	101	79	76	72	54
December	50	86	94	101	97	72	76	79	72	50	50	47	36
Annual total	3298	3604	3557	3326	2520	3449	3366	3154	2419	3100	2858	2581	2038
Monthly average	274	299	295	277	209	288	281	263	202	259	238	216	169

Based on BRE data derived from Kew average solar irradiation data for 1959–68 and reproduced by permission of the Director. It has been assumed that diffuse radiation does not change significantly with orientation; so data for direct radiation at various orientations has been combined with Courtney's data on diffuse radiation. Diffuse radiation includes ground reflections.

areas to become uncomfortably high in some circumstances. When placed in the living area itself, this can lead to low temperatures elsewhere when the living area is well populated. Some systems employ thermostats in each room and area. These are normally thermostatic radiator valves that act to restrict the flow of the heating water. Alternatively, the external temperature can be measured and used to boost the heating when it falls. This type of system is usually completely independent of internal conditions.

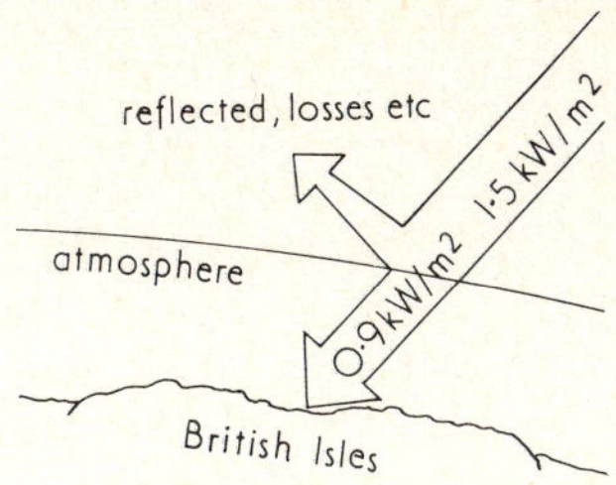

41.32 *Solar power available in the UK*

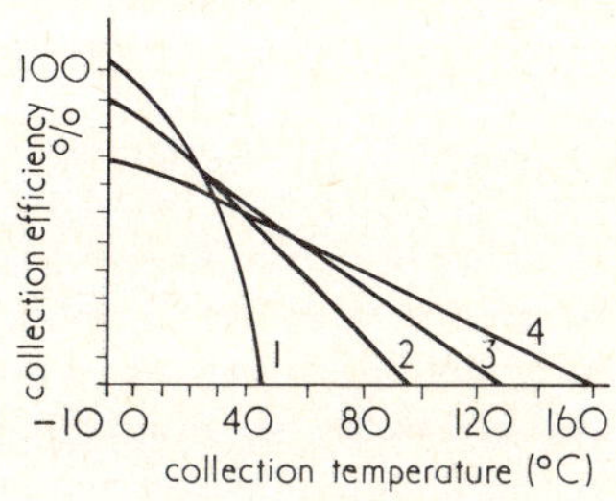

41.33 *Average outputs in the UK of four types of solar collector assuming ambient temperature about 10°C:*
curve 1—black plastic absorber uninsulated and unglazed
curve 2—black plastic absorber single glazed in insulated box
curve 3—good selective surface absorber single glazed in insulated box
curve 4—good selective surface absorber double glazed in insulated box

8.04 Frost-stats

In para **5.13** mention was made of thermal lag, and the encouragement this gives to condensation. To prevent this, many systems use a *frost-stat* control that overrides the time clock, etc when the temperature drops to near freezing point.

9 BIBLIOGRAPHY

Architects' Journal, Domestic heating, 17.8.77, 7.9.77 and 28.9.77
Architects' Journal, Handbook of Building Environment, published between 2.10.68 and 13.8.69
Architects' Journal, Solar Energy, 20.7.77
British Standard Code of Practice CP3: Chap II: 1970 *Thermal insulation in relation to the control of environment*
Brian Day and Peter Burberry, Condensation in buildings, *Architects' Journal* 19.5.71, 26.5.71, 2.6.71
Institution of Heating and Ventilating Engineers (now the Chartered Institution of Building Services) *IHVE Guide 1970*, London. New edition of this work being published in separate parts
D. Kut *Heating and hot water services in buildings*, Oxford, Pergamon, 1968
D. Kut, *Warm air heating*, Oxford, Pergamon, 1970

42 Tropical design

Martin Evans

CI/SfB (A3) (E7t)
CISfB (1976 revised) (A3) (H11)

Martin Evans is an architect in practice in the tropics

Contents

1 INTRODUCTION

The purpose of this section is to collect data and assemble checklists to assist readers designing for tropical conditions. There is a shortage of textbooks on tropical design both for designers trained and living in temperate climates, and for nationals of tropical climates who find they are aware of the problems, but do not have any realistic data.

This section concentrates on environmental comfort, giving general parameters, but these must be supplemented by local knowledge or research—the designer must ensure that he complies with local requirements, legislation and building controls, and considers the merits of traditional local solutions before adopting any of the recommendations made here.

2 CLIMATIC ZONES

2.01 Critical data

The main climatic zones of the tropics are shown in table I, together with the latitudes in which they are found, typical vegetation and climatic characteristics and the problems requiring special design solutions to ensure comfort.

The main climatic data required for tropical design will include the criteria listed in table II. Mean figures may be calculated over short periods (five years or more), but it can be dangerous to assume peaks from records of such a short period.

3 COMFORT AND REQUIREMENTS FOR COMFORT

3.01 Thermal comfort

Thermal comfort is dependent on temperature, humidity, radiation and air movement as well as type of activity, clothing and degree of acclimatisation. A guide to the range of bulb temperatures which are likely to be perceived as comfortable is given in table III.

These thermal comfort limits apply when there is no loss or gain of heat by radiation or air movement. The first step in the climatic design process is to compare the comfort limits with the meteorological data.

3.02 Meteorological data

Annual average temperature is the average of the mean monthly maximum and minimum temperatures. The average of the highest mean monthly maximum and lowest mean monthly minimum gives a close approximation.

A comparison of the monthly mean maximum temperature with the appropriate day comfort limit indicates whether days are hot, comfortable or cold.

Similarly a comparison of the mean monthly minimum temperature with the appropriate high comfort limits indicates if nights are hot, comfortable or cold. The degree of thermal stress can be used to indicate the design features required to achieve comfort during a particular month. These indicators are shown in table IV. The total number of indicators for each month for the whole year can be used in table V to find the requirements for plan form and building elements. This method takes into account the possibility of conflicting requirements over the year and the requirements shown in table V are optimised over the whole year. Some flexibility in interpretation may be required in borderline cases.

4 THERMAL PERFORMANCE REQUIREMENTS

4.01 Parameters

The thermal characteristics for walls and roofs in tropical climates should be defined not only by the insulation they provide (ie U value or air-to-air transmittance) but also by their ability to reflect solar radiation and to delay the flow of heat through the construction.

In addition to the U value, two other useful parameters of thermal performance are:

- *solar heat factor* The proportion of incident solar radiation transmitted through a construction
- *time lag* The response of a construction to temperature change. The requirements for U value, solar heat factor and time lag are given in table VI which uses the indicators from table IV. The following paragraphs explain the solar heat factor and time lag in greater detail.

4.02 Sol air temperature

Sol air temperature is defined as the temperature of the outside air which would give the same rate of heat transfer and the same distribution of temperature through a construction as the combined effects of solar radiation and air temperature. The sol air temperature will be higher than the air temperature when a surface is subject to solar radiation:

$$\theta sa = \frac{\alpha I}{fo} + \theta o$$

Where θsa = sol air temperature (°C)
α = absorptivity of surface to solar radiation
fo = outside surface conductance (W/m^2K)
I = intensity of solar radiation (W/m^2)
θo = outside air temperature (°C).

4.03 Solar heat factor

The solar heat factor is defined as the heat flow through the construction due to solar radiation expressed as a proportion of the total solar radiation incident on the surface of the construction. When a building has large openings and is well ventilated to the exterior (as is often the case in the tropics),

Table I Occurrence and characteristics of main climatic zones

Zone	Approximate latitude range	Natural vegetation	Typical cultivation	Climate	Problems	Requirements
Warm humid equatorial	7½N–7½S	Tropical rain forest	Banana, palm oil	Warm with high humidity and rainfall	Humidity prevents sweat evaporation; hot nights make sleep difficult; high rainfall and glare from overcast sky, sun on east and west facades	Air movement from fans or cross ventilation, low thermal capacity construction, sloping roofs and large overhangs, windows facing north and south
Tropical island	5–30°N 5–30°S	Rain forest	Sugar cane	Warm, humid but less cloud than warm humid zone	Similar to warm humid equatorial, but clear skies and bright sun more frequent	Similar to warm humid but with additional care in the design of shading the south facing windows in the northern hemisphere (vice versa in the southern)
Hot dry tropical	15–32°N 15–32°S	Desert, steppe	Palms, grazing (nomadic)	Hot and dry with high annual and daily variation of temperature	High diurnal range, very hot days in summer, cool winter days, low rainfall, very strong solar radiation and ground glare, sandy and dusty environment	High heat capacity construction, shading devices which allow solar heating in winter, small windows, flat roofs (often used for sleeping), small courtyards to give shade and protection
Maritime desert	15–30°N 15–30°S	Desert	Palms, grazing	Hot, humid with low rainfall	Similar to hot dry climates but with higher humidity causing discomfort by preventing sweat evaporation	Similar to hot dry but air movement is desirable at times
Intermediate composite or monsoon	5–20°N 5–20°S	Monsoon forest, dry tropical forest or scrub, savannah	Paddy rice, sugar cane, millet	Warm humid and hot dry seasons often with cool season	Combines the problems of warm humid and hot dry climate	Compromise between the requirements of warm humid and hot dry climate or ideally (but more expensively) two buildings or parts of buildings for use at different times of the year
Equatorial upland	10°N–10°S	Broadleaf forest, mountain vegetation	Millet	Temperate to cool depending on altitude	Combine the problems of warm humid and hot dry climates with those of a temperate or cold climate for all or part of the year	Design to take advantage of solar radiation when cool or cold. Heating and additional insulation may be required
Tropical upland	10–30°N 10–30°S	Steppe, cedars	Wheat	Hot summers, cold winters	Do	Do
Mediterranean	32–45°N 32–45°S	Mediterranean scrub	Vines, olives, citrus fruits	Hot dry summers, cool wet winters	Summers have some of the problems of a hot dry climate while winters are cold and humid with moderate rainfall	Design with high thermal capacity, medium to small openings, and courtyards to give shade and protection

Table II Climatic data*

Data required	Units	Relevance
Monthly mean max temperature	°C	thermal comfort analysis
Monthly mean min temperature	°C	
Monthly mean max humidity	%	
Monthly mean min humidity	%	
Monthly mean rainfall	mm	vegetation
Peak rainfall intensity and duration	mm/unit of time	storm damage
(Daily or hourly rainfall may be the only data available)	mm	rainwater drainage
Sunlight	hours	natural lighting
Cloud cover	oktas† or %	
Absolute max temperature	°C	thermal expansion
Absolute min temperature	°C	and effect on building materials
Frequency distribution of wind for different speeds and directions	% m/sec	siting and orientation
Frequency of special phenomena, ie sandstorms, fog, hail, thunder	days per year	provision of special precautions

* For a more detailed list see CIB Bulletin 3:1966
† 1 okta = 1 eighth of the sky

Table III Thermal comfort limits (°C)

Monthly average relative humidity	Annual average temperature over 20°C Day	Night	15–20°C Day	Night	under 15°C Day	Night
0–30%	26–34	17–25	23–32	15–23	21–30	14–21
30–50%	25–31	17–24	22–30	15–22	21–27	14–20
50–70%	23–29	17–23	21–28	15–21	19–26	14–19
70–100%	22–27	17–21	20–25	15–20	18–24	14–18

the solar heat factor is dependent on the U value and absorptivity the external surface conductance ('fo' can be taken as a constant):

$$\frac{q}{I} = \frac{U\alpha}{fo}$$

Surfaces heated by solar radiation will get hottest when wind velocities are low. Therefore external surface resistances for cold conditions should not be used. In hot conditions with low wind speeds a recommended value for fo is 20 W/(m^{2}K). If the solar heat factor is expressed as a percentage then:

$$\frac{q}{I} = \frac{U\alpha}{20} \times 100 \text{ or } 5\,U\alpha$$

Table IV Indicators of requirements for comfort for each month

Humid indicators		
H1	Air movement essential	mean monthly maximum temperature above the day comfort limits combined with humidity over 70% or humidity between 30–70% and a diurnal range of less than 10°C
H2	Air movement desirable	mean monthly maximum temperatures within the comfort limits combined with humidities over 70%
Arid indicators		
A1	Thermal storage required	diurnal range of temperatures over 10°C and humidity less than 70%
A2	Space required for outdoor sleeping	mean monthly minimum temperatures above the night comfort limits and humidity below 50%. Outdoor sleeping may also be indicated where maximum temperatures are above the day comfort limits and diurnal range is above 10°C with humidities less than 50%
Cold indicators		
C1	Solar radiation desirable	mean monthly maximum temperatures below day comfort limits
C2	Additional heating required	mean monthly maximum temperature below 15°C

Table V Building and planning requirements*

Requirements	Indicators (from table IV)
Layout alternatives	
1 Buildings orientated on E–W axis	A1 for up to 10 months and C1 for up to 2 months and/or H1 for over 6 months
2 Compact courtyard planning	A1 for 11 or 12 months and C1 for up to 4 months
3 Compact forms	All other cases
Spacing for breeze	
1 Open spacing of 5 times building height for breeze penetration (see **9**)	H1 for 11 or 12 months
2 Open spacing with some precautions against cold or hot dry winds	H1 for between 2–10 months
3 Compact planning	H1 for one month or less
Spacing for solar radiation	
1 Spacing not dependent on solar radiation	C1 for up to 3 months
2 Spacing to allow solar radiation (but high sun angles may still allow close spacing)	C1 for 4 months or more
Building form for air movement	
1 Rooms single banked with permanent provision for cross ventilation	H1 for over 3 months; or H1 for one or two months and A1 for 5 months or less
2 Cross ventilation not essential	Never H1 and H2 for one month or less
3 Double banked rooms with temporary provision for cross ventilation	All other cases
Outdoor sleeping	
Space required for outdoor sleeping	A2 for two months or more
Openings	
1 Large openings 40–80% of north and south walls	A1 one month or less and C1 never
2 Medium sized openings 20–40% of area of external walls	A1 one month or less and C1 one month or more or A1 for 2–5 months
3 Composite 15–30% of area of external walls	A1 for between 6 and 10 months
4 Small 10–20% of external wall area	A1 for 11 or 12 months C1 for up to 3 months
5 Medium 15–25% of external wall area	A1 for 11 or 12 months C1 for 4 months or over

* This table is intended for housing or similar buildings

U values may be increased if absorptivities are proportionately reduced, while still maintaining a constant solar heat factor. Most reflective surfaces require good maintenance if they are to remain effective. In many situations it is not prudent to rely on maintenance or repainting of surfaces and therefore U values should be decreased to obtain the required thermal performance standard.

Table VI Thermal performance recommendations

Structure	Performance	Indicator (table IV)
External walls		
1 Light walls	U value: 2·8 W/(m²K) max Solar heat factor: 4% max Time lag: 3 hours max	A1 for up to 2 months
2 Heavy walls	U value: 2·0 W/(m²K) max Solar heat factor: 4% max Time lag: 8 hours minimum	A1 for 3 months or more
Roofs		
1 Light roof	U value: 1·1 W/(m²K) max Solar heat factor: 4% max Time lag: 3 hours max	H1 for 10 months or over and A1 for up to 2 months
2 Heavy roofs	U value: 0·85 W/(m²K) max Solar heat factor: 3% max Time lag: 8 hours minimum	H1 for 9 months or less and A1 for 6 months and more
3 Well insulated roofs	U value: 0·85 W/(m²K) max Solar heat factor: 3% max Time lag: 3 hours max	All other cases

A solar heat factor of less than 4 per cent will ensure that ceiling temperatures will not be more than 5°C above air temperatures and will not add to discomfort.

4.04 Solar radiation

Wide variations may occur according to the colour and weathering of the surface. Surfaces exposed to solar radiation should have low absorptivities and high emissivities to reflect solar radiation and to re-radiate absorbed solar radiation. Surfaces facing cavities, however, will transmit less heat by radiation across the cavity if the 'low temperature' emissivity is low. Table VII gives the percentage absorptivity and emissivity of surfaces.

Table VII Percentage absorptivity and emissivity of surfaces

Material of surface	Absorptivity of solar radiation (Short wave 0.3–2.5 microns)	Emissivity of low temperature radiation (Long wave 5–20 microns)
Aluminium polished	5–15	5–10
Aluminium weathered	15–40	10–15
Whitewash new	10–15	80–90
Whitewash weathered	20–30	80–90
White paint	25–30	80–90
Aluminium paint	55	55
Pinewood	40	95
Asbestos cement weathered	70	95
Galvanised iron: new	25	25
Galvanised iron: rusty	90	28
White marble	50–60	95
Red brick	75–85	65–80
Black non-metalic surface ie bitumen	85–90	65–80

For all materials:
reflectivity = 1−absorptivity (for radiation of a given wavelength)
absorptivity = emissivity (for radiation of a given wavelength)

The approximate absorptivity of solar radiation of paints can be calculated if the Munsell value is known (as it is for colours in the BS range of paints for building purposes). The 'value' of the colour is given by the number which appears after the 'hue' letter in the Munsell number. This should be substituted for V in the formula:

Absorptivity = 100 − [V (V − 1)] (for solar radiation)

Example: Munsell number 6·25Y8·5/13 (Yellow)
V = 8·5 Absorptivity = 36 per cent.

At low temperatures most paints have an emissivity of 80–90 per cent.

4.05 Time lag

Time lag is defined as the phase difference (delay) between external periodic variations in temperatures and the resulting internal temperature variations. The period of variations is 24 hours and the lag is measured in hours. Table VIII gives time lag for homogeneous materials.

For non-homogeneous construction, the order in which different layers are placed can change the time lag considerably. If insulation is placed on the external surface of a dense material, the time lag is considerably increased. Table IX gives the time lag of constructions

Although the time lag indicates when the thermal impact of outside temperature swings will affect the interior, the actual internal conditions can only be calculated when the heat flow into and through all room surfaces is considered.

Table VIII Time lag for homogeneous materials (in hours)

Materials		**Thickness of material (mm)** 25	50	100	150	200	300
Dense concrete	min	—	1·5	3·0	4·4	6·1	9·2
	max	—	1·1	2·5	3·8	4·9	7·6
Brick	min	—	—	2·3	—	5·5	8·5
	max	—	—	3·2	—	6·6	10
Wood	min	0·4	1·3	3·0	—	—	—
	max	0·5	1·7	3·5	—	—	—
Fibre insulating board	ave	0·27	0·77	2·7	5·0	—	—
Concrete with foamed slag aggregate	ave	—	—	3·25	—	8	—
Stone	ave	—	—	—	—	5·5	8·0
Stabilised soil	ave	—	—	2·4	4·0	5·2	8·1

Table IX Time lag for composite roof constructions

Construction (described from the external surface inwards)		**Time lag (hours)**
40 mm	Mineral wool	
100 mm	Concrete	11·8
100 mm	Concrete	3
40 mm	Mineral wool	(same as concrete alone)
14 mm	Cement plaster	
165 mm	Concrete	
14 mm	Cement plaster	3·8
14 mm	Cement plaster	
50 mm	Vermiculite concrete	
115 mm	Concrete	
14 mm	Cement plaster	13
Any finish		
25 mm	Expanded polystyrene	
any structural concrete slab		over 8
Any finish		
75 mm	Lightweight concrete screed	
100 mm	Concrete slab	over 8
30 mm	Concrete tiles	
20 mm	Mortar bed waterproof membrane	
60 mm	Screed	
240 mm	Hollow pot slab	
14 mm	Render	12

5 SOLAR RADIATION

5.01 Criteria

The intensity of solar radiation on a surface depends on the altitude of the sun, the orientation of the surface in relation to the sun and the absorption of solar radiation by the atmosphere, pollution, cloud etc, **42.1**.

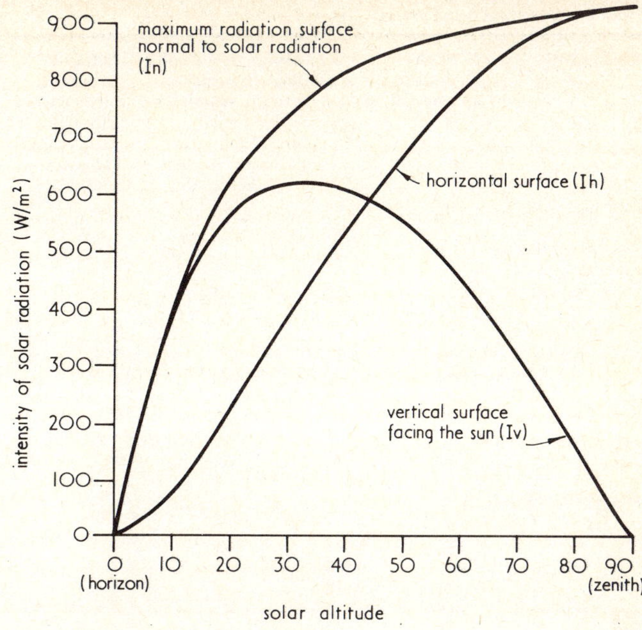

42.1 *Intensity of solar radiation*

5.02 Calculation

For vertical surfaces inclined at an angle θ to the azimuth (horizontal angle of sun on plan) the intensity of radiation on surface I will be:

$$Iv\theta = Iv \times \cos\theta$$

where Iv is taken from **42.1**.

The altitude and azimuth of the sun can be found from sunpath diagrams. A list of sources of these appears below.

The radiation from **42.1** should be multiplied by the values from tables X and XI to give total radiation at the appropriate altitude and/or for appropriate atmospheric conditions.

Table X Increase in solar radiation with altitude

Height above sea level	**Altitude of sun in degrees** 20°	30°	40°	60°	80°
900	1·14	1·12	1·10	1·08	1·08
1500	1·26	1·20	1·17	1·15	1·15
3000	1·30	1·31	1·28	1·25	1·23

Table XI Effect of cloud and atmospheric pollution on radiation*

Very low humidities and clear skies	1·1 (increase)
High humidities and pollution 'clear sky'	0·9 (decrease)
Overcast sky	0·1–0·3 (decrease)

* Varies greatly with cloud and solar altitude

6 SHADING AND GLAZING

6.01 Solar control glass

Special glasses for solar control absorb a considerable proportion of solar radiation. This heats the glass and special precautions must be taken to avoid problems resulting from thermal movement. About a third of the absorbed heat is transmitted indirectly to the interior and two-thirds to the exterior, so the total proportion of transmitted solar radiation may be considerably greater than the directly transmitted solar radiation. The high temperature of the glass may also cause discomfort. For comparison, a sheet of aluminium or an uninsulated concrete slab has been shown in table XII. Almost all special glasses will get as hot as these materials.

Table XII Solar heat gains through glass*

Fenestration	Visible radiation transmitted %	Direct solar radiation transmitted %	Total solar radiation transmitted %	Index of increase of surface temp above air temp (Clear glass = 1)
Clear float glass	85	80	84	1
Glass with reflective polyester film	18	17	25	2
	33	31	39	2
Solar energy reflecting glass	42	47	52	2·5
	58	59	62	4
Surface modified heat absorbing glass	50	56	67	3
	50	48	62	4
Tinted solar control glass				
grey	42	45	62	4
	42	40	58	4
	24	22	47	4
	19	16	43	4·5
green	76	52	66	4
	74	45	61	4
	62	30	51	4
Clear glass with open weave curtain internally	40	70	82	1
Corrugated aluminium (new)	0	0	9	2
100 mm concrete	0	0	15	4

Sources: Manufacturers' data; heat gains through fenestration F. J. Lotz and J. F. van Straaten, CSIR: R/Bov 223.
* Ranges of products are given. Consult manufacturers for specific data.

6.02 Reducing solar gain

The best way to reduce solar radiation heat gain is to reduce window size (or provide external shading) for windows receiving direct solar radiation, though precautions may be required against glare from bright cloudy sky or the ground.

6.03 External shading

If a view is not required a coat of white paint on glass will give reduced light and solar radiation transmission at considerably less cost than special glasses. The best way to control solar radiation, however, is to use external shading design with the aid of a sun path diagram. When the sunpath diagrams for the northern hemisphere are used for the southern hemisphere, changes should be made to the time, month, azimuth, direction as shown in table XIV. Shading coefficients are given in table XIII.

Internal shades are comparable with special glasses. Most forms of external shading are better than special glasses or internal shades.

Table XIII Shading coefficients: the quantity of solar radiation transmitted as a proportion of that transmitted through clear glass

Fenestration	Shading co-efficient
Clear 6 mm glass	1·00
Glass with internal dark roller blind	0·70–0·80
Glass with internal dark venetian blind	0·75
Glass with internal medium venetian blind	0·55–0·65
Glass with internal white venetian blind	0·45–0·55
Glass with external miniature louvres	0·50–0·10 (depends on angle of incidence)
Glass with dark canvas external awning	0·20–0·28
Glass with dense trees providing shade	0·20–0·30
Glass with movable louvres	0·10–0·20
Heat absorbing glasses	0·45–0·80

Sources for sun path diagrams

1 *Manual of tropical housing and building* Part 1. Climate design. Koenigsberger, Ingersol, Mayhew, Szokolay. Longmans. 1974.

2 Sharma, M. R. and Rao, K. R. *Solar radiation protractors* Central Building Research Institute. Rorkee, India. Equidistant projection 15°N–15°S; 15°N–35°N at 5 degree intervals (reproduced in Givoni *Man, climate and architecture* Elsevier 1969).

3 *Solar charts and shadow angle protractor for daylight planning* Catalogue no 374. Henry Hope & Sons. London. 1969. Stereographic projection 32°N–28°S at 4 degree intervals.

4 Richards, S. J. *South African architectural record* Vol 36 No 11. Stereographic projection 20°S–34°S at 2 degree intervals (reprinted by South African Council for Scientific and Industrial Research, Pretoria 1952).

5 AJ Handbook of Building Environment. Information Sheet—Sunlight 5, 30.10.68 pp 1024–1035. Gnomic projection 0–60° (N or S) at 2 degree intervals.

All sunpath diagrams for the northern hemisphere can be used for the southern hemisphere by reversing the hours and months and rotating the azimuth scale by 180°.

Table XIV Changes for sunpath diagrams in southern latitudes

Time (solar time) North	→South	Date North	→South	Azimuth degrees clockwise North	→South	Direction North→South
4	20	28 Jan	30 July	0	180	
5	19	28 Feb	30 Aug	30	210	
6	18	21 Mar	23 Sept	60	240	North–South
7	17	15 April	15 Oct	90	270	
,,	,,	15 May	15 Nov	120	300	
11	13	22 June	22 Dec	150	330	East–West
12	12	30 July	28 Jan	180	360	
13	11	30 Aug	28 Feb	210	30	
,,	,,	23 Sept	21 Mar	240	60	South–North
17	7	15 Oct	15 April	270	90	
18	6	15 Nov	15 May	300	120	
19	5	22 Dec	22 June	330	150	West–East
20	4			360	180	

7 AIR MOVEMENT AND VENTILATION

7.01 Wind velocities

The conversion of wind velocities from imperial to metric with the Beaufort scale and the corresponding wind pressures on a flat plate normal to the wind are given in **42.2**.

7.02 Urban areas

Wind velocities in urban or suburban areas may be estimated in relation to wind velocities in open country, **42.3**.

7.03 Effect of wind speed

Table XV shows the effect of different wind speeds in warm humid climates.

Table XV Effect of internal wind speeds in warm humid climates

Range of speeds m/min	Effect
0–15	Not noticeable, less than 1 °C of apparent cooling as air passes over skin
15–30	Just noticeable cooling effect equivalent to 1–2 °C
30–60	Effective and pleasant cooling effect
60–100	Maximum windspeed for cooling without undesirable side effects
100–200	Too fast for desk work; papers start to blow around
over 200	Too fast and uncomfortable for internal conditions.

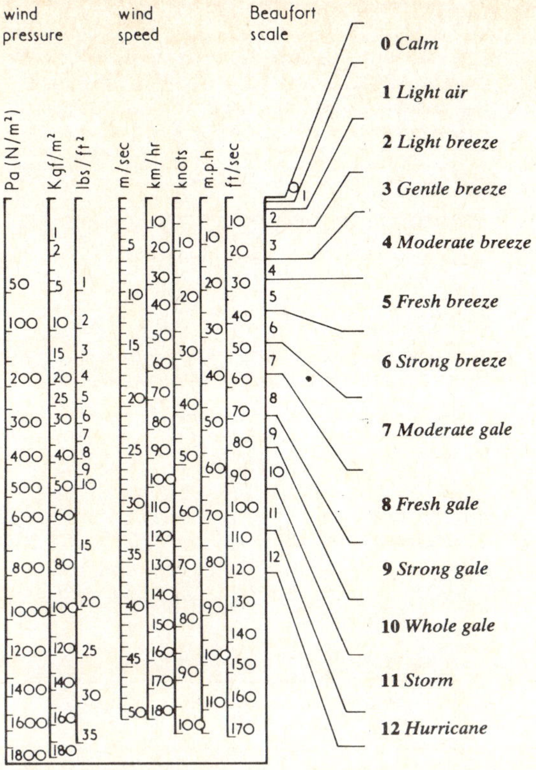

Key
0 *Calm* Smoke rises vertically. **1** *Light air* Direction shown by smoke but not wind vanes. **2** *Light breeze* Wind felt on face, leaves rustle. Vanes moved by wind. **3** *Gentle breeze* Wind extends light flags. Leaves in constant motion. **4** *Moderate breeze* Raises dust and loose paper. Small branches are moved. Onset of mechanical discomfort. **5** *Fresh breeze* Small trees in leaf begin to sway. Uncomfortable in urban areas. **6** *Strong breeze* Large branches in motion, telegraph wires whistle. Umbrellas difficult to use. **7** *Moderate gale* Whole trees in motion. Inconvenient to walk against wind. **8** *Fresh gale* Breaks twigs off trees, generally slows down walking. **9** *Strong gale* Slight structural damage occurs, tiles and slates dislodged. **10** *Whole gale* Seldom experienced inland. **11** *Storm* Trees uprooted. **12** *Hurricane* Rarely experienced. Accompanied by widespread damage.

42.2 *Conversion chart for wind velocities and the corresponding wind pressures. Reference: figure* 13, *BRS Digest no* 101. *January* 1969. *HMSO*

7.04 Wind shadow

The length of the wind shadow for various shapes of building is given in **42.4**.

Example: A two storey building 6 m high, 3 m wide and 24 m long (A = 3 m). The wind shadow will be $11\frac{3}{4} \times A$, that is 35 m long.

The variations in wind direction will alter the direction of the wind shadow, and allowance must be made for these variations. As a rough guide the wind shadow will be 5 × height of building (including the pitched roof).

8 MECHANICAL AIDS TO COMFORT

8.01 Ceiling fans

Ceiling fans give a wide distribution of blown air, **42.5**. Since the diameter of the fan is large, the fan can have a relatively low speed, reducing noise. Typical installation:

ceiling height	3·0 m min
blade height	2·5 m
fan diameter	1·0 m

8.02 Wall-mounted fans

These give higher air speeds and a more concentrated air stream, **42.5b**. They are often arranged to swing from side to side. Typical data:

mounting height	1·5–2·0 m
angle of swing	up to 60°
'reach' of blown air	3–4 m

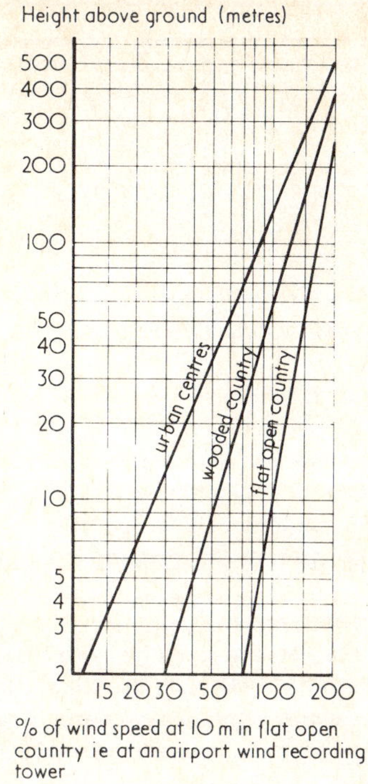

42.3 *Wind speed may be interpolated from the norm (*10*m height in flat open country) by reading across from the height in metres to the appropriate diagonal line for the types of surroundings (eg, if average velocity at the airport is* 3*m/sec, at the top of a building in an urban centre,* 35*m high, with a wind speed of* 50 *per cent, average speed will be* 1·5 *m/sec). Reference: figure* 5, *Paper* 2, *'Wind effects on building and structure'. HMSO,* 1965

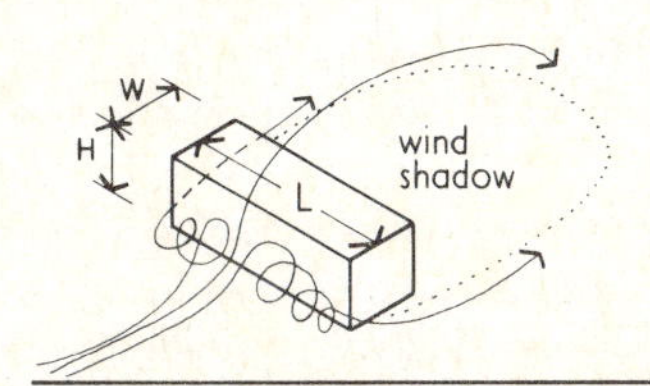

Wind shadow length chart

Section / Wind direction →	Length of building L: 2A	4A	8A	16A	24A
A x A	2½	3¾	5¼	8	8¾
H=1 W=2A	2	2¾	3¾	6	7
H=1 W=3A	2¼	3¼	4¼	5¾	5½
H=2A	5¼	8¼	11¾	16¼	18
H=3A	6¾	11½	16½	18¾	20¾
H=1 roof=45°	2¾	5¼	9¼	13¼	15
H=1 roof=30°	3	4	6¾	10	13
H=1 roof=15°	3	5¼	8¼	11½	14½
H=1 roof=15°	2½	4¼	6½	11	13¾

42.4 *Wind shadow length chart. Reference: B. H. Evans, Research report* 59. 1957. *Texas Engineering Station. See also Overseas Building Note* 112

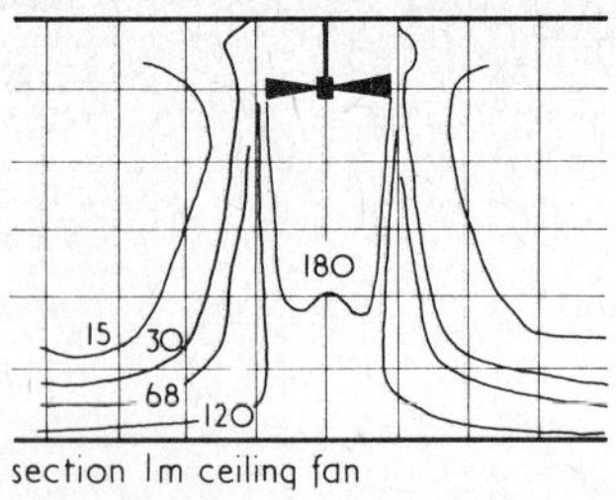

42.5a *Ceiling fans are fairly slow moving and quiet. Blades must be higher than an outstretched arm (2·4m)*

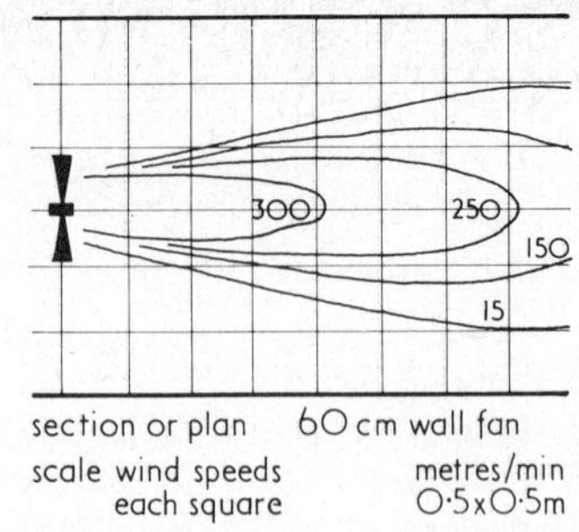

42.5b *Wall fans rotate at speed and may be noisy. Those swinging from side to side cover a larger area with less regular air movement. Reference: L. G. Wood, 'Performance of ceiling and desk fans'. Electrical Energy. March* 1957

8.03 Roof extract fans

With light roof constructions, extract fans are used to remove hot air from the ceiling void. Internal air temperatures may be reduced by 3 to 5°C in climates where high solar radiation is combined with moderate air temperatures. The extract fan does not give perceptible air movement within the building.

8.04 Unit air conditioners

Room air conditioners are available with a cooling capacity of from about 1·5 to 7 kW (Btus/hr are commonly used to measure capacity as many manufacturers and designs originate in the USA). Specifications vary considerably and may include additional features such as heaters. They are usually accommodated in walls or under windows (often as afterthoughts). Table XVI shows typical dimensions for a range of cooling capacities based on six widely available makes. Unit air conditioners require external air for removing excess heat, fresh air inlet and drainage for water removed from the air during cooling, **42.6**.

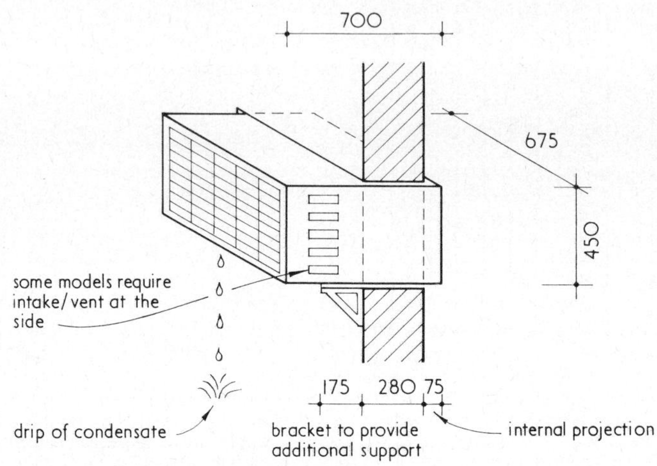

42.6 *A typical air conditioning unit*

8.05 Split air conditioning units

Split air conditioning units have the following advantages over unit (or 'room') air conditioners:

- minimal structural alterations if installed in an existing building or minimal requirements for new buildings: the two pipes connecting the condenser to the air handler will fit through a 100 mm hole
- greater security; important for banks, shops, etc
- reduced noise due to external position of condenser
- greater flexibility internally as air handler may be placed on internal wall or even ceiling
- improved external appearance as condenser may be placed on the roof, reducing problems of dripping condensate.

The air cooler and fan are not usually designed to provide ventilation by the exchange of internal air for external air.

8.06 Traditional aids to comfort

There are many traditional aids to comfort which may still be cheaper and as effective as mechanical aids. These include:

Cooling towers Towers which catch the breeze and bring it down a vertical shaft into the interior of the building, often passing an earthenware water jar which cools and humidifies the air. They are used in desert and maritime desert climates.

Air coolers Water is thrown at or dripped onto brushwood branches hung over a window to cool the air and filter out the dust. A modern version is the desert cooler—an electric fan

Table XVI Unit air conditioners: typical data

Capacity† BTUs per hour (or kilojoules/hour)	Capacity kW	Height mm	Width mm	Depth mm	Weight kg	Air handled m^3/min	Moisture removed litres/hour
Domestic and small rooms							
5 000	1·47	320	485	490	24	4·6	0·6
6 500	1·90	360	500	360–500	42–47	5·4	0·7
7 200	2·11	360	500	360–500	42–47	5·4	0·8
7 600	2·23	379	618	587	57	8·3	0·9
Small shops, small to medium offices, small apartments							
9 000	2·64	367–380	618	570	53	8·3	1·1
9 300	2·73	367–380	560–620	517–586	47–60	8·0–8·3	0·9–1·2
10 500	3·08	367–380	615	570	55	9·0–10·5	1·5
12 000	3·52	380–460	650–677	609–700	65–78	9·5–10·5	1·3–1·5
13 600	4·00	380–460	615	570	54–58	10·5	1·9
15 000	4·40	380–470	650–677	630–750	73–80	11·7	1·9
16 000	4·67	380–470	615	570–700	55–59	11·7	2·4
16 500		400–470	660	700	90–92	11·4–11·8	1·6–1·8
Larger offices, large apartments							
17 000	5·00	450	600	700–792	90–92	8·7–12·8	1·9–2·3
18 000	5·28	380	615	570	69	11	2·7
20 000	5·86	535	693	783	108	18·5	2·8
23 500	6·89	452	660	792	107	20·0	3·0

* Sources: manufacturers' data   † 1 BTU = 1·055 kJ

which blows air over a damp fabric cooling and humidifying as it passes.

Planting Many plants, climbers and trees provide shade during hot summer months and drop their leaves during the cool winter allowing the sun to provide welcome heating; an automatic adjustable shading device used in composite, desert, mediterranean and even temperate climates.
It should be realised, however, that some traditional aids have undesirable side effects (such as attracting as harbouring insects), high capital cost, as well as being regarded as symbols of the past (like thatched roofs in Northern Europe).

9 ILLUMINATION FROM SUN AND SKY IN THE TROPICS

9.01 Illumination at work surfaces

The illumination required at the work surface for a given task is the same regardless of latitude. However, since light is associated with heat (both physically and psychologically), there is a case for adopting slightly lower lighting standards to achieve higher levels of thermal comfort. In some countries, the cost of achieving high lighting standards may also be a factor.
The illumination from the sky is greater in tropical and subtropical regions so that a lower daylight factor can be used to achieve the same illumination at the work surface. The illumination from an overcast sky varies with latitude, altitude, degree of cloudiness and pollution. A guide is given in table XVII.

Table XVII Illumination from a design sky

Latitude (N or S)	Design sky illumination
0°	17 000 lux
10°	15 000 lux
20°	13 000 lux
30°	9 000 lux
40°	6 000 lux
50°	5 000 lux

Table XVIII Sun, sky and ground brightness

Hot dry desert conditions	Sky (away from sun)	3 000 lux
	Sun	50 000 lux
	Ground (20% reflectivity)	11 000 lux
Warm humid equatorial conditions	Sky (overcast)	10 000 lux
	Ground (20% reflectivity)	2 000 lux

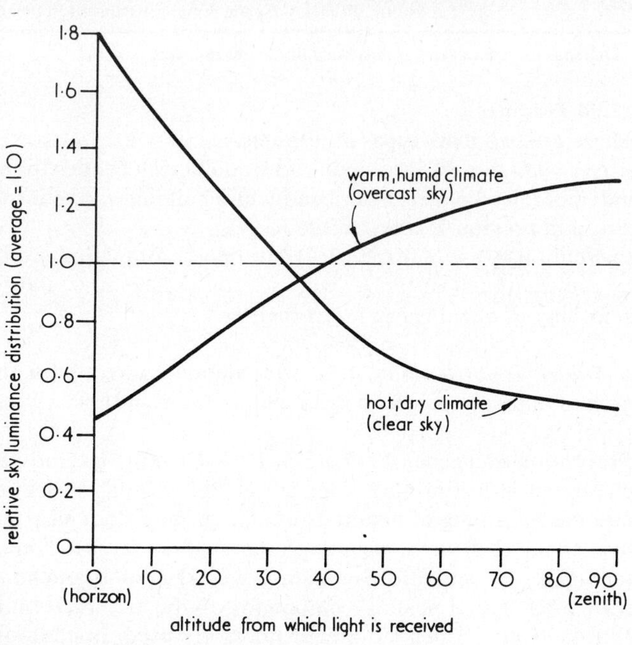

42.7 *Sky luminance distribution*

9.02 Position of windows

With an overcast sky, the sky below 35° has below average brightness and windows which allow a view of this part of the sky will avoid glare from the brighter overcast sky at higher angular altitudes, **42.7**.
Conversely, in hot dry conditions windows should allow a view of the deep blue sky at high altitudes rather than the horizon, as this may be a source of glare from haze on the horizon (and glare from the sunlit ground). In this case, the window must be protected from direct solar radiation.
In the equatorial regions, however, the warm humid climate is often associated with a bright but constant overcast sky with a luminance which may drop to 10 000 lux or below. The hot dry desert regions receive light mainly from direct sunlight and considerably less from the (usually) deep blue sky.

9.03 Incidence of light

Table XVIII gives a very rough comparison of the relative amount of light from different sources.
Figure **42.7** shows the relative sky luminance distribution in hot dry desert climates with a clear sky, and for warm humid climates with an overcast sky. (It is assumed that the warm, humid climate has a similar distribution to the CIE standard overcast sky.)

10 PUBLIC UTILITIES IN THE TROPICS

Principles of design of public utilities are similar to those used in temperate regions but there are a number of differences due to climate, economic conditions and social factors, some of which are mentioned below.

10.01 Water supply

Table XIX gives standards which have been recommended by international organisations. However, the standards actually adopted must be related to local resources and conditions. In some countries, brackish water for irrigation is supplied by an independent system to conserve drinking water. Table XX shows the large amounts of water required to maintain lawns and western style gardens.

Table XIX Daily domestic water supply standards (litres per capita)*

Distribution	Minimun or temporary	Normal	With wastage allowance
Standpipe for up to 100 persons	20	40	60
Single tap connection	120	160	180
Multiple tap connection	160	200	240
Multiple tap connection in areas of water shortage	100	150	—

* Sources WHO, World Bank

Table XX Water supply for irrigation in hot dry climates

Type of vegetation	Water supply requirement	
Private gardens	maximum	350 000 litres/hectare/day
	average	225 000 do
Private gardens without grass		170 000 do
Irrigated vegetables	average	80 000 do
Public parks		60–140 000 do
Tree plantations		2–7 000 do

10.02 Foul sewerage

The capacity of piped systems will be related to the water supply standard. Where sewers are laid with low falls and where high soil temperatures exist, sewage may become septic and attack asbestos cement and cement pipes. Sewage disposal may be by pit latrine, aqua privy or septic tank in low density development. Pit latrines require adequate space for replacements after the average life of five years, although this depends on soil conditions etc. They should be at least 7 m away from any building.

10.03 Surface water drainage

In many parts of the world there is inadequate data to establish design rainfall intensities. In some areas of high rainfall, open or openable monsoon drains are required to provide adequate capacity and ease of cleaning.
Where sand may enter piped systems, large sand traps should be provided at gulleys—150 mm minimum pipe size should be used (some authorities require 200 mm minimum) and gradients should achieve self cleansing velocities of at least 1 m per second. Hot dry desert conditions may pose severe rainwater drainage problems as rainfall, though low on average and infrequent, can occasionally be very heavy.

10.04 Electricity

Standards for voltage, demand, etc, vary widely. The design of the distribution system is as for temperate climates except that high soil temperatures may lead to a requirement for cable sizes to be larger to avoid over-heating.

11 HOUSING STANDARDS

Housing standards vary widely. Some may be unrealistically high in relation to national or family income. Others may be considered too low in relation to basic needs. Tables XXI and XXII give some guide to the range of plot and dwelling size. Table XXIII gives requirements affecting ceiling height.

12 HAZARDS

12.01 Earthquakes

Data on the location and severity of significant earthquakes between 1900 and the present day for all tropical regions can be obtained from the Institute of Geological Sciences, Edinburgh and other centres. The maps give data on date, magnitude (on the Richter scale) and depth of epicentre.

Table XXI Plot sizes and densities

Type of plot	Minimum frontage: m	Size: m²	Minimum net density: plots/ha
Site and service projects:			
minimum	5	60	100
recommended minimum	7·5	100	66
(USAID and UNRWA):			
low density	—	200	33
very low density	—	400	18
OPEC country standard:			
minimum	9	200	25
normal minimum	12	300	17
medium density	—	400	14
low density	—	1000	7

Table XXII Dwelling size

Type of dwelling	Area m² low	medium	high*
Core house (for self-build expansion)	8	18	—
Core house with one room	20	30	—
Bachelor units	22	—	—
Family dwellings:			
2 room units	36	50	—
3 room units	52	65	82
4 room units	65	85	100
5 room units	78	100	120
6 room units	—	—	140

* High, medium and low refers to the GNP of the countries selected

Table XXIII Floor to ceiling height*

To allow for ceiling beyond reach (in some regions 95 percentile reach is less than 2·264 found in Britain)	2·300 (ISO recommendation)
To allow for floor to ceiling as multiple of 200 (concrete block) or 300 (less usual vertical module)	2·400 (ISO recommendation)
To allow for ceiling mounted fan (2·400 + fan) (avoid mounting below light to avoid flicker)	3·000
Common requirement in tropical regions	3·000

* High ceiling heights have little effect on thermal comfort. See Givoni: *Man, climate and architecture*, Elsevier, 1969; and United Nations *Modular co-ordination of low cost housing*, New York, 1970; and BRE Overseas Building note 155

Structures must be designed to resist seismic forces. The severity of these forces will depend on the type of structure, the dead and live load, the location (seismic zone) and soil conditions. In general, building plans and massing should be simple, avoiding T, L, or similar forms. Expansion joints and adjoining structures should be avoided as buildings may knock one another during minor tremors.
For all but the simplest structures in minor seismic areas, a structural engineer's advice should be sought.

12.02 Hurricanes

In addition to the abnormally high wind loads, precautions are required to resist or reduce loads resulting from differences between internal and external pressures. Countries experiencing hurricanes and typhoons have appropriate building regulations concerning fixing for roofs, windows and structural design loadings.

12.03 Sandstorms

Table XXIV shows the effect of wind on sand movements. Tight closing of *all* openings is required to reduce nuisance from blown sand. Complete protection is not usually practical.

Table XXIV Effect of wind on sand

Windspeed m/min	Effect*
200	Sweeping sand. Visibility not impaired. Sand blown along the surface or up to 1 m from the ground
300	Driving sand. Visibility impaired. Sand rises up to 2 m high
600	Sandstorm. Particles of sand remain suspended in the air

* Depends on size of sand grains, humidity of ground, etc

12.04 Termites

There are two main types of termites:

- *Drywood termites* are similar to wood boring beetles found in temperate climates. They can fly into buildings or be introduced in previously infested timber.

Prevention by use of (expensive) naturally resistant timbers, or by pre-treatment of timber with a wood preservative and by screening of openings and roof cavities.

- *Subterranean termites* need to maintain contact with the ground and can survive in drier conditions than the drywood termites.

Prevention by a general soil poison over the entire ground slab combined with poisoning of soil around the building perimeter and the poisoning of materials used in ground floor slab and lower courses of walls, including hardcore, building sand, mortar and render and effective sealing of cracks, joints and holes with pitch based sealing compound, or the use of termite shields where suspended timber floors are used. In addition, regular inspection is required,**42.8 a, b**.

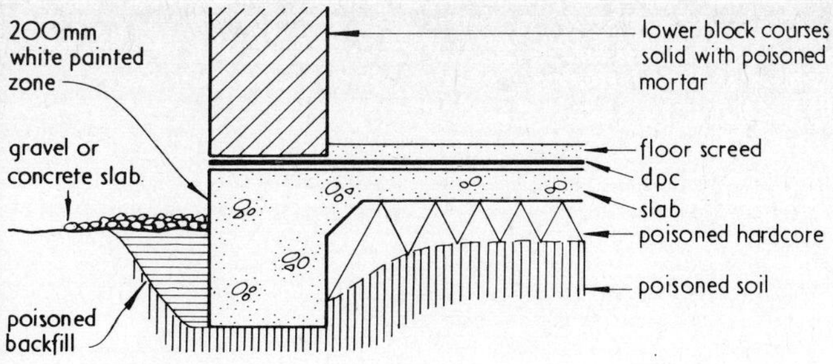

42.8a *Protection against termites when using a concrete slab floor*

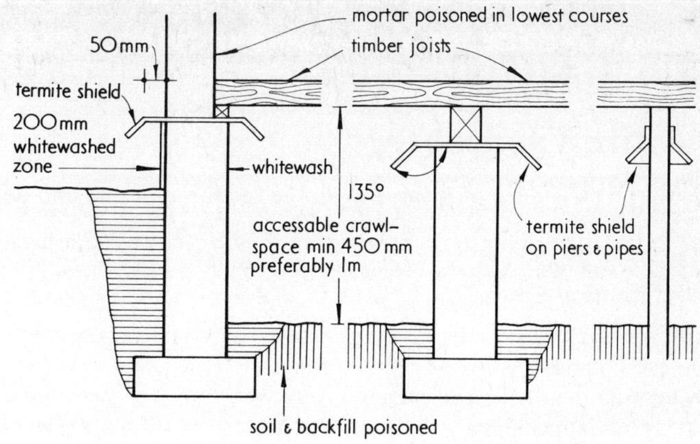

42.8b *Metal termite shields protecting a suspended timber floor over a crawl space*

12.05 Mosquitoes

Nuisance can be reduced by:

1 Cutting back undergrowth near buildings
2 Avoiding standing water on or near buildings, ie gutters
3 Screening of all openings with 16 mesh, 30 gauge wire screen or plastics mesh where windows open. Screening of bedrooms should have priority.
4 Avoid lighting which attracts mosquitoes (and other flying insects) indoors, especially over dining tables, beds, etc.
5 Use external lighting to divert mosquitoes and other insects away from openings.

12.06 Diseases

Bad housing conditions contribute to many diseases common in the tropics. The main factors required to avoid conditions which encourage the spread of disease are:

safe water supply
safe sewage treatment and disposal
easily cleaned wc and bathroom, **42.9**
sound construction without cracks or crevices
fly screening of kitchen and food storage areas
adequate natural light
adequate space standards
clearing of vegetation around the dwelling
avoiding standing water on badly drained surfaces, gutters and vegetation.

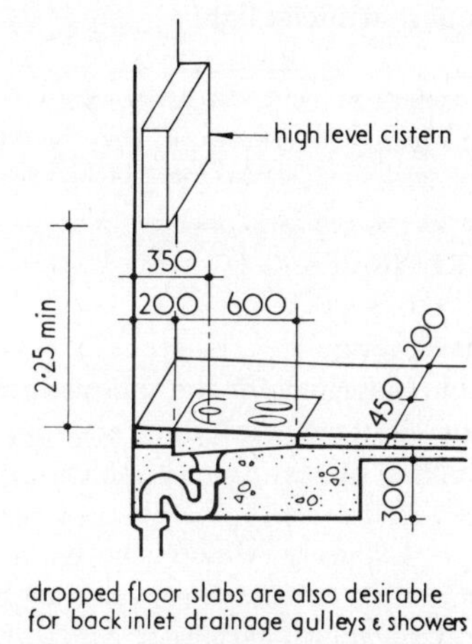

42.9 *Easily cleaned 'Asian' type* WC, *preferable in hot climates*

13 BIBLIOGRAPHY

Comfort and requirements for comfort
Design of low cost housing and community facilities, vol 1, *Climate and house design*, New York, United Nations, 1971

Thermal performance requirements
B. Givoni, ed, *Man, climate and architecture*, Elsevier, 1969
Institute of Heating and Ventilating Engineers, *Guide 1970*, Book A, London, 1970
D. H. Koenigsberger, T. G. Ingersol, A. Mayhew and S. V. Szokolay *Manual of tropical housing and design*, Part 1 Climatic design, Harlow, Essex, Longmans, 1973
A. G. Loudon *Summertime temperatures in buildings without air-conditioning*, BRS CP 47/68

Solar radiation
P. Petherbridge *Sunpath diagrams and overlays*, London, HMSO, 1969
W. H. Ransom Tropical Building Studies 3 *Solar radiation thermal effects on building materials*, London, HMSO, 1962

Illumination from sun and sky
C. G. H. Plant *Windows: design and function under tropical conditions*, Overseas Building Note 142, BRE, 1972
C. G. H. Plant, J. Longmore and R. G. Hopkinson, A study of interior illumination due to skylight and reflected sunlight under tropical condition, Proceedings of the CIE conference *Sunlight in buildings* (1965) Rotterdam, Bouwcentrum, 1967

Public utilities
B. H. Dietrich and J. M. Henderson *Urban water supply conditions in 75 developing countries*, WHO, 1963
E. G. Wagner and J. Lanoix *Water supply for rural areas and small communities*, WHO
E. G. Wagner and F. Lanoix *Excreta disposal for rural areas and small communities*, WHO, 1958

Hazards
Building Research Station *Tropical building legislation: model building regulations for small buildings in earthquake and hurricane areas*, 1966
W. Victor Harris *Termites: their recognition and control*, Longmans, 1971
Preservation of personal health in warm climates, Ross Institute of Tropical Hygiene, London, 1970
Termites and tropical building, Overseas Building Note 170, BRE, 1976

43 Light

Maritz Vandenberg

with the assistance of **Brian Day** in the section on units and definitions

Maritz Vandenberg is Technical Editor for books at the Architectural Press

CI/SfB (63)
CI/SfB (1976 revised) (63)

Contents

1 USER REQUIREMENTS

1.01 Functions of lighting

Light is required in buildings for the following purposes:

- For the carrying out of *tasks* (reading, writing, assembling or repairing instruments, operating mechanical equipment, and the like).
- For the creation of *amenity* (which is both a matter of pleasant general illumination of interiors, whether by daylight or electric light; and of adequate direct sunlight, which is an additional psychological requirement for most people).
- For the creation of *deliberate aesthetic effect* (such as display lighting; or dramatic lighting of certain kinds of interior, eg restaurants or night clubs).
- For the provision of *safety* in and around buildings.

These functions are briefly described in the paragraphs which follow; then practical design guidance is given in sections **2** to **7**.

1.02 Task lighting (electric light)

This is the aspect of lighting most easily described, specified and measured. The convention is simply to measure (or specify) the intensity of illumination on a hypothetical 'working plane'. This is normally taken to be a horizontal surface situated 0·85 m above floor level (unless the task is known to be carried out at a different level), **43.1**.

This intensity of illumination is known as 'illuminance', and is measured in lux—see section **8** for definitions.

The more detailed and exacting the work that is to be carried out in a working environment (such as office, laboratory, studio or workshop), the higher the level of illumination on the working plane needs to be, roughly speaking.

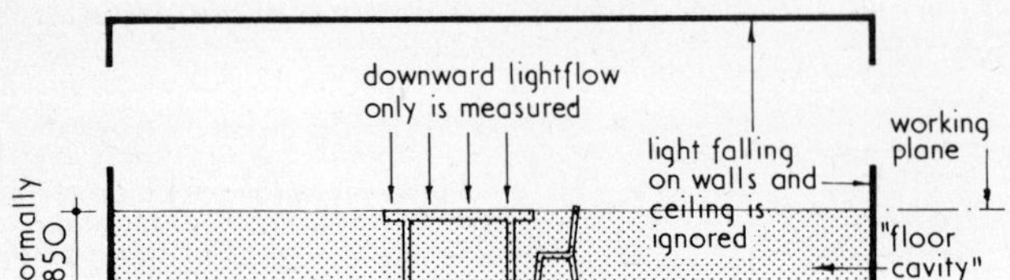

43.1 *The traditional method of measuring the effectiveness of a lighting scheme is in terms of illumination on the horizontal working plane*

There are accepted recommendations of illuminance on the horizontal plane for all work situations, in accordance with the above principle. These are given in tables I and II, for task lighting by *electric* light. In the case of rooms lit by *daylight*, the measure used is not illuminance (measured in lux), but Daylight Factor; and this is discussed in para **1.04**. Recommended daylight factors for various work situations are given in table III.

Recommended illuminances in electrically lit rooms vary from 150 lux for storage areas, to 3000 lux for work surfaces on which very fine assemblies (such as watches) are inspected and worked upon.

To give an appreciation of what these figures mean, the following are commonly encountered levels of illuminance on the horizontal plane:

Domestic living rooms	50 lux
Offices, laboratories, studios	500 lux
Out-of-doors, under overcast sky	5000 lux
Out-of-doors, under direct sun	50 000 lux

Most published recommendations for lighting levels inside buildings are on this simple basis of 'illuminance on the working plane' (together with limiting glare index—see explanation in para **1.03** below).

But it should be clearly understood that while task illumination, measured on the working plane, provides an important *minimum* lighting specification that must be met, other forms of measurement must be used in addition; and these are discussed in para **1.03** below.

1.03 Amenity lighting (electric light)

Illuminance on the horizontal working plane guarantees only that upper surfaces of desks and tables will be adequately lit. It does not necessarily ensure that any of the other criteria of a pleasantly and adequately lit room are met—enough light on walls and ceiling; enough contrast between light and shadow to ensure good visual modelling; avoidance of harshness; accurate colour rendering of objects and surfaces.

To ensure that a room is pleasantly lit in all the above respects, at least four further variables must receive attention:

1 relative brightness of surfaces
2 modelling characteristics of illumination
3 glare control
4 colour rendering.

These characteristics, and the methods of measuring and specifying them, are described below.

Relative brightness of surfaces

For all normal environments (exceptions are identified in para **1.05**), adequate light should fall not merely on the horizontal working plane, but also on walls and ceiling; or the room will look gloomy.

This requires a properly designed lighting system, directing light both horizontally and vertically; and also sensible colours for floor, wall and ceilings. Light-coloured room surfaces make it very much easier to achieve good light distribution than dark-coloured surfaces.

Table I Recommended illuminances, limiting glare indexes, and lamp colours for specific situations *(continued overleaf)*

Situation	Standard service illuminance (lux)	Position of measurement	Limiting glare index	Colour appearance of lamps	Reference letters of suitable lamps (see table VI)	Notes
Circulation areas						
Corridors, passageways	100 scalar	1·2 m above floor	22	Intermediate or warm	CDEFHIJLMQ	Scalar illuminance to be not less than $\frac{1}{8}$ horizontal planar illuminance in adjacent areas, and not less than 120 lux if there is no daylight
Lifts (passenger)	150	Floor	—	,,	CDEFHIJ	
Stairs	150	Treads	—	,,	CDEFHIJLMQ	Limit glare: see Section 3.5.3, recommendation d in IES Code
Escalators	150	Treads	—	,,	,,	Avoid specular reflections on treads
External covered ways	30	Ground	—	,,	CHIKLMNPQ	Illuminance should be compatible with adjacent lit areas
Entrances						
Entrance halls, lobbies, waiting rooms	150 scalar	1·2 m above floor	—	Intermediate or warm	CDEFHIJKL MPQ	
Enquiry desks	500	Desk	19	,,	,,	
Kitchens See SI No 1172: Food Hygiene (General) Regulations 1970						
Food stores	150	Floor	—	Intermediate or warm	DEFHIQ	
Working areas	500	Working surface	22	,,	,,	Position luminaires relative to working areas
Medical and first aid centres						
Consulting rooms, treatment areas	500	Desk or bed	—	Intermediate or warm	DEPQ	Examination lighting should be provided
Rest rooms	150	Bed	—	,,	DEQ	Restrict luminance seen by recumbent patient (glove)
Outdoor areas						
Controlled entrance or exit gates	150	Working plane	—	Intermediate or warm	CHIKLMNPQ	Working plane varies according to requirements
Entrances and exits	30	Ground	—	,,	,,	
Staff restaurants See SI No 1172: Food Hygiene (General) Regulations 1970						
Canteens, cafeterias, dining rooms	200	Tables	22	Intermediate or warm	DEFHIJQ	Arrange switching to allow variation of illuminance for social functions
Servery	300	Counter	22	,,	,,	
Staff rooms						
Changing, locker and cleaners' rooms, cloakrooms, lavatories	150	Floor	—	Intermediate or warm	DEFHIJQ	
Rest rooms	150	Table height	19	,,	,,	Change in character from general lighting desirable
Stores and stock rooms						
General	150	Vertical plane	—	Intermediate or warm	CHIKLQ	
Assembly shops						
Casual work	200	Working plane	25	Intermediate or warm	CHIKLMQ	
Rough work, eg frame and heavy machinery assembly	300	,,	25	,,	,,	
Medium work, eg engine assembly, vehicle body assembly	500	,,	22	,,	,,	
Fine work, eg electronic and office machinery assembly	1000	Bench	19	Cool, intermediate or warm	ACDEFHIKL MQ	Use A, D or E if colour rendering critical. Avoid specular reflections
Very fine work, eg instrument and small precision mechanism assembly	1500	Bench	16	,,	,,	Use A, D or E if colour rendering critical. Optical aids may be required
Car parks—indoor See BSCP 1004 Part 9 and IES Lighting Guide: The Outdoor Environment. Proof luminaires may be required						
Underground	30	Floor	22	Intermediate or warm	CHILMN	Vertical obstructions should be lit to a higher illuminance than floor, possibly by appropriate positioning of luminaires
Multi-storey						
parking floors	30	Floor	22	,,	,,	
ramps	50	Vertical sides	19	,,	,,	
Laboratories (general) For requirements of specialised laboratories see relevant building or process						
General	750	Bench	19	Cool, intermediate or warm	ACDEFHIQ	Position luminaires over benches. Proof luminaires may be required
Assembly and concert halls, cinemas and theatres See Theatres Act 1968 and Recommendations for safety in cinema						
Auditoria						
theatres and concert halls	100	Horizontal at seat level	—	Intermediate or warm	DEFJPQ	Dimming facilities required
cinemas	50	,,	—	,,	,,	,,

Table I (*continued*)

Situation	Standard service illuminance (lux)	Position of measurement	Limiting glare index	Colour appearance of lamps	Reference letters of suitable lamps (see table VI)	Notes
multi-purpose	100–500	,,	19	,,	DEFHIJKLM PQ	Allow for variation of luminance to suit function. Dimming facilities usually required
Foyers	75 scalar	1·2 m above floor	19	Intermediate or warm	DEFJPQ	
Libraries See IES Technical Report No 8: Lighting of Libraries						
Shelves, book stacks	150	Vertical at floor level	19	Intermediate or warm	CDEFHIKL	
Reading tables	300	Table	19	,,	,,	Low noise level required
Reading rooms						
newspapers and magazines	300	Desk	19	,,	,,	
reference libraries	500	,,	19	,,	,,	
Schools See SI No 890: Standards for School Premises Regulations 1959, amended 1970. See also *Guidelines for environmental design* (current edition). Where schools are also used for further education the more stringent requirements should be observed						
Assembly halls						
general	300	Working plane	19	Intermediate or warm	DEFHIKLQ	Provision for dimming should be included for stage and film use
platform and stage	Special lighting					
Teaching spaces						
general	300	Working plane	19	,,	DEFHIQ	Illuminance may be reduced to 150 lux in spaces lit by tungsten lamps, eg nursery and infant schools. Where main view is across space, eg to chalkboard, LGI should be 16 in that direction
Lecture theatres						
general	300	Desk	16	,,	,,	
chalkboard	500	Vertical plane	—	,,	,,	
demonstration benches	500	Bench	16	,,	,,	
Art rooms	500	Easel	19	Cool or intermediate	ADE	
Laboratories	500	Bench	19	Cool, intermediate or warm	ACDEFHIQ	Type of laboratory may dictate colour rendering requirements. Local lighting may be used
Workshops	300	Working plane	19	Intermediate or warm	CHIQ	Additional local lighting may be required
Homes						
Living rooms						
general	50	Working plane	—	Intermediate or warm	DEFGIJQ	In all home areas, attention should be given to the lighting of room surfaces. Luminaires should be selected and positioned to give occupants a compromise between attractive 'sparkle' and unwanted glare. Dimming is useful for changing atmosphere. Additional mirror lighting required in bedrooms
casual reading	150	Task	—	,,	,,	
sewing and darning	300	,,	—	,,	,,	
Studies						
desk and prolonged reading	300	,,	—	,,	,,	
Bedrooms						
general	50	Floor	—	,,	,,	
bedhead	150	Bed	—	,,	,,	
Kitchens						
working areas	300	Working surface	—	,,	,,	
Bathrooms	100	Floor	—	,,	,,	Additional mirror lighting required. Enclosed luminaires should be used
Halls and landings	150	Floor	—	,,	,,	High luminance areas should be screened from view when ascending or descending stairs
Stairs	100	Treads	—	,,	,,	
Workshops	300	Bench	—	,,	HIQ	
Garages	50	Floor	—	,,	,,	
Offices See Shops, Offices and Railways Premises Act 1963						
General offices with mainly clerical tasks and occasional typing	500	Desk	19	Intermediate or warm	CDEFHIKL	Minimise desktop reflections by suitable luminaire location
Deep-plan general offices	750	,,	19	,,	,,	,,
Business machine and typing offices	750	Copy	19	,,	,,	,,
Conference rooms	750	Tables	16	,,	DEFJQ	Consider variation of illuminance to suit different functions, eg conferences and lectures. Dimming may be required
Executive offices	500	Desk	16	,,	,,	Possible need to vary lighting using dimmers
Banking halls						
working spaces	500	,,	19	,,	DEFHIKL	
public spaces	300	Floor	19	,,	DEFHIKLMPQ	
Computer rooms	500	Working plane	19	,,	DEFHIKL	Avoid specular reflections in consoles. Limit illuminance on internally lit signals

Table I (*continued*)

Situation	Standard service illuminance (lux)	Position of measurement	Limiting glare index	Colour appearance of lamps	Reference letters of suitable lamps (see table VI)	Notes
Drawing offices						
drawing boards	750	Board	16	Cool, intermediate or warm	ACDEFHIKL	Where colour rendering is important use A, D or E. Boards may be vertical or inclined. Consider local lighting
Shops See Shops, Offices and Railway Premises Act 1963. For food shops see SI No 1172: Food Hygiene (General) Regulations 1970						
Conventional with counters	500	Counters—horizontal	19	Cool, intermediate or warm	ADEFGHIJK LPQ	Type of merchandise will dictate required colour rendering. Local or localised lighting needed to emphasise particular displays
Conventional with wall displays	500	Display—vertical	19	,,	,,	
Self-service	500	Vertical on displayed merchandise	19	Intermediate or warm	DEFGHIKLPQ	,,
Supermarkets	500	,,	22	,,	,,	,,
Hypermarkets	500	,,	22	,,	DEFGHIKLM PQ	In these very large areas, definition of perimeter walls by higher luminance on walls is desirable
	1000	Horizontal on working plane	22	,,		
Showrooms						
car	500	Vertical on cars	19	Cool, intermediate or warm	ACDFHIKLPQ	
general	500	Merchandise	19	,,	,,	Vertical surfaces may be important. Use A, D or E where colour rendering is important
Covered shopping precincts and arcades						
main circulation spaces	100–200 or 100 scalar	Floor	22	Intermediate or warm	DEFHIKLMQ	

Published by courtesy of the I.E.S.

Table II Conditions in which illuminances recommended in table I should be increased

Task group and typical task or interior	Standard service illuminance lux	Are reflectances or contrasts unusually low?	Will errors have serious consequences?	Is task of short duration?	Is area windowless?	Final service illuminance lux
Storage areas and plant rooms with no continuous work	150					**150**
Casual work	200	——	——	——	no —— 200 yes	**200**
Rough work Rough machining and assembly	300	no —— 300 yes	no —— 300 yes	—— 300	no —— 300 yes	**300**
Routine work Offices, control rooms, medium machining and assembly	500	no —— 500 yes	no —— 500 yes	yes no —— 500	—— 500	**500**
Demanding work Deep-plan, drawing or business machine offices. Inspection of medium machining	750	no —— 750 yes	no —— 750 yes	yes no —— 750	—— 750	**750**
Fine work Colour discrimination, textile processing, fine machining and assembly	1000	no —— 1000 yes	no —— 1000 yes	yes no —— 1000	—— 1000	**1000**
Very fine work Hand engraving, inspection of fine machining or assembly	1500	no —— 1500 yes	no —— 1500 yes	yes no —— 1500	—— 1500	**1500**
Minute work Inspection of very fine assembly	3000	—— 3000	—— 3000	yes no —— 3000	—— 3000	**3000**

Published by courtesy of the I.E.S.

Table III Recommended daylight factors and limiting glare indexes for specific situations

Situation	Average daylight factor (per cent)	Minimum daylight factor* (per cent)	Position of measurement	Limiting daylight glare index	Notes
Assembly and concert halls					
Foyers, auditoria	1	0·6	Working plane	24	
Corridors	2	0·6	Floor	—	
Stairs	2	0·6	Treads	—	
Drawing offices					
General	5	2·5	On boards	21	
General building areas					
Entrance halls and reception areas	2	0·6	Working plane	24	
Offices					
General offices	5	2	Desks	23	
Typing, business machines, manually operated computers	5	2·5	,,	23	
Schools and colleges See DES Building Bulletin No 33 and *Guidelines for environmental design*, DES (1972), and IES Lighting Guide: Lecture Theatres					
Assembly halls	1	0·3	Working plane	21	
Classrooms	5	2	Desks	21	
Art rooms	5	2	Easels	21	
Laboratories	5	2	Benches	21	
Staffrooms, common rooms	5	1·5	Working plane	23	
Sports halls Top lit. See IES Lighting Guide: Sports					
General	5	3·5	Working plane	21	
Surgeries (medical and dental)					
Waiting rooms	2	0·6	Working plane	24	
Surgeries	5	2·5	,,	21	
Laboratories	5	2	Benches	22	
Swimming pools See IES Lighting Guide: Sports					
Pool	5	2	Pool surface	23	Care should be taken to minimise glare and reflections from water surface
Surrounding areas	1	0·5	Working plane	23	

*In general all areas where the daylight factor is less than 1 per cent will require supplementary electric lighting
Published by courtesy of the I.E.S.

The problem is to find a technique of measurement which is suitable for specifying proper light distribution both horizontally and vertically.

The appropriate measure is that of *scalar* illuminance.

Unlike measuring illuminance on the working plane (discussed in para **1.02**) which is a *planar* measure, taking account only of light falling in one direction, scalar illuminance is defined as the average illuminance over the whole surface of a tiny sphere at a particular point in space, due to light arriving from all directions, **43.2.**

This is a reliable measure for the subjectively judged 'brightness' of a room. If the scalar illuminance (say 150 lux) is not less than half the horizontal illuminance (say 300 lux) as given, for instance, in table I, then it may be assumed that all surfaces of the room will appear adequately lit.

The calculation of the scalar illuminance in a room is too complex to be fully described here (see reference [10.6]); but it takes into account three sets of data:

1 The *proportions* of the room (length, width, and height), combined into a single figure called the 'room index'. The latter is easily determined using a nomogram [10.5].

2 The *reflectances* of the major surfaces of the room—floor, walls and ceiling. These are obtained from paint manufacturers' literature; or from BS 4800.

3 The pattern of downward light spread achieved by the electric light fittings—the so-called *BZ classification* of the fitting, listed in manufacturers' catalogues. A low BZ number means that downward light distribution tends to be *concentrated*; a high BZ number means that downward light distribution tends to be *dispersed*. The higher the BZ number, the more light will tend to fall on the walls (which is good); but the higher, too, will tend to be the risk of glare (which is bad)—see below under 'Glare index'.

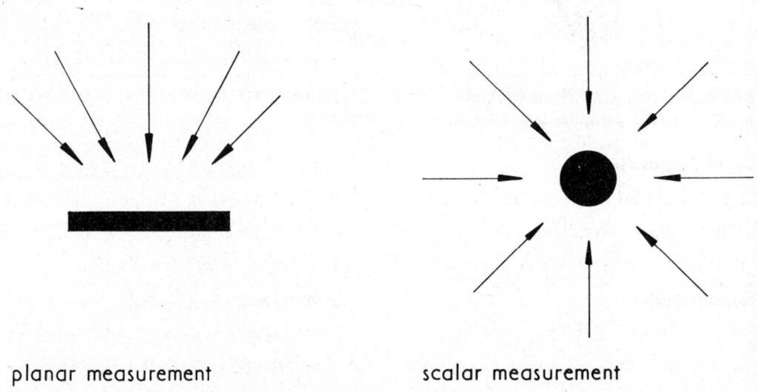

43.2 *Horizontal illumination, measured in lux, equals lumens per m^2 of horizontal surface; scalar illumination, also measured in lux, equals lumens per m^2 of spherical surface*

Rule of thumb

As a rule of thumb, light-coloured room surfaces (including the floor); side-windows, rather than rooflights; and electric light fittings of high BZ number (say between BZ 4 and BZ 10) will aid the achievement of good scalar illumination in most ordinary rooms.

Modelling characteristics of illumination

It is stated above that all room surfaces, both horizontal and vertical, should be adequately lit. But this can be overdone. If there is absolutely uniform distribution of light in all directions, the *modelling power* of the illumination may be inadequate. Ideally, there should be sufficient contrast between highlight and shadow to give interest and firm modelling to three-dimensional objects, without being harsh.

To measure this modelling quality, a further specifying concept is introduced—that of *illumination vector*.

Just because scalar illumination records all ambient illumination on the tiny notional sphere of measurement, irrespective of direction, it can tell us nothing about the direction of light flow in a room; it merely records how much light is flowing in all directions in that room.

If, however, the point of highest illumination, and the opposite point of lowest illumination, are identified on that tiny notional sphere (as a bright 'north pole' and a dark 'south pole'), then both the *direction* of light flow will be known, and the *illumination difference* between the two opposite poles, measured in lux, **43.3**.

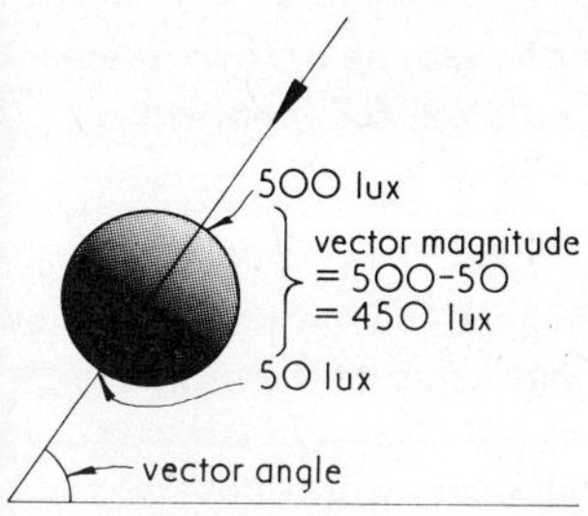

43.3 *Illumination vector has both direction and magnitude (the latter measured in lux)*

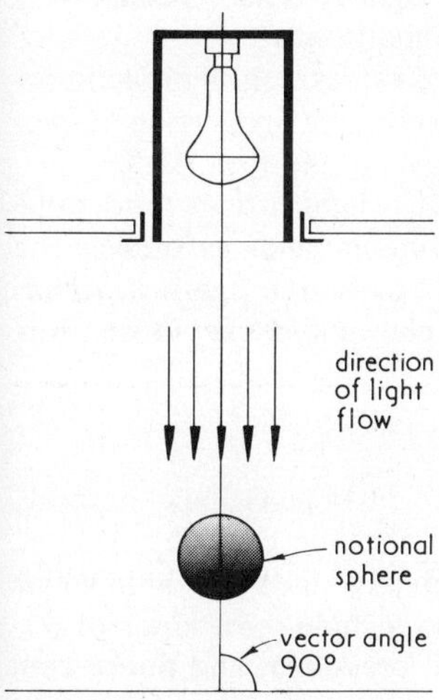

43.4 *Direction of illumination vector under a downlighter will be virtually vertical; most people find this unpleasant*

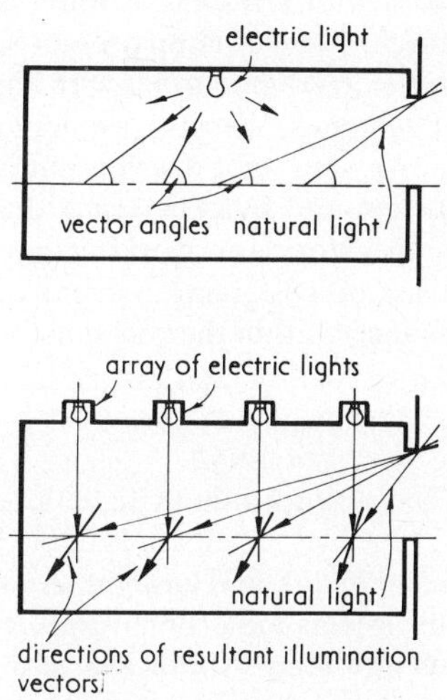

43.5 *A single overhead light fitting, in the middle of the room, gives good modelling (provided it emits light sideways as well as downwards). Daylight through side window also gives a satisfactory vector angle of about 45°; and the combination of overhead electric lights plus side windows gives excellent modelling qualities to light*

Taken together, the two characteristics give us the illumination vector of the lighting in a room, a vector comprising both magnitude and direction. For instance, a room lit exclusively by recessed downlighters will have a vector direction that must necessarily be vertical, **43.4**.

Vector magnitude will be affected not only by the lighting system, but also by room finishes. If the floor is very *dark*-coloured, little light will be reflected upward, and the underside of the notional tiny sphere will be much darker than the bright upper hemisphere, **43.6**. Modelling in the room will then be very contrasty, and unpleasantly harsh. If the floor is *light*, light will be reflected upward, and even with illumination by recessed downlighters, there may be sufficient upward light-flow to give pleasant, soft modelling.

The ratio between the magnitude of the illumination vector, and the scalar illumination, is known as the *vector/scalar ratio*; and this ratio gives a working measure of the modelling effect of lighting. If it is higher than about 1·8, then modelling will look harsh to most people. If it is less than about 1·2, then modelling will look indistinct and excessively soft. A figure of about 1·5 would be satisfactory in normal circumstances. Table IV gives vector/scalar ratios for various common situations, and the lighting effect that is achieved.

A calculation method is set out in *Interior lighting design* page 109, and in *The Architects' Journal* 9.7.1969 pp 61–64 [10.6]. It is too involved to give here.

Rule of thumb

As a rule of thumb, luminaires classified in manufacturers' literature as having low BZ number and low flux fraction ratio will provide good modelling in rooms with light-coloured floors. Luminaires with high BZ numbers and low flux fraction ratios will provide better modelling in rooms with dark-coloured floors, but in the latter case the requirements for good modelling may conflict with those of glare-prevention (see below).

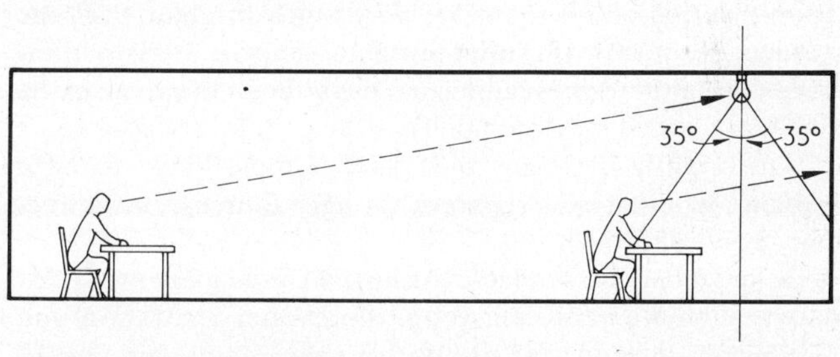

43.6 *Overhead light fittings are less likely to cause glare to occupants than more distant fittings; glare is therefore more likely in long rooms*

Table IV Vector/Scalar criteria for lighting design

Vector/scalar ratio	Strength of the flow of light	Typical situation	Typical appraisal
3·0	Very strong	Selective spotlighting. Direct sunlight	Strong contrasts: detail in shadow is not discernable
2·5	Strong	Low BZ*; low FFR*; dark floor. Windows on one side, dark wall surfaces	Noticeably strong directional effect: suitable for display but generally too harsh for human features
2·0	Moderately strong	Low BZ with medium or light floor. Medium or high BZ. Side windows with light surfaces PSALI†	Pleasant appearance of human features, for formal or distant communication
1·5	Moderately weak	Low BZ with medium or light floor. Medium or high BZ. Side windows with light surfaces PSALI	Pleasant appearance of human face, for informal or close communication
1·0	Weak	Medium or high BZ with light floor. Side windows in opposite walls	Soft lighting effect for subdued contrasts
0·5	Very weak	Luminous ceiling or indirect lighting with light room surfaces	Flat, shadow-free lighting; directional effect not discernible

*See para **7.03** †See para **2.01**

Published by courtesy of Pilkington Brothers Ltd

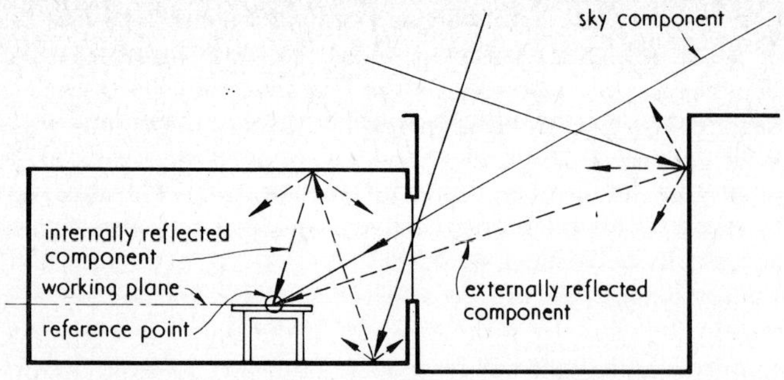

43.7 *Daylight reaches indoor working plane in three ways: 'daylight factor' calculation takes account of all three*

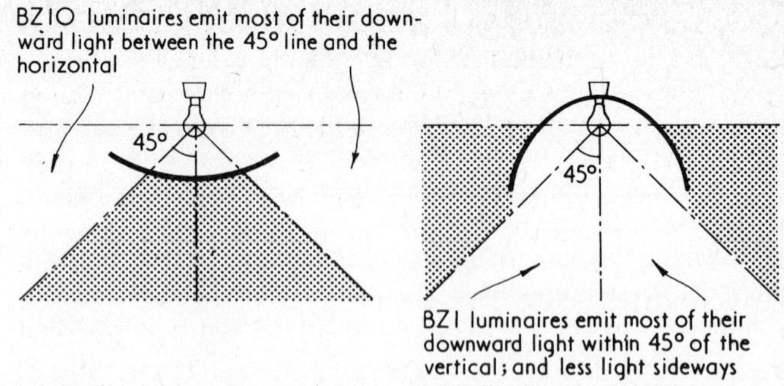

43.9 *The downward light-distribution patterns of light fittings with high BZ numbers, and those with low BZ numbers*

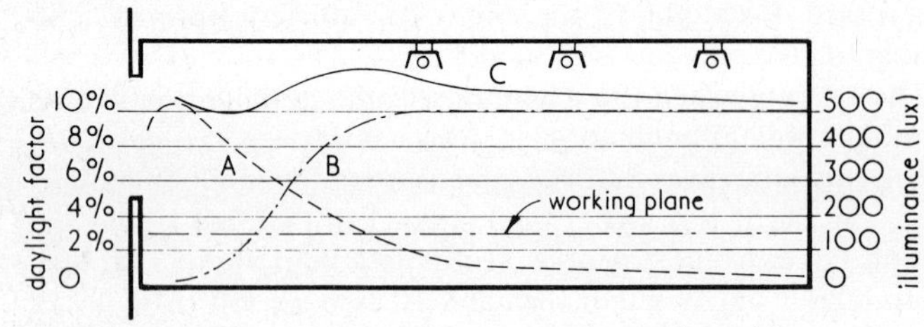

43.8 *The PSALI principle. Daylight near the window, and artificial light deeper in room, combine during daylight hours*

Taking the lighting environment as a whole, there should be a dominant direction of lighting in rooms (whether from windows/rooflights, or from electrical fittings, or both), rather than completely diffused light. But the directionality should not be so dominant as to cause harsh shadows. A generally well-liked angle of lighting (direction of illumination vector) is downward and from the side, at an angle of 15° to 45 ° to the horizontal plane. Side windows in conjunction with overhead electrical lights give satisfactory results, **43.5**. Directly vertical lighting gives the least popular modelling.

Glare control

If some surfaces in a room are very much brighter than the average, occupants will suffer visual discomfort or even 'disability' (the latter term meaning that it is difficult to carry out the visual task one is engaged upon).

It is in brightly lit rooms that glare is most likely to be a problem, whether direct glare from light sources, or indirect glare from reflections.

Published tables are available listing the 'initial glare index' for many combinations of room proportion, room decoration, and lighting fittings and layouts. With further data, this 'initial' index is then modified to a more precise figure. See *Interior lighting design*[2] pp 30–35; and 102–108.

The above calculation takes into account the brightness and size of each light source; the number of sources and their distance from the viewer; the direction of view in relation to the direction of the source; the brightness of the background; and the conditions to which the observer's eyes have become accustomed.

Establishing that a lighting layout is satisfactory from this point of view is important, because 'glare index' is one of the three criteria most commonly specified in Codes of Practice (or other official recommendations) for lighting installations, the other being illuminance on the working plane (para **1.02**) and daylight factor (para **1.04**).

Rule of thumb

In general, risk of glare is *increased* with:

- Long rooms (because distant light fittings come more easily within the normal angle of view than overhead fittings, **43.6**).
- Low ceilings in large rooms (because low-mounted fittings come more easily within the normal angle of view than high ones).
- Dark decorations (particularly dark ceiling finishes) because contrasts in lighting level are heightened.
- Lack of diffusers or louvres on light sources (because very bright parts of lamp are not screened from view).
- Electric light fittings with high BZ values (see manufacturers' catalogues); because a comparatively large proportion of their downward light distribution is directed sideways, **43.9**.

It is found by experience that BZ 3 light fittings tend to be satisfactory for working environments such as offices and classrooms, giving agreeable and economic lighting performance, within the recommended glare indices for those room types.

Colour rendering

There are kinds of light in which lit objects look 'natural'. These include all kinds of daylight, incandescent electric light, candlelight, and gaslight. And there are kinds of light in which lit objects look 'unnatural'. These include most kinds of discharge lamps (both low and high pressure); and fluorescent lamps.

However, some of the latter have been improved to the point where colour performance is satisfactory even for exacting situations (unfortunately at the expense of efficacy, which is one of the great advantages of discharge and fluorescent lamps); and these will be discussed in section **6**.

1.04 Task lighting (natural light)

If task lighting is to be provided by *electric* light, then intensity of illumination on the working plane is specified and measured in lux, as already explained; see para **1.02**.

But if workspaces are sufficiently close to a window (or rooflight) to be lit by *daylight*, then required intensity of illumination on the working plane are specified in terms of the 'daylight factor'. Table III lists the required daylight factors for various situations in buildings.

Daylight factor

Daylight factor expresses the illuminance on a horizontal working plane *inside* the building, as a ratio of the illuminance on a similar plane *outside* the building at the same instant, when the latter is lit by indirect natural light only (ie direct sunlight is excluded). If, for instance, only 2 per cent of the light falling on the outside surface happens to fall on the surface inside the building (via windows and skylights, and reflected off surrounding surfaces), then the daylight factor at that point inside the building is 2 per cent, **43.7**.

To standardise the measurement, the external sky is conventionally assumed to be heavily overcast, and to give an illumination, on the horizontal plane, of 5000 lux at roof level

outside the building. This ensures that specified illumination levels will be achieved within the building even on dull days. And it follows that a daylight factor of 10 per cent is equivalent to 500 lux; a daylight factor of 5 per cent is equivalent to 250 lux; and so on.
Whether task lighting in rooms should be provided by electric light (following the recommendations given in table I) or by natural light (following the recommendations given in table III) is one of the most fundamental questions the designer has to resolve at the early stages of tackling a lighting design. The decision affects the whole design of the building: the depths of rooms; the sizes of windows; the functions of windows.
This is a problem of design strategy, and it is more fully discussed in para **2.01** to **2.05**.

1.05 Amenity lighting (natural light)

Indirect sunlight
To ensure that a room is pleasantly lit by natural light requires attention to the same factors as those listed in para **1.03** for electric light. But the problems are easier with natural light.
The main points to remember are that:
• Moderately sized side windows will add much to the amenity of a room, by throwing light on wall surfaces (which overhead electric lights may leave underlit); and by providing pleasant modelling qualities to the essentially vertical flow of light from electric fittings.
• If side-windows are too large, problems both of glare, and heat-control are likely.
• Rooflights may provide very good task lighting, but are less likely to give good amenity lighting. A very pronounced vertical light direction is found unpleasant by most people, particularly if the floor is dark-coloured. Rooflights should therefore be used in moderation.

Direct sunlight
The penetration of direct sunlight into buildings (and the spaces surrounding buildings) is not really a matter of *lighting* design as such; it is more a matter of catering for a distinctly felt psychological need. However, as the control of *direct* sunlight by means of building orientation, and room and window design, is closely related to the decisions which control *indirect* sunlight (daylight), it is logical to deal with both aspects together.

Admittance of direct sunlight There are no mandatory rules in the UK (or most other countries) for the admittance of direct sunlight into buildings. But (particularly for homes, schools and hospitals) strong official recommendations have been published, and should be followed.
The more stringent recommendations in the UK are contained in CP3, chap I(B) *Sunlight: houses, flats and schools only*.
They require that:
a In living rooms (and, where possible, also kitchens and bedrooms) sunlight should be able to enter at some time of day for at least an hour, during not less than 10 months of the year from February to November. Preferably it should enter living rooms in the afternoon, and kitchens and bedrooms in the morning.
b In teaching rooms sunlight should be able to enter for about two hours in the morning throughout the year.
c Sunlight should have as much access as possible to all external façades of houses, flats and school buildings; and area of ground in permanent shadow should be reasonably small.
Under **a** and **b** above, sunlight counts as entering a room only if the 'grazing' angle between the sun's rays and the plane of the window is $22\frac{1}{2}°$ or greater.
There is also a less stringent set of recommendations, contained in *Sunlight and daylight*, a study published by the DOE in 1971. Being less stringent, these rules also more realistic, and more likely to be applied by designers. They are therefore described in Design techniques, para **3.01**.
It is necessary also to *exclude* unwanted direct sunlight at particular times of year. In countries with clear skies this may be the major purpose of sunlight penetration calculations; and even in Britain unwanted solar penetration can cause two kinds of nuisance.
The first is *glare*. Direct sunshine on working surfaces can give illumination levels of 60 000 lux compared with perhaps 500 lux on surrounding surfaces—an intolerable contrast in working environments.
The second is *heat* build-up. If windows are large and approximately south-facing and unshaded, heat gain on a sunny summer day can cause intolerably high temperatures even in Britain.
Sun control techniques for exclusion are outlined in para **3.02**.

1.06 Lighting for deliberate aesthetic effect

So far we have discussed *task* lighting (ensuring adequate light for working environments), and *amenity* lighting (creating pleasant general lighting conditions by bathing all room surfaces in relatively even light, of subdued contrast and correct colour, to promote easy seeing conditions).
But there is a third kind of purpose, which may apply particularly in entertainment and display buildings, which is to create dramatic and possibly deliberately misleading effects by means of selective lighting.
One example is the commercial display window, in which lighting is used to draw attention to selected objects at the expense of others. The aim may even be to change the true colour and appearance of the object by means of lighting, to enhance colour, to impart gloss, to make colour appear 'warmer' or 'cooler'.
Another example is the restaurant, cinema or night-club interior where, instead of promoting correct judgement of distances, shapes and colours, the intention might be to create illusory effects, even to disorient the occupant.
In a cinema, the intention might be to make side walls and ceiling 'disappear', to focus all attention on the screen. In a restaurant, lighting may be used to make the tables glow warmly and invitingly in an ambience of soft gloom; to make the walls appear more distant and more dramatic in colour and texture than they really are; to make the ceiling disappear altogether behind glittering or sculpturally dramatic light sources.
A final example of display lighting is the illuminated sign[10.23].
All these are matters of specialist skill. Detailed design guidance is beyond the scope of this brief treatment (see references AJ[10.1–10.3] and AJ[10.23]). What must be said is that the designer should be very clear in his mind what he is trying to achieve with his proposed lighting design. It cannot be both dramatic (in the sense explained above) *and* give good task and amenity lighting, unless unusual skill is brought to bear. Trying to create dramatic effects in a working environment is likely to cause unacceptable glare, or unpleasant contrasts between light and gloomy surfaces, rather than please the occupants.

1.07 Lighting for safety

This final aspect of lighting is to minimise the risks of tripping and falling (especially on stairs or ramps in places of assembly and crowding, in conditions of panic or rush); the risk of colliding with people or objects; and the risk of being attacked or assaulted in conditions of inadequate visibility.
It is also a matter of minimising risk of burglary, theft and pilferage.
Where lighting is adequate for the functions of task and amen-

ity, discussed in foregoing paragraphs, it will be adequate also for safety.

But in the following circumstances safety illumination requires special and separate consideration:

• In *outside* areas where task and amenity lighting recommendations do not apply, but which can be dangerous if underlit (alleys, courtyards, pedestrian areas generally). Make sure that these are well lit at all times, using fittings out of harm's way and not easily put out of action.

• In *internal* areas which would be dangerous if the normal lighting system failed (escape routes; public assembly areas; and such rooms as operating theatres). Battery-operated luminaires are usual for escape routes: only nominal illumination levels are required. For hospitals, alternative power plant capable of sustaining the full lighting system for operating theatres, etc, in case of supply failure, is required.

• In *all* areas where theft, break-in and pilferage may occur without adequate lighting. Power supply and light fittings should be protected from the possiblity of interruption.

Design guidance is given in para **5.04**.

1.08 Conclusion

At the outset of designing a lighting installation, user requirements should be checked using the following headings:

• Task lighting—identify rooms in which tasks are to be carried out; and identify category of task lighting for each room, using tables I, II and III.

• Amenity lighting—identify rooms which should be bright and cheerful at all times; rooms which need be so only occasionally; rooms which should be subdued. Identify rooms in which natural *daylighting* is very desirable, from amenity point of view; and rooms with distinct direct *sunlight* requirements.

• Special lighting effects—identify areas where lighting is required to create dramatic effects, or display objects, or attract people's attention, rather than provide task or amenity illumination. Make quite sure what overriding purpose is to be achieved in each case; and make sure requirements do not conflict.

• Safety—identify areas (especially circulation spaces) where battery-operated luminaires must be provided to assist safe movement in case of power failure.

• Security—ensure that all spaces where theft, burglary, attack or assault may occur if light is inadequate are identified.

• Emergency—identify areas where automatic alternative system must come into operation if power fails (eg operating theatres).

2 DESIGN STRATEGY

2.01 Daylight and electric light

The real cost of electric light has fallen over the past few decades to the point where it is of the same order (if high-efficacy discharge and fluorescent sources are used) as the real cost of daylighting.*

Also, fluorescent sources are now available which, in addition to being economical, are compatible with daylight in terms of colour, and do not generate a large heat-load.

Designers are therefore free to select one of three different strategies for lighting a building interior:

• By daylight alone (with artificial light used only at night, or when daylight conditions fall below tolerable levels).

• By daylight permanently supplemented by artificial light (known as PSALI, or Permanent Supplementary Artificial Lighting of Interiors).

• By artificial light alone, with windows used purely for view out and to provide interest (known as PAL, or Permanent Artificial Lighting).

Selecting from the above strategies is a matter of profound importance. The decision affects not only the lighting installation, but the building design as a whole.

If PAL is decided upon, the designer is left with very great freedom in deciding building form. Interiors, not being dependent on windows for their illumination, can be of any required depth; windows can be small (thus eliminating problems of solar heat gain; wintertime heat loss; sky glare; loss of privacy; and the like); and tinted glazing can be used. The deep, cube-form building which may result from such freedom is likely to be cheaper to construct and maintain than a building of equivalent volume, but shallower shape. However, in cases of power failure such an interior would possibly become unusable, unless full standby power is available.

If PSALI is chosen, most of the above advantages would still be possible, and room areas near the outside wall would have the benefit of natural lighting (which many people prefer to electric light). But as windows would then have to be fairly large and unobstructed to give the required illumination level in working environments, the familiar problems of summer heat gain, wintertime heat loss, and sky glare would arise and need to be solved.

If full daylighting is chosen as the method of lighting a working environment, all the problems of thermal control, glare control, and noise control, associated with large windows, must be effectively solved or the occupants will suffer.

Notes on each of the above design strategies follow.

2.02 Natural light alone

Daylight factor

The appropriate illumination standard in naturally lit rooms is the daylight factor (explained in para **1.04**). Table III gives recommended daylight factors which must be achieved by rooms for particular functions.

The recommended daylight factors will be more easily achieved with windows in two or more walls than with windows in one wall only.

Rooflights, too, will help achieve higher daylight factors. But if rooflights are used, then daylight factors should be not less than 5 per cent. The reason is that a vertical light flow (downward from a rooflight) registers more strongly on the working plane than a horizontal lightflow of the same strength (sideways from a window); therefore measurements taken on the working plane have an inbuilt bias in favour of rooflighting. A side-lit room giving a daylight factor reading of 2 per cent may look just as bright as a top-lit room giving a daylight factor reading of 5 per cent.

Large rooflights can cause great problems of solar heat gain in summer (unless they face north); and of heat loss in winter (whatever their orientation). A powerful lightflow from vertically overhead also tends to give an unpleasant quality of light (see under 'modelling characteristics', para **1.03**). Therefore rooflights should be used in moderation, and with caution.

Daylight glare index

Table III also gives a limiting daylight glare index for each kind of situation. Sky glare (and direct sun glare) are inherent problems with daylit rooms, because large window sizes are required.

Daylight glare indexes can be calculated using methods given in IES Technical Report no 4.[9]

* Real cost of daylighting includes the cost of providing large areas of glazing (especially expensive in the case of rooflights); the cost of keeping that glazing adequately clean for unobstructed light transmission (again, a problem in the case of rooflights); and the cost of dealing with increased heating and cooling loads owing to the large glazing areas. Winter heating costs can be very significantly higher in a building with large windows and rooflights than in a well-insulated enclosure.

But this is laborious, and most designers will prefer to provide the following commonsense methods.
• Provide curtains, blinds, and other sunshading devices; preferably under direct control of occupants.
• Provide windows in more than one wall in each room, so that brightnesses of framing wall-areas around windows are raised (by cross-lighting), thus reducing contrast between dark wall surface and bright sky.
• Provide white-painted deep splayed window reveals, as in traditional architectural detailing, to soften the contrast between wall and sky.
• Use tinted glass (provided this does not also reduce daylight factor to below recommended level).

2.03 Lighting by PSALI
Even in general office environments, people expect an illuminance on the working plane of 500 lux (see table I). This is equivalent to a daylight factor of 10 per cent. But in any largish room lit by side-windows (as must be the case with multistorey buildings) such a daylight factor is unobtainable over the whole of the working area.
Therefore the notion that daylight alone can be enough must be explicitly abandoned, and most working environments lit either by PSALI (if the designer decides that the side-window should still play a significant role in providing light), or PAL (if this role is abandoned altogether, and the window becomes merely a provider of view, amenity, and visual stimulus).
A common building arrangement suited to PSALI is the double-corridor layout. A pair of corridors surround a 'no-daylight' core, the latter comprising lifts, staircases, toilets, and other rooms not continuously occupied. The occupied, working environment then spreads outward from the corridors to the outside walls.
The working spaces are lit mainly by daylight, for desks near the windows; and mainly by artificial light for desks away from the windows. The two kinds of light combine in the way shown in **43.8**.
It is seen that there is a problem of *visual balance*, if two kinds of light are combined in this way. The problem is twofold.

Illuminance balance
If the artificial lighting system in a room of the above kind were simply to follow the standard recommendations as in table I, the room would be much brighter near the window than away from it. Eye adaptation, glancing from one zone to the other, would be troublesome. Therefore it would be better to upgrade the official recommendation to take account of this, and design the whole room for an illuminance on the working plane of, say, 750 lux.

Colour balance
Similarly, the colour of the artificial light should be brought in balance with that of the daylight range. Fluorescent lamps should be of the 'natural' or 'daylight' types, producing a light no cooler in colour than natural light (see para **6.02**).
The artificial lighting system should cover the whole of the room area, to the same illuminance standard; but have a switching arrangement allowing the lamps near the windows to be switched off while those in the deeper parts of the room are left on.

2.04 Lighting by PAL
Once the window as a provider of light begins to be replaced by an artificial system, as above, it might be better in many cases to abandon all compromise and go over fully to artificial lighting.
The advantages listed in para **2.01** can then be exploited to the full; the modern liking for high illumination levels (1000 or even 2000 lux, according to some authorities) can be satisfied; and the various powered environmental services fully integrated.

Integration of services
In a cube-shaped building with smallish windows and a high-illuminance lighting installation, the combination high heat-input and low heat-loss could make a separate heating system unnecessary. Provided the building is well-insulated, and provided services have been designed as a totality, waste heat from the lighting system can be drawn off to heat the building throughout the winter.
Special air-handling luminaires are available allowing this to be done. The logical extension of this principle is the integrated ceiling system, in which lighting, ventilation/airconditioning, acoustic control, and even the sprinkler system, are all provided in an all-embracing modular system making maximum use of the ceiling space.

Contract maintenance
The advantages of routine contract maintenance, in which a specialist outside firm calls regularly in order to clean, relamp, and replace defective components in a single operation become overwhelming in the case of sophisticated integrated systems of this kind.

2.05 Conclusion
Strategic decisions on the kind of lighting to be provided, involve much larger decisions on building shape, site coverage, and energy conservation.
If the building can be narrow, with shallow room spaces, daylighting alone (during daylight hours) may be best.
But this is an increasingly unlikely solution for working environments. PSALI is far more likely; and if the advantages of a deep, cube-shaped building with smallish windows and controlled internal environment are to be exploited, PAL will be the answer.

3 DESIGN TECHNIQUES: DIRECT SUNLIGHT

3.01 Site layout
There are no mandatory requirements for direct sunlight penetration, with regard to site layouts, in the UK or most other countries. But authoritative criteria for housing are given in *Sunlight and daylight*,[3] published by the Department of the Environment; and these should be followed wherever possible. They state:

> In proposed (residential) buildings, sides facing due south, or in any direction east or west of south (including due west), should have all points 2 m above ground level accessible to sunlight for 3 hours on March 1. Sunlight is only counted if the sun is 10° or more above the horizon, but sunlight at a bearing of less than 22½° to the side of the building is *not* excluded.

If buildings are set out on site so that their southerly faces accord with the above rules, then it should be possible, at a later design stage, to ensure that the rooms inside those buildings will receive the required amount of sunlight internally (see para **1.05**).

Techniques
The recommended method for checking that the above site layout criteria are met is the *Sunlight indicator* issued as a set of transparencies in the UK.[7]
Three latitudes are catered for (51°, 53°, and 55°); and all relate to the crucial date of March 1. Instructions for constructing sunlight indicators for other dates, and other latitudes, are given on p 37 of *Sunlight and daylight.*[3]

The above indicators will show for how long the sun will be visible at particular points on a site layout (taking account of obstacles); and where shadows will fall at any particular time. Other methods, some of them more comprehensive and precise than the above, include the following:

1 *Sunscan*—a small portable instrument used on the plans of buildings to provide information about sunlight, shadow, daylight, and solar heating. Contact Research Engineers Ltd, Orsman Road, Shoreditch, London N1.

2 *Waldram diagram*—a graphic method which will cope with the most complicated forms of construction. It is more complicated to use than the sunlight indicator. Waldram diagrams are not readily available to architects; but a set of calculation sheets meant particularly for carrying out the calculations for March 1, defined in the above paragraphs, has been published in *The Architects' Journal.*[11] They were specially developed by Professor Peter Burberry, of the University of Manchester Institute of Science and Technology.

3 *Pilkington Environmental Advisory Service Charts*—a graphic method consisting of charts and overlays used in conjunction with perspective views through windows, to establish sunpaths.[6] Contact the Environmental Advisory Service, Pilkington Brothers Ltd, St Helens, Lancs, UK.

4 *Models*—three-dimensional models can be used to simulate real conditions. The environmental science departments of most large schools of architecture have equipment of this kind, which may be available, by arrangement, to architectural practices. Contact the nearest university or polytechnic.

Example

Figure **43.10** shows an example of a site layout evaluation taken from *Sunlight and daylight.*[3]

a shows the sunlight indicator; and **b** its use on a housing layout plan. It is seen

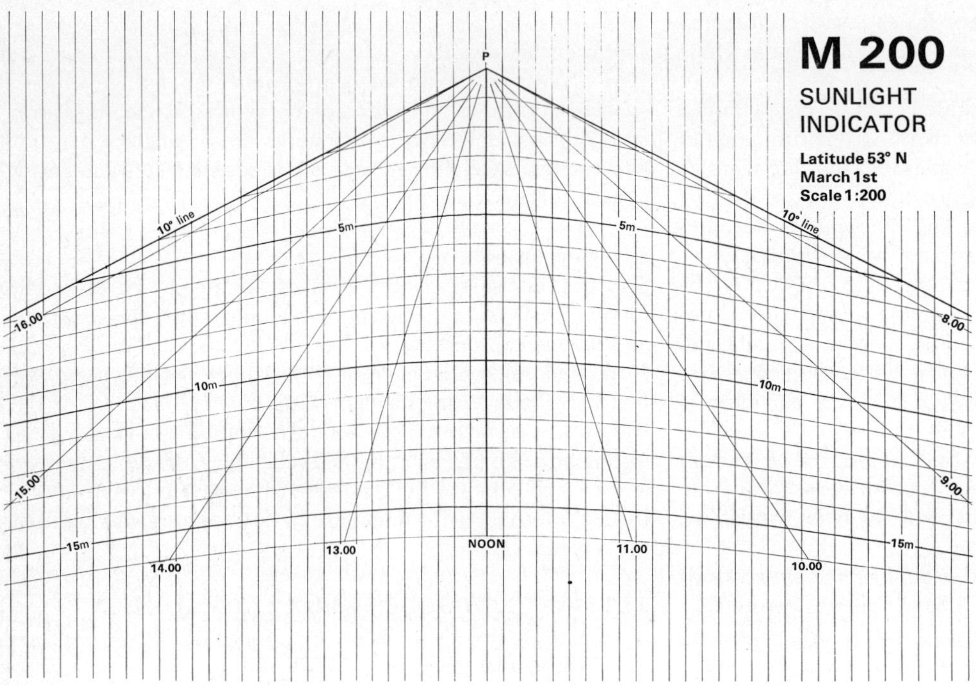

43.10a *Example of a DOE sunlight indicator (shown at reduced scale)*

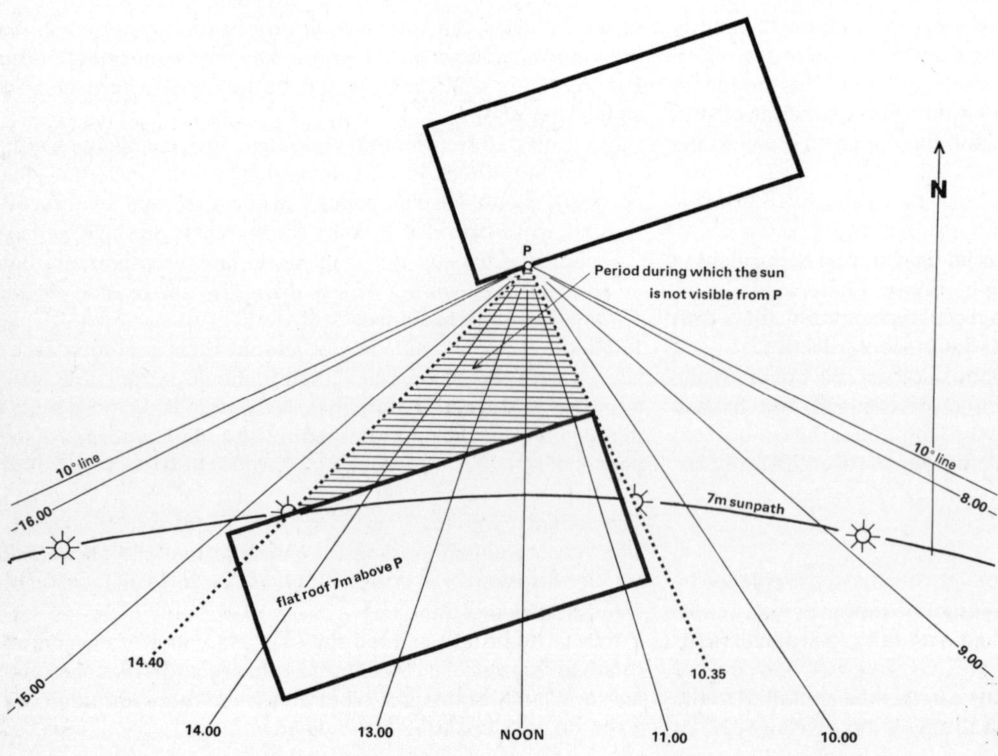

43.10b *The use of the indicator on a housing layout*

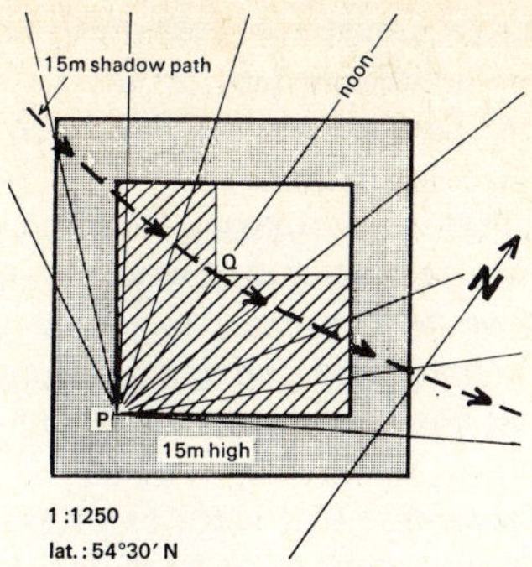

43.11 *Use of the daylight indicator to show the extent of shadow in a courtyard surrounded by buildings 15 m high*

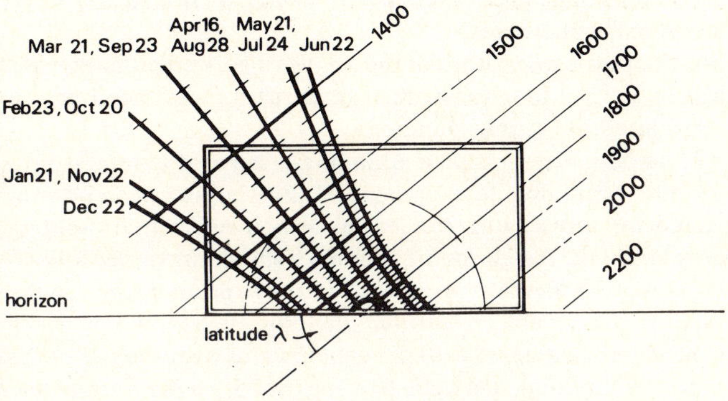

43.12 *Sunpaths as seen through a window (gnomic projection); dates and times of sun entry can be read off the curves*

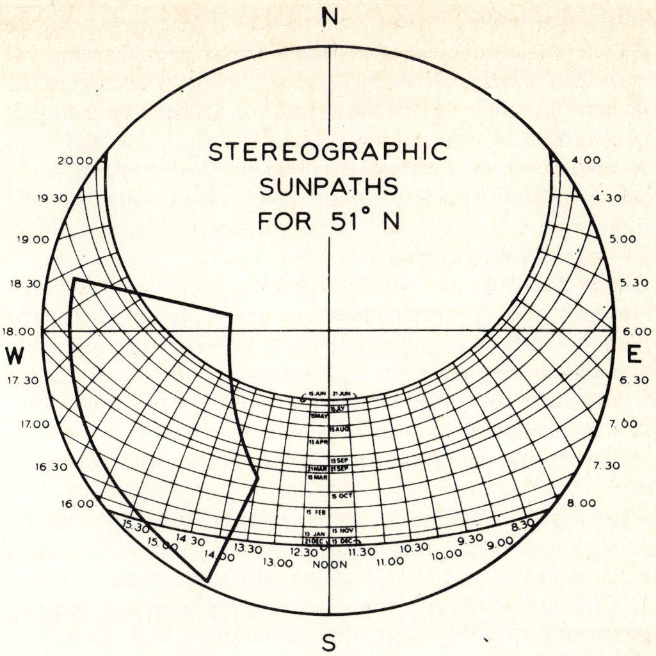

43.13 *Similar exercise to that carried out in* **43.12**, *but using stereographic projection*

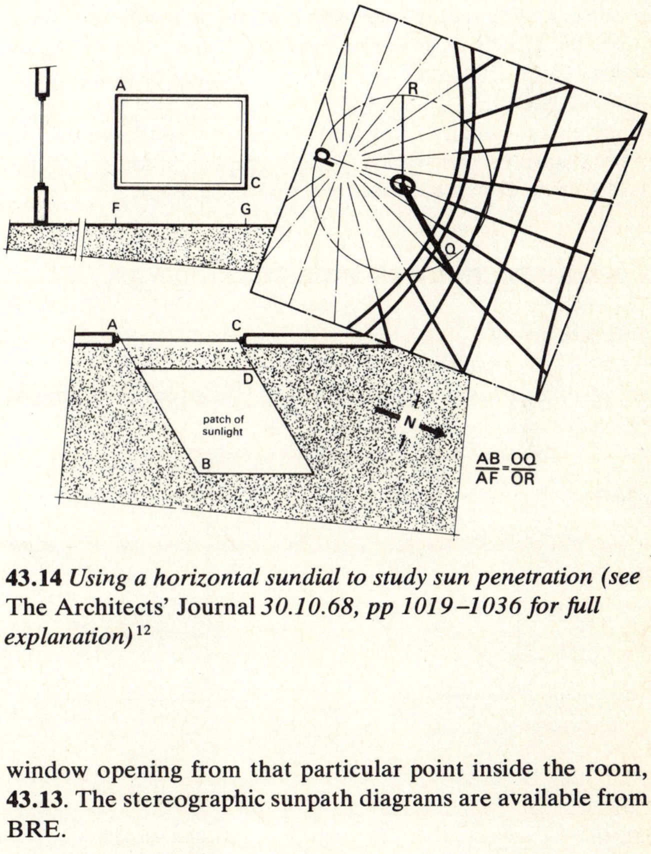

43.14 *Using a horizontal sundial to study sun penetration (see* The Architects' Journal *30.10.68, pp 1019–1036 for full explanation)*[12]

that from point P on the façade, the sun is obscured from 10·35 am to 4·40 pm, on the crucial date of March 1. On a clear day, therefore, the sun will be visible for 4¼ hours, and obscured for 4 hours (between the limits of the 10° lines specified in the criterion quoted in para **3.01** above).

Figure **43.11** shows an example of using the sunlight indicator to determine the extent of shadow at noon, March 1, in a residential courtyard surrounded by buildings 15 m high.

For the method itself, the *Sunlight indicator* set must be consulted.[3, 7] The above examples are merely intended to show the kind of graphic exercise that is involved; they cannot replace the original.

3.02 Room and window design

Penetration of direct sunlight

The principal recommendations for direct sunlight penetration in the UK were quoted in para **1.05**.

The necessary calculations are carried out graphically. A particular design point is chosen inside the room; the window is drawn as seen from that point; and the apparent sunpaths traced by the sun upon the skydome, superimposed upon the window outline, **43.12**.

The periods (times of day, for specific dates in the year) when the sun can be seen through the window, from that point in the interior, are then easily read off the diagram. This will indicate whether the room/window design satisfies the criteria laid down in CP 3.[13]

Three such graphic methods are briefly described below. For more detailed discussion, the quoted references must be consulted.

Stereographic projection

The window outline, as seen from the chosen point in the room, is projected onto the hemispherical skydome. Sunpaths for various dates are also shown, stereographically projected for a particular latitude; and it is easy to read off the times of day, and the dates, when the sun will be visible through the window opening from that particular point inside the room, **43.13**. The stereographic sunpath diagrams are available from BRE.

Gnomonic projection

The window outline, as seen from the chosen point in the room, is projected gnomonically (giving, unlike the previous method, a true rectangular shape). The curves representing sunpaths on specific dates are similarly projected, and the positions occupied by the sun at each time of day are indicated by the intersections between the sunpath curves, and the oblique straight lines, **43.12**. Again, it is easy to read off the times of day, for each particular date, when the sun will be visible through the window opening from a chosen point inside the room. The diagrams for the gnomonic method are available from Pilkington Brothers Ltd.[6]

Horizontal sundial
As a third alternative, the gnomonic sunpath diagram (as described in the previous paragraph) can be placed next to the plan of the room, and the patch of sunlight that would appear in the room on a specific time and date, plotted as shown in **43.14**. This method was fully explained in *The Architects' Journal* 30.10.1968 pp 1019–1036,[12] which should be consulted.

Exclusion of direct sunlight
It is probably unnecessary, in the UK, to carry out detailed graphic studies for this purpose provided window areas are of moderate size, and provided one or more of the following precautions are taken:
• Provide *external shading devices* to all south-facing windows (the most effective measure against overheating).
• Provide internal *cotton curtains* (next most effective measure).
• Use *heat-absorbing glazing* (less effective than either of the above methods; but still useful).
• Provide *venetian blinds*. The ordinary kind, on the inside of an ordinary window, is not a particularly effective form of preventing overheating (cotton curtains are better). But a venetian blind provided between the panes of a double-glazed window is effective.
• Provide the possibility of *high ventilation rates* during hot weather. This is the least effective of all the above measures, because the ventilation rates which would be needed to counteract the effects of excessive insolation on a hot day are so high as to be a nuisance to occupants (papers blowing away; dust entry; etc).
If, however, windows are both large and south-facing (or approximately south-facing), then graphic studies will be necessary to design adequate external sunshades.

4 DESIGN TECHNIQUES: INDIRECT SUNLIGHT

4.01 Site layout
As with *direct* sunlight on southerly façades (para **3.01**), there are no mandatory requirements in the UK for *indirect* sunlight

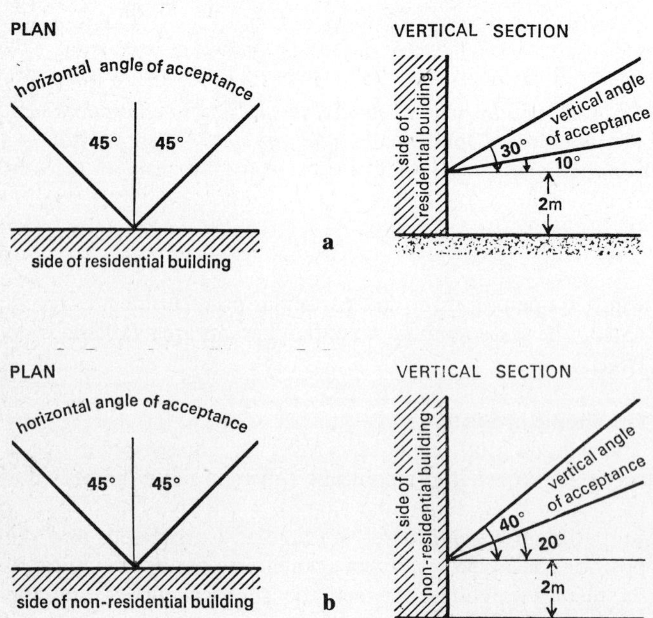

43.15 *Angles within which daylight is 'counted' by DOE criteria. Residential buildings are governed by* **a**; *non-residential by* **b**

on northerly façades; but there are authoritative recommendations which ought to be followed.
The DOE document *Sunlight and daylight*[3] recommends the following:
• *Residential buildings*—In proposed buildings, sides facing due north, or in any direction east or west of north (including due east), should have, at all points 2 m above ground level, a sky component of at least 0.84 per cent between the bearings of 45° to the normal and elevations of 10° and 30° upwards, **43.15a**.
• *Non-residential buildings*—All sides of proposed buildings should have a sky component of at least 0.97 per cent at all points 2 m above ground level and between bearings of 45° to the normal and elevations of 20° to 40° upwards, **43.15b**. This criterion, applying to the whole range of non-residential uses, is no more than a general indication and may not be appropriate in particular cases.
For the purposes of both of the above rules, end or flank walls less than 15 m long (as distinct from main faces) need not be considered as sides of buildings.
• *Plot boundaries*—If land adjoining or across the road from a proposed building development of any kind needs to be protected for foreseeable residential or non-residential building development, the common boundary should have the following sky components at all points 2 m above ground level. In the case of foreseeable *residential* building development, a sky component of at least 4.30 per cent inwards over the site of the proposed building, between bearings of 65° to the normal and elevations of 19° 26′ and 49° 6′, **43.16**. In the case of foreseeable *non-residential* a sky component of at least 2.90 per cent between bearings of 65° from the normal and elevations of 36° 3′ and 59° 13′ upwards.

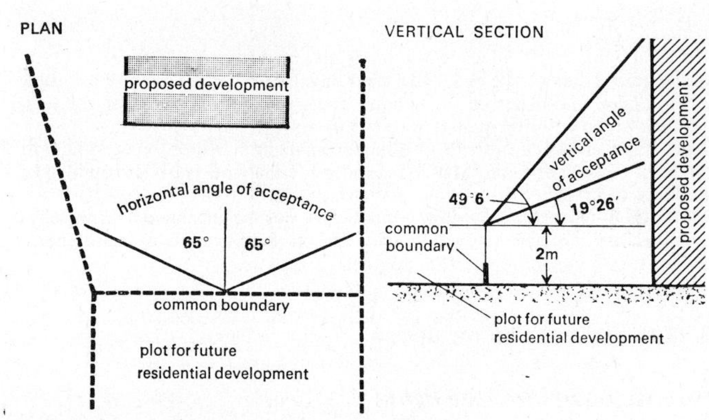

43.16 *Plot boundary criteria for daylight*

Techniques
The recommended method for checking that the above criteria have been met on site layouts is the *Daylight indicator*, issued as a set of 12 transparencies.[8]
Each transparency contains 3 indicators, **43.17**, and there are four transparencies at each of three scales—1:1250, 1:500, and 1:200.
The arcs in the indicators show the restriction on the height and the nearness of obstructions; and the central angle shows the horizontal sweep over which the restriction applies. An example is given below (see further down).
Other methods, some of them more comprehensive or precise than daylight indicators, include the following:
1 *Sunscan*—See under para **3.01** above.
2 *Waldram diagram*—For situations where close design is required, the daylight indicators described above give too limited a range of answers to be of maximum use; and the Waldram diagram (which is difficult) may be used instead. A

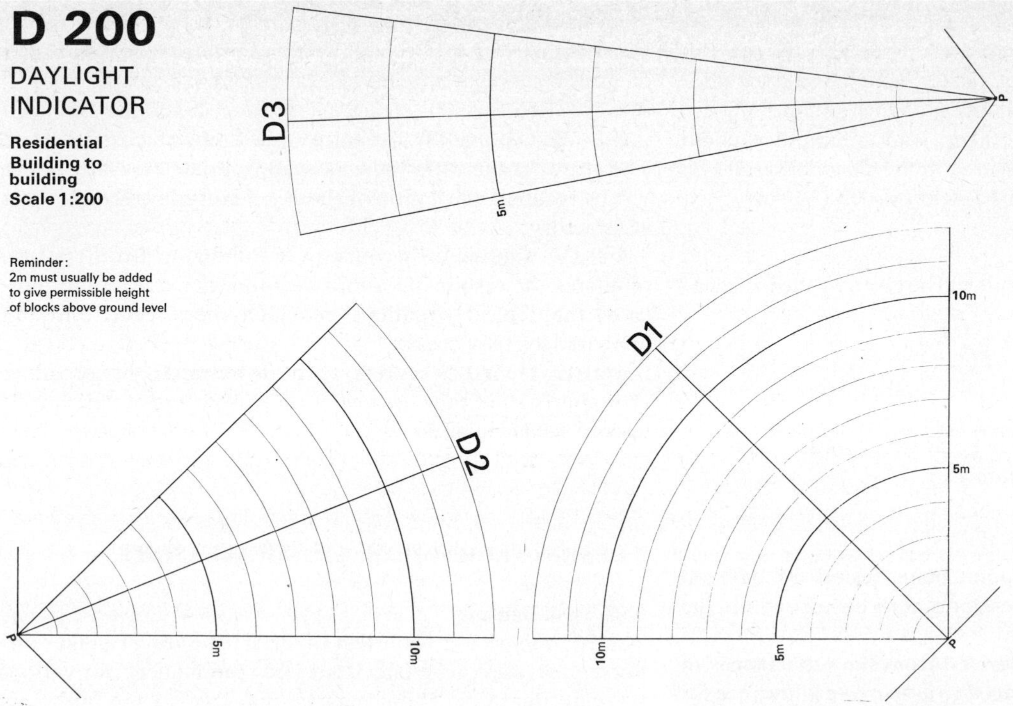

43.17 *Example of a DOE daylight indicator (shown at reduced scale)*

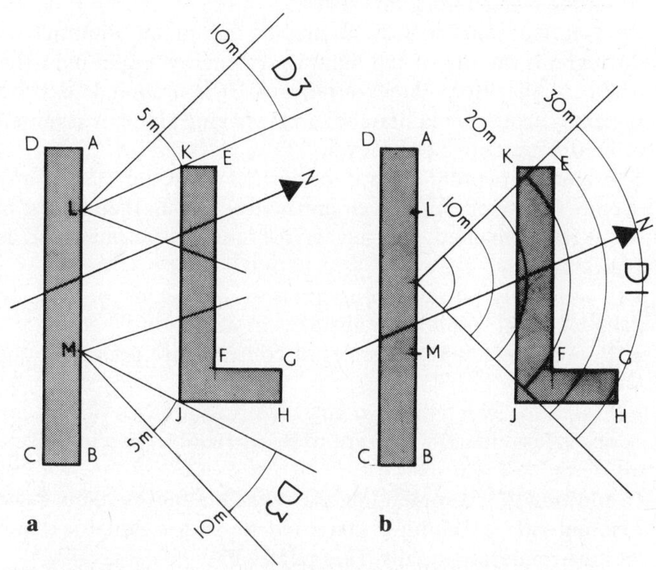

43.18 *Use of the DOE daylight indicator to test a housing layout*

set of sheets specially prepared by Professor Peter Burberry, too detailed to reproduce here, were published in *The Architects' Journal*.[11] They are the most accessible version.

3 *BRS Daylight Protractors*—Again, more useful than the daylight indicator where close design is required. Available from HMSO at 25p each.[4] Full instructions are contained in the separate publication *BRS daylight protractors* by J. Longmore, HMSO 1968[5].

4 *Pilkington Environmental Advisory Service Charts*[6]—See under para **3.01** above.

Example

Figures **43.18a** and **b** use two of the transparencies described under 'Techniques' above, at 1:1250 scale.

Figure **a** tests whether sections AL and BM of side AB (block ABCD) receive enough light round the side of the obstruction EFGHJK to satisfy the daylight criterion for residential buildings. It confirms that they do.

Figure **b** tests the remaining section LM. It is seen that 13 m would be the 'permissible height' for side KJ (the 13 m arc would just touch the façade) from the reference point chosen. If the site is flat, then according to the rules published in the set, 2 m are added to this permissible height, giving 15 m.

For the method itself, the *Daylight indicator*[3, 8] set must be consulted. The above example merely shows the kind of graphic exercise involved, and cannot replace the original.

4.02 Room and window design

Assuming that daylight plays a primary role in the illumination of any particular room, the following techniques apply.

4.03 Task lighting

Task illumination, as explained in para **1.04**, is measured on the horizontal working plane, and is specified in terms of daylight factor. Table III gives recommended daylight factors for various situations, and an explanation of how such daylight factors are calculated for a proposed room and window design follows.

Daylight factor

The DF for a particular point inside the building (normally a series of points on horizontal working planes, 0·85 m above floor level) is calculated by computing three 'components' **43.7**, and adding them together:

• Light falling directly from the *unobstructed sky* to the point under investigation (sky component).

• Light reflected onto the point under investigation from *external surfaces* beyond the window, such as the light-coloured walls of facing buildings (externally reflected component).

• Light reflected from *internal surfaces* within the room, to the point under investigation (internally reflected component).

As some of the light coming in through the window will be obstructed by the kind of glass used; by the presence of dirt on the glass; and by the window glazing bars, a correction factor must be applied to allow for this reduction in light. Conventional factors are recommended in each of the various calculation kits available.

All of the above calculations can be carried out by one or the other of two principal methods:

• BRS Daylight Protractors, published by BRE.[4, 5] Figure **43.19a** shows one stage of the calculation using this method—determination of the *sky component* at a chosen point in the room, with a particular window design.

• Dotted overlays, published by Pilkington Environmental Advisory Service.[6] Figure **43.19b** shows the dotted overlay method of computing *sky component*.

Rule of thumb 1

In the very early design stages, the area of glazing required to give a certain minimum daylight factor may be roughly estimated by the following formula (which is for side-lit rooms only):

P = 10 × D where P is area of glazing as percentage of floor area; and D is minimum daylight factor.

If, for instance, floor area is 100 m^2, and minimum daylight factor 2 per cent, then window area in the sidewall should be roughly 20 per cent (2 × 10) of 100 m^2, which is 20 m^2.

Rule of thumb 2

A more precise preliminary result will be given by the formula:

$$D = \frac{10WH^2}{L(L^2 + H^2)} + \frac{4GR}{F(1\text{-}R)}$$

where W is width of window
H is height of window
L is distance of reference point from window wall
G is area of glass in window
F is area of floor
R is average reflectance of walls.

The 'reference point' is the point being tested for daylight factor (probably a point on a desk or other worktop, within the room area that is to be daylit).

The 'average reflectance' is derived from the reflectances of the various wall surfaces, averaged so as to make allowance for the different wall areas. BS 5252:1976 gives reflectances for a number of sample colour patches; and for estimation purposes these may be simplified to a choice of three—0·5 (light colours); 0·3 (medium); and 0·1 (dark colours).

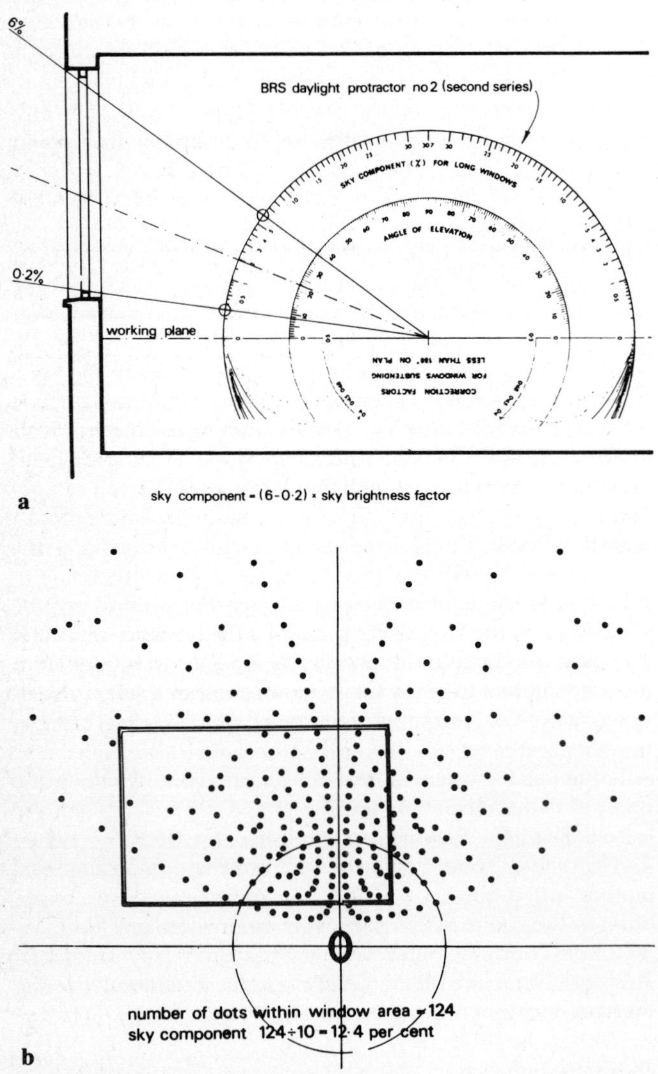

43.19 *Estimating sky component using 'protractor' method,* **a**; *and using 'dotted overlay' method,* **b**

4.04 Amenity lighting

It has been explained in para **1.03** that whereas *task* lighting is measured simply on the horizontal working plane, those qualities of light which create a pleasing luminous environment—that is, *amentiy*—require other forms of measurement. The vector/scalar ratio is the most appropriate measure; and it will be recalled that a ratio of about 1·5 is usually thought most satisfactory.

Table IV, published by courtesy of Pilkington Brothers Ltd, tabulates the results of various vector/scalar ratios; and indicates the typical situations in which they occur, and the appearances they create.

'Low BZ', High BZ', etc refer to the downward light-spreading characteristics of the electrical fittings used; and are explained in para **1.03** (see **43.9**).

5 DESIGN TECHNIQUES: ARTIFICIAL LIGHT

5.01 Task lighting

Table I, published by courtesy of the Illuminating Engineering Society, gives recommended 'service illuminances' for various working interiors. These may be regarded as the minimum conditions which must be satisfied.

By 'service illuminance' is meant the mean illuminance throughout the life of the lighting system, averaged over the whole of the illuminated room area. It is assumed that the system is properly maintained; and 'working plane' is assumed to be 0·85 m above floor level.

These recommendations are for normal situations only. If any of the following special circumstances exist, then table II should be consulted, and the service illuminance increased as indicated:

- If unusually serious consequences (in terms of cost or danger) could result from mistakes in perception.
- If unusually low reflectances or contrasts are present in the task.
- If tasks for which the standard recommended service illuminance is less than 500 lux are to be carried out in windowless interiors.

In addition to recommending a service illuminance, table I also recommends a 'limiting glare index' and a suitable lamp 'colour-rendering' quality (see para **1.03**).

Calculation

It is impossible to do more in the limited space available here than demonstrate the starting point of a lighting calculation for a working environment. The calculation shown is intended to determine the 'installed flux' that is necessary to produce the 'service illuminance' recommended in table I. This quick calculation will show whether *quantity* of light on the working plane is adequate; further checks will be required, perhaps at a later design stage, to ensure that *quality* of light too is satisfactory (see para **5.02**).

1 Establish required illuminance on working plane (from table I).

2 Establish 'utilisation factor' for light fitting to be employed (from tables published by manufacturer).

3 $\text{Installed flux} = \dfrac{\text{illuminance} \times \text{area to be illuminated}}{\text{utilisation factor}}$

Worked example 1

A factory production area measures 30 × 24 m. It is desired to light this area using lamps with an output of 16 000 Lighting Design Lumens each (manufacturers' catalogue will give LDL for particular fittings). How many lamps will be required?

1 From table I, illuminance needed for 'medium work' assembly area is 500 lux.

2 From manufacturers' catalogue, utilisation factor for provisionally chosen lamp fitting is 0·45.

3 Installed flux $= \frac{\text{illuminance} \times \text{area}}{\text{utilisation factor}}$

$= \frac{500 \times 720}{0{\cdot}45}$

$= \frac{36\,000}{0{\cdot}45}$

$= 480\,000$ lumens

4 Number of lamps $= \frac{\text{Installed flux}}{\text{Lighting Design Lumens per lamp}}$

$= \frac{480\,000}{16\,000}$

$= 30$ lamps.

If 30 lamps of the chosen type are required, how are they to be spaced, and at what height, to give uniform illumination over the working area?

While 100 per cent uniformity is impossible, the generally accepted rule is that *minimum* illuminance shall be not less than 70 per cent of *maximum* illuminance.

Assume that the 30 fittings are to be arranged in 5 rows of 6 each; and that mounting height is 2 m above working plane (ie 2·85 m above floor level).

1 Consult catalogue of luminaire manufacturer for maximum ratio of *spacing* to *height* of luminaires. Spacing is centre to centre; height is workplane to centre of lamp.

2 If this information is not given consult table V.

3 In the present case, the information is not given; therefore assume that light source is an unlouvred, plastic-trough fluorescent lamp, with a BZ classification (from manufacturers' catalogue) of BZ 5.

4 From table V, maximum ratio for BZ 5 fitting is 1·5:1.

5 Horizontal spacing has already been established as 5 m; vertical height as 2 m. This gives a ratio of 2·5:1, which is too large.

6 To give a ratio of 1·5:1, mounting height can be increased to 3·3 m. This is however not practical.

7 Alternatively, horizontal spacing can be reduced to 3 m. This would give a higher illuminance than that originally aimed for; but it might be an acceptable option.

8 As a third alternative, choose a different light fitting with lower output, thus allowing closer spacing while still aiming for the original illuminance.

9 Finally, a fitting with a different BZ classification might be tried, allowing a lower ratio of horizontal to vertical spacing.

Worked example 2

It is often more convenient to start with a layout, and find the illuminance this would produce, in order to judge whether the preferred layout will be acceptable.

Assume that two alternative layouts would be geometrically satisfactory, as ceiling layouts. One: two rows of four 85 W fittings. The other: three rows of six 40 W fittings. The space is 6 × 15 m.

1 Illuminance $= \frac{\text{installed flux} \times \text{utilisation factor}}{\text{floor area}}$

2 If the 85 W fittings have a utilisation factor of 0·5 and output of 6800 Lighting Design Lumens each (from manufacturers' catalogue), then

Illuminance $= \frac{2 \times 4 \times 6800 \times 0{\cdot}5}{6 \times 15}$

$= \frac{27\,200}{90}$

$= 322$ lux.

3 If the 40 W fittings have a utilisation factor of 0·5 and output of 2800 Lighting Design Lumens each (from manufacturers' catalogue), then

Illuminance $= \frac{3 \times 6 \times 2800 \times 0{\cdot}5}{6 \times 15}$

$= \frac{25\,200}{90}$

$= 280$ lux.

Table I will show whether these illuminances are acceptable for the kind of workspace being catered for.

Table V Ratios of horizontal spacing to vertical height

BZ classification of luminaire	Ratio of horizontal spacing (centre to centre) and vertical height (plane of work to lamp centre)
BZ 1 and BZ 2	1:1 maximum
BZ 3 and BZ 4	1·25:1
BZ 5 to BZ 10	1·5:1

Some fittings (particularly those using prismatic control) permit much wider spacing than shown above; therefore manufacturers' catalogues should always be consulted if possible. This table is for preliminary design only.

Rule of thumb

In the very early design stages, the following simple formula will enable required lamp wattage to be related to lighting level:

W = kL where W = installed wattage per m^2

L = required illuminance in lux

k = 0·6 (tungsten lamps in room with dark internal surfaces);

0·3 (tungsten lamps in room with light internal surfaces);

0·2 (fluorescent lamps in fittings with low BZ classifications);

0·05 (open fluorescent lamps in room with light coloured internal finishes).

5.02 Amenity lighting

Paragraph **5.01** dealt only with illumination on the working plane—a minimum condition for working situations; but not a full or adequate lighting specification from the point of view of amenity.

The requirements of amenity were discussed in para **1.03**; and there are two principal design methods which will take account of these requirements:

Luminance design method

This method enables two characteristics of light fittings to be established for a particular room—'BZ classification' (see para **1.03**); and 'Flux Fraction Ratio'—which will give the required level of illumination not only on the working plane, but also on wall and ceiling surfaces, thus producing a bright, cheerful atmosphere in the room.

It also acts as a check on the room colours selected. If the required illumination levels cannot be achieved on walls, floor and ceiling, following the method, then the room colours selected are too dark, and the architect is alerted to a decoration problem as well as a lighting one.

The method is too lengthy to be fully described here; designers should consult [10.5].

Integrated design criteria method

The requirements outlined in para **1.03** (good modelling qualities, glare control, and a proper relationship between scalar and horizontal illumination) are likely to be achieved if illumination levels on various surfaces in the room fall within the following ranges:

Task (eg work on desk top)	Immediate surround	Walls	Ceiling
1	1	0·5 to 0·8	0·3 to 0·9

A complete step-by-step procedure for calculation has been developed by Christopher Cuttle; starting with the room description and lighting objectives (step 1) it ends with detailed specifications of suitable lamps (step 15).

The method is too lengthy to be given here; designers should consult [10.8].

Exploitation of side-windows

Finally, it is worth repeating the point already made, that even if primary reliance is placed on artificial light to illuminate a room the horizontal light flow from windows in a side wall can improve the quality of light by adding modelling qualities to the essentially vertical (downward) flow of electrical installed light, **43.5**.

5.03 Lighting for special effect

The methods described so far are for the relatively even spread of light over a large area (such as a room). In certain cases, however, such as display areas, the requirement might be for light to be strongly concentrated on a particular display point. The point-by-point method allows direct illumination to be calculated for any position lit by a light fitting of known intensity at a known distance.

Some manufacturers publish tables or diagrams which can be used directly, for simple cases.

Both the calculation method, and a nomogram which simplifies

Table VI Properties of lamps. A general guide for initial design (for precise up-to-date data, consult manufacturers)

Lamp type	Lamp reference letter (see table I)	Lamp designation and code (capital letters denote BS colours)	Constructional and operational characteristics	Lamp colour appearance	Approximate correlated colour temperature (K)	CIE colour rendering group	Colour rendering characteristics (based on visual assessment)	Typical applications
Tubular fluorescent (MCF)	A	NORTHLIGHT; COLOUR MATCHING	Tubular fluorescent lamps are low pressure linear light sources with differences of colour appearance and colour rendering due to the use of different phosphor coatings. Generally, efficacy decreases as colour rendering improves. Immediate light output when switched on cold or hot. Light output affected by ambient temperature. Can be dimmed if used with special control gear. All lamps have universal operating positions. Smaller wattage lamps are available	Cool	6500	1	Similar to north skylight; emphasizes blues and, to a lesser extent, greens	Applications where colour rendering similar to north sky daylight is needed. Appears cold at low illuminances
	B	ARTIFICIAL DAYLIGHT		Cool	6500	1	Similar to Northlight/Colour Matching, but emits more ultra-violet radiation to conform with natural skylight	As above, but for use where critical colour matching to BS 950 Part 1 is required
	C	DAYLIGHT		Intermediate	4300	3	Emphasizes yellows strongly and, to a lesser extent, greens; subdues reds	Factories. Light blends well with daylight
	D	NATURAL		Intermediate	4000	2	Emphasizes yellows and, to a lesser extent, greens and blues. Reds shift slightly toward orange	Offices, department stores
	E	°Kolor-rite; Trucolor 37		Intermediate	4000	1	Equal emphasis given to all colours	Display lighting. Shops, hospitals (clinical areas), museums
	F	Colour 84		Intermediate	4000	2	Emphasizes oranges, greens and blue-violets, but subdues yellows and deep reds	Offices, department stores
		Plus-White			3600	2	Emphasizes yellows, oranges and greens; subdues deep reds. Blues shift towards violet	Offices, department stores
	G	de luxe Natural		Intermediate	3600	1	Equal emphasis given to blues, greens and yellows, but more to reds	Shops
	H	WHITE		Intermediate	3500	3	Emphasizes yellows and, to a lesser extent, greens. Subdues reds and, to some extent, blues which shift towards violet	General purpose use where high efficacy is primary requirement
	I	WARM WHITE		Warm	3000	3	Emphasizes yellows and, to a lesser extent, greens. Reds slightly subdued. Subdues blues which shift towards violet	General purpose use where high efficacy is primary requirement
	J	de luxe Warm White; Softone 32		Warm	2700 3000	2	Emphasizes yellows and, to a lesser extent, greens. Subdues blues which tend towards violet	Hotels, restaurants, homes
High pressure discharge	K	Mercury halide (MBI)	High pressure mercury lamp with metal halide additives in fused silica arc tube; clear outer bulb. Some ratings available in reflector form. Run-up period to full light output about 5 min. Re-ignition about 10 min; less if special circuits are used.	Intermediate	4200	2	Equal emphasis given to greens and blues; more to yellows. Red rendering variable according to lamp type and manufacturer	Industrial and commercial applications, eg high-bay factories, area lighting, shops and offices
		Mercury halide (MBIL)	MBI lamp in linear form, normally with outer bulb. Horizontal operation only		4200	2	As for MBI	As for MBI
		Mercury halide fluorescent (MBIF)	MBI lamp with fluorescent coating on inside of outer bulb		4000	2	As for MBI	As for MBI

Table VI Properties of lamps (*continued*)

Lamp type	Lamp reference letter (see table I)	Lamp designation and code (capital letters denote BS colours)	Constructional and operational characteristics	Lamp colour appearance	Approximate correlated colour temperature (K)	CIE colour rendering group	Colour rendering characteristics (based on visual assessment)	Typical applications
	L	Mercury fluorescent (MBF)	High pressure mercury lamp with fused silica arc tube and outer bulb with fluorescent coating to increase proportion of red light. Run-up and re-ignition times as for MBI	Intermediate	4000	3	Emphasizes yellows and blues which shift towards violet. Subdues reds	Industrial and commercial applications, eg shops, and offices, high-bay factories and area lighting, roadways
		Mercury fluorescent reflector (MBFR)	As for MBF but with internal reflector		4000	3	As for MBF	As for MBF
		Mercury-tungsten fluorescent (MBTF) (blended)	An MBF lamp with tungsten filament in series with arc tube to act as ballast and increase red light emitted		3600	3	Emphasizes yellows and blues which shift towards violet. Subdues reds to some extent	May be used as direct replacement for some GLS incandescent lamps, particularly where long life is important and access for maintenance is difficult
	M	High-pressure sodium (SON)	High pressure sodium lamp with arc tube in diffuse outer bulb. Run-up period to full light output 2 min. Re-ignition can take place within 1 min if external ignitor is used	Warm	2100	3	Emphasizes yellows strongly, reds to a lesser extent. Greens acceptable. Blues, which shift towards violet, strongly subdued	Industrial and commercial applications, eg public buildings, high-bay factories, area lighting, roadways
		High-pressure sodium (SONT)	SON lamp with clear tubular outer envelope		2100	3	As for SON	As for SON
		High-pressure sodium (SONL)	Linear SON lamp with clear slim tubular outer envelope. Enables higher degree of light control to be achieved if required		2100	3	As for SON	As for SON
		High-pressure sodium (SONR)	SON lamp with internal reflector bulb		2100	3	As for SON	As for SON
Incandescent	P	Tungsten halogen (T/H)	Compact tungsten filament lamps with clear or diffusing silica tube or bulb containing halides or halogens which prevent blackening and increase life and/or efficacy which is relatively low. Linear types restricted to horizontal operation only	Warm	2900	1	Emphasizes reds strongly, yellows and greens to lesser extent. Blues strongly subdued	Display and area lighting
	Q	Tungsten—General lighting service (GLS)	Tungsten filament in clear or diffusing glass bulb filled with inert gas. Relatively low efficacy. Immediate full light output. Light output sensitive to voltage variations. Easily switched and dimmed	Warm	2700	1	As for T/H	Hotels, restaurants, homes
		Tungsten reflector	Tungsten filament in internally metallized blown or pressed bulb. Available in a range of reflector contours		2700	1	As for T/H	Display and accent lighting

Published by courtesy of the I.E.S.

the process, are available from several sources, which should be consulted. There is insufficient space for reproducing it here. See reference [10.7], and appendix 6 of reference[2].

5.04 Lighting for safety

BS 5266 defines three functions for escape lighting:

- To indicate escape routes in buildings (in case of fire) clearly and unmistakably; and to indicate the exits.
- To illuminate such routes, thus allowing safe movement towards the exits.
- To illuminate fire alarm call points and firefighting equipment.

Emergency lighting for the above purposes may be of two kinds, a maintained system, or a non-maintained system.

With *maintained* systems, all emergency lamps are in operation at all times. They have the advantage that failures of either circuit or lamp will be immediately obvious (instead of becoming obvious only when the light is needed), and can be rectified before an emergency arises.

With *non-maintained* systems, the lamps come into operation only when the normal lighting system fails.

Sources of power may be of three kinds:

- From a central generator (unsuitable in many cases, because BS 5226 requires emergency lighting to come on within 5 seconds of mains failure, whereas most generators cannot be assumed to start in less than 12 seconds or more).
- From a central battery (which has the disadvantage, compared with option 3, of requiring an expensive central master unit with wiring to each individual lamp. The latter must be quite separate from, and additional to, the building's normal wiring circuits).
- From local battery systems, contained within self-powered individual luminaires. Self-contained luminaires are more expensive, individually, than the simple lamps associated with central power supplies; but they cut out the cost of the wiring. Therefore they will prove cheaper, overall, when installed in *existing* buildings (where new circuitry will be expensive to install); and also in *new* buildings where cabling involves long or complex runs.

It seems clear that for most purposes the self-contained luminaires will provide the best solution. Because they are of fairly recent origin, there is a lack of official standards; and of reliable information on length of battery life. The latter is thought to be perhaps 5 years, compared with perhaps 20 years for the battery types used in central systems.

Recommended illuminances

All exits and escape routes in places of public assembly must be clearly identified by signs reading 'Exit' or 'Emergency exit'—see BS 5266 and BS 2560 for details of lettering size etc. In a darkened auditorium, an illuminance of 0·02 lux is accepted, provided another 0·18 lux is added from a non-maintained system when the mains fail, to bring the illuminance up to 0·2 lux.

For other buildings the required illuminance is 0·2 lux; and where the general service illuminance in the room is more than 500 lux (the recommended figure for, eg, airport booking offices and reception areas) this should be increased to 0·5 lux.

There is no particular spacing laid down for luminaires along the exit route; experience, and an inspection of existing buildings, may provide the best guide. Dark areas must be avoided, particularly on stairs, landings and at emergency exit doors.

6 LIGHT SOURCES

6.01 Incandescent lamps

Consist of a coiled filament in a glass bulb; and come in various forms: the familiar pear shape (called gls, meaning 'general lighting service'); a tubular shape, with the filament strung out along the length of the lamp; and the 'reflector' type, which is mushroom-shaped and has a silvered inside to give a spotlight effect.

Rated life is standardised at a nominal 1000 hours—lower efficacy lamps could have an extended life; and conversely, longer-life lamps have the disadvantage of lower output. A more expensive type, the 'tungsten halide', has a rated life of 2000 hours.

Incandescent lamps give a warm, pleasant light which is accepted by the human eye as 'natural', even though in fact its spectrum is very different from that of sunlight.

Appropriate use

Incandescent lamps are cheap to install, and give a pleasant, warm light. But they are expensive to run, and generate a lot of heat relative to light output. They are appropriate for domestic use; and for display purposes, prestige interiors, and where 'sparkle' or other special effects are required. But they would not normally be sensible for ordinary working interiors such as offices, factories, or large shops.

6.02 Fluorescent tubes

Consist of a glass tube filled with mercury vapour at low pressure, through which an electric arc is struck. This produces some visible blue light, plus ultra-violet radiation. The latter is converted into visible light by the chemical coating on the inside of the glass tube.

Colour rendering depends on the type of chemical coating used in the tube; and there are two broad options. One is to go for *maximum efficacy*, at the expense of colour rendering. This produces the 'high-efficacy' class of fluorescent lamps, giving 50 to 80 lumens of light output per watt of electricity, but providing less than ideal colour rendering. The other is to go for *good colour*, at the expense of efficacy—this produces the 'de luxe' class of fluorescent lamps, giving only 30 to 55 lumens per watt. For comparison, incandescent lamps deliver 10 to 18 lumens per watt; and the tungsten halogen incandescent lamp (more expensive than the conventional type) up to 22 lumens per watt.

There are three kinds of 'high-efficacy' tube, in terms of colour-rendering—daylight (the coolest colour); white (intermediate); and warm white (the warmest).

There are four principal kinds of 'de luxe' tube, in terms of colour rendering. Arranged in sequence from the coolest colour to the warmest, they are Northlight and Colour-matching (both giving rather a cold light at low illuminances); Natural; and De luxe warm-white (the latter warm enough to be suitable for hotels, restaurants and homes).

Table VI lists all the main types, together with their characteristics and uses.

Lamp life is less important than with incandescent lamps, because while life expectancy is very long, light output falls progressively; and it is worth replacing the lamp well before the end of its lifetime, in the interests of light output. It is therefore 'replacement period' rather than 'expected life' that is important with fluorescent tubes.

6.03 Discharge lamps

Like fluorescent tubes, discharge lamps generate light by the passage of electricity through a gas; but in this case the electrical arc produces light directly, instead of producing ultra-violet radiation which is then converted to visible light by a fluorescent coating on the glass.

Colour rendering tends to be very 'unnatural'; and most of these lamps are suitable only for industrial and commercial use (see table VI). But the 'blended mercury-tungsten fluorescent' type, which is of intermediate colour temperature, blends with

incandescent lamps, while being more economical in use. Efficacies are high—between 35 (for high-pressure mercury lamps) to 100 lumens per watt (for high-pressure sodium).
Lamp life exceeds 'replacement period'; therefore the notes under para **6.02** apply here.

6.04 Choice between light sources

In general, choice should be clear from the above notes and table VI.
A final point is the relationship between fluorescent tube colour and illuminance level.
While cool-coloured lamps are satisfactory where illuminance levels are high, they give a depressing effect at low levels. Therefore Northlight and Colourmatching tubes (which are cool) should be used only at illuminances of 500 lux and more; and Natural (which is intermediate) only at 350 lux or more.

7 LUMINAIRES

7.01 Incandescent lamps

There are basically three luminaire types for incandescent lamps.
The first group of luminaires are intended to give a mainly diffused light, for general space illumination. The totally enclosed opal sphere gives a good glare-free light distribution, but tends to suck in dirt owing to the phenomenon known as 'breathing'. When hot, the air inside the globe expands and is forced out. When the light is switched off, the cooling air contracts, sucks in outside air (containing dust), and the dust remains inside. Therefore opal shades with open bottoms tend to be more common; but in this case the drawback is that of glare, owing to the bright filament being visible from certain angles of vision.
The second group of luminaires are highly directional. Spotlights are one example: they can be very effective, but must be situated in positions where they do not blind occupants. The recessed downlighter is another example. Unlike the spotlight it is fixed and unadjustable; and often its purpose is general illumination of the area below, rather than display spotlighting. But for general lighting the downlighter gives a harsh and unpleasant modelling quality, particularly of faces; and it should be used with care.
The third group of luminaires are deliberately intended to give 'sparkle' to a room: like the chandelier of earlier days. This should not be overdone, excessive brightness can cause severe visual distress.

7.02 Fluorescent fittings

The simplest type is the *batten* fitting: simply a metal channel housing the control gear, and with one or two tubes held on the lid. A metal trough may be added to the batten half enclosing the tubes; and this acts as a reflector. If the trough is made of translucent plastic, it also diffuses some of the light. All of these types are fairly rudimentary, and mainly for industrial and similar uses.
A second type is the *box* fitting: in this case the control gear is inside the box, together with the tubes. The underside of the box is generally covered with some diffusing element (opal sheet, louvres, etc), and this may be adjustable. The latter facility allows the box to be recessed above a ceiling, and the diffusing cover to be accurately adjusted to match ceiling plane. All 'recessed' fittings are of the box type.
A third arrangement is to do without either batten or box, and simply to mount the tube, on its own, in a concealed position (eg behind a pelmet). The control gear is then mounted separately, on a convenient surface, and connected to the tube caps by cable.

7.03 Characteristics of luminaires

Luminaires should not be selected from catalogues simply on a basis of appearance. They have a vital effect on all the aspects of light distribution pattern, glare, and light output, discussed in earlier paragraphs.
The following technical characteristics of luminaires are basic to the computations carried out in para **5.01**.

British Zonal classification

The distribution of light from a light source and its enclosing luminaire can be plotted graphically to give a 'polar curve', an example of which is shown in **43.20**.

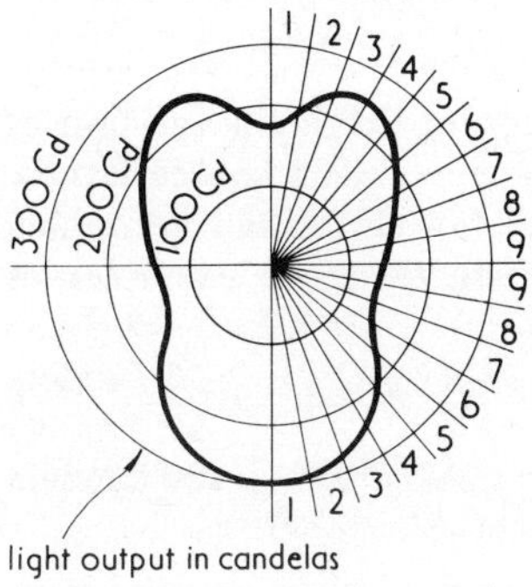

43.20 *Polar curve for one particular light fitting. Intensity of light output (in candelas) is plotted all round fitting, in vertical section, at 10° intervals. Each fitting type has its own directional characteristics*

The graph represents a vertical section through the light source. The heavy curve indicates intensity of luminous output: intensity is proportional to distance from the centre. Clearly, in the case illustrated, intensity of output in the *downward* direction is greater than in the *upward*; and the sideways output is less intense than either of the above.
A simple classification method, enabling luminaires to be specified in terms of this light distribution pattern, is provided by the British Zonal (BZ) system.
Put simply, it takes that proportion of light emitted *below* the horizontal (upward light is ignored), and indicates the pattern of downward distribution. A BZ 1 fitting gives out most of its downward light within 45° of the vertical; a BZ 10 fitting gives out most of its downward light between 45° and the horizontal. BZ 2 to BZ 9 form a series between these two extremes, **43.9**.
BZ classification has obvious importance for *glare prevention*—the greater the sideways light output, the greater the risk of glare to observers seeing the luminaire from a distance, **43.6**.
It also has significance for direction of light flow in a room, and on relative brightness of floors, worktops, and walls. The

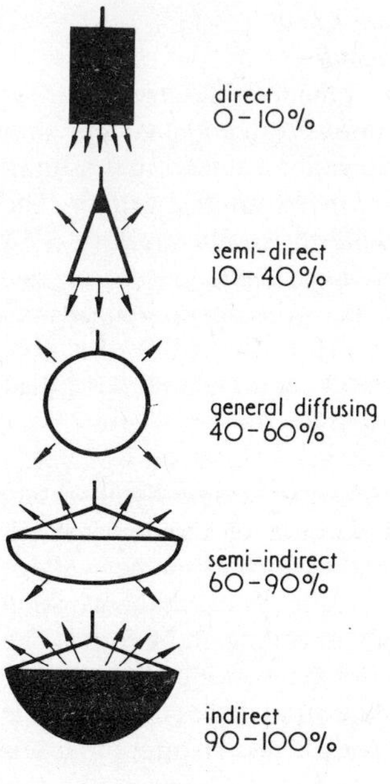

43.21 *Luminaires classified in terms of upward and downward light emission. Percentages of upward light output shown at right. A fitting with 'flux fractions' of 25 per cent upward, and 75 per cent downward, would have 'flux fraction ratio' of 0·33 (ratio of 25:75). Diagram from reference[2]*

greater the sideways light output, the brighter the walls will appear (see para **1.03**).

Flux fraction ratio
The is the ratio of the *upward* flux (light output) from a luminaire, to the *downward*. For instance a semi-direct luminaire of the type shown in **43.21** could have 'flux fractions' of 20 per cent up, and 80 per cent down (meaning that 20 per cent of its light output was upward, and 80 per cent downward). Its flux fraction ratio would then be 0·25 (because 20 is one quarter, or 0·25, of 80).

8 UNITS AND DEFINITIONS

Candela
The initial unit in the specification of lighting is the unit of *source intensity*, the *candela*. A one candela source is now defined in terms of a small white-hot cavity, but is roughly equal to a large domestic wax candle. Some examples of luminous intensity are:

Torch bulb	1 candela
60 W tungsten lamp	50 candela
Twin-tube fluorescent fitting	400 candela

Lumen
Quantity of light (*luminous flux*) is expressed in *lumens*. A one candela source continuously emits one lumen of light flux into each steradian (unit solid angle) around itself. The total flux emitted by a source is therefore 12·566 (4π) lumens per candela, assuming that it emits uniformly in all directions. Some examples of luminous flux are:

Bicycle lamp	10 lumens
150 W spot lamp (tungsten)	2 000 lumens
140 W sodium street lamp (emission)	13 000 lumens

Lux
The intensity with which a surface is illuminated (its *illuminance*) is measured in *lux*. An illuminance of 1 lux is provided if a luminous flux of 1 lumen falls on each square metre of the lit surface. This surface may be in any orientation to the incident but is usually taken as a plane. Paragraph **1.02** gives some examples of illuminance on horizontal planes.
Service illuminance (also measured in lux) is the mean illuminance throughout the life of an installation and averaged over the relevant area; this area may be the whole area of the working plane in an interior, or the area of the visual task and its immediate surround.

Candela per square metre and apostilbs
The brightness of a surface (its *luminance*) is measured in terms of its equivalent source intensity (candelas) per unit area. An alternative unit is the apostilb: a surface has a luminance of 1 apostilb if each square metre emits 1 lumen. The units are related so that 1 candela/m^2 is equivalent ot 3·14 apostilbs. Some examples of luminance are:

Dark ceiling in a 'gloomy' office	3 cd/m^2 (10 apostilbs)
White paper on a desk	130 cd/m^2 (400 apostilbs)
Bright, overcast sky	1600 cd/m^2 (5000 apostilbs)

Utilisation factor
The luminous flux emitted by a light source is partly absorbed within the light fitting (luminaire), and of the light which escapes some falls directly onto the working plane while a proportion of the remainder may eventually reach the working plane after reflection from the walls or ceiling of the room. The ratio between the flux which reaches the working plane (either directly or after reflection), and the output flux of the source, is the utilisation factor. Because of the reflected component, it is not just a function of the properties of the luminaire, but also depends on the proportions of the room and its surface reflectances.

9 REFERENCES

[1] Illuminating Engineering Society *The IES Code: recommendations for lighting building interiors,* London, 1977, The Society. (An essential reference for all lighting design)
[2] Electricity Council and Lighting Industry Federation Ltd *Interior lighting design*, London, 1977
[3] DOE and Welsh Office *Sunlight and daylight*, London, 1971, HMSO
[4] Building Research Station *Daylight protractors*, London, HMSO
[5] J. Longmore *BRS daylight protractors*, London, HMSO, 1968
(Gives full instructions for use of BRS daylight protractors)
[6] Environmental Advisory Service *Windows and environment*, Pilkington Brothers Ltd, 1969
[7] DOE and Welsh Office *Sunlight indicators*, London, HMSO, 1971
[8] DOE and Welsh Office *Daylight indicators,* London, HMSO, 1971
[9] Illuminating Engineering Society *Daytime lighting in buildings*, and supplement, London, 1972, The Society
Contains 'daylight glare' calculating procedure for use with daylight factor calculation
[10] *The Architects' Journal* AJ Handbook of Building Environment, section 9: Electrical lighting. Articles listed below were published on the following dates (they are *not* available in book form; nor as back numbers from the Architectural Press).

Technical studies:

[10.1]	TS 1: Light and vision	4.6.69, 1527–1544
[10.2]	TS 2: What sort of light?	11.6.69, 1589–1597
[10.3]	TS 3: Equipment and techniques	11.6.69, 1599–1611

Design guide:

[10.4]	Electrical lighting	18.6.69, 1665–1669

Information sheets:

[10.5]	IS 1: Luminance design method	18.6.69, 1671–1678
[10.6]	IS 2: Scalar illumination	9.7.69, 61– 64
[10.7]	IS 3: Point source calculations	9.7.69, 65– 67
[10.8]	IS 4: Integrated design method	30.7.69, 245–256
[10.9]	IS 5: Office lighting	16.7.69, 115–117
[10.10]	IS 6: Shop lighting	16.7.69, 119–120
[10.11]	IS 7: Factory lighting	16.7.69, 121
[10.12]	IS 8: Warehouse lighting	16.7.69, 123–124
[10.13]	IS 9: Indoor swimming pool lighting	16.7.69, 125–127
[10.14]	IS 10: Library lighting	16.7.69, 129–131
[10.15]	IS 11: School lighting	16.7.69, 133
[10.16]	IS 12: Church lighting	16.7.69, 135–136
[10.17]	IS 13: Hostel lighting	23.7.69, 185–186
[10.18]	IS 14: Hotel lighting	23.7.69, 187
[10.19]	IS 15: Lamp and tube data	23.7.69, 189–190
[10.20]	IS 16: Luminous ceilings	23.7.69, 191–192
[10.21]	IS 17: Cleaning and maintenance	23.7.69, 193
[10.22]	IS 18: Estimating annual costs	23.7.69, 195–196
[10.23]	IS 19: Illuminated signs	23.7.69, 197–202

[11] Peter Burberry An improved method of daylight prediction, *The Architects' Journal* 27.6.1973, pp 1573–1593
[12] Sunlight: Predicting the sun's track, *The Architects' Journal*, 30.10.1968, pp 1019–1036. Information sheet 5 of AJ Handbook of Building Environment, section 2: Sunlight.
[13] British Standards Institution BS CP 3 Chap 1 (B) *Sunlight. Houses, Flats and Schools*, London, HMSO

44 Sound

Neil Spring
Sandy Brown Associates

CI/SfB (M)
CI/SfB (1976 revised) (P)

Neil Spring is an acoustic consultant

Contents

1 INTRODUCTION

Sound affects the occupants of a building in two distinct ways:

- the quality of sounds generated within, eg a concert hall,
- annoyance with loud noise. (*Noise* is a term used to describe unwanted sound.)

The factors that determine sound quality are still imperfectly understood, despite much recent research and experience. The same is true to a lesser extent of the factors affecting noise. Part of the difficulty is that asking someone how he feels about his environment may itself modify his reactions.

Very loud prolonged sounds, such as occur in some industries, can result in permanent damage to the ear. The fear that they can damage buildings has been exaggerated, and need not normally be considered by the architect.

Many acoustical problems in buildings can be avoided by considering the broad requirements early in the design process. Later on, rectification is rarely satisfactory or economical.

2 FUNDAMENTAL ACOUSTICS

Sound is perceived when the ear-drum is set vibrating by variations in the air pressure just outside the ear. These pressure variations will have been caused by some vibrating object, said to radiate sound. The simplest kind of sound is a single pure tone, for which the graph of air pressure plotted against time produces a sine wave, **44.1**. The greater the amplitude A of the pressure variation, the louder the tone. The more rapid the variation (ie the higher the frequency), the higher the *pitch* of the tone.

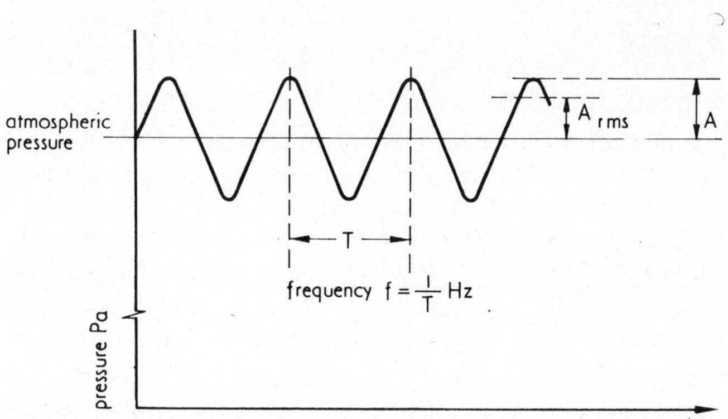

44.1 *Sinusoidal variation of air pressure at a point due to a pure tone*

2.01 Sound pressure level

The *sound pressure* of a pure tone generally means the root mean square value of the variation in pressure of the air due to the sound, not the amplitude or peak value A. This is because the rms value is the best measure of energy whether or not it is a pure tone. For a sine wave, the rms value is the amplitude divided by $\sqrt{2}$.

Audible sound pressures extend approximately from 2×10^{-5} Pa (the quietest sound that most people can hear—the threshold of hearing) to 100 Pa (sound so loud that it starts to be actually painful). This enormous range is telescoped by using a logarithmic notation employing the *decibel.*

The *sound power* of a pure tone is proportional to the square of the pressure. The ratio of the powers of two sounds is therefore the square of the ratios of the pressures. The *sound pressure level* of a given sound expressed in decibels is ten times the common logarithm of the ratio of its power and that of the internationally recognised threshold of hearing:

$$\text{ie SPL} = 10 \log_{10}\left\{\frac{p_1^2}{p_0^2}\right\} = 20 \log_{10}\left\{\frac{p_1}{2 \times 10^{-5}}\right\} \text{(decibels dB)}$$

Expressed in this way, the audible range extends roughly from 0 dB SPL to 134 DB SPL. A sound usually seems twice as loud when its pressure has trebled, ie it has increased by 10 dB.

2.02 Sound power level

The sound power of a source, measured in watts, is deduced from measurements of sound pressure in a well-defined acoustical environment, such as an anechoic chamber or a reverberation room. The range of powers being even greater than that of pressure levels, decibel notation is again used. The international reference level for sound power is 1 picowatt, so the *sound power level* is given by:

$$\text{SWL} = 10 \log_{10} \frac{W_1}{10^{-12}} \text{ db where } W_1 \text{ is the sound power in watts.}$$

The SPL produced at a particular place will depend on the distance, orientation and SWL of the source; and also on the amount of acoustical *absorption* present. It is important in any application to distinguish between SPL and SWL as they are usually numerically quite different.

2.03 Frequency

The frequency of a pure tone is measured in hertz (Hz), equal to and formerly called cycles per second, **44.1**. The human range of audible frequencies varies, but is roughly from 20 to 20 000 Hz. The ability to hear higher frequency sounds progressively deteriorates with age.

Any steady sound, however complex, can be reproduced by combining enough pure tones of the right amplitudes and frequencies. Thus, if we know the behaviour of a material, wall or room, etc with regard to the audible spectrum of pure tones, we can predict the behaviour with any steady sound.

The acoustical properties of materials or rooms are described in relation to contiguous frequency bands. Octave-band measurements are carried out where economy is needed. One-third octave bands or smaller are used when greater accuracy is required.

Octave-band centre frequencies: 63, 125, 250, 500, 1000, 2000, 4000 Hz

One-third-octave band centre frequencies: 50, 63, 80, 100, 125, 160, 200, 250, 315, 400, 500, 630, 800, 1000, 1250, 1600, 2000, 2500, 3150, 4000, 5000, 6300, 8000 Hz

2.04 Reverberation time

The *reverberation time* is defined as the time taken for an interrupted sound to fall in level by 60 dB. The reverberation time and its variation with frequency is probably the most significant measurable factor determining the acoustical character of a room, and it can be calculated from Sabine's formula:

$$T = \frac{0{\cdot}16V}{S\bar{a} + xV}$$

where V = room volume in m^3
$S\bar{a}$ = total surface absorption in m^2
x is a coefficient related to the sound attenuation of air.

The total surface absorption is obtained by adding together the separate areas of absorbent:

$$S\bar{a} = S_1 a_1 + S_2 a_2 + S_3 a_3 + \ldots + S_n a_n$$

where S_1 is the area in m^2 with absorption coefficient $\bar{a}_1$ etc.

Table I gives examples of typical absorption coefficients.

The Norris-Eyring formula is accepted as more accurate for rooms with a high average absorption coefficient:

$$T = \frac{0{\cdot}16V}{S[-2{\cdot}30 \log_{10}(1 - \bar{a})] + xV}$$

The range of T varies from 0·25 s for a small pop recording studio, to over 10 s for a large cathedral. For clear speech in a lecture hall or theatre, T should be between 0·7 and 1·0 s. Outside these limits, a specially designed speech reinforcement system will be required.

For most uses the RT (reverberation time) should be the same for low, middle and high frequencies. A moderate rise in the bass RT is acceptable for speech, and is often considered preferable for music; this is fortunate as sufficient low frequency absorption is seldom obtainable. In large auditoria the high frequency RT inevitably falls because the enclosed air has a high absorption value. In any case allowance must be made for the absorption of the audience, usually the greatest single component.

The acoustical shortcomings inevitable in multi-purpose auditoria can be alleviated by installing an electroacoustic artificial reverberation system such as the *assisted resonance* system at the Royal Festival Hall in London.

3 ROOM SHAPE AND QUALITY

3.01 Preferred dimensions—large auditoria

The size of a room is usually determined by factors other than the acoustics.

Very large auditoria are difficult to fill with sound. Most of the direct sound will never reach the remoter parts, and the large surface area produces a corresponding absorption resulting in a weak level of reverberant sound.

Little evidence for an ideal set of proportions exists for auditoria of conventional shape. Unconventional shapes should generally be avoided because there is a real danger of incorporating intolerable defects impossible to correct.

Successful concert halls are usually rectangular both in plan and section, eg Symphony Hall, Boston. This shape is convenient, and produces the reverberation time of up to two seconds preferred for symphonic music. Where a shorter reverberation time is desired, other shapes are satisfactory, such as the horseshoe of the traditional opera house, and the fan for a theatre.

Table I Absorption coefficients

	Frequency Hz						
	63	125	250	500	1000	2000	4000
Air, x (per m^3)	0	0	0	0	0·003	0·007	0·002
Audience seated in fully upholstered seats (per person) m^2	0·15	0·18	0·40	0·46	0·46	0·51	0·46
Orchestral player with instrument (average), m^2	0·18	0·37	0·8	1·1	1·3	1·2	1·1
Carpet, pile over thick felt on concrete floor	0·05	0·07	0·25	0·5	0·5	0·6	0·65
Plaster, on solid backing	0·05	0·03	0·03	0·02	0·03	0·04	0·05

The likelihood of audible echoes in large auditoria can usually be predicted, but conditions producing them are too various to summarise. Elimination of echoes is often expensive rather than technically difficult: an elegant example being the 'flying saucers' in the Royal Albert Hall, London.

Auditoria for speech and drama have clarity as a prime requirement. The shape should ensure that the audience receives strong sound reflections immediately after the direct sound.

For musical performances many prefer early sound reflections arriving at the listener from a lateral direction.

3.02 Preferred dimensions—small rooms

Small rooms can present serious acoustical problems, being often bedevilled with *colouration*, that is the excessive accentuation of one or more notes of particular pitches. We particularly notice this phenomenon in bathrooms and telephone boxes, and it comes from the fact that the dimensions of these small rooms are comparable with the wavelengths* of speech. Colourations are particularly evident in rectangular rooms where the length, breadth and height bear a simple numerical ratio to each other; the theory for this is well understood. However, the ideal proportions have still to be discovered.

3.03 Acoustical models

Recently it has become common to build scale models of proposed auditoria for acoustical testing, eg Sydney Opera House, De Doelen, Rotterdam. The scale factors range from 1:50 to 1:8. The 1:50 model gives only very crude results, but is much cheaper than the refined 1:8 model.

Computers have been used to predict some acoustical characteristics, but complete mathematical modelling by this means is still a long way off, and would probably be as expensive as a 1:8 analogue model.

4 NOISE

4.01 Noise criteria

As the human ear is most sensitive to frequencies between 1 and 3 kHz, a 1 kHz tone will sound much louder than a 100 hz tone of the same sound pressure level. Any measuring of loudness must take this frequency sensitivity into account, and the simplest of sound-level meters has a device which roughly compensates for it called the A-weighting network. Readings from this meter are designated dB(A) to distinguish them from plain unweighted dB SPL. Although the loudness and annoyance of a sound can depend on other factors than the dB(A), it is a useful measure for many of the sounds encoun-

* The wavelength λ in metres of a tone of frequency f in hertz is given by $\lambda = \frac{c}{f}$ where c is the velocity of sound (approximately 340 m/s in air).

tered in and around buildings.

A building performance specification could contain criteria expressed in dB(A) for various areas, but it is more usual for them to be given in Noise Criteria (NC) curves, or the similar Noise Rating (NR) curves, **44.2**. These plot octave-band SPL against frequency. If NR-35 is specified, for example for a private office, then the noise level in any octave band should not exceed that indicated for the NR-35 line on the graph.

The preference for the NC and NR criteria rather than the single-figure dB(A) arises from the octave-band data used in the design of mechanical services systems.

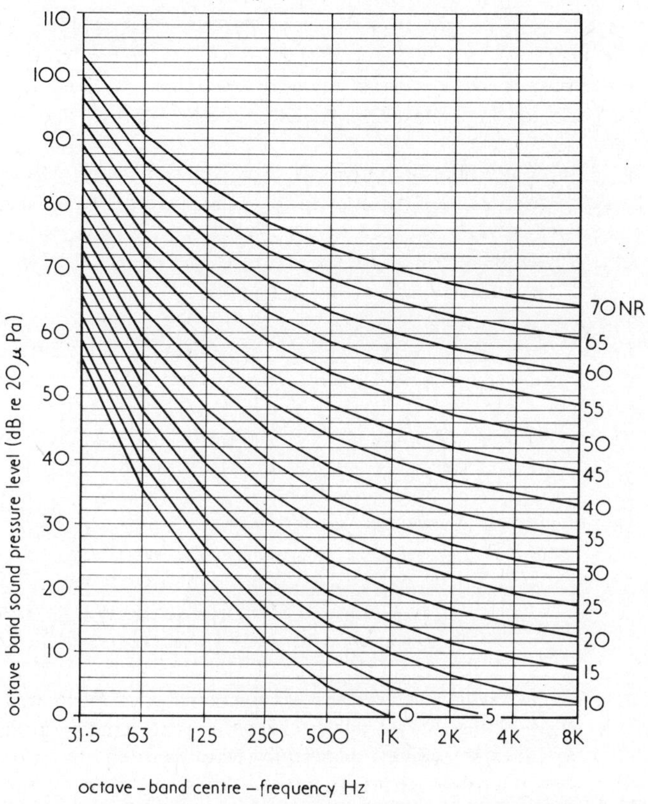

44.2 *Noise rating curves*

Noise criteria are commonly specified as maxima rather than optimal levels, and this is considered realistic in view of frequent failures in the past to meet the criteria for one reason or another. On the other hand, an office could be so quiet that private conversation can easily be overheard and occasional outside noises are distracting. In such a circumstance, a continuous masking (or *white*) noise can be deliberately introduced to drown the other sounds, using loudspeakers or by increasing the ventilation system noise.

Apart from dB(A), NC and NR ratings for noise, there are others the architect may encounter. Each has its merits, but most require rather more than a simple sound level meter for measurement. Some of these are:

L_{10} the level of noise in dB(A) exceeded for 10 per cent of the time.

L_{10} *(18 hour)* the average of the values of L_{10} measured hourly between 06.00 and 24.00 hrs on a normal working day. Used for planning and design as an index of road traffic noise.

L_{eq} *The equivalent continuous sound level.* This is the notional constant sound level which would give the same A-weighted sound energy as that of the actual sound over a period of time. Sometimes called *equivalent energy level* or *mean energy level.* L_{eq} is effectively an average sound level taken logarithmically, and may be calculated from:

$$L_{eq} = 10 \log_{10} \frac{(\Sigma t_i \, 10^{L_i/10})}{\Sigma t_i}$$

where t_i = the duration for which the sound level in dB(A) lies between L_{i-1} and L_{i+1}

Σ denotes summation over all values of i.

The period Σt_i over which the measurement is carried out depends on its purpose. L_{eq} is used in a number of countries for rating industrial and road traffic noises.

CNL *Corrected noise level.* An index of industrial noise based on dB(A) but corrected for tonal character, impulsive character, intermittency and duration.

L_{PN} *Perceived noise level*, or *PNdB*. A measure of aircraft noise. For subsonic jets the perceived noise level is about 13 times greater than the measured dB(A) value.

All the above are measured in decibels. Considerable confusion therefore arises if the particular index is not identified in each case.

The multiplicity of indices, not all of which have been described above, has led to the search for a single unified noise scale. Research is in progress to determine if L_{eq} which is the most likely candidate can adequately fill this role.

4.02 Internal noise sources

Because effective noise insulation is often impractical and usually costly, noise-producing areas should be sited away from noise-sensitive areas. In modern non-industrial buildings, the mechanical plant room is likely to be the noisiest area, especially if it contains heavy and inherently unbalanced plant. Chillers and large boilers can present severe noise problems particularly at low frequencies where curative measures are difficult. A buffer zone formed by a corridor or storage area around the plant room is a useful noise control measure.

The airborne noise radiated by industrial machinery can usually be calculated sufficiently accurately using the manufacturer's sound power level data. If the calculated noise level is too high, the reverberant noise level can be reduced by lining part of the plant room surfaces with an efficient acoustical absorbent; otherwise more costly measures may be necessary.

The noise transmitted into the structure of the building via the machinery mounts, etc is much more difficult to estimate. This is because the basic mechanical noise-generating characteristics of the machine are generally not known, and also because the ways in which sound propagates through building structures are imperfectly understood. In designing noise-isolation measures, rule-of-thumb methods are frequently used. They do not always work, and a long and expensive investigation is then needed to discover why; sound is easily propagated through structures with relatively little attenuation.

Many structure-borne noise problems are avoided by siting the plant room in a basement.

When specifying ventilation noise levels, etc, account should be taken of the noise arising from activities within the areas served, so that the criteria are compatible. In some circumstances, impact noise from footsteps can be a problem.

As well as the effects on people within the building, due consideration must nowadays be given to the effects on the neighbourhood. Roof-mounted cooling towers produce a noise problem in residential areas, nearly as severe as that from a poorly designed discotheque equipped with the latest in electronic amplifiers and loudspeaker technology.

4.03 External noise sources

The most important kind of external noise affecting buildings is transport noise from road traffic, aircraft and railways.

An essential characteristic of such sources is that they are not generally under the control of those affected by their noise. It is therefore essential to assess the likely level of external noise to which a proposed building is to be subjected. If this is done early enough in the design process, the scheme can be economically produced to alleviate the effects of the noise.

An example of planning against noise profoundly influencing the design is the five- to eight-storey Byker Wall in Newcastle-upon-Tyne. Here, a barrier against the noise from an adjacent motorway is formed by a long block of flats. The flats all have small windows facing the motorway and the noise-sensitive rooms on the quiet side of the barrier. The whole structure protects more conventional dwellings on the side remote from the traffic.

Where the source of noise is known it is often possible to calculate the likely noise level from its characteristics and the geometry of the site. In the UK there is now an official procedure for calculating the noise from motor vehicles, and this is recommended as preferable to measurement. At first sight this seems strange, but in practice it can be difficult to achieve a valid measurement. Some cynics have suggested that in the British Isles the prevailing wind direction and speed and the rainfall are such that the valid 18-hour traffic noise measurements are impossible except on a few days in the year!

Other sources of transport noise, such as railways, are not so well documented; direct site measurements may be the only course to take. For example, the sound of carriage doors being slammed shut can be a serious nuisance near a station.

Aircraft are a well-publicised source of serious noise, and there are usually severe planning restrictions on dwellings close to airports. However, hotels are often built here, and with care in siting and design the noise problems can be successfully overcome.

Public concern about jet engine noise has led to the development of quieter engines, and we can look forward to a diminution of the problem as the older aircraft are replaced; in spite of the exceptionally high noise levels generated by the present supersonic aircraft.

Other sources of noise that may have to be considered in particular situations include helicopters, hovercraft and industrial plant.

5 SOUND INSULATION

When a sound wave strikes a wall only a fraction of the incident sound energy is transmitted through the wall. The ratio of the incident to the transmitted sound energy, expressed in decibels, is called the *sound reduction index*. It can be properly measured only in a laboratory. The reduction in sound pressure level between adjacent rooms in an actual building depends not only on the sound reduction index of the separating wall, but also upon its area, the acoustic absorption present in the receiving room and the amount of transmission by *flanking paths* (see para **5.4**). Neglecting flanking transmission, the relation between the average sound level difference between two rooms and the sound reduction index of the separating wall is:

$$L_1 - L_2 = R + 10 \log_{10} \frac{A}{S}$$

where L_1 = sound pressure level averaged over the room containing the source

L_2 = sound pressure level averaged over the receiving room

R = sound reduction index of the separating wall

S = area of separating wall in m^2

A = acoustic absorption of receiving room in m^2 units.

Often A is comparable in size to S, making the level difference $L_1 - L_2$ vary little from the sound reduction index R so that it is commonly referred to simply as the 'sound insulation'.

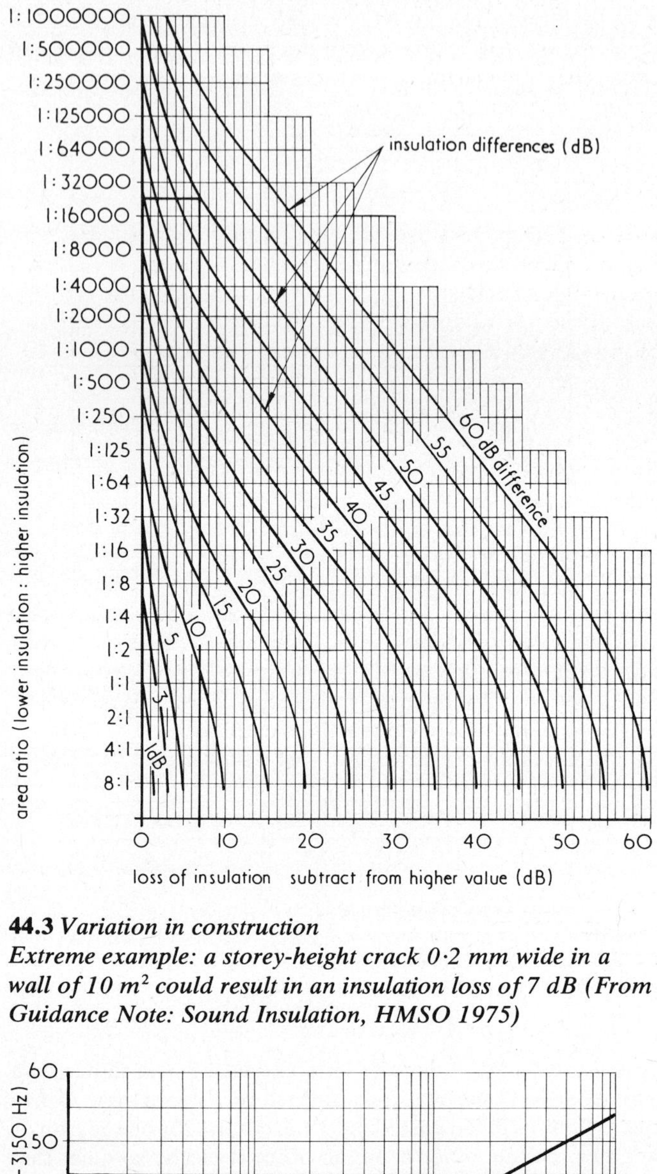

44.3 *Variation in construction*
Extreme example: a storey-height crack 0·2 mm wide in a wall of 10 m^2 could result in an insulation loss of 7 dB (From Guidance Note: Sound Insulation, HMSO 1975)

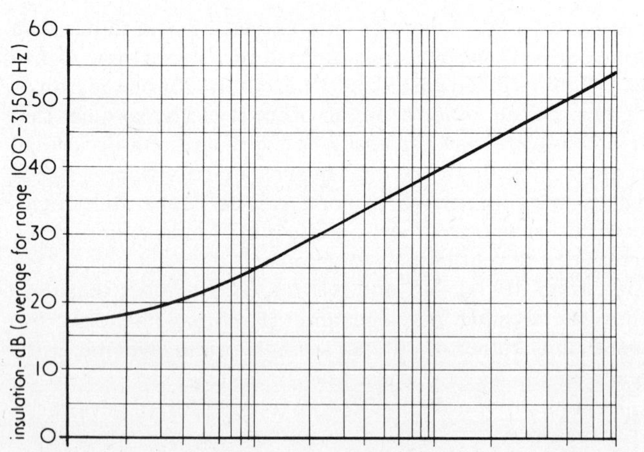

44.4 *Relationship of sound insulation to mass per unit area (From same as* **44.3***)*

5.01 Composite insulation

44.3 is an aid for calculating the sound insulation of a partition composed of two different materials. Consider a wall consisting of brickwork with insulation 45 dB and a glazed area amounting to $^1/_5$th the total wall area of insulation 25 dB. On the vertical axis of **44.3** the area ratio 1:4 meets the 20 dB difference curve at 13 dB on the horizontal axis. The composite insulation is therefore 45 − 13 = 32 dB.

5.02 Mass law

To a first approximation the insulation of a single leaf wall or floor depends on its mass per unit area. From **44.4** the insulation averaged over the frequency range 100 to 3150 Hz increases by about 5 dB for each doubling of mass (the *mass law*).

5.03 Coincidence effect

The sound insulation also increases by 5 to 6 dB for each doubling of frequency provided the partition is very limp, for example a lead sheet. However, most partitions are fairly stiff. With some materials the stiffness combines with the mass in such a way as to produce a resonance effect, seriously reducing the insulation below the mass-law value. This resonance, called the *coincidence effect*, is caused by flexural waves in the partition, and its significance depends on its nature and thickness. For a 215 mm brick wall the effect occurs at about 100 Hz which is generally too low to matter. Window glass has coincidence frequencies in the upper audible range.

5.04 Flanking transmission

In real buildings, sound is transmitted from one room to an adjacent room via many paths, **44.5**. Where the separating wall has a sound reduction index of 35 dB or less, most of the sound is transmitted through the wall. If, however, we try to improve the insulation by using a heavier wall, the *flanking paths*, or indirect transmission routes, become more important. In fact, it is difficult to achieve better than 60 dB sound level difference without special measures, such as carefully designed structural discontinuities, to reduce the flanking transmission.

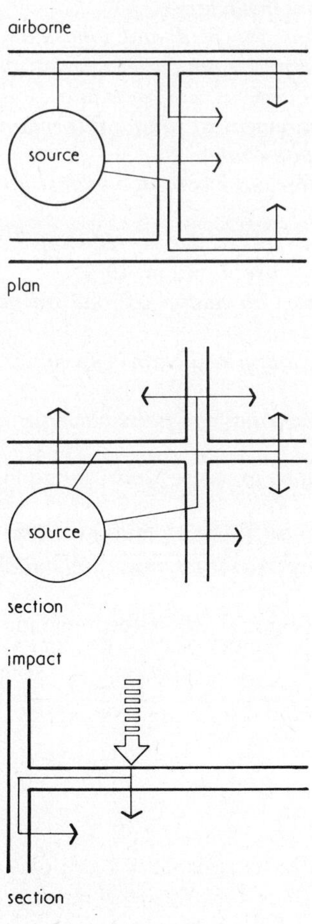

44.5 *Sound transmission paths*
(From same as **44.3***)*

5.05 Openings

Any kind of opening in a partition will seriously impair the sound insulation. The effect can be assessed by using **44.3**, and assigning a value of 0 dB for the insulation of the opening.

5.06 Double walls

Two single walls of 30 dB sound insulation combined would not produce 60 dB but only about 35 dB; the doubling of the mass per unit area adding 5 dB by the mass law. However, separating the two walls several metres apart would, if there were no flanking transmission, approach 60 dB. Practical double walls lie between these extremes, but it is difficult to theorise how a particular combination will actually behave.

Double walls and multi-leaf partitions are used where the required mass of a single leaf would be excessive. They improve dramatically on the mass law at middle and high frequencies, but at low frequencies the insulation is usually little better, and sometimes worse than a single leaf of the same mass per unit area.

5.07 Floors

The insulation of floors against airborne sound follows the same laws as for walls. An additional problem is the direct structural excitation of a hard-surfaced floor by footsteps, **44.5**. If a carpet is unacceptable, a floating floor must be used if footsteps noise is to be avoided in the room underneath. The resilient element for such a floor may be of mineral wool, slab or blankets, rubber, expanded polystyrene or springs. Care must be taken not to bridge this resilient element with any rigid connection.

The airborne sound insulation of a floor can be improved by suspending an impermeable ceiling below it, but the improvement is often limited by structural limitations on its weight and by height considerations.

5.08 Windows

The windows are usually the weakest part of the envelope where sound insulation is concerned. The mass per unit area of the glazing is generally small compared with the rest of the envelope. Table II is a guide to the sound insulation of different windows. In order to obtain the highest insulation double glazing will be required, at least one pane of which is sealed. Specially designed sound-attenuated ventilation will then be required.

5.09 Doors

Single doors having a sound insulation greater than 35 dB are expensive and difficult to install. Seals are required around the edge to prevent leakages, and where these are effective they make the door hard to open and close. Magnetic door seals similar to those on refrigerators are a small improvement. The most effective solution where the space is available is the use of two moderately insulating doors separated by an absorbent-lined 'sound lock'.

Table II Sound insulation of windows (from BRE Digest 140)

Description	Sound Reduction (av 100–3150 Hz)
Any type of window when open	about 10dB
Ordinary single openable window closed but not weather-stripped, any glass	up to 20dB
Single fixed or openable weather-stripped window, with 6 mm glass	up to 25dB
Fixed single window with 12 mm glass	up to 30dB
Fixed single window with 24 mm glass	up to 35dB
Double window, openable but weather-stripped, 150–200 mm air space, any glass	up to 40dB
Double window in separate frames, one fixed, 300–400 mm air space, 6–10 mm glass, sound-absorbent reveals	up to 45dB

5.10 Barriers

Barriers which intercept the line-of-sight between a sound source and the receiver are a common method of reducing a noise level. Outdoors they are used as a shield against traffic and aircraft noise, and some types of machinery. Indoors, they are used in open-plan offices and schools, and for altering the acoustics of broadcasting and recording studios.

The psychological effect of a visually opaque barrier can be very strong, giving a misleading impression of its acoustical effectiveness. For example, a line of trees is often proposed as a sound barrier although the measured acoustical effect is small.

There have been extensive theoretical studies on the effectiveness of barriers in idealised situations, but not much has resulted which can be applied in practice. Quite apart from the nature and geometry of the barrier itself, its performance can depend appreciably on the frequency of the sound, the weather and the nature of the ground between source and receiver.

As a rough guide, screen-type barriers 1 to 4 m high and mass about 10 kg/m^2 can give transmission losses of 5 to 20 dB.

5.11 Enclosures

Enclosures are used to suppress the noise from a stationary machine or item of plant such as a diesel generator. For them to be effective, they should have little or no openings. Where this is not possible because of the need to ventilate, properly designed attenuated air routes are required. The effectiveness of the enclosure is enhanced by lining internally with an acoustical absorbent.

Telephone hoods are an example of a partial enclosure, the effectiveness of which depends on how well the user obstructs the opening.

5.12 Cost and the designer

Structures designed to give a higher-than-average degree of sound insulation, such as broadcasting studios, are usually costly. The mass law, providing only a 5dB increase for doubling the material used illustrates how quickly the law of diminishing returns sets in. Attempts to beat the mass law by installing double or multi-leaf partitions incur the penalty of loss of usable space.

It is usually difficult to increase the sound insulation of an existing modern building. Older buildings of heavy construction have been successfully converted into broadcasting and recording studios. Several local radio stations in the UK can bear witness to this.

Because it is generally costly to increase sound insulation, the acoustical designer is rarely allowed the luxury of a safety margin. Unfortunately, the design of sound-insulating structures is still an imprecise science, and an economically designed building can fail to meet the expected performance. No reputable and knowledgeable acoustical consultant will guarantee the success of his design, any more than a doctor would guarantee to cure his patient.

The acoustical consultant's task is to achieve the right balance between a design that is too costly initially, and one for which failure would be disastrous.

6 ACOUSTICAL TEST DATA

One of the problems facing the designer is the lack of reliable acoustical data on materials and products. Some well-established manufacturers still supply information which is quite meaningless, although appearing authentic to the uninitiated. One must adopt a most sceptical attitude.

7 STANDARDS AND CODES OF PRACTICE

There are a growing number of national and international standards and codes of practice, many listed in the bibliography at the end of this chapter. These are invaluable in defining the methods used in testing acoustical materials, measuring traffic noise, etc. Before such standards existed, manufacturers would look for laboratories giving the most favourable results!

8 LEGISLATION

The architect has now to take into account a growing body of legislation concerned with reducing the objectionable effects of noise generated within his building, and noise sources externally affecting the occupants.

The Building Regulations 1976 for England and Wales lays down specific minimum sound insulation requirements for party walls and floors in dwellings. These requirements are couched in two alternative forms: either the known acoustical performance of the wall or floor conforms to the criteria in table III, or the construction conforms to one of the standard types in tables IV, V or VI.

In other countries other forms of legislative control apply. In the USA, the Environmental Protection Agency demands that a proposed development is preceded by an Environmental Impact Statement to show that no noise nuisance will result from it. European countries have specific noise limits for sanitary facilities such as wcs. Many countries have strict planning controls near known noise sources such as airports, and some provide financial assistance to insulate against sound in the proximity of motorways, etc.

9 BIBLIOGRAPHY

Acoustics—Assessment of occupational noise exposure for hearing conversation purposes, ISO Standard 1999, 1975

AJ Handbook of Building Environment, Section 5, Sound, *Architects' Journal* 22.1.69, 29.1.69, 5.2.69, 12.2.69

Assessment of noise with respect to community response, ISO Recommendation R 1996, May 1971

British Standard Code of Practice CP3: chapter III: part 2, 1972, *Code of basic data for the design of buildings: Sound insulation and noise reduction*

British Standard Code of Practice CP 153: part 3: 1972, *Windows and rooflights, part 3, Sound insulation*

BS 2750:1956 *Recommendations for field and laboratory measurement of airborne and impact sound transmission in buildings*

BS 3638:1963 *Method of measurement of sound absorption coefficients (ISO) in a reverberation room*

BS 5228:1975 *Code of Practice for noise control on construction and demolition sites*

Department of Employment *Code of Practice for reducing the exposure of employed persons to noise*, London, 1972

Department of the Environment *Calculation of road traffic noise*, London, 1975

Department of the Environment *Sound insulation in dwellings*, London, HMSO, 1971

W. Furner *Room and building acoustics and noise abatement*, London, Butterworths, 1964

Noise Advisory Council *A guide to noise units*, London, HMSO, 1973

Noise Advisory Council *Noise units*, London, HMSO, 1975

P. H. Parkin and H. R. Humphreys *Acoustics, noise and buildings*, London, Faber, 1969

Rating of sound insulation for dwellings, ISO Recommendation R 717, May 1968

Table III Sound reduction criteria to Building Regulations 1976

Frequency (in Hz) (1)	Walls Sound reduction (in dB) (2)	Transmission of airborne sound thro' floors Sound reduction (in dB) (3)	Transmission of impact sound thro' floors Octave band sound pressure level (in dB) (4)
100	40	36	63
125	41	38	64
160	43	39	65
200	44	41	66
250	45	43	66
315	47	44	66
400	48	46	66
500	49	48	66
630	51	49	65
800	52	51	64
1000	53	53	63
1250	55	54	61
1600	56	56	59
2000	56	56	57
2500	56	56	55
3150	56	56	53

Note The regulations (G2(1) and G4(1)) are satisfied if the sum of the deficiencies at each frequency does not exceed 23 dB.

Table IV Suitable wall constructions for sound insulation to Building Regulations 1976, G2(2)

Specification (1)	Construction of wall (2)
1.	A solid wall consisting of– (a) bricks or blocks with plaster not less than 12·5 mm thick on at least one face; or (b) dense concrete cast in situ or panels of dense concrete having all joints solidly grouted in mortar; or (c) lightweight concrete with plaster not less than 12·5 mm thick on both faces of the wall, in each case the average mass of the wall (calculated over any portion of the wall measuring 1 metre square and including the mass of any plaster) being not less than 415 kg/m^2
2.	A wall having a cavity not less than 50 mm wide constructed of two leaves each consisting of bricks, blocks or dense concrete with plaster not less than 12·5 mm thick on both faces of the wall, and having any wall ties of the butterfly wire type, the average mass of the wall (calculated over any portion measuring 1 metre square and including the mass of the plaster) being not less than 415 kg/m^2
3.	A wall having a cavity not less than 75 mm wide constructed of two leaves each consisting of lightweight concrete with plaster not less than 12·5 mm thick on both faces of the wall and having any wall ties of the butterfly wire type, the average mass of the wall (calculated over any portion of the wall measuring 1 metre square and including the mass of the plaster) being not less than 250 kg/m^2

Requirements

Must

(a) extend for a distance of at least 460 mm beyond an external flanking wall;

or

(b) be tied into or bonded to one leaf of an external flanking wall of bricks, blocks or concrete–
- (i) which is of a construction having an average mass (calculated over any portion of the leaf measuring 1 metre square) of not less than 120 kg/m^2; and
- (ii) in which any window or door opening on one side of the separating wall is not less than 690 mm measured horizontally, from any such opening on the other side of that wall unless the height of each opening does not exceed two-thirds of the height of the storey and the external flanking wall above and below the openings extends for a distance of not less than 3 m, measured horizontally, on both sides of the separating wall

or

(c) extend to the outer face of an external flanking wall of timber or other light construction other than tile hanging and at the top and bottom of each storey be tied into or bonded to–
- (i) a solid floor next to the ground; or
- (ii) a suspended concrete floor having an average mass (calculated over any portion of the floor measuring 1 metre square) of not less than 220 kg/m^2; or
- (iii) a concrete roof having an average mass (calculated over any portion of the roof measuring 1 metre square) of not less than 145 kg/m^2

Table V Suitable floor constructions for resistance to the transmission of airborne and impact sound to Building Regulations 1976, G4(2) and G5(2)

Specification (1)	Construction of floor (2)
1.	A floor consisting of– (a) a solid concrete slab; or (b) a slab of concrete beams and hollow infilling blocks of clay or concrete; or (c) a slab of hollow concrete beams, in each case having an average mass (calculated over any portion of the floor measuring 1 metre square and including the mass of any screed or ceiling plaster directly bonded to the slab but excluding the mass of any floating floor or suspended ceiling) of not less than 365 kg/m^2 and having either of the following laid upon it– (i) rubber on sponge rubber underlay having a total thickness of not less than 4·5 mm; or (ii) cork tiles not less than 8 mm thick
2.	A floor consisting of– (a) a solid concrete slab; or (b) a slab of concrete beams and hollow infilling blocks of clay or concrete; or (c) a slab of hollow concrete beams, in each case having an average mass (calculated over any portion of the floor measuring 1 metre square and including the mass of any screed or ceiling plaster directly bonded to the slab but excluding the mass of any floating floor or suspended ceiling) of not less than 220 kg/m^2 and having any of the following laid upon it– (i) boarding nailed to battens so laid as to float upon a layer of glass fibre or mineral wool quilt, in either case capable of retaining its resilience under imposed loading; or (ii) any covering directly applied to concrete or other cementitious screed, not less than 38 mm thick, so laid as to float upon a layer of glass fibre or mineral wool quilt, in either case capable of retaining its resilience under imposed loading; or (iii) rubber on sponge rubber underlay having a total thickness of not less than 4·5 mm or cork tiles not less than 8 mm thick, in either case laid upon a dense airtight sealing layer upon lightweight screed, not less than 50 mm thick, of a density of not more than 1100 kg/m^3
3.	Boarding nailed to battens laid to float upon a layer of glass fibre or mineral wool quilt, in either case capable of retaining its resilience under imposed loading, the layer being draped over wooden joists, beneath which a ceiling of lath and plaster or of plasterboard, in either case not less than 19 mm thick, has been constructed, with pugging on the ceiling such that the combined mass of the ceiling and pugging is not less than 120 kg/m^2

Table V Suitable floor constructions for resistance to the transmission of airborne and impact sound to Building Regulations 1976, G4(2) and G5(2) *(continued)*

Requirements

must

(a) in the case of a concrete floor, extend to the outer face of the inner leaf of any adjoining external wall and be tied into or bonded to every adjoining separating wall and every other internal wall which gives support to the floor

or

(b) in the case of a timber floor–
  (i) be bounded below on at least three sides by walls having an average mass (calculated over any portion of the wall measuring 1 metre square) of not less than 415 kg/m²; and
  (ii) every external flanking wall extends for not less than 600 mm, measured vertically from the underside of the floor, without any window or door opening therein other than a window or door opening above a balcony forming an extension to the floor

Table VI Suitable floor constructions providing resistance to the transmission of airborne sound only, to Building Regulations 1976, G5(2)

Specification (1)	**Construction of floor (2)**
1.	A floor consisting of a solid concrete slab having an average mass (calculated over any portion of the floor measuring 1 metre square and including the mass of any screed or ceiling plaster directly bonded to the slab but excluding the mass of any floating floor or suspended ceiling) of not less than 365 kg/m² and having any type of floor finish

Requirement

Subject to the same stipulations as in table V above

45 Structure

David Adler

CI/SfB (K)
CI/SfB (1976 revised) (J)

Contents

1 INTRODUCTION

1.01 Definition

Structural engineering is the technology of ensuring that loads are transmitted safely and economically to the ground, allowing for considerations of function, aesthetics, internal and external environment, and other restraints imposed by other members of the building team, legislation, etc.

The subject is compounded of roughly equal amounts of experience and investigation. Most of the latter involves the use of mathematics, although model testing is occasionally employed. Recent and future advances centre on more sophisticated techniques in the mathematics, particularly the use of computers.

The treatment of the subject in this section is necessarily brief, and the reader may have difficulty as some of the terms will only be sketchily explained. A fuller exposition will be found in the *AJ Handbook of Building Structure* (section 2, Structural analysis), Architectural Press, London.

1.02 Limit state

Modern techniques in structural engineering are based on the concept of the *limit state*. Since the new Codes of Practice use this concept freely, some explanation of it will be useful, even though the rest of this section is on traditional lines.

A building or structure may become unserviceable or unsatisfactory in a variety of ways. It may collapse completely, it may crack locally so badly as to let in the weather, it may deflect until the users feel unsafe. These are all limit states—in the case of the complete collapse the limit state is the condition immediately prior to the collapse! Obviously the task of the structural designer is to ensure that none of the possible limit states are ever reached. He does this by calculating, for example, the load at which the collapse limit state is reached; he then applies a load factor to determine the safe working load: usually about half the collapse load. A similar procedure is adopted for the other limit states. This method of structural design is not fundamentally different from traditional methods, it is simply a more organised way of approaching the problems.

Although the concepts of structural engineering are being reassessed in line with limit state design, the materials in which the engineer has been working remain almost unchanged. The basic palette consists of masonry (stone, brick and block), timber, steel and reinforced concrete. New materials such as plastics are unlikely to have more than a marginal effect in the near future.

1.03 Loads

In para **1.01** above, reference was made to loads. A load is an example of a force, and the term is usually used to describe those outside forces that act on a building structure. These loads are of three types:

- *dead loads:* from the weight of the structure itself and that of other fixed parts of the building such as cladding, finishes, partitions, etc
- *imposed loads:* of people, furniture and of materials stored in the building
- *dynamic loads:* these are of many origins. The commonest dynamic load on a building is that caused by wind, which can produce horizontal and vertical pressures, and also suctions. Other horizontal dynamic loads are produced by earthquakes, and by moving machinery such as the kinds of overhead cranes used in workshops. For the purposes of design, dynamic loads are transformed into equivalent static loadings.

1.04 Forces

Forces, including loads are measured in newtons (N). 1 N is the force required to give a mass of 1 kg an acceleration of 1 m/sec^2. A tip to remember is that a newton is about the weight of 1 apple.

Most forces in structural engineering are expressed in kN (kilonewtons), but MN (meganewtons) are often encountered and in very heavy work GN (giganewtons) which are N × 10^9 are used. Table I gives conversions from and to SI, MT and FPS units for loadings of all types.

1.05 Mass and weight

Confusion often arises between the terms *mass* and *weight*. Outside nuclear physics, the mass of an object is a fixed quantity which is a basic property of that object. Its weight will depend on the mass, but also on the value of the gravitational effect on the object. This is not a constant, but can vary, not only in extra-terrestial conditions, but even very slightly on different places on earth. However, for all practical purposes the acceleration of gravity is taken as 9·80665 m/sec^2, so that the weight of a kilogram mass is 9·80665 N.

Table II gives the loads of various materials which may comprise the fabric of a building, or be stored within it. The figures are given both in the usual mass density form: kg/m^3, and also in the more convenient weight density kN/m^3. Table III indicates the imposed loads that should be allowed for in designing buildings for various purposes.

Example 1

A concrete slab 200 mm in thickness forms a floor 6·5 m square in an office building. What would be the total load on this floor?

Slab self-weight: from table II concrete weighs about 24 kN/m^3 if it is reinforced to a normal 2% extent

Therefore weight of slab is 24 × 0·2 × 6·5 × 6·5 = 202·80 kN

Imposed load: from table III an office floor should be designed to take 2·5 kN/m^2

Therefore on this floor load is 2·5 × 6·5 × 6·5 = 105·62 kN

Total load = 308·42 kN

Table I Various conversions for loadings

Point loads

1 N = 0·102 kgf = 0·225 lbf
1 kN = 101·972 kgf = 224·81 lbf = 0·1004 tonf
1 MN = 101·972 tf = 224·81 kipf = 100·36 tonf
1 kgf = 9·807 N = 2·205 lbf
1 tf = 9·807 kN = 2·205 kipf = 0·9842 tonf
1 lbf = 4·448 N = 0·4536 kgf
1 kipf = 4·448 kN = 453·59 kgf = 0·4536 tf
1 tonf = 9·964 kN = 1·016 tf

Linearly distributed

1 N/m = 0·0685 lbf/ft = 0·206 lbf/yd
1 kN/m = 68·5 lbf/ft = 0·0306 tonf/ft
1 kgf/m = 9·807 N/m = 0·672 lbf/ft = 2·016 lbf/yd
1 tf/m = 9·807 kN/m = 0·672 kipf/ft = 2·016 kipf/yd = 0·3 tonf/ft = 0·9 tonf/yd
1 lbf/ft = 14·59 N/m = 1·488 kgf/m
1 kipf/ft = 14·58 kN/m = 1·488 tf/m
1 tonf/ft = 32·69 kN/m = 3·33 tf/m
1 tonf/yd = 10·90 kN/m = 1·11 tf/m

Superficially distributed

1 N/m^2 = 0·0209 lb/ft^2
1 kN/m^2 = 20·89 lb/ft^2
1 MN/m^2 = 9·324 tonf/ft^2
1 kgf/m^2 = 9·80665 N/m^2 = 0·2048 lbf/ft^2 = 1·843 lbf/yd^2
1 tf/m^2 = 9·80665 kN/m^2 = 0·2048 kipf/ft^2 = 0·0914 tonf/ft^2 = 0·823 tonf/yd^2
1 lb/ft^2 = 47·88 N/m^2 = 4·88 kgf/m^2
1 kipf/ft^2 = 47·88 kN/m^2 = 4·88 tf/m^2
1 tonf/ft^2 = 107·25 kN/m^2 = 10·93 tf/m^2
1 tonf/yd^2 = 11·92 kN/m^2 = 1·215 tf/m^2

Densities

1 N/m^3 = 0·00637 lbf/ft^3
1 kN/m^3 = 6·37 lbf/ft^3
1 MN/m^3 = 2·844 tonf/ft^3
1 kg/m^3 = 0·0624 lb/ft^3 (mass density)
1 t/m^3 = 62·4 lb/ft^3
1 lb/ft^3 = 16·02 kg/m^3 — 1 lbf/ft^3 = 157 N/m^3
1 ton/ft^3 = 35·88 t/m^3 — 1 tonf/ft^3 = 351·9 kN/m^3
1 ton/yd^3 = 1·33 t/m^3

Table II Densities of materials

Material	kg/m^3	kN/m^3
Adamantine clinkers, stacked	2082	20·4
Aerated concrete	801–961	7·8–9·4
Aggregates:		
coarse	1522	14·9
fine	850	8·3
Alabaster	2691	26·4
Alcohol:		
absolute	785	7·7
commercial proof spirit	913	9·0
wood—barrels	449	4·4
Alluvium, undisturbed	1602	15·7
Aluminium:		
cast	2771	27·2
rolled	2675	26·2
bronze	7545	74·0
DTD alloys	2675–2787	26·2–27·3
paint	1201	11·8
paste	1474	14·4
powder	721–801	7·0–7·9
sheet, per mm thickness	kg/m^2	kN/m^2
	2·8	0·027
	kg/m^3	kN/m^3
sulphate, bags	721	7·1
Ancaster stone	2499	24·5
Animal food, cases	400	3·9
Anthracite, broken	805	7·9
Antimony:		
pure	6680	65·5
ore, bags	1442	14·1
Apples, barrels	400	3·9
Argentine	7208	70·7
Asbestos:		
crude	897	8·8
fibre, cases	673	6·6
natural	3044	29·8
pressed	961	9·4
cement	1922–2082	18·8
sand	961	9·4
felt	150	1·5
Ash		
English	689	6·8
Canadian	737	7·2

Table II Densities of materials *(continued)*

Material	kg/m^3	kN/m^3
Ashes, dry	641	6·3
Asphalt:		
natural	1009	9·9
paving	2082	20·4
Automatic machines, cases	160	1·6
Automobiles, cases	128	1·2
Aviation spirit	753	7·4
Axles and wheels	513	5·0
Baggage	128	1·2
Ballast:		
loose, graded	1602	15·7
undisturbed	1922	18·8
Balsa wood	112	1·1
Bamboo	352	3·4
Barbed wire	384	3·7
Barium oxide, solid	4645–5446	45·5–53·4
Barley:		
grain	705	6·9
bags	593	5·8
ground	529	5·2
Barrels, empty	128	1·2
Bars, steel, bundles	2723	26·7
Basic slag, crushed	1794	17·6
Bath stone	2082	20·4
Baths, iron, cases	208	2·0
Bauxite	2563	25·1
crushed	1281	12·6
ore, bags	1201	11·8
Beech	769	7·5
Beer	1025	10·0
bottled, cases	449	4·4
barrels	529	5·2
Beeswax	961	9·4
Bell metal	8490	83·2
Benzene	881	8·6
Benzol	881	8·6
Bicycles, crates	128	1·2
Birch:		
American	641	6·3
logs	449	4·4
squares	625	6·1
yellow	705	6·9
Bitumen:		
natural	1089	10·7
prepared	1362	13·4
emulsion	1121	11·0
Blood	1057	10·4
dried, casks	561	5·5
Bolts and nuts, bags	1201	11·8
Bone	1762–2002	17·3–19·6
manure, bags	513	5·0
meal, bags	801	7·8
Books:		
on shelves	641	6·3
bulk	961	9·4
Boots and shoes, cases	384	3·8
Bottled goods, cases	897	8·8
Bottles, empty crates	416	4·1
Boxwood	929	9·1
Brass:		
cast	8330	81·7
rolled	8570	84·0
casks	721	7·1
tubes, bundles	897	8·8
Brewer's grains:		
wet	497	4·9
desiccated	256	2·5
Bricks:		
(common burnt clay)		
stacked	1602–1920	15·7–18·8
sand cement	1840	18·0
sand lime	2080	20·4
ballast	1200	11·8
brickwork	1920	18·8
British Columbia pine	529	5·2
Bronze:		
cast	8330	81·7
drawn, sheet	8794	86·2
Cadmium	8618	84·5
Calcium carbide, solid	2211	21·7
Canvas, bales	769	7·5
Carpets, rolls	256	2·5
Casein	1346	13·2
Casks, empty	128	1·2
Cedar, western red	384	3·8
Celluloid	1346–1602	13·2–15·7
Goods, cases	160	1·6

Table II Densities of materials (*continued*)

Material	kg/m³	kN/m³
Cement:		
bags	1281	12·6
bulk	1281–1442	12·6–14·2
casks	961	9·4
slurry	1442	14·7
Chalk	1602–2723	15·7–26·7
broken, barrels	961	9·4
Cheese, cases	513	5·0
Cherry wood	721	7·1
Chestnut:		
horse	513	5·0
sweet	561	5·5
Chromium	7096	69·6
Cigarettes, cases	240	2·4
Cinders	641	6·3
Clay: Fill:		
dry, lumps	1041	10·2
dry, compact	1442	14·1
damp, compact	1762	17·3
wet, compact	2082	20·4
undisturbed	1922	18·8
undisturbed, gravelly	2082	20·4
china, compact	2243	22·0
Clinker, furnace	1025	10·0
Coal:		
loose lumps	897	8·8
slurry	993	9·7
Cobalt	8586	84·2
Coke	481–561	4·7–5·5
Columbian pine	529	5·2
Concrete: cement, plain:		
aerated	961	9·4
brick aggregate	1840	18·0
clinker	1440	14·1
stone ballast	2240	22·0
Concrete: cement, reinforced:		
1 per cent steel	2370	23·2
2 per cent steel	2420	23·7
5 per cent steel	2580	25·3
Copper:		
cast	8762	85·9
drawn or sheet	8938	87·6
ingots	3588	35·2
Cork	128–240	1·2–2·4
bales	80	0·78
Corn, bulk	721	7·1
Cotton:		
raw, compressed	400–577	3·9–5·7
bales, American	272	2·7
pressed bales, Egyptian	529	5·2
Cupro-nickel (60 per cent to 80 per cent Cu)	8938	87·6
Cypress wood	593	5·8
Deal, yellow	432	4·2
Delta metal	8602	84·4
Diatomaceous brick	481	4·7
Diesel oil	881	8·6
Doors, crates	320	3·1
Douglas fir	529	5·2
Dry goods, average	481	4·7
Duralumin	2787	27·3
Dutch clinkers, stacked	1602	15·7
Dynamite	1233	12·1
Earth:		
dry, loose	1280	12·6
dry, compact	1550	15·2
moist, loose	1440–1600	14·1–15·7
moist, compact	1760–1840	17·2–18·1
Earthenware, packed	320	3·1
Ebonite	1201–1281	11·8–12·6
Ebony	1185–1330	11·6–13·1
Elm:		
American	673	6·6
Canadian	673	6·6
Dutch	577	5·6
English	577	5·6
wych	689	6·8
Felt:		
hair	272	2·7
roofing, rolls	593	5·8
Fibreboard	160–400	1·5–4·0
Files, etc, cases	897	8·8
Fir:		
Douglas	529	5·2
silver	481	4·7
Firebrick, Stourbridge	2002	19·6

Table II Densities of materials (*continued*)

Material	kg/m³	kN/m³
Fish, boxes	721	7·1
Flint	2563	25·1
Flour	705	6·9
sacks	641	6·3
barrels	545	5·3
Foam slag	700	6·9
Forest of Dean stone	2435	23·9
Freestone	2243–2483	22·0–24·4
masonry, dressed	2403	23·6
rubble	2243	22·0
Fuller's earth, natural	1762–2403	17·3–23·6
Galvanised sheets, bundles	897	8·8
Glass:		
bottle	2723	26·7
common green	2515	24·7
crown, extra white	2451	24·0
silicate	2195	21·5
flint, best	3076	30·2
heavy	4966–5927	48·7–58·1
optical	3524	34·6
plate	2787	27·3
crates	801	7·8
Pyrex	2243	22·0
bottles, crates	416	4·1
refuse (broken)	1522	14·9
silk	160–208	1·6–2·0
Gold	19318	189·4
Grain:		
barley	625	6·1
oats	416	4·1
rye	721	7·1
Granite	2643	25·9
chippings	1442	14·1
dressed, cases	2243	22·0
Granolithic	2243	22·0
Gravel:		
loose	1602	15·7
undisturbed	1922–2162	18·8–21·2
Gunmetal:		
cast	8458	82·9
rolled	8794	86·2
Gunpowder	897	8·8
Gypklith	449	4·4
Gypsum:		
crushed	1041–1602	10·2–45·7
bags	833	8·2
solid	2563	25·1
plaster	737	7·2
Hardcore	1922	18·8
Hemlock, western	497	4·9
Hiduminium	2803	27·5
Hoggin	1762	17·3
Hosiery, cased	224	2·2
Ice	913	9·0
Implements, agricultural bundles	256	2·5
Indiarubber	1121	11·0
Iroko	657	6·4
Iron:		
cast	7208	70·7
malleable cast	7368–7497	72·2–73·5
wrought	7689	75·4
corrugated, bundles	897	8·8
pig,		
random	2723	26·7
stacked	4485	44·0
pyrites,		
ground	2883	28·3
solid (60 per cent Fe)	4806–5128	47·1–50·3
sulphate, powdered	1121	11·0
wire, coils	897	8·8
Ironstone:		
Cleveland, lumps	2162	21·2
Spanish, lumps	2403	23·6
Swedish, lumps	3684	36·1
Ironmongery, packages	897	8·8
Ironwood	1137	11·2
Ivory	1842	18·1
Jointing compo, for tanks	801	7·8
Jute:		
bales	481	4·7
bales, compressed	641	6·3
Kentish rag	2675	26·2
crushed	1602	15·7
Kupfernickel	7208–7609	70·6–74·6

Table II Densities of materials (*continued*)

Material	kg/m³	kN/m³
Larch wood	593	5·8
Lead:		
cast or rolled	11325	111·1
pigs	3588	35·2
bronze (Cu 70 Pb 30)	9771	95·8
red, powder	2082	20·4
white, powder	1378	13·5
paste in drums	2787	27·3
Leather	961	9·4
hides, compressed	368	3·6
rolls	160	1·6
Lime:		
acetate of, bags	1281	12·6
Blue Lias,		
ground	849	8·3
lump	993	9·7
carbonate of, barrels	1281	12·6
chloride of, lead lined cases	449	4·4
grey chalk, lump	705	6·9
grey stone, lump	881	8·6
hydrate, bags	513	5·0
hydraulic	721	7·1
quick, ground	1025	10·0
slaked,		
ground, dry	561	5·5
ground, wet	1522	14·9
Lime mortar:		
dry	1650	16·2
wet	1746	17·1
Lime wood	561	5·5
American	416	4·1
Linoleum, rolls	481	4·7
Loam (sandy clay):		
dry, loose	1201	11·8
dry, compact	1602	15·7
wet, compact	1922	18·8
Logwood	913	9·0
Macadam	2082	20·4
Magnesia, solid	2403	23·6
Magnesite	3044	29·8
Magnesium	1730	17·0
alloys, about	1842	18·1
Magnetic oxide of iron	4966	48·7
Magnetite	4966	48·7
Mahogany:		
African	561	5·5
Honduras	545	5·3
Spanish	689	6·8
Manganese	7368	72·2
bronze	8602	84·4
Manganite	4325	42·4
Maple:		
Canadian	737	7·2
English	689	6·8
Marble	2595–2835	25·4–27·8
Mastic	1121	11·0
Mercury	13536	132·7
Mica	2723–3044	26·7–29·8
Millstone grit	2323	22·8
Molybdenum	9980	97·9
Mortar:		
cement, set	1922–2082	18·8–20·4
lime, set	1602–1762	15·7–17·3
Mud	1762–1922	17·3–18·8
Muntz metal:		
cast	8394	82·3
sheet	8922	87·5
Nails, wire, bags	1201	11·8
Neoprene	1201	11·8
Nickel	8810	86·4
silver	8730	85·6
Oak:		
African	961	9·4
American red	721	7·1
white	769	7·5
Austrian	721	7·1
English	801–881	7·8–8·6
Ore. See individual kinds		
Oregon pine	529	5·2
Padauk	785	7·7
Paint:		
aluminium	1201	11·8
bituminous emulsion	1121	11·0
red lead	3123	30·6
red lead dispersed	1522	14·9
white lead	2803	27·5
zinc	2403	23·6
Paper:		
blotting, bales	400	3·9
printing, reels	897	8·8
wall, rolls	384	3·8
writing	961	9·4
Paraffin:		
oil	801	7·8
wax	897	8·8
Peat:		
dry, stacked	561	5·5
sandy, compact	801	7·8
wet, compact	1362	13·4
Perspex	1346	13·2
Peruvian bark, bales	240	2·4
Petrol	689–769	6·7–7·5
cans or drums	721–801	7·0–7·8
Phosphates:		
ground	1201	11·8
bags	849	8·3
Phosphor-bronze:		
cast	8650	84·8
drawn	8810	86·4
Pine:		
American red	529	5·2
British Columbian	529	5·2
Christiania	689	6·8
Columbian	529	5·2
Dantzig	577	5·6
Memel	545	5·3
Kauri, Queensland	481	4·7
New Zealand	609	6·0
Oregon	529	5·2
pitch	657	6·4
Riga	545–753	5·3–7·4
Pipes:		
brass, bundles	897	8·8
cast iron, stacked	961–1281	9·4–12·6
earthenware, loose	320	3·1
salt-glazed, stacked	400	3·9
wrought iron,		
stacked $\frac{3}{8}$in (9·5 mm)	3204	31·4
3 in (76·2 mm)	1442	14·1
6 in (152·4 mm)	801	7·8
Pitch	1089	10·7
Plaster of Paris:		
loose	929	9·1
set	1281	12·6
Platinum	21465	210·5
Plywood	481–641	4·7–6·3
plastic-bonded	721–1442	7·0–14·2
Polystyrene	1057	10·4
Polvinyl chlor. acetate	1201–1346	11·8–13·2
Poplar	449	4·4
Porcelain	2323	22·8
Porphyry	2803	27·5
Portland cement:		
loose	1201–1362	11·8–13·4
bags	1121–1281	11·0–12·6
drums	1201	11·8
Portland stone	2243	22·0
Potatoes	641	6·3
Pulp, wood:		
dry	561	5·5
wet	721	7·1
Pumice stone	481–913	4·7–9·0
Purbeck stone	2707	26·5
Pyrites:		
iron,		
ground	2883	28·3
solid (60 per cent Fe)	4806–5126	47·1–50·3
copper, solid	4085–4325	40·0–42·4
Quartz	2643	25·9
loose	1442–1682	14·1–16·5
Quartzite	2723	26·7
Quicklime, ground, dry	1025	10·0
Quilt, eel grass	176	1·72
Ragstone	2403	23·6
Rails, railway	2403	23·6
Redwood:		
American	529	5·2
Baltic	497	4·9
non-graded	432	4·2
Rhodesian	913	

Table II Densities of materials (*continued*)

Material	kg/m³	kN/m³
Resin:		
lumps	1073	9·0
barrels	769	7·5
Resin bonded plywood	721–1362	7·0–13·4
Resin oil	993	9·7
Rubber:		
crepe, cases	400	3·9
processed sheet	1121	11·0
raw	929	9·1
sponge	48–160	0·47–1·57
vulcanised	1201	11·8
Salt, bulk	961	9·4
Salt-glazed ware	2243	22·0
Sand:		
saturated	1922	18·8
undisturbed dry	1682	16·5
saturated	2002	19·6
Satinwood	961	9·4
Sawdust	208	2·0
Screws, iron, packages	1602	15·7
Sea water	1009–1041	9·9–10·2
Shale	2563	25·1
granulated	1121	11·0
oil, Scottish	945	9·3
Silica, fused transparent	2211	21·7
translucent	2050	20·1
Silicon, pure	291	2·8
Silk, bales	352	3·4
Silver:		
cast	10444	102·4
pure	10492	102·9
glance	7208	70·7
Sirapite, powder	1025	10·0
Slag:		
coarse	1442	14·1
granulated	961	9·4
Slag wool	224–288	2·2–2·8
Slate:		
Welsh	2803	27·5
Westmorland	2995	29·4
Sludge cake, pressed, 50 per cent water	929	9·1
Snow:		
fresh	96	0·94
wet compact	320	3·1
Soap, boxed	913	9·0
Soapstone	2723	26·7
Soda, bags	657	6·4
Solder, pigs	2723	26·7
Soot	352	3·4
Spar:		
calcareous	2723	26·7
feld	2691	26·4
fluor	3204	31·4
Spirits of wine	785	7·7
Sponge rubber	48–160	0·47–1·57
Spruce		
Canadian	465	4·6
Norway	465	4·6
Sitka	449	4·4
Stationery cases	513	5·0
Steel:	7833	76·8
balls, barrels	1201	11·8
punchings	4806	47·1
Stone:		
Ancaster	2499	24·5
Bath	2082	20·4
Caen	2002	19·6
Darley Dane	2371	23·2
Forest of Dean	2439	23·9
freestone	2243–2483	22·0–24·3
granite	2643	25·9
Ham Hill	2162	21·2
Hopton Wood	2531	24·8
Kentish rag	2675	26·2
Mansfield	2259	22·2
marble	2723	26·7
millstone grit	2323	22·8
Portland	2243	22·0
Purbeck	2707	26·5
slate, Welsh	2803	27·5
Westmorland	2995	29·4
York	2243	22·0
Stoneware	2243	22·0
Straw:		
pressed	96	0·94
compressed bales	304	3·0
Strawboards, bundles	593	5·8

Table II Densities of materials (*continued*)

Material	kg/m³	kN/m³
Strontium white:		
solid	3844	37·7
ground	1762	17·3
Sulphate of:		
aluminium, bags	721	7·1
ammonia, bags	641	6·3
copper, cryst	1346	13·2
iron, powder	1121	11·0
Sulphur, pure solid	1922–2082	18·8–20·4
Sulphuric acid, 100 per cent	1970	19·3
commercial	1682–1794	16·5–17·6
Sycamore	609	6·0
Tar	1137–1233	11·2–12·1
barrels	801	7·8
Tarmacadam	2082	20·4
Tarpaulins, bundles	721	7·1
Teak, Burma African	657	6·4
Terracotta	1794	17·6
Tetraethyl lead	1602	15·7
Timbers, See individual kinds		
Tinned goods, cases	481–641	4·7–6·3
Tinplate, boxes	3204–4485	31·4–44·0
Tinstone	6407–7048	62·8–69·1
Tinware, cases	192	1·88
Titanium	4485	44·0
oxide, solid	3684	36·1
Tools, hand, cases	897	8·8
Treetex	208	2·0
Tubes, see 'Pipes'		
Tungsten	19222	188·5
Tyres, rubber	176–268	1·7–2·5
Vanadium	5991	58·8
Varnish:		
barrels	593	5·8
tins in cases	721	7·1
Walnut	657	6·4
Waste paper	352	3·4
pressed packed	449–513	4·4–5·0
Water:		
fresh	1001	9·8
salt	1009–1201	9·9–11·8
Wax:		
bees'	961	9·4
Brazil	993	9·7
cases of barrels	593	5·8
paraffin	897	8·8
White lead:		
powder	1378	13·5
paste in drums	2787	27·3
paint	2803	27·5
White metal	7368	72·2
Whitewood	465	4·6
Willow:		
American	577	5·6
English	449	4·4
Wine:		
bulk	977	9·6
bottles in cases	593	5·8
Wire		
iron, coils	1185	11·6
nails, bags	1201	11·8
rod, coils	801	7·8
rope, coils	1442	14·1
Wolfram (Wolframite)	7368	72·2
Wood block paving	897	8·8
Wool:		
compressed bales	769	7·5
uncompressed	208	2·0
Yew	673–801	6·6–7·8
York stone	2243	22·0
Zinc:		
cast	6804	66·7
rolled	7192	70·5
sheets packed	897	8·8
Zincblende	4085	40·1

Table III Design imposed loads on buildings for various uses
From CP3: Chap V: Part I: 1967

Use to which building or structure is to be put	Intensity of distributed load kN/m	Concentrated loads to be applied, unless otherwise stated, over any square with a 300 mm side kN
Art gallery (see 'Museum floors')		
Assembly buildings such as public halls and theatres, but excluding drill halls, places of worship, public lounges, schools and toilet rooms:		
with fixed seating*	**4·0**	—
without fixed seating	**5·0**	**3·6**
Balconies	Same as the rooms to which they give access	1·5 per metre run concentrated at the edge
Banking halls	**3·0**	—
Bedrooms:		
Domestic buildings	**1·5**	**1·4**
Hotels and motels	**2·0**	**1·8**
Institutional buildings	**1·5**	**1·8**
Billiard rooms	**2·0**	**2·7**
Boiler rooms	**7·5**	To be determined
Book stores	**2·4** for each metre of storage height	To be determined
Broadcasting studios:		
Corridors (see 'Corridors')		
Dressing-rooms	**2·0**	**1·8**
Fly galleries	**4·5kN** per metre run uniformly distributed over the width	—
Grids	**2·5**	—
Stages	**7·5**	**4·5**
Studios	**4·0**	—
Toilet rooms	**2·0**	—
Bungalows	**1·5**	**1·4**
Catwalks	Concentrated loads only	**1·0** at 1·0 m centres
Ceilings	Concentrated loads only	**0·9** on any joist
Chapels and churches	**3·0**	**2·7**
Cinemas (see 'Assembly buildings and 'Broadcasting studios')		
Classrooms	**3·0**	**2·7**
Clubs:		
Assembly areas with fixed seating*	**4·0**	—
Assembly areas without fixed seating	**5·0**	**3·6**
Bedrooms	**1·5**	**1·8**
Billiard rooms	**2·0**	**2·7**
Corridors (see 'Corridors')		
Dining-rooms	**2·0**	**2·7**
Kitchens	to be determined but not less than **3·0**	**4·5**
Lounges	**2·0**	**2·7**
Laundries	**3·0**	**4·5**
Toilet rooms	**2·0**	—
Cold storage	**5·0** for each metre of storage height, with a minimum of **15·0**	To be determined
Colleges:		
Assembly areas with fixed seating*	**4·0**	—
Assembly areas without fixed seating	**5·0**	**3·6**
Bedrooms	**1·5**	**1·8**
Classrooms	**3·0**	**2·7**
Corridors (see Corridors)		
Dining-rooms	**2·0**	**2·7**
Dormitories	**1·5**	**1·8**
Gymnasia	**5·0**	**3·6**

*Fixed seating implies that removal of the seating and use of the space for other purposes is improbable

Table III Design imposed loads on buildings for various uses (*continued*)
From CP3: Chap V: Part I: 1967

Use to which building or structure is to be put	Intensity of distributed load kN/m	Concentrated loads to be applied, unless otherwise stated, over any square with a 300 mm side kN
Kitchens	To be determined but not less than **3·0**	**4·5**
Laboratories, including equipment	To be determined but not less than **3·0**	To be determined but not less than **4·5**
Stages	**5·0**	**3·6**
Toilet rooms	**2·0**	—
Corridors, hallways passageways, aisles, public spaces and footbridges between buildings:		
Buildings subject to crowd loading, except grandstands	**4·0**	**4·5**
Buildings subject to loads greater than from crowds, including wheeled vehicles, trolleys, and the like	To be determined but not less than **5·0**	To be determined not less than **4·5**
All other buildings	Same as the rooms to which they give access	
Dance halls	**5·0**	**3·6**
Department stores:		
Shop floors for the display and sale of merchandise	**4·0**	**3·6**
Dormitories	**1·5**	**1·8**
Drill rooms and drill halls	**5·0**	To be determined but not less than **9·0**
Driveways and vehicle ramps other than in garages for the parking only of passenger vehicles and light vans not exceeding 2500 kg (2½ tons) gross weight	To be determined but not less than **5·0**	To be determined but not less than **9·0**
Dwellings	**1·5**	**1·4**
Factories and similar buildings	**5·0; 7·5** or **10·0** as appropriate	To be determined
File rooms in offices	**5·0**	To be determined
Flats	**1·5**	**1·4**
Footpaths, terraces and plazas leading from ground level:	To be determined but not less than	To be determined but not less than
No obstruction to vehicular traffic	**5·0**	**9·0**
Used only for pedestrian traffic	**4·0**	**4·5**
Foundries	To be determined but not less than **20**	—
Garages:		
Car parking only, for passenger vehicles and light vans not exceeding 2500 kg (2½ tons) gross weight including driveways and ramps	**2·5**	**9·0**
All repair workshops for all types of vehicles and parking for vehicles exceeding 2500 kg (2½ tons) gross weight, including driveways and ramps	To be determined but not less than **5·0**	Worst possible combination of wheel loads
Grandstands:		
Assembly areas with fixed seating*	**4·0**	—
Assembly areas without fixed seating	**5·0**	**3·6**
Corridors and passageways	**5·0**	**4·5**

Table III Design imposed loads on buildings for various uses (*continued*)
From CP3: Chap V: Part I: 1967

Use to which building or structure is to be put	Intensity of distributed load kN/m	Concentrated loads to be applied, unless otherwise stated, over any square with a 300mm side kN
Toilet rooms	2·0	—
Gymnasia	5·0	3·6
Halls:		
Corridors, hallways and passageways (see 'Corridors')		
Dressing-rooms	2·0	1·8
Fly-galleries	4·5kN per metre run uniformly distributed over the width	—
Grids	2·5	—
Projection rooms	5·0	—
Stages	5·0	3·6
Toilet rooms	2·0	—
Hospitals:		
Bedrooms and wards	2·0	1·8
Corridors, hallways and passageways (see 'Corridors')		
Dining-rooms	2·0	2·7
Kitchens	To be determined but not less than 3·0	4·5
Laundries	3·0	4·5
Toilet rooms	2·0	—
Utility rooms	2·0	4·5
X-ray rooms and operating theatres	2·0	4·5
Hotels and motels:		
Bars and vestibules	5·0	—
Bedrooms	2·0	1·8
Corridors, hallways and passageways (see 'Corridors')		
Dining-rooms	2·0	2·7
Kitchens	To be determined but not less than 3·0	4·5
Laundries	3·0	4·5
Lounges	2·0	2·7
Toilet rooms	2·0	—
Houses	1·5	1·4
Indoor sporting facilities:		
Areas for equipment	To be determined but not less than 2·0	To be determined
Assembly areas with fixed seating*	4·0	—
Assembly areas without fixed seating	5·0	3·6
Corridors (see 'Corridors')		
Dressing-rooms	2·0	1·8
Gymnasia	5·0	3·6
Toilet rooms	2·0	—
Institutional buildings:		
Bedrooms	1·5	1·8
Communal kitchens	To be determined but not less than 3·0	4·5
Corridors, hallways and passageways (see 'Corridors')		
Dining-rooms	2·0	2·7
Dormitories	1·5	1·8
Laundries	3·0	4·5
Lounges	2·0	2·7
Toilet rooms	2·0	—
Kitchens other than in domestic buildings, including normal equipment	To be determined but not less than 3·0	4·5
Laboratories, including equipment	To be determined but not less than 3·0	4·5
Landings	Same as the floors to which they give access	
Laundries other than in domestic buildings, excluding equipment	To be determined but not less than 3·0	4·5
Libraries:		
Reading-rooms without book storage	2·5	4·5
Rooms with book storage (eg public lending libraries)	4·0	4·5
Stack rooms	2·4 for each metre of room height with a minimum of 6·5	To be determined
Dense mobile stacking on mobile trucks	To manufacturer's recommendations	
Corridors	4·0	4·5
Toilet rooms	2·0	—
Machinery halls circulation spaces therein	4·0	To be determined
Maisonettes	1·5	1·4
Motor rooms, fan rooms and the like, including weight of machinery	To be determined but not less than 7·5	To be determined
Museum floors and art galleries for exhibition purposes	To be determined but not less than 4·0	To be determined
Offices:		
Corridors and public spaces (see 'Corridors')		
Filing and storage spaces	5·0	To be determined
Offices for general use	2·5	2·7
Offices with computing, data processing and similar equipment	3·5	To be determined
Toilet rooms	2·0	—
Pavement lights	To be determined but not less than 5·0	1½ times the wheel load but not less than 9·0
Places of worship	3·0	2·7
Printing plants:		
Paper storage	To be determined but not less than 4·0 for each metre of storage height	To be determined
Type storage and other areas	To be determined but not less than 12·5	To be determined
Public halls (see 'Halls')		
Public lounges	2·0	2·7
Residential buildings such as apartment houses, boarding houses, guest houses, hostels, lodging houses and residential clubs, but excluding hotels and motels:		
Bedrooms	1·5	1·8
Communal kitchens	To be determined but not less than 3·0	4·5
Corridors, hallways and passageways (see 'Corridors')		
Dining-rooms and public rooms	2·0	2·7
Dormitories	1·5	1·8
Laundries	3·0	4·5
Toilet rooms	2·0	—
Schools (see 'Colleges')		
Shop floors for the display and sale of merchandise	4·0	3·6
Stairs:		
Dwellings not over three-storey	1·5	1·8
All other buildings	Same as the floors to which they give access, but not less than 3·0	Same as the floors to which they give access

Table III Design imposed loads on buildings for various uses (*continued*)
From CP3: Chap V: Part I: 1967

Table III Design imposed loads on buildings for various uses (*continued*)
From CP3: Chap V: Part I: 1967

Use to which building or structure is to be put	Intensity of distributed load kN/m	Concentrated loads to be applied, unless otherwise stated, over any square with a 300mm side kN
	and not more than **5·0**	
Stationery stores	**4·0** for each metre of storage height	To be determined
Storage other than types listed separately	To be determined but not less than **2·4** for each metre of storage height	To be determined
Television studios (see 'Broadcasting studios')		
Theatres (see 'Assembly buildings' and 'Broadcasting studios')		
Universities (see 'Colleges' and 'Libraries')		
Warehouses (see 'Storage')		
Workrooms, light, without storage	**2·5**	**1·8**
Workshops (see 'Factories')		

1.06 Structural elements

For convenience, large structures are broken up into *elements*. These are of different types according to the function they perform in the building. Before describing each type, it will be necessary to go deeper into the forces that are found internally in the materials of the structure.

1.07 Stress and strain

If a bar of uniform cross-section has a force applied at each end, **45.1**, it will stretch slightly. This stretch is called the *strain* in the bar, and is defined as $\frac{\text{change in length}}{\text{original length}}$. Now for elastic materials, which all normal structural materials can be assumed to be, Hooke's Law applies. This states that $\frac{\text{stress}}{\text{strain}}$ is a constant, Young's Modulus. The *stress* is the force on the cross-section divided by its area. From Hooke's Law it can be seen that there can be no stress without strain, and no strain without stress.

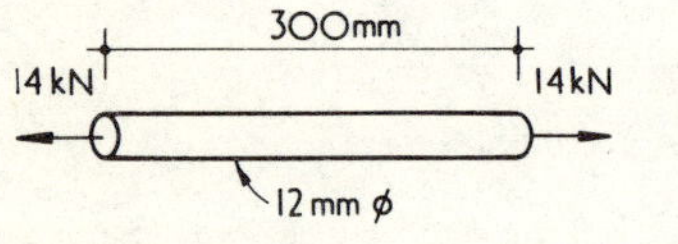

45.1 *A bar of uniform cross-section under a tensile force (Example 2)*

Example 2

A round steel bar 12 mm in diameter and 300 mm long, has a load of 14 kN applied at each end as in **45.1**. *How much will this bar stretch?*

The area of the cross-section is $\pi \times 6^2 = 113\,\text{mm}^2$

Therefore the stress on the cross-section is $\frac{14 \times 10^3}{113} = 123{\cdot}9\ \text{N/mm}^2$

Young's Modulus for steel is $207\,000\ \text{N/mm}^2$

Now strain $= \frac{\text{stress}}{\text{Young's Modulus}} = \frac{123{\cdot}9}{207 \times 10^3} = 0{\cdot}599 \times 10^{-3}$

Therefore, the stretch in the bar is strain × original length $= 0{\cdot}599 \times 10^{-3} \times 300 = \underline{0{\cdot}180\ \text{mm}}$

1.08 Units of stress

The basic SI unit of stress is the N/m^2, but this is a very small unit for practical purposes. The correct form for the normal unit is MN/m^2 but this is often expressed as N/mm^2, and this form is allowable. The codes for steel, concrete and timber use N/mm^2, while the code for masonry uses MN/m^2; this section will follow the use in the appropriate code. Eventually, it is intended to replace both forms with MPa, but as yet the pascal is not in general use in Britain.

1.09 Tension members

If the stress in the member is tending to lengthen it, it is said to be in *tension*. Elements in tension are called *ties*. In many ways, this is the simplest kind of stress. Some materials are ideal for resisting it: steel in particular. Cables, wires and chains can be used to carry tension, but no other kind of force. Some materials have little or no resistance to tension: stone, cast iron and unreinforced concrete fall into this category.

1.10 Compression members

If the stress in the member is tending to shorten it, it is said to be in *compression*. Elements in compression are called *struts* if they are small; or sometimes *columns, piers* or *stanchions*. The term used depends on the position and the material of which the member is composed.

Most materials other than cables, wires and chains can be used to carry compression. However, there is a special phenomenon that occurs with compression, and this is called *buckling*. For certain types of member, particularly those that are slender in comparison with their length, increase of compressive load will cause the strut to bend until failure occurs in tension on one side. In the case of masonry construction, it is this buckling effect that causes the collapse of towers and high walls.

The ultimate load that can be carried by a compression member before it buckles is given by the Euler formula:

$$P_e = \frac{\pi^2 EI}{l^2}$$

where P_e is the maximum load before buckling occurs
E is Young's Modulus
l is the effective length of the strut
and I is a quantity called loosely the moment of inertia of the section, but should be referred to as 'the second moment of area'.

1.11 The second moment of area

Since this quantity crops up frequently in structural design, it is important to grasp its meaning, and be able to find the correct value when required.

In **45.2a** a cross-section of a structural member is shown, together with a reference axis ox. Consider a small area a of the section, which is distance y from ox. We use the symbol Σ to mean 'the sum of all the different quantities' so that $\Sigma a = A$ the total area of the cross-section. Similarly we can calculate Σay, which is called the first moment of area of the section about ox. Now, it is possible to find an axis OX for which Σ aY is zero; this is called the *neutral axis* and passes through the *centre of area* of the section. In fact, the distance between ox and OX is given by $\frac{\Sigma\ ay}{A}$. There are many axes for which the first moment of area is zero, but they all pass through the centre of area at different angles. Two of them, OX and OY, are called the *principal axes* and are at right angles to each other, **45.2b**.

If OX is now used as the reference axis, consider the quantity aY^2. If this quantity is summed for the whole section: $\Sigma\ aY^2$, the second moment of area is obtained about axis OX. This is the quantity often called the moment of inertia of the section, and is abbreviated I_x. Second moments may be obtained for all the axes through the centre of area. Conventionally I_x is the maximum value and I_y is the minimum, and these are the principal axes. If the section is symmetrical about one or two axes, these will coincide with the principal axes. For obtaining I_x table IV gives formulae for common section shapes.

In the case of the Euler formula, since the strut could buckle in any direction, the I value that must be substituted in the formula is the minimum value I_y.

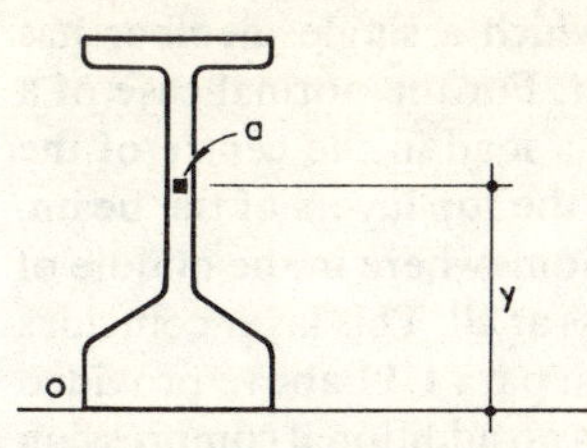

45.2a *Cross-section of a structural member with an arbitrary reference axis*

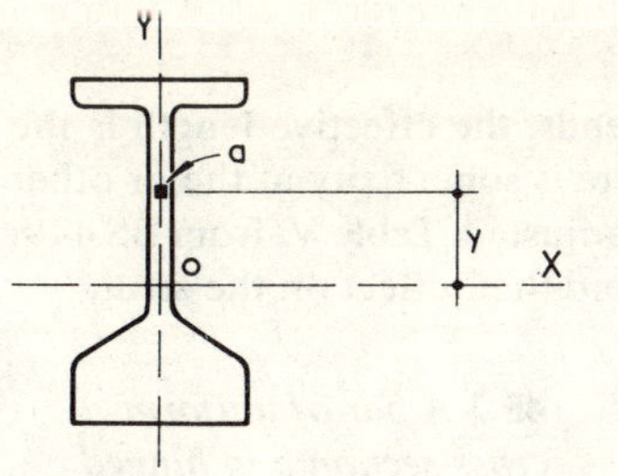

b *The same member with principle axes through the centre of area*

Table IV Properties of sections

Section shape	Area of section A	Distance (y_1) of extremity of section from neutral axis	Moment of inertia about neutral axis X × (I_x)	Modulus $Z_x = \left(\frac{I_x}{y_1}\right)$	Radius of gyration $k = \sqrt{\frac{I_x}{A}}$
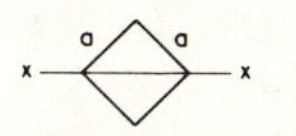	a^2	$\frac{a}{2}$	$\frac{a^4}{12}$	$\frac{a^3}{6}$	$\frac{a}{\sqrt{12}} = 0{\cdot}289a$
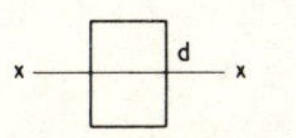	bd	$\frac{d}{2}$	$\frac{1}{12}bd^3$	$\frac{1}{6}bd^2$	$\frac{d}{\sqrt{12}} = 0{\cdot}289d$
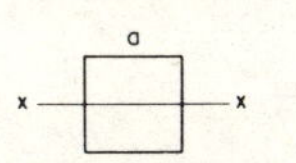	a^2	$\frac{a}{\sqrt{2}} = 0{\cdot}707a$	$\frac{a^4}{12}$	$\frac{\sqrt{2}}{12}a^3 = 0{\cdot}118a^3$	$\frac{a}{\sqrt{12}} = 0{\cdot}289a$
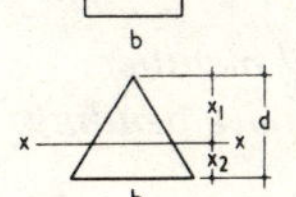	$\frac{bd}{2}$	$\frac{d}{3}$	$\frac{bd^3}{36}$	$\frac{bd^2}{24}$	$\frac{d}{\sqrt{18}} = 0{\cdot}236d$
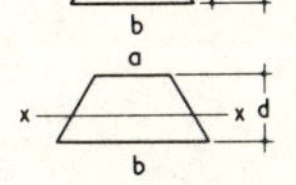	$\frac{a+b}{2}d$	$\frac{a+2b}{a+b}\frac{d}{3}$	$\frac{a^2+4ab+b^2}{36(a+b)}d^3$	$\frac{a^2+4ab+b^2}{12(a+2b)}d^2$	$d\sqrt{\frac{a^2+4ab+b^2}{18(a+b)^2}}$
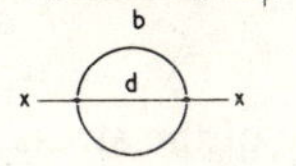	$\frac{\pi d^2}{4} = 0{\cdot}7854d^2$	$\frac{d}{2}$	$\frac{\pi d^4}{64} = 0{\cdot}0491d^4$	$\frac{\pi d^3}{32} = 0{\cdot}0982d^3$	$\frac{d}{4}$
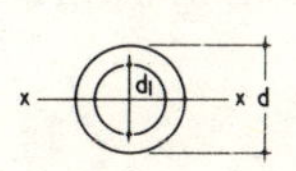	$\frac{\pi}{4}(d^2 - d_1^{\,2})$	$\frac{d}{2}$	$\frac{\pi}{64}(d^4 - d_1^{\,4})$	$\frac{\pi}{32}\frac{d^4 - d_1^{\,4}}{d}$	$\frac{\sqrt{d^2 + d_1^{\,2}}}{4}$
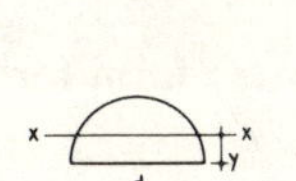	$\frac{\pi d^2}{8} = 0{\cdot}3927d^2$	$\frac{2d}{3\pi} = 0{\cdot}212d$	$\frac{9\pi^2 - 64}{1152\pi}d^4 = 0{\cdot}007d^4$	$\frac{(9\pi^2 - 64)d^3}{192(3\pi - 4)} = 0{\cdot}024d^3$	$\frac{\sqrt{9\pi^2 - 64}d}{12\pi} = 0{\cdot}132d$
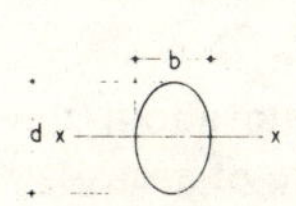	$\frac{\pi bd}{4} = 0{\cdot}7854bd$	$\frac{d}{2}$	$\frac{\pi bd^3}{64} = 0{\cdot}0491bd^3$	$\frac{\pi bd^2}{32} = 0{\cdot}0982bd^2$	$\frac{d}{4}$
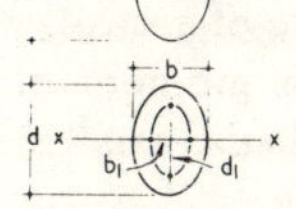	$\frac{\pi}{4}(bd - b_1d_1)$	$\frac{d}{2}$	$\frac{\pi}{64}(bd^3 - b_1d_1^{\,3})$	$\frac{\pi}{32}\frac{bd^3 - b_1d_1^{\,3}}{d}$	$\frac{1}{4}\sqrt{\frac{bd^3 - b_1d_1^{\,3}}{bd - b_1d_1}}$
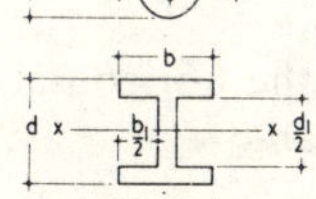	$(bd - b_1d_1)$	$\frac{d}{2}$	$\frac{1}{12}(bd^3 - b_1d_1^{\,3})$	$\frac{bd^3 - b_1d_1^{\,3}}{6d}$	$\sqrt{\frac{bd^3 - b_1d_1^{\,3}}{12(bd - b_1d_1)}}$
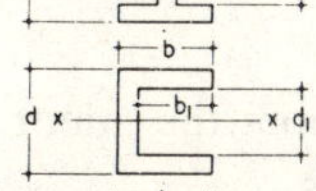	$(bd - b_1d_1)$	$\frac{d}{2}$	$\frac{1}{12}(bd^3 - b_1d_1^{\,3})$	$\frac{bd^3 - b_1d_1^{\,3}}{6d}$	$\sqrt{\frac{bd^3 - b_1d_1^{\,3}}{12(bd - b_1d_1)}}$
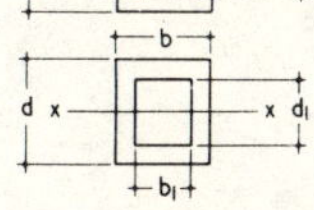	$(bd - b_1d_1)$	$\frac{d}{2}$	$\frac{1}{12}(bd^3 - b_1d_1^{\,3})$	$\frac{bd^3 - b_1d_1^{\,3}}{6d}$	$\sqrt{\frac{bd^3 - b_1d_1^{\,3}}{12(bd - b_1d_1)}}$
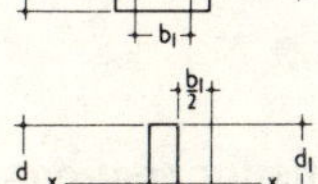	$(bd - b_1d_1)$	$\frac{bd^2 - 2b_1d_1d + b_1d_1^{\,2}}{2(bd - b_1d_1)}$	$\frac{(bd^2 - b_1d_1^{\,2})^2 - 4bdb_1d_1(d - d_1)^2}{12(bd - b_1d_1)}$	$\frac{(bd^2 - b_1d_1^{\,2})^2 - 4bdb_1d_1(d - d_1)^2}{6(bd^2 - 2bdd_1 + b_1d_1^{\,2})}$	–
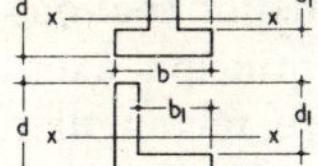	$(bd - b_1d_1)$	$\frac{bd^2 - 2b_1d_1d + b_1d_1^{\,2}}{2(bd - b_1d_1)}$	$\frac{(bd^2 - b_1d_1^{\,2})^2 - 4bdb_1d_1(d - d_1)^2}{12(bd - b_1d_1)}$	$\frac{(bd^2 - b_1d_1^{\,2})^2 - 4bdb_1d_1(d - d_1)^2}{6(bd^2 - 2bdd_1 + b_1d_1^{\,2})}$	–
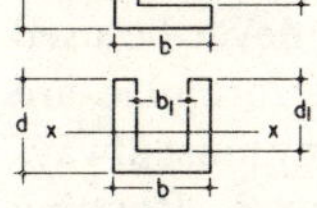	$(bd - b_1d_1)$	$\frac{bd^2 - 2b_1d_1d + b_1d_1^{\,2}}{2(bd - b_1d_1)}$	$\frac{(bd^2 - b_1d_1^{\,2})^2 - 4bdb_1d_1(d - d_1)^2}{12(bd - b_1d_1)}$	$\frac{(bd^2 - b_1d_1^{\,2})^2 - 4bdb_1d_1(d - d_1)^2}{6(bd^2 - 2bdd_1 + b_1d_1^{\,2})}$	–
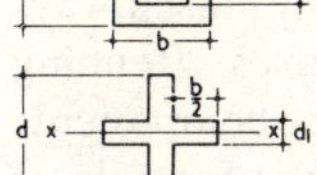	$(bd_1 + b_1d)$	$\frac{d}{2}$	$\frac{1}{12}(bd_1^{\,3} + b_1d^3)$	$\frac{b_1d^3 + bd_1^{\,3}}{6d}$	$\sqrt{\frac{bd_1^{\,3} + b_1d^3}{12(bd_1 + b_1d)}}$
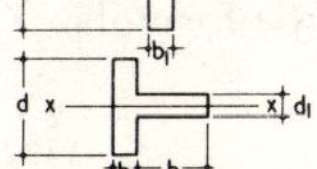	$(bd_1 + b_1d)$	$\frac{d}{2}$	$\frac{1}{12}(bd_1^{\,3} + b_1d^3)$	$\frac{b_1d^3 + bd_1^{\,3}}{6d}$	$\sqrt{\frac{bd_1^{\,3} + b_1d^3}{12(bd_1 + b_1d)}}$
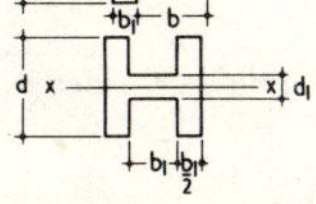	$(bd_1 + b_1d)$	$\frac{d}{2}$	$\frac{1}{12}(bd_1^{\,3} + b_1d^3)$	$\frac{b_1d^3 + bd_1^{\,3}}{6d}$	$\sqrt{\frac{bd_1^{\,3} + b_1d^3}{12(bd_1 + b_1d)}}$

1.12 A strut

If the strut is hinged at both ends, the effective length is the actual length. If, however, there is some fixity at the or other end, or both, this value can be adjusted. Table V, from BS 449, indicates various conditions, and their effect on the strut.

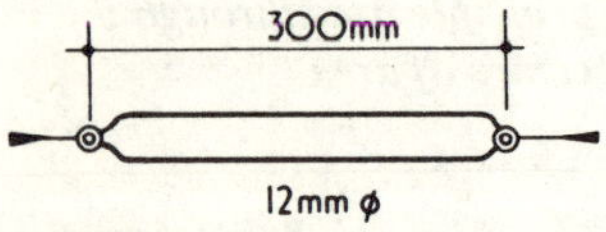

45.3 *A bar of uniform cross-section and hinged ends under a compressive force (Example 3)*

Example 3

The bar in example 2 is hinged at each end and placed in compression. What stress can it carry before it buckles? **45.3**

For a circular cross-section, the second moment of area about all axes is the same, $\frac{\pi d^4}{64} = 0{\cdot}0491 \times 12^4 = 1018\ \text{mm}^4$ from the Euler formula:

$$P_e = \frac{\pi^2 \times 207 \times 10^3 \times 1018}{300^2} N$$

$$= 23{\cdot}11\ \text{kN}$$

Hence, the stress immediately before buckling is $\frac{\text{Load}}{\text{Area}}$

$$= \frac{23{\cdot}11}{113}\ \text{kN/mm}^2$$

$$= \underline{\underline{204{\cdot}5\ \text{N/mm}^2}}$$

Table V Effective length of struts

For the purpose of calculating the Euler load for struts (see 1.11) the effective length *l* shall be taken as follows:	
1. Effectively held in position and restrained in direction at both ends	$l = 0{\cdot}7L$
2. Effectively held in position at both ends and restrained in direction at one end	$l = 0{\cdot}85L$
3. Effectively held in position at both ends, but not restrained in direction	$l = L$
4. Effectively held in position and restrained in direction at one end, and at the other partially restrained in direction but not held in position	$l = 1{\cdot}5L$
5. Effectively held in position and restrained in direction at one end, but not held in position or restrained in direction at the other end	$l = 2{\cdot}0L$
where L = Length of strut from centre-to-centre of intersections with supporting members.	

It should not be thought that the structural member in the example above would be allowed to carry a load as great as the ultimate load calculated. A *factor of safety*, or a *load factor* of about 2 is normally used to cover for all the unknown situations that might arise during the life of the element. These include accidental overloading, reduction of the member due to corrosion or damage, and for inaccurate assumptions in the design process itself.

1.13 Pin-jointed frames

Some structures are designed and constructed completely of members that are in tension or compression. The familiar roof truss is of this type; the generic term for which is *pin-jointed frames*. Methods for finding the magnitude of the forces in each member can be found in the standard text-books.

1.14 Bending

Struts and ties transmit forces along their length without changing the magnitude or direction of those forces. However, the object of a structure is almost always to make some change in the disposition of the forces carrying a load: as for example, to provide a large open space beneath it. This change can be accomplished by using a pin-jointed frame, but most commonly a *bending* member is used.

Bending is the phenomenon by which a single member has both compression and tension in it. For the normal case of a beam supported at each end with a load in the centre of the span, there will be compression in the top layers of the beam, and tension in the bottom layers. Somewhere in the middle of the beam will be layer with no stress at all. This layer coincides with the neutral axis as described in para **1.11** above provided the beam is in pure bending and has no additional compression or tension, **45.4**.

A cross-section of a beam will have imposed on it a combination of forces known as a *bending moment*. A bending moment is to a beam what a load is to a column, and a method of calculating the bending moment will be given in para **1.17**.

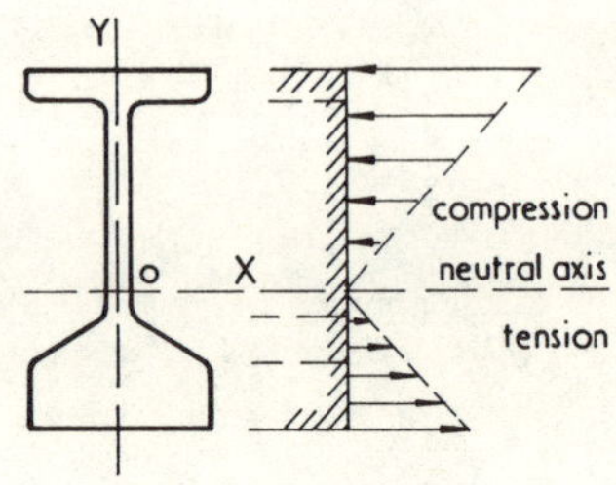

45.4 *Structural member in* **45.2** *under pure bending*

1.15 Bending stress

The basic bending formula, giving the magnitude of the stress in any layer of the beam once the bending moment is known, is expressed as follows:

$$\frac{M}{I} = \frac{f}{y} \text{ or alternatively } M = f\,Z$$

In fact this formula is slightly deficient. The correct form for bending about the X direction is:

$$\frac{M_x}{I_x} = \frac{f}{y} \text{ or } M_x = f\,Z_x$$

where M_x = the bending moment in the beam about the horizontal axis through the centre of area

I_x = the second moment of area (or moment of inertia) of the section about the same axis

f = the stress in the layer

y = the distance of the layer from the horizontal axis through the centre of area and

Z_x = the section modulus $= \frac{I_x}{y}$

There is an equivalent formula for bending about the minor principal axis, ie conventionally the vertical one:

$$\frac{M_y}{I_y} = \frac{f}{x} \text{ or } M_y = f\,Z_y$$

Note: These formulae only apply in these forms for sections that are symmetrical about at least one of the principal axes. They may be used for I-sections, channels, square, rectangular or round solid or hollow sections. They should never be used for angle sections or Z-sections, for which the formulae are more complex. Where the formulae are used, they must always be applied to bending about the principal axes. Bending moments about other axes should be resolved into components about the principal axes before substitution in the formulae.

Example 4

What is the stress in the extreme fibre of a 16 mm square solid bar under a bending moment of 60 Nm about axis OX in **45.5***?*

For a square section $I = \frac{bd^3}{12}$ from table IV

$$= \frac{16 \times 16^3}{12} = 5461 \text{ mm}^4$$

The centre of area of the section is at the geometric centre of the square. The extreme fibres are therefore 8 mm from the bending axis.

Hence $\frac{M}{I} = \frac{f}{y}$, $f = \frac{M}{I}y = \frac{60 \times 10^3}{5461} \times 8 = \underline{\underline{88 \text{ N/mm}^2}}$

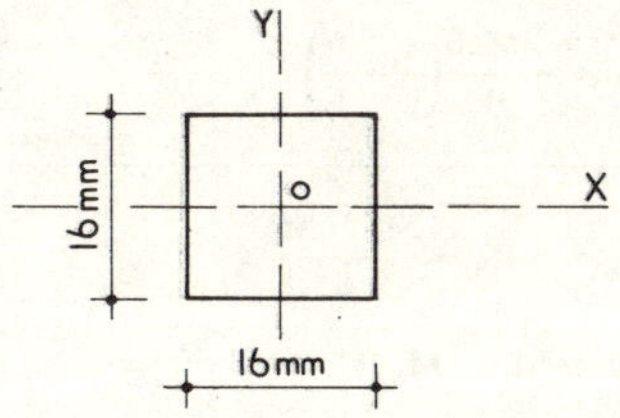

45.5 *Square solid bar (Example 4)*

1.16 Materials in bending

Since bending includes both compression and tension, only materials that are strong in both are suitable for beams. Steel and timber are good examples of such materials. Stone, being weak in tension, makes poor bending members. This is the reason why the ancient Greeks had to build their columns so close together. Only when the arch had been invented could the spans be increased, because the arch is wholly in compression. Cast iron is another material weak in tension, although not as weak as stone. Beams of this material have lower flanges larger than the upper to allow for this. Concrete is also poor under tensile force, so steel is used to reinforce the bottom of concrete beams.

The stress in the extreme fibre of a beam depends on the second moment of area of the cross-section. The larger this is, the smaller will the maximum stress be. It is therefore beneficial to choose cross-section shapes that have large I-values for the given area of material. The most common shape of cross-section for a steel beam is an I for this reason. The further away from the neutral axis that the flanges can be, the less they need be in area to give a required second moment of area.

There is a limit to this, however. Since the top flange is in compression, if it becomes too small in itself, and too divorced from the rest of the beam, it can buckle. This is particularly significant in the design of steelwork, and is the reason that bending stresses in the extreme fibres of steel beams are reduced well below the ultimate strength of the material.

1.17 Bending moments

In order to calculate the stresses on a section in bending, the bending moment has first to be found. For many simple beams, only a knowledge of elementary statics is necessary. (Such knowledge is assumed in the following, but a résumé will be found in the *AJ Handbook of Building Structure*.) The principles of the method will be found illustrated in the examples

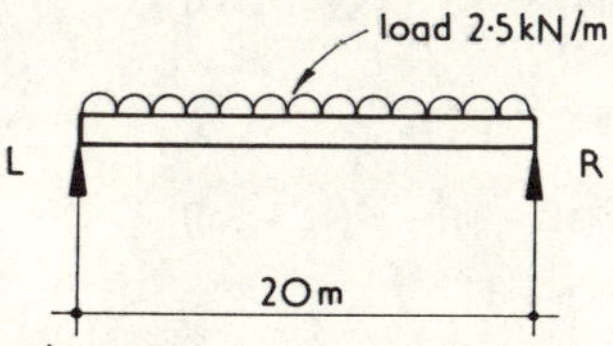

45.6a *Beam with uniformly distributed load (Example 5)*

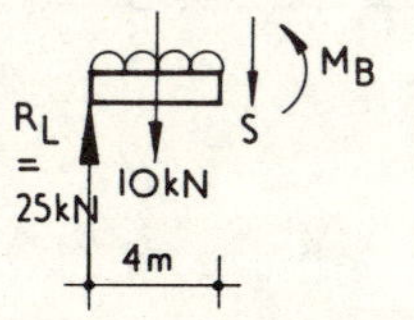

b *Forces acting on the left-hand 4 m length of beam*

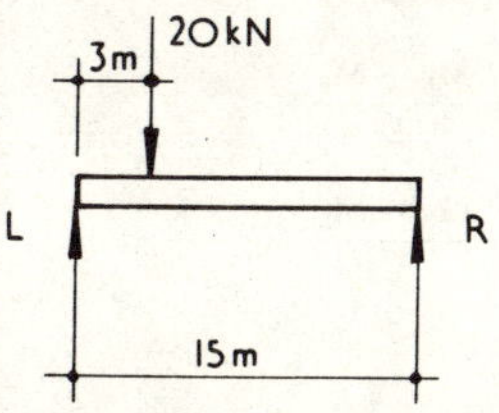

45.7a *Beam with point load (Example 6)*

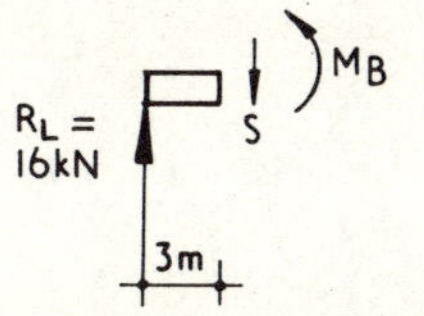

b *Forces acting on the left-hand length of beam up to the load*

Example 5

A beam of 20 m span has a load uniformly distributed along its length of 2·5 kN per metre run. What is the bending moment 4 m from the end? **(45.6a)**
The total load on the beam is $2{\cdot}5 \times 20\text{kN} = 50\text{kN}$. Since the system is symmetrical, the reaction at each end is 25 kN.
For the length of beam 4 m long, take moments about the cut end **(45.6b)**:

$$25 \times 4 - 2{\cdot}5 \times \frac{4}{2} \times 4 = M_B = \underline{\underline{80 \text{ kNm}}}$$

Example 6

A beam of 15 m span has a single point load of 20 kN 3 m from the left-hand end. What is the moment at the load position? **(45.7a)**
This time, more work is needed to find the value of the end reactions. Take moments about the right-hand end for the whole beam **(45.7a)**:
$20 \times 12 - R_L \times 15 = 0$ Hence $R_L = 16\text{kN}$
Now take moments for the left-hand end of the beam about a point immediately to the left of the load position, **45.7b**:
$R_L \times 3 = M_B$ from which $M_B = \underline{\underline{48 \text{kNm}}}$

1.18 Beams

The bending moment at any section for such simple beams (often called statically determinant) may be found by similar methods, further details of which can be found in the standard text-books. These beam types include simply-supported spans, cantilevers and beams with overhanging ends.

Other types of beams, such as those where the ends are rigidly held (as in most reinforced concrete), beams that are continuous over three or more supports, and those forming parts of complete frames are not susceptible to this method. Such beams are described as *statically indeterminant*, and the methods of dealing with them, although more complicated, may also be found in the standard texts, including the *AJ Handbook of Building Structure.*

Formulae for calculating the bending moments of a number of simple beams, and also of the common types of fixed-ended beams (sometimes called encastré or built-in) are given in table VI.

Beams that are supported on brickwork, such as lintols over doorways and windows, are not considered structurally encastré, and are normally designed as simply-supported.

Table VI Standard beam conditions

1 Cantilevers

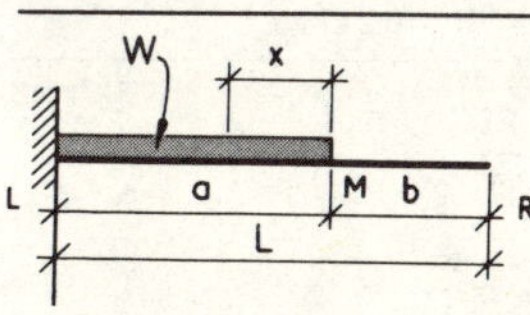	$M_x = \frac{Wx^2}{2a}$ $M_{max} = \frac{Wa}{2}$ $S_{max} = R_L = W$ $\delta_M = \frac{Wa^3}{8EI}$ $\delta_{max} = \delta_R = \frac{Wa^3}{8EI} \times \left(1 + \frac{4b}{3a}\right)$
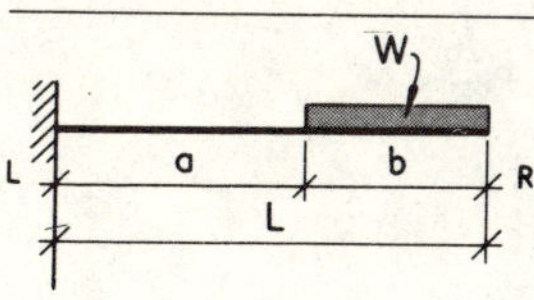	$M_{max} = W\left(a + \frac{b}{2}\right)$ $S_{max} = R_L = W$ $\delta_{max} = \delta_R$ $= \frac{W}{24EI}(8a^3 + 18a^2b + 12ab^2 + 3b^3)$
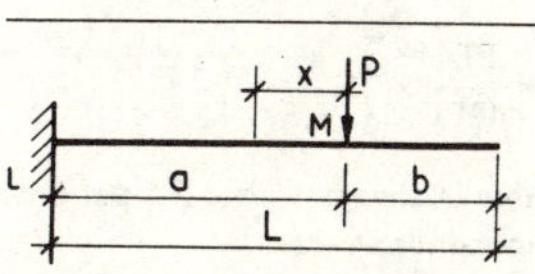	$M_x = P_x$ $M_{max} = Pa$ $S_{max} = R_L = P$ $\delta_M = \frac{Pa^3}{3EI}$ $\delta_{max} = \delta_R = \frac{Pa^3}{3EI} \times \left(1 + \frac{3b}{2a}\right)$

Table VI Standard beam conditions *(continued)*

2 Free support beams

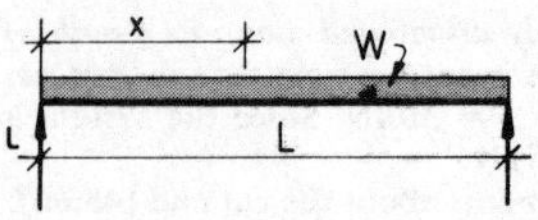

$$M_x = \frac{Wx}{2}\left(1 - \frac{x}{L}\right)$$

$$M_{max} = \frac{WL}{8}$$

$$R_L = R_R = \frac{W}{2}$$

$$\delta_{max} \text{ at centre} = \frac{5}{384} \times \frac{WL^3}{EI}$$

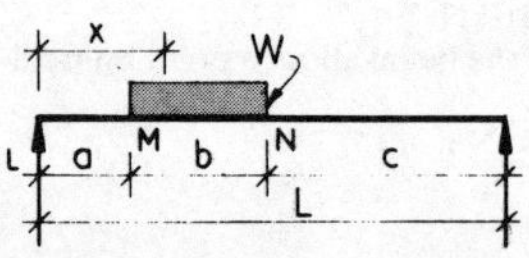

$$M_{max} = \frac{W}{b}\left(\frac{x^2 - a^2}{2}\right)$$

when

$$x = a + R_L \times \frac{b}{W}$$

$$R_L = \frac{W}{L}\left(\frac{b}{2} + c\right)$$

$$R_R = \frac{W}{L}\left(\frac{b}{2} + a\right)$$

if $a = c$

$$M = \frac{W}{8}(L + 2a)$$

$$\delta_{max} = \frac{W}{384EI} \times (8L^3 - 4Lb^2 + b^3)$$

$$M_x = W_x\left(\frac{1}{2} - \frac{2x^2}{3L^2}\right)$$

$$M_{max} = \frac{WL}{6}$$

$$R_L = R_R = \frac{W}{2}$$

$$\delta_{max} = \frac{WL^3}{60EI}$$

If $\phi = 60°$ $M = 0{\cdot}0725\,wL^3$
$R = 0{\cdot}217\,wL^2$

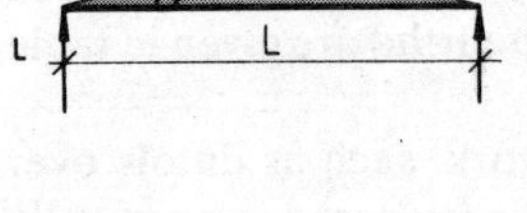

$$M_{max} = \frac{PL}{4}$$

$$R_L = R_R = \frac{P}{2}$$

$$\delta_{max} = \frac{PL^3}{48EI}$$

$$M_{max} = P\frac{ab}{L} = M_p$$

$$R_L = \frac{Pb}{L} \qquad R_R = \frac{Pa}{L}$$

δ_{max} always occurs within 0·0774L of the centre of the beam.
When $b > a$

$$\delta_{centre} = \frac{PL^3}{48EI} \times \left[3\frac{a}{L} - 4\left(\frac{a}{L}\right)^3\right]$$

This value is always within 2·5 per cent of the maximum value.

$$\delta_p = \frac{PL^3}{3EI}\left(\frac{a}{L}\right)^2\left(1 - \frac{a}{L}\right)^2$$

Table VI Standard beam conditions *(continued)*

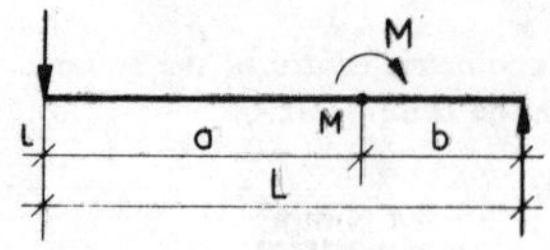

$$M_{ML} = M\frac{a}{L} \qquad M_{MR} = M\frac{b}{L}$$

$$R_A = R_B = \frac{M}{L}$$

when $a > b$

$$\delta_M = -\frac{Mab}{3EI}\left(\frac{a}{L} - \frac{b}{L}\right)$$

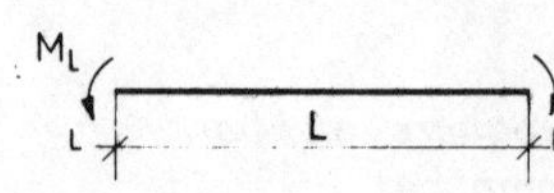

$$R_L = -R_R = \frac{M_L - M_R}{L}$$

when $M_L = M_R$,

$$\delta_{max} = -\frac{ML^2}{8EI}$$

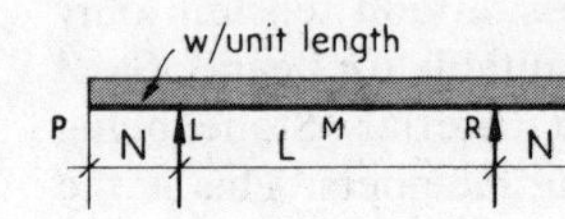

$$M_L = M_R = -\frac{wN^2}{2}$$

$$M_{max} = \frac{WL^2}{8} + M_L$$

$$R_L = R_R = w\left(N + \frac{L}{2}\right)$$

$$\delta_p = \delta_q = \frac{wL^3N}{24EI} \times (1 - 6n^2 - 3n^3)$$

$$\delta_{max} = \frac{wL^4}{384EI}(5 - 24n^2)$$

$$n = \frac{N}{L}$$

3 Fixed-end beams

$$M_L = M_R = -\frac{WL}{12}$$

$$M_M = \frac{WL}{24}$$

$$R_L = R_R = \frac{W}{2}$$

points of contraflexure 0·21L from each end

$$\delta_{max} = \frac{WL^3}{384EI}$$

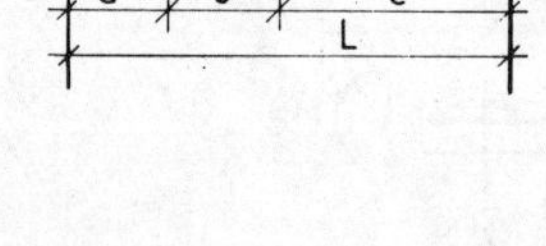

$$M_L = -\frac{W}{12L^2b}$$

$[c^3(4L - 3c) - c^3(4L - 3c)]$

$$M_R = -\frac{W}{12L^2b}$$

$[d^3(4L - 3d) - a^3(4L - 3a)]$

$a + b = d$

$b + c = e$

when r = reaction if the beam were simply supported.

$$R_L = r_L + \frac{M_L - M_R}{L}$$

$$R_R = r_R + \frac{M_R - M_L}{L}$$

when $a = c$,

$$\delta_{max} = \frac{W}{384EI} \times (L^3 + 2L^2a + 4La^2 - 8a^3)$$

Table VI Standard beam conditions (*continued*)

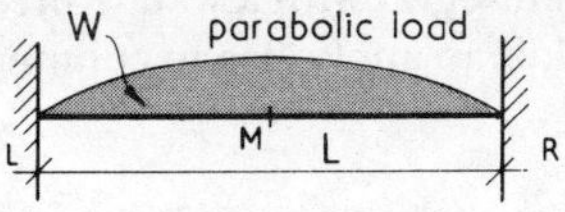

$$M_L = M_R = -\frac{WL}{10}$$

$$M_M = \frac{5WL}{32} - \frac{WL}{10} = \frac{9WL}{160}$$

$$R_L = R_B = \frac{W}{2}$$

$$\delta_{max} = \frac{1{\cdot}3WL^3}{384EI}$$

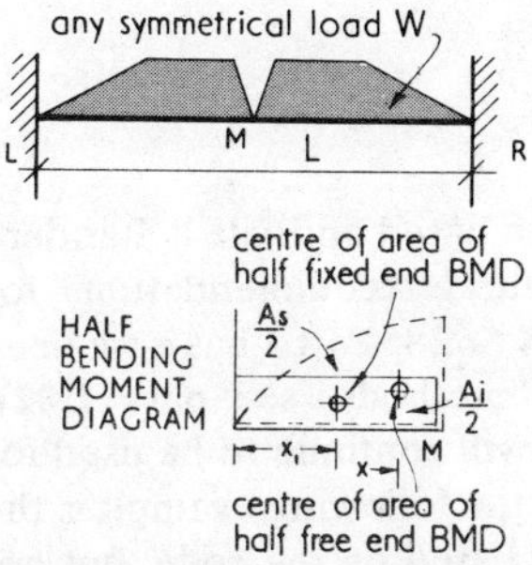

$$M_L = M_R = -\frac{A_S}{L}$$

where A_S is the area of the free bending moment diagram

$$R_L = R_R = \frac{W}{2}$$

$$\delta_{max} = \frac{A_S x - A_1 x_1}{2EI}$$

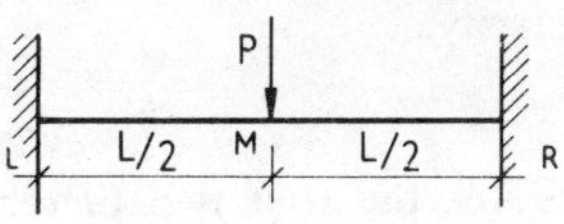

$$M_L = M_R = -\frac{PL}{8}$$

$$M_M = \frac{PL}{8}$$

$$R_L = R_R = \frac{P}{2}$$

$$\delta_{max} = \frac{PL^3}{192EI}$$

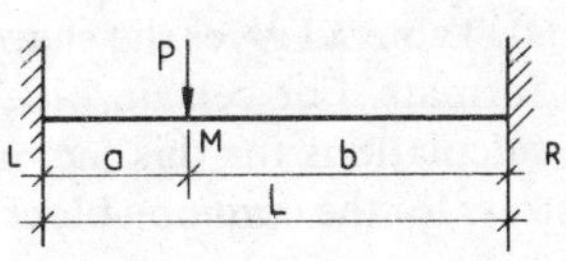

$$M_L = -\frac{Pab^2}{L^2}$$

$$M_R = \frac{Pba^2}{L^2}$$

$$M_M = \frac{2Pa^2b^2}{L^3}$$

$$R_L = P\frac{b^2}{L^2}\left(1 + 2\frac{a}{L}\right)$$

$$R_R = P\frac{a^2}{L^2}\left(1 + 2\frac{b}{L}\right)$$

$$\delta_M = \frac{Pa^3b^3}{3EIL^3}$$

$$\delta_{max} = \frac{2Pa^2b^3}{3EI(3L - 2a)^2}$$

$$\text{at } x = \frac{L^2}{3L - 2a}$$

4 Propped cantilevers

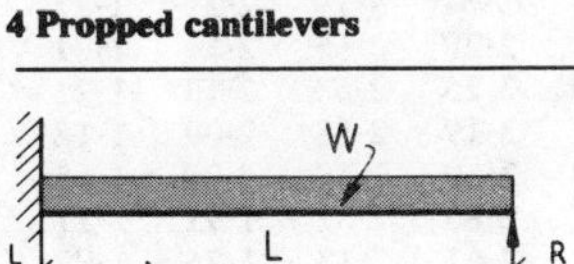

$$M_L = -\frac{WL}{8}$$

$$M_{max} = \frac{9WL}{128} \text{ at } x' = \tfrac{5}{8}$$

$M = 0$ at $x' = \frac{1}{4}$
$R_L = \frac{5}{8}W$
$R_R = \frac{3}{8}W$
if $m = 1 - x'$

$$\delta = \frac{WL^3}{48EI} \times (m - 3m^3 + 2m^4)$$

$$\delta_{max} = \frac{WL^3}{185EI}$$

at $x' = 0{\cdot}5785$

Table VI Standard beam conditions (*continued*)

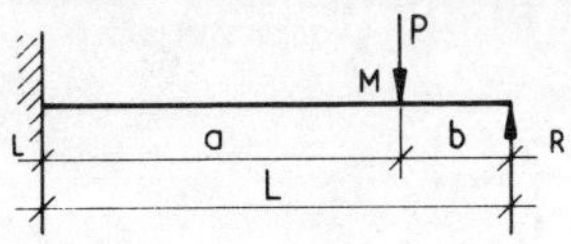

$$M_L = -\frac{Pb}{2}(1 - b'^2)$$

(maximum 0·193PL if $b' = 0{\cdot}577$)

$$M_M = \frac{Pb}{2}(2 - 3b' + b'^3)$$

(maximum 0·174PL if $b' = 0{\cdot}366$)

$$R_R = \tfrac{1}{2}Pa'^2(b' + 2)$$

$$\delta_m = \frac{Pa^3b^2}{12EIL^3} - (4L - a)$$

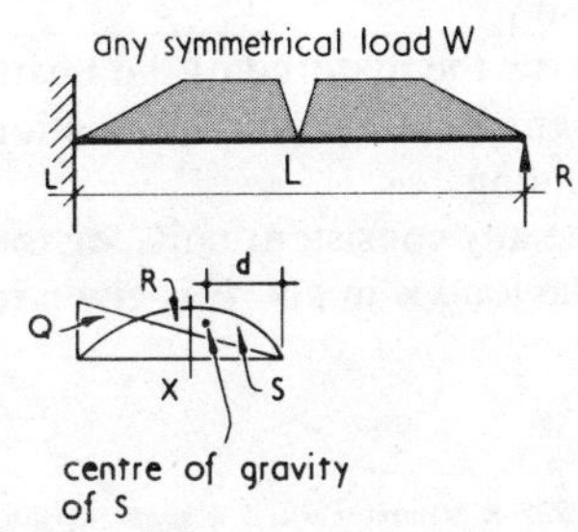

If A_S = area of free bending moment diagram

$$M_L = \frac{3A_S}{2L}$$

$$R_L = \frac{W}{2} + \frac{M_L}{L}$$

$$R_R = \frac{W}{2} - \frac{M_L}{L}$$

δ_{max} at X where area Q = area R

$$\delta_{max} = \frac{\text{area S} \times X \times d}{EI}$$

1.19 Shear

Referring back to example 5 it will be noted that by taking moments about the cut end the forces through this section could be ignored. If the vertical forces on the left-hand portion of the beam are resolved, it will be seen that a downward force of 16 kN must act on the cut section. This force is a *shear* force acting in the same plane as the section.

In the case of an I-section, the shear force acts mainly within the web connecting together the two flanges that are in compression and tension. If the web becomes too slender, it can indeed buckle under the influence of the shear force, **45.8**.

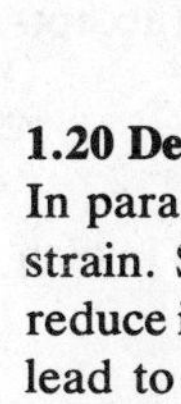

45.8 *Buckling of web of an I-section beam*

1.20 Deflection

In para **1.07** it was said that there could be no stress without strain. Since the top of the beam is in compression, it must reduce in length; the bottom, in tension, must stretch. This will lead to the beam taking up a curved form: in the case of a simply-supported beam with vertical loading, it will sag. Excessive sagging is not only unsightly, it may cause damage to finishes, or even cause load to be transferred onto elements of construction, such as partitions, that are not designed to carry such load. Formulae are published giving the deflections of various kinds of beams under different loadings (some included in table VI), the commonest of which is for a simply-supported beam under a uniformly distributed load which is:

$$\Delta = \frac{5}{384} \times \frac{WL^3}{EI} \qquad (45.9)$$

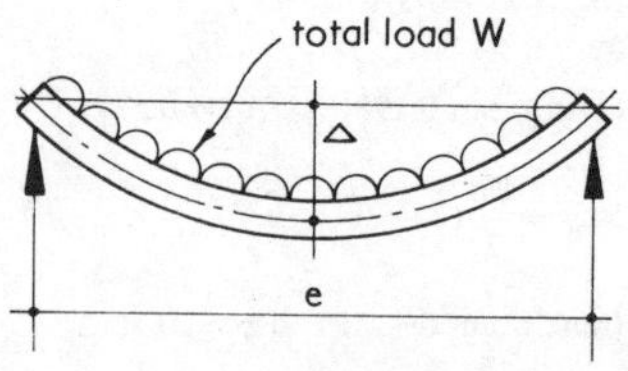

45.9 Deflection of a uniformly loaded, simply-supported beam

where Δ is the deflection of the centre of the beam below its unloaded position
W is the total load on the beam
L is the span of the beam
E is Young's Modulus for the material of the beam
I is the second moment of area of cross-section about the axis of bending.

The units of this formula could be any consistent units, eg the deflection will be in mm when the load is in kN, E is given in kN/mm^2, and I is mm^4.

Example 7

What is the deflection of a timber joist 225 × 50 mm over a 5 m span carrying a load of 0·75 kN/m? (Young's Modulus for timber to be taken as 6900 N/mm².)

For a rectangular section $I = \frac{bd^3}{12} = \frac{50 \times 225^3}{12} = 47{\cdot}5 \times 10^6 \text{ mm}^4$

Hence $\Delta = \frac{5}{384} \times \frac{0{\cdot}75 \times 10^3 \times 5 \times 5^3 \times 10^9}{6900 \times 47{\cdot}5 \times 10^6} = \underline{\underline{18{\cdot}6 \text{ mm}}}$

This is $\frac{18{\cdot}6}{5000} = \frac{1}{268}$ of the span of the beam, and is therefore a little excessive.

A limiting figure of $\frac{1}{325}$ is normally assumed.

As a guide to probable deflection characteristics, rule-of-thumb span to depth ratios are often used. Provided the use in confined to preliminary investigation, and the actual deflections are later checked, the ratios in table VII will be found of value.

Table VII Maximum span/depth ratios (rule-of thumb)

Concrete beams	20
Concrete slabs	30
Steel beams (I-section)	25
Timber joists	20

2 MASONRY

Masonry is the term that covers load-bearing construction in brick, block and stone. Since these materials and the mortar that is used to fill the gaps between the elements of the materials are both weak in tension, such construction is normally used to carry only simple compressive forces: usually in vertical elements such as walls and piers, sometimes in arches. Nevertheless, tensile stresses up to $0{\cdot}1 \text{ N/mm}^2$ may be acceptable in certain circumstances.

2.01 Walls

A wall is a vertical load-carrying element whose length in plan is at least four times its width, **45.10a**; otherwise it is a column, **45.10b**. A pier, **45.10c** is a column integral with a wall. References to walls in the text can be taken to apply also to columns and piers unless stated otherwise.

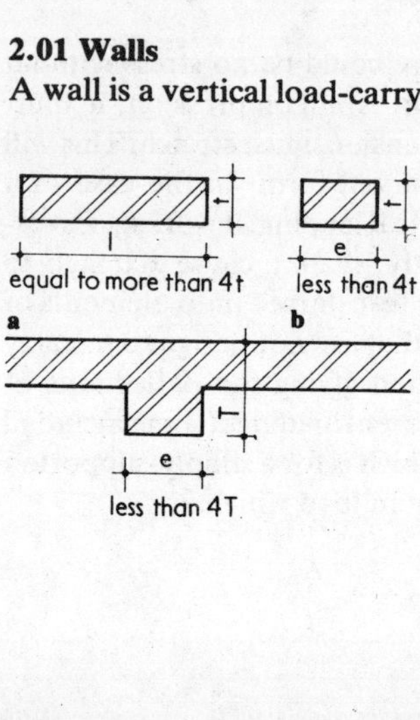

45.10a *A masonry wall*
b *a column*
c *a pier*

The load-carrying capacity of a wall depends on:

- the crushing strength of the brick, block or stone
- the composition of the mortar
- the size and shape of the brick, block or stone
- the height of the wall relative to its width—its slenderness ratio
- the eccentricity of the loading

2.02 Design method

Masonry design in Britain has been based on British Standard Code of Practice CP111, Structural recommendations for loadbearing walls. A new code BS 5628: Part I has now been produced embodying limit state methods (see para **1.02**). However, the methods in CP111 will continue to be used for simple design problems such as in the following examples; the methods given and the tables are based on the code, but are slightly simplified for the sake of clarity.

2.03 Permissible stress

The working basic stress incorporates the first two factors above. Table VIII gives the working basic stress for various brick and block classes for each grade of mortar. The mortar grade system is explained in table IX.

The shape factor modifies the working basic stress which applies directly to the standard brick format (ie a shape factor of 1). Table X gives the British Standard brick and block sizes **45.11** (see also section 39, Materials). Table XI gives the shape modification factors for common formats. For certain block types, the code requires complex calculations for this factor: table XII gives the modified basic stress for the common blocks of this type.

2.04 Slenderness ratio and eccentricity

These are covered by the stress reduction factor which is given in table XIII, from the code. The assessment of the eccentricity of loading is not easy. The self-weight of the wall, and any walls above resting directly on it can be taken as acting axially. So

Table VIII Working basic stresses for units of standard brick format

$\frac{\textbf{Height}}{\textbf{Thickness}}$ **= or less than 0·75**

Brick class	Strength	Mortar grade I	II	III	IV	V	VI
14	96·5 MN/m² and over	5·85	4·50	3·80	3·10	2·40	1·40
12	82·5	5·19	4·04	3·44	2·79	2·22	1·27
11	76	4·88	3·83	3·28	2·65	2·14	1·21
	75	4·83	3·80	3·25	2·63	2·13	1·21
10·5	72·5	4·72	3·71	3·19	2·58	2·09	1·18
10	69	4·55	3·60	3·10	2·50	2·05	1·15
9	62	4·12	3·27	2·85	2·31	1·91	1·11
8	55	3·69	2·94	2·61	2·13	1·76	1·07
7·5	52	3·50	2·80	2·50	2·05	1·70	1·05
7	48	3·27	2·63	2·35	1·96	1·63	1·00
6	41	2·87	2·33	2·09	1·80	1·51	0·92
–	35	2·53	2·07	1·87	1·66	1·41	0·86
5	34·5	2·50	2·05	1·85	1·65	1·40	0·85
4	27·5	2·05	1·70	1·60	1·45	1·15	0·75
3·5	24	1·85	1·58	1·45	1·30	1·05	0·72
3	20·5	1·65	1·45	1·30	1·15	0·95	0·70
2·5	17·5	1·47	1·30	1·20	1·06	0·88	0·66
	17	1·44	1·28	1·18	1·05	0·86	0·65
2	14	1·25	1·15	1·10	1·00	0·80	0·60
1·5	10·5	1·05	0·95	0·95	0·85	0·70	0·55
1	7	0·70	0·70	0·70	0·55	0·49	0·42
	5·5	0·55	0·55	0·55	0·46	0·38	0·34
0·5	3·5	0·35	0·35	0·35	0·35	0·23	0·23
	2·8	0·28	0·28	0·28	0·28	0·21	0·21

Table IX Recommended mortar mixes (proportions by volume)

Direction of change in properties	Mortar Grade*	Hydraulic lime/sand	Cement/ lime/sand	Masonry/ cement/sand	Cement/sand with plasticiser	Required mortar grade strengths (MN/m²) as BS 4551 7-day	28-day
Increasing strength	I	—	1:0–¼:3	—	—	11·0	16·0
but decreasing ability	II	—	1:½:3½ to 4½	1:3 to 3½	1:3½ to 4	5·5	8·0
to accommodate	III	—	1:1:5 to 6	1:4 to 5	1:5 to 6	2·75	4·0
movements due to settlement, shrinkage,	IV	1:2:3	1:2:8 to 9	1:5½ to 6½	1:7 to 8	1·0	4·5
etc.	V	1:3	1:3:10 to 12	1:6½ to 7	1 to 8	1·0	1·5

Changing characteristics ⟶ Increasing resistance to damage by freezing after hardening
within any one
mortar designation ⟵ Improvement in bond and consequent resistance to rain penetration

* Grades I and II should only be used with high strength blocks in walls subject to high loading. Grade III should normally be used in external walls subject to severe exposure, eg retaining walls, walls below dpc, parapets and free standing walls, or in walls not subject to severe exposure if there is a danger of early frost action. Grade IV should be used in walls not subject to severe exposure, eg external walls between dpc and eaves except in exposed situations and internal walls, provided no early frost action is possible. Grade V may be used in internal walls and partitions provided no early frost action is possible.

Table X BS Brick and block sizes

Unit	Length × height Co-ordinating size (mm)	Work size (mm)	Thickness work size (mm)
Standard Brick	225 × 75	215 × 65	102·5
Metric modular bricks	200 × 100	190 × 90	90
	200 × 75	190 × 65	90
	300 × 100	290 × 90	90
	300 × 75	290 × 65	90
Type A Blocks	400 × 100	390 × 90	75, 90, 100
	400 × 200	390 × 190	140 and 190
	450 × 225	440 × 215	75, 90, 100 140, 190 and 215
Type B Blocks	400 × 100	390 × 90	75, 90, 100
	400 × 200	390 × 190	140 and 190
	450 × 200	440 × 190	75, 90, 100 140, 190 and 215
	450 × 225	440 × 215	
	450 × 300	440 × 290	
	600 × 200	590 × 190	
	600 × 225	590 × 215	
Type C Blocks	400 × 200	390 × 190	60 and 75
	450 × 200	440 × 190	
	450 × 225	440 × 215	
	450 × 300	440 × 290	
	600 × 200	590 × 190	
	600 × 225	590 × 215	

Note

Blocks of work size 448 mm × 219 mm × 51, 64, 76, 102, 152 or 219 mm thick, and 397 mm × 194 mm × 75, 92, 102, 143 and 194 mm thick will be produced as long as they are required.

If blocks of entirely non-standard dimensions or design are required the limits of size or the design shall be agreed. Such blocks shall then be deemed to comply with this standard provided they comply with the other requirements. Types A, B and C are as described under table III.

Table XI Shape modification factors

Unit width	Unit heights 50	65	90	190	215
75				2	2
90		1	1·20	2	2
100				1·92	2
102·5	1	1			
140				1·49	1·63
150				1·41	1·55
170		1			
190				1·20	1·31
215				1·11	1·20
220				1·09	1·18
230				1·06	1·15
250				1·01	1·09
300				1	1

Note

These shape modification factors apply only to units:

a of concrete, with no voids, of strength not greater than 35 MN/m²

b of other types of strength not greater that 5·5 MN/m².

For units of concrete with no voids, of strength greater than 35 MN/m² use the greater of:

Table XI Shape modification factors (*continued*)

the basic working stress from table VIII for the actual strength, or the basic working stress from table VIII for a block of strength 35 MN/m² × the appropriate shape modification factor.

For other units of strength greater than 21 MN/m², use the basic working stress from table VIII unmodified.

For such units of strengths between 5·5 and 21 MN/m², the formula is complicated. Table XII gives values of modified working stress for a range of common block sizes and strengths.

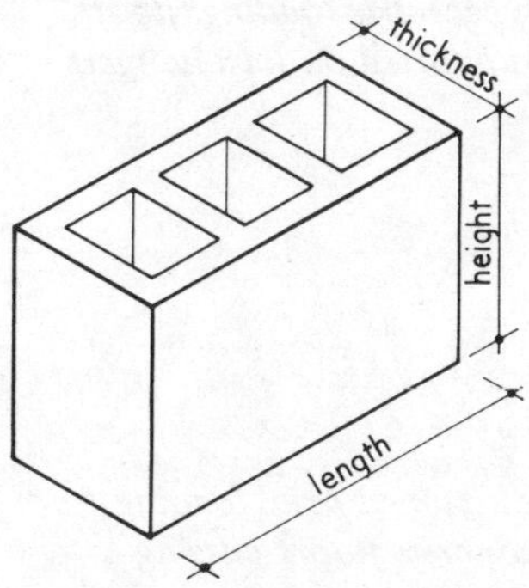

45.11 *Dimensional terms used for a building block*

can timber floor beams with roughly equal spans either side of the wall, and passing over the top, or through it, **45.12a**. In respect of other cases there is some difference of opinion. Timber beams on joist hangers should be taken as applying their loads 25 mm from the outer face of the wall, **45.12b**. Reinforced concrete slabs passing through the wall are normally assumed to be acting axially, **45.12c**.

2.05 Maximum permissible stress

Once the modification and reduction factors have been found, max permissible stress = working basic stress × shape modification factor × reduction factor for slenderness and eccentricity.

This is compared with the actual working stress under load to see that the latter does not exceed the permissible stress. When calculating the actual stress, the effect of eccentricity of loading must be taken into account.

Example 8

A long brick wall at ground floor level in a multi-storey load-bearing brick building carries a load of 600 kN/m. The storey height floor to floor is 3 m, and the floors are of reinforced concrete. If the wall is nominal 225 mm thick, and built of class 7·5 bricks in grade 1 mortar, will this be satisfactory? **(45.13)**

All clause references are to CP 111: Part 2: 1970 as amended June 1971.

Compressive stress in brickwork $= \frac{W}{A}$

where W = the load

and A = the area over which the load acts.

Consider 1 m run of wall:

W = 600 kN = 0·06 MN

A = 215 (actual thickness of wall) × 1000 = 0·215 m²

compressive stress $= \frac{0{\cdot}06}{0{\cdot}215} = \underline{3{\cdot}02\,\mathrm{MN/m^2}}$

Units of class 7·5 have a crushing strength of 52 MN/m². When used with a mortar grade 1 (1:¼:3), the working basic stress is 3·50 MN/m² (table VIII)

Using standard bricks, there is no shape modification factor.

The effective height of the wall is 0·75 × 3 m (cl 305a(i))
= 2·25 m
The effective thickness of the wall (cl 307a) = 0·215 m
slenderness ratio = $\frac{2{\cdot}25}{0{\cdot}215}$ = 10·47
for an axially loaded wall with this slenderness ratio, the reduction factor from table XIII (table 4 in the code) would be 0·88
the max permissible stress is 3·50 × 1 × 0·88 = 3·08 MN/m²
This is greater than the actual stress; the design is satisfactory.

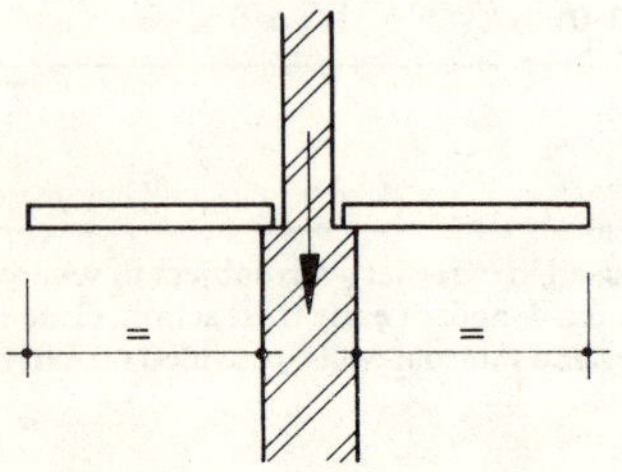

45.12a *axial loading: walls above, and equal floor spans on either side*

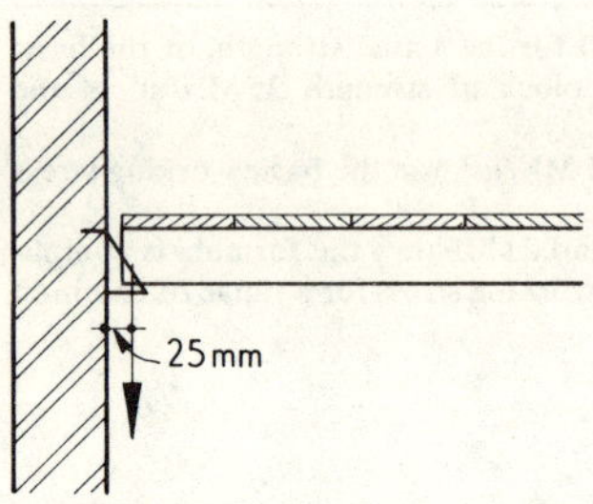

b *eccentric loading: floors supported on joist hangers*

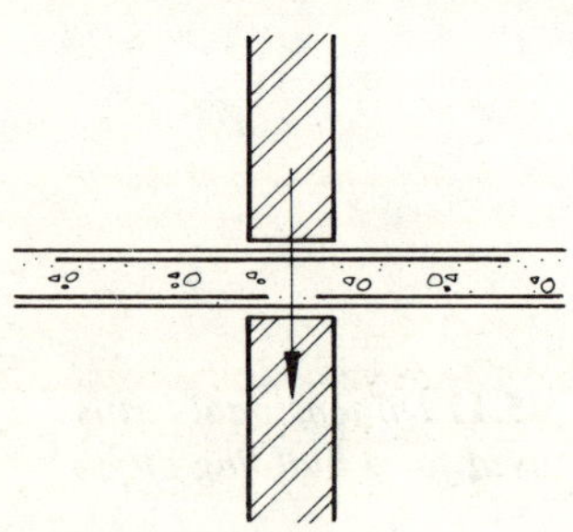

c *assumed axial loading from concrete floors passing through the wall*

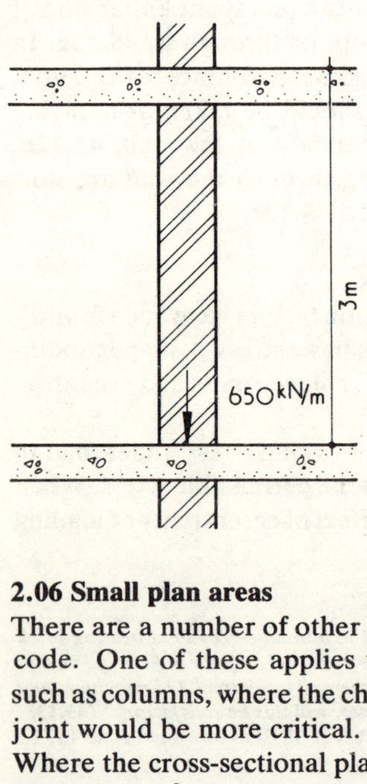

45.13 *Design of brick wall (Example 8)*

2.06 Small plan areas

There are a number of other modification factors given in the code. One of these applies to small plan areas of masonry, such as columns, where the chance of an inferior unit or mortar joint would be more critical.

Where the cross-sectional plan area A of an element does not exceed 0·3 m² the reduction factor is

$$0{\cdot}75 + \frac{A}{1{\cdot}2}$$

Example 9

An isolated brick column nominally 338 mm square carries a steel stanchion base 135 mm square on its top which is 2·75 m above its base. The stanchion centre-line is 55 mm eccentric from the centre-line of the brick column, and the load carried is 60 kN. What bricks in grade I mortar will do for the column shaft, and what bricks in the same mortar would be required instead of a padstone at the top of the column? (Ignore column self-weight) **45.14**

Actual stress
If the stanchion base is not to rotate, the reaction generated by the stress diagram must be equal to the load applied in the opposite direction in the same line. The effective eccentricity of the loading is

$$\frac{55}{327{\cdot}5} = \frac{1}{6}$$

With this eccentricity, the stress diagram will be a triangle, and the maximum stress will be twice the average value,

Maximum stress = $\frac{2 \times W}{A}$
load W = 60 kN = 0·06 MN
area A = 327·5 × 327·5 mm² = 0·107 256 m²
Clause 307a lays down that the thickness of brickwork used in calculation should be the actual thickness. A nominal pier of 338 mm side will be composed of alternate headers and stretchers on each face. A normal brick header is 102·5 mm, a stretcher 215 mm, and a joint 10 mm. The side of the pier is therefore actually 102·5 + 215 + 10 = 327·5 mm
It will be seen that nominal dimensions are 10 mm larger than actual, rounded off.

Maximum stress = 2× $\frac{0{\cdot}06}{0{\cdot}107}$ = 1·12 MN/m²

Permissible stress
Reduction factor for slenderness and eccentricity:
effective height of column (cl 305a(iii)) = 2 × 2·75 = 5·5 m
effective thickness of column (cl 307a) = 0·3275 m

slenderness ratio = $\frac{5{\cdot}5}{0{\cdot}3275}$ = 16·8

eccentricity of loading (as above) = $\frac{1}{6}$

from table XIII reduction factor = 0·595
However, from clause 315d, as the load is eccentric and the magnitude of the maximum stress is due to this eccentricity, the permissible stress can be increased by a factor of 1·25.
Shape factor:
As we are using bricks, the shape factor is unity (1).
Small area reduction factor:
The area of the column at 0·107 m² is less than 0·3 m². The stress reduction factor is therefore

$$0{\cdot}75 + \frac{0{\cdot}107}{1{\cdot}2} = 0{\cdot}84$$

The formula for calculating the permissible stress in this case would be:
maximum permissible stress = working basic stress × shape factor × reduction factor for slenderness and eccentricity × small area factor × additional factor,
that is: working basic stress × 1 × 0·595 × 0·84 × 1·25
= working basic stress × 0·623
Design
The actual stress is 1·12 MN/m², so the working basic stress must be at least

$$\frac{1{\cdot}12}{0{\cdot}623} = 1{\cdot}80\,MN/m^2$$

From table VIII, class 3·5 bricks in grade I mortar have a working basic stress of 1·85 MN/m².
The actual stress on the underside of the stanchion baseplate is $\frac{W}{A}$
where W = 60 kN = 0·06 MN
A = 135 × 135 mm² = 0·0182 m²

stress = $\frac{0{\cdot}06}{0{\cdot}0182}$ = 3·29 MN/m²

Cl 315e provides that underneath a concentrated load the stress can be increased by a factor of 1·5. The reduction factors for slenderness, eccentricity and small area still apply.
The same formula as above can be used giving:
maximum permissible stress = working basic stress × 1 × 0·595 × 0·84 × 1·5
= 0·75 × working basic stress

required working basic stress = $\frac{3{\cdot}29}{0{\cdot}75}$ = 4·39 MN/m²

Use class 10 bricks in grade 1 mortar at column top in lieu of padstone —say, three courses. Basic stress = 4·55 MN/m² (table VIII).

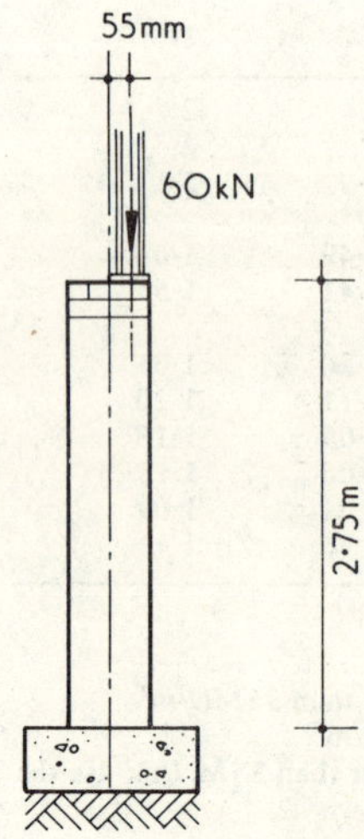

45.14 *Design of brick column (Example 9)*

Table XII Allowable stress on certain units of strengths 7, 9 and 11 MN/m²

Mortar grade	75 × 190 90 × 190 75 × 215 90 × 215 100 × 215	100 × 190	140 × 215	150 × 215	140 × 190	150 × 190	190 × 215	190 × 190 215 × 215 220 × 215 90 × 90 (metric format brick)	230 × 215	215 × 190 220 × 190 250 × 215	230 × 190	250 × 190 300 × 190 300 × 215
Unit formats thickness × height												
7 MN/m²												
I	1·16	1·12	0·98	0·93	0·91	0·87	0·81	0·75	0·73	0·71	0·69	0·66
II	1·13	1·10	0·95	0·91	0·88	0·84	0·79	0·73	0·71	0·68	0·66	0·64
III	1·12	1·09	0·94	0·90	0·87	0·84	0·78	0·72	0·70	0·67	0·65	0·63
IV	0·95	0·91	0·79	0·76	0·74	0·70	0·66	0·60	0·59	0·57	0·56	0·53
V	0·78	0·76	0·66	0·63	0·61	0·58	0·55	0·50	0·49	0·47	0·46	0·44
VI	0·68	0·65	0·56	0·55	0·53	0·50	0·47	0·43	0·42	0·40	0·41	0·37
9 MN/m²												
I	1·24	1·20	1·08	1·04	1·02	0·99	0·94	0·89	0·87	0·85	0·83	0·81
II	1·18	1·15	1·02	0·99	0·96	0·93	0·88	0·83	0·82	0·79	0·78	0·75
III	1·16	1·13	1·00	0·96	0·94	0·91	0·86	0·81	0·79	0·77	0·75	0·73
IV	0·98	0·95	0·85	0·82	0·81	0·77	0·74	0·69	0·68	0·66	0·65	0·63
V	0·81	0·79	0·71	0·68	0·67	0·64	0·61	0·57	0·57	0·54	0·54	0·52
VI	0·68	0·66	0·58	0·57	0·55	0·53	0·51	0·47	0·46	0·44	0·44	0·42
11 MN/m²												
I	1·31	1·29	1·18	1·15	1·13	1·11	1·07	1·02	1·01	0·99	0·98	0·96
II	1·22	1·20	1·09	1·06	1·04	1·02	0·98	0·93	0·92	0·90	0·89	0·87
III	1·19	1·16	1·06	1·03	1·01	0·98	0·94	0·90	0·89	0·87	0·85	0·83
IV	1·02	0·99	0·91	0·88	0·87	0·84	0·81	0·77	0·77	0·75	0·74	0·72
V	0·84	0·83	0·75	0·74	0·72	0·70	0·68	0·64	0·64	0·62	0·61	0·60
VI	0·69	0·67	0·60	0·59	0·58	0·56	0·54	0·51	0·50	0·49	0·49	0·47

Table XIII Stress reduction factors for slenderness and eccentricity
From CPIII: 1970

Slenderness ratio	Stress reduction factor: Axially loaded	Eccentricity of vertical loading as a proportion of the thickness of the member: 1/6	1/4	1/3†
6	1·00	1·00	1·00	1·00
8	0·95	0·93	0·92	0·91
10	0·89	0·85	0·83	0·81
12	0·84	0·78	0·75	0·72
14	0·78	0·70	0·66	0·62
16	0·73	0·63	0·58	0·53
18	0·67	0·55	0·49	0·43
20	0·62	0·48	0·41	0·34
22	0·56	0·40	0·32	0·24
24	0·51	0·33	0·24	
26	0·45	0·25		
27	0·43	0·22		

3 TIMBER

3.01 Structure of timber

Timber is probably the oldest building material used by mankind. It is composed of hollow tubular fibres of cellulose impregnated with the resin lignin, packed closely together not unlike a bundle of drinking straws. The result is that the material is strong in the longitudinal direction—in tension and compression; but weak in the interface between the fibres.

3.02 Advantages of timber

Consequently, timber has the supreme virtue of 'toughness'. It usually gives fore-warning of imminent failure, as the weakness between the fibres inhibits the progress of transverse cracks. Even when failure has occurred, there is often enough residual strength to carry a substantial load. Its principal drawbacks are susceptibility to insect and fungal attack, and vulnerability to fire. Biological resistance can be fortified by treatment, and fire-resistance, particularly in the larger section sizes, is greater than generally realised.

Timber is one of that minority of materials that is almost equally strong in tension and compression. This strength is such that buckling of the compression flange of bending members is rarely a problem. Rectangular sections are easily formed and used for this purpose. Timber is easily worked by hand and machine tools, and it is simple to connect with other members, both other timber members and those of steel, masonry and concrete.

3.03 Design methods

The advanced technology of timber started in the railway era when it was used for elaborate viaducts and bridges. These were generally constructed by trial and error, calculation methods being developed later. In recent years these methods of calculation have been taken to the point where it has become an extremely specialised field. In the United Kingdom, they have been set down in CP 112, Structural Use of Timber (part 2, Metric units, and part 3, Roof trusses). The use of this code by non-specialists is not recommended. In complicated timber structures, for example, the sizes of the members can depend more on the design of the connections than on the internal stresses.

3.04 Roof trusses

The average architect meets timber in two common places: roof trusses and floor joists. Nowadays, most trusses are of the gang-nail type supplied to order for the required conditions. The manufacturer will supply calculations based on part 3 of the code for submission to the local building inspector. For preliminary design purposes, the data in tables XIV–XVI, reprinted from part 3, will be found of value for trusses of the shapes in **45.15** and **16**.

The calculation of floor joists is within the competence of a non-expert. However, before proceeding to the subject, it will be necessary to deal with certain aspects of the material.

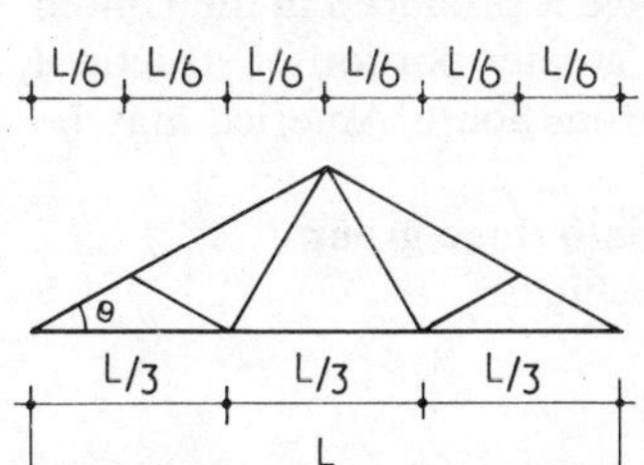

45.15 *Diagram of fan trussed rafter*

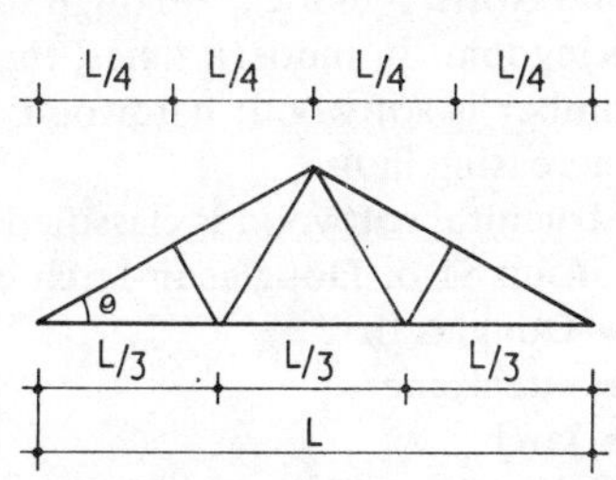

45.16 *Diagram of fink or 'W' trussed rafter*

Trussed rafter tables
Timber must be of one of the following species: Western Hemlock (commercial), including western balsam and some mountain hemlock, European redwood or white-wood, Canadian spruce, Douglas fir.

Table XIV Maximum permissible spans for rafters for Fan trussed rafters (timber to BS 4471 and of composite grade to CP 112) (see 45.15)

Basic size	Actual size	Pitch (degrees)								
		15	17½	20	22½	25	27½	30	32½	35
mm	mm	m	m	m	m	m	m	m	m	m
38 × 75	35 × 72	8·03	8·38	8·64	8·87	9·08	9·27	9·46	9·65	9·85
38 × 100	35 × 97	9·89	10·37	10·67	10·96	11·00	11·00	11·00	11·00	11·00
38 × 125	35 × 120	11·00	11·00	11·00	11·00	–	–	–	–	–
44 × 75	41 × 72	8·65	9·00	9·25	9·48	9·73	9·89	10·08	10·30	10·48
44 × 100	41 × 97	10·71	11·00	11·00	11·00	11·00	11·00	11·00	11·00	11·00
44 × 125	41 × 120	11·00	–	–	–	–	–	–	–	–
50 × 75	47 × 72	9·26	9·62	9·86	10·10	10·36	10·53	10·70	10·93	11·00
50 × 100	47 × 97	11·00	11·00	11·00	11·00	11·00	11·00	11·00	11·00	–

Table XV Maximum permissible spans for rafters for Fink trussed rafters (timber to BS 4471 and of composite grade to CP 112) (see 45.16)
Timber must be of one of the following species: Western Hemlock (commercial), including western balsam and some mountain hemlock, European redwood or white-wood, Canadian spruce, Douglas fir.

Basic size	Actual size	Pitch (degrees)								
		15	17½	20	22½	25	27½	30	32½	35
mm	mm	m	m	m	m	m	m	m	m	m
38 × 75	35 × 72	6·03	6·16	6·29	6·41	6·51	6·60	6·70	6·80	6·90
38 × 100	35 × 97	7·48	7·67	7·83	7·97	8·10	8·22	8·34	8·47	8·61
38 × 125	35 × 120	8·80	9·00	9·20	9·37	9·54	9·68	9·82	9·98	10·16
44 × 75	41 × 72	6·45	6·59	6·71	6·83	6·93	7·03	7·14	7·24	7·35
44 × 100	41 × 97	8·05	8·23	8·40	8·55	8·68	8·81	8·93	9·09	9·22
44 × 125	41 × 120	9·38	9·60	9·81	9·99	10·15	10·31	10·45	10·64	10·81
50 × 75	47 × 72	6·87	7·01	7·13	7·25	7·35	7·45	7·53	7·67	7·78
50 × 100	47 × 97	8·62	8·80	8·97	9·12	9·25	9·38	9·50	9·66	9·80
50 × 125	47 × 120	10·01	10·24	10·44	10·62	10·77	10·94	11·00	11·00	11·00

Table XVI Maximum permissible spans for ceiling ties for Fink and Fan trussed rafters (timber to BS 4471 and of composite grade to CP 112) (see 45.15 and 16)
Timber must be of one of the following species: Western Hemlock (commercial), including western balsam and some mountain hemlock, European redwood or white-wood, Canadian spruce, Douglas fir.

Basic size	Actual size	Pitch (degrees)								
		15	17½	20	22½	25	27½	30	32½	35
mm	mm	m	m	m	m	m	m	m	m	m
38 × 75	35 × 72	5·07	5·31	5·53	5·74	5·94	6·12	6·31	6·50	6·67
38 × 100	35 × 97	7·03	7·36	7·68	7·99	8·27	8·54	8·81	9·06	9·33
38 × 125	35 × 120	8·66	9·10	9·49	9·88	10·24	10·59	10·93	11·00	11·00
38 × 150	35 × 145	10·17	10·71	11·00	11·00	11·00	11·00	11·00	–	–
44 × 75	41 × 72	5·53	5·78	6·03	6·26	6·48	6·69	6·89	7·08	7·28
44 × 100	41 × 97	7·53	7·90	8·25	8·59	8·90	9·19	9·48	9·75	10·04
44 × 125	41 × 120	9·13	9·60	10·04	10·46	10·86	11·00	11·00	11·00	11·00
44 × 150	41 × 145	10·52	11·00	11·00	11·00	11·00	–	–	–	–
50 × 75	47 × 72	5·92	6·20	6·46	6·72	6·94	7·17	7·39	7·60	7·81
50 × 100	47 × 97	7·93	8·33	8·71	9·06	9·38	9·70	10·02	10·32	10·62
50 × 125	47 × 120	9·42	9·94	10·40	10·86	11·00	11·00	11·00	11·00	11·00
50 × 150	47 × 145	10·59	11·00	11·00	11·00	–	–	–	–	–

3.05 Origins

Most structural timber is imported, mainly from Scandinavia and North America, although some is produced in the United Kingdom. In modern times, the greater portion of structural timber is softwood; hardwood from South America may be increasing in use,

Structural softwood is classified into three groups:

Group S1 or Douglas fir-larch includes:

- Douglas fir
- pitch-pine
- larch

Group S2 or Hem-fir includes:

- western hemlock
- parana pine
- redwood
- whitewood
- Canadian spruce
- Scots pine
- pricess spruce
- western white spruce

Group S3 or Spruce-pine-fir includes:

- European spruce
- Sitka spruce
- western red cedar.

Each species of timber is graded according to its structural quality. There are two methods of doing this: visual grading and machine grading.

3.06 Machine grading
Machine grading is relatively new. The machine measures the modulus of elasticity of the timber and its strength properties are derived from the known behaviour of the species relative to that modulus. The four gradings used are: M75, MSS (machine special structural), M50 and MGS (machine general structural). Machine grading is carried out in the United Kingdom on home-produced and on imported timber. Some timber from Scandinavia is now imported ready graded to BS 4978.

3.07 Visual grading
Visual grading depends on the skill of the grader. The two grades now becoming universal are SS (special structural) and GS (general structural). However, hardwood is still graded using the old numbered grades: Basic, 75, 65, 50 and 40.

3.08 Permissible stresses
Having established the grading of the timber from the marking that should be on it, the permissible stresses can be determined from table XVII. Some timber is imported from Canada and the US with their own grading system, and the table gives permissible stresses for this system as well.

3.09 Moisture content
The behaviour of most timbers vary with the moisture content. For convenience, only two conditions are specified: dry (for moisture contents less than 18 per cent), and green. Dry stresses may be used for timbers wholly enclosed within buildings that may be expected to be more or less heated all the time. Green stresses should be used for external timber work. Exposure during the construction period need not be taken into account.
Timber behaviour depends greatly on time. This effect can be beneficial: short-term overloads can be carried, but with time gradual failure occurs. This effect is due to creep as the fibres of the material slowly stretch. Consequently, grade stresses are multiplied by a modification factor, table XVIII, based on the duration of loading on bending and tension members.

3.10 Load-sharing
Joists in floors and roofs are rarely used singly. They are usually placed at constant centres, and some form of boarding is laid across them. Under these conditions the loading tends to be shared by a number of the joists, and the effect of one weak piece of timber among them is reduced. Hence, the permissible stress in the timber may be increased by 10 per cent in these circumstances; and in deflection calculations the mean modulus of elasticity used rather than the minimum. The latter is not permitted if the stress induced by the long-term loads is more than 60 per cent of the permissible stress.

3.11 Beam design
The following symbols will be used, (**45.17**):

- b — for the breadth of a timber beam section in metres
- d — for its depth
- r — for the spacing centre-to-centre of the joists
- l — for the span
- q — for the total loading on the floor in MN/m^2, assumed uniformly distributed
- p — for the maximum permissible bending stress in the timber in N/mm^2 (see para **1.08**)
- σ — for the maximum permissible shear stress
- E — for the modulus of elasticity in N/mm^2
- E_{min} — for its minimum value, used where there is little or no load-sharing between a number of joists
- E_{mean} — for the mean value, used when the load is shared between at least four joists (but see para **3.10** above).

For western hemlock (commercial), grade MGS, dry, the value of E_{mean} is $9800\,N/mm^2$ and this is a convenient figure to use in general situations.

3.12 Bending stress
From table VI it is seen that the bending moment in a free-support, uniformly loaded beam is:

$$\frac{WL}{8}$$

Substituting the appropriate symbols above:

$$M = \frac{q \times r \times 1 \times 1}{8} = \tfrac{1}{8} qrl^2$$

Now the bending stress from para **1.15** is given by:

$$M = f\,Z$$

From table IV, the Z for a rectangular section is $^1/_6\ bd^2$

Hence $$\tfrac{1}{8} qrl^2 = \tfrac{1}{6} pbd^2 \text{ or } \frac{b}{qr} = \frac{3}{4p}\left(\frac{l}{d}\right)^2 \qquad \text{equation (i)}$$

3.13 Deflection
Again from table VI the formula for deflection is:

$$\Delta = \frac{5}{384} \times \frac{W\,l^3}{E\,I}$$

A deflection greater than about 0·003 of the span has been found to cause damage to finishes (plaster ceilings etc), and so this value is normally taken as a safe limit.
Again substituting:

$$0{\cdot}003 \times 1 = \frac{5}{384} \times \frac{qrl^4}{E \times \dfrac{bd^3}{12}}$$

hence $$\frac{b}{qr} = \frac{5000}{96\,E}\left(\frac{l}{d}\right)^3 \qquad \text{equation (ii)}$$

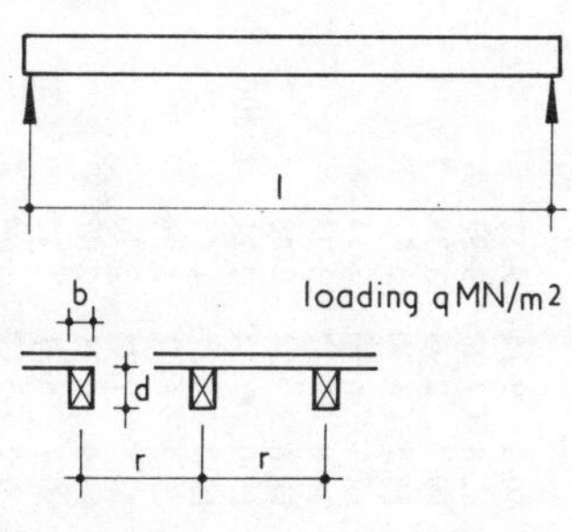

45.17 *Design of timber joist floors:*
a *elevation of joist*
b *section showing spacing*

3.14 Shear
By similar methods the shear equation is set up:

$$\text{Shear force} = \frac{W}{2} = \tfrac{1}{2}\,q\,r\,l = \frac{2^*}{3}\,b\,d$$

$$\text{hence } \frac{b}{qr} = \frac{3}{4}\left(\frac{l}{d}\right) \qquad \text{equation (iii)}$$

* For an explanation for the presence of this $^2/_3$ factor, see the *AJ Handbook of Building Structure*, Analysis 1, para 3.59 et seq.

3.15 Beam characteristic graphs
It is seen that each of the equations (i), (ii) and (iii) relate the quantity $\frac{b}{qr}$ against the span/depth ratio $\frac{l}{d}$. In **45.18** the equations are plotted on a horizontal axis representing the span/depth ratio and a vertical axis representing the quantity $\frac{b}{qr}$. The graph is for western hemlock (commercial), grade MGS.

Table XVII Permissible stresses

Number	Species	Origin	Grade	Dry stresses (Moisture content not exceeding 18%) N/mm² Bending	Tension	Compression parallel to grain	Compression perpendic. to grain	Shear parallel to grain	Modulus of Elasticity Mean	Minimum	Loading sharing medium term Shear limit	Bending limit	Single memb. long term Shear limit	Bending limit	Green stresses (moisture content exceeding 18%) N/mm² Bending	Tension	Compression parallel to grain	Compression perpendic. to grain	Shear parallel grain	Modulus of elasticity Mean	Minimum	Load sharing medium term Shear limit	Bending limit	Single memb long term Shear limit	Bending limit	Number
	Hardwoods																									
1	Greenheart	I	Basic	41·4	41·4	30·3	9·31	5·52			5·9	*	5·9	*	37·9	37·9	27·6	6·20	4·83			6·1	*	6·1	*	1
2			75	31·0	31·0	22·8	7·93	3·93			7·0	*	7·0	*	28·3	28·3	20·7	5·38	3·59			7·1	*	7·1	*	2
3			65	26·9	26·9	19·7	7·93	3·38	18600	13400	7·6	*	7·6	*	24·1	24·1	17·9	5·38	3·10	17200	12400	7·6	*	7·6	*	3
4			50	20·7	20·7	15·2	6·90	2·62			7·9	9·4	7·9	9·3	19·0	19·0	13·8	4·62	2·41			7·9	9·5	7·9	9·4	4
5			40	16·5	16·5	12·1	6·90	2·14			7·7	11·8	7·7	11·7	15·2	15·2	11·0	4·62	1·93			7·9	11·9	7·9	11·7	5
6	Opepe	I	Basic	29·0	29·0	24·8	8·27	3·72			6·2	*	6·0	*	25·5	25·5	22·1	5·52	3·10			6·5	*	5·9	*	6
7			75	22·4	22·4	18·6	7·24	2·48			7·6	*	7·3	*	19·3	19·3	16·5	4·83	2·28			7·5	*	6·9	*	7
8			65	18·6	18·6	15·9	7·24	2·21	13800	9300	8·1	*	7·8	*	16·5	16·5	14·1	4·83	2·00	12400	7600	8·1	*	7·4	*	8
9			50	14·5	14·5	12·4	6·21	1·65			8·8	10·0	8·8	9·2	12·8	12·8	11·0	4·14	1·52			8·4	10·1	8·4	8·6	9
10			40	11·7	11·7	9·7	6·21	1·34			8·7	12·4	8·7	11·4	10·3	10·3	8·6	4·14	1·24			8·3	12·6	8·3	10·6	10
11	Karri	I	Basic	26·2	26·2	22·1	7·24	2·76			7·7	*	7·1	*	22·1	22·1	16·5	4·83	2·48			9·9	*	9·0	*	11
12			75	19·3	19·3	15·9	6·21	2·07			8·9	*	8·2	*	16·5	16·5	12·4	4·14	1·86			8·8	*	8·0	*	12
13			65	15·9	15·9	13·1	6·21	1·72	15500	9700	9·2	10·2	9·0	*	14·1	14·1	10·7	4·14	1·59	13800	8300	8·9	10·3	8·7	*	13
14			50	12·4	12·4	9·7	5·17	1·34			9·3	13·1	9·3	11·3	11·0	11·0	8·3	3·59	1·24			8·9	13·1	8·9	10·9	14
15			40	9·7	9·7	7·9	5·17	1·10			8·8	16·7	8·8	14·4	8·6	8·6	6·6	3·59	0·97			8·9	16·8	8·9	13·9	15
16	Teak	I	Basic	26·2	26·2	22·1	6·21	2·62			7·0	*	6·6	*	22·1	22·1	16·5	4·14	2·34			7·0	*	6·5	*	16
17			75	19·3	19·3	15·9	5·17	1·86			8·4	*	7·8	*	16·5	16·5	12·4	3·59	1·72			8·2	*	7·6	*	17
18			65	15·9	15·9	13·1	5·17	1·65	12400	7900	8·9	*	8·3	*	14·1	14·1	10·7	3·59	1·52	11000	6900	8·7	*	8·1	*	18
19			50	12·4	12·4	9·7	4·48	1·28			9·7	10·5	9·4	*	11·0	11·0	8·3	3·10	1·17			9·4	10·5	9·2	*	19
20			40	9·7	7·9	7·9	4·48	0·97			10·0	13·4	10·0	11·7	8·6	8·6	6·6	3·10	0·90			9·6	13·4	9·6	11·6	20
21	Afrormosia	I	Basic	26·2	26·2	22·1	6·21	2·76			6·8	*	6·4	*	22·1	22·1	15·9	4·14	2·62			6·4	*	6·2	*	21
22			75	19·3	19·3	15·2	5·17	2·07			7·8	*	7·4	*	16·5	16·5	11·7	3·59	1·93			7·5	*	7·2	*	22
23			65	15·9	15·9	12·4	5·17	1·79	12100	7900	8·4	*	8·0	*	14·1	14·1	10·3	3·59	1·65	10300	6900	8·1	*	7·8	*	23
24			50	12·4	12·4	9·3	4·48	1·38			9·0	10·2	9·0	9·2	11·0	11·0	7·9	3·10	1·31			8·4	9·8	8·4	9·0	24
25			40	9·7	9·7	7·6	4·48	1·10			8·8	13·1	8·8	11·7	8·6	8·6	6·2	3·10	1·03			8·3	12·5	8·3	11·6	25
26	Jarrah	I	Basic	23·4	23·4	20·7	6·21	2·62			7·0	*	6·6	*	19·3	19·3	15·9	4·14	2·34			6·8	*	6·5	*	26
27			75	16·9	16·9	15·2	5·17	1·86			8·3	*	7·8	*	14·5	14·5	11·7	3·59	1·72			7·9	*	7·6	*	27
28			65	14·1	14·1	12·4	5·17	1·65	12100	7900	8·5	9·0	8·3	*	12·4	12·4	10·3	3·59	1·52	10300	6900	8·2	8·7	8·1	*	28
29			50	11·0	11·0	9·3	4·48	1·28			8·6	11·5	8·6	10·3	9·7	9·7	7·9	3·10	1·17			8·3	11·1	8·3	10·2	29
30			40	8·6	8·6	7·6	4·48	0·97			8·9	14·7	8·9	13·2	7·6	7·6	6·2	3·10	0·90			8·4	14·2	8·4	13·1	30
31	Sapele	I	Basic	23·4	23·4	20·7	6·21	2·76			6·5	*	6·0	*	19·3	19·3	15·9	4·14	2·34			6·6	*	6·2	*	31
32			75	16·9	16·9	15·2	5·17	1·86			7·9	*	7·3	*	14·5	14·5	11·7	3·59	1·72			7·7	*	7·2	*	32
33			65	14·1	14·1	12·4	5·17	1·65	11000	6900	8·4	*	7·8	*	12·4	12·4	10·3	3·59	1·52	9700	6200	8·2	*	7·7	*	33
34			50	11·0	11·0	9·3	4·48	1·28			8·6	10·5	8·6	99·0	9·7	9·7	7·9	3·10	1·17			8·3	10·5	8·3	9·2	34
35			40	8·6	8·6	7·6	4·48	0·97			8·9	13·4	8·9	11·6	7·6	7·6	6·2	3·10	0·90			8·4	13·4	8·4	11·7	35
36	Iroko	I	Basic	23·4	23·4	19·3	6·21	2·62			6·4	*	6·2	*	20·7	20·7	15·2	4·14	2·34			6·3	*	6·0	*	36
37			75	17·6	17·6	14·5	5·17	1·86			7·6	*	7·3	*	15·5	15·5	11·4	3·59	1·72			7·4	*	7·0	*	37
38			65	15·2	15·2	12·1	5·17	1·65	10300	6900	8·1	*	7·8	*	13·4	13·4	9·7	3·59	1·52	9000	5900	7·9	*	7·5	*	38
39			50	11·7	11·7	9·0	4·48	1·28			9·1	9·2	8·8	*	10·3	10·3	7·6	3·10	1·17			8·8	9·2	8·5	*	39
40			40	9·3	9·3	7·2	4·48	0·97			9·6	11·6	9·6	10·7	8·3	8·3	5·9	3·10	0·90			9·2	11·4	9·2	10·2	40
41	Gurjun/keruing	I	Basic	22·8	22·8	19·3	4·48	2·62			7·4	*	7·1	*	17·2	17·2	13·8	3·10	2·34			7·4	7·6	7·1	*	41
42			75	14·8	14·8	13·1	3·79	1·86			8·00	9·8	8·0	9·0	12·8	12·8	10·3	2·76	1·72			7·4	10·1	7·4	9·3	42
43			65	12·4	12·4	11·0	3·79	1·65	13800	9300	7·5	11·7	7·5	10·8	11·0	11·0	9·0	2·76	1·52	12400	8300	7·2	11·8	7·2	10·9	43
44			50	9·7	9·7	8·3	3·45	1·28			7·6	14·9	7·6	13·8	8·6	8·6	6·9	2·34	1·17			7·4	15·1	7·4	13·9	44
45			40	7·9	7·9	6·6	3·45	0·97			8·1	18·3	8·1	17·0	6·9	6·9	5·5	2·34	0·90			7·7	18·8	7·7	17·3	45

Table XVII Permissible stresses (*continued*)

Number	Species	Origin	Grade	Dry stresses (moisture content not exceeding 18%) N/mm² Bending	Tension	Compression parallel to grain	Compression perpendic. to grain	Shear parallel to grain	Modulus of Elasticity Mean	Minimum	Loading sharing medium term Shear limit	Bending limit	Single memb. long term Shear limit	Bending limit	Green stress (moisture content exceeding 18%) N/mm² Bending	Tension	Compression parallel to grain	Compression perpendic. to grain	Shear parallel grain	Modulus of elasticity Mean	Minimum	Load sharing medium term Shear limit	Bending limit	Single memb. long term Shear limit	Bending limit	
46	European ash & European beech	H	Basic	22·8	22·8	15·2	4·48	3·10			6·2	*	5·8	*	17·2	17·2	11·0	3·10	2·76			6·2	*	5·9	*	46
47			75	14·8	14·8	10·3	3·79	2·28			6·5	8·1	6·5	7·0	12·8	12·8	8·3	2·76	2·07			6·2	8·2	6·2	7·4	47
48			65	12·4	12·4	8·6	3·79	2·00	11400	7200	6·2	9·6	6·2	8·4	11·0	11·0	6·9	2·76	1·79	10000	6·000	6·1	9·5	6·1	8·6	48
49			50	9·7	9·7	6·6	3·45	1·52			6·4	12·3	6·4	10·7	8·6	8·6	5·5	2·34	1·38			6·2	12·2	6·2	11·1	49
50			40	7·9	7·9	5·2	3·45	1·24			6·4	15·1	6·4	13·1	6·9	6·9	4·5	2·34	1·10			6·3	15·2	6·2	13·8	50
51	European oak	H	Basic	20·7	20·7	15·2	4·48	3·10			5·7	*	4·9	*	15·9	15·9	11·0	3·10	2·48			6·0	*	5·1	*	51
52			75	13·8	13·8	10·3	3·79	2·07			6·7	7·4	6·0	*	11·7	11·7	8·3	2·76	1·86			6·3	7·7	5·9	*	52
53			65	11·7	11·7	8·6	3·79	1·72	9700	5200	6·8	8·7	6·6	*	10·3	10·3	6·9	2·76	1·59	8600	4500	6·5	8·7	6·4	*	53
54			50	9·0	9·0	6·6	3·45	1·34			6·7	11·3	6·7	8·3	7·9	7·9	5·5	2·34	1·24			6·4	11·4	6·4	8·2	54
55			40	7·2	7·2	5·2	3·45	1·10			6·5	14·1	6·5	10·4	6·2	6·2	4·5	2·34	0·97			6·4	14·5	6·4	10·5	55
56	Abura	I	Basic	16·5	16·5	13·8	3·45	2·41			6·4	*	5·4	*	13·8	13·8	10·3	2·34	2·07			6·5	*	5·6	*	56
57			75	12·1	12·1	10·0	3·10	1·65			7·3	8·0	6·9	*	10·3	10·3	7·6	2·07	1·52			6·8	8·4	6·5	*	57
58			65	10·3	10·3	8·3	3·10	1·45	9300	4800	7·1	9·5	6·9	*	9·0	9·0	6·6	2·07	1·31	8300	4500	6·9	9·7	6·9	7·2	58
59			50	7·9	7·9	6·2	2·48	1·14			6·9	12·3	6·9	8·7	6·9	6·9	5·2	1·72	1·03			6·7	12·6	6·7	9·4	59
60			40	6·2	6·2	4·8	2·48	0·90			6·9	15·7	6·9	11·1	5·5	5·5	4·1	1·72	0·83			6·6	15·8	6·6	11·8	60
61	African mahogany	H	Basic	15·2	15·2	13·1	3·10	1·93			6·8	*	5·8	*	12·4	12·4	9·7	2·07	1·72			6·9	*	5·9	*	61
62			75	10·7	10·7	9·3	2·62	1·34			8·0	8·4	7·0	*	9·3	9·3	7·2	1·79	1·24			7·5	8·9	6·9	*	62
63			65	9·0	9·0	7·6	2·62	1·21	8600	4500	7·4	10·0	7·3	*	7·9	7·9	6·2	1·79	1·10	7900	4100	7·2	10·5	7·2	7·5	63
64			50	7·2	7·2	5·5	2·21	0·90			8·0	12·5	8·0	9·0	6·2	6·2	4·8	1·52	0·83			7·5	13·3	7·5	9·5	64
65			40	5·5	5·5	4·5	2·21	0·76			7·2	16·4	7·2	11·8	4·8	4·8	3·8	1·52	0·69			7·0	17·2	7·0	12·3	65
66	Red meranti/ red seraya	I	Basic	15·2	15·2	13·1	2·62	1·72			7·1	*	6·1	*	12·4	12·4	9·7	1·79	1·52			7·2	*	6·2	*	66
67			75	10·7	10·7	9·3	2·34	1·21			8·5	*	7·3	*	9·3	9·3	7·2	1·59	1·10			8·5	8·6	7·3	*	67
68			65	9·0	9·0	7·6	2·34	1·07	8300	4500	8·4	9·7	7·8	*	7·9	7·9	6·2	1·59	0·97	7600	4100	8·1	10·1	7·8	*	68
69			50	7·2	7·2	5·5	1·93	0·83			8·7	12·1	8·7	9·0	6·2	6·2	4·8	1·31	0·76			8·2	12·8	8·2	9·5	69
70			40	5·5	5·5	4·5	1·93	0·66			8·3	15·8	8·3	11·8	4·8	4·8	3·8	1·31	0·62			7·7	16·6	7·7	12·3	70
	Softwoods— Group S1																									
			Machine																							
101	Douglas fir	H	M75	12·4	8·7	12·4	2·25	1·30	12500	8000	9·5	10·6	9·4	*	10·0	7·0	9·2	1·50	1·03	11300	7100	9·7	11·8	9·7	10·2	101
102			MSS	9·0	6·3	9·0	1·93	0·87	10800	7200	10·3	12·6	10·3	11·5	7·3	5·1	6·7	1·29	0·70	9700	6400	10·4	13·9	10·4	12·6	102
103			M50	8·3	5·8	8·3	1·93	0·87	10500	7000	9·5	13·2	9·5	12·1	6·7	4·7	6·1	1·29	0·70	9500	6200	9·6	14·8	9·6	13·3	103
104			MGS	6·3	4·4	6·3	1·72	0·87	9500	6500	7·2	15·8	7·2	14·9	5·1	3·6	4·6	1·15	0·70	8600	5800	7·3	17·7	7·3	16·4	104
			Visual																							
105	Douglas fir-larch	I	SS	9·3	6·5	9·6	2·00	0·96	12500	7300	9·7	14·1	9·7	11·3	7·6	5·3	7·2	1·35	0·86	11000	6600	8·8	15·2	8·8	12·5	105
106			GS	6·5	4·6	6·7	1·8	0·96	11300	6600	6·8	18·2	6·8	14·6	5·3	3·7	5·0	1·20	0·86	10200	6000	6·2	20·2	6·2	16·3	106
107	Pitch-pine	I	SS	9·3	6·5	9·1	2·00	0·94	12900	8500	9·9	14·5	9·9	13·2	7·6	5·3	7·2	1·35	0·86	11900	7800	8·8	16·4	8·8	14·8	107
108			GS	6·5	4·6	6·3	1·8	0·94	11600	7600	6·9	18·7	6·9	16·8	5·3	3·7	5·0	1·20	0·86	10700	7000	6·2	21·1	6·2	19·0	108
109	Douglas fir	H	SS	9·0	6·3	9·0	1·93	0·87	10700	5200	10·3	12·5	9·3	*	7·3	5·1	6·7	1·29	0·70	9800	4700	10·4	14·1	9·8	*	109
110			GS	6·3	4·4	6·3	1·72	0·87	9800	4700	7·2	16·3	7·2	10·7	5·1	3·6	4·6	1·15	0·70	8800	4200	7·3	18·1	7·3	11·9	110
111	Larch	H	SS	8·6	6·0	8·6	2·00	1·00	10600	4600	8·6	12·9	8·1	*	6·9	4·8	6·3	1·35	0·86	9200	4400	8·0	14·0	8·0	9·2	111
112			GS	6·0	4·2	6·0	1·80	1·00	9500	4600	6·0	16·6	6·0	11·0	4·8	3·4	4·4	1·20	0·86	8200	4000	5·6	17·9	5·6	12·0	112
113	Group S1		SS	8·6	6·0	8·6	1·93	0·87	10600	5100	9·9	12·9	9·2	*	6·9	4·8	6·3	1·29	0·70	9200	4400	9·9	14·0	9·5	*	113
114			GS	6·0	4·2	6·0	1·72	0·87	9500	4600	6·9	16·6	6·9	11·0	4·8	3·4	4·4	1·15	0·70	8200	4000	6·9	17·9	6·9	12·0	114

Table XVII Permissible stresses (*continued*)

Number	Species	Origin	Grade	Dry stresses (moisture content not exceeding 18%) N/mm²											Green stress (moisture content exceeding 18%) N/mm²											Number
									Modulus of Elasticity		Loading sharing medium term		Single memb. long term							Modulus of elasticity		Load sharing medium term		Single memb. long term		
				Bending	Tension	Compression parallel to grain	Compression perpendic. to grain	Shear parallel to grain	Mean	Minimum	Shear limit	Bending limit	Shear limit	Bending limit	Bending	Tension	Compression parallel to grain	Compression perpendic. to grain	Shear parallel grain	Mean	Minimum	Shear limit	Bending limit	Shear limit	Bending limit	
			Canadian																							
115	Douglas fir-larch Joists and planks	I	Select struc.	12·3	8·6	10·2	2·00	0·96	13800	8100	12·3	*	11·0	*	10·0	7·0	7·6	1·35	0·86	12500	7300	11·6	13·1	11·1	*	115
116			No 1	10·4	7·3	9·2	2·00	0·96	13000	7600	10·8	13·1	10·7	*	8·5	6·0	6·8	1·35	0·86	11700	6900	9·9	14·4	9·9	11·7	116
117			No 2	8·4	5·9	7·7	1·80	0·96	12100	7100	8·8	15·1	8·8	12·2	6·8	4·8	5·7	1·20	0·86	10900	6400	7·9	16·8	7·9	13·6	117
118			No 3	4·8	3·4	4·9	1·35	0·86	10700	6300	5·6	23·3	5·6	18·9	4·0	2·8	3·6	0·90	0·77	9700	5700	5·2	25·4	5·2	20·5	118
119	Light framing*		Construction	7·2	4·4	8·3	2·00	0·96	11100	6500					6·0	3·6	6·2	1·35	0·86	10100	5900					119
120			Standard	4·0	2·5	6·8	1·80	0·96	10000	5900					3·3	2·0	5·1	1·20	0·86	9000	5300					120
121			Utility	1·9	1·2	4·4	1·35	0·86	9200	5400					1·6	1·0	3·3	0·90	0·77	8300	4900					121
122	Stud grading			5·6	3·4	4·4	1·35	0·86	10700	6300					4·5	2·8	3·3	0·90	0·77	9700	5700					122
	Softwoods—Group S2																									
			Machine																							
201	Western hemlock (commercial)	I	M75	10·0	7·0	10·8	1·80	1·19	12000	8000	8·4	12·6	8·4	11·5	8·1	5·7	7·9	1·20	1·03	10600	7100	7·9	13·7	7·9	12·6	201
202			MSS	7·3	5·1	7·9	1·55	0·80	10800	7200	9·1	15·5	9·1	14·2	5·9	4·0	5·9	1·03	0·70	9600	6400	8·4	17·0	8·4	15·6	202
203			M50	6·6	4·6	7·1	1·55	0·80	10500	7000	8·2	16·7	8·2	15·3	5·3	3·7	5·2	1·03	0·70	9300	6200	7·6	18·4	7·6	16·8	203
204			MGS	5·1	3·5	5·5	1·38	0·80	9800	6500	6·4	20·1	6·4	18·4	4·0	2·9	4·0	0·92	0·70	8700	5800	5·7	22·8	5·7	20·9	204
205	Redwood	I	M75	10·0	7·0	10·8	1·98	1·28	10700	6700	7·8	11·2	7·8	9·6	8·1	5·7	7·3	1·32	1·03	8700	5500	7·9	11·2	7·9	9·8	205
206			MSS	7·3	5·1	8·0	1·70	0·86	10200	6400	8·5	14·6	8·5	12·6	5·9	4·0	5·4	1·13	0·70	8300	5200	8·4	14·7	8·4	12·6	206
207			M50	6·6	4·6	7·1	1·70	0·86	9000	5500	7·7	14·3	7·7	12·0	5·3	3·7	4·8	1·13	0·70	7300	4500	7·6	14·4	7·6	12·2	207
208			MGS	5·1	3·5	5·6	1·52	0·86	8800	5400	5·9	18·1	5·9	15·2	4·0	2·9	3·7	1·01	0·70	7200	4400	5·7	18·9	5·7	15·8	208
209	Whitewood	I	M75	10·0	7·0	10·8	1·80	1·28	10700	6700	7·8	11·2	7·8	9·6	8·1	5·7	7·3	1·20	1·03	8700	5500	7·9	11·2	7·9	9·8	209
210			MSS	7·3	5·1	8·0	1·55	0·86	10200	6400	8·5	14·6	8·5	12·6	5·9	4·0	5·4	1·03	0·70	8300	5200	8·4	14·7	8·4	12·7	210
211			M50	6·6	4·6	7·1	1·55	0·86	9000	5500	7·7	14·3	7·7	12·0	5·3	3·7	4·8	1·03	0·70	7300	4500	7·6	14·4	7·6	12·2	211
212			MGS	5·1	3·5	5·6	1·38	0·86	8800	5400	5·9	18·1	5·9	15·2	4·0	2·9	3·7	0·92	0·70	7200	4400	5·7	18·9	5·7	15·8	212
213	Western whitespruce	I	M75	9·7	6·7	9·5	1·80	1·19	10200	6400	8·2	11·0	8·2	9·5	7·6	5·3	7·3	1·20	1·03	9200	5800	7·4	12·7	7·4	11·0	213
214			MSS	6·9	4·8	7·0	1·55	0·79	9200	5800	8·7	14·0	8·7	12·1	5·5	3·8	5·4	1·03	0·70	8300	5200	7·9	15·8	7·9	13·6	214
215			M50	6·2	4·3	6·3	1·55	0·79	8600	5400	7·8	14·5	7·8	12·5	5·0	3·5	4·8	1·03	0·70	7700	4900	7·1	16·1	7·1	14·1	215
216			MGS	4·8	3·3	4·8	1·38	0·79	8200	5100	6·1	17·9	6·1	15·3	3·8	2·7	3·7	0·92	0·70	7400	4600	5·4	20·4	5·4	17·4	216
217	Scots pine	H	M75	9·6	6·7	10·0	2·25	1·28	12000	8000	7·5	13·1	7·5	12·0	7·0	4·8	6·7	1·50	1·03	10300	6800	6·8	15·4	6·8	14·0	217
218			MSS	7·6	5·3	8·0	1·93	0·86	11200	7500	8·8	15·4	8·8	14·2	5·5	3·8	5·4	1·29	0·70	9600	6400	7·9	18·3	7·9	16·8	218
219			M50	6·2	4·3	6·5	1·93	0·86	10500	7000	7·2	17·7	7·2	16·3	4·5	3·1	4·3	1·29	0·70	9000	6000	6·4	20·9	6·4	19·2	219
220			MGS	5·3	3·7	5·6	1·72	0·86	10200	6800	6·2	20·2	6·2	18·5	3·8	2·7	3·7	1·15	0·70	8700	5800	5·4	24·0	5·4	22·0	220
			Visual																							
221	Scots pine	H	SS	7·6	5·3	8·0	1·93	0·86	10600	5100	8·8	14·6	8·8	9·7	5·5	3·8	5·4	1·29	0·70	9000	4300	7·9	17·1	7·9	11·3	221
222			GS	5·3	3·7	5·6	1·72	0·86	9500	4600	6·2	18·8	6·2	12·5	3·8	2·7	3·7	1·15	0·70	8100	3900	5·4	22·3	5·4	22·3	222
223	Hem-fir	I	SS	7·3	5·1	7·9	1·55	0·80	10700	6600	9·1	15·4	9·1	13·0	5·9	4·0	5·9	1·03	0·70	9500	5900	8·4	16·9	8·4	14·4	223
224			GS	5·1	3·5	5·5	1·38	0·80	9600	5900	6·4	19·7	6·4	16·7	4·0	2·9	4·0	0·92	0·70	8500	5200	5·7	22·3	5·7	18·7	224
225	Parana pine	I	SS	7·3	5·1	9·1	1·71	0·90	10500	6000	8·1	15·1	8·1	11·8	5·9	4·0	6·7	1·14	0·76	9500	5400	7·8	16·9	7·8	13·2	225
226			GS	5·1	3·5	6·3	1·52	0·90	9400	5300	5·7	19·3	5·7	15·0	4·0	2·9	4·6	1·01	0·76	8500	4800	5·3	22·3	5·3	17·3	226
227	Redwood	I	SS	7·3	5·1	8·0	1·71	0·86	10000	5700	8·5	14·3	8·5	11·2	5·9	4·0	5·4	1·14	0·70	8200	4700	8·4	14·6	8·4	11·5	227
228			GS	5·1	3·5	5·6	1·52	0·86	8600	4900	5·9	17·7	5·9	13·8	4·0	2·9	3·7	1·01	0·70	7000	4000	5·7	18·3	5·7	14·4	228
229	Whitewood	I	SS	7·3	5·1	8·0	1·55	0·86	10000	5700	8·5	14·3	8·5	11·2	5·9	4·0	5·4	1·03	0·70	8200	4700	8·4	14·6	8·4	11·5	229
230			GS	5·1	3·5	5·6	1·38	0·86	8600	4900	5·9	17·7	5·9	13·8	4·0	2·9	3·7	0·92	0·70	7000	4000	5·7	18·3	5·7	14·4	230

Table XVII Permissible stresses (*continued*)

Number	Species	Origin	Grade	Dry stresses (moisture content not exceeding 18%) N/mm² Bending	Dry Tension	Dry Compression parallel to grain	Dry Compression perpendic. to grain	Dry Shear parallel to grain	Dry Modulus of Elasticity Mean	Dry Modulus of Elasticity Minimum	Dry Loading sharing medium term Shear limit	Dry Loading sharing medium term Bending limit	Dry Single memb. long term Shear limit	Dry Single memb. long term Bending limit	Green stress (moisture content exceeding 18%) N/mm² Bending	Green Tension	Green Compression parallel to grain	Green Compression perpendic. to grain	Green Shear parallel grain	Green Modulus of elasticity Mean	Green Modulus of elasticity Minimum	Green Load sharing medium term Shear limit	Green Load sharing medium term Bending limit	Green Single memb. long term Shear limit	Green Single memb. long term Bending limit	Number
231	Spruce-pine-fir	I	SS	6·9	4·8	7·0	1·55	0·79	8900	4800	8·7	13·5	8·7	10·0	5·5	3·8	5·4	1·03	0·70	8000	4300	7·9	15·2	7·9	11·3	231
232			GS	4·8	3·3	4·8	1·38	0·79	8000	4300	6·1	17·4	6·1	12·9	3·8	2·7	3·7	0·92	0·70	7200	3900	5·4	19·8	5·4	14·8	232
233	Group S2		SS	6·9	4·8	7·0	1·55	0·79	8900	4800	8·7	13·5	8·7	10·0	5·5	3·8	5·4	1·03	0·70	8000	4300	7·9	15·2	7·9	11·3	233
234			GS	4·8	3·3	4·8	1·38	0·79	8000	4300	6·1	17·4	6·1	12·9	3·8	2·7	3·7	0·92	0·70	7000	3900	5·4	19·3	5·4	14·8	234
	Hem-fir	I	Canadian																							
235	Joists and planks		Select struc.	9·6	6·7	8·4	1·55	0·80	11800	7300	12·0	12·9	11·5	*	7·7	5·4	6·2	1·03	0·70	10400	6400	11·0	14·1	11·0	12·0	235
236			No 1	8·1	5·7	7·5	1·55	0·80	11100	6800	10·1	14·4	12·1	12·1	6·6	4·6	5·6	1·03	0·70	9800	6000	9·4	15·6	9·4	13·1	236
237			No 2	6·5	4·6	6·4	1·38	0·80	10300	6400	8·1	16·6	8·1	14·2	5·3	3·7	4·7	0·92	0·70	9100	5600	7·6	18·0	7·6	15·2	237
238			No 3	3·8	2·6	4·0	1·03	0·70	9100	5600	5·4	25·1	5·4	21·2	3·0	2·1	3·0	0·69	0·61	8100	5000	4·9	28·3	4·9	24·0	238
239	Light framing*		Construction	5·7	3·4	6·8	1·55	0·80	9500	5900					4·6	2·8	5·0	1·03	0·70	8400	5200					239
240			Standard	3·2	1·9	5·6	1·38	0·80	8600	5300					2·5	1·5	4·1	0·92	0·70	7600	4700					240
241			Utility	1·5	0·9	3·7	1·03	0·70	7800	4800					1·2	0·7	2·7	0·69	0·61	7000	4300					241
242	Stud grading*			4·1	2·5	3·2	1·03	0·63	7600	4100					3·5	2·1	2·7	0·69	0·61	8100	5000					242
	Softwoods— Group S3																									
			Machine																							
301	Sitka spruce	H	M75	6·6	4·6	6·4	1·45	0·98	9000	5500	6·7	14·3	6·7	12·0	4·9	3·4	4·6	0·97	0·82	7500	4600	6·0	16·0	6·0	13·5	301
302			MSS	5·2	3·6	5·0	1·24	0·66	8200	5100	7·9	16·5	7·9	14·1	3·8	2·7	3·6	0·82	0·55	6800	4200	6·9	18·7	6·9	15·9	302
303			M50	4·5	3·2	4·3	1·24	0·66	7700	4800	6·8	17·9	6·8	15·3	3·3	2·3	3·1	0·82	0·55	6400	4000	6·0	20·3	6·0	17·5	303
304			MGS	3·6	2·5	3·5	1·10	0·66	7200	4500	5·5	20·9	5·5	18·0	2·7	1·9	2·5	0·73	0·55	6000	3700	4·9	23·3	4·9	19·7	304
			Visual																							
305	European spruce	H	SS	5·5	3·9	5·6	1·24	0·71	7500	4000	7·7	14·3	7·7	10·5	4·2	2·9	4·0	0·82	0·55	6100	3200	7·6	15·2	7·6	11·0	305
306			GS	3·9	2·7	3·9	1·10	0·71	6800	4900	5·5	18·3	5·5	18·1	2·9	2·0	2·8	0·73	0·55	5500	2900	5·3	19·9	5·3	14·4	306
307	Western red cedar	I	SS	5·2	3·6	5·9	1·16	0·70	8000	5500	7·4	16·1	7·4	15·2	4·5	3·2	4·0	0·77	0·62	7400	5000	7·3	17·2	7·3	16·0	307
308			GS	3·6	2·5	4·1	1·03	0·70	7200	4900	5·1	20·9	5·1	19·6	3·2	2·2	2·8	0·69	0·62	6700	4600	5·2	21·9	5·2	20·7	308
309	Sitka spruce	H	SS	5·2	3·6	5·0	1·24	0·66	7800	3700	7·9	15·7	7·9	10·2	3·8	2·7	3·6	0·82	0·55	6500	3100	6·9	17·9	6·9	11·7	309
310			GS	3·6	2·5	3·5	1·10	0·66	7000	3300	5·5	20·4	5·5	13·2	2·7	1·9	2·5	0·73	0·55	5900	2800	4·9	22·9	4·9	14·9	310
311	Group S3		SS	5·2	3·6	5·0	1·16	0·66	7500	3700	7·9	15·1	7·9	10·2	3·8	2·7	3·6	0·77	0·55	6100	3100	6·9	16·8	6·9	11·7	311
312			GS	3·6	2·5	3·5	1·03	0·66	6800	3300	5·5	19·8	5·5	13·2	2·7	1·9	2·5	0·69	0·55	5500	2800	4·9	21·3	4·9	14·9	312
	Spruce-pine-fir		Canadian																							
313	Joists and planks		Select struc.	9·1	6·4	7·4	1·55	0·79	9800	5300	11·4	*	9·8	*	7·3	5·1	5·7	1·03	0·70	8800	4800	10·4	12·6	9·9	*	313
314			No 1	7·7	5·4	6·7	1·55	0·79	9300	5000	9·7	12·6	9·5	*	6·2	4·3	5·1	1·03	0·70	8200	4400	8·9	13·9	8·9	10·2	314
315			No 2	6·2	4·3	5·6	1·38	0·79	8600	4700	7·8	14·5	7·8	10·9	5·0	3·5	4·3	0·91	0·70	7700	4200	7·1	16·1	7·1	12·1	315
316			No 3	3·6	2·5	3·6	1·03	0·63	7600	4100	5·7	22·1	5·7	16·4	2·9	2·0	2·7	0·69	0·55	6800	3700	5·3	24·6	5·3	18·4	316
317	Light framing*		Construction	5·4	3·1	6·0	1·55	0·79	7900	4300					4·3	2·6	4·6	1·03	0·70	7100	3900					317
318			Standard	3·0	1·8	5·0	1·38	0·79	7100	3900					2·4	1·5	3·8	0·91	0·70	6300	3400					318
319			Utility	1·4	0·9	3·2	1·03	0·63	6600	3600					1·1	0·7	2·5	0·69	0·55	5800	3100					319
320	Stud grading*			4·1	2·5	3·2	1·03	0·63	7600	4100					3·3	2·0	2·5	0·69	0·55	6800	3700					320

***See explanation in para 3.16**

Table XVIII Stress modification factors for duration of loading

Long-term loads (dead + permanent imposed)	× 1·00
Medium-term loads (dead + snow, dead + temporary* loads)	× 1·25
Short-term loads (dead + imposed + wind)	× 1·50
Transitory loads (wind gusting)	× 1·75

* The use of the building will usually determine whether the live loads are transient (such as people), semi-permanent or permanent. In ordinary residential buildings, it has been found that the loading due to furniture (permanent) amounts to only 0·25 to 0·5 kN/m². However, in office buildings the permitted imposed loadings are often found to be exceeded by the furniture and its contents without even allowing for the occupants. In all such buildings the imposed loads should be taken as long term.

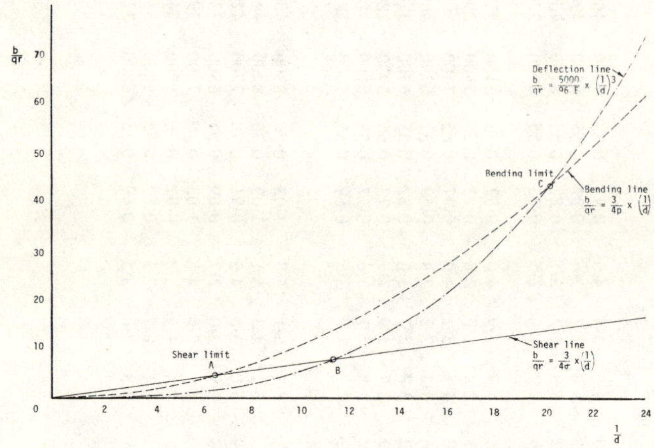

45.18 *Beam characteristics for Western Hemlock (commercial) Grade MGS, Load sharing, medium-term loads*

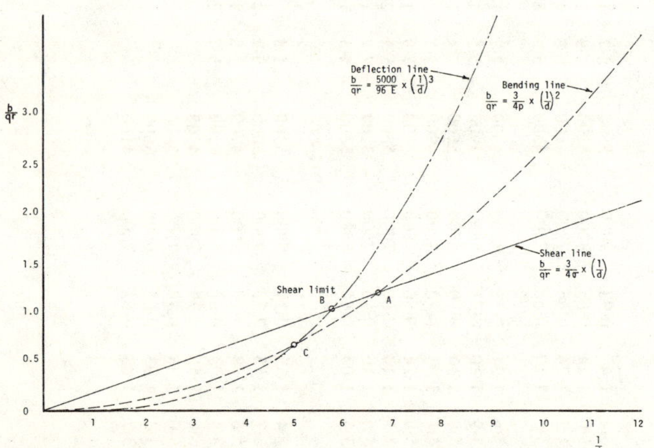

45.19 *Beam characteristics for European Oak, basic grade, Load sharing, medium-term loads*

The value of $\frac{b}{qr}$ is proportional to the *inverse* of the loading on the floor; the greater this quantity is, the less load the joist can carry. It is not therefore surprising that $\frac{b}{qr}$ increases with increasing $\frac{1}{d}$ for the shear, the bending and the deflection, although at different rates.

Since the load capacity is less the higher the value of $\frac{b}{qr}$, it follows that the uppermost line on the graph for any given $\frac{1}{d}$ represents the first mode of failure that will be reached as the load increases. From $\frac{1}{d} = 0$ to 6·4 the shear line (equation (iii), full line) is the highest, and the beam will fail by shear in this range. Point A at 6·4 is called the *shear limit*: where the shear line crosses the bending line. Similarly, the *bending limit* at C where $\frac{1}{d} = 20·1$ marks the point at which the bending ceases to be critical, and deflection becomes the controlling factor. C is the point where the bending line (equation (i), dashed line) crosses the deflection line (equation (ii), chain-dotted line). Similar graphs can be constructed for each species and grade of timber. **45.19** shows the characteristics for a hardwood, European oak, basic grade. In this case the bending is never critical, the shear limit occurring at B where the shear line cuts the deflection line.

3.16 Shear and bending limits

The shear and bending limits for each species and grade of timber are given in table XVII. They are shown for two cases which cover the possible extremes:

- medium-term load (table XVIII) with load-sharing (para 3.10)
- long-term loading with no loadsharing (single member).

The use of these limits during the design process will save time, for it will be seen that in the majority of cases it will be the deflection criterion that will control the design. However, it should be noted that there are cases where a deflection exceeding 0·003 of the span would be acceptable, for example where there is no plaster ceiling under the floor.

Where in table XVII there is a star * in the bending limit column, this indicates that, as in European oak above, bending is never a critical factor.

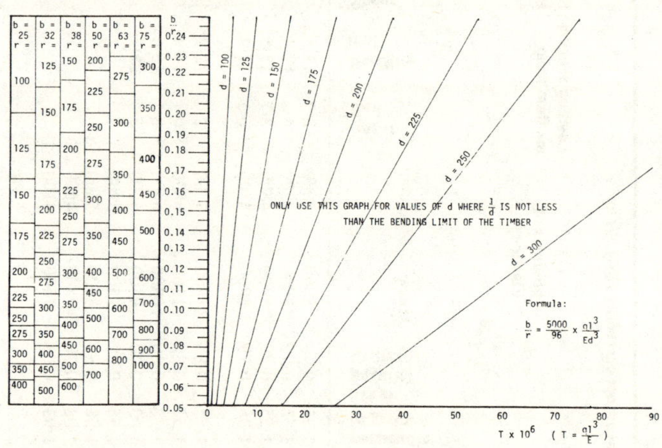

45.20 *Design chart for deflection criterion*

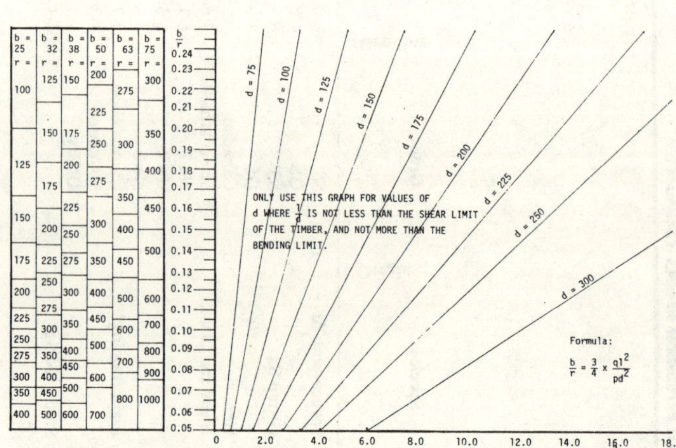

45.21 *Design chart for bending criterion*

3.17 Design graphs

Once having established from table XVII whether bending or deflection is critical in any specific case (shear will almost never be critical in an ordinary floor joist), **45.20** or **45.21** can be used to find a convenient size and spacing for the joists. The method is best described by means of an example.

Example 10

What sizes and spacings of joists in western hemlock (commercial) grade MGS, could be used for a floor in a heated building spanning 3·75 m, and carrying a medium-term load of 2·7 kN/m² total? The maximum spacing is to be 600 mm.

Answer: From table XVII obtain the bending limit for the timber under load-sharing and medium-term load conditions: $\frac{1}{d} = 20{\cdot}1$

For span 3·75 m, limiting d is $\frac{3750}{20{\cdot}1} = 187$ mm, sections below this depth will be critical in deflection, above it in bending.

Use **45.20**, deflection criterion:

$$T = \frac{ql^3}{E} = \frac{2{\cdot}7 \times 10^{-3} \times 3{\cdot}75^3}{9800} = 14{\cdot}53 \times 10^{-6}\ \mathrm{m}^3$$ giving a position along the x-axis of the graph.

Using the line on the graph for d = 175 (mm), $\frac{b}{r} = 0{\cdot}1425$ and from the chart on the left-hand side of the graph, the following alternatives can be used:

175 × 25 @ 175, not really satisfactory owing to lack of section width for nailing the floorboards down to (t' = 25 mm)
175 × 32 @ 200 (t' = 28)
175 × 38 @ 250 (t' = 26·6)
175 × 50 @ 350 (t' = 25)

The quantity t' is the effective thickness of the floor in mm. It is derived from the formula $t' = \frac{bd}{r}$ where the breadth, depth and spacing are all expressed in mm. The effective thickness is an indication of the cost of the structure, as it represents the volume of timber used. The design with minimum t' (within limits) is likely to be the most economical. In addition, 6t' will give the approximate loading of the weight of the structural timber in N/m².

Now, it may be necessary for architectural reasons to use a 150 mm deep joist. From the chart $\frac{b}{r} = 0{\cdot}225$ and we can use:

150 × 50 @ 200 (t' = 37·5)

Now consider joists deeper than 175 mm. For these joists bending will be the critical item.

Use **45.21**, bending criterion:

$$U = \frac{ql^2}{p} = \frac{2{\cdot}7 \times 10^{-3} \times 3{\cdot}75^2}{1{\cdot}25 \times 1{\cdot}1 \times 5{\cdot}1} = 5{\cdot}42 \times 10^{-3}\ \mathrm{m}^2$$ giving the appropriate position along the x-axis.

Note the modification factors on the grade stress for load-sharing and medium-term load. Compressive stress of 5·1 N/mm² is found in table XVII.

Using the line on the chart for d = 225 (mm), $\frac{b}{r} = 0{\cdot}08$

225 × 32 @ 400 (t' = 18—most economical solution)
225 × 38 @ 450 (t' = 19)
225 × 50 @ 600 (t' = 18·8)

4 REINFORCED CONCRETE

4.01 Composition

Reinforced concrete is probably the most prolific and versatile structural material in use today. It is composed of two distinct materials: concrete and reinforcement, each of which can be varied in quality, disposition and quantity to fulfil almost any requirement.

The concrete component is itself an amalgam of at least three constituents: aggregate, cement and water. These are mixed together into a homogenous mass, and are then put in place and left for the chemical and physical changes to occur that result in a hard and durable material. The strength and durability will depend on the quality and quantity of each of the constituents; and whether any other material, such as an admixture, has been added to the wet mix. Since the mixing of concrete has now in the majority of cases been taken off the building site and is done by the ready-mix companies, control over this aspect, as far as the architect is concerned, has become less direct. The strength of the hardened material as demonstrated by the ubiquitous cube crushing test, will not necessarily indicate that sufficient cement has been included to fulfil the requirement for long-term durability. Sometimes admixtures are included in the mix to promote workability and early strength. Deterioration in the material due to these may not become evident for some years, but may then be disastrous.

4.02 Specification

Provided a clear specification is laid down and checked by site staff, the concrete should fulfil its function indefinitely. This specification should now be in accordance with BS 5328: 1976, Methods for specifying concrete, which includes not only for strength, but also for minimum cement content, aggregate size, cement type and other relevant aspects. For most architectural purposes ordinary prescribed mixes should be sufficient; the composition of these is given in tables XIX and XX.

4.03 Reinforcement

Reinforcement is rarely other than steel, although other materials such as glass fibre have been tried (mainly in cladding units). The steel may be smooth round steel, in which case the bars are referred to as R25, etc, the numbers indicating the diameter in millimetres. Only those sizes in tables XXI and XXII are available.

The other type of bar reinforcement is a high-yield deformed bar referred to as Y25 etc. In this case the numbers refer to the plain bar diameter of equivalent cross-sectional area. The actual bar size will be about 10 per cent greater than this due to the deformation. The two kinds of high-yield bar used are hot-rolled and cold-worked, but for practical purposes the difference is rarely significant.

4.04 Mesh reinforcement

For many positions, such as in slabs and walls, it is convenient to use reinforcement in the shape of a sheet or mesh composed of bars in both directions. Table XXIII gives the standard sizes of available meshes.

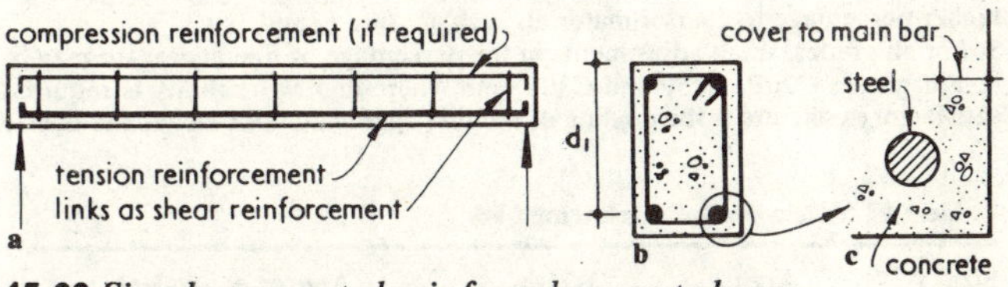

45.22 *Simply supported reinforced concrete beam*
a *elevation showing reinforcement conventionally*
b *section*
c *detail of section showing cover to main reinforcement*

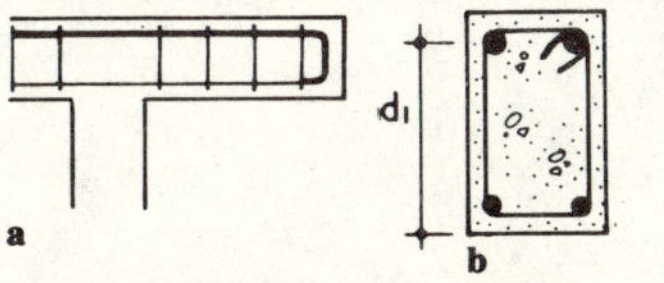

45.23 *Cantilever in reinforced concrete:*
a *elevation showing reinforcement conventionally*
b *section*

4.05 Reinforcement position

Concrete is strong in compression, but very weak in tension. The reinforcement is used to compensate for this weakness. There must be some reinforcement wherever tension is likely to occur, and sufficient at places of maximum tension. Simply spanning beams are reinforced near the bottom, with most reinforcement at midspan, **45.22**. Shear forces can also produce tensile stresses: links or stirrups are used to reinforce the concrete against the effects of these stresses. Sometimes the compressive strength of the concrete is insufficient for the loading, in this case reinforcing bars can be used to help take the compression as well. Such use for reinforcement is expensive, and is only used when increasing the size of the beam is not possible.

In cantilever beams the tension occurs near the top. These beams have their heaviest reinforcement at the top, with most near the root, **45.23**.

Table XIX Ordinary prescribed concrete mixes (BS 5328: 1976): Weights of dry aggregate to be used with 100 kg of cement

Grade of concrete	Nominal maximum size of aggregate (mm)	40		20		14		10	
	Workability	Medium	High	Medium	High	Medium	High	Medium	High
	Range of slump (mm)	50–100	80–170	25–75	65–135	5–55	50–100	0–45	15–65
		kg	kg	kg	kg	kg	kg	kg	kg
C7P		1080	920	900	780	N/A	N/A	N/A	N/A
C10P		900	800	770	690	N/A	N/A	N/A	N/A
C15P	Total	790	690	680	580	N/A	N/A	N/A	N/A
C20P	aggregate	660	600	600	530	560	470	510	420
C25P		560	510	510	460	490	410	450	370
C30P		510	460	460	400	410	360	380	320

N/A not applicable

Table XX Percentage by weight of fine aggregate to total aggregate

Grade of concrete	Nominal maximum size of aggregate (mm)	40		20		14		10	
	Workability	Medium	High	Medium	High	Medium	High	Medium	High
C7P, C10P, C15P		30–45		35–50		N/A		N/A	
C20P, C25P, C30P	Grading zone 1	35	40	40	45	45	50	50	55
	2	30	35	35	40	40	45	45	50
	3	30	30	30	35	35	40	40	45
	4	25	25	25	30	30	35	35	40

N/A not applicable

Notes

1. The proportions given in the tables will normally provide concrete of the strength in N/mm² indicated by the grade except where poor control is allied with the use of poor materials.
2. For grades C7P, C10P and C15P a range of fine-aggregate percentages is given; the lower percentage is applicable to finer materials such as zone 4 sand and the higher percentage to coarser materials such as zone 1 sand.
3. For all grades, small adjustments in the percentage of fine aggregate may be required depending on the properties of the particular aggregates being used.
4. For grades C20P, C25P and C30P, and where high workability is required, it is advisable to check that the percentage of fine aggregate stated will produce satisfactory concrete if the grading of the fine aggregate approaches the coarser limits of zone 1 or the finer limits of zone 4.

Table XXI Metric sizes of reinforcing bars

Size (mm)	Area (mm²)	Weight (kg/m)	Approx equivalent inch size
6	28	0·222	1/4
8	50	0·395	5/16
10	79	0·617	3/8
12	113	0·888	1/2
16	201	1·58	5/8
20	314	2·47	3/4
25	491	3·86	1
32	804	6·31	1 1/4
40	1260	9·87	1 1/2
50	1964	15·41	2

4.06 Effective depth

The effective depth of the beam is the distance from its top (or compression flange) to the centroid of area of the tensile steel reinforcement. It is indicated by the symbol d_1. **45.22** and **45.23**.

4.07 Minimum reinforcement

Stresses arise in concrete not only from the applied loads but also from a variety of other causes. For example, when concrete dries and sets it tends to shrink slightly. If it cannot move it will tend to crack. Similar cracking will occur if movement induced by thermal expansion and contraction is inhibited. Consequently, to reduce the tendency to form large cracks, a modicum of reinforcement is used throughout: this causes the formation of a multitude of fine cracks instead.

4.08 Deflection

In addition to limiting the working stresses below the permissible values, reinforced concrete must possess sufficient stiffness to prevent deflection or deformation which might impair the strength or efficiency of the structure, or produce cracks in finishes or partitions. For all normal cases, it may be assumed that the stiffness will be satisfactory if the ratio of span to overall depth does not exceed the appropriate value from table XXIV, **45.24**.

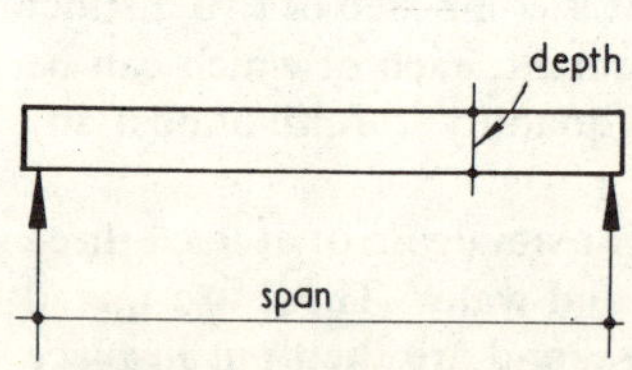

45.24 *Deflection rules for reinforced concrete beams and slabs by span/depth ratios*

4.09 Concrete cover to reinforcement

In all cases there must be sufficient concrete cover to reinforcement, both to preserve it from corrosion and to ensure an adequate bond with the concrete, **45.22c**. CP 114:1969 (see para **4.11** below) lays down certain rules which may be summarised as follows. The thickness of concrete cover exclusive of plaster or other decorative finish should be:

1 for each end of a reinforcing bar, not less than 25 mm nor less than twice the diameter of such bar;

2 for a longitudinal reinforcing bar in a column, not less than 40 mm nor less than the diameter of such bar. In the case of columns with a minimum dimension of 200 mm or under,

Table XXII Areas of round bar reinforcement (mm²)

Diam (mm)	Weight (kg/m)	Areas in mm² for numbers of bars 1	2	3	4	5	6	7	8	9	10	11	12
6	0·222	28	57	85	113	142	170	198	226	255	283	311	340
8	0·395	50	101	151	201	252	302	352	402	453	502	552	604
10	0·617	79	157	236	314	393	471	550	628	707	785	864	942
12	0·888	113	226	339	452	565	678	791	904	1017	1130	1243	1356
16	1·58	201	402	603	804	1005	1206	1407	1608	1809	2010	2211	2412
20	2·47	314	628	942	1256	1570	1884	2198	2512	2826	3140	3454	3768
25	3·86	491	983	1474	1966	2457	2948	3439	3932	4423	4915	5406	5896
32	6·31	804	1608	2412	3216	4020	4824	5628	6432	7236	8040	8844	9648
40	9·87	1260	2520	3780	5040	6300	7560	8820	10080	11340	12600	13860	15120

Diam (mm)	Areas in mm²/m for spacings in mm 50	75	100	125	150	175	200	250	300
6	566	376	283	226	188	162	141	113	94
8	1006	670	503	402	335	287	251	201	168
10	1570	1048	785	628	524	449	393	314	262
12	2262	1508	1131	904	754	646	565	452	377
16	4020	2680	2010	1608	1340	1149	1005	804	670
20	6284	4190	3142	2514	2095	1795	1571	1257	1047
25	9830	6552	4915	3932	3276	2809	2457	1966	1638
32		10720	8040	6432	5360	4594	4020	3216	2680
40			12600	10080	8400	7200	6300	5040	4200

Table XXIII Sizes of reinforcing meshes

BS reference	Mesh sizes Nominal pitch of wires (mm)		Size of wires (mm)		Cross sectional area (mm²) per metre width		Nominal mass kg per m²
Square mesh fabric							
	Main	*Cross*	*Main*	*Cross*	*Main*	*Cross*	
A 393	200	200	10	10	393	393	6·16
A 252	200	200	8	8	252	252	3·95
A 193	200	200	7	7	193	193	3·02
A 142	200	200	6	6	142	142	2·22
A 98	200	200	5	5	98	98	1·54
Structural fabric							
	Main	*Cross*	*Main*	*Cross*	*Main*	*Cross*	
B 1131	100	200	12	8	1131	252	10·90
B 785	100	200	10	8	785	252	8·14
B 503	100	200	8	8	503	252	5·93
B 385	100	200	7	7	385	193	4·53
B 283	100	200	6	7	283	193	3·73
B 196	100	200	5	7	196	193	3·05
Long mesh fabric*							
	Main	*Cross*	*Main*	*Cross*	*Main*	*Cross*	
C 785	100	400	10	6	785	70·8	6·72
C 503	100	400	8	5	503	49·0	4·34
C 385	100	400	7	5	385	49·0	3·41
C 283	100	400	6	5	283	49·0	2·61
Wrapping fabric							
	Main	*Cross*	*Main*	*Cross*	*Main*	*Cross*	
D 49	100	100	2·5	2·5	49·0	49·0	0·760
D 31	100	100	2·0	2·0	31·0	31·0	0·480
Recommended fabric for carriageways							
	Main	*Cross*	*Main*	*Cross*	*Main*	*Cross*	
C 636	80–130	400	8–10	6	636	70·8	5·55

* Cross wires for all types of long mesh may be of plain hard drawn steel wire

Table XXIV Span to depth ratios (see 45.24)

	A	B	C
Beams			
Simply supported beams	20	18	17
Continuous beams	25	23	21
Cantilever beams	10	9	8
Slabs			
Slabs spanning in one direction, simply supported	30	27	25
Slabs spanning in one direction, continuous	35	31	30
Slabs spanning in two directions, simply supported	35	31	30
Slabs spanning in two directions, continuous	40	36	34
Cantilever slabs	12	11	10

Notes

Column A Members with steel stresses not more than 140 N/mm² and concrete stresses not more than 10·0 N/mm²

Column B Members with either steel stresses greater than in column A or concrete stresses greater, but not both

Column C Members with both steel and concrete stresses greater than in column A

whose bars do not exceed 12 mm diameter, 25 mm cover may be used;
3 for a longitudinal reinforcing bar in a beam, not less than 25 mm nor less than the diameter of such bar;
4 for tensile, compressive, shear or other reinforcements in a slab, not less than 15 mm nor less than the diameter of such reinforcement;
5 for any other reinforcement not less than 15 mm nor less than the diameter of such reinforcement.
For all external work, for work against earth faces and also for internal work where there are particularly corrosive conditions, the cover of the concrete should not be less than 40 mm for all steel, including stirrups, links and so on, except where the face of the concrete is adequately protected by a suitable cladding or by a protective coating, which may indeed be advisable where the corrosive conditions are unduly severe.
Additional cover may be necessary where lightweight or porous aggregates are used, or to comply with fire resistance requirements.

4.10 Distance between bars

The horizontal distance between two parallel steel reinforcements in reinforced concrete should usually, except at splices, be not less than the greatest of the three following distances:
1 the diameter of either bar if their diameters be equal;
2 the diameter of the larger bar if the diameters be unequal;
3 5 mm more than the nominal maximum size of the coarse aggregate used in the concrete.
A greater distance should be provided where convenient. Where immersion vibrators are intended to be used, however, the horizontal distance between bars of a group may be reduced to two-thirds of the nominal maximum size of the coarse aggregate provided that enough space is left between groups of bars to enable the vibrator to be inserted; this would normally be a space of 75 mm.
The vertical distance between two horizontal main steel reinforcements, or the corresponding distance at right-angles to two inclined main steel reinforcements, should be not less than 15 mm or the nominal maximum size of aggregate, whichever is the greater, except at splices or where one of such reinforcements is transverse to the other.
The pitch of the main bars in a reinforced concrete solid slab should be not more than three times the effective depth of such slab.
The pitch of distributing bars in a reinforced concrete slab should not be more than five times the effective depth of such slab.

4.11 Design

The design of reinforced concrete structures is covered by two Codes of Practice: CP 114:1969 and CP 110:1972. The latter is based on the concept of limit state design which has been discussed in para **1.02**. For the purposes of the simple examples given here the methods of CP 114 are sufficient and appropriate. These methods consist of calculating the working stresses in the structure: those are the stresses under maximum working load conditions. The designer then checks that these stresses are less than the permissible stresses given in the code for the particular material. Tables XXV and XXVI give permissible stresses for concrete and steel respectively.

4.12 Beam stress calculation

The method of calculating the working stresses called *the load factor method* takes into account the behaviour of the materials when approaching ultimate failure. This may appear to be illogical, but it is a method that has been evolved over many years, and combines safety with economy. It may be accepted for all normal structures, but its application to an unusual design should be treated with reservations.
In a beam, the concrete and the steel are both assumed to have reached their ultimate stress, **45.25**. The concrete stress will be a rectangle from the top of the beam down to the neutral axis, and the depth of the neutral axis will be only enough to generate a sufficiently large bending moment. The first equation is derived by resolving horizontally:

$$p_c \times b \times n = p_s \times A_s \qquad \text{———————(i)}$$

where:

p_c is the ultimate stress in the concrete (compressive)
p_s is the ultimate stress in the steel (tensile)
b is the breadth of the beam
n is the depth of the neutral axis
A_s is the area of the steel reinforcement.

The second equation comes from taking moments about a point at the position of the reinforcement:

$$p_c \times b \times n \times 1_a = M \qquad \text{———————(ii)}$$

where:

1_a is the lever arm between the centre of area of the stress block and the reinforcement
M is the bending moment at that section of the beam.

Now when failure is imminent, it is found that the neutral axis

Table XXV Permissible concrete stresses (compare with table 6 in BS CP 114: 1969)

Mix	28 day works cube strength	Compressive stresses Direct	Bending	Shear stress	Bond stresses average	local
Nominal mixes						
1:1:2	30	7·6	10·0	0·9	1·0	1·5
1:1½:3	25·5	6·5	8·5	0·8	0·93	1·4
1:2:4	21	5·3	7·0	0·7	0·83	1·25
Ordinary prescribed mixes						
C30P	30	7·5	10·0	0·87	1·0	1·47
C25P	25	6·2	8·3	0·77	0·9	1·33
C20P	20	5·0	6·7	0·67	0·8	1·20
C15P	15	3·75	5·0	0·50	0·6	0·9

Table XXVI Permissible stresses in steel reinforcement (compare with table XI in BS CP 114: 1969)

	Tensile stress in other than shear reinforcement N/mm^2	Tensile stress in shear reinforcement N/mm^2	Compressive stress N/mm^2
Mild steel, diameter not exceeding 40 mm	140	140	125
Mild steel, diameter exceeding 40 mm	125	125	110
High yield bars f_y not less than $418 N/mm^2$, and diameter not exceeding 20 mm	230	175	175
High yield bars f_y not less than $418 N/mm^2$, and diameter exceeding 20 mm	210	175	175

occurs at half the effective depth of the beam d_1. At this stage The bending moment is called the *moment of resistance of the concrete, M_r*.

so, from (ii) with $n = \frac{1}{2} d_1$ and $l_a = \frac{3}{4} d_1$

$$M_r = \tfrac{3}{8} p_c \times b \times d_1^2$$

in fact, this is not quite correct as the stress block is not a perfect rectangle. To allow for this the equation becomes:

$$M_r = \tfrac{1}{4} p_c b d_1^2 \qquad \text{(iii)}$$

Combining equations (i) and (ii):

$$p_s \times A_s \times l_a = M$$

therefore

$$A_s = \frac{M}{p_s l_a} \qquad \text{(iv)}$$

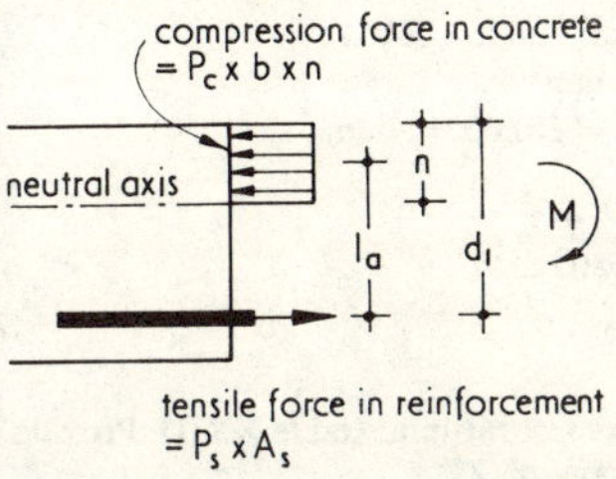

45.25 *Forces on a reinforced concrete beam section (as seen in elevation)*

4.13 Calculation procedure

a First, find the moment of resistance

Use the following form of equation (iii)

$$M_r = \tfrac{1}{4} p_{cb} b d_1^2$$

p_{cb} is the allowable compressive stress *in bending* as given in table XXV.

b Next ensure that the applied moment M is less than M_r

If it is not, increase the size of the beam until it is. Alternatively, compression reinforcement can be used, but it is desirable that this is done only under the supervision of a structural engineer.

c Now calculate the amount of tensile reinforcement

Use equation (iv):

$$A_s = \frac{M}{p_{st}\, l_a}$$

where p_{st} is the allowable tensile stress in the steel from table XXVI.

l_a, the lever arm is calculated from the ratio of the applied and resistance moments:

$\frac{M}{M_r}$	1	0·9	0·8	0·7	0·6	0·5 and below
$\frac{l_a}{d_1}$	0·75	0·78	0·81	0·84	0·87	0·9

d The shear stress is now calculated

$$q = \frac{Q}{b l_a} \qquad \text{(v)}$$

where q is the shear stress

Q is the shear force

If this stress is greater than the permissible shear stress in table XXV shear reinforcement must be used. This can be in the form of bent-up bars or in the form of stirrups, **45.24**. The shear values for single bars bent-up at an angle of 45 deg are given in table XXVII. When using stirrups the following formula can be used:

$$A_{sh} = \frac{hQ}{p_{ss}\, l_a\, N} \qquad \text{(vi)}$$

where A_{sh} is the area of shear steel in length 'h'

p_{ss} is the permissible tensile stress in shear reinforcement from table XXVI

N is the number of 'arms' in the links.

4.14 Simply supported slab

Example 11

Design a simply supported slab spanning 5 m to carry a super-imposed load (including finishes) of 14 kN/m². In this and the following examples assume all the concrete to be C25P as table XIX.

If high yield reinforcement is to be used the span: depth ratio must not exceed 27 (table XXIV)

$$\therefore \text{depth} = \frac{\text{span}}{27}$$

$$\text{span} = 5\,\text{m} = 5000\,\text{mm}$$

$$\therefore \text{depth} = \frac{5000}{27} = 185\,\text{mm}$$

assume slab is 200 mm deep density of concrete is 24 kN/m³ (table II)

$$\therefore \text{self-weight of slab} = 0{\cdot}2 \times 24 = 4{\cdot}8\,\text{kN/m}^2$$

$$\text{superload} = 14{\cdot}0\,\text{kN/m}^2$$

$$\therefore \text{total load} = 18{\cdot}8\,\text{kN/m}^2$$

$$M = \frac{WL}{8} \quad \text{for a simply supported beam (table VI)}$$

Consider 1 m width of slab as a beam

$$W = 18{\cdot}8 \times 5\,\text{kN}$$
$$L = 5\,\text{m}$$
$$\therefore M = \frac{18{\cdot}8 \times 5^2}{8} = 58{\cdot}75\,\text{kNm/m width}$$

assume the reinforcement to be in the form of 20 mm diameter bars

cover required = 20 mm (para **4.09**)

$$d_1 = \text{effective depth of slab} = \text{total depth} - \text{cover} - \tfrac{1}{2}\,\text{bar diam} = 200 - 20 - 10 = 170\,\text{mm}$$

now the concrete moment of resistance

$$M_r = \frac{p_{cb}}{4} b d_1^2 \quad \text{(para **4.13a**)}$$

$$p_{cb} = 8{\cdot}3\,\text{N/mm}^2 \quad \text{(table XXV)}$$

$$\therefore \frac{p_{cb}}{4} = 2{\cdot}08\,\text{N/mm}^2$$

this value may be retained and substituted directly in future calculations.

$$b = 1000\,\text{mm}$$
$$d_1 = 170\,\text{mm}$$
$$\therefore M_r = 2{\cdot}08 \times 1000 \times 170^2 = 60\,100\,000\,\text{Nmm} = 60{\cdot}1\,\text{kNm/m width}$$

The concrete moment of resistance is greater than the applied moment (para **4.13b**). The depth of the slab is therefore satisfactory.

$$\frac{M}{M_r} = \frac{58{\cdot}75}{60{\cdot}1} = 0{\cdot}98$$

from this the lever arm is found (para **4.13c**)

$$\frac{l_a}{d_1} = 0{\cdot}75$$
$$\therefore l_a = 0{\cdot}75 \times 170 = 127\,\text{mm}$$

the area of steel is now calculated (para **4.13c**)

$$A_t = \frac{M}{p_{st} l_a}$$
$$p_{st} = 230\,\text{N/mm}^2 \quad \text{(table XXVI)}$$
$$M = 58{\cdot}75\,\text{kNm/m width} = 58{\cdot}75 \times 10^6\,\text{Nmm/m}$$
$$\therefore A_t = \frac{58{\cdot}75 \times 10^6}{230 \times 127} = 2011\,\text{mm}^2\text{/m width}$$

Provide Y20 (20 mm diameter high yield) bars at 150 mm centres = 2095 mm²/m (from table XXII), **45.26**.

Deflection should be satisfactory if the span/depth ratio is within the limit laid down in table XXIV.

In solid slabs (ie those that are not ribbed) shear is never a problem, and the calculation is not required.

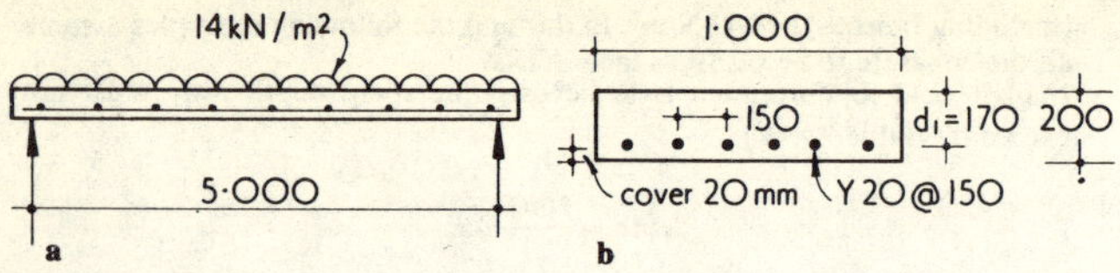

45.26 *Simply supported reinforced concrete slab (Example 11)*
a *elevation showing reinforcement conventionally*
b *section across main reinforcement*

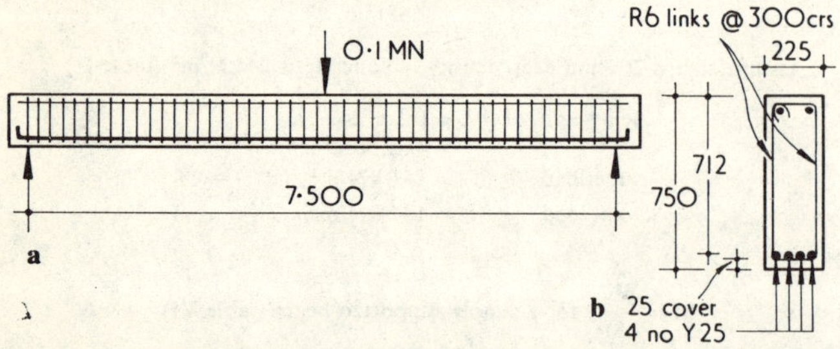

45.27 *Rectangular section reinforced concrete beam (Example 12)*
a *elevation showing reinforcement conventionally*
b *section*

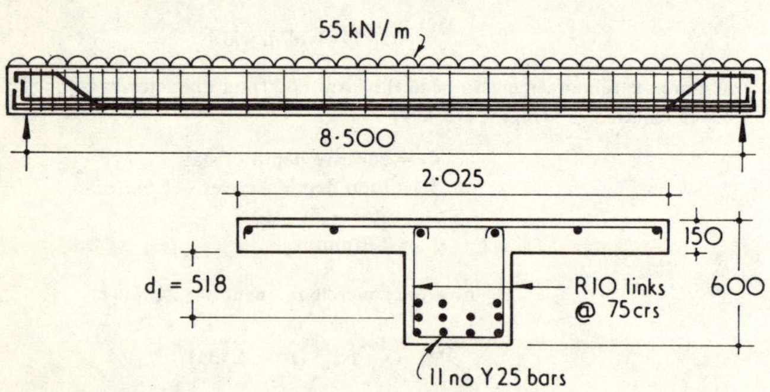

45.28 *Reinforced concrete T-beam (Example 13)*
a *elevation showing reinforcement conventionally*
b *section showing reinforcement*

Table XXVII Bent-up bars as shear reinforcement. Singles bars bent-up at 45 deg. in single system: shear value

Size mm	Mild steel kN	High yield steel kN
12	11·2	14·0
16	19·9	24·9
20	31·1	38·9
25	48·6	60·8
32	79·6	99·5
	25·0	156·0

4.15 Rectangular section reinforced concrete beam

Example 12

Design a rectangular reinforced concrete beam 750 × 225 mm simply supported with a span of 7·5 m carrying a central point load weighing 0·1 MN. (Self-weight may be neglected.)

First find the applied moment

$$M = \frac{WL}{4} \quad \text{for a central point load on a simply supported beam (table VI)}$$

$$W = 100\,kN$$
$$L = 7{\cdot}5\,m$$
$$\therefore M = \frac{100 \times 7{\cdot}5}{4} = 187\,kNm$$

If the reinforcement is to be 25 mm diam bars

$$\text{cover} = 25\,mm \text{ (para } \mathbf{4.09})$$
$$\text{effective depth} = \text{total depth} - \text{cover} - \tfrac{1}{2}\text{bar diameter}$$
$$\therefore d_1 = 750 - 25 - 13 = 712\,mm$$
$$\text{beam breadth } b = 225\,mm$$

now the concrete moment of resistance

$$M_r = \tfrac{1}{4}p_{cb}\,bd_1^2 \text{ (para } \mathbf{4.13a})$$
$$= 2{\cdot}08 \times 225 \times 712^2$$
$$= 237\,000\,000\,Nmm$$
$$= 237\,kNm$$

$$\therefore \frac{M}{M_r} = \frac{187}{237} = 0{\cdot}79$$

$\therefore$ from para **4.13c**

$$l_a = 0{\cdot}81 \times d_1 = 0{\cdot}81 \times 712 = 580\,mm$$

$$\text{now } A_t = \frac{M}{p_{st}l_a} \text{ (para } \mathbf{4.13c})$$

$$p_{st} = 210\,N/mm^2 \text{ (table XXVI)} = 0{\cdot}210\,kN/mm^2$$
$$M = 187\,kNm = 187\,000\,kNmm$$

$$\therefore A_t = \frac{187\,000}{0{\cdot}210 \times 580} = 1535\,mm^2$$

Four Y25 (25 mm diameter high yield) bars = 1966 mm² (table XXII). Provide this reinforcement at the bottom of the beam, **45.27**.

$$\text{Shear force } Q = \frac{W}{2} = 50\,kN = 50\,000\,N$$

(from para **4.13d**)

$$\text{shear stress } q = \frac{Q}{bl_a} = \frac{50\,000}{225 \times 580} = 0{\cdot}383\,N/mm^2$$

permissible shear stress in the concrete (table XXV) = 0·77 N/mm²

As shear stress is less than permitted amount, nominal shear reinforcement only is required.

This is taken as 0·15 per cent of the horizontal section of beam that would be required to take shear force on the concrete alone, and the spacing of the links must not exceed the lever arm l_a. If b_c is the breadth of this required beam and q_p is the maximum permitted shear stress on the concrete alone, then

$$b_c = \frac{Q}{q_pl_a}$$

$$\text{now } Q = 50\,000\,N \text{ (above)}$$
$$q_p = 0{\cdot}77\,N/mm^2 \text{ (table XXV)}$$
$$l_a = 580\,mm \text{ (above)}$$
$$\therefore b_c = \frac{50\,000}{0{\cdot}77 \times 580} = 112\,mm$$

the horizontal area of 1 m run of the assumed beam is then

$$112 \times 1000 = 112\,000\,mm^2$$

the required area of reinforcement is

$$\frac{0{\cdot}15}{100} \times 112{\cdot}000 = 168\,mm^2/m$$

each link has two arms

$$\frac{168}{2} = 84\,mm^2/m$$

6 mm diameter bars at 300 mm = 94 mm²/m (table XXII) hence links at 300 mm = 188 mm²/m

The maximum permitted spacing is l_a = 580 mm.

4.16 Reinforced concrete T-beam

Example 13

Design a simply supported T-beam spanning 8·5 m; consisting of a 150 mm slab with downstand 450 × 225 mm width. The beam is to carry a total distributed load of 55 kN/m including self-weight.

Applied bending moment

$$M = \frac{WL}{8} \text{ for a simply supported beam under a distributed load (table VI)}$$

load W = 55 × 8·5 kN
span L = 8·5 m

$$\therefore M = \frac{55 \times 8{\cdot}5^2}{8} = 497\,\text{kNm}$$

from BS CP 114:1969 clause 311e the maximum breadth of a T-beam flange is the least of the following three figures:

1 span ÷ 3 = $\frac{8{\cdot}5}{3}$ = 2·8 m

2 distance between centres of beams (not applicable in this case)
3 breadth of rib + (12 × depth of flange) = 225 + (12 × 150) = 2025 mm
this last figure 2025 mm is the maximum permitted flange breadth

Assume reinforcement to consist of Y25 (25 mm diameter high yield) bars in three layers at the bottom of the rib,
then effective depth d_l = total depth − cover − bar diameter − vertical spacing − half bar diameter (see **45.28**).
from para **4.09** cover = 25 mm
from para **4.10** vertical spacing = 12 mm or nominal max size of aggregate, if larger
nominal max aggregate size = 20 mm
∴ vertical spacing = 20 mm
∴ d_l = (150 + 450) − 25 − 25 − 20 − 12 = 518 mm

Now in para **4.12** it was stated that at ultimate load the stress in the concrete would be the ultimate stress, but the depth of the neutral axis would be only that sufficient to produce a large enough stress block to generate the required ultimate moment. If the stress block remains entirely within the top flange of the T-beam, its behaviour as far as bending is concerned would be the same as that of a rectangular beam of breadth the same as the breadth of the flange. The areas of section cut away in the T-beam, being wholly in tension, make no contribution to the beam strength.
Assume this being the case, a rectangular beam with breadth = 2025 mm and d_l = 518 mm

$$\text{moment of resistance} = M_r = \tfrac{1}{4} p \frac{bd\,2}{cb\,1} \text{ (para 4.13a)}$$
$$= 2{\cdot}08 \times 2025 \times 518^2$$
$$= 1\,130\,000\,000\,\text{Nmm}$$
$$= 1130\,\text{kNm}$$
$$\therefore \frac{M}{M_r} = \frac{497}{1130} = 0{\cdot}44$$

Now, if n is the depth of the neutral axis in a rectangular beam below the top of the beam, the following formula relating n to the bending moment can be derived:

$$\frac{n}{d_1} = 1 - \sqrt{1 - \frac{3M}{4M_r}}$$

$$\text{in this case } \frac{n}{518} = 1 - \sqrt{1 - (0{\cdot}75 \times 0{\cdot}44)}$$
$$= 0{\cdot}181$$
$$\therefore n = 94\,\text{mm}$$

This means that the neutral axis is within the 150 mm depth of the flange of the beam, and justifies the assumption made above.

Now referring back to para **4.13c**

$$\frac{M}{M_r} = 0{\cdot}44$$
$$\therefore l_a = 0{\cdot}9 \times d_1$$
$$= 0{\cdot}9 \times 518 = 465\,\text{mm}$$

$$\text{now } A_t = \frac{M}{p_{st}\,l_a}$$

p_{st} = 210 N/mm² (table XXVI)
= 0·210 kN/mm²
M = 497 kNm (above)

$$\therefore A_t = \frac{497 \times 10^6}{210 \times 465} = 5090\,\text{mm}^2$$

Use eleven Y25 bars = 5406 mm² (See table XXII, **45.28a**)

$$\text{Shear force } Q = \frac{W}{2}$$

W = total load on span
= 55 × 8·5 = 468 kN
∴ Q = 234 kN = 234 000 N

then shear stress (para **4.13d**)

$$q = \frac{Q}{bl_a}$$

now in this case b = breadth of rib

= 225 mm
l_a = 465 mm

$$\therefore q = \frac{234\,000}{225 \times 465} = 2{\cdot}24\ \text{N/mm}^2$$

permitted shear stress on concrete alone (table XXV) = 0·77 N/mm²
∴ shear reinforcement is required

First some bent-up bars will be used. Bend-up two Y25 bars in single system.
shear force carried (table XXVII)

= 122 kN

shear force left to be carried by links

= Q − 122
= 234 − 122 = 112 kN
= 112 000 N

from para **4.13**

$$A_s = \frac{hQ}{p_{ss} l_a N}$$

take

h = 1 m = 1000 mm
p_{ss} = 140 N/mm² (table XXVI)
l_a = 465 mm (above)
N = 2 (each link has two arms)

$$\therefore A_s = \frac{1000 \times 112\,000}{140 \times 465 \times 2}$$
$$= 860\,\text{mm}^2/\text{m}$$

use R10 links at 75 mm

= 1048 mm²/m (table XXII, **45.28a**)

4.17 Reinforced concrete column

The principles of column design in reinforced concrete are similar to those for beams given in para **4.12**: both the steel and the concrete are assumed to have reached their full permissible stress under the working load.
Buckling, as described in para **1.10**, is rarely a factor with concrete columns. This is because the actual height/least lateral dimension of a concrete column would normally have to exceed 20 before it would be considered 'slender', and such a column would be extremely difficult to pour and compact without causing segregation in the wet mix.

Example 14

What reinforcement is required in a concrete column of section 300 mm square carrying a load of 750 kN?

The maximum permissible stress in the concrete from table XXV = 6·2 N/mm²
∴ the load carried by the concrete will be 6·2 × 300² N = 558 kN
this leaves 750 − 558 kN = 192 kN to be carried by the steel.
The maximum permissible stress in the steel, if high yield deformed bars are used is 175 N/mm².
But these bars will occupy an area of the column for which a concrete stress of 6·2 N/mm² has already been taken. Therefore, the effective stress of the steel is 175 − 6 = 169 N/mm².

$$\text{Area of steel required} = A_{sc} = \frac{192\,000}{169} = 1136\,\text{mm}^2$$

Use four Y20 bars = 1256 mm², **45.29**.
Links are needed in all columns to provide restraint for the main reinforcement which might otherwise buckle.
CP 114, cl 321b lays down that the diameter of the link bars should not be

less than ¼ the diameter of the main bars, and the pitch of the links should not be more than the least of the following:

1 the least lateral dimension of the column
2 12 × the diameter of the smallest main reinforcement
3 300 mm

Thus in this column, R6 links at 240 mm should be provided.

Note also that the percentage of main reinforcement in a column should lie between 0·8 and 8 per cent (cl 321a). In this example, the steel area should not be less than 720 mm², nor more than 7200 mm². (It is usually found very difficult physically to exceed the upper limit!)

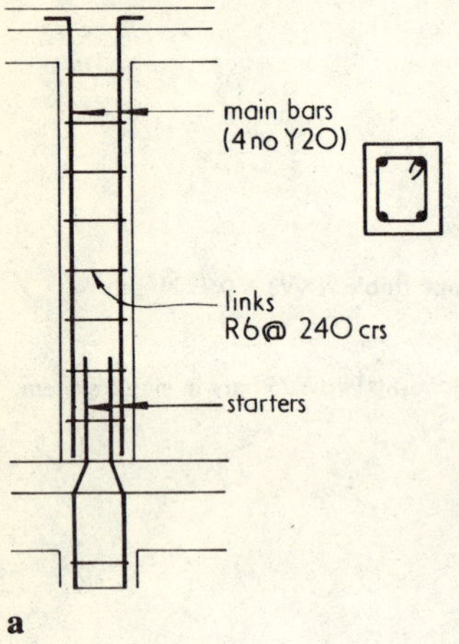

45.29 *Reinforced concrete column (Example 14)*
a *elevation showing reinforcement conventionally*
b *section*

5 STRUCTURAL STEELWORK

5.01 Grades

Steel for structural purposes is available in the United Kingdom in three grades increasing in strength: grade 43, which corresponds to the previous description of 'mild steel', grade 50 and grade 55. Design is currently carried out in conformity with BS 449:1969—The use of structural steel in building (as amended up to July 1975). A new design code based on limit state methods (see para **1.02**) is due to be published soon, but for most architectural work is not expected to supersede BS 449 for some time.

5.02 Handbooks

Books of design tables are issued by the Constructional Steel Research and Development Organisation (Constrado), which is a branch of the British Steel Corporation. These give the properties of the various sections available from the Corporation.

These books are as follows:

- *Structural Steel Handbook* (published jointly with BCSA) contains dimensional details of all BSC sections and safe load tables for beams and H stanchions of all grades.
- *Structural Steelwork Handbook—Metric angles to BS 4848: 1972* contains dimensions of all angles (also in the Handbook above) but also safe load tables for single and compound ties and struts of grades 43 and 50.
- *Structural Steelwork Handbook—Structural Hollow Sections to BS: 4848: part 2: 1975* contains dimensions of structural hollow sections (also in full Handbook above); also safe load tables for the use of these sections as struts (all grades).

For further details of available steel products, see section 39, Materials.

5.03 Beams

For bending purposes I-sections and channels are generally used, although the latter may cause problems due to torsion unless used in pairs. The Handbook mentioned above gives tables of safe load for most of these sections, but not for universal column sections used as beams. This is a pity, as these sections are valuable, especially in conversion work where the wide flanges assist in carrying brickwork, and the shallow depth helps where headroom is limited. Safe load tables for some of these sections are given in table XXX and extracts from the Handbook for commonly used sections in tables XXVIII, XXIX and XXXI.

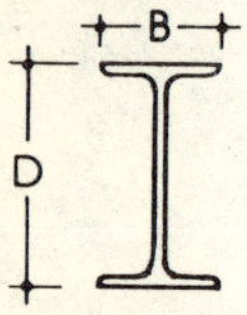

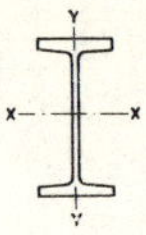

Table XXVIII Universal beams
1 Dimensions and properties

Serial size mm	Mass per metre kg	Depth of section D mm	Width of section B mm	Area of section cm²	Moment of inertia Axis x–x Gross cm⁴	Moment of inertia Axis x–x Net cm⁴	Moment of inertia Axis y–y cm⁴	Radius of gyration Axis x–x cm	Radius of gyration Axis y–y cm	Elastic modulus Axis x–x cm³	Elastic modulus Axis y–y cm³	Ratio D/T
305 × 165	54	310·9	166·8	68·4	11710	10134	1061	13·09	3·94	753·3	127·3	22·7
	46	307·1	165·7	58·9	9948	8609	897	13·00	3·90	647·9	108·3	26·0
	40	303·8	165·1	51·5	8523	7384	763	12·86	3·85	561·2	92·4	29·9
305 × 127	48	310·4	125·2	60·8	9504	8643	460	12·50	2·75	612·4	73·5	22·2
	42	306·6	124·3	53·2	8143	7409	388	12·37	2·70	531·2	62·5	25·4
	37	303·8	123·5	47·5	7162	6519	337	12·28	2·67	471·5	54·6	28·4
305 × 102	33	312·7	102·4	40·8	6487	5800	193	12·46	2·15	415·0	37·8	29·0
	28	308·9	101·9	36·3	5421	4862	157	12·22	2·08	351·0	30·8	34·8
	25	304·8	101·6	31·4	4387	3962	120	11·82	1·96	287·9	23·6	44·6
254 × 146	43	259·6	147·3	55·1	6558	5706	677	10·91	3·51	505·3	92·0	20·4
	37	256·0	146·4	47·5	5556	4834	571	10·82	3·47	434·0	78·1	23·4
	31	251·5	146·1	40·0	4439	3879	449	10·53	3·35	353·1	61·5	29·1
254 × 102	28	260·4	102·1	36·2	4008	3569	178	10·52	2·22	307·9	34·9	26·0
	25	257·0	101·9	32·2	3408	3046	148	10·29	2·14	265·2	29·0	30·8
	22	254·0	101·6	28·4	2867	2575	120	10·04	2·05	225·7	23·6	37·2
203 × 133	30	206·8	133·8	38·0	2887	2476	384	8·72	3·18	279·3	57·4	21·5
	25	203·2	133·4	32·3	2356	2027	310	8·54	3·10	231·9	46·4	26·0

Note
In calculating the net amount of inertia, each flange is reduced by one hole.

2 Safe loads for grade 43 steel

Serial size	**Mass per metre**	**Safe distributed loads in kilonewtons for spans in metres and deflection coefficients**													**Critical span L_c**
		2·00	**2·50**	**3·00**	**3·50**	**4·00**	**4·50**	**5·00**	**5·50**	**6·00**	**7·00**	**8·00**	**9·00**	**10·00**	
mm	**kg**	**112·0**	**71·68**	**49·78**	**36·57**	**28·00**	**22·12**	**17·92**	**14·81**	**12·44**	**9·143**	**7·000**	**5·531**	**4·480**	**m**
305 × 165	54	**479***	**398**	**331**	*284*	*249*	*221*	*199*	181	166	142	124	110	99	3·69
	46	**412***	**342**	**285**	**244**	*214*	*190*	*171*	155	143	122	107	95	86	3·53
	40	**370**	**296**	**247**	**212**	*185*	*165*	*148*	135	123	106	93	82	74	3·38
305 × 127	48	**404**	*323*	*269*	*231*	*202*	*180*	*162*	147	135	115	101	90	81	2·59
	42	**351**	*280*	*234*	*200*	*175*	*156*	*140*	127	117	100	88	78	70	2·45
	37	**311**	*249*	*207*	*178*	*156*	*138*	*124*	113	104	89	78	69	62	2·37
305 × 102	33	**274**	*219*	*183*	*158*	*137*	*122*	*110*	100	91	78	68	61	55	1·90
	28	**232**	*185*	*154*	*132*	*116*	*103*	*93*	84	77	66	58	51	46	1·79
	25	**190**	*152*	*127*	*109*	*95*	*84*	*76*	69	63	54	47	42	38	1·64
254 × 146	43	**333**	**267**	*222*	*191*	*167*	148	133	121	111	95	83			3·41
	37	**286**	**229**	*191*	*164*	*143*	127	115	104	95	82	72			3·22
	31	**233**	*186*	*155*	*133*	*117*	104	93	85	78	67	58			2·96
254 × 102	28	*203*	*163*	*135*	*116*	*102*	90	81	74	68	58	51			2·01
	25	*175*	*140*	*117*	*100*	*88*	78	70	64	58	50	37			1·87
	22	*149*	*119*	*99*	*85*	*74*	66	60	54	50	43	37			1·75
203 × 133	30	**184**	*147*	*123*	*105*	92	82	74	67	61	53				3·03
	25	*153*	*122*	*102*	87	77	68	61	56	51					2·80

Notes
Loads printed in *italic* type do not cause overloading of the unstiffened web, and do not cause deflection exceeding span/360.
Loads printed in ordinary type should be checked for deflection, see example 16.
* Load is based on allowable shear of web and is less than allowable load in bending.

3 Safe loads for grade 50 steel

Serial size	**Mass per metre**	**Safe distributed loads in kilonewtons for spans in metres and deflection coefficients**													**Critical span L_c**
		2·00	**2·50**	**3·00**	**3·50**	**4·00**	**4·50**	**5·00**	**5·50**	**6·00**	**7·00**	**8·00**	**9·00**	**10·00**	
mm	**kg**	**112·0**	**71·68**	**49·78**	**36·57**	**28·00**	**22·12**	**17·92**	**14·81**	**12·44**	**9·143**	**7·000**	**5·531**	**4·480**	**m**
305 × 165	54	**670***	**554**	**462**	*396*	347	308	277	252	231	198				3·27
	46	**576***	**477**	**397**	**341**	298	265	238	217	199	170				3·14
	40	**516**	**413**	**344**	**295**	258	229	207	188	172	148				3·03
305 × 127	48	**563**	*451*	*376*	*322*	282	250	225	205	188	161				2·29
	42	**489**	*391*	*326*	*279*	244	217	195	178	163	140				2·10
	37	**434**	*347*	*289*	*248*	217	193	174	158	145	124				2·12
305 × 102	33	**382**	**305**	*255*	*218*	191	170	153	139	127	109				1·70
	28	**323**	**258**	*215*	*185*	161	144	129	117	108	92				1·60
	25	**265**	*212*	*177*	*151*	132	118	106	96	88	76				1·48
254 × 146	43	**465**	**372**	*310*	266	232	207	186	169	155					3·00
	37	**399**	**319**	*266*	228	200	177	160	145	133					2·86
	31	**325**	**260**	*217*	186	162	144	130	118	108					2·64
254 × 102	28	**283**	*227*	*189*	162	142	126	113	103	94					1·79
	25	*244*	*195*	*163*	139	122	108	98	89	81					1·68
	22	*208*	*166*	*138*	119	104	92	83	76	69					1·57
203 × 133	30	**257**	*206*	171	147	128	114	103							2·68
	25	*213*	171	142	122	107	95								2·50

See notes to section 2 of this table.

Example 15
A wall carries a load of 24 kN/m run, mostly dead load. An opening 3·2 m wide is to be cut in it. What size beam should be used to carry the load:
(a) in grade 43 steel, minimum weight
(b) in grade 43 steel, minimum depth
(c) in grade 50 steel, minimum weight
(d) in grade 50 steel, minimum depth?
The span = opening + a bearing length (say 300 mm)
ie span = 3·2 + 0·3 = 3·5 m
Total load on span = 3·5 × 24 = 84 kN
Refer to tables:
Grade 43, UB 203 × 135 @ 25 over a span of 3·5 m carries 87 kN (a)
Grade 50, RSJ 178 × 102 @ 21·54 over a span 3·5 m carries 90 kN (c)
Grade 43, UC 152 × 152 @ 37 over a span of 3·5 m carries 103 kN (b)
Grade 50, UC 152 × 152 @ 23 over a span of 3·5 m carries 87 kN (d)

Example 16
If in the example above the load contained a live load of 20 kN/m, what beam sizes would then be required?
The significance of live load is in its effect on deflections.
At the head of each column in tables XXVIII to XXXI will be seen a 'deflection coefficient'. The formula for using this is:

$$W_L \text{ (live load in kN)} = \text{coefficient} \times \frac{I}{1000}$$

where I is in cm^4 units as tabulated.

In this case $20 \times 3{\cdot}5 = 36{\cdot}57 \times \frac{I}{1000}$

whence $I = 1{\cdot}914 \times 1000\,cm^4 = 1914\,cm^4$

This will give the minimum value of I that will deflect less than 1/360 of the span as required by BS 449. It can be seen that there would be no advantage in this case of using any higher grade of steel than 43.
RSJ 203 × 102 @ 25·33: I = 2294 cm^4
UB 203 × 133 @ 25 I = 2348 cm^4 (a)
UC 152 × 152 @ 37 I = 2218 cm^4 (b)

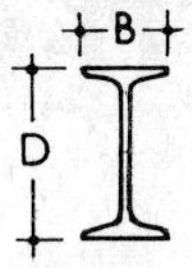

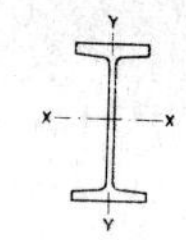

Table XXIX Joists
1 Dimensions and properties

Nominal size	Mass per metre	Depth of section D	Width of section B	Area of section	Moment of inertia Axis x–x Gross	Moment of inertia Axis x–x Net	Moment of inertia Axis y–y	Radius of gyration Axis x–x	Radius of gyration Axis y–y	Elastic modulus Axis x–x	Elastic modulus Axis y–y	Ratio D/T
mm	kg	mm	mm	cm²	cm⁴	cm⁴	cm⁴	cm	cm	cm³	cm³	
254 × 203	*81·85*	254·0	203·2	104·4	12016	10527	2278	10·7	4·67	946·2	224·3	12·8
254 × 114	*37·20*	254·0	114·3	47·4	5092	4243	270·1	10·4	2·39	401·0	47·19	19·8
203 × 152	*52·09*	203·2	152·4	66·4	4789	4177	813·3	8·48	3·51	471·3	106·7	12·3
203 × 102	25·33	203·2	101·6	32·3	2294	2024	162·6	8·43	2·25	225·8	32·02	19·5
178 × 102	21·54	177·8	101·6	27·4	1519	1339	139·2	7·44	2·25	170·9	27·41	19·8
152 × 127	*37·20*	152·4	127·0	47·5	1818	1627	378·8	6·20	2·82	238·6	59·65	11·5
152 × 89	17·09	152·4	88·9	21·8	881·1	762·6	85·98	6·36	1·99	115·6	19·34	18·4
152 × 76	*17·86*	152·4	76·2	22·8	873·7	736·2	60·77	6·20	1·63	114·7	15·90	15·9
127 × 114	*29·76*	127·0	114·3	37·3	979·0	866·9	241·9	5·12	2·55	154·2	42·32	11·0
127 × 114	*26·79*	127·0	114·3	34·1	944·8	834·6	235·4	5·26	2·63	148·8	41·19	11·1
127 × 76	*16·37*	127·0	76·2	21·0	569·4	476·1	60·35	5·21	1·70	89·64	15·90	13·2
127 × 76	13·36	127·0	76·2	17·0	475·9	400·0	50·18	5·29	1·72	74·94	13·17	16·7
114 × 114	*26·79*	114·3	114·3	34·4	735·4	651·2	223·1	4·62	2·54	128·6	39·00	10·7
102 × 102	*23·07*	101·6	101·6	29·4	486·1	425·1	154·4	4·06	2·29	95·70	30·32	9·9
102 × 64	9·65	101·6	63·5	12·3	217·6	182·2	25·30	4·21	1·43	42·84	7·97	15·4
102 × 44	*7·44*	101·6	44·4	9·5	152·3	126·9	7·91	4·01	0·91	29·99	3·44	16·7
89 × 89	*19·35*	88·9	88·9	24·9	306·7	263·7	101·1	3·51	2·01	68·99	22·78	9·0
76 × 76	*14·67*	76·2	80·0	19·1	171·9	144·1	60·77	3·00	1·78	45·06	15·24	9·1
76 × 76	*12·65*	76·2	76·2	16·3	158·6	130·7	52·03	3·12	1·78	41·62	13·60	9·1

Note
In calculating the net moment of inertia, one hole is deducted from each flange.

2 Safe loads for grade 43 steel

Nominal size	Mass per metre	Safe distributed loads in kilonewtons for spans in metres and deflection coefficients 1·00	1·25	1·50	1·75	2·00	2·25	2·50	2·75	3·00	3·25	3·50	4·00	4·25	Critical span Lc
mm	kg	**448**	**287**	**199**	**146**	**112**	**88·5**	**71·7**	**59·2**	**49·8**	**42·4**	**36·6**	**28·0**	**24·8**	m
254 × 203	*81·85*						**518***	**500**	**454**	**416**	**304**	*357*	*312*	*294*	5·80
254 × 114	*37·20*		**386***	**353**	**302**	**265**	*235*	*212*	*192*	*176*	*163*	*151*	*132*	*125*	2·35
203 × 152	*52·09*			**362***	**356**	**311**	**277**	*249*	*226*	*207*	*191*	178	156	146	4·47
203 × 102	25·33		**236***	**199**	**170**	*149*	*132*	*119*	*108*	*99*	*92*	85	75	70	2·22
178 × 102	21·54	**188***	**180**	**150**	**129**	*113*	*100*	*90*	*82*	*75*	69	64	56	53	2·21
152 × 127	*37·20*	**315**	**252**	*210*	*180*	*158*	*140*	*126*	115	105	97	90	79	74	3·79
152 × 89	17·09	**149***	**122**	*102*	*87*	*76*	*68*	*61*	56	51	47	44	38	36	2·01
152 × 76	*17·86*	**151**	*121*	*101*	*86*	*76*	*67*	*61*	55	50	47	43	38	36	1·77
127 × 114	*29·76*	**204**	*163*	*136*	*116*	*102*	90	81	74	68	63	58	51	48	3·55
127 × 114	*26·79*	**188***	**157**	*131*	*112*	*98*	87	79	71	65	60	56	49	46	3·61
127 × 76	*16·37*	**118**	*95*	*79*	*68*	*59*	53	47	43	39	36	34	30	28	2·06
127 × 76	13·36	**99**	*79*	*66*	*57*	*49*	44	40	36	33	30	28	25	23	1·82
114 × 114	*26·79*	**170**	*136*	*113*	97	85	75	68	62	57	52	49	42	40	3·63
102 × 102	*23·07*	*126*	*101*	*84*	72	63	56	51	46	42	39	36	32	30	3·48
102 × 64	9·65	*57*	*45*	*38*	32	28	25	23	21	19	17	16	14	13	1·58
102 × 44	*7·44*	*40*	*32*	*26*	23	20	18	16	14	13	12	11	10	9	0·95
89 × 89	*19·35*	*91*	*73*	*61*	52	46	41	36	33	30	28	26	23	21	3·33
76 × 76	*14·67*	*60*	*48*	40	34	30	26	24	22	20	18	17	15	14	2·91
76 × 76	*12·65*	*55*	*44*	37	31	27	24	22	20	18	17	16	14	13	2·95

Notes
Sections with mass shown in italics are, although frequently rolled, not in BS 4. Availability should be checked with BSC Sections Product Unit. Flanges of BS 4 joists have a 5° taper; all others taper at 8°.
The loads listed are based on bending stresses of $165\,N/mm^2$ in section 2 and $230\,N/mm^2$ in section 3 and assume adequate lateral support. Without such support the span must not exceed L_c unless the compressive stress is reduced in accordance with clause 19a of BS 449.
Loads printed in bold type may cause overloading of the unstiffened web, the capacity of which should be checked.
Loads printed in italic type do not cause overloading of the unstiffened web, and do not cause deflection exceeding span/360.
Loads printed in ordinary type should be checked for deflection (see example 16).
* Load is based on allowable shear of web and is less than allowable load in bending.

3 Safe loads for grade 50 steel

Nominal size	Mass per metre	Safe distributed loads in kilonewtons for spans in metres and deflection coefficients													Critical span L_c
		1·00	1·25	1·50	1·75	2·00	2·25	2·50	2·75	3·00	3·25	3·50	4·00	4·25	
mm	kg	**448**	**287**	**199**	**146**	**112**	**88·5**	**71·7**	**59·2**	**49·8**	**42·4**	**36·6**	**28·0**	**24·8**	m
254 × 203	*81·85*						**725***	**696**	**633**	**580**	**536**	497	435	410	4·94
254 × 114	*37·20*		**541***	**492**	**422**	**369**	*328*	*295*	*268*	*246*	227	211	184	174	2·06
203 × 152	*52·09*			**506***	**496**	**434**	**385**	347	315	289	267	248	217	204	3·80
203 × 102	25·33		**330***	**277**	**237**	*208*	*185*	166	151	138	128	119	104	98	1·95
178 × 102	21·54	**264***	**252**	**210**	**180**	*157*	140	126	114	105	97	90	79	74	1·94
152 × 127	*37·20*	**439**	**351**	*293*	*251*	220	195	176	160	146	135	125	110	103	3·21
152 × 89	17·09	**209***	**170**	*142*	*122*	106	95	85	77	71	65	61	53	50	1·76
152 × 76	*17·86*	**211**	*169*	*141*	*121*	*105*	94	84	77	70	65	60	53	50	1·53
127 × 114	*29·76*	**284**	*227*	*189*	162	142	126	113	103	95	87	81	71	67	2·99
127 × 114	*26·79*	**263***	**219**	*183*	156	137	122	110	100	91	84	78	68	64	3·05
127 × 76	*16·37*	**165**	*132*	*110*	94	82	73	66	60	55	51	47	41	39	1·76
127 × 76	13·36	**138**	*110*	*92*	79	69	61	55	50	46	42	39	34	32	1·58
114 × 114	*26·79*	**237**	*189*	158	135	118	105	95	86	79	73	68	59	56	3·06
102 × 102	*23·07*	*176*	141	117	101	88	78	70	64	59	54	50	44	41	2·92
102 × 64	9·65	*79*	63	53	45	39	35	32	29	26	24	23	20	19	1·36
102 × 44	*7·44*	*55*	44	37	32	28	25	22	20	18	17	16	14	13	0·83
89 × 89	*19·35*	*127*	102	85	73	64	56	51	46	42	39	36	32	30	2·79
76 × 76	*14·67*	83	66	55	47	42	37	33	30	28	26	24	21	20	2·43
76 × 76	*12·65*	77	61	51	44	38	34	31	28	26	24	22	19	18	2·47

See notes to section 2 of this table.

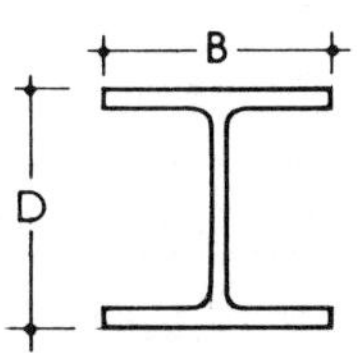

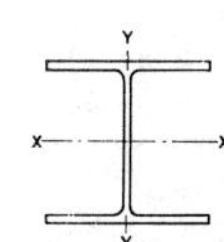

Table XXX Universal columns used as beams
1 Dimensions and properties

Serial size	Mass per metre	Depth of section	Width of section	Area of section	Moment of inertia			Radius of gyration		Elastic modulus		
					Axis x–x		Axis	Axis	Axis	Axis	Axis	Ratio
		D	B		Gross	Net	y–y	x–x	y–y	x–x	y–y	D/T
mm	kg	mm	mm	cm^2	cm^4	cm^4	cm^4	cm	cm	cm^3	cm^3	
254 × 254	167	289·1	264·5	212·4	29914	27171	9796	11·9	6·79	2070	740·6	9·1
	132	276·4	261·0	167·7	22416	20350	7444	11·6	6·66	1622	570·4	11·0
	107	266·7	258·3	136·6	17510	15889	5901	11·3	6·57	1313	456·9	13·0
	89	260·4	255·9	114·0	14307	12973	4849	11·2	6·52	1099	378·9	15·0
	73	254·0	254·0	92·9	11360	10299	3873	11·1	6·46	894·5	305·0	17·9
203 × 203	86	222·3	208·8	110·1	9462	8373	3119	9·27	5·32	851·5	298·7	10·8
	71	215·9	206·2	91·1	7647	6756	2536	9·16	5·28	708·4	246·0	12·4
	60	209·6	205·2	75·8	6088	5383	2041	8·96	5·19	581·1	199·0	14·8
	52	206·2	203·9	66·4	5263	4651	1770	8·90	5·16	510·4	173·6	16·5
	46	203·2	203·2	58·8	4564	4035	1539	8·81	5·11	449·2	151·5	18·5
152 × 152	37	161·8	154·4	47·4	2218	1931	709	6·84	3·87	274·2	91·78	14·0
	30	157·5	152·9	38·2	1742	1516	558	6·75	3·82	221·2	73·06	16·8
	23	152·4	152·4	29·8	1263	1104	403	6·51	3·68	165·7	52·95	22·3

In calculating the moment of inertia, each flange is reduced by one hole.

2 Safe loads for grade 43 steel

Serial size	Mass per metre	Safe distributed loads in kilonewtons for spans in metres and deflection coefficients														Critical span L_c
		1·5	2·0	2·5	3·0	3·5	4·0	4·5	5·0	5·5	6·0	6·5	7·0	7·5	8·0	
m	kg	**199·1**	**112·0**	**71·68**	**49·78**	**36·57**	**28·00**	**22·12**	**17·92**	**14·81**	**12·44**	**10·60**	**9·143**	**7·964**	**7·000**	m
254 × 254	167		**1554***	**1524**	**1270**	*1088*	952	846	762	692	635	586	544	508	476	11·094
	132		**1208***	**1193**	**994**	*852*	746	663	597	542	497	459	426	398	373	9·265
	107		**970***	**966**	**805**	*690*	604	537	483	439	402	372	345	322	302	8·068
	89			**706***	**674**	**578**	505	449	404	368	337	311	289	270	253	7·280
	73			**612***	**530**	**454**	397	353	318	289	265	244	227	212	199	6·611

2 Safe loads for grade 43 steel (*continued*)

Serial size	Mass per metre	Safe distributed loads in kilonewtons for spans in metres and deflection coefficients														Critical span
		1·5	2·0	2·5	3·0	3·5	4·0	4·5	5·0	5·5	6·0	6·5	7·0	7·5	8·0	L_c
m	kg	199·1	112·0	71·68	49·78	36·57	28·00	22·12	17·92	14·81	12·44	10·60	9·143	7·964	7·000	m
203 × 203	86	**810***	**783**	**627**	522	448	392	348	313	285	261	241	224	209	196	7·511
	71		**622***	**521**	434	372	325	289	260	237	217	200	186	174	163	6·552
	60	**546***	**534**	**427**	356	305	267	237	213	194	178	164	152	142	133	5·864
	52		**462***	**375**	313	268	235	209	188	171	156	144	134	125	117	5·491
	46	**416***	**413**	**330**	275	236	206	183	165	150	138	127	118	110	103	5·154
152 × 152	37	**336**	252	201	168	144	126	112	101	92	84	77	72	67	63	4·504
	30	**270**	202	162	135	116	101	90	81	74	67	62	58	54	51	4·028
	23	**202**	152	121	101	87	76	67	61	55	51	47	43	40	38	3·462

Notes
The loads listed are based on bending stresses of 165 N/mm² in section 2 and 230 N/mm² in section 3 and assume adequate lateral support. Without such support the span must not exceed L_c unless the compressive stress is reduced in accordance with clause 19a of BS 449.
Loads printed in **bold** type may cause overloading of the unstiffened web, the capacity of which should be checked.
Loads printed in *italic* type do not cause overloading of the unstiffened web, and do not cause deflection exceeding span/360.
Loads printed in ordinary type should be checked for deflection. See example 16.
* Load is based on allowable shear of web and is less than allowable load in bending.

3 Safe loads for grade 50 steel

Serial	Mass per metre	Safe distributed loads in kilonewtons for spans in metres and deflection coefficients														Critical span
		1·5	2·0	2·5	3·0	3·5	4·0	4·5	5·0	5·5	6·0	6·5	7·0	7·5	8·0	L_c
mm	kg	199·1	112·0	71·68	49·78	36·57	28·00	22·12	17·92	14·81	12·14	10·60	9·143	7·964	7·000	m
254 × 254	167		**1110***	**1093**	**911**	*781*	*683*	*607*	546	497	455	420	390	364	342	9·277
	132		**862***	**856**	**714**	*612*	*535*	*476*	428	389	359	329	306	285	268	7·822
	107		**694***	**693**	**578**	*495*	*433*	*385*	347	315	289	267	248	231	217	6·886
	89			**546***	**484**	**414**	*363*	322	290	264	242	223	207	193	181	6·283
	73			**436***	**394**	**337**	*295*	262	236	215	197	182	169	157	148	5·773
203 × 203	86	**578***	**562**	**450**	*375*	*321*	281	250	225	204	187	173	161	150	140	6·334
	71		**444***	**374**	*312*	*267*	234	208	187	170	156	144	134	125	117	5·673
	60	**390***	**384**	**307**	*256*	*219*	192	170	153	139	126	118	110	102	96	5·052
	52		**330**	**269**	*225*	*192*	168	150	135	122	112	104	96	90	84	4·768
	46		**296**	**237**	*197*	169	148	132	119	108	99	91	85	79	74	4·511
152 × 152	37	**241**	*181*	*145*	121	103	90	80	72	66	60	56	52	48	45	3·867
	30	**195**	*146*	*117*	97	83	73	65	58	53	49	45	42	39	36	3·502
	23	**146**	*109*	*87*	73	62	55	49	44	40	36	34	31	29	27	3·063

See notes to section 2 of this table.

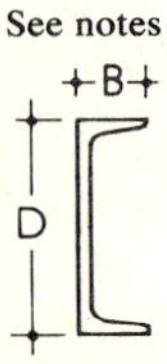

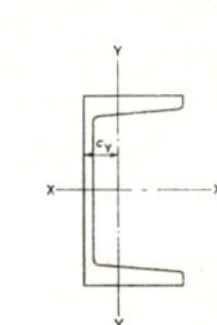

Table XXXI Channels
1 Dimensions and properties

Nominal size	Mass per metre	Depth of section D	Width of section B	Area of section	Dimension Cy	Moment of inertia Axis x–x Gross	Moment of inertia Axis x–x Net	Moment of inertia Axis y–y	Radius of gyration Axis x–x	Radius of gyration Axis y–y	Elastic modulus Axis x–x	Elastic modulus Axis y–y	Ratio D/T
mm	kg	mm	mm	cm²	cm	cm⁴	cm⁴	cm⁴	cm	cm	cm³	cm³	
305 × 102	46·18	304·8	101·6	58·83	2·66	8214	6583	499·5	11·8	2·91	539·0	66·59	20·5
305 × 89	41·69	304·8	88·9	53·11	2·18	7061	5826	325·4	11·5	2·48	463·3	48·49	22·3
254 × 89	35·74	254·0	88·9	45·52	2·42	4448	3612	302·4	9·88	2·58	350·2	46·70	18·7
254 × 76	28·29	254·0	76·2	36·03	1·86	3367	2669	162·6	9·67	2·12	265·1	28·21	23·2
229 × 89	32·76	228·6	88·9	41·73	2·53	3387	2732	285·0	9·01	2·61	296·4	44·82	17·2
229 × 76	26·06	228·6	76·2	33·20	2·00	2610	2041	158·7	8·87	2·19	228·3	28·22	20·5
203 × 89	29·78	203·2	88·9	37·94	2·65	2491	1996	164·4	8·10	2·64	245·2	42·34	15·8
203 × 76	23·82	203·2	76·2	30·34	2·13	1950	1508	151·3	8·02	2·23	192·0	27·59	18·2
178 × 89	26·81	177·8	88·9	34·15	2·76	1753	1397	241·0	7·16	2·66	197·2	39·29	14·5
178 × 76	20·84	177·8	76·2	26·54	2·20	1337	1028	134·0	7·10	2·25	150·4	24·72	17·3
152 × 89	23·84	152·4	88·9	30·36	2·86	1166	924·1	215·1	6·20	2·66	153·0	35·70	13·2
152 × 76	17·88	152·4	76·2	22·77	2·21	851·5	653·6	113·8	6·12	2·24	111·8	21·05	16·9
127 × 64	14·90	127·0	63·5	18·98	1·94	482·5	367·1	67·23	5·04	1·88	75·99	15·25	13·8
102 × 51	10·42	101·6	50·8	13·28	1·51	207·7	167·8	29·10	3·95	1·48	40·89	8·16	13·3
76 × 38	6·70	76·2	38·1	8·53	1·19	74·14	54·48	10·66	2·95	1·12	19·46	4·07	11·2

Note In calculating the moment of inertia, each flange is reduced by one hole.

2 Safe loads for grade 43 steel

Nominal size	Mass per metre	Safe distributed loads in kilonewtons for spans in metres and deflection coefficients													Critical span L_c
		1·50	**2·00**	**2·50**	**3·00**	**3·50**	**4·00**	**4·50**	**5·00**	**5·50**	**6·00**	**7·00**	**8·00**	**9·00**	
mm	kg	**199·1**	**112·0**	**71·68**	**49·78**	**36·57**	**28·00**	**22·12**	**17·92**	**14·81**	**12·44**	**9·143**	**7·000**	**5·531**	m
305 × 102	46·18	*474*	*356*	*285*	*237*	*203*	*178*	*158*	*142*	129	119	102	89	79	3·077
304 × 89	41·69	*408*	*306*	*245*	*204*	*175*	*153*	*136*	*122*	111	102	87	76	68	2·537
254 × 89	35·74	*308*	*231*	*185*	*154*	*132*	*116*	103	92	84	77	66	58		2·842
254 × 76	28·29	*233*	*175*	*140*	*117*	*100*	*87*	78	70	64	58	50	44		2·142
229 × 89	32.76	*261*	*196*	*156*	*130*	*112*	98	87	78	71	65	56			3·005
229 × 76	26·06	*201*	*151*	*121*	*100*	86	75	67	60	55	50	43			2·317
203 × 89	29·78	*216*	*162*	*129*	*108*	92	81	72	65	59	54				3·190
203 × 76	23·82	*169*	*127*	*101*	*84*	72	63	56	51	46	42				2·497
178 × 89	26·81	*174*	*130*	*104*	*87*	74	65	58	52	47	43				3·388
178 × 76	20·84	*132*	*99*	*79*	*66*	57	50	44	40	36	33				2·577
152 × 89	23·84	*135*	*101*	*81*	67	58	50	45	40						3·629
152 × 76	17·88	*98*	*74*	*59*	49	42	37	33	30						2·591
127 × 64	14·90	*67*	*50*	40	33	29	25								2·477
102 × 51	10·42	*36*	*27*	22	18										1·993
76 × 38	6·70	17	13	10											1·724

Notes
The loads listed are based on bending stresses of 165 N/mm² in section 2 and 230 N/mm² in section 3 and assume adequate lateral support. Without such support the span must not exceed L_c unless the compressive stress is reduced in accordance with clause 19 *a* (ii) of BS 449.
Loads printed in bold type may cause overloading of the unstiffened web, the capacity of which should be checked.
Loads printed in italic type do not cause overloading of the unstiffened web, and do not cause deflection exceeding span/360.
Loads printed in ordinary type should be checked for deflection. See example 16.

3 Safe loads for grade 50 steel

Nominal size	Mass per metre	Safe distributed loads in kilonewtons for spans in metres and deflection coefficients											Critical span L_c
		1·50	**2·00**	**2·50**	**3·00**	**3·50**	**4·00**	**4·50**	**5·00**	**5·50**	**6·00**	**7·00**	
mm	kg	**199·1**	**112·0**	**71·68**	**49·78**	**36·57**	**28·00**	**22·12**	**17·92**	**14·81**	**12·44**	**9·143**	m
305 × 102	46·18	*661*	*496*	*397*	*331*	*283*	248	220	198	180	165	142	2·557
305 × 89	41·69	*568*	*426*	*341*	*284*	*244*	213	189	170	155	142	122	2·123
254 × 89	35·74	*430*	*322*	*258*	*215*	184	161	143	129	117	107		2·340
254 × 76	28·29	*325*	*244*	*195*	*163*	139	122	108	98	89	81		1·799
229 × 89	32·76	*364*	*273*	*218*	182	156	136	121	109	99			2·454
229 × 76	26·06	*280*	*210*	*168*	140	120	105	93	84	76			1·924
203 × 89	29·78	*301*	*226*	180	150	129	113	100					2·581
203 × 76	23·82	*235*	*177*	141	118	101	88	78					2·050
178 × 89	26·81	*242*	*181*	145	121	104	91						2·717
178 × 76	20·84	*185*	*138*	111	92	79	69						2·105
152 × 89	23·84	*188*	141	113	94	80							2·881
152 × 76	17·88	*137*	103	82	69	59							2·113
127 × 64	14·90	*93*	70	56	47								1·977
102 × 51	10·42	50	38										1·585
76 × 38	6·70	24											1·350

See notes to section 2 of this table.

6 FOUNDATIONS

6.01 Nature

The purpose of a foundation is to transmit the loads of and contained by a building structure to the ground. The nature of the foundation will depend on:

- the characteristics of the soil
- the magnitude of the loads of the structure
- the nature of the loads of the structure.

In the majority of buildings, the loads transmitted to the ground will arrive either as point loads down columns, or line loads down walls. For the type of building with which these notes deal, the magnitudes of these loads will not be so great as to significantly effect the choice of foundation system.

6.02 Soil

This will basically depend on the strength of the soil to carry the load. The term 'soil' in this context means not vegetable material suitable for growing crops, but the material forming the surface of the earth to a depth of about 100 m, which is not so hard in nature as to be classified as a 'rock'.
The technology of the physical properties of soil is called *soil mechanics*. It is not appropriate to deal in depth with this subject, but some simple principles are necessary to understand the design of foundations.

6.03 Bearing pressure

The bearing pressure that can be carried by the soil is the *additional* load that can be carried on unit area. A soil stratum

Table XXXII Safe bearing capacities of soils from British Standard Code of Practice CP2004: 1972, table 1

Group	Class	Types of rocks and soils	Presumed bearing value kN/m²	Remarks
I Rocks	1	Hard igneous and gneissic rocks in sound condition	10 000	These values are based on the assumption that the foundations are carried down to unweathered rock
	2	Hard limestones and hard sandstones	4 000	
	3	Schists and slates	3 000	
	4	Hard shales, hard mudstones and soft sandstones	2 000	
	5	Soft shales and soft mudstones	600 to 1 000	
	6	Hard sound chalk, soft limestone	600	
	7	Thinly bedded limestones, sandstones, shales	To be assessed after inspection	
	8	Heavily shattered rocks		
II Non-cohesive soils	9	Compact gravel, or compact sand and gravel	>600	Width of foundation (*B*) not less than 1 m (3 ft). Ground-water level assumed to be a depth not less than *B* below the base of the foundation
	10	Medium dense gravel, or medium dense sand and gravel	200 to 600	
	11	Loose gravel, or loose sand and gravel	<200	
	12	Compact sand	>300	
	13	Medium dense sand	100 to 300	
	14	Loose sand	<100	
III Cohesive soils	15	Very stiff boulder clays and hard clays	300 to 600	Group III is susceptible to long-term consolidation settlement
	16	Stiff clays	150 to 300	
	17	Firm clays	75 to 150	
	18	Soft clays and silts	<75	
	19	Very soft clays and silts	Not applicable	
IV	20	Peat and organic soils	Refer to engineer	
V	21	Made ground or fill	Refer to engineer	

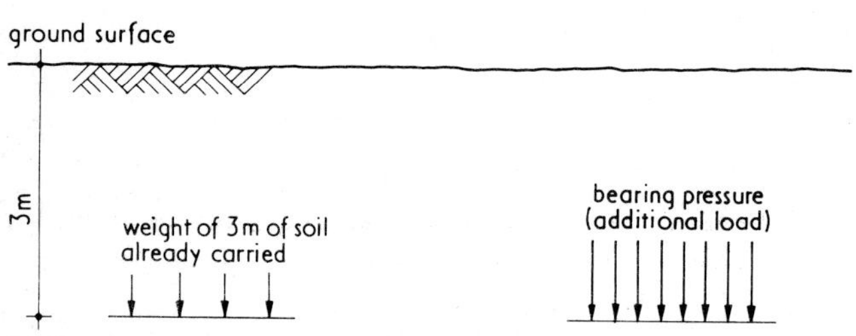

45.30 *Bearing pressure on formation level of soil*

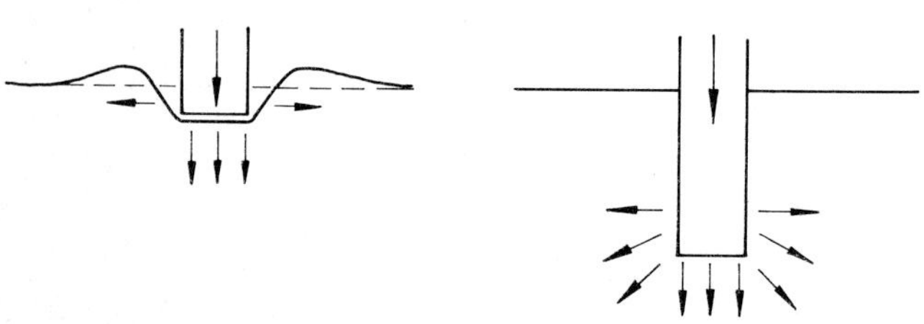

45.31 *How the soil increases in strength with depth:*
a *shallow foundation can cause heave*
b *deeper foundation is restrained*

at a depth of, say, three metres is already carrying the weight of that three metres of soil **45.30**. In fact, the bearing pressure of many soils increase substantially with depth. This is because a common mode of failure under excessive load is sideways spillage of the soil, often accompanied by upward heave of the material around the area of application. Obviously this is much less likely where the load is carried at some depth, **45.31**. Since the bearing capacity represents the additional load the soil can carry, the greater the depth, the smaller proportion of the total pressure (or gross pressure) this will form. In fact, it is even possible to produce zero or negative net pressure by removing the overburden, and replacing it with something weighing much less, for example, a hollow box. This is the principle by which loads can be carried on soft marshy soil; in fact, the extreme case is a boat! In many cases, the architect will be told what the bearing capacity of the soil is at normal foundation depth—about 1 m. Table XXXII gives figures for common soils, but should be used with caution.

6.04 Pad and strip foundations

The design of simple pad and strip foundations is best shown from example.

Example 17

A brick wall forming the outside of a house carries a load of 85 kN/m. What width of foundation will be required at a depth of 1·5 m, given a bearing capacity of the soil at that depth of 60 kN/m²? Ignore the weight of the foundation itself.

If B = the width of the foundation in metres,

The pressure transmitted to the soil will be $\frac{85}{B}$ kN/m²

Put this equal to the capacity to obtain the minimum allowable value of B:

$$\frac{85}{B} = 60 \qquad \text{hence } B = 1{\cdot}417\,\text{m}$$

The practical width of foundation therefore 1·5 m **45.32**.

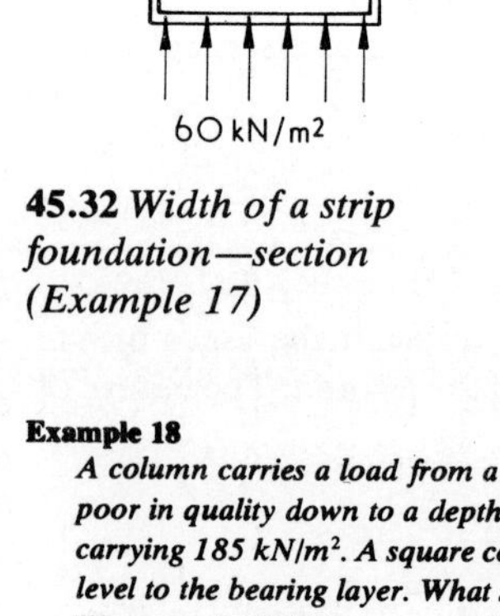

45.32 *Width of a strip foundation—section (Example 17)*

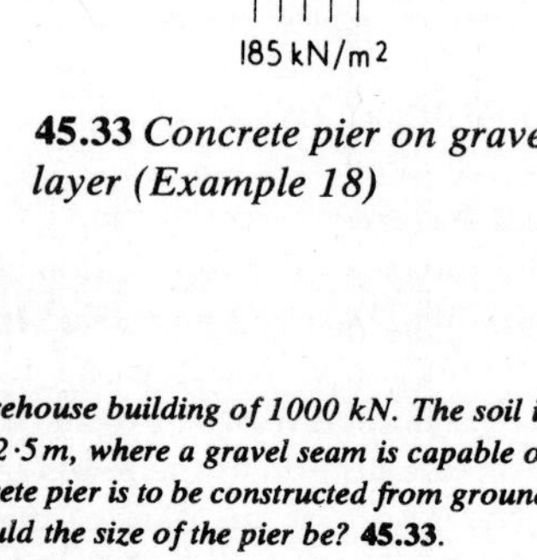

45.33 *Concrete pier on gravel layer (Example 18)*

Example 18

A column carries a load from a warehouse building of 1000 kN. The soil is poor in quality down to a depth of 2·5 m, where a gravel seam is capable of carrying 185 kN/m². A square concrete pier is to be constructed from ground level to the bearing layer. What should the size of the pier be? **45.33**.

The gravel will have to carry the weight of the concrete pier as well as the column load. The load imposed by any material will be

density × height

The density of concrete is assumed to be 24 kN/m³ (table II). The pressure at the gravel due to this concrete will therefore be

$$24 \times 2{\cdot}5 = 60\ \text{kN/m}^2$$

However, the same depth of soil will have been removed to construct the pier. The pressure this exerted on the gravel layer was $2{\cdot}5 \times 16$ kN/m^2 (generally assumed density of soil)

$= 40$ kN/m^2.

Consequently the pressure capacity available to carry the load of the column will be:

bearing capacity	+	soil pressure	−	pier weight
185	+	40	−	60
				= 165 kN/m^2

The required pier area is hence $\frac{1000}{165} = 6{\cdot}06\,\text{m}^2$

The side of the pier will be $\sqrt{6{\cdot}06} = 2{\cdot}46$ m say $2{\cdot}5$ m

6.05 Other foundation types

It is frequently found that the loads of the building are so large, or the bearing capacity of the soil is so poor that suitable pad or strip foundations will be either very deep, or required to be so large that adjoining bases impinge on one another. The two common solutions to this problem are

- raft foundations **45.34**, where the bases are combined together to form one large base. This has to be so reinforced as to allow for the stresses induced by inequalities of loading and bearing capacity; or
- piles **45.35**, which are devices for carrying loads down to deeper levels than would otherwise be practical.

Raft foundations are beyond the scope of this section, but some knowledge of piles will be useful.

6.06 End bearing piles

Piles can be divided into those that carry their loads into the soil mainly by the pressure of their lower ends **45.36a**, and those that act by virtue of the friction between the soil and their shaft lengths, **45.36b**. End-bearing piles normally sit on rock or gravel strata with high bearing capacities. They may consist of a precast concrete shaft, driven into place with a large mechanical hammer. Alternatively a hollow shell is driven and afterwards filled with wet concrete to form the pile. In either of these cases, the amount of penetration achieved at each hammer blow is an indication of the carrying capacity of the pile. A design method for simple end-bearing piles in the shape of the short-bored variety sometimes used in housing is illustrated in the following example:

Example 19

The brick wall in example 17 above is to be carried on short-bored piles taken down to a gravel seam 4 m deep. The gravel has a capacity of 400 kN/m², and the piles are 650 mm diameter. What is the required spacing of the piles? **45.37**

The area of the pile base is $\pi \times \left(\frac{0{\cdot}65}{2}\right)^2 \text{m}^2 = 0{\cdot}332\,\text{m}^2$

The load that can be carried by each is therefore

$400 \times 0{\cdot}332 = 133$ kN

The length of wall loaded at 85 kN/m carried on one pile is hence

$\frac{133}{85} = 1{\cdot}56$ m

Therefore, piles should be at, say, $1{\cdot}5$ m centres.

Minimum permissible spacing is $3 \times$ diameter or $1{\cdot}35$ m in this case.

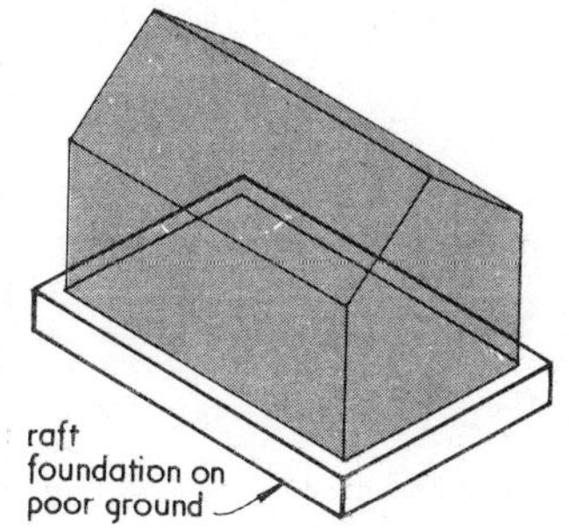

45.34 *Raft foundation as used on poor ground*

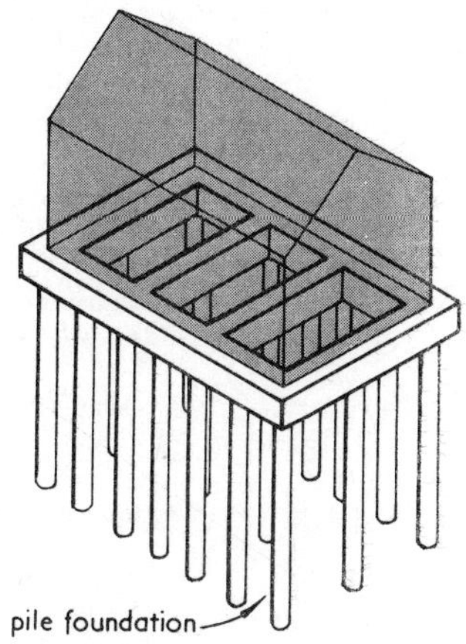

45.35 *Alternative pile foundation*

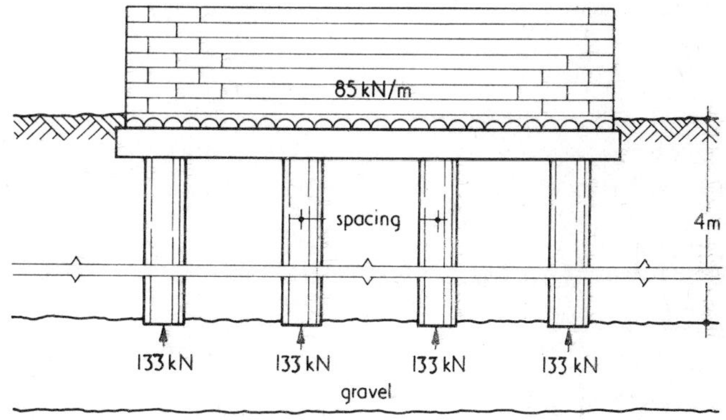

45.37 *Short-bored piles (Example 19)*

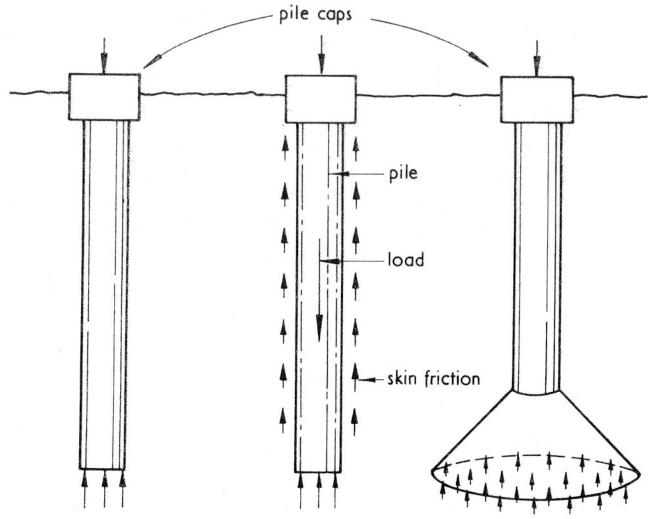

45.36 *Types of piles:* **a** *end-bearing pile* **b** *friction pile* **c** *under-reamed pile*

6.07 Skin-friction piles

Skin-friction piles, mostly appropriate for cohesive soils such as clays are usually bored. In this method a hole is constructed in the ground by augering or other methods. If necessary, the sides of the hole are temporarily sleeved. When the necessary depth has been reached, reinforcement is lowered into the hole and concrete is poured in. Some method of compacting the concrete is employed, so that all cavities in the ground are properly filled. The sleeve is withdrawn as the concrete goes in, to ensure intimate contact between pile and soil, **45.36(b)**.

The capacity of this type of pile is not self-evident as in the case of driven piles. Calculation is used to determine the length of pile required, based on the shear strength of the clay at various depths below the ground. This information is gained by carrying out a site investigation prior to the design.

While the calculations themselves are not particularly difficult, the subject is one where engineering judgement is required in their interpretation. It is recommended that this type of pile should be installed either under the supervision of a consulting engineer, or to a design guaranteed by the piling firm.

6.08 Pile testing

In both driven and bored piles it is usual to carry out one or more tests on the actual piles on each contract. On very large contracts additional piles are tested to failure, but normally one of the working piles is simply loaded to 1½ times the working load to prove the efficacy of the design.

6.09 Under-reamed piles

There is a third type of pile which is a bored end-bearing pile. The shaft is augured in the usual way, but when the required depth has been reached a special tool is used to enlarge the base to a bell-shape, **45.36(c)**. These piles are substantial in size, and men are lowered to clean out the base prior to concreting, and also to inspect and approve the soil on which the load will rest. It is not often possible to test this type of pile in view of the magnitude of the loads carried.

7 Bibliography

Allan Hodgkinson (ed) *AJ Handbook of Building Structure*, London, Architectural Press, 1974.

46 Lifts

Frank Williams

CI/SfB (66)
CI/SfB (1976 revised) (66)

1 INTRODUCTION

1.01 Definitions

For the purpose of this general guide the following definitions apply:

- *passenger lift:* a lift intended primarily for carrying people and having automatic sliding doors
- *group:* two to six passenger lifts with an integrated control system and common call buttons on the landings
- *goods lift:* a lift intended primarily for carrying goods accompanied by personnel and having manually operated collapsible doors
- *service lift:* a lift for transporting goods only (eg documents, food) and having a car not exceeding 1400 mm height and 1·20 m^2 floor area.

1.02 Location

Passenger lifts should be within a reasonable walking distance from the furthest part of the floor areas served (say 70 m maximum) and, where they are the only or main lifts, near an entrance but with the stairs nearer to the entrance. A shorter walking distance (say 50 m maximum) is desirable in an office building where interfloor journeys are to be catered for. The location of goods and service lifts will depend on their function, but they should not open into passenger lift lobbies or public areas.

2 PASSENGER LIFTS

2.01 Selecting type, size, speed and number

Details of the most widely used types of passenger lift are shown in table I and **46.1**. The first type is inappropriate to the main lift service of such as an office building, especially in the single speed motor form (speed 0·25 or 0·50 m/s), but their robust single panel doors and associated specification package makes them particularly suitable for municipal housing and hostels. The doors of types 1 and 5 (which close relatively slowly from the side) are particularly suited to the infirm or for goods traffic. The quicker doors of types 3 and 4 maximise handling capacity and service quality, by minimising journey and waiting times.

The quality/quantity of the lift service needed by buildings other than offices is essentially a matter for individual assessment with the assistance of specialist advice and comparison with similar cases. However, for preliminary purposes it may be assumed that passenger lifts suited to an office building can be stretched (in terms of population and visitors served) up to 100 per cent for a hospital, hotel or shop. The passenger lift service (if any is required) for a two or three storey building, including offices but excluding hospitals and the like, will usually need to be only nominal, ie not related to the population figures. It is particularly relevant to such cases to note that a load capacity smaller than eight persons is unsuitable for a disabled person in a wheelchair and of limited usefulness for furniture removal.

2.02 Lifts in offices

For lift types 3 and 4, table II gives the configurations for a good service when dealing with typical office building traffic. Table III gives single lift schemes for situations where the previous table cannot be applied; as, for example, where only two lifts can be justified and have to be separated for functional reasons instead of being placed in the more efficient two-lift group.

The second column in tables II and III gives the population per floor (averaged) for which the lift or group of lifts is suitable. For example, the second scheme in table II is suitable for a

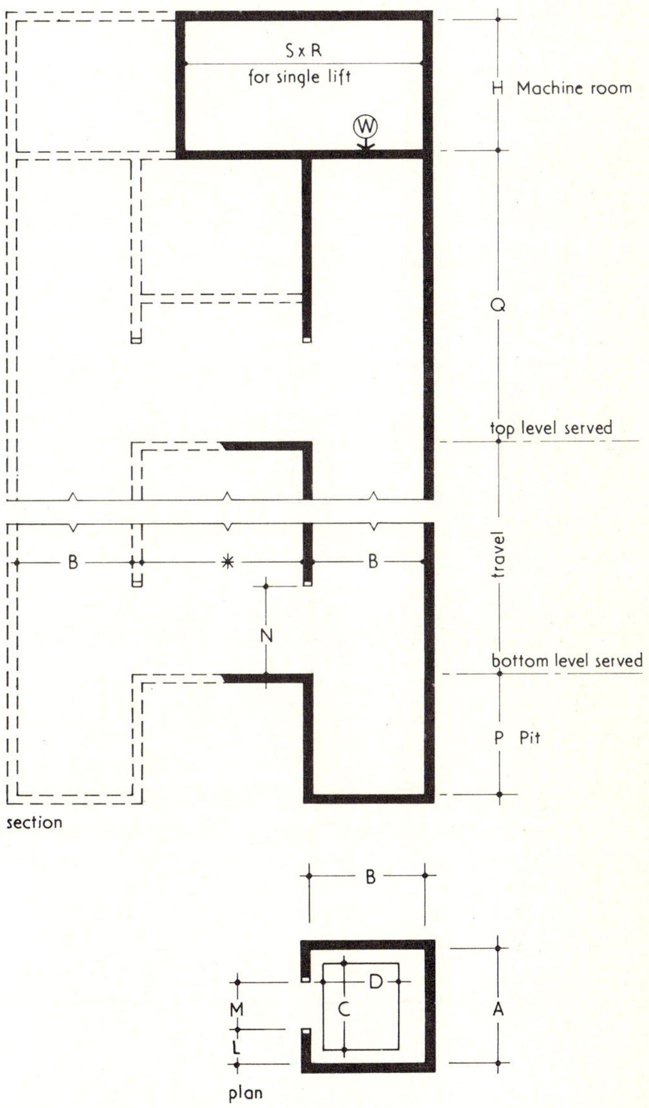

46.1 *Passenger lifts. (Dimensions as table I)*
See **46.2 for multi-lift machine rooms and lobby arrangements*
Prefer machine room floor on one level with lift-well capping

Table I Passenger lift dimensions
Based generally on BS 2655: part 3‡

Type	Load capacity persons/kg	Speed m/s	Well (nominal)* A	B	Car (Internal) C	D	Doors M	N	L	Pit P	Machine room Q	H	R	S	W†
1 Light traffic	8/600	0·25 or 0·50	2050	1950	1100	1400	800	2000	275	1400	4200	2600	2050	4400	72
doors: single panel		0·75								1500					
side-opening		1·00								1500					
	10/750	0·25 or 0·50	2050	1950	1300	1400	800	2000	275	1500	4200	2600	2050	4400	90
		0·75								1600					
		1·00								1600					
3 General purpose	8/600	0·75	1850	1950	1100	1400	800	2000	525	1500	4200	2600	2050	4400	72
(speed up to 1·50 m/s)		1·00								1700	4200	2600	2050	4400	
		1·50								1700	4400	2600	3100	4800	
4 Intensive traffic	10/750	0·75	2050	1950	1300	1400	800	2000	625	1600	4200	2600	2050	4400	90
(speed of 2·50 m/s		1·00								1700	4200	2600	2050	4700	
or more)		1·50								1700	4400	2600	3100	5000	
doors: two panel	12/900	0·75	2150	2200	1600	1400	800	2000	675	1600	4200	2700	2150	4700	108
centre-opening		1·00								1800	4400	2700	2150	5100	
		1·50								1800	4400	2700	3300	5100	
		2·50								2800	6400	3000	3200	7500	
	16/1200	0·75	2650	2300	2000	1400	1100	2000	775	1700	4300	2700	2650	4900	144
		1·00								1900	4400	2700	2650	5300	
		1·50								1900	4500	2700	3500	5300	
		2·50								2800	6500	3000	3200	8000	
		3·50								3400	7400	3000	3200	8000	
	20/1500	0·75	2650	2600	2000	1700	1100	2000	775	1700	4300	2700	2650	5300	180
		1·00								1900	4400	2700	2650	5600	
		1·50								1900	4500	2700	3500	5600	
		2·50								2800	6700	3000	3200	8300	
		3·50								3400	7600	3000	3200	8300	
	24/1800	2·50	2650	2800	2000	1900	1100	2000	775	2800	6700	3000	3200	8600	216
		3·50								3400	7600	3000	3200	8600	
5 Hospital	20/1500	0·50	2250	3050	1400	2400	1200	2000	275	1700	4500	2700	2250	5700	180
doors: two panel		0·75								1700	4600	2700	2250	6000	
side-opening		1·00								1700	4700	2700	3500	6000	
		1·50								1800	4700	2700	3500	6000	
	24/1800	0·50	2450	3050	1600	2400	1300	2200	275	1700	4700	2900	2450	5800	216
		0·75								1700	4800	2900	2450	6100	
		1·00								1700	4800	2900	3500	6100	
		1·50								1800	4800	2900	3500	6100	
	28/2100	0·50	2450	3350	1600	2700	1300	2200	275	1700	4700	2900	2450	6100	252
		0·75								1700	4800	2900	2450	6400	
		1·00								1700	4800	2900	3500	6400	
		1·50								1800	4800	2900	3500	6400	

* Includes 50 mm allowance for out of plumb
† Approx maximum force imposed by lift (kN)

‡BS 2655 has recently been superseded by BS 5655, which is being issued in parts. Dimensional information has only been published so far for lifts in residential buildings.

Table II Selection table for a good passenger lift service
Based on the handling capacity of lift types 3 and 4 (table I) for typical officer traffic

Levels served	Population per floor	Net area served m²	No of lifts in group	Load persons	Speed m/s	Cost of* group £
4	60	2400	1	8	0·75	8250
	69	2760	1	10		8800
	76	3040	1	12		9350
	147	5880	2	8		16500
	176	7040	2	10		17600
	202	8080	2	12		18700
5	33	1650	1	8	1·00	8800
	37	1850	1	10		9350
	88	4400	2	8		17600
	105	5250	2	10		18700
	147	7350	2	10	1·50	22000
	168	8400	2	12		23100
	199	9950	2	16		24200
6	51	3060	2	8	1·00	18700
	58	3480	2	10		19800
	63	3780	2	12		20900

* Approx cost for comparison purposes (mid 1973) excluding builder's work

Table II Selection table for a good passenger lift service (*continued*)

Levels served	Population per floor	Net area served m²	No of lifts in group	Load persons	Speed m/s	Cost of* group £
6(*cont.*)	86	5160	2	8	1·50	22000
	100	6000	2	10		23100
	109	6540	2	12		24200
	121	7260	2	16		25300
	151	9060	3	10		34600
	168	10080	3	12		36300
	199	11940	3	16		38000
7	69	4830	2	10	1·50	24200
	75	5250	2	12		25300
	109	7630	3	10		36300
	122	8540	3	12		38000
	138	9660	3	16		39600
	152	10640	3	20		41900
	163	11410	4	12		50600
	184	12880	4	16		52800
	207	14490	4	20		55900
8	51	4080	2	10	1·50	26400
	84	6720	3	10		39600
	94	7520	3	12		41300
	109	8720	3	16		42900
	119	9520	3	20		45200

Table II Selection table for a good passenger lift service (*continued*)

Levels served	Population per floor	Net area served m²	No of lifts in group	Load persons	Speed m/s	Cost of* group £
8(*cont.*)	125	10000	4	12		55000
	146	11680	4	16		57200
	165	13200	4	20		60300
	183	14640	5	16		71500
	206	16480	5	20		75400
9	75	6750	3	12	1·50	42900
	85	7650	3	16		44600
	92	8280	3	20		46900
	100	9000	4	12		57200
	116	10440	4	16		59400
	131	11790	4	20		62500
	146	13140	5	16		74300
	164	14760	5	20		78100
	175	15750	6	16		89100
	197	17730	6	20		93700
10	61	6100	3	12	1·50	44600
	68	6800	3	16		46200
	73	7300	3	20		48500
	82	8200	4	12		59400
	96	9600	4	16		61600
	107	10700	4	20		64700
	120	12000	5	16		77000
	134	13400	5	20		80900
	144	14400	6	16		92400
	161	16100	6	20		97000
11	51	5610	3	12	1·50	46200
	56	6160	3	16		47900
	61	6710	3	12	2·50	61100
	69	7590	3	16		64300
	73	8030	3	20		72600
	81	8910	4	12		81400
	93	10230	4	16		85800
	103	11330	4	20		96800
	112	12320	4	24		99000
	129	14190	5	20		121000
	140	15400	5	24		123700
	155	17050	6	20		145200
	168	18480	6	24		148500
12	53	6360	3	12	2·50	62700
	58	6960	3	16		66000
	61	7320	3	20		74300
	71	8520	4	12		83600
	80	9600	4	16		88000
	89	10680	4	20		99000
	96	11520	4	24		101200
	111	13320	5	20		123800
	120	14400	5	24		126500
	134	16080	6	20		148500
	144	17280	6	24		151800
13	46	5980	3	12	2·50	64400
	52	6760	3	16		67700
	61	7930	4	12		85800
	70	9100	4	16		90200
	77	10010	4	20		101200
	82	10660	4	24		103400
	97	12610	5	20		126500
	105	13650	5	24		129200
	116	15080	6	20		151800
	126	16380	6	24		155100
14	40	5600	3	12	2·50	66000
	54	7560	4	12		88000
	61	8540	4	16		92400
	68	9520	4	20		103400
	71	9940	4	24		105600
	77	10780	5	16		115500
	85	11900	5	20		129200
	92	12880	5	24		132000
	102	14280	6	20		155100
	110	15400	6	24		158400
15	48	7200	4	12	2·50	90200
	54	8100	4	16		94600
	60	9000	4	20		105600
	68	10200	5	16		118300
	76	11400	5	20		132000
	82	12300	5	24		134700
	91	13650	6	20		158400
	98	14700	6	24		161700
16	44	7040	4	12	2·50	92400
	50	8000	4	16		96800
	53	8480	4	20		107800
	62	9920	5	16		121000
	68	10880	5	20		134700
	74	11840	5	24		137500
	82	13120	6	20		161700
	89	14240	6	24		165000
	91	14560	6	24	3·50	184800
17	40	6800	4	12	2·50	94600
	45	7650	4	16		99000
	47	7990	4	20		110000
	56	9520	5	16		123800
	62	10540	5	20		137500
17	64	10880	5	20	3·50	154000
	69	11730	5	24		156700
	77	13090	6	20		184800
	83	14110	6	24		188100
18	42	7560	4	16	3·50	114400
	45	8100	4	20		125400
	53	9540	5	16		143000
	59	10620	5	20		156700
	63	11340	5	24		159500
	71	12780	6	20		188100
	76	13680	6	24		191400

* Approx cost for comparison purposes (mid 1973) excluding builder's work

Table III Selection table for separated lifts
Based on estimated handling capacity of lift types 3 and 4 in table I

Levels served	Population per floor	Net area served m²	Load persons	Speed m/s	Cost* £
5	64	3200	8	1·50	10450
	70	3500	10		11000
	77	3850	12		11550
6	40	2400	8	1·50	11000
	44	2640	10		11550
	47	2820	12		12100
7	29	2030	10	1·50	12100
	31	2170	12		12650

population averaging over 60 per floor, and up to and including an average of 69 people per floor. The third column is an extension of the previous column, allowing 10 m² per person, and therefore gives some indication of the total net area for which the lift or group is suitable. Dividing the area figure by 10 will given an indication of the total population for which the lift or group is suitable.

It is advisable to work within the tabulated population figures until any conditions that would reduce the suitability of the lifts can be assessed with specialist advice. These conditions include situations where:

- the floor to floor heights average more than 3·3 m
- there is more than one main floor, ie the lifts populate the building from the ground floor and another level or levels
- there are levels below the ground floor requiring normal service for passengers
- the distance from ground floor to first floor exceeds 3·3 m and/or the stairs are located so as to be unattractive to people entering the building
- there is a canteen above the ground floor, especially where it is used by people not counted in the population served by the affected lift or group
- there are large numbers of visitors
- there is a significant amount of goods traffic or other 'requisitioning' which restricts availability for normal passenger traffic.

Firemen's lift service, where required (unusual under seven storeys), can usually be provided by the most suitably located passenger lift of at least eight persons capacity.

2.03 Lifts in residential accommodation

Standard public housing
Lift access is provided to dwellings where there is a climb of more than two storeys to reach the front door. The climb is measured from the ground, or from a main pedestrian deck. Where there are dwelling entrances on storeys up to the sixth storey (counting the ground or pedestrian deck as the first) one lift is usually sufficient. Where there are dwelling entrances above the sixth storey, two lifts are provided to serve not less than 20 dwellings each, or more than 50. These numbers include the numbers of dwellings at ground or deck level.
In order to avoid the provision of lifts in three storey blocks of flats, or four storey maisonettes, it is frequent practice to have a further storey-height climb beyond the dwelling entrance.
There should be no direct lift access to enclosed underground or under-deck garage spaces, for safety reasons. Where direct access is necessary, an open area should be provided between lift and garage; otherwise stair access from the lift is normally provided.
For blocks not higher than 11 storeys including ground or deck level, the type of lift normally used is the eight-passenger (630 kg) lift with standard speed 0·5 m/s (table I). For higher blocks a faster speed may be necessary. For example, in a twenty storey block, a person on the top floor calling the lift from the bottom and returning there, assuming four other 15 second stops en route, will take approximately 5 minutes at a speed of 0·5 m/s, 4 minutes at 0·75 m/s and 3 minutes at 1 m/s. In large blocks the slow speed may well encourage overloading. If our passenger had just missed the lift, his trip times will be increased by half as much again.
Lifts should be located in the block so that the walk to the dwelling entrance does not exceed 60 m. Habitable rooms, particularly bedrooms, should not abut liftshafts or machinery rooms.
The standard lift car size is 1·1 × 1·4 × 2·2 m, and the door width is 0·8 m. It will be seen that this will accommodate a wheelchair and most items of furniture except the largest. It will not accommodate a stretcher or a coffin unless this can be stood on end. For this reason a stretcher recess is sometimes provided in one lift of a tall block. This projects into the shaft for only part of the car height, but usually for the full width. This means the shaft is increased in size, but there is still a saving over a larger car because this would have to be designed for the greater loading that could be accommodated. However, stretcher recesses can be a problem where vandalism is concerned.

Private sector housing
The standards of lift accommodation in the private sector are not usually markedly more generous than in the public. A faster speed might be provided, and the standard of finishing will probably be more luxurious and less vandal-resistant.

Housing for elderly and handicapped persons
There is some difference of opinion on the desirable scale of provision in these cases. The Greater London Council does not accommodate these people above the fifth storey from the normal pedestrian or vehicular level, whichever is the lower. However, the Department of the Environment only permits one lift up to four storeys, and two lifts for five and six storey buildings only if there are at least 12 dwellings above fourth storey level.
No reduction in the standard lift size can be allowed, as it is necessary to have sufficient door width to pass self-propelled wheelchairs. Passenger-operated controls should be low enough to be operated from a wheelchair.

Hostels
Where students and nurses are accommodated it may be assumed that they are generally younger and fitter than the general population. It is therefore uncommon to install lifts unless more than four storey-heights have to be climbed. However, a goods hoist should be provided for linen etc over two storeys. For blocks higher than four storeys one lift will probably be sufficient. Provided no communal facilities are provided on upper floors there should be no need for access by students in wheelchairs, who will be accommodated at ground level.

Hotels
Provision in hotels should be on the scale laid down for office buildings. Again, no hotel rooms should abut shafts or machinery rooms.

2.04 Planning passenger lifts
Cul de sac or recessed lobbies, **46.2**, are essential to get the maximum performance from lift groups. Clearly, having called

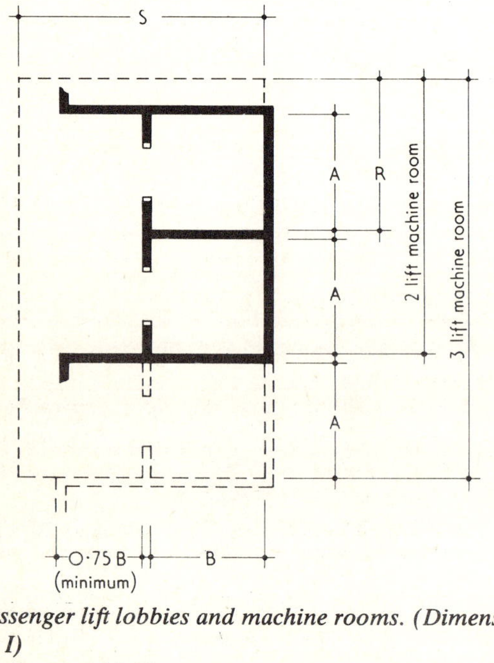

46.2 *Passenger lift lobbies and machine rooms. (Dimensions as table I)*
a *Plan of recessed lobby*

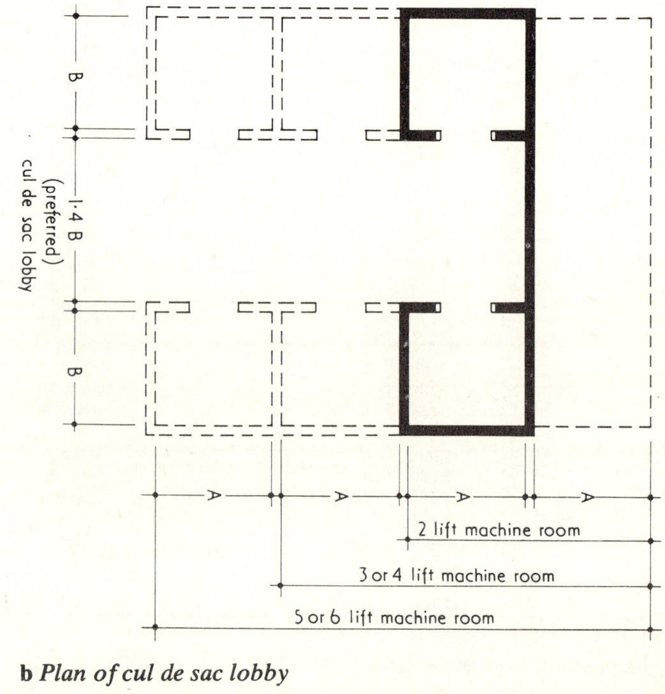

b *Plan of cul de sac lobby*

for lift service the waiting passengers should not have to walk further than necessary to the responding lift car, and should not be obstructed by passers-by. The lobbies for separate lifts or groups should be adequately separated in order to promote the channelling of their respective traffic and to discourage the duplication of calls on the landing buttons.

Within a group it is preferable for all the lifts to serve all levels in order to avoid a particularly annoying inconvenience to the users. If only one or some lifts of a group serve a basement car park, for example, a normal push button system cannot ensure (at a higher floor) that the responding lift car will be one that is able to travel to the basement.

3 GOODS LIFTS

Except for heavy industrial use (eg self-propelled trucks), goods lifts as in table IV and **46.3** are usually satisfactory. The selected lift car size should allow for a person to accompany the largest item or batch of goods to be catered for in normal use (to operate the controls at the side of the car).

For over eight storeys, a requirement for a separate goods lift service is usually met by a lift of the passenger type having automatic side-opening doors, eg type 5 in table I (which could accommodate a stretcher) or a pro-rata special, or type 1.

4 OTHER PRELIMINARY PLANNING REQUIREMENTS

Below the lift well, a space used by people should be avoided wherever possible. Where there must be such a space, even a large duct, the necessary safety provisions will involve a wider or deeper well, additional structural support and/or additional lift equipment.

Lift wells for passenger and goods lifts have to be vented at the top, directly or by duct (not into machine room), to external air for smoke dispersal purposes. Safe and convenient access to the machine room has to be arranged for lift personnel and the handling of replacement assemblies. Personnel access via a ladder and trapdoor should be avoided wherever possible. No services installation or access route, other than those provided for lift equipment and lift personnel, should share or pass through the machine room or lift well.

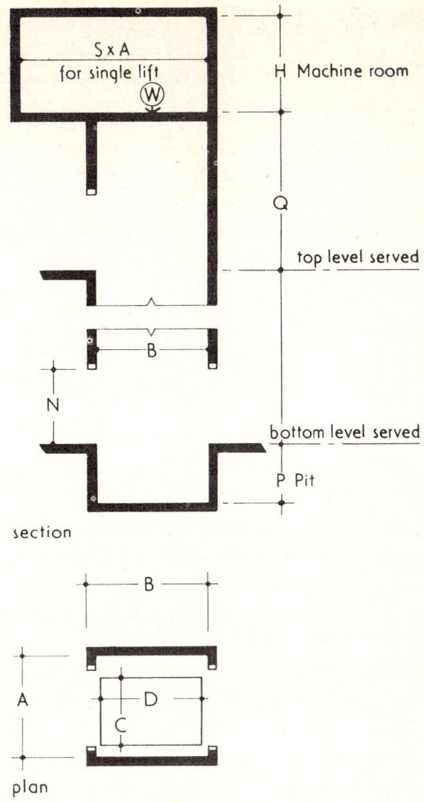

46.3 *Goods lifts. (Dimensions as table IV)*
Entrances can be: front only
or front and back
Prefer machine room floor on one level with lift-well capping

Table IV Goods lift dimensions
Based generally on BS 2655: part 3

Load Capacity kg	Speed m/s	Well (nominal)* A	B	Car and doors (internal/clear) C	D	N	Pit P	Machine room Q	H	S	W†
500	0·38 or 0·50	1850	1550	1100	1050	2000	1400	4000	2400	3700	60
	0·75						1500				
	1·00						1500				
1000	0·25	2150	2150	1400	1650	2000	1500	4000	2400	4300	120
	0·38 or 0·50						1500				
	0·75						1500				
	1·00						1500				
1500	0·25	2550	2350	1700	1850	2300	1500	4200	2700	4500	180
	0·50						1700	4300			
	0·75						1800	4400			
	1·00						1800	4400			
2000	0·25	2550	2850	1700	2350	2300	1500	4300	2900	5100	240
	0·50						1700	4500			
	0·75						1800	4700			
	1·00						1800	4700			
	0·25	2850	2450	2000	1950	2300	1500	4300	2900	4700	240
	0·50						1700	4500			
	0·75						1800	4700			
	1·00						1800	4700			
3000	0·25	3050	3350	2000	2850	2300	1500	4400	2900	5600	360
	0·50						1700	4600			
	0·75						1800	4700			
	0·25	3550	2750	2500	2250	2300	1500	4400	2900	5000	360
	0·50						1700	4600			
	0·75						1800	4700			

* Includes 50 mm allowance for out of plumb
† Approx maximum force imposed by lift (kN)

5 SERVICE LIFTS

Unfortunately there is no standard range from which to select service lifts, and makers' preferences on general arrangement, size and speed vary considerably; but most makers can provide lifts similar to those shown in **46.4** and **46.5**. Where there are other lifts in the building or associated buildings, a service lift by their maker would assist management and minimise maintenance cost.

Table V Dimensions for medium size service lifts

Load capicity kg	Well (nominal)* A	B	Car and doors (internal/clear) C	D	N
50	1000	800	550	550	700
100	1200	1000	750	750	800

Note: speed up to 0·5 m/s preferred 0·38 m/s
* Includes 25 mm allowance for out of plumb

Table VI Dimensions for large size service lifts suitable for trolleys

Load capicity kg	Well (nominal)* A	B	Car and doors (internal/clear) C	D	N
150 or 250	1300	1350	750	1075	1200
250	1450	1550	900	1275	1400

Note speed 0·25 m/s or 0·38 m/s
* Includes 25 mm allowance for out of plumb

A maker's own standard pre-assembled and clad unit might be advantageous for such as an existing building where minimum builder's work and quick delivery are overriding factors. These packages do have disadvantages for some applications, however, eg insufficient fire resistance of cladding, unsafe for location above a space used by people.

6 ESCALATORS

Dimensions, speed and finishes vary but a 30° incline is available from all makers, **46.6**. For preliminary purposes or an approximate comparison with lifts' performance (tables II and III), allow a handling capacity of 1600 people in 30 minutes per 600 mm of step width. In many buildings of course, a lift or lifts will also be needed for the infirm, wheelchairs, prams and /or goods traffic.

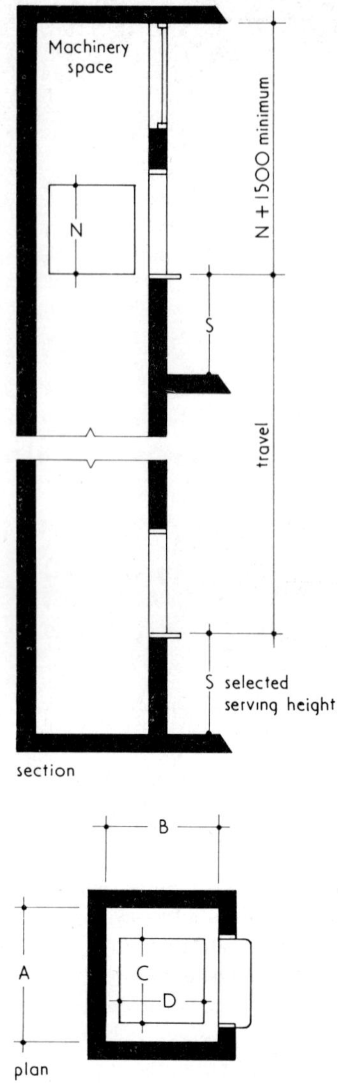

46.4

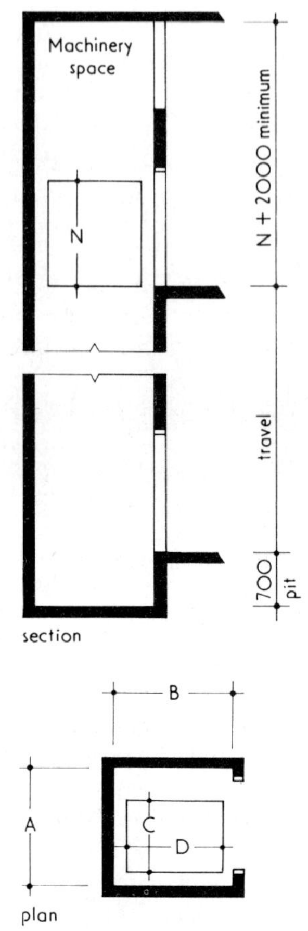

46.5

46.4 *Medium size service lift. (Dimensions as table V)*
Note: Speed up to 0·50 m/s, but prefer 0·38 m/s
Well size includes 25 mm allowance for out of plumb
Openings can be front and rear as required
Car can be fitted with shelves

46.5 *Large service lifts suitable for trolleys. (Dimensions as table VI)*
Note: Speed 0·25 or 0·38 m/s
Well size includes 25 mm allowance for out of plumb
Openings can be front and rear as required

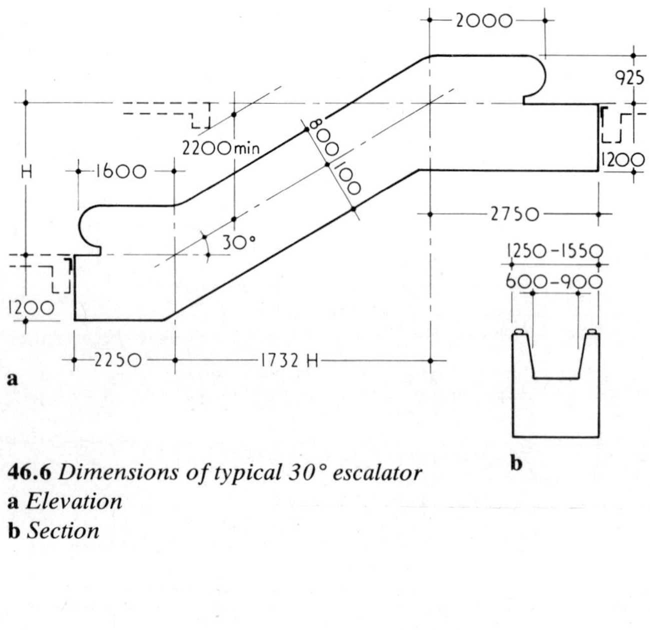

46.6 *Dimensions of typical 30° escalator*
a *Elevation*
b *Section*

7 BIBLIOGRAPHY

The Architects' Journal Selection of passenger lifts for office buildings, 9.8.72

The Architects' Journal Selection of passenger lifts for tall office buildings, 4.12.74

BS 2655 *Lifts, escalators, passenger conveyors and paternosters*, Part 3 Arrangements of standard electric lifts, Part 6 Building construction requirements

BS CP 407 *Electric, hydraulic and hand-powered lifts*

PAS 20* General purpose passenger lifts

PAS 22* General purpose goods lifts

* Purchasing specifications published by BSI for the Public Sector Standardization Team.

47 Security

David Adler

CI/SfB (68)
CI/SfB (68)

Crimes
The main types of crime are:

1 pilferage by staff or other insiders
2 pilferage by public (shoplifting)
3 vandalism without gain to the perpetrator
4 casual break-ins
5 planned break-ins
6 attacks on persons for immediate gain (muggings, etc)
7 various types of kidnapping, hi-jacking, etc, mainly for ransom
8 bomb attacks, etc, for political objectives.

Table I indicates which particular building types are subject to these crimes and suggests suitable precautionary measures. Most combative measures are designed either to slow up the criminal until he feels vulnerable to discovery or else makes it difficult for him to remain unobserved.

Fences
Where a site perimeter protection is indicated a fence is better than a wall as one can see through it. It should be not less than 2·5 m high and topped with two strands of barbed wire. If a wall has to be used, it should have barred openings in it.

Windows
Window panes less than 0·05 m^2 in area cannot be climbed through. Larger panes should be as large as possible. For more security windows should be barred. Vertical bars are more effective than horizontal. Ideally, the bars should be of square cross-section minimum 20 mm at a maximum spacing of 125 mm and built in 75 mm. Transverse tie-bars should be provided at 600 mm centres.

Doors
The 'last door out' (which cannot be barred or bolted from inside) should be as stout as possible. It should be solid, not of hollow construction, and the hinge should be internal. When opening in, the stop should be formed by rebating the solid and not planted; otherwise the tongue of the lock is readily accessible. The lock should be a mortice lock to BS 3621: 1963 or equivalent, unless the door is less than 45 mm thick. In this case the mortice will weaken the door, and an automatically deadlocking rim latch should be fitted. No lock should be fitted that can be opened from inside without the key—this will hinder an intruder's escape. A letter plate in the door should be sized and positioned so that it cannot be used to gain entrance. The minimum distance between it and the lock should be more than 400 mm. Unless there is another method of identifying callers, a lensed spyhole should be provided.

Table I Relationship between crime and building type

Type of building	Principal risks	Vulnerable points	Design solutions
Single person/ family dwellings	4	Ground floor doors and windows	Doors and windows fitted with security locks that cannot be opened by merely breaking the glass Overlooking of all doors and windows from neighbouring properties, with adequate lighting from street lights, etc
		Upper windows near to low-level roofs, drainpipes, etc	Anti-climb paint or barbed wire on drainpipes. Locks and visibility as above
Flats	3, 4, 5, 6	Door to flat, particularly where this opens off internal lobby, as no window is allowed by fire regulations. This means that criminals can often proceed without the possibility of being seen from a neighbouring flat	Solid door with bolts at hinge side in addition to security locks, bolts and spyhole Good lighting proof against interference Minimum length of corridors, few corners
Multi occupancy dwellings	3, 4, 6	Generally as flats	Generally as flats
Hostels	1, 4	Communal areas Rooms	No architectural measures other than ensuring that fire exits cannot be used for unauthorised ingress
Hotels	1, 4, 5	Kitchens, linen stores, rooms	Ensure that all exits are under constant casual observation at all times. This makes it difficult for staff or intruders to remove their booty. A substantial safe should be provided near the reception for guests' valuables. Safe less than 600 kg weight must be secured against bodily removal. Master key system for rooms under good control
Shops	1, 2, 4, 5, 8	Ground floor doors and windows	Security locks and easy observation
		Back-up stores and rear corridors Unfrequented areas of sales floor, fitting rooms in garment shops	Should be designed to be under constant casual observation. If not architecturally possible closed-circuit television can be used, although this may affect trade by alienating customers
Offices	1, 4, 5, 8	Ground floor doors and windows, particularly rear fire escapes	Panic-type locks on rear escape doors. All areas under constant casual observation. Supply all staff with lockable furniture for personal valuables. Consider a secure store for expensive items such as typewriters

Table I Relationship between crime and building type (*continued*)

Type of building	Principal risks	Vulnerable points	Design solutions
Factories and storage buildings	1, 4, 5, 8	Ground floor doors and windows, lorry loading banks	Doors barred with heavy duty locks—in many cases machinery for cutting through such devices will be to hand. Constant casual observation, including security patrolling at night. Good fencing around the site, with permanent lighting of the area between fence and building
Sports buildings	1, 3, 8	Changing rooms, cash desk	Stout lockers, good observation, substantial safe for takings, if must be left on premises
Restaurants	1, 8	Kitchens, stores, cash desk	Constant casual observation
Banks	5, 7, 8	Almost everything	Sophisticated security measures that are not generally known outside the particular organisation
Car parks	1, 2, 3, 4, 8		Observation at all times, including the use of lighting and closed circuit television. In this case public acceptance is universal

48 Window cleaning

CI/SfB (75)
CI/SfB (1976 revised) (75)

Contents

1 METHOD AND FREQUENCY OF CLEANING

1.01 Method

The methods by which windows and façades are to be regularly cleaned and maintained must be considered at an early stage of design, and the necessary equipment incorporated into the structure.

Regular washing with cold or warm water (sometimes with a mild detergent) is normally adequate, applied either by swab with chamois leather to dry and scrim to polish; or by squeegee, which is much quicker over large areas and when used from cradles. Between 400 m^2 and 500 m^2 in eight hours is average, using a squeegee in ideal conditions.

1.02 Frequency

Table I shows DOE recommendations for frequency of washing according to locality, and table II gives frequency of washing particular building types in non-industrial areas. For industrial areas and cities the interval between cleans should be halved.

Table I Recommended frequency of cleaning per year for DOE buildings

Location	Ground floor facing street	Other windows	Rooflights
London postal area and smoky industrial areas of large cities	16	8	2
Semi-industrial towns	12	6	2
Non-industrial towns	8	4	1

Table II Recommendations for frequency of washing of particular building types

Type of building	Side windows	Rooflights
Offices	Every 3 months*	every 12 months
Public offices, banks, etc	2 weeks	3 months
Shops	Outside every week Inside every 2 weeks	6 months
Shops (in main streets)	Outside daily inside every week	3 months
Hospitals	3 months	6 months
Schools	3–4 months	12 months
Hotels (first class)	2 weeks	3 months
Factories (precision)	4 weeks	3 months
Factories (heavy work)	2 months	6 months
Domestic (by contract)	4–6 weeks	—

* Ground floor windows facing streets should be cleaned at twice this frequency

2 ACCESS

2.01 Internal or external access?

Type of access is decided by:

- method and frequency of cleaning
- capital and running costs
- whether cleaned by tenants or professional window cleaners
- safety requirements
- appearance of equipment when not in use.

A general guide to selecting external types of access is shown in **48.1**.

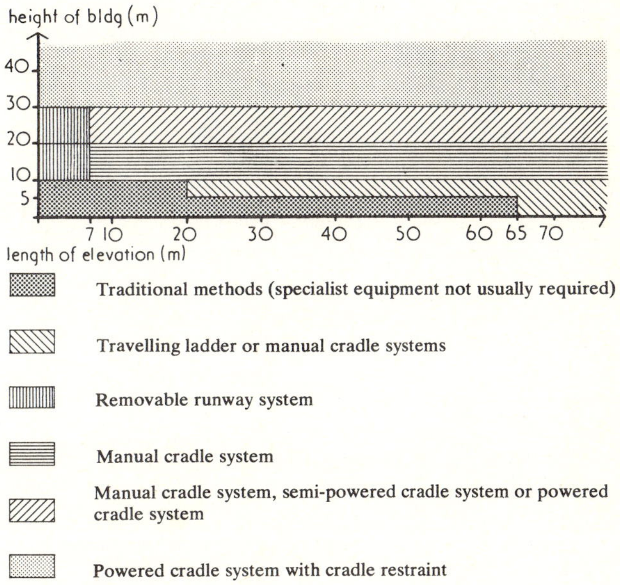

48.1 *Chart for selecting system for external access*

When selecting internal types of access, take into account possible problems:

- type of window (especially high-rise housing)
- method of cleaning adjacent exterior cladding
- freak draughts and disruption to air-conditioning when opening windows
- disruption to furniture and activities; and damage to property
- relative cost of providing opening windows (for cleaning from inside) against cost of cradle (for cleaning from outside)
- safety (beware cleaners, especially tenants, having to lean out to clean adjacent fixed lights).

Cleaning the internal glass face is usually no problem unless inaccessible.

Often two separate contractors are given the work of cleaning the inside and outside faces. Cleaning the outside *from* the outside will usually give better results, and can effect long-term savings over the extra cost of providing opening windows to allow cleaning from inside.

3 INTERNAL ACCESS

3.01 Anthropometrics

Human dimensions related to window cleaning are shown in **48.2, 48.3, 48.4, 48.5** and table III.

3.02 Types of window

- *Side hung casements* should have offset pivot hinges to give

Table III Access to external faces from the inside
Note: shaded area indicates glass face

	Good	Satisfactory	Bad
Casement	 (1) Inward opening	 (2) Outward opening with extending hinges	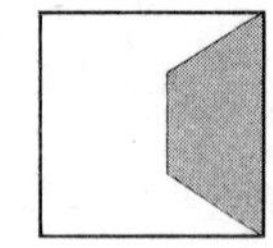 (3) Outward opening
Double opening hopper	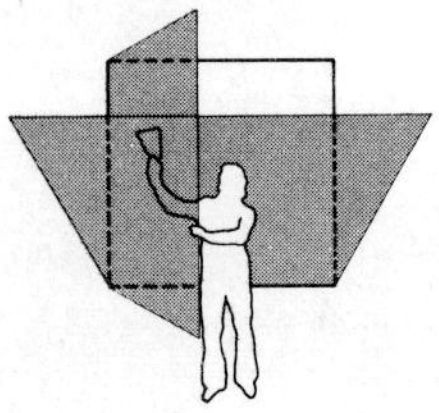 (4)	 (5) Inward opening 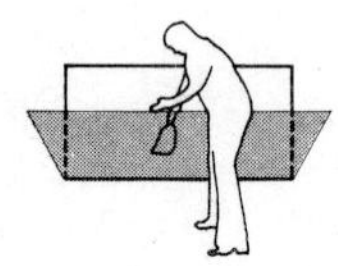 (6)	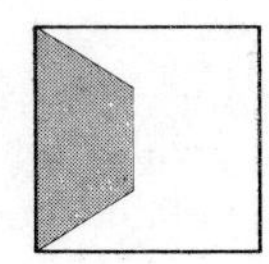 (7) Outward opening
Vertical slide			 (8)  (9)
Horizontal slide		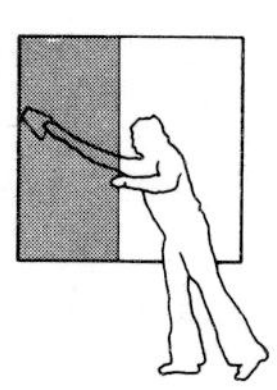 (10) Top corner reach possible (see **48.2**)	 (11) Corner reach not possible
Horizontal pivot	 (12) Completely reversible		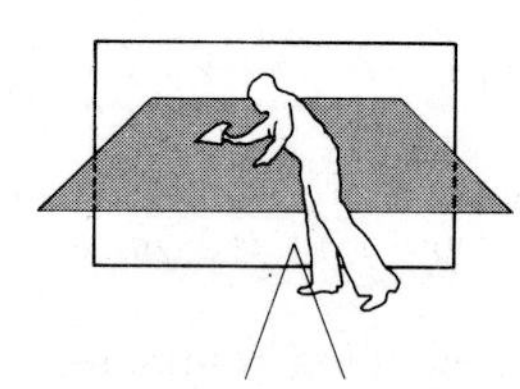 (13) Not completely reversible and too high (see 3)
Vertical pivot	 (14) Completely reversible	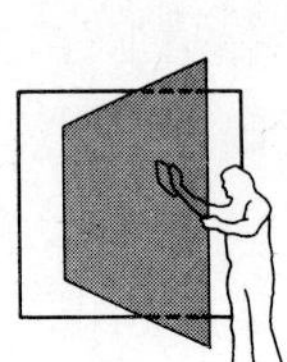 (15) Not reversible but at correct height (see **48.3**)	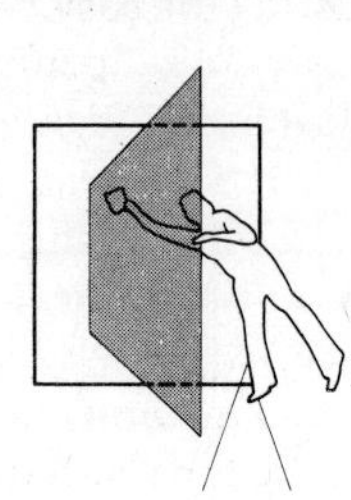 (16) Not reversible and too high (see **48.3**)

Table III (*continued*)

Good	Satisfactory	Bad
Top hung	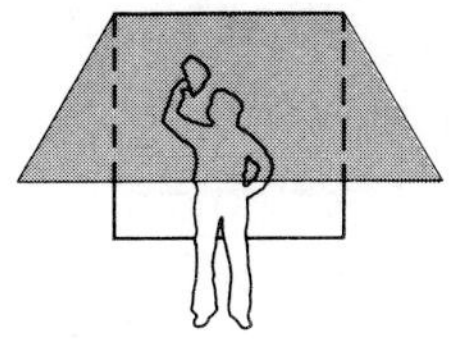 (17) Top-hung opening in	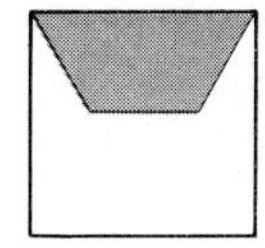 (18) Top-hung opening out is impossible to clean
Fixed adjacent  (19) Corner reach possible		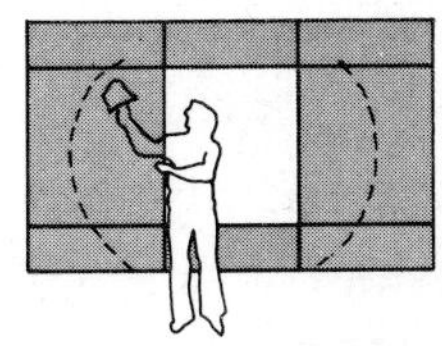 (20) Corner reach impossible
	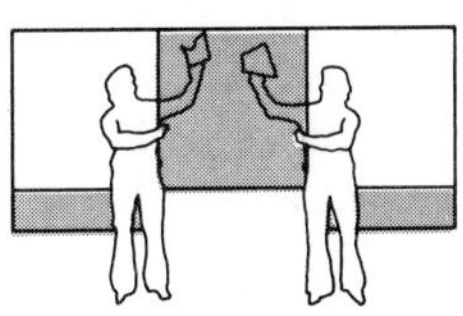 (21) Centre reach possible	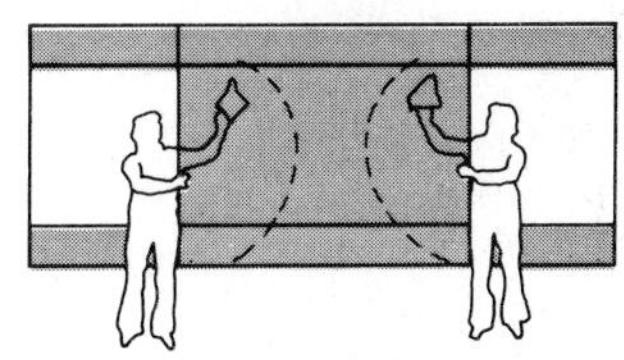 (22) Centre reach impossible
Access to internal faces Double glazing 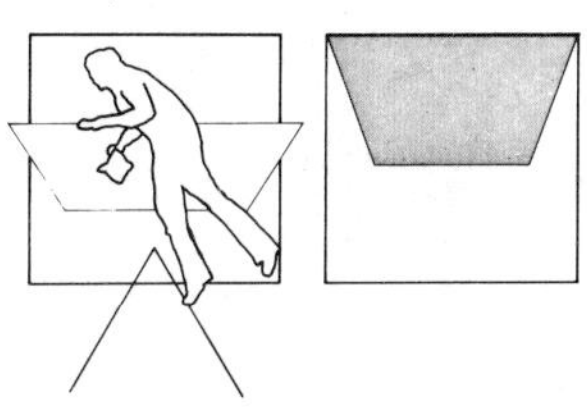 (23) Fully reversible pivot	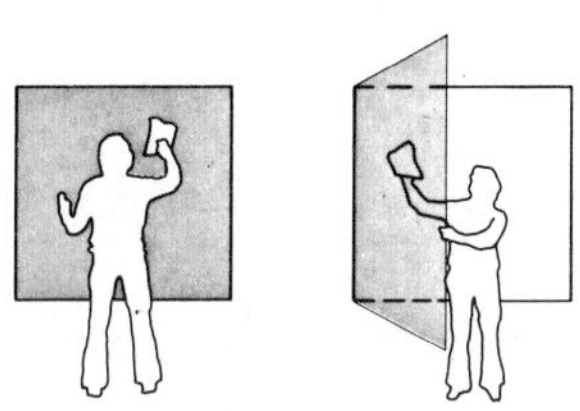 (24) Inward opening casement	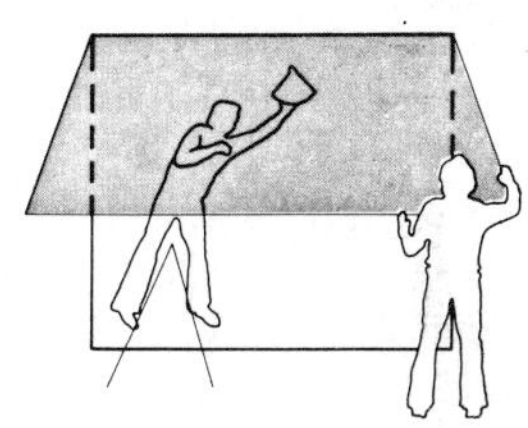 (25) Top-hung opening. Too large and distorts when held by corner

minimum 100 mm gap, set well forward of the frame which should not be fixed more than 100 mm in from the external face.

Consider using continental type *inward* opening casements which solve most window cleaning problems.

- *Double opening windows* have both side hinges and hopper hinges allowing for easy cleaning and safety.
- *Hopper windows* opening inwards must be low and narrow for easy cleaning. If high and large, they can be dangerous.
- *Vertical and horizontal sliding sash windows* should not be used for internal cleaning.
- *Horizontal and vertical pivot windows* are satisfactory for internal cleaning if they can be fully reversed and securely fixed with locking bolts both when reversed for cleaning and open normally for ventilation.
- *Sliding projecting windows* can be dangerous for internal access cleaning unless maximum depth is 750 mm, but even then cleaning can be hazardous.

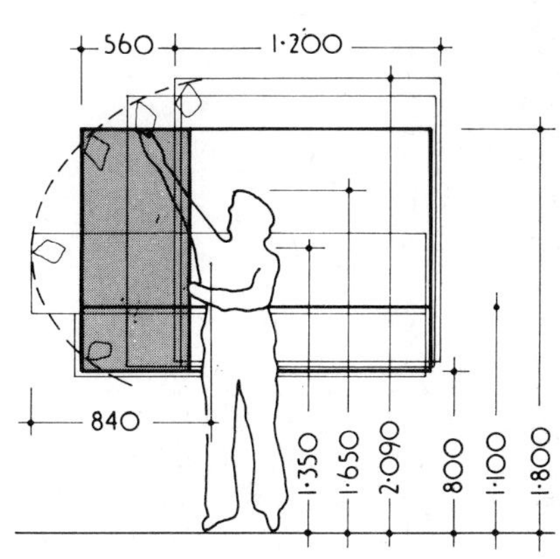

48.2 *Exterior reach to adjacent fixed light through opening light*
Shaded area is average acceptable size for ease of cleaning

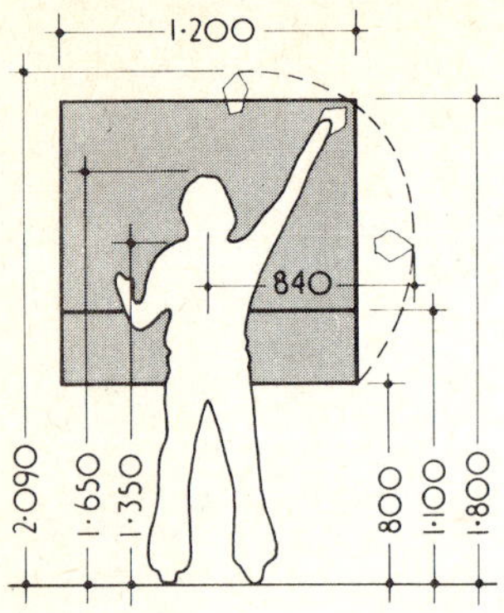

48.3 *Interior reach to fixed, reversible or pivot window*

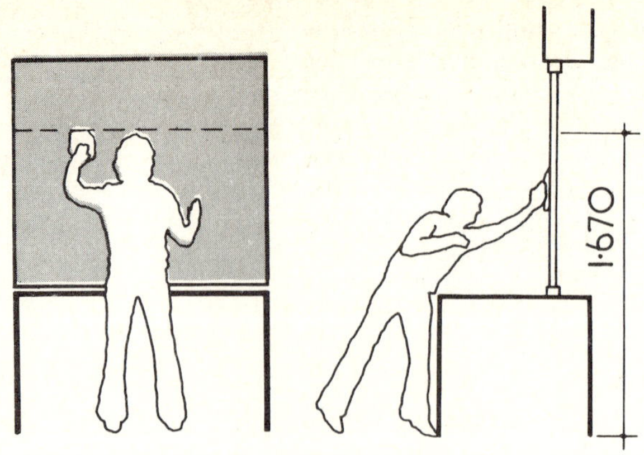

48.4 *Reach becomes less over bench or worktop*

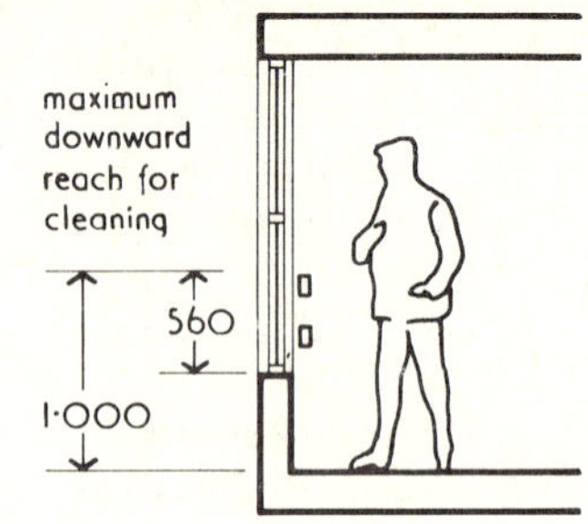

48.5 *Dimensions of fixed light heights and guard rails for domestic buildings*

4 EXTERNAL ACCESS

4.01 Manual cleaning: access from ground

Type of access can be initially assessed from **48.1**. Manual cleaning methods with access from the ground include:

- *On foot*—maximum window height 1·8 m providing there are no awkward projections.
- *Single part ladder*—up to 3 m, but awkward with long horizontal windows (use *travelling ladders*—see below **4.02**). Long-handled squeegee can sometimes be used instead.
- *Ladders over 3 m.* Usually less than 9 to 11 m; maximum 15 m, but can be difficult and dangerous. Safe inclination: 83°.
- *Mechanical ladder* on mobile chassis. Can be either free-standing or leant against a wall. More rigid than simple ladders but still only gives access to limited areas.
- *Single step ladder* in the form of a mobile trestle. Maximum height is 5·4 m.
- *Lightweight portable scaffolding.* Height is maximum three times least base dimension unless weighted, tied back to building, or outriggers fitted. Special scaffolds can be made to suit building design. Provides safe, rigid platform leaving both hands free.
- *Zip-up staging* in light, hinged aluminium alloy sections each 2·14 m high × 1·6 m long × 1·35 m wide. Height is maximum four times least base dimension, but outriggers and restraint can increase this ratio. Again, variations are possible to suit building design.
- *Mobile folding and telescopic platforms*, only for use as secondary access for difficult areas. Generally of fixed height between 12 m and 15 m.

4.02 Manual cleaning, permanent access

There are five main possibilities:

- *Balconies*: but only if *all* windows can be reached, otherwise some other forms of access will be needed.
- *Sills and ledges*: if continuous, more useful to a professional window cleaner than a balcony. Construction regulations 1966 suggest 630 mm as minimum width, but some cleaners will accept 300 to 500 mm width. But all are unsafe without permanent anchorage points for safety belts. Continuous safety belt anchorage should be provided 900 mm above the side or at the top of the window, provided it can be reached by the cleaner.
- *Catwalks*: mainly for lateral movement. Must be level and non-slip. Maximum gradient of 20° with regularly spaced stepping laths for sloping roofs; above 20° needs steps. Internal catwalks need 2 to 2·15 m headroom.

Minimum footing width 630 mm (870 mm if materials put on gangway): guard rails between 900 mm and 1150 mm above platform when more than 2 m above ground; toe boards

48.6a *Fixed ladder leading to interior catwalk*

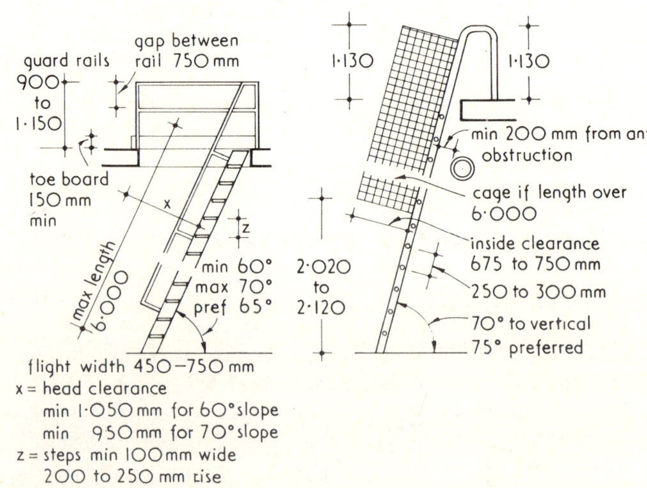

48.6b and c *Recommended dimensions for fixed ladders and landings, based on Construction Regulations 1966 (with additional information from Industrial Data Sheet 53) Australian Department of Labour*

150 mm deep with maximum distance of 750 mm between the board and lowest guard rail.

• *Fixed ladders*: use steps up to 70°, rungs over 70° pitch. Vertical ladders not recommended. Use landings every 6 m height positioned to break fall, or use metal mesh safety cage over the ladder, **48.6**.

• *Travelling ladders*: with top and bottom fixings on continuous rail or channel to allow ladder to slide along and round the façade. Useful for long bands of glazing up to 4·5 m high; can be fixed at almost any angle.

5 ROOF SUSPENSION SYSTEMS

5.01 Temporary systems

These are usually hired and erected and dismantled each time. There are two systems:

• *Counterweighted system* as shown in **48.7**. Rather unwieldy and limited; roof structure and parapet must be capable of taking load.

• *Fixed davits* as shown in **49.8**. Same problems as the counterweight system but safer, although horizontal traverse is more difficult.

There are also a few proprietary portable gantry systems.

5.02 Permanent systems: trolley units

A permanent system is usually desirable and for frequent cleaning soon covers the extra initial cost; but unless carefully designed and integrated with the structure and facade it can look very unsightly. In all cases roof structure and finishes must be able to carry the imposed loads. There are two elements to consider: the roof trolley system and the suspended chair or cradle (see **5.03**) There are two trolley systems:

• *Manual roof trolley* consisting of a continuous rail, often RSJ, positioned about 450 mm in front of the wall face, to which the cradle is attached by ropes and castors.

The most common is a pair of continuous rails, fixed to the roof about 750 mm apart, on which runs a cantilevered trolley, **48.9**.

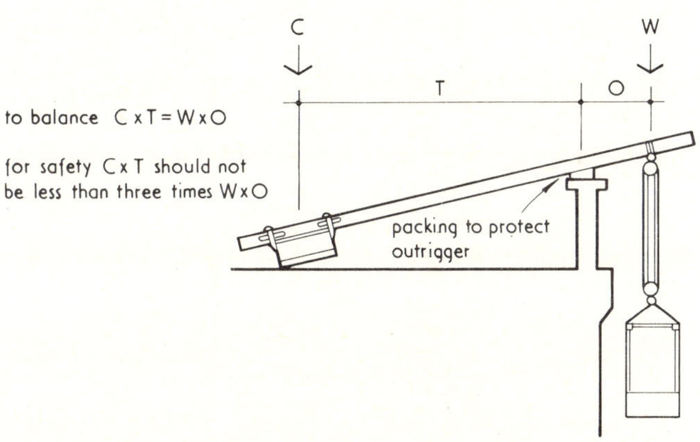

48.7 *Cradle using typical counterweight system*

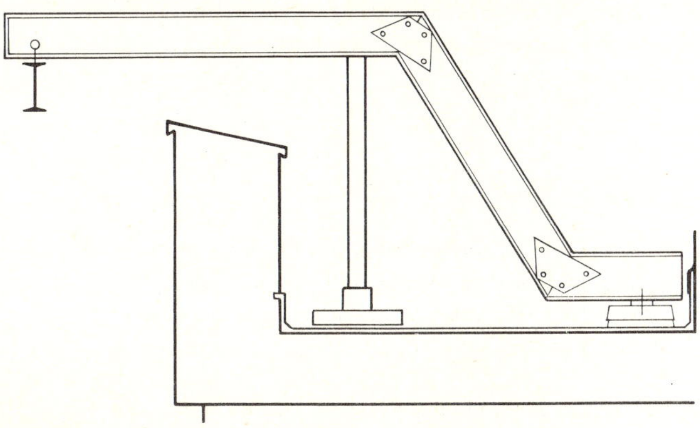

48.8 *Fixed davits on a roof with parapet*

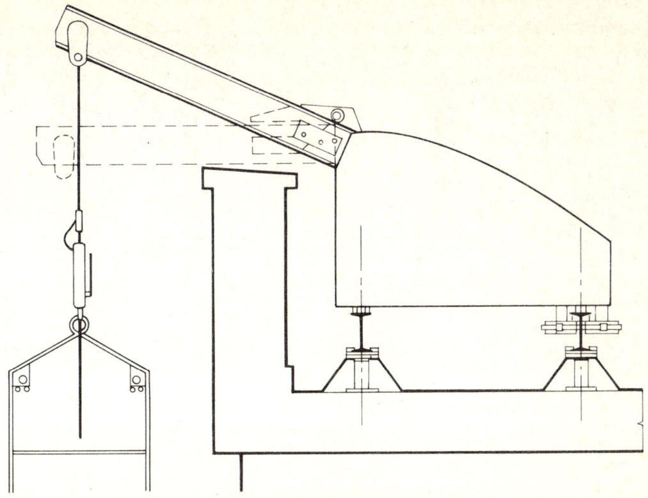

48.9 *Typical hand-operated roof trolley travelling on twin track. The boom can be lowered to the horizontal to deal with projections on the face of the building*

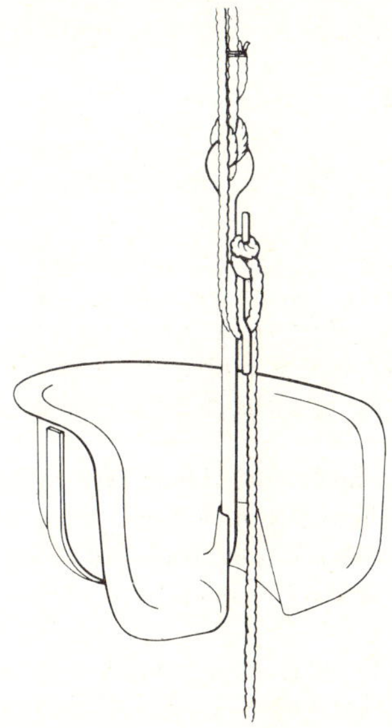

48.10 *Preformed plastic bosun's chair*

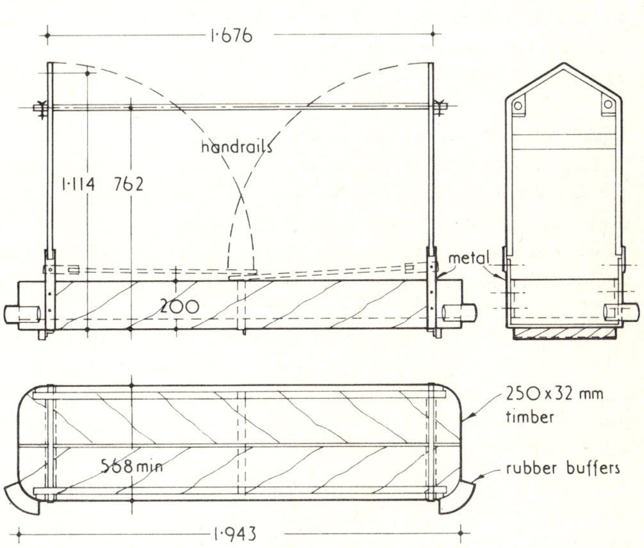

48.11 *Typical standard timber cradle*
a *Side elevation*
b *End elevation*
c *Plan*

• *Powered roof trolley* is the most efficient and safest and is essential for heights of over 45 m. It is also the most expensive, but can be relatively cheap for large buildings. It must be considered at the very earliest design stages.
The general principle is the same as the manual trolley except that the unit is powered. Power supply needed is 440 v 3 phase.

5.03 Suspended units
There are two basic types—chairs and cradles:
• *Bosun's chair*, **48.10**, extensively used for awkward areas and always used with manual gantries. A modern version is the *facing bicycle*, with pedals to work the winch.
• *Manually operated cradle.* A typical standard timber cradle is shown in **48.11**. Not recommended for heights over 30 to 45 m.
• *Power operated cradle.* Sizes range from 1·8 to 9 m width; materials can be steel, aluminium or GRP.
In all cases some form of manual or mechanical chair or cradle restraint, and of independent safety harness for the occupants, must be provided. The only method to provide continuous restraint is a *mullion guide*, **48.12**.

6 BIBLIOGRAPHY

Handbooks
AJ Handbook: Design and cleaning of windows and facades. Published in *The Architects' Journal* from 7 March 1973 to 2 May 1973 (AJ, 7.3.73; 4.4.73; 18.4.73; 2.5.73)

General
BS CP 153: Part 1: 1969 *Windows and rooflights: Part 1 Cleaning and safety*
Department of Employment Construction (Working Places) Regulations 1966
DOE *Bibliography: Current information on maintenance. Cleaning of buildings*, London, HMSO, 1972
Ministry of Public Building and Works *Window cleaning provision of means and access*, London, HMSO, 1970
R. M. Rostron *Light cladding*, London, Architectural Press, 1964
L. Wright, The cleaning of large buildings, *The Architectural Review*, April 1960, pp 283–6

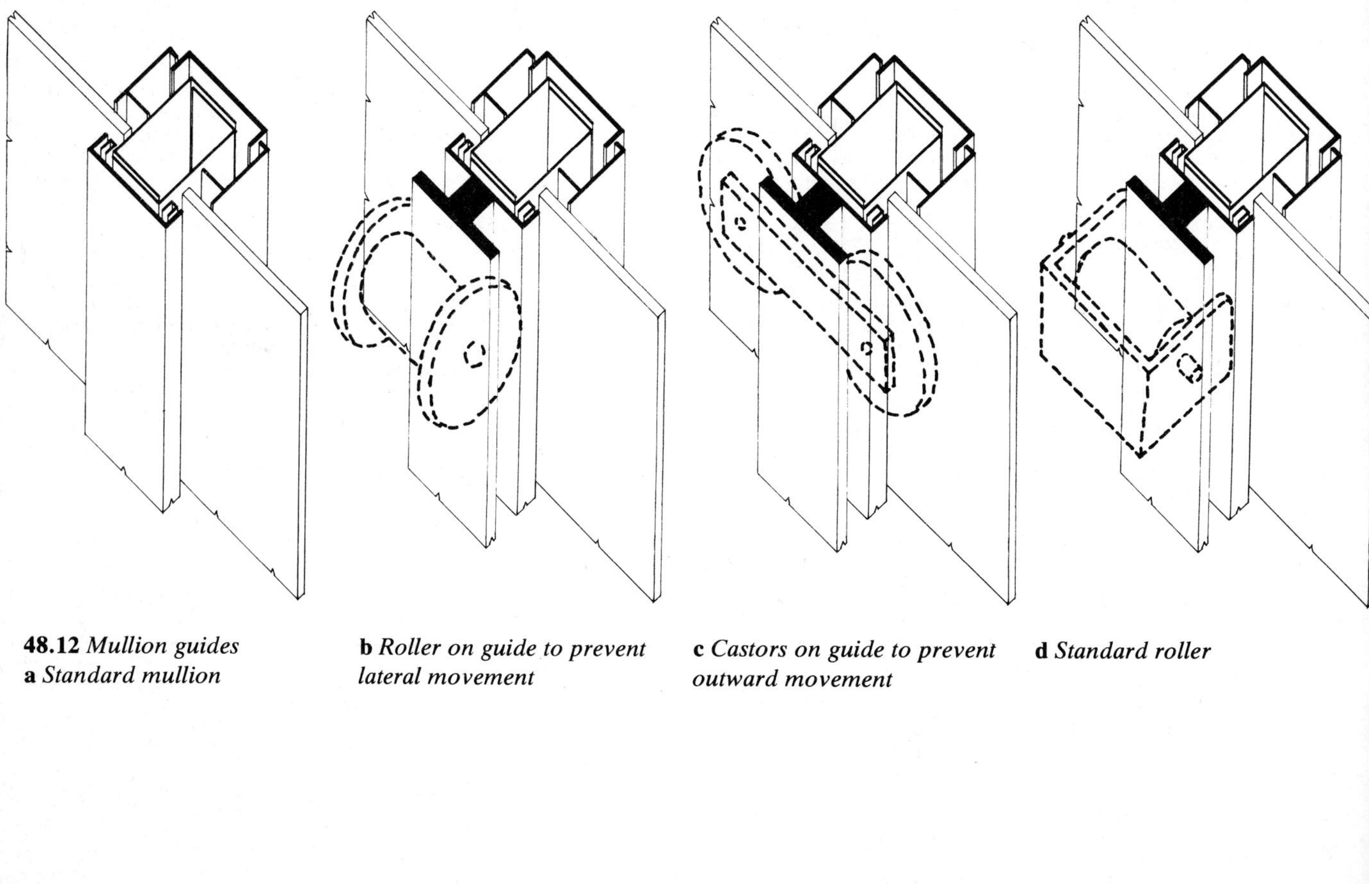

48.12 *Mullion guides*
a *Standard mullion*
b *Roller on guide to prevent lateral movement*
c *Castors on guide to prevent outward movement*
d *Standard roller*

Appendix
Conversion factors and tables

Table 1 Conversion factors

Bold type indicates exact conversions. Otherwise four or five significant figures are given.

Quantity	Conversion factors	
General purposes		
Length	1 mile	= 1·609 km
	1 chain	= **20·1168** m
	1 yard	= **0·9144** m
	1 foot	= **0·3048** m = **304·8** mm
	1 inch	= **25·4** mm = **2·54** cm
Area	1 square mile	= 2·590 km^2 = 259·0 ha
	1 hectare	= **10000** m^2
	1 acre	= 4046·9 m^2 = 0·40469 ha
	1 square yard	= 0·8361 m^2
	1 square foot	= 0·09290 m^2 = 929·03 cm^2
	1 square inch	= 645·2 mm^2 = 6·452 cm^2
Volume	1 cubic yard	= 0·7646 m^3
	1 litre	= **1** dm^3 = **1000** cm^3
	1 millilitre	= **1** cm^3 = **1000** mm^3
	1 cubic foot	= 0·02832 m^3 = 28·32 litre
	1 petrograd standard	= 4·672 m^3
	1 cubic inch	= 16387 mm^3 = 16·387 cm^3 = 16·387 ml = 0·016387 litre
Capacity	1 UK gallon	= 4·546 litre
	1 UK quart	= 1·137 litre
	1 UK pint	= 0·5683
	1 UK fluid ounce	= 28·413 cm^3
	1 US barrel (for petroleum)	= 159·0 litre
	1 US gallon	= 3·785 litre
	1 US liquid quart	= 0·9464 litre
	1 US dry quart	= 1·101 litre
	1 US liquid pint	= 0·4732 litre
	1 US dry pint	= 0·5506 litre
	1 US liquid ounce	= 29·574 cm^3
Mass	1 UK ton	= 1·016 tonne = 1016·05 kg
	1 US (or short) ton	= 0·9072 tonne = 907·2 kg
	1 kip (1000 lb)	= 453·59 kg
	1 UK hundredweight	= 50·80 kg
	1 short (US) hundredweight	= 100 lb = 45·36 kg
	1 pound	= 0·4536 kg
	1 ounce avoirdupois	= 28·35 g
	1 ounce troy	= 31·10 g
Mass per unit length	1 UK ton per mile	= 0·6313 kg/m = 0·6313 t/km
	1 lb per yard	= 0·4961 kg/m
	1 lb per foot	= 1·4882 kg/m
	1 lb per inch	= 17·86 kg/m
	1 oz per inch	= 1·1161 kg/m
Length per unit mass	1 yd per lb	= 2·016 m/kg
Mass per unit area	1 ton per square mile	= 392·3 kg/km^2 = 0·3923 g/m^2 = 3·923 kg/ha
	1 ton per acre	= 0·2511 kg/m^2
	1 hundredweight per acre	= 0·01255 kg/m^2
	1 lb per square foot	= 4·882 kg/m^2
	1 lb per square inch	= 703·07 kg/m^2
	1 oz per square yard	= 33·91 g/m^2
	1 oz per square foot	= 305·15 g/m^2
	1 kg/cm^2	= 10 t/m^2
Mass density (mass per unit volume)	1 ton per cubic yard	= 1329 kg/m^3 = 1·3289 t/m^3
	1 lb per cubic yard	= 0·5933 kg/m^3
	1 lb per cubic foot	= 16·02 kg/m^3
	1 lb per cubic inch	= 27·68 g/cm^3 = 27·68 t/m^3

Table I (*continued*)

Area coverage	x square yards per ton	$=\frac{1}{x}\times$ 1215 kg/m^2
	x square yards per gallon	$=\frac{1}{x}\times$ 5·437 $litre/m^2$
Volume rate of flow	1 cubic feet per minute	= 0·4719 litre/s = 471·9 cm^3/s = 0·000 4719 m^3/s
	1 cusec (cu ft per sec)	= 0·02832 m^3/s ('cumec')
	1 cu ft per thousand acres	= 0·06997 litre/ha = 0·006997 m^3/km^2 = 6997 cm^3/km^2
	1 cubic inch per second	= 16·39 ml/s
	1 gallon per year	= 4546 cm^3/a* = 0·004546 m^3/a
	1 gallon per day	= 4546 cm^3/d
	1 litre/s	= **86·4** m^3/d
	1 million gallons per day	= 0·05262 m^3/s
	1 gallon per person per day	= 4·546 litre/(person day)
	1 gallon per sq yd per day	= 0·005437 $m^3/(m^2\cdot d)$ = 0·000062928 mm/s
	1 gallon per cu yd per day	= 0·005946 $m^3/(m^3\cdot d)$
	1 gallon per hour	= 4·5461 litre/h
	1 gallon per minute	= 0·07577 litre/s
	1 gallon per second	= 4·5461 litre/s
Fuel consumption	1 gallon per mile	= 2·825 litre/km
	1 mile per gallon	= 0·354 km/litre
	x miles per gallon	$=\frac{1}{x}\times$ 282·5 litre 100km
Velocity	1 mile per hour	= 1·609 km/h = **0·44704** m/s
	1 foot per minute	= **0·3048** m/min = 0·0051 m/s
	1 foot per second	= **0·3048** m/s
	1 inch per second	= **25·4** mm/s
	1 UK knot	= 0·5148 m/s = 1·853 km/h = 1·00064 international knot
Acceleration	1 foot per sec per sec	= **0·3048** m/s^2
	1 mile per hr per sec	= **0·44704** m/s^2
	1 g (standard gravity)	= **9·80665** m/s^2
Heating		
Temperature	x° Fahrenheit	$=\frac{5}{9}\times(x-32)$ ° Celsius
Temperature interval	1° F	= 0·5556 K = 0·5556 °C
Energy (heat)	1 British thermal unit	= 1055 J = 1·055 kJ
	1 Therm	= 105·5 MJ
	1 calorie	= **4·1868** J
	1 kilowatt-hour	= **3·6** MJ
	1 foot pound-force	= 1·356 J
	1 kilogram force-metre	= **9·806 65** J
Power (also heat flow rate)	1 Btu per hour	= 0·293 07 W
	1 horsepower	= 745·70 W
	1 ft-lbf per second	= 1·356 W
	1 kgf-metre per second	= **9·806 65** W
	1 calorie per second	= **4·1868** W
	1 kilocalorie per hour	= **1·163** W
	1 metric horsepower	= 735·5 W
Density of heat flow rate	1 Btu per square foot hour	= 3·155 W/m^2
Thermal conductivity k value	1 Btu inch per square foot hour degree Fahrenheit	= 0·1442 W/(m.K)
Thermal transmittance or coefficient of heat transfer or thermal conductance or U value	1 Btu per square foot hour degree Fahrenheit	= 5·678 $W/(m^2K)$

*a (for annum) is the symbol for year.

Table I Conversion factors *(continued)*

Thermal resistivity $\frac{1}{k}$ value	1 sq ft hr °F per Btu inch	= 6·933 m.K/W
Thermal or specific heat capacity	1 Btu per lb °F	= 4·187 kJ/(kg.K)
	1 Btu per cu ft °F	= 67·07 kJ/(m^3.K)
Calorific value	1 Btu per pound	= **2·326** kJ/kg
	1 Btu per cubic foot	= 37·26 kJ/m^3 = 37·26 J/litre
	1 Btu per gallon	= 232·1 J/litre
Refrigeration	1 ton	= 3517 W
Lighting Illumination	1 foot-candle	= 10·76 lx
	1 lumen per sq ft	= 10·76 lx
Luminance	1 candela per square inch	= 1550 cd/m^2
	1 candela per square foot	= 10·76 cd/m^2
	1 apostilb	= $\frac{1}{\pi}$ cd/m^2 = 0·3183 cd/m^2
Structural design (All tons are UK tons)		
Force	1 pound-force	= 4·448 N
	1 kip-force	= 4·448 kN
	1 ton-force	= 9·964 kN
	1 kilogram-force	= 9·807 N
	1 kilopond	= 9·807 N

Table I *(continued)*

Force per unit length	1 pound-force per foot	= 14·59 N/m
	1 pound-force per inch	= 175·1 kN/m = 175·1 N/mm
	1 ton-force per foot	= 32·69 kN/m
	1 kilogram-force per metre	= 9·807 N/m
	1 kilogram-force per centimetre	= 0·9807 kN/m
Force per unit area or Stress or Pressure	1 lbf per square foot	= 47·88 N/m^2 = 47·88 Pa = 0·04788 kN/m^2
	1 lbf per square inch	= 6·895 kN/m^2 = 6·895 kPa
	1 tonf per square foot	= 107·3 kN/m^2 = 107·3 kPa
	1 tonf per square inch	= 15·44 MN/m^2 = 15·44 N/mm^2 = 15·44 MPa
	1 kgf per square metre	= 9·807 N/m^2 = 9·807 Pa
	1 kgf per sq centimetre	= 98·07 kN/m^2 = 98·07 kPa
	1 bar	= **100** kN/m^2 = **100** kPa
	1 millibar	= **100** N/m^2 = **100** Pa
	1 standard atmosphere	= **101·325** kPa
	1 inch of mercury	= 3·386 kPa
	1 foot of water	= 2·989 kPa
Bending moment of torque	1 pound-force foot	= 1·356 Nm
	1 pound-force inch	= 0·1130 Nm = 113·0 Nmm
	1 kip-force foot	= 1·356 kNm
	1 kip-force inch	= 0·1130 kNm = 113·0 Nm
	1 ton-force foot	= 3·037 kNm
	1 ton-force inch	= 0·2531 kNm = 253·1 Nm
	1 kilogram-force metre	= 9·807 Nm

Table II Inches and fractions of an inch to millimetres (1/16 in increments up to 11 15/16 in)

	Inches	1/16	1/8	3/16	1/4	5/16	3/8	7/16	1/2	9/16	5/8	11/16	3/4	13/16	7/8	15/16
		millimetres														
		1·6	3·2	4·8	6·4	7·9	9·5	11·1	12·7	14·3	15·9	17·5	19·1	20·6	22·2	23·8
1	25·4	27·0	28·6	30·2	31·8	33·3	34·9	36·5	38·1	39·7	41·3	42·9	44·5	46·0	47·6	49·2
2	50·8	52·4	54·0	55·6	57·2	58·7	60·3	61·9	63·5	65·1	66·7	68·3	69·9	71·4	73·0	74·6
3	76·2	77·8	79·4	81·0	82·6	84·1	85·7	87·3	88·9	90·5	92·1	93·7	95·3	96·8	98·4	100·0
4	101·6	103·2	104·8	106·4	108·0	109·5	111·1	112·7	114·3	115·9	117·5	119·1	120·7	122·2	123·8	125·4
5	127·0	128·6	130·2	131·8	133·4	134·9	136·5	138·1	139·7	141·3	142·9	144·5	146·1	147·6	149·2	150·8
6	152·4	154·0	155·6	157·2	158·8	160·3	161·9	163·5	165·1	166·7	168·3	169·9	171·5	173·0	174·6	176·2
7	177·8	179·4	181·0	182·6	184·2	185·7	187·3	188·9	190·5	192·1	193·7	195·3	196·9	198·4	200·0	201·6
8	203·2	204·8	206·4	208·0	209·6	211·1	212·7	214·3	215·9	217·5	219·1	220·7	222·3	223·8	225·4	227·0
9	228·6	230·2	231·8	233·4	235·0	236·5	238·1	239·7	241·3	242·9	244·5	246·1	247·7	249·2	250·8	252·4
10	254·0	255·6	257·2	258·8	260·4	261·9	263·5	265·1	266·7	268·3	269·9	271·5	273·1	274·6	276·2	277·8
11	279·4	281·0	282·6	284·2	285·8	287·3	288·9	290·5	292·1	293·7	295·3	296·9	298·5	300·0	301·6	303·2

Table III Feet to millimetres (up to 200 ft)

feet	0	1	2	3	4	5	6	7	8	9
	Millimetres									
		304·8	609·6	914·4	1219·2	1524·0	1828·8	2133·6	2438·4	2743·2
10	3048·0	3352·8	3657·6	3962·4	4267·2	4572·0	4876·8	5181·6	5486·4	5791·2
20	6096·0	6400·8	6705·6	7010·4	7315·2	7620·0	7924·8	8229·6	8534·4	8839·2
30	9144·0	9448·8	9753·6	10058·4	10363·2	10668·0	10972·8	11277·6	11582·4	11887·2
40	12192·0	12496·8	12801·6	13106·4	13411·2	13716·0	14020·8	14325·6	14630·4	14935·2
50	15240·0	15544·8	15849·6	16154·4	16459·2	16764·0	17068·8	17373·6	17678·4	17983·2
60	18288·0	18592·8	18897·6	19202·4	19507·2	19812·0	20116·8	20421·6	20726·4	21031·2
70	21336·0	21640·8	21945·6	22250·4	22555·2	22860·0	23164·8	23469·6	23774·4	24079·2
80	24384·0	24688·8	24993·6	25298·4	25603·2	25908·0	26212·8	26517·6	26822·4	27127·2
90	27432·0	27736·8	28041·6	28346·4	28651·2	28956·0	29260·8	29565·6	29870·4	30175·2
100	30480·0	30784·8	31089·6	31394·4	31699·2	32004·0	32308·8	32613·6	32918·4	33223·2
110	33528·0	33832·8	34137·6	34442·4	34747·2	35052·0	35356·8	35661·6	35966·4	36271·2
120	36576·0	36880·8	37185·6	37490·4	37785·2	38100·0	38404·8	38709·6	39014·4	39319·2
130	39624·0	39928·8	40233·6	40538·4	40843·2	41148·0	41452·8	41757·6	42062·4	42367·2
140	42672·0	42976·8	43281·6	43586·4	43891·2	44196·0	44500·8	44805·6	45110·4	45415·2
150	45720·0	46024·8	46329·6	46634·4	46939·2	47244·0	47548·8	47853·6	48158·4	48463·2
160	48768·0	49072·8	49377·6	49682·4	49987·2	50292·0	50596·8	50901·6	51206·4	51511·2
170	51816·0	52120·8	52425·6	52730·4	53035·2	53340·0	53644·8	53949·6	54254·4	54559·2
180	54864·0	55168·8	55473·6	55778·4	56083·2	56388·0	56692·8	56997·6	57302·4	57607·2
190	57912·0	58216·8	58521·6	58826·4	59131·2	59436·0	59740·8	60045·6	60350·4	60655·2
200	60960·0									

Note: use tables II and III together to obtain the metric equivalent of any dimension up to 200 ft. For example

56 ft 3¾ in: 56 ft = 17068·8
3¾ in = 95·3

TOTAL = 17164·1 mm = 17·164 m

Table IV Miles (up to 100 miles) to kilometres (to two places of decimals)

Miles	0	1	2	3	4	5	6	7	8	9
	Kilometres									
0	—	1·61	3·22	4·83	6·44	8·05	9·66	11·27	12·87	14·48
10	16·09	17·70	19·31	20·92	22·53	24·14	25·75	27·36	28·97	30·58
20	32·19	33·80	35·41	37·01	38·62	40·23	41·84	43·45	45·06	46·67
30	48·28	49·89	51·50	53·11	54·72	56·33	57·94	59·55	61·16	62·76
40	64·37	65·98	67·59	69·20	70·81	72·42	74·03	75·64	77·25	78·86
50	80·47	82·08	83·69	85·30	86·90	88·51	90·12	91·73	93·34	94·95
60	96·56	98·17	99·78	101·39	103·00	104·61	106·22	107·83	109·44	111·05
70	112·65	114·26	115·87	117·48	119·09	120·70	122·31	123·92	125·53	127·14
80	128·75	130·36	131·97	133·58	135·19	136·79	138·40	140·01	141·62	143·23
90	144·84	146·45	148·06	149·67	151·28	152·89	154·50	156·11	57·72	159·33
100	160·93	—	—	—	—	—	—	—	—	—

Table V Square inches (up to 100 sq in) to square millimetres (to one place of decimals)

Square inches	0	1	2	3	4	5	6	7	8	9
	Square millimetres (mm²)									
0	—	645·2	1 290·3	1 935·5	2 580·6	3 225·8	3 871·0	4 516·1	5 161·3	5 806·4
10	6 451·6	7 096·8	7 741·9	8 387·1	9 032·2	9 677·4	10 322·6	10 967·7	11 612·9	12 258·0
20	12 903·2	13 548·4	14 193·5	14 838·7	15 483·8	16 129·0	16 774·2	17 419·3	18 064·5	18 709·6
30	193 54·8	20 000·0	20 645·1	21 290·3	21 935·4	22 580·6	23 225·8	23 870·9	24 516·1	25 161·2
40	25 806·4	26 451·6	27 096·7	27 741·9	28 387·0	29 032·2	29 677·4	30 322·5	30 967·7	31 612·8
50	32 258·0	32 903·2	33 548·3	34 193·5	34 838·6	35 483·8	36 129·0	36 774·1	37 419·3	38 064·4
60	38 709·6	39 354·8	39 999·9	40 645·1	41 290·2	41 935·4	42 580·6	43 225·7	43 870·9	44 516·0
70	45 161·2	45 806·4	46 451·5	47 096·7	47 741·8	48 387·0	49 032·2	49 677·3	50 322·5	50 967·6
80	51 612·8	52 258·0	52 903·1	53 548·3	54 193·4	54 838·6	55 483·8	56 128·9	56 774·1	57 419·2
90	58 064·4	58 709·6	59 354·7	59 999·9	60 645·0	61 290·2	61 935·4	62 580·5	63 225·7	63 870·8
100	64 516·0									

Table VI Square feet (up to 500 ft²) to square metres (to two places of decimals)

Square feet	0	1	2	3	4	5	6	7	8	9
	Square metres (m²)									
0	—	0·09	0·19	0·28	0·37	0·46	0·56	0·65	0·74	0·84
10	0·93	1·02	1·11	1·21	1·30	1·39	1·49	1·58	1·67	1·77
20	1·86	1·95	2·04	2·14	2·23	2·32	2·42	2·51	2·60	2·69
30	2·79	2·88	2·97	3·07	3·16	3·25	3·34	3·44	3·53	3·62
40	3·72	3·81	3·90	3·99	4·09	4·18	4·27	4·37	4·46	4·55
50	4·65	4·74	4·83	4·92	5·02	5·11	5·20	5·30	5·39	5·48
60	5·57	5·67	5·76	5·85	5·95	6·04	6·13	6·22	6·32	6·41
70	6·50	6·60	6·69	6·78	6·87	6·97	7·06	7·15	7·25	7·34
80	7·43	7·53	7·62	7·71	7·80	7·90	7·99	8·08	8·18	8·27
90	8·36	8·45	8·55	8·64	8·73	8·83	8·92	9·01	9·10	9·20
100	9·29	9·38	9·48	9·57	9·66	9·75	9·85	9·94	10·03	10·13
110	10·22	10·31	10·41	10·50	10·59	10·68	10·78	10·87	10·96	11·06
120	11·15	11·24	11·33	11·43	11·52	11·61	11·71	11·80	11·89	11·98
130	12·08	12·17	12·26	12·36	12·45	12·54	12·63	12·73	12·82	12·91
140	13·01	13·10	13·19	13·29	13·38	13·47	13·56	13·66	13·75	13·84
150	13·94	14·03	14·12	14·21	14·31	14·40	14·49	14·59	14·68	14·77
160	14·86	14·96	15·05	15·14	15·24	15·33	15·42	15·51	15·61	15·70
170	15·79	15·89	15·98	16·07	16·17	16·26	16·35	16·44	16·54	16·63
180	16·72	16·82	16·91	17·00	17·09	17·19	17·28	17·37	17·47	17·56
190	17·65	17·74	17·84	17·93	18·02	18·12	18·21	18·30	18·39	18·49
200	18·58	18·67	18·77	18·86	18·95	19·05	19·14	19·23	19·32	19·42
210	19·51	19·60	19·70	19·79	19·88	19·97	20·07	20·16	20·25	20·35
220	20·44	20·53	20·62	20·72	20·81	20·90	21·00	21·09	21·18	21·27
230	21·37	21·46	21·55	21·65	21·74	21·83	21·93	22·02	22·11	22·20
240	22·30	22·39	22·48	22·58	22·67	22·76	22·85	22·95	23·04	23·13
250	23·23	23·32	23·41	23·50	23·60	23·69	23·78	23·88	23·97	24·06
260	24·15	24·25	24·34	24·43	24·53	24·62	24·71	24·81	24·90	24·99
270	25·08	25·18	25·27	25·36	25·46	25·55	25·64	25·73	25·83	25·92
280	26·01	26·11	26·20	26·29	26·38	26·48	26·57	26·66	26·76	26·85
290	26·94	27·03	27·13	27·22	27·31	27·41	27·50	27·59	27·69	27·78
300	27·87	27·96	28·06	28·15	28·24	28·34	28·43	28·52	28·61	28·71
310	28·80	28·89	28·99	29·08	29·17	29·26	29·36	29·45	29·54	29·64
320	29·73	29·82	29·91	30·01	30·10	30·19	30·29	30·38	30·47	30·57
330	30·66	30·75	30·84	30·94	31·03	31·12	31·22	31·31	31·40	31·49
340	31·59	31·68	31·77	31·87	31·96	32·05	32·14	32·24	32·33	32·42

Table VI Square feet (up to 500 ft)²) to square metres (to two places of decimals) (*continued*)

Square feet	0	1	2	3	4	5	6	7	8	9
	Square metres (m²)									
350	32·52	32·61	32·70	32·79	32·89	32·98	33·07	33·17	33·26	33·35
360	33·45	33·54	33·63	33·72	33·82	33·91	34·00	34·10	34·19	34·28
370	34·37	34·47	34·56	34·65	34·75	34·84	34·93	35·02	35·12	35·21
380	35·30	35·40	35·49	35·58	35·67	35·77	35·86	35·95	36·05	36·14
390	36·23	36·33	36·42	36·51	36·60	36·70	36·79	36·88	36·98	37·07
400	37·16	37·25	37·35	37·44	37·53	37·63	37·72	37·81	37·90	38·00
410	38·09	38·18	38·28	38·37	38·46	38·55	38·65	38·74	38·83	38·93
420	39·02	39·11	39·21	39·30	39·39	39·48	39·58	39·67	39·76	39·86
430	39·95	40·04	40·13	40·23	40·32	40·41	40·51	40·60	40·69	40·78
440	40·88	40·97	41·06	41·16	41·25	41·34	41·43	41·53	41·62	41·71
450	41·81	41·90	41·99	42·09	42·18	42·27	42·36	42·46	42·55	42·64
460	42·74	42·83	42·92	43·01	43·11	43·20	43·29	43·39	43·48	43·57
470	43·66	43·76	43·85	43·94	44·04	44·13	44·22	44·31	44·41	44·50
480	44·59	44·69	44·78	44·87	44·97	45·06	45·15	45·24	45·34	45·43
490	45·52	45·62	45·71	45·80	45·89	45·99	46·08	46·17	46·27	46·36
500	46·45									

Table VII Cubic feet (up to 100 ft³) to cubic metres (to two places of decimals)

Cubic feet	0	1	2	3	4	5	6	7	8	9
	Cubic metres (m³)									
0	—	0·03	0·06	0·08	0·11	0·14	0·17	0·20	0·23	0·25
10	0·28	0·31	0·34	0·37	0·40	0·42	0·45	0·48	0·51	0·54
20	0·57	0·59	0·62	0·65	0·68	0·71	0·73	0·76	0·79	0·82
30	0·85	0·88	0·91	0·93	0·96	0·99	1·02	1·05	1·08	1·10
40	1·13	1·16	1·19	1·22	1·25	1·27	1·30	1·33	1·36	1·39
50	1·42	1·44	1·47	1·50	1·53	1·56	1·59	1·61	1·64	1·67
60	1·70	1·73	1·76	1·78	1·81	1·84	1·87	1·90	1·93	1·95
70	1·98	2·01	2·04	2·07	2·10	2·12	2·15	2·18	2·21	2·24
80	2·27	2·29	2·32	2·35	2·38	2·41	2·44	2·46	2·49	2·52
90	2·55	2·58	2·61	2·63	2·66	2·69	2·72	2·75	2·78	2·80
100	2·83	—	—	—	—	—	—	—	—	—

Table VIII Pounds (up to 500 lb) to kilogrammes (to two places of decimals)

Pounds	0	1	2	3	4	5	6	7	8	9
	Kilogrammes (kg)									
0	—	0·45	0·91	1·36	1·81	2·27	2·72	3·18	3·63	4·08
10	4·54	4·99	5·44	5·90	6·35	6·80	7·26	7·71	8·16	8·62
20	9·07	9·53	9·98	10·43	10·89	11·34	11·79	12·25	12·70	13·15
30	13·61	14·06	14·52	14·97	15·42	15·88	16·33	16·78	17·24	17·69
40	18·14	18·60	19·05	19·50	19·96	20·41	20·87	21·32	21·77	22·23
50	22·68	23·13	23·59	24·04	24·49	24·95	25·40	25·85	26·31	26·76
60	27·22	27·67	28·12	28·58	29·03	29·48	29·94	30·39	30·84	31·30
70	31·75	32·21	32·66	33·11	33·57	34·02	34·47	34·93	35·38	35·83
80	36·29	36·74	37·19	37·65	38·10	38·56	39·01	39·46	39·92	40·37
90	40·82	41·28	41·73	42·18	42·64	43·09	43·54	44·00	44·45	44·91
100	45·36	45·81	46·27	46·72	47·17	47·63	48·08	48·53	48·99	49·44
110	49·90	50·35	50·80	51·26	51·71	52·16	52·62	53·07	53·52	53·98
120	54·43	54·88	55·34	55·79	56·25	56·70	57·15	57·61	58·06	58·51
130	58·97	59·42	59·87	60·33	60·78	61·24	61·69	62·14	62·60	63·05
140	63·50	63·96	64·41	64·86	65·32	65·77	66·22	66·68	67·13	67·59
150	68·04	68·49	68·95	69·40	69·85	70·31	70·76	71·21	71·67	72·12
160	72·57	73·03	73·48	73·94	74·39	74·84	75·30	75·75	76·20	76·66
170	77·11	77·56	78·02	78·47	78·93	79·38	79·83	80·29	80·74	81·19
180	81·65	82·10	82·55	83·01	83·46	83·91	84·37	84·82	85·28	85·73
190	86·18	86·64	87·09	87·54	88·00	88·45	88·90	89·36	89·81	90·26
200	90·72	91·17	91·63	92·08	92·53	92·99	93·44	93·89	94·35	94·80
210	95·25	95·71	96·16	96·62	97·07	97·52	97·98	98·43	98·88	99·34
220	99·79	100·24	100·70	101·15	101·61	102·06	102·51	102·97	103·42	103·87
230	104·33	104·78	105·23	105·69	106·14	106·59	107·05	107·50	107·96	108·41
240	108·86	109·32	109·77	110·22	110·68	111·13	111·58	112·04	112·49	112·95
250	113·40	113·85	114·31	114·76	115·21	115·67	116·12	116·57	117·03	117·48
260	117·93	118·39	118·84	119·30	119·75	120·20	120·66	121·11	121·56	122·02
270	122·47	122·92	123·38	123·83	124·28	124·74	125·19	125·65	126·10	126·55
280	127·01	127·46	127·91	128·37	128·82	129·27	129·73	130·18	130·64	131·09
290	131·54	132·00	132·45	132·90	133·36	133·81	134·26	134·72	135·17	135·62

Table VIII Pounds (up to 500 lb) to kilogrammes (to two places of decimals) *(continued)*

Pounds	0	1	2	3	4	5	6	7	8	9
	Kilogrammes (kg)									
300	136·08	136·53	136·99	137·44	137·89	138·35	138·80	139·25	139·71	140·16
310	140·61	141·07	141·52	141·97	142·43	142·88	143·34	143·79	144·24	144·70
320	145·15	145·60	146·06	146·51	146·96	147·42	147·87	148·33	148·78	149·23
330	149·69	150·14	150·59	151·05	151·50	151·95	152·41	152·86	153·31	153·77
340	154·22	154·68	155·13	155·58	156·04	156·49	156·94	157·40	157·85	158·30
350	158·76	159·21	159·67	160·12	160·57	161·03	161·48	161·93	162·39	162·84
360	163·29	163·75	164·20	164·65	165·11	165·56	166·02	166·47	166·92	167·38
370	167·83	168·28	168·74	169·10	169·64	170·10	170·55	171·00	171·46	171·91
380	172·37	172·82	173·27	173·73	174·18	174·63	175·09	175·54	175·99	176·45
390	176·90	177·36	177·81	178·26	178·72	179·17	179·62	180·08	180·53	180·98
400	181·44	181·89	183·34	182·80	183·25	183·71	184·16	184·61	185·07	185·52
410	185·97	186·43	186·88	187·33	187·79	188·24	188·69	189·15	189·60	190·06
420	190·51	190·96	191·42	191·87	192·32	192·78	193·23	193·68	194·14	194·59
430	195·05	195·50	195·95	196·41	196·86	197·31	197·77	198·22	198·67	199·13
440	199·58	200·03	200·49	200·94	201·40	201·85	202·30	202·76	203·21	203·66
450	204·12	204·57	205·02	205·48	205·93	206·39	206·84	207·29	207·75	208·20
460	208·65	209·11	209·56	210·01	210·47	210·92	211·37	211·83	212·28	212·74
470	213·19	213·64	214·10	214·55	215·00	215·46	215·91	216·36	216·82	217·27
480	217·72	218·18	218·63	219·09	219·54	219·99	220·45	220·90	221·35	221·81
490	222·26	222·71	223·17	223·62	224·08	224·53	224·98	225·44	225·89	226·34
500	226·80	—	—	—	—	—	—	—	—	—

Table IX Pounds per cubic foot to kilogrammes per cubic metre (to one place of decimals)

Pounds per cubic foot	0	1	2	3	4	5	6	7	8	9
	Kilogrammes per cubic metre (kg/m³)									
0	—	16·0	32·0	48·1	64·1	80·1	96·1	112·1	128·1	144·2
10	160·2	176·2	192·2	208·2	224·3	240·3	256·3	272·3	288·3	304·4
20	320·4	336·4	352·4	368·4	384·4	400·5	416·5	432·5	448·5	464·5
30	480·6	496·6	512·6	528·6	544·6	560·6	576·7	592·7	608·7	624·7
40	640·7	656·8	672·8	688·8	704·8	720·8	736·8	752·9	768·9	784·9
50	800·9	816·9	833·0	849·0	865·0	881·0	897·0	913·1	929·1	945·1
60	961·1	977·1	993·1	1009·2	1025·2	1041·2	1057·2	1073·2	1089·3	1105·3
70	1121·3	1137·3	1153·3	1169·4	1185·4	1201·4	1217·4	1233·4	1249·4	1265·5
80	1281·5	1297·5	1313·5	1329·5	1345·6	1361·6	1377·6	1393·6	1409·6	1425·6
90	1441·7	1457·7	1473·7	1489·7	1505·7	1521·8	1537·8	1553·8	1569·8	1585·8
100	1601·9	—	—	—	—	—	—	—	—	

Table X UK gallons (up to 100 galls) to litres (to two places of decimals)

UK gallons	0	1	2	3	4	5	6	7	8	9
	Litres									
0	—	4·55	9·09	13·64	18·18	22·73	27·28	31·82	36·37	40·91
10	45·46	50·01	54·55	59·10	63·64	68·19	72·74	77·28	81·83	86·37
20	90·92	95·47	100·01	104·56	109·10	113·65	118·20	122·74	127·29	131·83
30	136·38	140·93	145·47	150·02	154·56	159·11	163·66	168·20	172·75	177·29
40	181·84	186·38	190·93	195·48	200·02	204·57	209·11	213·66	218·21	222·75
50	227·30	231·84	236·39	240·94	245·48	250·03	254·57	259·12	263·67	268·21
60	272·76	277·30	281·85	286·40	290·94	295·49	300·03	304·58	309·13	313·67
70	318·22	322·76	327·31	331·86	336·40	340·95	345·49	350·04	354·59	359·13
80	363·68	368·22	372·77	377·32	381·86	386·41	390·95	395·50	400·04	404·59
90	409·14	413·68	418·23	422·77	427·32	431·87	436·41	440·96	445·50	450·05
100	454·60	—	—	—	—	—	—	—	—	—

Table XI Acres (up to 1000 acres) to hectares (to two places of decimals)

Acres	0	1	2	3	4	5	6	7	8	9
	Hectares									
	—	0·40	0·81	1·21	1·62	2·02	2·43	2·83	3·24	3·64

Acres	0	10	20	30	40	50	60	70	80	90
	Hectares									
	—	4·05	8·09	12·14	16·19	20·23	24·28	28·33	32·37	36·42
100	40·47	44·52	48·56	52·61	56·66	60·70	64·75	68·80	72·84	76·89
200	80·94	84·98	89·03	93·08	97·12	101·17	105·22	109·27	113·31	117·36
300	121·41	125·45	129·50	133·55	137·59	141·64	145·69	149·73	153·78	157·83
400	161·87	165·92	169·97	174·02	178·06	182·11	186·16	190·20	194·25	198·30
500	202·34	206·39	210·44	214·48	218·53	222·58	226·62	230·67	234·72	238·77
600	242·81	246·86	250·91	254·95	259·00	263·05	267·09	271·14	275·19	279·23
700	283·28	287·33	291·37	295·42	299·47	303·51	307·56	311·61	315·66	319·70
800	323·75	327·80	331·84	335·89	339·94	343·98	348·03	352·08	356·12	360·17
900	364·22	368·26	372·31	376·36	380·41	384·45	388·50	392·55	396·59	400·64
1000	404·69	—	—	—	—	—	—	—	—	—

Table XII Miles per hour (up to 100mph) to metres per second (to two places of decimals)

Miles per hour	0	1	2	3	4	5	6	7	8	9
	Metres per second									
0	—	0·45	0·89	1·34	1·79	2·24	2·68	3·13	3·58	4·02
10	4·47	4·92	5·36	5·81	6·26	6·71	7·15	7·60	8·05	8·49
20	8·94	9·39	9·83	10·28	10·73	11·18	11·62	12·07	12·52	12·96
30	13·41	13·86	14·31	14·75	15·20	15·65	16·09	16·54	16·99	17·43
40	17·88	18·33	18·78	19·22	19·67	20·12	20·56	21·01	21·46	21·91
50	22·35	22·80	23·25	23·69	24·14	24·59	25·03	25·48	25·93	26·38
60	26·82	27·27	27·72	28·16	28·61	29·06	29·50	29·95	30·40	30·85
70	31·29	31·74	32·19	32·63	33·08	33·53	33·98	34·42	34·87	35·32
80	35·76	36·21	36·66	37·10	37·55	38·00	38·45	38·89	39·34	39·79
90	40·23	40·68	41·13	41·57	42·02	42·47	42·92	43·36	43·81	44·26
100	44·70	—	—	—	—	—	—	—	—	—

Table XIII £sd to £p in 1d increments

Shillings	Pence											
	0	1	2	3	4	5	6	7	8	9	10	11
	£p equivalents											
0	0·00	0·00	0·01	0·01	0·02	0·02	0·03	0·03	0·03	0·04	0·04	0·05
1	0·05	0·05	0·06	0·06	0·07	0·07	0·07	0·08	0·08	0·09	0·09	0·10
2	0·10	0·10	0·11	0·11	0·12	0·12	0·13	0·13	0·13	0·14	0·14	0·15
3	0·15	0·15	0·16	0·16	0·17	0·17	0·17	0·18	0·18	0·19	0·19	0·20
4	0·20	0·20	0·21	0·21	0·22	0·22	0·23	0·23	0·23	0·24	0·24	0·25
5	0·25	0·25	0·26	0·26	0·27	0·27	0·27	0·28	0·28	0·29	0·29	0·30
6	0·30	0·30	0·31	0·31	0·32	0·32	0·33	0·33	0·33	0·34	0·34	0·35
7	0·35	0·35	0·36	0·36	0·37	0·37	0·37	0·38	0·38	0·39	0·39	0·40
8	0·40	0·40	0·41	0·41	0·42	0·42	0·43	0·43	0·43	0·44	0·44	0·45
9	0·45	0·45	0·46	0·46	0·47	0·47	0·47	0·48	0·48	0·49	0·49	0·50
10	0·50	0·50	0·51	0·51	0·52	0·52	0·53	0·53	0·53	0·54	0·54	0·55
11	0·55	0·55	0·56	0·56	0·57	0·57	0·57	0·58	0·58	0·59	0·59	0·60
12	0·60	0·60	0·61	0·61	0·62	0·62	0·63	0·63	0·63	0·64	0·64	0·65
13	0·65	0·65	0·66	0·66	0·67	0·67	0·67	0·68	0·68	0·69	0·69	0·70
14	0·70	0·70	0·71	0·71	0·72	0·72	0·73	0·73	0·73	0·74	0·74	0·75
15	0·75	0·75	0·76	0·76	0·77	0·77	0·77	0·78	0·78	0·79	0·79	0·80
16	0·80	0·80	0·81	0·81	0·82	0·82	0·83	0·83	0·83	0·84	0·84	0·85
17	0·85	0·85	0·86	0·86	0·87	0·87	0·87	0·88	0·88	0·89	0·89	0·90
18	0·90	0·90	0·91	0·91	0·92	0·92	0·93	0·93	0·93	0·94	0·94	0·95
19	0·95	0·95	0·96	0·96	0·97	0·97	0·97	0·98	0·98	0·99	0·99	1·00
20	1·00	—	—	—	—	—	—	—	—	—	—	—

This table has been compiled to conform to banking and accounting requirements

Table XIV Pressure and stress. Pounds-force per square inch to kilonewtons per square metre (to two places of decimals)

lbf per sq in	kN/m²	1 or kPa	2	3	4	5	6	7	8	9
		6·90	13·79	20·68	27·58	34·48	41·37	48·26	55·16	62·06
10	68·95	75·84	82·74	89·64	96·53	103·42	110·32	117·22	124·11	131·00
20	137·90	144·80	151·69	158·58	165·48	172·38	179·27	186·16	193·06	199·96
30	206·85	213·74	220·64	227·54	234·43	241·32	248·22	255·12	262·01	268·90
40	275·80	282·70	289·59	296·48	303·38	310·28	317·17	324·06	330·96	337·86
50	344·75	351·64	358·54	365·44	372·33	379·22	386·12	393·02	399·91	406·80
60	413·70	420·60	427·49	434·38	441·28	448·18	455·07	461·96	468·86	475·76
70	482·65	489·54	496·44	503·34	510·23	517·12	524·02	530·92	537·81	544·70
80	551·60	558·50	565·39	572·28	579·18	586·08	592·97	599·86	606·76	613·66
90	620·55	627·44	634·34	641·24	648·13	655·02	661·92	668·82	675·71	682·60
100	689·50	—	—	—	—	—	—	—	—	—

Note: the same table will convert kipf per sq in to MN/m² or MPa

Table XV British thermal units per hour to watts

Btu per hr	W	1	2	3	4	5	6	7	8	9
		0·29	0·59	0·88	1·17	1·47	1·76	2·05	2·34	2·64
10	2·93	3·22	3·52	3·81	4·10	4·40	4·69	4·98	5·28	5·57
20	5·86	6·16	6·45	6·74	7·03	7·33	7·62	7·91	8·21	8·50
30	8·79	9·09	9·38	9·67	9·97	10·26	10·55	10·84	11·14	11·43
40	11·72	12·02	12·31	12·60	12·90	13·19	13·48	13·78	14·07	14·36
50	14·66	14·95	15·24	15·53	15·83	16·12	16·41	16·71	17·00	17·29
60	17·59	17·88	18·17	18·47	18·76	19·05	19·34	19·64	19·93	20·22
70	20·52	20·81	21·10	21·40	21·69	21·98	22·28	22·57	22·86	23·15
80	23·45	23·74	24·03	24·33	24·62	24·91	25·21	25·50	25·79	26·09
90	26·38	26·67	26·97	27·26	27·55	27·84	28·14	28·43	28·72	29·02
100	29·31	—	—	—	—	—	—	—	—	—

Table XVI U value: British thermal units per square foot per hour per degree Fahrenheit to watts per square metre per kelvin

Btu per sq ft hr °F		0·01 W/(m²K)	0·02	0·03	0·04	0·05	0·06	0·07	0·08	0·09
		0·057	0·114	0·170	0·227	0·284	0·341	0·397	0·454	0·511
0·1	0·568	0·624	0·681	0·738	0·795	0·852	0·908	0·965	1·022	1·079
0·2	1·136	1·192	1·249	1·306	1·363	1·420	1·476	1·533	1·590	1·647
0·3	1·703	1·760	1·817	1·874	1·931	1·987	2·044	2·101	2·158	2·214
0·4	2·271	2·328	2·385	2·442	2·498	2·555	2·612	2·669	2·725	2·782
0·5	2·839	2·896	2·953	3·009	3·066	3·123	3·180	3·236	3·293	3·350
0·6	3·407	3·464	3·520	3·577	3·634	3·691	3·747	3·804	3·861	3·918
0·7	3·975	4·031	4·088	4·145	4·202	4·258	4·315	4·372	4·429	4·486
0·8	4·542	4·599	4·656	4·713	4·770	4·826	4·883	4·940	4·997	5·053
0·9	5·110	5·167	5·224	5·281	5·337	5·394	5·451	5·508	5·564	5·621
1·0	5·678	—	—	—	—	—	—	—	—	—

Index